SEACOAST DEFENSES
A REFERENCE GUIDE

EDITED BY
MARK A. BERHOW

Published by the CDSG Press
Third Edition, revised

September 2024 Revision

THE COAST DEFENSE STUDY GROUP, INC.

CDSG.ORG

The Coast Defense Study Group, Inc. (CDSG) is a tax-exempt corporation dedicated to the study of seacoast fortifications. The purposes of the CDSG include educational research and documentation, preservation and interpretation of historic sites, and assistance to other organizations dedicated to the preservation and interpretation of coast defense sites. Membership is open to any person or organization interested in the study or history of coast defenses and fortifications. Membership in the CDSG will allow you to attend annual conferences, special tours, and receive quarterly newsletter and journal. To find our more about the CDSG, please visit the CDSG website at **cdsg.org.**

The CDSG Press

The **CDSG Press**, the publishing arm of the CDSG, offers for sale back issues of the CDSG's quarterly publication, *Coast Defense Journal/CDSG Journal/CDSG News* (first issued in1985), both as hard copy and in electronic PDF format. Also available are the comprehensive *Notes from the CDSG Conferences and Special Tours* (from 1987) to various United States harbor defenses. These notes include summary histories, site plans, engineering drawings, Reports of Completed Works, etc. The Press offers for sale several reprints of coast defense history:

Notes on Seacoast Fortification Construction by Col. Eben E. Winslow (1920, 428-pages, hard-cover).
Seacoast Artillery Weapons by U.S. War Department (1944, 202-pages, hard-cover).
The Service of Coast Artillery by Frank Hines and Franklin Ward (1910, 736-pages, hard-cover).
Permanent Fortifications and Sea-Coast Defences by the Committee on Military Affairs, U.S. House of Representatives
  Report No. 56 (1862, 544-pages, hard-cover).
American Coast Artillery Matériel, by Office of the Chief of Ordnance, ODD #2042, (1922, hard-cover).
The Endicott and Taft Board Reports to the U.S. Congress (1886 & 1906, hard-cover).
Artillerists & Engineers: The Beginings of American Seacoast Fortifications 1794-1815, by Arthur P. Wade (1971, 2011)
WW II Harbor Defenses of San Diego, by H.R. Everett (2021)

The CDSG Press also offers an ever-expanding number of key reprints reports and manuals in electronic PDF format on compact disks. To order these books and other **CDSG Press** publications, please access the **CDSG Press** pages on the **CDSG web site** at **cdsg.org.**

The CDSG Press
1717 Forest Lane
McLean, VA 22101-3322 USA

All original compositions are © copyright 1999, 2004, 2015, 2023 by the individual authors
& the Coast Defense Study Group, Inc.

Library of Congress Cataloging-in-Publication data
American Seacoast Defenses: A Reference Guide/ Mark A. Berhow, Editor—3rd Edition
p.   cm.
Includes bibliographical references and index.
Library of Congress Control Number: 2004100338
ISBN 978-0-9748167-3-9 (h.c.)
1. Military History, 2. Artillery I. Mark A. Berhow

Preface to the Revised Third Edition

This guide is a compilation of selected information and illustrations intended to help interested researchers get started in the detailed study of American seacoast artillery and seacoast defenses. Research efforts in this area of history have most often centered on the extraction of information from difficult to find primary sources. This guide is intended as a starting point to acquaint the reader with the terms, nomenclature, and data available on this fascinating subject. The intent was to create a series of documents which can be used to bridge the gap between E.R. Lewis' fine introductory book *Seacoast Fortifications of the United States, an Introductory History* and the more detailed reference materials concerning American seacoast defenses—the weapons, structures, terminology, and original documents— that this kind of study entails.

Since the founding of the Coast Defenses Study Group by a small group of enthusiasts to exchange information on American seacoast fortifications in1978, it has grown slowly and steadily in both membership and in the production of published information, both in the form of reprints and original articles. This reference guide was first published in 1999. The 2004 second edition marked the 25th anniversary of the CDSG and added seven significant new sections of material to that of the 1999 edition. It featured an extensive selection of new photographs in all sections, as well as correcting a number of errors and omissions found in the first edition. The 2015 third edition featured a number of new illustrations and corrections and the first publication of a hard cover edition. The 2023 revised edition features additional revisions and more corrections, with more changes to the text and more new photographs and illustrations.

Since the publication of the first edition a number of source documents have been made available by the CDSG Press at cdsg.org. In addition to reprints, the CDSG ePress also features several collections of digital PDF files, including a number of primary reference documents, manuals, lists, articles, maps, battery drawings, reports, and memos cited here, mostly reproduced from originals in the holdings of the National Archives. This electronic document collection has grown with the reproduction of the series of "Reports of Completed Works," "Annexes to Harbor Defense Projects," "Harbor Defense Engineer Notebooks," "Quartermaster Records" for harbor defenses, and other key documents for various defended harbor locations.

Acknowledgments

A number of people have contributed to the sections contained in this guide. Most of these people were members of the Coast Defense Study Group, which has facilitated my study of this subject. The membership of the CDSG has contributed encouragement, information, and helpful criticism and corrections. I would like to thank in particular Matt Adams, Tom Batha, Gordon Bliss, Charles Bogart, Roger Davis, Elliot Deutsch, Joel Eastman, Dale Floyd, William Gaines, Alvin Grobmeier, Greg Hagge, Milton (Bud) Halsey, Alan Hardey, Gale Hemmen, David Hansen, Alex Holder, Michael Kea, David Kirchner, Nelson Lawry, Danny Malone, John Martini, Terry McGovern, Gary Paliwoda, Leo Polaski, Bolling Smith, Terry Sofian, Chris Sterling, Debbie Steitz, Sam Stokes, Tom Thomas, Thomas Vaughan, John Weaver, Glen Williford, and Robert Zink.

The information contained in this guide was culled from many sources, manuals, documents, and site visits, much of which are not generally available to a researcher in a given city. It has taken years to locate and assemble this material into the condensed format presented here. This material is as accurate as it was possible to make it, but due to the necessity of keeping it brief and introductory in nature, there will no doubt be some errors, if nothing else in my interpretation. As compiler of this material I take full responsibility for any errors that appear here. Any corrections and suggestions readers may have would be greatly appreciated. Corrections and changes will continued to be updated in the electronic edition.

Mark A. Berhow

AMERICAN SEACOAST DEFENSES: A REFERENCE GUIDE
REVISED THIRD EDITION

Table of Contents

12-inch Seacoast artillery mortars at Battery Laidley, Fort De Soto Park, St. Petersburg, FL, in 2010

Introductory Remarks

by Mark A. Berhow

I visited my first seacoast defense site in the mid-1960s. It was a deserted concrete complex with the name Battery David Russell on it, located in Oregon's Fort Stevens State Park at the entrance to the Columbia River. The clean lines of the concrete work was captivating, the massive size was impressive. I became interested in the battery and the park's namesake fort. Why was it built? When was it built? What was it used for? The park rangers at that time could only answer a few of my questions. In the mid-1970s I discovered a small pamphlet, entitled *The Cape Forts,* written by Marshall Hanft and published by the Oregon Historical Society, which gave precious few historical photographs and some background on the U.S. military reservations of Fort Stevens, on the Oregon side of the Columbia, and Forts Canby and Columbia on the Washington side of the river. I began to look for information on the other coastal defense forts around the nation. Several years later I stumbled on Ian Hogg's short book *Fortress* and E.R. Lewis' excellent book *Seacoast Fortifications of the United States.* Both books had a wealth of good general information, but were frustratingly short on specific details.

For several years I tried to hunt down more specific information on US seacoast defense forts. I eventually made contact with the members of CAMP, the Council on Abandoned Military Posts (now the Council on America's Military Past) and the members of the CDSG, the Coast Defense Study Group. I began accumulating very specific information on guns, carriages, batteries, forts, mines, etc., etc. It was a long and often frustrating process. Most of the technical data I wanted was long out of print and it was difficult to find copies of the original sources. There are a number of current publications available on the history of many of the individual forts, especially those that were involved in the American Civil War, but precious little available in print on the overall history and equipment of the US Army coast artillery. I accumulated a great collection of photocopied documents, drawings, and other minutia from my fellow members in these organizations and slowly learned the terms, definitions, and lingo of the field. This guide is an effort to help others bypass some of that long process.

Many of these old forts and military reservations have become parks or public lands today. I would especially hope that members of the staff of these parks will be able to use this guide in preparing interpretive signs, brochures, publications, and guided talks for the visitors to these parks, so that others may come to find and appreciate this fascinating part of our historical past.

1. Starting by reading *Seacoast Fortifications of the United States: an Introductory History*, by E. R. Lewis (Naval Institute Press, Annapolis, MD, current edition copyright 1992). This is the one basic introductory text to the American seacoast defenses. This fine work of 145 pages is an excellent short history with a good discussion of the development of the seacoast fortifications of the United States from the colonial times through their demise in the years following the Second World War. Originally published by the Smithsonian Institution in 1970, this work has been reprinted several times. Currently, it is being published by the Naval Institute Press, Annapolis, MD, and can be ordered by any bookseller. This reference guide will make more sense AFTER one has read Dr. Lewis' book. A second excellent introductory work, *Two if by Sea, the Development of American Coast Defense Policy* by Robert S. Browning III (Greenwood Press, Westport, CT 1983), is a brief, but well referenced discussion of the development of US coast defense policies from the colonial period until the beginning of World War I. A third good introductory work is *American Coastal Defenses 1885-1950* by Terrance McGovern and Bolling Smith, which is part of Osprey Publication's Fortress Series FOR #44 published in 2006. For the earlier U.S. fortifications I recommend *Artillerists and Engineers: The Beginnings of American Seacoast Fortifications, 1794-1815* by Arthur P. Wade, which is the best source of information on the First and Second Systems of American fortifications, published by CDSG Press and available from lulu.com; and *A Legacy in Brick and Stone: American Coastal Defense Forts of the Third System 1816-1867, Second Edition,* by John R. Weaver II, published by Redoubt Press (2018), which is a great introduction to the brick and stone forts of the American Third System harbor defense fortifications built between 1816 and 1867.

2. Join the Coast Defense Study Group (cdsg.org). The CDSG publishes a newsletter and a journal four times a year, which are a great source of information on this subject area. The back issues of the *Coast Defense Journal/CDSG News/Journal* are treasure troves of specific information on a number of different seacoast artillery topics and can be used to supplement the information found in this guide. The CDSG is committed to making available rare reference material on this subject area, including the publication of this reference guide. The CDSG Press has reprinted several rare primary source books. The CDSG now has a large collection of electronically scanned documents which provide detailed collection of maps, records, manuals, reports and other reference materials, mostly of original army records and documents, on general coast artillery subjects and on each of the harbor locations where it has held meetings. Visit (cdsg.org) for more information.

3. Look for books, manuals, and documents on seacoast defense history. Major metropolitan libraries and major university libraries are repositories for federal records. Many have collections of old army manuals from their ROTC programs, especially the land grant universities. You can use the interlibrary loan system to get books from other libraries, and you can browse catalogs on the Internet. One of the best libraries to go to if you can is the Military History Institute Library, located at Carlisle Barracks, Carlisle, PA. This library has the best and most complete collection of Army publications anywhere in the United States, save the Library of Congress, and certainly has one of the premier published military history collections in the world. As noted above, you can browse their catalog on the internet and order books published after 1939 by interlibrary loan. However, many of the items of interest to the coast artillery researcher were published before 1939, hence a visit to Carlisle should be considered by the serious student (http://Carlisle-www.army.mil/usamhi/). The Internet is a storehouse of information on seacoast artillery, especially at sites related to the army, parks, and museums. Finally, many out of print military manuals can be found by browsing the used book stores (both in person and on the Internet) and at swap meets specializing in military items. Some publishers and organizations, including the CDSG, offer photocopies of a few documents related to seacoast fortifications (for example www.military-info.com).

4. If one is interested in primary documents, records, and photographs concerning the army's seacoast defenses, THE place to go is the National Archives. However, there is an art to getting the information you want, even if you are there in person, let alone try to get material by mail. The best way to get material from the Archives is indeed to go there yourself, use the finding aids and the technical staff to get an idea of what is in the record groups and boxes so that you can make the best use of your time there. Further information on getting material from the National Archives can be found online at www.nara.gov. An excellent guide to doing research in the National Archives and Records Administration holdings is *Environmental Cleanup at Former and Current Military Sites: A Guide to Research* (EP-870-1-64), USACE, Historical Division, Office of the Chief of Engineers, GPO, Washington, DC, 2001) which can be downloaded as a PDF file from the Office's publications page—www.hq.usace.army.mil/history/pubs.htm. The National Archives cartographic holdings include a large collection of maps and drawings on all aspects of American seacoast defenses.

Visit as many of the parks and museums at old seacoast fortifications as you can. There are outstanding parks on old harbor defense sites literally all around the country. Besides being able to actually see these remains (and occasionally an actual artillery piece!), the bookstores and information centers are good sources of currently published material, and their archives and holdings can be a good source of primary documents and photographs.

Now that you have an interest in this subject, get involved with a park, museum or organization that is working to preserve these sites. This is a critical time for historic preservation in the United States. Government budgets are being cut—parks and museums are relying more and more on volunteer hours and fund raising to make ends meet. Old seacoast fortification structures are especially vulnerable. Located on prime coastal real estate, often in need of expensive repairs and expensive regular maintenance, these structures are vulnerable to modern pressures. Organizations dedicated to the preservation and interpretation of these sites are in desperate need of your help. If you can give volunteer time, in whatever way you can contribute, it will be appreciated. If you can contribute to the fund raising efforts, that will be appreciated as well. But, what ever you do, get involved!

United States Seacoast Defense Construction 1781-1948:
A Brief History and Guide
By Mark A. Berhow (All photos by author, except where indicated)

The First and Second Systems, 1794-1815

When the United States gained its independence in 1783, its surviving seacoast defenses were in poor condition. The 1794 and 1807 war scares caused Congress to appropriate money for fortifications to guard key harbors. These programs are called the First and Second Systems of American seacoast fortification. However, interest in building fortifications subsided once the threat of attack disappeared and the completed works deteriorated. Built mostly of earth with some masonry backing and designed to hold smoothbore cannons, the structures were neither uniform nor durable. Subsequent construction and erosion have modified or destroyed many of these works.

From the colonial era, the Spanish-built Castillo de San Marcos in St. Augustine FL is now a national monument and remains today pretty much unmodified. Fort Niagara State Park, near Niagara Beach NY, was originally built by the French, though extensively modified in latter years. No unmodified First System forts remain today, though Fort Mifflin Historic Site, Philadelphia PA, was originally built during the Revolutionary War and modified somewhat over the following years. Several fine Second System forts remain including Fort James Jackson Historical Park, Savannah GA; Fort McHenry National Monument, Baltimore MD; Fort Washington National Park, a "transitional work" south of Washington D.C. in VA; Fort Jay (Columbus) and Castle Williams in Governors Island National Monument New York NY; Fort Edgecomb State Park, Wiscasset ME; and Fort Moultrie National Monument on Sullivans Island SC.

Fort Mifflin, a First System work near Philadephia, PA (1996)

Castle Williams, a Second System work on Governor's Island, New York harbor (1997)

The Third System, 1816-1860

In 1816 Congress appropriated over $800,000 for a new fortification program, which became the Third System, which was the most ambitious American fortification construction program to date. Begun under peaceful conditions, the works were built more methodically and were permanent in nature. President James Madison appointed a Board of Engineers for Fortifications, which visited potential sites and prepared plans for the new works. Its first report in 1821 was the basis for a fortification program that remained the backbone of American coastal defense until well into the latter part of the 19th Century. The original report suggested 50 sites, but by 1850 the board had recommended nearly 150 more be built. In all, the board suggested building at 200 Atlantic and Gulf coast sites and 20 Pacific coast sites. However, fortifications were only actually built at about one-fourth of these sites. The construction of these works, as were all subsequent seacoast fortification construction projects, was overseen by officers of Corps of Engineers. The mainstays of the defensive works were the large masonry structures built to house many guns in their vertical faces. Smaller works were built to guard less important harbors. The larger works, principally around the major harbors, were largely replacements of earlier works.

Fort Monroe, a large Third System work near Hampton, VA (MAB 2019)

Fort Delaware, a Third System work on Pea Patch Island near Delaware City, DE (1996)

Fort Point, a late Third System Fort in San Francisco, CA (1992)

A large number of these works remain today. Relatively unmodified works include Fort Knox State Park, Bucksport ME; Fort Independence State Park, Boston, MA; Fort Taber (Rodman) Park, New Bedford MA; Fort Adams State Park, Newport RI; Fort Trumbull State Park, New London CT; Fort Schuyler, New York NY (now SUNY Maritime College campus); Fort Richmond (Battery Weed) and Fort Tomkins in the Fort Wadsworth Unit of Gateway National Recreation Area, Staten Island NY; Fort Ontario State Park, Ontario NY; Fort Wayne Historic Site, Detroit MI; Fort Monroe National Monument, Hampton, VA; Fort Macon State Park, Atlantic Beach, NC; Fort Pulaski National Monument, Tybee Island SC; Fort Clinch State Park, Fernandia Beach FL; Fort Jefferson National Monument, Dry Tortugas FL; Fort Barrancas, a unit of the Gulf Shores National Seashore, Pensacola FL; and Fort Point National Monument, San Francisco CA. Slightly modified Third System forts include Fort Warren, Boston Harbor Islands NRA MA; Fort Delaware State Park, Delaware City DE; Fort Pickens, a unit of the Gulf Shores National Seashore Gulf Breeze FL; Fort Morgan State Park, Mobile Point AL; and Fort Gaines Historic Park, Dauphine Island AL.

The Civil War Era, 1861-1865

Black Point Battery, a Civil-War era battery in San Francisco, CA (1994)

The urgencies of the Civil War required that fortifications be constructed rapidly and at minimal cost. Forts begun in the Third System were completed, but new construction was primarily of wood-revetted earthworks. Sometimes these earthworks were constructed near a Third System fort – supplementing the firepower of the fort – and sometimes they were stand-alone fortifications. Often, they were constructed to protect harbors or close channels to ports that had increased in importance during the war. They were, due to the expediency of war, temporary in nature. In addition, the war saw the first use of underwater mines as a planned part of seacoast defenses.

A number of Civil War era earthworks forts remain, including remains of the Civil War defenses of Washington D.C. which includes Fort Foote, VA, a great example of coast defense post of that era. The Black Point Battery at Fort Mason, now a part of the Golden Gate National Recreation Area, is another restored example.

The Post-Civil War Era, 1865-1875

Following the war, construction began on several Third System forts in New England. Built of stone, and designed to accommodate the large-bore cannon developed during the war, these massive forts would have been formidable works. In 1867, however, money for masonry fortifications was halted. A combination of controversy over the vulnerability of masonry to rifled cannon and large-caliber smoothbore cannon and a lessening of concerns regarding land-based attack led to the construction of masonry-revetted earthen fortifications. During the 1870s, a number of these new works were begun which were to include large caliber mortars and submarine mines. While the cannon emplacements with their brick-lined magazines were constructed and armed, facilities for the mortars and mines were not completed. Most of the defenses were essentially abandoned by the early 1880s.

Most of these defenses were built at existing Second and Third System forts or reservations and many remain today. Outstanding examples include Battery Cavello and other batteries at Fort Baker, as well as the East Battery at Fort Winfield Scott, now all part of the Golden Gate National Recreation Area, San Francisco CA; and a section of the 1870s batteries at Fort Wadsworth, Gateway National Recreation Area, Staten Island NY.

An 1870's-era battery at Fort Wadsworth in New York, NY (1997)

Modern U.S. Harbor Defense Construction, 1886-1917; The Endicott and Taft Boards

During the years following the termination of the harbor defense construction in the 1870s, several advances took place in the design and construction of heavy ordnance, including the development of breech-loading, longer-ranged cannon. Coupled with these developments was a growing alarm over the obsolescence of existing seacoast defenses. In 1883, the navy began a new construction program for the first time since the Civil War. The navy's new ships were to be used in offensive roles rather than defensively. This change in naval policy, along with the advances in weapon technology, required a new system of seacoast defenses which would safeguard the harbors and free the navy for its new role.

In 1885 President Cleveland appointed a joint army, navy and civilian board, headed by Secretary of War William C. Endicott, to evaluate proposals for new defenses. The board painted a grim picture of existing defenses in its 1886 report and recommended a massive $127 million construction program of breech-loading cannons and mortars, floating batteries, and submarine mines for some 29 locations around the country. In 1888 Congress created the Board of Ordnance and Fortifications to test weapons and implement the new program. Funding for the actual construction of a more modest building program began in 1890, under the direction of the U.S. Army Corps of Engineers. The new guns were dispersed over a large area in widely separated concrete emplacements, having underground magazines and earthen and concrete parapets that were designed to blend in with their surroundings. A key part of these new defenses were electrically controlled mine fields. Facilities for planting, retrieving, storing and controlling these mines were installed at many locations.

In 1898 the outbreak of the Spanish-American War necessitated the use of some of the older muzzle-loading ordnance and other guns in hurriedly-built "temporary" defenses around the United States. Mine fields were planted and maintained during the time of that conflict.

A 1898 emergency battery built next to Fort Popham, ME (2005)

Outstanding accessible examples of the Endicott-era posts (with both gun batteries and garrison buildings remaining) include the WA state park Forts Worden, Casey, and Flagler, at the entrance to the Puget Sound; Fort Columbia State Park, Chinook WA; Forts Winfield Scott, Baker, and Barry, all now part of the Golden Gate National Recreation Area CA; Fort Mott State Park, Salem NY; and Fort Hancock, now the Sandy Hook unit of Gateway National Recreation Area NJ. It should be noted that rare examples of guns from this era remain at Batteries Worth (10-inch DC) and Trevor (3-inch)-Fort Casey WA; Battery Cooper (6-inch DC)- Fort Pickens FL; Battery Irwin (3-inch)- Fort Monroe VA; Battery Gunnison (6 inch)- Fort Hancock NJ; Batteries Bingham (3-inch) and McCorkle (4.7 inch)- Fort Moultrie SC; Battery Chamberlin (6-inch DC)- Winfield Scott CA; and Battery Laidley (12-inch mortar & 6-inch guns)- Fort De Soto County Park near St. Petersburg FL).

A early Endicott battery—Battery Godfrey at Fort Winfield Scott, San Francisco, CA, (2015)

Battery Worth with one of its 10 inch guns on a disappearing carriage, Fort Casey, Whidby Island, WA (2018)

A late Endicott-era battery—Battery David Russell at Fort Stevens, near Astoria, OR. (2018)

In 1905, President Theodore Roosevelt convened another board, this one under his secretary of war William H. Taft, to update and review the progress of the earlier board's program. Most of the changes recommended by this board were technical; such as adding more searchlights, electrification including lighting, communications, and projectile handling, and a more sophisticated optical aiming technique. The board also recommended the fortification of key harbors in the newly acquired territories of Cuba, the Philippines, Hawaii, Panama, and a few other sites. The Taft program fortifications differed slightly in battery construction and had fewer numbers of guns at a given location than those of the Endicott program. These two modern programs, although not fully realized, gave the United States a coastal defense system that was equal to any other nation by the beginning of World War I.

The only Taft-era fort built in the CONUS was Fort MacArthur in San Pedro, CA. The lower reservation (garrison area) is now a US Air Force reservation, and the upper reservation (gun battery area) is now Angels Gate Park. The other outstanding example of a fort built during this time frame is Fort Mills on Corregidor Island in Manila Bay, the Philippines. Corregidor was the site of fierce combat during WWII. It has an outstanding collection of structural remains and gun armament. Although open to tourists previously, it is currently closed to general visitors.

A Taft-era battery—Battery Osgood-Farley at Fort MacArthur in Los Angeles, CA (1992)

Plans and Projects Between the Wars, 1917-1940

The years after the turn of the century were very productive in the design and building of naval guns and gunnery. By 1915, several foreign battleships could out range any harbor defense weapon in the United States. Moreover, the high firing angles of naval guns generally nullified the advantages of the disappearing carriages then used by the United States. During WW I, many guns were removed from the existing seacoast defenses to be used as field artillery overseas. Many coast artillery units manned these and other field pieces in Europe. At home the coast artillery posts served as enlistment and training centers for those going overseas.

The U.S. Army entered a period of austerity following the end of the war in 1918. Many of the coast defense forts were put on caretaker status, maintained by a small number of soldiers, and used as summer training camps for Reserve, National Guard, Reserve Officers Training Corps (ROTC), and Civilian Military Training Camp (CMTC). New long-range 12-inch and 16-inch army ordnance for seacoast armament were built, supplemented by 16-inch naval guns made available as the result of naval reductions due to the Washington Naval Treaty of 1922. A number of new harbor defense construction plans were designed, but only 15 new two-gun 12 inch batteries and 8 new two-gun 16-inch batteries were actually completed between 1917 and 1930.

Mobile guns left over from the war were utilized as supplemental firepower to support the older guns. 12-inch mortars and 8-inch guns and a few 14-inch guns on railroad carriages were stationed at seacoast defense sites. During WW I, the United States had purchased from France a number of 155 mm GPF tractor drawn guns, which were later made in the U.S. as the M1918. In the late 1920s, these were pressed into a coastal defense role. To increase the accuracy of these guns, they were mounted on easily built circular concrete "Panama" mounts, which were named after the area in which they were first used. The growing importance of aircraft as an offensive weapon resulted in the formation and training of specialized Coast Artillery Corps antiaircraft artillery units during this period. A number of antiaircraft guns were installed at all harbor defense reservations during WWI and years thereafter.

Most of the large batteries built during this time were modified during the WWII years. The only remaining publicly accessible unmodified battery is Battery Kimble at Fort Travis County Park, on the Bolivar Peninsula near Galveston, TX.

Post World War I era battery—Battery Kimble, Fort Travis, TX (2017, Ralph Stenzel, Galveston County Hist. Assn)

A 155 mmm GPF gun in travel arrangment (MAB) 2000

The World War II era, 1940-1950

Rearming the American coastline with long range 16-inch weapons from existing army and navy stocks began in the late 1930s. These guns were emplaced in positions with substantial overhead protection. Two 16-inch prototype batteries were constructed at San Francisco from 1937 to 1940, and a few other batteries were started during this time. A full construction program was authorized by Congress in September of 1940. The program planned for new defenses at 18 harbors along both coasts of the United States. The fortifications were built using two standardized designs, a two-gun 16-inch batteries (or in some cases new or updated 12-inch batteries) and a two-gun 6-inch batteries (or in some cases new 8-inch batteries), along with their supporting command and observing stations. When America entered the war in December of 1941, a large number of mobile weapons were rushed to both coasts. A number of other "temporary" seacoast defenses were built using old naval weapons and relocated Army seacoast weapons.

A 16-inch WWII-era battery—Battery Issac N. Lewis (BCN 116) Highlands, NJ (2017)

The seacoast defense construction program went into high gear in early 1942, with priority for the sites along the Pacific coast. Batteries of new 90 mm guns were added to the program as anti-motor torpedo boat (AMTB) units. By that year permanent defenses were planned for 33 harbor areas. However, after the Battle of Midway in June of 1942, the possibility of major hostile attacks on the American mainland diminished. As a result the construction program was curtailed in late 1942 and halted altogether by 1946.

A WW II-era 6-inch battery—BCN 236 at Fort Travis near Galveston, TX.(2017, Ralph Stenzel)

Fort Wool, the Harbor Defenses of the Chesapeake Bay, 2000, showing three phases of American seacoast defense works. The unique WW II-era Battery Gates (#229) is in front, the turn of the century Endicott-era batteries are lined along the middle of the island, and the remains of the uncompleted Third System fort is at the very back of the island, (photo by Terry McGovern, 2000)

Most of the fortifications built during the WWII years remain today, and many are on older era fort reservations. A great example is Fort Cronkite in the Marin Headlands Unit of the Golden Gate National Recreation Area with a collection of WWII-era barracks and administration buildings and the early design 16-inch Battery Townsley. Casemated batteries can be seen at the Salt Creek Recreation Area, Crescent Bay WA, the location of WWII-era Camp Hayden; Odiornes Point State Park, Portsmouth NH, the location of WWII-era Fort Dearborn; Cape Henlopen State Park, Lewes DE, the location of WWII-era Fort Miles; and Hartshorne Woods County Park, Highlands NJ, the site of the WWII-era Navesink Military Reservation. Sites with accessible 6-inch batteries include Fort Ebey State Park, Whidbey Island, WA; the Columbia River defenses—Fort Stevens State Park, OR, Fort Columbia State Park, WA (with two WWII era 6-inch guns), and Cape Disappointment State Park (Fort Canby). The Fort Pickens unit of Gulf Shores National Seashore, Gulf Breeze FL, also has two 6-inch guns at Battery 234.

By 1950 the day of seacoast artillery was past, outdated by the airplane, the missile, and new amphibious landing techniques. Nearly all of the big guns were scrapped, the harbor defense commands were dismantled, and the Coast Artillery Corps abolished as a separate branch of the army. The old coast defense reservations were either converted to other uses by the military or declared surplus.

Many of these old reservations around the U.S. coastline have become public parks and are accessible to visitors today. A complete listing of Third System forts is found on pages 52-53 and a listing of the modern era forts is found on pages 201-231. These lists note the status and current ownership as of the date of this book. Privately owned and military properties are not open to the public.

The first post-seacoast artillery American continental defense system was the Nike surface to air missiles (1954-1974). This is Site SF-88L at Fort Barry, CA, the only restored Nike site in the U.S, now a part of the Golden Gate National Recreation Area in the Marin Headlands Unit. Adjacent to this site are a number of Endicott-era and WW II-era seacoast artillery structures.(1994)

Fixed Armament in American Harbor Defenses, 1946
Compiled by Mark Berhow, 1995

Abbreviations are as follows: A = Army gun tube on open barbette mount, N = Navy gun tube on open barbette mount, H = Army howitzers on open barbette mounts, CB = Navy 16-inch gun tube on barbette mount or Army 12-inch in casemate, BC = Army 12-inch gun or Navy 8-inch gun on barbette carriages. SB = Army gun tube on barbette mount with shield. All batteries consisted of two guns each.

Batteries built prior to 1940 Program[1]

	16"A	16"H	16"N	16"CB	12"BC	8"BC
retained, but not rebuilt	1	2	1	2	2	3
casemated 1941-44	2	-	2	-	10	-
disarmed	-	-	-	-	2	-

New batteries built under 1940 Program

	---------CONUS--------			-------------Other Sites----------				-AMTB-
	16"CB	12"CB	6"SB	16"CB	12"CB	8"BC	6"SB	90mm
planned[2]	34	2	51	5	4	9	38	127
built	21	2	49	0	1	8	25	89+13
armed	15	2	34	0	1	5	16	89+13

Notes—

1. The Hawaiian 16-inch Army gun battery (Williston) and the two 16-inch howitzer batteries (Pennington and Walke) at Cheapeake Bay were not casemated, but were modernized with protective shields. In Panama, both 16-inch Navy gun batteries (Murray and Haan) and both 12-inch batteries (Pratt and MacKenzie) were to be casemated but the casemating was only carried out on one battery of each type (Murray and Pratt). Three other existing 12-inch batteries were not casemated, one at Galveston (later disarmed) and two at the Delaware River (one was disarmed, the other never reactivated). The two singl- gun Philippines 12-inch batteries (disabled by their crews before surrender in May 1942) are not included in this table.

2. In the continental United States, twenty-seven new 16-inch batteries and fifty new 6-inch batteries were proposed in 1940 in addition to five 16-inch batteries and one 6-inch battery that had been previously authorized (and later added to the new program), while five 16-inch, nine 8-inch and twenty nine 6-inch batteries were proposed for locations outside the continental limits. Additional proposed batteries were added in 1941-42 to give the totals shown in this table. The Hawaiian turret batteries (built 1942-44) are not included in this table.

Harbor Defense Works of the First and Second Systems

Derived from the article "Early Coast Fortification," *Coast Artillery Journal* 70 (1929), pp. 134-144.

The entrance to Fort McHenry, Baltimore, Maryland (Mark Berhow 2009)

Of all the plans submitted to Congress at the close of the Revolutionary War looking to the organization of a peacetime military establishment, not one took into consideration the necessity of providing for the defense of the maritime frontier. During the war, coast defense had been a function of the several States, the Government finding it necessary to devote its entire attention to defeating the enemy in the field. It is probable that the proponents of the peace measures considered—if they thought of the matter at all— that the States could continue to furnish their own coast fortifications, but if so they neglected the obvious fact that the States had not theretofore provided effective fortifications. Even during the colonial period, the defenses had almost invariably been inadequate to the requirements; and at the close of the Revolution there were few coastal works not in ruins, and none in serviceable condition.

In the years immediately following the disbanding of the Continental troops the entire force—too small to be called an army—in the service of the United States was employed along the land frontiers. The artillery was armed as infantry and served as infantry. The only difference between the two branches was that the artillery also served the guns in the frontier forts and those taken on expeditions against the Indians. Properly speaking, they were artificers rather than artillerymen, and when the time came to take up their duties in coast defense they were unprepared.

The threat of war with Great Britain, growing out of disputes over unsettled boundaries and over British treatment of American seamen, turned the eyes of the infant nation from the depths of the backwoods to the undefended seaboard. Here was opening up an entirely new field of service for the artillery, which brought about the reorganization and expansion of that branch of service under the act of May 9, 1794.

Before this date, however, the fortification of the coast had been begun. On February 27, 1794, a committee had recommended to Congress the fortification of sixteen points along the Atlantic shore line—Portland, Portsmouth, Cape Ann, Salem, Marblehead, Boston, Newport, New London, New York, Philadelphia, Baltimore, Norfolk, Wilmington (N.C.), Ocracoke Inlet, Charleston, and Savannah, to which Wilmington (Del), Annapolis, Alexandria, and Georgetown (S.C.) were subsequently added. On March 20, Congress appropriated the necessary funds, and by the end of the year the project was near completion save in Boston Harbor and at one or two other points.

This first project contemplated the erection of earthen batteries, faced with timbers at such places where earth of an adhesive quality could not be obtained. The strictest economy was necessary, and it was felt that a tenacious earth, properly sloped, sodded, and seeded with knot-grass, would be durable and would afford sufficient protection so far as naval attacks were concerned.

Naval science had not then developed to a point where landings in force on an open beach were considered practicable, and the coast batteries were therefore required only to prevent the use of harbors and wharves by the enemy and to protect communities from bombardment. Small landings on beaches were, nevertheless, practicable, and the batteries themselves required protection from land attack or raids. This introduced into coast defense a conception from which the Coast Artillery is still suffering (in 1929)—local defense by the artillerymen themselves.

In the immediate vicinity of each battery, or group of batteries, particularly where the battery occupied an exposed position at a distance from the town it defended, on a point of land, or on an island, there was to be erected a strong redoubt or other inclosed work (or a blockhouse for batteries of lesser importance), in which one or two pieces of light artillery would be mounted. This redoubt, or blockhouse, thus became a barrack for the garrison and a citadel protecting the battery from attack from the landward side. In case of such an attack, the apparent idea was that the gunners would retire to the citadel, take up the small arms with which they were provided, and become infantry for the time being—an idea which the Artillery accepted without protest until within very recent years.

The weapons best suited for the coast forts were considered to be the 24 and 32-pounders, of which the entire project called for about 450. Of these, it was thought that 150 could be obtained from materiel on hand and 150 from guns in possession of the States, leaving about 150 to be manufactured. To allow for possible shortages, the purchase of one hundred of each of these heavier calibers was authorized by Congress.

At that time there was on hand a great variety of calibers remaining from the Revolutionary War. The return of ordnance, arms, and implements, of December 14, 1793, shows 214 iron guns, 49 iron howitzers, 2 iron mortars, 2 iron cohorns, 153 brass guns, 43 brass howitzers, 63 brass mortars, and 1 brass cohorn. The calibers included: Iron, 1, 2, 3, 4, 6, 9, 12, 18, and 24-pounder cannon, 3-1/2 and 5-1/2-inch howitzers, 13-inch mortars, and 18-pounder carronades; brass, 2, 3, 4, 6, 8, 12, and 24-pounder guns, 2-3/4, 4-1/2, 5-1/2, and 8-inch howitzers, and 4, 4-1/2, 5-1/2, 8, 10, 13, and 16-inch mortars. The following were available among the heavier types:

	Iron	Brass
24-pounder	12	3
18-pounder	36	—
12-pounder	49	11
8-inch howitzer	—	18
5-1/2-inch howitzer	2	17
16-inch mortar	—	1
13-inch mortar	2	4
10-inch mortar	—	19
8-inch mortar	—	3
5-1/2-inch mortar	—	19
	103	92

Prior to 1800 there was no noteworthy change in the calibers of artillery constructed for seacoast artillery. The 42-pounder was added in 1801 and the 50-pounder Columbiad in 1811. In the project of 1818, these, as well as a 100-pounder, formed a part of the seacoast materiel.

The guns available in 1794 were of both brass and cast iron. Though more expensive than cast iron, brass cannon were favored because there was less danger of bursting. The Revolution had practically compelled the colonists to use the iron and thus demonstrate its possibilities, and there ensued a long contest between the two metals (the brass being substantially what was afterwards known as bronze), with cast iron steadily growing in favor. In the end it displaced brass, only itself to be superseded at about the opening of the Civil War. In the heavier guns for coast defense, the project of 1794 established cast iron as the metal to be used, and from that time until wrought iron appeared, no other metal was used for the heavy coast cannon.

The multiplicity of calibers was not, of itself, a great inconvenience, but there were many varieties of each caliber, owing to the fact that each foundry cast its guns according to its own plans. This led to great confusion in the manufacture of gun carriages. These carriages were, as a rule, wooden frames, although there were also carriages made in two parts—a chassis and a top carriage. In 1818 cast-iron carriages were adopted to replace those made wood, but in 1839 a reactionary spirit brought the wooden carriage again into favor, where it held its place for fifteen years before being definitely and finally displaced.

The project of 1794 contemplated the use of two kinds of carriages for seacoast armament— "coast" carriages (which might be casemate or barbette) and "traveling" carriages. These latter, which must not be confused with the "light field" carriages, are particularly worthy of note in view of the use of mobile artillery in coast defense today.

The term "traveling carriage" was not applied to the carriages of any particular calibers. There were "heavy" and "light" guns for every caliber in the service. Light field carriages were used with the light guns of whatever calibers constituted the field artillery of a force in the field. The traveling carriage, less mobile and more rugged in construction, was used to transport every type of "heavy" gun, and was therefore as necessary with the heavy 48-pounder as with the heavy 24-pounder. Guns mounted on traveling carriage were employed as siege or garrison artillery or, in battle, as guns of position. In coast defense they were, as a rule, held in reserve, to be moved into position when and where danger threatened.

The construction and occupation of the works of 1794 demanded both engineers and artillerists, of which the Army possessed neither. Pending the organization of the Corps of Artillerists and Engineers, the Government employed a number of civilians as temporary engineers to put up the necessary works. Stephen Rochefontaine, assigned to the New England coast from New London north, was the most capable of the engineers so employed, and by the end of the year had his works practically completed, except at Boston, where the Governor would not approve the plans without the sanction of the Legislature, which delayed taking action. Charles Vincent, appointed engineer for New York; John Jacob Ulrick Rivardi, for Baltimore and Norfolk; and Paul Hyacinte Perrault, for South Carolina and Georgia, had their portions of the project well under way by December. Charles L'Enfant, engineer for Philadelphia and Wilmington; John Vermonnet, for Annapolis and Alexandria; and Nicholas Francis Martinon, for North Carolina, accomplished little.

The project called for a battery, a redoubt, and a blockhouse each at Portland, Portsmouth, Governor's Island (Boston), New London, Groton, Governor's Island (New York), Paulus [Sandy?] Hook, Baltimore, Norfolk, Wilmington (N. C.), Charleston (three sets), and Savannah; a battery and a blockhouse each at Gloucester (Cape Ann), Salem, Marblehead, in New York City (several sets), and Ocracoke Inlet; traveling carriages, with no battery, at Newport; and repair of works only at Castle Island (Boston), Goat Island (Newport), and Mud Island (Delaware). The total estimated cost was $76,053.62 for the fortifications, and $96,645.00 for the manufacture of two hundred cannon.

With the dissipation of the war clouds there was a relaxation in the matter of coast defense, although some work continued. The first project may be considered to have been complete by the end of 1795, but almost at once preparations on a second project became necessary, for war with France appeared to threaten. The earthen works of 1794 had deteriorated rapidly and large appropriations were necessary to effect repairs. Philadelphia, New York, Newport, Baltimore, and Charleston were considered inadequately defended and large sums were spent at these points in new construction. No new places appear[ed] in the project of 1798, but Cape Ann, Wilmington (Del), Annapolis, Alexandria, and Georgetown (S. C.) disappear[ed]. At a few of the other harbors no funds were spent, but at most of them some repairs were found necessary. Later, the Louisiana purchase brought New Orleans into the program.

At this time the artillery was scattered in many small detachments along the seacoast and on the land frontier. The largest detachment, in December 1802, consisted of 118 officers and men at New Orleans; and no other exceeded seventy-five. Ten stations were garrisoned by from fifty to seventy-five officers and men; twelve had from twenty-five to fifty; and four numbered less than twenty-five. It was therefore impracticable to keep the coast forts in good repair, especially those not garrisoned. In 1807, under the stress of imminent war with Great Britain, the necessity for the repair of the coast defenses brought out an entirely new project.

In December, 1807, the Government, in preparing this new program, classified the harbors into the more important ports and those of minor importance. In the two groups it listed practically all the ports and harbors of the Atlantic and Gulf seaboards, and then, from fear that some might have been overlooked, it made provision for other places that might be found to require defense. Work was undertaken promptly and was advanced rapidly. By February, 1810, $640,000 had been expended. When the war actually broke out, the project was essentially complete; at which time the results of the three programs— 1794, 1798, and 1807—were about as follows, all works being in good condition unless otherwise stated:

Passamaquoddy: *Fort Sullivan,* erected on Moose Island in 1808-1809, was a circular battery of stone, mounting four heavy guns, covered by a blockhouse.

Machias: Under the project of 1807 there was erected a circular battery of stone, mounting four heavy guns, covered by a blockhouse.

Penobscot: Under the project of 1807 there was erected a small inclosed battery, mounting four heavy guns.

Fort St. Georges, at Robinson's Point, on the east side of St. Georges River, erected in 1808-1809, was a small inclosed battery, mounting three heavy guns.

Damariscotta: On the southeastern angle of Narrow Island, and in the town of Boothbay, on the Damariscotta River, there was erected, under the project of 1807, a small inclosed battery, mounting three heavy guns, covered by a blockhouse.

Edgecomb: On Davis' Point, on the east side of Sheepscot River, there was erected a small inclosed battery, with six heavy guns, covered by a blockhouse, as a part of the project of 1807.

Georgetown: On Shaw's Point, on the west side of the mouth of Kennebec River, there was erected in 1808 an inclosed work, with a battery of six heavy guns.

Portland: *Fort Sumner,* authorized in 1794, was built on the hill formerly occupied by Fort Allen as a small inclosed work with parapets supported by stone walls and sod; largely rebuilt in 1798-1799, and kept in repair until 1802; comprised also a blockhouse and a detached battery for heavy cannon near the water; rebuilt in 1808 as a battery of five guns, with a brick gunhouse containing four and eighteen-pounders on traveling carriages. *Fort Preble* (1808), on Spring Point, at the entrance to the harbor, was an inclosed star fort of stone and brick masonry, with a circular battery with flanks, mounting fourteen heavy guns. *Fort Scammel* (1808), on House Island, opposite Fort Preble, was a circular battery of masonry, mounting fifteen heavy guns covered in the rear with a wooden blockhouse mounting six guns.

Portsmouth: *Fort Constitution,* on the eastern point of Newcastle Island, at the entrance to Piscataqua River, three miles below Portsmouth, was begun in 1794 as a fort of masonry and sods, with a citadel; practically rebuilt in 1800-1801, it was completed under the project of 1807 as an irregular work of masonry, mounting thirty-six heavy guns. *Fort McClary* (1808), on Kittery Point, opposite Fort Constitution, was a circular battery of masonry, inclosed by earth and palisades, mounting ten heavy guns. In *Portsmouth,* a brick arsenal (1808) contained three 24-pounders and three 18-pounders on field carriages.

Newburyport: On the east point of Plum Island, at the mouth of Merrimac River, an inclosed battery of timber and earth, mounting five heavy guns, was built as part of the project of 1807.

Gloucester (Cape Ann): In 1794 a battery and a blockhouse were erected at the head of the harbor on the site of an old fort. Omitted from the project of 1798. An inclosed battery, mounting seven heavy guns, covered with a blockhouse, was erected under the project of 1807.

Salem: *Fort Pickering,* situated on the west side of the harbor entrance, was erected in 1794 on the site of old Fort William as an inclosed work of masonry and sods; repaired in 1800 and improved in 1808 to mount six heavy guns.

Marblehead: *Fort Sewall,* situated on the west point of the entrance to the harbor, erected in 1794 on the site of an old fort, was an inclosed work of masonry and sods, covered with a blockhouse; rebuilt in 1799 and improved in 1808 to mount eight heavy guns.

Boston: Boston Harbor was included in the project of 1794, but delay in securing State approval of the plans prevented any work except a limited amount of repairs among the ruins of *Castle William,* on Castle Island, on the south side of the inner harbor. *Fort Independence,* a regular pentagon, with five bastions of masonry, mounting forty-two heavy guns, and two batteries for six guns, was begun in 1800, practically completed in 1803, and extensively repaired under the project of 1807. *Fort Warren,* on the summit of Governor's Island, opposite Fort Independence, a star fort of masonry, mounting twelve guns, was erected under the project of 1807. On the south point and the west bead of the island, circular batteries of masorny, mounting ten guns each, were also constructed.

Charlestown: Near the Navy Yard, on the point formed by Charles and Mystic Rivers, a circular battery of earth, on a stone foundation, mounting eight heavy guns, was erected in 1808.

Plymouth: On Gurnet Point, at the entrance to the harbor, an old inclosed fort, mounting five guns, was repaired with stone and sod in 1808

New Bedford: On Eldridge Point, at the entrance to the inner harbor, an inclosed work of masonery, mounting six guns, was erected in 1808.

Newport: In 1794, a fort on *Goat Island,* a guard house on Tammany Hill, and a battery at Howland's Ferry were erected. *Fort Adams,* on Briton (Brenton) Point, on the east side of the entrance to the harbor, was an irregular fort of masonry, with an irregular indented work of masonry adjoining it, mounting seventeen heavy guns, begun in 1798 and repaired and extended in 1808. *Fort Wolcott,* on Goat Island, in the center of the harbor, was a small inclosed irregular work, with open batteries, extending from two opposite flanks, of stone and earth, mounting thirty-eight heavy guns; principally built in 1798 on the site of the 1794 fort, and repaired and extended in 1808. On *Rose Island,* situated to defend the north and south passages of the harbor, a regular work of masonry with four bastions (two of them circular), to mount sixty guns, was begun in 1798, but was left unfinished. On a bluff of rocks called the *Dumplins,* on Conanicut Island, nearly opposite Fort Adams, a circular tower of stone, with casemates, was begun in 1798, but was left unfinished. On *Eaton's Point,* at the north point of the town, an elliptical stone battery had been erected, but was in ruins by the end of 1811. In Newport were some guns on traveling carriages.

Bristol: Ten guns on traveling carriages protected this town under the project of 1807.

Stonington: A brick arsenal, with four 18-pounders on traveling carriages, was provided by the project of 1807.

New London: *Fort Trumbull,* situated on the west side of the harbor, was an inclosed irregular work of masonry and sod, mounting eighteen heavy guns, erected during the Revolutionary War, repaired in 1794-1795, restored in 1799, and further improved in 1808.

Groton: A fort of earth and sods was begun in 1794, but was left unfinished.

New Haven: *Fort Hale,* on the eastern side of the harbor, was an elliptical inclosed battery, mounting six heavy guns, erected in 1808-1809.

New York: *Fort Jay,* on Governor's Island, within half a mile of the city, was a regular inclosed work, with detached batteries for heavy cannon and mortars. The first fort, of earth, with two detached batteries, which had been built in 1794-1795, was rebuilt in 1798-1801 at considerable expense; but in 1806 the whole was demolished except walled counterscarp, grate, sallyport, magazine, and two barracks) and removed as rubbish to make room for a new work of the same shape. *Fort Columbus,* built on the site of Fort Jay, was a regular inclosed pentagonal work of masonry, with four bastions and a ravelin, mounting sixty heavy guns. *Castle William,* on a projecting point of rocks at the western extremity of the island, begun in 1808, was a stone tower, with fifty-two 42 and 32-pounders, mounted in two tiers, under a bomb-proof roof with a terrace above intended to mount twenty-six 50-pounder Columbiads. *Bedloe's Island,* nearly opposite Governor's Island, was provided with a battery in 1794. *Fort Wood,* a star fort of masonry, mounting twenty-four heavy guns, with a brick arsenal, was erected in 1809-1810. *Ellis (Oyster) Island,* opposite Fort Columbus, was also provided with a battery in 1794-1795. *Fort Gibson,* an inclosed circular battery of masonry, mounting fourteen heavy guns, was erected in 1809 to cover the entrance to North River. In *New York,* a formidable battery of heavy cannon and mortars, erected at the southwest point of city in 1794-1795, was in ruins by 1806. *Castle Clinton,* an inclosed circular battery of stone, mounting twenty-eight heavy guns, was erected in 1809 about a hundred yards in front of the west head of the grand battery. *Humbert Battery,* an inclosed circular stone battery, mounting sixteen heavy guns, was built in 1809 one mile up North River. Within the city was a brick arsenal, with one brass 24-pounder, seven 12-pounders, 4 brass howitzers, and twenty-two iron 18-pounders, all on traveling carriages; and three miles above the city was a brick arsenal and laboratory.

Sagg Harbor: Under the project of 1807, a brick arsenal, with four 18-pounders on field carriages, was provided.

West Point: *Fort Putnam* was repaired and altered in 1794-1795.

Philadelphia: A fort on *Mud Island,* seven miles below Philadelphia, was begun in 1794, and a large pier, as a foundation for a battery, was laid on a sand bar opposite the island. *Fort Mifflin,* principally built in 1798-1800 and extensively repaired in 1808-1809, was an irregular inclosed work of masonry, defended by bastions, demi-bastions, etc., mounting twenty-nine heavy guns, with a water battery without the works, mounting eight heavy guns.

Wilmington, Del.: A site was selected and surveyed in 1794, but no works were erected. A brick arsenal, with four 12-pounders on field carriages, was built in 1809.

Newcastle: A brick arsenal, with four heavy guns on field carriages, was built in 1809.

Baltimore: Under the project of 1794, a battery was erected and some guns mounted. *Fort McHenry,* at the entrance to the harbor, erected principally in 1798-1800, was a regular pentagon of masonry, mounting thirty guns, with a water battery, mounting ten heavy guns.

Annapolis: A site was selected and surveyed in 1795 and some preliminary work was done, but an unfavorable report caused the project to be abandoned. *Fort Madison,* at the western entrance to the harbor, erected in 1809, was an inclosed work of masonry comprised of a semi-elliptical face, with circular flanks, mounting thirteen guns. *Fort Severn,* on Windmill Point, a circular battery of masonry, mounting eight heavy guns, was erected in 1809.

Washington: *Fort Washington,* at Warburton, on the east side of Potomac River, between Alexandria and Mount Vernon, erected in 1808-1809, was an inclosed work of masonry, comprehending a semi-elliptical face, with circular flanks, mounting thirteen heavy guns, defended in the rear by an octagon tower of masonry, mounting six guns.

Alexandria: Some progress had been made in the construction of works in 1795, but an unfavorable report upon the plans caused the project to be abandoned.

Norfolk: *Fort Nelson,* on the western side of Elizabeth River, begun in 1794, extensively repaired and improved in 1802-1804, and again repaired in 1808, was an irregular work, defended by whole and half bastions, built of brick and sods, closed in the rear by a brick parapet, mounting thirty-seven guns. *Fort Norfolk,* on the northeastern side of Elizabeth River, a thousand yards distant from Fort Nelson, erected in 1794-1795 and rebuilt in 1808-1809, was an irregular inclosed work of masonry, comprehending a semi-elliptical battery defended on the flanks and rear by irregular bastions, mounting thirty heavy guns.

Hood's Point: *Fort Powhatan,* on James River, begun in 1808, was a strong battery of masonry, intended for thirteen guns, but unfinished in 1811.

Ocracoke Inlet: The foundation of a fort was laid on Beacon Island in 1794, but no further work was done; in 1799 an inclosed work was ordered on the ruins of the former work, but none was erected.

Wilmington, N. C.: *Fort Johnston,* on the right bank of Cape Fear River, twenty-eight miles below Wilmington, was originally a colonial fort. In 1794, a battery was erected on the site of the old fort, and in 1799-1800 some progress was made in constructing new works. Delays prevented the completion of the fort until after 1806. As finished, it was a flank battery of tapia, mounting eight heavy guns.

Beaufort: *Fort Hampton,* on Old Topsail Inlet, erected in 1808-1809, was a small inclosed work, mounting five guns.

Georgetown, S. C.: A battery was begun in 1794, but was abandoned because of the unhealthfulness of the site. *Fort Wingaw,* a small battery and blockhouse, was erected in 1809.

Charleston: *Charleston* was included in the projects of 1794 and 1798, but, since the State had not then ceded any sites to the United States, little was accomplished until the project of 1807. *Fort Johnson,* on James Island, *Fort Moultrie,* on Sullivan's Island, at the entrance to the harbor, and *Fort Pinckney* were colonial or Revolutionary War forts. In 1794 Fort Johnson was ordered repaired and foundations for forts were laid at Forts Moultrie and Pinckney. Work was soon suspended, except for a battery *(Fort Mechanic)* in Charleston which was completed by the mechanics. In 1798-1799 the old works were repaired and improved but were practically demolished by an unusual storm it 1804. As rebuilt under the 1807 project, the new *Fort Johnson* was a marine battery of irregular form, built of brick and wood, mounting sixteen guns; the new *Fort Moultrie* was a brick work of irregular form, presenting a battery on the sea front, with the whole inclosed with ramparts, parapets, etc., mounting forty guns; *Castle Pinckney* was a brick work of elliptical form, with two tiers, mounting thirty guns; the new *Fort Mechanic (Mechonric),* on the point of the city, crossing its fire with that of the Castle at nine hundred yards, was a temporary masonry battery, falling into decay; in *Charleston* was a brick arsenal.

Beaufort, S. C: *Fort Marion,* a work of tapia, circular of form in front and straight line in rear, was begun in 1809 but was unfinished in 1811.

Savannah: *Fort Green,* on Cockspur Island, near the mouth of Savannah River, erected in 1794-1796, was an irregular work, with a battery. In 1804 the works were totally destroyed and a part of the garrison drowned in an unusually severe storm. *Fort Jackson,* at Five Fathom Hole, in a marsh on the west side of Savannah River, three miles below the town and twelve hundred yards from the nearest dry land, begun in 1808, was an inclosed work of masonry and mud, mounting six heavy guns.

St. Mary's (Point Petre): A battery of timbers, filled with earth and in-closed with pickets, was erected in 1799-1801 but was abandoned before 1804. Included in the 1807 project, no work had been accomplished because no site had been secured.

New Orleans: *Fort St. Philip,* at Plaquemines, on the eastern side of Mississippi River, thirty-two nautical miles from the mouth, an irregular work of brick built by Governor Carondelet in 1793, was acquired in 1803 in poor repair and rebuilt as an inclosed work of masonry and wood, mounting twenty guns. At *English Turn,* on the ruins of some French works, an inclosed work, with two bastions and a battery of masonry, for nine guns, was built in 1809-1811. When acquired in 1803, *New Orleans* was surrounded by five redoubts— Forts Burgundy, St. John, and St. Ferdinand in the rear, and Forts St. Louis and St. Charles in front, all dilapidated—connected by a line of ditches. *Fort St. Charles,* immediately below and at the northeast corner of the city, was restored as an inclosed redoubt of five sides, of masonry and earth, mounting nineteen guns. On the site of the Spanish *Fort St. John,* on Lake Ponchartrain, at the mouth of Bayou St. John, a strong battery of six guns, commanding the approach to New Orleans by way of the lake, was erected under the 1807 project.

The war with England brought about additional construction, and the acquisition of Florida in 1819 added to the ports and harbors to be defended. As a result the following new fortifications appear in the war and post-war years: Fort Lewis, New York; Craney Island, Virginia; Fort Scott, Point Petre, Georgia; Fort Marion (Castillo de San Marco, or St. Mark's fort), Florida; Fort Barrancas, Florida; Fort Bowyer, Mobile Point; Pass Christian, and a number of lake and river forts.

In 1819, while a new coast project was in process of formation, the coast and inland forts were manned by the following garrisons:

Station	Guns	Commanding Officer	Organization	Aggregate
Fort Sullivan, Maine	4	Lieut. Merchant		
Machias, Maine	4		Det., Corps of Arty.	39
Fort St. Georges, Maine	9	Capt. Leonard	1 Co., Lt. Arty.	70
Damariscotta Maine	3			
Edgecomb, Maine	6			
Georgetown, Maine	6			
Fort Preble, Maine	14			
Fort Scammel, Maine	15	Bvt. Major Crane	1 Co.. C. of Arty.	98
Old Fort Sumner, Maine	5			
Fort McClary, Maine	10	Bvt. Lt. Col. Walbach	2 Cos., C. of Arty.	195
Fort Constitution, N. H.	36			
Fort Pickering, Mass	6			
Gloucester, Mass.	6			
Fort Sewall, Mass.	8	Bvt. Lt. Col. Harris	1 Co., Lt. Arty.	70
Fort Independence, Mass.	42			
Fort Warren, Mass	12	Bvt. Lt. Col. Eustis	5 Cos., Lt. Arty.	390
Boston, Mass. (2 batteries)	14			
Plymouth, Mass.	5			
New Bedford, Mass	6			
Fort Wolcott, R. I	28			
Fort Adams, R. I	17			
Fort Hamilton, R. I.		Bvt. Lt. Col. Towson	2 Cos., Lt. Arty.	146
Fort Green, R. I	6			
Dumplins, R. I	10			
Fort Griswold, Conn.	12	Capt. McDowell	1 Co., Lt. Arty.	53
Fort Trumbull, Conn.	18			
Fort Hale, Conn.	6			
Fort Columbus, N.Y.	60			
Castle William, N. Y.	102			
Fort Lewis, New York				
Fort Wood, New York	24	Lt. Col. House	Corps of Arty.	345
Fort Gibson, New York	14			
Castle Clinton, New York	28			
Humbert [North] Battery, NY	16			
Fort Gansevoort, N. Y.	12			
Sagg Harbor, New York	6			
Fort Mifflin, Pa.	37	Major Biddle	1 Co., C. of Arty.	121
Fort McHenry, Md.	30	Col. Hindman	1 Co., C. of Arty.	118
Fort Madison, Md.	13	Capt. Reed	1 Co., C. of Arty.	103
Fort Severn, Md	6			
Fort Washington, Md.	19	Lt. Col. Jones	2 Cos., C. of Arty.	123
Fort Nelson, Va.	37	Lt. Col. McRea	Corps of Arty.	88
Fort Norfolk, Va.	30	Lieut. McIlvain	Corps of Arty.	50
Craney Island, Va	20			
Fort Powhattan, Va.	13			

Station	Guns	Commanding Officer	Organization	Aggregate
Fort Johnston, N. C.	9	Lieut. N. G. Wilkinson	Corps of Arty.	10
Fort Hampton, N. C.	5			
Fort Wingaw, S. C.	6			
Fort Johnson, S. C.	16			
Castle Pinckney, S. C.	30	Lieut. Washington	Small det., C. of A.	
Fort Mechanic, S. C.	7			
Fort Moultrie, S. C.	40			
Fort Marion, S. C.	6	[?] McCall		
Fort Jackson, Georgia	6			
Fernandina, Amelia Island		Capt. Payne	Corps of Arty.	222
Fort St. Mark's, Florida		Major Fanning	C. of A. and Inf'y.	108
Fort San Carlos de Barancas		Major Brook	C. of A. and Inf'y.	77
Fort Charlotte, Alabama				
Fort Bowyer, Alabama			C. of A. and Inf'y.	46
Fort St. Philip, La.	20	Major Humphreys	1 Co., C. of Arty.	85
Fort Petit Coquille, Lake Ponchartrain			1 Co., C. of Arty.	34
Bayou St. John, La.		Major Maney	1 Co., C. of Arty.	57
Fort St. Charles, La.				
Sacketts Harbor, New York		Col. Brady	Inf. & 1 Co., C. of A.	432
Greenbush, New York		Capt. Worth	Inf. & 1 Co., C. of A.	99
Fort Niagara, New York		Lt. Col. Pinkney	Inf. & Corps of Arty.	4
Detroit, Michigan Ter.		Major Marston	Inf. & 1 Co., C. of A.	169
Mackinac, Michigan Ter.		Capt. Pierce	1 Co., C. of A., & Inf.	131
Fort Scott, Georgia		Capt. Donoho	C. of Arty. and Inf.	75
Fort Gaines, Georgia			G. of Arty. and Inf.	13
Newport, Kentucky		Capt. L. Scott	1 Co., C. of Arty.	31

(Above) The exposed casemates of Fort Pickens, a Third System fortification near Penasacola, FL. The missing section was a bastion destroyed by a magazine explosion in the 1880s. (photo by Mark Berhow, 2003)

(Right) The scarp wall of Fort Pulaski, GA showing the damage from the bombardment in 1863. The major breech in the center of the photo was repaired after it was recaptured by Union forces. (Mark Berhow 2021)

Introduction to the Architecture & Weapons of the Third System of American Coastal Fortification 1816 – 1867

John R. Weaver II

The Third System of American Coastal Defense was constructed in response to the War of 1812, in anticipation of further threats to our shores. The 51-year program resulted in the construction of 42 new forts, several other defensive structures, and the refitting of numerous older forts.

Following is a look at the architecture of fortifications constructed during the Third System. It is excerpted from *A Legacy in Brick and Stone: American Coastal Defense Forts of the Third System, 1816 – 1867* (Redoubt Press, Pictorial Histories Publishing Company, 2001), which provides more information on the Third System forts.

The nomenclature applied to the elements of American forts is almost exclusively French. This was not due to Bernard's influence, but predates his coming to the United States by many years. The terminology is derived from the French influence during the infant years of the American military in general, and of fort design in particular. French engineers designed our earliest coastal forts after we became a nation and this became the First System of coastal defense. In addition, the textbooks on fort design initially used at the United States Military Academy were written in French (1); it was not until the middle nineteenth century that instructors at West Point translated some of the texts into English and eventually wrote original works based on the French texts.(2) By this time, the French names for most elements of a fort had been ingrained in the minds of the American engineers.

Elements of Nineteenth Century Fortifications

A fort, like any structure, can be viewed three ways: in plan, or as it would be seen when looking straight down on it; in section, or from the side as if a portion had been sliced off; or in elevation, looking at the entire structure from the side. This analysis of military architecture during the Third System will start by analyzing the plan of a fort.

The line that defines the outer wall of a fort is called the **trace** of the fort, also referred to as the magistral line. The trace of a fort is said to be **regular** if it forms a regular (equal-sided) geometric figure, otherwise the trace is irregular. Many times the trace of a fort was a **truncated polygon**, a regular polygon that was "chopped off" at less than the complete figure.

All the structures defined by the trace of the fort are considered the **enceinte**, or body of the work. All components of the fort outside of the trace are **outworks**. The perimeter of the enceinte is faced by a surrounding wall, known as the **scarp** or **scarp wall.** This wall is considered to be part of the enceinte, and provides the division between the enceinte and the outworks.

The scarp wall of all Third System forts was constructed of either brick, stone, or a combination of the two materials. Some forts in the latter part of the Third System used masonry faces on the scarp wall, and filled the interior with materials such as concrete or ground-shell formulas. The scarp wall was a minimum of five feet thick at its junction with the parapet in early Third System works. This was increased to eight feet thick in later works to compensate for the technological advances that allowed the construction of more powerful artillery.(3) The scarp wall gradually increased in thickness as it progressed down toward the foundation. Near the foundation, it sometimes tapered sharply outward to its thickest point, often more than twenty feet.

Truncated Hexagon

Third System forts were often built on the plan of a partial regular polygon. The most common of these geometric forms was the truncated hexagon. Four sides of the fort were on the plan of a regular hexagon, while the fifth side closed the polygon.

In most cases, the mass of the scarp was reinforced with a very thick (approximately 25 feet) arched structure of brick or stone masonry. In several Third System forts, the arched structure behind the landward scarp was replaced with approximately 30 feet of earth to provide additional resistance to siege cannon.

While it is obvious that this formidable barrier was designed for self-protection, that protection addresses two distinct types of opponent. Seaward, the primary purpose of the scarp was to protect against cannon fire from enemy ships; landward, the purpose was to prevent entry of the fort by the enemy, by siege or by storm. For this reason, the scarp wall on the **gorge**, or landward side, was often somewhat different from the scarp wall on the **faces**, or seaward sides of a fort.

It was commonly believed throughout the early and middle nineteenth century that shipboard cannon, due to the movement of the deck of a ship even in relatively calm waters, could not consistently strike the same area of a masonry fort and create a breach in the wall.(4) For this reason, the seaward faces of a fort were generally left as unprotected masonry with the scarp walls placed close to the shoreline and devoid of outworks. This allowed maximum offensive and defensive viability through increased visibility, better angles of fire, and more tiers of guns.

The seaward side of a scarp wall was generally penetrated by openings, called **embrasures**, for cannon to fire on the enemy. These embrasures may be simple openings in the wall, as in early Third System forts, or may be protected by complex mechanisms such as the Totten Shutters used in later Third System forts. Embrasures were tapered inward from both the inner and outer surface of the scarp. The narrowest point of the embrasure is called the **throat**, and provided the smallest opening possible to enemy fire while still allowing an adequate angle of traverse for the cannon.

The embrasures were the weakest part of the scarp wall. For this reason, a great deal of research was conducted in this area.(5) In the early years, cannon were less powerful, and fewer precautions were required. Granite forts required no special treatment of the embrasure, as the granite provided sufficient strength. Brick forts used either special, hardened brick or granite for the embrasures.

Totten initiated the use of iron in embrasures, using 4-inch-thick pieces of cast iron at the throat and thinner plates around the outer surface of the embrasure to reinforce the stone or brick masonry. These "Totten Embrasures" significantly strengthened the thin portions of the masonry, but created problems due to the vulnerability of iron to the salt air. Not only did the iron weaken when rusted, it expanded and broke loose large masonry sections of the embrasures.(6)

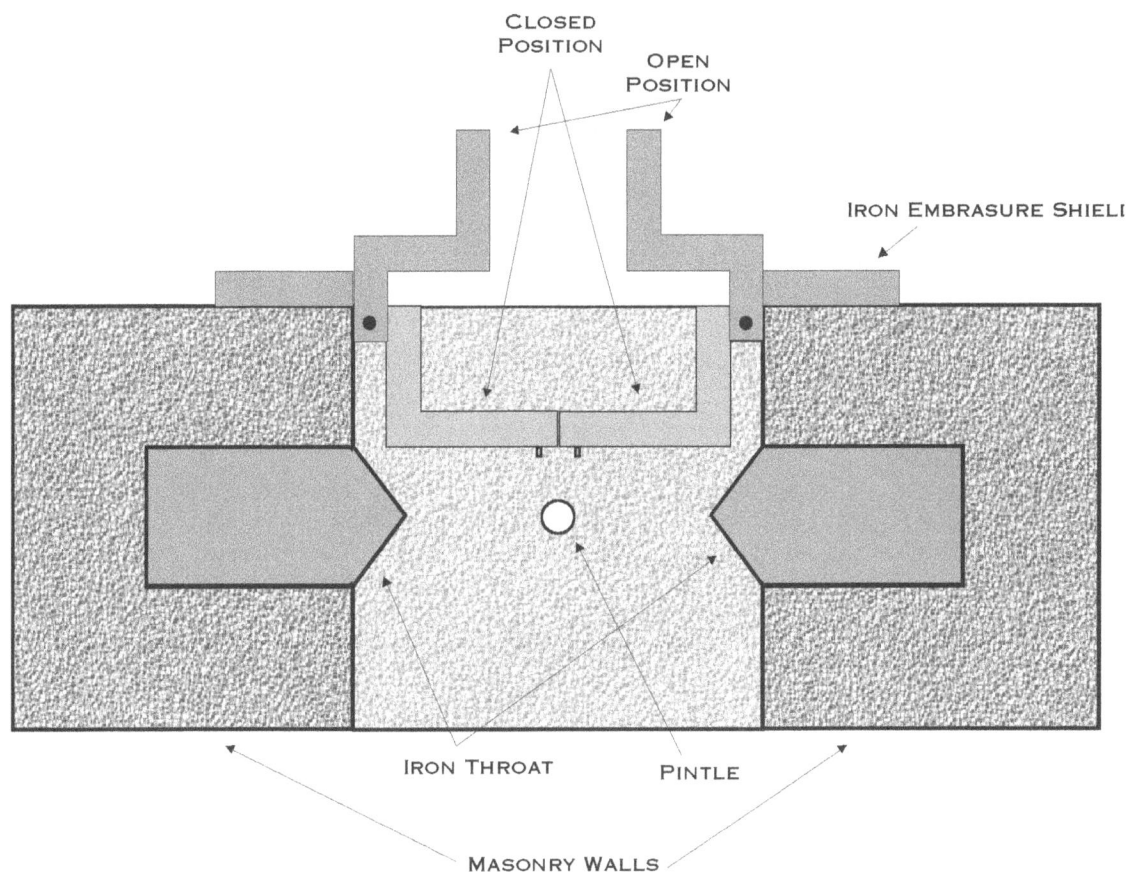

CLOSED
POSITION
OPEN
POSITION

IRON EMBRASURE SHIELL

IRON THROAT   PINTLE

MASONRY WALLS

The Totten Embrasure System consisted of iron components inserted into, and attached to the face of, a masonry embrasure. This system protected gunners between firings, and reinforced the most vulnerable portion of the scarp - an embrasure.

A further innovation by Totten was the development of the so-called **Totten Shutters**. These iron plates, installed with a complex, self-closing mechanism, were designed to protect cannon and gunners during loading and aiming operations. The shutters automatically opened when the cannon was run out and closed when the cannon recoiled after firing. Their angular shape allowed them to open wide enough to allow the full traverse of the cannon. If not properly maintained, these shutters were vulnerable to the salt air as well. In the worst case, the shutters tended to rust closed. An unsubstantiated tradition has it that attempts to fire through "frozen" Totten shutters caused entire embrasures to be blown from the scarp wall.

On the landward side, the threat was somewhat different. Emplaced siege cannon could consistently strike a small area of the scarp, eventually creating a breach. For this reason, the scarp wall was generally protected from direct cannon fire by outworks. Also of concern was the ability of an attacker to storm the walls. Cannon embrasures were often traded for musketry **loopholes**, or rifle slits. These tapered, narrow openings allowed a rifleman to command a relatively wide field of fire, while enjoying the protection of a relatively small opening. Most forts mixed loopholes with embrasures on the landward side, usually mounting howitzers (short-barreled cannon generally used for antipersonnel missions) in the embrasures. These howitzers acted as large shotguns, firing canisters filled with shot varying in size from three-fourths inch to two inches in diameter.

The primary location for howitzer embrasures was in the flank of the bastions, but howitzer embrasures could also be found in the landward curtains of some forts and the faces of the bastions of others. Another favored location was at the junction of two counterscarp walls. Some forts have a howitzer embrasure immediately above the main sally port.

To support the incredible weight of a Third System fort and its artillery, the arched form was used. In addition to the magnificent arches seen when touring these forts, hidden arches below the floor distribute the load more evenly on the foundation. This inverted arch is in the seacoast front of Fort Pickens, guarding Pensacola, Florida. (MAB)

A typical feature found in all Third System forts was the **casemate**. A casemate is an arched gunroom or bombproof shelter, generally located along the curtain wall of the fort and set in the flanks and sometimes faces of the bastions.(7)

The casemate was developed in Europe, long before the American Third System, and was applied in several Second System forts. Castle Clinton in New York Harbor was the first American fort to make use of the casemate in its design, though casemates had been used extensively in Europe by this time. A most impressive use of the casemate prior to the Third System was Fort Lafayette, also in New York Harbor. It boasted two casemated tiers of cannon with a third tier of barbette guns. This transitional work was completed in 1819, and "modernized" later. Also of note is Castle Williams with 78 guns.

In several instances, casemates were constructed external to the fort. Fort Monroe had a casemated Water Battery immediately across the moat from the fort, and several forts had casemated counterscarp batteries that fired the length of the ditch. A few had casemated caponiers and demilunes, while Fort Adams had a casemated crownwork and tenailles.

In most cases, embrasures provided the capability for guns to fire out of these protected areas, and in many forts housed a major portion of the fort's armament. In addition to their principal function as a gun room, casemates also served as barracks, kitchens, powder magazines, storage rooms, and for other miscellaneous uses. This minimized the number of separate, freestanding structures required in the fort and provided a more protected location for the fort's supplies and quarters.

An important goal of the casemate was to provide overhead protection for gunners, partially due to the early nineteenth century practice of placing snipers in the rigging of attacking ships. These snipers were charged with the elimination of the gun crews in the fort under attack. Additionally, casemates were considered "bombproof," or resistant to exploding or solid shells hitting their tops.(8)

Early Third System forts used long passageways between the gun position and the parade to provide additional protection for the gunners. Shown above is a "tunnel casemate" in Fort Wood, defending New Orleans, LA. (MAB)

To expand this protection to the gunners' rear, casemates of early Third System forts – Forts Jackson, Pike, and Wood defending New Orleans, and Monroe defending Hampton Roads – communicated with the parade through long tunnels. While this was effective in protection, the amount of smoke and noise generated from the firing of the large weapons made life nearly intolerable. Smoke vents were provided in the casemates, but these proved inadequate. Throughout the duration of the system, experimentation continued regarding the placement of these vents. Unfortunately, no adequate solution was forthcoming. Most Third System forts solved the problem through the design of casemates that were relatively shallow, and open to the parade in the rear.(9)

Casemate of a Third System fort. This casemate at Fort Knox near Buckport, ME, has positions for two guns instead of the more standard design with one gun per casemate (Photo by Mark Berhow, 2005)

Another major benefit of the casemate, rapidly put into effect, was to provide multiple tiers of guns, greatly increasing the firepower of a fort. As the Third System progressed, this became the primary motivation for building casemated forts. Powerful forts could be built with smaller traces, and large forts could amass tremendous firepower. Several forts in the later portion of the Third System utilized three tiers of guns, with two forts boasting four tiers.(10)

Casemates were constructed as a masonry web, with piers supporting arches in both axes. These arches, constructed without keystones, were masterpieces of brickwork and, in a few cases, granite work. The scarp wall closed the face of a casemate, but the scarp was not structurally joined to the casemate. This architectural separation allowed the destruction of the scarp without bringing down the casemates, and also allowed for differential settling of the casemates and scarp.(11)

The casemates, fronted with the scarp wall, comprised the basic mass of the work, referred to as the **ramparts**. The term ramparts is derived from earthwork forts where the mass of the work was an earthen mound excavated from the ditch. In the Third System, the mass of the work was usually made up of casemates. The top of the casemates contained drainage piping that fed the cisterns of the fort, and the entire area was covered with earth. This earthen area, called the **terreplein** (12), formed the top of the ramparts.

Some Third System forts had non-casemated walls, usually on a landward face. These walls used an earthen rampart that either sloped to the parade or had a masonry parade wall. This earthen rampart was more resistant to the bombardment of siege cannon, and was used where casemates were not needed.

In all cases, the earthen terreplein provided an inexpensive, easily leveled surface. Being a compressible material, it could absorb the impact of shells striking the rampart and the vibration of the firing of heavy cannon without damaging the masonry below. In some cases this earthen terreplein was paved with brick or slate, but was generally planted with grass.(13)

The front of the terreplein was the **parapet**, the highest part of the defensive structure. In Third System forts, the parapet was revetted with masonry on the rear, and the front was usually made up of two sloped, earthen masses. The entire structure, usually breast high, was often topped with a stone coping. In some forts the parapet was an extension of the scarp, with no sloped surfaces. In others, only one slope ran from the top of the scarp to the coping of the parapet. The classic shape, however, included a steep **exterior slope** joining a flatter **superior slope.**

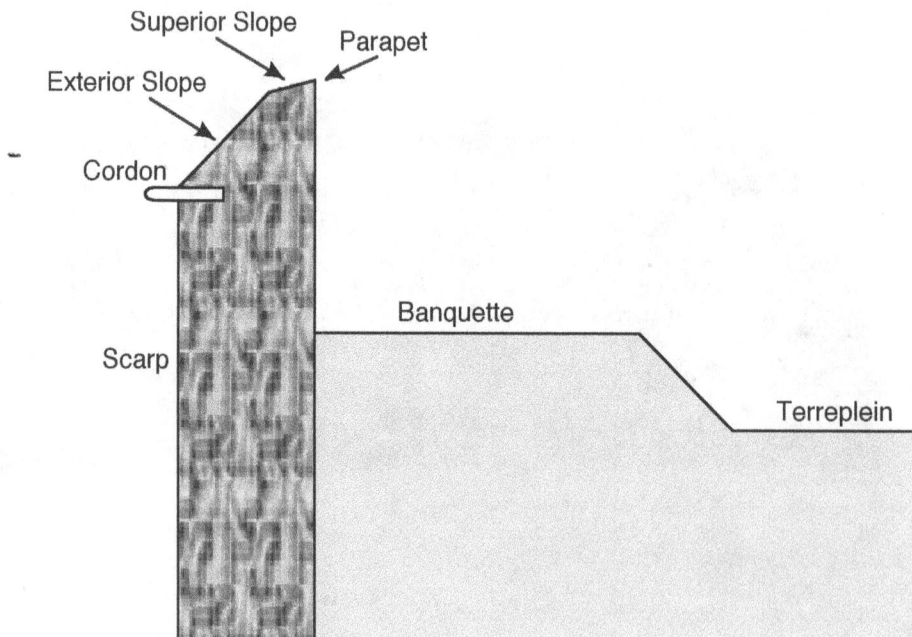

This sketch shows the typical elements in the enciente, or body, of a Third System fort. In the drawing, the attackers would be approaching from the left, with the parade of the fort to the right.

Immediately behind the parapet, on top of the terreplein, a wide, continuous step, or **banquette**, was generally provided. This platform allowed small arms to be fired over the top of the parapet, while a person not standing on the banquette was shielded from enemy fire. Also provided behind the parapet were elevated mounts for cannon, referred to as **barbettes** or barbette platforms. Cannon mounted on this level were said to be mounted en barbette. In rare cases, open-topped embrasures called **crenels** penetrated the parapet and provided more protection to the gun crews than the open barbette mounts.(14)

While the scarp wall formed a difficult obstacle to forcing entry into the fort, it also provided shelter for an attacker who came close enough to it. By standing very close to the scarp wall, the attacker was out of the trajectory covered by embrasures and loopholes as well as being masked from view from the parapet. **Bastions**, or projections from the fort walls, were designed to eliminate this avenue of shelter. These bastions, generally—but not exclusively—located at the corners of the fort, provided a view along the straight wall, or **curtain**, of the fort. The curtain is the portion of the scarp wall between the bastions.

When a bastion protruded from only one wall of a fort at a corner, it was called a **demibastion**, or "half bastion." Bastions and demibastions were always casemated, and were designed to mount the small, short-barreled howitzers mentioned previously. These howitzers fired canister shot along the walls to "sweep" them of an enemy who was seeking refuge against the scarp. Canister shot consisted of a container filled with small balls, ranging in size from three-quarters inch to balls about two inches in diameter. On firing, the canister ruptured, discharging the balls in a pattern similar to a shotgun.

The design of the bastions of a fort was a lesson in geometry. The sides of the bastion that attached to the curtain were called the **flanks** of the bastion. These flanks were at an angle of at least 90 degrees to each curtain, depending on the angle formed by the projection of the two curtains.

The **faces** of a bastion were the other two walls, which formed the point or **salient** of the bastion. The intersection of the face and the flank of a bastion is known as the **shoulder** of the bastion. The angle of the face of a bastion, if projected, met the curtain at the flank of the opposite bastion. This ensured that the cannon in the flank of the bastion could sweep the salient of the opposite bastion.

Since a bastion projected outward from the fort, it was the most vulnerable part of a fort. If overrun, it would provide an area for the attacker to assemble that was level with the terreplein. To control this area, a **cavalier** is a wall that divides the bastion from the remainder of the fort. It generally forms a continuation of the parapet as if the bastion were not present. In the Third System, Fort Jackson, on the Mississippi River south of New Orleans, had cavaliers at each bastion.

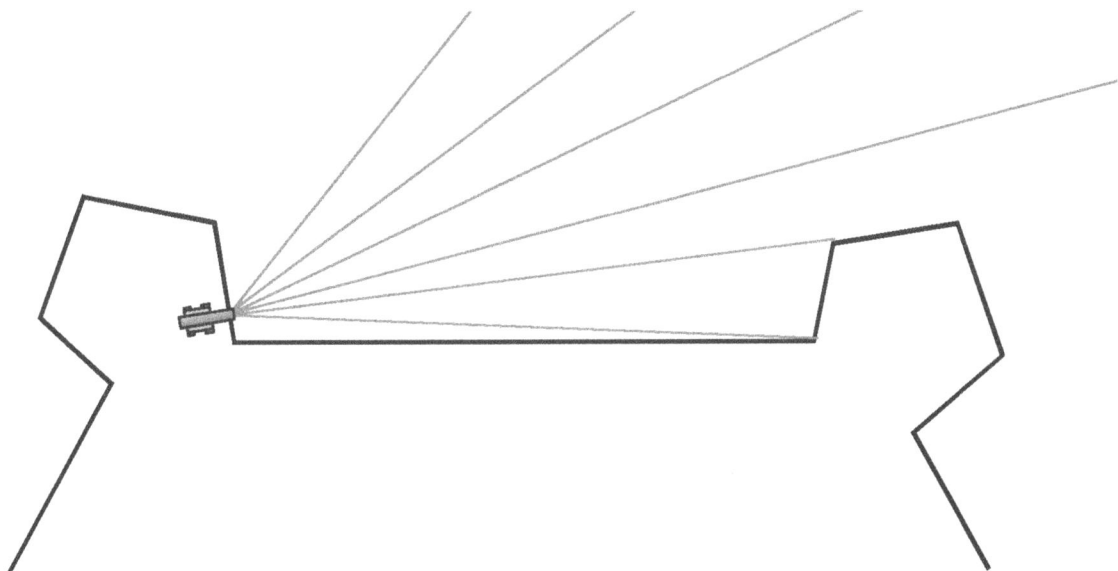

Bastions are designed to eliminate any points of refuge for an attacker, especially against the scarp of the fort. The geometric form of a bastion allows the opposite bastion to sweep its face as well as the curtain that joins the two bastions.

The citadel at Fort Pike, guarding New Orleans, Louisiana, is one of the few remaining citadels in the Third System. Its role as a barracks was less than successful because of the poor ventilation provided by the loopholes, and additional barracks were constructed elsewhere on the site. (MAB)

Inside the ramparts was a flat area, generally level with the lowest level of casemates.(15) This area, called the **parade**, was the primary place for the training and assembly of troops. The parade often housed barracks, and sometimes a strong defensive barracks called a **citadel**. These citadels generally had loopholes in the walls, and some were designed to mount cannon on top of them, en barbette. The citadel was considered to be the last line of defense if the remainder of the fort fell.

Around the outside of the scarp was usually a **ditch** or **moat**. While, by technical definition, a ditch can refer to either a dry ditch or a mud- or water-filled ditch, the term ditch generally is applied to a dry ditch, and moat is generally applied to a wet ditch. Both served as a physical obstacle to deter the approach of an attacker to the scarp wall. A dry ditch provided an obstruction to attackers descending into it, provided the scarp additional effective height, and increased the effectiveness of canister shot by providing surfaces from which it could ricochet. A wet moat would prevent (if deep) or slow (if shallow) an attacker wading across it, giving defenders more time to fire on him. A wet moat also inhibited mining operations against the fort. In both cases, earth from the ditch was used to form the glacis and build the outworks, as well as to level the top of the ramparts.

Some Third System forts immediately adjacent to the water provided a small sea wall around the perimeter of the fort to define the perimeter of a wet ditch. This sea wall would also serve the structural function of protecting the fort from wave action. A well-preserved example of such a sea wall is at Fort Jefferson in the Dry Tortugas.(16)

Defensive Outworks

Elements of a fortification in advance of the scarp wall are referred to as **outworks**. These took many forms, from simple earthworks to elaborate earth-and-masonry designs. Outworks had four principal functions: 1) to provide multiple lines of defense and fall-back points for defenders; 2) to provide additional firepower on the seaward side of the fort; 3) to guard the enceinte and ditch from attackers; and, 4) to protect the scarp from direct cannon fire.

The concept of *defense in depth* was strongly supported by most engineers of the smoothbore-artillery period, and was well exemplified by the works of the famous French engineer, the Marquis de Vauban (1633-1707). A major part of his theories involved the principle of multiple lines of defense. He believed that siege warfare was a battle of attrition, and that defenses should consist of progressive layers of fortification which would have to be overrun one at a time, at a high cost to the attacker. While it was understood at the time of the Third System that **any** fort would eventually fall to a protracted siege, fort designers used progressive layers of fortification as a delaying tactic until the siege could be lifted through outside reinforcements, or the price to the attacker became too high.

The outworks of a fort included all elements from the scarp wall to the structures furthest from the scarp wall. Many of these structures were located in the ditch, others were well beyond the ditch, up to the range of the main guns of the fort.(17)

The outworks provided additional landward defenses and protection from direct fire by artillery. At Fort Pulaski, (near Savannah, GA) a large ravelin surrounded by a moat covers the gorge scarp wall of the fort.

In this treatment, we begin with the outworks located the maximum distance from the fort and proceed back toward the landward facing scarp wall.

The detached, self-contained works generally located most distant from the enceinte were **redoubts**. A redoubt was simply a small, fortified work that had been constructed in advance of the main fort. This meant that an attacking force would come under fire from the redoubt before reaching the main fort. Classically, the defense of the redoubt raised the alarm that an attack was imminent. Scouting patrols could not reconnoiter the outworks or the enceinte of the fort without the capture of the redoubt, which would generally require more than a small force. In the Third System, the redoubt generally controlled topographical features – high ground – that could expose the main work. In both cases, a redoubt served as a first line of defense in the "defense in depth" concept. When the attack on the redoubt began, the war of attrition also began. The attacker must expend significant resources in the reduction of the redoubt, but possession of the redoubt does not give him any advantage in his assault on the main works.

The redoubt at Fort Adams, in Newport Rhode Island, represents the classic purpose of a redoubt. It was located on high ground on the peninsula that led to the main work, requiring the attacker to first reduce this work before proceeding to the other levels of defense. Two proposed redoubts at Fort Point in San Francisco were designed in a similar manner. They were not constructed.

The Advanced Redoubt at Pensacola Florida was built for an entirely different purpose. The Advanced Redoubt was designed to seal off one end of the land approach to the Navy Yard, with Fort Barrancas designed to seal off the other end. Rather than sitting in advance of the main work, Fort Barrancas, it was to cooperate with it. This factor, as well as it being similar in size to Fort Barrancas, has caused the Advanced Redoubt to be treated as a separate fort by this author.

An important feature in a redoubt was the design of its gorge – the wall facing the main work. To deny the attacker the establishment of a strong point in the captured redoubt, the gorge was designed such that guns could not be mounted on that face. The defenses of the gorge of a redoubt were limited to those that would prevent a coup de main, generally a ditch and demibastions.

The approach to the fortifications, redoubt, outworks, or main work, was the **glacis**, a gentle slope cleared of all trees and obstacles, thus providing a clear area for fire from the fort's guns. The slope of the glacis was carefully calculated to provide maximum protection to the scarp wall while providing an area that was raked by the barbette guns of the fort. A line projected from the top (superior) of the parapet, across the ditch to the crest of the glacis, would generally define the slope of the glacis. From a distance, this would make the earthen parapet of the fort appear to be an extension of the slope of the glacis.

In addition to providing a clear field of fire for the guns of the fort, the glacis also provided a significant amount of protection for the walls of the fort. The earthen slope hid almost the entire scarp of the fort from land-based cannon, making it very difficult for shot to strike the more vulnerable masonry walls. Instead, most of the shot was absorbed in the soft earth of the glacis or parapet.

Immediately toward the fort from the glacis outer slope was the parapet revetment and sometimes a banquette. The revetment was generally masonry, but earth or earth-and-wood construction was also used. Some forts had barbette mounts for cannon in addition to the banquette for small-arms fire. Behind (and lower than) the banquette was a large pathway, called the **covert way** or covered way. This pathway was hidden from the attackers, but covered by the guns of the fort. It allowed communication along the perimeter of the outwork, and often opened into a **place d'armes** or place of arms. A place d'armes was a gathering place for a counterattacking force, and was referred to as a salient place d'armes if at an outward-pointing angle in the outworks or a reentering place d'armes if at an inward-pointing angle in the outworks. The places d'armes were fronted by the same banquette, parapet, and glacis as the covert way.

Immediately behind the covert way was the **counterscarp wall** or slope that formed the side of the ditch opposite the fort. Communication between the banquette and the ditch was provided by stairways and/or ramps that led from the ditch to the covert way.

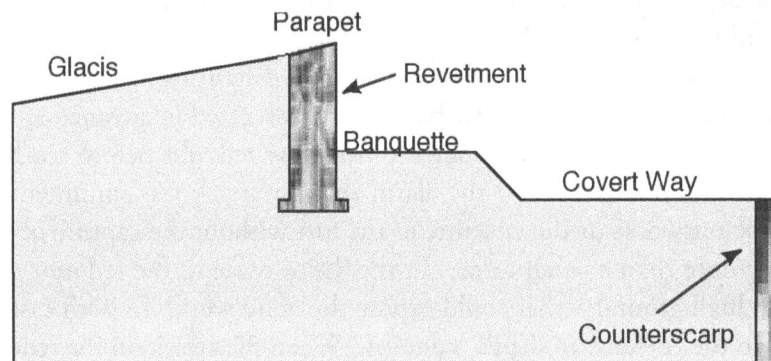

This sketch shows the typical elements in the outworks of a Third System fort. The glacis gently slopes toward the route of attack, while the ditch is to the immediate right of the drawing. Defense of the ditch was provided for by additional outworks, by elements of the main fort, or both.

While in most Third System forts the ditch was either empty or filled with water, some of the forts provided a further level of defense within the ditch. Large detached **crownworks** or **hornworks** were positioned in advance of the walls of the fort. A crownwork consisted of a central bastion with two flanking demibastions, joined by masonry curtain walls.(18) A hornwork consisted of two demibastions joined by a masonry curtain. Both provided flanking fire as well as forward fire from casemates in the bastions and demibastions. In addition, they often had cannon mounted en barbette for forward fire.

Tenailles, small detached works within the ditch and between the bastions of a fort, protect the curtain of the fort from direct fire. In the Third System, the only tenailles also provided flanking fire in the ditch and forward fire against an attacker who had crossed the parapet of the outworks. These tenailles consisted of two casemated demibastions joined by a masonry-and-earthen curtain. Fort Adams, Newport, Rhode Island, is a classic example of the use of a crownwork with tenailles for land defense, while Fort Schuyler, New York City, has a classic hornwork.

Another type of outwork designed to control the ditch through the use of enfilading (or flanking) fire was the **caponier.** A caponier was a masonry gunroom running perpendicular (or nearly perpendicular) to the ditch, with loopholes and/or embrasures in each side. It was often a projection from the scarp wall, and could be accessed directly from the fort. Detached caponiers could be accessed either by a tunnel from the fort or by a separate entrance from the ditch. Caponiers were generally used to defend the ditch of unbastioned forts or the unbastioned sides of forts.

Main Fort

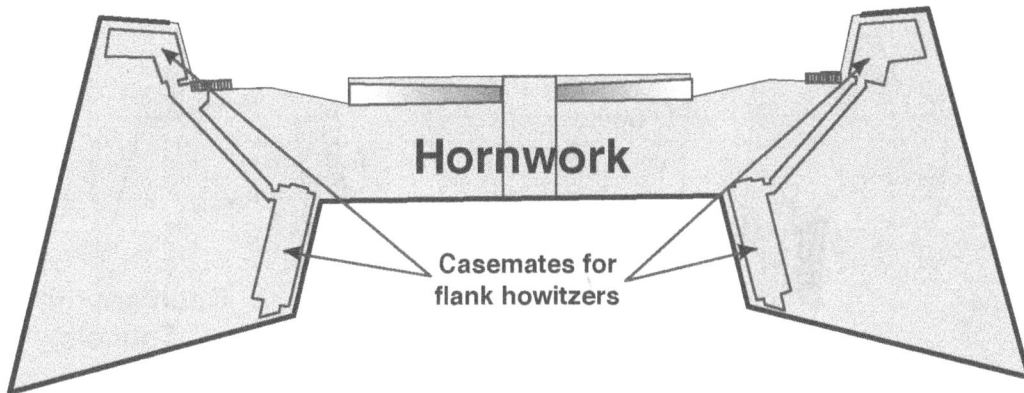

Hornwork

Casemates for flank howitzers

A hornwork is an elaborate land defense that divides the ditch into two distinct parts—an inner ditch and an outer ditch. Its major role is defense in depth, and - because of its massive size - the assault of a hornwork is very similar to the assault of the main work. The above sketch is based on the hornwork at Fort Schuyler, guarding the northern approaches to New York City.

The most extensive type of outworks designed to guard the scarp wall and control the ditch was a **counterscarp gallery**. A counterscarp gallery, also called a counterfire room, was a masonry vault behind the counterscarp wall equipped with loopholes and, often, casemates for howitzers. These galleries, completely protected from artillery fire and any other forward fire against the fort, provided a crossfire with the fort on an attacker who has entered the ditch. At the salient angles of the counterscarp were generally casemates with howitzer embrasures, which provided enfilading fire down the ditch.(19) Counterscarp galleries and casemates were entered either from the ditch or from the fort through a tunnel under the ditch.

An additional outwork seen in some Third System forts were **listening galleries** or **countermining tunnels**. These consisted of masonry-revetted tunnels that extended beyond the scarp, and often beyond the ditch. They were designed to allow defenders to detect mining operations by listening. If mining operations were detected, openings in the walls allowed charges to be placed in the direction of the mining operation. These charges would be detonated, bringing down the mine. Additional chambers under the bastions would allow the bastion to be reduced if it fell to the enemy, killing the enemy emplaced there. Fort Adams in Newport Rhode Island and Fort Pickens in Pensacola Florida have excellent examples of these chambers.

The **demilune** and **ravelin** are members of another family of outworks that may be located in the ditch or beyond the ditch. In classic fortifications, the terms are synonymous, but Third System engineers distinguished between them. They were both freestanding, open-backed works consisting of a parapet with banquette and/or barbette platforms.(20) The demilune, literally translated as "half moon" from French, was curved while the ravelin was an open-backed triangle.

Third System ravelins were generally constructed opposite a front, with the open end of the ravelin approximating the length of the curtain. They usually had cannon positions on both fronts of the ravelin.

Third System demilunes were generally smaller, with the smallest one consisting of a curved rifle gallery with no cannon positions. Larger demilunes included cannon positions.

Ravelins and demilunes were commonly located opposite the gorge wall of the fort, providing defense of the sally port, but were also on seacoast fronts to provide additional cannon for seacoast defense.(21) Prior to the Third System, ravelins were often as tall as the scarp of the fort, masking a curtain wall. Access to these ravelins was via bridges across the ditch. By the Third System, most ravelins and demilunes were constructed at an elevation even with the parade.

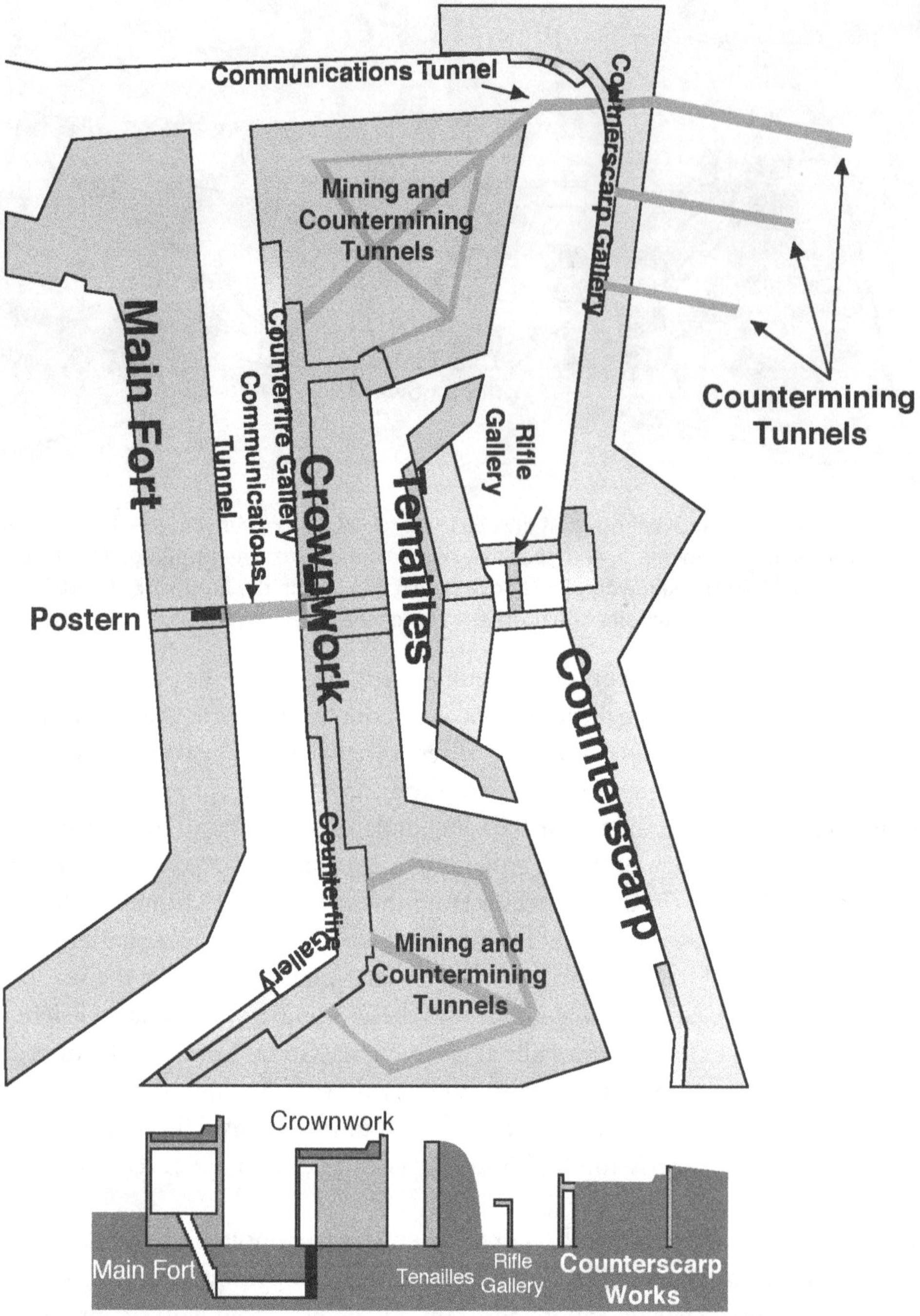
Communications Tunnel
Counterscarp Gallery
Mining and Countermining Tunnels
Main Fort
Counterfire Gallery
Communications Tunnel
Crownwork
Tenailles
Rifle Gallery
Countermining Tunnels
Postern
Counterscarp
Counterfire Gallery
Mining and Countermining Tunnels
Crownwork
Main Fort
Tenailles
Rifle Gallery
Counterscarp Works

The amazing land defenses of Fort Adams, guarding Newport, Rhode Isand, were the most extensive in the Third System. Most impressive to today's visitor is the network of tunnels that provide communication between defensive elements and provide mining and countermining capability. Also note that each outwork is progressively lower in height, allowing all elements to be used simultaneously.

Outside the ditch of a fort were another group of works, constructed on the channel-bearing front and designed entirely for seacoast defense. These works were designed to mount additional seacoast guns to supplement the firepower of a fort. **Water batteries** were freestanding works designed to add additional cannon to the seacoast fronts of a fort, while **coverfaces** were works that protected the scarp of the fort while providing additional cannon emplacements. These works were either of masonry, masonry-revetted earth, or all-earth construction. Two of these structures remain, the coverface of Fort Warren in Boston Harbor and the water battery of Fort Barrancas in Pensacola, Florida. Unfortunately, the two most unique water batteries have been lost. The fully casemated water battery at Fort Monroe was removed during later periods of construction, and the detached-scarp (Carnot wall) water battery at Fort McRee was lost with the fort due to coastal erosion.

Integration of the Design

During the Third System, basic design elements were integrated in many unique combinations to provide for the defense of particular locations. While the elements themselves evolved during the period, so did the relative importance of the functions that drove their implementation.

Third System coastal fortifications were charged with the six functions outlined in the 1821 Report of the Board of Engineers,(22) but these can be summed up in one goal: to secure a particular harbor or waterway against a hostile naval force. In order to meet this goal, a fort had three basic defensive missions: 1) to guard against attack by ships at sea; 2) to preclude the possibility of a coup de main; and 3) to provide a resistance to a land-based siege. The structural response to these three missions manifested itself in the implementation of the different architectural elements that were described in the previous portion of this chapter.

Since the primary purpose of the Third System forts was to prevent passage of ships into a harbor or near a city, the forts were generally located at a narrows in the channel; this is why so many Third System forts now lie in the shadows of a major bridge.(23) A narrow channel forced ships closer to the guns of the fort where they *literally* had more impact, gave the ships less opportunity to maneuver, and caused the ships to be in the range of the guns of the fort for a longer time. To take full advantage of this effect and therefore maximize their ability to deny passage to ships, the seaward faces of the fort would mount a maximum number of cannon. For most forts, this was achieved through stacking cannon tier over casemated tier, in a similar manner to the way a ship-of-the-line stacks guns deck over deck.

Tiers of Cannon in Third System Forts

One Tier	15 Forts
Two Tiers	15 Forts
Three Tiers	10 Forts
Four Tiers	2 Forts

The front of a fort that most directly bore on the channel was referred to as the **primary seacoast front** of the fort. Depending on the location of the channel in relation to the fort – and whether multiple channels were involved – there could be one or more primary seacoast fronts.

Supporting the primary seacoast front were fronts that bore on the channel at a less-ideal angle. These were referred to as **secondary seacoast fronts**. They would generally provide fire on a ship before or after the ship passed the fort, increasing the amount of time that a ship was under fire from the fort.

As the Third System progressed, so did the technology of warship construction. At the beginning of the Third System, capital ships were constructed of wood, powered by sail, and contained multiple decks of light, smoothbore cannon.

By the end of the system, however, all these changed. The wooden ships were armored with iron, sails gave way to steam-driven engines, and multiple tiers of small cannon were replaced by small numbers of large-bore, rifled guns. These were often placed in turrets to allow firing in all directions.

To combat these developments, the designs of the forts also evolved. The number of tiers of guns increased, making it possible to keep a high rate of fire on increasingly faster steam-powered ships. At the beginning of the Third System forts, especially when coupled with floating obstructions and torpedoes (mines), could deny passage through a narrows. The Civil War, however, proved that forts alone could no longer deny passage to fast-moving ships.(24)

Land defenses for Third System forts lessened as the period progressed. It was found that the basic design of the enceinte was sufficient to prevent a coup de main, and that even relatively simple land defenses would require the attackers to mount a protracted siege, and reduce the work through artillery fire. This reduced the need for extensive land defenses designed to repel infantry attacks, requiring only outworks sufficient to shield the scarp from artillery.

The front of the fort that provided the primary land defense of the fort was called the **gorge**. This front, often containing the main sally port, was the most vulnerable front in a siege, and required the most extensive protection of the masonry.

Early Third System land defenses were quite extensive. Defending New Orleans, both Forts Pike and Wood sported two moats, with a glacis, parapet and banquette between them.(25) At Newport, Rhode Island, Fort Adams boasted a major crownwork, with two tenailles, in addition to counterscarp galleries and the conventional glacis, parapet, and banquette. Fort Schuyler, defending the northern approach to New York City, had an impressive hornwork, complete with casemates. Fort Caswell, in North Carolina, had six caponiers, and the original design of Fort Morgan, Mobile, Alabama, included three caponiers in addition to the five bastions of the enceinte, though none were built. Fort Tompkins (Staten Island), Fort Macon (North Carolina), and Fort Barrancas and the Advanced Redoubt (Pensacola, Florida) had full counterscarp galleries, including emplacements for howitzers in the counterscarp. Both Fort Barrancas and the Advanced Redoubt communicated with the galleries through tunnels under the ditch, as did Fort Adams.

A classic, but unique to the Third System, application of separate land and water defenses was at The Narrows in New York Harbor. Topographical limitations drove the fort designers to build separate land-defense and water-defense forts, rather than combine both elements in a single structure.

Fort Lafayette, a transitional work, was a casemated water-defense fort on a shoal in The Narrows, but it had bare masonry walls that would be vulnerable to land-based batteries on the shore. Fort Hamilton was constructed immediately ashore of Fort Lafayette, and while it had several guns which would be used for water defense; it was primarily a land-defense fort in support of Fort Lafayette. Likewise across the channel, Fort Richmond was built as a tower battery with four tiers of cannon to seaward but minimal defense to landward. Fort Tompkins was built on the high ground above Fort Richmond to provide land defense. In this way, two land defense forts and two water defense forts could be designed and constructed with a purity of purpose and not have to compromise their design to a mixed mission.

An interesting note is the counterpoint between the Corps of Engineers and the leadership of the soldiers who were to man the forts. While the Corps of Engineers worked very hard to design structures that were readily defensible, in many cases they did not expend a commensurate effort on living quarters. The garrisons often complained that the citadels and the casemate quarters were very inhospitable.(26) In some cases, the living conditions were blamed for sickness and death among the garrison.(27) While some concessions were made, engineer reports indicated that the Corps lacked interest in dealing with these problems and the troops were frequently forced to construct temporary quarters outside the confines of the fort. These structures were usually wooden, and were to be burned upon attack of the fort by an enemy. Ironically, at Fort Pike these "temporary" buildings got out of hand, with gun emplacements on the ramparts of the fort being taken up by a commandant's quarters, and fields of fire completely blocked by barracks structures.

In some forts however, comfortable and sometimes elaborate barracks buildings were constructed. These engineer-designed buildings, generally masonry, were designed specifically as quarters and served that function well.

Later in the Third System, loopholes were often broken out to form window-like openings that would provide better ventilation. In the citadel of Fort Alcatraz in San Francisco Bay, the openings in the wall were made at full window width, with bricks stockpiled to close them to loopholes should a threat present itself. At Fort Warren, Boston Harbor, and Fort Gorges, Portland Harbor (see descriptions), extensive efforts to reduce dampness were undertaken by constructing double walls of masonry with heat from the fireplace ducted between them. While these efforts made some progress, the healthiest living conditions throughout the period were in separate barracks buildings rather than dual-purpose structures.

Location of Third System Forts

Four general locations were chosen for Third System forts: islands, shoals, shorelines or riverbanks, and hilltops. Which of these locations were used depended on the topography of the land and on the characteristics of the waterway they were designed to secure. Shorelines and riverbanks were the most common locations, and were the site of nearly half of the Third System forts. These twenty forts were divided between six riverbank locations, thirteen shore sites and one inland location.

Of nearly equal proportion were islands. Eighteen Third System forts occupied island locations, with some dominating a small island while others were located at one end of a large island. Shoals came next, with four forts built on moles (artificial islands) in these shallow areas. Finally, only three forts were located on hilltops. Two of these forts were discussed previously, Forts Hamilton and Tompkins, with the third being Fort Trumbull in New London, Connecticut. These three forts also meet the shoreline or riverbank criteria, and are therefore double-counted.

There is a tactical reason why Third System forts were usually not located on "high ground." This reason was the practice, common in this period, of skipping cannon balls along the water surface en route to a target. It allowed for significant error in the calculation of the distance to a ship, as the ball would seek the first projection over the water it found. To accomplish this, the cannon had to be fairly close to water level.

A second item in regard to the location of Third System forts is that they were generally some distance from the city they protected. This was due to the desire to allow as much time as possible to muster a secondary defense of the city. The longer a distance an invader had to travel after laying siege to a fort, the more time was allowed for the transportation of more defensive forces and/or the "calling out" of a militia defensive force. This drove the Third System designers to locate the forts farther from the population centers when geography allowed.

An additional factor was that the longer range of the artillery by the time the Third System had begun. A fort had to be located farther from a city for that city to be beyond the range of a ships guns, and the longer-ranged guns of a fort allowed a wider channel to be protected than was previously possible. This wider channel was generally further from a population center and closer to the sea.(28)

Finally, Third System forts were usually located very close to the water. Engineering had sufficiently matured by this time to allow the construction of these massive forts on less-than-ideal soil systems.(29) By taking advantage of this technology, designers were able to maximize the effective range of their cannon to block a passageway, and to provide fire on an attacking vessel for a longer period of time.

The decision of what places needed to be defended on the nation's long coastline was laid out in the 1821 report, but "pork barrel politicize" played a role in modifying this philosophy. The dominant factor, however, was when the Navy was involved. The Navy had significant backing in Congress, and when a fort or series of forts were needed to protect a naval installation, the funds were always forthcoming. Both Bernard and Totten were astute enough to pick up on this fact and were careful to support the desires of the Navy in the formation of projects for forts.

By the close of the Third System, a very distinct "clustering" of new-construction forts could be noted. The coast from just west of the mouth of the Mississippi to Pensacola, Florida, had the largest cluster, with eleven forts filling this rather small area of coastline. Two forts guarded the Florida Strait at Key West and the Dry Tortugas, and forts from the Florida-Georgia border were fairly evenly spaced up the long coastline to New York City. New York had a cluster of six forts in a very small space, with four more filling the relatively short distance between New York and Boston. Five more forts were clustered on the coast of

New Hampshire and Maine, from Portsmouth to Prospect. Finally, two closely spaced forts guarded the harbor at San Francisco.

Note that the forts cited were new construction only. These were supplemented by renovation of existing forts, sometimes very substantial, by defensive towers, and by smaller batteries. Together this formidable line of defense incorporated a set of works that numbered more than 150.

Sizes of Third System Forts

Certainly the size of a fort can be considered proportional to the perceived importance of the harbor that it protected, but many other factors were considered in the sizing of the forts of the Third System.

These forts, as discussed earlier, were to be garrisoned by only small peacetime units until a war was imminent, then the garrisons reinforced with local militia artillery. This meant that fort size would depend on the number of militia artillery available from the populace to garrison the fort in time of war. Forts Monroe, Adams, and Pickens were designed to hold relatively large permanent garrisons. These "headquarters" forts had many functions other than defense that were implemented within the enceinte, and they required a large area to carry out these activities. Additionally, these forts were near significant population centers that could supplement the permanent garrison in time of war.

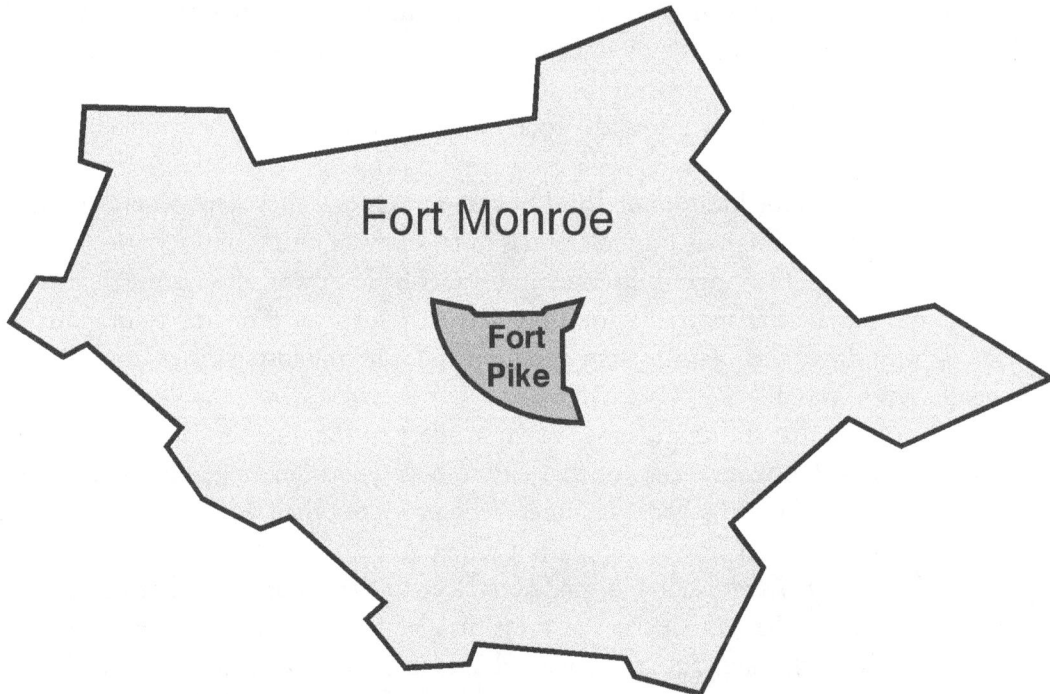

As shown in this scale drawing, the sizes of Third System forts varied dramatically. Fort Monroe, the largest of the system, is shown here with Fort Pike, which began construction in the same year. While most Third System forts fell between these two in size, there was a very wide variety in area, armament, and garrison.

Also of consideration was the number of guns that would be needed to defend the site. Fort Jefferson, designed to be the most heavily armed fort in the history of the United States, mounted three tiers of cannon around the perimeter of its very large trace. Away from any population centers, it relied on a large permanent garrison to control this crucial location in the Florida Strait.

As construction technology improved through the course of the Third System, multiple levels of casemates were used, allowing forts of smaller trace to mount the same number of guns. Fort Richmond with four tiers of guns and Forts Taber and Popham(30) with three tiers were of relatively small trace, but sported large numbers of cannon. The local soil, however, played a major role in the number of casemates that could be stacked. This was a difficult lesson learned at both Fort Calhoun (located on a shoal) and Fort Pulaski (located on a mud island). Although designed for multiple levels of casemates, neither foundation was able to support more than one casemate tier and one barbette tier of cannon. Many other forts, however, were able to support many tiers of guns through proper foundation design. Fort Delaware, for example, boasted four tiers of cannon on a foundation of timber piles and grillage.(31)

Shapes of Third System Forts

The forts of the Third System were the last closed forts in the United States. Their complex geometrical shapes were the accumulation of fortification knowledge spanning at least two centuries, with contributions from engineers from several countries. Subsequent fortification efforts (32) involved the use of multiple detached batteries, not at all "forts" in the classical sense.

The first consideration involving the design of a fort was the location it was to defend, driving its overall shape. The major factors in this involved the position of the fort in relation to the channel or channels it was to defend, and the maximum arc of the guns mounted *en casemate*. While center-pintled barbette guns had very large angles of rotation, the shape of the embrasure and the relationship of the fore-pintled carriage to that embrasure governed the maximum traverse of a gun. With the goal of the fort to apply as much firepower to the channel as possible, it was desirable to have multiple fronts bearing on a given channel.

A counterpoint to this idea was the size of the fort. The larger the angle between walls, the larger the interior of the fort became, and the area consumed and the cost of construction went up proportionately. Several early Third System forts were built on the trace of a regular pentagon, the classic shape of a fort of the French school. This was a compromise of size and angle of the fort wall. Later, the designs shifted to truncated shapes, with the gorge wall on the landward side treated differently from the seacoast fronts. This changed the rules: in an unrestricted setting, a regular figure of shallower angle could be used without increasing the size of the fort. A 60° angle of traverse of the guns was common, which promoted the use of 120° as the angle between adjacent faces of the fort. This is the angle of a regular hexagon; therefore many Third System forts were built on the trace of a truncated hexagon.

The amount of truncation of the polygon varied according the desired size of the fort and the number and position of channels to defend. A five-sided plan was most common, with a single gorge wall connecting the two secondary seacoast fronts.

In significantly smaller forts, however, a four-sided plan was employed. In this design, one primary front faced the channel and two secondary fronts came off at 120° angles. The gorge wall then connected the ends of these secondary fronts. This left a very small parade, and made the use of conventional bastions guarding the gorge inconvenient. The size and angles of a conventional bastion would be very awkward, and demibastions were used. Also, the importance of a bastion guarding a seacoast front in the latter Third System was much less and may have contributed to the decision to use demibastions. The demibastion would continue the seacoast front beyond the gorge wall, then turn nearly parallel to the gorge. A flank of the demibastion, perpendicular to the gorge wall, would provide defense of the gorge wall and sally port through the use of howitzer emplacements.

In rare cases, a fort was designed with all sides acting as fronts, with no gorge. Fort Jefferson, located on an island with deep water surrounding it, was a classic example of this design, having six faces and no gorge. Other forts, especially those forming a regular geometric figure, would have multiple landward faces. The two regular pentagons, Fort Jackson and Fort Morgan, each had two landward faces.

The true exercise in geometry came about with the design of the bastions and their relationship to the connecting curtain wall. The purpose of a bastion was to provide flanking fire down the length of the curtain and the face of the opposite bastion, thus leaving no area around the perimeter of the fort uncovered. To achieve this purpose, the flank of the bastion was formed at greater than 90° to the curtain and provided embrasures for howitzers and sometimes loopholes for rifles.(33) When constructed, the face of a bastion is on a line that terminates at the point where the flank of the opposite bastion meets the curtain. This geometric exercise assured that no area on the face of a bastion was outside the fire of the howitzers of the opposing bastion.

In Third System forts of very large trace, the design of the bastions changed. In conventional Third System forts, the bastions were closed-back structures with relatively narrow throats protruding from the salient angles of the trace. The bastions were entered from the adjacent casemates or from the parade, but usually with a clear separation from the parade.

In three very large Third System forts, Fort Monroe at Hampton Roads, Virginia, Fort Adams at Newport, Rhode Island, and Fort Warren in Boston harbor, the bastions were open structures with parade-like areas in the middle. These areas were open to the sky, unlike the interior of conventional Third System bastions.

Viewed from the parade, these bastions appeared to be angles in the fort wall. A row of casemates followed the scarp wall along the flank of the bastion. In the case of Fort Adams and Fort Warren, the faces of the bastion also contained casemates and embrasures. In Fort Adams, these embrasures were for seacoast cannon to increase the firepower of the seaward front. Fort Warren had seacoast cannon embrasures in its channel-bearing faces and loopholes in the bastion faces that secured the coverface, ditch, and rear glacis.

Although the appearance of these very large bastions varied from the smaller bastions employed in most Third System forts, the function was the same. The flanks of these bastions were pierced with howitzer embrasures, and Fort Adams actually had howitzers mounted en barbette atop the gorge demibastion. The faces of these bastions also followed the geometric rule mentioned previously, allowing them to be washed by howitzers. They also served as extensions of the curtain wall to increase the complement of barbette guns.(34)

Another major consideration relative to shape of a Third System fort was the available area and its physical features. Fort Point in San Francisco was shaped to fit on the very small amount of land available near the water. In fact, when it was found that shallows existed on either end of the point, the trace of the fort was modified to use these areas as well. Fort Warren in Boston Harbor was designed as a regular pentagon "squashed" to fit on Georges Island. Fort Jefferson on Garden Key in the Dry Tortugas was built in the general shape of the island on which it was located. This resulted in a regular hexagon that was shortened on two parallel sides. Fort McRee, on Foster's Bank near Pensacola, Florida, was built very long and narrow to fit on the narrow island.

Fort Alcatraz was the ultimate in regard to adapting a fort to its physical surroundings. It was built using the physical features of the land, with some blasting to make some cliffs steeper and the addition of masonry walls in some locations, for its enceinte. In this way the island and the fort were inseparable.

Finally, the unknown prevails. Why did some forts have sharp angles between adjoining seaward faces while others were rounded? Why were some forts built as truncated hexagons and others of the same period as truncated octagons?

It is interesting to note that there were two pair of forts with a shared design, and one set of three forts shared a common design.(35) Only two of the three forts sharing the same design were constructed as such, Fort Pike and Fort Wood. Fort Livingston was to be the third of that design, but delays in construction allowed new technologies to be employed, and it was completely redesigned prior to its construction. The two pair of forts coincidentally includes Fort Gaines in both. Originally designed as a twin to Fort Morgan, construction was halted after only foundation work had begun. When construction was resumed many years later, the revised design was the same as the more modern Fort Clinch.(36) They shared a unique design, utilizing the "detached scarp" concept. This left the Third System with 40 individual designs, each with its distinctive touches.

Other Structures

While the heart of the Third System was in the group of 42 newly constructed forts, the success of the system depended on the integration of these forts with other subsidiary structures. These subsidiary structures were used to protect lesser channels, to assist the newly constructed forts in the protection of major channels, and to provide a second line of defense for important harbors.

These structures fell into three primary categories: older forts modernized and/or preserved as part of the system, defensive towers, and masonry-revetted earthen batteries. The size and shape of these structures varied greatly, from very large, relatively modern forts to very small Martello Towers designed to mount one or two cannon. As with the newly constructed forts, their designs were appropriate to their function in the overall protection of a harbor.

The modernization of existing forts consisted of changes needed to accommodate the larger cannon prevalent in the Third System, and the modification of existing designs that were not up to Bernard's (and later Totten's) standards. A classic example of the latter took place at Fort Washington, guarding the Potomac River south of Washington, D. C. This transitional fort had been designed and construction began before Bernard came to the United States. When inspecting the fort, Bernard was concerned about the absence of defenses on the rear of the fort, facing the deep ravine that separated the fort's promontory from the Maryland countryside. To remedy the shortcomings, Bernard modified the trace of the fort to include two bastions and added a caponier at the midpoint between the bastions.

At Fort McHenry, Bernard was concerned about the rather weak sally port and the design and locations of the supporting ravelins. He modified the fort with a conventional sally port and rearranged the ravelins and supporting batteries.

Defensive Towers

Defensive Towers became popular following the successful defense of the Mortella Point in Corsica by three small cannon[37] in a defensive tower. It is believed that the popular name Martello Tower is a corruption of the Mortella Tower that held off the British Navy.

Seven defensive towers were constructed during the Third System. Some of these towers, such as Tower Dupré, were very similar in design to classic Martello Towers and were referred to by that name. Other towers, such as Fort Proctor and Fort Winthrop, were much larger structures that reflected the general principles of a Martello Tower but departed from the classic design. These were generally referred to by the more general term towers or sometimes defensive towers.

The original Castle Pinckney at Charleston, South Carolina, the Martello Tower on Tybee Island, Georgia, and Tower Duprè near New Orleans, were the three towers that approximated the design of the classic Martello Towers of Europe. The towers at Key West were square towers that departed from the design of these classic Martello Towers, but were referred to by that name. They also departed significantly from the Martello Tower concept with the addition of a surrounding scarp, ditch, and external batteries. The large, square tower of Fort Winthrop differed most from the Martello Tower design and concept in size, strength, and the presence of external batteries. Fort Proctor, near New Orleans, was very similar in design to Fort Winthrop, but lacked the external batteries. Several other towers were planned, and some were designed, but they were not constructed.

Batteries

The general term batteries was used to describe a number of structures that differed greatly in size and complexity. The more general term works also referred to these structures.

The batteries referred to here are stand-alone defensive structures, not the external batteries under the protection of the main fort. Those batteries were described earlier in this section.

Batteries in their simplest form were masonry-revetted earthworks. These consisted of a glacis terminating in a parapet, with barbettes located behind the parapet. These were either linear, or more often V-shaped, essentially designed as a ravelin. The angle of the V was, however, usually much less acute than a ravelin.

Where a battery extended over a projection of land – such as at Lime Point near the northern terminus of the Golden Gate Bridge – it might form a multiple-sided, open-backed structure.

In there most complex form, batteries consisted of closed geometric forms, even incorporating a sally port. A battery at the mouth of the Columbia River, now part of Fort Stevens, was a truncated hexagon with a sally port in the "gorge" wall. It was also protected by a surrounding ditch.

In the Third System, batteries were masonry-revetted earth[38] and were not casemated. Usually, the only interior structures in these batteries were earthen-covered magazines. Battery Bienvenue outside New Orleans – the most impressive Third System battery – had a closed form with barracks inside the enceinte.

Cannon at Third System Forts

While the purpose of this section is to discuss the design and construction of Third System forts, a better understanding of the forts can be achieved by understanding the weapons these forts housed. This is in no way intended to be a detailed look at the artillery of the period, but a brief overview of the cannon typical in Third System forts.

Seacoast Cannon

The primary armament at Third System forts was the seacoast cannon. These cannon were the weapons called upon to control the waterways adjacent to the fort, to provide a defense against ship borne attack, and to combat a siege. They were cast iron muzzle-loading artillery – both the powder and the ball were loaded through the muzzle of the cannon.

During most of the Third System – until the early 1860s – all cannon were smoothbores. This means that the cannon fired a spherical ball and that the interior of the gun tube had no rifling. The introduction of rifled cannon added lands, or ridges, to the inside of the barrel that caused the projectile – now cylindrical – to spin.

This modification served two purposes. First, the spinning cylindrical projectile had significantly greater accuracy than the spherical ball fired from a smoothbore cannon. Second, the windage – the difference between the diameter of the projectile and the bore of the cannon – was reduced. This caused more of the force of the powder explosion to drive the projectile, increasing both the range of the cannon and the size of the projectile.

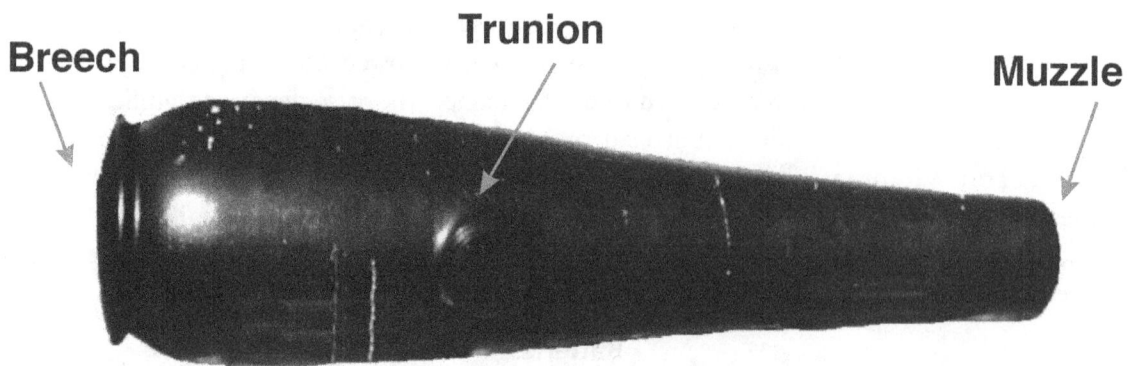

The above diagram shows the basic parts of a cannon from this era. The majority of the cannon used in Third System forts were iron - falling between the older brass cannon and the yet-to-be-developed steel cannon.
(Photo by author)

A Columbiad cannon on a fore-pintle mount. This type of carriage and mount were very common in the Third System, used both en barbette—as shown here—and en casemate. (MAB)

A Parrott Rifle is shown here on a forepintle mount

A 10-inch Rodman cannon on a center-pintle barbette mount. Note the iron carriage used for this massive gun, as opposed to the wooden carriage on the Columbiad shown above. (MAB)

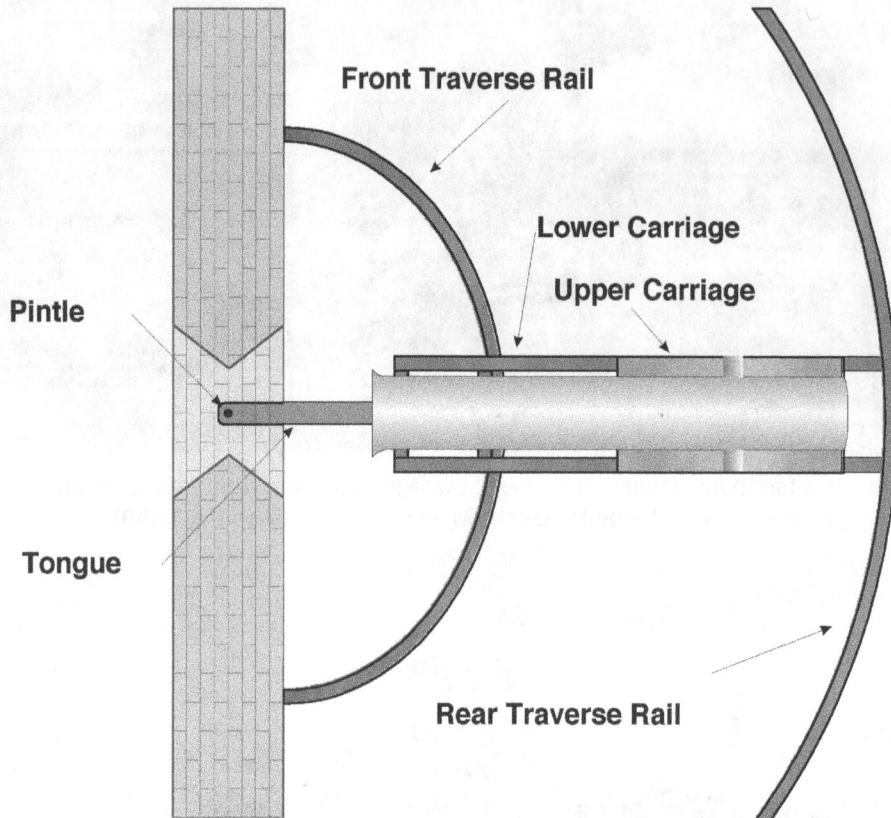

The fore-pintle mount, later referred to as front pintie, was designed to give a maximum angle of traverse in a minimum area. Placing the pivot point at the narrowest point of the embrasure accomplished this goal.

About the same time that rifled cannon were being introduced, Thomas J. Rodman was making a major improvement in the methods of designing and manufacturing smoothbore cannon. The result of his development, the Rodman cannon, was a series of cannon that had larger bores and longer range than the previous generation of seacoast guns.

Rodman's method was essentially the solving of a physics problem. When traditional cannon were cast, the tube cooled from the outside surface to the inside of the casting, thus creating an expansive (outward) stress in the metal. Thus the stress was in the same direction as the forces encountered when firing the cannon. Over time, the metal would fatigue and the gun would explode on firing. Adding additional thickness to the metal increased the amount of stress, providing little benefit.

Rodman developed a method of cooling the bore of the cannon during the casting process, thus creating a compressive (inward) stress. The stresses in the metal were now opposed to the stresses encountered during firing. Less metal fatigue resulted, and made feasible the casting of thicker, and therefore larger bore, cannon.

The second step in Rodman's approach was to calculate the force vectors present when the cannon fired. If these vectors were plotted with the base of each vector along the bore of the cannon, the tip of the vector arrows defines a smoothly curved tapered form. This form is the exterior of a Rodman cannon.

Both types of smoothbore cannon – the pre-Rodman design and the Rodman cannon – fired cast iron spherical shot, solid iron balls, and shell, hollow iron balls filled with an explosive charge.

Both types of cannon went under the general name *Columbiad.* The pre-Rodman design was generally classified by the weight of the solid shot fired from the cannon. Rodman cannon were generally classified by the diameter of the bore. Thus the pre-Rodman seacoast cannon present in Third System forts were generally 32-pounder and 42-pounder cannon, and the Rodman cannon were generally 8-inch, 10-inch,

The most impressive weapon used in Third System forts was the 15-inch Rodman cannon. Weighing up to 66,000 pounds - not including the massive iron carriage - these weapons were generally mounted on center pintles. This provided maximum flexibility in the use of the weapon by allowing a full 360-degree traverse.
(MAB)

Two of the common types of shot used in howitzers are grape-shot, shown on the left, and solid shot, shown on the right. This round of solid shot is shown attached to a sabot, a wooden "donut" that fixes the position of the shot in the bore of the cannon.
(Photos by author)

and 15-inch cannon. A 32-pounder had a bore just under 6-1/2 inches and a 42-pounder had a bore of just over seven inches. An 8-inch shot weighed approximately 45 pounds, a 10-inch shot weighed approximately 90 pounds and a 15-inch shot weighed approximately 300 pounds.

Rifled cannon increased the range and weight of the projectile for a given bore size. An 8-inch Parrott rifle fired a 200-pound shot, more than twice the weight of the shot fired by a 10-inch Rodman gun. A 10-inch Parrott rifle fired a shot approximately the same weight as that fired by a 15-inch Rodman gun.

Howitzers

The primary purpose of howitzers – smoothbore cannon with significantly shorter barrels than other cannon – in Third System forts was for defense against infantry. The workhorse weapon for this purpose was the 24-pounder flank howitzer.

While rated a 24-pounder, these cannon were generally not used for shot or shell, but for canister rounds. A canister consisted of a thin metal casing – similar to a coffee can – that contained a large number of iron balls. On firing, the casing would disintegrate spraying the balls in a pattern similar to that of a shotgun.

A 24-pounder flank howitzer would fire a canister with 48 iron balls approximately $1^{1}/_{3}$ inches in diameter. As can be easily imagined, this would be a formidable weapon against massed infantry, especially in a confined area such as a ditch.

Relative Sizes of Shot

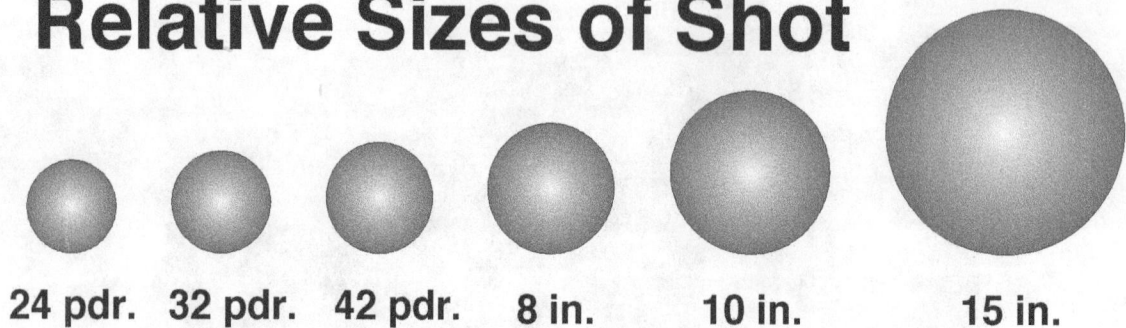

24 pdr.   32 pdr.   42 pdr.   8 in.   10 in.   15 in.

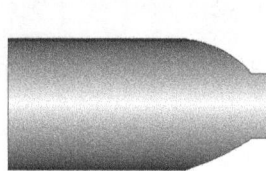

8 in. Rifle   10 in. Rifle

The above scale drawing shows the relative sizes of shot used in the artillery emplaced in Third System forts. The increased mass of a bolt used in a rifled cannon over the round shot used in a smoothbore cannon can be clearly seen in the diagram.

Mortars

While not the primary armament at Third System forts, most forts had one or more seacoast mortars assigned.

Mortars were large-bore, short-barreled weapons that would fire shot or shell on a very high trajectory, resulting in what is generally referred to as plunging fire. These had two purposes during this period. First, the high trajectory would allow shot and shell to penetrate the lightly armored decking of a ship. Shot would use its heavy mass to cause significant damage below decks, while shell would penetrate the upper deck and explode below decks. Shell could also be fused to explode over the upper deck of a ship, causing significant damage and/or loss of life on this deck.

The second use of mortars was to hurl shells over obstacles – natural or man-made – that provided cover to besieging troops and artillery.

The most common seacoast mortars used during the Third System were 10-inch and 13-inch models. 10-inch mortars weighed between five- and eight-thousand pounds, and 13-inch mortars weighed over seventeen thousand pounds.

Carriages

A carriage is the often-complex mechanism that supports a cannon or mortar. Carriages are designed for particular applications, by function, by location within a fort, and by the size of the cannon being supported.

The carriages used for both seacoast guns and howitzers consisted of two parts – an upper carriage and a lower carriage. The upper carriage was firmly attached to the cannon, and moved with the cannon during recoil. It also provided the mechanism for adjusting the elevation of the cannon.

The lower carriage provided the support on which the upper carriage rode, and provided the traverse – the side-to-side aiming of the cannon.

The most common carriage design is the fore-pintle carriage, later called the front-pintle carriage. Within this category are two types of carriage, a barbette carriage that is designed to allow the gun to fire over a parapet and a casemate carriage that is designed to allow the gun to fire through an embrasure.

Both of these designs use a similar upper carriage – it is the lower carriage that separates the two applications. The upper carriage consisted of a framework – metal, wood, or a combination of both – that supported the gun barrel by the trunions. This framework was supported by two axles, each with two eccentric cams and two wheels.

When the gun was fired, the axle would turn such that it would rotate off the eccentric cam and the bottom of the upper carriage would contact the top of the lower carriage. The friction caused by these two flat surfaces sliding against each other would dissipate the energy of the recoil. To return the gun to its firing position, the axel would be rotated – using long poles placed in holes in the surface of the wheels – such that the upper carriage was supported by the wheels. The upper carriage and gun would then be rolled forward into firing position.

The lower carriage of a fore-pintle casemate cannon consisted of a four-wheel framework that supported the upper platform. The wheels of this framework were mounted transverse to the length of the carriage, allowing the carriage to be rotated left and right. Protruding from the front of this framework was a long metal tongue that extended through a slot beneath the embrasure. This tongue had a circular hole near its front that was engaged by the pintle, a long large-diameter cylinder that extended from the bottom of the embrasure to the floor of the casemate. The pintle was located at the throat – or narrowest part – of the embrasure.

The wheels of the lower carriage rode on two sets of flat, semicircular plates attached to the floor of the casemate, called traverse rails. The angle of traverse of this type of carriage was approximately 120 degrees.

The lower carriage for a barbette mount consisted of a similar framework as the casemate mount, but with only two wheels. The lower front member of the lower carriage rested on a fixed pintle set in a barbette, or gun platform (hence the name barbette mount). The rear wheels again ran on a traverse rail attached to the terreplein.

A center-pintle barbette carriage again used the same upper carriage, but with a unique lower carriage. The lower carriage had wheels fore and aft, and the center member attached to a pintle mounted in the barbette. The traverse rails extended to form a full circle, allowing a full 360-degree rotation of the gun.

The most common center-pintle barbette mounts were for 15-inch Rodman cannon. Because of the weight of these massive cannon –over 50,000 pounds – three pair of wheels on the front and three pair on the rear supported the lower framework. This necessitated three concentric circles of traverse rails.

The carriage for the flank howitzer was much smaller than the carriages for the larger guns, as there was far less weight to support. The upper carriage looked like a miniature version of the seacoast carriage, but the lower carriage was much lighter in design. It consisted of a upper framework that supported the upper carriage. This framework connected directly to a pintle located in the throat of the embrasure. The rear of the framework was supported on a vertical frame that terminated in two wheels at floor level. These wheels rode on a single traverse rail.

Mortars had the simplest carriages of all. They consisted of a framework that connected to the trunions of the mortar and terminated in a sled that was manually turned for right-left movement. Elevation was set by a mechanism built into the framework of the carriage.

While the style of each of these carriages is consistent, the size of the carriages varied greatly. The massive iron carriage of a 15-inch Rodman gun was of a similar design to the wooden carriage of a 32-pounder, but the difference in size is dramatic.

Time and technology were major factors in the manner in which guns were mounted. As with fort design and artillery design, evolutionary changes were taking place in the design of carriages. The early wooden carriages used to mount cannon at Fort Pike in the 1820s looked quite different from the iron carriages in use late in the Third System. Necessity and availability also played a role – identical guns were sometimes mounted on different carriages, depending on what could be allocated to the fort at that time.

In general, one can also identify the size of the gun last mounted in a casemate by looking at the floor of the casemate – providing that the floor has not been altered. A single traverse rail and a pintle located above the sill of the embrasure indicate that a flank howitzer was emplaced at that location. Two traverse rails indicate a seacoast gun, with the distance between the embrasure and the traverse rails in the casemate floor giving an indication of its size.

On the terreplein, a semicircular traverse rail truncated by the scarp is indicative of a fore-pintle mount. A full-circle traverse rail (usually multiple concentric circles) was used for a center-pintle mount.

For a more in-depth review of nineteenth-century artillery, there are many books on the subject. (39) Several forts have artillery emplaced, some actual artillery that has been restored and some replica hardware. In both cases, the emplacements aid the visitor in understanding the appearance of the forts when armed.

Manning the Forts

Private contractors, under the leadership of the Corps of Engineers, performed the construction of the Third System forts. After an early construction debacle at Fort Delaware, a superintending engineer was assigned to each fort-construction site. His role was to watch over the government's interest during the construction process, modify the designs as appropriate, to manage the contracts for materials, and to control the budget for that project.

The engineers who performed this function were some of the best of the Corps of Engineers. Such household names as Joseph Totten, Robert E. Lee, P. G. T. Beauregard, and William Chase served in the role of superintending engineer during the construction of various Third System forts.

The ideal process for the transfer of a fort from the construction phase to the operational phase was for the Corps of Engineers to complete the construction, supervise the mounting of the guns, then turn the fort over to the Regular Army unit that was to man it. In the absence of a threat to our shores, that is the procedure that was used. For many of the forts, however, it was necessary to mount some of the guns and man the fort prior to the completion of construction. In these cases, the Engineers were still building the fort while the soldiers were training on the emplaced cannon. Many forts of the Third System were never officially completed, although all of them received armament.

With the arrival of the troops came the materiel to sustain them – quartermaster supplies, individual weapons and ammunition, etc. Information regarding the arming of individual forts is beyond the scope of this work, but is available in the books and guidebooks related to each particular fort.

The seaward face of Fort Pickens, Florida (MAB)

Notes

1. The Academy itself was patterned after the École de Polytechnique in France.

2. The first text, Gay de Vernon's *A Treatise on the Science of War and Fortification,* was the standard text at the Military Academy prior to works by Henry Wager Halleck and Dennis Hart Mahan, both professors at the Academy.

3. These advances included improved methods for casting cannon and the development of rifled cannon. Both increased the size and velocity of the shot and shell that could be brought to bear on the scarp wall.

4. Following the Civil War, this subject again became a subject of debate, but the majority opinion still held that even rifled cannon mounted on an ironclad ship was not adequate to maintain a prolonged attack on a masonry fort.

5. Totten gained prominence through the studies he conducted at Fort Adams (and elsewhere) regarding the firing of ordnance against various types of embrasures. This work led to many of the innovations he developed for various latter Third System forts.

6. Several forts undergoing restoration have had the iron removed to minimize further damage to the fort. If the iron were properly maintained, it would remain free of rust and therefore be sound. The poor maintenance of Third System forts, due primarily to lack of allocated funds and manpower, caused these iron components to become vulnerable to the salt air and the problems, now irreversible, began.

7. Fort Livingston and twin Forts Gaines and Clinch had interior casemates which served only as bombproof shelters.

8. Fort Macon did not have casemated cannon emplacements, all cannon were mounted en barbette. When the Union siege of the fort took place, the gunners drove the defenders from the ramparts and forced the surrender of the fort. It can be speculated that casemated cannon positions would have allowed the fort to mount a longer defense.

9. Many forts had a few casemates that were closed in the rear because of the design of the fort. The casemates in the curved portions of Fort Calhoun were closed, as were the casemates in the bastions and demibastions of most of the forts.

10. Ten forts had three tiers of guns; Forts Richmond (NYC) and Point (San Francisco) had four tiers.

11. The separation of the scarp and the casemates at Fort Point in San Francisco is very dramatic. After the 1990 earthquake, the scarp angled outward, leaving a large gap in the upper tiers. Work is planned to stabilize the wall and attempt to draw it back to its previous position.

12. Many documents of the period used two words, terra plein, to define this area. The single word, terreplein, was coming into use at that time and is used here.

13. Fort Morgan, at the entrance to Mobile Bay, Alabama, is an excellent example of a fort with a brick-paved terraplein.

14. In Fort Adams, Newport R.I., and the Advanced Redoubt, Pensacola Florida, crenellated embrasures were used to protect the main sally port.

15. Fort Adams has a unique "internal ditch" which provides access to the lower-tier casemates on the seaward face. The second level of casemates was at the level of the parade and was accessible from the parade through a series of bridges over this ditch. Fort Warren has an analogous access ditch to the lower level of casemates guarding the gorge. Fort Totten was to have a parade even with the highest level of casemates on the seaward fronts, but this was the level of the only casemates on the remaining sides. The fort was designed this way because of the sloping terrain on which it was constructed. This portion of the fort was not constructed.

16. Fort Taylor on Key West, and Forts Pike, Wood, and Jackson defending New Orleans had similar seawalls, but these are no longer intact. The absence of the seawall at Fort Pike and on the marina side of Fort Wood demonstrates the devastating effects of marine and riverine erosion on the foundation of a masonry fort. Both forts have suffered structural damage and will eventually be destroyed if the erosion is not abated and the foundation repaired.

17. In Third System forts, this maximum distance is approximately one mile. While the larger guns of the fort had considerably more range than this, this distance was adopted as the maximum distance to a redoubt in advance of the fort.

18. Mahan, in his *Attack and Defence of Permanent Works* defines a crownwork in the more general case. He described it as having "two or more bastioned fronts" rather than the more classical definition used here. Only one crownwork was constructed in the Third System, and that followed the definition in the text. An earthen crownwork was constructed at Yorktown, Virginia during the Revolutionary War and reconstructed during the Civil War. This also followed the definition in the text. The author is not aware of any larger crownworks in the United States.

19. Fort Adams, Newport, RI, had both counterscarp galleries and counterfire rooms. The counterfire rooms were located in the rear of the crownwork, so cannot be called "counterscarp" galleries though they were identical in design to counterscarp galleries.

20. While open-topped, open-backed works defines this family of outworks, there are exceptions. The demilune at Fort Warren, Boston Harbor, was a masonry vault with both a roof and a rear wall. A tunnel from the covert way provided access.

21. Fort Warren, Boston Harbor, had both a demilune and a ravelin. The ravelin was an earthwork seacoast battery, while the demilune was an all-masonry casemated rifle gallery set in an earthen coverface guarding the gate.

22. See Chapter One of *A Legacy in Brick and Stone,* by John Weaver, for the definition of these six goals.

23. Fort Point actually lies below the Golden Gate Bridge. The architect of the bridge saw the historical significance of the fort and provided a special arch to allow construction of the bridge without demolition of the fort.

24. At Fort Jackson in the Mississippi River and at Forts Gaines and Morgan in Mobile Bay, Federal ships ran past the guns of the fort to enter the guarded area. These forts, however, had a significantly smaller concentration of armament than the "tower forts" of the latter Third System.

25. Fort Wood was renamed Fort Macomb, the name it carries today. In this text, however, it is referred to by its original name.

26. A note from J. M. Scarritt to J. G. Totten on 20 March 1849 states "...in time of peace the citadel be not used as quarters for officers & men. In summer the rooms were warm, damp, and wanting in ventilation. We have also myriads of mosquitoes and it requires a thorough draft of air and that moving with all the force of the sea breeze to make the rooms habitable. The soldiers & laborers I have here are forced to sleep on top of the parapets."

27. See especially the construction and early garrison records of the Louisiana forts of the Third System. Fort Jackson had special problems in this regard, but the other Louisiana forts had near-mutinies of the garrison due to living conditions inside the fort.

28. *Seacoast Fortifications of the United States* by E.R. Lewis (Naval Institute Press, Annapolis, MD, 1972) p. 12.

29. The major exception to this progress in engineering is the ever-sinking Fort Calhoun. This fort was never completed due to the constant settling of its foundations into the bottom of Hampton Roads. Only one and one-half tiers of what was to be a three-tiered fort were ever completed. The fort site is still sinking today.

30. Fort Popham was planned for three tiers, but only the first casemated tier was completed and the second tier partially completed.

31. *The District: A History of the Philadelphia District, U. S. Army Corps of Engineers 1866-1971,* (Snyder and Guss, 1974): p. 46-51.

32. These efforts began even before the close of the Third System. The subsequent work, sometimes referred to as the Fourth System, began with masonry-revetted earthworks built from 1866-1876. In 1885, the Endicott System used structural concrete for detached batteries, a practice that, with refinements, continued through World War II.

33. This angle varied from 90° to 135°, though rarely did it hit the larger figure. Using a larger angle between the flank and the curtain made the bastion much smaller, a necessity if the curtain wall was relatively short or the enceinte angle relatively small.

34. There was a tendency to reduce the size of the bastions in Third System forts as the period progressed. This was not a strong trend, however, as Fort Totten was designed late in the system with very large bastions. Fort Hancock, also a later fort, also was designed with large bastions. As a general trend, however, smaller bastions and even the unbastioned shoal forts like Carroll and Sumter became prevalent.

35. Fort Pike and Fort Wood were of nearly identical design, with Fort Wood being slightly smaller than Fort Pike. An additional difference in the two was due to the fact that they were both constructed by the same team. Lessons learned in the construction of the former had an impact on the way the latter was finally built.

36. In actuality, Fort Clinch was an insignificant 25 feet shorter than Fort Gaines on each of its curtains. The bastions were the same size and all other proportions of the fort were identical.

37. The armament of the tower consisted of one six-pounder and two eighteen-pounder cannon.

38. Fort Stevens, at the mouth of the Columbia River, was wood-revetted earth. This causes a controversy on whether it was part of the "permanent" system.

39. Works by Edwin Olmstead and Wayne Stark are highly recommend, as is a very nice web page by Chuck TenBrink at www.cwartillery.org .

The seaward front of Fort Adams, Newport, Rhode Island (Mark Berhow, 2011)

the interior of Fort Adams, Newport, Rhode Island (Mark Berhow, 2011)

Practice battery at West Point (Ordnance Museum, 1880s?)

COASTAL DEFENSES OF THE THIRD SYSTEM, 1819-1860

Name	Location	Notes	current status
Fort Sullivan	Eastport, ME	Second System blockhouse & battery	buildings only
Fort Edgecomb	Wiscasset, ME	Second System blockhouse & battery	state park
Fort Knox	Bucksport, ME	Third System fort	state park
Fort Popham	Popham Beach, ME	Third System fort	state park
Fort Gorges	Portland, ME	Third System fort	city park
Fort Scammel	Portland, ME	Rebuilt Second System fort	private
Fort Preble	Portland, ME	Rebuilt Second System fort	college campus
Fort Constitution	Portsmouth, NH	Rebuilt Second System fort	state park
Fort McClary	Portsmouth, NH	Rebuilt Second System fort	state park
Unnamed (Glover)	Gloucester, MA	First System fort	city park
Fort Pickering	Salem, MA	First System fort	ruins, city park
Fort Lee	Salem, MA	First System earthworks	ruins, city park
Fort Sewall	Marblehead, MA	First System fort	city park
Fort Warren	Georges Island, MA	Third System fort	city park
Fort Winthrop	Governor's Island, MA	Rebuilt Second System fort	destroyed
Fort Independence	Boston, MA	Third System fort, replaced earlier work	city park?
Fort at Clark's Point	New Bedford, MA	Third System fort	city park?
Fort Phoenix	New Bedford, MA	battery added to Second System fort	county/city pk
Fort Adams	Newport, RI	Third System fort	state park
Fort Wolcott	Goat Island, RI	First System fort, batteries upgraded	destroyed
Fort Greene	Newport, RI	First System fort, batteries upgraded	destroyed
Barnett Point	Barnett Point, RI	Second System batteries upgraded	status unknown
Fort Hamilton	Rose Island, RI	Second System batteries upgraded	partial, navy land
Fort Trumbull	New London, CT	Third System fort	state park
Fort Griswold	New London, CT	battery replaced earlier work	state park
Fort Schuyler	Throngs Neck, NY	Third System fort	college campus
Fort Totten	Willet's Point, NY	Third System fort, incomplete	City of NY park
Fort Hamilton	N. Narrows, NY	Third System fort	US Army
Fort Lafayette	N. Narrows, NY	transitional fort	destroyed
Fort Tomkins	Staten Island, NY	Third System fort, replaced earlier work	national park
Fort Richmond	Staten Island, NY	Third System fort, replaced earlier work	national park
Fort Wood	Bedlow's Island, NY	Second System fort	national park
Castle Clinton	Manhattan Island, NY	Second System fort	national park
Castle Williams	Governor's Island, NY	Second System fort	city property
Fort Columbus (Jay)	Governor's Island, NY	Second System fort	city property
Fort Gibson	Ellis Island, NY	Second System fort	destroyed
South Battery	Governor's Island, NY	First System fort	city property
Battery Morton	Staten Island, NY	Second System Battery	destroyed
Battery Hudson	Staten Island, NY	Second System Battery	destroyed
Fort Ganesvoort	Manhattan Island, NY	Second System fort	destroyed
North Battery	Hubert Island, NY	Second System battery	destroyed
Fort at Sandy Hook	Sandy Hook, NJ	Third System fort, incomplete	national park
Fort Delaware	Pea Patch Island, DE	Third System fort	state park
Fort Mifflin	Philadelphia, PA	transitional early system fort	city park
Fort Carroll	Baltimore Harbor, MD	Third System fort	private
Fort McHenry	Fort Point, MD	Second System fort	national park
Fort Severn	Annapolis, MD	Second System fort	destroyed
Fort Madison	Annapolis, MD	Third System battery	ruins

Name	Location	Notes	current status
Fort Washington	Potomac River, MD	transitional work	national park
Fort Monroe	Hampton, VA	Third System fort	US Army
Fort Calhoun (Wool)	Hampton, VA	Third System fort	city park
Fort Norfolk	Norfolk, VA	First & Second System fort	army property
Fort Nelson	Norfolk, VA	First & Second System Fort	destroyed
Fort Macon	New Bern, NC	Third System fort	state park
Fort Hampton	Beaufort Inlet, NC	Second System fort	destroyed
Fort Caswell	Oak Island, NC	Third System fort	ruins, private
Fort Johnson	Cape Fear River, NC	Second System fort	ruins, city park
Fort Winyaw	Georgetown, SC	Second System fort	destroyed
Fort Sumter	Charleston, SC	Third System fort	national perk
Fort Moultrie	Charleston, SC	Second System fort	national park
Castle Pickney	Charleston, SC	Second System tower	private
Fort Johnson	Charleston, SC	First, Second System fort	ruins
Beaufort Battery	Port Royal, SC	First System battery	destroyed
Fort Pulaski	Savannah, GA	Third System fort	national park
Fort Jackson	Savannah, GA	Second System fort	state park
Fort Clinch	Jacksonville, FL	Third System fort	state park
Fort Marion	St. Augustine, FL	Spanish fort	national park
Fort Taylor	Key West, FL	Third System fort	state park
East Martello Tower	Key West, FL	Third System tower	priv. museum
West Martello Tower	Key West, FL	Third System tower	private
Fort Jefferson	Dry Tortugas, Fl	Third System fort	national park
Fort Pickens	Pensacola, FL	Third System fort	national park
Fort Barrancas	Pensacola, FL	Third System fort, with earlier works	national park
Advanced Redoubt	Pensacola, FL	Third System fort	national park
Fort McRee	Pensacola, FL	Third System fort	destroyed
Fort Morgan	Mobile, AL	Third System fort	state park
Fort Gaines	Mobile, AL	Third System fort	county park
Fort Massachucsetts	Ship Island, MS	Third System fort	national park
Fort Pike	Rigolet, LA	Third System fort	state park
Fort Wood (Macomb)	Chef Menteur, LA	Third System fort	state property
Battery Bienvenue	New Orleans, LA	Third System battery	private
Tower Dúpre	New Orleans, LA	Third System tower	ruins
Tower at Proctor's Ld	New Orleans, LA	Third System tower	county property
Fort Jackson	Plaquemines, LA	Third System fort	county park
Fort St. Phillip	Plaquemines, LA	Spanish, Second System fort	private
Fort Livingston	Grand Terra Island, LA	Third System fort	county property
Fort on Alcatraz Is.	San Francisco, CA	Third System fort	national park
Fort Point	San Francisco, CA	Third System fort	national park

Great Lakes Forts

Name	Location	Notes	current status
Fort Mackinac	Mackinac Island, MI	Second System fort	state park
Fort Wayne	Detroit, MI	Third System fort	city property
Fort Niagara	Lewiston, NY	First, Second, Third System fort	state park
Fort Ontario	Oswego, NY	Third System fort, replaced earlier work	state/county
Fort Montgomery	Rouses Point, NY	Third System fort	ruins, private
Fort Porter	Buffalo, NY	Third System tower	destroyed

UNITED STATES COAST DEFENSE FORTIFICATIONS 1870-1876

Joel W. Eastman, Bolling W. Smith and Glen M. Williford

At the end of the Civil War, the U.S. Army Corps of Engineers was forced to re-evaluate the Third System of fortifications in light of the experience of combat and of the technological advances made during the war. Steam-powered ironclad naval vessels, mounting powerful smoothbore and rifled guns, posed a serious threat to the masonry Third-System forts, and land-based, large rifled siege guns were an even greater danger.

The engineers struggled to find an answer that would minimize the necessity for a war-weary country to rebuild its coastal defenses. Iron sheathing for masonry forts seemed promising, but was found to be ineffective. On the other hand, earthen forts had served well during the war, even against heavy naval bombardment, and parapets, traverses, and parados of earth and sand had long been used to protect masonry fortifications. Earth and sand were relatively cheap and plentiful, and damage to these fortifications could be easily repaired. Traverses and parados were also logical locations for powder magazines.

The ordnance was also obvious. Ten and 15-inch smoothbore, muzzle-loading Rodman guns had been manufactured in great quantities for seacoast defense during the war, and few had seen action. Thus, large numbers were available. Large Parrott rifled guns were also on hand, although their tendency to burst made them suspect. While 10-inch Rodman guns were ineffective against the latest ironclads, they were useful against the wooden ships that still constituted the bulk of the world's navies, and 15-inch Rodmans were capable of destroying ironclads at close range.

A number of batteries were proposed and a few were built at new sites, but most were adjacent to or incorporated into Third System forts. Many masonry forts were modified to take the new, larger guns on their barbette tiers. The new batteries required major alterations of the old forts, including new, heavier emplacements, removing structures from the interior of the fort, covering exposed masonry with earth and sand, and constructing powder magazines to store the large amounts of powder required by the Rodman guns.

Other batteries were built outside the old forts. The standard plan for separate batteries included pairs of gun platforms protected by traverse magazines. The platforms were of granite, with stepped granite interior walls to aid in loading. The interior walls were extended above the granite by pieces of slate secured by iron brackets, allowing a higher earthen parapet, to better protect the guns and crews.

The traverse magazines featured a new innovation–concrete rather than brick construction. The interior walls of some batteries were also built of concrete, and the engineers experimented with concrete gun blocks and bases for traversing rails. At batteries subject to fire from the rear, parados were built behind the batteries. In addition to gun batteries, sunken batteries were also built for 13-inch smoothbore seacoast mortars.

Finally, the engineers incorporated submarine mining after its successful debut during the Civil War. The first four mining casemates to be built by the United States were constructed for this system, even though technical issues with the mines, and particularly with cables and firing techniques, posed difficult problems.

Based on appropriations actually made and work actually undertaken, 18 ports were defended by the new batteries, at 45 "separate" fortification reservations. Of these 45 reservations, 17 were Second and Third-System works modified for new gun batteries, another 15 were fort reservations dating from the Third System, but having new exterior batteries and not utilizing the older masonry works, and 14 were new reservations– bought for this purpose or dating from Civil War temporary position acquisitions. One of the latter was later canceled after appropriations were approved, but prior to any construction (Point Lazaretto in Baltimore Harbor). The engineers prepared plans for additional forts and for defending additional ports, including several in the Great Lakes, but work had not started before funding was halted by Congress for the 1876 fiscal year.

15-inch Rodman smoothbore in the 1870s-era South Battery of Fort Preble. (courtesy Joel Eastman)

Thus, many of the new gun batteries remained uncompleted, and almost all unarmed. If the 1870s fortifications had been finished and armed, they might have been known as the "Fourth System." However, this system was stillborn, although the batteries provided a model for a number of the temporary works built during the war with Spain.

Some of the best surviving examples of the 1870s fortifications are at Fort Wadsworth in New York City, Fort McHenry in Baltimore, Fort Pulaski near Savannah, and Battery Cavallo at Fort Baker in San Francisco.

U.S. Coast Defense Fortifications, 1870-1876

(Based on sites with provided appropriations)

Eighteen Ports Received Some Funding:

Portland	Portsmouth	Boston	Newport	New London
New York	Delaware	Baltimore	Washington	Hampton Roads
Charleston	Savannah	Key West	Pensacola	Mobile
New Orleans	San Diego	San Francisco		

Construction was planned or initiated at Forty-five Distinct Military Reservations/Sites:

Significant Remains:

Fort Gorges	Fort Scammell	Fort Wadsworth	Fort Mifflin	Fort McHenry
Fort Foote	Fort Moultrie*	Fort Jackson, GA	Fort Pulaski	Fort Jefferson
Fort Jackson, LA	Fort W. Scott	Lime Point	Cavallo Point	

Partial Remains:

FortWilliams	Fort Preble	Ft. Independence	Fort Strong	Fort Warren
Fort Adams	Fort Greble	Fort Trumbull*	Fort Schuyler	Fort Totten
Fort Delaware	Fort Mott	Fort Dupont	Fort Washington	Fort Monroe
Fort Sumter	Fort Pickens	Fort Morgan	Fort St. Philip	Gravelly Beach

Very Little or Nothing Remains:

Fort Foster	Fort Stark	Fort Winthrop	Fort Columbus	Fort Wood
Fort Hamilton	Lazaretto Point	Fort Taylor	Fort Rosecrans	Alcatraz Island
Yellow Bluff				

Note: Many of these "fort" or location names were not in use in the 1870s, but are later, mostly Endicott names. They are used because they will be familiar to the reader and serve as an aid to help place and locate the sites today.

*Fort Moultrie's battery is a restored/reconstructed battery, but historically accurate. Fort Trumbull's battery is also mostly rebuilt for display purposes.

For more detailed information please consult: Glen M. Williford, The Transitional Coast Defense Generation, American Seacoast Defenses of the 1870s, *Coast Defense Journal* (2007) Vol. 21, Issue 2, p. 51, Issue 3, p. 51, Issue 4, p. 59.

A Guide to the Weapons and Batteries of Modern American Harbor Defenses

Prepared by Mark Berhow and Glen Williford
with assistance from Alex Holder, Danny R. Malone, and Gregory J. Hagge

The permanent concrete batteries and temporary batteries of the early modern era (Endicott & Taft), the post First World War period, and Second World War programs (1885-1945).

This section provides a general overview of the seacoast artillery weapons and their supporting defensive structures built by the United States during the modern or concrete era. The weapons are illustrated by photographs and drawings. Unfortunately, most of these weapons are now long gone and all that remains are the concrete emplacements. The major fixed coast artillery emplacements of the modern era can generally be easily identified by several methods. Obviously name plates on the batteries, or clearly identified locations on historical maps or plans is a virtually foolproof method. Sometimes, these are not available or incomplete with regard to earlier or later works. Also at times identification aids are needed to place photos or diagrams correctly.

There is enough standardization in recognizable features of the gun emplacements for a particular carriage model to allow positive identification. Each type of major carriage was often unique in terms of anchoring bolt pattern, and size and positioning of the surrounding parapet and loading platform structures. From the Endicott period through the Second World War, the design of emplacements was fairly consistent (particularly with regard to the actual gun mounting structures). The Ordnance Department provided the engineers with the plan for the mounting. In turn the chief of engineers provided the local office with type plans and approved designs or deviations submitted by these offices. Thus most batteries of a "type" are very consistent in many design features. Temporary, mobile, antiaircraft, and obsolete carriage types are covered, but not as extensively as the textual references were not currently available to the authors.

Battery Chamberlin, Fort Winfield Scott, Harbor Defenses of San Francisco—an excellent example of an early 1900s American seacoast artillery battery. A six-inch gun on a M1903 disappearing carriage was re-installed in the early 1970s. The fort is now part of the Golden Gate National Recreation Area. (Mark Berhow, 2006).

The weapons are listed from small and medium calibur (3 to 8-inch) (usually classified as secondary armament) to large calibur (10 to 16-inch) (the primary armament). The basic sources used were the Ordnance Department descriptive pamphlet series, issued for each of the major carriage types. Later versions (the Technical Manuals) were also of use. A variety of secondary Ordnance and Technical Engineering documents were used to fill in necessary details. While carriage models are extensively illustrated here, the gun barrel models are not generally identified.

More information on most of the Army seacoast artillery gun and carriage models can be found in: *The Service of Coast Artillery,* by Frank T. Hines and Franklin W. Ward (Goodenough & Woglom Co., New York, 1910); *American Seacoast Material,* Ordnance Department Document #2042, GPO, Washington, DC (June 1922); and *Seacoast Artillery Weapons,* Army Technical Manual (Army Coast Artillery) TM-210, GPO, Washington D.C., (1944). General information on the design and construction of the early modern American seacoast fortifications can be found in *Notes on Seacoast Fortification Construction* by Col. Eben E. Winslow (Occasional Papers No. 61, Engineer School, U.S. Army, G.P.O., 1920). All four of these books have been reprinted by the CDSG Press (www.cdsg.org).

For more specific information on seacoast artilery weapons, see these *CDSG Journal* articles by Greg Hagge: "Mortar Carriage Notes" Vol. 6, Iss. 2 (May 1992) pp. 23-31; "Installations and accessories for gun carriages" Vol. 13, Iss. 4 (Nov. 1999) pp. 23-35; and by Bolling W. Smith: "12 inch breechloading rifles" Vol. 7, Iss. 4 (Nov. 1993) pp. 57-58; "Coast Artillery Projectiles 1892-1915" Vol. 9, Iss. 1 (Feb. 1995) pp. 16-24; "Seacoast Artillery Gun Sights." Vol. 9, Iss. 2 (May. 1995) pp. 25-49; "Charcoal Powder Propellants for the Coast Artillery." Vol. 10, Iss. 4 (Nov. 1996) pp. 20-29; "Primers and Firing Mechanisms." Vol. 12, Iss. 1 (Feb. 1998) pp. 67-88; "Smokeless Powder." Vol. 12, Iss. 2 (May 1998) pp. 52-62; "Seacoast Artillery Fuzes." Vol. 12, Iss. 3 (Aug. 1998) pp. 32-54; "Seacoast Guns and Carriages." Vol. 12, Iss. 4 (Nov. 1998) pp. 73-118 & Vol. 13, Iss. 1 (Aug. 1993) p. 90; "The 16-inch Batteries at San Francisco and the Evolution of the Casemated 16-inch Battery," Vol. 15, Iss. 1 (Feb. 2001) pp. 16-83.

NOMENCLATURE OF US SEACOAST ARTILLERY CANNON AND CARRIAGES
Derived and updated from "Seacoast Defense Terminology and Abbreviations"
by Robert D. Zink and E. R. Lewis, *CDSG News* Vol. 1, No. 3 (May 1986) pp. 15-16.

Cannon: The modern (or breech-loading) era of American artillery began with the development of new rifled breech-loading guns in the late 1800s. The guns used to defend the American coastline were designed by the Ordnance Department of the U.S. Army. These cannon were termed either mortars, howitzers, or guns, depending on their tube length and a few other parameters. During the years 1883-1930, the weapons were designated by caliber, the year they were approved (the model or M-year), and any minor modifications thereof (modification [M] 1, etc.) Most American-built cannon had this information stamped on the face of the muzzle, along with the serial number of individual piece and the place of manufacture. Example: 14-inch gun M1910MI (model 1910 modification 1). For gun models designed and built after 1930, the year appellation was dropped, the weapons were designated by their caliber and a model number (M#). Example: the 6-inch gun M1.

In times of crisis, the United States was often caught in short supply of artillery pieces. During the Spanish-American War, a number of rifled muzzle-loading cannon were quickly mounted on modified cast iron and steel carriages at several locations along the Eastern and Gulf coasts. The U.S. also purchased a number of British Armstrong guns and carriages which were mounted in the same areas. During World War I, the U.S. purchased a number of French 155 mm tractor-drawn artillery guns. The Ordnance Department produced the carriage and a large number of American manufactured guns. Some of these guns were used in harbor defenses during the 1930s and 40s.

Carriages were either fixed (permanently emplaced in a prepared position), or mobile (i.e. had wheels or tracks). Early modern era carriages (1883-1930) were designated like the cannons, that is by gun caliber, carriage type, model approval year, and any modifications thereof, for example: 12-inch mortar carriage M1896MIII (model 1896 modification 3). After 1930, as noted above for the cannon, the year appellation was dropped for the new carriage models which were designated as M# (see drawings and photographs which follow in main body of the guide).

Fixed Barbette carriages—

Barbette carriage (BC): A carriage arrangement in which the gun was fired over a parapet. The carriage was comprised of a set of frames or cheeks which supported the gun tube in a manner that allowed for elevation and depression, on a base that accommodated the firing recoil and that could be moved horizontally. Only a few of these early designs were made for the 8, 10, and 12-inch guns by the early 1890s when the disappearing carriage design was adopted for the heavy seacoast artillery guns.

Barbette carriage (long-range) (BCLR): A new barbette design of the World War I period allowed for greater firing elevations and ranges. It eventually became the standard American seacoast artillery carriage, and these carriages have collectively become known today as barbette carriages (long-range). During World War II the new 16-inch BCLRs were protected by modern concrete casemates, while the new 6-inch BCLRs were protected by thick steel shields.

Pedestal mount (Ped.): Carriages designed for guns of 7-inch or lesser caliber. This type of mount consisted of a fixed cylindrical base on top of which rotated the yoke which held the cannon in a cradle equipped with recoil absorbing cylinders. Most pedestal mounts had a frontal shield.

Antiaircraft mount (AA): When antiaircraft gun carriages were developed following World War I, they were technically pedestal mounts, but were always were referred to as AA mounts.

Turret mount (T): At Fort Drum, in the entrance to Manila Bay, the Philippines, a pair of Army designed twin-gun turrets was installed during 1909-1918.

Navy gun mount (NT): Two salvaged three-gun 14-inch navy gun mounts and eight surplused two-gun 8-inch navy gun mounts were installed as seacoast defenses on the Island of Oahu, Hawaii, during WW II. The army referred to these as turret mounts.

Howitzer mount (How): Four fixed 16-inch howitzers, M1920, were installed at the entrance to Chesapeake Bay. The carriages were BCLR's.

Mortar Carriage (MC): Mortar mounts for firing the short weapons in high arcs.

Fixed Retractable Carriages—

Gun-Lift Carriage (GLC): An early American retractable design which was essentially a barbette carriage atop an elevator platform. The entire gun platform was lowered into the emplacement between rounds for reloading. Five of these gun carriages were built, but only a single two-gun battery of this design was completed, due to the success of the disappearing carriage design.

Altered Gun-Lift Carriage (AGLC): The three additional GLC's were modified and eventually installed as standard BC's, but were known as AGL carriages.

Disappearing Carriage (DC): The prevalent carriage design for heavy American seacoast guns in the period between 1890 and 1917, of which large numbers were constructed and installed. These were further designated, as were all carriages, as either "limited fire" (LF) or "all around fire" (ARF) with regard to their traversing abilities. The guns were mounted on one end of a pair of swiveling arms which were counterweighted at the other end. When the catch was released after loading, the falling counterweights would raise the gun to firing position, and the recoil energy of firing would push the gun back down to the loading position.

Balanced Pillar Mount (BPM): A trade name for an early modern era retractable carriage for the 5-inch gun which allowed the weapon to be lowered into its emplacement between drills or engagements. The gun and gunner's platform rested on a telescoping tube which could be manually cranked up or down.

Masking Parapet Mount (MPM): Somewhat similar to the balance pillar mount, this mount for the 3-inch gun was designated by the trade name used by the manufacturer.

Over 100 of these 3 and 5-inch guns and carriages were installed around the United States at the turn of the century. They were permanently locked in the up position around 1914 and all were removed from service by 1920.

Mobile Carriages—

The mobile coast defense weapons of the post-World War I era were from stocks of weapons originally designed and built for use in European Theater. The availability of these weapons, the new emphasis on mobility in seacoast defense, and the lack of funding for new fixed coast artillery guns, brought a "mobile era" to coast artillery in the period between the wars.

Railway mount (RY): Several American railway mount designs were developed during World War I, most utilizing weapons removed from the existing coast defenses. The most numerous of these remounted weapons were 8-inch guns (on a M1918 BC) and 12-inch mortars, both of which were eventually used for coast defense. A few designs for new 14-inch guns were developed and built by the Navy for use in Europe. The Army took one of these experimental designs, improved it, and built four 14-inch M1920 railway mounts which were used for coast defense in Panama and Los Angeles. The RM guns could be fired from anywhere along existing or specially built railroad tracks, but thyey were not very versatile. This problem was alleviated somewhat by the construction of simple circular concrete emplacements at specific sites. The guns were brought to the platform by rail, the carriage lowered onto the platform, after the wheels were removed. The gun mount was then essentially a BC which had a 360 degree firing arc.

Tractor-drawn mount (TD): Also left over from the war were a great number of truck (initially tractor)-drawn artillery pieces which could be moved along roads. The most numerous of these was a 155 millimeter gun, an American-built (M1918) version of a French 1917 design. This became the standard tractor-drawn coast defense weapon, and with some modifications to the carriage and the addition of pneumatic tires, remained in service through World War II. The guns could be set up to fire at any location by preparing an earthen Field Mount (FM). During the 1920s a simple circular concrete platform was developed in Panama for this gun, and these were called Panama Mounts (PM). Many PMs were built for 155s which were used as temporary defenses at the beginning of World War II.

240 mm Howitzer mount (How): Twelve 240 mm howitzers on modified M1918 mobile mounts were sent to Hawaii in 1920s. A number of concrete emplacement platforms were built to hold the mobile mounts, which were stored in warehouses at various military bases.

At times some field artillery pieces (such as the 75 mm) and mobile antiaircraft guns were used for coast defense in the early stages of World War II.

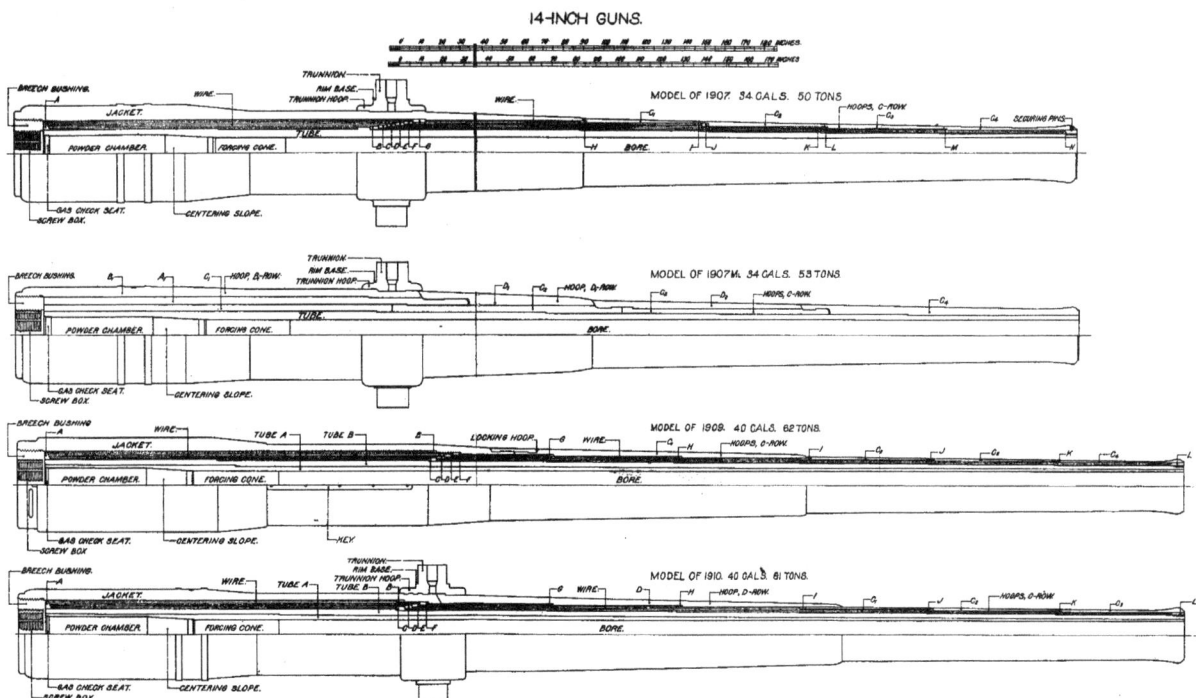

Four 14-inch gun models

12 INCH   MORTARS.

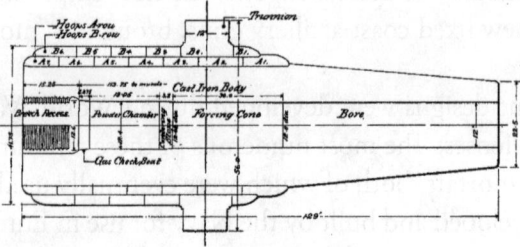

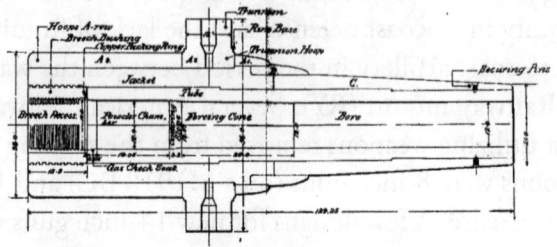

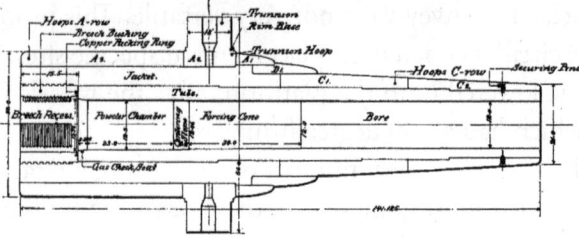

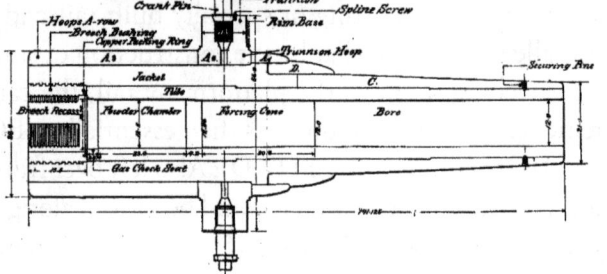

Four 12-inch mortar models

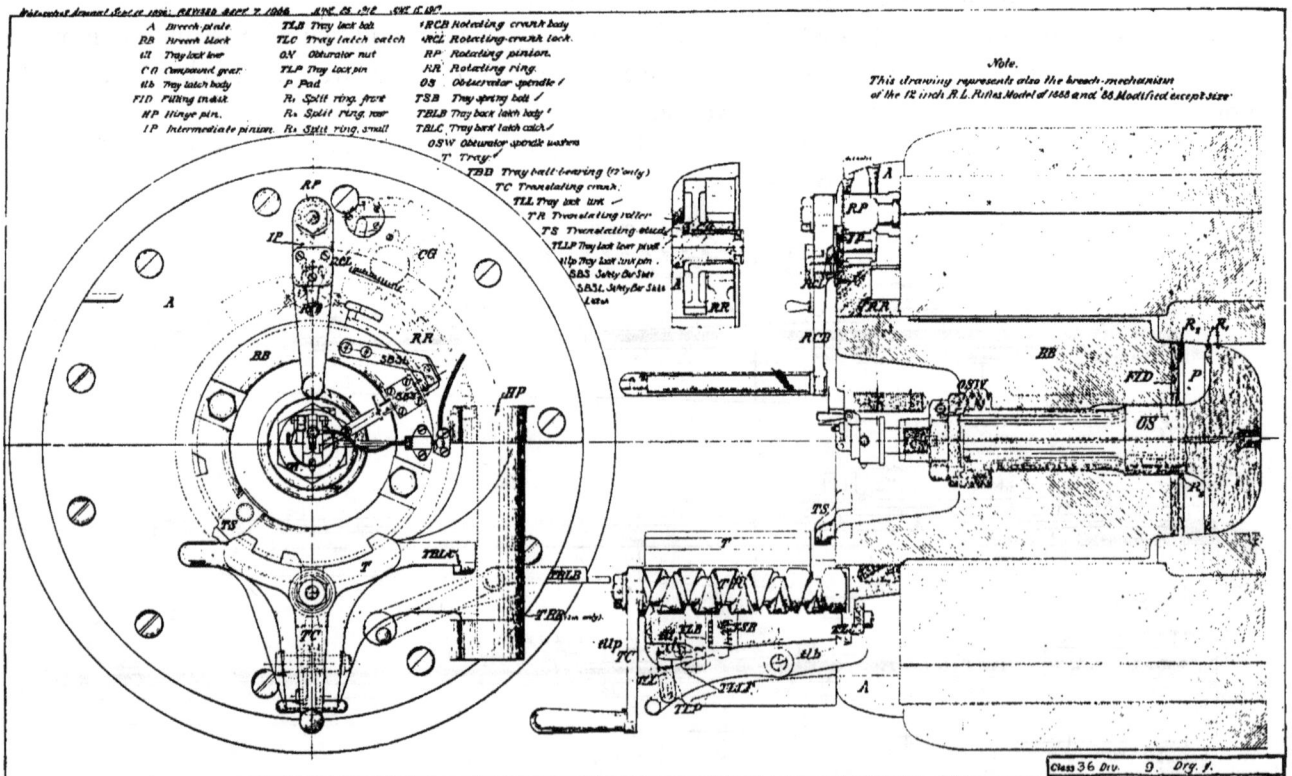

TYPE OF BREECH MECHANISM USED IN THE 10 AND 12-INCH GUNS, MODEL OF 1888.

Breech mechanism used in 10- and 12-inch guns

Characteristics of the Principal Types of US Army Seacoast Artillery

Coast Artillery Field Manual (1930s)

Cannon Caliber & Type	Carriage type	Elevation (degrees)	Traverse (degrees)	Projectile type	Projectile weight (pounds)	Range yards (miles) (maximum)	rate of fire (rounds per minute)
16-in gun	barbette	65 (55)	360	AP proj.	2,340	44,700 (25)	> 1
				AP proj.	2,100	49,100 (28)	
16-in howitzer	barbette	65	360	AP proj.	2,100	24,500 (14)	> 1
14-in gun	disappearing	20	170	AP S&S	1,515	25,000 (14)	> 1
	railway	35	7	HE shell	1,215	48,200 (27)	> 1
			(360)	AP Proj.	1,560	42,200 (24)	> 1
12-in gun	disappearing	15	170	AP Proj.	975	18,400 (10)	< 1
				AP S&S	1,070	18,000 (10)	
	barbette	35	360	AP Proj.	975	30,100 (17)	< 1
				AP S&S	1,070	27,500 (15)	
	railway	38	10	AP proj.	975	30,100 (17)	> 1
			360	AP S&S	1,070	27,500 (15)	
12-in mortar	fixed	65	360	DP shell	1,046	11,345 (6)	< 1
	railway				824	12,280 (7)	> 1
					700	14,650 (8)	
10-in gun	disappearing	12	170	AP S&S	617	14,100 (8)	> 1
				HE shell	510	14,700 (8)	
8-in gun	disappearing	12	120	AP S&S	323	12,900 (7)	< 1
				AP shell	260	14,200 (8)	
				HE shell	200	12,400 (7)	
	railway	42	360	HE shell	200	21,300 (12)	< 1
				AP shell	260	23,900 (13)	
6-in gun	disappearing	15	170	AP S&S	108	14,600 (8)	4
				HE shell	90	12,900 (7)	
	barbette (00)	20	360	AP S&S	108	16,000 (9)	5
				HE shell	90	14,800 (8)	
3-in gun	barbette	16	360	HE shell	15	11,000 (6)	12
155-mm gun	tractor-drawn	35	360	HE shell	95	17,400 (10)	3

War Dept. Technical Manual TM 4-210 Seacoast Artillery Weapons (Oct. 15, 1944)

Cannon Caliber & Type	Carriage type	Elevation (degrees)	Traverse (degrees)	Projectile weight (pounds)	Range yards (miles) (maximum)	rate of fire (rounds per minute)
16-in	howitzer M1920	65	360	2100	24,500 (14)	1
16-in	barbette M1919	65	360	2340	49,100 (28)	1
	barbette M4	47	145		45,150 (25)	
14-in	railway M1920	50	7 (360)	1560	48,200 (27)	
12-in	barbette M1917	35	145	975	29,300 (16)	2
8-in	railway M1918	42	360		24,900 (14)	2
	railway M1	45	360	260	35,300 (20)	
	barbette M1A1	45	360		35,300 (20)	
6-in	barbette M1900	20	360	105	17,000 (9)	4
M1903A2	barbette M1	47.5	360		27,100 (15)	
M1	barbette M3, M4	47	360		27,500 (15)	
155-mm	tractor-drawn M1917M2	35	60 (360)		19,100 (11)	3
	tractor-drawn M1918M3	35	60 (360)		19,100 (11)	
	tractor-drawn M1	63	60 (360)		27,500 (15)	
90-mm	fixed M3	80	360	23.5	19,500 (11)	30
3-in	pedestal M1902	15	360		11,000 (6)	12
	pedestal M1903	16	360		12,000 (7)	

Note: ranges and cannon performance varied considerably over the years depending on projectile used, degree of rifling in the gun, and variations in quality, type, and amount of the powder used. These tables reflect general information on performance of these weapons specific for the year noted.

Abbreviations: DC- disappearing carriage; trav.-traverse; deg.-degrees; bar.-barbette, RY-railway mount; TD-tractor-drawn; how.-howitzer; AP-armor-piercing; S&S- shot and shell; proj.-projectile; HE-high explosive; DP-deck-piercing; mor.-mortar

US SEACOAST ARTILLERY GUN EMPLACEMENT NOMENCLATURE

The term "battery" as used in articles concerning harbor defenses unfortunately now has several technical meanings. The term originated from the tactical organization of artillery units. Its simplest and most accurate definition is one or more pieces of artillery which are under the command of a single individual and intended to concentrate their fire on a single target. In the American Army, the term also applied to the artillery corps unit that was the equivalent of an infantry company. An artillery battery would ideally be the complement of men required to man a designated gun battery. The term was not generally applied to the large multi-storied forts built by the United States for seacoast defense before the Civil War, however, it was used when referring to detached works containing smaller numbers of weapons. Following the Civil War, seacoast defense guns were emplaced almost exclusively in these detached batteries. By the turn of the century, in the modern era harbor defense construction programs, the term "battery" was used to describe a set of guns under a single commander together with the entire structure erected for the emplacement, protection, and service of those guns. Today we apply the term "battery" to those structures which were built to hold seacoast artillery, even though their guns (and thereby the original definition) have long since been removed. The modern era seacoast gun emplacements were fairly complex structures that had specialized terminology that was all its own.

Battery Names: Many of the detached gun batteries built in the 1860s & 70s received some sort of a name to help tell them apart. Thus, when a pair of detached gun batteries were built near San Francisco's Fort Point, one was named West Battery and the other East Battery. The large number of detached gun batteries built during the 1890s quickly became difficult to keep track of. Therefore, beginning in 1902, the War Department assigned names to each battery. The batteries were most often named to honor deceased individuals who had served the nation with distinction, though there were many exceptions. Many of the batteries were named after army officers who had been killed in action or who had died recently, but a few were named after famous Indian chiefs, governors, and localities. Some structures, while physically a single unit, were divided into two (or more) tactical batteries at one time or another, each with its own official name. Many mobile gun batteries were designated by their locality or the name of their battery commander.

For more details on official fort and battery names, see *Designating US Seacoast Fortifications, War Department General Orders and Letters from the Adjutant General 1809-1950,* compiled by Matthew Adams, privately published by the compiler, Australia, 2000.

Modern harbor defense battery structures were built to perform three basic functions: anchor (emplace) the guns to a stable platform for accurate firing, protect the gun and its crew from enemy fire, and provide protected space for the storage of ammunition. By 1900, a typical harbor defense gun battery was composed of a maze of concrete rooms and walls, each with a specialized name and function. Some of the terms for the parts of concrete batteries were defined in the following mimeograph.

```
            Office of the Chief of Engineers United States Army
                          Mimeograph No. 60
            Subject: System of nomenclature of parts of modern batteries.

March 17, 1903

    The following report of the Board of Engineers submitting a list of definitions
for the various parts of modern seacoast batteries is published for the information
of all Engineer officers.

    By command of Brig. Gen. Gillespie:

                                        Fredrick V. Abbot
                                        Major, Corps of Engineers
```

                    The Board of Engineers
                       Army Building
                       New York City
                      March 11, 1903

Brig. Gen. G.L. Gillespie,
Chief of Engineers,
Washington, D.C.

General:

    Complying with Department indorsement of Dec. 23, 1902 (45488), I have the hon-
or to forward the accompanying list of definitions of the various parts of modern
seacoast batteries as recommended by the Board of Engineers. It is further recommended that these terms be used in future on all maps and in all written matter
where they may properly apply. * * * It is not desirable to use terms relating to
the old fortifications which will not readily apply to similar parts of a modern
battery.   Other terms and definitions may be required and these can be added
from time to time as the necessity arises.

    For the Board:
    Very respectfully,
    Your obedient servant,
    Chas. R. Suter
    Colonel, Corps of Engineers,
    President of the Board

                          Nomenclature
         List of Terms and Definitions Relating to Modern Seacoast Batteries.

Battery— The entire structure erected for the emplacing, protection and service
of one or more cannon.

Emplacement— That part of a battery pertaining to the position and service of one
gun or mortar, or a group of mortars.

Pit— That part of a mortar emplacement designed for mounting one or more mortars.

Parapet— That part of the battery which gives protection to the armament and personnel from front fire.

Interior Crest— The line of intersection of the interior wall or slope with the
superior slope.

Traverse— The structure perpendicular or oblique to the parapet wall, protecting
the armament and personnel from flank fire.

Gun Platform— That part of the battery on which the gun carriage rests.

Counterweight Well— The pit in the gun platform for the reception of the counterweight of a disappearing carriage.

Loading Platform— The surface upon which the cannoneers stand while loading the piece.

Truck Platform— If the ammunition trucks run on a different surface from that of the loading platform, this surface is called the truck platform.

Banquette— The step between the truck and loading platforms used in the transfer of ammunition.

Superior Slope— The top of the parapet or traverse.

Apron— That portion of the superior slope of a parapet and the interior slope of a pit designed to protect against blast.

Exterior Slope— The outer slope of the battery.

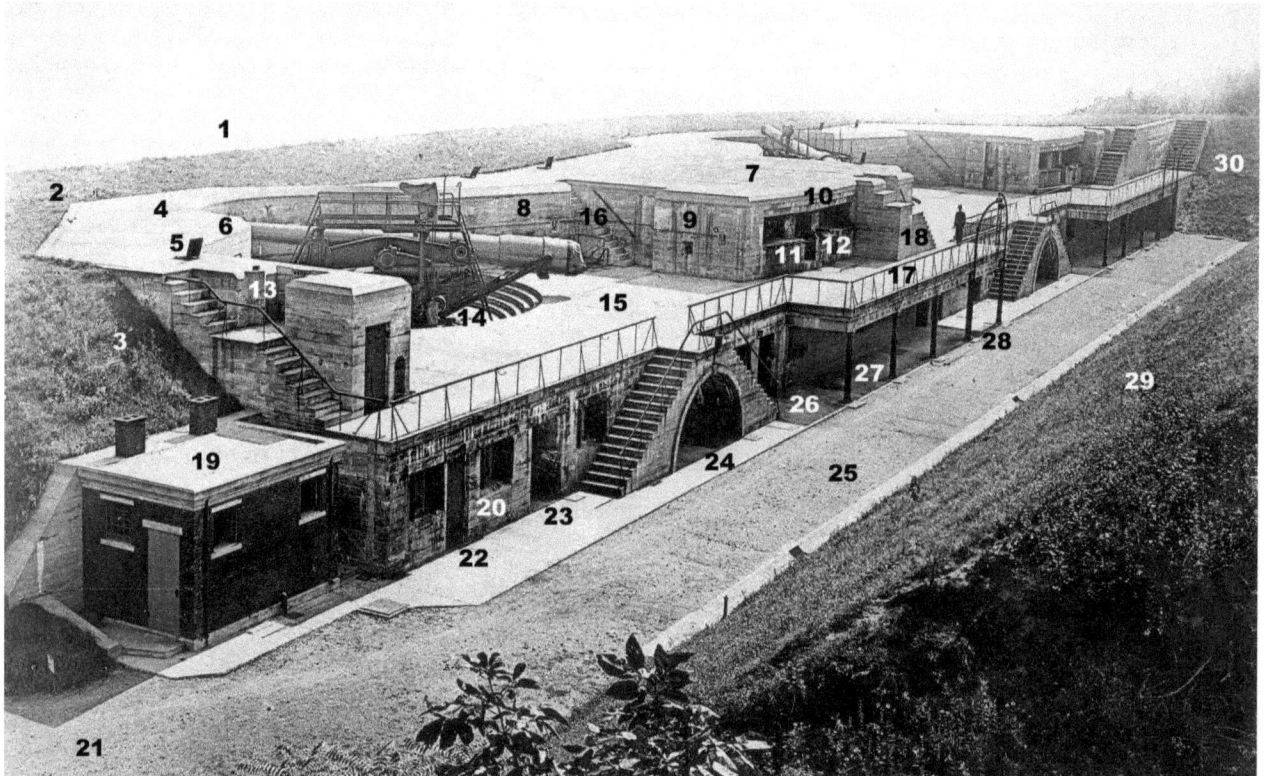

A Seacoast Gun Battery (Battery Dix, Fort Wadsworth, NY)
Adapted from: Frank T. Hines and Franklin W. Ward, *Service of Coast Artillery,* NY 1910

1. Exterior Slope of Parapet	16. Platform Stairs
2. Superior Slope of Parapet	17. Corridor
3. Interior Slope of Parapet	18. Corridor Wall
4. Blast Slope or Apron	19. Latrine
5. Blast Ventilator	20. Parade Wall
6. Interior Crest	21. Approach
7. Traverse	22. To Oil and Tool Room
8. Interior Wall	23. To Shell Room
9. Traverse Wall	24. To Magazine
10. Canopy	25. Battery Parade
11. Reserve Table	26. Office
12. Delivery Table	27. Gallery
13. Observing Station (Crow's Nest)	28. Crane
14. Gun Platform	29. Interior Slope of Parados
15. Loading Platform	30. Traverse Slope of Parados

Interior Slope or Wall— The inner slope or wall of gun parapets or mortar pits.

Traverse Slope or Wall— The side slope or wall of the traverse.

Rear Slope— The rear slope to the parade.

Corridor— The elevated passageway, in the rear of a traverse, connecting adjacent gun positions.

Corridor Wall— The traverse wall along the corridor.

Gallery— Any passageway covered overhead and at the sides.

Parade Slope or Wall— The rear slope or wall of the emplacement.

Battery Parade— The place in the rear of the emplacements where the detachments form.

Parados— A structure in rear of the battery for protection against fire from the rear. It may have an interior, superior and exterior slope.

Ramp— An inclined plane serving as a means of communication from one level to another.

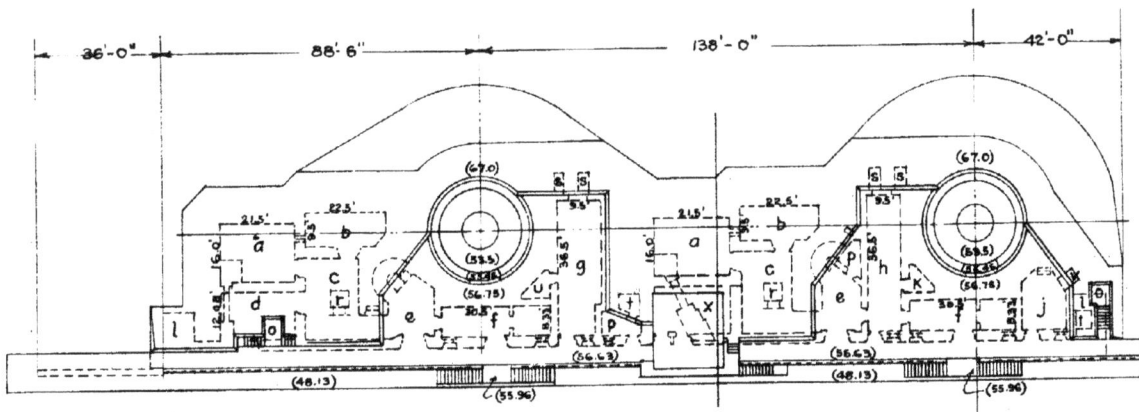

PLAN

Layout of an American seacoast 10-inch gun battery built 1905 (Battery Russell, Fort Stevens, OR)

Legend:
a- magazine
b- shell room
c- lobby
d- shot room
e- tool room
f- guard room
g- office & relocator room
h- sleeping room for guards
j- officer's room
k- store room
l- latrine
o- observing station
p- oil room
r- shell/shot hoist
s- shot trucks
t- telephone rooms
u- map vault
x- plotting room
y- powder hoist

Approaches— Roadways entering the battery parade.

Observing Station— A protected position constructed in the parapet or traverse for the purpose of observation.

Booth— Any recess or construction for the accommodation of teleautograph, telephone, etc.

Magazine— The rooms and galleries for the storage of ammunition.

Cartridge Room— The room of the magazine for the storage of cartridges.

Shell Room— The room of the magazine for the storage of projectiles.

Shot Gallery— The gallery of the magazine for the storage of projectiles.

Ammunition Hoist— The device by means of which the ammunition is raised from the magazine to the loading platform.

Receiving Table— The hoist table on which the ammunition is placed preparatory to raising.

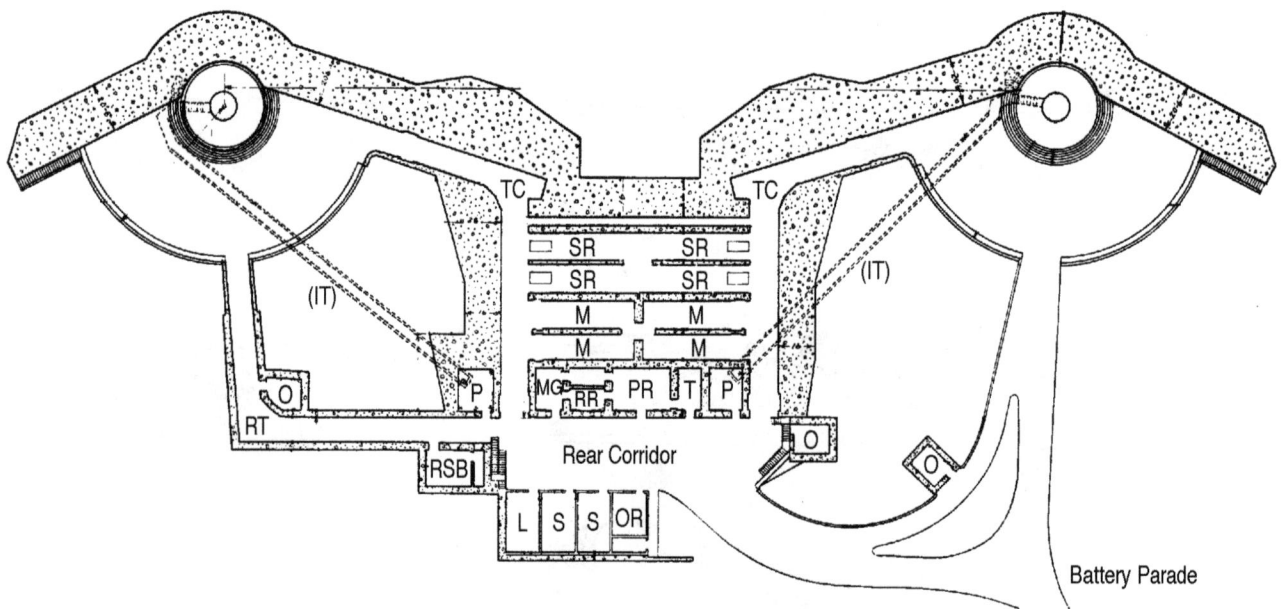

Layout of an American seacoast 14-inch gun battery built 1915-19
(Battery Osgood-Farley, Fort MacArthur, CA)

SR	shell room	MG	motor generator room
M	powder room	P	plotting room
IT	indicator tunnel	RSB	radio & swtitchboard room
TC	truck corridor	S	store room
PR	power room	L	latrine
T	transformer room	OR	officer's room
RR	radiator room	RT	rear tunnel

Delivery Table— The table on which ammunition is delivered from the hoist.

Reserve Table— A table in a sheltered position for reserve ammunition.

Trolley— A mechanical device for transporting projectiles on horizontally suspended tracks.

Crane— A mechanical device for raising ammunition by means of differential or other blocks.

In addition to the above there will always be the various rooms of a battery designated by their uses, as Store room, Guard room, Tool room, Accumulator room, Power room, etc.

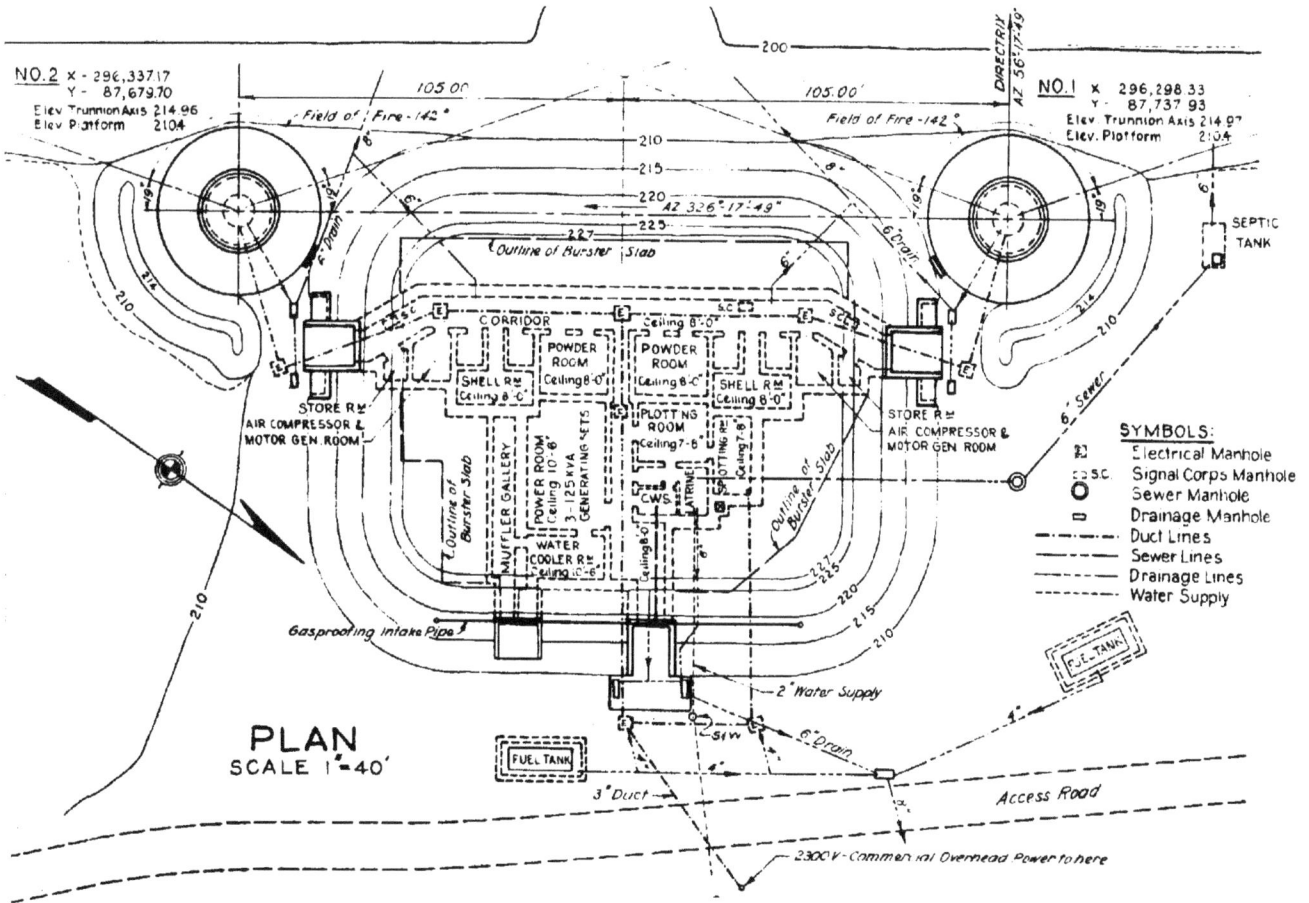

Battery plan of a World War II-era 6-inch battery (BCN 248, Fort Ebey, WA)

NOTES ON EMPLACEMENTS

An understanding of what is required to securely bolt a gun to the emplacement is necessary to interpret visible remains.

The **gun block** is the mass required to provide a solid foundation for the gun carriage. The anchor bolts are cast into this mass of concrete along with all provisions set forward by the Ordnance Department (such as clearance for recoil, counterweights, electrical boxes and conduits, drain holes, gutters etc.). The gun block is cast independently from the rest of the emplacement platform, to prevent it from being influenced by any general settling of the battery structure.

The **base ring** of the carriage is the foundation and lower roller bearing path for the gun carriage. It is vitally important that the base ring be perfectly level. The base ring is lowered over the anchor bolts, and another set of bolts is threaded into the flange of the ring to adjust the level against the thrust plates.

Thrust plates: These are the small steel plates used to support the leveling bolts of the carriage's base ring. They can be square or rectangular, and are often found still in place embedded in the fine white grout used to fill the space under the base ring. When the steel plates are gone their imprint is usually clearly visible along with the rust stain on the surface.

Anchor bolts: Bolts that are intact indicate that the carriage was removed with an eye toward using it elsewhere or the possibility that the emplacement would be rearmed at some point in time. For example those batteries that lost their guns for another site during WW I or WW II and those scrapped in the early 1920s tend to have left their bolts intact, often with nuts attached. Bolts cut off about 2-3 inches above the deck indicate that the nuts were cut off above the base ring flange and that the carriage was scrapped on site. In this case sometimes evidence of the work (scoring in the deck, pitting due to dropping molten slag from torches, and gouges resulting from dragging hunks of steel across the deck) can be readily seen. Bolts cut off flush with the deck generally indicate removal of protruding metal for safety reasons, typically done at active park sites.

Parapet wall: Generally the fewer the features on walls the earlier the emplacement. Light niches and exposed wire conduit paths indicate later construction. Because of the workings of disappearing guns and the fixed height of barbette guns, the height of the wall directly in front of the gun has a practical maximum and can usually be measured. Likewise the anchor bolt circle is a predictable distance from the wall directly in front in most types of gun emplacements and can also be measured.

Loading platform: Most older construction disappearing type batteries had their loading platforms extended to provide for more space for the crew. Changes in color of concrete and remnants of old design features (former stairways now blocked or evidence of ammunition davits) can be a sign of these changes. On the deck of the loading platform note the finish of the surface. Older construction often has a "diamond plate" nonskid finish. On disappearing batteries the gun carriage deck was at a lower level than the loading platform. The gun sat in a sort of well, semicircular in shape. As the carriage and breech always retracted to the same height above the loading platform (thus making possible standard size ammunition trucks), the depth of the well was "standard." To make servicing the carriage possible, concentric "steps" were built into the drop-off between the loading platform and the carriage deck. At a normal step size, it is possible on many batteries to count the steps and identify the carriage type. Exceptions do exist, particularly on the early M1894 10-inch carriage emplacements; but the technique is really quite useful for field and photographic identification.

Wall details: On earlier batteries many telephone and telautograph niches (along with most electrical fixtures) were retrofitted onto walls. Ammunition hoists were also added or upgraded after construction, explaining odd or awkward placements. Masking parapet guns (15-pounder and 5-inch) emplacements had barrel niches built into flank walls to allow a place for the gun barrel to rest when the carriage was retracted.

GUIDE NOTES

Usual gun model: The common model number of guns usually mounted on this particular carriage. Usually for all "M" variants of the year and model number (but not specifically discribed in this guide). Only occasionally useful in photo identification, and of course not at all of help in empty emplacements.

Carriages built: Number of this type of carriage physically fabricated, or acquired by the army if built by another service or nation. Records are not at hand on the more modern types, so approximations are given for an idea of commonness.

Original emplacements: Number of physical gun sites built for the original "run" of this type of carriage. That might count some relocations during the early phases of use, but generally not major efforts to move during future years (as happened to many 3-inch and 6-inch guns as well as 12-inch mortars). Also included as a guide to frequentness and likelihood of encountering.

Time emplaced: General time frame of when these types of carriages were installed in their structural emplacements. Approximate only, major relocations generally are also noted.

Number of bolts: Actual number of anchor foundation bolts set in concrete on the gun block. Usually arranged in a regular circle, and countable in existing emplacements as an aid to identification. For some carriages both an interior and exterior ring of bolts existed.

Bolt circle diameter: The diameter described by the anchor bolt circle. Some of the data comes from field measurement and projections from drawings and might be close but not exact.

Parapet height: Measured height from base of deck directly in front of the gun mounting and crest of the parapet. Some variation existed, obviously there was a maximum but no minimum height.

Carriage center-to-parapet wall: The measured distance from the center of the bolt circle to the edge of the parapet wall directly in front of the gun at deck level (some walls had upper lips or were sloped). Can be converted to distance from bolt circle line with measurement from diameter halved for radius. While not physically demanded, except to allow for clearance of retracting guns, this distance is usually dependable.

Notes: Additional comments relative to carriage type identification. Includes presence of collars, barrel niches, locations of unique items, hoist uniqueness, traversing rings, and step number analysis.

Gun and carriage information & drawings: Material was derived from *The Service of Coast Artillery*, by Frank T. Hines and Franklin W. Ward (Goodenough & Woglom Co., New York, 1910); *American Seacoast Material*, Ordnance Department Document #2042, GPO, Washington, DC (June 1922); and *Seacoast Artillery Weapons*, Army Technical Manual (Army Coast Artillery) TM-210, GPO, Washington D.C., (1944), and/or the individual Ordnance Department Documents (*Instructions for Mounting, Using and Caring for . . .)* specific for the carriage model illustrated. All battery drawings are from the Army Corps of Engineers "Report of Completed Works, Form 7," (National Archives) for the battery illustrated.

Photographs: GMW: Glen M. Williford, MAB: Mark A. Berhow, TM: Terrance McGovern, AMH: Alex M. Holder, Jr., NA: The National Archives/ U.S. Army Photographs, other photograph sources are fully noted in the captions.

M1898 & M1898MI Masking Parapet Mount for 3-inch Gun

Usual guns: M1898, M1898M1
Carriages Built: 120
Original Emplacements: 111
Time Emplaced: 1899-1905

Number of Bolts: 8
Circle Diameter: 3' 5"
Parapet Height: 3' 6"
Center-to-parapet: 2' 9"
Notes: generally collar base is present with filled center well. Barrel niches in flank walls. Later, M1898 carriages were modified to M1898MI standard by removing counterweight and filling well below pedestal. The carriage was then fixed in raised position. No new M1898MI carriages were built.

Total weight: 3 tons
Max./min. elevation: +15 / –5 degrees
Turning radius: 360 degrees

M1898 gun
Weight: 1,782 lbs
Length: 50 calibers

Photos:
Upper: As installed in a battery, 1900s (NA).

Middle: Battery Crenshaw, Fort Columbia, HD Columbia River, 1987 (AMH).

Lower: Battery Irwin, Fort Monroe, HD Chesapeake Bay, 2019 (MAB).

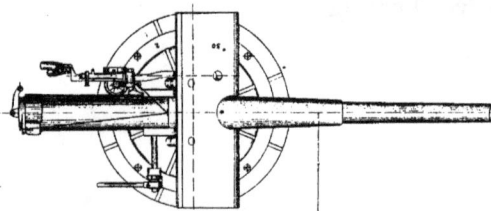
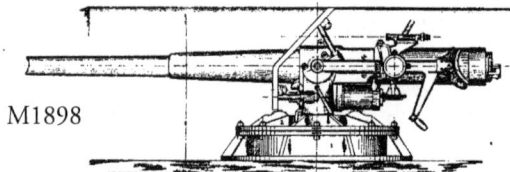

M1898

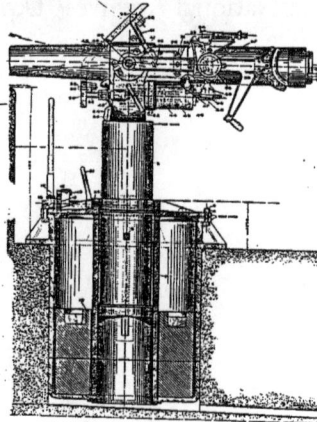
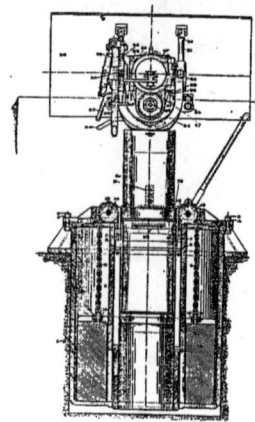

M1898 & M1898MI Masking Parapet Mount for 3-inch Gun

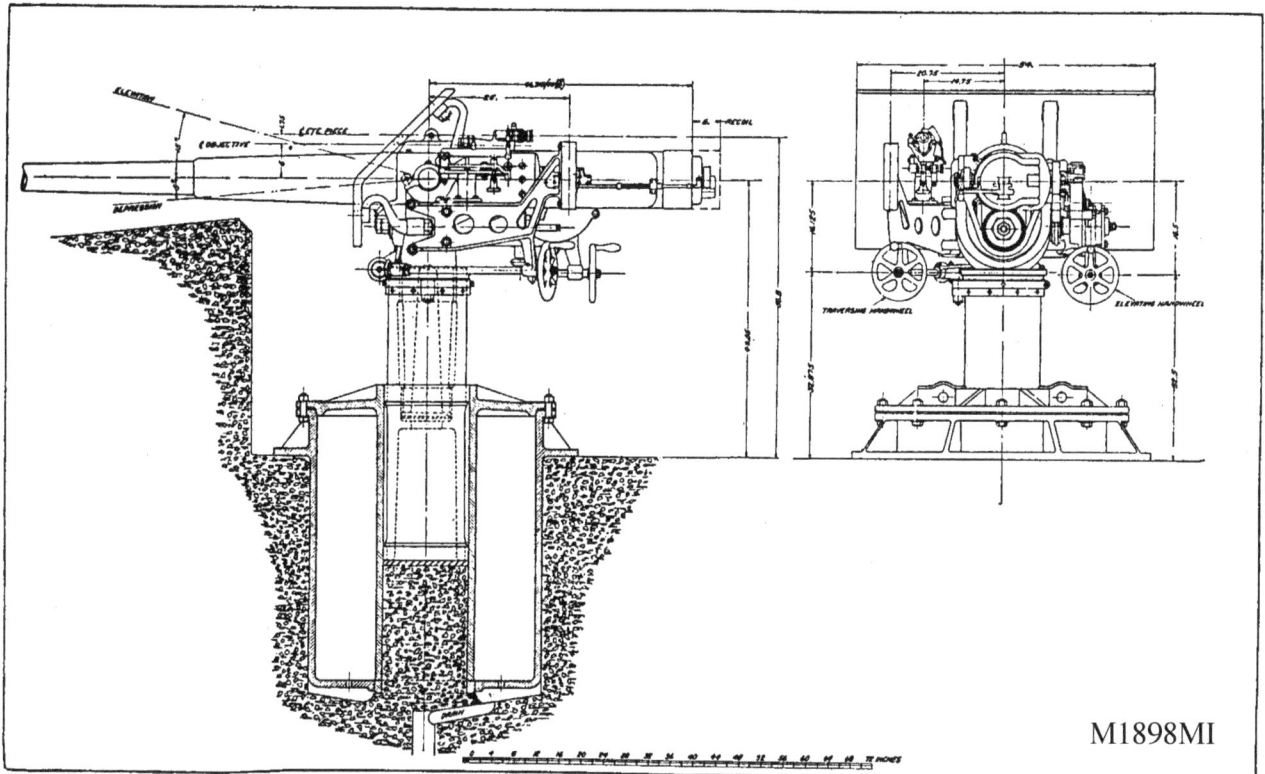

3-INCH (15 PDR.) BARBETTE CARRIAGE, MODEL OF 1898MI.

SECTION B-B.

SECTION A-A.

REAR ELEVATION
from
SECTION AT EDGE OF SIDEWALK.

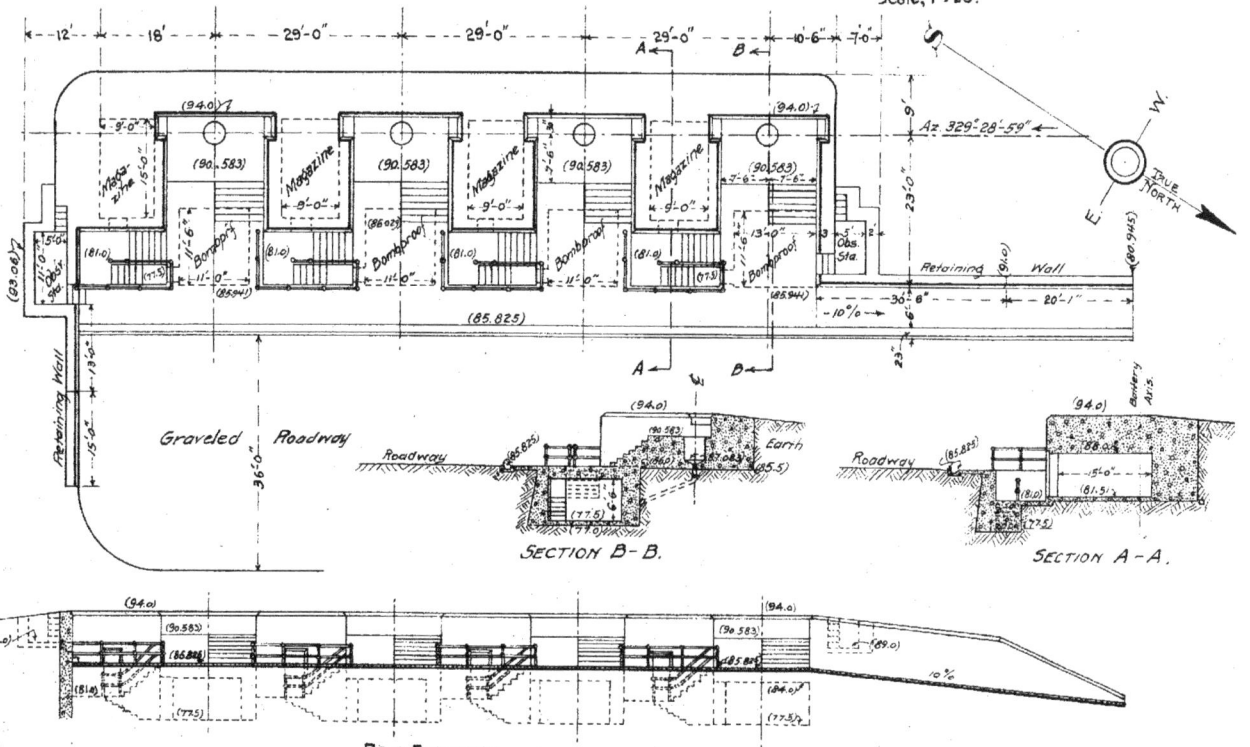

Battery Thornburgh, Fort Ward, HD Puget Sound

M1902 Pedestal Mount for 3-inch Gun

Usual guns: M1902
Carriages Built: 60
Original Emplacements: 60
Time Emplaced: 1903-1910, relocations until 1942

Number of Bolts: 10
Circle Diameter: 3' 1"
Parapet Height: 3'
Center-to-parapet: variable
Notes: uneven bolt distribution around circle, often a center pipe for electrical conduit. (See also entry for casemate mount).

Carriage weight: 4,075 lbs
Max/min elevation: +15 / –10 degrees
Turning radius: 360 degrees

Gun: M1902
Weight: 1,950 lbs.
Length: 50 calibers

3" (15 PDR) GUN AND MOUNT
DESIGNED AND BUILT BY
BETHLEHEM STEEL COMPANY
SO. BETHLEHEM, PA., U.S.A.

Photos:
Upper: from Bethlehem Steel Archives (GW).

Middle: Battery Irwin, Fort Monroe, HD Cheaspeake Bay (M1902A1 as modified during WW II), guns still in place, 2010 (John Stanton).

Lower: Battery Yates, Fort Baker, HD San Francisco, 1992 (MAB).

M1902 Pedestal Mount for 3-inch Gun

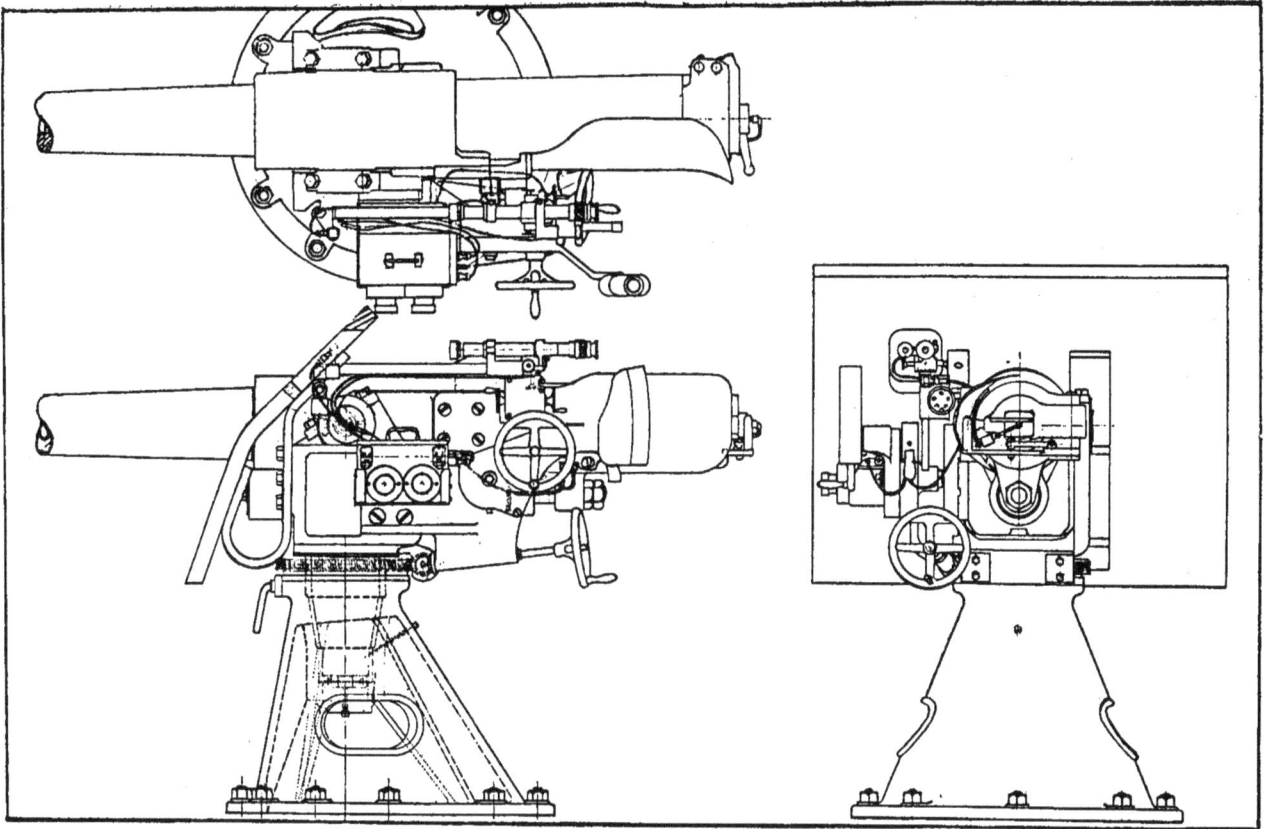

3-INCH BETHLEHEM BARBETTE CARRIAGE.

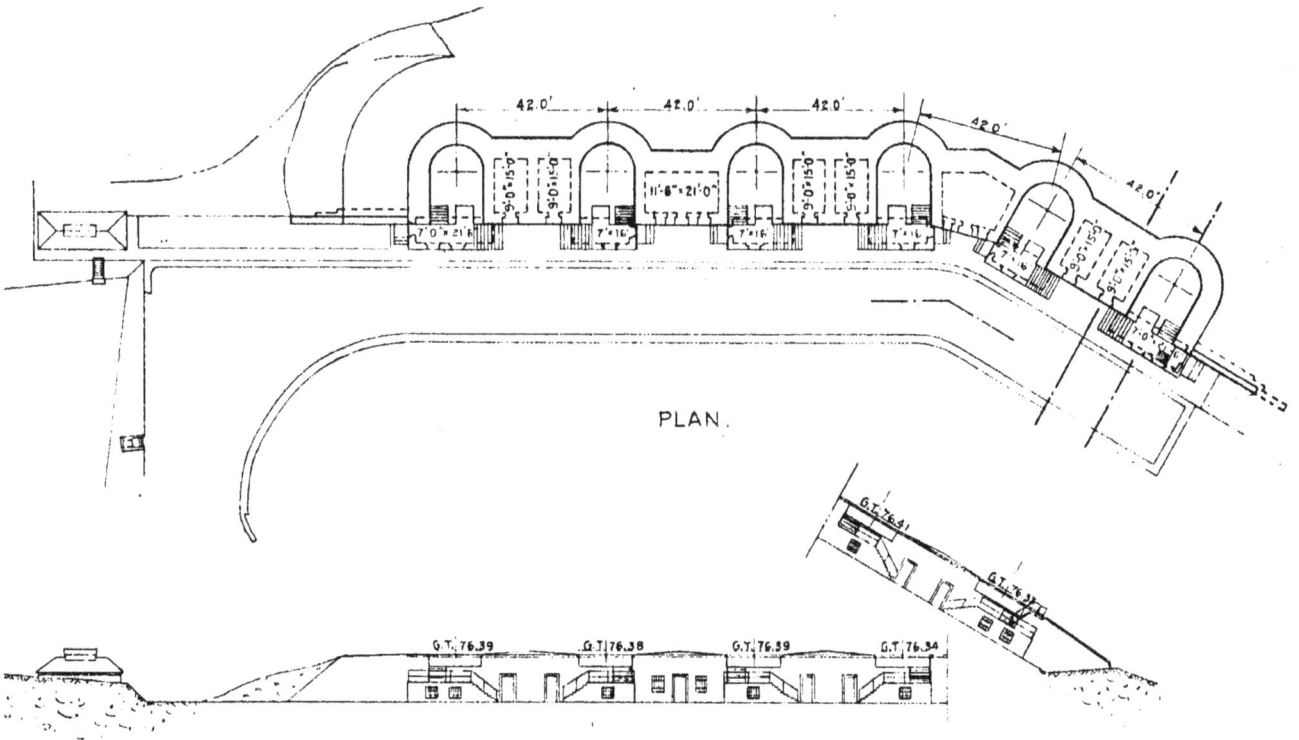

PLAN.

REAR ELEVATION.

Battery Yates, Fort Baker, HD San Francisco

M1903 Pedestal Mount for 3-inch Gun

Usual guns: M1903
Carriages Built: 107
Original Emplacements: 79 US,
22 Terr.
Time Emplaced: 1904-1917,
relocations until 1942

Number of Bolts: 12
Circle Diameter: 3' 1"
Parapet Height: 3'
Center-to-parapet: variable

Notes: even distribution of 12
bolts, often with center pipe for
electrical conduit

Carriage weight: 3,310 lbs
Max/min elevation: +16 / −10
degrees
Turning radius: 360 degrees

Gun: M1903
Weight: 2,690
Length: 55 calibers

Battery Trevor, Fort Casey, HD Puget Sound, gun brought over from the Phillipines in the early 1960s by the Washington State Parks Dept. 2006 (MAB).

Photos:

Middle left: Battery
Van Horne, Fort Casey,
HD Puget Sound , 1995
(MAB).

Middle Right:: Battery
Belton, Fort Adams,
HD Narrangansett Bay,
1987 (AMH).

Below: Fort Casey,
1920s (Dan Kerlee
Collection)

M1903 Pedestal Mount for 3-inch Gun

3-INCH BARBETTE CARRIAGE, MODEL OF 1903.

PLAN

REAR ELEVATION
WITH LONGITUDINAL SECTION THRU GUTTER

Battery Samuel Walker, Fort Worden, HD Puget Sound

M1917 Pedestal Mount for 3-inch AA Gun

Usual guns: M1917 & modifications
Carriages Built: over 200
Original Emplacements: about 170
Time Emplaced: 1918-1939, relocations 1942

Number of Bolts: 16
Circle Diameter: 5' 1.5"
Parapet Height: none
Center-to-parapet:
Notes: "keyhole" indented shape in concrete, bolts recessed below platform.

The M3 mount for 3-inch AA guns is not shown in this guide.

Photos:
Above: On Corregidor Island, Manila Bay, the Philippines, 1922 (NA)

Middle: Sand Island, Oahu, December, 1941, manning detachment from Battery F, 55th CA (TD) Regiment. (NA)

Below: emplacement at Fort Cronkhite, HDSF, just before gun installation, 1939. (NA)

Due to lack of complete information, no listing is provided for these sites.

M1917 Pedestal Mount for 3-inch AA Gun

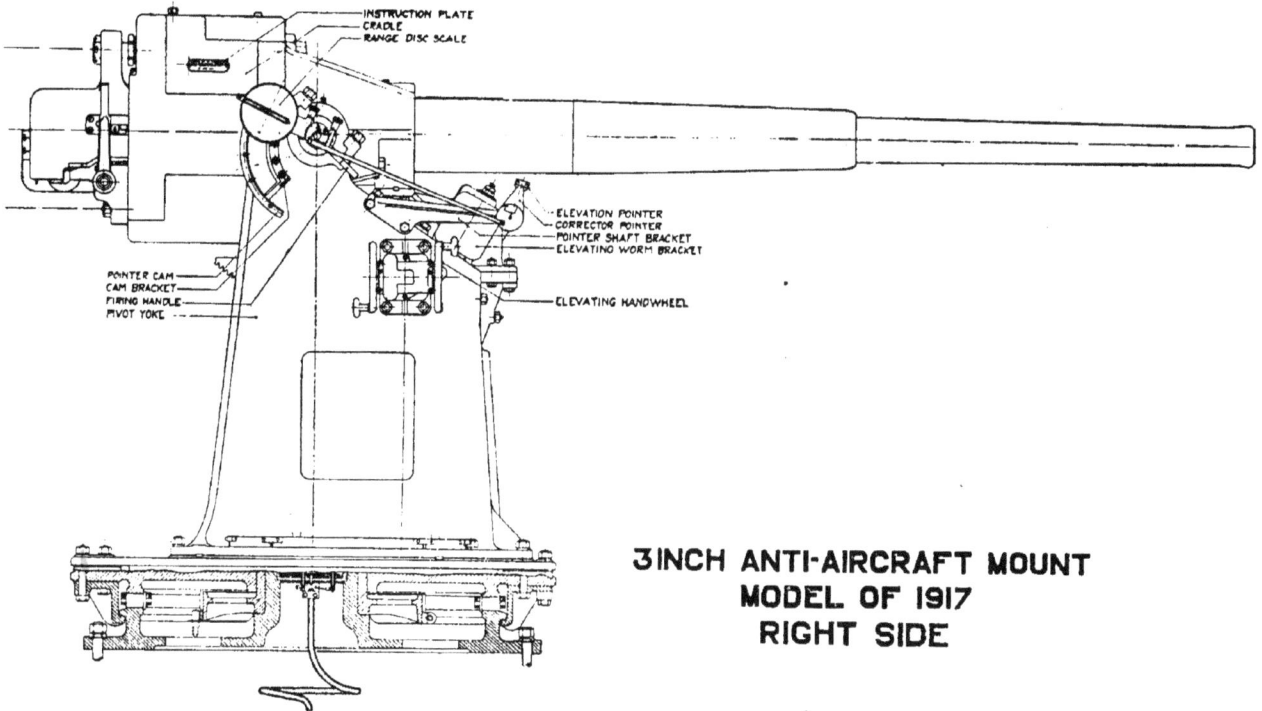

INSTRUCTION PLATE
CRADLE
RANGE DISC SCALE

ELEVATION POINTER
CORRECTOR POINTER
POINTER SHAFT BRACKET
ELEVATING WORM BRACKET

ELEVATING HANDWHEEL

POINTER CAM
CAM BRACKET
FIRING HANDLE
PIVOT YOKE

3 INCH ANTI-AIRCRAFT MOUNT
MODEL OF 1917
RIGHT SIDE

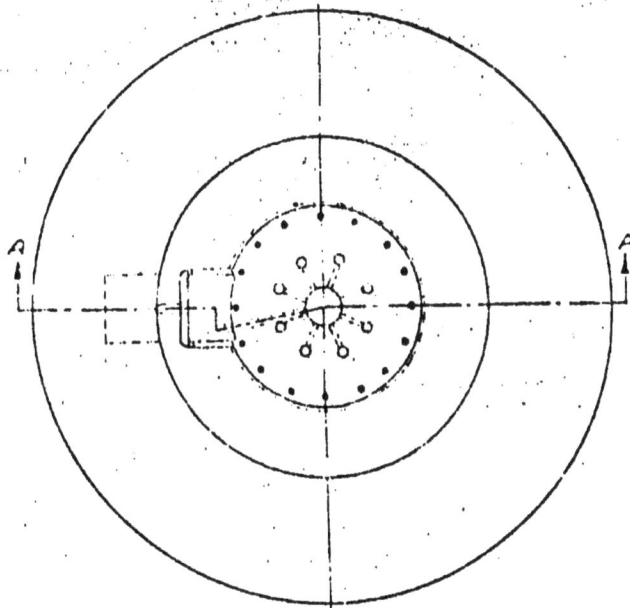

PLAN

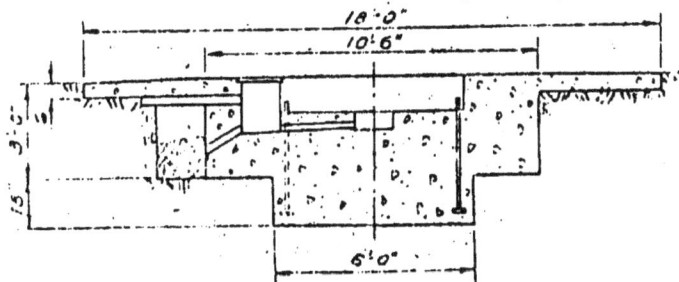

18'-0"

10'-6"

3'-0"

1'-5"

6'-0"

SECTION A-A

M1915 Pedestal Mount for 4.7-inch Howitzer & 75 mm gun

Usual Tubes: M1917 & Mod. G
Carriages Built:
Original Emplacements:
Time Emplaced: 1920s?

Number of Bolts: 12
Circle Diameter: 37 inches
Parapet Height: none
Center-to-parapet:

Approximately two-dozen 4.7-inch M1906 field howitzers were emplaced on permanent gunblocks (on fixed M1915 pedestals) for land defense in the Panama Canal Zone. The bolt pattern was purposely designed to be interchangeable with the M1903 pedestal for 3-inch guns (12 bolts, 37" diameter). These served from 1918 to 1926, when the tubes were removed and replaced with 75 mm M1897 field guns.

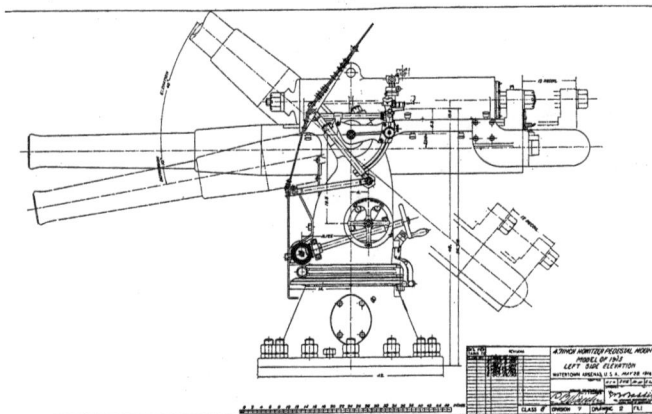

Photos:
Upper: Panama Canal Zone site, 1993 (AMH)

Middle: Panama Mark Berhow Collection

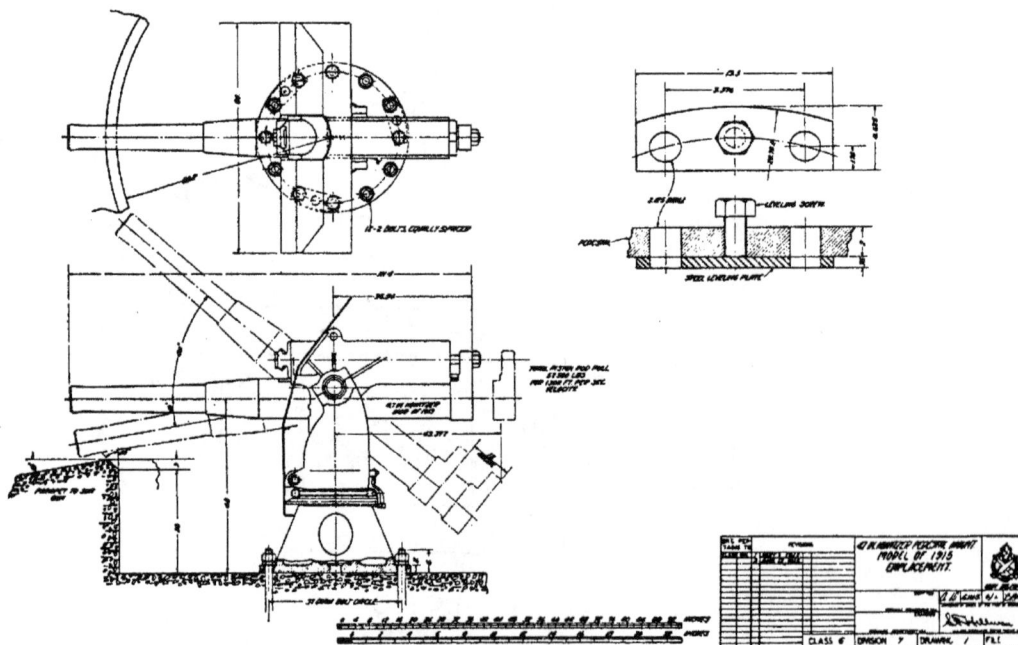

M1918 MOBILE MOUNT FOR 3-INCH AA GUN

PLATFORM FOR M1918 3"AA GUN.

10'-6"

Hole for lifting jackscrews

A

10'-6"

A

DETAIL PLAN

Concrete
oak sill
Anchor bolts for gun carriage

2'-0"

Pipe for electric wires
Anchor bolts
Reinforcement

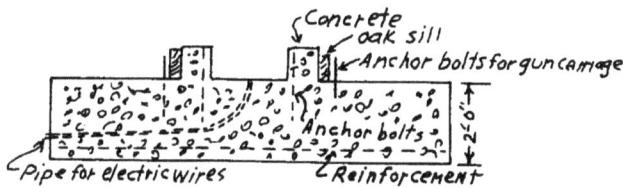

SECTION A-A
Scale 1/4 inch = 1 ft.

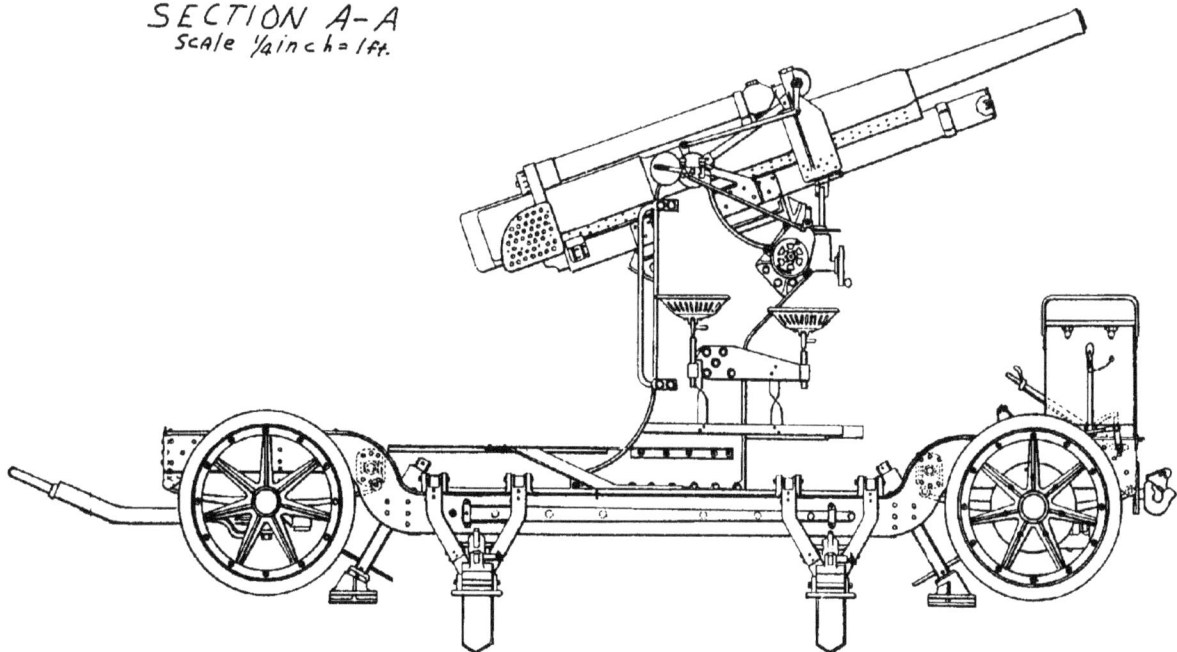

Photo: Puget Sound Coast Artillery Museum Collection.

Notes: Found at some locations, notably Puget Sound, San Francisco and Portland, ME.

Due to lack of complete information, no listing is provided for these sites.

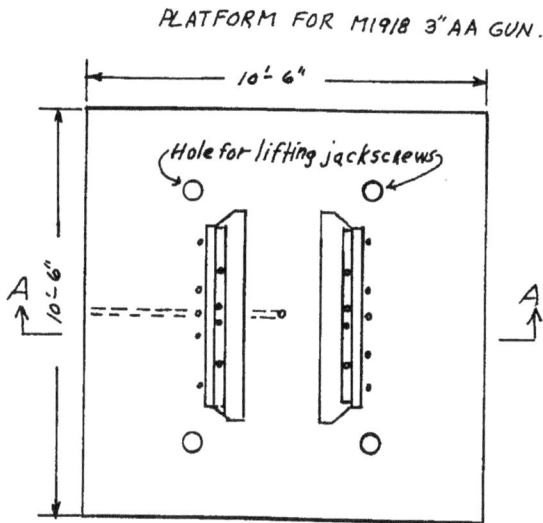

SIDE ELEVATION

M3 FIXED MOUNT FOR 90 MM GUN

Usual guns: M1
Carriages Built: 248?
Original Emplacements: over 200
Time Emplaced: 1943-1946

Number of Bolts: 16
Circle Diameter: 3' 10"
Notes: concrete block 14 feet in diameter, often with collar remaining, and a rectangular conduit hatch adjacent to bolt circle. Complete AMTB battery consisted of two 90 mm guns on fixed mounts, two 90 mm gun on mobile (M1A1) mounts and two 37 mm or 40 mm guns on mobile mounts. Some AMTB batteries featured only mobile armament. The mobile mounts M1, M2 and M1A1 are discussed in *Antiaircraft Artillery used by the Coast Artillery Corps.*

Carriage weight: 4,110 lbs
Maximum/minimum elevation: +80 / −10 degrees
Turning radius: 360 degrees

gun: M1
Weight: 1,465 lbs
Length: 50 calibers

Photos:
Top: AMTB Battery Fort Williams, HD Portland, ME, ca. 1944 (James Piper Collection, courtesy Joel Eastman)

Middle: AMTB #962, Fort Levett, HD Portland, ME, 1986. (AMH)

Bottom: In Battery Parrot, at Fort Monroe, 2010 (John Stanton)

M3 FIXED MOUNT FOR 90 MM GUN

90mm Gun M1 - Mount M3

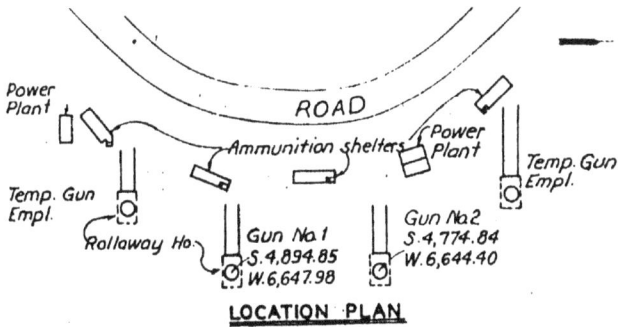

LOCATION PLAN

HARBOR DEFENSES OF LOS ANGELES, CALIF.
GAFFEY BULGE
90MM-A.M.T.B. BATTERIES
U.S. ENGINEER OFFICE, LOS ANGELES, CALIF.

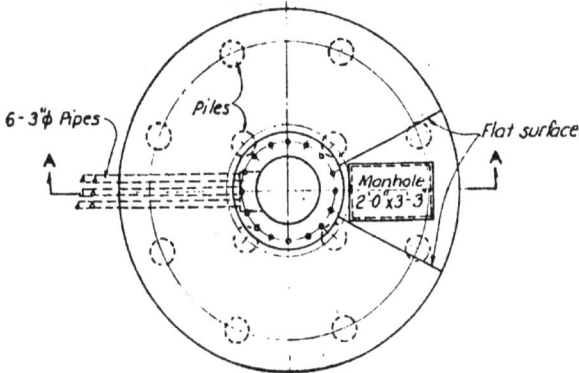

FIRING PLATFORM

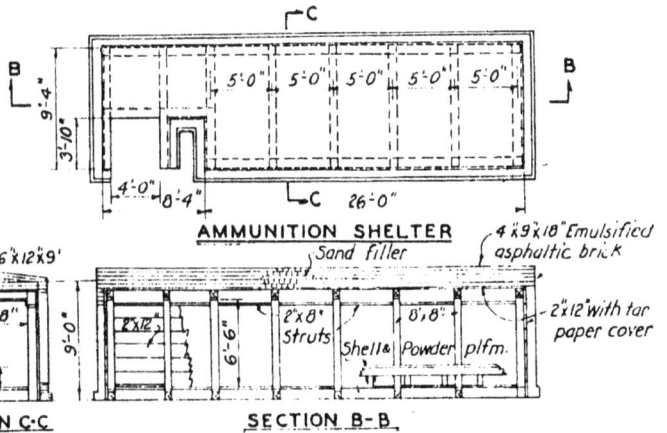

AMMUNITION SHELTER

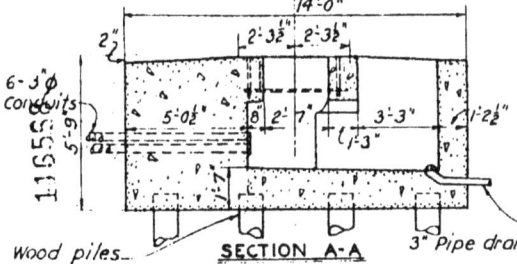

SECTION A-A

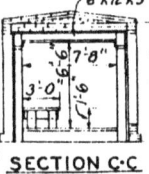

SECTION C-C

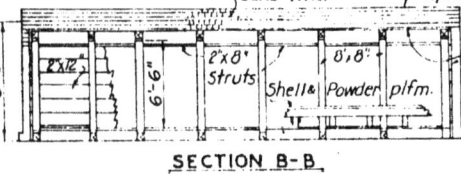

SECTION B-B

AMTB battery at Gaffey Street Bulge, Fort MacArthur, HD Los Angeles

M1 BARBETTE MOUNT FOR 105 MM AA GUN

Another unique Panama Canal Zone mount. The M3 gun on M1 barbette (only 12 produced and emplaced) were re-emplaced in four three-gun batteries in the Zone around 1938.

THIS IS THE STANDARD 105mm ANTIAIRCRAFT WEAPON, FOR FIXED MOUNTS. (1940)

105mm GUN M3 AND MOUNT M1 MANNED, RIGHT HAND VIEW

'LETED WORKS SEACOAST DEFENSES ANTI-AIRCRAFT DEFENSES OF THE PANAMA CANAL.
ABOUT 600 YARDS EAST OF MADDEN ROAD

BATTERY No 14 (ANTI-AIRCRAFT)

No. of Guns 3, Caliber 105mm Carriage

Corrected to Oct. 13, 1934.   Scale:

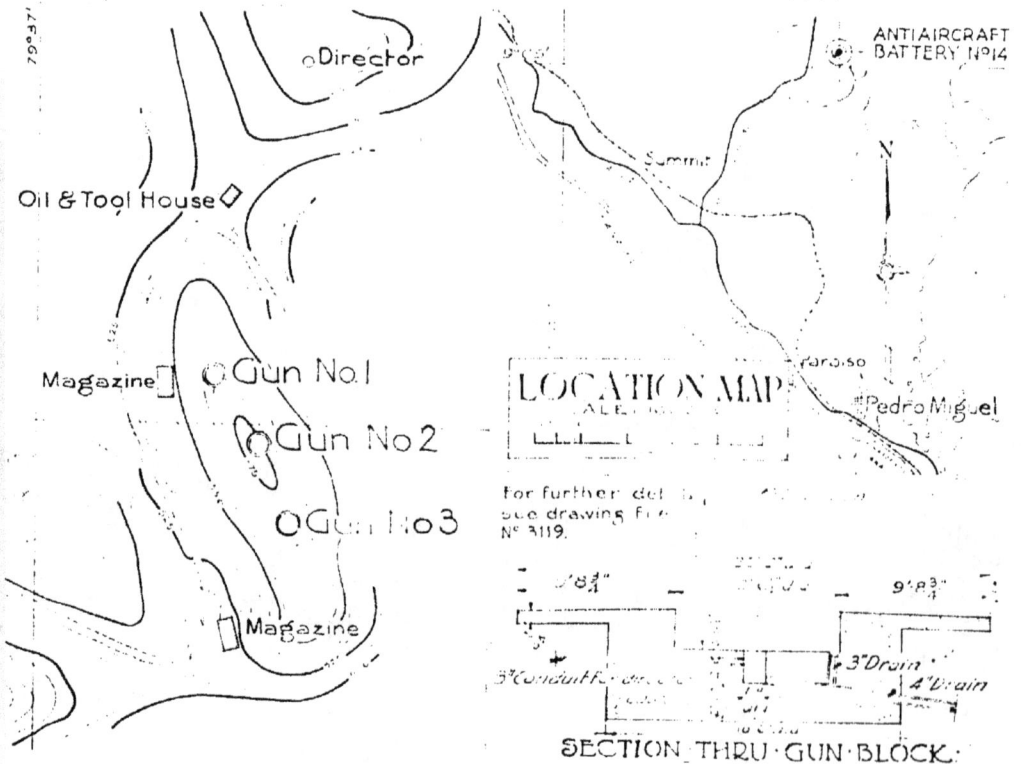

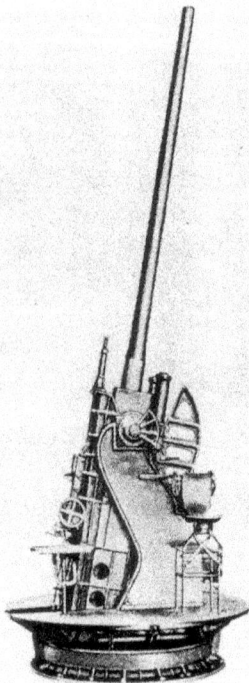

oDirector

Oil & Tool House ◇

Magazine □ o Gun No.1

o Gun No 2

o Gun No 3

□ Magazine

ANTIAIRCRAFT BATTERY No 14

Summit

LOCATION MAP

Paraiso

Pedro Miguel

For further det..
see drawing f..
No 3119.

3"Drain
4" Drain

3" Conduit

SECTION THRU GUN BLOCK

M1 BARBETTE MOUNT FOR 105 MM AA GUN

105 mm AA mount in Panamanian national park, Pacific side, 1999 (AMH)

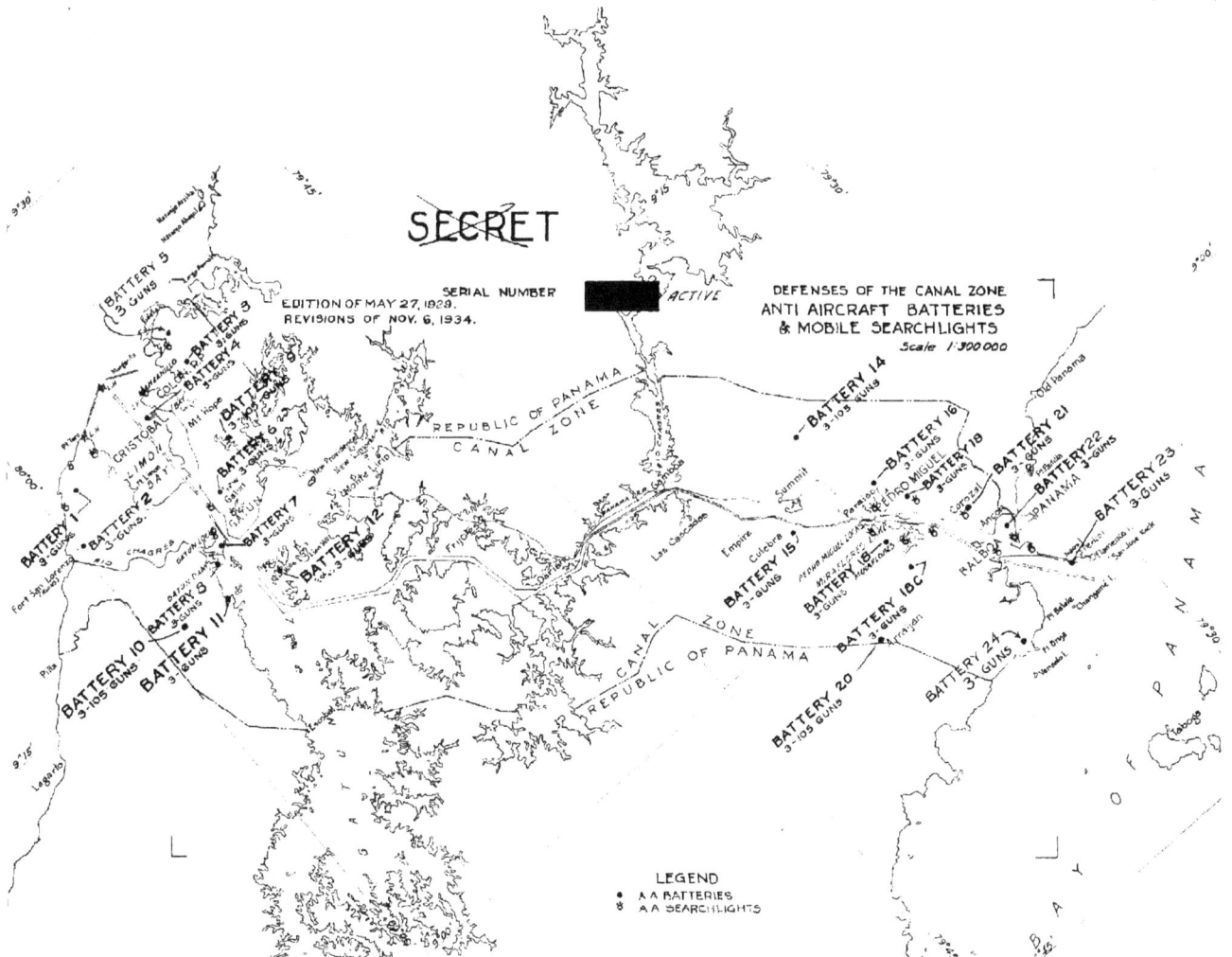

ARMY PEDESTAL MOUNT FOR 4-INCH NAVY GUN

Usual Tubes: 4-inch navy gun
Carriages Built: 4 by Army
Original Emplacements: 4
Time Emplaced: 1898

Number of Bolts: 14
Circle Diameter: not circular
turning radius: 360 degrees
Notes: "D" shaped bolt pattern
Two batteries: Battery Plunkett, Fort Warren, HD
Boston, and Battery White, Fort Washington, HD
Potomac River. The guns were manufactured by
Driggs-Schroeder Company, the carriages by William
Cramp & Sons. Plunkett's guns were later moved to
another location at Fort Warren.

Gun: Driggs Schroeder
gun weight: 3,613 lbs
length: 40 calibers

Photos:
Top: Battery Plunkett, Fort Warren, HD Boston, 1910s. (NA)

Middle: as display pieces at Fort Warren, 1941 (MHI; Al Schroeder Collection)

Bottom: Battery Plunkett, 1985 (AMH).

ARMY PEDESTAL MOUNT FOR 4-INCH NAVY GUN

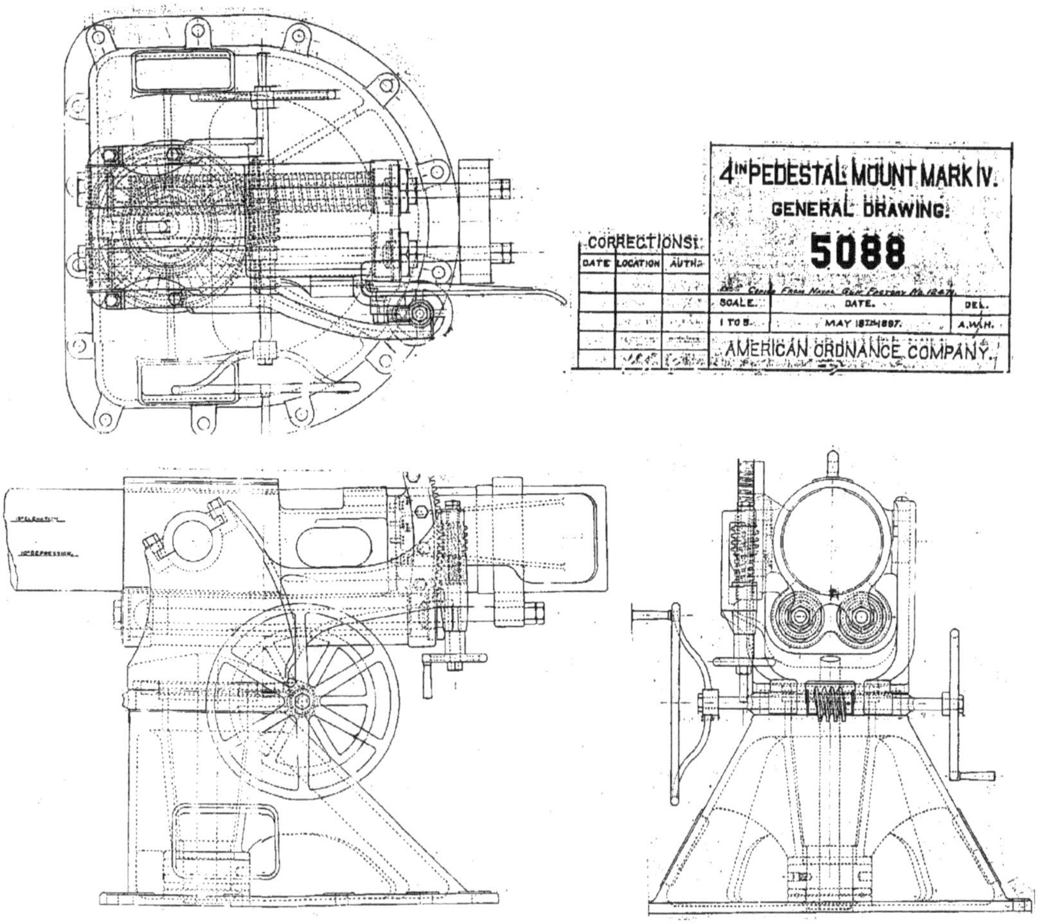

4 IN PEDESTAL MOUNT MARK IV.
GENERAL DRAWING
5088

CORRECTIONS:		
DATE	LOCATION	AUTH:

SCALE. | DATE. | DEL.
1 TO 5. | MAY 18TH 1897. | A.W.H.

AMERICAN ORDNANCE COMPANY.

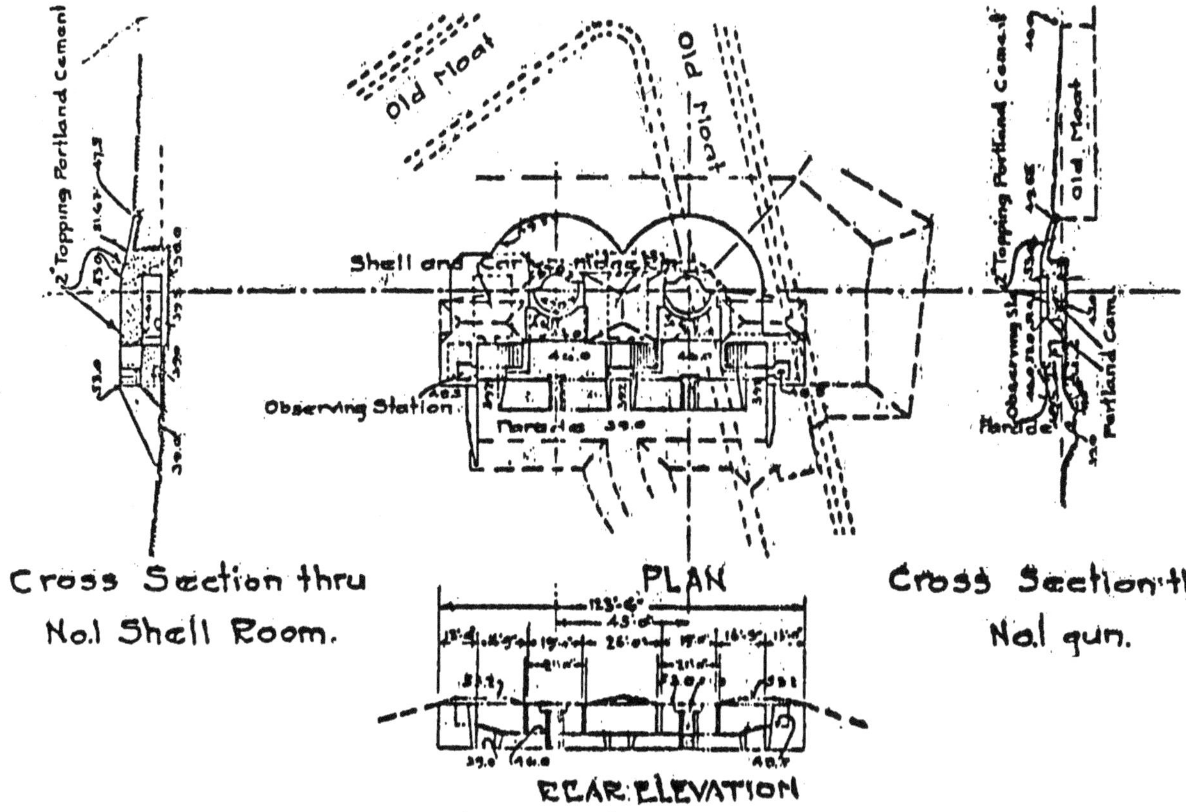

Cross Section thru
No.1 Shell Room.

PLAN

Cross Section thru
No.1 gun.

REAR ELEVATION

Battery White, Fort Washington, HD Potomac River

ARMSTRONG PEDESTAL MOUNT FOR 4.72-INCH GUNS

Usual Tubes: 40, 45, & 50 calibers
Carriages Bought: 34
Original Emplacements: 34
Time Emplaced: 1898-1899,
relocations until 1917

Number of Bolts: 18
Circle Diameter: 2' 9"
Parapet Height: 1' 10"
Center-to-parapet: 4' 10"
Notes: Central pivot mount, often
on elevated concrete pedestal

Total weight:
Max./min. elevation:
Turning radius: 360 degrees

gun: Armstrong
Weight: 4,648 lbs
Length: 40 calibers

gun: Armstrong
Weight: 5,958 lbs
Length: 45 calibers

gun: Armstrong
Weight: 6,160 lbs
Length: 50 calibers

Photos:
Above: Newport, RI, 2011
(MAB)

Middle: Ansonia, CT, 2003
(MAB)

Below: Battery Talbot, Fort
Adams, HD Narragansett Bay,
2011 (MAB)

ARMSTRONG PEDESTAL MOUNT FOR 4.72-INCH GUNS

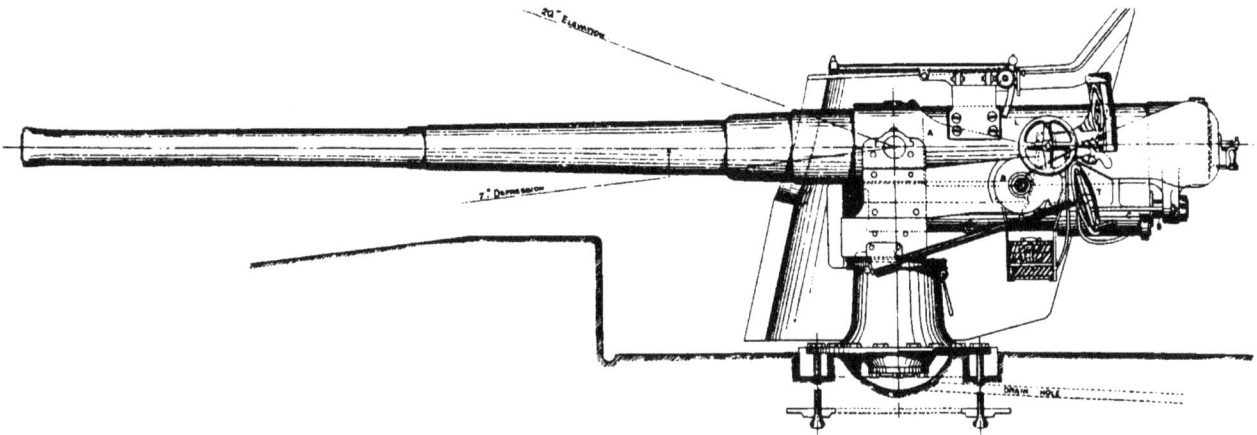

4.7 INCH BARBETTE CARRIAGE. ARMSTRONG

50 Calibre Gun

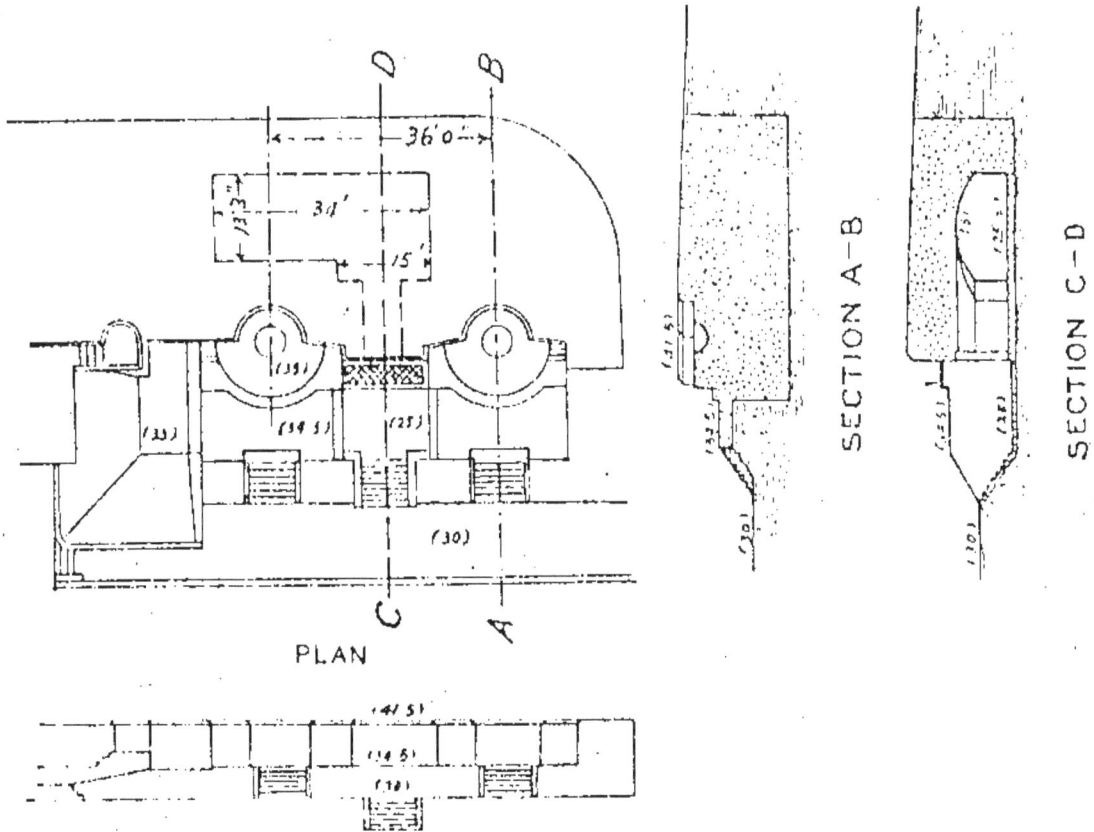

PLAN

REAR ELEVATION

SECTION A-B

SECTION C-D

Battery Irons, Fort Armistead, HD Baltimore

M1896 BALANCED PILLAR MOUNT FOR 5-INCH GUNS

Usual Tubes: M1897
Carriages Built: 32
Original Emplacements: 32
Time Emplaced: 1897-1903,
relocations until 1917

Number of Bolts: 24 usually covered
by collar & plate
Circle Diameter: 4' 9" collar diameter
Parapet Height: 6' 8"
Center-to-parapet: 10' 6"
Notes: collar on 12" elevated concrete
"cone," barrel niches in emplacement
flanks. During WWI some M1896
weapons were removed from the
balanced pillar and re-emplaced on
simple concrete pedestals.
Number of bolts: 24; circle diameter:
43.5 inches. An
example remains
at Westport (Grays
Harbor), WA

carriage weight: 48,809
lbs
Max./min. elevation:
Turning radius: 360
degrees

M1897 gun
Weight: 7,583 lbs
Length: 45 calibers

Photos:
Top: Battery Vicars,
Fort Worden, HD Puget
Sound, 1995 (AMH).

Upper middle: Coutesy
Micheal Kea.

Lower Middle: Battery
Vicars, Fort Worden,
HD Puget Sound, 1910s
(Dan Kerlee Collection)

Below: Battery Orlando
Wagner, Fort Baker, HD
San Francisco. 1994
(MAB).

M1896 BALANCED PILLAR MOUNT FOR 5-INCH GUNS

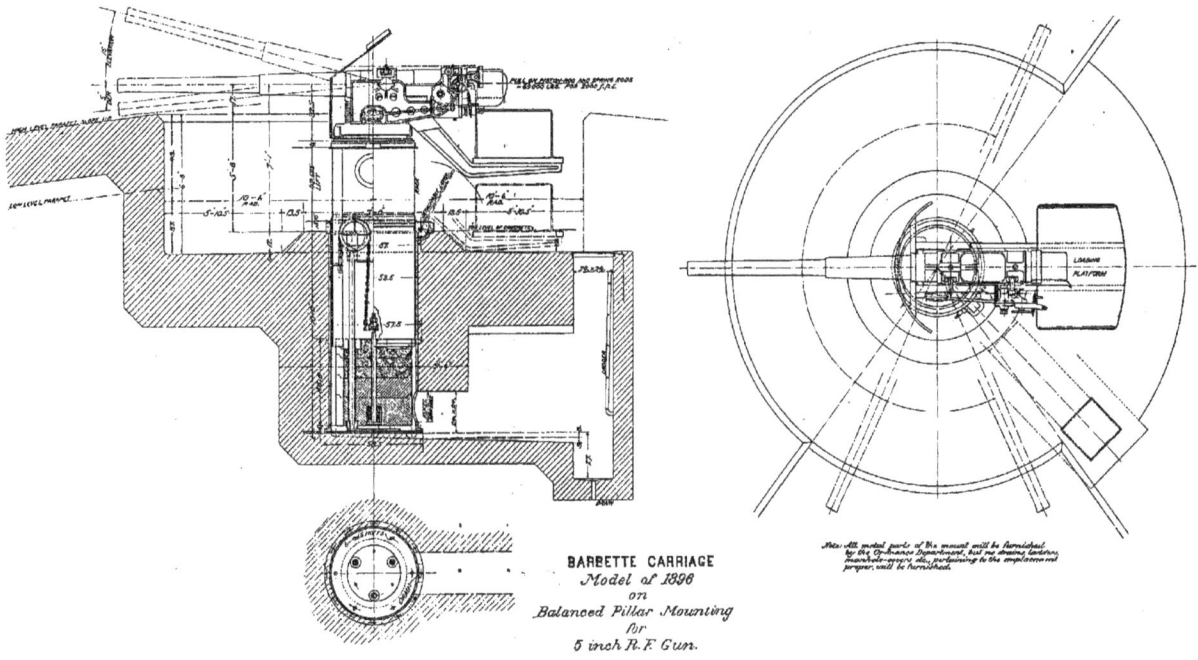

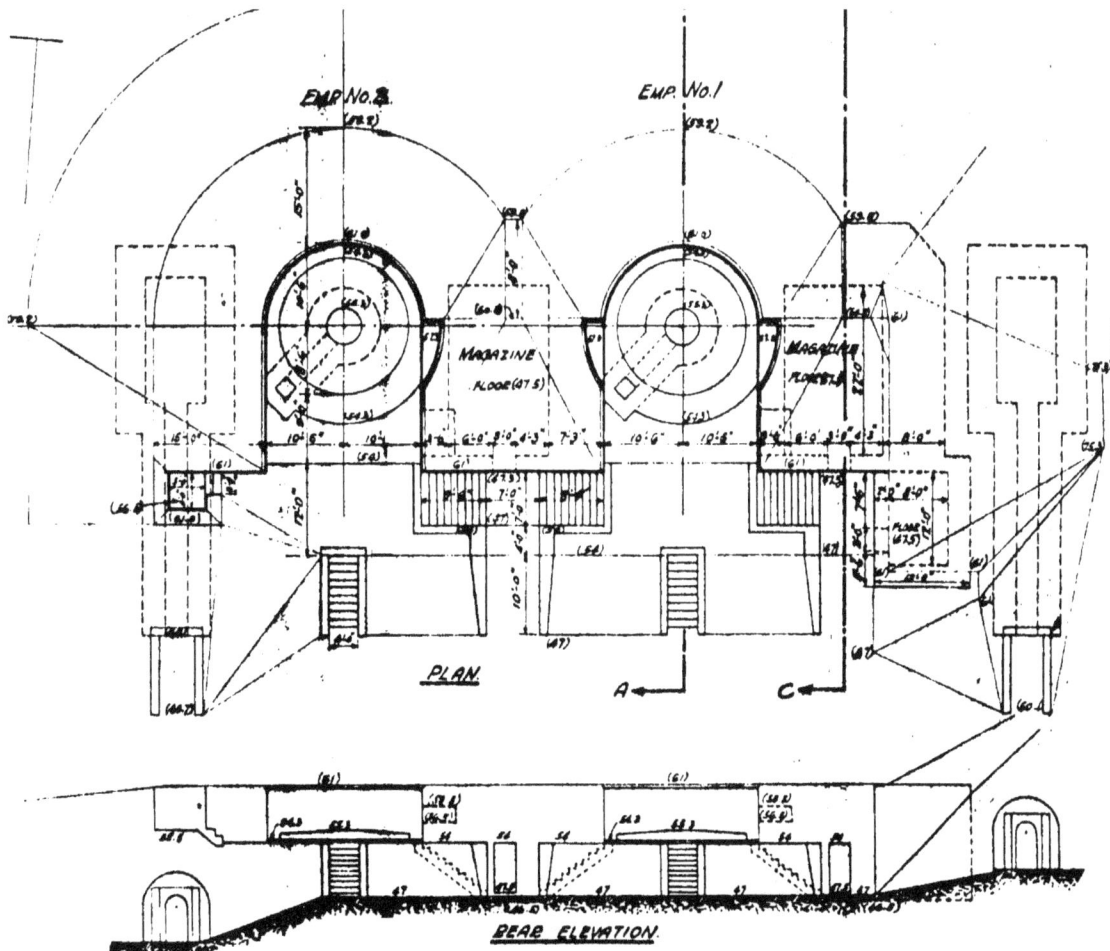

BARBETTE CARRIAGE
Model of 1896
on
Balanced Pillar Mounting
for
5 inch R.F. Gun.

PLAN.

REAR ELEVATION.

Battery Stuart, Fort Totten, HD Eastern New York

M1903 Pedestal Mount for 5-inch Guns

Usual Tubes: M1900
Carriages Built: 21
Original Emplacements: 20
Time Emplaced: 1904-1909,
relocations until 1919

Number of Bolts: 16
Circle Diameter: 4'1-5/16"
Parapet Height: 2' 9"
Center-to-parapet: 3' 2"
Notes: often on elevated
pedestal

Total weight:
Max./min. elevation:
Turning radius: 360 degrees

M1900 gun
Weight: 11,120 lbs
Length: 50 calibers

Photos:

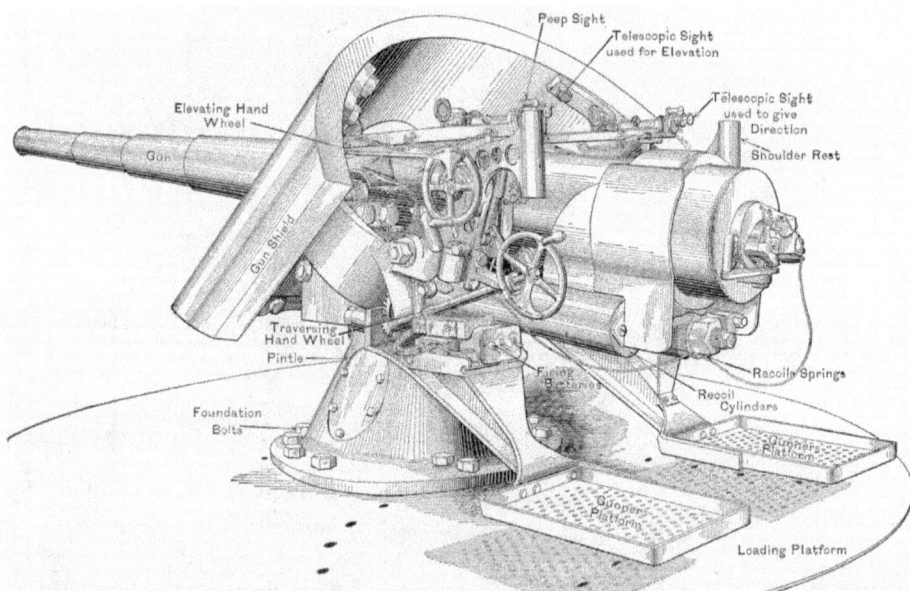

Top: Sandy Hook, NJ 1900s?
(NA—courtesy GMW)

Below: Battery Ledyard,
Fort McDowell, HD San
Francisco,1994 (MAB).

M1903 PEDESTAL MOUNT FOR 5-INCH GUNS

Battery Warner, Fort Ward, HD Puget Sound

ARMSTRONG PEDESTAL MOUNT FOR 6-INCH VICKERS GUN

Usual guns: Vickers
Carriages Bought: 8
Original Emplacements: 11 US, 4 Terr.
Time Emplaced: 1898-1900, relocations until 1916

Number of Bolts: 24
Circle Diameter: 2' 10"
Parapet Height: 2' 3"
Center-to-parapet: 5'
Notes: no ammunition hoists

Total weight:
Maximum/minimum elevation: +16 / −7 degrees
Turning radius: 360 degrees

Gun: Vickers
Weight: 14,784 lbs.
Length: 40 calibers

Photos:
Top: Battery Burchsted, Fort Dade, HD Tampa Bay, 1910s.(NA)

Middle: on display at Fort De Soto County Park, St. Petersburg, FL, 2010 (MAB).

Bottom: Battery Logan, Fort Moultrie, HD Charleston, 1990s (GMW).

ARMSTRONG PEDESTAL MOUNT FOR 6-INCH GUN

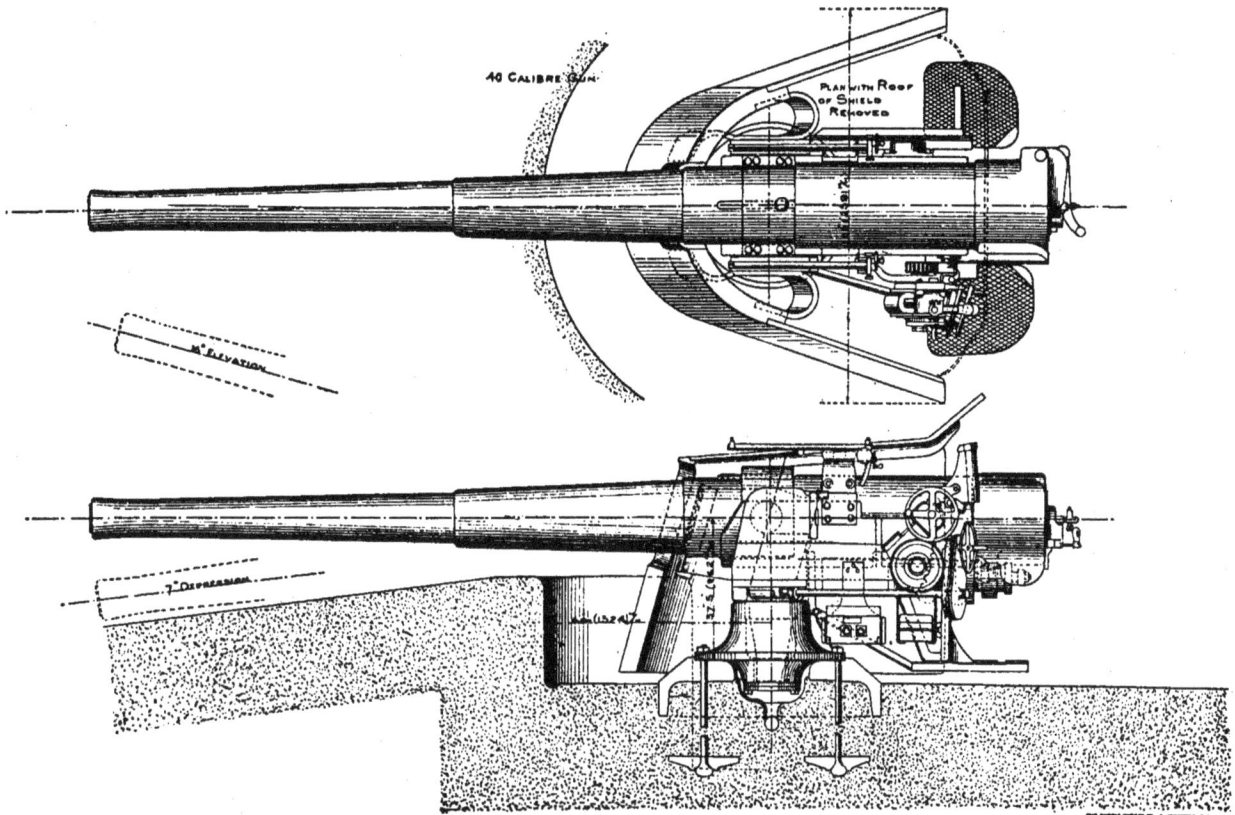

6-INCH ARMSTRONG CARRIAGE.

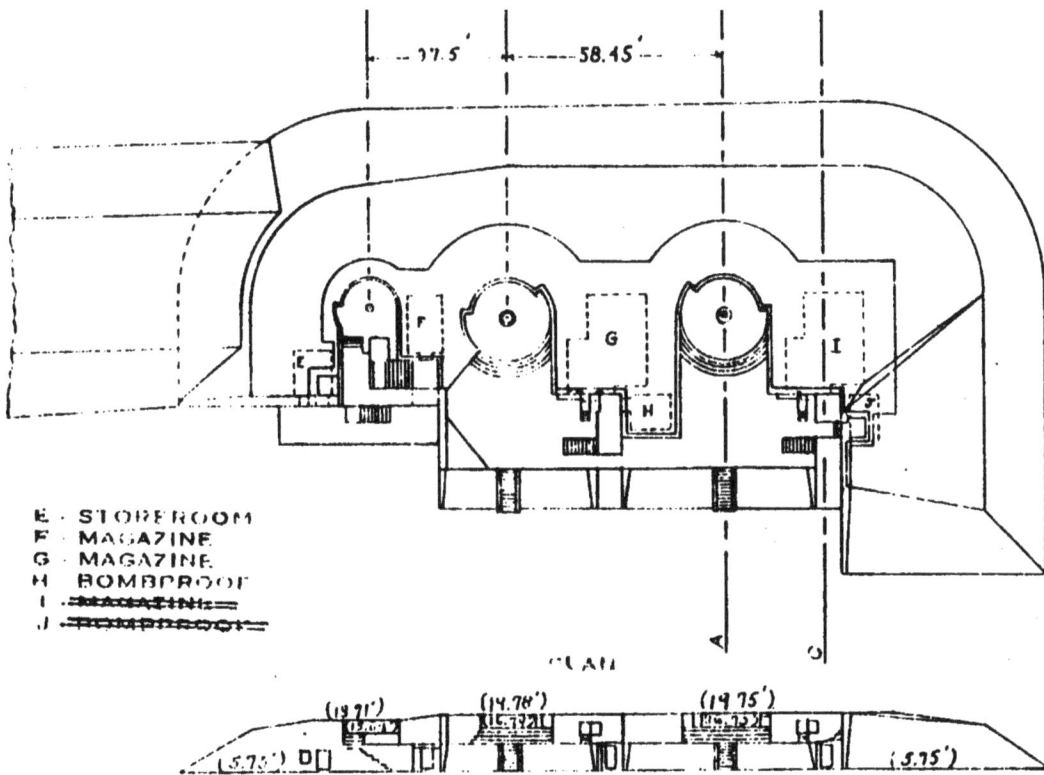

E · STOREROOM
F · MAGAZINE
G · MAGAZINE
H · BOMBPROOF
I · MAGAZINE
J · BOMBPROOF

PLAN

HEAD ELEVATION

Battery Burchstead, Fort Dade, HD Tampa Bay

M1898 Disappearing Carriage for 6-inch Gun

Usual Guns: M1897M1
Carriages Built: 29
Original Emplacements: 29
Time Emplaced: 1899-1903

Number of Bolts: 12
Circle Diameter: 9' 3.5"
Parapet Height: 8' 8"
Center-to-parapet: 8' 10"
Notes: ammunition hoists present,
1 step

Total weight: 33 tons
Maximum/minimum elevation:
+15 / –5 degrees
Turning radius: 360 degrees
(limited by emplacement)

Gun: M1897MI
Weight: 16,216 lbs
Length: 44.58 calibers

Photos:
Top: Battery Dutton, Fort
H.G. Wright, HD Long Island
Sound, 1930s (photo by
Adeland J. LeGere, Pierce
Rafferty Collection)

Upper middle: Battery
Dutton, 1930s (photo by
Adeland J. LeGere, Matt
Edwards Collection)

Lower middle: Battery Pratt,
Fort Stevens, HD Columbia
River with full scale replica
of a 6-inch M1898 DC, 2016
(MAB)

Below: Battery Crosby,
Fort Winfield Scott, HD San
Francisco 1992 (MAB)

M 1898 Disappearing Carriage for 6-inch Gun

6-INCH DISAPPEARING CARRIAGE, MODEL OF 1898, SIDE ELEVATION.

PLAN

REAR ELEVATION

Battery Crosby, Fort Winfield Scott, HD San Francisco

M1900 PEDESTAL MOUNT FOR 6-INCH GUN

Usual guns: M1900
Carriages Built: 45
Original Emplacements: 44
Time Emplaced: 1902-1906, frequent relocations until 1943

Number of Bolts: 16
Circle Diameter: 5'
Parapet Height: 2' 9"
Center-to-parapet: 3' 6"

Notes: some emplacements have hoists, most have "v" groove in platform behind the carriage for recoil at increased elevation. The battery designs for this model of gun varied considerably over the years. Also this gun model was relocated quite frequently, often into modified batteries which originally had 6-inch disappearing carriages.

Total weight: 20 tons, carriage: 24,800 lbs
Maximum/minimum elevation: +15 / −5 degrees (+20) (1944)
Turning radius: 360 degrees

Gun: M1900
Weight: 19,968 lbs.
Length: 50 calibers

Photos:
Top: 6-in model 1900 pedestal carriage of the original Battery Peck, Fort Hancock (NA, GMW collection)

Middle: Battery Smith, Fort Barry, HD San Francisco 1996 (AMH).

Bottom: One of two guns currently in place in Battery "New" Peck (ex-Battery Gunnison), Sandy Hook Unit, Gateway National Recreation Area, NJ, 2019 (MAB).

M1900 PEDESTAL MOUNT FOR 6-INCH GUN

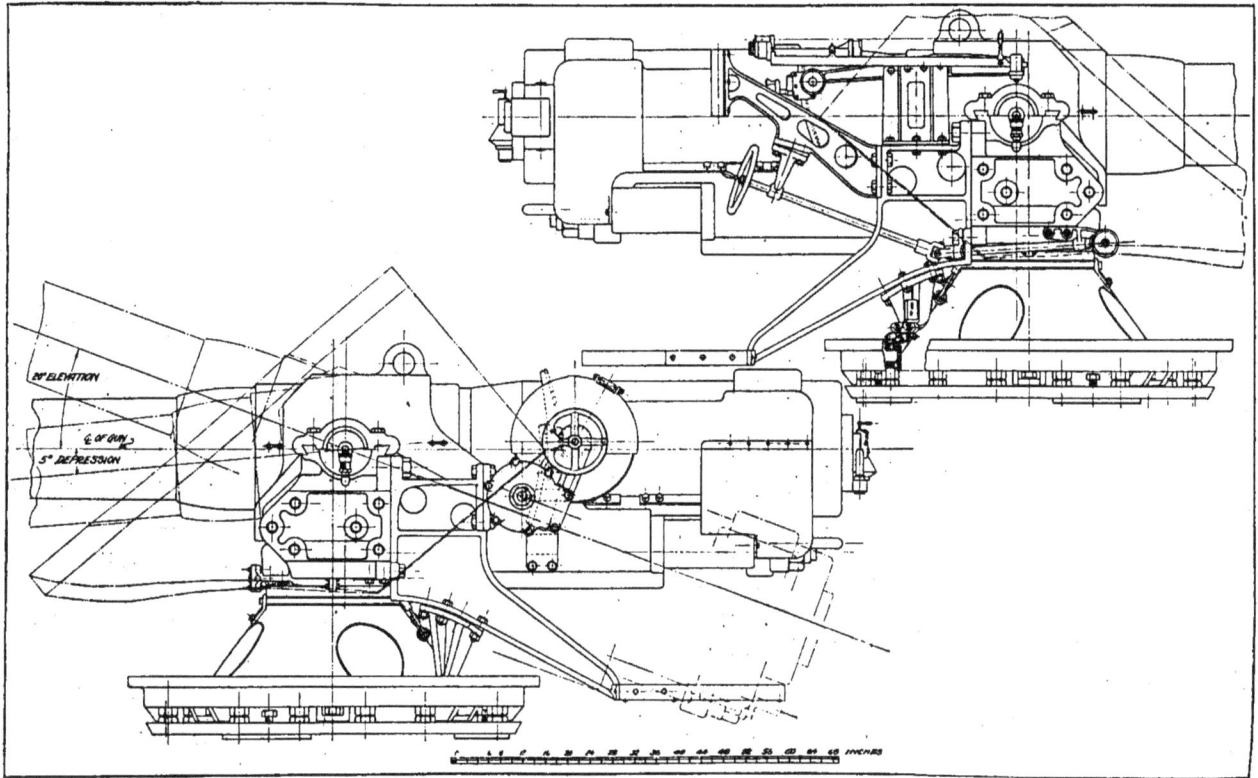

6-INCH BARBETTE CARRIAGE, MODEL OF 1900.

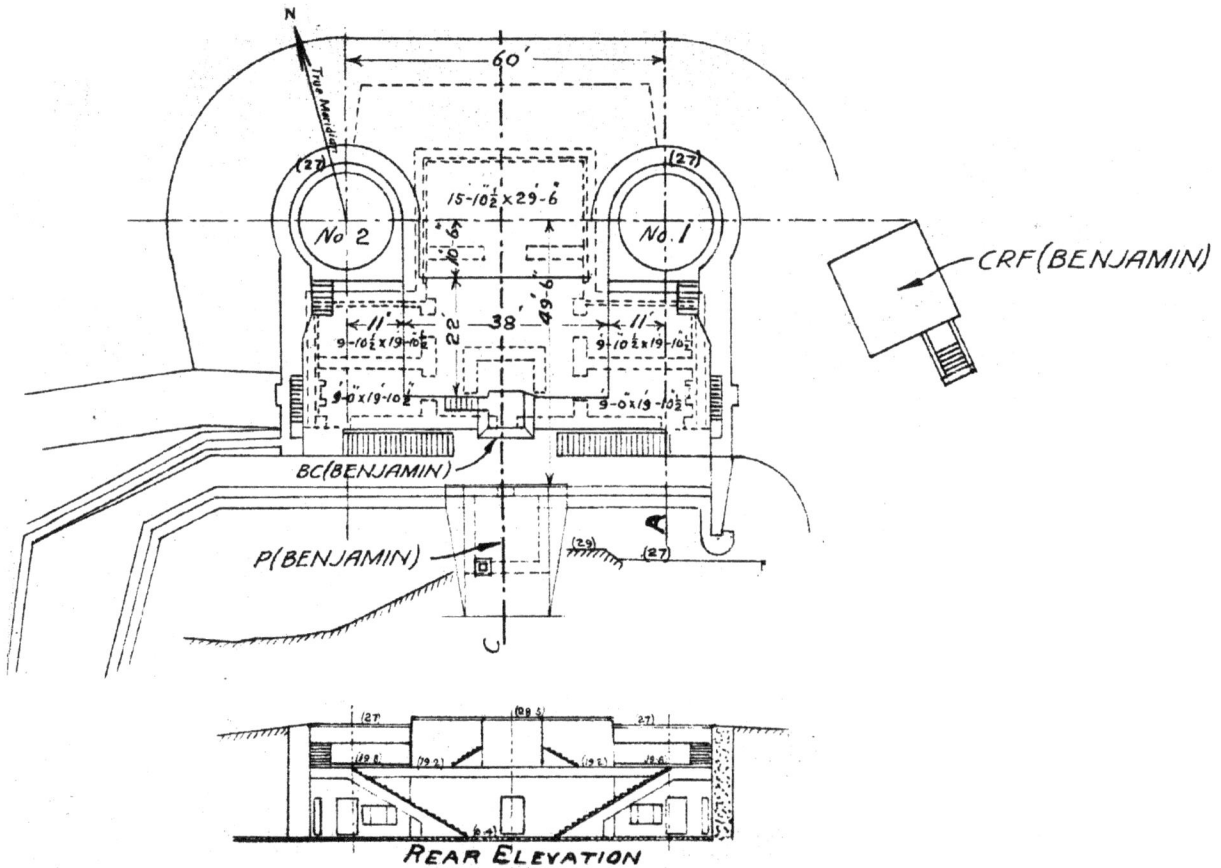

Battery Benjamin, Fort Michie, HD Long Island Sound

M1903 Disappearing Carriage for 6-inch Gun

Usual guns: M1903, M1905, M1900
Carriages Built: 90
Original Emplacements: 90 US,
2 relocated
Time Emplaced: 1904-1908

Number of Bolts: 12
Circle Diameter: 9' 6.5"
Parapet Height: 10' 3"
Center-to-parapet: 10' 5"
Notes: 3 steps, rear of loading
platform is half-circle shape

M1903 6-DC
Total weight: 48 tons
Maximum/minimum elevation: +15 /
−5 degrees
Turning radius: 360 degrees
(limited by emplacement)

Gun: M1900
Weight: 19,869 lbs
Length: 50 calibers

Gun: M1903
Weight: 19,990
Length: 50 calibers

Photos:
Top: Battery Marcy, Fort H.G. Wright,
HD Long Island Sound, 1911 (Photo
by Quimby, Jim Diaz Collection)

Upper Middle: Gun and carriage now
in Battery Chamberlin, Fort Winfield
Scott, Golden Gate National Rec Area,
CA 2006 (MAB).

Lower Middle:
Battery
Valleau, Fort
Casey, HD
Puget Sound
1996 (AMH).

Bottom:
Battery
Stoddard, Fort
Worden, HD
Puget Sound,
1910.

M1903 DISAPPEARING CARRIAGE FOR 6-INCH GUN

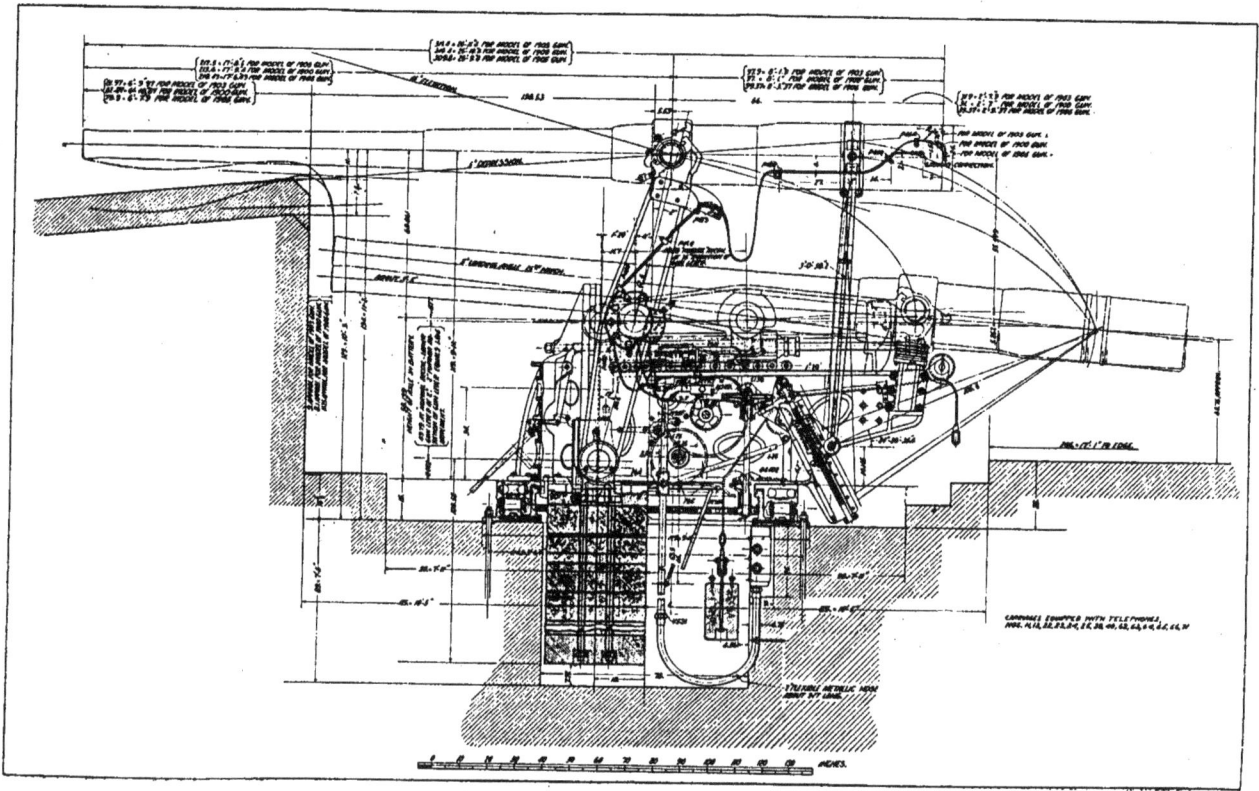

6-INCH DISAPPEARING CARRIAGE, MODEL OF 1903.

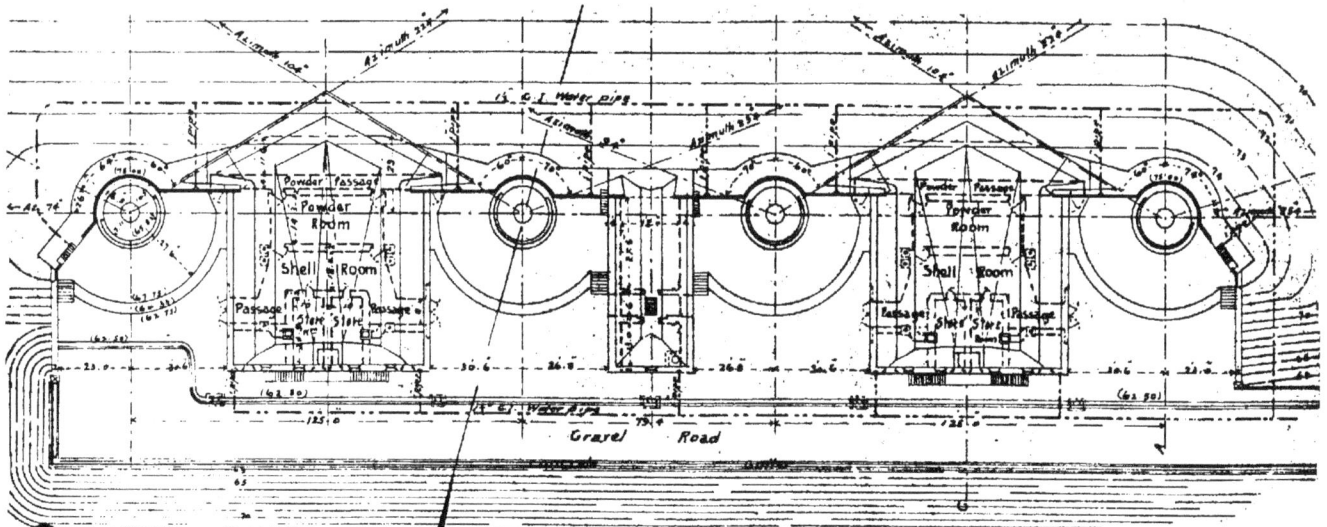

PLAN

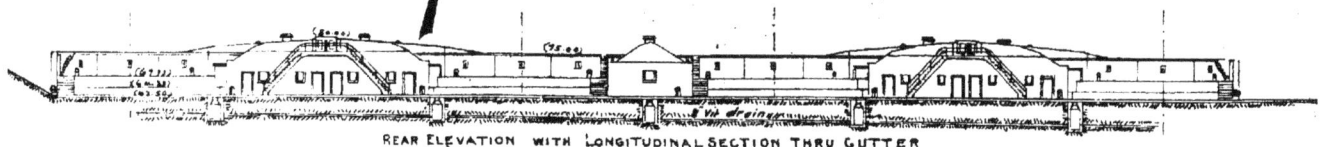

REAR ELEVATION WITH LONGITUDINAL SECTION THRU GUTTER

Battery Caldwell, Fort Flagler, HD Puget Sound

M1905, MI, MII DISAPPEARING CARRIAGES FOR 6-INCH GUN

Usual Tubes: M1905, M1908, M1908M1
Carriages Built: 33
    M1905: 10
    M1905MI: 9
    M1905MII: 14
Original Emplacements: 4 US, 29 Terr.
Time Emplaced: 1910-1917

Number of Bolts: 12
Circle Diameter: 10' 7.5"
Parapet Height: 9' 2"
Center-to-parapet: 10' 5"
Notes: no ammunition hoists, 3 steps

Gun: M1905
Weight: 21,148 lbs
Length: 50 calibers

Gun: M1908
Weight: 12,500
Length: 44.58 calibers

M1905 carriage
M1905MI carriage
M1905MII carriage
Total weight: 56 tons
Max/min elevation: +15 / −5 degrees
Turning radius: 360 degr. (limited by empl)

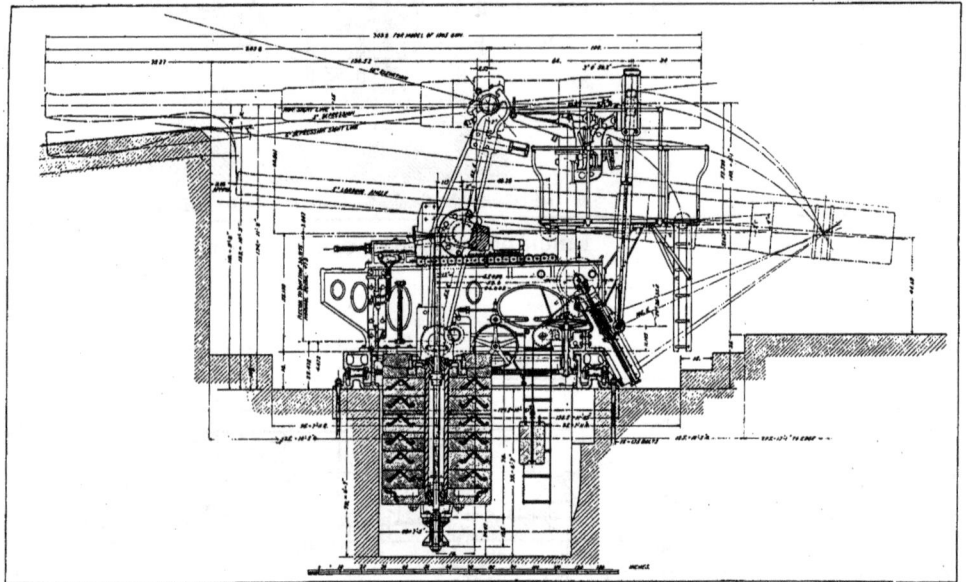

6-INCH DISAPPEARING CARRIAGE, MODEL OF 1905, SHOWING PARAPET.

Photo:
Top: Battery Hall (M1905), Fort Wint, HD Subic Bay, The Philippines, 2006 (MAB).

(right) Battery Harrison (plan) Fort Whitman, HD Puget Sound (M1905MI)

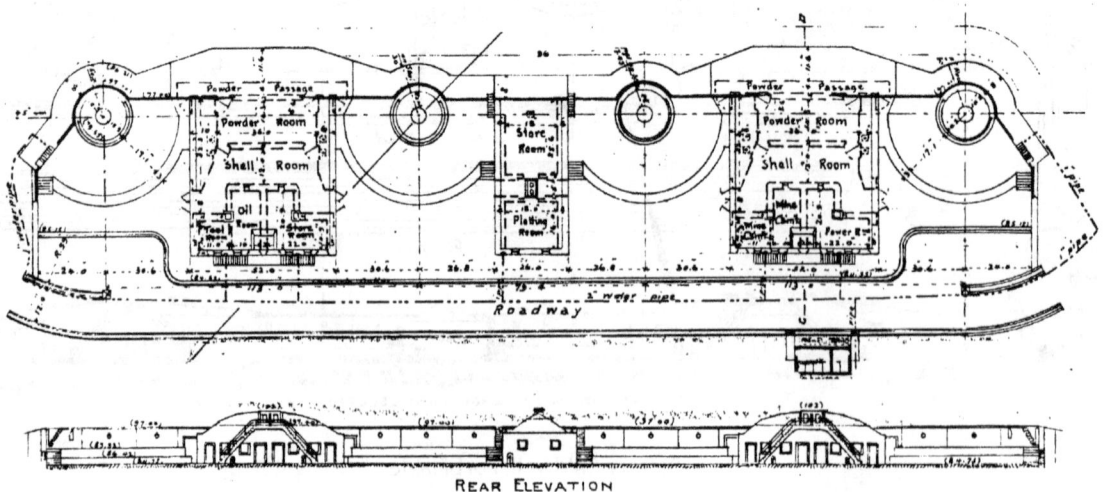

REAR ELEVATION

M1905, MI, MII Disappearing Carriages for 6-inch Gun

6-INCH DISAPPEARING CARRIAGE, MODEL OF 1905 MI, SHOWING PARAPET.

Model of 1905 MII

M1910 PEDESTAL MOUNT FOR 6-INCH GUN

Usual guns: M1908M2
Carriages Built: 6
Original Emplacements: 6 Terr.
Time Emplaced: 1915-1917

Number of Bolts: 16
Circle Diameter: 4' 10"
Parapet Height: 2' 11"
Center-to-parapet: 4' 2"

Notes: Unique circular shield, 4 were casemated (Batteries Roberts and McCrea, Ft. Drum), 2 pedestal (Battery Morgan, Ft. DeLesseps)

Carriage weight: 30,000 lbs
Max/min elevation: +12 / –3 degrees
Turning radius: 360 degrees (127 degrees for casemate mounts)

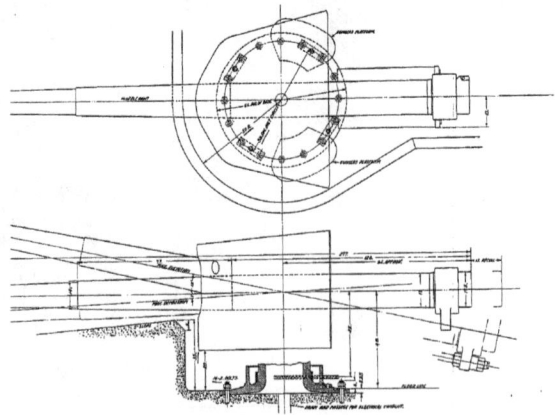

Gun: M1908 MII
Weight: 12,500 lbs
Length: 44.58 calibers

Photos:
Top: Battery Morgan, Fort Delesseps, HD Cristabol, Panama Canal Zone,1939. (LIFE)

Middle: Battery Morgan, 1993 (AMH).

Lower Left: Battery Roberts, Fort Drum, HD Manila Bay, 1930s (Bolling Smith Collection).

Lower right: Battery Roberts, Fort Drum, HD Manila Bay, 1995 (AMH).

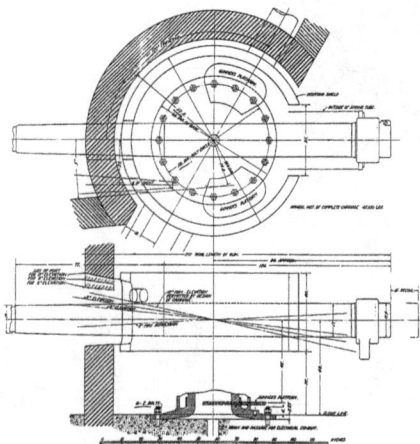

for casemate mount

For pedestal mount

M1910 Pedestal Mount for 6-inch Gun

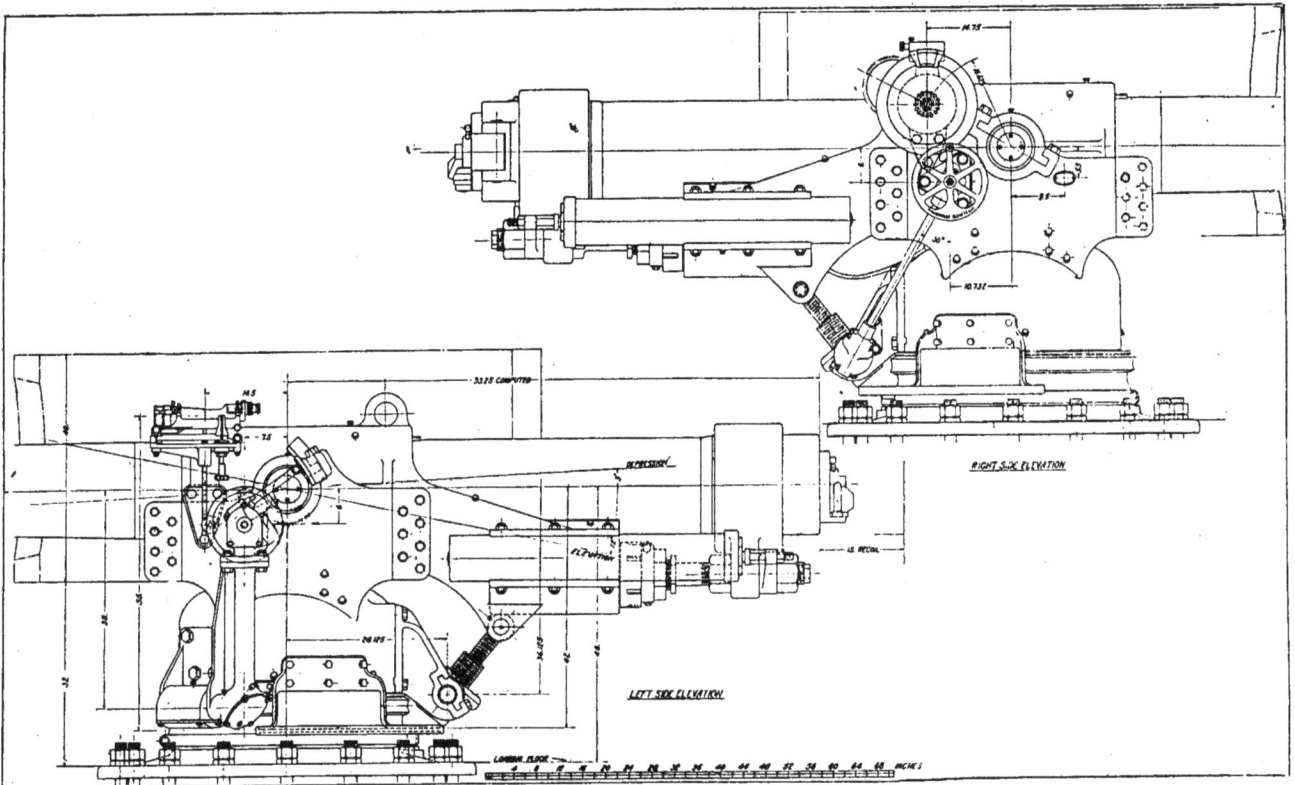

6-INCH BARBETTE CARRIAGE, MODEL OF 1910, SIDE ELEVATIONS.

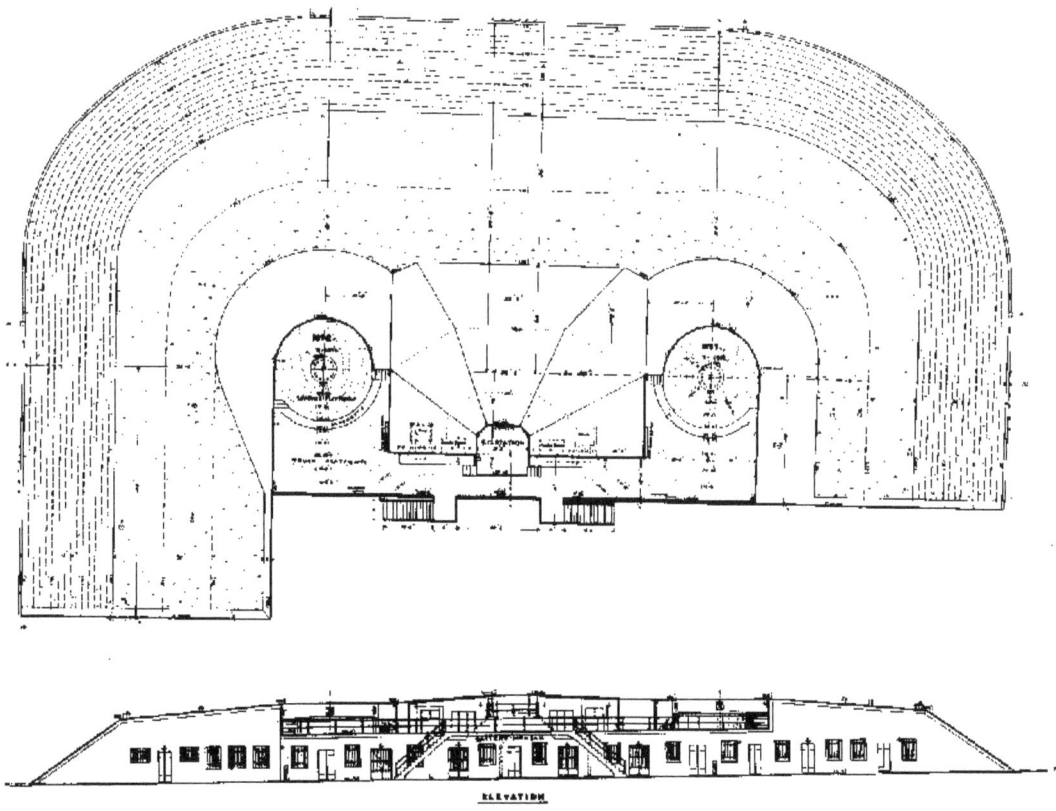

Battery Morgan, Fort DeLessups, HD Cristobal, Panama Canal Zone

M1, M2, M3 & M4 Barbette Carriages for 6-inch Guns

Guns: M1903A2, M1905A2,
　　M1(T2)
Carriages built:　Approx. 143
M1: 68　　　　M2: 2
M3: 15　　　　M4: 58
Batteries planned: 89
Batteries built: 48 CONUS,
18 Terr., 4 overseas
Batteries armed: ~ 60
(Carriages emplaced: ~ 120)
Time Emplaced: 1941-1946

Number of Bolts: 16
Circle Diameter: 11' 4"
Parapet Height: none
Center-to-parapet:
Notes: base ring in depressed
circular pit. For more discussion
and illustrations see: "The Six-
Inch Part of the Modernization
Program of 1940," by Robert
Zink, *CDSG Journal* 1994 Vol. 8,
Issue 2, pp. 21-38.

M1Carriage
Carriage weight: 75,300 lbs
Maximum/minimum elevation:
+47.5 / –5 degrees
Turning radius: 360 degrees

M3, M4 Carriages
Carriage weight: 67,450 lbs
Maximum/minimum elevation:
+47 / –5 degrees
Turning radius: 360 degrees

Guns: M1903A2, M1905A2
Weight: 20,700 lbs
Length: 50 calibers

Gun: M1
Weight: 20,550 lbs
Length: 50 calibers

Photos:
Top: Battery 249, Camp
Hayden, HD Puget Sound
1944 (Puget Sound CA
Museum)

Others: Battery 246, Fort
Columbia, HD Columbia
River (2008 MAB).

Pensacola, FL.

M1, M2, M3 & M4 Barbette Carriages for 6-inch Guns

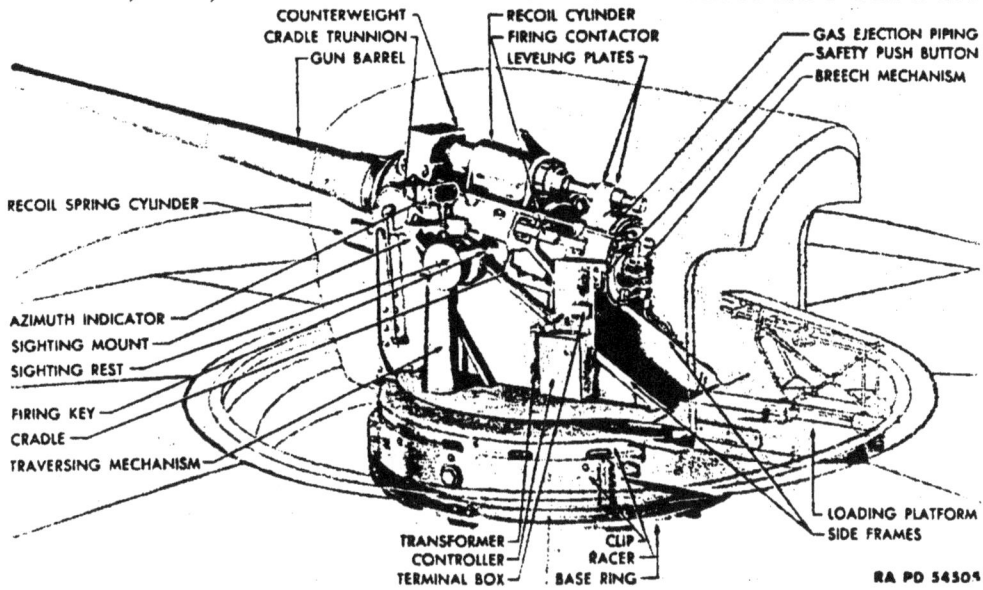

FIGURE 8A.—6-inch gun, M1903A2, and Barbette Carriage, M1, left rear view.

In 1993 the State of Washington placed two 6-inch guns and carriages in BCN 246, originally from BCN 281, Fort McAndrew, Argentia, Newfoundland, Canada. Two guns and carriages remain at BCN 282 at Argentia and two guns and carriages are on display at Battery 234, Fort Pickens, HD

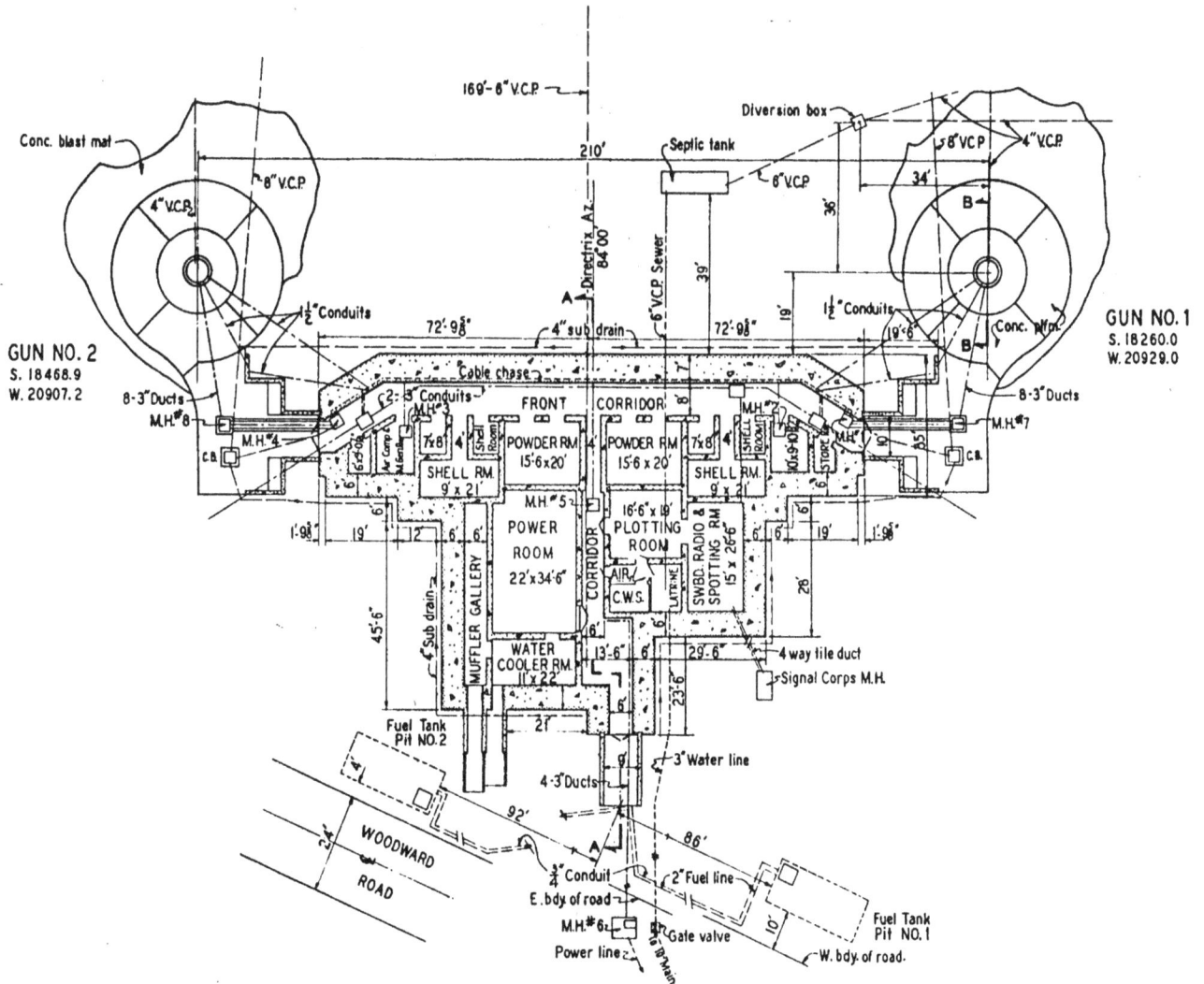

BCN 237, Fort Rosecrans, HD San Diego.

M1892 BARBETTE CARRIAGE FOR 8-INCH GUN

Usual guns: M1888
Carriages Built: 9
Original Emplacements: 9
Time Emplaced: 1898-1900

Number of Bolts: 16
Circle Diameter: 9' 10"
Parapet Height: 6"
Center-to-parapet:
Notes: most in temporary emplacements; there were only two designed batteries: Drew (1 emplacement) and Duncan (2 emplacements) in the San Francisco defenses. The West Point Military Academy had one installed as a practice weapon.

Total weight: 42 tons
Max/min elevation: +18 / −7 degrees
Turning radius: 320 degrees

Gun: M1888
Weight: 32,218 lbs
Length: 32 calibers

Photos:
Top: Battery Drew, Fort McDowell, HD San Francisco, 1910s (Fort Hulan Museum Collection)

Middle: West Point, 1900s. (NA)

Bottom: Battery Duncan, 1983 (AMH).

M1892 BARBETTE CARRIAGE FOR 8-INCH GUN

8 inch B.L.Rifle.– Steel 14½ tons.
Barbette Carriage

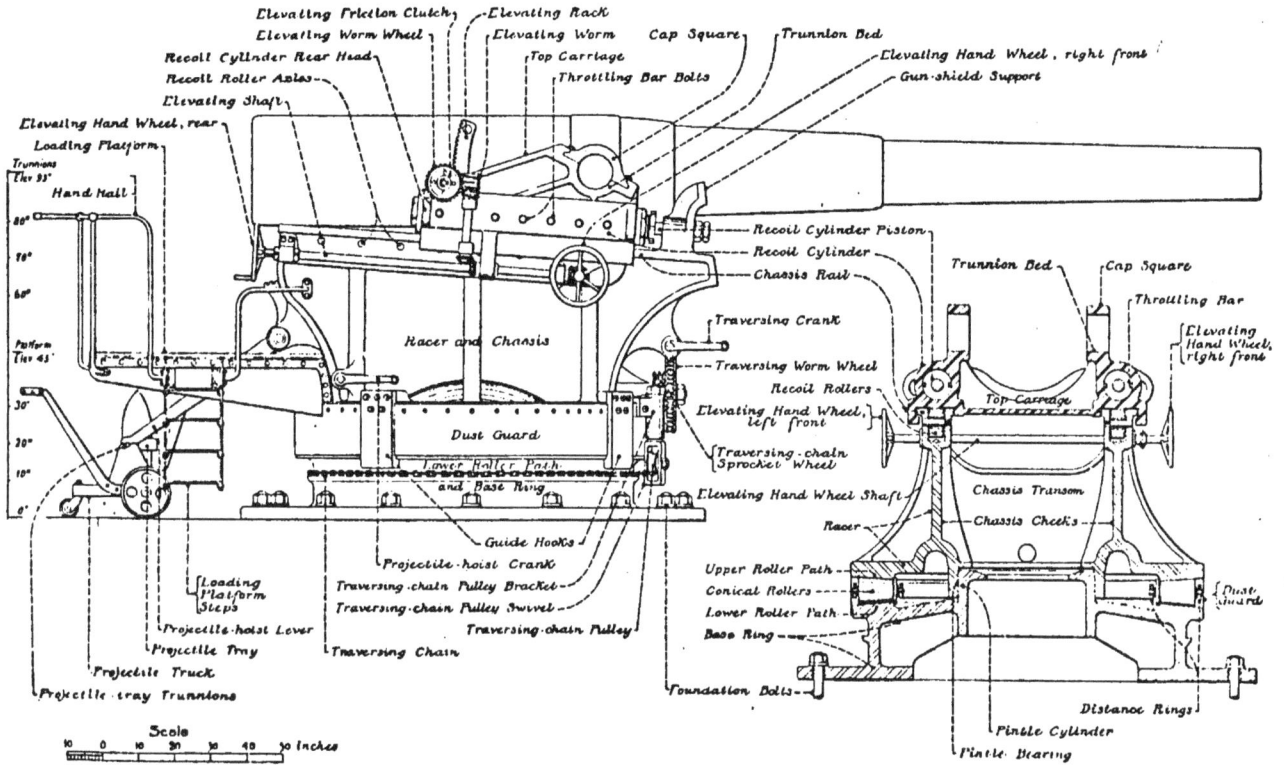

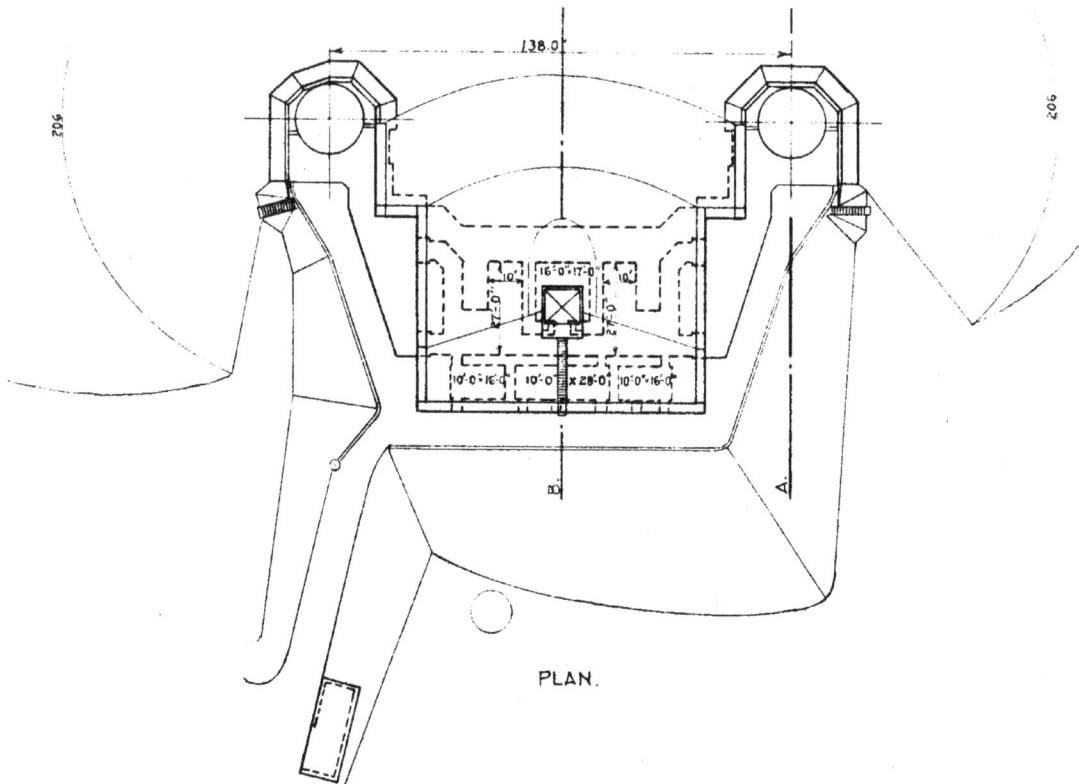

Elevating Friction Clutch
Elevating Worm Wheel
Recoil Cylinder Rear Head
Recoil Roller Axles
Elevating Shaft
Elevating Hand Wheel, rear
Loading Platform
Trunnions Thr 95°
Hand Rail
Elevating Rack
Elevating Worm
Top Carriage
Throttling Bar Bolts
Cap Square
Trunnion Bed
Elevating Hand Wheel, right front
Gun shield Support

Recoil Cylinder Piston
Recoil Cylinder
Chassis Rail

Racer and Chassis

Traversing Crank
Traversing Worm Wheel
Recoil Rollers
Elevating Hand Wheel, left front
Traversing chain Sprocket Wheel
Elevating Hand Wheel Shaft

Dust Guard

Lower Roller Path and Base Ring

Guide Hooks
Projectile hoist Crank
Traversing chain Pulley Bracket
Traversing chain Pulley Swivel
Traversing chain Pulley

Loading Platform Steps
Projectile hoist Lever
Projectile Tray
Projectile Truck
Projectile tray Trunnions

Traversing Chain

Trunnion Bed
Cap Square
Throttling Bar
Elevating Hand Wheel, right front

Top Carriage

Chassis Transom
Chassis Cheeks
Racer
Upper Roller Path
Conical Rollers
Lower Roller Path
Base Ring
Foundation Bolts
Dust Guard
Distance Rings
Pintle Cylinder
Pintle Bearing

Scale
10 0 10 20 30 40 50 Inches

138.0

PLAN.

Battery Duncan, Fort Baker, HD San Francisco

M1894 Disappearing Carriage for 8-inch Gun

Usual guns: M1888
Carriages Built: 26
Original Emplacements: 29
Time Emplaced: 1896-1908

Number of Bolts: 16
Circle Diameter: 8' 4"
Parapet Height: 10' 5"
Center-to-parapet: 5'
Notes: 4 steps (except Battery Duane, Ft. Wadsworth, HDNY, which has 3 steps), rear traversing roller path

Carriage weight: 94,000 lbs
Max./min. elevation:
Turning radius:

M1888 gun
Weight: 32,218 lbs
Length: 32 calibers

Photos:
Above: As installed at Battery Arrowsmith, Fort Hancock, HD Sandy Hook (Gateway Natl. Rec. Area collection)
Middle: Battery Bowyer, Fort Morgan, HD Mobile Bay 2003, (AMH). Left Inset Battery Duane (3 steps), 2003 (AMH).
Below: From Scientific American Special Edition 1898.

M1894 DISAPPEARING CARRIAGE FOR 8-INCH GUN

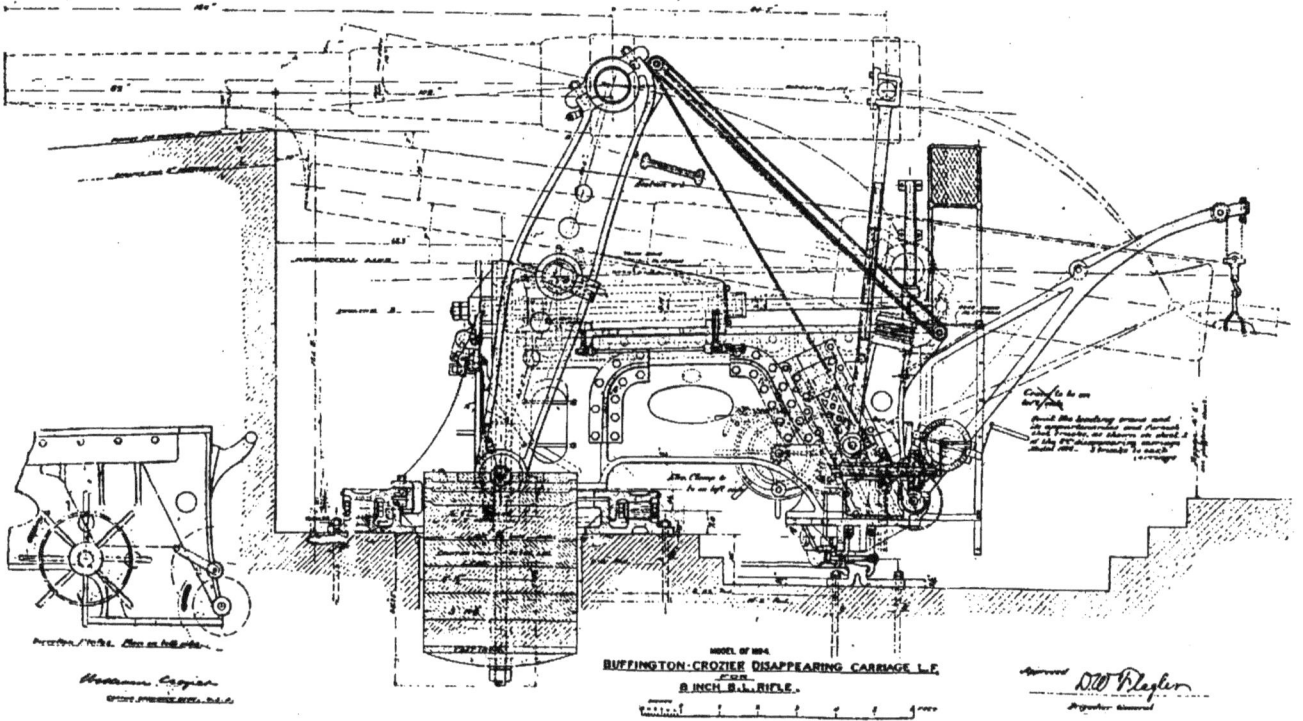

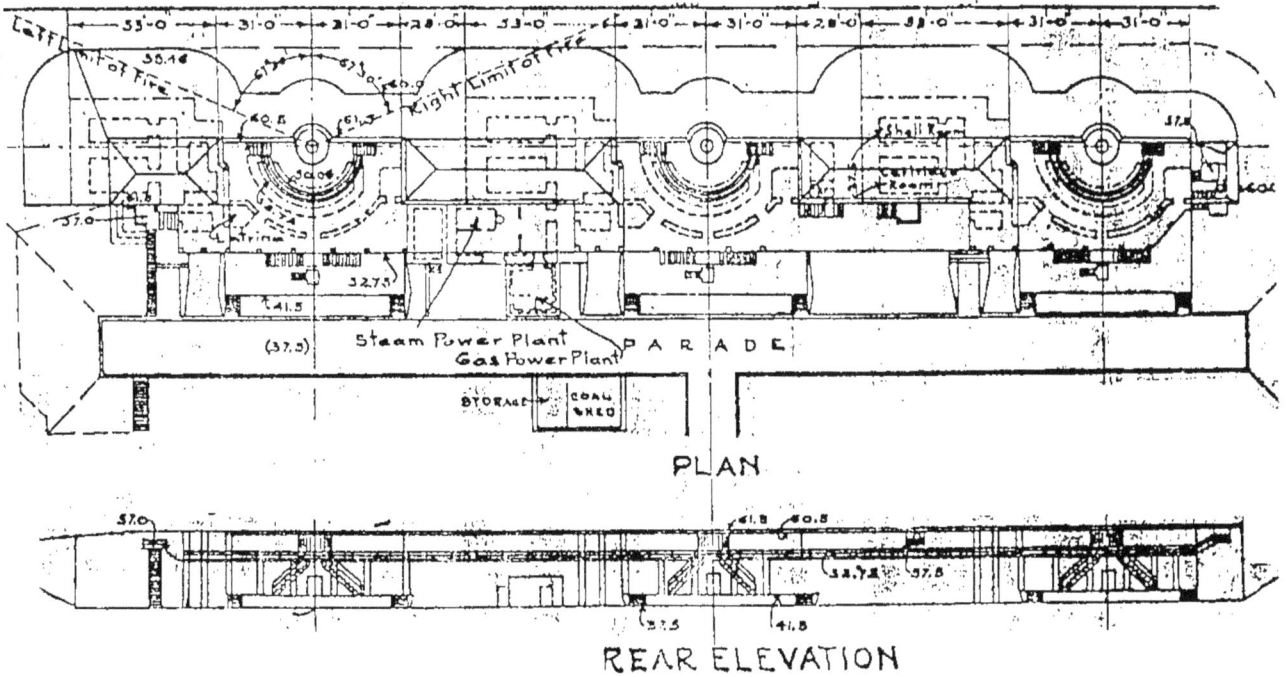

Battery Mt. Vernon, Fort Hunt, HD Potomac River

M1896 DISAPPEARING CARRIAGE FOR 8-INCH GUN

Usual guns: M1888
Carriages Built: 38
Original Emplacements: 39
Time Emplaced: 1897-1901

Number of Bolts: 12
Circle Diameter: 12' 1"
Parapet Height: 10' 3"
Center-to-parapet: 8' 6.5"
Notes: 2 steps

Total weight: 104 tons
Max/min elevation: +12 / –5 deg.
Turning radius: 360 degrees
(limited by emplacement)

Gun: M1888MI, MII
Weight: 32,218 lbs
Length: 32 calibers

Photos:
Above: Battery Nash, Fort Ward, HDPS, 1900s
(Friends of Fort Ward, WA)
Middle: Battery McIntosh, Fort Dade, HD
Tampa Bay, 1994 (AMH).
Below right: Battery Jules Ord, Fort Columbia,
HD Columbia River, 1999 (AMH).

Below Left: Gun and carriage, originally
from Fort Dade, HD Tampa Bay, on display in
monument in Plant Park, University of Tampa,
1930s, This gun was scrapped during WW II
and replaced by a 8-inch gun on a M1918 BC.

M1896 DISAPPEARING CARRIAGE FOR 8-INCH GUN

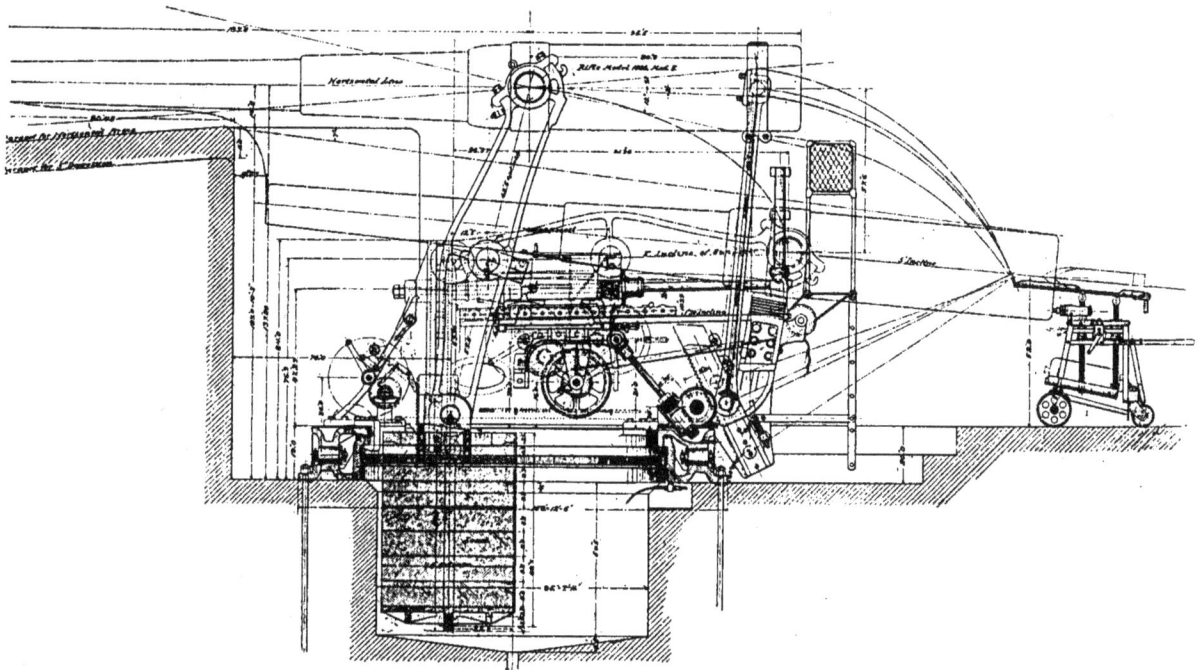

8-INCH DISAPPEARING CARRIAGE, MODEL OF 1896.

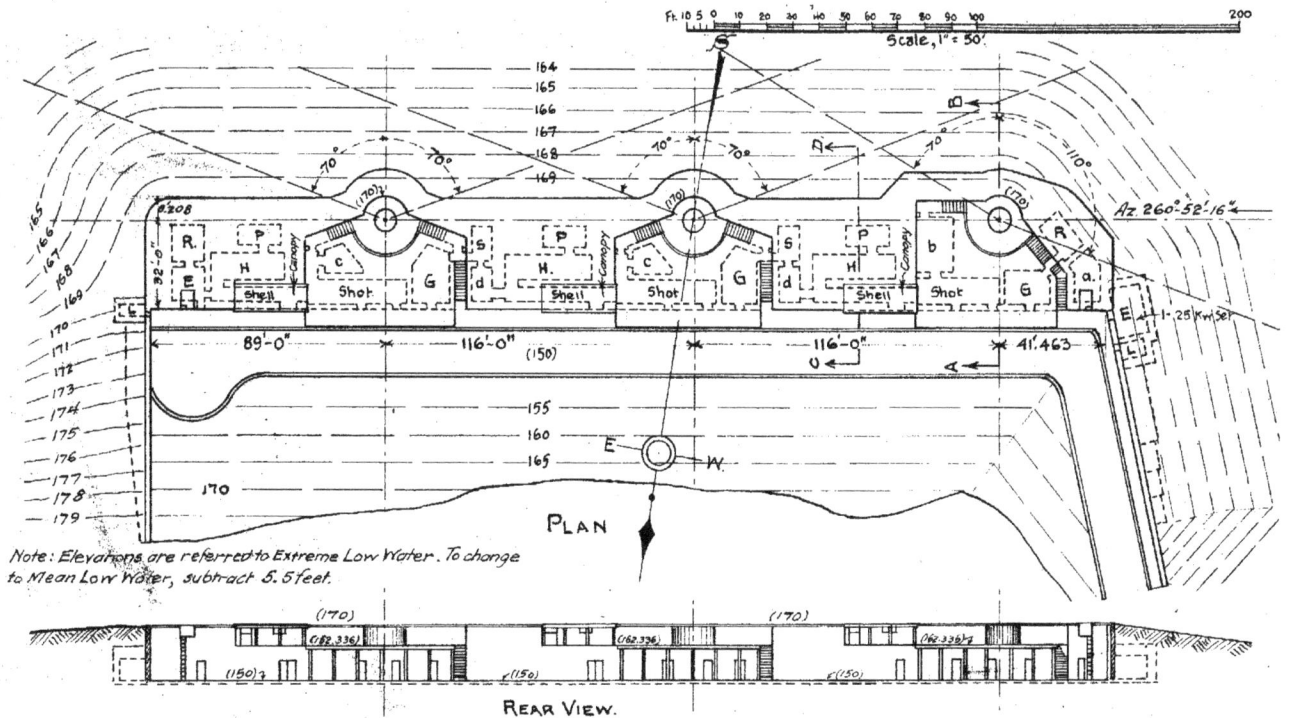

Note: Elevations are referred to Extreme Low Water. To change to Mean Low Water, subtract 5.5 feet.

PLAN

REAR VIEW.

Battery Nash, Fort Ward, HD Puget Sound

M1918 BARBETTE CARRIAGE FOR 8-INCH GUN

Usual guns: M1888MIA1
Carriages Built: 47
Original Emplacements: about 12
Time Emplaced: 1935-1942

Number of Bolts: 36
Circle Diameter: flattened circle
Parapet Height: none
Center-to-parapet:
Notes: carriage designed for mounting on a M1918MI railway car. Also emplaced as a barbette.

Carriage weight: 148,000 lbs
Max/min elevation: +42 / 0 degrees
Turning radius: 360 degrees

Gun: M1888MII
Weight: 33,200 lbs
Length: 32 calibers

Photos:
Above: Platform for Battery RJ-43, Corregidor Island, The Philippines, 1996 (AMH).

Below: monument in Plant Park, University of Tampa, 1992 (MAB).

M1918 Barbette Carriage for 8-inch Gun

PLATE 267.

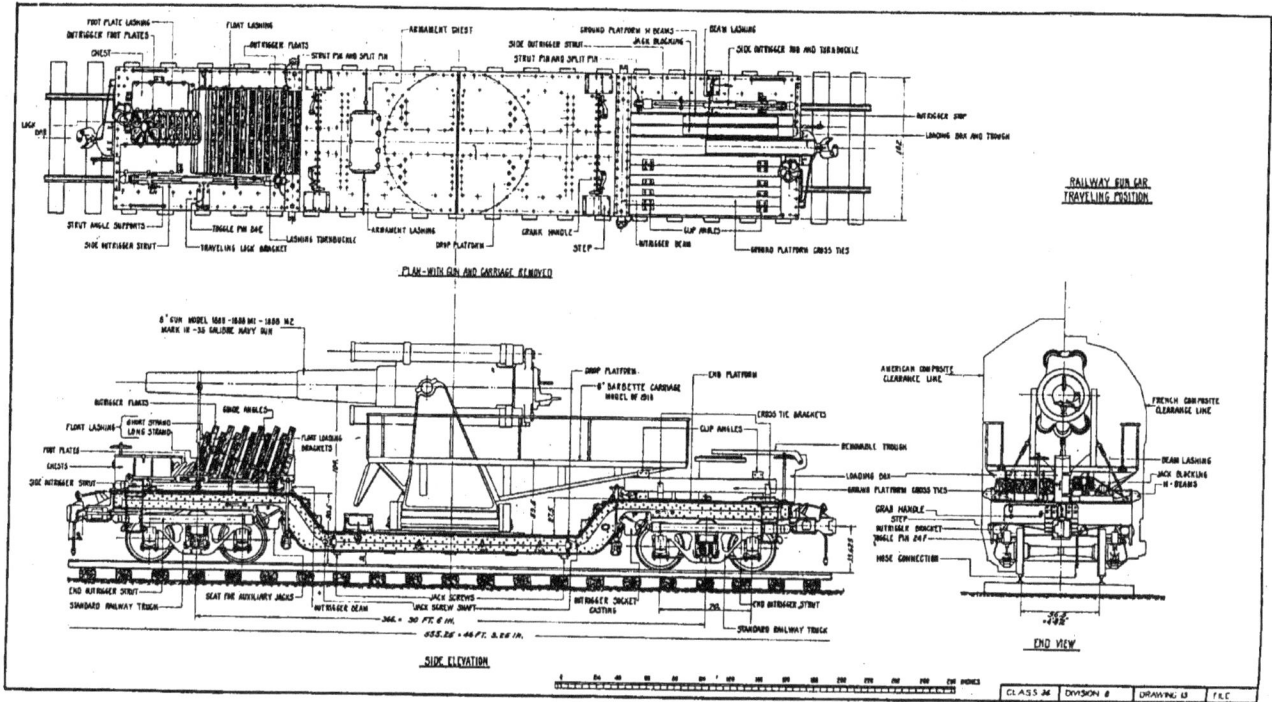

PLAN—WITH GUN AND CARRIAGE REMOVED

RAILWAY GUN CAR TRAVELING POSITION

SIDE ELEVATION

END VIEW

8-INCH RAILWAY MOUNT, MODEL OF 1918, IN TRAVELING POSITION.

Aberdeen Proving Grounds, MD (Ordnance Museum)

M1 Barbette Carriage for 8-inch Navy Gun

Guns: Navy MkVIM3A2
Carriages Built: 24
Original Emplacements: 16
Time Emplaced: 1937-1943

Number of Bolts:
Circle Diameter:
Notes: The first battery using this carriage model was Battery Strong, built at Fort Rosecrans in 1937. A single casemated battery, Reilly, was built at Fort Church in 1940. The 400-series batteries built during the 1940s in Hawaii, Alaska, and the Caribbean were similar in design to Battery Strong. They were to be protected by a shield like that made for the WW II 6-inch batteries, but the shields were never made.

Carriage weight: 58,470 lbs
Max/min elevation: +45 / –5 degrees
Turning radius: 360 degrees
(145 degrees in casemate)

Gun: Navy MkVIM3A2
Weight: 42,000 lbs
Length: 45 calibers

Photos:
Above: BCN 402, Fort Schwatka, HD Dutch Harbor, AK, 1997 (AMH).
Middle: Battery Reilly, Fort Church, HD Narragansett Bay, 1988 (AMH).
Below: The casemated Battery Reilly, as installed (TM 4-210).

M1 Barbette Carriage for 8-inch Navy Gun

Battery Strong, Fort Rosecrans, HD San Diego.

M1 RAILWAY MOUNT FOR 8-INCH NAVY GUN

Guns: Navy MkVIM3A2
Carriages Built: about 24
Original Emplacements:
Time Emplaced: 1942-1945

Notes: Two units were assigned to
Los Angeles, two to Puget Sound, the
remainder left on the east coast mainly
at Fort Hancock, HD New York; Fort
Miles, HD Delaware River; and Fort John
Custis, HD Cheaspeake Bay.

Carriage weight: 188,000 lbs
Max/min elevation: +47 / –5 degrees
Turning radius: 360 degrees

Gun: MkVIM3A2
Weight: 42,000 lbs
Length: 45 calibers

Photos:
Above: Photo taken at the Aberdeen
Proving Ground in 1938
(U.S. Army Ordnance Department)

Middle: Baldwin Locomotive Works
Photograph

Below: US Army manual

Figure 156. 8-inch gun Mk. VI Mod. 3A2 on 8-inch gun railway mount M1A1, traveling position.

M1 Railway Mount for 8-inch Navy Gun

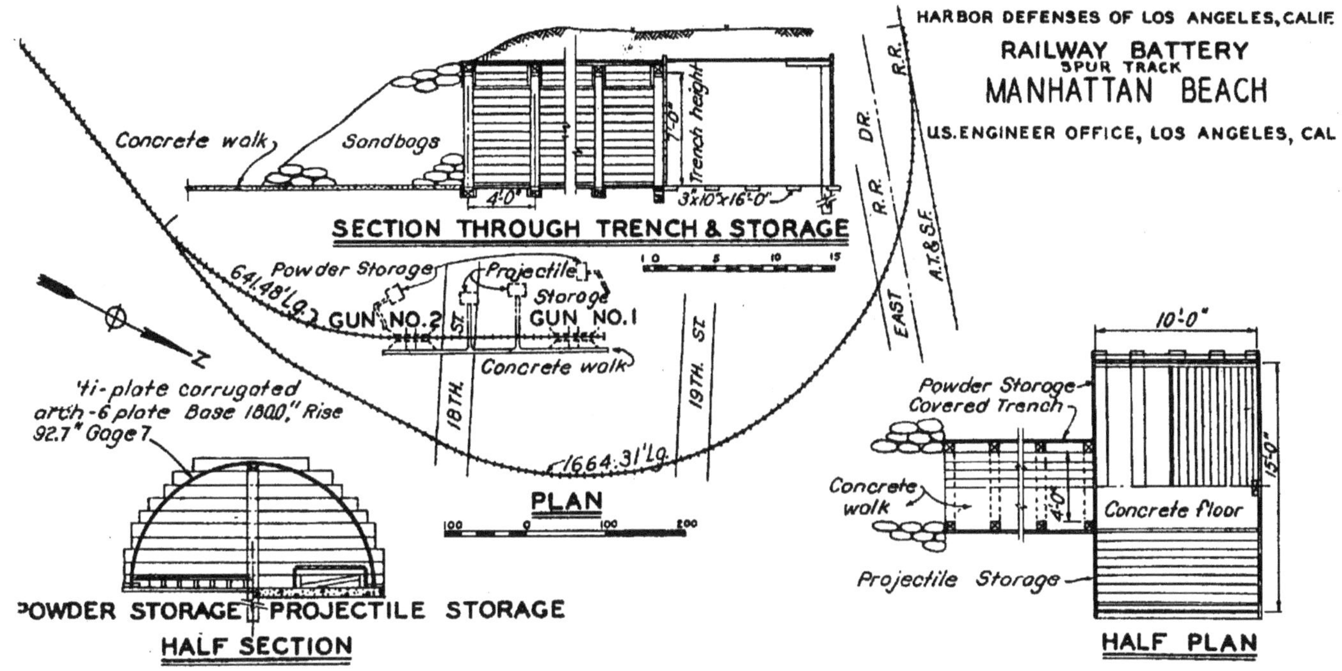

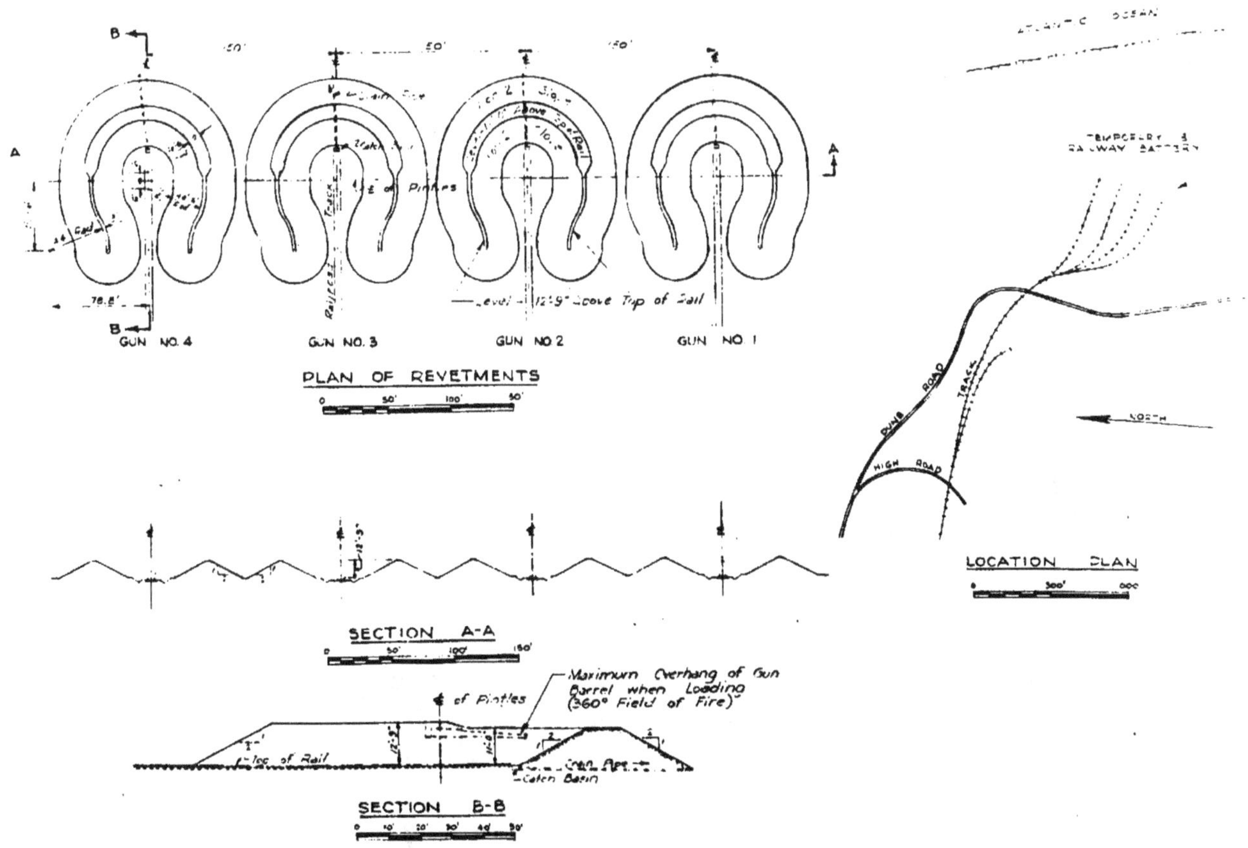

SECTION THROUGH TRENCH & STORAGE

Concrete walk, Sandbags

Trench height

4'-0" 3"x10"x16'-0"

HARBOR DEFENSES OF LOS ANGELES, CALIF.

RAILWAY BATTERY
SPUR TRACK
MANHATTAN BEACH

U.S. ENGINEER OFFICE, LOS ANGELES, CAL

Powder Storage, Projectile Storage
64.48' Lg. GUN NO.2   GUN NO.1
Concrete walk

PLAN

'ti-plate corrugated arch-6 plate Base 1800," Rise 92.7" Gage 7.

POWDER STORAGE   PROJECTILE STORAGE
HALF SECTION

10'-0"
Powder Storage
Covered Trench
Concrete walk
Projectile Storage
Concrete floor
15'-0"
4'-0"
HALF PLAN

PLAN OF REVETMENTS

GUN NO. 4   GUN NO. 3   GUN NO 2   GUN NO I

SECTION A-A

SECTION B-B

Maximum Overhang of Gun Barrel when Loading (360° Field of Fire)

LOCATION PLAN

Emplacements at Fort Miles, HD Delaware River

NAVAL TURRET FOR 8-INCH 55-CALIBER GUNS

Tubes:Mk ??
turrets used: 8
Original Emplacements: 8
Time Emplaced: 1942

Notes: Twin gun turrets removed from the USS Lexington and USS Saratoga during refit in 1942. Batteries Kilpatrick, Burgess, Riggs, and Ricker, Oahu Island, Hawaii

Reference: Excellent drawings and photographs of the Hawaii turret batteries can be found in the article "The Oahu Turrets" by E. R. Lewis and D. P. Kirchner, printed in the magazine Warship International, 1992 Issue No. 3, pp. 273-301.

Photos:
Above: Battery Ricker No. 4 (Brodie Camp) in the summer of 1942, showing the battery prior to placement of camouflage netting. Personnel are most likely members of the 809th CA Battalion (Separate). (US Army Museum of Hawaii)

Middle: Battery Kilpatrick, Wilwilinui Ridge Military Reservation, Oahu Island, HI, 1994 (TM).

Below: Battery at Opaeula (BatteryRiggs) William Gaines Collection)

NAVAL TURRET FOR 8-INCH 55-CALIBER GUNS

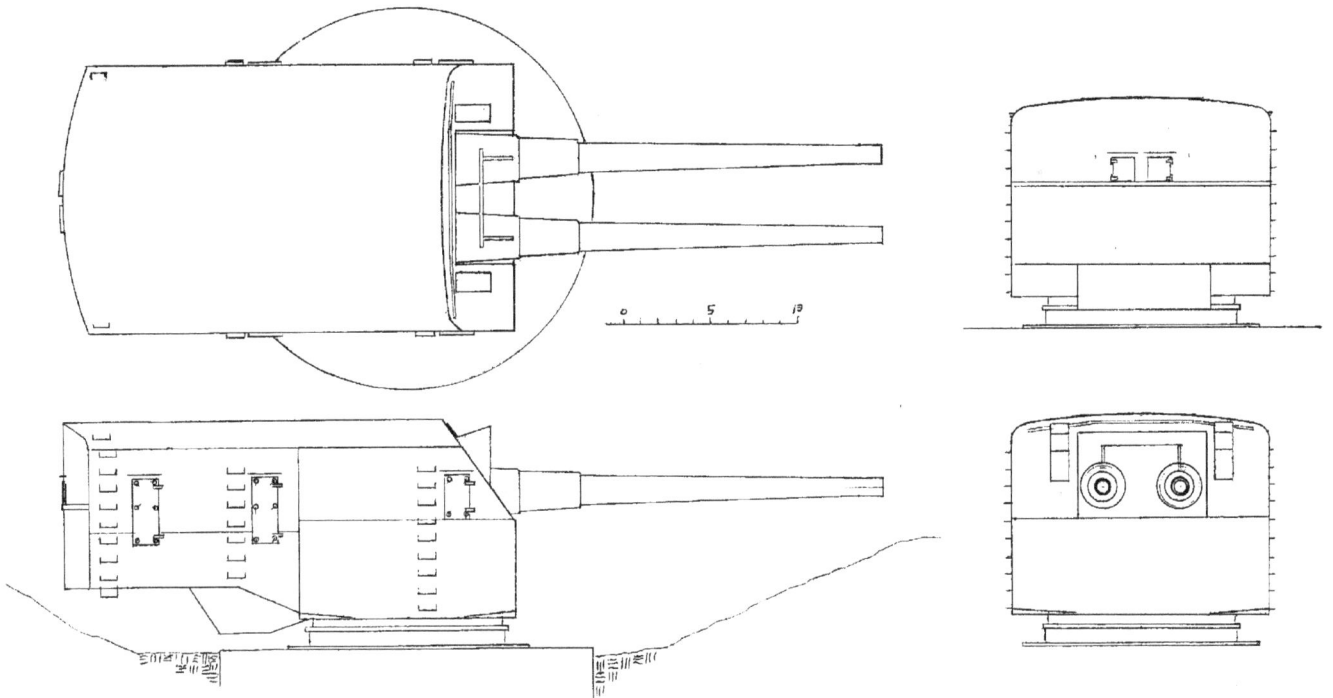

DAN
ROWBOTTOM
1-9-1998

Drawing by Dan Rowbottom

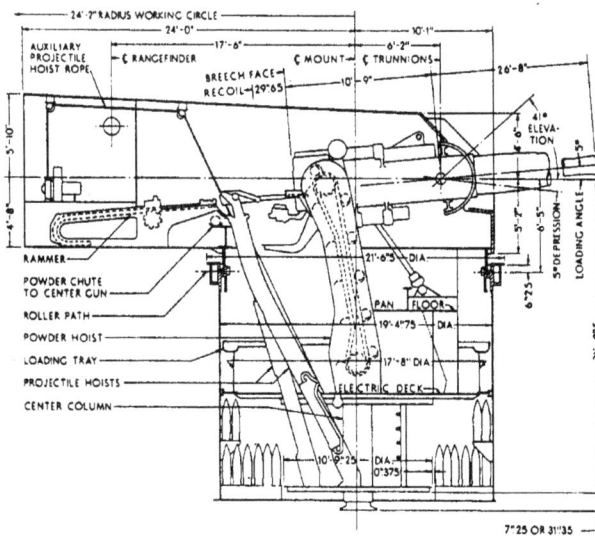

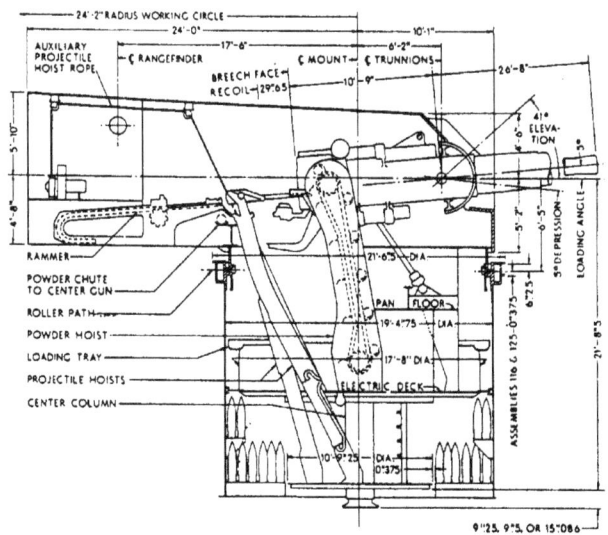

24'-7" RADIUS WORKING CIRCLE
24'-0"
17'-6"
₡ RANGEFINDER
₡ MOUNT
6'-2"
10'-1"
₡ TRUNNIONS
AUXILIARY PROJECTILE HOIST ROPE
BREECH FACE RECOIL 29"65
26'-8"
10' 9"
41° ELEVATION
5'-10"
5°
5'-7"
6'-5"
4'-8"
6'-0"
5° DEPRESSION
LOADING ANGLE
RAMMER
POWDER CHUTE TO CENTER GUN
21'-6"5 DIA
PAN FLOOR
6"725
ROLLER PATH
POWDER HOIST
19'-4"75 DIA
21'-8"5
LOADING TRAY
17'-8" DIA
PROJECTILE HOISTS
ELECTRIC DECK
CENTER COLUMN
10'-9"25 DIA
0"375
7"25 OR 31"35

24'-7" RADIUS WORKING CIRCLE
24'-0"
17'-6"
₡ RANGEFINDER
₡ MOUNT
6'-2"
10'-1"
₡ TRUNNIONS
AUXILIARY PROJECTILE HOIST ROPE
BREECH FACE RECOIL 29"65
26'-8"
10' 9"
41° ELEVATION
5'-10"
5°
5'-7"
6'-5"
4'-8"
6'-0"
5° DEPRESSION
LOADING ANGLE
RAMMER
POWDER CHUTE TO CENTER GUN
21'-6"5 DIA
PAN FLOOR
6"725
ROLLER PATH
POWDER HOIST
19'-4"75 DIA
21'-8"3
LOADING TRAY
17'-8" DIA
ASSEMBLES 116 G. 125-0"375
PROJECTILE HOISTS
ELECTRIC DECK
CENTER COLUMN
10'-9"25 DIA
0"375
9"25, 9"5, OR 15"086

M1893 Barbette Carriage for 10-inch Gun

Usual guns: M1888
Carriages Built: 11
Original Emplacements: 9
Time Emplaced: 1898-1900

Number of Bolts: 16
Circle Diameter: 9' 9"
Parapet Height: 6' 6"
Center-to-parapet: 10' 10"

Notes: Battery Randol (2), and Battery Quarles (3) at Ft. Worden, Battery Rawlins (2) and Battery Revere (2), at Ft. Flagler, all HD Puget Sound. In 1940 five of these weapons were declared surplus and sold to Canada where they were installed in 1941, where one relativel intact gun and carriage remains.

Total weight: 63 tons
Max/min elevation: +15 / −7 degrees
Turning radius: 320 degrees

Gun: M1888
Weight: 67,183 lbs
Length: 34 calibers

Photos:
Top: Battery Revere, Fort Flagler, HD Puget Sound. (NA, n.d.)

Middle: McNutt Island, Newfoundland, Canada, 2011 (Charles Bogart).

Below: Battery Revere, Fort Flagler 2003 (Dan Rowbottom).

M1893 Barbette Carriage for 10-inch Gun

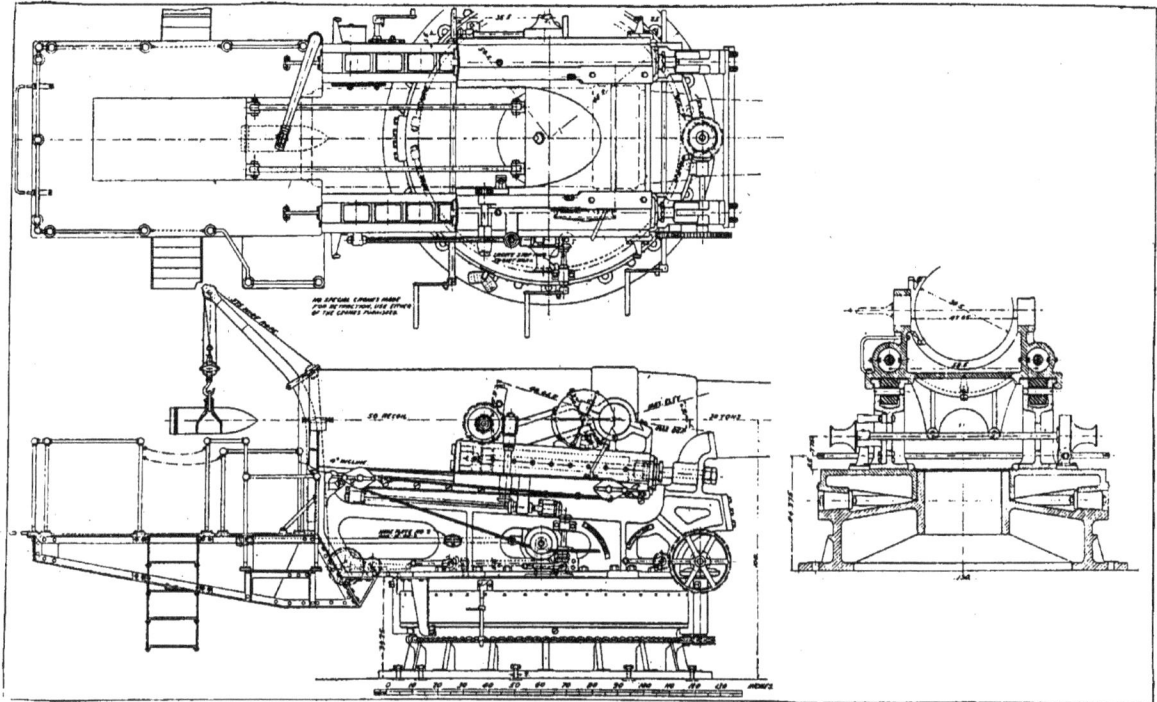

10-INCH BARBETTE CARRIAGE, MODEL OF 1893.

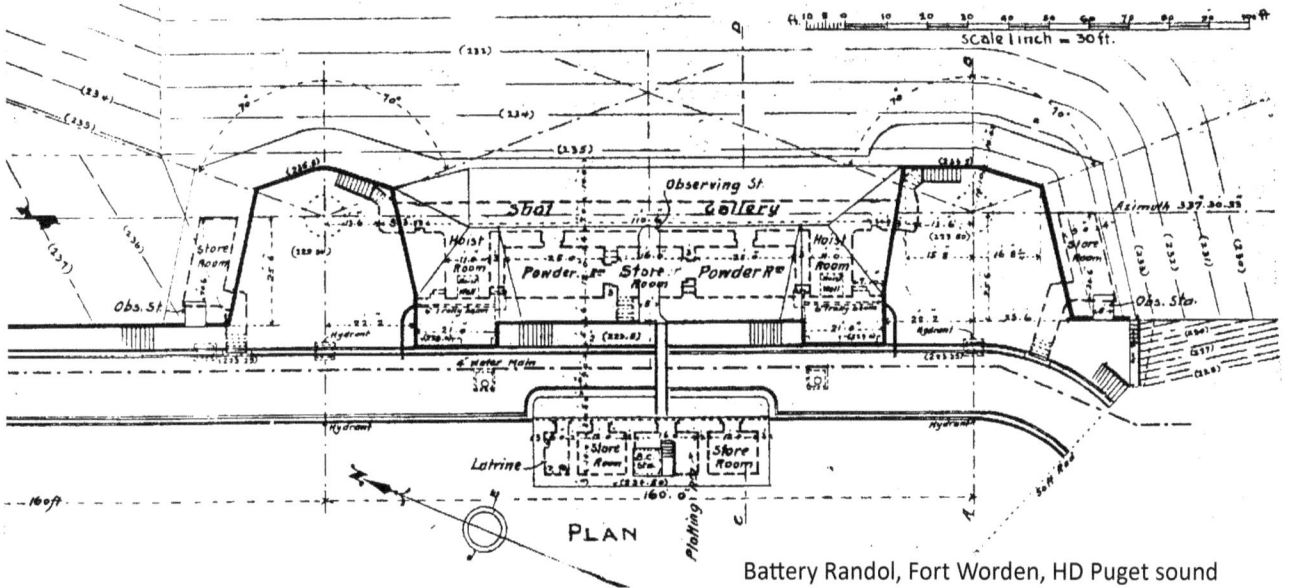

Battery Randol, Fort Worden, HD Puget sound

Battery Quarles, Fort Worden, 1995 (AHM)

M1894 & M1894MI Disappearing Carriage for 10-inch Gun

Usual Guns: M1888
Carriages Built: 35
Original Emplacements: 35
Time Emplaced: 1896-1900

Number of Bolts: 16
Circle Diameter: 10' 6"
Parapet Height: 12' 1"
Center-to-parapet: 8' 8"
Notes: 5 "giant" steps (sometimes 6 steps), has traverse rail 24' 7"diameter

Notes: The M1894MI upgrade was made to existing M1894 carriages. No new M1894MI carriages were built. Three of these weapons were sent to Canada via lend-lease in 1941. Two barrels still remain in Canada.

Total weight: 131 tons
Max/min elevation: +12 / −5 degrees
Turning radius: 140 degrees

Gun: M1888
Weight: 67,183 lbs
Length: 34 calibers

Photos:
Above: Battery Sevier, Fort Pickens, HD Pensacola, circa 1910 (Gulf Island Natl. Seashore Collection). Middle: Battery Walker (5 steps), Fort Stevens, HD Columbia River, 1987 (AMH)
Below (left): Battery Jack Adams, Fort Warren, HD Boston, 2003 (AHM); (Right) Battery Bartlett, (6 steps) Fort Warren, 1985 (AMH).

M1894, M1894MI DISAPPEARING CARRIAGE FOR 10-INCH GUN

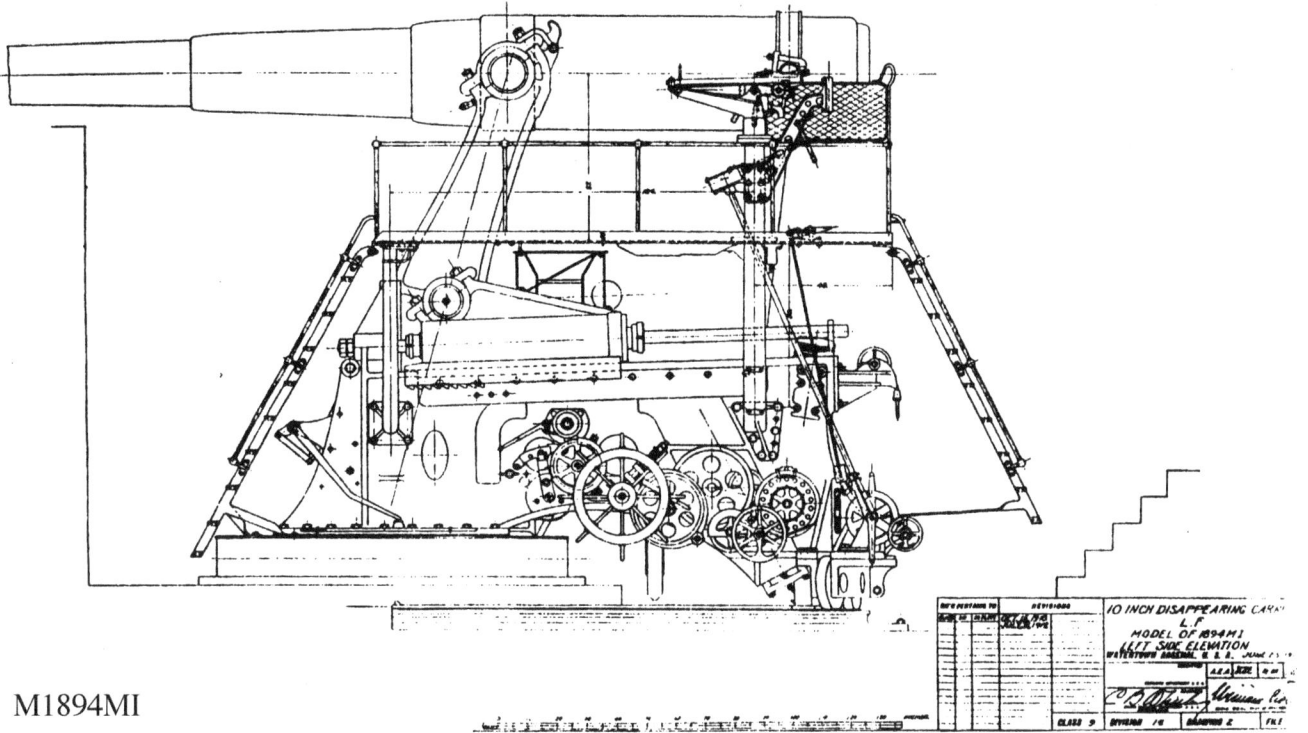

M1894MI

M1894

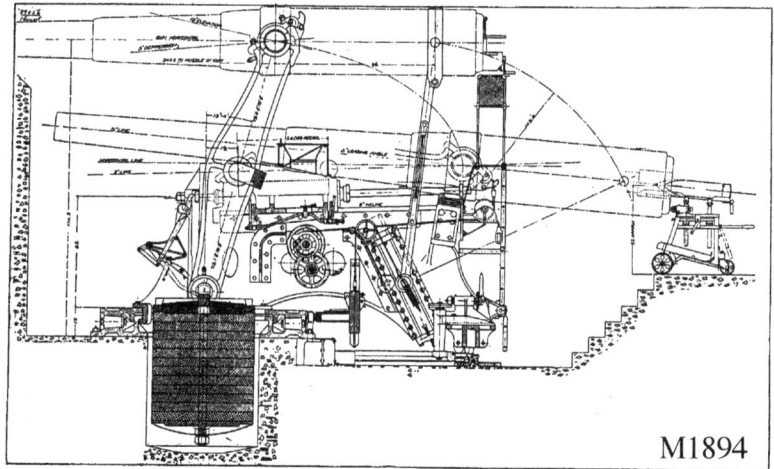

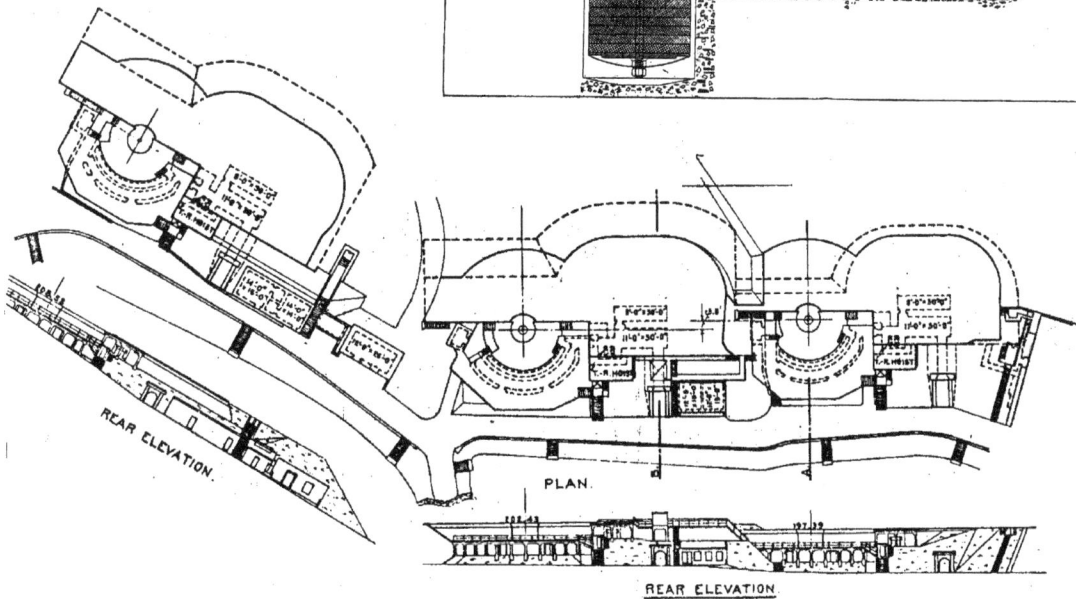

Battery Marcus Miller, Fort Winfield Scott, HD San Francisco

M1896 DISAPPEARING CARRIAGE FOR 10-INCH GUN

Usual Guns: M1888, M1895
Carriages Built: 74
Original Emplacements: 74
Time Emplaced: 1898-1902

Number of Bolts: 16
Circle Diameter: 15' 4"
Parapet Height: 12' 10.5"
Center-to-parapet: 12'
Notes: 4 steps

Total weight: 252 tons
Max/min elevation: +12 / –5 degrees
Turning radius: 360 degrees (limited by emplacement)

Gun: M1888
Weight: 67,183 lbs
Length: 34 calibers

Gun: M1895
Weight: 66,700 lbs
Length: 35 calibers

Photos
Top: Battery Barlow, Fort H.G. Wright, HD Long Island Sound, circa 1916 (Pierce Rafferty Collection)

Middle: Bottom: Battery Jesup, Fort Frémont, HD Port Royal Sound, 1990s,(GMW).

Bottom:Battery Cullem, Fort Pickens, HD Pensacola. (Note this 5 step emplacement was designed for the earlier M1894 carriage) Stillions Collection Gulf Shores Natl Seashore.

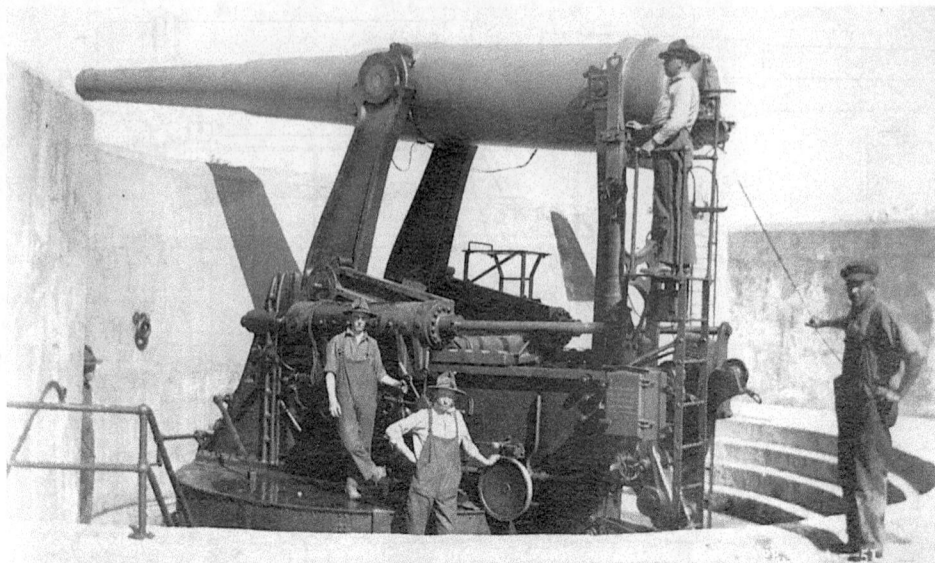

M1896 DISAPPEARING CARRIAGE FOR 10-INCH GUN

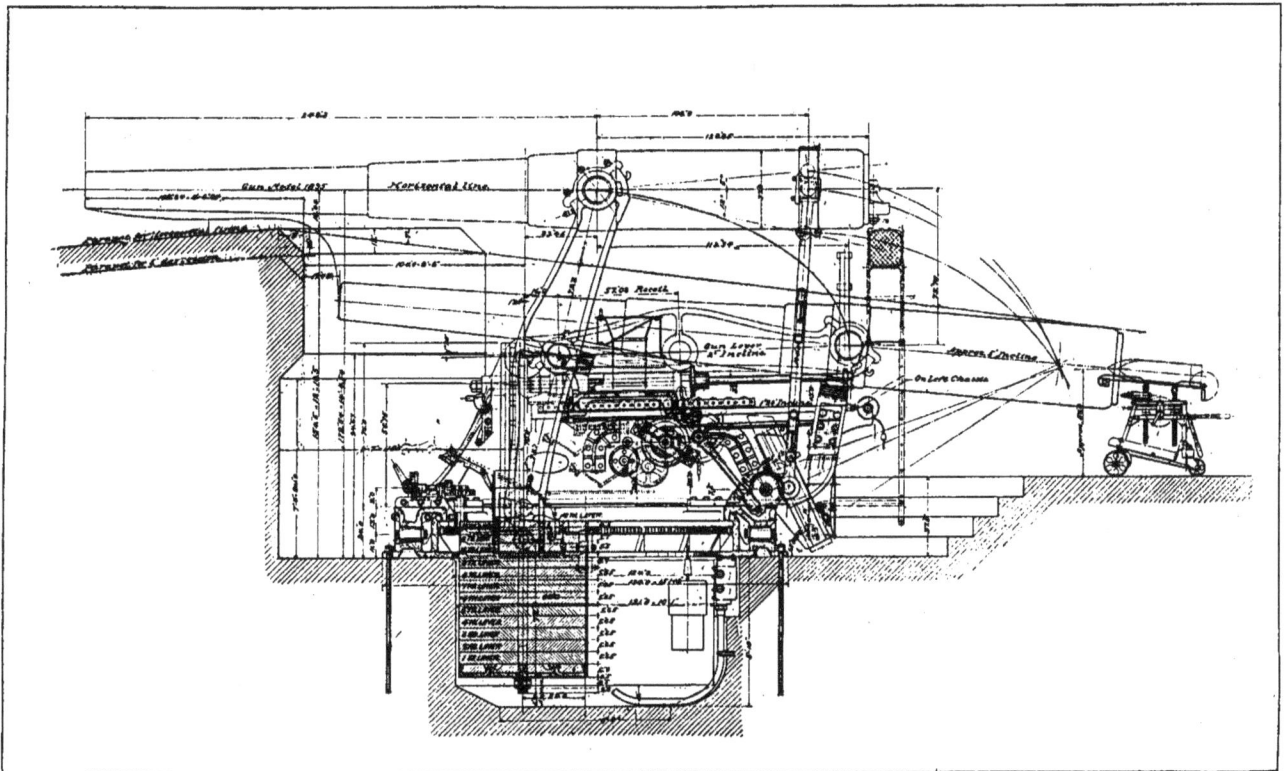

LIMITED-FIRE DISAPPEARING CARRIAGE, MODEL OF 1896, MOUNTING 10-INCH GUNS.

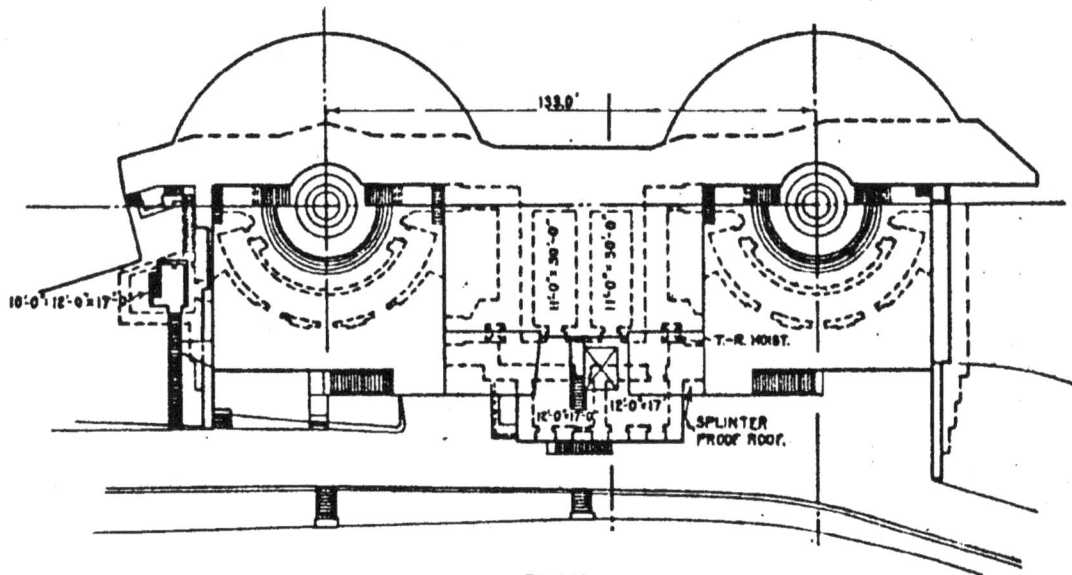

PLAN.

REAR ELEVATION.

Battery Cranston, Fort Winfield Scott, HD San Francisco

M1896 DISAPPEARING CARRIAGE ARF FOR 10-INCH GUN

Usual Guns: M1888
Carriages Built: 3
Original Emplacements: 3
Time Emplaced: 1898-1899

Number of Bolts: 12 inner, 44 outer
Circle Diameter: 7' 10" inner, 25' 8" outer
Parapet Height: 9' 10"
Center-to-parapet: 8' 8" to first well
Notes: 2 emplacements were Battery Mishler, Fort Stevens; HD Columbia River. The third emplacement was half of Battery Heileman, Fort San Jacinto, HD Galveston. Battery Mishler was covered with a concrete roof during 1948, but is otherwise intact; Battery Heileman has been destroyed.

Total weight: 122 tons
Max/min elevation: +12 / −5 degrees
Turning radius: 360 degrees

Gun: M1888
Weight: 67,183 lbs
Length: 34 calibers

BATTERY MISHLER—FORT STEVENS, ORE.
SOUTH JETTY IN DISTANCE.

Photos: Battery Mishler, Fort Stevens, HD Columbia River, 1930s (from the collection of the Fort Stevens Museum holdings).

Plan for Battery Mishler, Fort Stevens, HD Columbia River

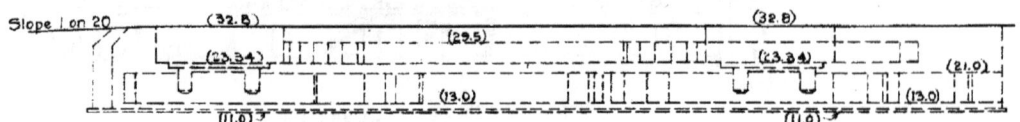

SECTIONAL PLAN

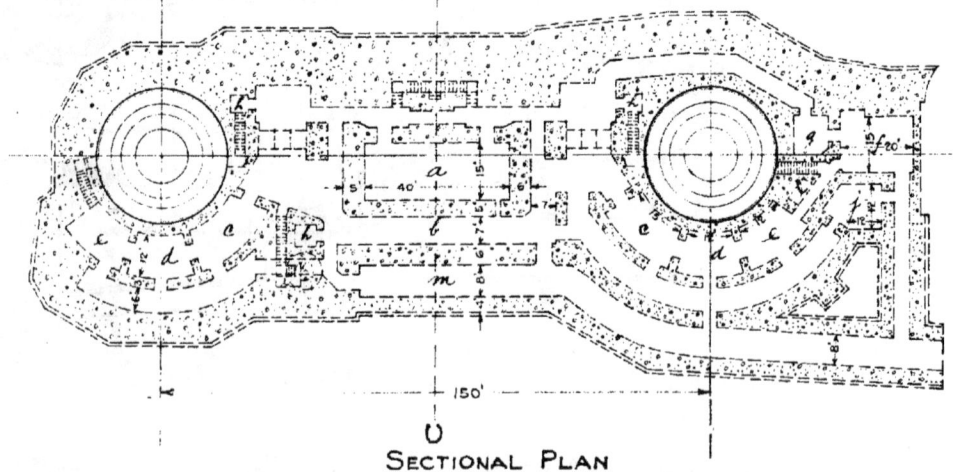

M1896 DISAPPEARING CARRIAGE ARF FOR 10-INCH GUN

Battery Heileman, Fort San Jacinto, HD Galveston. The left emplacement was for the 10-inch M1896 DC ARF carriage. This battery has been destroyed.

M1901 DISAPPEARING CARRIAGE FOR 10-INCH GUN

Usual Guns: M1900, M1895
Carriages Built: 16
Original Emplacements: 12 US, 4 Terr.
Time Emplaced: 1904-1910

Number of Bolts: 16
Circle Diameter: 15' 4"
Parapet Height: 11' 6.5"
Center-to-parapet: 13' 9"
Notes: 1 large or 2 small steps

Total weight: 167 tons
Max/min elevation: +12 / −5 degrees
Turning radius: 360 degrees (limited by emplacement)

Gun: M1900
Weight: 76,830 lbs
Length: 40 calibers

Note: Two 10-inch guns on M1901 DCs currently at Fort Casey State Park, Whidbey Island, Washington. The guns and carriages were removed from Battery Warwick, Fort Wint, HD Subic Bay, the Philippines, and installed in 1967 by the Washington State Park Service in Battery Worth (which originally had M1896 10-inch DCs). See section on fort buildings for photos.

Photos:
Top: Battery David Russell, Fort Stevens, HD Columbia River, 1930s (Marshall Hanft Collection).

Middle: Battery Thomson (2 steps), Fort Moultrie, HD Charleston, 1986 (AMH)

Below: Battery Grubbs (1 step), Fort Mills, HD Manila Bay, 2006 (MAB),

M1901 Disappearing Carriage for 10-inch Gun

10-INCH GUN ON DISAPPEARING CARRIAGE, MODEL OF 1901.

PLAN

REAR ELEVATION

Battery Benson, Fort Worden, HD Puget Sound

M1891 Gun Lift Carriage for 12-inch Gun

Usual Guns: M1888
Carriages Built: 5
Original Emplacements: 2
Time Emplaced: 1892-1896

Number of Bolts:
Circle Diameter:
Parapet Height: 5'
Center-to-parapet:
Notes: Battery Potter, Fort Hancock, was the only GL battery built. Reference: "The Gun-Lift Battery in the Defenses of the United States," CDSG Journal 1996 Vol. 10, No. 4, pp. 4-16.

Gun: M1888
Weight: 117,127 lbs
Length: 34 calibers

Photos:
Top: Battery Potter entrance 1994 (MAB)

Upper Middle: Battery Potter gun platform 1999 (AMH)

Middle lower: Battery Potter test firing, 1892 (Gateway Natl. Rec. Area)

Bottom: Looking southeast at the Sandy Hook gun-lift battery ca. 1894. National Archives, No. 77-F-45-90-22.

M1891 Gun Lift Carriage for 12-inch Gun

Top: Looking at the recently mounted north gun from the still-unfinished surface of the south emplacement. The gun-lift carriage is in the recoiled and locked position, ready for lowering in to the battery interior. National Archives, No. 77-F-45-90-14.

Middle: The Sandy Hook gun-lift battery housed an impressive collection of equipment to power the lifts; two of the pumps are shown here. National Archives, No. 77-F-45-90-20.

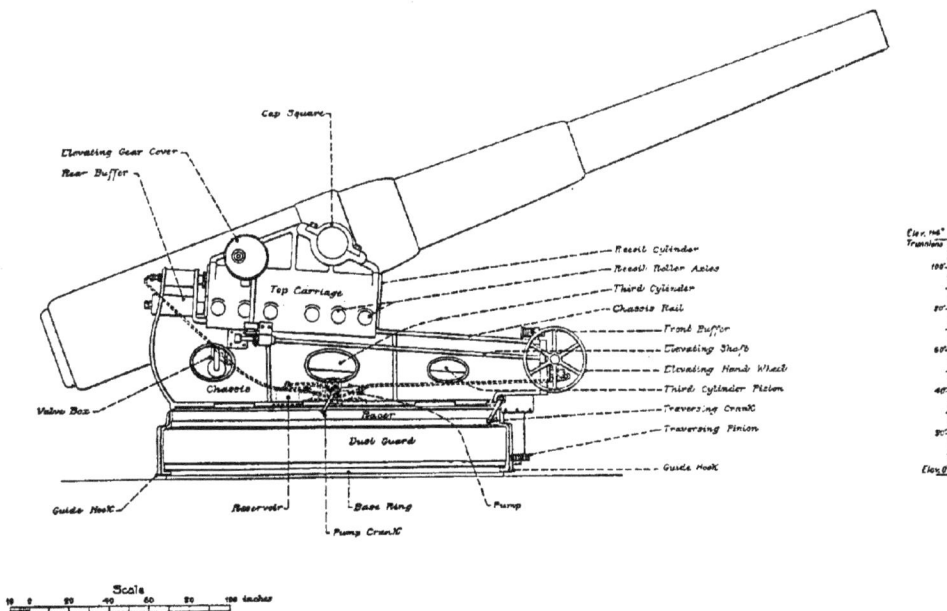

M1891 Gun Lift Carriage for 12-inch Gun

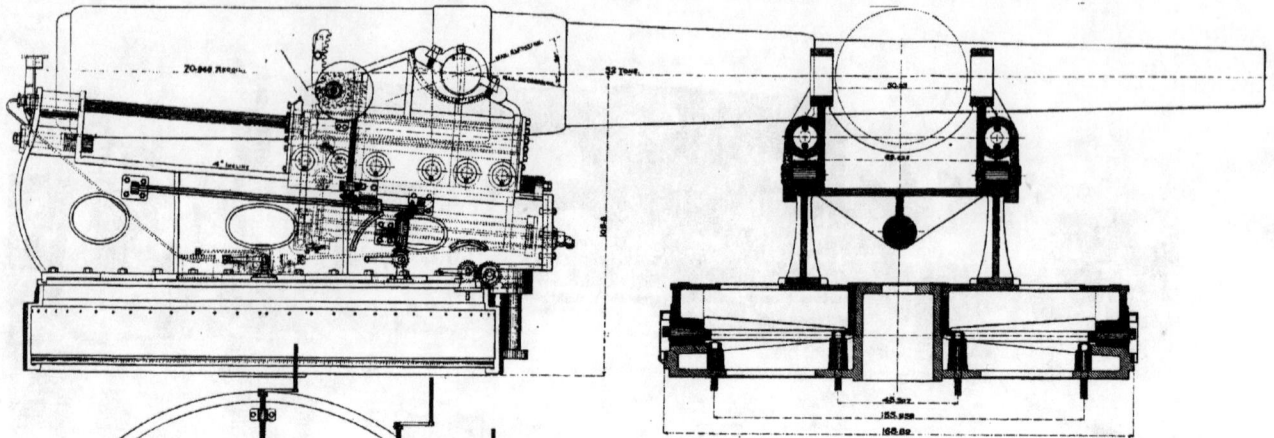

GUNLIFT CARRIAGE FOR 12 INCH B. L. RIFLE.

Key to drawings on next page:
Interior layout of gun lift battery at Fort Hancock
First Floor A: Accumulator Room A, L: Ammunition Lift B: Boiler Room L: Lift, P: Pumps, Pr: Projectile Storage, Pw: Powder Magazine
Second Floor A: Accumulator Room (open from first to second floor), B: Boiler Room (open from first to second floor, C: Casemate (intended use not known), L: Lifts, R: Rammer Gallery
The unidentified spaces to the left and right of the lifts, and in front of the battery, provide access to the ends of the threaded rods holding the lift rails in place. A wooden bridge connected the passageways at the rear of the second story.

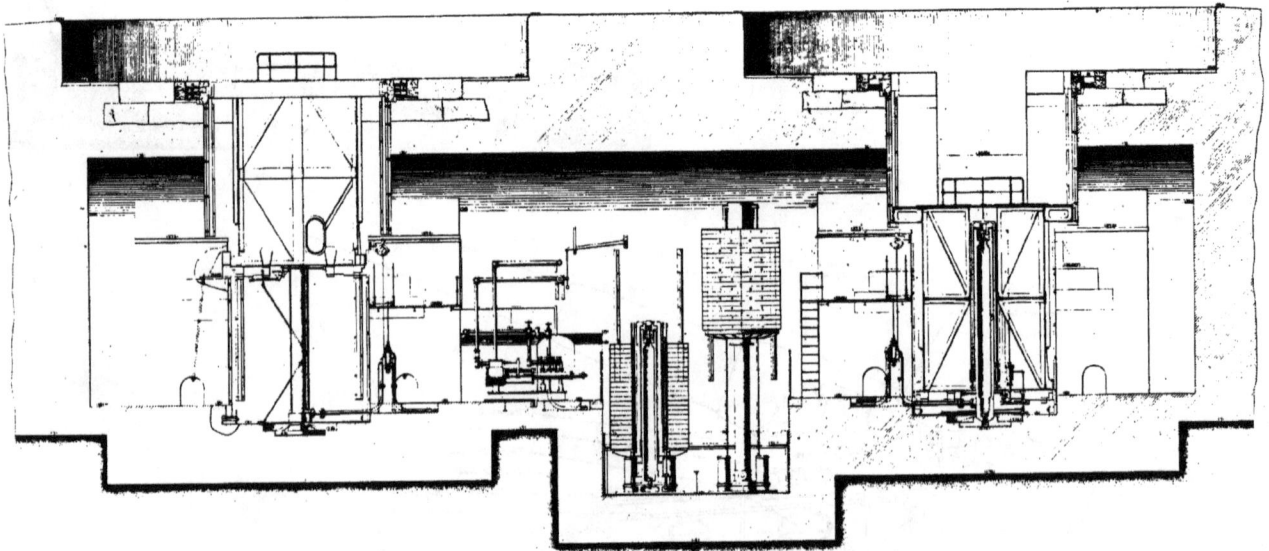

Section through centre line of Accumulator Room
showing
Arrangement of Mechanism.

Interior cross section, Battery Potter, Fort Hancock, HD New York (drawing from the National Archives)

M1891 Gun Lift Carriage for 12-inch Gun

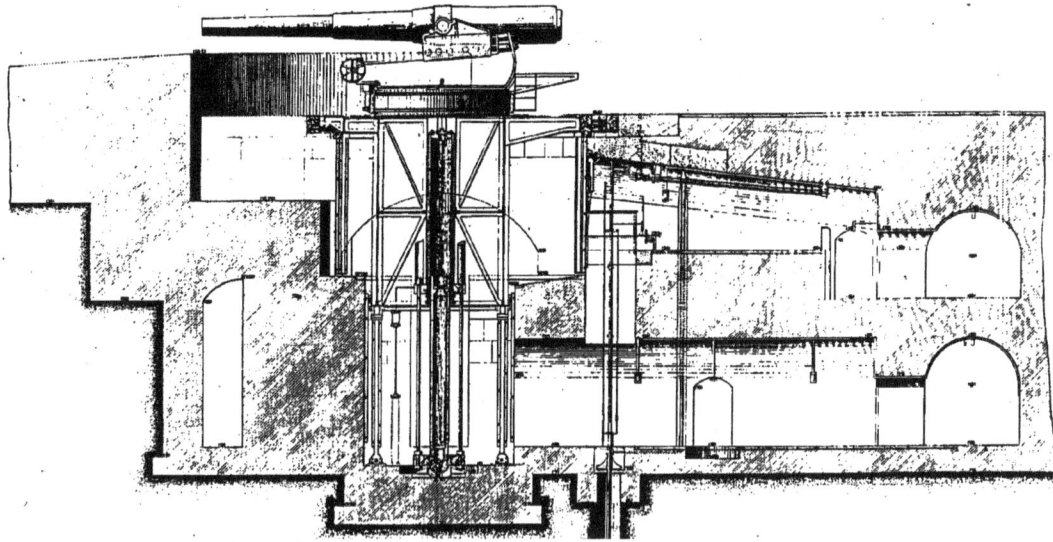

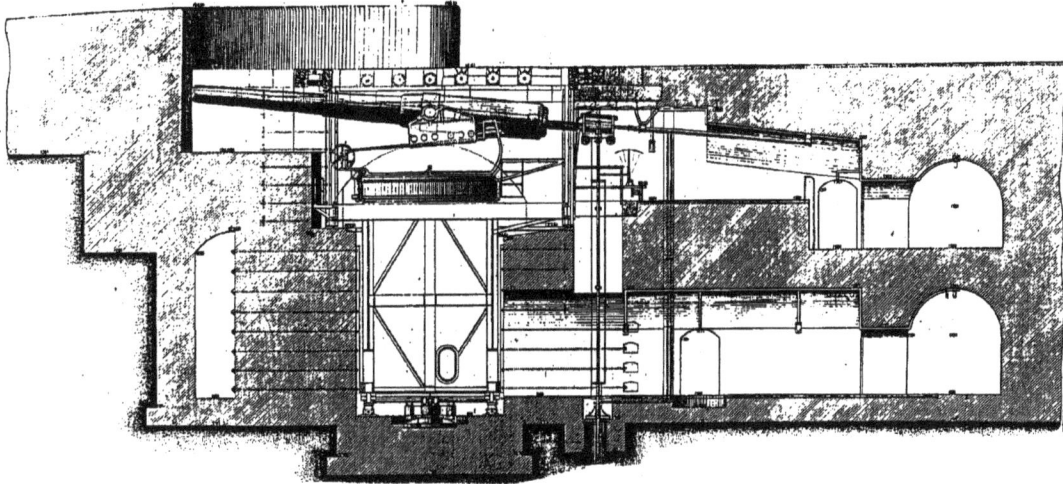

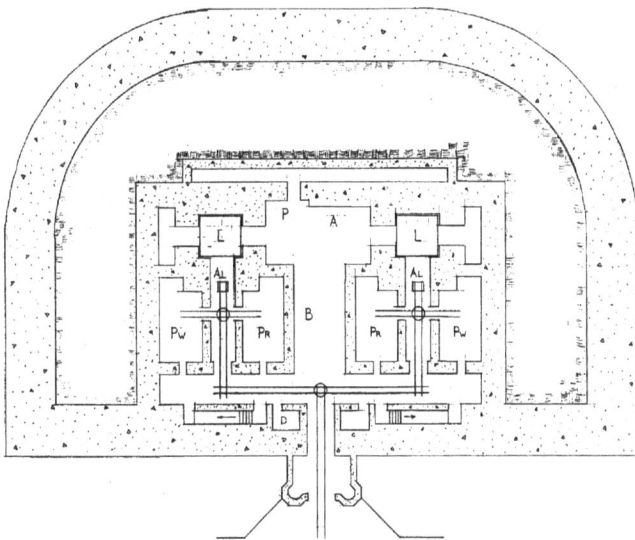

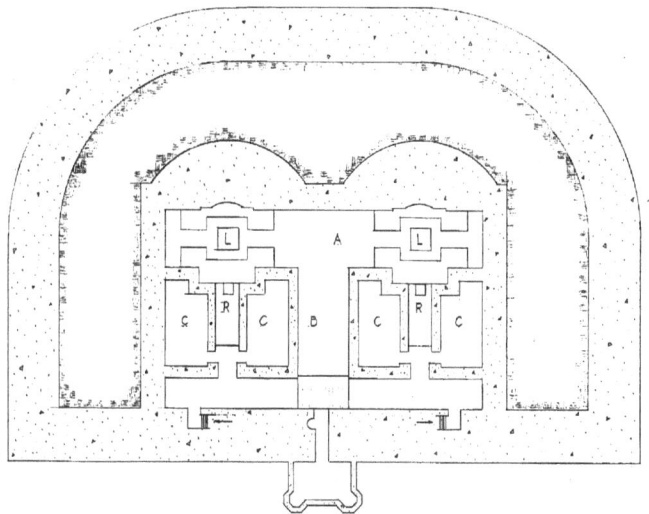

Battery Potter: upper & lower levels (drawings by David Hansen)

M1891 CARRIAGE FOR 12-INCH MORTAR

Mortars: M1886, M1886M,
M1886-90M1
Carriages Built: 86
Original Emplacements: 84
Time Emplaced: 1893-1898

Number of Bolts: 12 inner, 12 outer
Circle Diameter: 13' 7"
Notes: all were replaced with M1896MI,
or removed from service by 1920

Carriage weight: 79,714 lbs
Max./min. elevation:
Turning radius: 360 degrees
M1886 mortar
Weight: 31,920 lbs
Length: 10 calibers

Photos:
Top: Battery Bagley, HD Cape Fear
River, 1900s, M1886 cast iron tube.

Middle: Battery McCook or
Reynolds, Fort Hancock, HD New
York, 1900s.

Bottom: Battery McCook-Reynolds
1997 (MAB).

M1891 CARRIAGE FOR 12-INCH MORTAR

Spring Return Carriage for 12 inch B.L. Steel Mortar.

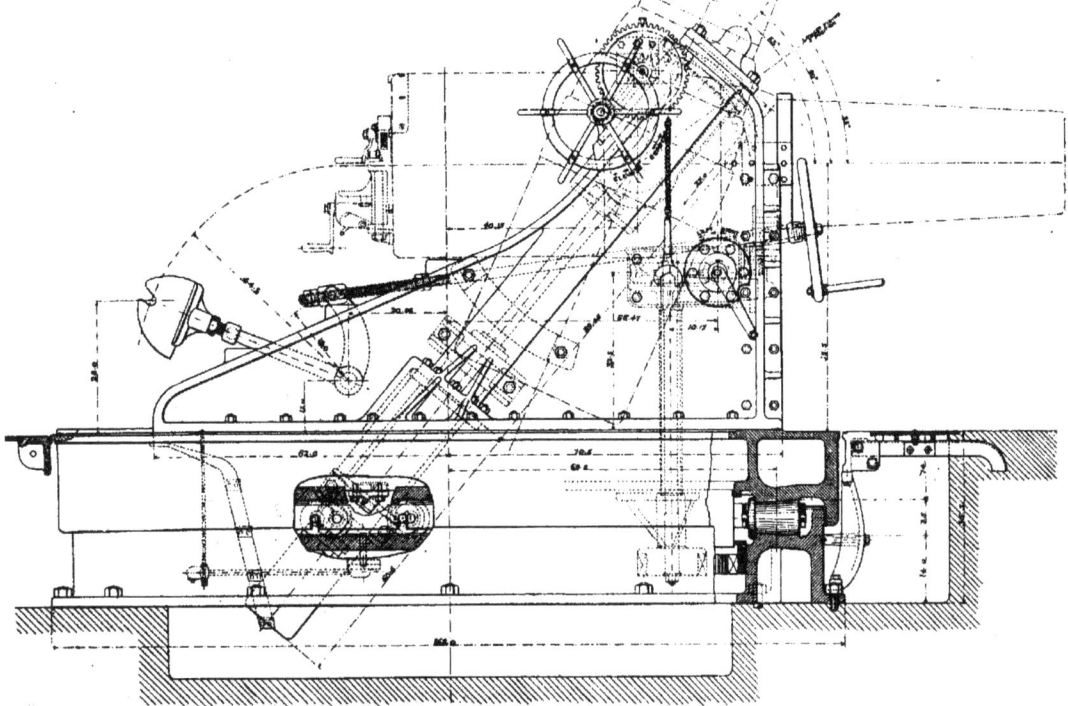

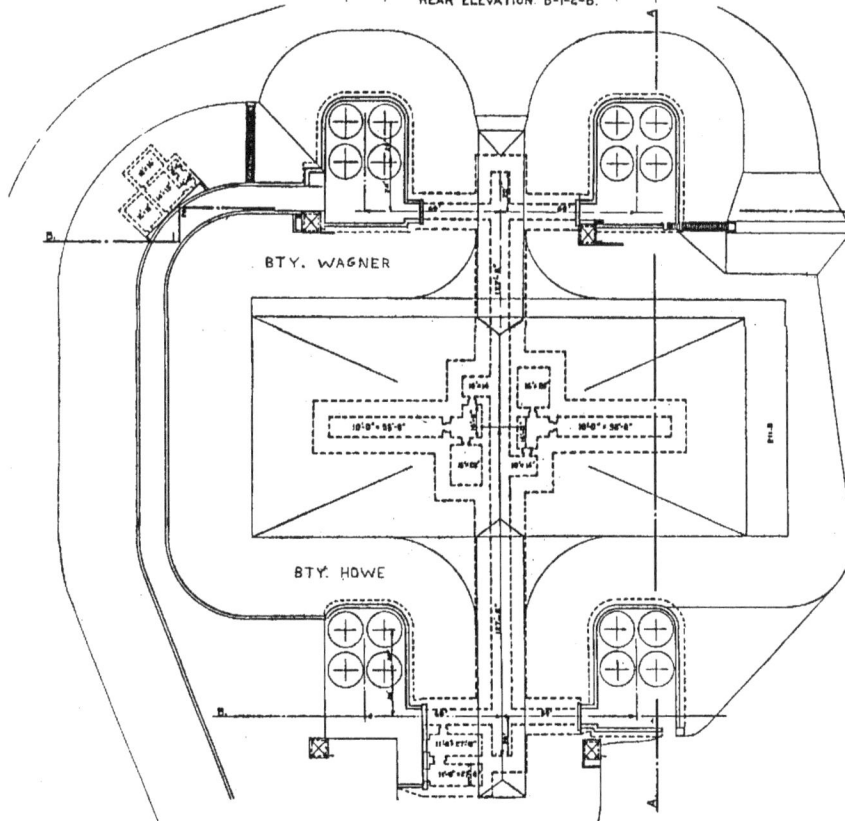

Model 1891.

REAR ELEVATION B-1-2-B.

BTY. WAGNER

BTY. HOWE

Batteries Howe and Wagner, Fort Winfield Scott, HD San Francisco (from Winslow)

M1892 Barbette Carriage for 12-inch Gun

Usual Guns: M1888
Carriages Built: 28
Original Emplacements: 28
Time Emplaced: 1895-1900

Number of Bolts: 28
Circle Diameter: 11' 8"
Parapet Height: 6' 8"
Center-to-parapet:
Notes: often on elevated
concrete pedestal.

Total weight: 118 tons
Max/min elevation: +15 / −7
degrees
Radius: 360 degrees

Gun: M1888MI, MII
Weight: 117,127 lbs
Length: 34 calibers

Photos:
Top:Battery Neary, Fort
Hamilton, HD New York (NA,
GMW Collection)

Upper Middle: Battery
Godfrey, Fort Winfield Scott,
HD San Francisco, 1992 (MAB).

Lower Middle: Battery Read,
Fort DuPont, HD Delaware
River, 1996 (AMH)

Bottom: Battery Huger, Fort
Sumter, HD Charleston, 1986
(AMH).

M1892 Barbette Carriage for 12-inch Gun

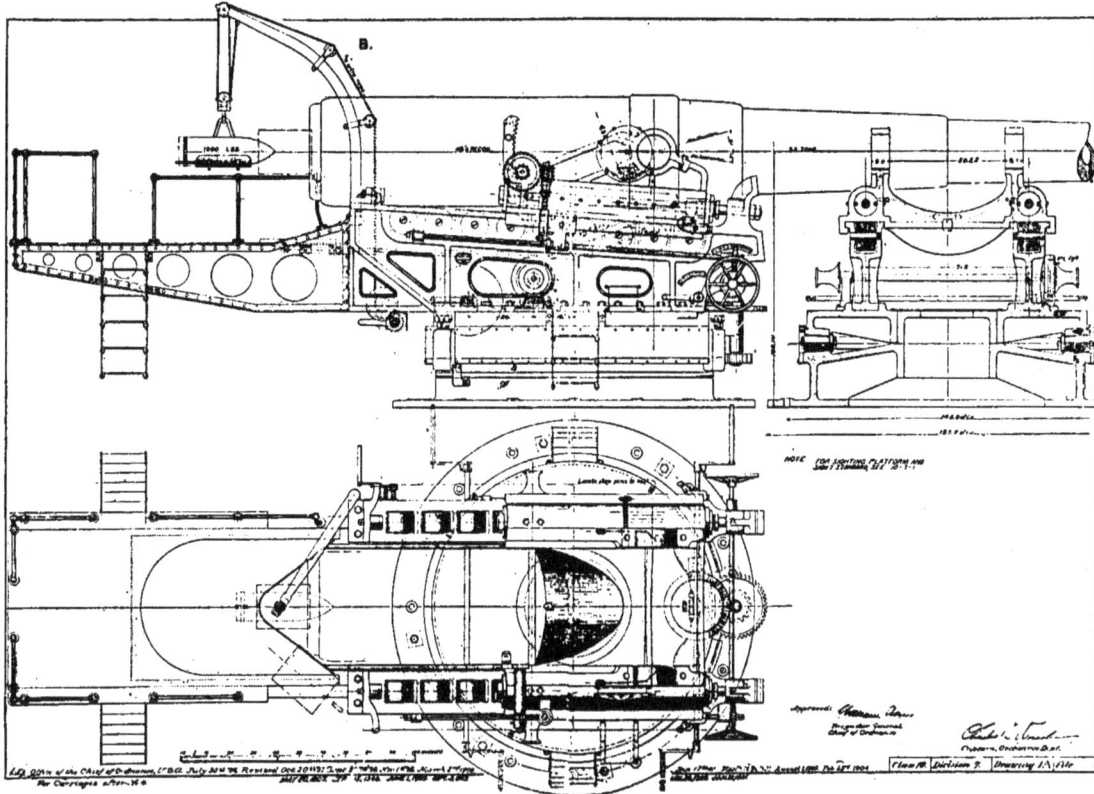

TOP CARRIAGE RECOIL TYPE OF BARBETTE CARRIAGE, MODEL OF 1892, FOR 12-INCH GUNS.

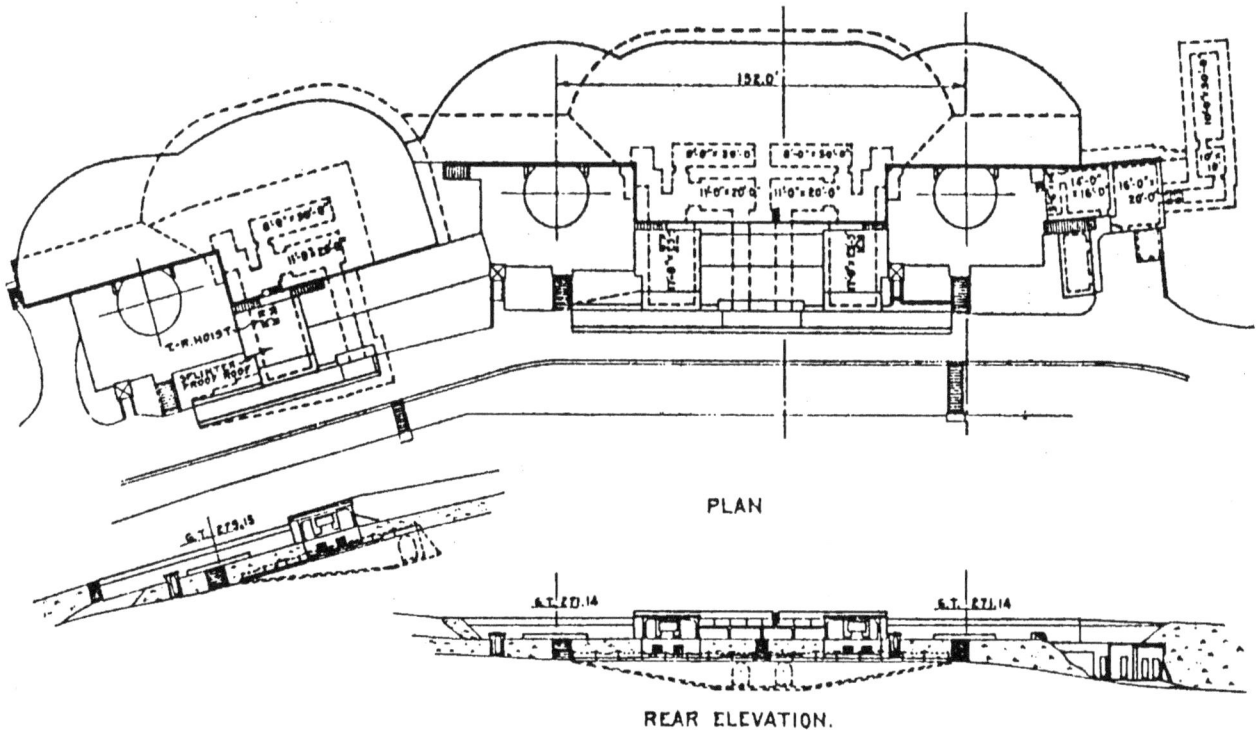

PLAN

REAR ELEVATION.

Battery Godfrey, Fort Winfield Scott, HD San Francisco

M1896 Disappearing Carriage for 12-inch Gun

Usual Guns: M1888, M1895
Carriages Built: 27
Original Emplacements: 27
Time Emplaced: 1898-1900

Number of Bolts: 14 inner, 12 outer
Circle Diameter: 14' 2" inner, 18' 4" outer
Parapet Height: 17' 4.5"
Center-to-parapet: 13' 5"
Notes: deep well usually 9 steps but some 7 or 8 steps

Total weight: 186 tons
Max/min elevation: +10 / −5 degrees
Radius: 360 degrees (limited by emplacement)

Gun: M1888
Weight: 117,127 lbs
Length: 34 calibers

Photos:
Top: Either Battery Bloomfield or Alexander, Fort Hancock, HD Sandy Hook (NA)

Upper Middle: Battery Stricker, Fort Howard, HD Baltimore, 1990 (AMH)

Lower Middle: Battery Duportail (8 steps), Fort Morgan, HD Mobile Bay, 2002 AMH).

Bottom: Battery Torbert (7 steps), Fort Delaware, HD Delaware River, 1996 (AMH)

M1896 DISAPPEARING CARRIAGE FOR 12-INCH GUN

UNITED STATES DISAPPEARING CARRIAGE.
L.F., MODEL OF 1896
FOR
12 INCH B.L. RIFLE.

PLAN

REAR ELEVATION

Battery Duportail, Fort Morgan, HD Mobile Bay

M1896 & M1896MI Carriages for 12-inch Mortar

Mortars: M1890, M1890M1
Carriages Built: 310
Original Emplacements: 308
Time Emplaced: 1897-1910, various relocations & partial removals 1917-1921
Notes: The M1896MI upgrade was made to existing M1896 carriages. No new M1896MI carriages were built.

Number of Bolts: 12 inner, 12 outer
Circle Diameter: 13' 7" (inside bolt ring omitted on later constructions)

Total weight: 64 tons
Max/min elevation: +70 / +45 degrees
Radius: 360 degrees

Mortar: M1890MI
Weight: 29,120 lbs
Length: 10 calibers

Photos:
Top: Battery Way, Fort Mills, HD Manila Bay, the Philippines, 2006 (MAB).

Middle: Battery Seminole, Fort Taylor,, HD Key West, 1910s (Signal Corps).

Bottom: Battery Alexander, Fort Barry, HD San Francisco, 1994 (MAB).

M1896 & M1896MI Carriages for 12-inch Mortar

UNITED STATES CARRIAGE.
Model of 1896
for
12 Inch B.L. Mortar.
[Steel]

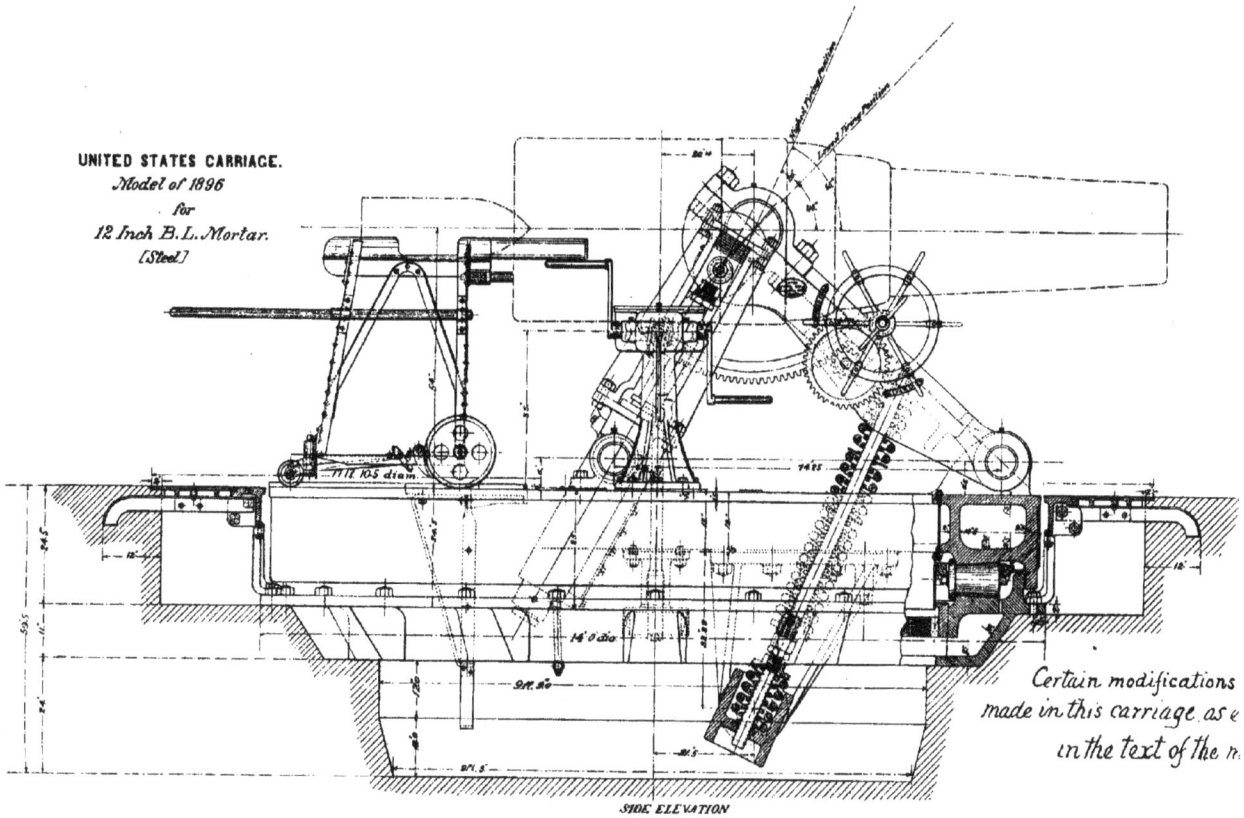

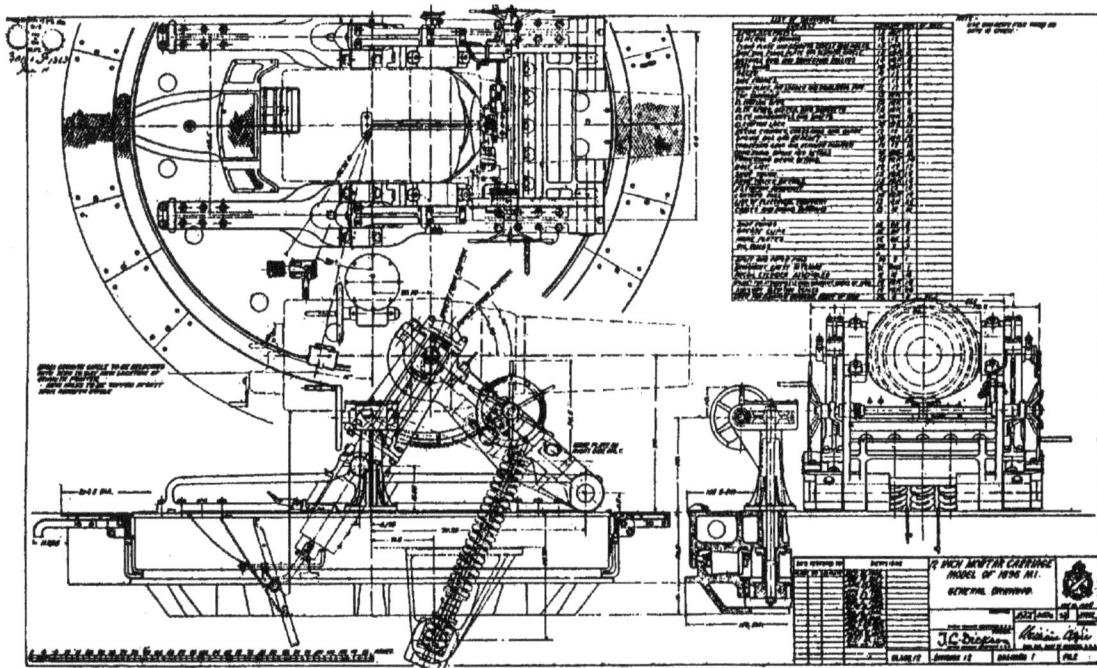

Certain modifications
made in this carriage as e
in the text of the n

SIDE ELEVATION

M1896MI

M1896 & M1896MI Carriages for 12-inch Mortar

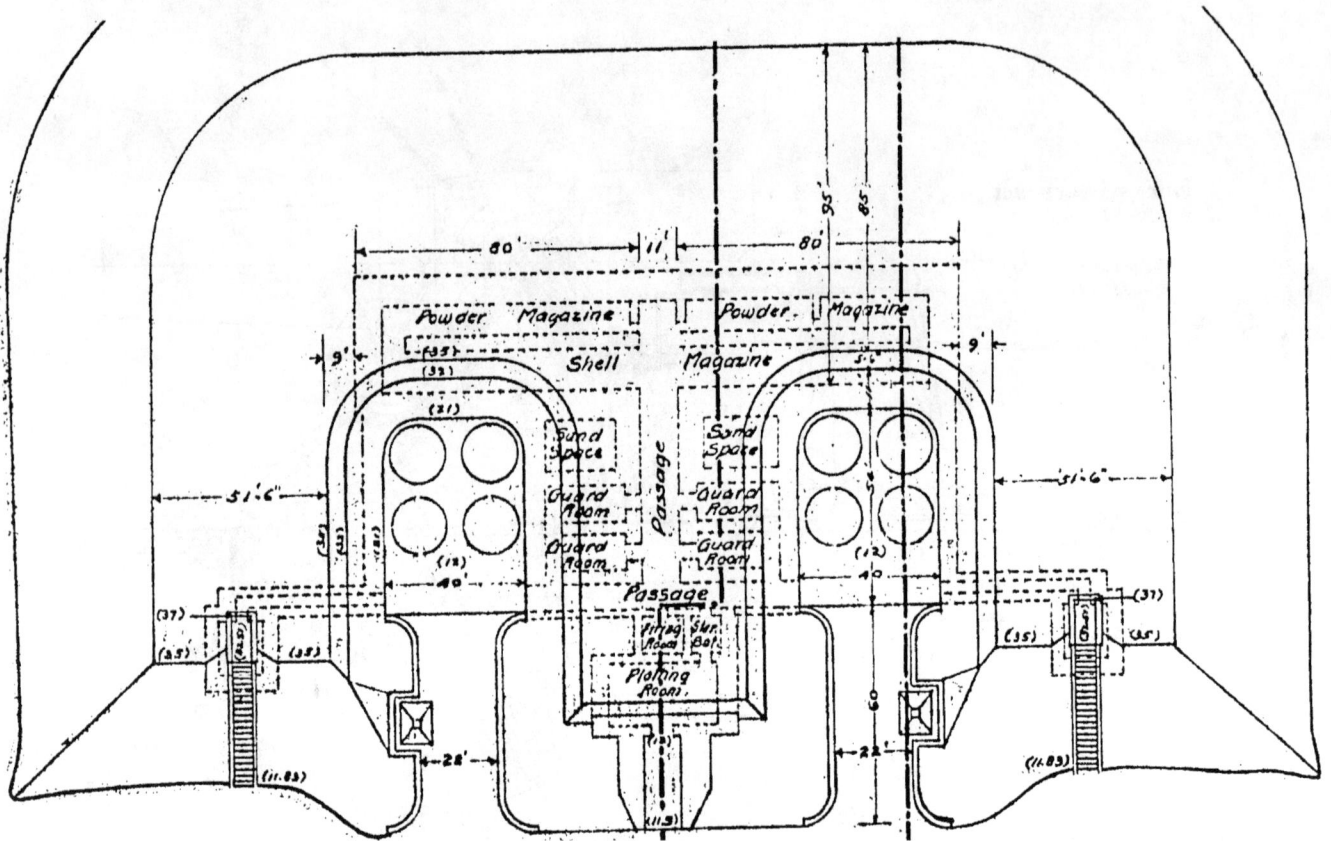

Battery Worth, Fort Pickens, HD Pensacola

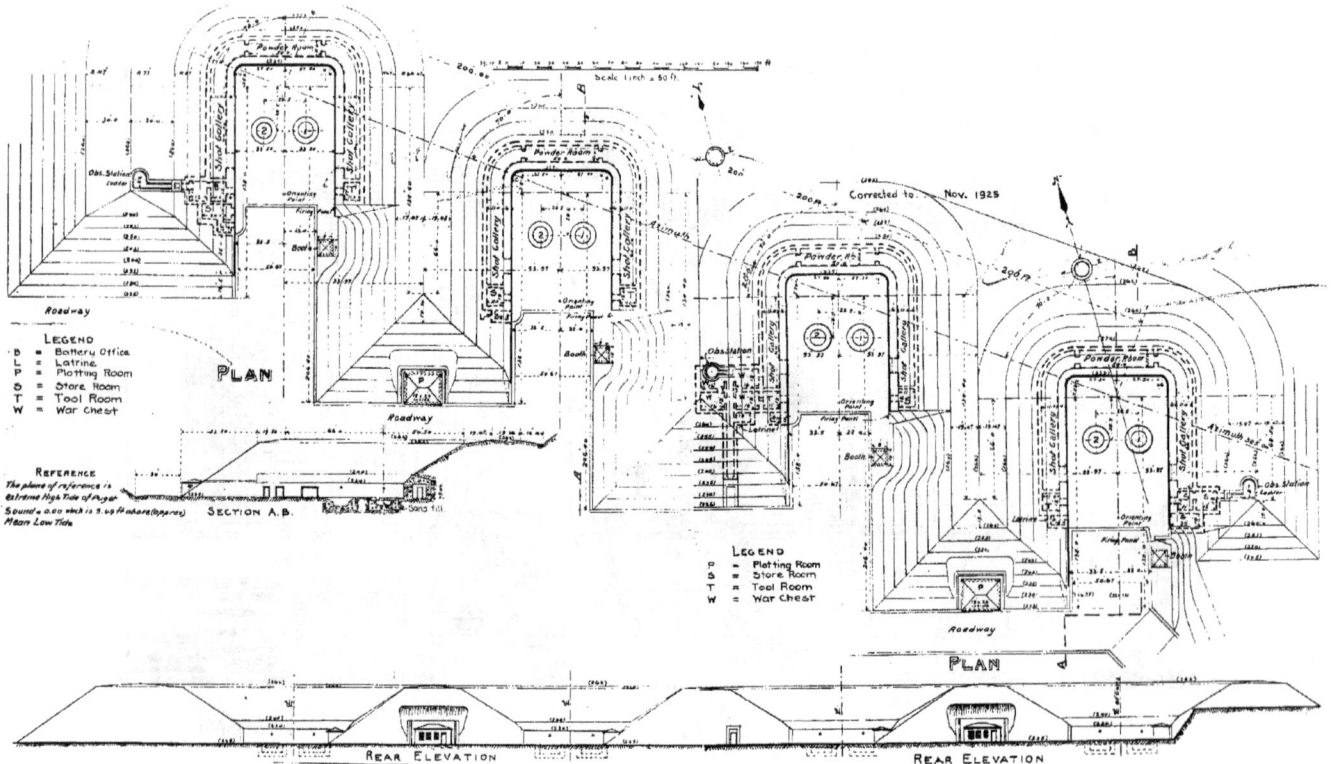

Batteries Powell and Brannon, Fort Worden, Harbor Defenses of Puget Sound

M1896MII Carriage for 12-inch Mortar

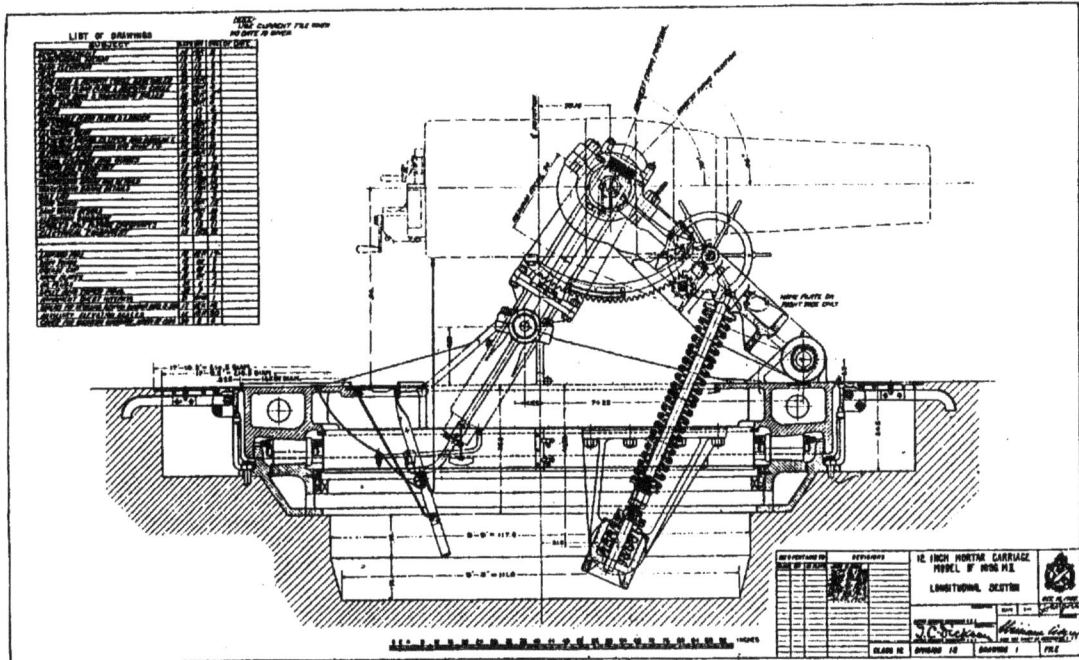

Only four 12-inch mortars on M1896MII carriages were built. They were emplaced in Pit B of Battery Harlow, Fort Ruger, HD Honolulu, which shown in this photograph (US Army Museum of Hawaii).

M1896MIII Carriage for 12-inch Mortar

Mortar: M1912
Carriages Built: 41
Original Emplacements: 8 US, 32 Terr.
Time Emplaced: 1914-1917

Number of Bolts: 8 inner, 12 outer
Circle Diameter: 10' 10" inner, 13' 10" outer
Notes: designed for two mortars per pit—Batteries Barlow-Saxton (Los Angeles), Prince, Merritt, Carr, Tidball, Zalinski, Baird, Howard (Panama) & Craighill (Philippines).

Total weight: 66 tons
Max/min elevation: +70 / 0 degrees
Turning radius: 360 degrees

Mortar: M1912
Weight: 33,854 lbs
Length: 15 calibers

Photos:
Above: Battery Barlow-Saxton, Fort MacArthur, HD Los Angeles, 1919 (USACE)

Middle: Battery Craighill, Fort Hughes, HD Manila Bay, the Philippines, 2006 (MAB).

Below: Battery Barlow-Saxton, Fort MacArthur, 1994 (MAB).

M1896MIII Carriage for 12-inch Mortar

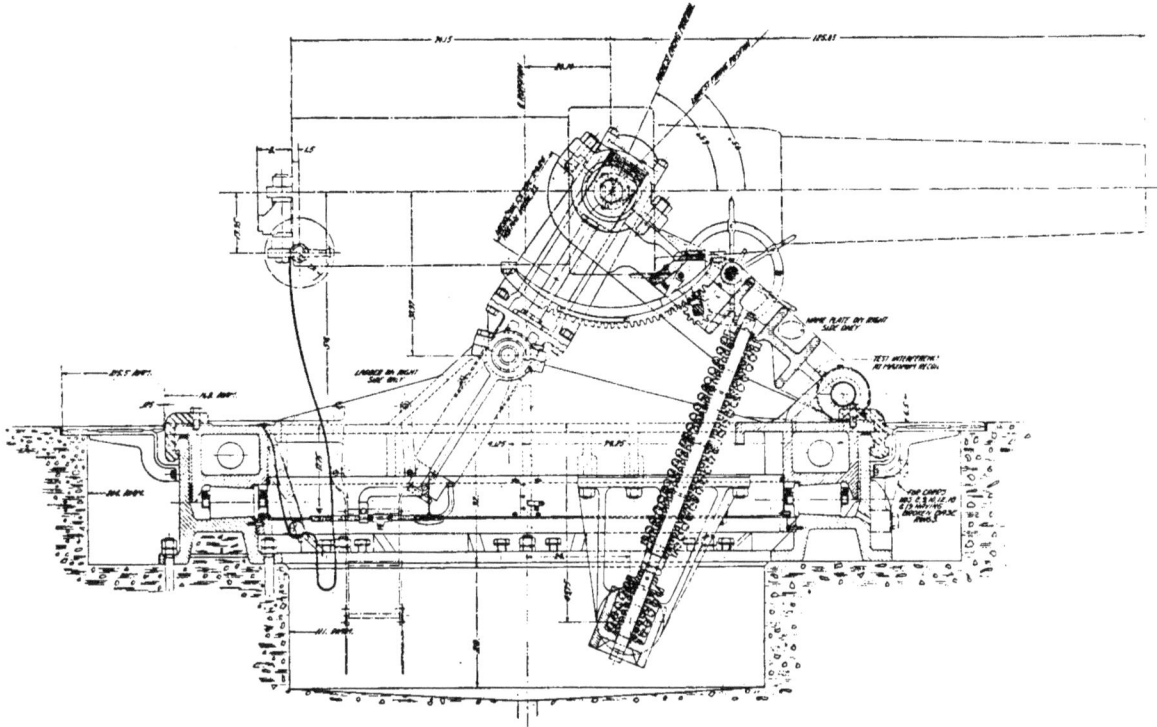

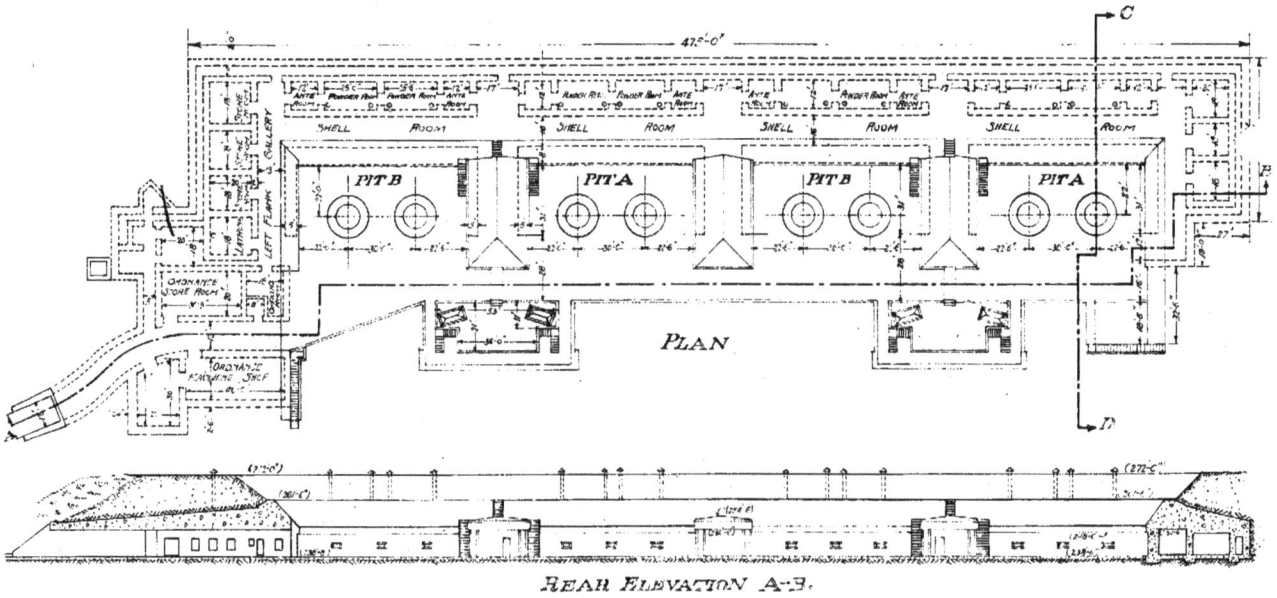

REAR ELEVATION A-B.

Batteries John Barlow and Saxton, Fort MacArthur, HD Los Angeles

M1897 ALTERED GUN LIFT CARRIAGE FOR 12-INCH GUN

Photos: Battery Wilhelm, Fort Flagler, HD Puget Sound
top circa 1916, middle 2003 (Dan Rowbottom),
bottom 1995 (AHM).

Usual Guns M1888
Carriages Built: 3
Original Emplacements: 3
Time Emplaced: 1900

Number of Bolts: 8 inner, 16 outer
Circle Diameter: 3' 7" inner, 11' 2.5"
Parapet Height: 5'
Center-to-parapet: 10' 11"

Three other gun lift carriages were built in addition to the two purchased from France that were used in Battery Potter. After the decision was made not to use the gun lift design, these carriages were modified for use as barbette mounts and redesignated Altered Gun Lift carriages. The three carriages were again modified and modernized after installation, two at Fort Flagler and one at Fort Worden.

Note: The third AGL emplacement at Fort Worden initially was emplaced in position #5 in a line of 7 emplacements with the other barbette carriage guns and was named Battery Powell. Between 1905 and 1909 the gun line was completely rebuilt and modernized, including the addition of the new Taylor-Raymond hoists. After much debate about replacing the AGL, it was decided to move it to the #7 emplacement and move the 10-inch carriage mounted there to the #5 emplacement. This would give both 12-inch guns a seaward field of fire. The #7 emplacement was then rebuilt to hold the 12-inch AGL with a double bolt ring and larger dimensions to accommodate it. Soon after this, this AGL was dismounted and used as parts to rebuild the two AGLs at Fort Flagler. The Worden AGL emplacement was then modified to mount a M1892 BC, formerly the proof mount at Sandy Hook Proving Ground. The inner bolt ring was cut away, the mounting holes of the M1892 carriage were redrilled to fit the AGL foot print, and the excess mounting bolt holes were filled in. The two 12-inch emplacements (#6 & #7) became Battery Ash. Emplacement #5 with a 10-inch barbette was added to Battery Quarles (emplacements #4 and #3). The excess name of Battery Powell was eventually transferred to two pits of the mortar battery.

M1897 ALTERED GUN LIFT CARRIAGE FOR 12-INCH GUN

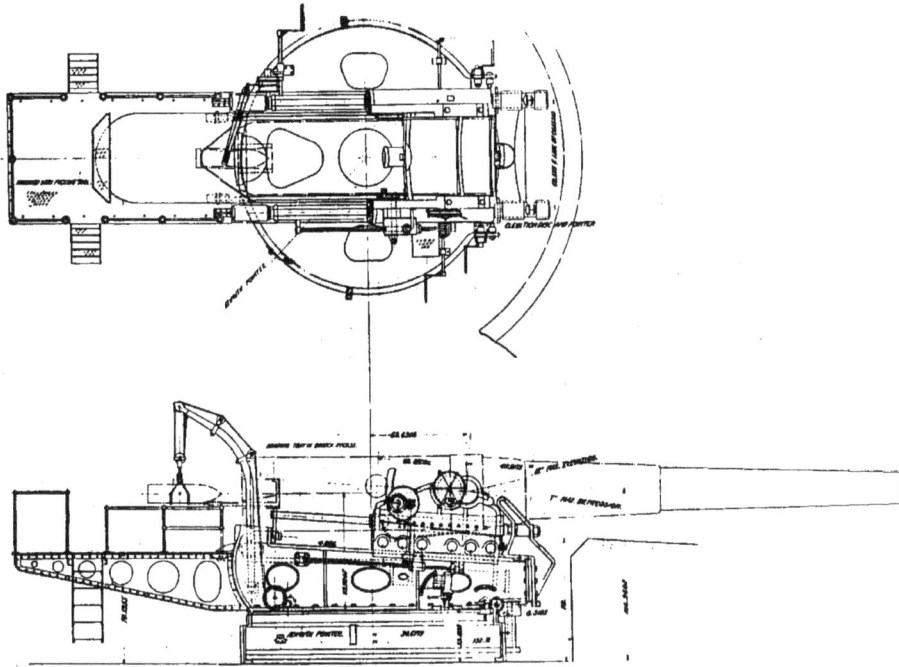

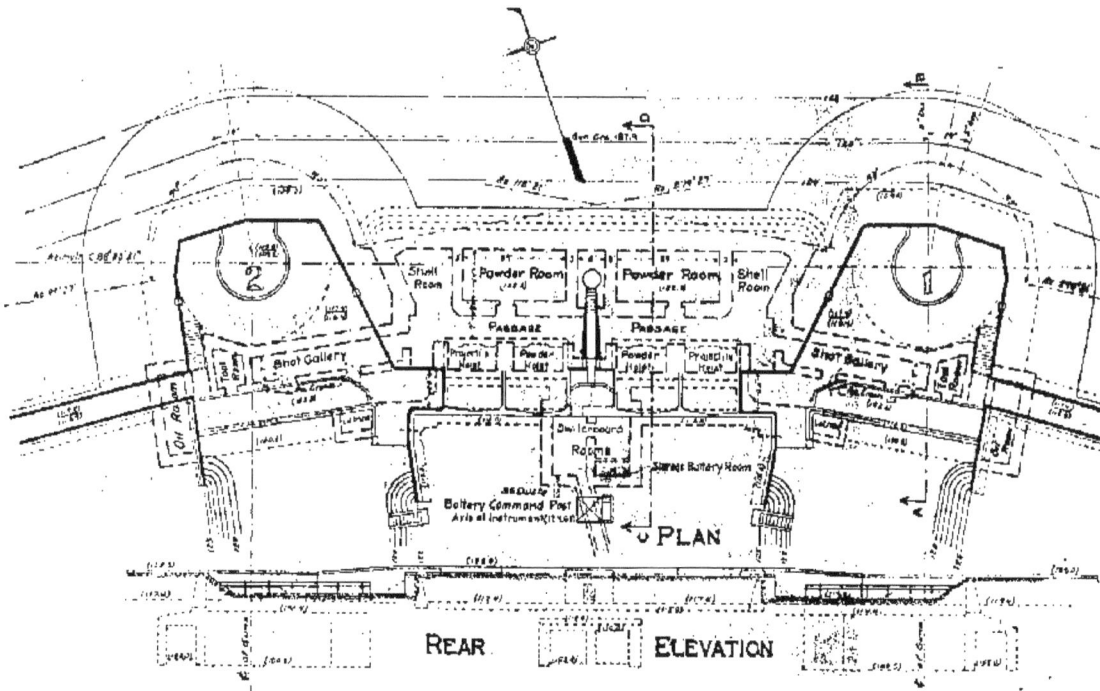

Battery Wilhelm, Fort Flagler, HD Puget Sound

M1897 DISAPPEARING CARRIAGE FOR 12-INCH GUN

Usual Guns: M1895
Carriages Built: 35
Original Emplacements: 35
Time Emplaced: 1899-1904

Number of Bolts: 14 inner, 12 outer
Circle Diameter: 14' 2" inner, 18' 4" outer
Parapet Height: 13' 11.5"
Center-to-parapet: 13' 5"
Notes: 5 steps

Total weight: 243 tons
Max/min elevation: +10 / −5 degrees
Radius: 360 degrees (limited by emplacement)

Gun: M1895
Weight: 115,000 lbs
Length: 40 calibers

12" B.L. Rifle

Photos:
Top: Battery Butterfield, Fort H.G. Wright, HD Long Island Sound, circa 1912 (Sandy Esser Collection).

Middle: Battery DeRussy, Fort Monroe, HD Chesapeake Bay. (The Casemate Museum).

Bottom: Battery Stevenson, Fort Warren, HD Boston 1988 (AMH).

M1897 DISAPPEARING CARRIAGE FOR 12-INCH GUN

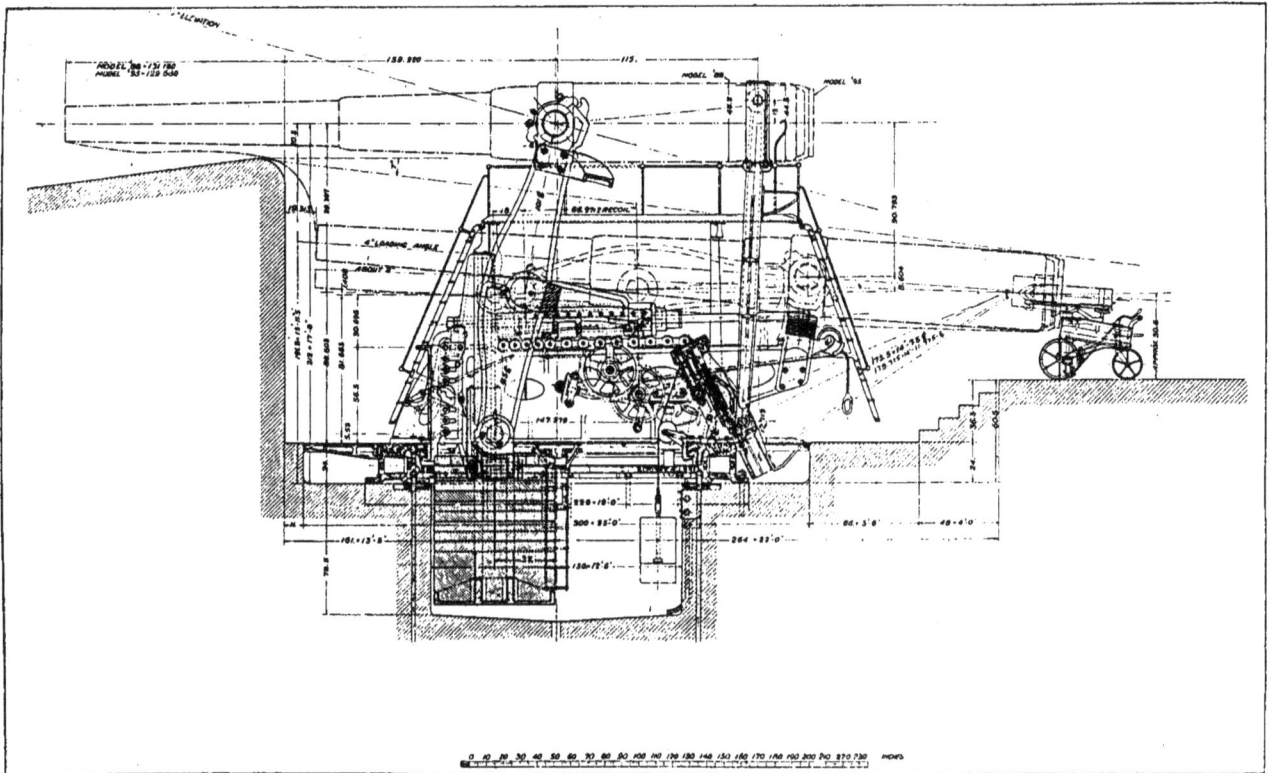

MODEL OF 1897 DISAPPEARING CARRIAGE FOR 12-INCH GUNS.

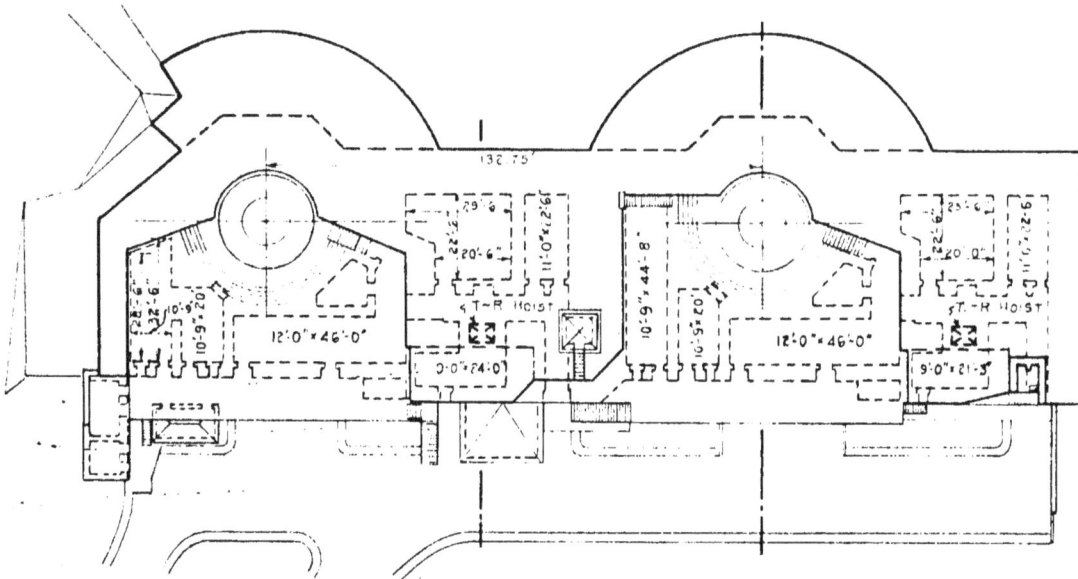

PLAN.

REAR ELEVATION.

Battery Mendell, Fort Barry, HD San Francisco

M1901 DISAPPEARING CARRIAGE FOR 12-INCH GUN

Usual Guns: M1900, M1895
Carriages Built: 21
Original Emplacements: 13 US, 8 Terr.
Time Emplaced: 1904-1910

Number of Bolts: 12
Circle Diameter: 18' 4"
Parapet Height: 14' 7"
Center-to-parapet: 17'
Notes: 5 steps

Total weight: 251 tons
Max/min elevation: +10 / −5 degrees
Radius: 360 degrees (limited by emplacement)

Gun: M1900
Weight: 132,380 lbs
Length: 40 calibers

Photos:
Top: Battery Richardson, Fort Hancock, HD Sandy Hook (Gateway NRA).

Upper Middle: Battery Crockett, Fort Mills, HD Manila Bay, the Philippines, 2006 (MAB).

Lower Middle: Battery Wheaton, Fort Wetherill, HD Narragansett Bay, 1984 (AMH).

Bottom: Battery Kinzie, Fort Worden, HD Puget Sound, 1995 (MAB).

M1901 Disappearing Carriage for 12-inch Gun

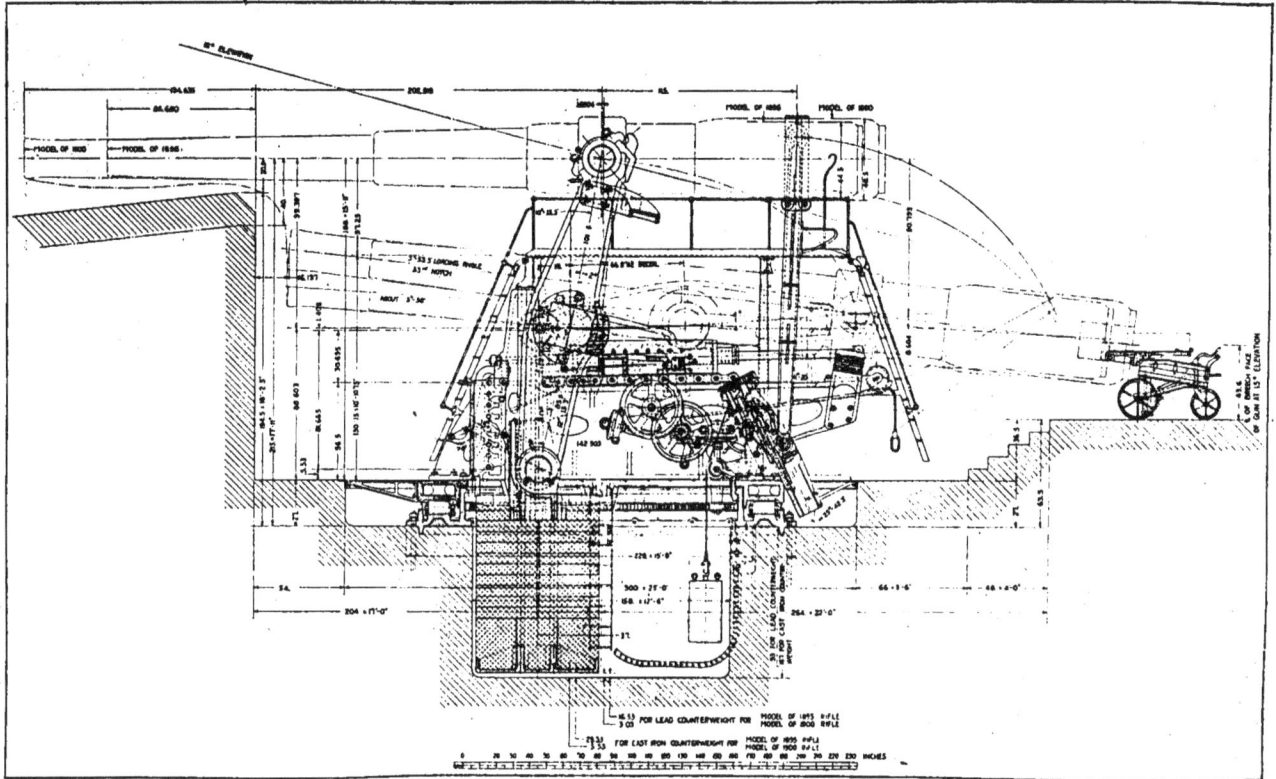

12-INCH DISAPPEARING CARRIAGE, MODEL OF 1901.

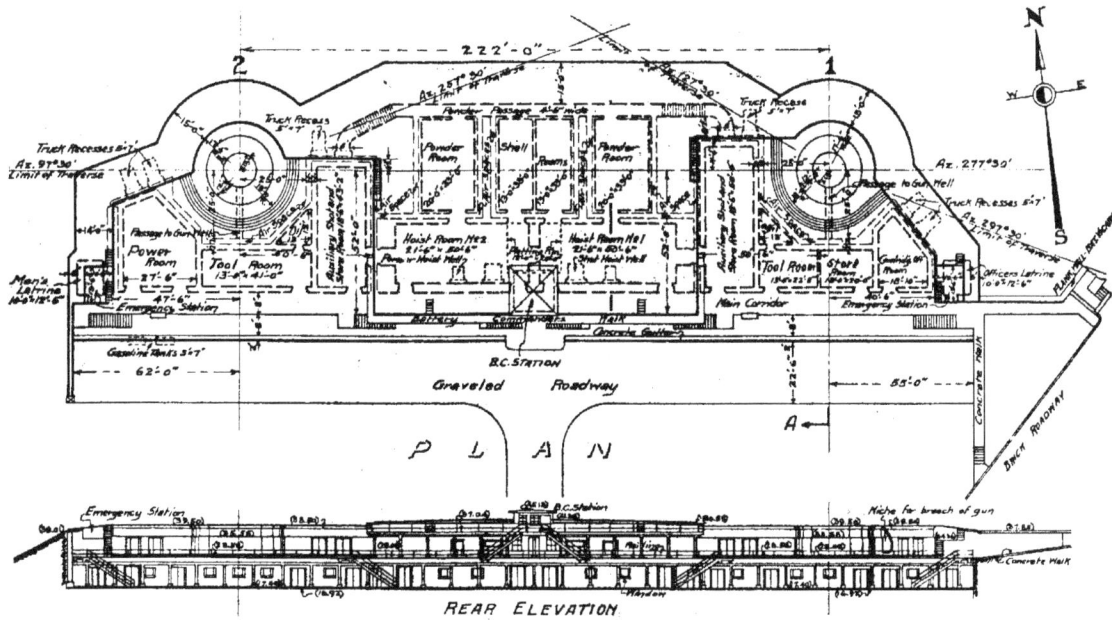

Battery Kinzie, Fort Worden, HD Puget Sound

M1908 CARRIAGE FOR 12-INCH MORTAR

Mortar: M1908
Carriages Built: 24
Original Emplacements: 20 territorial, 4 US
Time Emplaced: 1911-1916

Number of Bolts: 16
Circle Diameter: 15' 10"

Notes: Batteries Hasbrouck (8), Geary (4), Koehler (8). The remaining four mortars and carriages went to Boston. In 1910, four mortars from pit A, Battery Frank Whitman, Fort Andrews, HD Boston, were removed and sent to the Philippines and emplaced in Geary's second pit. The four M1908s were then mounted at Battery Frank Whitman.

Total weight: 63 tons
Max/min elevation: +65 / 0 degrees
Radius: 360 degrees

Mortar: M1908
Weight: 18,200 lbs
Length: 10 calibers

Photos:
Top: Battery Hasbrouck, Fort Kamahameha, HD Pearl Harbor, 1930s.

Middle: Pit A, Battery Frank Whitman, Fort Andrews, HD Boston (NA; Al Schroder Collection).

Below: Battery Hasbrouck, 1992 (GMW).

M1908 CARRIAGE FOR 12-INCH MORTAR

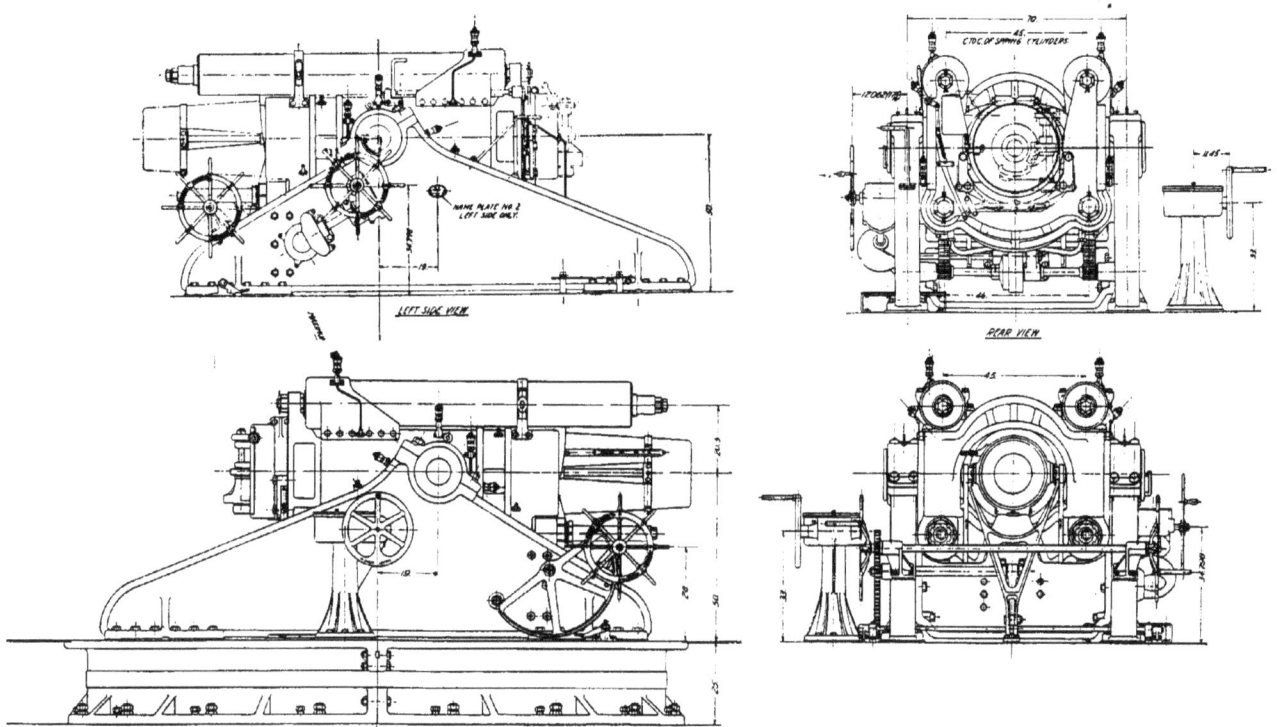

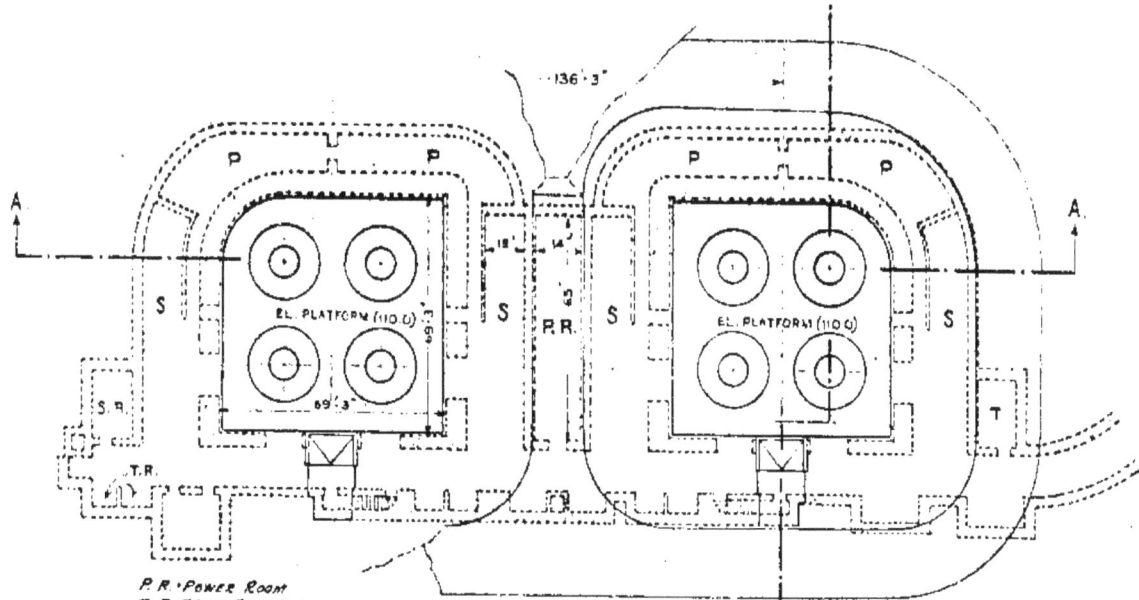

P. R. · POWER ROOM
T. R. · TRUCK RECESSES
S. R. · STORE ROOM
S · SHELL ROOM
P · POWDER MAGAZINE
T · TANK ROOM

PLAN
Scale: 1in · 50 ft.

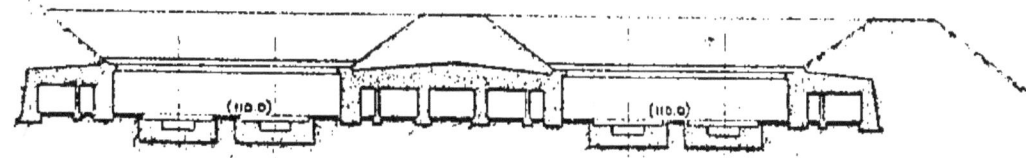

SECT. A.·A.

Battery Koehler, Fort Frank, HD Manila Bay, the Philippines

M1917 Barbette Carriage for 12-inch Gun

Usual Tubes: M1895A2
Carriages Built: 32
Original Emplacements: 22 US, 8 Terr. 4 relocated
Time Emplaced: 1917-1924, relocation 1940-1943

Number of Bolts: 12 inner, 24 outer
Circle Diameter: 12' 1" inner, 17' 0.5" outer
Parapet: none

Notes: The original mount was ARF, in a pair of flat concrete circles. Only Batteries Kimble (Galveston), Pratt (Panama), Hall, Haslet (Delaware R.) Smith and Hearn (Philippines) remain in more or less their original configuration. The other 11 batteries and 3 new batteries were casemated in the 1940s.

Total weight: 151 tons, carriage: 302,000 lbs
Max/min elev: +35 / 0 degrees
Radius: 360 degrees
(145 degrees casemated)

Gun: M1895MIA2
Weight: 114,700 lbs
Length: 35 calibers
Photos:

Top: Battery Hearn, Fort Mills, HD Manila Bay, 2006 (MAB)

Upper Middle: Battery Hearn, Fort Mills, HD Manila Bay, in the 1930s (NA).

Lower Middle: Battery Mills, Fort Hancock, HD New York, 1946 (Courtesy Gateway NRA)

Bottom: Battery Millikin, Fort Rodman, HD New Bedford, 1986 (AMH).

M1917 BARBETTE CARRIAGE FOR 12-INCH GUN

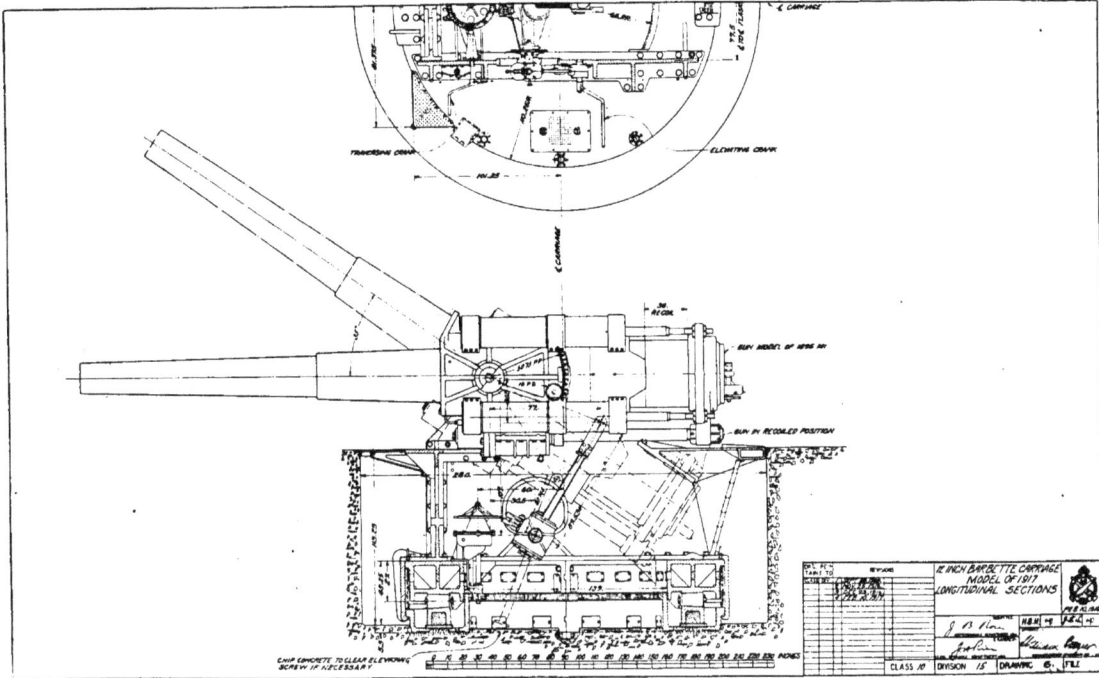

12-INCH GUN ON BARBETTE CARRIAGE, MODEL OF 1917.

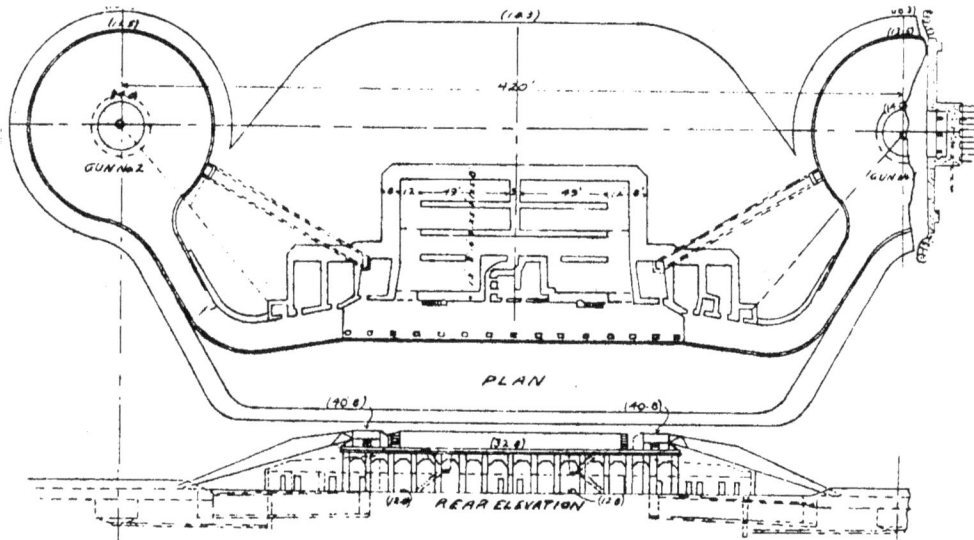

Battery Kimble, Fort Travis, HD Galveston

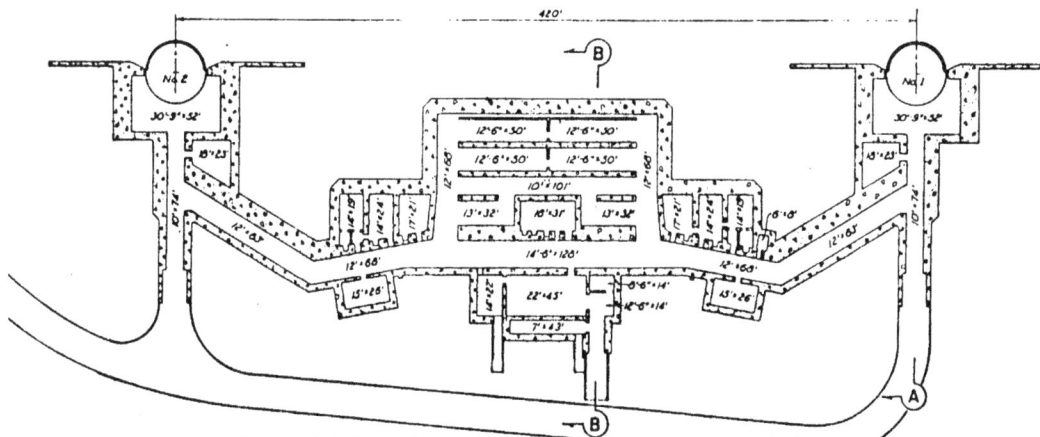

Battery Wallace, Fort Barry, HD San Francisco, as rebuilt

M1918 CARRIAGE (RAILWAY) FOR 12-INCH MORTAR

Usual Guns: M1890
(modified)
Carriages Built:
Original Emplacements:
Time Emplaced:
Parapet: none

Total weight: 88 tons
Max/min elev:
+60 / -5 degrees
radius: 360 degrees

Mortar: M1890MI
Weight: 29,120 lbs
Length: 10 calibers

Photos:
Aberdeen Proving Ground,
MD (Ordnance Museum)

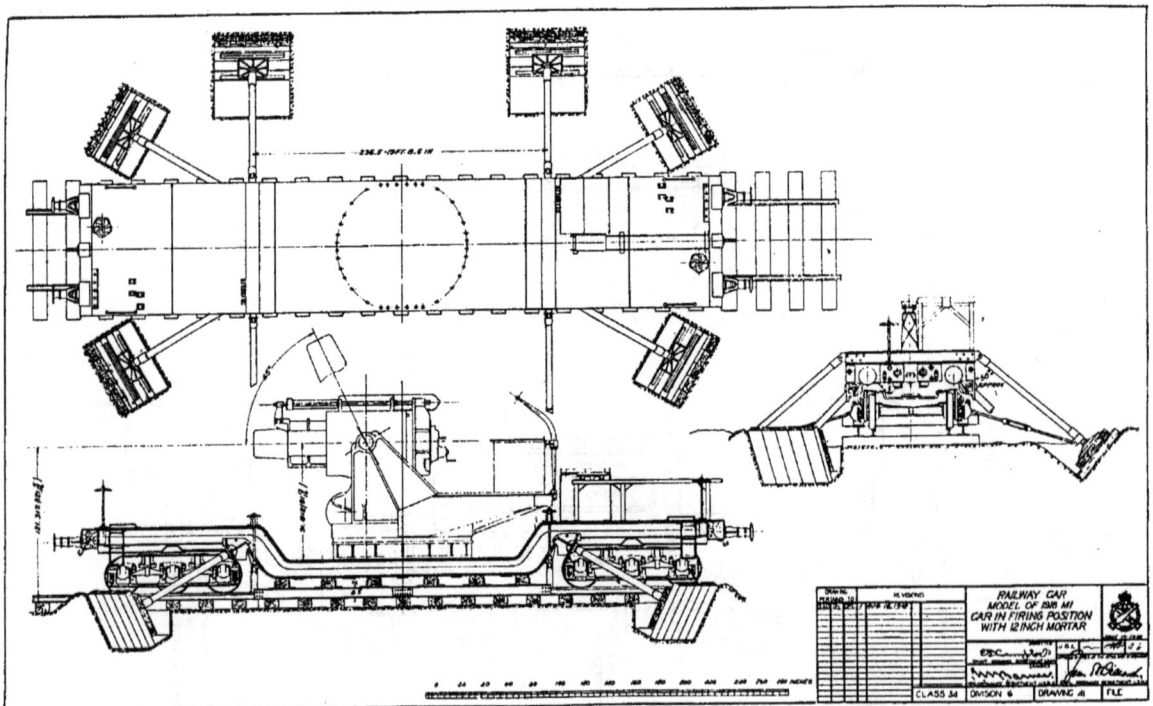

12-INCH MORTAR CARRIAGE, MODEL OF 1918, ON RAILWAY CAR, MODEL OF 1918 MI, FIRING POSITION WITH OUTRIGGERS
AND RAIL CLAMPS IN PLACE.

M1918 Carriage (Railway) for 12-inch Mortar

PLATE 241.

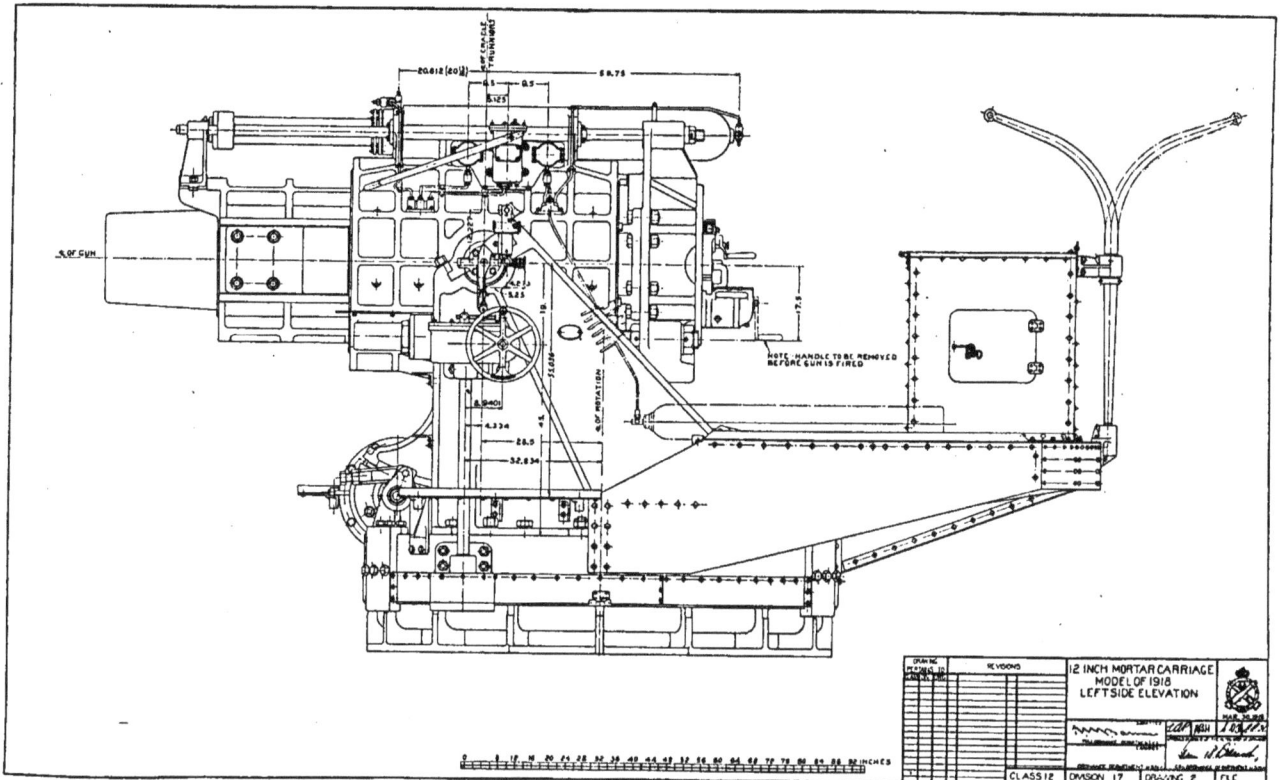

12 INCH MORTAR CARRIAGE
MODEL OF 1918
LEFT SIDE ELEVATION

CLASS 12   DMSON 17   DRAWING 2   FLC

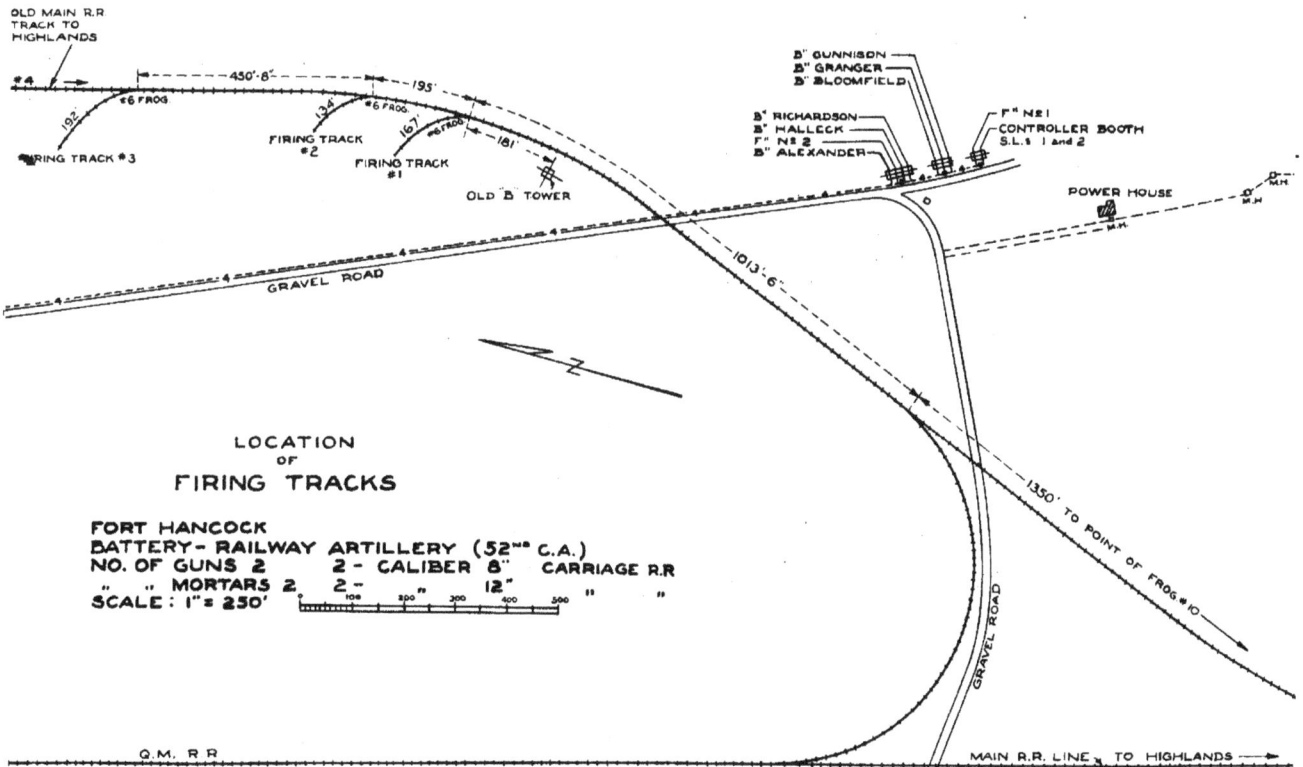

LOCATION
OF
FIRING TRACKS

FORT HANCOCK
BATTERY- RAILWAY ARTILLERY (52ND C.A.)
NO. OF GUNS 2   2- CALIBER 8"   CARRIAGE R.R
   "   " MORTARS 2   2-   "   12"   "   "
SCALE: 1"= 250'

M1907 & M1907MI DISAPPEARING CARRIAGES FOR 14-INCH GUN

Usual Guns:
M1907, M1907MI, M1910, M1910MI

Carriages Built: 21
M1907: 5 M1907MI: 16
Original Emplacements: 4 US, 16 Terr.
Time Emplaced: 1910-1917

Number of Bolts: 14 inner, 12 outer
Circle Diameter: 14' 2" inner, 18' 4" outer
Parapet Height: 15' 1" from gun plate
Center-to-parapet: 15'

M1907 carriage   (four emplacements)
Total weight: 318 tons

M1907MI carriage
Total weight: 341 tons

Max/min elev: +15 / –5 degrees
Radius: 170 degrees

Gun: M1907MI
Weight: 118,700 lbs
Length: 34 calibers

Gun: M1910
Weight: 138,675
Length: 40 calibers

Photos:
Top: Battery Osgood, Fort MacArthur, 1919.

Upper Middle: Battery Woodruff, Fort Hughes, HC Manila Bay, the Philippines 2006 (MAB).

Lower Middle: Battery Greer (M1907 emplacement), Fort Frank, Manila Bay, PI, 2003, (AMH)

Bottom: Battery Buell (M1907MI emplacement), Fort Grant, Panama Canal Zone, 1993 (AMH)

M1907 & M1907MI DISAPPEARING CARRIAGES FOR 14-INCH GUNS

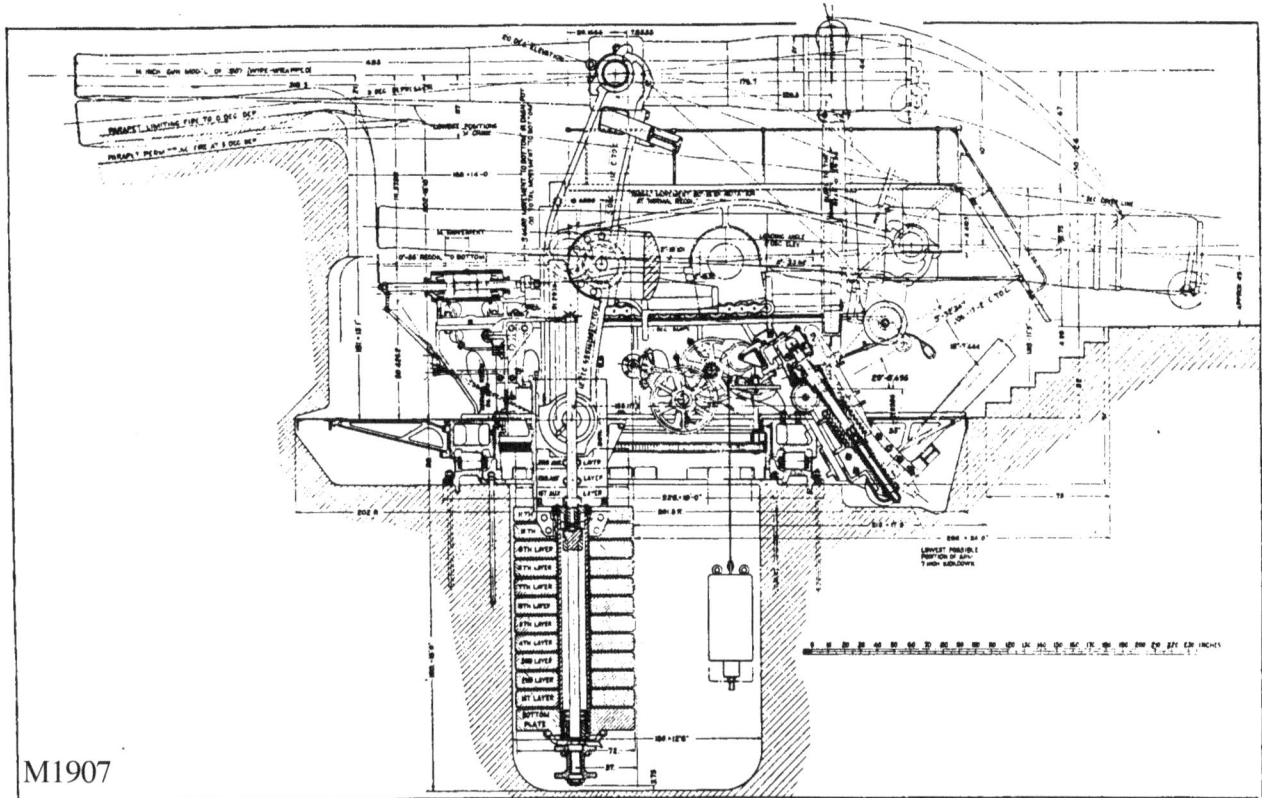

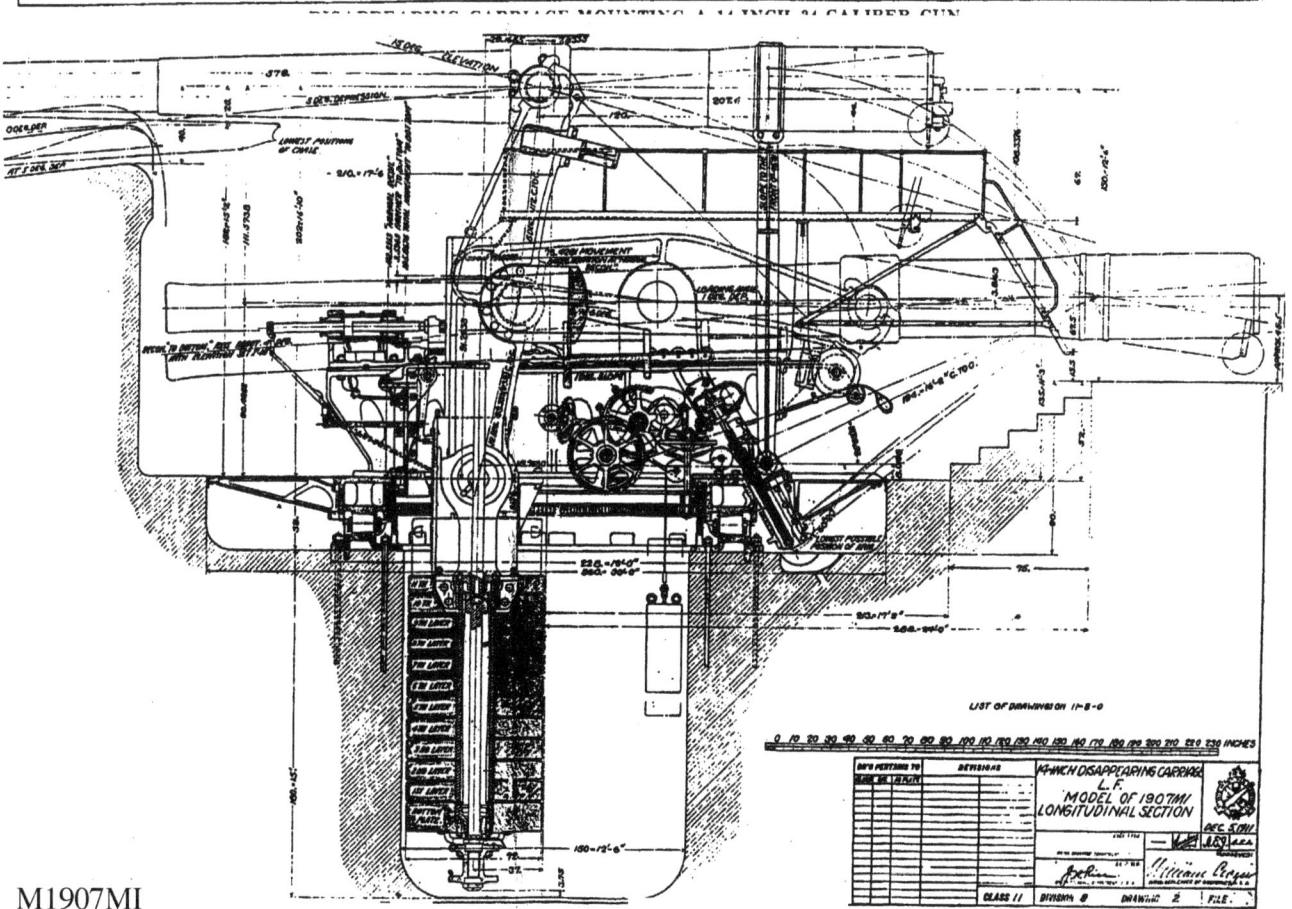

M1907

M1907MI

for emplacement drawing see page 66 (Battery Osgood-Farley (M1907MI)

M1909 TURRET MOUNT FOR 14-INCH GUNS

Guns: M1909
Original Emplacements: 2
(2 guns per turret)
Time Emplaced: 1917

Number of Bolts: 82 (in two
rings)
Circle Diameter: 31' 7"
Parapet: none

Notes: Batteries Wilson &
Marshall, Fort Drum,
HD Manila Bay, the
Philippines.

Total weight:
Max/min elevation:
+15 / –0.09 degrees
Radius: 360 degrees
(limited by surrounding
structures)

Gun: M1909
Weight: 139,210 lbs
Length: 40 calibers

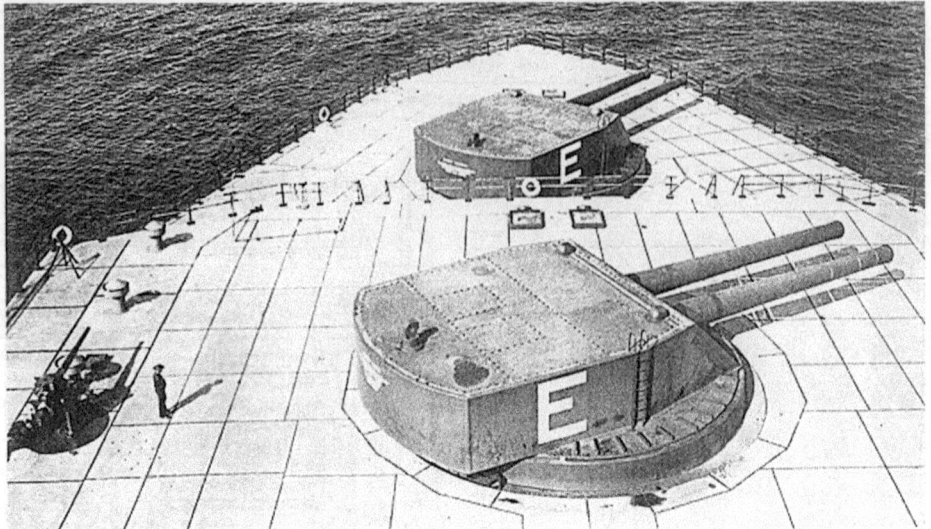
Fort Drum, 1930s (NA).

Lower: Assaulting Fort Drum, 1945 (NA)

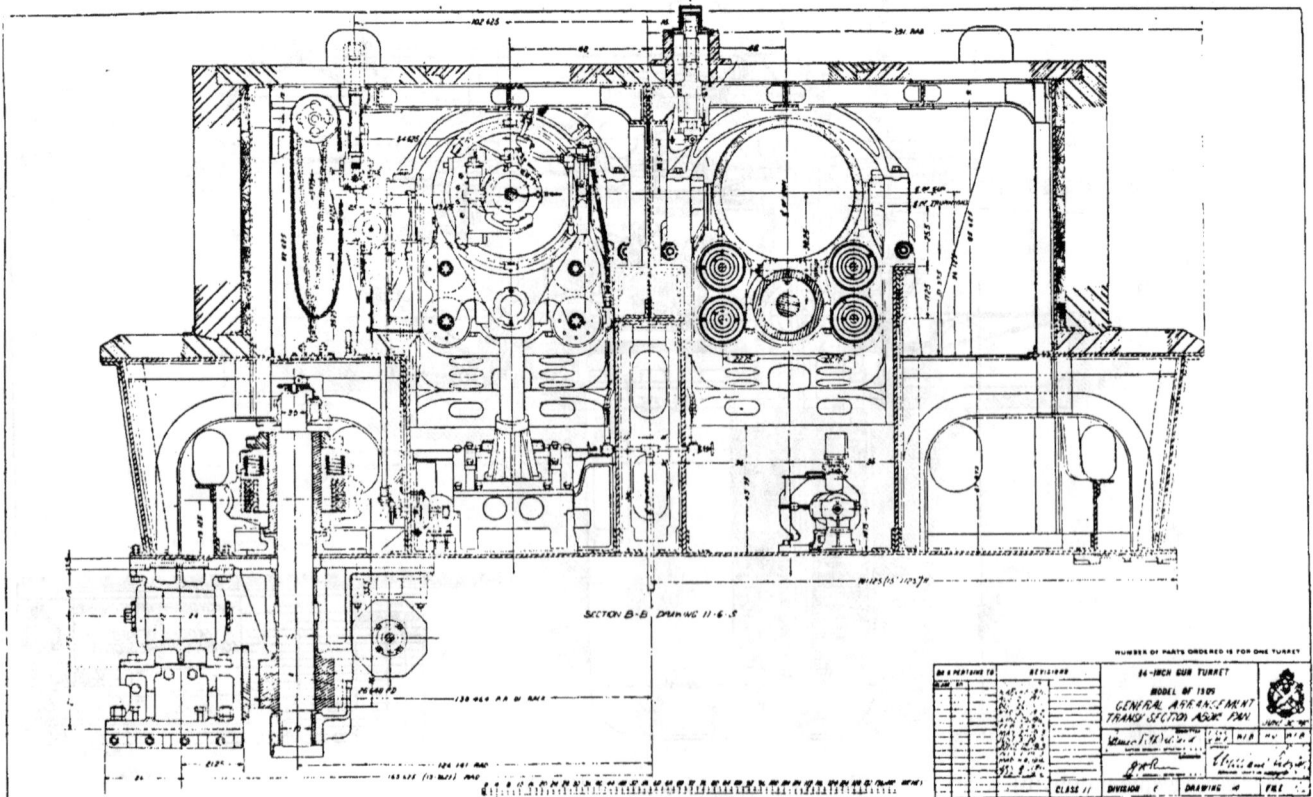

M1909 Turret Mount for 14-inch Guns

M1909 Turret Mount for 14-inch Guns

Fort Drum, 1930s (Karl Schmidt collection)

Fort Drum 2006 (MAB)

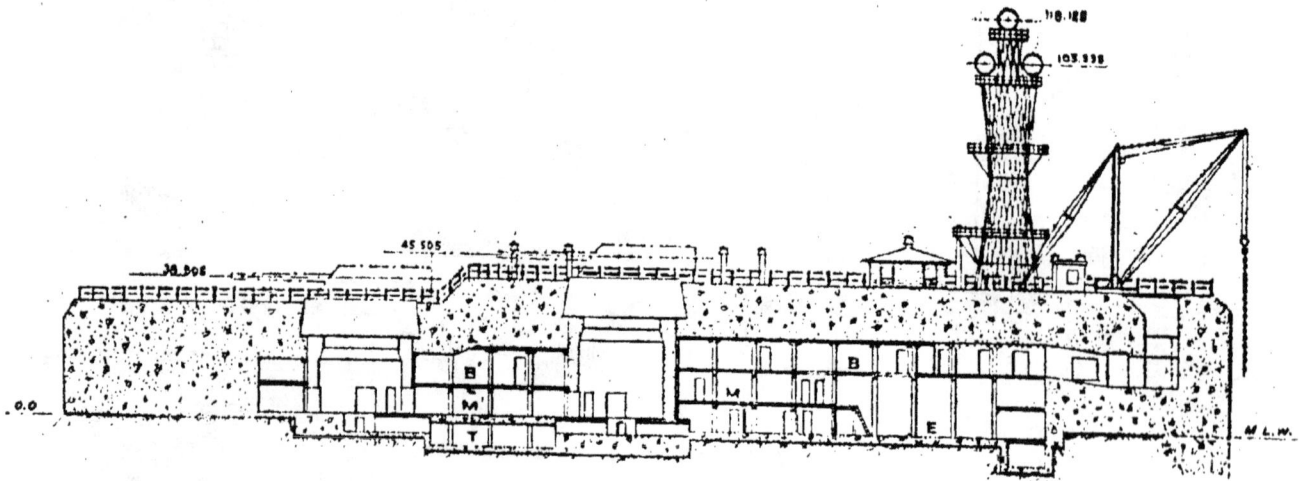

M1909 Turret Mount for 14-inch Guns

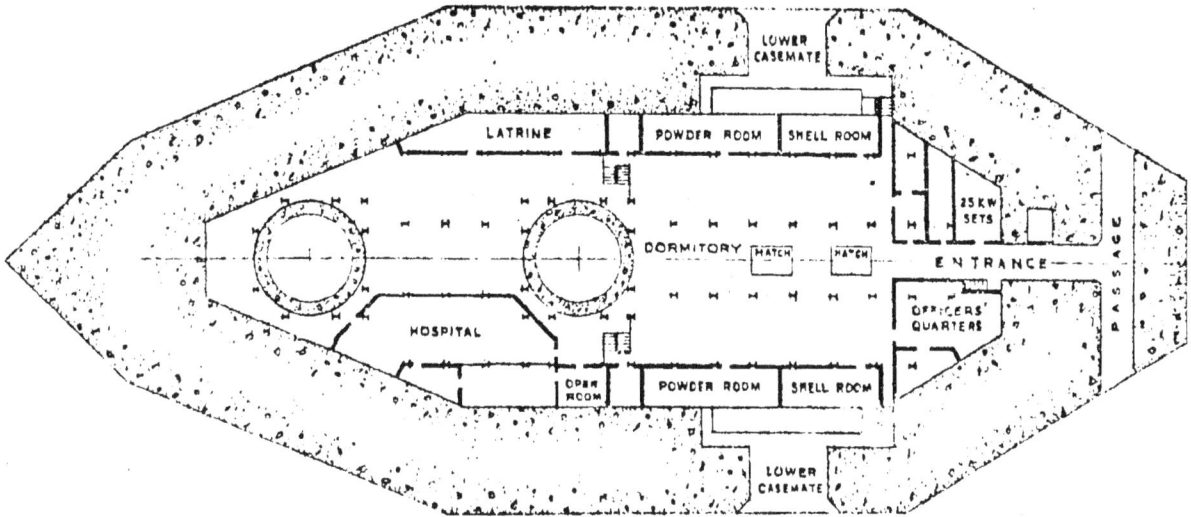

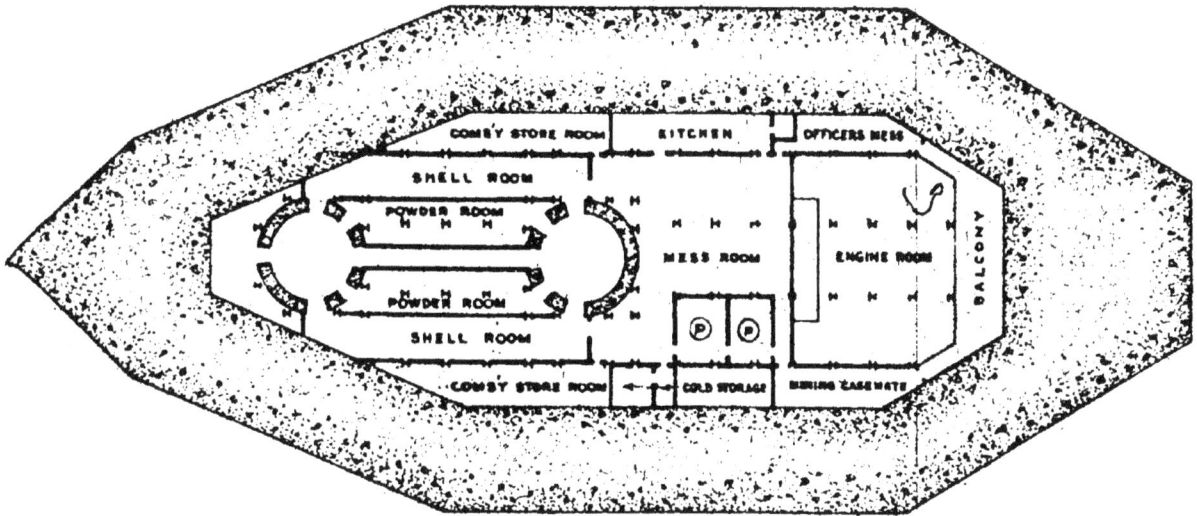

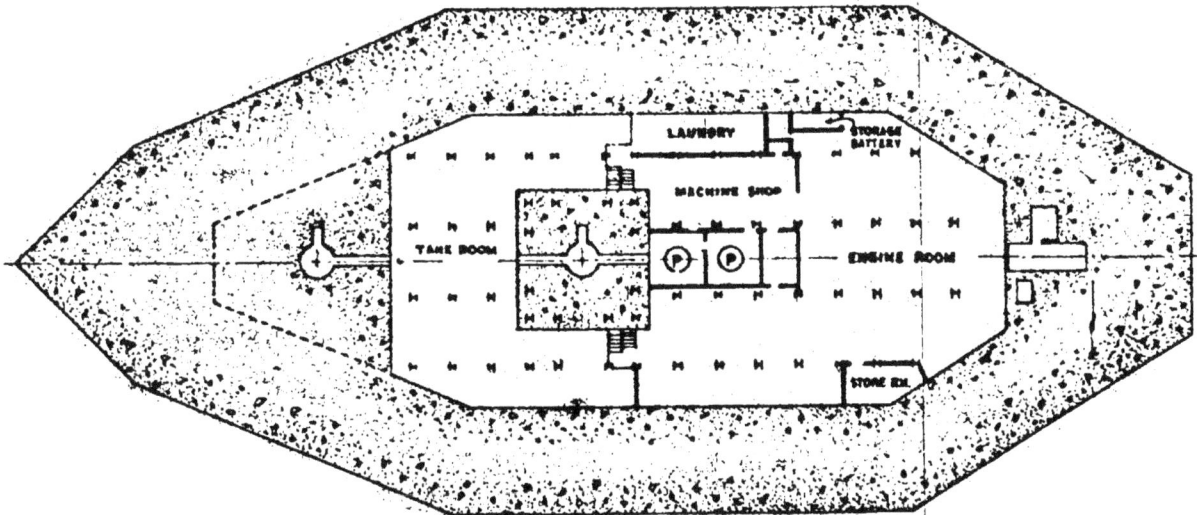

LOWER CASEMATE

LATRINE | POWDER ROOM | SHELL ROOM

25 KW SETS

DORMITORY | HATCH | HATCH

ENTRANCE

HOSPITAL

OFFICERS' QUARTERS

PASSAGE

OPER ROOM | POWDER ROOM | SHELL ROOM

LOWER CASEMATE

COM'SY STORE ROOM | KITCHEN | OFFICERS MESS

SHELL ROOM

POWDER ROOM

MESS ROOM | ENGINE ROOM

BALCONY

POWDER ROOM

SHELL ROOM

COM'SY STORE ROOM | COLD STORAGE | FIRING CASEMATE

LAUNDRY | STORAGE BATTERY

MACHINE SHOP

TANK ROOM | ENGINE ROOM

STORE RM.

M1920 RAILWAY MOUNT FOR 14-INCH GUN

Usual Tubes: M1920
Carriages Built: 4
Original
Emplacements: 1 Los
Angeles 1925, 2 more
built in LA 1938, 4
Panama.
Time Emplaced: 1926-
1930.
Parapet: none

Total weight: 218 tons,
carriage: 496,200 lbs
Max/min elevation:
+50 / −7 degrees
Turning radius:
360 degrees (on
emplacement) 7
degrees on tracks

Gun: M1920
Weight: 192,500 lbs
Length: 50 calibers

Photos:
Top and Upper Middle: Fort MacArthur, HD Los
Angeles, 1930s (Fort MacArthur Museum, Los
Angeles Times).

Lower Middle: Emplacement at Fort Randolph,
Panama Canal Zone, 1993 (AMH).

Bottom: Target practice at Camp Pendleton, CA,
1937 (Geoge Ruhlen Collection, Fort MacArthur
Museum).

M1920 Railway Mount for 14-inch Gun

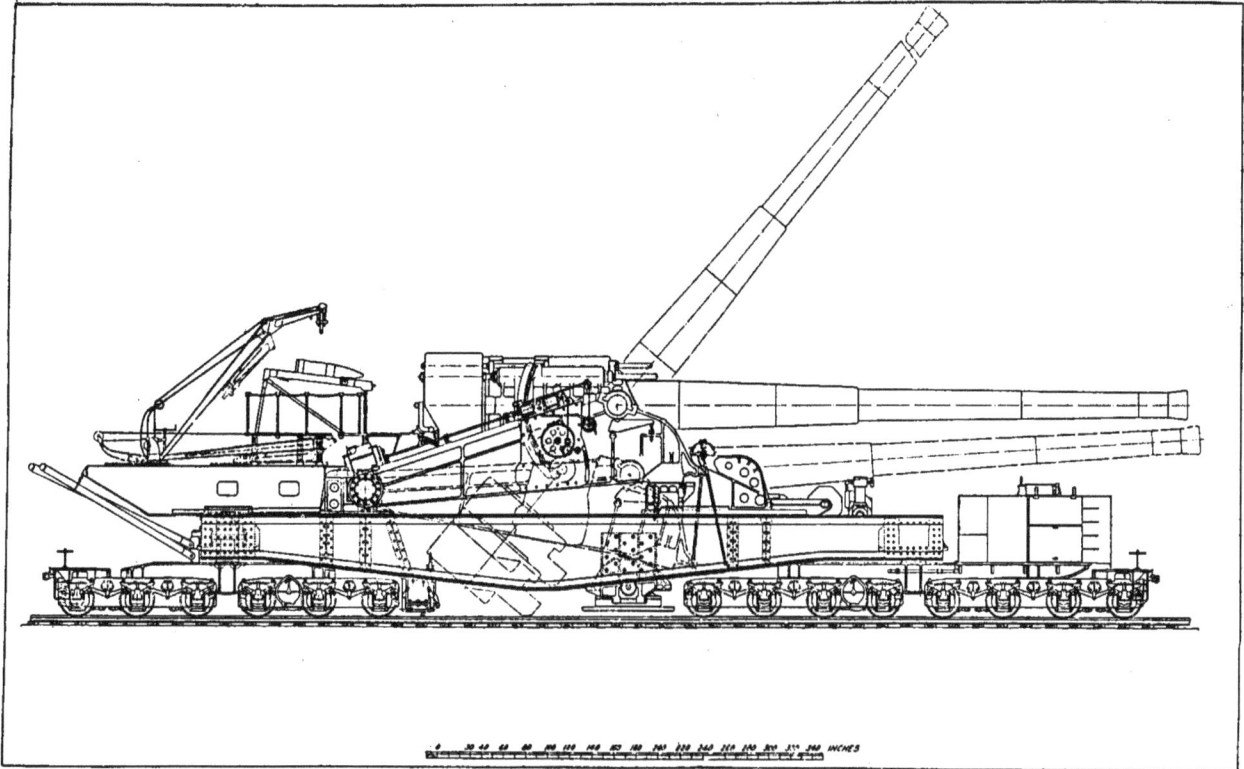

14-INCH 50-CALIBER GUN ON RAILWAY MOUNT, MODEL OF 1920.

REPORT OF COMPLETED WORKS – SEACOAST FORTIFICATIONS.
BATTERY PLAN.

FORM 7.   Corrected to July, 20, 1942

HARBOR DEFENSES OF LOS ANGELES, CALIFORNIA.
FORT MAC ARTHUR
FIRING PLATFORM
No. of guns-TWO. Caliber-14 INCH. Carriage-RY.

Note:
Each firing platform enclosed by No. 6 ga. woven wire mesh fence 84" high. 14–150 watt lights mounted on top.

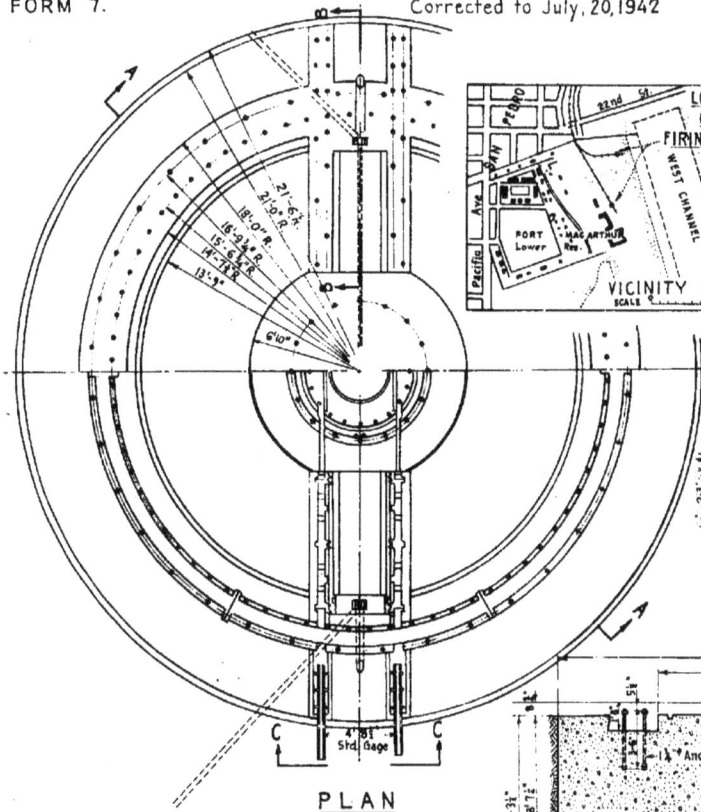

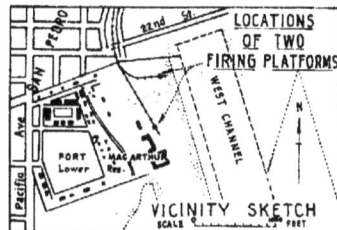

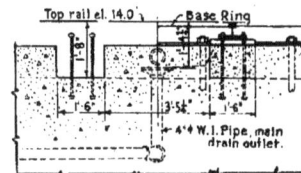

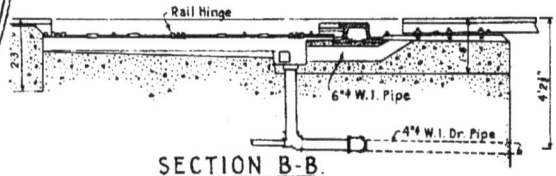

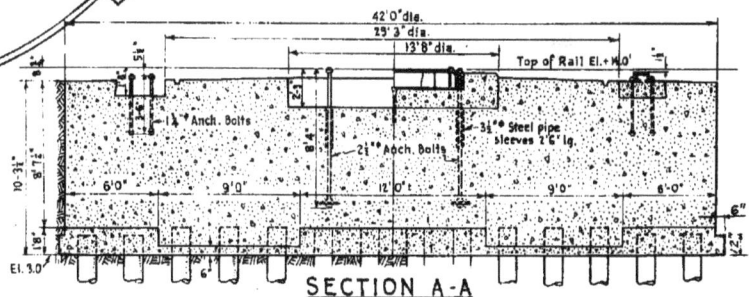

LOCATIONS OF TWO FIRING PLATFORMS

VICINITY SKETCH

ELEVATION C-C.

SECTION B-B.

PLAN

SECTION A-A.

NAVAL TURRET FOR 14-INCH GUNS

Turrets used: 2
Original Emplacements: 2
Time Emplaced: 1943-1945
Parapet: none

Notes: Batteries Pennsylvania &
Arizona, Oahu Island, Hawaii. The two
aft triple-gun turrets salvaged from the
USS Arizona.

Reference: Excellent photographs
and drawings of the Hawaii turret
batteries can be found in the article
"The Oahu Turrets" by E. R. Lewis and
D. P. Kirchner in the magazine Warship
International 1992, issue No. 3, pp.
273-301.

Photos:
Top: Battery Pennsylvania proof firing,
1945 (National Archives)

Middle: Looking up from inside Battery
Pennsylvania, 1992 (GMW)

Bottom: Battery Arizona, 1992 (GMW)

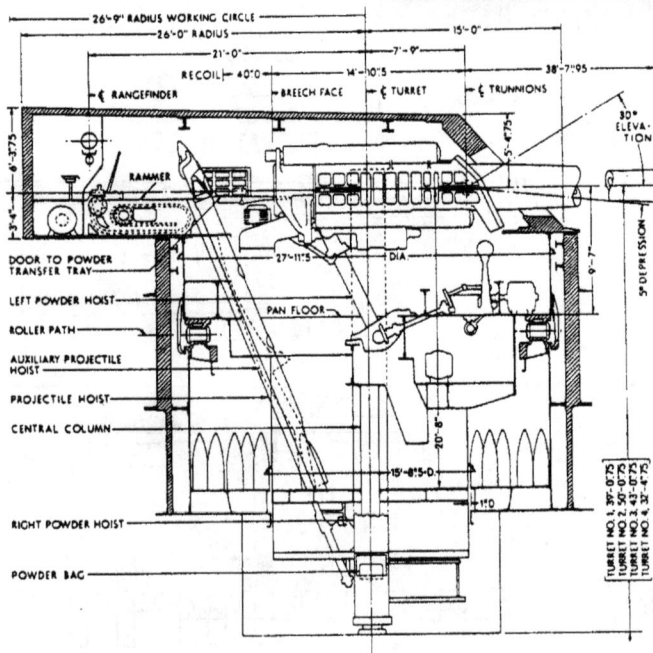

Naval Turret for 14-inch Guns

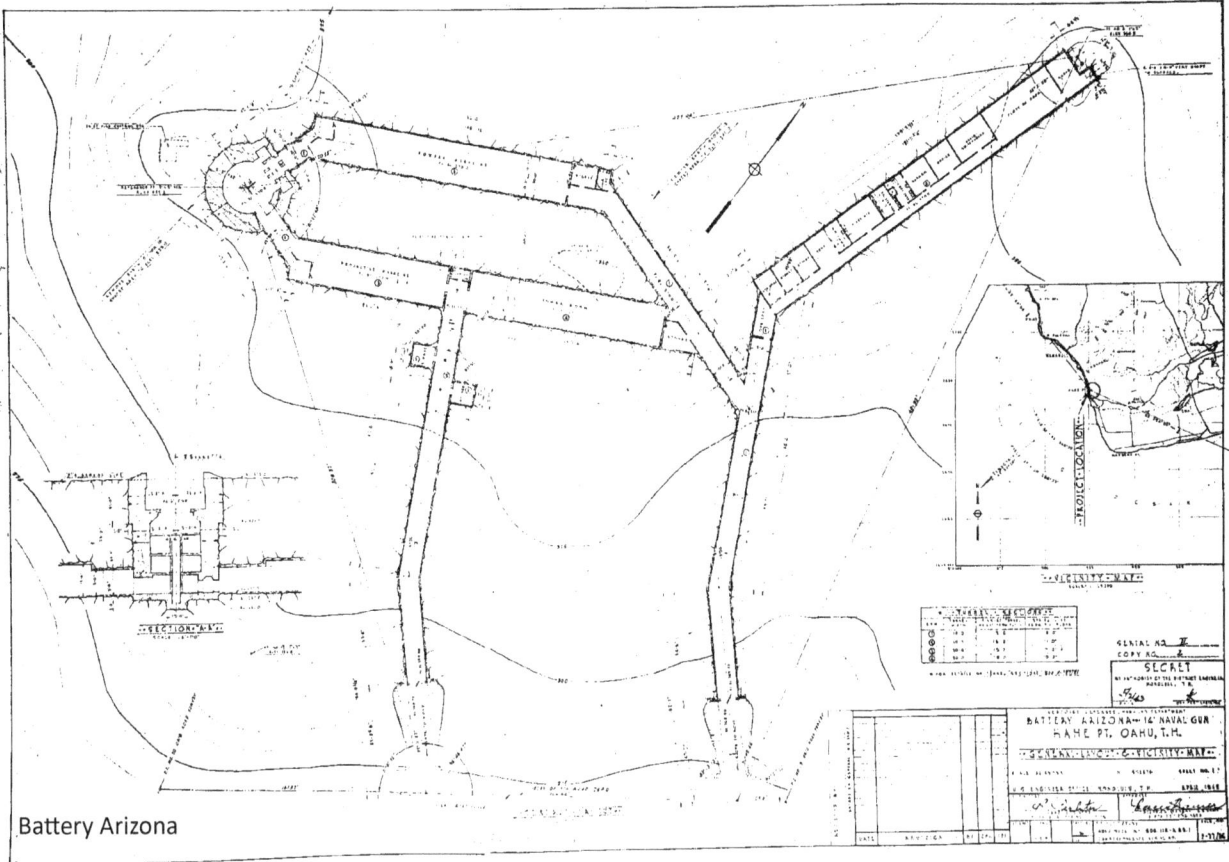

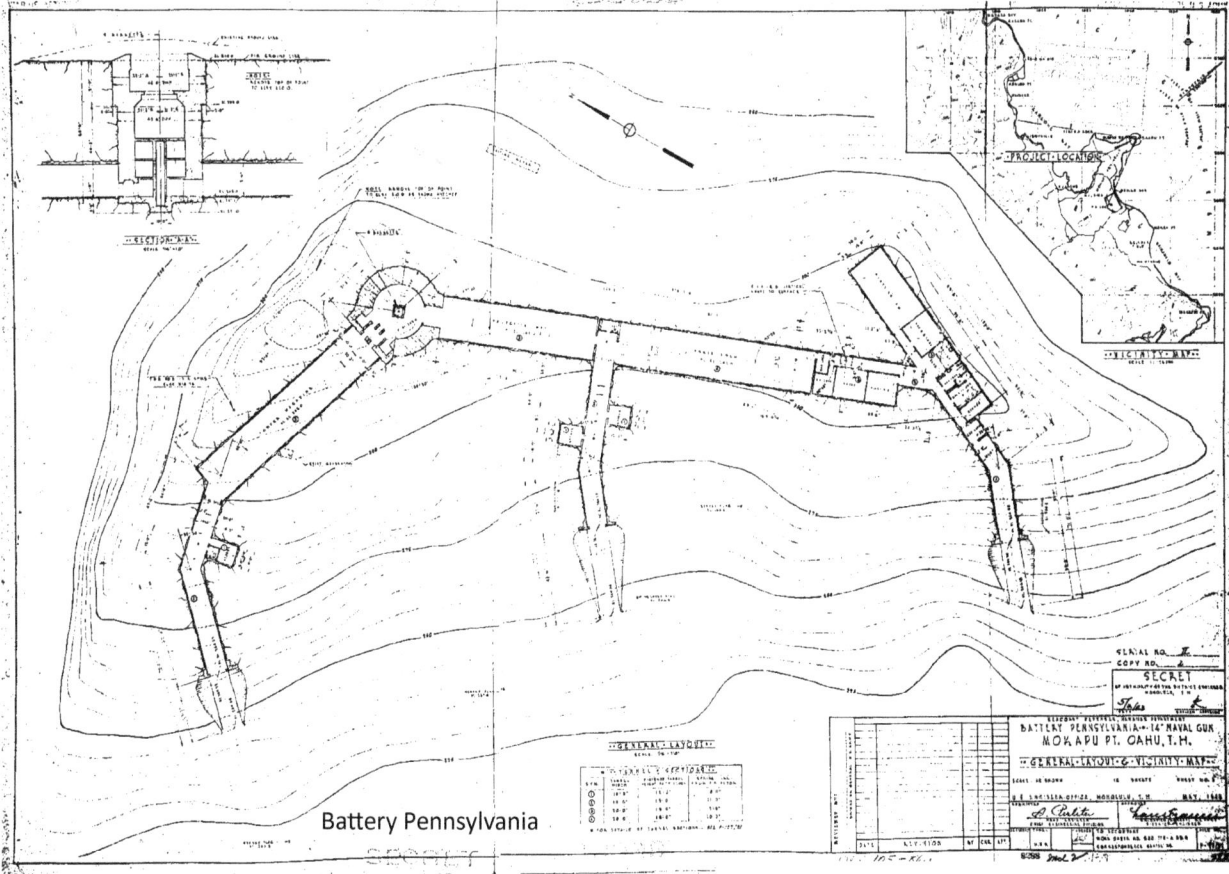

Battery Arizona

Battery Pennsylvania

M1912 DISAPPEARING CARRIAGE FOR 16-INCH GUN

Gun:: M1895
Carriages Built: 1
Original Emplacements: 1
Time Emplaced: 1917
Battery Newton, Fort Grant, PCZ

Number of Bolts: 16 inner, 14 outer
Circle Diameter: 17' 9" inner, 23' 2" outer
Parapet Height: 19' 4"
Center-to-parapet: 19' 10"
Notes: 8 steps

Total weight: 637 tons
Max/min elevation: +20 / −5 degrees
Radius: 360 degrees
(limited by emplacement)

Gun: M1895
Weight: 284,500 lbs
Length: 35 calibers

Photos:
Top: Battery Newton, Fort Grant, Panama Canal Zone, 1995 (GMW)

Middle: Sandy Hook Proving Ground (Gateway NRA)

Bottom: Battery Newton, 1920s?

M1912 DISAPPEARING CARRIAGE FOR 16-INCH GUN

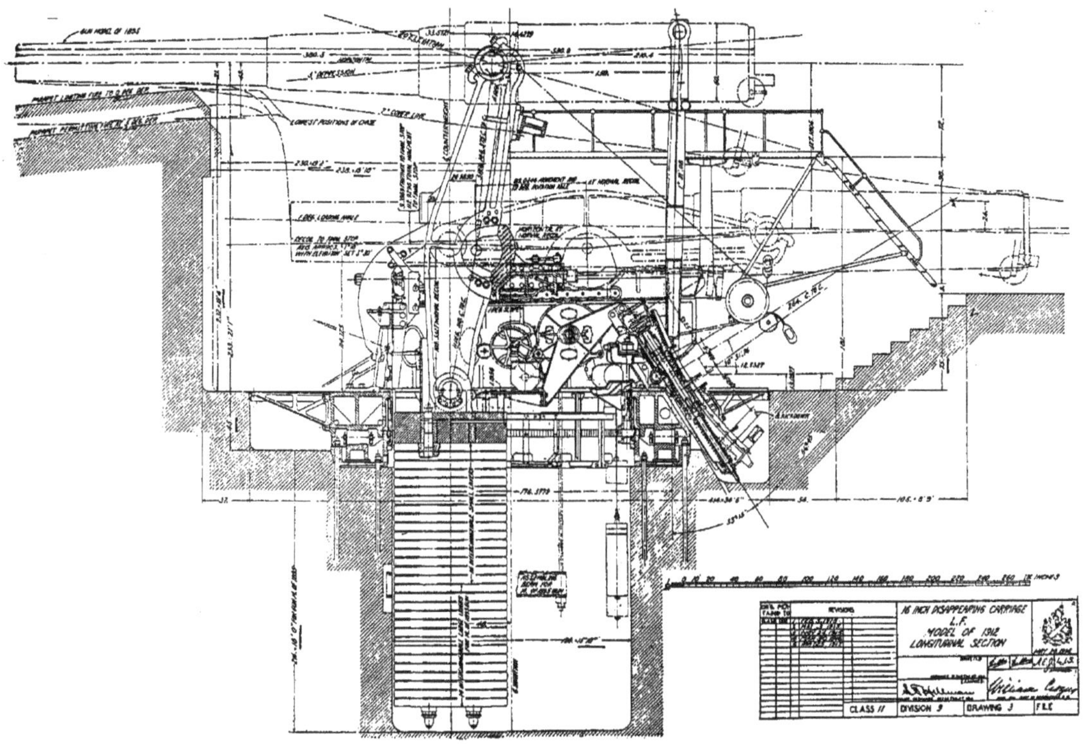

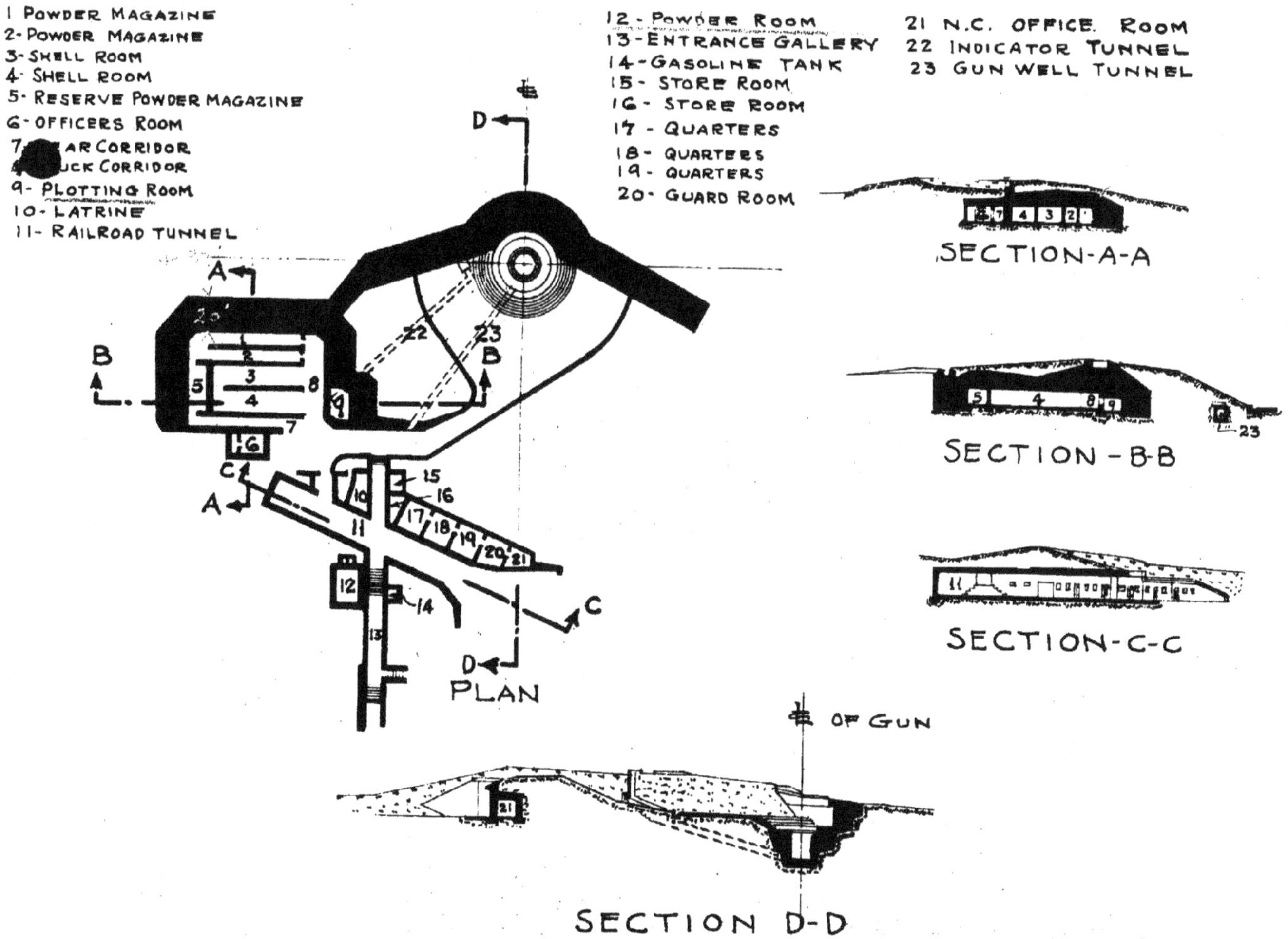

1 POWDER MAGAZINE
2- POWDER MAGAZINE
3- SHELL ROOM
4- SHELL ROOM
5- RESERVE POWDER MAGAZINE
6- OFFICERS ROOM
7- ᴀᴿ CORRIDOR
8- ᴜᴄᴋ CORRIDOR
9- PLOTTING ROOM
10- LATRINE
11- RAILROAD TUNNEL

12- POWDER ROOM
13- ENTRANCE GALLERY
14- GASOLINE TANK
15- STORE ROOM
16- STORE ROOM
17- QUARTERS
18- QUARTERS
19- QUARTERS
20- GUARD ROOM

21 N.C. OFFICE ROOM
22 INDICATOR TUNNEL
23 GUN WELL TUNNEL

PLAN

SECTION-A-A

SECTION-B-B

SECTION-C-C

SECTION D-D

Battery Newton, Panama Canal Zone

M1917 Disappearing Carriage for 16-inch Gun

Gun: M1919
Carriages Built: 1
Original Emplacements: 1
Time Emplaced: 1921
Battery J.M.K. Davis, Fort Michie, HD
Long Island Sound

Number of Bolts: 16 inner, 14 outer
Circle Diameter: 17' 9" inner, 23' 2"
outer
Parapet Height: 15'
Center-to-parapet: 29' 8"
Notes: 10 steps,

Total weight: 835 tons
Max/min elevation: +30 / –5 degrees
Radius: 360 degrees

Gun: M1919
Weight: 385,600 lbs
Length: 50 calibers

Photos:
Top: Battery J.M.K.
Davis, Fort Michie, H.D.
Long Island Sound, circa
1933 (Pierce Rafferty
Collection).

Middle: Watertown
Arsenal 1920s (NA).

Bottom: Battery J.M.K.
Davis, Fort Michie,
HD long Island Sound,
(AMH).

M1917 Disappearing Carriage for 16-inch Gun

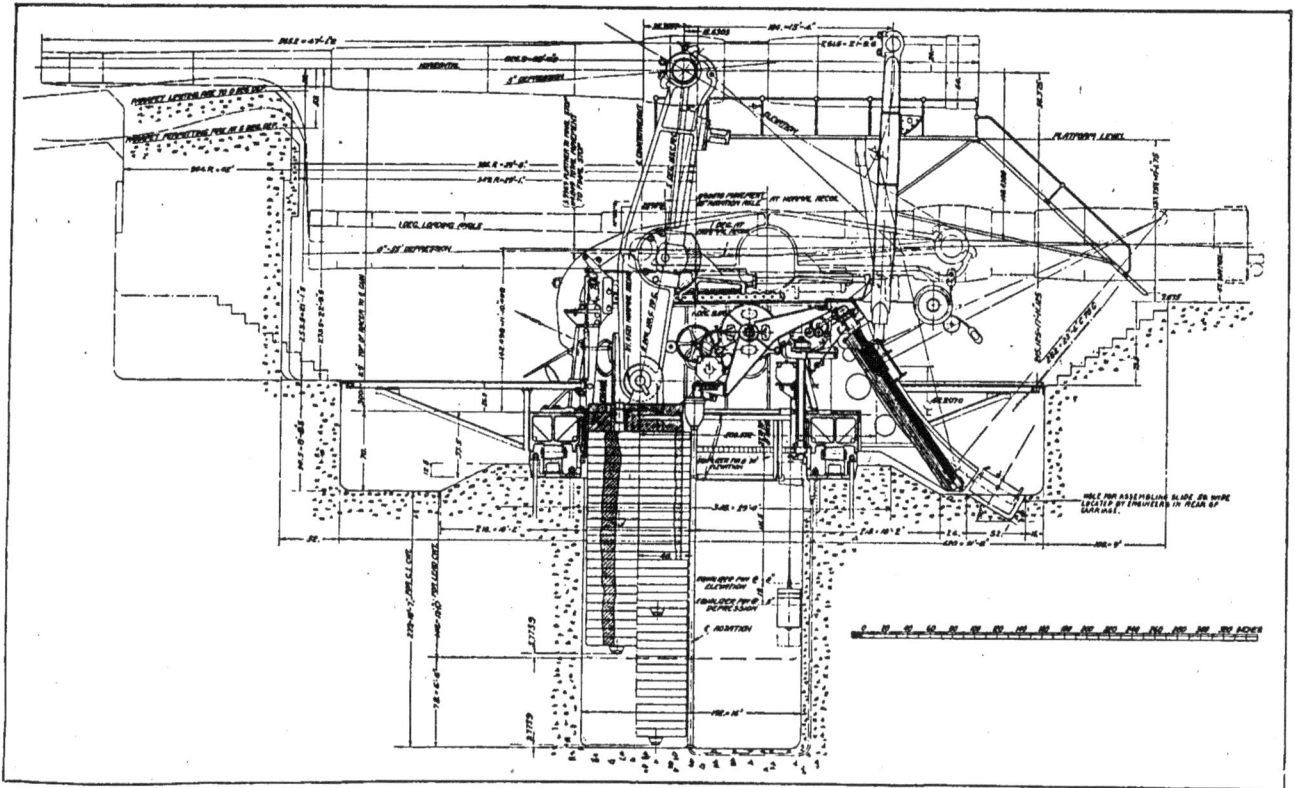

DISAPPEARING CARRIAGE MODEL OF 1917, MOUNTING 16-INCH 50-CALIBER GUN.

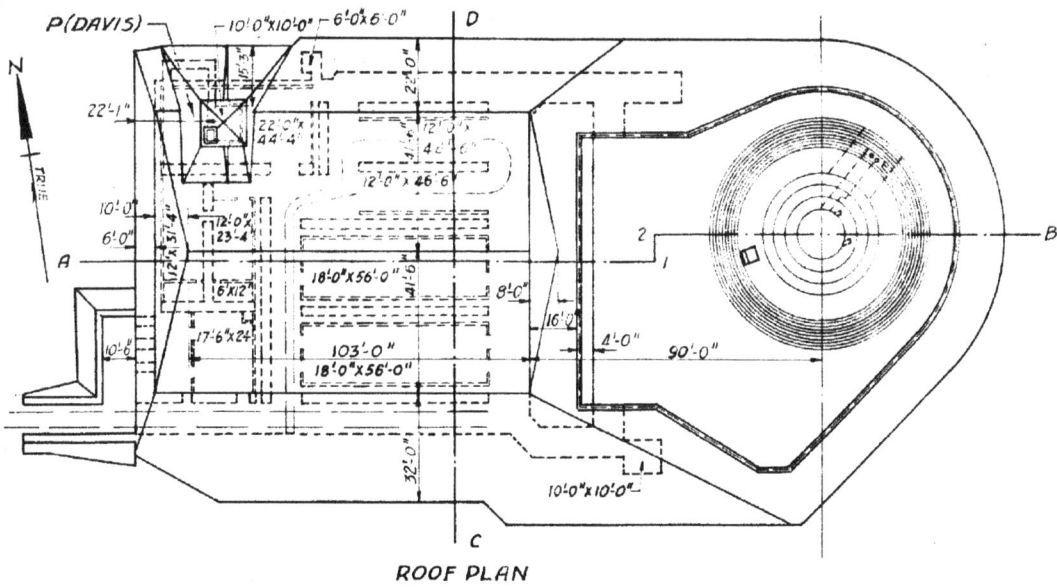

ROOF PLAN

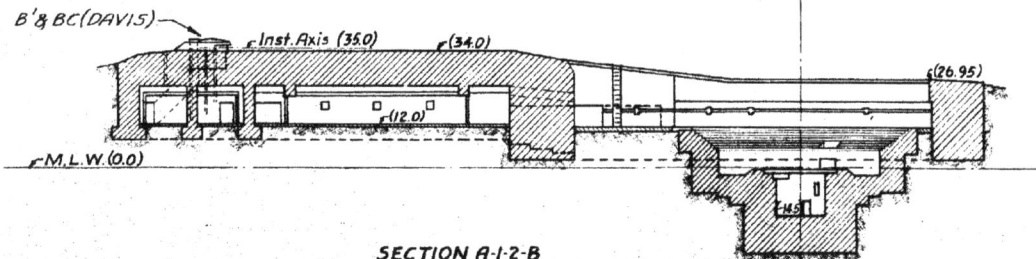

SECTION A-1-2-B

Battery J.M.K. Davis, Fort Michie, HD Long Island Sound, NY

M1919 BARBETTE CARRIAGE FOR 16-INCH GUN

Usual guns: M1919 Army
Carriages Built: 6
Original Emplacements: 6
Time Emplaced: 1922-1930

Number of Bolts: 16 inner,
20 outer
Circle Diameter: 24' 10 inner,
31' 8" outer
Parapet: none
Notes: deep circular pit 23'
8" diameter.

Batteries Long, Harris &
Williston. Long and Harris
were casemated 1940s
Total weight: 500 tons,
carriage: 699,153 lbs
Max/min elev: +65 / −7
degrees
Radius: 360 degrees (145
degrees in casemate)

Gun: M1919MII, MIII
Weight: 385,600 ilbs
Length: 50 calibers

Battery Williston, Fort Weaver, HD Pearl Harbor, HI, 1930s?

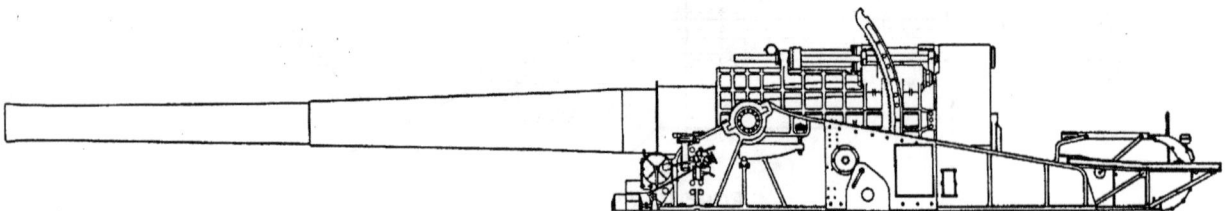

Drawing © Robert D. Fritz, used by permission.

M1919 BARBETTE CARRIAGE FOR 16-INCH GUN

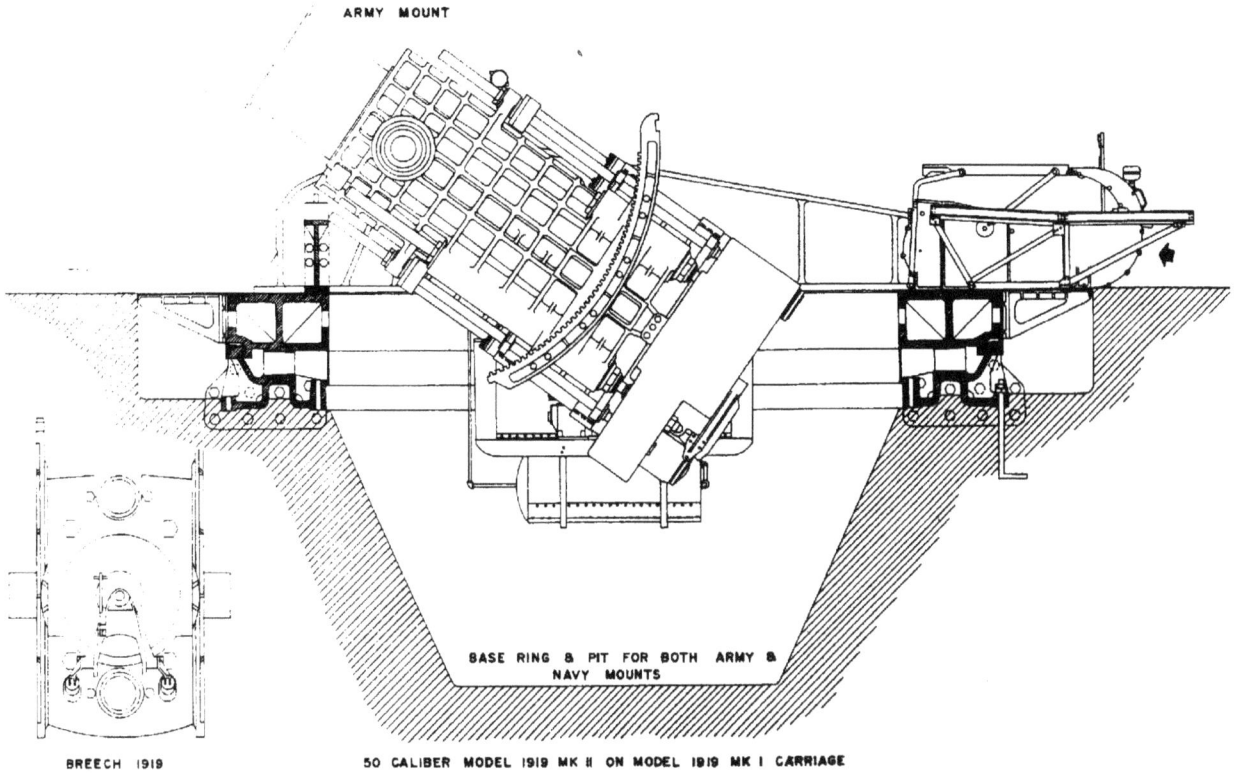

ARMY MOUNT

BASE RING & PIT FOR BOTH ARMY & NAVY MOUNTS

BREECH 1919

50 CALIBER MODEL 1919 MK II ON MODEL 1919 MK I CARRIAGE

Drawing © Robert D. Fritz, used by permission.

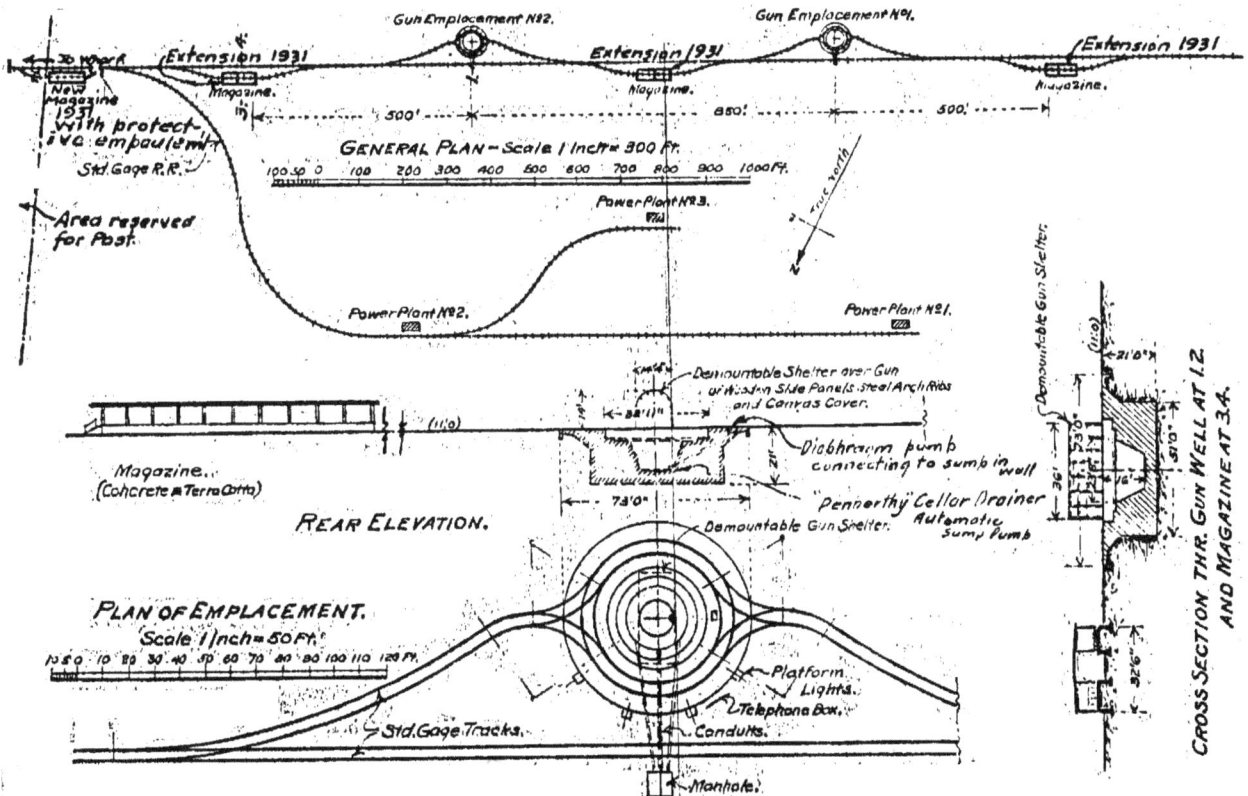

GUN EMPLACEMENT Nº2. GUN EMPLACEMENT Nº1.

GENERAL PLAN - Scale 1 inch = 300 ft.

REAR ELEVATION.

PLAN OF EMPLACEMENT.
Scale 1 inch = 50 ft.

CROSS SECTION THR. GUN WELL AT 1.2 AND MAGAZINE AT 3.4.

Battery Harris, Fort Tilden, HD New York

M1919M1 Barbette Carriage for 16-inch Navy Gun

Usual guns: Navy MkIIM1
Carriages Built: 6
Original Emplacements: 6
Time Emplaced: 1922-1934

Number of Bolts: 16 inner, 20 outer
Circle Diameter: 24' 10 inner, 31' 8" outer
Parapet: none
Notes: deep circular pit 23' 8" diameter,
Batteries Murray, Haan & Hatch; Murray & Hatch casemated 1940s.

Gun: Navy MkIIMI
Weight: 307,185 lbs
Length: 50 calibers

Photos:
Either Battery Murray or Haan, Fort Kobbe, Panama Canal Zone (NA)

M1919M1 Barbette Carriage for 16-inch Navy Gun

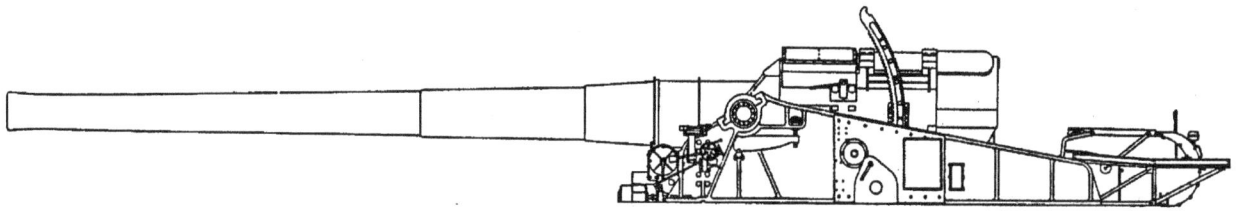

Drawing © Robert D. Fritz, used by permission.

Battery Haan, 1995 (GMW)

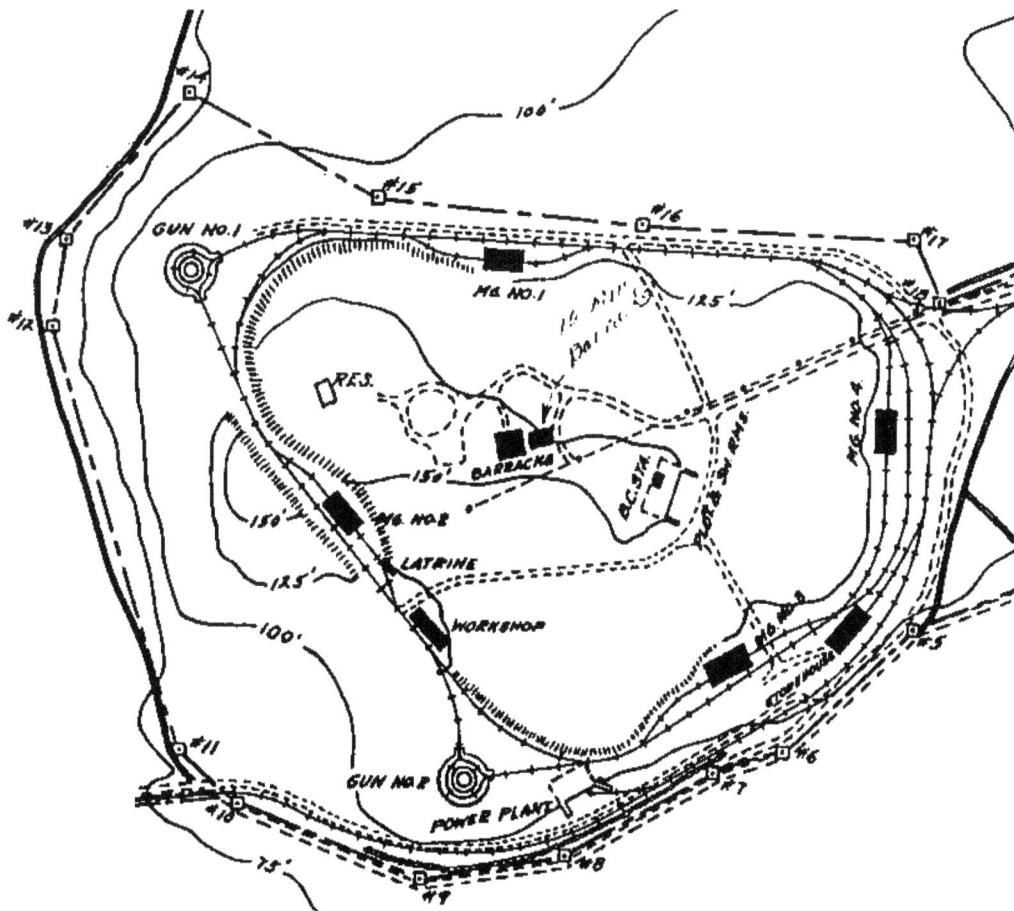

Battery Hatch, Fort Barrette, HD Pearl Harbor, as built, 1935

M2, M3, M4, M5 Barbette Carriages for 16-inch Navy Guns

Usual guns: Navy MkIIM1
Carriages Built: 46-49?
M2: 4   M3: 2
M4: 27   M5:13?
Original Emplacements: 40
Time Emplaced: 1938-1945

Number of Bolts: 16 inner, 20 outer
Circle Diameter: 24' 10" inner, 31' 8" outer
Parapet: none
Notes: deep circular pit 38' 11" diameter.
Notes: Of some 41 planned new batteries in 1937-40 only 23 were completed and only 17 of these actually armed.

M4 Carriage
Carriage weight: 665,315 lbs
Max/min elevation: +47 / −3 degrees
Radius: 180 degrees (145 degrees in casemate)

Gun: MkIIMI
Weight: 307,185 lbs
Length: 50 calibers

Photos:
Top: Fort Story, HD Chesapeake Bay, 1942

Middle: Battery Steele, Peaks Island Military Reservation, ca.1946 (5th Maine Regiment Museum Collection)

Bottom: Battery Paul D. Bunker (BCN 127), Fort MacArthur-White Point MR, HD Los Angeles (1993 MAB).

M2, M3, M4, M5 Barbette Carriages for 16-inch Navy Guns

16-inch MkIII gun on Army proof mount, Aberdeen Proving Ground (2009 MAB)

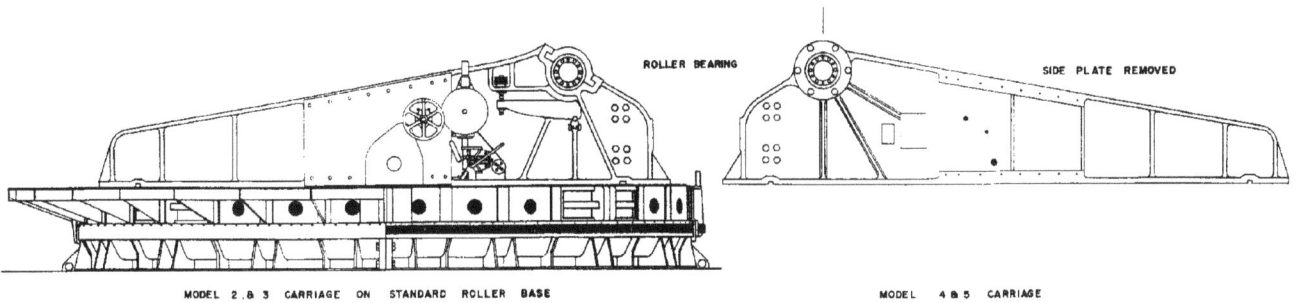

MODEL 2 & 3 CARRIAGE ON STANDARD ROLLER BASE

MODEL 4 & 5 CARRIAGE

Drawing © Robert D. Fritz, used by permission.

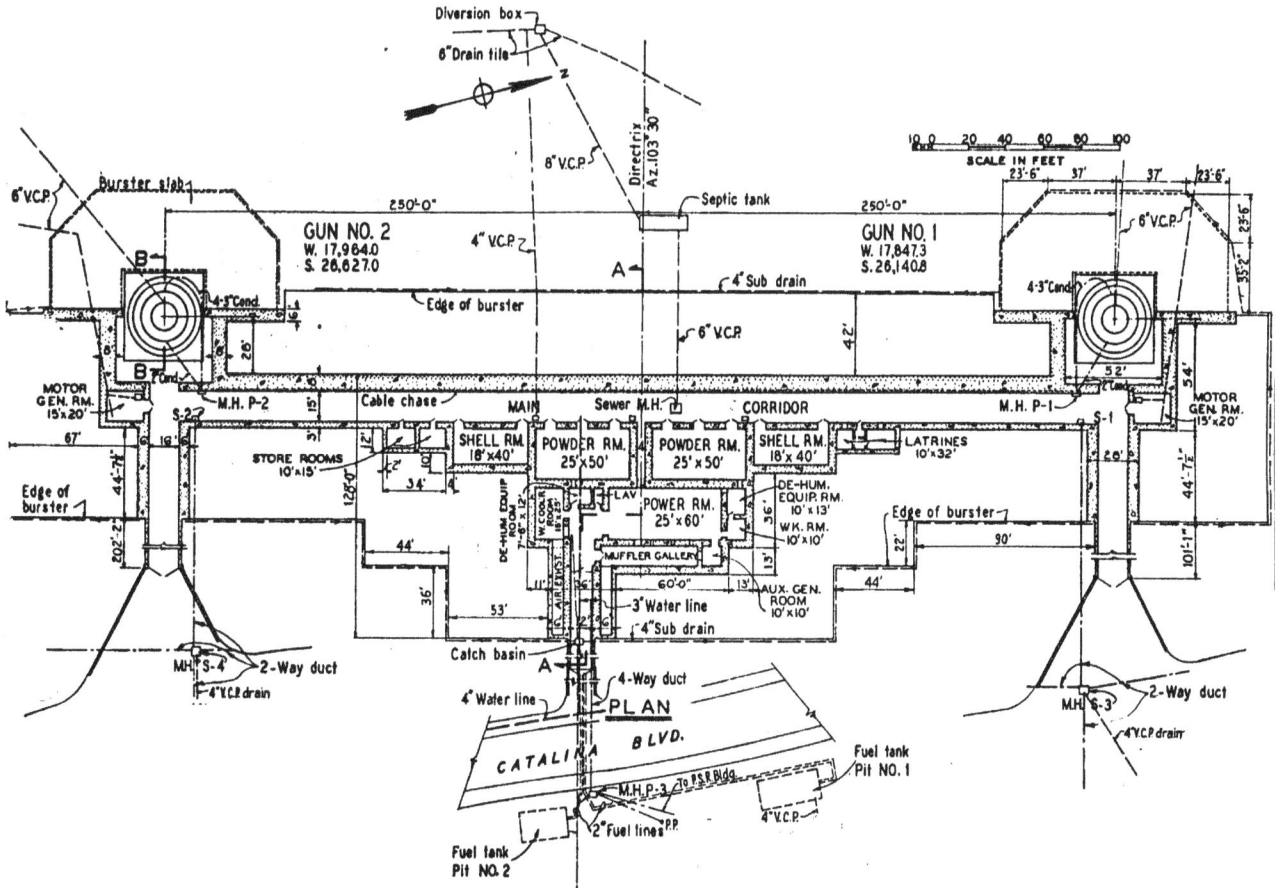

BCN 126, Battery Thomas Q. Ashburn, Fort Rosecrans, HD San Diego

M1920 Barbette Carriage for 16-inch Howitzer

Howitzer: M1920
Carriages Built: 4
Original Emplacements: 4
Time Emplaced: 1922

Number of Bolts: 16 inner, 20 outer
Circle Diameter: 24' 10" inner, 31' 8" outer
Parapet: none
Notes: Battery Pennington (1940 divided into Batteries Pennington & Walke), Ft. Story, HD Chesapeake Bay. Shields added during WW II.

Carriage wt: 705,000 lbs
Max/min elev: +65 / −7 d.
Radius: 360 degrees

Gun: M1920
Weight: 195,300 lbs
Length: 25 calibers

Photos: Top: M1920 16-inch howitzer on M1920 barbette

carriage, Battery Pennington, Fort Story, VA, May 1933. (Casemate Museum)

Middle: Battery Pennington or Walke, 1946 (Master Sgt. Herbert F. Markland)

Below: Battery Pennington, Fort Story, HD Chesapeake Bay, 1985, (AMH),

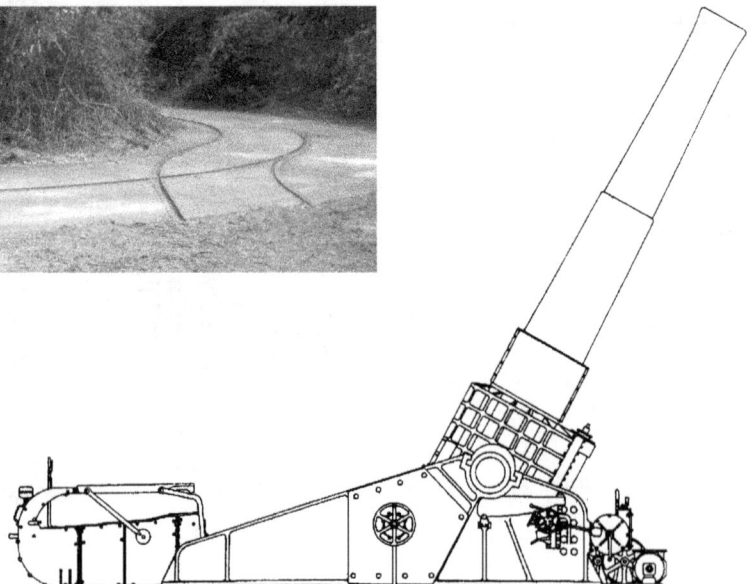

25 CALIBER HOWITZER ON BARBETTE CARRIAGE MODEL 1920

Drawing © Robert D. Fritz, used by permission.

M1920 Barbette Carriage for 16-inch Howitzer

LOADER

HOWITZER BASE RING & PIT

25 CALIBER MODEL 1920

BREECH OF HOWITZER

Drawing © Robert D. Fritz, used by permission.

SHELL ROOM

TELEPHONE

Scale - 1in.= 20ft.

TERMINAL STRIP

T.I.BELL

C. SPOTTER

C.PLOTTER

OBSERVER

SWITCH
KEY SETS

B.C. STATION
Scale 1in.= 10ft.

POST PHONES

SHEET 1-B
FORT STORY, A.
FIRE CONTROL INSTALLATION
FOR
BATTERY PENNINGTON
GENERAL LAYOUT OF CABLES, PHONES, ETC.,
ON RESERVATION
Scale-1in.= 400 ft.

TELEPHONE
IN BOOTH

POWER HOUSE
Scale-1in.= 20ft.

U.S. ENGINEER DEPARTMENT RAILROAD

25 PAIR L.C.A.S.
TO GRANITE

TRUE NORTH

NOTE:- THIS STATION TO
BE REMOVED.

NORFOLK AND SOUTHERN RAILROAD

DEAD ENDED

POWER HOUSE No.2

1-PAIR L.C.A.

No4

SHELL
ROOM

MAGAZINE

No3

SHELL
ROOM

MAGAZINE

No2

POWER HOUSE No.1

25-PAIR L.C.A.S.
TO EMERSON, HOLLIES
AND RIFLE RANGE

1-PAIR L.C.A.

MAGAZINE

1-PAIR L.C.A.

MAGAZINE

SHELL ROOM

MAGAZINE

1-PAIR L.C.A.

MAGAZINE

1-PAIR L.C.A.

MAGAZINE

25-PAIR L.C.A.S.

MAGAZINE

No1

SHELL ROOM

1-PAIR L.C.A.

SHELL ROOM

SHELL ROOM

25-PAIR L.C.A.S.

SHELL ROOM

1-PAIR L.C.A.

MAGAZINE

B.C. STATION

25-PAIR L.C.A.S.

18-DUCTS

PLOTTING ROOMS &
SWITCHBOARD ROOM

SHELL ROOM

MAGAZINE
TELEPHONE

Scale-1in.= 20ft.

AIR SPACE

DISTRIBUTION BOX

STORAGE

BATTERY

MOTOR
GEN.

DYNA
MOTOR

T.I.AP
PARATUS

F.C.SWBD

SPOT

CHECK

SPOT

CHECK

PLOT.

DATA

PLOT

DATA

SWBD
ROOM

PLOTTING
ROOM NO.2

PLOTTING
ROOM NO.1

PLOTTING ROOMS-SWITCHBOARD ROOM
Scale-1in.= 20ft.

1890s "Rebuilt" and "Emergency" Batteries
for Civil War Era Muzzleloading Barbette Mounts

From 1896 to 1898 "new" emplacements were built for older Civil-war era artillery at a variety of US ports. Some of the earlier sites were virtual reconstructions of older emplacements built from preservation and repair funding by the Corps of Engineers. Additional new emplacements were built as Spanish-American War emergency emplacements. Many other sites simply used repaired older mounts. New authorized gunblocks were built for (approximately):

 8-15-inch Rodman guns
16-10-inch Rodman guns
30-8-inch converted rifles (10-inch
    Rodmans converted to muzzle
    loading rifles)
 8-10-inch(?) Parrott rifles

The emplacements were usually gunblocks built on the older pattern with pintles (4 inch diameter for all but the 15-inch Rodman, which had a 6 inch diameter pintle) and iron racers for the traversing wheels. A few were quite elaborate stand-alone concrete emplacements (like Ward's Point, Staten Island)

Photos:
Top: 15-inch Rodman smoothbore on a center pintle mount, Fort Pickens, Gulf Islands Natl. Seashore, Pensacola, FL (MAB)

Middle right: Spanish-American War batterieries at Fort Marion, FL (GMW)

Lower right: Ridge Battery, Fort Baker, HDSF (GMW)

Left: at Fort Wadsworth, HDNY (GMW)

1890s "Rebuilt" and "Emergency" Batteries
for Civil War Era Muzzleloading Barbette Mounts

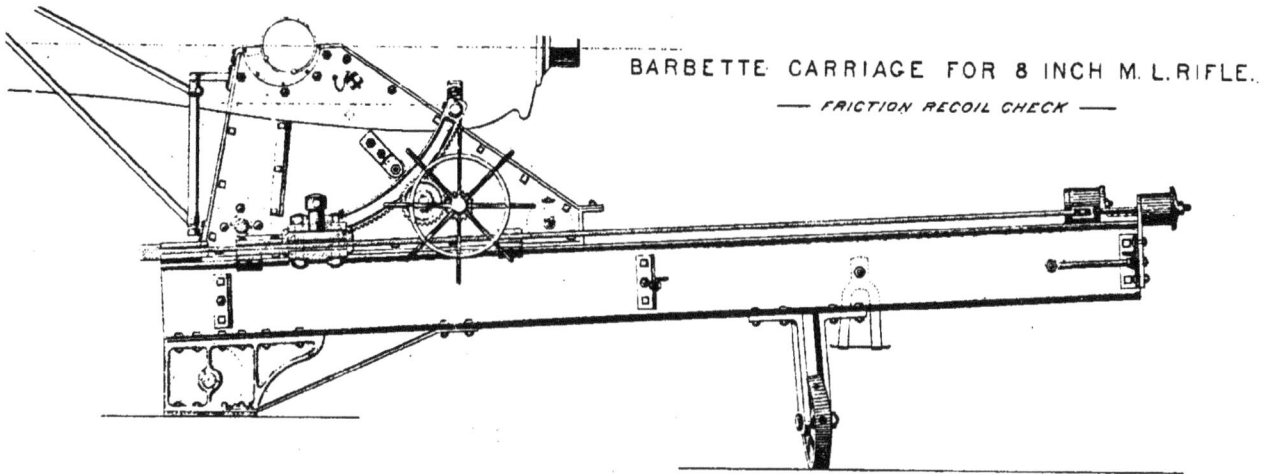

BARBETTE CARRIAGE FOR 8 INCH M. L. RIFLE.
— *FRICTION RECOIL CHECK* —

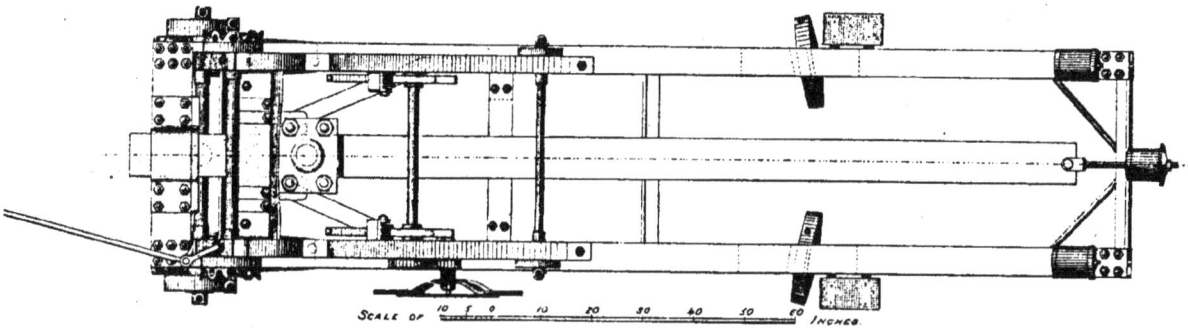

SCALE OF   10 5 0   10   20   30   40   50   60   INCHES.

Ord. Mem. 24.

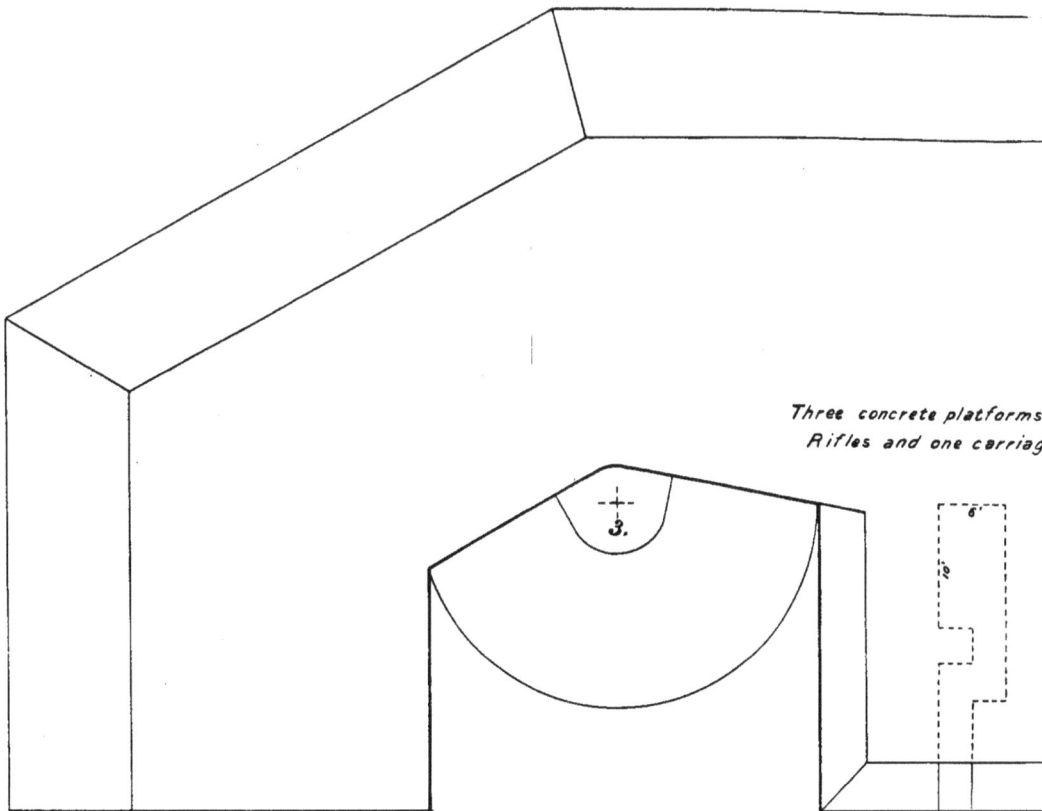

Three concrete platforms
Rifles and one carriag

MODIFIED RODMAN CARRIAGE FOR 8-INCH BL GUN

Usual Tubes: M1888, M1888MI
Carriages Built: 22
Original Emplacements: 21
Time Emplaced: 1898-1900

Total weight:
Max./min. elevation:
Turning radius: 360 degrees, limited by emplacement

Gun: M1888
Weight: 32,218 lbs
Length: 32 calibers

Photos:

Top: Fort Tyler, 1911 (Bolling Smith Collection)

Upper Middle: St. Johns Bluff Battery, 1986 (AMH)

Lower Middle: Fort Clinch, FL, 1998 (AMH)

Bottom Right: Fort Popham, ME, 1986 (AMH)

Bottom Left: Jerry's Point, NH circa 1900

MODIFIED RODMAN CARRIAGE FOR 8-INCH BL GUN

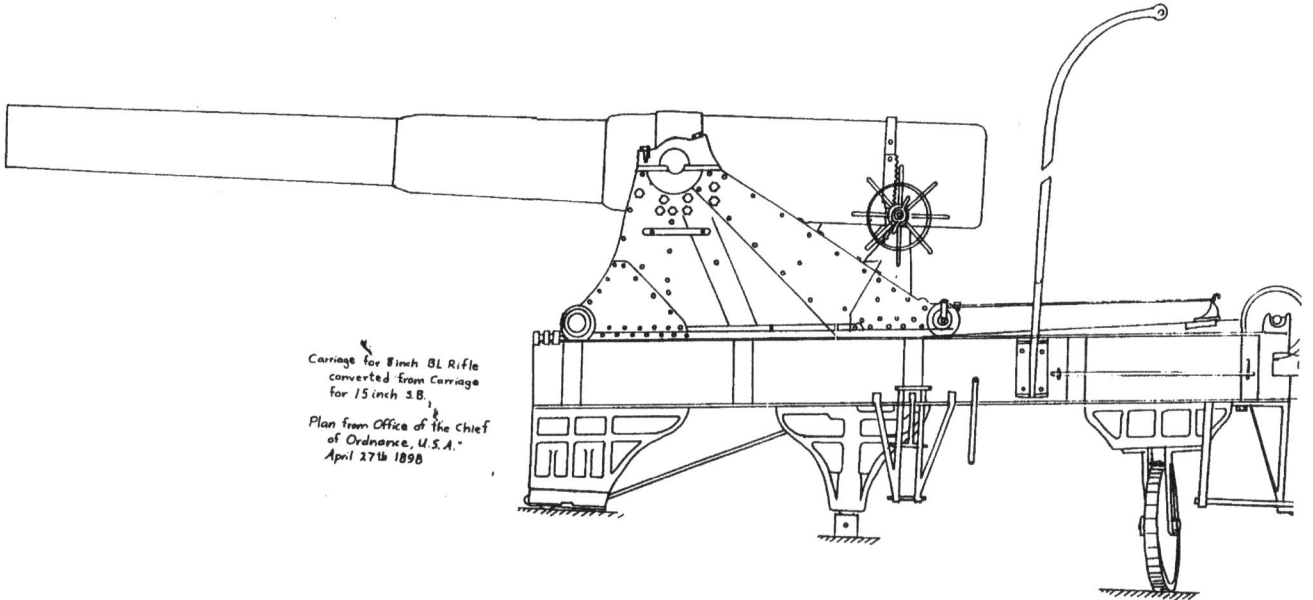

Carriage for 8 inch BL Rifle
converted from Carriage
for 15 inch S.B.

Plan from Office of the Chief
of Ordnance, U.S.A.
April 27th 1898

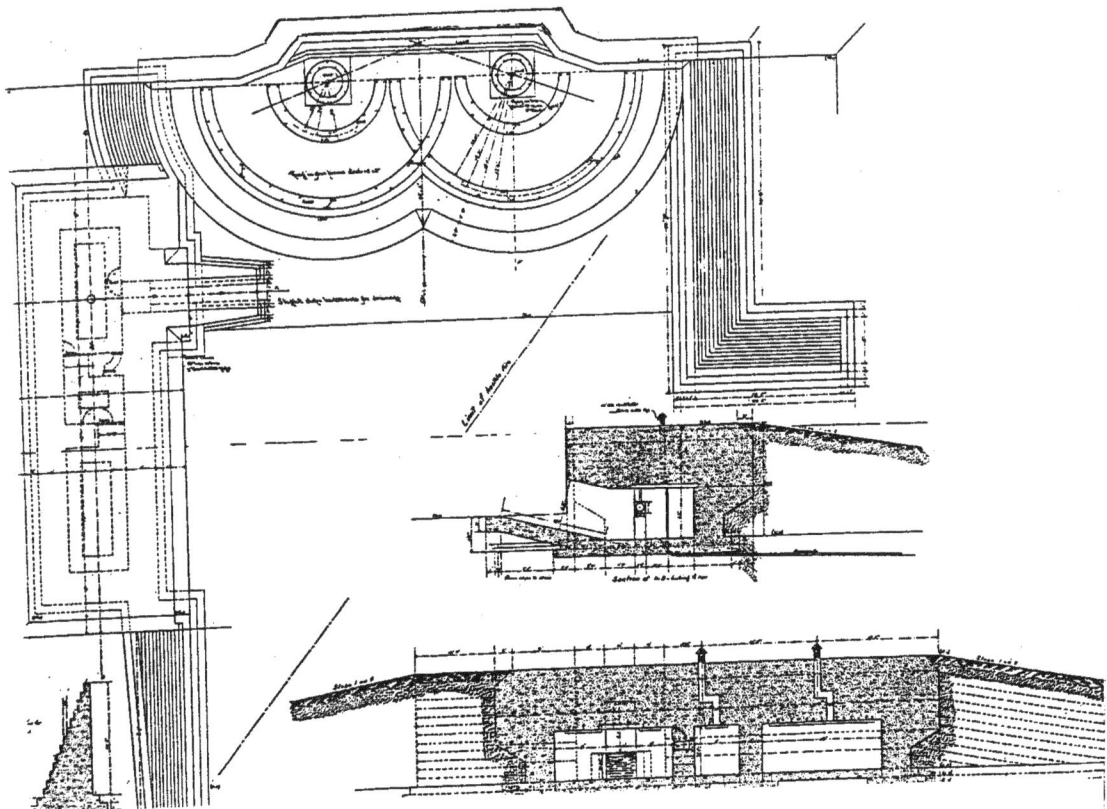

Egmont Key, HD Tampa Bay, FL

MOUNT FOR PNEUMATIC (DYNAMITE) GUN

Usual Tubes: 15-in Pneumatic M1886(?), 8-in Pneumatic M1886(?)

Carriages used: M1886 (15-in), M1886 (8-in) M1890 (15-in)

Original Emplacements: Concrete structures built for 2 x 15 inch, 1 x 8 inch at Fort Hancock, HD New York, 3 x 15-inch at Fort Winfield Scott, HD San Francisco, 1 x 15 inch at Fort H.G. Wright, HD Long Island Sound, and 1 x 15 inch at Hilton Head, SC.

Time Emplaced: 1890s
Parapet: variable
Notes: Discontinued by 1900

Photos:
Above: At Fort Winfield Scott, HD San Francisco, 1890s.

Middle: Lithograph from Scientific American Special Edition of 1898 of the battery at Fort Hancock, HD New York

Bottom: Emplacement for 15-inch Dynamite gun at Fort H.G. Wright, HD Long Island Sound, 1994 (AMH)

15-INCH PNEUMATIC DYNAMITE GUN, AS INSTALLED AT SANDY HOOK, NEW YORK, AND THE ENTRANCE TO SAN FRANCISCO HARBOR.
Range, with 1,000-pound shell, 2,400 yards; with 240-pound shell, 6,000 yards.

MOUNT FOR PNEUMATIC (DYNAMITE) GUN

15 inch Coast Defence Gun

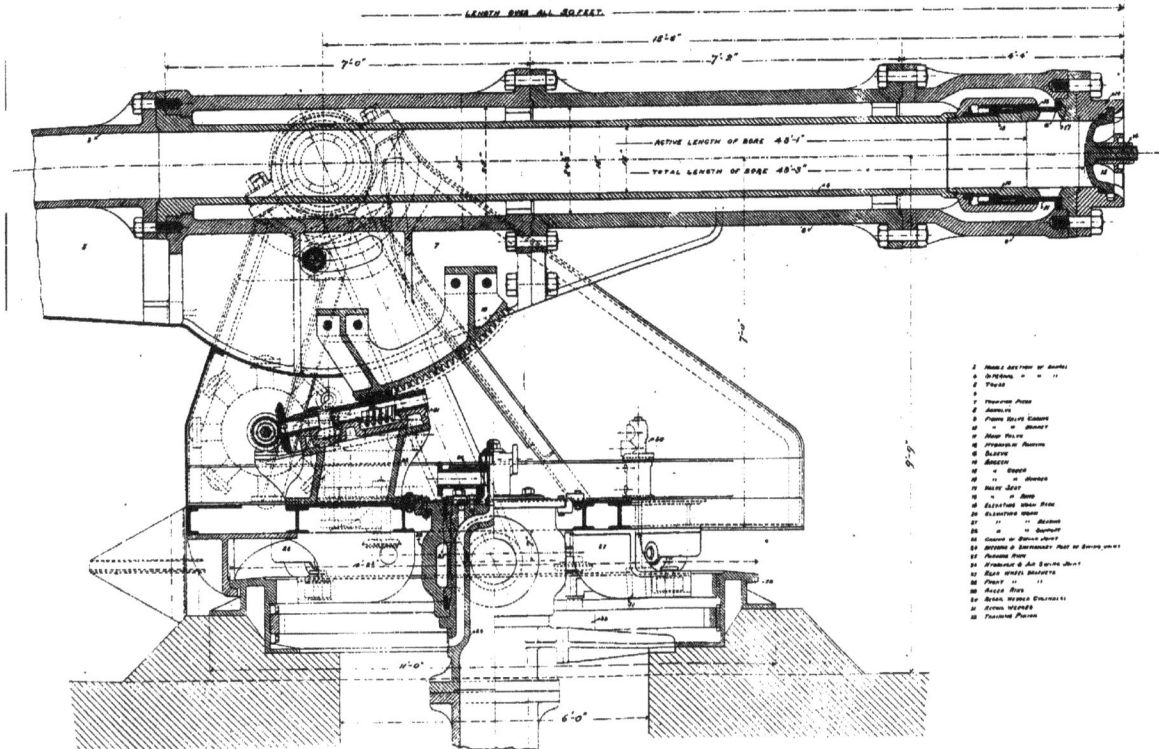

15-inch Pneumatic Gun (M1890)

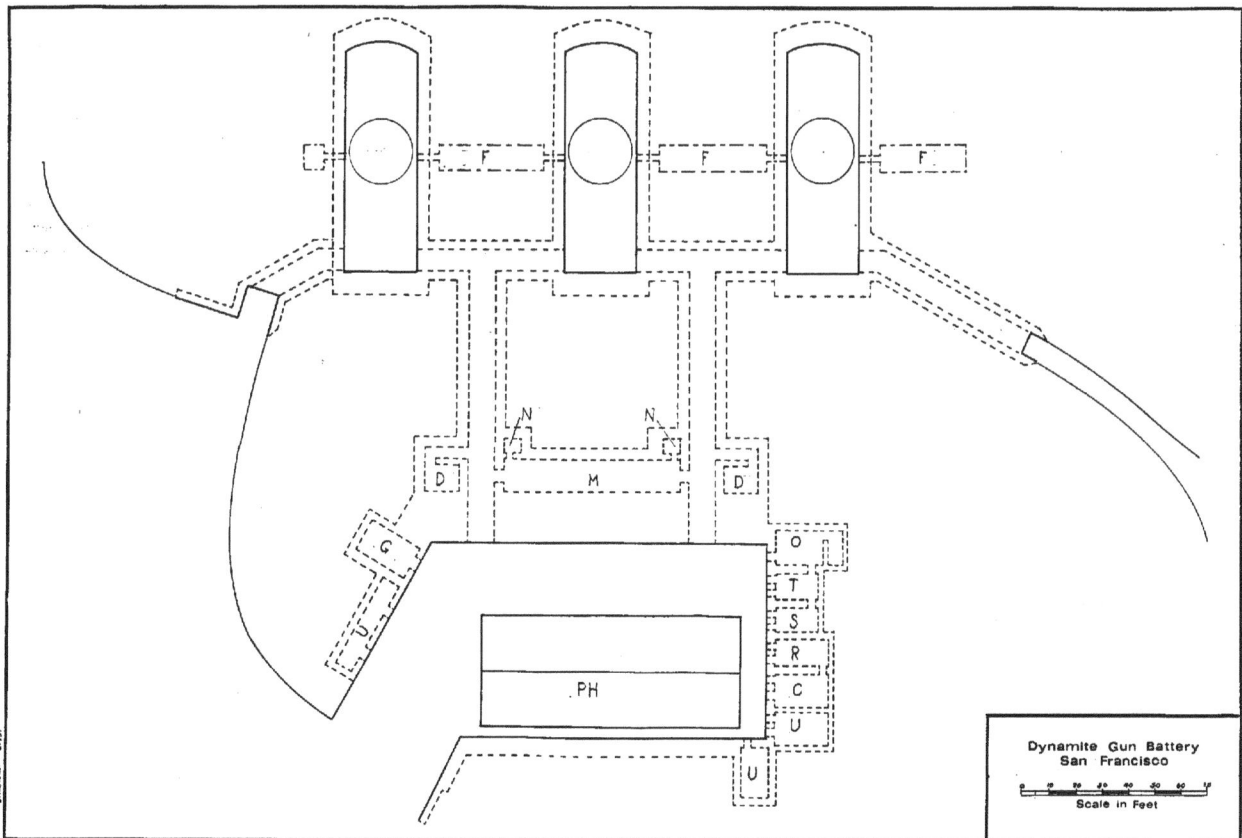

Dynamite Battery, Fort Winfield Scott, HD San Francisco, as completed

MOUNT FOR PNEUMATIC (DYNAMITE) GUN

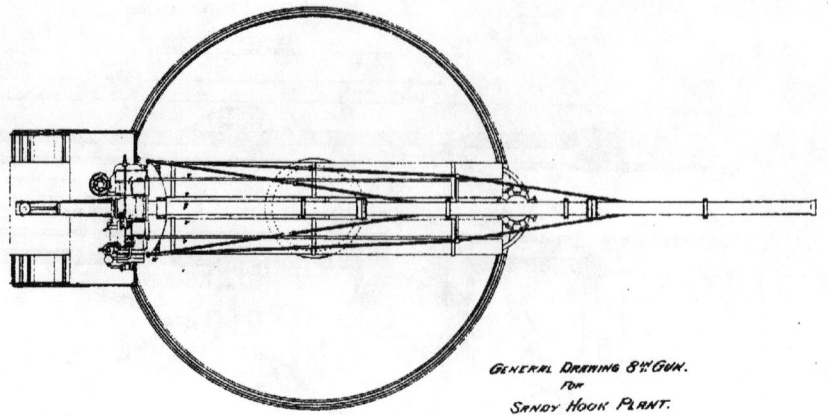

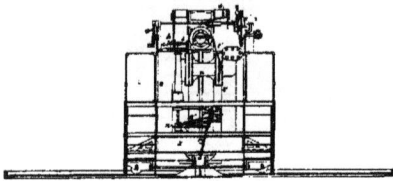

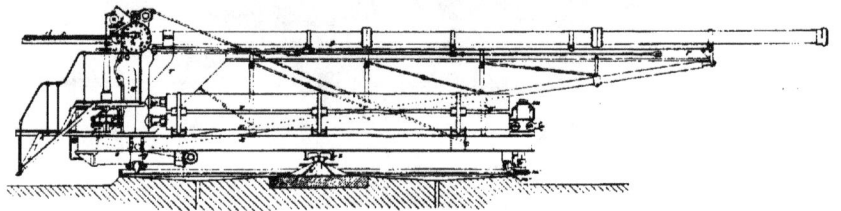

M1886 8 inch dynamite gun (Fort Hancock)

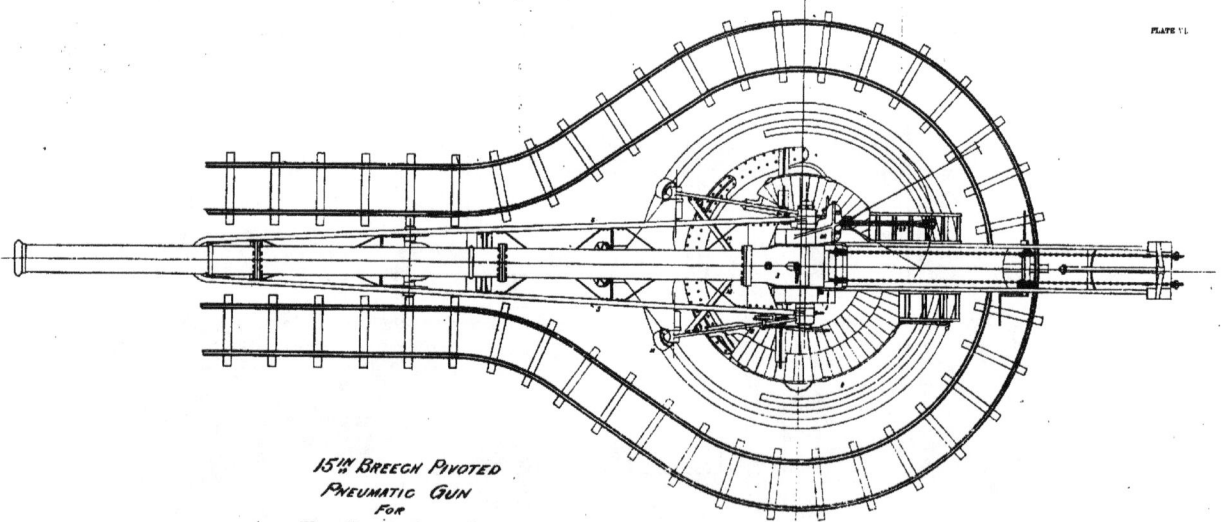

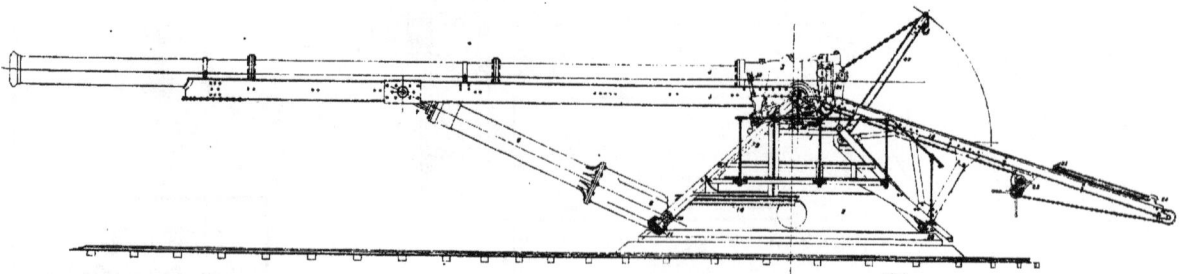

M1886 15 inch dynamite gun (Fort Hancock)

MOUNT FOR PNEUMATIC (DYNAMITE) GUN

~ 15˚ PNEUMATIC DYNAMITE SEA COAST GUN ~

— LENGTH OVER ALL 50 FT —

M1890 15 inch dynamite gun

15 inch dynamite gun in action at Sandy Hook? (Ordnance Museum)

Emplacement for a 15-inch dynamite gun at Hilton, Head, SC, 1995 (AMH)

WHEELED CARRIAGE PARAPET MOUNTS FOR 6 POUNDER GUNS

6 pdrs. (2.24") Guns M1898, M1900, and Mk III on wheeled carriage M1898, M1900 and American Ordnance Company carriage

Generally used a mobile defense and saluting weapons mostly on Coast Artillery posts. Sometimes positioned in forts (always still on wheeled carriage) on older gun pintles. A very few experimental emplacements were built in the continental US at about the turn of the century, and a dozen fixed emplacements at Fort Ruger, Oahu, emplaced as part of the Land Defense Project of 1915.

Photos:
Upper: emplacement on east rim of Diamond Head, Fort Ruger, HD Honolulu, HI, 1996 (GFW)

Lower: Postcard of Fort Monroe, 1900s

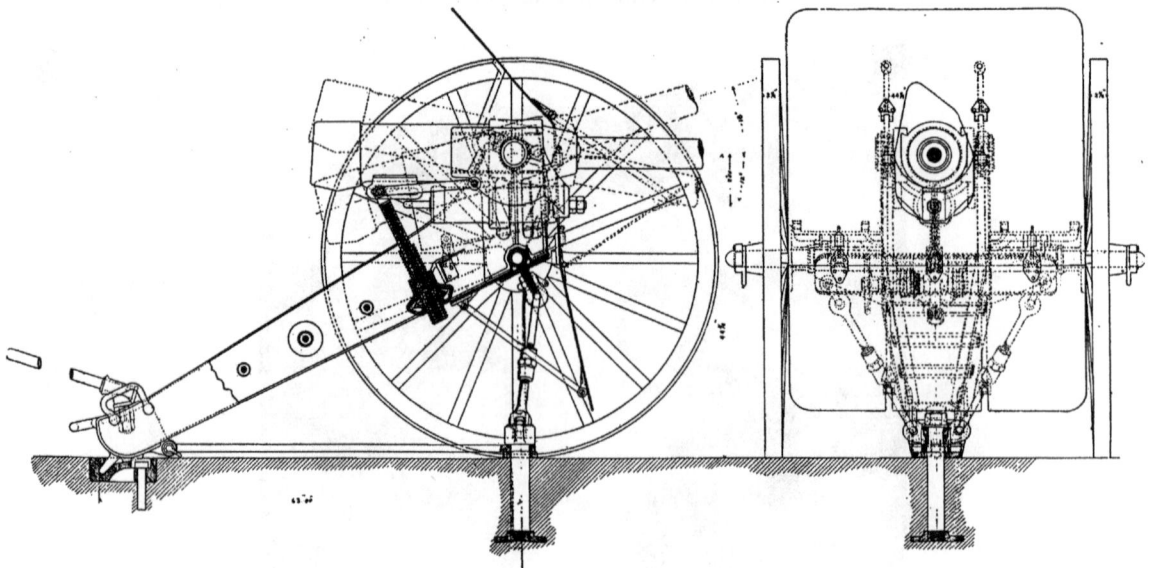

American Ordnance Company 6-pounder parapet mount

WHEELED CARRIAGE PARAPET MOUNTS FOR 6 POUNDER GUNS

M1898

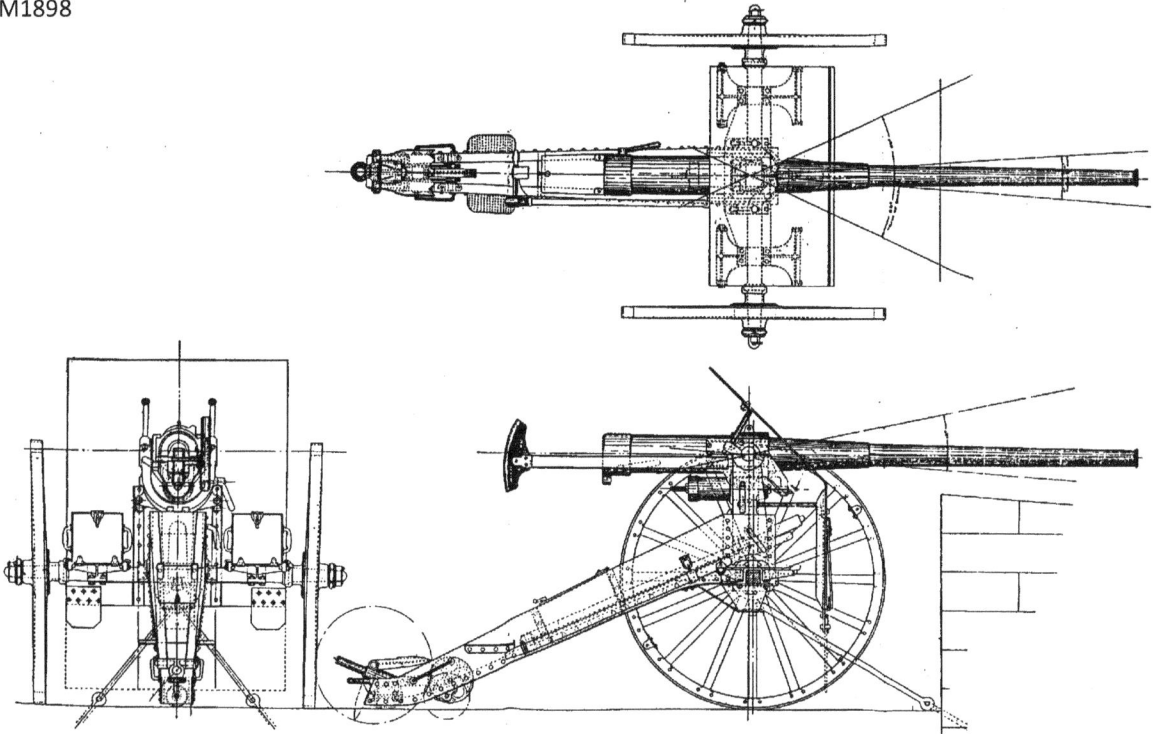

6 pounders most probably located behind Battery North, Fort Michie, HD Long Island Sound,1910s (NA)

M1917 & M1918 TRACTOR-DRAWN MOUNTS FOR 155 MM GUNS

Usual guns: M1917, M1917M1, M1918, M1918M1
Carriages Built: over 2000
Original Emplacements: over 300
Time Emplaced: 1928-1943
"Grand Puissance Filloux" (G.P.F.) originally bought from France during WWI, also built in the US.

Number of Bolts: none
Circle Diameter: 17' 8" radius center pintle to racer ring
Notes: Center concrete base usually 10' in diameter with raised center piece. A fairly standardized emplacement dating from the late 1930s. Approximately three hundred concrete emplacements were built in 180, 270, and 360-degree varieties. Some were later converted for the new 155 mm M1, which also had a metal-frame ring ("Kelly" mount) for use as a fixed emplacement, primarily used in overseas temporary defenses.

M1917 & M1918 carriages
Carriage weight: 14,500 lbs
M1917AI & M1918AI modified wheel bearings, M1918M2, M3, modified with pneumatic wheels
Maximum/minimum elevation: +35 / 0 degrees
Turning radius: 360 degrees (on Panama mount)

Guns: M1917 & M1918
Weight: 8,713 lbs
Length: 37 calibers

Fort Weaver, HD Pearl Harbor, HI, circa 1935 (NA)

155 mm GPF on M1918M3 carriage, Presidio of San Francisco, 2011 (MAB)

Fort MacArthur, CA, 1992 (AMH)

(left) Fort Baldwin, ME, 1994, (right) 180 degree mount at Harmon Field, Newfoundland, Canada, 1985 (AMH)

M 1917 & M 1918 Tractor-Drawn Mounts for 155 mm Guns

M1918A1 carriage with limber In traveling arrangement, Aberdeen Proving Grounds, 2009 (MB)

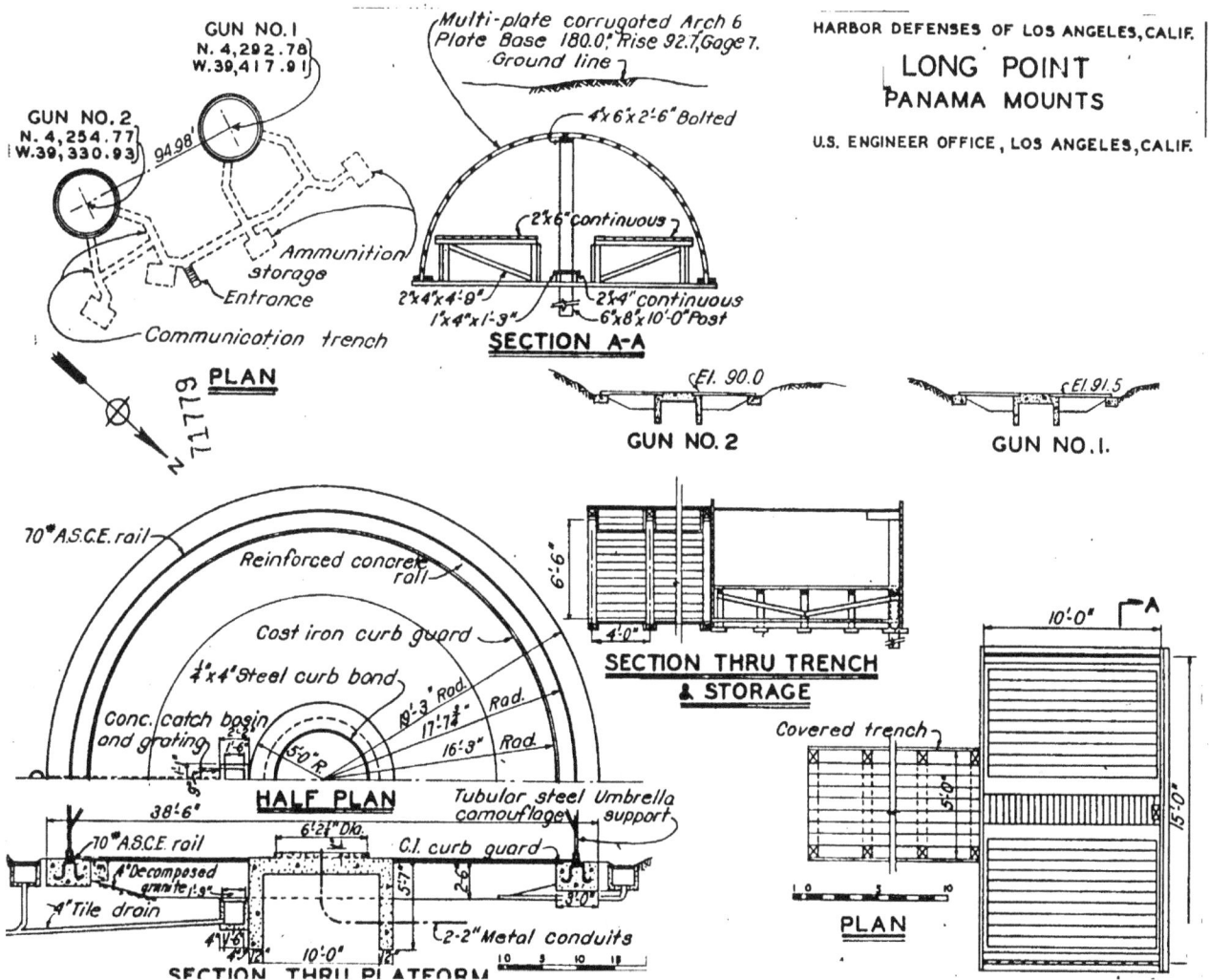

PHILIPPINE TRIAL MOUNT FOR 155 MM GUN

Guns: GPF guns and mounts without trails
Original Emplacements: 1
Time Emplaced: 1930s

Number of Bolts: 4
Trapezoid Dimensions:
24 in x 33 in x 24 in x 55 in

Notes: Battery West, Fort Mills, the Philippines
Fixed emplacement for gun no trails.

PLAN
Scale: ½" = 1'

ELEVATION
Scale: ½" = 1'

Copy of C of
662 B Manila
Files
BW

MODIFIED M1918 MOUNT FOR 240 MM HOWITZER

Usual Tubes:
Carriages used: 12
Original Emplacements:
Time Emplaced: 1920s, 1940s

Number of Bolts:
Circle Diameter:
Parapet: none

Notes: Twelve 240 mm M1918 howitzers were diverted to Hawaii from the Philippines at the signing of the Washington Treaty. They were converted from mobile mounts and used by the coast artillery on Oahu on a central concrete pintle with an outer ring for traversing. As the sites were changed over time, the number of emplacements exceeds the actual total of weapons ever present.

Photos:
Upper: Battery C, 90th FA Battalion firing at the Anahulu Flats emplacements, circa 1940s. (NA, courtesy William Gaines

Lower: Somewhere on Oahu Island. (NA)

M1918 MOUNT FOR 240 MM HOWITZER

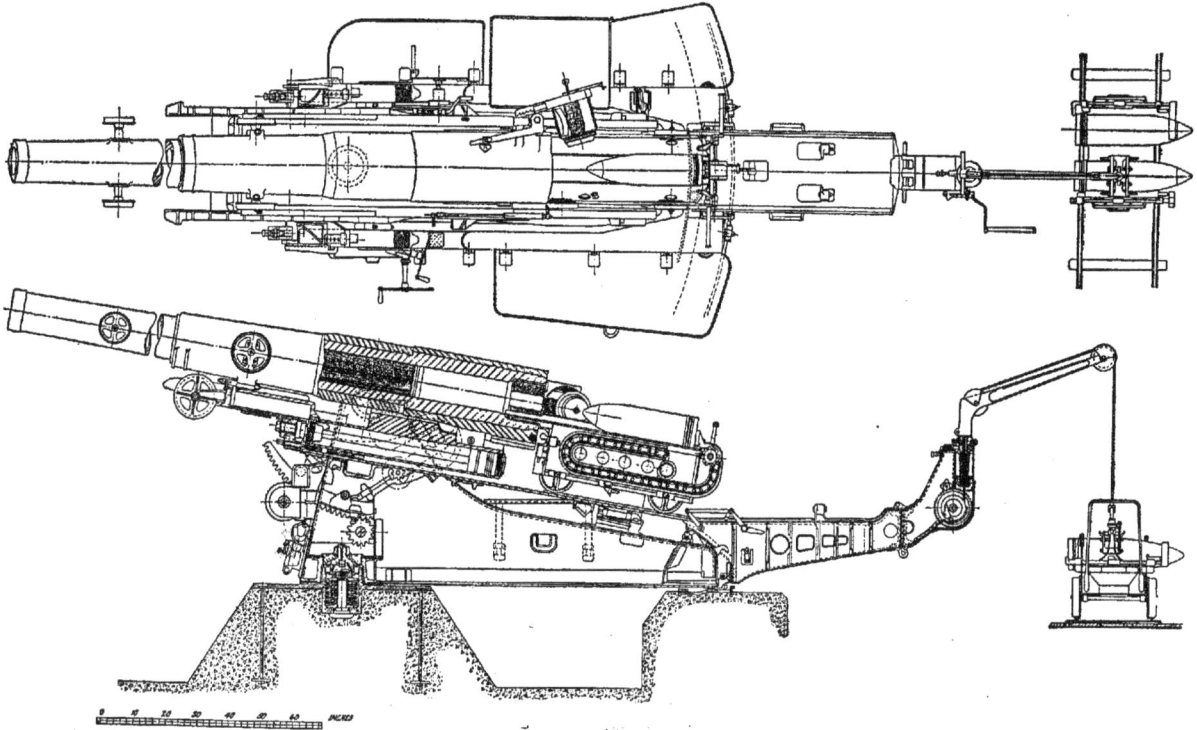

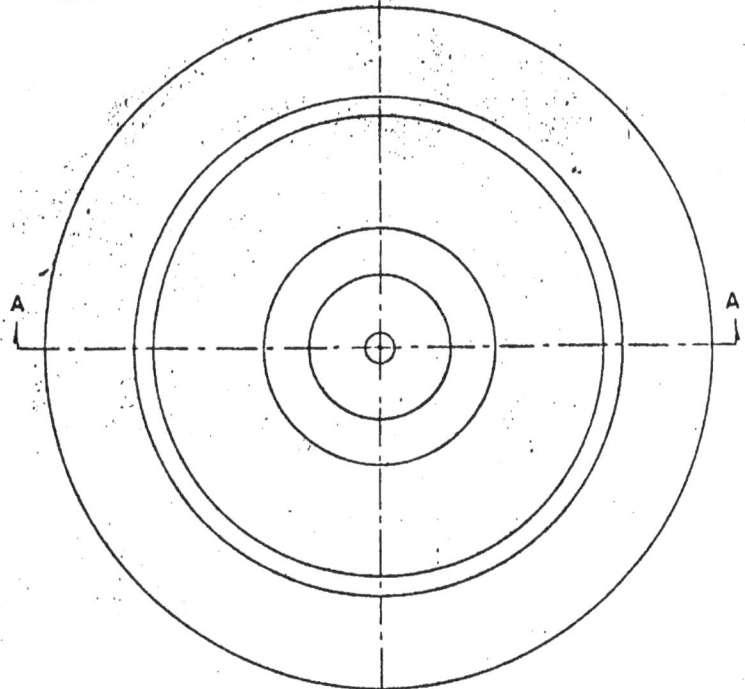

PLAN

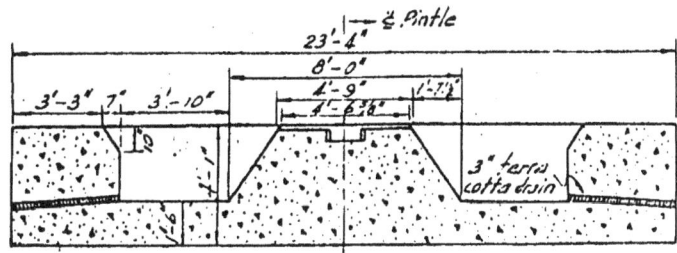

SECTION A-A

emplacement for 240 mm howitzer on modified mount

WW II Emergency Batteries with Naval Carriages & Guns

3-inch Naval Mark 6 on Mark 7 M4 pedestals. At least one battery in Hawaii was armed with this arrangement. (no illustrations)

4-inch Naval Mk 9 (with some Mk 8 and Mk 12) on Mk 12 M1 pedestals were used in temporary emplacements in the Hawaiian Islands. Approximately 26 emplacements documented. (no illustrations)

5-inch Naval Mk 8 and Mk 15 on Mk 13 M4 or M8 pedestals were used at several places. Approximately 17 emplacements (Hawaii, San Diego, North Carolina). Additionally several dozen served with the Marine Defense Battalions on naval base duty (Wake I., Midway I., Johnston I., Palmyra, Guantanamo Bay, etc.). (Drawings of gun only)

6-inch Naval Mk 8/Mk 10 pedestals were also used in temporary batteries. About 28 accounted for (Los Angeles, San Francisco, Western Washington, North Carolina, Aleutian Islands, Argentia, Espiritu Santo). (pictures of emplacements)

7-inch Naval Mk 2 M2 pedestals also used in temporary batteries. 22 accounted for (10 in Hawaiian Islands, 3 in San Diego, 8 in Bora Bora, Samoa). (gun drawing and pictures of emplacements)

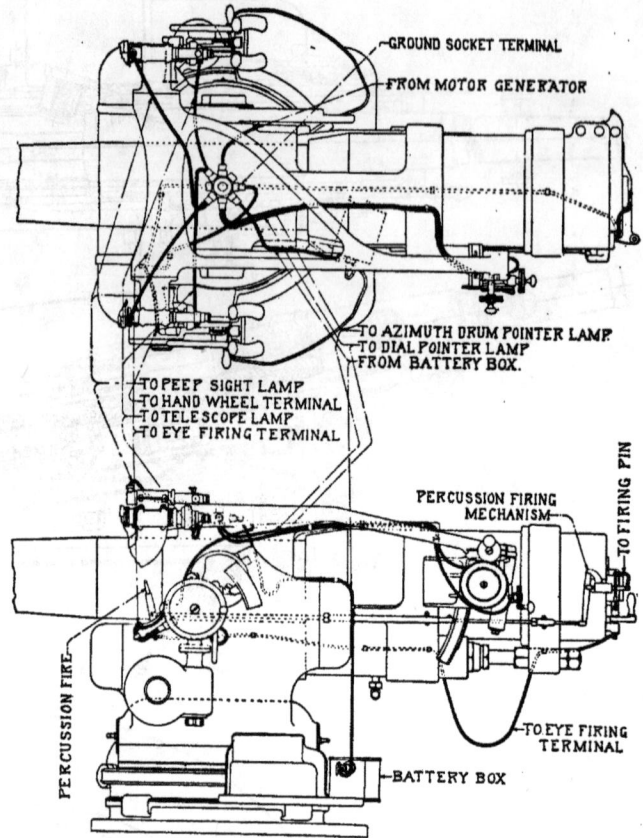

5" Naval Mk 15 on Mk 13 M4 pedestals.

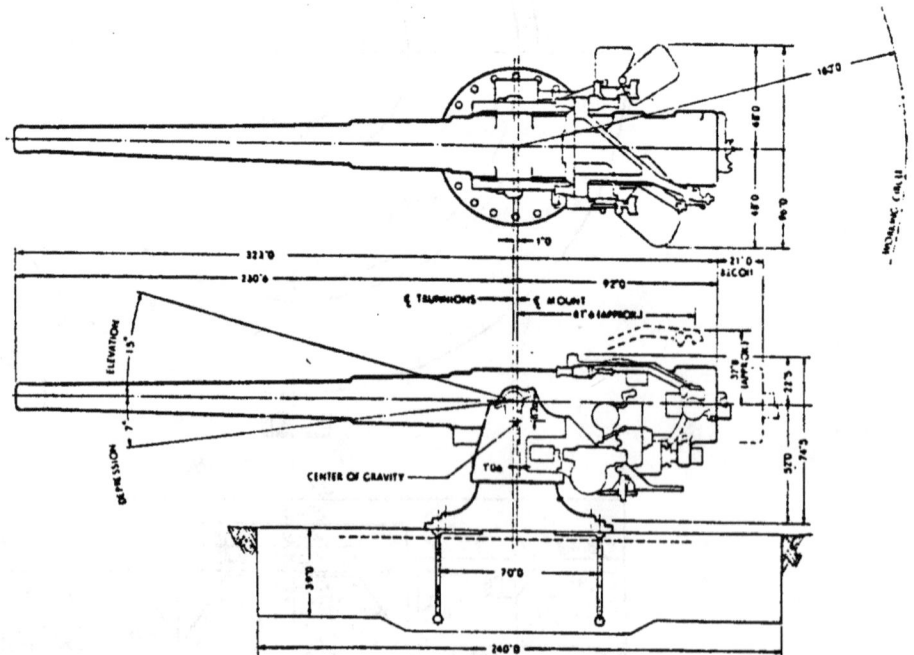

Note: Not all these batteries (for instance the Alaskan WW II emergency 6-inch gun batteries) are listed in the battery list.

7-inch naval Mk2M3 pedestal mount

WW II Emergency Batteries with Naval Carriages & Guns

photos (left to right from top):

Top: Two photos of an Alaskan 6-inch naval gun positions (1943?) (GMW collection).

Upper Middle left: 7-inch naval pedestal emplacements: Sand Island MR, Honululu, HI (GMW)

Upper Middle Right: 6-inch naval pedestal emplacement Argentia, Newfoundland, Canada, 1980s (GMW)

Lower Middle: 5-inch naval pedestal mount, Rose Island, Narragansett Bay, RI 1994 (AMH)

Bottom: 7-inch naval pedestal mount on display at Hawaii Army Museum, Honolulu, HI, 2002 (Photo by John D. Bennett)

4.7 INCH SCHNEIDER GUN AND MOUNT

Usual guns: Schneider
Carriages purchased: ?
Original Emplacements: 1
Time Emplaced:1890s

Notes: A number of different 4.7 in guns were purchased for testing during the mid-1890s. A single Schneider 4.7 in gun was tactically emplaced at Fort Hancock during the Spanish-American War.

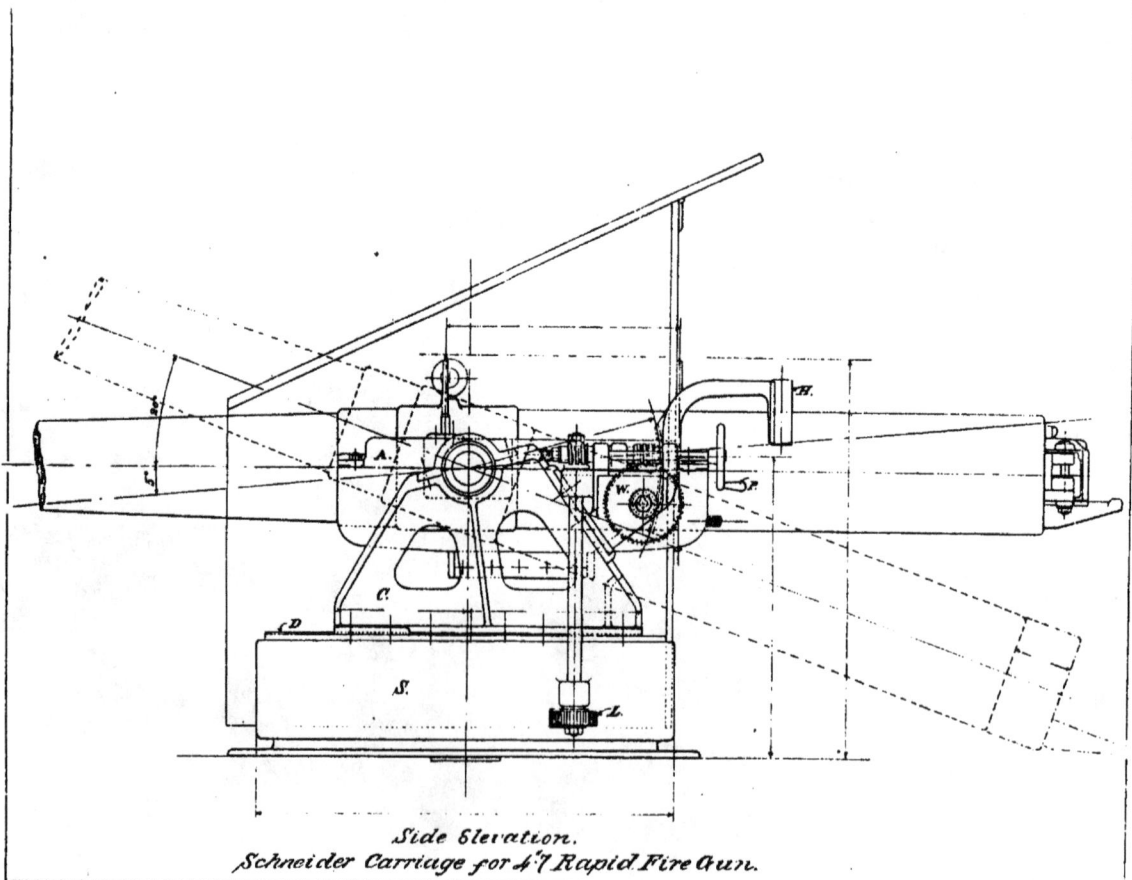

Side Elevation.
Schneider Carriage for 4.7 Rapid Fire Gun.

Appendix 26, 1896.
Ord 54 2

4.7 INCH SCHNEIDER GUN AND MOUNT

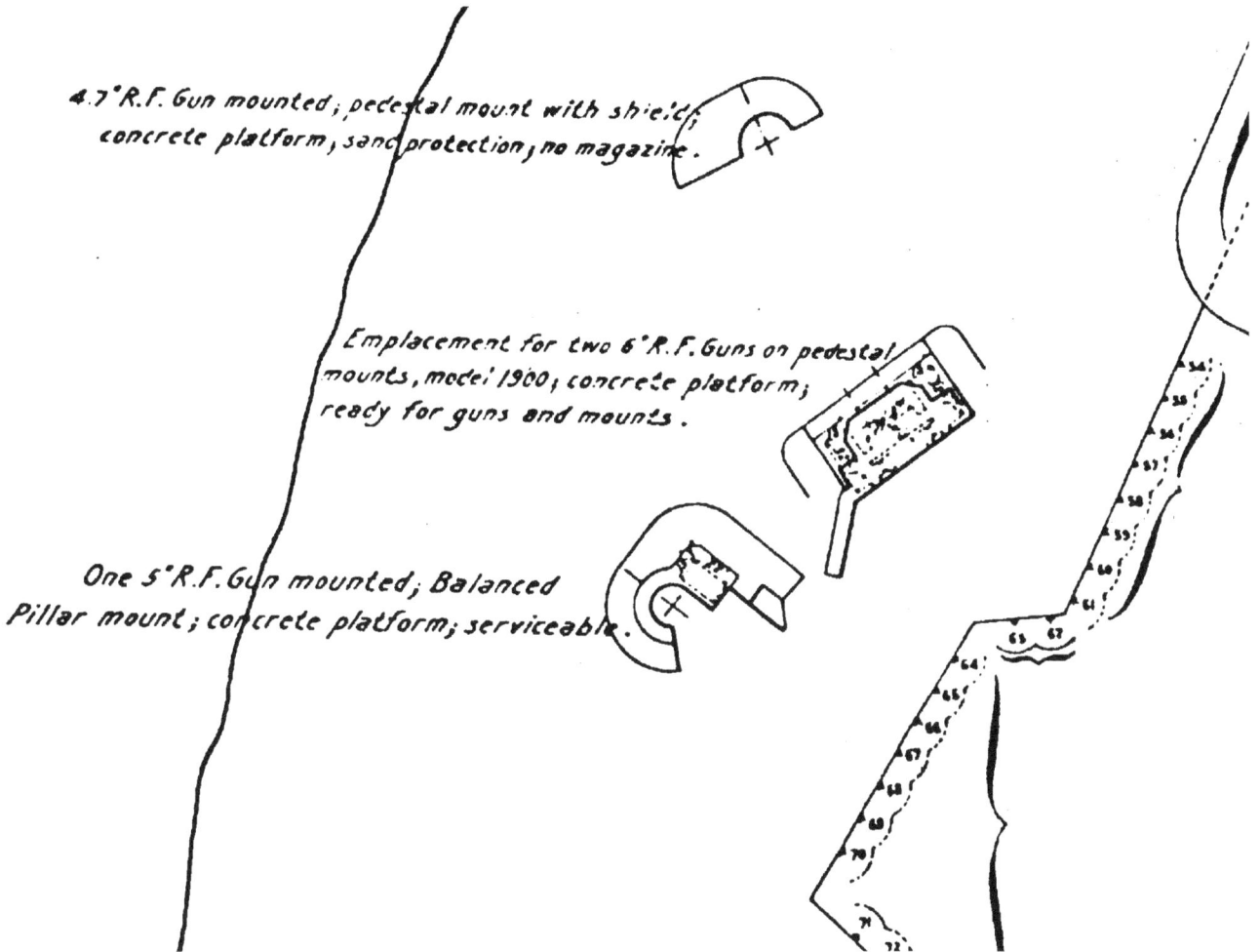

4.7"R.F. Gun mounted; pedestal mount with shield; concrete platform, sand protection, no magazine.

Emplacement for two 6"R.F. Guns on pedestal mounts, model 1900; concrete platform; ready for guns and mounts.

One 5"R.F. Gun mounted; Balanced Pillar mount; concrete platform; serviceable.

Layout at Fort Hancock, 1899

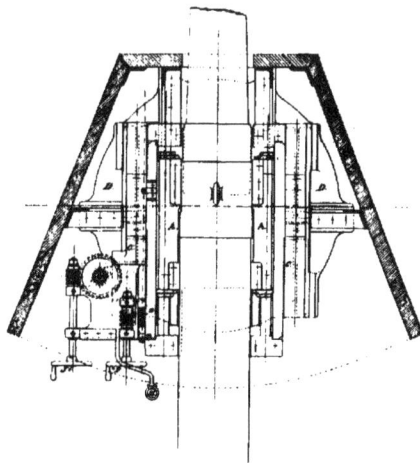

Plan
Schneider Carriage for 4"7 Rapid Fire Gun

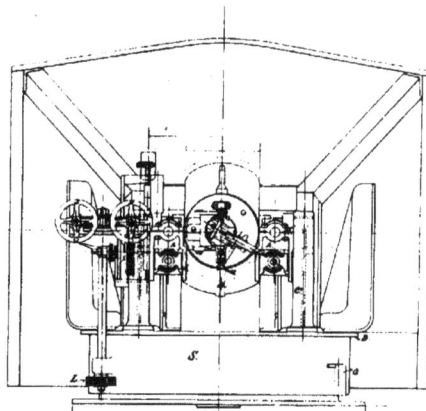

Rear Elevation.
Schneider Carriage for 4"7 Rapid Fire Gun.

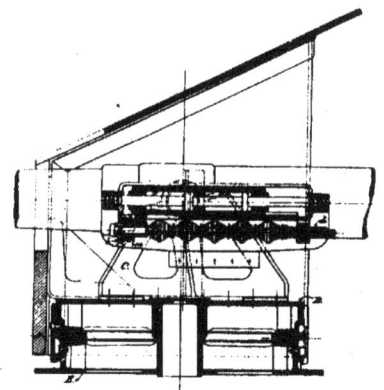

Sectional Side Elevation.
Schneider Carriage for 4"7 Rapid Fire Gun.

4.7 in Schneider gun mount and shield

3 INCH GUN CASEMATE MOUNT

Guns: M1898
Carriages used: 2
Original Emplacements: 2
Time Emplaced: 1901

Number of Bolts: 10
Circle Diameter: 3' 1"
Notes: uneven bolt distribution around circle, with a center pipe for electrical conduit. The carriages were modified at Watertown Arsenal as Pedestal Mounts, Driggs-Seabury, Casemate Modified.

Battery Edwards, Fort Mott, HD Delaware River. Casemate openings were below Battery Krayenbuhl (2 x 5-in BP) which sat above the corridors that made up Battery Edwards.

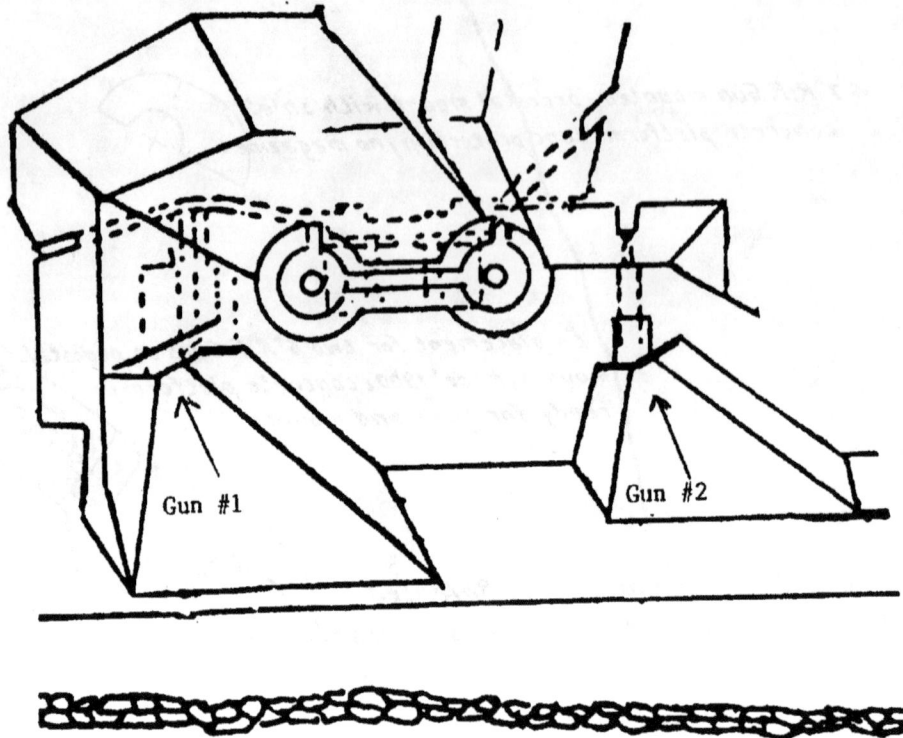

Gun #1

Gun #2

Casemates Nos. 1 & 2: Interiors and embrasures completed, Two (2) Driggs-Seabury 15-pdr. R.F. Guns (Nos. 62 & 63) on Casemate mounts with shields, mounted, ready for service.

Battery Edwards, Fort Mott, HD Delaware
left, exterior view (MAB); right, interior view, (AMH) 1996

MODERN AMERICAN SEACOAST DEFENSES

A LIST OF MILITARY RESERVATIONS
AND CONCRETE GUN BATTERIES

1890-1950

Compiled by Mark A. Berhow
© 1995-2024, Mark Berhow

Revision Date:
January 18, 2025

Fort Michie (Great Gull Island, NY)
Once part of the Harbor Defenses of Long Island Sound, NY, the island is now a national wildlife refuge.
The large concrete structure in the forground is Battery J.M.K. Davis, built for a single 16-inch gun on a disappearing
carriage. Photograph by Terry McGovern, 2003.

AMERICAN SEACOAST DEFENSES
A LIST OF MILITARY RESERVATIONS AND CONCRETE GUN BATTERIES 1890-1945

This is an attempt to list all the concrete emplacements built by the U.S. Army Corps of Engineers to hold seacoast armament of the "Modern era" (1890-1950). It includes four major generations of American coast defense construction—the Early Modern Program batteries (the "Endicott Board" and the "Taft Board"), the post-World War I batteries, and the WW II Modernization Program batteries—as well as those batteries built during emergency situations. Every effort has been made to make this list as accurate as possible, but it will most likely contain a number of errors and omissions. The author would greatly appreciate being contacted about any corrections.

Fort and battery names used in this list are those perceived by the author as *being the last official designation*. Named batteries are *listed by surname only*, even though as many were designated by the full name of the person they were named after. See *Designating US Seacoast Fortifications, War Department General Orders and Letters from the Adjutant General 1809-1950*, compiled by Matthew Adams (privately published by the compiler, Australia, 2000) for more information on fort and battery names. The battery service years listed here are generally from the year in which the battery was transferred to the Artillery or Coast Artillery Corps (or the year the battery was completed) to the year the battery was ordered removed from service (or the year the last gun was removed). Where possible, the information in a listing was confirmed from a report of completed batteries or report of completed works; otherwise information from other published sources was used.

Gun and carriage year models (or M# after 1935) are given where known.

Sample Entry:
Harbor Defenses of . . .

FORT NAME	Location	service years (if known*)	current ownership	MC**, MF**	rating
Battery name	/ # of guns	/ caliber & model #	/ carriage type & model #	/ service years***/	notes

     * Several coast artillery forts were officially abandoned as harbor defense posts by 1928, all by 1950.

     ** Mine Casemate (MC) or Mine Shore Facilities (MF), see next page for explanation

     *** Batteries whose exact service years are not known are designated by an era, such as WWII.

Abbreviations:

MC**	mine casemate (see note next page)
MF**	Mine Facilities: mine wharf & shore buildings (see note next page)
destroyed	emplacement destroyed
buried	emplacement buried
empl	emplacement
repl	replaced
rem	removed
ARF	carriage designed for 360 degree fire.
number (#101)	1940 Project battery construction number (used for battery name in some cases)
NB	emplacement not built
NC	emplacement built, but not completed
NA	emplacement completed, but not armed (gun tube missing or not installed, carraige at site or installed)
Still Emplaced (SE)	Original (or appropriately replaced) guns in the battery today.

Carriage Abbreviations (N = Navy gun)

A	British Armstrong guns on pedestal mounts
AGL	altered gun-lift carriage
B	barbette carriage
Rod	breechloading gun on altered 15-inch Rodman carriage
BL	long range barbette carriage, Army gun
BN	long range barbette carriage, Navy gun
BP	balanced pillar mount
CB	long range barbette carriage in casemate, Navy gun (16" 1940 Program)
CM	casemated mount
D	disappearing carriage
F	fixed pedestal mount (anti-motor torpedo battery (AMTB) mount)
GL	gun-lift carriage
H	long range howitzer carriage
M	mortar carriage
MP	masking pedestal mount
NT	turret mount—Navy
NC	casemated mount, Navy gun and carriage
NP	Navy gun on pedestal mount
P	pedestal mount
PM	155 mm GPF gun on tractor-drawn carriage with concrete "Panama" mount
Pne	pneumatic (dynamite) gun and carriage
RM	railway mount—mortar
RY	railway mount—gun
SB	long range barbette carriage with shield (6" 1940 Program)
TM	turreted mount—Army

Mine defense facilities are indicated in the fort name entry: Controlled mine fields were an integral part of the modern American harbor defenses. "MF" indicates that there were mine loading and storage facilities at the reservation for storing the mines and their cables and for deploying the mines for planting by the mine planters. These shore facilities usually included a mine wharf, mine loading rooms, magazines, cable tanks, torpedo storehouses, and a rail tramway system connecting these structures. "MC" indicates that there was one or more mine casemates, the protected structure which housed the actual firing circuits for the deployed submerged mine groups, on the reservation. Mine facilities were built during all major construction program eras. The mine defenses of some harbors were discontinued long before the harbor defenses themselves were abandoned. Other harbors had a major update of their mine facilities in the late 1930s and early 1940s.

ARF Disappearing Gun Carriages: There were three 10" DC ARF installed in circular concrete emplacements (one at Fort San Jacinto, two at Fort Stevens) and one 16" DC ARF was emplaced at Fort Michie. All 12" and 16" BCLR emplacements *built prior to 1936* were ARF until they were casemated.

Mortar Pits: Many (but not all) of the original 4-mortar pits had 2 mortars removed during the years 1905-1920. This is not necessarily noted in the list.

Prepared 240 mm Howitzer Positions in Hawaii: Twelve emplacements for 240mm howitzers on modified mobile M1918 carriages were prepared in 1920. Ten more emplacements were built during 1938-1945 to replace the original 12 emplacements.

155 mm GPFs on Panama mounts: These guns were used in harbor defenses beginning in the 1920s. A concrete platform was designed for a permanent emplacement. Only those batteries which had concrete Panama mounts constructed are listed here. Due to lack of information, the 155 Panama mount batteries in the Caribbean are NOT all listed.

Modernization Project Battery Constructions: For the sake of completeness, *all* Modernization Project batteries (6", 8", 12" and 16") are listed here, including those not actually built *[indicated by italics and brackets]*.

AMTB (Anti-Motor Torpedo Boat) Batteries: Only the fixed emplacements are listed here. Complete AMTB batteries were composed of two 90 mm M1 guns on fixed M3 mounts, two 90 mm M1 guns on mobile M1 mounts and two 37 mm (later 40 mm) automatic guns. Some of the AMTBs listed here were not completely armed with full complement of the mobile guns. Many other positions not listed here were armed only with mobile guns (some 90 mm, but mostly 37 mm sections). Earlier (1942) "AMTB" batteries (repositioned M1903 3" pedestal mounts) are listed.

Coast Artillery railway artillery: Two 14" RY guns were at Fort MacArthur, CA and two were in the Panama Canal Zone from the late 1920s - 1940s. Some 8" RY and 12" RM railway carriage guns parked at sites are not specifically listed unless emplacements were built, and not all the positions prepared in Hawaii during WWII may be listed. Known locations for parked 12" RY and 8" RY guns during the 1920s and 1930s includes Fort Hancock, NJ, Cape Henlopen, DE, Camp Eustis, VA, Fort Story, VA, Camp Pendleton, VA, Fort MacArthur, CA, Fort Stevens, OR, Oahu, HI. Plans included preparing positions for railway artillery at Fort Stevens, OR (never started) Grays Harbor, WA (positions prepared) and Cape George, WA (positions prepared). 8" RY artillery were later deployed at postions in Canada and near Port Angeles along the Strait of Juan de Fuca, as well at Los Angeles, CA, Fort John Custis, VA, Fort Miles, DE, Fort Hancock, NJ, and Oahu, HI during WWII.

Not Listed: Coast Artillery Corps troops manned several other types of harbor defense weapons and sites which are *not* listed here due to incomplete information held by the compiler. This includes: all fixed (emplaced) antiaircraft (AA) guns; some of the WW II emergency Navy guns & mounts, the 75mm howitzer mounts used in Panama, field mounted 155mm GPF batteries, fire control stations; and searchlights.

Current Disposition of Military Reservations and Batteries: This list contains the current information (as of the date of this revision) on the ownership of the coast defense military reservations. This, unfortunately, is subject to change. Emplacements known to have been modified, buried, or destroyed are so noted. The fort rating system is an arbitrary device used by the author to give the reader some overall idea of what to expect to find at the sites today.

Fort Rating System

★ ★ ★ ★ ★	all or most emplacements intact, many or most buildings remain
★ ★ ★ ★	all or most emplacements intact, several buildings remain, reservation(s) may be divided up
★ ★ ★	all or most emplacements intact, some buildings remain
★ ★	all or most emplacements intact, few or no buildings remain
★	some emplacements intact, few or no buildings
X	nothing remains at the site

Sources: Much of the information tabulated in this list came from a variety of original documents, lists, books, and correspondence. As far as possible the data listed in this table came from the reports of completed works filed by the Corps of Engineers for each gun battery built. Additional information was obtained from some of the various annexes to harbor defense projects. More information was gleaned from the various articles, books, brochures and other publications both of a general nature and on specific forts or harbor defenses by various agencies such as the National Park Service, state and local historical agencies, and private publishing companies, which are too numerous to list here. Additional help and information was obtained from Alvin Grobmeier, Terrance McGovern, Bob Zink, Gregg Hagge, Joel Eastman, Alex M. Holder, Bill Dorrance, John D. Bennett, Hans Neuhauser, Luis Ramos, and others.

The Harbor Defenses of the Kennebec

FORT BALDWIN			Sabino Head 1905 - 1928		state park	MC (improvised)	* *
unnamed	1	8"		Rod	1899-1910	at Fort Popham	
Hardman	1	6" 1905		D 1903	1908-1917		
Hawley	2	6" 1900		P 1900	1908-1924	modified for 155 mm gun emplacements, WWII	
Cogan	2	3" 1903		P 1903	1908-1924		
unnamed	4	155 mm		PM		two empl. on Hawley	

The Harbor Defenses of Portland

Long Island Military Reservation					private		* *
unnamed	2	3" 1903?		P 1903?	1942-1943	gunblocks covered	
AMTB 965	2	90 mm M1		F M3	1943-1946	one gun block built over	
AMTB 966	2	90 mm M1		F M3	1943-1946		
			Other locations				* *
unnamed	2	3" 1903?		P 1903?	1942-1944	Great Chebeague Island	
AMTB 969	2	90 mm M1		F M3	1943-1946	Great Chebeague Island, buried	
AMTB 970	2	90 mm M1		F M3	1943-1946	Bailey Island, one gun block covered	
FORT LYON			Cow Island 1896		private		* *
Bayard	3	6" 1903		D 1903	1907-1917		
Abbot	3	3" 1903		P 1903	1909-1946		
FORT McKINLEY			Great Diamond Island 1896		private	MF, MC	* * * * *
Ingalls	8	12" 1890MI	M 1896MI		1904-1942		
Berry	2	12" 1888MI	D 1896		1901-1943		
Thompson	3	8" 1888MII	D 1896		1902-1942		
Weymouth	3	8" 1888MII	D 1896		1901-1942		
Honeycutt	2	8" 1888MII	D 1896		1901-1942		
Acker	2	6" 1897MI	D 1898		1902-1943		
Carpenter	2	6" 1900	P 1900		1906-1947		
Farry	2	3" 1898	MP 1898MI		1902-1920		
Ramsey	2	3" 1898	MP 1898MI		1902-1920		
Peak's Island Military Reservation					private	MC	* *
Steele (#102)	2	16" MkIIMI	CB M4/M5		1945-1948		
Cravens (#203)	2	6" 1905A2	SB M1		1945-1948	built on	
unnamed	2	3" 1903?	P 1903?		1942-1943	one gun block covered	
AMTB 963	2	90 mm M1	F M3		1943-1946		
AMTB 964	2	90 mm M1	F M3		1943-1946	buried/destroyed	
Jewell Island Military Reservation					state prop		* *
#202	2	6" M1	SB M3		1944 NC		
AMTB 967	2	90 mm M1	F M3		1943-1946		
AMTB 968	2	90 mm M1	F M3		1943-1946		
FORT LEVETT			Cushing Island 1903		private		* * * *
Foote	2	12" 1895MI	B 1917		1924-1948	casemated-WWII	
Bowdoin	3	12" 1895	D 1897		1903-1943		
Kendrick	2	10" 1895	D 1896		1903-1942		
Ferguson	2	6" 1900	P 1900		1906-1947		
Daniels	3	3" 1898	MP 1898MI		1903-1920		
AMTB 962	2	90 mm M1	F M3		1943-1946		
FORT PREBLE			Spring Point 1808		state community college		* * *
Kearny	8	12" 1890MI	M 1896MI		1901-1942	2 mortars moved to West Point 1911, buried	
Chase	8	12" 1890MI	M 1896MI		1901-1942	buried	
Rivardi	2	6" 1903	D 1903		1906-1918		
Mason	1	3" 1902MI	P 1902		1906-1942	gun moved to new position 1942	
Mason II	1	3" 1902MI	P 1902		1942-1946	destroyed	
FORT WILLIAMS			Cape Cottage 1898		city park	MF, MC	* * *
Blair	2	12" 1895	D 1897		1903-1943	partially buried	
Sullivan	1	10" 1888MII	D 1896		1898-1938	buried	
&	2	10" 1888MI	D 1894			buried	
DeHart	2	10" 1888MI	D 1894MI		1898-1942	1 gun M1888MII, buried	
Garesche	2	6" 1900	D 1903		1906-1917	guns & carr. orig on display St. Louis Expo. 1903	
Hobart	1	6" Armstr	Armstrong		1900-1913		
Keyes	2	3" 1902MI	P 1902		1905-1946		
AMTB 961	2	90 mm M1	F M3		1943-1946	one gun block covered	

Cape Elizabeth Military Reservation				state park		**
[#101]	2	16"	CB	*Not Built*	location not in park but nearby	
#201	2	6"	SB M4	NA		
Biddleford Military Reservation				private?		*
unnamed	4	155 mm	PM			

The Harbor Defenses of Portsmouth

FORT FOSTER	Gerrish Island	1900		Kittery ME town park		**
Bohlen	3	10" 1895	D 1896	1901-1943	buried to loading platform	
Chapin tactical #5	2	3" 1902MI	P 1902	1904-1943	sod and sand cover removed	
#205 tactical #6	2	6"	SB M4	1944 NC		
AMTB 952	2	90 mm M1	F M3	1943-1946	one block in floor of shelter	
FORT CONSTITUTION	Newcastle Is.	1798		NH state hist. site & USCG	MF, MC	*
Farnsworth	2	8" 1888MII	DC 1894	1898-1917		
Hackleman tactical #4	2	3" 1903	P 1903	1904-1942	destroyed, guns to H.G. Wright	
FORT STARK	Jerry's Point	1902		state hist. site	MC	**
Hunter	2	12" 1895MI	D 1897	1904-1945		
unnamed	2	8"	Rod	1898-1900	buried	
Kirk	2	6" 1903	D 1903	1904-1917	modified for HECP-HDCP	
Hays	2	3" 1902MI	P 1902	1905-1942		
Lytle	2	3" 1902MI	P 1902	1905-1942		
Lytle II tactical #3	2	3" 1902MI	P 1902	1942-1945	guns from Lytle	
FORT DEARBORN	Odiorne Point, Rye NH	1942-1948		Odiorne State Park		**
Seaman (#103) tact #2	2	16" MkIIMI	CB M5	1942-1948		
#204 tactical #1	2	6" T2	SB M3	1943-1948		
unnamed	4	155 mm	PM	1942-?	1 mount covered	
AMTB 951	2	90 mm M1	F M3	1943-1946	Pulpit Rock, one gun block dest.	
Salisbury Beach Military Reservation				NH state park		*
unnamed	4	155 mm	PM		1 mount buried	

The Harbor Defenses of Boston

East Point Military Reservation				municipal park, Northeastern Univ. Marine Lab		**
Murphy (#104)	2	16" MkIIMI	CB M4	1944-1948		
#206	2	6" 1903A2	SB M1	1943-1948		
unnamed	2	155 mm	PM		buried	
FORT RUCKMAN	Nahant			municipal park		**
Gardner	2	12" 1895MI	B 1917	1924-1946	casemated- WWII, partially buried	
FORT BANKS	Winthrop 1899			municipal park/private		*
Lincoln	8	12" 1890MI	M 1896MI	1896-1943	magazines covered, M1890 mortars?	
Kellogg	8	12" 1890MI	M 1896MI	1896-1943	magazines covered, 1 pit buried	
FORT HEATH	Grover's cliff			private & municipal park		X
Winthrop	3	12" 1888	D 1896	1901-1945	destroyed	
AMTB 945	2	90 mm M1	F M3	1943-1946	buried	
FORT DAWES	Deer Island			Massachusetts Water Resources Authority	MF, MC	X
#105	2	16"	CB	1944 NC	(guns were on site) now destroyed	
#207	2	6"	SB M4	1944 NA	destroyed	
Taylor	2	3" 1902MI	P 1902	1942-1944	from Strong, destroyed	
AMTB 944	2	90 mm M1	F M3	1943-1946	destroyed	
FORT WARREN	Georges Island			Metropolitan District Commission park	MF, MC	****
Stevenson	2	12" 1895	D 1897	1903-1945		
Bartlett	4	10" 1888MII	D 1894	1899-1942	1 emplacement destroyed	
Adams	1	10" 1888	D 1894	1899-1914		
Plunkett	2	4" 1896 DS	P 1896 DS	1899-1925	guns later moved to saluting position	
Lowell	3	3" 1898	MP1898MI	1900-1920		
FORT STANDISH	Lovell's Island			Metropolitan District Commission park		**
Morris	2	10" 1900	D 1901	1907-1942		
Burbeck	2	10" 1900	D 1901	1907-1942		
Terrill	3	6" 1897MI	D 1898	1902-1943	erosion encroachment	
Whipple	2	6" 1900	P 1900	1904-1947	modified, WWII	
Vincent	4	3" 1898	MP1898MI	1904-1920	#1 & #4 mod. to 3" AA- WWII, partially buried	
Weir	2	3" 1902MI	P 1902	1906-1926	destroyed	
Williams	3	3" 1903	P 1903	1904-1946		
AMTB 943	2	90 mm M1	F M3	1943-1946		

FORT STRONG	Long Island 1870			city hospital	MF, MC	* *
Hitchcock	2	10" 1888MI	D 1894	1899-1939		
&	1	10" 1888MI	D 1896			
Ward	2	10" 1888MI	D 1894	1899-1939		
Drum	2	4.7" Armstr	Armstrong	1899-1917		
Smyth	2	3" 1902	P 1902	1906-1921	guns to Basinger	
Stevens	2	3" 1902	P 1902	1906-1946		
Taylor	2	3" 1902	P 1902	1906-1942	guns to Ft. Dawes, WWII	
Basinger	2	3" 1898	MP 1898MI	1901-1947	2-3" P 1902 guns from Smyth after 1921	
FORT ANDREWS	Peddocks Island			Metropolitan District Commission park		* * *
Cushing	8	12" 1890MI	M 1896MI	1904-1942		
Whitman	8	12" 1890MI	M 1896MI	1904-1942	4 Mor. rem. 1910, repl with M1908 mortars	
McCook	2	6" 1900	P 1900	1904-1947	rebuilt WWII	
Rice	2	5" 1900	P 1903	1904-1917	CRF on #1	
Bumpus	2	3" 1902	P 1902	1904-1946		
FORT DUVALL	Hog Island 1920-1944			private (Renamed Spinnaker Is)		*
Long	2	16" 1919	B 1919	1927-1948	casemated- WWII, built on	
FORT REVERE	Telegraph Hill, Hull			Metropolitan District Commission , private		*
Ripley	2	12" 1888MI	B 1892	1901-1943	buried, built on	
Field	2	5" 1897MI	BP 1896	1901-1917	part. dest. and buried	
Sanders	4	6" 1903	D 1903	1906-1917	2 guns rem.	
Pope	2	6" 1903	D 1903	1906-1943	later Pope inc. 3rd gun from Sanders	
AMTB 941	2	90 mm M1	F M3	1943-1946	dest./buried	
	Outer Brewster Island			Massachuseltts Dept. Environmental Management		* *
Jewell (#209)	2	6" M1	SB M3	1943-1948		
	Great Brewster Island			Massachuseltts Dept. Environmental Management		* *
AMTB 942	2	90 mm M1	F M3	1943-1946		
Fourth Cliff Military Reservation Scituate				Air Force recreation area		* * *
[#106]	2	16"	CB	*Not Built*	no final location (Flowers Hill)	
#208	2	6" T2	SB M4	1944-1948	Generators still in place	
Sagamore Hill Military Reservation						*
unnamed	2	155 mm	PM			
Buzzards Bay						X
AMTBs	8	90 mm	F M3		planned, not built?	

The Harbor Defenses of New Bedford

FORT RODMAN	Clarke Point 1840-(1947)			waste water treatment plant/ city park	MF, MC	* *
Milliken	2	12" 1895MI	B 1917	1924-1946	casemated-WWII	
Walcott	1	8" 1888MII	D 1896	1899-1942		
Barton	1	8" 1888MII	D 1896	1899-1942		
Cross	2	5" 1900	P 1903	1902-1920		
Craig	2	3" 1898	MP 1898MI	1902-1920	#1 mod. to 3-in AA, WW II	
Gaston	2	3" 1898	MP 1898MI	1902-1920		
unnamed	2	155 mm	PM			
Mishaum Point Military Reservation						*
#210	2	6" M1	SB M3	1945-1947	private, built on	
unnamed	2	155 mm	PM		covered	
	Other locations					*
AMTB 931	2	90 mm M1	F M3	1943-1946	Barney's Joy Point	
AMTB 932	2	90 mm M1	F M3	1943-1946	Cuttyhunk Island	
AMTB 933	2	90 mm M1	F M3	1943-1946	Nashawena Island	
AMTB 934	2	90 mm M1	F M3	1943-1946	Butler's Point	
unnamed	2	155 mm	PM		Butler's Point	

The Harbor Defenses of Narragansett Bay

FORT CHURCH	Sakonnet Point (3 sections)			private club/private/private		*
Gray (#107)	2	16" MkIIMI	CB M3	1942-1948	early design, West Res. partially buried, Sakonnet Golf Club	
Reilly	2	8" Mk3A2	CB M1	1942-1947	East Res., completely buried	
#212	2	6" 1903A2	SB M1	1943-1948	South Res., Warren Point, private, built on	
unnamed	2	155 mm	PM			

FORT ADAMS	Newport 1824			state park, Navy housing	MF, MC	* * * *
Greene	8	12" 1890MI	M 1896MI	1898-1943	renamed Gilmore in 1940	
Edgerton	8	12" 1890MI	M 1896MI	1898-1943	One M1890 mortar	
Reilly	2	10" 1888MII	D 1896	1899-1917		
unnamed	1	8"	Rod	1898-		
Bankhead	3	6" Armstr	Armstrong	1907-1913	guns from Greble, Screven, Moultrie	
Talbot	2	4.7" Armstr	Armstrong	1899-1917	one 4.7 gun is in Equality Park, Newport	
Belton	2	3" 1903	P 1903	1907-1925		
FORT WETHERILL	Conanicut Is. 1900-1945			state park	MF, MC	* *
Varnum	2	12" 1888	B 1892	1901-1943		
Wheaton	2	12" 1900	D 1901	1908-1945		
Walbach	3	10" 1888MII	D 1896	1908-1942		
Zook	3	6" 1903	D 1903	1908-1918		
Dickenson	2	6" 1900	P 1900	1908-1947	modified WW II	
Crittenden	2	3" 1902	P 1902	1908-1946		
Cooke	2	3" 1898	MP 1898MI	1901-1920		
AMTB 923	2	90 mm M1	F M3	194? NC	modified old 3-in AA blocks	
FORT GETTY	Conanicut Island 1903			Jamestown city park		*
Tousard	3	12" 1900	D 1901	1910-1942	partially buried	
House	2	6" 1900	P 1900	1910-1942	guns & carr. & name to Ft. Varnum, now mostly buried	
Whiting	2	3" 1903	P 1903	1910-1942	guns & carr. & name to Ft. Burnside	
Armistead (3)	2	3" 1903	P 1903	1943	Whiting empl., guns & carr from Varnum	
AMTB 922	2	90 mm M1	F M3	1943-1946	#1 block buried	
FORT BURNSIDE	Beavertail 1940-1948			state park		* * *
#110	2	16"	CB	*Not Built*		
#213	2	6" 1905A2	SB M1	1943-1948		
Whiting (2)	2	3" 1903	P 1903	1942-1946		
FORT GREBLE	Dutch Island 1863			state land/park	MF, MC	* * *
Sedgwick	8	12" 1890MI	M 1896MI	1901-1942		
Hale	3	10" 1898MII	D 1896	1898-1942		
Mitchell	3	6" 1903	P 1903	1905-1917		
	1	6" Armstr	Armstrong	1898 NC	dest. replaced by new 6" battery, gun to Ft. Adams	
Ogden	2	3" 1898	MP 1898MI	1900-1920		
FORT KEARNY	Saunderstown 1904			state university campus		* *
French	4	6" 1905	D 1903	1908-1917	research facility built on middle of battery	
Cram	2	6" 1905	D 1903	1908-1943	buildings in emplacements	
Armistead	2	3" 1903	P 1903	1908-1942	guns & carr & name to Ft. Varnum, now built on	
FORT VARNUM	Boston Neck 1941			National Guard		* * * *
House	2	6" 1900	P 1900	1942-1947	guns from Ft. Getty	
Armistead (2)	2	3" 1903	P 1903	1942-1943	guns from Ft. Kearny, then sent to Getty	
AMTB 921	2	90 mm M1	F M3	1943-1946		
FORT GREENE	Point Judith (3 sections) 1941					* *
Hamilton (#108)	2	16" MkIIMI	CB M4	1943-1948	early design, East Res., National Guard/Reserve	
#109	2	16"	CB M4/M5	1944 NC	West Res., guns were on site, state park	
#211	2	6" 1903A2	SB M1	1945-1948	South Res., city park, entrances buried, erosion threat	
unnamed	4	155 mm	PM		mostly buried	
	Other locations					X
unnamed?	2	4.7" Armstr	Armstrong	1917-1919	Sachuest Point?	
AMTB 924		90 mm	F	*Not Built?*		
AMTB 925	2	90 mm M1	F M3	1943?	Brenton Point, moved to Ft. Wetherill?	
unnamed	4	155 mm	PM		Brenton Point, 3 mounts destroyed	

The Harbor Defenses of Long Island Sound

FORT MANSFIELD	Napatree Point, R.I. 1900			state park		* *
[#114]	2	16"	CB	*Not Built*	near Watch Hill, original site at Ft. Terry	
Wooster	2	8" 1888	D 1896	1901-1917		
Crawford	2	5" 1897	BP 1896	1901-1917		
Connell	2	5" 1900	P 1903	1901-1917	mostly destroyed, in surf	
	other Connecticut sites					X
AMTB 915	2	90 mm M1	F M3	1943-1946	Pine Island, state land	
AMTB 914	2	90 mm M1	F M3	1943-1946	Goshen Point, covered, Harkness State Park	
unnamed	4	155 mm	PM		Oak's Inn Military Reservation	

FORT H.G. WRIGHT	Fisher's Island 1899		private	MF, MC		* * *
Dynamite	1	15"	Pne	?	Race Point	
Clinton	8	12" 1890MI	M 1896MI	1902-1943	partially buried	
Butterfield	2	12" 1895	D 1897	1901-1945	partially. buried	
Barlow	2	10" 1888MII	D 1896	1901-1939	partially buried	
Dutton	3	6" 1897MI	D 1898	1901-1945	partially buried	
Hamilton	2	6" 1903	D 1903	1905-1917		
Marcy	2	6" 1903	D 1903	1906-1917		
#215	2	6" 1903A2	SB M1	1943-1946	Race Point	
Hoffman	2	3" 1902	P 1902	1906-1946		
Hoppock	2	3" 1903	P 1903	1905-1946	guns replaced once	
AMTB 913	2	90 mm M1	F M3	1943-1946	buried	
	Fisher's Island (other sites)		private			* *
#111	2	16"	CB M4/M5	194? NC	guns were on site, Wilderness Pt. (Mt. Prospect)	
#214	2	6"	SB M4	194? NC	Wilderness Point, built on	
unnamed	4	5"	P + BP	1917-1919	NB?, North Hill	
New Hackleman	2	3" 1903	P 1903	1944-1946	North Hill	
AMTB 916	2	90 mm M1	F M3	*Not Built?*	East Point	
FORT MICHIE	Great Gull Island 1896-1948		bird sanctuary	MC		* * *
J.M.K. Davis	1	16" 1919	D 1917	1923-1945	ARF	
Palmer	2	12" 1895	D 1897	1900-1945		
North	2	10" 1888MII	D 1896	1900-1918	destroyed for Davis	
Benjamin	2	6" 1900	P 1900	1908-1947		
Maitland	2	6" 1900	P 1900	1908-1947		
Pasco	2	3" 1903	P 1903	1905-1934		
AMTB 912	2	90mm M1	F M3	1943-1946		
FORT TERRY	Plum Island		USDA Animal Disease Lab	MF, MC		* * * *
Stoneman	8	12" 1890MI	M 1890MI	1901-1943		
Steele	2	10" 1888MII	D 1896	1900-1942		
Bradford	2	6" 1897MI	D 1898	1901-1944		
Floyd	2	6" 1903	D 1903	1906-1917		
Dimick	2	6" 1903	D 1903	1905-1917		
#217	2	6"	SB M4	194? NC		
Kelly	1	5" 1900	P 1903	1900-1917	partially buried	
&	1	4.7" Armstr	Armstrong		later replaced by 5" P 1903	
Hagner	2	3" 1903	P 1903	1906-1932	mostly destroyed	
Eldridge	2	3" 1903	P 1903	1906-1946		
Dalliba	2	3" 1903	P 1903	1905-1946	broken up in surf	
Greble	2	3" 1902	P 1902	1905-1932		
Campbell	2	3" 1903	P 1903	1905-1934	partially destroyed	
AMTB 911	2	90 mm M1	F M3	1943-1946	Plum Island	
unnamed	4	155 mm	PM			
FORT TYLER	Gardiner's Point Island		former Navy target range			*
unnamed	2	8"	Rod	1898 NA	entire fort partially destroyed	
Edmund Smith	2	8" 1888	D 1894	1898 NA		
&	2	5"	P	1898 NA		
CAMP HERO	Montauk Point 1929		FAA radar sta., state park			*
Dunn (#113)	2	16" MkIIMI	CB M4	1944-1948	entrances covered	
#112	2	16" MkIIMI	CB M4	1944-1948	entrances covered	
#216	2	6" 1903A2	SB M1	1944-1947	entrances covered	

The Harbor Defenses of Eastern New York

FORT SLOCUM	Davids Island 1863-(1928)		city property, leveled for redevelopment		X?
Haskin	8	12" 1886	M 1891	1897-1919	
Overton	8	12" 1886	M 1891	1897-1919	Pit A destroyed
"Practice"	2	8"	Rod	1896-1899	1 orig. 15" SB, partially buried
Kinney	2	6" 1900	P 1900	1904-1917	destroyed, guns to Tilden
Fraser	2	5" 1900	P 1903	1901-1917	destroyed

FORT SCHUYLER	Throg's Neck 1833-(1928)		New York State Maritime Acadamy	MC	*
Gansevoort	1	12" 1888MI D 1896	1899-1935	1 empl. destr., 1 empl. buried	
&	1	12" 1895 D1897			
Hazzard	2	10" 1888MI D 1896	1898-1930	destroyed	
Bell	2	5" 1897 BP 1896	1900-1917	destroyed	
Beecher	2	3" 1898 MP 1898MI	1900-1920	destroyed	
FORT TOTTEN	Willet's Point 1862-(1928)		city park	MF, MC	* * * * *
King	8	12" 1890MI M 1896MI	1900-1935	buried	
Mahan	2	12" 1895 D 1897	1900-1918		
Graham	2	10" 1888 D 1894MI	1897-1918		
Sumner	1	8" 1888MI D 1894	1899-1918		
&	1	8" 1888MII D 1896			
Stuart	2	5" 1897 BP 1896	1899-1917		
Baker	4	3" 1898 MP 1898MI	1900-194?	2 P M1903 replaced during 1920s?	
Burnes	2	3" P	1904-1946?		

The Harbor Defenses of Southern New York

FORT TILDEN	Rockaway 1926-1945		Gateway National Recreation Area		* * *
Harris	2	16" 1919 B 1919	1924-1948	casemated WWII	
unnamed	4	12" 1896MI M 1896MI	1917-1919	at NAS Rockaway	
Kessler (West)	2	6" 1900 P 1900	1917-1947	from Kinney, Slocum, rebuilt 1941-42	
Ferguson (East)	2	6" 1900 P 1900	1917-1942	from Burke, Hamilton, b uried	
#220	2	6" SB M1	1942 NA	part. buried	
FORT HAMILTON	East Narrows, Brooklyn 1831		Army Post		X
Piper	8	12" 1890MI M 1896MI	1901-1942	(all Hamilton emplacements have been-	
Brown	2	12" 1895 D 1897	1902-	destroyed or buried)	
Doubleday	2	12" 1895 D 1897	1900-1943		
Neary	2	12" 1888MII B 1892	1900-1937		
Gillmore	4	10" 1888/95 D 1896	1899-1942	originally 7 empl., 3 later designated Spear	
Spear	3	10" 1888 D 1894	1898-1917		
Burke	4	6" 1900 P 1900	1903-1917?	2 guns to Tilden1917	
Livingston	2	6" 1903 D 1903	1905-1907	guns to Btty Schofield at West Point, repl with 05 tubes 1919	
&	2	6" 1900 P 1900	1905-1947	gun tubes to Btty Peck, Ft Hancock, 1948	
Johnston	2	6" 1897/MI D 1898	1902-1943	one gun M1897, one gun M1897MI	
Mendenhall	4	6" 1903 D 1903	1905-1917		
Griffin	2	4.7" Armstr Armstrong	1899-1913		
&	2	3" 1898 MP 1898MI	1902-1920		
&	2	3" 1903 P 1903	1903-1946		
FORT WADSWORTH	West Narrows, Staten Is. 1821		Gateway NRA, Federal Empl. Housing MF, MC		* * * *
[#115]	2	16" CB	*Not Built*		
Ayres	2	12" 1895 D 1897	1901-1942	partially buried	
Dix	2	12" 1900 D 1901	1902-1944		
Hudson	2	12" 1888MII D 1896	1899-1944	partially buried	
Richmond	2	12" 1888 D 1896	1899-1942	partially buried	
Barry	2	10" 1888 D 1896	1899-1918		
Upton	2	10" 1888 D 1896	1899-1925	partially buried	
unnamed	2	8" Rod	1898		
Duane	5	8" 1888/MII D 1894	1897-1915	magazines removed, 1 empl. destroyed	
Barbour	2	6" Armstr Armstrong	1898-1920	1 empl. buried	
&	2	4.7" Armstr Armstrong	1898-1920	1 empl. buried, 1 partially buried	
Mills	2	6" 1897MI D 1898	1900-1943	both emplacements partially buried	
#218	2	6" SB M1	1943 NC	-1958?	
Catlin	6	3" 1903 P 1903	1904-1942		
Bacon	2	3" 1898 MP 1898MI	1899-1918		
Turnbull	6	3" 1902 P 1902	1903-1944	later 2 guns removed, partially buried	
New Turnbull (#14)	4	3" 1902 P 1902	1942-1946	covered	

FORT HANCOCK		Sandy Hook, N.J. 1885		Gateway National Recreation Area, USCG	MF, MC * * * * *
Proving Ground	many - various		various	1890-1920	Proving Ground moved to Aberdeen, 1920
Dynamite	2	15"	Pne	1896-1902	USCG shooting range
&	1	8"	Pne	1896-1902	mine casemate 1921
Potter	2	12" 1888	GLC 1891	1890-1907	
McCook	8	12" 1886	M 1891	1898-1923	advanced HECP-HEDP post 1943
Reynolds	8	12" 1886	M 1891	1898-1918	
Alexander	2	12" 1888/MII	D 1896	1899-1943	
Bloomfield	2	12" 1888M1	D 1896	1899-1944	1 gun M1888MI-1/2
Richardson	2	12" 1900	D 1901	1904-1944	
Kingman	2	12" 1895MI	B 1917	1922-1946	casemated 1943
Mills	2	12" 1895MI	B 1917	1922-1946	casemated 1943
Halleck	3	10" 1888/MII	D 1896	1898-1944	one gun removed 1910s
Granger	2	10" 1888	D 1896	1898-1940s	
Arrowsmith	3	8" 1888MI	D 1894	1909-1921	partially destroyed
Peck	2	6" 1900	P 1900	1903-1943	guns and carr. relocated '43 to Gunnison
Gunnison	2	6" 1903	D 1903	1905-1943	extensively rebuilt for Peck's guns
Peck "II" /New Peck	2	6" 1903	D 1903	1943-1964	(ex-Gunnison), guns returned 1976 Still Emplaced restored
Engle	1	5" 1897	BP 1896	1898-1917	CRF built on empl.
unnamed	1	4.7"	P	1898	temporary (Schneider)
Urmston	4	3" 1898	MP 1898MI	1903-1920	
&	2	3" 1903	P 1903	1904-1946	built between original two emplacement sets
Morris	4	3" 1903	P 1903	1904-1946	
New Urmston (#6)	2	3" 1903	P 1903	1942-1946	partially covered
AMTB #7	2	90 mm M1	F M3	1943-1946	
AMTB #8	2	90 mm M1	F M3	1943-1946	on ex-Peck
Highlands Military Reservation, New Jersey (Navesink)		Hartshorne Woods Monmouth County Park			* *
Lewis (#116)	2	16" MkIIM1	CB M4	1944-1948	renovated 2018, 16" navy MkVII barrel in emplacement
unnamed	4	12" 1890MI	M 1896MI	1917-1920	only one empl. remains uncovered
#219	2	6" 1903A2	SB M1	1944-1949	
		other sites			* ?
[#117]	2	16"	CB	*Not Built*	Nigger Pt. (now JFK Airport)
#20	2	3" 1902	P 1902	1942-1946	Rockaway Point
New Catlin (#18)	4	3" 1903	P 1903	1942-1946	Norton Point, buried
AMTB	2	90 mm M1	F M3	1943-1946	Rockaway Point
AMTB #19	2	90 mm M1	F M3	1943-1946	Norton Point, buried
AMTB	2	90 mm M1	F M3	1943-1946	Miller Field
AMTB #12	2	90 mm M1	F M3	1943-1946	Swinburne Island

The Harbor Defenses of the Delaware

FORT MOTT		Finns Point, N.J. 1900		state park	* * * *
Arnold	3	12" 1888MII	D 1896	1899-1943	
Harker	3	10" 1888	D 1894	1899-1941	guns and carr to Canada, barrels remain in Canada
Gregg	2	5" 1900	P 1903	1901-1910	
Krayenbuhl	2	5" 1897	BP 1896	1900-1918	
Edwards	2	3" 1900 DS	CM 1902	1902-1920	
FORT DELAWARE		Pea Patch Island 1847		state park	MC * * *
Torbert	3	12" 1895	D 1896	1901-194?	guns to Battery Reed, Puerto Rico
Dodd	2	4.7" Armstr	Armstrong	1899-1918	
Hentig	2	3" 1903	P 1903	1901-1942	
Alburtis	2	3" 1898	MP 1898MI	1901-1920	
Allen	2	3" 1898	MP 1898MI	1901-1920	
FORT DuPONT		Delaware City, Del. 1896		assorted public agencies, state park	MF, MC * * * *
Rodney	8	12" 1890MI	M 1896MI	1900-1941	
Best	8	12" 1890MI	M 1896MI	1900-1941	
Read	2	12" 1888	B 1892	1899-1918	emplacements split, Btty Gibson located between
Gibson	2	8" 1888	D 1896	1899-1917	one structure with Read
Ritchie	2	5" 1900	P 1903	1900-1917	destroyed
Elder	2	3" 1903	P 1903	1904-1942	relocated 1922 to Delaware Beach
Elder II	2	3" 1903	P 1903	1942-1942	Liston front range light
FORT SAULSBURY		Slaughter Beach, Del. 1902		private, Wildlife Refuge	* *
Hall	2	12" 1895MI	B 1917	1918-1945	
Haslet	2	12" 1895MI	B 1917	1918-1942	guns to Ft. Miles

Cape May M. R	Cape May, NJ			state park		*
#223	2	6" M1	SB M3	1944-1947	now in surf	
AMTB #7	2	90 mm M1	F M3	1943-1946	now in surf	
unnamed	4	155 mm	PM		in surf	
		Other locations				X
temporary	1	6"	P	1917-1919	C.G. base east of town, destroyed	
FORT MILES	Cape Henlopen 1940			state park	MF, MC	***
Smith (#118)	2	16" MkIIMI	CB M4	1943-1948	early design	
[#119]	2	16"	CB	*Not Built*		
#519	2	12" 1895MI	CB 1917	1944-1948	guns from Haslet (replaced #119), Museum restoration 12 in navy barrel in casemate 2010	
unnamed	4	8"	RY		covered/buried	
unnamed	4	8"	RY		covered/buried	
temporary	1	6" 1900	P 1900	1917-1918	Cape Henlopen M.R. from Ft. St. Phillip	
Herring (#221)	2	6" 1903A2	SB M1	1944-1948	earthen cover removed	
Hunter (#222)	2	6" 1903A2	SB M1	1943-1947		
Exam	4	3" 1903	P 1903	1942-1946	mostly buried	
AMTB #5A	2	90 mm M1	F M3	1943-1946	mostly buried	
AMTB #5B	2	90 mm M1	F M3	1943-1946	1 block covered by parking lot	
unnamed	4	155 mm	PM			
					one 16" navy MkVII barrel on display 2012	

The Harbor Defenses of Baltimore

FORT HOWARD	North Point 1900-(1928)			VA hospital, county park	MF, MC	****
Key	8	12" 1890MI	M 1896MI	1900-1927		
Stricker	2	12" 1888	D 1896	1899-1918		
Nicholson	2	6" 1897	D 1898	1900-1927		
Harris	2	5"1897	BP 1896	1900-1917		
Clagett	2	3" 1898	MP 1898MI	1901-1920		
Lazear	2	3" 1898	MP 1898MI	1900-1920	destroyed	
FORT CARROLL	Soller's Point Flats 1848-(1928)			private		**
Towson	2	12" 1888	B 1892	1900-1918		
Heart	2	5" 1897	BP 1896	1900-1917		
Augustin	2	3" 1898	MP 1898MI	1900-1920		
FORT ARMISTEAD	Hawkins Point 1896-(1928)			city park	MC	**
Winchester	1	12" 1888	D 1896	1900-1918		
McFarland	3	8" 1888	D 1894	1900-1917		
Irons	2	4.7"Armstr	Armstrong	1900-1913		
Mudge	2	3" 1898	MP 1898MI	1901-1920	partially destroyed	
FORT SMALLWOOD	Rock Point 1890-(1928)			city park		*
Hartshorne	2	6" 1897	D 1898	1900-1927		
Sykes	2	3" 1902	P 1902	1905-1927	destroyed	

The Harbor Defenses of the Potomac

FORT WASHINGTON	east bank, MD 1814-(1928) 1947			Capital National Park Area	MF, MC	***
Meigs	8	12" 1890MI	M1896MI	1902-1914		
Decatur	2	10" 1888	D 1894	1899-1918		
Emory	2	10" 1888	D 1894	1898-1929		
Humphreys	2	10" 1888MII	D 1896	1899-1929		
Water	1	10" 1888	B 1893	1898-1898		
Wilkin	2	6" 1897	D 1898	1902-1928		
White	2	4" Dri-Schr	P Crampton	1899-1921		
Smith	2	3" 1898	MP 1898MI	1903-1920		
Many	2	3" 1902	P 1902	1905-1928		
FORT HUNT	west bank, VA 1899-(1928)1933			Capital National Park Area		**
Mount Vernon	3	8" 1888MI	D 1894	1898-1917		
Porter	1	5" 1897	BP 1896	1901-1917		
Robinson	1	5" 1897	BP 1896	1901-1917		
Sater	3	3" 1898	MP 1898MI	1904-1920		

The Harbor Defenses of Chesapeake Bay

FORT MONROE   Old Point Comfort 1819 (C.A. School 1907-1948) NPS State Park-Property MF, MC   * * * *

Battery	#	Caliber	Mount	Years	Notes
[#124]	2	16"	CB	*Not Built*	
Anderson	8	12" 1890MI	M 1896MI	1898-1943	two mortar tubes were M1890
Ruggles	8	12" 1890MI	M 1896MI	1898-1943	
DeRussy	3	12" 1895	D 1897	1904-1944	
Parrott	2	12" 1900	D 1901	1906-1943	emplacements used for AMTB #23
Humphreys	1	10" 1888	D 1894	1897-1910	destroyed
Eustis	2	10" 1888	D 1896	1901-1942	destroyed
Church	2	10" 1888	D 1896	1901-1942	
Bomford	2	10" 1888	D 1894	1897-1940	destroyed
N.E. bastion	1	10" 1888	D 1894	1900-1908	
Barber	1	8" 1888	B 1892	1898-1915	destroyed
Parapet	4	8" 1888	B 1892	1898-1915	mostly buried
Water	1	8" 1888	B 1892	1897-1898	
Montgomery	2	6" 1900	P 1900	1904-1947	guns removed '20s, guns repl. '41, destroyed
Gatewood	4	4.7" Armstr	Armstrong	1898-1914	50 cal. mostly buried
Irwin	4	3" 1898	MP 1898MI	1903-curr.	all Gs rem. '20s, 2-3" P 1902 repl. '46 <-Lee/Wool, SE
AMTB #23	2	90 mm M1	F M3	1943-1946	1 gun (from Fisherman's Is) Still emplaced

FORT WOOL   Rip Raps, Hampton Roads 1830 Bird Sanctuary: No Public Access   * *

Battery	#	Caliber	Mount	Years	Notes
Claiborne	2	6" 1903	D 1903	1908-1918	
Dyer	2	6" 1903	D 1903	1908-1917	
Gates	2	6" 1903	D 1903	1908-1942	converted to #229
Gates (#229)	2	6"	SB M4	1944 NA	
Lee	4	3" 1902	P 1902	1905-1946	2 guns to Ft. J. Custis, 2 guns to Irwin, Ft. Monroe
Hindman	2	3" 1902	P 1902	1905-1946	

FORT JOHN CUSTIS   Cape Charles 1940   Fish and Wildlife Reserve   * *

Battery	#	Caliber	Mount	Years	Notes
Winslow (#122)	2	16" MkIIMI	CB M4	1943-1948	one 16" navy MkVII barrel in casemate 2013
[#123]	2	16"	CB	*Not Built*	
Unnamed	2	8"	RY	1942-1943	
#228	2	6"	SB M4	1943 NC	buried

Fisherman Island   Fish and Wildlife Reserve   * *

Battery	#	Caliber	Mount	Years	Notes
#227	2	6" 1905A2	SB M1	1943-1965	guns to Smithsonian, Power equip. still in place
unnamed	4	5"	P	1917-1919	2-5"/2-6"?
Lee (AMTB #20)	2	3" 1902	P 1902	1942-1944	guns from Lee, Ft. Wool, partially buried
AMTB #24	2	90 mm M1	F M3	1943-1946	
unnamed	4?	155 mm	PM		

FORT STORY   Cape Henry 1917   Army post   MC (MF at Little Creek)   * * *

Battery	#	Caliber	Mount	Years	Notes
Pennington	2	16" 1920	H 1920	1922-1947	original name for all 4 guns
Walke	2	16" 1920	H 1920	1922-1947	named as separate battery in 1941, empl. covered
Ketcham (#120)	2	16" MkIIMI	CB M4	1943-1948	early design
#121	2	16" MkIIMI	CB M4	1943-1948	
Worcester (#224)	2	6" 1900	P 1900	1942-1947	prototype for the WW II 6-in construction
Cramer (#225)	2	6" 1903A2	SB T2(M2)	1943-1949	
#226	2	6" M1	SB M4	1943-1949	
unnamed	2	6" 1900	P 1900	1917-1919	broken up, in surf
unnamed	2	5" 1900	P 1903	1917-1919	broken up, in surf
Exam (AMTB #19)	2	3" 1902	P 1902	1942-1945	guns from Irwin, Ft. Monroe, buried
AMTB #21	2	90 mm M1	F M3	1942-	buried
AMTB #22	2	90 mm M1	F M3	1942-	awash
unnamed	4?	155 mm	PM		awash
unnamed	4?	155 mm	PM		

Temporary Harbor Defenses of Beaufort, NC   X

Battery	#	Caliber	Mount	Years	Notes
unnamed	2	6" Navy	Navy P	1942-1945	near Fort Macon, buried
Lookout	2	5" Navy	Navy P	1942-1945	Cape Lookout –in surf

The Harbor Defenses of Cape Fear River

FORT CASWELL		Oak Island	(-1928?)	church camp	MF, MC	* * * * *
Bagley	8	12" 1886	M 1891	1903-1925	1 gun replaced 1912	
Caswell	2	12" 1888	B 1892	1899-1925	modified for swimming pool	
Swift	3	8" 1888	D 1894	1898-1920	2 guns removed 1917	
&	1	8" 1888	D 1896			
New Madison	2	6" 1903	D 1903	1905-1917		
McDonough	1	5" 1897	BP 1896	1902-1905	gun to Shipp	
Shipp	2	5" 1897	BP 1896	1901-1919	originally 1 emplacement	
Madison	1	4.7" Armstr	Armstrong	1899-1904	guns to Screven	
McKavett	2	3" 1898	MP 1898MI	1903-1920		
New McDonough	2	3" 1902	P 1902	1904-1925		
unnamed	4	155mm	FM	1942-1944	Kure Beach, Temp. HD Wilmington	

The Harbor Defenses of Charleston

Marshall Military Reservation		Sullivan's Island	(-1947)	private		*
#520	2	12" 1895MI	CB 1917	1944-1947	guns from Kimble, Travis, modified for homes	
unnamed	4	155 mm	PM			
FORT MOULTRIE		Sullivan's Island 1776-(1947)		Fort Sumter N.H.M. & private	MF, MC	* * * *
[#125]	2	16"	CB	Not Built	James Island	
Capron	8	12" 1886	M 1891	1898-1942		
Butler	8	12" 1886	M 1891	1898-1942		
Jasper	4	10" 1888MI	D 1896	1898-1942	One gun M1888MII	
Thomson	2	10" 1900	D 1901	1906-1945	used by Fire Department	
Gadsen	4	6" 1903	D 1903	1906-1917	magazine area used as library	
Logan	1	6" Armstr	Armstrong	1898-1904	removed 1904	
&	1	6" 1897MI	D 1898	1906-1944		
#230	2	6"	SB M4	1943 NC		
Bingham	2	4.7" Armstr	Armstrong	1899-1918	1 gun replaced 1980 Still Emplaced	
McCorkle	2	3" 1898	MP 1898MI	1899-1943	1 gun replaced 1980 Still Emplaced	
Lord	2	3" 1902	P 1902	1899-1946	destroyed	
AMTB #2	2	90 mm M1	F M3	1943-1946	on Jasper's parapet	
FORT SUMTER		artifical island 1842-(1947)		Fort Sumpter National Historic Monument		* *
Huger	1	12" 1888MII	D 1896	1906-1943	built over	
&	1	12" 1888MII	B 1892	1906-1943		
AMTB #1	2	90 mm M1	F M3	1943-1946	covered	

The Harbor Defenses of Port Royal Sound

FORT FREMONT		St. Helena (now Oak?) Island	(-1928)	county nature preserve		* *
Jesup	3	10" 1888MII	D 1896	1899-1914	1 gun M1895	
Fornance	2	4.7" Armstr	Armstrong	1899-1913	built on	
Ft. Welles		Hilton Head Island				*
Dynamite	2	15"	Pne	1897 NA	partially destroyed	
unnamed	2	8"	Rod	1898	partially destroyed	

The Harbor Defenses of Savannah

FORT SCREVEN		Tybee Island 1897-(1928)		private	MF, MC	* * *
Habersham	8	12" 1890MI	M 1896MI	1900-1928		
Garland	1	12" 1888	B 1892	1899-1920	now a museum	
Fenwick	1	12" 1888	B 1892	1898-1920	home built on	
Brumby	4	8" 1888	D 1894	1899-1917	home built on (#3 intact)	
Backus	1	6" Armsrtr	Armstrong	1899-1920	later rearmed with 4.7-in	
&	2	4.7" Armstr	Armstrong	1899-1920	home built on	
Gantt	2	3" 1898	MP 1898MI	1904-1920	home built on 2003	
unnamed	4	155mm	FM	1942-1944	Temp. HD of Savannah	
Fort Pulaski				National Historic Monument		* *
unnamed	1	8" 1888	B 1892	1898-1899	located on demilune, + 2-15 in Rodmans	
Hambright	2	3"	MP	1903 NA		
		Other location				X
unnamed	2	4.7" Armstr	Armstrong	1898-1899	Wassaw Island, guns to Screven, parially dest. named Battery Henry Sims Morgan by GA in 2006	

The Harbor Defenses of Jacksonville * *

unnamed	2	8"	Rod	1898-1899	St. John's Bluff
unnamed	4	155mm	FM	1942-1944	Mayport

The Harbor Defenses on the St. Mary's River * *

unnamed	1	8"	Rod	1898-1900	Fort Clinch

Temporary Harbor Defenses of Fort Lauderdale ?

unnamed	4	155mm	FM	1942-1944	

Temporary Harbor Defenses of Miami Beach ?

unnamed	4	155mm	FM	1942-1944	

The Harbor Defenses of Key West

FORT TAYLOR   Key West 1846-(1947)   state park, Naval Reservation   *

Seminole	8	12" 1890MI	M 1896MI	1904-1943	
Osceola	2	12" 1888	B 1892	1900-1944	
DeLeon	4	10" 1888	D 1896	1904-1940	destroyed 1962
Covington	2	8" 1888	D 1894	1904-1917	destroyed 1962
DeKalb	2	6" 1900	P 1900	1906-1917	destroyed 1950s
#231	2	6"	SB M4	1942 NA	
Gardiner	2	4.7" Armstr	Armstrong	1898-1913	destroyed 1962
Ford	2	3" 1903	P 1903	1906-1946	destroyed 1964
Dilworth	2	3" 1898	MP 1898MI	1901-1920	destroyed 1970s
Adair	4	3" 1898	MP 1898MI	1901-1920	
AMTB #5	2	90mm M1	F M3	1943-1946	1 empl. on Adair
unnamed	4	155 mm	PM	1940-1942	2 guns to Miami Beach, destroyed

Other locations   *

#232	2	6" M1	SB M3	1944-1946	east of Martello Tower, Salt Ponds M.R. (airport)
Inman	2	3" 1903	P 1903	1906-1946	inside West Martello Tower
AMTB #6	2	90 mm M1	F M3	1943-1946	near West Martello Tower
unnamed	4	155mm	PM	1942-1942	West Martello Tower
unnamed	4	155mm	PM	1943-1944	near East Martello Tower

The Harbor Defenses of Tampa Bay

FORT DADE   Egmont Key 1899-(1928)   state park (1989)   MF, MC   *

unnamed	2	8"	Rod	1899-1900	guns to McIntosh, Mellon built over
McIntosh	2	8" 1888	D 1896	1900-1923	damaged by surf
Howard	2	6" 1903	D 1903	190?-1926	damaged by surf erosion
Burchsted	2	6" Armstr	Armstrong	1899-1919	guns & carr. to Ft. DeSoto 1980, broken up in the gulf
&	1	3" 1898	MP 1898MI	1904-1920	broken up in the gulf
Mellon	3	3" 1898	MP 1898MI	1904-1920	
Page	2	3" 1903	P 1903	1910-1919	broken up in the gulf

FORT DE SOTO   Mullet Key 1901-(1928)   County Park (1963   * *

Laidley	8	12" 1890MI	M 1896MI	1902-1921	4 M to Rosecrans 1918, 4 mortars Still Emplaced
Bigelow	2	3" 1898	MP 1898MI	1904-1920	broken up in the gulf
"modern display"					2 - 6" Armstrong guns from Ft. Dade, Burchsted
unnamed	4	155mm	FM	1942-1944	Passa Grill Beach, Temp. HD of Tampa

Other Florida sites   X

unnamed	2	155 mm	PM	1943-1944	Temp. HD of Panama City, FL
unnamed	2	155 mm	PM	NA	Port St.. Joe, FL

The Harbor Defenses of Pensacola

FORT PICKENS   Santa Rosa Island 1845-(1947) Gulf Islands National Seashore   MF, MC   * * * *

Langdon	2	12" 1895MI	B 1917	1923-1947	casemated-WWII
Worth	8	12" 1890MI	M 1896MI	1899-1942	4 guns removed 1918, HECP-HDCP 1943
Pensacola	2	12" 1895	D 1897	1899-1935	
Cullum	2	10" 1888	D 1894	1898-1918	modified for 2 - 3" from Trueman 1942
Sevier	2	10" 1888	D 1894	1898-1934	
Cooper	2	6" 1903	D 1903	1906-1917	1 gun repl. 1976, (from West Point) Still Emplaced
#234	2	6"	SB M4	1943 NC	2 guns repl. 1976 (from Ft. J. Custis) Still Emplaced
Van Swearingen	2	4.7" Armstr	Armstrong	1898-1921	one gun & carr at Danielsville, GA
Trueman	2	3" 1902	P 1902	1905-1942	guns to Cullum
Payne	2	3" 1902	P 1902	1904-1946	
AMTB	2	90 mm M1	F M3	1943-1946	
unnamed	4	155 mm	PM	1937-1945	

FORT McREE		Perdido Key	1843-(1947)	Gulf Islands National Seashore		* *
Slemmer	2	8" 1888	D 1896	1900-1918	partially buried	
#233	2	6"	SB M4	1943 NC		
Center	4	3" 1898	MP 1898MI	1901-1920	partially buried	

The Harbor Defenses of Mobile

FORT MORGAN		Mobile Point	1833-(1928)	state park	MF, MC	* * *
Dearborn	8	12" 1890MI	M 1896MI	1901-1928		
Duportail	2	12" 1888	D 1896	1900-1928		
Test	1	10" 1888	D 1896	1916-1918	mostly buried	
Bowyer	2	8" 1888	D 1894	1898-1918	1 gun in M1918 BC at University of Tampa, FL	
&	2	8" 1888	D 1896			
Thomas	2	4.7" Armstr	Armstrong	1899-1918		
Schenck	2	3" 1898	MP 1898MI	1900-1923		
&	1	3" 1902	P 1902	1900-1923		
unnamed	2	155 mm	PM	1942-1944	1 covered	
FORT GAINES		Dauphin Island	1821-(1928)	county park		* * *
unnamed	2	8"	Rod	1898-1899	covered by Stanton #1	
Stanton	2	6" 1897	D 1898	1901-1928		
&	1	6" 1903	D 1903			
Terrett	3	3" 1898	MP 1898	1901-1923		
						*
unnamed	2	155 mm	PM		Pacagoula, MS	

The Harbor Defenses of the Mississippi

FORT ST. PHILIP		north bank	1843-(1928) 1943	private	MF, MC	* *
Pike	2	10" 1888	D 1896	1898-1919		
unnamed	2	8"	Rod	1898-1899	gun tubes to Ransom, Jackson, 1 built over	
Forse	2	8" 1888	D 1896	1899-1918		
Merrill	4	6" 1900	P 1900	1907-1920	later only 2 guns, 2 guns to Delaware R.	
Ridgely	2	4.7" Armstr	Armstrong	1899-1913		
Scott	2	3" 1898	MP 1898MI	1901-1920		
Brooke	2	3" 1898	MP 1898MI	1904-1920		
FORT JACKSON		south bank	1832-(1928)	parish park		* *
Ransom	2	8" 1888	D 1896	1899-1918		
Millar	2	3" 1898	MP 1898MI	1901-1920		
						* ?
unnamed	2	155 mm	PM	1942-	Calcasieu Pass, LA	
unnamed	2	155 mm	PM	1942-1945	Port Eads, LA, Temp. HD of New Orleans	
unnamed	2	155 mm	PM	1942-1945	Burrwood, LA, Temp. HD of New Orleans	
unnamed	1	8"	Rod	1898-1899	Sabine Pass, TX, buried or destroyed?,	
unnamed	2	155 mm	PM	1942-	Sabine Pass, TX	

The Harbor Defenses of Galveston

FORT TRAVIS		Bolivar Point	(-1947)	county park		* *
Kimble	2	12" 1895MI	B 1917	1922-1943	guns to Moultrie	
Davis	2	8" 1888	D 1896	1900-1918	deteriorating	
#236	2	6"	SB M1	194? NC		
Ernst	3	3" 1898	MP 1898MI	1900-1946	converted to 3-in P	
FORT SAN JACINTO		Fort Point	(-1947)	Corps of Eng. prop	MF, MC	*
Mercer	8	12" 1890MI	M 1896MI	1898-1943	partially buried in mud	
Heileman	1	10" 1888	D 1896 ARF	1899-1943	mostly destroyed & buried	
&	1	10" 1888	D 1896			
#235	2	6" 1905A2	SB M1	1944-1946	partially buried	
Hogan	2	4.7" Armstr	Armstrong	1898-1917	partially destroyed & buried	
Croghan	2	3" 1898	MP 1898MI	1911-1946	converted to 3-in P	
AMTB #4	2	90 mm M1	F M3	1943-1946	buried	
FORT CROCKETT		Galveston	(-1947)	private, federal prop		* * *
Hoskins	2	12" 1895MI	B 1917	1924-1946	casemated WWII, now built on	
Izard	8	12" 1890MI	M 1896MI	1902-1943	destroyed	
Hampton	2	10" 1895	D 1896	1899-1943	destroyed	
Laval	2	3" 1898	MP 1898	1902-1946?	converted to 3-in P, partially covered	

	2	6" Navy	Navy P		Freeport, TX, WWII temp?
unnamed	2	155 mm	PM	1942-	Freeport, TX
unnamed	2	155 mm	PM	1942-	Port Aransas, TX

The Harbor Defenses of San Diego

FORT EMORY	Coronado Heights			Former Navy radio facility, current training site	**
#134	2	16"	CB	1944 NC	unofficially named Gatchell, destroyed 2016
Grant (#239)	2	6" 1905A2	SB M1	1943-1946	
Imperial	4	155 mm	PM		
AMTB Cortez	2	90 mm M1	F M3	1943-1946	Coronado Beach MR, Silver Strand, destroyed

FORT PIO PICO	North Island 1906-1919 (to Navy 1935) Naval Air Station				X
Meed	2	3" 1903	P 1903	1902-1919	guns to Rosecrans, destroyed
AMTB Pio Pico	2	90 mm M1	F M3	1942-1943	guns later moved to AMTB Fetterman

FORT ROSECRANS	Point Loma 1899-1958		Navy Base, Natl. Cem., Cabrillo N.M.		MF, MC ****
Ashburn (#126)	2	16" MkIIMI	CB M4	1943-1948	modified
Whistler	4	12" 1890MI	M 1896MI	1916-1942	guns from DeSoto, extensively modified
White	4	12" 1890MI	M 1896MI	1916-1942	
Wilkeson	2	10" 1888	D 1896	1900-1943	
Calef	2	10" 1888	D 1896	1900-1943	1 gun M1895
Strong	2	8" MkVI3A2	B M1	1941-1946	
#237	2	6" 1903A2	SB M1	1943-1946	unofficially "Woodward," Generators still in place
Humphreys (#238)	2	6" 1903A2	SB M1	1943-1946	
Zeilin	2	7" Navy	Navy P	1937-1943	buried
Gillespie	3	5" Navy	Navy P	1937-1943	No. 1 empl dest., 2 & 3 remain overgrown
McGrath	2	5" 1897	BP 1896	1900-1943	repl. by 2-3" guns from Pio Pico, 1919
Fetterman	2	3" 1898	MP 1898MI	1900-1917	destroyed July 1940
AMTB Fetterman	2	90 mm M1	F M3	1943-1946	guns from Pio Pico, destroyed
AMTB Cabrillo	2	90 mm M1	F M3	1943-1946	destroyed
Point Loma	4	155 mm	PM		

The Harbor Defenses of Los Angeles

Bolsa Chica Military Reservation	north of Huntington Beach			private, state reserve	X
#128	2	16"	CB	1944 NC	destroyed
Harrison (#242)	2	6" M1	SB M4	1944-1948	destroyed
unnamed	2	155 mm	PM	1942	

FORT MacARTHUR (Upper & Lower) San Pedro 1885-1975 (M) Air Force housing, city park					*****
Osgood	1	14" 1910	D 1907MI	1917-1944	gun remodeled 1923, being restored
Farley	1	14" 1910	D 1907MI	1917-1944	
Leary	1	14" 1910MI	D 1907MI	1917-1944	converted to HECP, 1944
Merriam	1	14" 1910MI	D 1907MI	1917-1944	converted to HECP, 1944
"Erwin"	2	14" 1920MII	RY 1920	1926-1946	1 empl. L.R. (covered), 2 empl. L.R. (destroyed)
Barlow	4	12" 1912	M 1896MIII	1917-1944	
Saxton	4	12" 1912	M 1896MIII	1917-1944	
#241	2	6" M1	SB M4	1948 NA	armed post-war to 1956, Generators still in place
JAAN #1	2	3" 1903	P 1903	1942-1946	L. R. (Cabrillo Beach), guns from Lodor
AMTB Gaffey Bulge	2	90 mm M1	F M3	1943-1946	
AMTB Navy Field	2	90 mm M1	F M3	1943-1946	L. R., buried?
unnamed	2	155 mm	PM	1942	

White Point Military Reservation	San Pedro		Air Force housing, city property		**
Bunker (#127)	2	16" MkIIMI	CB M4	1944-1948	

Long Point Military Reservation	Pt. Vicente (Rancho Palos Verdes) city park				**
Barnes (#240)	2	6" 1903A2	SB M1	1944-195?	Pt. Vicente
unnamed	2	155 mm	PM	1942	Long Point, dest.

	Other locations				*
"Eubanks"	2	8"	RY	1942-1945	Manhattan Beach, destroyed
Lodor	4	3" 1903	P 1903	1919-1927	Deadman's Island, destroyed
JAAN #2	2	3" 1903	P 1903	1942-1946	Bluff Park, Long Beach, guns from Lodor
AMTB Terminal Is.	2	90 mm M1	F M3	1943-1946	Terminal Island, covered
AMTB Bluff Park	2	90 mm M1	F M3	1943-1946	Bluff Park, Long Beach, buried?
unnamed	3	155 mm	PM	1942	Costa Mesa (only two guns assigned), destroyed
unnamed	3	155 mm	PM	1942	Rocky Point (only two guns were assigned), dest
unnamed	2	155 mm	PM	1942	El Segundo/Hyperion, originally 2 6-in NP, dest.
unnamed	2	155 mm	PM	1942	Playa Del Rey, destoyed

unnamed	2	155 mm	PM	1942	Pacific Palisades, destroyed
unnamed	2	155 mm	PM	1942	Port Hueneme, buried/destroyed
unnamed	2	155 mm	PM	1942	Oxnard, destroyed
unnamed	2	155 mm	PM	1942	Ventura, in surf 1992
unnamed	2	155 mm	PM	1942	Santa Barbara, destroyed

The Harbor Defenses of San Francisco

Milagra Ridge Military Reservation			Pacifica		Golden Gate National Recreation Area	* *
[#130]	2	16"	CB	*Not Built*		
#244	2	6" M1	SB M4	1945 NA	armed post war, guns from Ft. Columbia	

FORT FUNSTON		Lake Merced	1917-1975		Golden Gate National Recreation Area	* *
Davis	2	16" MkIIMI	CB M2	1939-1948	prototype	
Howe	4	12" 1890MI	M 1896MI	1919-1945	guns from Ft. W. Scott, destroyed	
Bruff	2	5" 1900	P 1903	1919-1919	guns from Ft. W. Scott, destroyed	
unnamed	4	155 mm	PM	1930s	destroyed by erosion	
unnamed	4	155 mm	PM	1942	buried	

FORT MILEY		Lands End	1900-1948		VA Hospital, Golden Gate National Recreation Area	* *
Livingston	8	12" 1890MI	M 1890MI	1900-1943		
Springer	8	12" 1890MI	M 1890MI	1900-1943		
Chester	2	12" 1895	D 1897	1898-1943	2 guns replaced 1920s	
&	1	12" 1888	B 1892	1898-1943		
Lobos	2	6" Navy	Navy P	1943-1945	Navy guns	
#243	2	6" M1	SB M3	1944 NA	armed post war?	
Call	2	5" 1900	P 1903	1915-1921	guns from Ft. McDowell, destroyed	
AMTB Lands End	2	90 mm M1	F M3	1943-1948		

FORT WINFIELD SCOTT		Presidio	(1852) 1885-1995		Golden Gate National Recreation Area MF, MC	* * * * *
Dynamite	3	15"	Pne	1894-1904	converted to HECP, WWII	
Lancaster	3	12" 1895	D 1897	1899-1918	partially buried for bridge	
&	1	12" 1888MI	D 1896		gun M1888MI-1/2	
Godfrey	3	12" 1888	B 1892	1895-1943		
Saffold	2	12" 1888MII	B 1892	1898-1943		
Howe	8	12" 1886	M 1891	1895-1920	1 empl. remains, others buried	
Wagner	8	12" 1886	M 1891	1895-1920	buried	
Stotsenburg	8	12" 1890MI	M 1896MI	1900-1943	4 guns removed 1918 to Ft. Funston	
McKinnon	8	12" 1890MI	M 1896MI	1900-1943		
Cranston	2	10" 1888MII	D 1896	1898-1943	built on	
Miller	3	10" 1888	D 1894	1898-1920		
East	2	8"	B	1897-1915	converted Rodman rifles	
Slaughter	3	8" 1888	D 1896	1899-1917	partially buried	
Crosby	2	6" 1897MI	D 1898	1902-1943		
Chamberlin	4	6" 1903	D 1903	1904-1918	1-6" DC installed by NPS 1976 <u>Still emplaced</u>, being restored	
Chamberlin	2	6" 1900	P 1900	1920-1947	in modified DC emplacement	
Boutelle	3	5" 1897	BP 1896	1898-1918		
Sherwood	2	5" 1900	P 1903	1900-1917	guns to Ft. Funston	
Baldwin	2	3" 1898	MP 1898MI	1901-1920	partially buried	
Blaney	4	3" 1898	MP 1898MI	1907-1920		
Point	2	3" 1902	P 1902	1944-1945	guns from Yates, Ft. Baker	
Gate	2	3" 1902	P 1902	1942-1945	guns from Gravelly Beach, Ft. Baker	
AMTB Baker Beach	2	90 mm M1	F M3	1943-1946	buried	

FORT MASON		Black Point	1852-1975		Golden Gate National Recreation Area MF, MC	* * * * *
Burnham	1	8" 1888	D 1896	1900-1909	gun & carr to Ft. Columbia, built on	

FORT McDOWELL		Angel Island	1852-(1928)1947	state park	MC	* * * * *
Drew	1	8" 1888	B 1892	1898-1915		
Wallace	1	8" 1888	D 1896	1901-1915		
Ledyard	2	5" 1900	P 1903	1901-1915	guns to Miley	
AMTB		90 mm M1	F M3	*Not Built?*		

FORT BAKER		Lime Point, 1867-			Golden Gate National Recreation Area MF	* * * * *
Spencer	3	12" 1888	B 1892	1893-1943	1 gun removed 1918	
Kirby	2	12" 1895	D 1897	1900-1941	1 gun replaced 1933	
Duncan	2	8" 1888	B 1892	1898-1917		
Wagner	2	5" 1897	BP 1896	1901-1917		
Yates	6	3" 1902	P 1902	1904-1943	4 guns removed 1941	
Gravelly Beach	2	3" 1902	P 1902	1942-1943	guns from Yates, part. buried	
AMTB Gravelly Beach	2	90 mm M1	F M3	1943-1946		

FORT BARRY	Point Bonita 1905-1975			Golden Gate National Recreation Area	MC	* * * * *
#129	2	16"	CB M4	1943 NC	guns were at emplacement	
Alexander	8	12" 1890MI	M 1896MI	1901-1943		
Mendell	2	12" 1895	DC 1897	1901-1943		
Wallace	2	12" 1895MI	B 1917	1919-1947	casemated WWII	
Guthrie	2	6" 1900	P 1900	1905-1946		
Smith	2	6" 1900	P 1900	1905-1946		
Rathbone	2	6" 1900	P 1900	1905-1947		
McIndoe	2	6" 1900	P 1900	1905-1947		
O'Rorke	4	3" 1903	P 1903	1905-1945		
FORT CRONKHITE	Tennessee Point 1937-1975			Golden Gate National Recreation Area		* * * * *
Townsley	2	16" MkIIMI	CB M2	1940-1948	restoration, one 16" navy MkVII barrel on display 2012	
unnamed	2	3" 1903	P 1903	1940s	training battery	

MD, Yerba Buena Island
MD, Alcatraz Is.

The Harbor Defenses of the Columbia

FORT STEVENS	Point Adams 1852-(1947)			state park, private	MF, MC	* * * *
Clark	8	12" 1890MI	M 1896MI	1899-1942	4 guns to Canby 1918	
Mishler	2	10" 1888MII	D 1896ARF	1897-1941	ARF, covered for HECP 1941	
Lewis	1	10" 1888MII	D 1894	1897-1920		
&	1	10" 1888MII	D 1896			
Walker	2	10" 1888MII	D 1894	1897-1920		
Russell	2	10" 1900	D 1901	1904-1944		
Pratt	2	6" 1897MI	D 1898	1902-1943	battery being restored, full scale replica gun & carriage	
Freeman	2	6" 1900	P 1900	1902-1920	6" guns to Willapa Bay, empl. destroyed	
&	1	3" 1898	MP 1898MI	1902-1920		
#245	2	6" M1	SB M4	1944-1947	2-5" naval gun mounts in empl. as display 1981	
Smur	2	3" 1898	MP 1898MI	1902-1920		
AMTB #1	2	90 mm M1	F M4	1943-1946	Clatsop Spit	
FORT COLUMBIA	Chinook Point 1864-(1947)			state park	MC	* * * * *
Ord	2	8" 1888	D 1896	1898-1917		
&	1	8" 1888	D 1894	1898-1917	Originally named Neary, now buried	
Murphy	2	6" 1897MI	D 1898	1900-1945		
#246	2	6"	SB M4	1944 NC	2 guns & carr. from Ft. McAndrew installed 1994	
Crenshaw	3	3" 1898	MP 1898MI	1900-1920		
FORT CANBY	Cape Disappointment 1852-(1947)			state park, USCG		* *
Guenther	4	12" 1890MI	M 1896MI	1921-1942	guns from Stevens	
Allen	3	6" 1903	D 1903	1906-1945	3 guns removed 1917, 2 guns replaced 1918	
O'Flyng	2	6" 1903	D 1903	1906-1918	guns to Allen	
#247	2	6" M1	SB M4	1944-1947		
AMTB #2	2	90 mm M1	F M3	1943-1946	Jetty	

The Harbor Defenses of Grays Harbor (temporary)

	North Cove (Willapa Bay)					*
unnamed	4	12"	M	1919-1919	NA	
unnamed	2	6" 1900	P 1900	1919-1932	guns from Ft. Stevens, later moved to Ft. Worden, under water	
unnamed	2	155 mm	PM		destroyed	
	Westport , private					*
unnamed	4	12"	M	1919	NA, covered	
#1	4	12"	RM	1942	at Pt. Brown- unknown	
#2	2	6" Navy	Navy P	1942-1944	one empl remains near park	
	4	6" navy	Navy P	1942	at Point Brown, one empl remains in water	
unnamed	2	5" 1897MI	BP 1896	1919-1919	guns from Lee, Ft. Flagler, one empl remains	
North	2	155 mm	PM	1942	one empl remains in park	
South	2	155 mm	PM	1942	destroyed	
Markham	2	155 mm	PM	1942	on school property, extant	

The Harbor Defenses of Puget Sound

FORT WARD	Bean Point 1903-(1928)1938			private, metro park	MF, MC	* * *
Nash	3	8" 1888	D 1896	1903-1918	private	1 gun M1888MII
Warner	2	5" 1900	P 1903	1903-1925	private	
Thornburgh	4	3" 1898	MP 1898MI	1903-1920		
Vinton	2	3" 1898	MP 1898MI	1903-1920	lower rooms filled	

Middle Point Military Reservation				state park		MF, MC	* * *
Mitchell	2	3"		MP	1903 NA	across channel from Ft. Ward	
FORT WHITMAN		Goat Island	1909-(1947)	state game refuge		MC	* *
Harrison	4	6" 1908	D 1905MI		1911-1943		
FORT FLAGLER		Marrowstone Island 1898-1954		state park			* * * *
Bankhead	8	12" 1890MI	M 1896MI		1902-1942		
Wilhelm	2	12" 1888MII	AGL 1891?		1898-1942		
Rawlins	2	10" 1888MII	B 1893		1899-1918	both empl. mod. for 3-in AA (3rd AA empl. separate)	
Revere	2	10" 1888MII	B 1893		1899-1941	guns and carriages to Canada	
Calwell	4	6" 1903	D 1903		1904-1918		
Grattan	2	6" 1903	D 1903		1905-1918		
Lee	2	5" 1897MI	BP 1896		1901-1918	guns to Grays Harbor	
Downes	2	3" 1903	P 1903		1905-1946		
Wansboro	2	3" 1903	P 1903		1906-1946	2 guns repl. 1963 from Ft. Wint for display	
AMTB Marrowstone	2	90 mm M1	F M3		1943-1946	in surf	
FORT CASEY		Admiralty Head	1897-1953	state park, college extention campus			* * * * *
Schenck	8	12" 1890MI	M 1896MI		1899-1942	no mortars removed, two tubes were M1890	
Seymour	8	12" 1890MI	M 1896MI		1899-1942	Schneck mortars held in reserve for temp. batteries	
Kingsbury	2	10" 1895	D 1896		1902-1942	2 guns rem. 1918, 1 repl 1920, 2 empl. mod. for 3-in AA	
&	1	10" 1900	D 1901				
Moore	3	10" 1895	D 1896		1904-1942	1 empl. mod. for 3-in AA	
Worth	2	10" 1895	D 1896		1898-1942	M1901 guns and carr. repl 1963 from Warwick, Ft. Wint	
Parker	2	6" 1903	D 1903		1905-1918		
Valleau	4	6" 1903	D 1903		1907-1918		
Turman	2	5"1897MI	BP 1896		1901-1918		
Trevor	2	3" 1903	P 1903		1905-1933	2 guns repl 1963 for display	
Van Horne	2	3" 1903	P 1903		1905-1945		
FORT WORDEN		Point Wilson	1898-1954	state park		MF?	* * * * *
Brannan	8	12" 1890MI	M 1896MI		1901-1943		
Powell	8	12" 1890MI	M 1896MI		1901-1943		
Ash	2	12" 1888MII	B 1892		1900-1942		
Kinzie	2	12" 1895MI	D 1901		1910-1944		
Benson	2	10" 1900	D 1901		1907-1943		
Quarles	3	10" 1888MII	B 1893		1900-1941	guns & carr. to Canada, one gun SE at Ft. McNutt	
Randol	2	10" 1888MI	B 1893		1900-1918		
Stoddard	4	6" 1903	D 1903		1906-1918		
Tolles	4	6" 1903	D 1903		1905-1943	2 guns rem '18	
"Tolles B"	2	6" 1900	P 1900		1937-1947	in empty Tolles empl, 2-6" P from Willapa Bay	
Vicars	2	5" 1897MI	BP 1896		1902-1918		
Putnam	2	3" 1903	P 1903		1907-1945		
Walker	2	3" 1903	P 1903		1907-1946		
AMTB Pt. Wilson	2	90 mm M1	F M3		1943-1946	In surf	
		Ebey's Landing		private			X
AMTB Ebey's Lndg	2	90 mm M1	F M3		1943-1946	destroyed	
FORT EBEY		Partridge Point		state park			* *
#248	2	6" 1905A2	SB M1		1943-1946		
		Deception Pass		state park			X
North & South	4	3"		P	*Not Built*		
		Cape George					X
unnamed	4	12"		RM	193?	NA, empl cuts remain	
unnamed	4	8"		RY		Tibbels Bluff	
		Angeles Point					X
unnamed	4	155 mm		PM	1942		
CAMP HAYDEN		Striped Peak		county park/state land			* *
#131	2	16" MkIIMI	CB M4		1944-1948		
#249	2	6" M1	SB M4		1945-1948		
FORT HAYDEN (Cape Flattery Military Res.)			Makah Indian Res				X
[#132]	2	16"		CB	*Not Built*	Construction initiated on all sites, ended 1943	
[#133]	2	16"		CB	*Not Built*		
[#250]	2	6"		SB	*Not Built*		
[#251]	2	6"		SB	*Not Built*		
		Canadian sites, British Columbia					*
	2	8"		RY	1941-1943	Christopher Point, Vancouver Is.	
	2	8"		RY	1941-1943	Fairview, Prince Rupert	

The Harbor Defenses of Honolulu, Hawaii

FORT RUGER		Diamond Head	1906-now	Natl. Guard , State Dept. Land Nat. Res. State Civil Def.	* * *
Harlow	4	12" 1890MI	M 1896MI	1910-1943	
&	4	12" 1890MI	M 1896MII		
Birkhimer	4	12" 1890MI	M 1896MI	1916-1943	rebuilt, site modified
Granger Adams	2	8" 1888MII	B 1918	1935-1946	Black Pt., destroyed
#407	2	8"	B M1	1942 NA	tunneled into rock
S.C. Mills	2	5" 1900/05	BP 1903/07	1916-1925	Black Pt., guns from Mott, destroyed
Dodge	2	4.7" Armstr	Armstrong	1915-1925	2-4" NP WWII, orig. barrells at NG Armory Wahiawa
Hulings	2	4.7" Armstr	Armstrong	1915-1925	
unnamed	12	2.24"	Ped.	1915-1919	emplacements for 6 pdr. parapit mounts
Ruger	4	155 mm	PM	1942-1943	destroyed
Wiliwilinui Ridge M.R.				private	*
Kirkpatrick	4	8"	2 x Navy T	1942-1948	
Wili	4	155 mm	PM		destroyed
FORT DeRUSSY		Waikiki Beach	1908- now	Army recreation center , Army Museum	*
Randolph	2	14" 1907	D 1907	1913-1944	2 - 7" guns now in emplacements, being restored
Dudley	2	6" 1908	D 1905MI	1913-1946	destroyed '69
AMTB No. 5	2	90 mm M1	F M3	1943-1946	destroyed
FORT ARMSTRONG		Kaakaukukui Reef 1899		private MF, MC	X
Tiernon	2	3" 1903	P 1903	1911-1943	destroyed
Sand Island Military Reservation					*
Sand	4	155 mm	PM	1937-1943	destroyed
Harbor	4	7" Navy	Navy P	1942-1944	
AMTB No. 4 (Sand)	2	90 mm M1	F M3	1943-1946	destroyed
Ala Moana AMTB	2	90 mm M1	F M3	1943-1945	Ala Moana Beach Park, destroyed
Other locations- Honolulu					?
#304	2	6"	SB	1942 NC	Punchbowl
[#305]	2	6"	SB	*Not Built*	Koko Saddle
unnamed	2	240 mm	H 1918	1929-1941	Waimanalo
unnamed	2	240 mm	H 1918	1927-1941	Kaaawa, buried
unnamed	2	240 mm	H 1918	1927-1941	Ulupau, destroyed 1992
Punchbowl	4	155 mm	PM	1943-1944	Punchbowl, 1 empl. remains
Koko Head	2	155 mm	PM	1941-1942	Koko Head
School	4	155mm	PM	1942-1944	Kamehameha School

The Harbor Defenses of Pearl Harbor, Hawaii

FORT BARRETTE		Kapolei 1934		city/county park	* * *
Hatch	2	16" MkIIMI	B 1919M1	1934-1948	casemated-WWII
FORT WEAVER		Puuloa 1899 (1922)		Navy Housing	*
Williston	2	16" 1919	B 1919	1924-1948	ARF, destroyed
AMTB No. 1	2	90 mm M1	F M3	1943-1945	
Weaver	4	155 mm	PM		
FORT KAMEHAMEHA		Queen Emma Point	1908-1949	Hickam AFB MF?, MC	* * *
Hasbrouck	8	12" 1908	M 1908	1914-1943	
Closson	2	12" 1895MI	B 1917	1924-1948	casemated 1942
Selfridge	2	12" 1895MII	D 1901	1913-1945	
Jackson	2	6" 1908	D 1905MI	1913-1943	
Ahua	3	5" Navy	Navy P	1942-1944	Ahua Pt.
Barri	2	4.7" Armstr	Armstr-CM	1915-1924	Bishop Pt., destroyed
Chandler	2	3" 1903	P 1903-CM	1915-1942	Bishop Pt., destroyed
Hawkins	2	3" 1903	P 1903	1914-1943	
AMTB No. 2	2	90 mm M1	F M3	1943-1946	
AMTB No. 3	2	90 mm M1	F M3		location?
AMTB No. 6	2	90 mm M1	F M3		location?
Kam	4	155 mm	PM		
Puu O Hulu M.R.					* *
"Hulu"	2	7"	Navy CM	1942-1944	incorporated into BCN #303
#303	2	6"	SB	1942 NC	tunneled into rock

Other locations- Pearl Harbor   *

Name					
Arizona	3	14"	Navy Turr	1945 NC	Kahe Pt., from *USS Arizona*
Burgess	4	8"	2 x Navy T	1942-1948	(Salt Lake) near Aliamanu Crater, destroyed
Brown's Camp	4	8"	RY	1937-1944	Browns Camp, Kahe Point
"Homestead"	3	7" Navy	Navy P	1942	Keaau Homesteads
Adair	2	6" Armstr	Armstr-CM	1917-1925	Ford Island, empl. elsewhere earlier ?
Boyd	2	6" Armstr	Armstr-CM	1917-1925	Ford Island
Nanakuli	2	5" Navy	Navy P	1941-1943	and 1-3" NP, destroyed
Oneula	2	5" Navy	Navy P	1942-1944	Oneula Beach (Ewa) destroyed
unnamed	2	240 mm	H 1918	1931-1941	Laie
unnamed	1	240 mm	H 1918	1927-1941	Pupukea
unnamed	3	240 mm	H 1918	1930-1941	Makua
Awanui	4	155 mm	PM	1942-1944	Brown's Camp
Barber's Point	4	155 mm	PM	1937-1942	Barber's Point, destroyed
"X-Ray"	4	155 mm	PM	1942-1944	Oneula Beach, destroyed
Homestead	4	155 mm	PM	1942-1944	Makua Military Reservation
Kahe Point	4	155 mm	PM	1942-1944	Kahe Point

The Harbor Defenses of Kaneohe Bay and the North Shore of Oahu, Hawaii

FORT HASE		Mokapu Point 1942		Marine Corps Base, Hawaii		* *
Pennsylvania	3	14"	Navy Turr	1945-1948	Ulupau Head, from *USS Arizona*	
Demerritt (#405)	2	8" MkVI3A2	B M1	1944-1948	Puu Papaa, tunneled into rock	
Sylvester	4	8" 1890MI	RY 1918	1942-1944	RY guns on specially built tracks, later dismounted	
French (#301)	2	6" 1903A2	SBC M1	1944-194?	Pyramid Rock	
East Beach	4	155 mm	PM			
North Beach	4	155 mm	PM			
Pyramid Rock	4	155 mm	PM			
Paumalu M.R.		Waialee				X
[#408]	2	8"	B	*Not Built*		
[unnamed]	4	8"	RY	*Not Built*		
Kawailoa M.R.						X
unnamed	4	8"	RY	1940-1942		
unnamed	4	155 mm	PM			
Opaeula M.R.			private			*
Riggs	4	8"	2 x Navy T	1942-1948		
Brodie Camp M.R.		Helemano	private			*
Ricker	4	8"	2 x Navy T	1942-1948		
Kaena Point M.R.						* *
#409	2	8"	B	1942 NC	tunneled into rock	
Other locations- North Shore						*
Kahuku	4	8" 1888MII	RY 1918	1940-1944	dismounted RY with shields, 1 empl. remains	
Cooper (#302)	2	6" 1903A2	SB M1	1944-194?	Lae o ka oio (opposite Kaioi Pt.), tunnels	
Kahana	2	5" Navy	Navy P	1942-1943	Kahana Bay	
Wailia	2	3" 1903	P 1903	1942-1943	Wailia Pt., guns from Chandler, repl by 90m F?	
AMTB No. 7	2	90 mm M1	F M3	1943-1945	2-90 mm M/ 2-37 mm M, S. Pyramid Rock, 1 empl. rem.	
AMTB No. 8	2	90 mm M1	F M3	1943-1945	2-90 mm M/ 2-37 mm M, destroyed	
Kahuku/Ranch	4	155 mm	PM	1941-1945	Kahuku	
Loko	4	155 mm	PM	1942-1944	Kualoa Ranch	
Ashley	4	155 mm	PM	1939-1944	Ashley Military Reservation	
Kaena	2	4" Navy	Navy P	1942	later converted to 155 mm site	
Kalihi	2	4" Navy	Navy P	1942	Mokuoeo Is?	
Dillingham	2	4" Navy	Navy P	1942	Mokuleia	
Kaneohe	2	4" Navy	Navy P	1942	Kaneohe Bay	
Oahu howitzer positions manned by field artillery units						?
Kalihi	2	240 mm	H 1918	1941-1944	Kalihi, atop Koalau Range	
"Quadropod"	2	240 mm	H 1918	1941-1944	Paalaa	
unnamed	2	240 mm	H 1918	1941-1944	Kole Kole Pass	
unnamed	2	240 mm	H 1918	1941-1944	Kunia	
Other locations- other islands						?
unnamed	2	7" Navy	Navy CM	1942	Ahukini, Kauai Is.	
unnamed	2	7" Navy	Navy CM	1942	Monument, Kauai Is.	

The Harbor Defenses of Manila and Subic Bays, the Philippines

FORT MILLS		Corregidor Island		Philippine National Monument	2-MF, MC	* * * * *
Hearn	1	12" 1895MI	B 1917	1921	Still Emplaced + spare barrel	
Smith	1	12" 1895MI	B 1917	1921	Still Emplaced	
Way	4	12" 1890MI	M 1896MI	1910	Still Emplaced	
Geary	4	12" 1890MI	M 1896MI	1910	4 guns from Ft. Andrews, part. dest., 6 tubes at site	
&	4	12" 1908	M 1908			
Cheney	2	12" 1895/MI	D 1901	1910	Still Emplaced	
Wheeler	2	12" 1895/MI	D 1901	1910	1 Gun Still Emplaced + spare barrel	
Crockett	2	12" 1895MI	D 1901	1910	Still Emplaced + spare barrel	
Grubbs	2	10" 1895MI	D 1901	1911	Still Emplaced + spare barrel	
RJ 43	1	8" 1888MII	B 1918	1942	gun barrel at wharf (also 1-8" BC near Bagac, Bataan)	
Morrison	2	6" 1905/08	D 1905MI	1910	Still Emplaced + spare barrel	
Ramsey	3	6" 1905	D 1905/MI	1911	1 gun Still Emplaced, 2 barrels outside emplacement	
James	4	3" 1903	P 1903	1910		
Keyes	2	3" 1903	P 1903	1913		
Cushing	2	3" 1903	P 1903	1919	reachable only by boat	
Hanna	2	3" 1903	P 1903	1919	empl #1 intact, empl # 2 buried	
Martin	2	155 mm	PM			
Hamilton (South)	3	155 mm	PM			
Kysor (North)	2	155 mm	PM			
Rock Point	2	155 mm	PM			
Sunset	4	155 mm	PM			
Stockade	2	155 mm	PM			
Monja	2	155 mm	PM		one emplacement was casemated	
Concepcion	3	155 mm	PM			
Levagood	2	155 mm	PM			
FORT FRANK		Carabao Island		abandoned		* *
Greer	1	14" 1907MI	D 1907	1913	partially destroyed	
Crofton	1	14" 1907MI	D 1907	1913	partially destroyed	
Koehler	8	12" 1908	M 1908	1913	partially destroyed	
Hoyle	2	3" 1903	P 1903	1913	removed before WWII, partially destroyed	
Frank	4	155 mm	PM			
FORT DRUM		El Fraile Island		abandoned		* * +
Wilson	2	14" 1909	TM 1910	1918	Still Emplaced	
Marshall	2	14" 1909	TM 1910	1918	Still Emplaced	
Roberts	2	6" 1908MII	CM 1910	1918	1 gun Still Emplaced	
McCrea	2	6" 1908MII	CM 1910	1918	partially destroyed	
New Hoyle	1	3"1903	P 1903		installed 1941, gun ped. remains	
FORT HUGHES		Caballo Island		Philippine Naval Station	MF, MC	* * +
Woodruff	1	14" 1910	D 1907MI	1914	Still Emplaced	
Gillespie	1	14" 1910MI	D 1907MI	1914	Still Emplaced	
Craighill	4	12" 1912	M 1896MIII	1919	Still Emplaced	
Leach	2	6" 1908	D 1905MII	1914	empl. destroyed, 1 barrel present	
Fuger	2	3" 1903	P 1903	1914	buried	
Willaims	2	155 mm	PM			
Hooker	1	155 mm	PM			
FORT WINT		Grande Island, Subic Bay		to Philippines 1992, radar site	MF, MC	* *
Warwick	2	10" 1895MI	D 1901	1910	guns to Casey 1967, built on	
Woodruff	2	6" 1905	D 1905	1910	partially destroyed by magazine explosion after WW II	
Hall	2	6" 1905	D 1905	1910	Mag dest. post war, guns & carr. Still Emplaced	
Flake	4	3" 1903	P 1903	1910	2 guns to Casey 1963	
Jewell	4	3" 1903	P 1903	1910	2 guns repl Navy 3", 2 guns to Flagler 1963	
unnamed	4	155 mm	PM		Ogonbolo, Bataan	

The Harbor Defenses of Cristobal, Panama (Panama Canal Zone, Atlantic side)

FORT RANDOLPH		Margarita Island	1911		to Panama, 1979; commercial development	* *
Webb	2	14" 1910	D 1907MI	1912-1948		
#1	2	14" 1920	RY 1920	1928-1948	2 guns for Panama, 4 empl. (#1 & #8)	
Tidball	4	12" 1912	M 1896MIII	1912-1943		
Zalinski	4	12" 1912	M 1896MIII	1912-1943		
Weed	2	6" 1908MII	D 1905MII	1912-1946		
X(4A)	4	155 mm	PM	1940		
2C	4	155 mm	PM			
5A	4	155 mm	PM			
FORT DeLESSEPS		Colon	1911		to Panama, 1950s	* *
Morgan	2	6" 1908MII	P 1910	1913-1944	modified casemate mounts M1910	
AMTB #3b	4	90 mm M1	F M3	1943-1948	Cristobal mole, built over	
FORT SHERMAN		Toro Point	1911	MF, MC to Panama 1999		* * * *
[#151]	2	16"	CB	*Not Built*		
Mower	1	14" 1910	D 1907MI	1912-1948		
Stanley	1	14" 1910	D 1907MI	1912-1948		
Howard	4	12" 1912	M 1896MIII	1912-1943		
Baird	4	12" 1912	M 1896MIII	1912-1943		
Pratt	2	12" 1895MI	B 1917	1924-1948	Iglesia Pt., casemated-WWII	
MacKenzie	2	12" 1895MI	B 1917	1924-1948	Iglesia Pt., not rebuilt	
Kilpatrick	2	6" 1908	D 1905MII	1913-1946		
W	4	155 mm	PM	1940		
Other sites						?
U	4	155 mm	PM	1918	Tortuguilla Point (the original "Panama" mounts)	
V	4	155 mm	PM	1940	Naranjitos Point	
Y	4	155 mm	PM	1940	Palma Media Island	
Z(1A)	4	155 mm	PM	1940	Galetta Is.	
1B	4	155 mm	PM		Galetta Is.	

The Harbor Defenses of Balboa, Panama (Panama Canal Zone, Pacific Side)

FORT KOBBE (ex-Ft. Bruja)		Bruja Point			to Panama 1999	* *
Murray	2	16" MkIIMI	B 1919MI	1926-1948	Bruja Pt., casemated-WWII	
Haan	2	16" MkIIMI	B 1919MI	1926-1948	Batele Pt., not casemated, empl. buried	
AMTB #6	4	90 mm M1	F M3	1943-1948		
Z (3A)	2	155 mm	PM			
FORT AMADOR		Balboa			to Panama, 1997; commercial development	*
Birney	2	6" 1908	D 1905MII	1913-1943	buried	
Smith	2	6" 1908	D 1905MII	1913-1943	buried	
FORT GRANT		Balboa			to Panama, 1979 MF, MC	* *
Newton	1	16" 1895	D 1912	1914-1943	Perico Is., filled to loading platform level	
Buell	2	14" 1910	DC 1907MI	1912-1948	Naos Is., one emplacement filled	
Burnside	2	14" 1910	DC 1907MI	1912-1948	Naos Is., emplacements filled	
Warren	2	14" 1910	DC 1907MI	1912-1948	Flaminco Is., empls. filled to parapit edge	
Prince	4	12" 1912	M 1896MIII	1912-1943	Flaminco Is., modified for shopping center	
Merritt	4	12" 1912	M 1896MIII	1912-1943	Flaminco Is., modified for shopping center	
Carr	4	12" 1912	M 1896MIII	1912-1943	Flaminco Is.	
Parke	2	6" 1908	D 1905MII	1912-1948	Naos Is., destroyed for condominiums	
#8	2	14" 1920	RY 1920	1928-1948	Culebra Is., empl (see #1, Randolph), covered	
T	2	155 mm	PM		Flamenco Is.	
U (10A)	2	155 mm	PM		Flamenco Is.	
V (10B)	2	155 mm	PM		Culebra Is.	
Other sites						?
W (1B)	4	155 mm	PM		Taboquilla Is.	
2B	2	155 mm	PM		Taboquilla Is.	
	4	155 mm	PM		Paitilla Pt.	
X	2	155 mm	PM		Urara Is.	
Y (1A)	4	155 mm	PM		Taboga Is.	

The Harbor Defenses of Sitka, Alaska

FORT BABCOCK	Shoals Pt., Kruzof Is.			Tongass Natl. Forest			* *
#290	2	6"		SB	NC	88% complete	
	2	6" Navy		Navy P	1942	named Allen? repl by BCN 290	
FORT PEIRCE	Biorka Is.			FAA and USCG			* *
#291	2	6" 1903A2		SB M1	NC	98% complete	
FORT ROUSSEAU	Makhnati Is.			State of Alaska			* *
#292	2	6" 1903A2		SB M1	1944-1950	completed	
	4	155 mm		PM			
	Other location						?
AMTB 1	2	90 mm M1		F M3	WWII	Watson Pt., buried? Church school	
AMTB 2	2	90 mm M1		F M3	WWII	Whale Is.	

The Harbor Defenses of Seward, Alaska

FORT McGILVRAY	Caine's Head			state recreation area			* *
#293	2	6" 1903A2		SB M1	NC	99% complete	
AMTB	2	90 mm M1		F M3	WWII	Lowell Point, moved to Whittier	
Rocky Point	4	155 mm		PM			
FORT BULKLEY	Rugged Is			State of Alaska			* *
#294	2	6" 1903A2		SB M1	NC	90% complete	

The Harbor Defenses of Kodiak, Alaska

FORT J.H. SMITH	St. Peters Head, Chiniak			Lesnoi Corp.			* *
#403	2	8" MkVI3A2	B M1		1944-1950	1 gun tube at Buskin River Inn near airport	
[#295]	2	6"		SB	Not Built		
Chiniak	4	155 mm		PM			
FORT TIDBALL	Castle Bluffs, Long Island			Lesnoi Corp			* *
#296	2	6" 1903A2		SB M1	1944-1950	2 shields remain (no carr)	
Deer Point	4	155 mm		PM		Deer Point	
FORT ABERCROMBIE	Miller Point			state park			* *
#404	2	8" 8" MkVI3A2		B M1	1944-1950	2 gun tubes displayed at battery	
AMTB	2	90 mm		F	WWII	Spruce Cape	
	Other Kodiak locations						?
[#297]	2	6"		SB	Not Built	East Cape, Spruce Is.?	
AMTB	2	90 mm M1		F M3	WWII	Puffin Is.	
Artillery Hill	4	155 mm		PM		Bushkin Hill	

The Harbor Defenses of Dutch Harbor, Alaska

FORT LEARNARD	Eider Point, Unalaska			Unalaska Corp.			* *
#298	2	6" 1903A2		SB M1	1944-1950	2 shields (no carr.) remain at site	
AMTB 3a	2	90 mm M1		F M3	WWII		
FORT SCHWATKA	Ulakta Head, Amaknak Is.			Aleutian WWII Natl. Hist. Area			* *
#402	2	8" MkVI3A2	B M1		1944-1950		
[#299]	2	6"		SB	Not Built		
AMTB	2	90 mm M1		F M3	1944-1950	on spit at Dutch Harbor	
Mt. Ballyhoo	4	155 mm		PM			
Hill 400	4	155 mm		PM		Fort Mears, Amaknak Is.	
FORT BRUMBACK	Summer Bay						* *
Summer Bay	4	155 mm		PM			

other Alaskan positions

							?
	2	6" Navy		Navy P		Cold Bay, Mortensen Pt.	
	2	6" Navy		Navy P		Chernofski	
	2	6" Navy		Navy P		Umnak, Sheep Pt.	
	1	6" Navy		Navy P		George Is., still emplaced	
	2	6" Navy		Navy P		Yakutat, Point Carrew, still emplaced	
	2	6" Navy		Navy P		Nome	
	2	6" Navy		Navy P		Annette Is.	
	2	6" Navy		Navy P		Adak Is.	
	2	6" Navy		Navy P		Shemya Is.	
	2	6" Navy		Navy P		Popof Is., Sand Point	
	4	155 mm		PM		Cold Bay, Mortensen Pt.	
	2	155 mm		PM		Umnak, Umnak Pass	

	7	155 mm	PM		Umnak?
	4	155 mm	PM		Yakutat, Point Carrew
	4	155 mm	PM		Annette Is.
	?	155 mm	PM		Adak Is.
	?	155 mm	PM		Shemya Is.
	2	155 mm	PM		Attu Is., Murder Point
	2	155 mm	PM		Attu Is., Chichagot Pt.
	8	155 mm	PM		Amchitka Is.
AMTB	2	90 mm M1	F M3		Whittier, from Seward defenses
AMTBs	6	90 mm M1	F M3		Amchitka Is.
AMTBs	8	90 mm M1	F M3		Adak Is.
AMTB	2	90 mm M1	F M3		Shemya Is., 1 gun remains, relocated from?

Defenses in Newfoundland, Canada
FORT McANDREW   Argentia, St. Johns, Stephenville, other locations; commercial development, private, parkland   * * *

	2	10"	B	1942-44	M1888 guns, St. Johns
	2	10"	B	1941-44	Fort McNutt, Shelburne
	2	10"	B/D	1942-44	Fort Prevel, Gaspe, Quebec
	2	10"	D	1942-43	Wiseman Cove, Botwood
#T2212	2	8"	RY	1941-44	Signal Hill, later Red Cliff Head, St. Johns
#281	2	6" 1903A2	SB M1	WWII	Argentia North, guns to Ft. Columbia 1993
#282	2	6" 1903A2	SB M1	WWII	Still Emplaced, Argentia Hill 195
#954	2	6" Navy	Navy P	WWII	Shalloway Pt., Argentia
#604	2	6" Navy	Navy P	WWII	Latine Point, Argentia
	2	3"	P	1942	Isaac Point, Argentia (relocated to Airport)
#955	2	3"	P	1942-44	Airport, Argentia
#T8503	2	155 mm	PM	WWII	Harmon Field, Stephenville
#T1653	2	155 mm	PM	WWII	St. Johns
AMTB	2	90 mm M1	F M3	WWII	Roche Pt., Argentia, covered
AMTB	2	90 mm M1	F M3	WWII	Ship Harbor Pt. ? Argentia
AMTB	2	90 mm M1	F M3		not built?

Defenses in Bermuda
* *

	2	8"	RY	1941-44	Scaur Hill
	2	8"	RY	1941-44	Fort Victoria
#283	2	6" 1903A2	SB M1	1943-46	Stone (Tudor) Hill, Southhampton Parish
#284	2	6" 1903A2	SB M1	1943-46	Fort Victoria, St. Georges Is. covered
	2	155 mm	PM	1942-46	Turtle Hill
	2	155 mm	PM	1942-46	Cooper's Island
AMTB	4	90 mm	F	?	Fort Victoria

Defenses in Jamaica (planned)
Portland Bight, Heathshire   X

#285	2	6"	SB	*Not Built*	battery location moved to Vieques Is.

Defenses in Guantanamo Bay, Cuba
U.S. Naval Station   never armed by Army   ?

	4	6"	D	191?-NA	Conde' Bluff Reservation Taft-era, Carr. mounted
	4	3"	P	NA	Cuzco Hill Reservation

The Harbor Defenses of Vieques Sound, Puerto Rico, Virgin Islands (Roosevelt Roads)
FORT CHARLES W. BUNDY, Roosevelt Roads Naval Station, Puerto Rico   now private & government   *

[#152]	2	16"	CB	*Not Built*	Punta Mata Redonda
[#155]	2	16"	CB	*Not Built*	Punta Yeguas
#406	2	8" MkVI3A2	B M1	NC?-1948	Punta Mata Redonda, magazine not built
#265	2	6" 1903A2	SB M1	WWII	Isla Pineros
#268	2	6" 1903A2	SB M1	WWII	Punta Lima, now a prision
[#311]	2	6"	SB	*Not Built*	Cabo San Juan
AMTB	2	90 mm M1	F M3	WWII	Punta Algodones, HECP-HDCP
[AMTB]	4	90 mm	F	*Not Built?*	
	2	155 mm	PM		Punta Yeguas

		Vieques Island sites				X
[#153]	2	16"	CBC	*Not Built*	Mt. Pirata, Vieques Is.	
[#154]	2	16"	CBC	*Not Built*	Cerro Matias Jalobre, Vieques Is. (orig. Culebra Is.)	
[#266]	2	6"	SBC	*Not Built*	Cerro Martineau or Punta Mulas, Vieques Is.	
[#267]	2	6"	SBC	*Not Built*	Punta Arenas or Mt. Pirata, Vieques Is.	
[#285]	2	6"	SBC	*Not Built*	East End, Vieques Is.	
		Culebra Island sites (planned)				X
[#312]	2	6"	SBC	*Not Built*	North Pt.	
[#313]	2	6"	SBC	*Not Built*	Dolphin Hd.	
FORT SEGARRA		St. Thomas Island sites 1944				*
#401	2	8" MkVI3A2 B M1		NC -1948	Fortuna Hill, gunblocks only	
#314	2	6"	SB M4	NA (97%)	Flamingo Pt., Water Is.	
[#315]	2	6"	SBC	*Not Built*	Hill 411	
AMTB	2	90 mm	F	WWII	Water Is. (Ft. Sagarra)	
AMTB	4	90 mm	F	WWII	Water Is.	

The Harbor Defenses of San Juan, Puerto Rico

FORT MASCARO		Punta Salinas		National Guard		* *
Buckey (#261)	2	6"	SB M4	WWII	NA? Punta Salinas	
Pence (#262)	2	6" 1903A2	SB M1	WWII	East Salinas Island	
FORT AMEZQUITA		Las Cabras Is.		San Juan park		* *
Reed	2	12" 1895MI	CB 1917	1941-1948	guns from Torbert, Ft. Delaware, new carr.	
Cabras Island	4	155 mm	PM	1941-		
AMTB	2	90 mm M1	F M3		mobile only?	
		Punta Escambron				X
Schwan (#263)	2	6" 1903A2	SB M1	WWII	dest., 1965	
		Punta Cangrejos				*
Lancaster (#264)	2	6" 1903A2	SB M1	WWII	modified	
FORT BROOKE		El Morro		National Park Service		* *
"Point"	1	3"	P	WWII	HECP-HDCP, exam battery	
Morro	1	4.7" Armstr	Armstrong	1917-1921	gun & mount orig. from Gatewood, Monroe	
Princesa	1	4.7" Armstr	Armstrong	1917-1921	gun & mount orig. from Gatewood, Monroe	
St. Elena	1	4.7" Armstr	Armstrong	1917-1921	gun & mount orig. from Gatewood, Monroe	
La Princesa	4	155 mm	PM	1941-		
		Other Puerto Rico 155 mm batteries				?
Acuda (Aguada)	4	155 mm	PM	1941-	Aguadilla	
Algarrobo	4	155 mm	PM	1941-	Mayaguez	
Ponce	4	155 mm	PM	1941-	Ponce	
Borinquen East	2	155 mm	PM	1941-	Borinquen	
Borinquen West	2	155 mm	PM	1941-	Borinquen	

Planned defenses in Trinidad

Fort Read (planned)						X
	2	12"	CBC	*Not Built*	Corozol Pt.	
	2	12"	CBC	*Not Built*	Chacachara Is.	
[#271]	2	6"	SBC	*Not Built*	Corozal Pt.	
[#272]	2	6"	SBC	*Not Built*	north Monos Is.	
[#273]	2	6"	SBC	*Not Built*	south Monos Is.	
[#274]	2	6"	SBC	*Not Built*	Chacachara Is.	
[#275]	2	6"	SBC	*Not Built*	Green Hill	
	24	155 mm	PM	WW II	various locations	
AMTBs	8	90 mm	F	*Not Built*	later reduced to 4 planned	
		Other AMTB sites				?
Bluie West 1	2	90 mm M1	F M3	1942-44	Narsarsuaq, Greenland	
Bluie West 7	2	90 mm M1	F M3	1942-44	Kangilinnguit, Greenland	
AMTBs	16	90 mm	F	*Not Built?*	Iceland	

United States Major Gun Seacoast Batteries Built or Modified
During the Years 1918-1945

by Mark A. Berhow

Originally published in the *CDSG Journal* Vol. 8, Iss. 4 (Nov. 1994) pp. 34-38. List updated January 2020. Additional information provided by Alvin C. Grobmeier, Robert D. Zink, Glen Williford, Mark Henkiel, Dave Kirchner, and Charles B. Robbins

Names: Battery names are denoted by the full name of the person they were named for; the bracketts enclose the part of that person's name not included in the official battery name. In 1940 a battery construction number (BCN) was assigned for identification purposes to each battery being built. The BCN was to be dropped when the battery was offically named, but as many of the 1940-45 batteries were not named by the Army, the BCNs have been retained as unoffical names.

Structure Status: Most of the batteries still remain and are unmodified except those denoted in the BCN or name columns as either:

  * structure destroyed   § in danger of destruction   † structure modified and/or built on   Δ current restoration efforts, see note [17]

Locations: Ft. stands for Fort, MR stands for Military Reservation.

Battery Status in 1946: The following abbreviations are used in this column: CM—original open emplacements casemated during 1942-44; DA— disarmed 1944; S—battery retained inactive; RM—battery not casemated but was to have been modernized with a shield for the the protection of the gun crew which retained the 360 degree firing arc; RC— battery to have been casemated but project canceled; C— completed, NC— not completed; NA— not armed (no gun tube); NB— not built[1] (also indicated by italicized brackett *[###]* BCN); PC— partially completed (gun emplacements built, guns mounted but magazines not completed).

Batteries built prior to 1940 still retained in mid-1945

16-inch M1919MII & MIII Army guns on BCLR batteries

Name	Harbor Defense	Location	Built	Status in 46	carriage model
Edward B. Williston*	Pearl Harbor	Ft. Weaver	1921-24	RM	M1919
Henry L. Harris	New York	Ft. Tilden	1921-23	CM	M1919
Frank S. Long †	Boston	Ft. Duvall	1921-27	CM	M1919

16-inch M1920 Army howitzer on BCLR batteries built 1921-22

Name	Harbor Defense	Location	Built	Status in 46	carriage model
Alexander C. M. Pennington [2]	Chesapeake Bay	Ft. Story	1921-22	RM	M1920
(Willoughby) Walke [2]*	Chesapeake Bay	Ft. Story	1921-22	RM	M1920

Early 16-inch MkIIMI Navy gun on BCLR batteries

Name	Harbor Defense	Location	Built	Status in 46	carriage model
Henry J. Hatch	Pearl Harbor	Ft. Barrette	1931-35	CM	M1919M1
(William G.) Haan	Balboa, Panama	Ft. Kobbe	1928-29	RC	M1919M1
(Arthur) Murray	Balboa, Panama	Ft. Kobbe	1924-29	CM	M1919M1

16-inch MkIIMI Navy gun on BCLR in casemated batteries

Name	Harbor Defense	Location	Built	Status in 46	carr. model
Richmond P. Davis	San Francisco	Ft. Funston	1936-40	C	M2
(Clarence P.) Townsley [17] Δ	San Francisco	Ft. Cronkhite	1938-40	C	M2

12-inch M1895/M1895MI gun on BCLR batteries

Name	Harbor Defense	Location	Built	Status in 46	carriage model
Stephen M. Foote	Portland	Ft. Levett	1917-24	CM	M1917
Augustus P. Gardner	Boston	Ft. Ruckman	1917-24	CM	M1917
Alfred S. Milliken	New Bedford	Ft. Rodman	1917-24	CM	M1917
(D.C.) Kingman	New York	Ft. Hancock	1918-22	CM	M1917
(Albert L.) Mills	New York	Ft. Hancock	1918-22	CM	M1917
(David) Hall	Delaware	Ft. Saulsbury	1917-24	S	M1917
(John) Haslet	Delaware	Ft. Saulsbury	1917-24	DA (to #519)	M1917
Loomis L. Langdon	Pensacola	Ft. Pickens	1917-23	CM	M1917
Leonard C. Hoskins †	Galveston	Ft. Crockett	1917-21	CM	M1917
Edwin R. Kimble	Galveston	Ft. Travis	1917-22	DA (to #520)	M1917
(Elmer J.) Wallace	San Francisco	Ft. Barry	1917-28	CM	M1917

12-inch M1895/M1895MI gun on BCLR batteries (cont.)

Name	Harbor Defense	Location	Built	Status in 46	carriage model
Henry W. Closson	Pearl Harbor	Ft. Kamehameha	1917-20	CM	M1917
Alexander MacKenzie	Cristobal, Panama	Ft. Sherman	1916-23	RC	M1917
Sedgwick Pratt	Cristobal, Panama	Ft. Sherman	1916-23	CM	M1917
Frank G. Smith [3]	Manila Bay	Ft. Mills	1919-21	-	M1917
[Clint C.] Hearn [3]	Manila Bay	Ft. Mills	1918-21	-	M1917

8-inch batteries[4]

Name	Harbor Defense	Location	Built	Status in 46	carriage model
Granger Adams * [4]	Honolulu	Ft. Ruger	1933-35	C	M1918
(Fredrick S.) Strong	San Diego	Ft. Rosecrans	1937	C	M1
(Henry J.) Reilly† [4]	Narrangansett Bay	Ft. Church	1938?	C	M1

Batteries built or planned during the years 1940-1945

16-inch MkIIMI Navy gun casemated batteries[5]

BCN	Name	Harbor Defense	Location	Status in 46	carriage model	BES
[101]		Portland	Cape Elizabeth[1]	NB	—	
102	(Harry L.) Steele	Portland	Peake Island MR	C	M4/M5	9
103	(Claudius M.) Seaman	Portsmouth	Ft. Dearborn	C	M5	14
104	John B. Murphy	Boston	East Point MR	C	M4	10
105 *		Boston	Ft. Dawes [6]	NC (deferred)	M5	9
[106]		Boston	Flowers Hill, Scituate[1]	NB	—	
107	(Quinn) Gray [5]	Narragansett Bay	Ft. Church	C	M3	10
108	(Alston) Hamilton [5]	Narragansett Bay	Ft. Greene	C	M4	12
109		Narragansett Bay	Ft. Greene	NC (deferred)	M4/M5	10
[110]		Narragansett Bay	Ft. Burnside[1]	NB	—	
111		Long Island Sound	Wilderness Pt., Fishers Is.	NC (deferred)	M4/M5	14
112 [5]		Long Island Sound	Camp Hero	C	M4	5
113	(John M.) Dunn	Long Island Sound	Camp Hero	C	M4	6
[114]		Long Island Sound	Ft. Terry ->Watch Hill, RI[1]	NB	—	
[115]		New York	Ft. Wadsworth[1]	NB	—	
116	Isaac N. Lewis [17] Δ	New York	Navesink MR	C	M4	8
[117]		New York	Nigger Pt. ->Jamaica Bay[1]	NB	—	
118	William R. Smith [5]	Delaware	Ft. Miles	C	M4	7
[119]		Delaware	Ft. Miles	NB (replaced by #519)		
120	(Daniel W.) Ketcham [5]	Chesapeake Bay	Ft. Story	C	M4	8
121		Chesapeake Bay	Ft. Story	C	M4	8
122	(Eben E.) Winslow [5, 17] Δ	Chesapeake Bay	Ft. John Custis	C	M4	8
[123]		Chesapeake Bay	Ft. John Custis[1]	NB	—	
[124]		Chesapeake Bay	Ft. Monroe[1]	NB	—	
[125]		Charleston	James Island	NB (replaced by #520)		
126 †	(Thomas Q.) Ashburn	San Diego	Ft. Rosecrans	C	M4	6
127	Paul D. Bunker	Los Angeles	White Point MR	C	M4	9
128 *		Los Angeles	Bolsa Chica MR [12]	NC (deferred)	M4	7
129		San Francisco	Ft. Barry	NC (deferred)	M4	12
[130]		San Francisco	Milagra Ridge MR	NB	—	
131		Puget Sound	Camp Hayden	C	M4	10
[132]		Cape Flattery	Ft. Hayden	NB	—	
[133]		Cape Flattery	Ft. Hayden	NB	—	
134 *		San Diego	Ft. Emory [16]	NC (deferred)	M4	5
[151]		Cristobal, Panama	Ft. Sherman[1]	NB	—	
[152]		Roosevelt Roads	Punta Mata Redondo, P.R.[1]	NB	—	
[153]		Roosevelt Roads	Mt. Pirata, Vieques Is.[1]	NB	—	
[154]		Roosevelt Roads	Mt. Jalobre, Vieques Is.[1]	NB	—	
[155]		Roosevelt Roads	Punta Yeguas, P.R.[1]	NB	—	

12-inch M1895/M1895M1 gun on BCLR casemated batteries

BCN	Name	Harbor Defense	Location	Status in 46	carriage model
	(Henry A.) Reed	San Juan	Ft. Amezquita, Cabras Is.	C	M1917
		Trinidad	Chacachacare Is.[1]	NB	—
		Trinidad	Corozal Pt.[1]	NB	—
519 [17] Δ		Delaware	Ft. Miles	C	M1917
520 †		Charleston	Marshall MR, Sullivan's Is.	C	M1917

8-inch Navy gun on BCLR batteries[4]

BCN	Name	Harbor Defense	Location	Status in 46	carriage model
401		Roosevelt Roads	Fortuna Hill, St. Thomas I.	PC	M1
402		Dutch Harbor	Ft. Schwatka	C	M1
403		Kodiak	Ft. J.H. Smith	C (two story)	M1
404		Kodiak	Ft. Abercrombie	C	M1
405 [8]	Robert E. DeMerritt	Kaneohe Bay	near PuuPapaa	C	M1
406		Roosevelt Roads	Punta Mata Redonda, P.R.	PC	M1
407 [8]		Honolulu	Ft. Ruger	NC	M1
[408]		North Shore, Oahu	Paumalu MR[1]	NB	—
409 [8]		North Shore, Oahu	Kaena Pt.	NC (tunnels excavated)	

6-inch gun BCLR shielded batteries

BCN	Name	Harbor Defense	Location	Status/1946	guns/carr.	# BES
201		Portland	Cape Elizabeth MR	NA	—/M4	5
202		Portland	Jewell Is. MR	C	M1/M3	5
203 †	(Richard K.) Cravens	Portland	Peake Is.	C	05A2/M1	5
204		Portsmouth	Ft. Dearborn	C	M1/M3	6
205		Portsmouth	Ft. Foster	NA	—/M4	6
206		Boston	East Point, Nahant	C	03A2/M1	5
207 *		Boston	Ft. Dawes [6]	NA	—/M3	5
208 [7]		Boston	Fourth Cliff	C	M1/M4	4
209	(Frank C.) Jewell	Boston	Outer Brewster Is.	C	M1/M3	6
210 †		New Bedford	Mishaum Pt.	C	M1/M3	5
211 §		Narrangansett Bay	Ft. Greene	C	03A2/M1	6
212 †		Narrangansett Bay	Warren Pt.	C	03A2/M1	6
213		Narrangansett Bay	Ft. Burnside	C	05A2/M1	5
214 †		Long Island Sound	Wilderness Pt., Fishers Is.	NC	—/M4	6
215		Long Island Sound	Ft. H.G. Wright	C	03A2/M1	5
216		Long Island Sound	Camp Hero	C	03A2/M1	4
217		Long Island Sound	Ft. Terry	NA	(M1)/M4	6
218		New York	Ft. Wadsworth	NC	—/M1	4
219		New York	Navesink	C	03A2/M1	5
220		New York	Ft. Tilden	NA	—/M1	4
221	(Ralph E.) Herring	Delaware Bay	Ft. Miles	C	03A2/M1	6
222	(Charles H.) Hunter	Delaware Bay	Ft. Miles	C	03A2/M1	6
223 §		Delaware Bay	Cape May MR	NC	M1/M3	4
224	(Phillip) Worcester [9]	Chesapeake Bay	Ft. Story	C	00/00 Ped.	2
225	(Raymond V.) Cramer	Chesapeake Bay	Ft. Story	C	03A2/M2	3
226		Chesapeake Bay	Ft. Story	C	M1/M4	3
227 [7, 10]		Chesapeake Bay	Fishermans Is.	C	05A2/M1	3
228		Chesapeake Bay	Ft. John Custis	NC	—/M4	4
229	(Horatio) Gates [11]	Chesapeake Bay	Ft. Wool	NA	—/M4	2
230		Charleston	Ft. Moultrie	NC	—/M4	5
231		Key West	Ft. Taylor	NA	—/M4	3
232		Key West	Salt Ponds MR	C	M1/M3	3
233		Pensacola	Ft. McRee	NC	—/M4	3
234 [10] Δ		Pensacola	Ft. Pickens	NC	—/M4	3
235		Galveston	Ft. San Jacinto	C	05A2/M1	3
236 Δ		Galveston	Ft. Travis	NC	—/M1	3
237 [7]		San Diego	Ft. Rosecrans	C	03A2/M1	5
238	(Charles) Humphreys	San Diego	Ft. Rosecrans	C	03A2/M1	4
239	(Homer B.) Grant	San Diego	Ft. Emory	C	05A2/M1	5

6-inch gun BCLR shielded batteries (cont.)

BCN	Name	Harbor Defense	Location	Status/1946	guns/carr.	# BES
240	Harry C. Barnes	Los Angeles	Point Vicente MR	C	03A2/M1	4
241 † [7, 12]		Los Angeles	Ft. MacArthur	NA	(M1)/M4	6
242 * [12]	Harry J. Harrison	Los Angeles	Bolsa Chica MR	C	M1/M4	6
243 [13]		San Francisco	Ft. Miley	NA	—/M4	6
244 [13]		San Francisco	Milagra Ridge MR	NA	(M1)/M4	5
245 Δ		Columbia River	Ft. Stevens	C	M1/M4	7
246 [13] Δ		Columbia River	Ft. Columbia	NC	—/M4	7
247 [13]		Columbia River	Ft. Canby	C	M1/M4	6
248 Δ		Puget Sound	Ft. Ebey	C	05A2/M1	5
249		Puget Sound	Camp Hayden	C	M1/M4	7
[250]		Cape Flattery	Ft. Hayden, Neah Bay	NB	—	
[251]		Cape Flattery	Ft. Hayden, Ocean Creek	NB	—	
261	(Mervyn C.) Buckey	San Juan	Punta Salinas	NA	—/M4	5
262	(William P.) Pence	San Juan	East Salinas Is.	C	03A2/M1	5
263 *	(Theodore) Schwan	San Juan	Punta Escambron	C	03A2/M1	4
264	(James W.) Lancaster	San Juan	Punta Cangrejos	C	03A2/M1	4
265		Roosevelt Roads	Pineros Is., P.R.	C	03A2/M1	
[266]		Roosevelt Roads	Martineau Hill, Vieques Is.[1]	NB	—	
[267]		Roosevelt Roads	Punta Arenas, Vieques Is.[1]	NB	—	
268		Roosevelt Roads	Punta Lima, P.R.	NA	—/M4	
[271]		Trinidad	Ft. Read, Corozal Pt.[1]	NB	—	
[272]		Trinidad	Ft. Read, Monos Is. N.[1]	NB	—	
[273]		Trinidad	Ft. Read, Monos Is. S.[1]	NB	—	
[274]		Trinidad	Ft. Read, Chacachacare Is.[1]	NB	—	
[275]		Trinidad	Ft. Read, Green Hill[1]	NB	—	
281 [13]		Argentia	Ft. McAndrew (North)	C	03A2/M1	
282 [14] Δ		Argentia	Ft. McAndrew, Hill 195	C	03A2/M1	
283		Bermuda	Stone Hill	C	03A2/M1	5
284 †		Bermuda	Ft. Victoria	C	03A2/M1	4
[285]	Jamaica -> Roosevelt Roads	Portland Bight-> Vieques I.[1]	NB	—	
290		Sitka	Ft. Babcock	NC(88%)	—/—	2
291		Sitka	Ft. Peirce	NC(98%)	03A2/M1	4
292		Sitka	Ft. Rousseau	C	03A2/M1	5
293		Seward	Ft. McGilvray	C (99%)	03A2/M1	4
294		Seward	Ft. Bulkley	NC(90%)	03A2/M1	3
[295]		Kodiak	Ft. J. H. Smith[1]	NB	—	
296		Kodiak	Ft. Tidball	C	03A2/M1	
[297]		Kodiak	Ft. Abercrombie[1]	NB	—	
298		Dutch Harbor	Ft. Learnard	C	03A2/M1	
[299]		Dutch Harbor	Ft. Schwatka[1]	NB	—	
301	Forrest J. French	Kanoehe Bay	Ft. Hase	C	03A2/M1	3
302 [8]	Avery J. Cooper	Kanoehe Bay	Lae-o-ka-oio	C/NA?	03A2/M1	3
303 [8]		Pearl Harbor	Puo-O-Hulu	NC	—/—	
304		Honolulu	Punchbowl	NC	—/—	
[305]		Pearl Harbor	Koko Saddle[1]	NB	—	
[311]		Roosevelt Roads	Cabo San Juan, P.R.[1]	NB	—	
[312]		Roosevelt Roads	Culebra Is., North Pt.[1]	NB	—	
[313]		Roosevelt Roads	Culebra Is., Dolphin Head[1]	NB	—	
314		Roosevelt Roads	Water Is., St. Thomas	NA	—/M4	
[315]		Roosevelt Roads	Hill 411, St. Thomas[1]	NB	—	

1942-1945 Hawaiian Naval turret batteries

Name	Armament	Location	Status in 1946
Lewis S. Kirkpatrick	two twin 8" turrets	Wiliwilinui Ridge	C
Louis S. Burgess *	two twin 8" turrets	Salt Lake MR	C
Carrol G. Riggs	two twin 8" turrets	Opaeula MR	C
George W. Ricker	two twin 8" turrets	Browns Camp MR	C
Pennsylvania	one triple 14" turret	Ft. Hase, Ulupau Head	C
Arizona [15]	one triple 14" turret	Kahe Pt.	NC

NOTES:

1. The given locations of the batteries that were not actually built (NB) on this list are the last ones proposed for that battery before they were officially canceled. Many of these had several different proposed locations during the years 1940-1942. Battery locations that are known to have been moved are indicated by the symbol -> between locations. The given locations are usually the first assigned location and the last assigned location for the indicated battery.

2. Battery Pennington was the name originally given to all four guns of the 16-inch howitzer battery. Later the battery was divided into two batteries, Pennington 'A' and 'B'. In 1941 Battery 'B' (gun positions 3 and 4) were renamed Battery Walke, while positions 1 and 2 retained the name Pennington.

3. The two 12-inch LRBC guns built at Fort Mills were originally designated Battery Smith. However, the magazines were actually built as two separate structures to service the guns which were located on adjacent ridges. In the 1920s, the battery was redesignated as two single-gun batteries, with the left gun receiving the name of Battery Hearn. These guns were rendered inoperable by their crews before Corregidor was surrendered to the Japanese. They are still in place.

4. Battery Granger Adams was built for two Army M1888 8-inch guns on M1918 barbette carriages. These guns were emplaced on simple concrete platforms by 1935 with a protected magazine between the two gun positions. This battery served as the model for Battery Strong built two years later, which utilized two 8-inch Navy guns on newly designed Army M1 barbette carriages. Battery Reilly was a unique 8-inch casemated battery. The new 6-inch and 8-inch batteries were designed with open emplacements, but the gun crews were to be protected from shrapnel by metal shields which surrounded the gun carriage. The shields were developed and manufactured for the 6-inch guns, but not for the 8-inch guns. Battery Reilly was buried in 1995.

5. The 16-inch battery construction program had an interesting progression of events from 1940 to 1945. In addition to the 27 new 1940 Program batteries authorized that year, six 16-inch batteries previously authorized for construction in 1939 were also incorporated—four then under construction and two not yet started: underway were two at Narragansett Bay, RI, later designated as Battery Gray-BCN 107 and Battery Hamilton-BCN 108; one underway at Cape Henlopen, DE, later designated Battery Smith-BCN 118; and one underway at Cape Henry (Fort Story), VA, later designated Battery Ketcham-BCN 120. These four had a slightly different design than the rest of 1940 program batteries. A fifth 16-inch battery had been assigned to Montauk Point, NY, (eventually designated BCN 112) and a sixth 16-inch battery had been assigned to Cape Charles, VA (eventually designated BCN 122). This resulted in a total of 33 new 16-inch batteries to be built or finished when the project officially started in late 1940. Once underway, priorities shifted resulting in the cancellation of a number of the planned 16-inch batteries. BCN 115 was dropped from the construction program in late 1941 and BCN 134 was added for San Diego. The construction of fourteen 16-inch batteries which had not been started were deferred by September 1942 (BCNs 101, 106, 110, 114, 117, 119, 123, 124, 125, 128, 130, 133, and 134) as well as the Carribean 16-inch batteries (BCNs 151, 152, 153, 154, 155). In November 1942 construction on BCNs 128, 133, and 134 were allowed to proceed, the remaining deferred batteries were eliminated. BCNs 132 and 133 were cancelled in October 1943. Work on BCNs 105, 109, 111, 128, 129, and 134 was suspended in November 1943—the concrete work was completed, but the armament and the power plants were not installed, resulting in designations of not completed (NC) or not armed (NA).

6. All emplacements at Fort Dawes on Deer Island were destroyed in the mid-1980s for a Boston municipal water treatment plant.

7. The power equipment, including the diesel generators, still remain in BCNs 208, 223, 237, and 241.

8. These batteries were tunneled in the hillsides of Oahu. The guns were protected from rockfall by by thin conrete aprons.

9. Battery Worcester was built for a pair of Model 1900 6-inch pedestal mount guns. The design of this battery was much simpler than the 6-inch design finally adopted. The construction was given BCN 224 to fit it into the program. A second battery built under this particular design was Battery Kessler at Fort Tilden which was totally rebuilt during 1941-1942, but not as a part of the modernization program.

10. The guns of BCN 227 were salvaged by the Smithsonian. In the early 1970s, they were installed in BCN 234 at Fort Pickens. Now part of the Gulf Shores National Seashore, the guns are one of three pairs of this model that remain today.

11. Battery Gates (BCN 229) was a conversion of an existing 6 inch battery, there was only 105 ft. between the guns.

12. Battery Harry C. Harrison (BCN 242) was armed, but not considered complete in 1944. In 1946, Battery Harrison's guns were transferred to BCN 241, which was not armed by late 1945. The guns remained at BCN 241 until the mid-1950s. BCN 241 still retains its power equipment and generators. Both Bolsa Chica batteries were demolished during 1993-95.

13. The carriages were installed for BCN 246 and the gun tubes were delivered to the site but not mounted. The guns of BCN 246 and those of BCN 247 were transferred to San Francisco and mounted in BCN 243 and BCN 244 sometime during 1945-47. In 1993, a pair of 6-in guns and carriages that remained at BCN 281 were removed, refurbished, and installed in BCN 246 at Fort Columbia State Park by the State of Washington. BCN 247 at Fort Stevens State Park has two 5-inch navy gun turrets in its emplacements.

14. The guns and carriages still remain at BCN 282.

15. The turret and its guns was mounted at Battery Arizona, but it was not considered complete nor were its guns proof-fired.

16. BCN 134 was destroyed by the Navy in 2016 to make room for a new SEAL training complex.

17. Modern restoration efforts. Battery 520 has a 12-inch navy barrel remounted in one casemate, the battery is home to the Fort Miles Museum, open regular hours for tours. A 16-inch navy MkVII barrel is on display nearby. Battery Lewis has been rennovated and restored by Monmouth County Parks and Recreation and is open on scheduled weekends and has a 16-inch MkVII Navy gun barrel mounted in one casemate. At Battery Winslow a 16-inch MkVII Navy gun barrel is mounted in one casemate. Battery Townsley is being restored and is open for tours once a month, with a 16-inch MkVII Navy gun barrel on display. BCN 234 and BCN 246 have had guns replaced in 1976 and 1992 respectively. BCN 248 at Fort Ebey, BCN 247 at Fort Stevens, and BCN 245 at Cape Disappointment are clean and open to the public.

SURVIVING AMERICAN SEACOAST ARTILLERY WEAPONS
December 2025

Compiled: Lists in various CDSG publications prepared by C.L. Kimbell (1985), R.D. Zink (1989), T.C. McGovern (1992 and 1996) and Tom Batha (2014-16).

General Note: This is an attempt to list surviving weapons (or the same model/type) that were used by the American armed forces, either in the U.S. and overseas, in a coast defense role in the "modern era" (1890 to 1950). Items to be included in this list must retain at least the whole gun/mortar/howitzer. Surviving weapons from earlier periods, muzzleloading cannon (rifled and smoothbore), field artillery (except for 155mm), mobile anti-aircraft guns, and British, Canadian, or Mexican coast artillery are excluded from this list, even if used in North America. In a few cases, weapons have been included because they represent weapons similar to those used for coast defense, and are sufficiently interesting to warrant inclusion. Every effort has been made to make this list as accurate as possible, but it will probably contain a number of errors and omissions. Corrections and additions can be sent to Tom Batha (tbatha@primelink1.net), Mark Berhow (berhowma@cdsg.org) or Terry McGovern (tcmcgovern@att.net).

The list is arranged by caliber (largest to smallest). The first line contains data about the weapon: the quantity at the site, the caliber of weapon (inches or millimeters), the model, serial number, place of manufacture, and carriage information, if known. The next lines contain information on where the weapon was previously located as coast defense weapon. The last line contains data about the location of the weapon: battery or site were the weapon is currently located, fort or installation, geographic location, and city and/or country.

Sources: Matt Adams, Tom Batha, Mark Berhow, Roger Davis, Colt Denfeld, Elliott Deutsch, Bill Dorrance, Clarence Drennon, Al Grobmeier, Greg Hagge, Ed Jerue, Nelson Lawry, Ray Lewis, Charles Kimball, David Kirchner, Danny Malone, John Martini, Dean Mayhew, Terry McGovern, Jim Parker, Dan Rowbottom, Bob Rutherford, Phil Sims, Peter Smith, Dick Weinert, Glen Williford, David Yeager, and Bob Zink.

16 Inch Guns

1. 16"/50 Mk III Mod 1, _____ No. 138 on M1919 Barbette Carriage Watertown Arsenal No. 1
   Location: Ordnance Museum, Aberdeen Proving Ground, MD
2. 16"/50 Mk II, Naval Gun Factory No. 111
   Location: U.S. Navy Museum, Washington Navy Yard, Washington, DC
3. 16"/50 Mk II _____ No. 96
   Location: Naval Surface Warfare Center, Dahlgren, VA
4. 16"/50 Mk II_____ No. 100
   Location: Naval Surface Warfare Center, Dahlgren, VA

14 Inch Guns

1. M1909, Watervliet No. 4 on M1909 Turret Mount, Newport News No. 1
   Location: Battery John M. Wilson, Fort Drum, El Fraile Island, PI
2. M1909, Watervliet No. 3 on M1909 Turret Mount, Newport News No. 1
   Location: Battery John M. Wilson, Fort Drum, El Fraile Island, PI
3. M1909, Watervliet No. 2 on M1909 Turret Mount, Newport News, No. 2
   Location: Battery William L. Marshall, Fort Drum, El Fraile Island, PI
4. M1909, Watervliet No. 1 on M1909 Turret Mount, Newport News, No. 2
   Location: Battery William L. Marshall, Fort Drum, El Fraile Island, PI
5. M1910, Watervliet No. 15 on M1907M1 Disappearing Carriage, Watervliet No. 20
   Location: Battery Gillespie, Fort Hughes, Caballo Island, PI
6. M1910, Watervliet No. 8 on M1907M1 Disappearing Carriage, Watervliet No. 17
   Location: Battery Woodruff, Fort Hughes, Caballo Island, PI
7. 14"/50 Mk II, _____ No. 119L2 on Mk 1 Railway Mount No. 148
   Location: U.S. Navy Museum, Washington Navy Yard, Washington, DC

12 Inch Guns

1. M1895M1A4, Watervliet No. 1 on M1917 Barbette Carriage, Eng. Machine No. 31
   Location: Battery Smith, Fort Mills, Corregidor, PI
2. M1895M1A4, Watervliet No. 6 on M1917 Barbette Carriage, Eng. Machine No. 30
   Location: Battery Hearn, Fort Mills, Corregidor, PI
3. M1895M1A4, Watervliet No. 8 (spare tube) Battery Hearn, Fort Mills, Corregidor, PI
   Location: Baywalk, Dolomite Beach, Manila for mock turret display
4. M1895, Watervliet No. 13 on M1901 Disappearing Carriage, Watertown No. 14
   Location: Battery Crockett, Fort Mills, Corregidor, PI
5. M1895, Watervliet No. 27 on M1901 Disappearing Carriage, Watertown No. 15
   Location: Battery Crockett, Fort Mills, Corregidor, PI
6. M1895, Bethlehem No. 9 (spare tube) Battery Crockett, Fort Mills, Corregidor, PI,
   Location: Baywalk, Dolomite Beach, Manila for mock turret display
7. M1895, Watervliet No. 37 on M1901 Disappearing Carriage, Watertown No. 16
   Location: Battery Cheney, Fort Mills, Corregidor, PI
8. M1895, Watervliet No. 12 on M1901 Disappearing Carriage, Watertown No. 17
   Location: Battery Cheney, Fort Mills, Corregidor, PI
9. M1895, Watervliet No. 16 (spare tube) Battery Cheney, Fort Mills, Corregidor, PI,
   Location: Bottomside Area, Fort Mills, Corregidor, PI
10. M1895, Watervliet No. 36, with partially scrapped Disappearing Carriage
    Location: Battery Wheeler, Fort Mills, Corregidor, PI
11. M1895, Bethlehem No. 7 on M1901 Disappearing Carriage, Watertown No. 2
    Location: Battery Wheeler, Fort Mills, Corregidor, PI
12. M1895, Bethlehem No. 10 (spare tube)
    Location: Battery Wheeler, Fort Mills, Corregidor, PI
13. M1895 Watervliet No. 19 on M1918 Railway Mount, Marion Steam Shovel No. 9
    Location: US Army Ordnance School, Fort Lee, VA

12 Inch Mortars

1. M1890M1, Watervliet No. 173 on M1896M1 Carriage, Watertown No. 158
   Location: Battery Way, Fort Mills, Corregidor Island, PI
2. M1890M1, Watervliet No. 174 on M1896M1 Carriage, Watertown No. 151
   Location: Battery Way, Fort Mills, Corregidor Island, PI
3. M1890M1, Watervliet No. 170 on M1896M1 Carriage, Watertown No. 241
   Location: Battery Way, Fort Mills, Corregidor Island, PI
4. M1890M1, Watervliet No. 172 on M1896M1 Carriage, Watertown No. 150
   Location: Battery Way, Fort Mills, Corregidor Island, PI
5. M1890M1, Bethlehem No. 31 on M1896M1 Carriage, _____ No. 104
   Location: Battery Geary Pit A, Fort Mills, Corregidor Island, PI
6. M1890M1, Bethlehem No. 40 on M1896M1 Carriage, _____ No. 210
   Location: Battery Geary Pit A, Fort Mills, Corregidor Island, PI
7. M1890M1, Watervliet No. 22 on M1896M1 Carriage, American Hoist No. 183
   Location: Battery Laidley Pit A, Fort Desoto, Mullet Key, FL
8. M1890M1, Watervliet No. 86 on M1896M1 Carriage, American Hoist No. 184
   Location: Battery Laidley Pit A, Fort Desoto, Mullet Key, FL
9. M1890M1, Watervliet No. 132 on M1896M1 Carriage, American Hoist No. 185
   Location: Battery Laidley Pit B, Fort Desoto, Mullet Key, FL
10. M1890M1, Watervliet No. 135 on M1896M1 Carriage, American Hoist No. 187
    Location: Battery Laidley Pit B, Fort Desoto, Mullet Key, FL
11. M1908, Watervliet No. 2 on M1908 Carriage, No. 17
    Location: Battery Geary Pit B, Fort Mills, Corregidor Island, PI
12. M1908, Watervliet No. 22 on M1908 Carriage No. 20
    Location: Battery Geary Pit B, Fort Mills, Corregidor Island, PI
13. M1908, Watervliet No. 13 on M1908 Carriage No. 19
    Location: Battery Geary Pit B, Fort Mills, Corregidor Island, PI
14. M1908, Watervliet No. 12on M1908 Carriage No. 18
    Location: Battery Geary Pit B, Fort Mills, Corregidor Island, PI

15. M1912, Watervliet No. 40 on M1896MIII Carriage, Watertown No. 41
    Location: Battery Craighill Pit A, Fort Mills, Corregidor Island, PI
16. M1912, Watervliet No. 39 on M1896MIII Carriage, Watertown No. 40
    Location: Battery Craighill Pit A, Fort Mills, Corregidor Island, PI
17. M1912, Watervliet No. 41 on M1896MIII Carriage, Watertown No. 38
    Location: Battery Craighill Pit B, Fort Hughes, Caballo Island, PI
18. M1912, Watervliet No. 38 on M1896MIII Carriage, Watertown No. 39
    Location: Battery Craighill Pit B, Fort Hughes, Caballo Island, PI
19. M1911, ____ No. 8, no carriage (non- standard training dummy)
    Location: Visitor Center, Fort Wadsworth, Staten Island, NY

10 Inch Guns

1. M1895M1, Watervliet No. 25 on M1901 Disappearing Carriage, Watertown No. 14
   Location: Battery Grubbs, Fort Mills, Corregidor Island, PI
2. M1895M1, Watervliet No. 22 on M1901 Disappearing Carriage, Watertown No. 16
   Location: Battery Grubbs, Fort Mills, Corregidor Island, PI
3. M1895M1, Watervliet No. 20 (spare tube only)
   Location: Battery Grubbs, Fort Mills, Corregidor Island, PI
4. M1895M1, Watervliet No. 26 on M1901 Disappearing Carriage, Watertown No. 13
   Formerly emplaced at Battery Warwick, Fort Wint, PI
   Location: Battery Worth, Fort Casey, Coupeville, WA
5. M1895M1, Watervliet No. 28 on M1901 Disappearing Carriage, Watertown No. 15
   Formerly emplaced at Battery Warwick, Fort Wint, PI
   Location: Battery Worth, Fort Casey, Coupeville, WA
6. M1888, Watervliet No. 41
   Formerly emplaced at Battery Harker, Fort Mott, NJ
   Location: Fort Cape Spear, St John's, Newfoundland
7. M1888, Watervliet No. 3
   Formerly emplaced at Battery Harker, Fort Mott, NJ
   Location: Fort Cape Spear, St John's, Newfoundland
8. M1888, Watervliet No. 12, on M1893 Barbette Carriage, Watertown No. 11
   Formerly emplaced at Battery Quarles, Fort Worden, WA
   Location: Fort McNutt, McNutt Island, Nova Scotia
9. M1888, Watervliet No. 37, on M1893 Barbette Carriage, Watertown No. 1
   Formerly emplaced at Battery Quarles, Fort Worden, WA
   Location: Fort McNutt, McNutt island, Nova Scotia (Note: Partially scrapped)
10. M1888, Bethlehem No. 9
    Location??

8 Inch Guns

1. M1888M1A1 Watervliet No. 32 on M1918M1 Barbette Carriage, Morgan Engineering No. 9
   Formerly emplaced at Battery Bowyer, Fort Morgan, AL
   Location: Plant Park, Tampa, FL
2. Navy 8"/45 Mk VI M3A2 _____No. 160L2
   Location: Battery 404, Fort Abercrombie, Kodiak, AK
3. Navy 8"/45 Mk VI M3A2_____No. 154L2
   Location: Battery 404, Fort Abercrombie, Kodiak, AK
4. Navy 8"/45 Mk VI M3A2_____No. 134L2
   Formerly emplaced at Battery 403, Fort J.H. Smith, Kodiak, AK
   Location: Kodiak Airport, Kodiak, AK
5. M1888MII, Bethlehem No. 8
   Location: Bottomside artillery park, Fort Mills, Corregidor Island, PI
6. Navy 8"/45 Mk VI M3A2_____No._____on M1 Railway Proof Mount Baldwin No.____
   Note: bored out to 9.12" for experimental purposes, missing 8 ft. from muzzle
   Location: Fort Miles Museum, Cape Henlopen State Park, DE

7 Inch Naval Guns

1. 7"/45 Mk II, Midvale No. 90 on Mk II Mod 4 Pedestal Mount No. ___
   Formerly armed USS *New Hampshire* (BB-25)
   Formerly emplaced at Harbor Battery, Sand Island, HI
   Location: Battery Randolph, Fort DeRussey, Oahu, HI
2. 7"/45 Mk II, Midvale No. 92 on Mk II Mod 4 Pedestal Mount No. ___
   Formerly armed USS *New Hampshire* (BB-25)
   Formerly emplaced at Harbor Battery, Sand Island, HI
   Location: Battery Randolph, Fort DeRussey, Oahu, HI
3. 7"/45 Mk II, Naval Gun Factory No. 100 on Mk II Mod 4 Pedestal Mount No. 43
   Location: Battery North, Bora Bora, Society Island, French Polynesia
4. 7"/45 Mk II, Bethlehem No. 97 on Mk II Mod 4 Pedestal Mount No. 55
   Formerly emplaced at Battery North, Bora Bora, Society Island, French Polynesia
   Location: private yard in Bora Bora.
5. 7"/45 Mk II, Watervliet No. 57 on Mk II Mod 4 Pedestal Mount No. 23
   Formerly armed USS *Louisiana* (BB-19).
   Location: Battery West, Bora Bora, Society Island, French Polynesia
6. 7"/45 Mk II, Naval Gun Factory No. 82 on Mk II Mod 4 Pedestal Mount No. 30
   Location: Battery West, Bora Bora, Society Island, French Polynesia
7. 7"/45 Mk II, Naval Gun Factory No. 81 on Mk II Mod 4 Pedestal Mount No. 19
   Location: Battery South, Bora Bora, Society Island, French Polynesia
8. 7"/45 Mk II, Naval Gun Factory No. 96 on Mk II Mod 4 Pedestal Mount No. 18 (?)
   Location: Battery South, Bora Bora, Society Island, French Polynesia
9. 7"/45 Mk II, Watervliet No. 60 on Mk II Mod 4 Pedestal Mount No. 13
   Formerly armed USS *Louisiana* (BB-19)
   Location: Battery East, Bora Bora, Society Island, French Polynesia
10. 7"/45 Mk II, Midvale No. 69 on Mk II Mod 4 Pedestal Mount No. 14
    Formerly armed USS *Kansas* (BB-21)
    Location: Battery East, Bora Bora, Society Island, French Polynesia

6 Inch Guns

1. M1905 Watervliet No. 30 on M1903 Disappearing Carriage, Watertown No. 1
   Formerly emplaced at Battery Schofield, West Point Military Academy, NY
   Location: Battery Cooper, Fort Pickens, Pensacola, FL
2. M1905 Watervliet No. 9 on M1903 Disappearing Carriage, Watertown No. 2
   Formerly emplaced at Battery Schofield, West Point Military Academy, NY
   Location: Battery Chamberlin, Fort Scott, San Francisco, CA
3. M1905 Watervliet No. 31 on M1905M1 Disappearing Carriage, Watertown No. 12
   Location: Battery Morrison, Fort Mills, Corregidor Island, PI
4. M1905 Watervliet No. 32 on M1905M1 Disappearing Carriage, Watertown No. 13
   Location: Battery Morrison, Fort Mills, Corregidor Island, PI
5. M1905 Watervliet No. 4 on M1905M1 Disappearing Carriage, Watertown No. 9
   Note: gun is in poor condition, cut up
   Location: Battery Ramsay, Fort Mills, Corregidor Island, PI
6. M1905 Watervliet No. 33 on M1905M1 Disappearing Carriage, Watertown No. 11
   Note: gun is in poor condition, cut up
   Location: Battery Ramsey, Fort Mills, Corregidor Island, PI
7. M1905 Watervliet No. 27
   Note: spare tube, no carriage
   Location: Battery Morrison, Fort Mills, Corregidor Island, PI
8. M1905 Watervliet No. 6 on M1905M1 Disappearing Carriage, Bethlehem No. 6
   Location: Battery Hall, Fort Wint, Grand Island, Subic Bay, PI
9. M1905 Watervliet No. 7 on M1905M1 Disappearing Carriage, Bethlehem No. 7
   Location: Battery Hall, Fort Wint, Grande Island, Subic Bay, PI
10. M1908 Watervliet No. 6
    Location: Battery Leach, Fort Hughes, Caballo Island, PI

11. M1908MII Watervliet No. 4
    Note: with shield, no carriage
    Location: Battery Roberts, Fort Drum, El Fraile Island, PI
12. M1900 Watervliet No. 22 on M1900 Barbette Carriage, Rock Island No. 12
    Formerly emplaced at Battery Peck, Fort Hancock, Sandy Hook, NJ
    Location: Battery Gunnison (Peck 2), Fort Hancock, Sandy Hook, NJ
13. M1900 Watervliet No. 23 on M1900 Barbette Carriage, Rock Island No. 17
    Formerly emplaced at Battery Peck, Fort Hancock, Sandy Hook, NJ
    Location: Battery Gunnison (Peck 2), Fort Hancock, Sandy Hook, NJ
14. M1905A2 Watervliet No. 16 on M1 Barbette Carriage, _____No. 58
    Formerly emplaced at Battery 227, Fort John Custis, VA
    Location: Battery 234, Fort Pickens, Pensacola, FL
15. M1905A2 Watervliet No. 21 on M1 Barbette Carriage, _____ No. 59
    Formerly emplaced at Battery 227, Fort John Custis, VA
    Location: Battery 234, Fort Pickens, Pensacola, FL
16. M1903A2 Watervliet No. 30 on M1 Barbette Carriage, _____No. 9
    Formerly emplaced at Battery 281, Fort McAndrew, Argentia, Newfoundland
    Location: Battery 246, Fort Columbia, Chinook Point, WA
17. M1903A2 Watervliet No. 61 on M1 Barbette Carriage, _____No. 10
    Formerly emplaced at Battery 281, Fort McAndrew, Argentia, Newfoundland
    Location: Battery 246, Fort Columbia, Chinook Point, WA
18. M1903A2 Watervliet No. 13 on M1 Barbette Carriage, _____No. 44
    Location: Battery 282, Fort McAndrew, Argentia, Newfoundland
19. M1903A2 Watervliet No. 8 on M1 Barbette Carriage, _____No. 45
    Location: Battery 282, Fort McAndrew, Argentia, Newfoundland
20. M1905A1 Watervliet No. 12 on M1917 carriage, Morgan Engineering No. 22
    Formerly emplaced at Battery Parker, Fort Casey, WA, Note: with limber, Morgan No. 82
    Location: U.S. Army Ordnance Training Support Facility, Fort Lee, VA
21. Rapid Fire BL Gun Mk 7, Armstrong No. 12139 on Mk 2 Barbette Carriage No. 11162
    Formerly emplaced at Battery Burchsted, Fort Dade, FL
    Location: Fort Desoto, Mullet Key, FL
22. Rapid Fire BL Gun Mk 7, Armstrong No. 12140 on Mk 2 Barbette Carriage No. 11157
    Formerly emplaced at Battery Burchsted, Fort Dade, FL
    Location: Fort Desoto, Mullet Key, FL

6 Inch Naval Guns

1. Navy 6"/50 Mk VIII Naval Gun Factory No. 368
    Location: Near airport, Nome, AK
2. Navy 6"/50_____ on Pedestal Mount
     Location: Cannon Beach, Yakutat, AK
3. Navy 6"/50_____ on Pedestal Mount
    Location: Cannon Beach, Yakutat, AK
4. Navy 6"/50_____
    Location: City dump, Cold Bay, AK
5. Navy 6"/50 _____ on Mk X Pedestal Mount No.____
    Location: George Island, Cross Sound, AK
6. Navy 6"/50 Mk VIII Mod II Midvale No. 550 on Mk X Pedestal Mount
    Location: Blunts Point, Pago Pago, American Samoa
7. Navy 6"/50 Mk VIII Mod II Midvale No. 554 on Mk X Pedestal Mount
    Location: Blunts Point, Pago Pago, American Samoa
8. Navy 6"/50 Mk VIII Mod II _____No.____ on Mk X Pedestal Mount
    Location: Breakers Point, Pago Pago, American Samoa
9. Navy 6"/50 Mk VIII Mod II _____No.____ on ___ Pedestal Mount
    Location: Breakers Point, Pago Pago, American Samoa
10. Navy 6"/50 Mk VI Mod 1 No. 309L on Pedestal Mount
    Note: Ex-navy gun emplaced in 1942 at Fort New Amsterdam
    Location: Paramaribo, Suriname (formerly Dutch Guiana)

11. Navy 6"/50
    Location: Paramaribo, Suriname
12. Navy 6"/50 Mk VI Mod 2 No. 314L on Pedestal Mount
    Location: Paramaribo, Suriname
13. Navy 6"/50
    Location: Paramaribo, Suriname

155mm GPF Guns

1. M1918M1, Bullard Eng. Works No. 161; Carriage M3 No.____
   Location: Presidio Army Museum, Fort Scott, San Francisco, CA
2. M1918M1, Watervliet Arsenal No. 906; Carriage M2 No. 82
   Location: Fort Stevens State Park, Astoria, OR
3. M1917, Puteaux No. 629; Carriage M1917 No. 222, with limber
   Location: U.S. Army Ordnance Training Support Facility, Fort Lee, VA
4. M1918M1, Watervliet Arsenal No. 1073, Carriage M3   Minneapolis Steel Co. No.X
   Location: Fort Morgan, Mobile Bay, AL
5. M1917, Puteaux No.918; Carriage M3 No.523
   Location: Fort MacArthur Military Museum, San Pedro, CA
6. M1918_____No. ___; Carriage M1918 No.____
   Location: Shrine of Valor Memorial, Bataan, Philippines
7. M1918_____No. 810; Carriage M1918 No.____
   Location: Pulaski Park, Pulaski, VA
8. M1918_____No. 75; Carriage M1918 No.____
   Location: Odlin County Park, Lopez Island, WA
9. M1917, Puteaux No. 1263; Carriage M1917 No. X
   Location: Washelli-Evergreens Cemetery, North Aurora St., Seattle, WA
10. M1918M1, Bullard Eng. Works No. 314; Carriage M1918
    Location: Memorial Park, Lansdale, PA
11. M1918M1, Bullard Eng. Works No. 209; Carriage M1918
    Location: Memorial Park, Lansdale, PA
12. M1918M1, Bullard Eng. Works No. 149; Carriage M1918 No.____, with limber
    Location: Barberton Veterans Memorial, Lake Anna, Barberton, OH
13. M1918M1, Watervliet Arsenal No. 218; Carriage M1918 No.____
    Location: Cemetery on Rte. 40, Indianapolis, IN
14. M1917M1, Puteaux No.____; Carriage M3 No.473
    Location: Fort Sill, OK
15. M1918M1, Watervliet Arsenal No. 920; Carriage M1918
    Location: Lake St., Lancaster, NY
16. M1917, Puteaux No. 959; Carriage M1918 No. 208 (missing wheels & axel)
    Location: Fort Miles Museum, Cape Henlopen State Park, Lewes, DE
17. M1917, Puteaux No.____; Carriage M1918 No.____
    Location: Camp Shelby, MS
18. M1918M1, Bullard Eng. Works No. 376; Carriage M1918 No.____
    Location: Cemetery on David Highway, Saranac, MI
19. M1917, Puteaux No. ___; Carriage M1918 No.____
    Location: Fort Macon State Park, Atlantic Beach, NC
20. M1818M1, Watervliet Arsenal No. 806; Carriage _____No.____
    Location: Cherryville, KS
21. M1918M1, Watervliet No. 989; Carriage M1918 No. X, with limber
    Location: Edgemont, SD
22. M1918M1, Watervliet No. 988; Carriage M1918 No. X, with limber
    Location: Edgemont, SD
23. M917, Bullard No. 427; Carriage M1918 No. X
    Location: Courthouse, Forsyth, MT
24. M1918M1, Bullard Eng. Works No. 171; Carriage _____No.____
    Location: Mechanicsville, IA

25. M1918, _____No. 967; Carriage _____No.___
    Location: Marine Base, Quantico, VA
26. M1918M1, Watervliet Arsenal No. 976; Carriage M1918 No.___
    Location: Quincy, IL
27. M1917A1, Puteaux No.765/USA No. 1158; Carriage M1917 No. 1158
    Location: Scotland, SD
28. M1917, Puteaux No. 103; Carriage M3 No. X
    Location: American Legion Post, Seattle, WA
29. M1918,_____No.___; Carriage _____No.___
    Location: St. Louis, MO (possibly from Jefferson Barracks)
30. M1918M1, Puteaux No. 764; Carriage M1918A1 Rock Island Arsenal No. 633 (missing wheels & axel)
    Location: Lockport, NY
31. M1917, Puteaux No. 602; Carriage M1917 No. X Note: Nos. 31 & 32 believed never used as CA
    Location: Fort Park, Walla Walla, WA
32. M1917, Puteaux No. 740; Carriage M1917 No. X Note: Nos. 31 & 32 believed never used as CA
    Location: Fort Park, Walla Walla, WA
33. M1918M1,_____No.____; Carriage _____No.__
    Location: Grand Rapids, WI
34. M1917, Puteaux No. ___; Carriage M1918 No.___
    Location: Fort Macon State Park, Atlantic Beach, NC

5 Inch Guns

1. M1897 Watervliet No. 6, Formerly emplaced at Battery Krayenbuhl, Fort Mott, NJ
   Location: Chapman Park, Stanley, WI
2. M1897 Bethlehem No. 16, Formerly emplaced at Battery Shipp, Fort Caswell, NC
   Location: Memorial Park, Woodstown, NJ
3. M1897 Bethlehem No. 21, Formerly emplaced at Battery Wagner, Fort Baker, San Francisco, CA
   Location: Veterans Memorial Park, Mahoning Ave, Warren, OH
4. M1897 Bethlehem No. 13, Formerly emplaced at Battery Bell, Fort Schuyler, NY
   Location: Valley Forge Military Academy, Wayne, PA
5. M1897 Bethlehem No. 14, Formerly emplaced at Battery McGrath, Fort Rosencrans, CA
   Location: VFW Post 1989, Indiana, PA
6. M1897 Bethlehem No. 3, Formerly emplaced at Battery Boutelle, Fort Winfield Scott
   Location: Virginia War Museum, Newport News, VA
7. M1897 Bethlehem No. 22, Formerly emplaced at Battery Vicars, Fort Worden, WA
   Location: Chewsville, MD
8. M1897 Bethlehem No.____ Note: muzzle cut off, data inferred
   Location: Boonsboro, MD

5 inch/51 Navy Guns

1. 5"/51 Mk VIII Four Lakes No. 1205
   Location: NAS Trumbo Point, Key West, FL
2. 5"/51 _____No. 1093L2
   Served on USS Indiana, USS Arizona
   Location: Ropkey Museum, Crawfordville, IN
3. 5"/51 Mk VII Watervliet No. 774
   Location: NAS Midway Island
4. 5"/51 Mk VII Watervliet No. _____
   Location: NAS Midway Island
5. 5"/51 Mk VIII_____No. 415L
   Location: NROTC, Tulane University, New Orleans, LA
6. 5"/51 Mk XV Naval Gun Factory No. 736L
   Served on USS Idaho
   Location: Naval Gun Factory, Washington, DC
7. 5"/51 _____No.__
   Location: Fort Schuyler, Kings Point, NY

8. 5"/51_____
   Location: Behind Museum, Treasure Island, CA
9. 5"/51 Mk XV _____No.___
   Location: Park, Lewsiton, ME
10. 5"/51 Mk XV Bethlehem No.____
    Location: NAS Brunswick, ME
11. 5"/51 _____No. ___
    Location: Mitchell, IN
12. 5"/51 Mk IX Mod 3 Naval Gun Factory Mod. 3 No. 938L
    Location: American Military Museum, S. El Monte, CA
13. 5"/51 _____ No. _____
    Served on USS Colorado, Museum of Science and Industry, University of Washington, Seattle, WA

4.7 Inch Guns

1. QF Mk IV Armstrong No. 12123 on Central Pivot Mount Mk 1 No. 10981
   Formerly emplaced at Battery Talbot, Fort Adams, RI
   Location: Equality Park, Newport, RI
2. QF Mk IV Armstrong No. 12124 on Central Pivot Mount Mk 1 No. 10982
   Formerly emplaced at Battery Talbot, Fort Adams, RI
   Location: Battery Bingham, Fort Moultrie, Sullivan's Island, SC (shield missing)
3. QF Mk 3 Armstrong No. 11933
   Formerly emplaced at Battery Dodge, Fort Ruger, HI
   Location: State Armory, Wahiawa, Oahu, HI
4. QF Mk 3 Armstrong No. 11009
   Formerly emplaced at Battery Dodge, Fort Ruger, HI
   Location: State Armory, Wahiawa, Oahu, HI
5. QF Mk IV Armstrong No. 11856 on Central Pivot Mount Mk 1 No. 10842
   Formerly emplaced at Battery Drum, Fort Strong, MA
   Location: State Armory, Main Street, Ansonia, CT
6. QF Mk IV Armstrong No. 9718 on Central Pivot Mount Mk 1 No. 10841
   Formerly emplaced at Battery Van Swearingen, Fort Pickens, FL
   Location: County Courthouse, Danielsville, GA
7. QF Mk 3 Armstrong No. 11855 on Central Pivot Mount Mk 1 No. 10843
   Formerly emplaced at Battery Drum, Fort Strong, MA
   Location: American Legion Concord Post 431, Springville, NY
8. QF Mk 3 Armstrong No. 9724 on Central Pivot Mount Mk 1 No. 10885
   Formerly emplaced at Battery Dodd, Fort Delaware, DE
   Location: American Legion Post, Dravosburg, PA
9. QF Mk 3 40-caliber Armstrong No. 11857 on Central Pivot Mount Mk 1 No. 10835
   Formerly emplaced at Battery Backus, Fort Screven, GA
   Location: East Park, Giradville, PA

3 Inch Guns

1. M1898M1, Driggs-Seabury No. 11
   Formerly emplaced at Battery Center, Fort McRee
   Location: Schell Memorial Cemetery, Boyertown, PA
2. M1898M1 Driggs-Seabury No. 118
   Formerly emplaced at Battery Terrett, Fort Gaines, AL
   Locations: Schell Memorial Cemetery, Boyerstown, PA
3. M1898M1, Driggs-Seabury No. 25
   Formerly emplaced at Battery Center, Fort McRee, FL
   Location: Mount Pleasant Cemetery, Elbridge, NY
4. M1898M1, Drigggs Seabury No. 27
   Formerly emplaced at Battery Beecher, Fort Schuyler, NY
   Location: Valley Forge Military Academy, Wayne, PA
5. M1898M1, Driggs Seabury No. 28
   Formerly emplaced at Battery Baker, Fort Totten, NY
   Location: Valley Forge Military Academy, Wayne, PA

6. M1898M1, Driggs-Seabury No. 60
   Formerly emplaced at Battery Burchsted, Fort Dade
   Location: Rte 22 Harrisburg, PA

7. M1898M1, Driggs-Seabury No. 75
   Formerly emplaced at Battery Blaney, Fort Scott, CA
   Location: Town Hall, Forked River, NJ

8. M1898M1, Driggs-Seabury No. 85
   Formerly emplaced at Battery Irwin, Fort Monroe, VA
   Location: Valley Forge Military Academy, Wayne, PA

9. M1898M1, Driggs-Seabury No. 88
   Formerly emplaced at Battery Griffin, Fort Hamilton, NY
   Location: Orange, MA

10. M1898M1, Driggs-Seabury No. 119
    Formerly emplaced at Battery Terrett, Fort Gaines, FL
    Location: Copper Street Cemetery, Vernon, NY

11. M1898M1, Driggs-Seabury No. 120
    Formerly emplaced at Battery Adair, Fort Taylor, Key West, FL
    Location: Central Park, West Rutland, VT

12. M1898M1, Driggs-Seabury No. 112
    Formerly emplaced at Battery Mellon, Fort Dade, FL
    Location: Pulaski Square, Cleveland, OH

13. M1898M1, Driggs-Seabury No.___
    Formerly emplaced at Battery_____
    Location: Valhalla Firehouse, Valhalla, NY

14. M1898MI, _____, No. _____
    Location: Oregon National Guard Museum storage, Camp Withycombe, Clackamas, OR

15. M1898M1, Driggs-Seabury No. 39
    Formerly emplaced at Battery Dilworth Fort Taylor
    Location: American Legion Post 397, Shrewsbury, MA

16. M1898M1, Driggs-Seabury No. 92
    Formerly emplaced at Battery Irwin, Fort Monroe, VA
    Location: Brighton Town Hall, State Route441, Brighton, N.Y.

17. M1898M1, Driggs-Seabury No. 37
    Formerly emplaced at Battery Laval, Fort Crockett, TX
    Location: American Legion Post, Penfield, N.Y.

18. M1898M1, Driggs-Seabury No. 38
    Formerly emplaced at Battery Laval, Fort Crockett, TX
    Location: American Legion Post, Penfield, N.Y.

19. M1898 Masking Parapet mounts (4ea)
    Formerly emplaced at Battery Adair, Fort Taylor, Key West, FL
    Location: Florida Historical Resources Conservation Laboratory, Tallahassee, FL and Fort Taylor, Key West, FL
    Note: cast-iron counter weights located in abandoned Fort St. Philip, New Orleans, LA

20. M1902M1, Bethlehem No. 6 on M1902 Pedestal Mount No. 6
    Formerly emplaced at Battery Hindman, Fort Wool, VA
    Location: Battery Irwin, Fort Monroe, VA

21. M1902M1, Bethlehem No. 7 on M1902 Pedestal Mount No. 7
    Formerly emplaced at Battery Hindman, Fort Wool, VA
    Location: Battery Irwin, Fort Monroe, VA

22. M1903, _____ No. 11 on M1903 Pedestal Mount No. 6
    Formerly emplaced at Battery Flake, Fort Wint, Philippine Is.
    Location: Battery Trevor, Fort Casey, Coupeville, WA

23. M1903, _____ No. 12 on M1903 Pedestal Mount No. 7
    Formerly emplaced at Battery Flake, Fort Wint, Philippine Is.
    Location: Battery Trevor, Fort Casey, Coupeville, WA

24. M1903, _____ No. 17 on M1903 Barbette Carriage No. 10
    Formerly emplaced at Battery Jewell, Fort Wint, Philippine Is.
    Location: Battery Wansboro, Fort Flagler, WA
25. 3"/50 Mk VI, American Radiator Co. No. 4954, on American Ordnance Pedestal Mount Mk II
    Formerly emplaced at Battery Jewell, Fort Wint PI as a Philippine Army practice battery
    Location: Battery Wansboro, Fort Flagler, Nordland, WA
26. M1911 Watervliet No. 2 on M1912 Cowdrey Machine Co. Barbette Carriage No. 1
    Note: training dummy (formerly at Aberdeen Proving Ground Museum)
    Location: US Army Air Defense Training Support Facility Fort Sill, OK
27. M1911 Watervliet No._____ on M1912 Cowdrey Machine Co. Barbette Carriage No.__
    Note: Training dummy
    Location: Battery McCorkle, Fort Moultrie, Sullivan's Island, SC

90 mm Guns (Only Fixed Emplacement Carriages)

1. 90mm M1, Chevrolet No. 6931 on Carriage T3, Fisher Body No. 123
   Formerly emplaced at Battery TAC 24, Fisherman Island, VA
   Location: Battery Parrott, Fort Monroe, VA
2. 90 mm M1,_____ No.___ on Carriage T3, _____ No. ____
   Location: Outside Building 600, Shemya Air Force Base, Shemya Island, AK (shield scrapped)
3. 90 mm Carriage T3,_____No. ____
   Formerly emplaced at Battery TAC 24, Fisherman Island, VA
   Location: Fort MacArthur Military Museum, San Pedro, CA

6 Pdr Field Mount

1. One 6 pounder (2.24-inch) Driggs-Seabury Gun M1900 No. 25, on Wheeled Carriage M1898
   Location: Prescott, AZ
2. One 6 pounder (2.24-inch) Driggs-Seabury Gun M1900 No. 41, on Wheeled Carriage M1898
   Location: Maquoketa, IA
3. One 6 pounder (2.24-inch) Driggs-Seabury Gun M1900 No. 58, on Wheeled Carriage M1898
   Location: Plymouth, MI
4. One 6 pounder (2.24-inch) Driggs-Seabury Gun M1900 No. 47, on Wheeled Carriage M1898
   Location: (Formerly at Mattituck, NY), currently unknown location
5. Two 6 pounder (2.24-inch) American Ordnance Gun M1900 No. ?, on Wheeled Carriage M1898
   Location: Sitka, AK
6. One 6 pounder (2.24-inch) American Ordnance Gun M1900 No. 19, on Wheeled Carriage M1898 No. 10
   Location: Gillman, IL
7. One 6 pounder (2.24-inch) American Ordnance Gun M1900 No. 17, on Wheeled Carriage M1898 No. 6
   Location: Murray, UT
8. One 6 pounder (2.24-inch) American Ordnance Gun M1900 No. 36, on Wheeled Carriage M1898
   Location: Virginia War Memorial, Norfolk, VA
9. Two 6 pounder (2.24-inch) Driggs-Seabury Gun M1898 Nos. 5 & 6, on navy M1898 pedestal mounts
   Location: Sparkill, NY
10. One 6 pounder (2.24-inch) Driggs-Seabury Gun M1900 No. 51, on Wheeled Carriage M1898
    Location: private collection Somerset County, NJ (formerly at Fort Williams, ME, then Jewitt City, CT)
11. One 6 pounder (2.24-inch) Driggs-Seabury Gun M1900 No. 54, on Wheeled Carriage M1898
    Location: private collection Riverside, CA (?) (formerly at Fort Williams, ME, then Jewitt City, CT)
12. One 6 pounder (2.24-inch) Driggs-Seabury Gun M1900
    Location: Mahan Foundation Collection, Basking Ridge, NJ

Other Surviving North American Seacoast Artillery Weapons
December 2025

General Note: This is an attempt to list surviving weapons (or the same model/type) that were used by British, Canadian, or Mexican armed forces, only in North America, in a coast defense role in the "modern era" (1890 to 1950). Items to be included in this list must retain at least the whole gun/mortar/ howitzer. Surviving weapons from earlier periods, muzzleloading cannon (rifled and smoothbore), field artillery (except for 155mm), mobile anti-aircraft guns, and coast artillery manned by American armed forces (see other list) are excluded from this list, even if used in North America. In a few cases, weapons have been included because they represent weapons similar to those used for coast defense. Every effort has been made to make this list as accurate as possible, but it will probably contain a number of errors and omissions. Terrance McGovern would greatly appreciate being contacted about any corrections at 1717 Forest Lane, McLean, VA 22101 USA or called at (703) 538-5403 (home).

The list is arranged by caliber (largest to smallest). The *first line* contains data about the weapon: the quantity at the site, the caliber of weapon (inches or millimeters), the model, serial number, and place of manufacture, if known. This information is repeated for the carriage. The *last line* contains data about the location of the weapon: battery or site were the weapon is currently located, fort or installation, geographic location, and city and/or country.

10 Inch
1. One 10-inch BLR Gun Mk (#)
   York Redoubt Ordnance Park, Halifax, Nova Scotia, Canada

9.2 Inch
1. Two 9.2-inch BL Guns Mk X (# & # Vickers) on Barbette Carriages Mk V (# & # Vickers)
   St. David's Battery, St. David's Head, Bérmuda
2. One 9.2-inch BL Gun Mk X (#272 Vickers) on Barbette Carriage Mk V (# Vickers)
   Dockyard Keep, Ireland Island, Bermuda

150 mm
1. One 150 mm Hooped Iron Gun Model 1885 on Seacoast Artillery Carriage (Ordoñez)
   Main Parade Ground, Presidio of San Francisco, Golden Gate Natl. Rec. Area, San Francisco, CA

6 Inch
1. Two 6-inch BL Guns Mk VII (# & # Vickers) on Central Pivot Mk II (# & # Vickers)
   Alexandria Battery, St. Georige Island, Bermuda (from Warwick Battery)
2. Two 6-inch BL Guns Mk VII (# & # Vickers) on Central Pivot Mk II (# & # Vickers)
   St. David's Battery, St. David's Head, Bermuda
3. One 6-inch BL Gun Mk VII Gun (# Vickers) on Central Pivot Mk II (# Vickers)
   Fort St. Catherine, St. George Island, Bermuda
4. Two 6-inch BL Guns Mk VII (#L/1029 & # RGF) on Central Pivot Mk II (# & # Vickers)
   (spare barrels from Warwick Bty and St. David Bty)
   Bastions D & C, Dockyard Keep, Ireland Island, Bermuda
5. One 6-inch BL Guns Mk II (80 cwt #85 1882) (from the Old Machinery and Casting Company, St. George)
   Bastion E Royal Dockyard Keep, Ireland Island, Bermuda
6. One 6-inch BL Guns Mk IV? (unreadable markings)
   (from the Old Machinery and Casting Company, St. George)
   Bastion E Royal Dockyard Keep, Ireland Island, Bermuda
7. One 6-inch BL Gun Mk VI (#841) on display base
   Upper Battery, Fort Rodd Hill, Victoria, British Columbia
8. One 6-inch BL Gun Mk III/IV (#302) on display base
   Lower Battery, Fort Rodd Hill, Victoria, British Columbia
9. One 6-inch BL Gun Mk VII (#) on Central Pivot Mk II (#)
   Fort Ogilvie, Point Pleasant Park, Halifax, Nova Scotia

4.7 Inch
1. Two 4.7-inch Armstrong BL Guns Mk III (# & #) on Central Pivot Mount Mk I (# & #)
   Fort Amherst, St. John's, Newfoundland
2. Two 4.7-inch Armstrong BL Guns Mk III (# & #) on Central Pivot Mount Mk I (# & #)
   Fort Peninsula, Forillon National Park, Gaspe, Quebec
3. Two 4.7-inch Armstrong BL Guns Mk III (# & #) on Central Pivot Mount Mk I (# & #)
   Bell Island, Newfoundland

12 Pounder
1. One 12 pounder QF Gun Mk I (2767) on Pedestal Mount W.G. Armstrong Whitworth Mk I (11798)
   Belmont Battery, Fort Rodd Hill, Victoria, British Columbia

75mm
1. One 75mm Gun M1917 (British) (#) on Pedestal Mount Model (#)
   Chain Rock Battery, Signal Hill National Park, St. John's, Newfoundland
2. One 75mm Gun M1917 (British) (#) on Pedestal Mount Model (#)
   St. John's, Newfoundland

6 Pounder
1. Twin 6 pounder QF Guns Mk I (# & #) on Pedestal Mount Mk I (#)
   Belmont Battery, Fort Rodd Hill, Victoria, British Columbia
2. One 6 pounder QF Gun Mk I (#502 Hotchkiss) on Garrison Carriage Mk I** (# Hotchkiss Cone Mount)
   Fort Rodd Hill, Victoria, British Columbia
3. One 6 pounder QF Gun Mk I (#) (tube only)
   Fort Rodd Hill, Victoria, British Columbia
4. One 6 pounder QF Gun Mk I (13980 & N1584) on Pedestal Mount Mk I (5110)
   Bay Street Armory, Victoria, British Columbia

Pacific Islands with major collections of artillery used for coast defense:
1. Viti Levu, Fiji Islands
2. Midway Island
3. Kiska and Attu Islands, Alaska

ANTIAIRCRAFT WEAPONS OF THE COAST ARTILLERY CORPS

Bolling W. Smith

The first army efforts toward antiaircraft defense, in 1913, involved modifying the trail and recoil system of field guns to allow them be fired against airplanes.(1) By 1916, however, specific antiaircraft guns were on the drawing board. What follows is a summary of the principal American antiaircraft weapons used by the Coast Artillery Corps. Omitted are a number of foreign weapons used during WWI, as well as the multitude of machine guns used occasionally in AA roles.

The data given must be read with care. Antiaircraft guns were only part of a system, other parts being the fire control system and the ammunition. Frequently the figures given for velocity and range were based on the performance of the ammunition, and more specifically, the fuzes. Changes in ammunition could make significant changes in the performance of the guns without any actual change in the guns. Therefore, the performance data is only accurate for the date indicated.

.30 Caliber Browning Machine Gun, M1917

During WWI, the United States used a wide variety of .30 caliber machine guns for close-in antiaircraft protection. Types used included American Colt and Lewis guns, British Vickers guns, and French Chauchats. Eventually, relative standardization was achieved by the use of the water-cooled, belt-fed M1917 Browning .30 caliber machine gun [fig 1] on improvised mounts. These guns, while adequate considering the state of aircraft development and the allied air supremacy, would nonetheless clearly be unsatisfactory against the new generation of aircraft. While the mounts were often unstable, the ammunition was the principal problem. The wartime tracers were very inaccurate, with a large percentage of blinds, and tracers did not follow the trajectory of the ball ammunition. Adding insult to injury, the tracers fouled the guns badly. By 1925, the ammunition had been completely reworked, with flatter trajectories for both ball and tracer, while the tracers were brighter and longer lasting, and no longer fouled the guns.

By 1926, the .30 caliber machine gun had been slated for replacement, although some .30 caliber guns remained in service into WWII; they were used in a dual role on Corregidor in 1942. The .30 caliber MG weighed 37 lbs, and had an vertical range 4,000 yds, 800 yds to tracer burnout, and an estimated effective range of 500 yds, with a muzzle velocity of 2,700 fps. The cyclic rate of fire was 525 rpm (1940).(2)

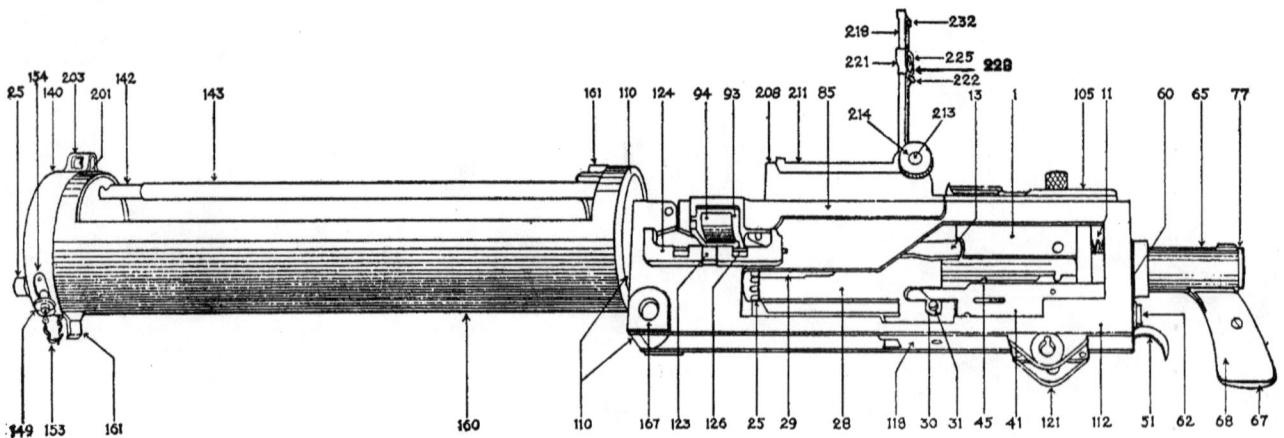

.30 cal. MG, M1917, Antiaircraft Defense (Harrisburg: Military Service Publishing Co., 1940), p. 149.

.50 Caliber Browning Machine Guns

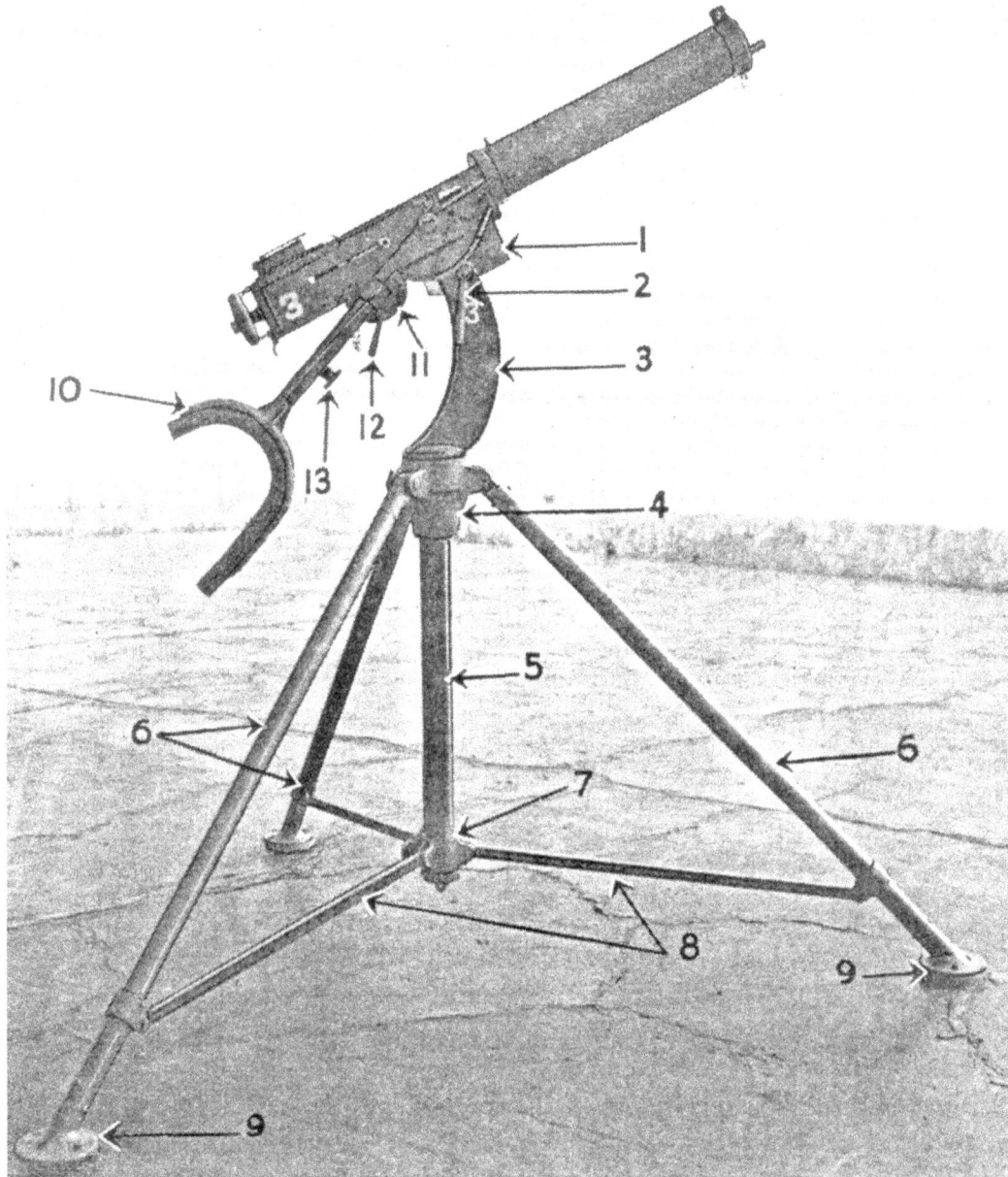

Browning M1921

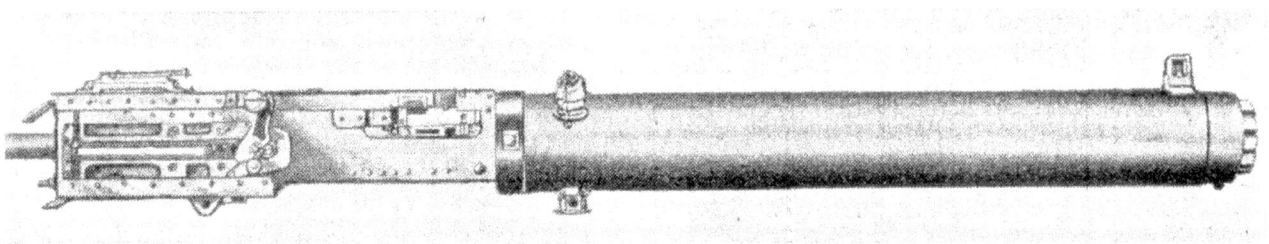

Browning M2 (water cooled)

In 1922 the army adopted the water-cooled, belt-fed Browning M1921 machine gun for both ground and antiaircraft use. On a high tripod mount, it served in an antiaircraft role into early WWII. By 1926, it had supplanted the .30 caliber gun as the standard machine gun for antiaircraft defense.

In 1932, the army adopted an improved version, the M2. This multipurpose weapon gradually replaced the M1921. It was available in three forms: aircraft gun, water-cooled antiaircraft gun, and M2 HB (heavy barrel) air-cooled ground mount (still in service today). The water-cooled gun was designed as the anti-aircraft weapon, with its water jacket allowing more sustained fire, but actual use showed that this feature was unnecessary, since aircraft targets were too fleeting to need the sustained fire, making the water-cooled version obsolescent by 1943.

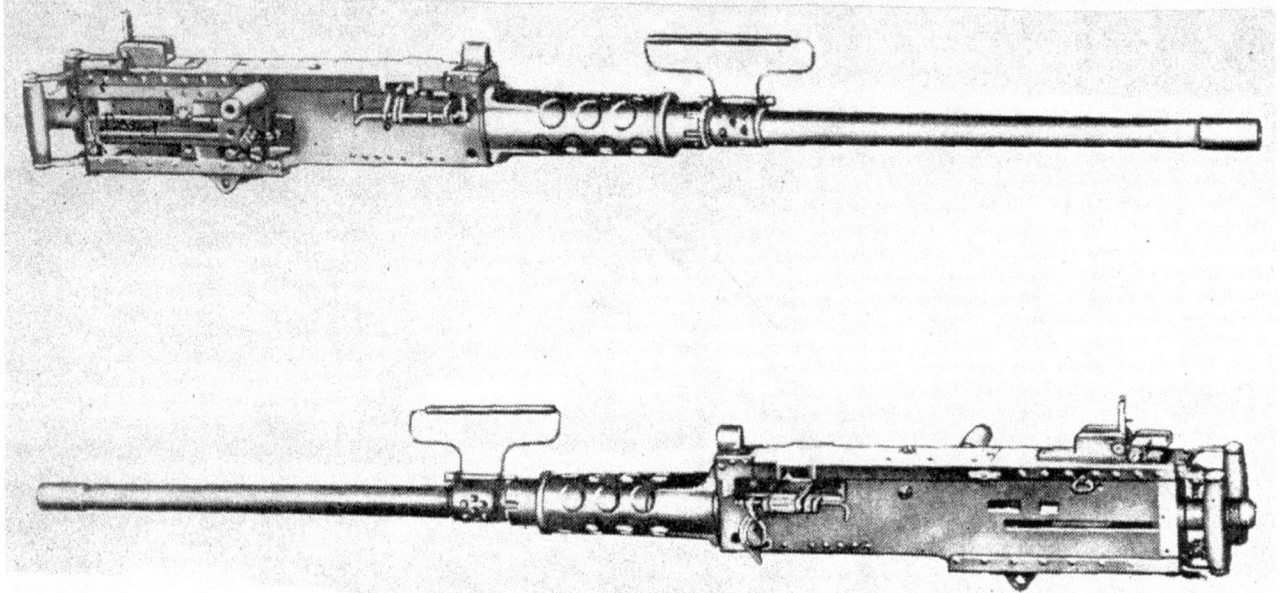

Browning M2 (heavy barrel)

The M2 HB was issued for antiaircraft service before WWII, and proved an excellent weapon. Single M2 HBs were mounted on army vehicles, providing most their only AA self-protection. In addition, the M@ HB wasmounted in multiple arrangements on a variety of mounts. In late 1941, the army adopted the multiple machine gun mount M33 for two .50 caliber guns, mounted on half tracks as multiple gun motor carriages M13 and M14, depending on the model of half track. By 1944, it had been largely re-placed by the multiple machine gun mount M45, with four .50 caliber guns and a shield to protect the gunner. Mounted on half tracks, this was known as the multiple gun motor carriage M16 and M17. A trailer mounted model was designated the multiple .50 cal. machine gun carriage M51. Airborne troops were issued the lighter M45C quad mount, without a shield, on a light-weight trailer, as multiple .50 cal. machine gun trailer mount M55. In all its roles, the multiple .50 caliber M2 HB gun was highly effective, and when the threat of low-flying enemy air attack diminished toward the end of the war, they were used with devastating effectiveness against ground targets. During the Korean War, this role was repeated.

The water-cooled M2 weighed 81 lbs, the M2 HB 121 pounds. They had a cyclic rate of fire of 450-600 rounds per minute, giving a muzzle velocity of 2,800 (M1 ammunition) to 2,900 (M2 ammunition) fps with a 750 grain, non-explosive projectile, effective out to 2,000 ft.(3)

37 mm AA Automatic Gun

In response to army requirements growing out of WWI, in 1922 John Browning began development of a 37 mm AA gun. Following his death in 1926, the development was continued by Colt until it was finally adopted in 1938. The gun was mounted on the M3 two-axle trailer, was manually loaded with ten round clips, and was manually aimed. The M1 gun was used in 1941-42, including the defense of the Philippines. By the invasion of North Africa in November 1942, the original M1 model had been upgraded to the M1A1, with new sights and power traverse and elevation for off-carriage fire control. Although it had been decided to replace the 37 mm gun with the 40 mm Bofors, the inability to produce sufficient 40 mm guns kept the 37 mm gun in service. As modified, the M1A1 and M1A2 models remained in use throughout WWII, although production ended in 1944.

The most effective use of the 37 mm gun was probably in combination with a pair of .50 cal. M2 HB machine guns. The original 37 mm combination gun mount M42 mounted two water-cooled M2 MGs above a single 37 mm gun. Subsequent improvements placed the 37 mm gun over the .50 cal. guns, and replaced the water-cooled MGs with air-cooled M2 HBs. Mounted on half tracks as combination gun motor carriage M15, the M42 mount and its successor, the M54 mount, performed well against both air and ground targets.

The 37 mm gun fired a 1.34 lb explosive shell at a cyclic rate of 120 rpm, with a muzzle velocity of 2,600 fps and a vertical range of 4,000 yards. Elevation was from -5° to over 90°.(4)

37-MM AA GUN MATERIEL

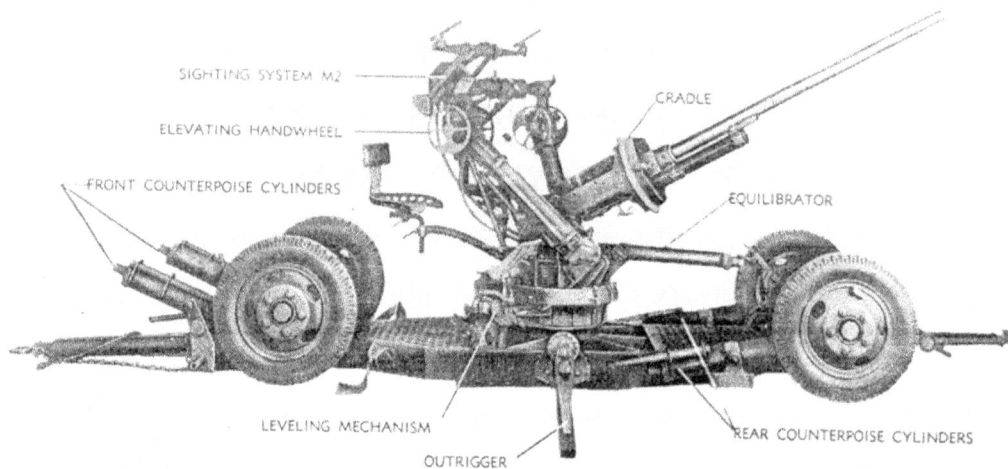

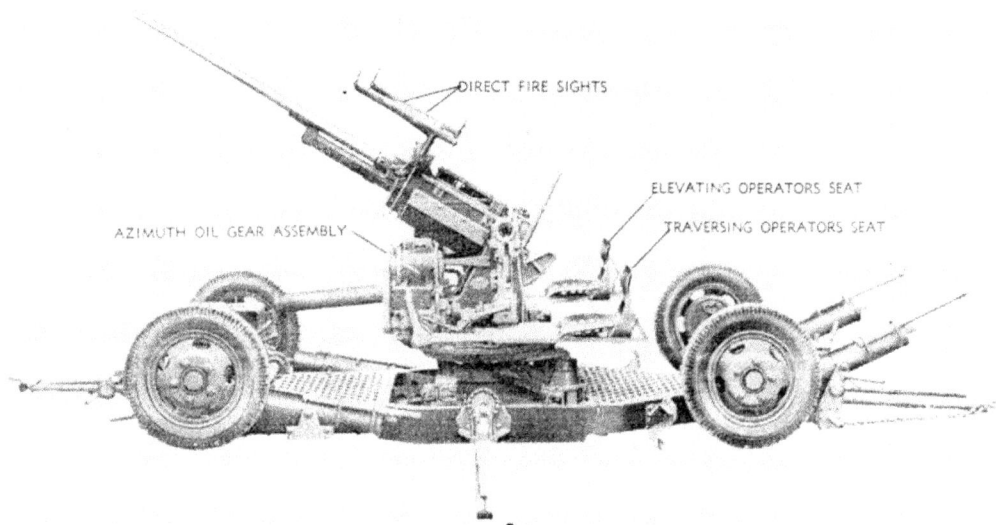

37-MM ANTIAIRCRAFT GUN M1A2

CALIBER .50 MACHINE GUN, M2

CALIBER .50 AMMUNITION CHEST, M2

TOP CARRIAGE

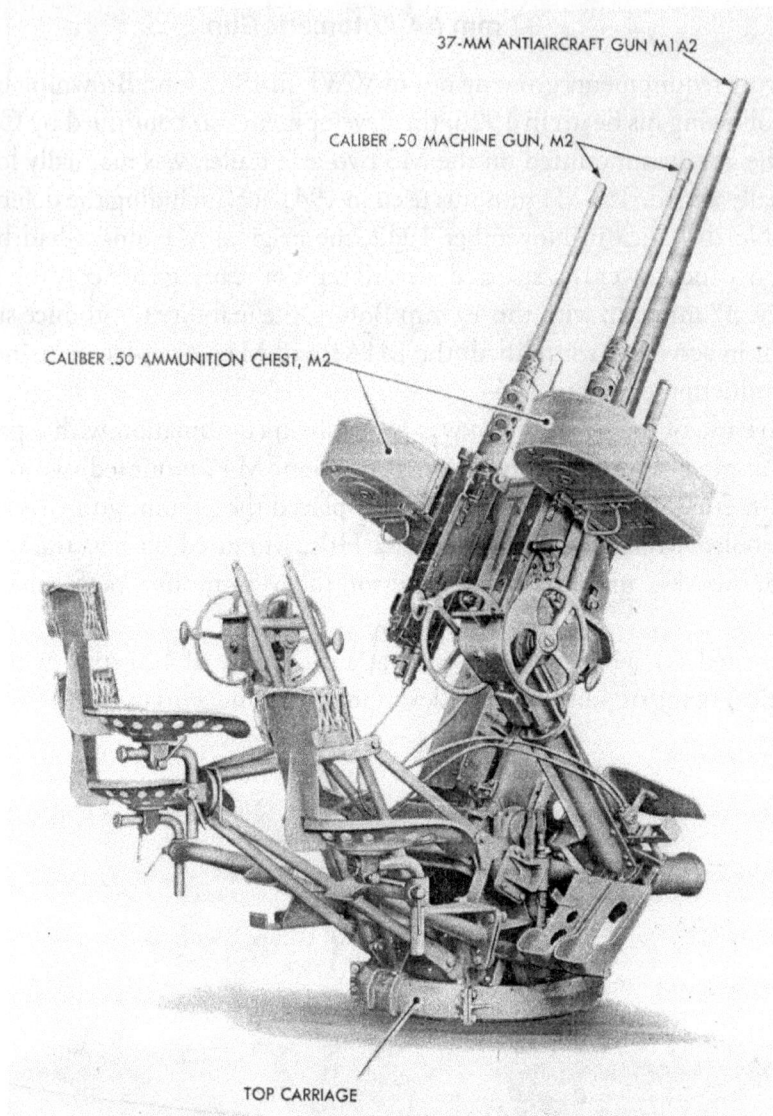

37 mm combination gun mount M42

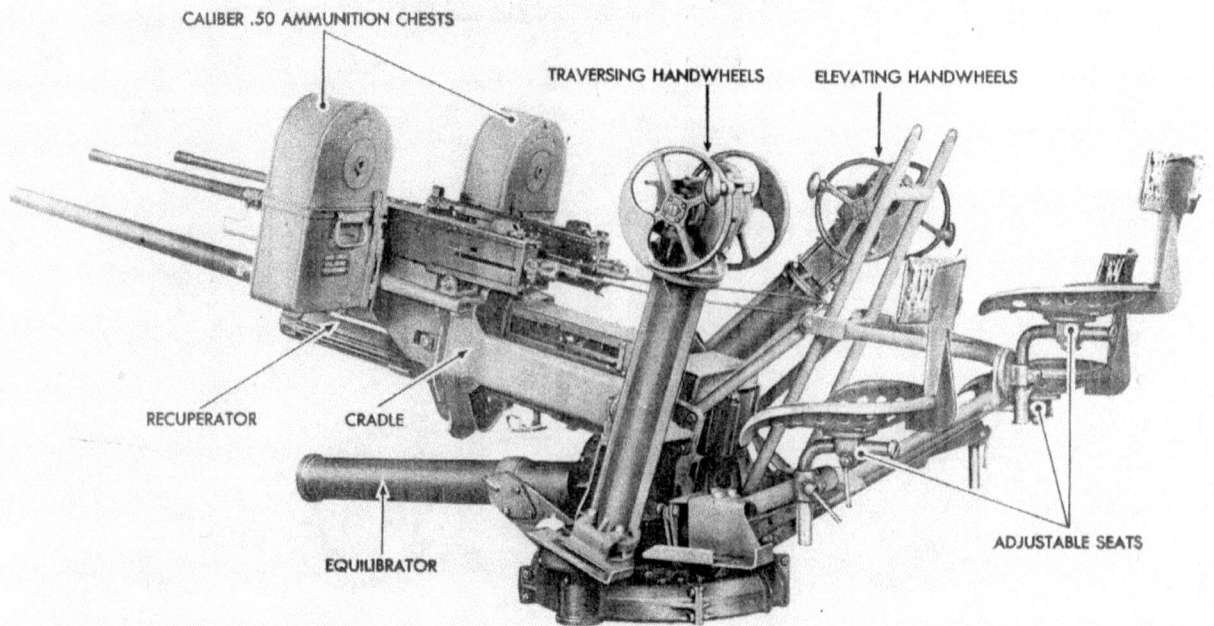

CALIBER .50 AMMUNITION CHESTS

TRAVERSING **HANDWHEELS**

ELEVATING HANDWHEELS

RECUPERATOR

CRADLE

EQUILIBRATOR

ADJUSTABLE SEATS

40 mm Automatic Gun

The 37 mm gun was superseded by a better weapon, the Bofors 40 mm automatic gun. This gun was developed in Sweden in the 1930s, and after a good showing in the Spanish Civil War, it was adopted by most combatants on both sides. By the time Britain had arranged for it to be manufactured in this country in 1940 under Lend-Lease, the army had already evaluated it and expressed their interest. The gun was re-engineered to American standards and adopted as the M1 in December 1940. Navy needs took priority, however, and the army received only a limited number of the guns before 1943.

The 40 mm gun was larger and heavier than the 37 mm gun, with a higher muzzle velocity and longer range. Due to its longer range, it worked best when coupled with off-carriage fire control. Because of the navy needs, production was high, and all anticipated army needs had been filled by early 1944. By the time the army received large numbers of this weapon in late 1943, the threat of enemy air attack had declined sharply, and the guns, like the 37 mm, were used extensively against ground targets.

The standard version was mounted on a two-axle trailer, the M2, while the light-weight, single-axle airborne trailer was the M5. The navy twin-40 mm mount was mounted on the chassis of the army's Chaffee light tank and termed the gun motor carriage M19. Few, if any, saw service in WWII, but both the M1 gun and the M19 GMC were retained long after WWII.

The 40 mm gun had a cyclic rate of fire of 120 rpm, with a 2 lb explosive projectile and a muzzle velocity of 2,960 fps. Its AA range was 3,500 yds (1942) and 4,200 yds (1944), limited by tracer burnout, increased to 5,100 yds (1949). The elevation was from -6° to 90°. The ammunition was fed in four-round clips.(5)

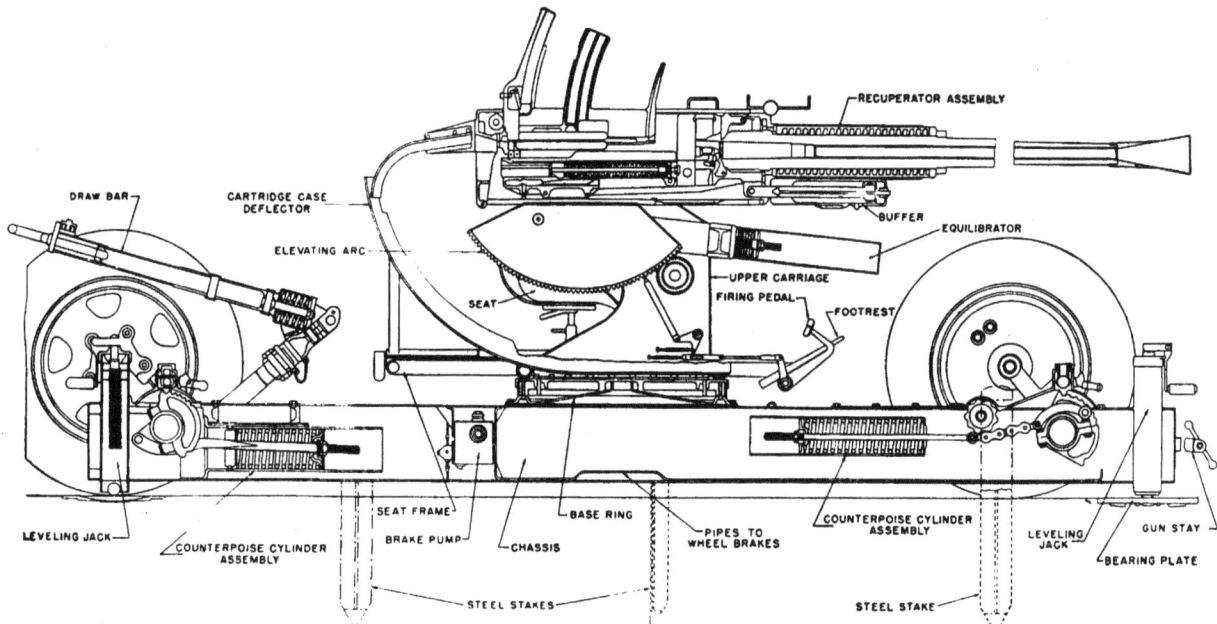

75 mm AA Gun M1916, Mobile

In 1917, the army was faced with the need to provide mobile antiaircraft guns for the field army. In the rush, there was insufficient time to develop and test a new design, so plans were made to use a field gun that was already in production, mounting it on a truck body. The design was completed on May 1, 1917, and production began immediately. It was recognized at the time that the gun was inadequate, but there seemed no alternative if the mobile forces were to be equipped at all. The American M1916 field gun was used because in July 1917, the recuperators for the French 75 mm gun had not yet been adapted to American standards. As a stop gap, 50 of these carriages were sent to France immediately upon completion, where they were mounted with French 75 mm guns. By the end of 1918, 51 of the M1916 75 mm AA guns were completed. It was the only American-made AA gun to reach France, but it did not see active service during the war. All American AA equipment used against the enemy was secured from our allies, primarily France.

The American gun was a makeshift job, an M1916 75 mm field gun with a hydro-spring recoil mechanism, mounted on $2\frac{1}{2}$ ton White trucks, designated AA truck mount M1917. The barrel was only 28.4 calibers long, giving a velocity of 1830 fps (1920) with a 14.7 lb shell. At 82° it had a maximum height of 5,500 yds, limited by the 20 sec fuze (1920). Its limited elevation (31° to 82°) and low muzzle velocity severely limited its effectiveness.

The gun was fired from the ground, behind the truck, and was limited to 240° in azimuth. The elevating and traversing mechanism jammed frequently, and although the truck body was equipped with jacks to relieve the strains of firing and to prevent the truck from overturning when firing at low elevation, it was so encumbered with heavy iron that it was scarcely mobile. By 1940, it was no longer in service, although a few were still in use for training.(6)

3-inch AA Gun, Fixed, M1917

Before 1917, the U.S. Army had concentrated its antiaircraft efforts on the protection of its coast defenses. By April 1916, the Ordnance Department had designed a 3-inch gun for fixed emplacements, and by mid-1916 Watertown Arsenal had undertaken the manufacture of one 3-inch AA mount, model E. Between May 1916 and June 18, 1917, 160 of these guns were ordered from Watervliet Arsenal and Bethlehem Steel, of which 116 were delivered by April 10, 1919, and sent to the fortifications. The M1917 gun, improved and upgraded, lasted through WWII.

The M1917 gun was 55 calibers long, with a drop-block breech mechanism. The firing mechanism was of the continuous-pull type, permitting a repetition of blows to the primer without opening the breech, in the case of a misfire. The gun was fired by means of a firing handle. It fired from 0° to 90° elevation, with 360° traverse, and threw a 15 lb projectile, high explosive or shrapnel, at a muzzle velocity of 2,600 fps (1920). The M1917 gun, along with the later M2 and M4 fixed 3-inch AA guns, had the same chamber dimensions and used the same cartridge case as the M1903 3-inch (15-Pdr) seacoast gun.

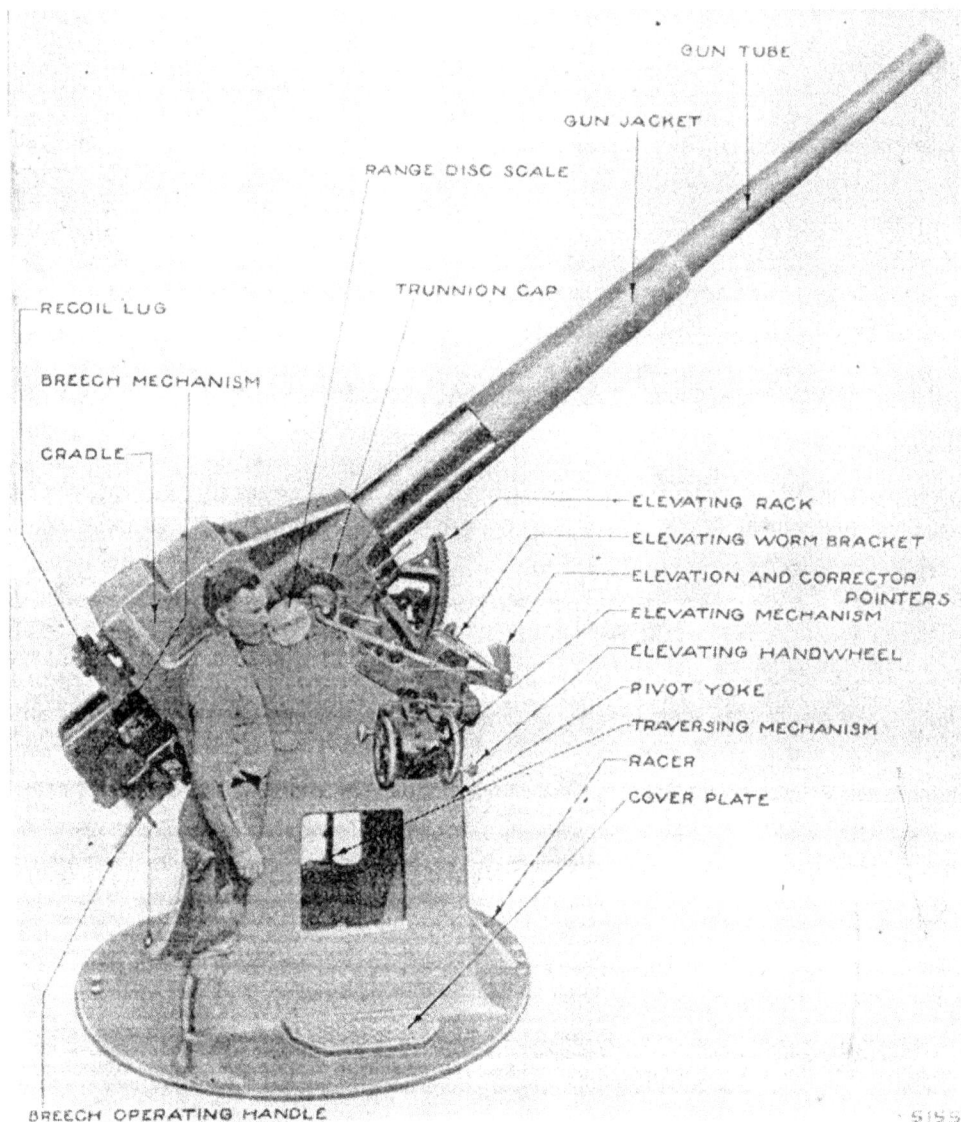

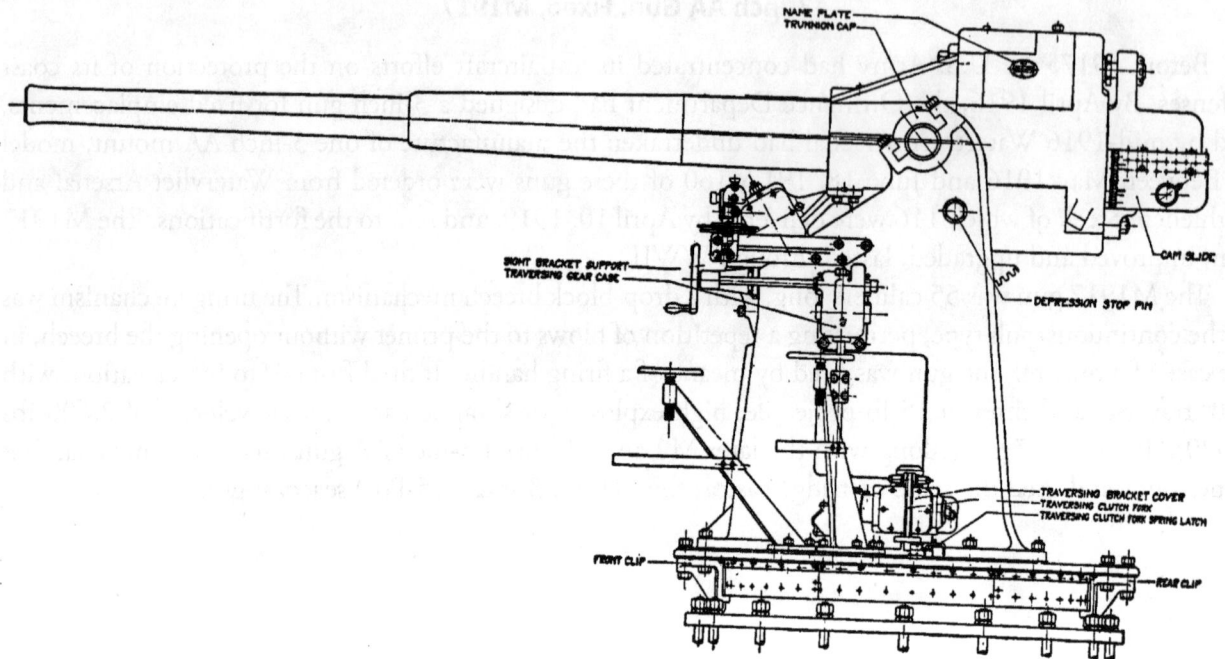

The recoil mechanism was of the hydro-spring type, with a single recoil cylinder above and two counterrecoil cylinders below. The M1917 MI mount added a second recoil cylinder above. The M1917 gun was trunnioned near the breech, allowing it to recoil when firing at high elevation. The barrel was balanced with a breech weight, avoiding the equilibrator used on later M1 and M3 mobile 3-inch guns. The M1917 3-inch AA gun mount was designed to be mounted, using 16 bolts, on a concrete base 30 inches thick and 18 feet in diameter. Emplacements for the M1917 3-inch AA gun remain at many of our coastal forts.(7)

3-inch AA Gun, Mobile, M1918

The truck-mounted 75 mm gun was clearly inadequate from inception. The Ordnance Department immediately began to develop a 3-inch replacement, which was to be trailer-mounted. The gun was "of the ballistic design" of the M1898 3-inch (15-Pdr) seacoast gun, which had become obsolescent, if not yet obsolete. The extent to which actual M1898 tubes were equipped with new breech mechanisms and

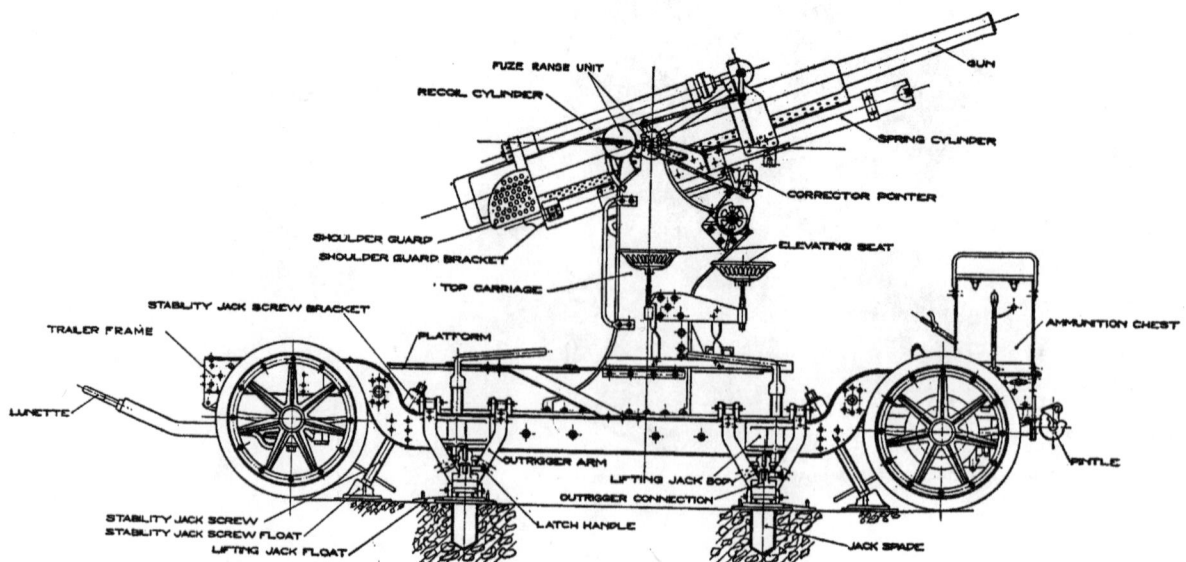

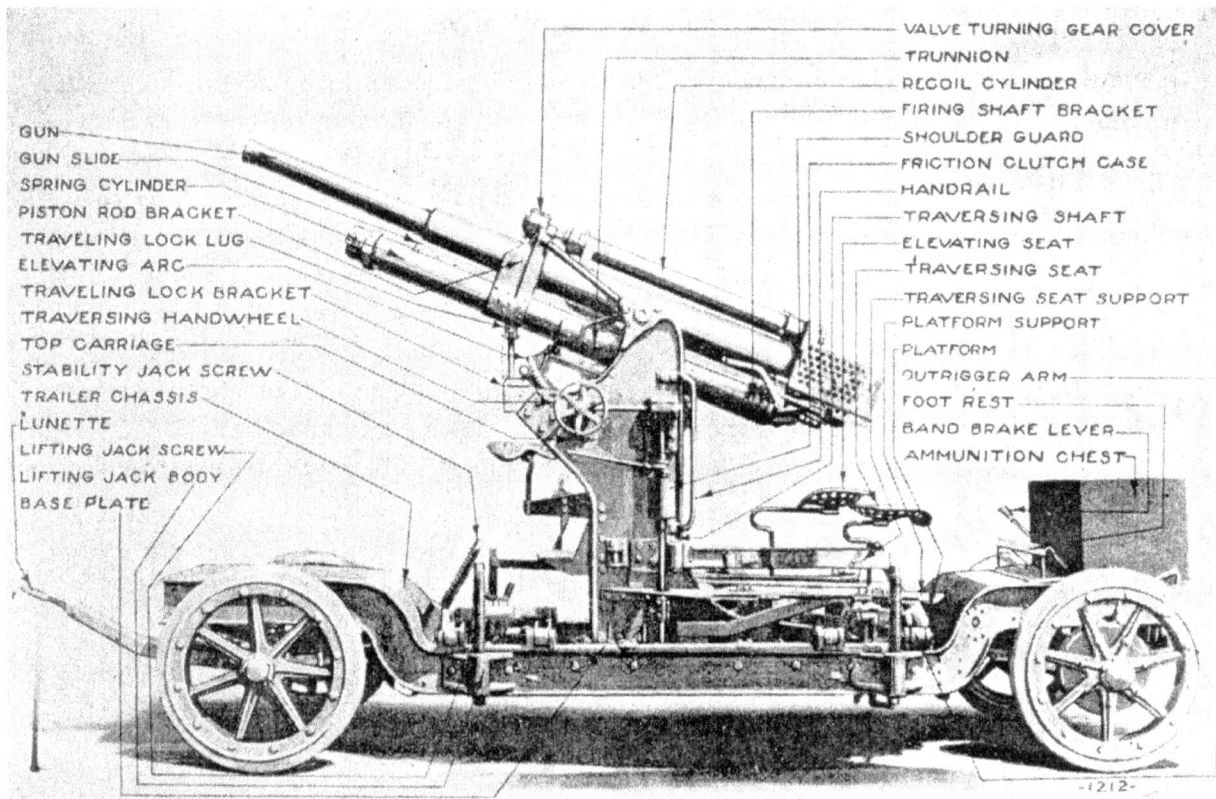

GUN
GUN SLIDE
SPRING CYLINDER
PISTON ROD BRACKET
TRAVELING LOCK LUG
ELEVATING ARC
TRAVELING LOCK BRACKET
TRAVERSING HANDWHEEL
TOP CARRIAGE
STABILITY JACK SCREW
TRAILER CHASSIS
LUNETTE
LIFTING JACK SCREW
LIFTING JACK BODY
BASE PLATE

VALVE TURNING GEAR COVER
TRUNNION
RECOIL CYLINDER
FIRING SHAFT BRACKET
SHOULDER GUARD
FRICTION CLUTCH CASE
HANDRAIL
TRAVERSING SHAFT
ELEVATING SEAT
TRAVERSING SEAT
TRAVERSING SEAT SUPPORT
PLATFORM SUPPORT
PLATFORM
OUTRIGGER ARM
FOOT REST
BAND BRAKE LEVER
AMMUNITION CHEST

-1212-

recoil systems to be reused as AA guns remains under dispute, but the M1918 gun, along with the later M1 and M3 mobile 3-inch AA guns, had the same chamber dimensions and used the same cartridge case as the M1898 and M1902 3-inch (15-Pdr) seacoast guns. The new semi-automatic breech mechanism was a drop-breech design, practically identical to that on the M1916 75 mm gun, and was operated by a lever on the right side of the gun. Pulling that back and down opened the breech. The barrel was only 40 calibers long, and gave a muzzle velocity of 2,400 fps (1920).

The gun was mounted on the 3-inch auto-trailer carriage, M1917. The mount allowed an elevation range of 10° to 85°, giving a maximum vertical altitude of 8,600 yds, limited by the fuze (1920). The recoil varied from 16 inches at 85° to 40 inches at 10°.

The gun and carriage were in turn mounted on the four-wheeled M1917 trailer. The drop-center trailer could be pulled by a tractor, or could be steered by the rear wheels if the rear-wheel lock was disengaged and a steering bar inserted. It was equipped with outriggers and jacks resting on detachable floats on the ground, which stabilized the unit and prevented it from overturning when firing at low elevations.

In actual operation, the gun was less than perfect. Despite the outriggers and jacks, the mount was unsteady during firing. The counterrecoil mechanism functioned irregularly, and loading at high elevations was difficult. Despite this, the gun remained a significant part of the American arsenal until the beginning of WWII, especially for National Guard regiments. A few were even assigned to the defense of permanent fortifications, and a mount for this purpose survives at Fort Worden, Washington.

On December 1, 1924, a change of nomenclature was approved, and the 3-inch AA gun mount was henceforth known as the M1918, and the trailer was known as the 3-ton AA gun trailer, M1918.(8)

3-inch AA gun, Mobile, M1 & M3

After WWI, the army moved forward with the development of a new 3-inch AA gun. The M1923-E 3-inch AA gun was standardized in 1927 as the M1. The guns were produced by the new method of cold-working, or autofrettage, and featured removable liners. The M1 and the M3 guns were very similar, the principal difference being that the M3 gun was larger, allowing for a larger liner. There were relatively few M1 guns produced; firing tables as early as 1928 showed the M3 & M4 guns. Both the M1 and the M3 were 50 calibers long and had a muzzle velocity of 2,600 fps with shrapnel (1941) and 2,800 fps with high explosive. The breech was a drop-block type, with the M14 continuous pull firing mechanism, fired by a short lanyard.

The gun recoiled in a cradle, which was attached to the mount by a hydro-pneumatic, constant recoil system. The barrel was trunnioned relatively near to the breech, allowing the gun to be mounted closer to the ground and still clear the platform when fired at high elevations. To balance the barrel, equilibrators were provided.

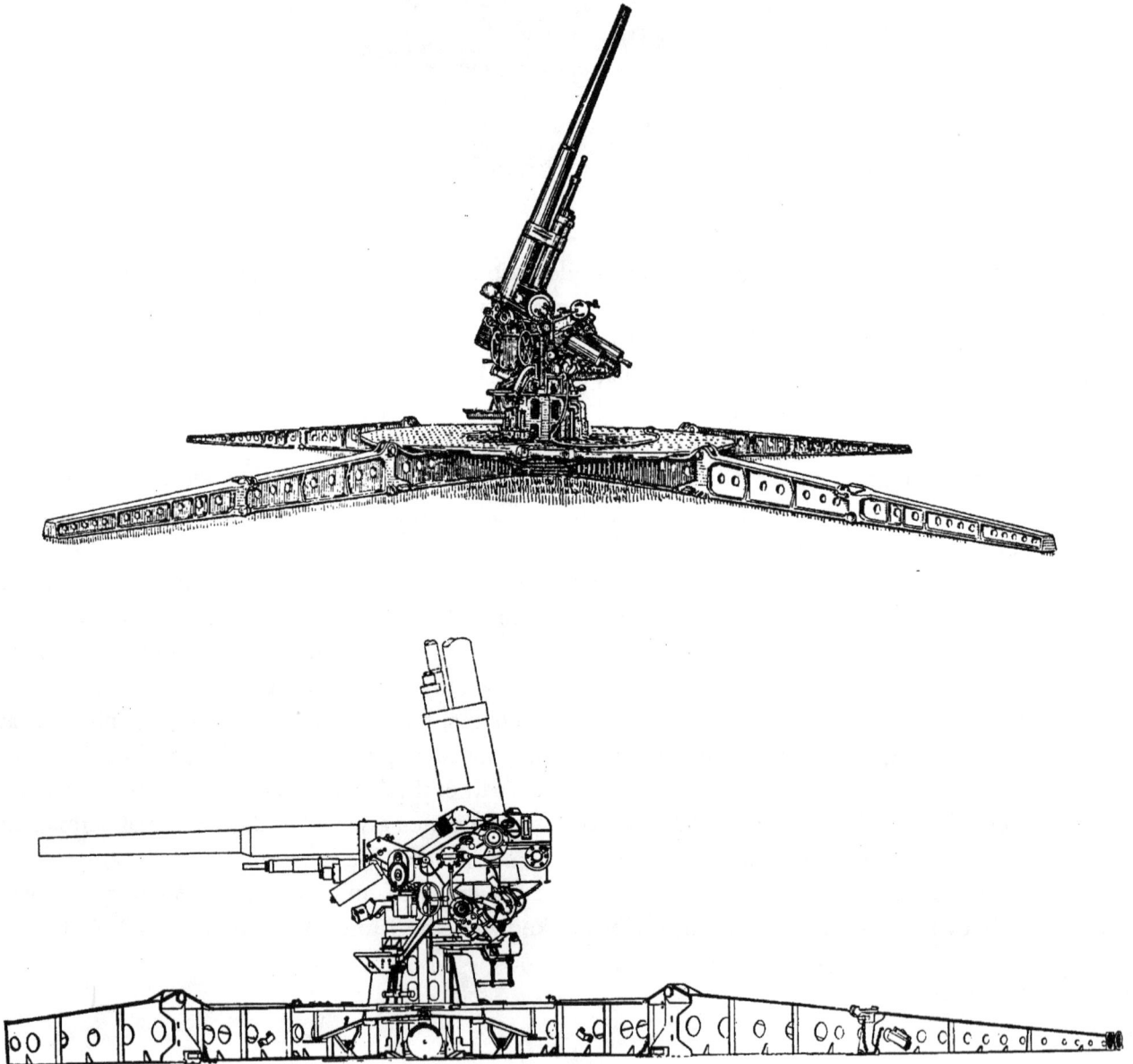

The M1 gun was mounted on the M1 mount, the M3 gun on the M2 mount. Like the guns, the differences between the mounts were only in their fabricated parts and the type of commercial brakes. The most common mount was the M2A2, a mobile trailer mount with balloon tires, capable of high speeds on good roads and good speed over irregular terrain. The most distinctive features were the perforated steel plate platform and four large outriggers. The gun could be elevated from -1° to 80°, and traversed through 360°.

The 3-inch gun M3 gun was the mainstay of the AA regiments at the beginning of WWII, until replaced by the 90 mm gun. Although eventually giving way to a larger, more powerful gun, the M3 was an excellent weapon.(9)

3-inch Guns, Fixed, M2 & M4

At about the same time the as the adoption of the M1 gun, the army adopted a new fixed 3-inch gun, the M2. Like the M1, this was rapidly superseded by the M4, which was virtually identical except for being a little larger because of a larger liner. The documentary evidence of the M2/M4 is relatively scanty, especially compared with the M1917 fixed and the M3 mobile guns, apparently because these guns were produced only in small numbers. As noted above, the M4 gun was in production by 1928. The author's files of Reports of Completed Works for AA batteries contain only one for M2 and M4 guns. In 1936, one battery at Fort Monroe, Virginia, showed one M2 gun and two M4 guns. This M2 gun, if correctly listed,

had several interesting characteristics. The serial number was 170, obviously a continuation of the M1917 serial numbers. The mount, meanwhile, was a M1917 MI. The M4 guns were serial number 1 and 2, on mounts 1 and 2. Whether more M2 and M4 guns were produced is unclear, but it is clear that the coast artillery relied on the M1917 guns to protect its fixed fortifications until they were eventually replaced by 90 mm guns, in some cases not until the end of WWII.

The M2/M4 guns were improved versions of the M1917, which they resembled. They did not improve on the M1917 as much as the M1/M3 guns improved on the M1918. They were also cold-worked, with removable liners, and used the M14 continuous pull firing mechanism. Apparently, the M2 gun was mounted on the M1917 MI mount, while the M4 gun was mounted on the M3 mount, which was similar in all respects to the M1917 MI mount, except for larger and heavier cradle and bearings to accommodate the heavier M4 gun.(10)

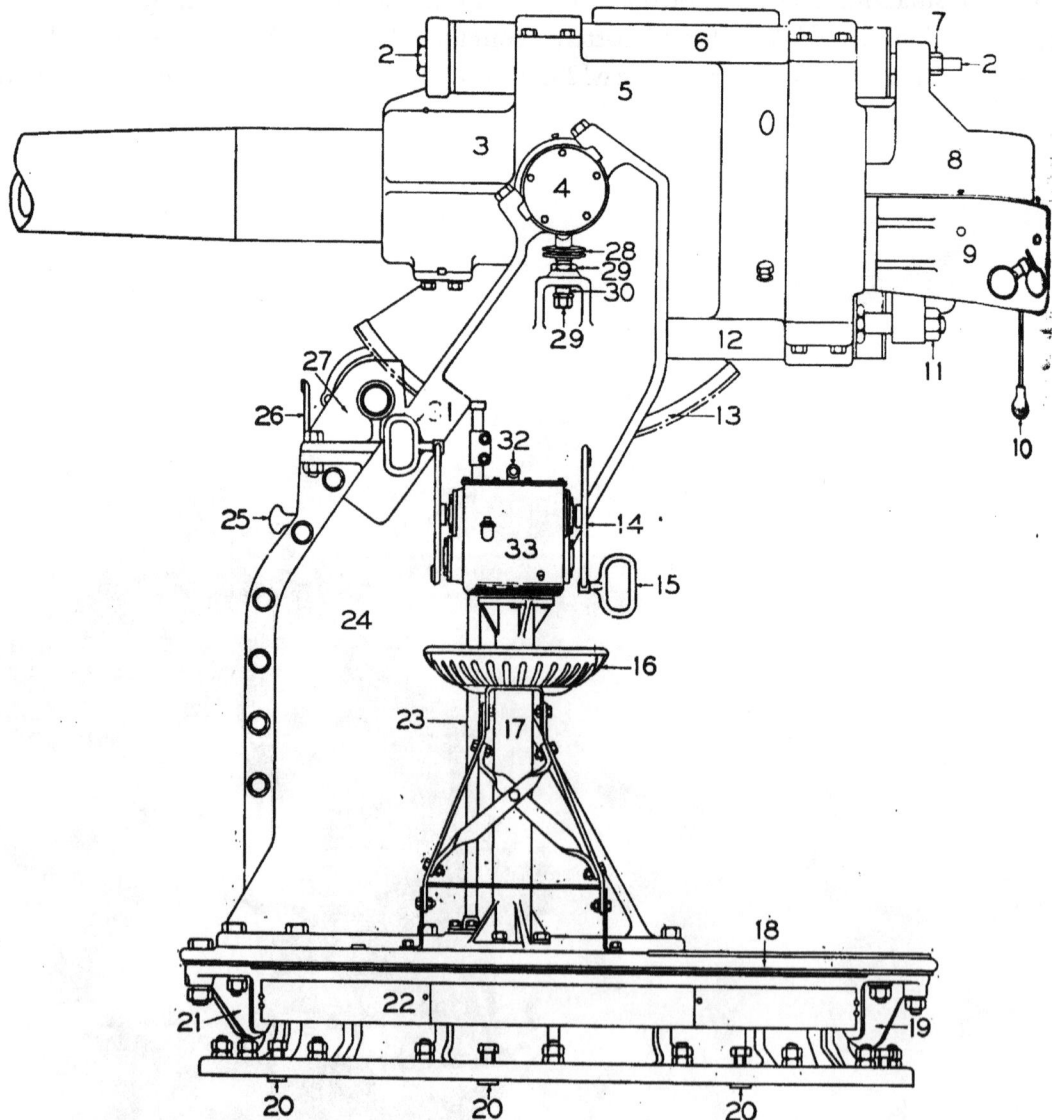

90 mm AA Gun

In 1938, the U.S. Army issued requirements for the development of antiaircraft guns to counter continued improvements in aircraft size, speed, and altitude. The 90 mm model T-1 was tested that same year, and modifications and improvements suggested by the testing produced the T-2, which was adopted and put into production as the M1 in 1940. By the time the United States entered the war in December 1941, the army had 171 M1 90 mm guns, and the very high-priority production of M1 guns continued into 1942. The M1 gun was a powerful weapon, with a 50 caliber barrel and a drop-block breech mechanism, using an inertia-type firing mechanism, a modification of the continuous pull type. All models of the 90 mm gun fired a 23 lb HE shell at 2,700 fps, to a range of 19,000 yds, limited by the proximity fuze (1944), or 12,000 yds, limited by the 30 sec time fuze (1949). Manual loading was standard for the M1 gun, but a rammer was designed for it. The rammer worked on the basis of recoil, so one round had to be loaded and fired in the normal manner before the rammer was operational.

The recoil was controlled by a hydro-pneumatic system, which used a recoil cylinder, a gas cylinder, and a floating piston cylinder, all placed under the barrel. The recoil was variable, ranging from 40-44 inches at $0°$ elevation to $24^1/_2$-26 inches at $80°$. A single equilibrator balanced the weight of the barrel, allowing the gun to be trunnioned near the breech.

The M1 and M1A1 mounts differed in that the M1A1 mount was designed to work with the M2 remote control system, while the M1 mount operated manually only. The gun sat, unprotected, on a single-axle trailer mount, with three outriggers for stability.

The M1 gun was also mounted on the M3 (originally the T-3) mount. This fixed mount was mounted on a concrete emplacement, and featured a shield to protect the gun and crew. This manually-loaded mount was used in seacoast role, but could also be used against aircraft. Like the M1A1 mount, it used a remote control system, in this case, the M13. The 90 mm gun on the M-3 mount had a maximum rate of fire of 30 rpm, and a normal rate of 25 rpm. On the M1 mount, the gun had a rate of fire of 22 rpm.

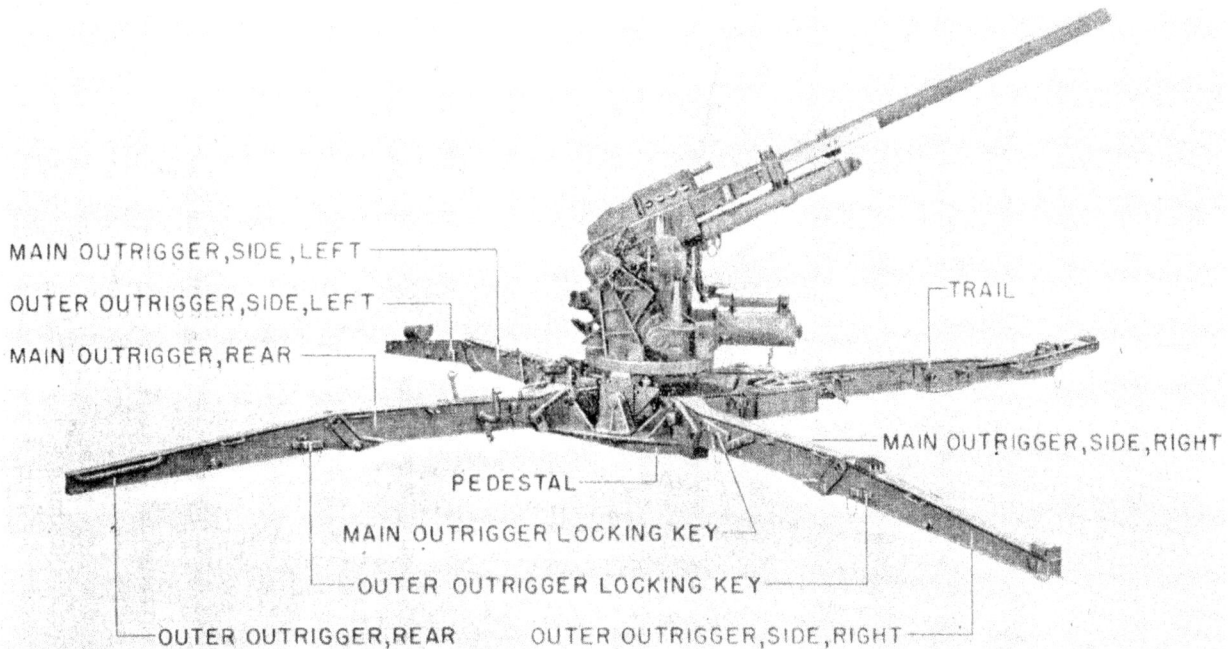

RA PD 2971

90 mm M1A1 mount

By mid-1941, the United States had seen the Germans use their 88 mm guns in both antiaircraft and antitank roles, and wanted a mount for the 90 mm gun which could be used against tanks. Although the production of the improved guns was delayed by the high priority assigned to the M1 gun, in May 1943 the army adopted the 90 mm gun M2, with the M2 AA mount. The gun was normally lowered to ground for firing, but could be fired while still on its wheels in an emergency. The mount included a small shield for use in the ground role, as well as an automatic fuze setter-rammer. Extremely versatile, the gun had all the M1 gun's antiaircraft equipment, along with sights and a mount that maximized its effectiveness as a field gun. The M2 had a rate of fire of 20-24 rpm. Production of M2 guns wound down in 1944, but both the M1 and M2 guns served on for years, in both antiaircraft and seacoast roles.(11)

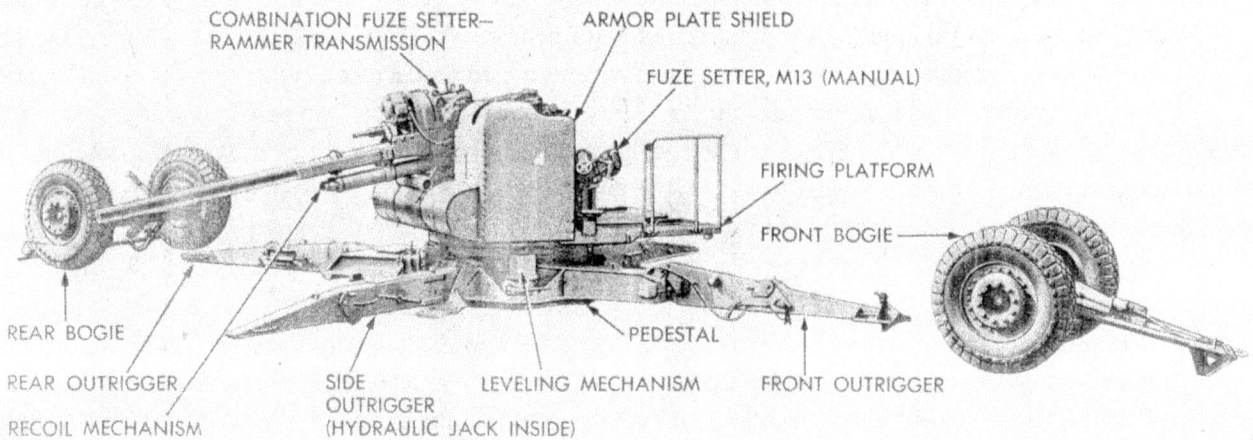

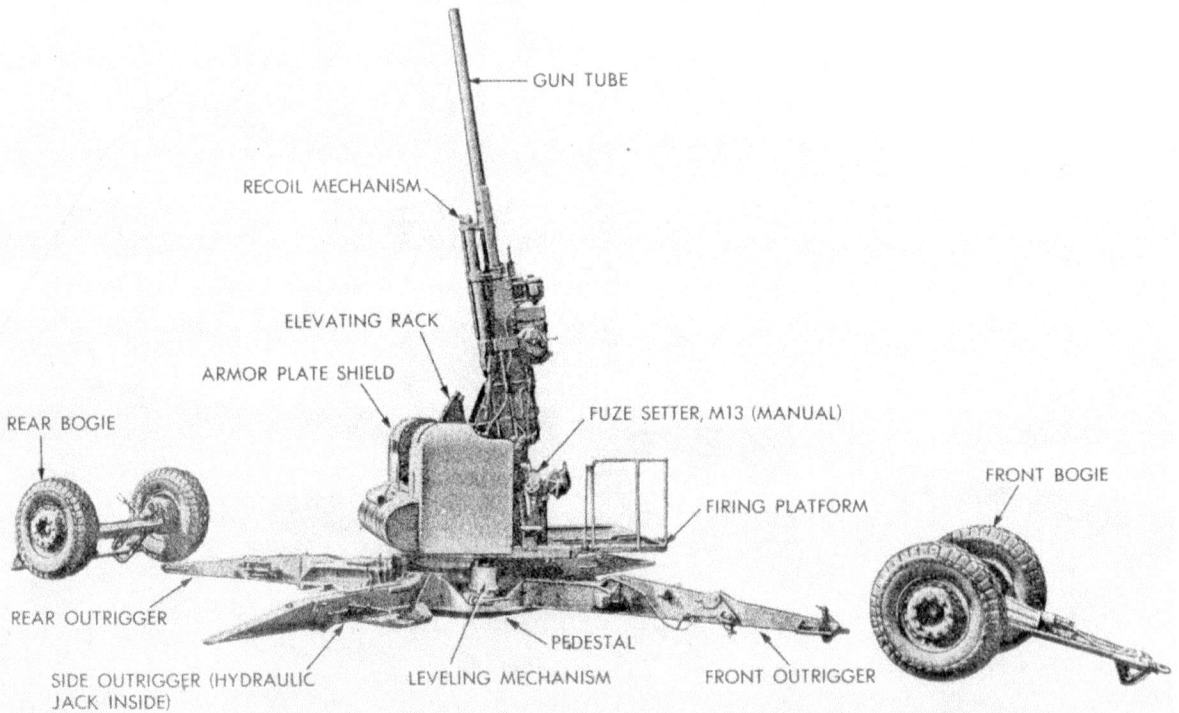

90 mm M2 mount

Figure 3—90-mm Gun M1 and 90-mm Gun Mount T3—Front View, Emplaced

90 mm T3 (M3) Fixed Mount

105 mm Antiaircraft Gun

In 1928, the Ordnance Department announced the adoption of a new gun, 105 mm, intended to be the standard equipment for fixed AA defenses. The weapon adopted was 60 calibers long, with a 4.1-inch bore. It fired a 33 lb projectile to an altitude of over 12,000 yds, at a muzzle velocity of 3,000 fps (1928). Using fixed ammunition and a compressed air rammer, the gun could fire 15 rpm. Congress, however, declined to fund the gun (or much of anything else), and it remained dormant until 1938.

When finally approved in 1938, the 105 mm gun barrel was of one piece, cold worked, monotube construction. Although the muzzle velocity was reduced to 2,800 fps, the altitude had increased to 14,000 yds (1938). The M1 mount was very similar to the M3 3-inch mounts, the principal differences being the larger dimensions to accept the larger barrel and the addition of a power rammer. Using the rammer, which was actuated by air compressed during the movement of the gun in recoil, a sustained rate of fire of 15-20 rpm was achieved, with a maximum of 30 rpm for short periods of time. A single hydraulic recoil cylinder was provided under the barrel, with two spring-cylinder recuperators.

Few 105 mm guns were built, and those actually constructed were sent to Panama to defend the canal. By 1944, it no longer appeared in listings of standard artillery.(12)

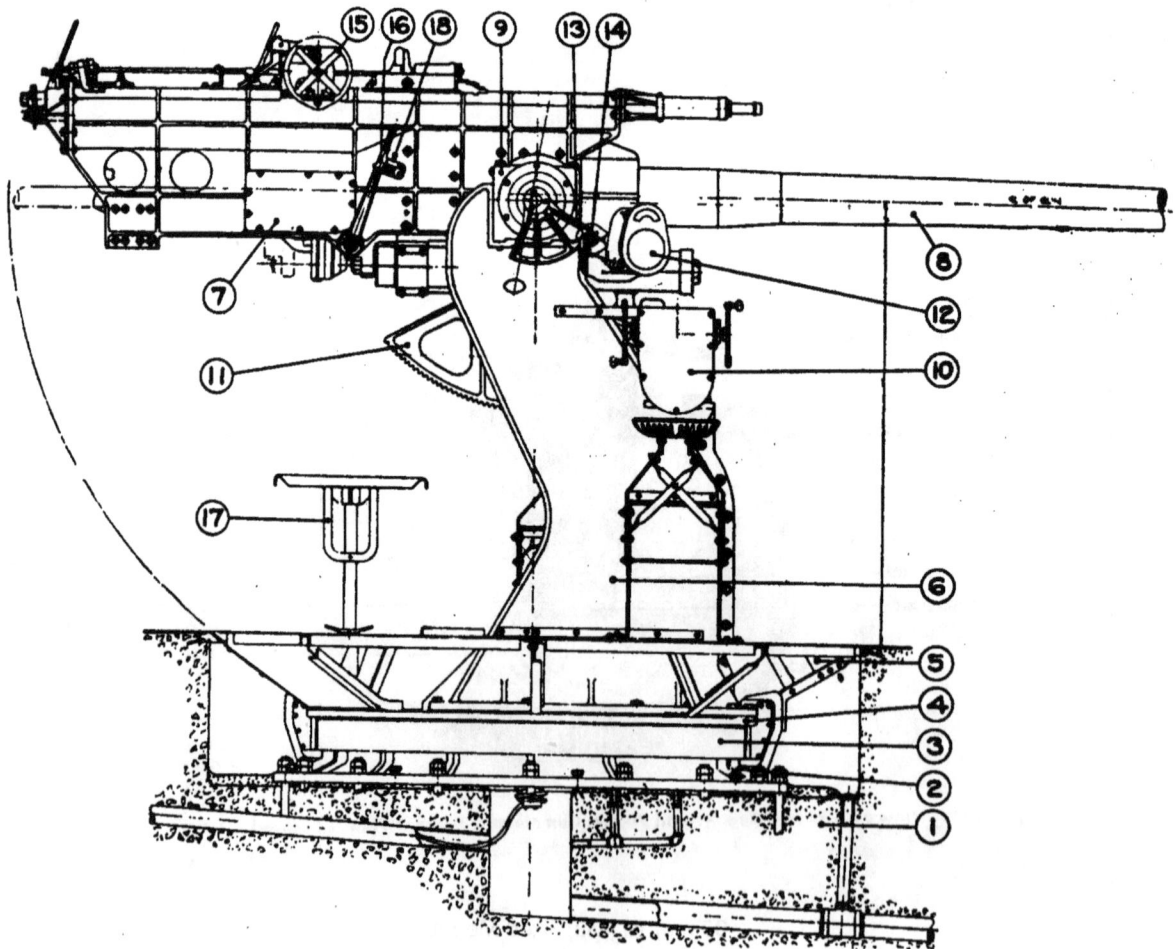

120 mm (4.7-inch) Antiaircraft Gun

The largest American antiaircraft gun was the 4.7-inch, later termed the 120 mm, or "stratosphere" gun. The concept for the gun originated during WWI, based on suggestions from General Pershing. Some development was carried out in the early 1920s, but on January 15, 1925, the Ordnance Department suspended work on the 4.7-inch AA gun M1920E, which had been under test since the latter part of 1922, concluding that there was no need for a gun this large. At the same time, the department announced 40% completion of AA fire control data for the gun.

In 1938, as funding began to become available, the army issued requirements for both 90 mm and 120 mm AA guns, in expectation of future aircraft developments. The production of 90 mm guns was a very high priority, while the 120 mm gun remained a low priority until after 1942. A handful of prototype guns were built in 1942, but the gun was not put into production until the USAAF adopted the high-flying B-29 bomber. Fearful that the Axis might develop a similar weapon, the army adopted the 120 mm M1 AA gun early in 1944, and by the end of the year, finished guns were being issued to the Coast Artillery Corps.

By the time the weapon was available, it had become clear that the enemy could not produce the type of aircraft it was designed to defend against. As a result, only 550 of these guns were built, many after the end of the war, and only a few were deployed to Hawaii and on Iwo Jima by the 752nd Gun Battalion, Batteries C and D. The gun remained an important element in the American arsenal for many years.

Twice the size of the 90 mm gun, the 120 mm gun was a massive piece of equipment. The gun traveled on a twin bogie mount, with dual tires. Both bogies were removed when the gun was emplaced, a process that only took 25 minutes. When traveling, the gun was racked 70 inches out of battery, in order to distribute its weight over the two bogies, which prevented the gun from firing off its wheels. When emplaced, the mount was supported by the pedestal and stabilized by four outriggers. The mount allowed fire up to 80° elevation. The hydropneumatic recoil mechanism was "somewhat temperamental," and required special attention, while as of 1950, the fuze setter-power rammer was mechanically complex, and "a source of malfunctions in the field." It was, however, being constantly improved. The 120 mm battalions were semi-mobile, due to insufficient organic transportation. The M6 tractor, 38 tons loaded, could tow the gun at 15 mph on good primary roads, while 4 mph across firm terrain was considered good. On road marches, weight limits on bridges were a constant problem.

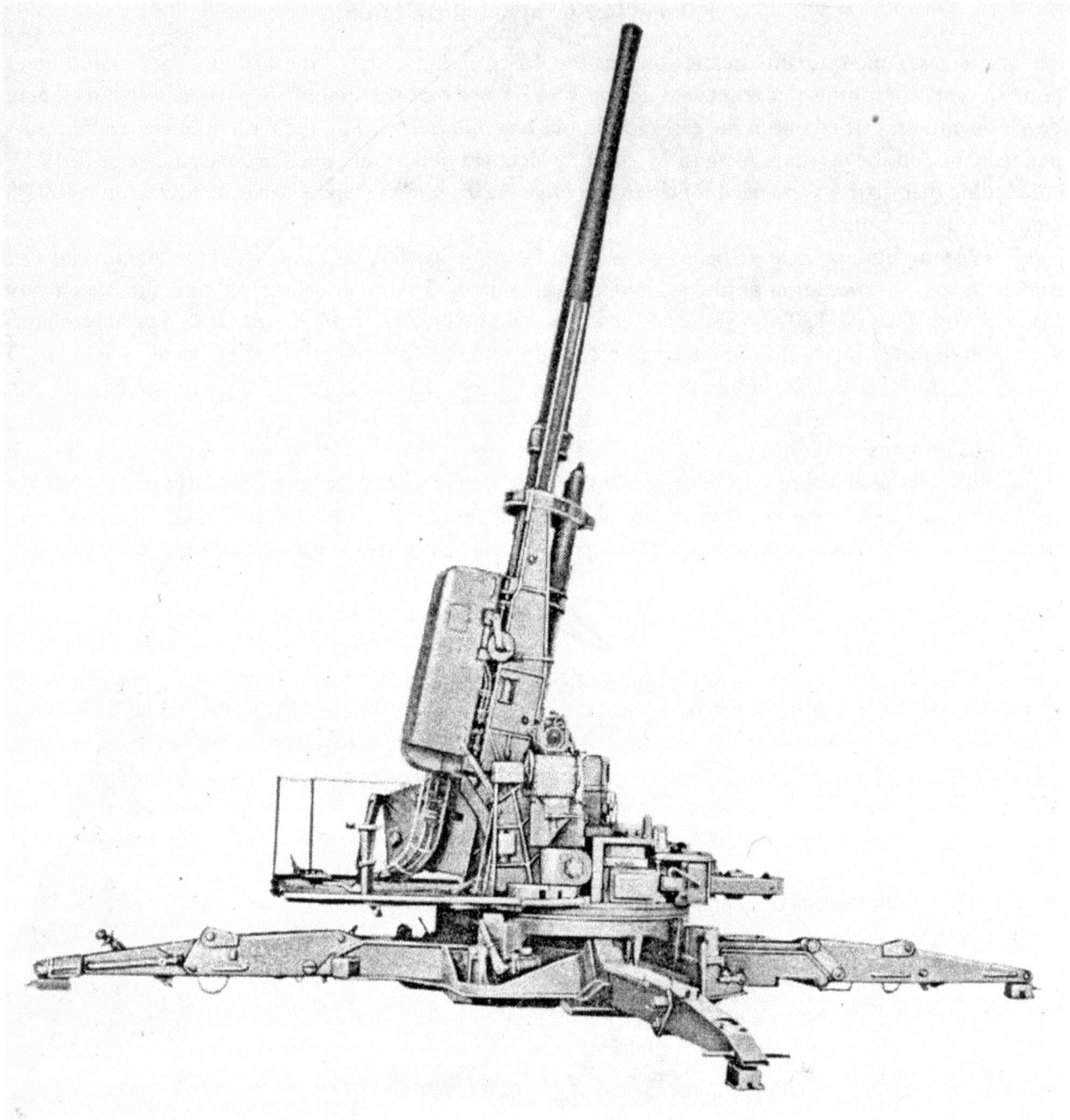

The 60-caliber barrel alone weighed 7,856 lbs, while the entire gun, mount, rammer, and remote control system weighed over 30 tons. The breech was of the drop-block variety, and the gun had a normal rate of fire of 10 rpm using the power rammer, although a well trained crew using proximity fuzes could reach 15 rpm. Due to the gun size, the ammunition was semi-fixed, that is, the 50 lb projectile was rammed first, then the cartridge case with the powder. The muzzle velocity was 3,100 fps in 1945, giving a maximum altitude of 19,150 yds (1950), or 15,800 (1944, limited by 30 sec. fuze). The 1950 altitude exceeded the capabilities of the fire control equipment.

Despite the 120 mm's higher velocity, it was slightly less accurate than the 90 mm, within the 90 mm's limited range, for reasons that were not entirely known.(13)

Sources

1. "Professional Notes," *The Journal of the United States Artillery*, Vol. 40, No. 2, (Sept.-Oct. 1913), pp. 254-55.

2. "Artillery Ordnance Development," *The Coast Artillery Journal* (hereafter *CAJ*), Vol 62, No. 5 (May 1925), p. 435. *Antiaircraft Defense*, (Harrisburg: Military Service Publishing Co., 1940), pp. 148-154. Herbert F. Markland, "A Coast Artilleryman's Experience on Fort Mills: Part Two," *Coast Defense Study Group Journal*, Vol. 9, No. 2 (May 1995), p. 9.

3. "Report of the Chief of Coast Artillery," *CAJ*, Vol. 64, No. 1 (Jan. 1926), p. 53. Konrad F. Schreier Jr., *Standard Guide to U.S. World War II Tanks and Artillery*, (Iola, WI: Krause Publications, 1994), pp. 211-214. War Department, *Service of the Piece, Caliber .50 AA Machine Gun*, Antiaircraft Artillery Field Manual 4-155, 1943, pp. 30-31, 83.

4. Schreier, *Standard Guide*, pp. 214-19. War Department, *Standard Artillery and Fire Control Matériel*, Technical Manual 9-2300, 1944, pp. 20-21, 86-87. War Department, *37-mm AA Gun Matériel*, Technical Manual 9-235, 1944, p. 9.

5. Schreier, *Standard Guide*, pp. 22-25. TM 9-2300, 1944, pp. 22- 25. War Department, *Artillery Matériel and Associated Equipment*, Technical Manual 9-2300, 1949, pp. 10-11, 30-31. War Department, *Service of the Piece, 40-mm Antiaircraft Gun*, Coast Artillery Field Manual 4-141, 1942, p. 5.

6. Benedict Crowell, *America's Munitions, 1917-1918*, GPO, 1919, pp. 87-90. *Basic Coast Artillery*, (Annapolis: National Service Publishing Co., 1928), p. 129-40. War Department, *Handbook of Artillery, Including Mobile, Anti-Aircraft and Trench Matériel, May, 1920*, Ordnance Department Document No. 2033, GPO, 1920, pp. 354-64. *Antiaircraft Defense*, pp. 9-10.

7. *Handbook of Artillery*, pp. 340-53. War Department, *3-inch Seacoast Gun Materiel*, Technical Manual 9-421, 1942, pp. 80- 81.

8. William J. Wuest, "The Development of Heavy Antiaircraft Artillery, Part 2," *Antiaircraft Journal* (hereafter *AAJ*), Vol. 95, No. 5 (Sept.-Oct 1952) p. 33. *Handbook of Artillery*, pp. 331- 39. "Professional Notes," *CAJ*, Vol. 62, No. 4 (April 1925), p. 332. War Department, *3-inch Seacoast Gun Materiel*, Technical Manual 9-421, 1942, pp. 78-79.

9. *Military Science and Tactics, Coast Artillery, Senior Division, Basic Course*, (Washington: P.S. Bond Pub. Co., 1939), pp. 114-117. War Department, *3-inch Gun Antiaircraft Gun Matériel (Mobile)*, Technical Manual 9-360, 1940, pp. 3-24. U.S. Army. Ordnance Department, *Firing Tables for 3-inch Antiaircraft Gun Models of 1917, 1917 MI, 1917 MII, 1925 MI, and MI, M2, M3 and M4, Firing A.A. Shrapnel, Mark I*, FT 3AA-J-2, (Washington: Engineer Reproduction Plant, 1928).

10. Firing Table 3AA-J-2, 1928. War Department, Report of Completed Works-Seacoast Fortifications, Harbor Defenses of Chesapeake Bay, Fort Monroe, Virginia, Form 1, Antiaircraft Battery, Corrected to May 27, 1936. *Military Science and Tactics, Coast Artillery, Senior Division, Basic Course*, 1939, pp. 112-114.

11. TM 9-2300, 1944, pp. 26-29, 128-29. War Department, *90-mm Antiaircraft Gun Matériel M1 and M1A1*, TM 9-370, 1942, pp. 5-45. War Department, *90-mm Gun M2 and Antiaircraft Mount M2*, TM 9-372, 1944, pp. 1- 20. Schreier, *Standard Guide*, pp. 235-245.

12. "Professional Notes," *CAJ*, Vol. 68, No. 4 (April 1928), pp. 263-64. James A. Sawiki, *Antiaircraft Artillery Battalions of the US Army*, Vol. 1, Wyvern Publications, 1991, p. 8. Thomas J. Hayes, *Elements of Ordnance*, (John Wiley: New York, 1938) pp. 329-332. TM 9-2300, 1944.

13. "Professional Notes," *CAJ*, Vol. 62, No. 5 (May 1925), pp. 438, 440. Schreier, *Standard Guide*, pp. 237-238, 246. Elmo E. Cunningham, "The One-Twenty Millimeter Gun," *AAJ*, Vol. 93, No. 5 (Sept.-Oct. 1950), pp. 7-10. TM 9-2300, 1944, pp. 18-20. War Department, *120-mm Gun M1 and Antiaircraft Mount M1*, Technical Manual 9-380, 1945.

U.S. ARMY COAST ARTILLERY
FIRE CONTROL, POSITION FINDING, AND GUN DRILL

Edited by Mark A. Berhow
Sections by Bolling W. Smith, John A. Martini, and Fred M. Baldwin

Battery DeRussy, Fort Monroe, HD Chesapeake Bay. (The Casemate Museum).

This section explains the basic details of the procedures used to fire the seacoast guns of the US Army Coast Artillery Corps. The procedures described here are similar those used to fire all the major guns of the "concrete artillery" during the years 1900-1945, from 3-inch to 16-inch. However, the specific details described in this section will be those used to fire the 12-inch and 14-inch disappearing guns and, in general, follow the fire control and position finding methods used in 1940. The firing procedures used during the earlier periods and for the other caliber and types of weapons were similar to those outlined in this section, but differed in a number of the specific details and in the numbers of men needed. For further specific information on the other big guns, please refer to the references listed at the end of this article.

The target range and azimuth for seacoast artillery guns were determined using command and equipment systems collectively referred to as *fire control and position finding.* Fire control was defined as the exercise of those functions of command connected with the concentration and distribution of fire, including the assignment and identification of targets. Once the battery received its target designation, it was responsible for determining the range and direction from its guns to the target. Position finding systems were used to determine the range and direction of a target from a battery directing point.

Fire Control and Position Finding: Background

Bolling W. Smith

In the early years of America's seacoast defenses, when gun ranges were only a few miles at most, aiming techniques were simple. A gunner pointed his cannon in the direction of a target, estimated its speed and course, fired a round or two and corrected his aim after each shot. This "point and pray" school of fire control continued to nearly the end of the 19th century. Each gunner made his own calculations and aimed his own gun, and no special structures were needed for fire control. By the last decade of the 19th century, though, improvements in guns and propellants were increasing ranges to almost ten miles, and with these increased ranges came the need to locate targets with increasing precision. While the gunners could still largely aim the guns at the targets, the precise determination of the range was the key problem.

Essentially, until the development of radar, all techniques involved the use of triangles. With one side and two angles known, the remaining angle and the length of the other two sides of the triangle could be calculated. The key choice was between vertical and horizontal triangles. A vertical triangle relied on an elevated observing instrument, or depression position finder (DPF). The angle at which the instrument was pointed down to see the target was measured, and from this the range to the target was calculated.

A horizontal triangle, on the other hand, used two instruments in separate stations with a carefully measured base line between them, both pointed at the same target. Each instrument measured the angle between the base line and the target, and this information, combined with the length and azimuth of the base line, established the location of the target. Single-station self-contained range finders were a further development of the horizontal-base method, but since accuracy was proportional to the length of the baseline, self-contained range finders were inherently less accurate, especially at longer ranges. Both vertical and horizontal methods involved advantages and drawbacks.

The vertical-base systems required only one station with two soldiers to observe and transmit the information. The instrument could be located above the plotting room, eliminating the need for long communications links. Also, officers and NCOs could directly supervise the observers, contributing to their diligence and accuracy.

The disadvantages, however, were considerable. The instrument had to be elevated. DPFs lower than 125 feet above sea level were inaccurate beyond 10,000 yards, which was less than the maximum range of the major-caliber disappearing guns, and much less than the maximum range of the long-range guns introduced during and after WWI. This height requirement posed little problem where cliffs and mountains abutted the seacoast, but the mid-Atlantic and Southern seashores were low-lying which meant expensive and vulnerable towers.

DPF observers had to carefully set their telescope crosshairs on the waterline of the target, since errors in discerning the waterline of a moving ship seriously distorted the range. The height necessary to waterline a target remained a serious limiting factor; an instrument 100 feet high could waterline a ship 23,253 yards away, but to waterline a ship at 45,000 yards required an instrument almost 400 feet above the water level. Great mechanical precision was required to measure the minute angles accurately, and DPFs were four times as expensive to install as azimuth instruments. A system stations to measure the tide was also required.

Lastly, even the air itself caused problems. Light is bent by refraction, and due to the small angles being measured, refraction at times caused angular changes several times greater than that due to distance. DPFs had to be pointed at datum points and adjusted to give the known distance. When this was done for several points at different ranges, the effects of refraction were largely eliminated, but this required a system of datum points visible from every DPF.

The horizontal-base system was the reverse of the vertical-base system. Unlike the vertical-base system, horizontal-base stations could be miles apart, and therefore much more accurate, especially at longer ranges. In addition, the instruments were much simpler, needing only to measure horizontal angles, without allowing for tide or refraction.

There were drawbacks, however. Horizontal-base systems required reliable communications. Even if the primary base end station was above the plotting room, efficient communications were essential between the secondary station and the plotting room. Before improvements in the telephone during the first decade of the century, this was the critical problem. Secondly, more men and stations were needed. Two stations were needed for every observation, doubling the manpower required. Thirdly, both stations had to track the same target. Fog, smoke, haze, or any other factor that prevented both stations from seeing the same target would thwart the system. When targets were numerous, there was the real danger that the stations would track different targets. Further, both stations had to read the azimuth to the target at precisely the same instant. Readings taken on a rapidly moving target several seconds apart would be worse than useless. To solve this problem, time interval systems were necessary. These systems were effective, but they and their required communications added another major complication, especially for mobile coast artillery. Lastly, horizontal-base stations could not effectively track targets on the flanks of the baseline.

While base end stations for the horizontal system did not need expensive and vulnerable towers at moderate ranges, the curvature of the earth also limited the range at which low-lying horizontal-base stations could track targets. Since base end stations used in the horizontal system did not have to waterline their targets, their range was less limited than vertical-base stations, but as gun ranges increased after World War I, towers or other high-sites frequently became necessary even for horizontal-base stations.

The army initially envisioned a horizontal-base system, at least in theory. During the waning years of the muzzle-loading era, the vague fire control system prescribed presumed the use of plane tables or the like to locate the target. The communications problems were left to the individual post commanders, which is to say they were mostly ignored. In any event, this theoretical system was virtually never attempted in practice. Where it was attempted, it was defeated by poor communications. Through the end of the 19th century, visual signaling was too slow and telephones were insufficiently reliable, while the telegraph was both too slow and required trained operators.

As the first breech-loading guns began to come into service in 1896, the army adopted a depression position finder for use with the vertical-base system, which the artillery favored "… on account of its apparent simplicity, and also on account of the length and difficulties of rapid and accurate communication involved in the horizontal base system." In addition, the relatively short ranges then in use favored the vertical-base system. Early DPFs seemed adequate for the needs of the service at the time, although there were dissenters.

By 1899, a type tower for range finders was constructed at Fort Hancock, NJ, mounting a DPF at a height of 60 feet. The chief of engineers considered high towers vulnerable aiming points for enemy ships, but a board of officers evaluated the tower and found it "admirably suited to its purpose in every respect."

Another board, convened in 1901 to evaluate the relative merits of the horizontal and vertical systems, subsequently reported "A system of position finding based purely on a vertical base will not meet the requirements of the service, in as much as it may often happen that the water-line of the vessel cannot be seen, due to the interference of other vessels, or from other causes. Therefore in all cases a depression system should be augmented by an auxiliary system with horizontal base."

The chief of artillery concluded that horizontal base lines were necessary for all batteries and fire commands, but they should be equipped with DPFs for target identification and other purposes. For instance, DPFs in the extreme flank stations could extend coverage beyond that of the horizontal-base system. These conclusions represented the final answer, at least for an era. With the installation of longer-range guns during and after World War I, the vertical-base system became less and less practical.

By the 1920s, the horizontal system was only marginally effective at the maximum range of the newer guns. Spurred by experience in the World War, considerable energy was devoted to aerial methods. Balloons and airplanes were tried, but the problems were never really solved. In the end, the final answer was radar. By the end of World War II it was effective far beyond both vision and the ranges of even the most powerful guns.

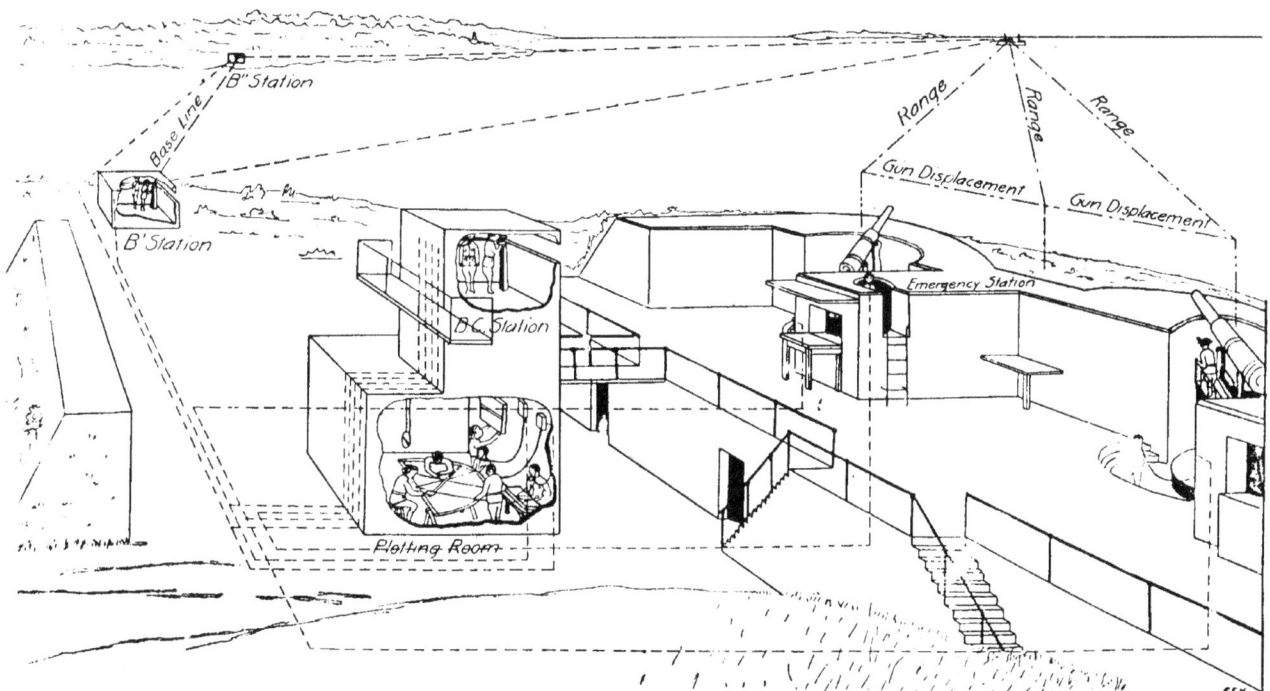

Elements of the horizontal position finding system

Specific Principles of Position Finding (1940)

Position finding equipment was designed to collect accurate and rapid data to:
- Correctly identify the target.
- Determine the speed and direction of the target.
- Determine the true range of the target from the directing point of the battery at any instant.
- Predict the position of the target at any instant in advance of its present position.
- Make ballistic corrections due to atmospheric or other conditions existing at the moment of firing.
- Make corrections due to range difference between directing point and location of the gun.
- Make adjustments to place center of impact of fire from guns on target.
- Determine deflection to be set on gun so that when fired, it will allow for the travel of the target during the time of flight of the projectile, and effect of wind and drift on the flight of the projectile.

In summary, to furnish accurate ranges and azimuths at which the guns of the battery were to be laid to insure hitting the target upon firing.

The function of a position finding system was to furnish data in the proper form for use in pointing the guns of a battery for firing at a target. In seacoast artillery, the guns had to be pointed at a moving target. There was a lapse of time between the instant an observation was taken on a target and the instant the guns were fired with the firing data that were calculated as a result of that observation. That interval was called the "dead time." Its length depended on the time necessary to calculate the firing data with the desired accuracy and apply them to the guns.

DEFINITION OF POINTING.-The operations of the range section culminated in the determination of the elevation (range) and the azimuth (direction) at which the gun had to be set for firing at a particular instant. In other words the determination of the setting of the axis of the bore of the cannon in elevation and in azimuth for the proper vertical angle and the proper horizontal angle. This setting operation was called "pointing."

CASES OF POINTING.-There were three cases of pointing, which were defined according to the combination of pointing methods used. The definitions were as follows:

Case I.-Pointing in which both direction and elevation were given at the gun by means of a sight pointed at the target. All firing data was determined by the firing section (the gun crew).

Case II.-Pointing in which direction was given at the gun by means of a sight pointed at the target, and elevation by means of an elevation quadrant or a range disk with all range data provided by the range section.

Case III.-Pointing in which direction was given the gun by means of an azimuth circle, or of a sight pointed at an aiming point other than the target, and elevation by means of an elevation quadrant or a range disk. All range and direction data was provided by the range section.

In a 3-inch rapid fire battery, case I or case II pointing was used. The ranges and times of flight were short and the dead time was negligible. The problem of determining firing data is comparatively simple, a self-contained range finder and a gun sight was used.

For a battery of 6-inch caliber or larger the operation of determining firing data for a moving target was divided into the following steps:

(1) Tracking-which included observing and plotting successive positions of the target.
(2) Determination of the set-forward point-which predicted the future position of the target, that is, its predicted position at the end of the projectile's time of flight.
(3) Relocation-which determined the range and direction of the future position of the target from the directing point.
(4) Calculation of firing data-which consisted of converting the relocation data into firing data for use in pointing the guns.

The three standard systems of position finding used by seacoast artillery were the 1) horizontal base, 2) vertical base, and 3) self-contained base systems. In all of these systems the procedures were similar. They differed only in the method of locating the target in tracking.

TRACKING PRINCIPLES COMMON TO ALL SYSTEMS.-

The first step in all position finding systems was the identification and location of the position of the assigned target with respect to the observation stations (base end stations) of the battery. This operation was called "tracking" and consisted of locating, at regular intervals of time by observation from one or more stations, successive positions of the target, and plotting those positions on a plotting board. The time interval between successive observations was called the "observing interval" and was 15 to 30 seconds in length (this varied from battery to battery). The observing intervals were indicated by TI (time interval) bells or buzzers which sounded simultaneously in all stations of the battery.

HORIZONTAL BASE SYSTEM

Description- In the horizontal base system, the target was located by the method of intersection used in surveying in which the direction of the target from two known points was determined. In the triangle involved, one side and the two adjacent angles were known. The solution was made graphically on the plotting board. The system required a base line on the ground, the azimuth and length of which had been accurately determined by surveying; two observation stations, one at each end of the base line, in each of which was mounted an instrument for measuring azimuths; a plotting room with plotting board; and the necessary communication lines. By 1944 most major caliber batteries had six to ten different possible baselines.

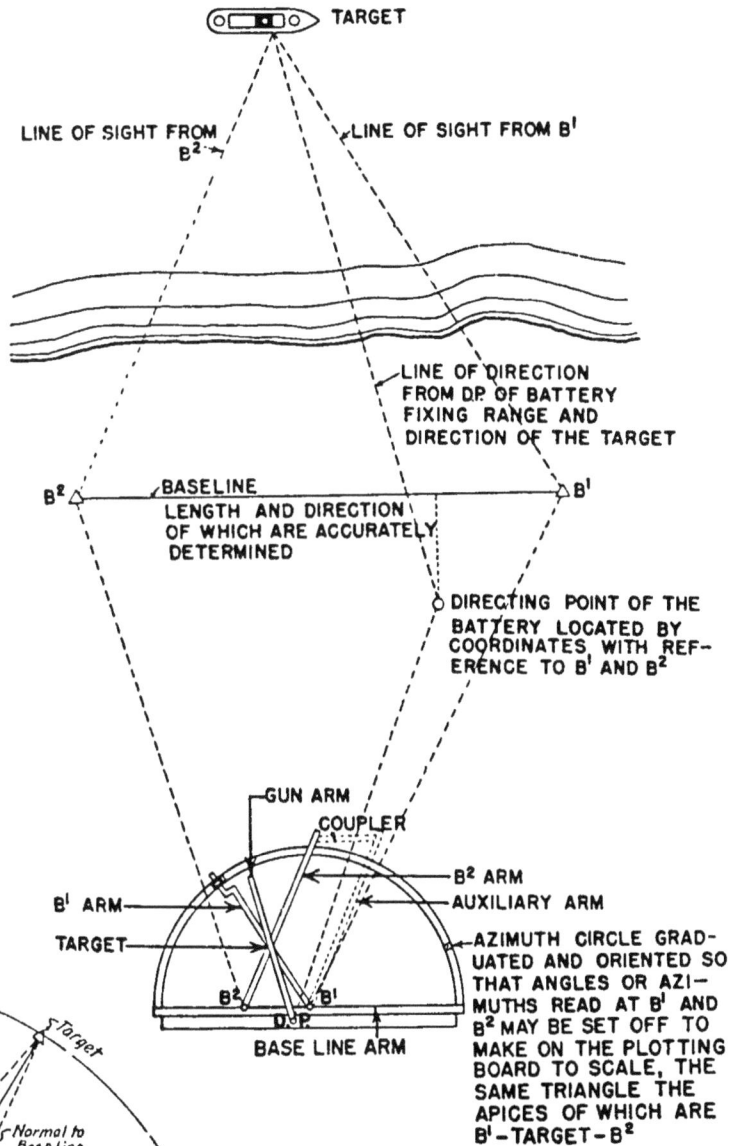

Elements of a horizontal base system: Base line orientation and the relationship of the directing point, the baseline, and target to the plotting board.

The horizontal angle, measured in a clockwise direction, was from the reference line in use to the line joining the observer and the objective. In the case of fixed harbor defense artillery, the reference line was a horizontal line running parallel to true SOUTH from the origin of the harbor defense coordinates; in the case of mobile artillery, the reference line is usually a line running grid NORTH.

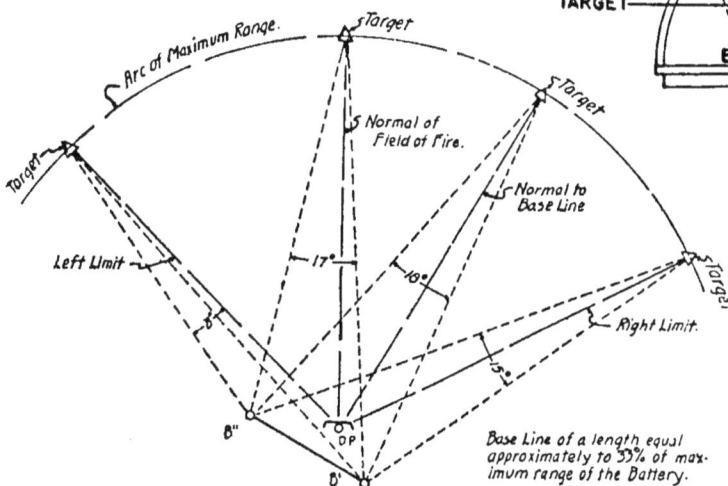

The plotting board represented to scale the field of fire of the battery. On it was located to scale—in their proper relation to each other—the observation stations, the base line, and the directing point (the point for which the firing data are to be determined). The figure on the previous page illustrates the relation between the installations in the field and the set-up on the plotting board.

The observation station nearest the directing point was usually called the primary station. The station at the other end of the base line was called the secondary station. The base line of a horizontal base system was called "righthanded" or "left-handed," according to whether the secondary station was to the right or to the left of the primary station, as viewed from behind the base line facing the field of fire. The base line for a horizontal base system conformed to the following principles:

(a) Its length was from one-fourth to one-third of the maximum range to be measured by the base line.

(b) Its direction was approximately perpendicular to the center line of the field of fire to be covered by the base line, wherever possible.

(c) The base end stations were of sufficient height above sea level to afford a field of view to seaward beyond the maximum range to be measured, wherever possible.

Operation.-The *observers* at the base end stations sighted and followed the target with the vertical cross wires of their instruments on a designated point on the target assigned by the *battery commander*. At the sounding of the TI bell, the *observers* stopped following the target with their instruments while the *readers* read the azimuths and then resumed tracking. Each *reader* was equipped with a telephone head set connecting him to an operator, called an *arm setter*, in the plotting room. There the successive observations were plotted on the plotting board. The plotting board had an arm corresponding to each of two observation stations with a means of setting each arm in azimuth. Each *arm setter* set his arm to the azimuth read by the corresponding *reader*. The point of intersection of the arms represented the position of the target at the instant the observations were taken. This point was marked by the *plotter*. The operation was repeated at the sounding of each successive TI bell. The points were called "plotted points." A line joining the plotted points represented the track or path of the target.

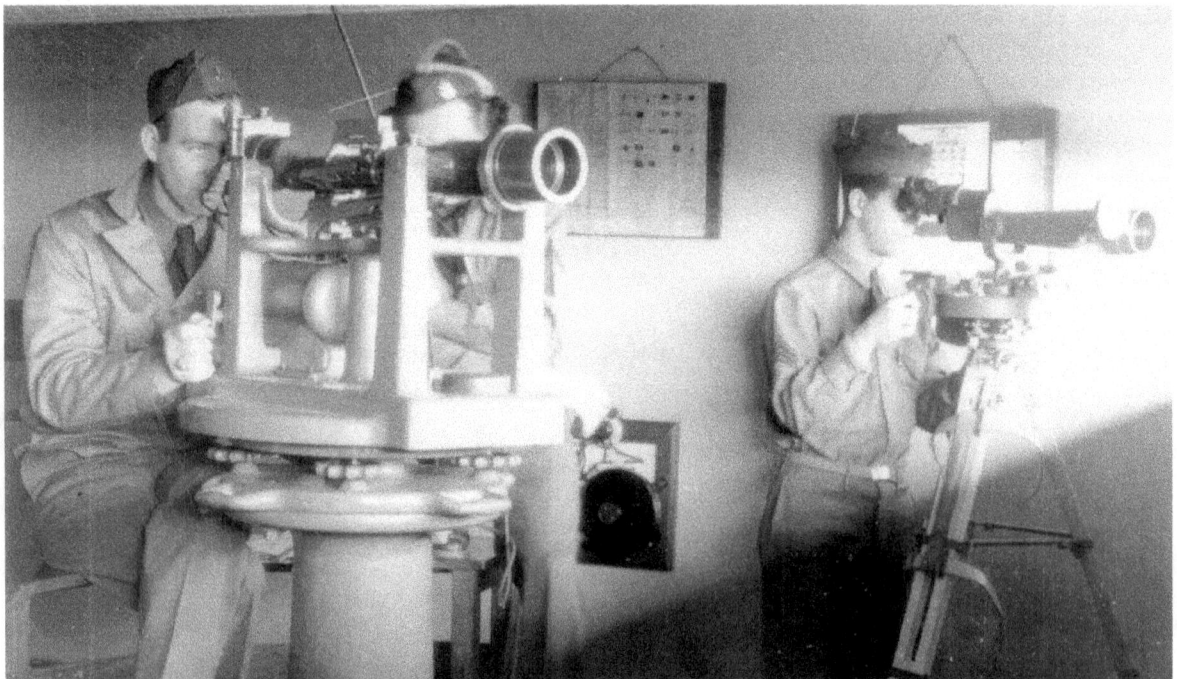

Interior of an observation station at Fort Standish, HD Boston, with a depression position finder (DPF) (left) and an azimuth scope (right). (NA; Al Schroeder Collection, Inst. of Mil. History, Carlisle, PA)

VERTICAL BASE SYSTEM

Description.-In the vertical base system, the target was located by the offset method used in surveying, in which the direction and distance of the target from a known point were determined. The direction was determined by reading the azimuth as in the horizontal base system. The distance was determined by the depression angle method which involved the solution of a vertical right triangle of which the known leg was the effective height of the observation instrument above sea level, the hypotenuse was the line of sight from the *observer* to the target, and the unknown leg was the desired range to the target. The known angle was the complement of the angle between the hypotenuse and the known side, corrected for refraction. It was the angle through which the line of sight was depressed from the horizontal to intersect the target and was called the depression angle. The triangle was solved mechanically by the observation instrument called a "depression position finder" (DPF). This system required but one observation station, the azimuth and range to the target being read from the same instrument.

Operation.-The *observer* tracked the target in azimuth with the vertical cross wire as in the horizontal base system. At the same time he tracked the target in range with the horizontal cross wire. In the plotting room, only one arm of the plotting board was used. The azimuth and range was received from the *reader* at each sounding of the TI bell. The *arm setter* set the arm in azimuth and repeated the range to the *plotter* who marked the point at that range by means of range graduations along the edge of the arm.

SELF-CONTAINED BASE SYSTEM

Description.-In the self-contained base system, the target was located by the offset method as in the vertical base system. The azimuth (direction) and range were determined by means of a self-contained range finder of either the coincidence (coincidence range finder or C.R.F.) or the stereoscopic type.

Operation.-The operation of tracking with this system was similar to that with the vertical base system except that azimuths may have been read from a separate instrument.

Cooincidence Range Finder at the Harbor Defenses of Portland, Maine (Courtesy Joel Eastman)

ALTERNATE BASE LINES AND ALTERNATE STATIONS

For batteries employing the horizontal base system, several alternate base lines frequently were provided in order that use might be made of the base line allowing the greatest accuracy under existing conditions of visibility, target position, and target course. Those stations of the horizontal base system which had sufficient height of site were usually provided with depression position finders for use in a vertical base system, thus offering a choice of the most advantageous system and stations.

EMERGENCY SYSTEMS

Emergency systems possessed features of reduced accuracy that were acceptable only under emergency conditions and were for use when all the normal systems broke down or were put out of action. Possible emergency methods included use of data determined from a station outside the battery—either a group command station or the directing point of an adjacent battery-and their conversion to suitable firing data by means of range difference and azimuth difference charts; use of aerial observation to determine initial data and to determine adjustment corrections thereto; and estimation of data from the guns by means of comparison with the known ranges and azimuths of reference points, such as buoys, in the field of fire, with subsequent adjustment as a result of observation of fire.

GUN DATA COMPUTERS

During World War II the US Army began using electro-mechanical gun data computers for determining firing solutions calculated in the plotting rooms and transmitting this information to the guns in both the coastal artillery and the antiaircraft artillery. In antiaircraft applications they are used in conjunction with a director. The M1 Gun Data Computer was used for major caliber seacoast guns, it computed continuous firing data for a battery of two guns that are separated by not more than 1000 feet. It utilized input data furnished by the range section. The M8 Gun Data Computer was used with medium caliber guns (up to 8-inches). The M8 series used electrical methods for computing firing data, and made corrections for wind, drift, earth's rotation, muzzle velocity, air density, height of site, and spot corrections.

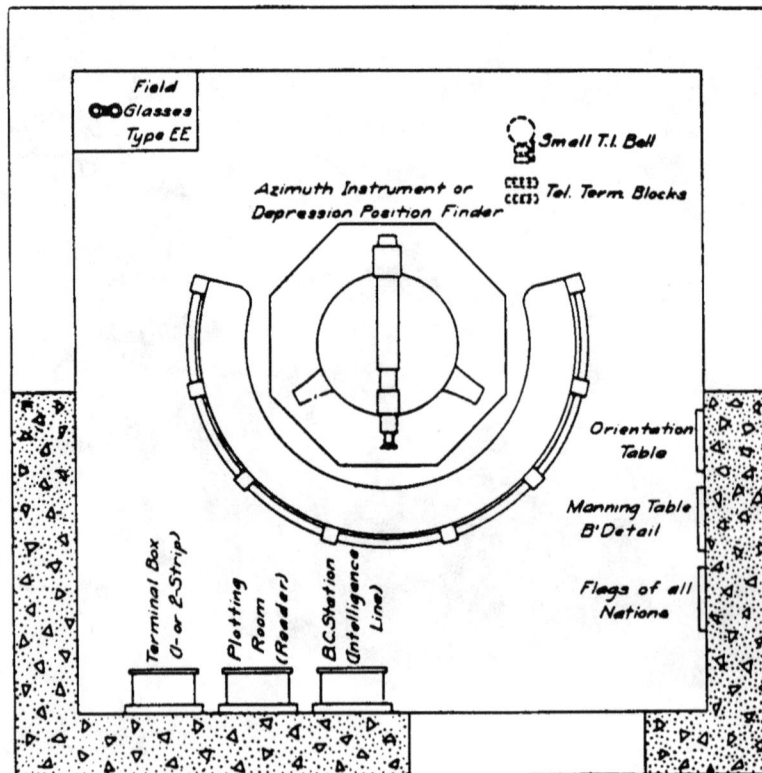

Interior arrangement of a base end station 1910s

Fire Control Observation Stations

John Martini

Fire control stations dot the coastline near many major American harbors. People coming across these nondescript military structures frequently, if erroneously, call them "pill boxes," "bunkers," "lookout posts," or "machine gun nests." In actuality these stations are remnants of the once-extensive network of army seacoast defenses that protected the United States during the first half of the 20th Century. These stations served as the eyes for seacoast guns that surrounded the harbors, taking measurements to enable gunners to precisely locate enemy ships, and controlling the submarine mines that guarded the harbors.

Fire control primary stations, Fort Casey State Park, WA (photo by Mark Berhow, 1996)

Technically, battery fire control stations for use with vertical-base systems were not base end stations, but the term was generally used for all position-finding stations. These stations came in several styles, usually reflecting the eras when they were constructed and the location where they were built. The first fire control stations, "crows nests," were small stations set into the battery parapets. The first separate stations at most harbor defenses began to appear just after the turn of the century. Army engineers frequently grouped five or six stations within a few dozen yards of each other, or several stations were located in large, multi-story frame buildings. Many clusters of fire control stations, especially primary stations, were within existing army forts. Secondary (and higher) stations were usually located some distance away. When possible, they too were located on government land, such as another fort, but often, especially during World War II, they were on small, military reservations purchased or leased for the purpose.

Where high land was close to the water, stations were often originally built on the reverse slope of hills; only their tops high enough to see or be seen. Later, simple rectangular structures were buried with only the top two or three feet exposed. A vision slit, usually with steel shutters, faced the sea. Interiors allowed just enough space for a cramped crew of soldiers to operate their telescopes and communications equipment. At other coastal locations, such as along much of the Gulf and South Atlantic Coasts, the land was relatively low and base end stations had to be elevated. Most of these stations consisted of a one, two, or three-floor house atop or otherwise integrated into a wood, metal, or concrete tower.

In the early 1900s, common station styles included concrete-walled, wood-frame, plaster-on-lath, and brick stations, with wood and tarpaper roofs. Other stations had glass circles embedded in the ceiling for light, while a few even had steel roofs. The 1930s brought more consistent concrete ground stations, either square or hexagonal, and in early World War II the style became the concrete "pillbox." Although these were somewhat standardized, they did vary from harbor to harbor. Around San Francisco, some base end stations had curved, dome-like steel roofs and counter-balanced visors that closed and covered the vision slit, while Los Angeles stations had simple or counter-balanced entry hatches and up to six separate metal shutters to cover and secure the observation windows. Exteriors tended to be plain concrete until wartime, when they were draped with camouflage netting or elaborately disguised with swirling paint schemes and carefully applied rockwork.

The "Barrancas"System and "Sewell" Construction.

An early system of base end station configurations was known as the Barrancas System, named after having been first tested at Fort Barrancas near Pensacola, Florida, in 1902-03. The Barrancas stations consisted of a series of stations for several batteries massed in one location. The stations at one location were generally housed in a single building, which had only one level, but with interior partition walls between each separate station. Most of these early Barrancas stations were stucco-covered wood frame buildings on a concrete foundation similar to the popular housing construction method in that same time period. This type of observation station construction became known in the Army as "Sewell" construction after the officer who proposed its use. Many harbor defenses received these types of stations before the system was discontinued in favor of more widely dispersed stations. Sadly today only a few of these structures, or their series of closely positioned instrument piers, remain. The photograph above is of the South Secondary Stations for Fort Worden, WA. (photo by Alex Holder 1980s).

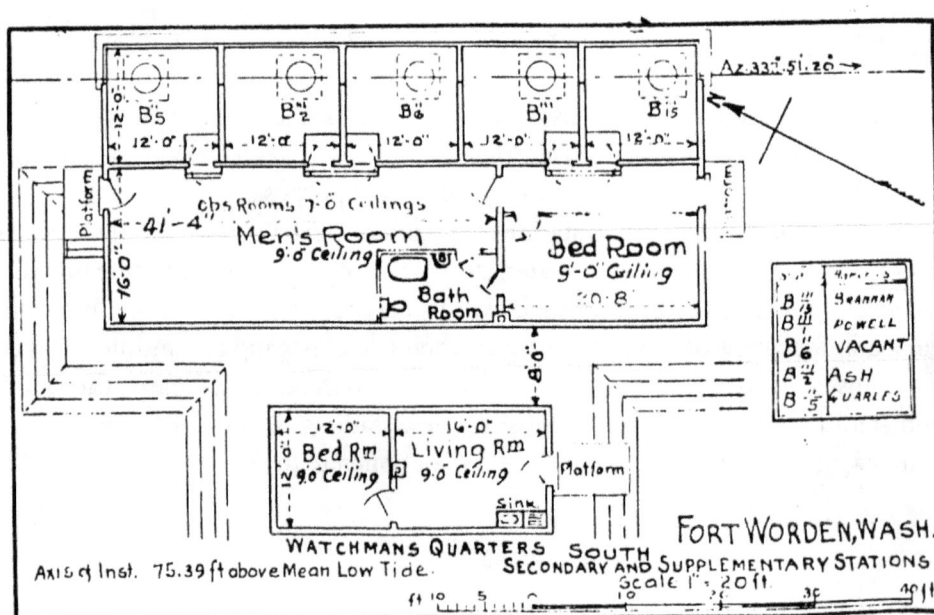

Base End Station (very early type) Fort Wadsworth,
HD New York (Alex Holder, 1997)

Base End Station, Fort Warren,
HD Boston (Alex Holder, 1995)

Most towers incorporated central instrument piers, isolated from the station floor, to provide vibration-free support for the DPF. These piers, at least three feet in upper diameter, were usually concrete, but occasionally a riveted steel lattice was used. Open-frame steel towers occasionally incorporated metal windshields around the central pier. Reinforced-concrete single or multi-station towers, of round or square section, were largely built from early 1939 through 1943. In addition to the DPFs, most 1940s stations normally contained azimuth instruments, and some concrete towers contained as many as nine observation instruments, requiring these concrete towers to substitute floor and wall rigidity for the stability of a central pier.

A few additional styles of base end station designs are worthy of mention. Some stations were disguised as other structures. In New England in particular, stations resembled common "Cape Cod cottages," while other stations appeared to be lighthouses, water towers, or church steeples, and especially around the flat beaches of Southern California, some resembled oil derricks. A few stations were built on or in the tops of existing structures, such as hotels, office buildings, and bridge towers.

Battery Commander's Station and CRF,
Battery Hackleman, Fort Constitution,
HD Portsmouth
(Alex Holder, 1990)

BES, Cape George, HD Puget Sound (Alex Holder, 1999)

Coincidence Range Finder (CRF) station, Fort Totten, NY (photo by Mark Berhow, 1997)

Open-top CRF Station, Battery Ernst, HD Galveston
(Alex Holder, 1989)

Base end station, Long Point Military Reservation
(Point Vicente), CA (Mark Berhow, 1992)

Supplementary Station B''''2
Long Point Reservation
Harbor Defenses of Los Angeles

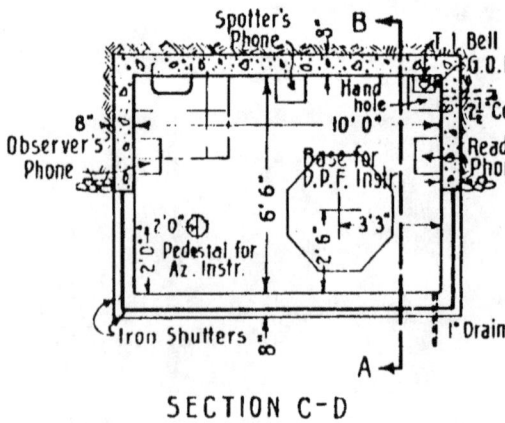

SECTION C-D

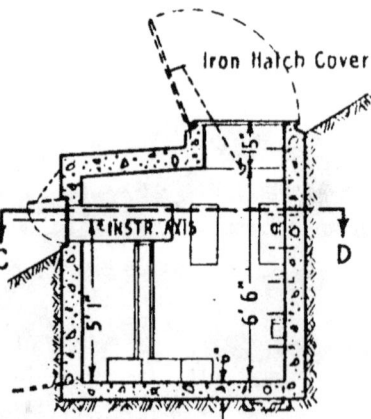

SECTION A-B

U.S. Engineer Office, Los Angeles, Calif.
Scale in feet

Fire control structures—clockwise from the upper left:

Ruins of double mine prime at Fort Wadsworth, NY HD New York (Alex Holder, 1997);

BES, Nashon Island, MA, HD New Bedford (Alex Holder 1988)

BES, Wolf Ridge, Fort Chronkhite, HD San Francisco (Alex Holder, 2001)

BES/Fire Command Station, Fort Washington, MD, HD Potomac River (Alex Holder 1990);

BES, Devils Slide, HD San Francisco (Alex Holder 2001)

BES/Mine/Fire Command Station, Fort Michie, HD Long Island Sound (Alex Holder, 1998)

Disguised fire control structures (clockwise from upper left)—Breakers Hotel (originally the Hilton Hotel), Long Beach, CA—note the tell-tale slitted window in "cupula" above sign(Mark Berhow 1991); CampHero, HD Long Island Sound (Alex Holder 1994); Fort Varnum, HD Narrangansett Bay (Alex Holder 1991) Charleston, RI (Narragansett Bay) (Alex Holder 1988); Warren Point, HD Narrangansett Bay (Alex Holder 1987); Marblehead Neck, HD Boston (Alex Holder 1991); Fourth Cliff, Scituate, MA, HD Boston (Alex Holder 1987)

Most in-ground base end stations contained only a single room, accessed through a top hatch or side door. By the mid 1920s, these single-room stations additionally served as "spotting" stations for observers to track the splash of the projectiles and phone in corrections. From then on the base end stations were generally designated "BS" stations for "base end/spotting."

Soldiers manning stations had two basic duties: watching for enemy ships and tracking assigned targets. When tracking they telephoned the azimuths of the target back to the plotting room. By late in WW2, some newer instruments did this automatically, on a continual basis. A typical base end station required two to four soldiers. Furnishings were primitive - two wall bunks, some footlockers, telephones, a couple of light fixtures, a stove, and two telescopes filled the interior; identification charts of enemy ships and aircraft decorated the walls. Contrary to modern misconception, the stations never served as "machine gun nests"; their only weapons were individual pistols and rifles for security.

Everything (and everyone) in the station became subservient to the telescopes, and a typical station housed two scopes. The larger DPF had greater magnification, up to 30x power. While either DPFs or azimuth instruments could locate a target along with a second station, the DPFs could locate a target by themselves if necessary, and so they were the primary instruments if the station was high enough to have one. The DPF could measure an extremely precise bearing to a target and it could measure the angle of depression to the target and convert that angle into the range.

The soldier on the DPF carefully tracked the ship by turning a hand wheel on the instrument's base, keeping the crosshairs aligned on the ship's waterline at either its bow or forward smokestack, as directed. Two observing stations sighting on different points of a ship could throw off the plotting-room's range calculations.

The second telescope was the azimuth instrument, with up to 15x power. In conjunction with another station, the azimuth instrument could track targets as well as the DPF, but when there was a DPF, the additional versatility of the DPF was preferable and the azimuth instrument was normally relegated to spotting the fall of the shots to enable the battery to correct its fire.

The time-interval bell rang simultaneously in every base end station observing the target every 30 seconds. The bell rang three dings, each one second apart, and azimuth readings were taken on the exact moment of the third ding. Both stations recorded the azimuth at the same instant. During wartime, a four-man crew would rotate shifts sweeping the horizon with their scopes, 24 hours a day, seven days a week, season in and season out, as they maintained their watch. While one man sat on the bench around the observation scope, the other three slept, performed maintenance, walked sentry, or scanned the skies for enemy aircraft. If a target was assigned, though, one soldier tracked the target with the DPF, a second read the range and azimuth from dials on the instrument and relayed them to the plotting room. The third soldier manned the azimuth instrument, spotting the splashes when the shells landed, and the fourth man telephoned the bearings from this scope to the plotting room.

Early batteries usually had only two stations, a primary and a secondary. By WWII, however, guns had such long range that multiple stations were necessary because of the long coastline covered by each battery. In the 1940s, each station had a specific designation, such as "$B^2_3 S^2_3$," (battery base end station No. 2 for tactical battery No. 3, and spotting station No. 2 for tactical battery No. 3). The plotting boards (and later computers) in the battery plotting room could choose from a variety of stations for triangulating the target location. The choice of stations provided options for finding an optimal base line regardless of the weather or target location. The ideal baseline for range finding was about one third the distance to the target, so WWII 16-inch gun batteries had as many as nine base end stations spread along 40 miles of coastline.

Fire control stations were not normally manned during peacetime. For target practice, sub-caliber drills, or similar exercises, soldiers were sent to man the stations, often taking their telephones and some fire control instruments with them. When harbor defenses were placed on wartime footing, one of the first steps was to man the fire control stations.

Wartime duty in these stations became crushingly boring. Along many areas of the Pacific Coast, no buildings were allowed above ground, so as not to draw attention to the stations. As the war progressed, though, soldiers constructed timber-lined underground "gopher holes," near many of the stations. These were roomier and had more storage space. There was no running water or sanitation in the stations, so the men constructed field latrines nearby. Rations were either cooked atop the small stove that doubled as a heater, or eaten cold. This sufficed until the supply truck arrived with hot(?) food, clean clothes, or even better, a relief crew.

Usually a sergeant was assigned to take supplies and food to the far-flung stations. The commanding officer issued him a truck, a handful of maps, and a government checkbook, and then set him loose on his own devices. He frequently became remarkably resourceful, acquiring needed (or at least desirable) supplies through normal army channels, purchasing from local stores, or by bartering with local ranchers. Rumor has it that some on occasion simply "borrowed" items from other units that his unit had more need of.

Fire control towers—
upper photos left to right—Fort Mott, NJ (Alex Holder 1996); South Fort Miles, DE (Alex Holder1992); Fourth Cliff, MA (Alex Holder 1998); Prouts Neck, ME (Alex Holder 1998);

Fire control towers—
Upper Middle from left to right—Steel tower at Fort Story, VA, HD Chesapeake Bay (Alex Holder 1996): Cape Elizabeth, ME, HD Portland (Alex Holder 1986). Appledore Island, NH (Alex Holder 1998); Fort Greble, RI (Alex Holder 1998)

Lower Left—Triple BES at Seabench Military Reservation, CA, HD Los Angeles (Alex Holder, 1992)

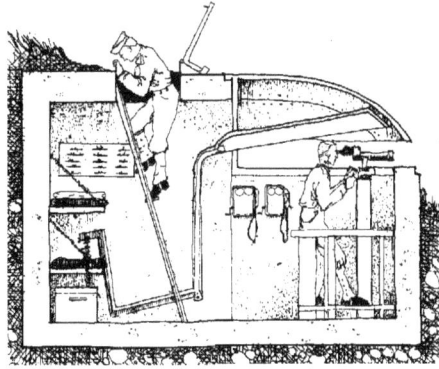

A west coast-style WW II-era base end station (drawing courtesy the Golden Gate National Park Association)

A WW II BES at Hill 640 Military Reservation near Stinson Beach, CA, Golden Gate National Recreation Area
(photo by John Martini, 1976)

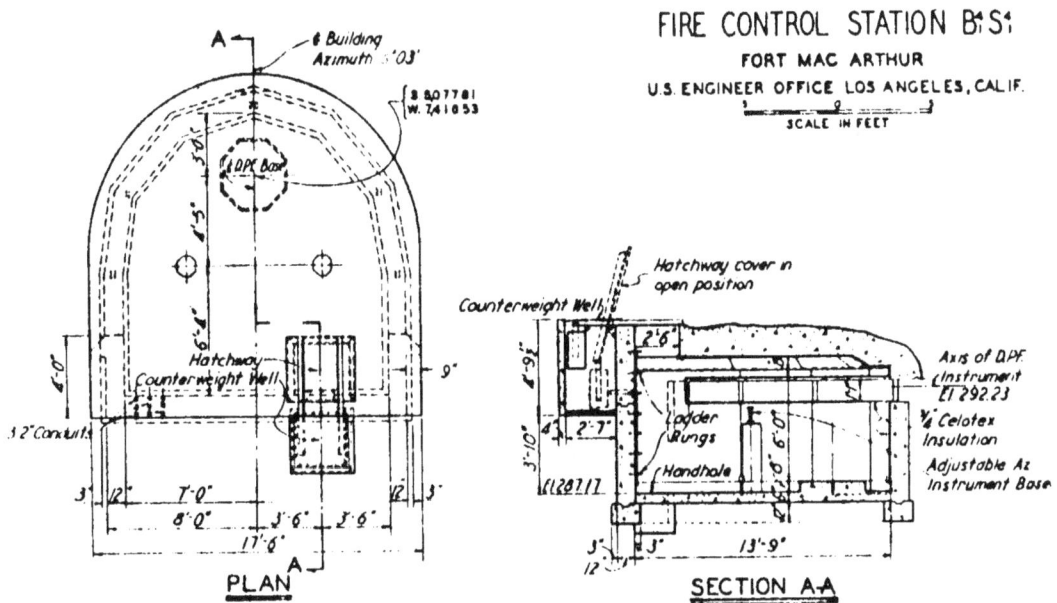

Plan for a WW II base end station (National Archives)

Above ground and tower stations were generally more comfortable than the buried stations, at least along the West Coast. Generally, there was no problem constructing barracks and latrines, or mess halls nearby the fire control structures if there were enough men. But, where stations were disguised as other structures, barracks were frequently included in the scheme. Men assigned to hotels or office buildings could probably use facilities in the buildings. Less user-friendly sites might include bridge towers. Some isolated stations on high ridges required mountain climbing skills. The famed five-story fire control station atop Diamond Head crater on Oahu, Hawaii, was a brisk 45-minute climb along "improved" trails, and mules were used to resupply some particularly remote stations.

During WWII, soldiers spent anywhere from a few days to several weeks in the stations before being rotated back to their barracks. Return to camp came as a much-needed respite, allowing the men to eat three hot meals daily, shower, occasionally go to town, and generally feel human again. Some of the worst duty stations in the defenses of San Francisco were in the wilderness area of Wildcat on Point Reyes, miles from nowhere and perpetually short of rations and supplies. By contrast, the best San Francisco stations were probably the pair in Sutro Heights within a city park, just across the street from the Cliff House, Sutro Baths, and a streetcar station providing direct service to Market Street and all its attractions. In contrast, isolated New England stations in winter were not desirable duty.

Mine and Command Stations

In addition to the fire control stations for the gun batteries, there were other stations for the mine defense and for the higher levels of command starting with the battery commander's station for the battery.

The controlled submarine mines, like the guns, depended on the fire control system to precisely locate their targets. Primary and secondary stations similar to those for the gun batteries were built for the mine units. The principal difference was that mine stations were normally paired, two connecting rooms enabling them to cover two targets.

The names for the units above the level of the batteries changed several times during the half century the fire control system was in place, but the essential organization remained unchanged. Batteries were grouped into higher commands, with the number of levels of command dependant on the size of the defense and the geographic complexity of the area to be covered. Each higher unit had a command post with an observation room and normally a secondary station. Command posts had numerous booths for

Battery commander's station, Battery Haslet, Fort Saulsbury, DE (photo by Mark Berhow, 1995)

Double mine primary station, Fort Wetherill, HD Narragansett Bay, RI (1980s, Alex Holder.)
This fine example of Sewell-style construction has since been destroyed.

Fort commander's station at Mount Prospect, Fort
H.G. Wright, HD Long Island Sound
(Alex Holder 1987)

Combined HDCP and HECP command center for the Harbor
Defenses of Narragansert Bay, Fort Burnside, RI
(Alex Holder 1989)

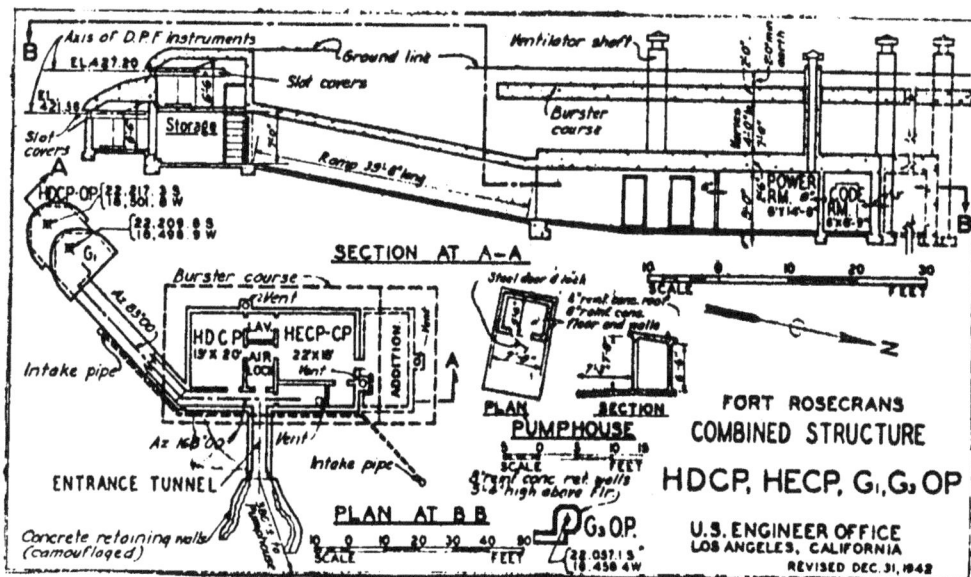

drawing
NA

telephones and operators to maintain communications, and some also had plotting rooms, offices, and other miscellaneous rooms.

At the start of WWII, harbor entrance command posts (HECPs) were activated. These, jointly manned by army and navy officers, controlled shipping entering and leaving the harbor and coordinated the harbor defense. The army, in turn, established harbor defense command posts (HDCPs) as their headquarters in the harbor defense. HDCPs were sometimes, but not always, located with the HECP. In some locations, additional observations stations for the HECP and the HDCP were established.

While early frame and plaster stations have largely disappeared, a number of the early brick stations survive, and many 1930s and 1940s stations still exist, their concrete and steel nearly impervious to time and weather. A few stations have been demolished or pushed from their former locations to make way for new development. The greatest loss has been the tower stations. A few steel towers dating back to the first decade of the century still stand, along with some 1930s steel towers. The steel towers, however, are rapidly showing their age, and except where strenuous preservation efforts have been made (notably at Fort Mott, NJ), those that have not been removed are on borrowed time. Most concrete towers along the East Coast, newer and stronger, still survive and some have even been converted into homes.

Many fire control stations are preserved within national parks, such as the Golden Gate National Recreation Area, Cabrillo National Monument, and Gateway National Recreation Area, as well as state and local parks. Current plans by the National Parks call for the preservation of many of these stations as remnants of this nationally significant system of America's seacoast defenses.

Gun Battery Organization and Duties I: HQ and Range Sections (1940)

The men of a gun battery were divided into a headquarters section, a range section, as many firing sections as were needed for the guns in the battery; and a maintenance section. The battery was under the over-all command of the battery commander, a commissioned officer who was usually a captain in rank. Each section contained a number of squads and details. All sections were commanded by a commissioned officer; and each section, squad and detail had a non-commissioned officer as its chief.

Headquarters Section (Battery Commander)

The headquarters section was the administrative unit that ran the day to day affairs of the battery at the fort. During a firing practice, most of these men were assigned duties in the other sections, leaving only the *battery commander,* the first sergeant of the battery, and the battery commander's (BC) detail. This section was charged with the fire direction and fire control of this battery, overseeing the communications network of the chain of command, and the collection, preparation, and dissemination of intelligence data.

Battery Commander's Detail

The BC detail was composed of those members of the battery who assisted the *battery commander* in the conduct of fire. The detail was posted in the battery commander's station and was composed of:

BC observer
Recorder
Azimuth reader
3 *Telephone operators*- intelligence, group, guns

The Range Section (Range Officer)

The range section of a battery was that subdivision of the personnel of the battery in which was centered the function of position finding. It was under the immediate command of the *range officer* and generally consisted of the observing details, the spotting details, and the plotting room detail.

The *range officer* commanded the battery range section and was responsible to the *battery commander* for the condition, adjustment, and use of the battery position finding equipment, for the training and efficiency of the battery range section, for the serviceability of the battery communication system, and for the policing of the stations pertaining to the battery position finding system. When the battery was firing, his station was in the battery plotting room. Before drill, practice, or action he made a careful examination of the plotting room equipment and apparatus. He verified the adjustment of all position finding equipment and apparatus as often as was necessary to insure their proper operation and their readiness for service at all times. During drill, practice, or action he maintained constant supervision over the functioning of the plotting room detail and, insofar as he was able to do so from his station in the plotting room, over the entire battery position finding service. During firing he supervised the adjustment of fire in range and direction.

Observing details.- Each observing detail was assigned to a particular observation station (also known as a base end station (BES) and labeled B', B'', etc., later B^1, B^2, etc.) and was composed of the *observer* and the *reader.*

The *observer* was responsible to the *range officer* for the care, adjustment, and use of his instrument, for the policing of his station, and for the functioning of his detail. Upon arrival at his station he made a careful inspection and examination, oriented his instrument, tested the means of communication, and reported to the *plotter,* "B^1 (or B^2, etc.) in order," or reported such defects as he was unable to remedy without delay. When the target had been indicated by the *battery commander* and identified by the *observer,* the latter reported, "B^1 (or B^2, etc.) on target"; and when the *battery commander* gave the command "track," the *observer* followed the target with his instrument, keeping the cross hairss accurately centered on the observing point of the target. When the third bell of each time interval signal was struck, he stopped following the target long enough to permit his *reader* to read and transmit to the plotting room the required data.

The *reader* functioned under the direction of his *observer.* He assisted in the care, adjustment, and orientation of the observation instrument and in the policing of the station. Upon arrival at his station he performs such duties as were directed by the *observer,* tested the functioning of his communication with the plotting room, and reported to his *observer.* When the target was assigned and tracking was started, the *reader* read from the observation instrument at each time interval the azimuth, or the azimuth and the range, and transmitted those data to the proper *arm setter* in the plotting room and recorded the data.

Spotting details.- Each spotting detail was assigned to a particular spotting station (S^1, S^2, etc.) and was composed of the *spotting observer* and the *spotting reader.* The *reader* was omitted when the *observer* read deviations from an internal scale.

The *spotting observer* was responsible for the care, adjustment, and use of his observation instrument, for the functioning of his detail, and for the police of his station. Upon arrival at his station he made an inspection of the station and equipment, tested the means of communication, oriented and adjusted his instrument, and reported to the *plotter,* "S^1 (or S^2, etc.) in order." The detailed duties of the *observer* depended upon whether observations of impacts were being reported as deviations from the target (or other point) or as azimuths. He identified the target when assigned, reported to the *battery commander,* "S^1 (or S^2, etc.) on target", and thereafter followed the target with the vertical hair of his instrument. When the splash occurred, he accurately adjusted his instrument on the splash (that is, the splash pointer if reading deviations, or the vertical cross hair if azimuths are to be read), and either transmitted the deviation thereof

Cover
Prism Retaining
Spring
Yoke Cap Clamp Screw
Yoke Cap
Depression Arm Clamp
Screw

Lower Focusing
Sleeve
Prism

Lower Power Field Lens
Protector Holder Shaft
Focusing Nut

Eye Piece Adapter
Protector Holder
Field Lens Cell
Eye Protector
Amber glass Disk
Eye Lens Cover
Lower Power Eye Lens
Eye piece Draw Tube

Objective Shutter
Objective Cell Holder   Cell Ring

Objective

Telescope Tube

Yoke Cap Pin
Pivot Master Screw
Depression Slow-motion
Plunger
Depression Plunger Spg

Yoke

Focusing
Ring

Prism holder   Prism

Draw Tube

Cross Wire
Adj. Screw

Depression Plunger
Spring Plug

Slow-motion Adj Screw Knob

Heart Surface Switch

Az. Stop
Az. Stop Stud

Depression Slow-motion
Screw
Oiler

Depression Slow-motion
Screw
Az. drum Lamp Shield

Az. Slow-motion Arm
Worm Box Crank Taper Pin
Crank Stop Pin
Worm Box Crank

Leveling Screw
Leveling Plate
Swivel Stud
Swivel Wing Nut
Left Swivel
Swivel Screw

Level Shoe

Plumb Bob Hook
Tripod Head

Right Swivel

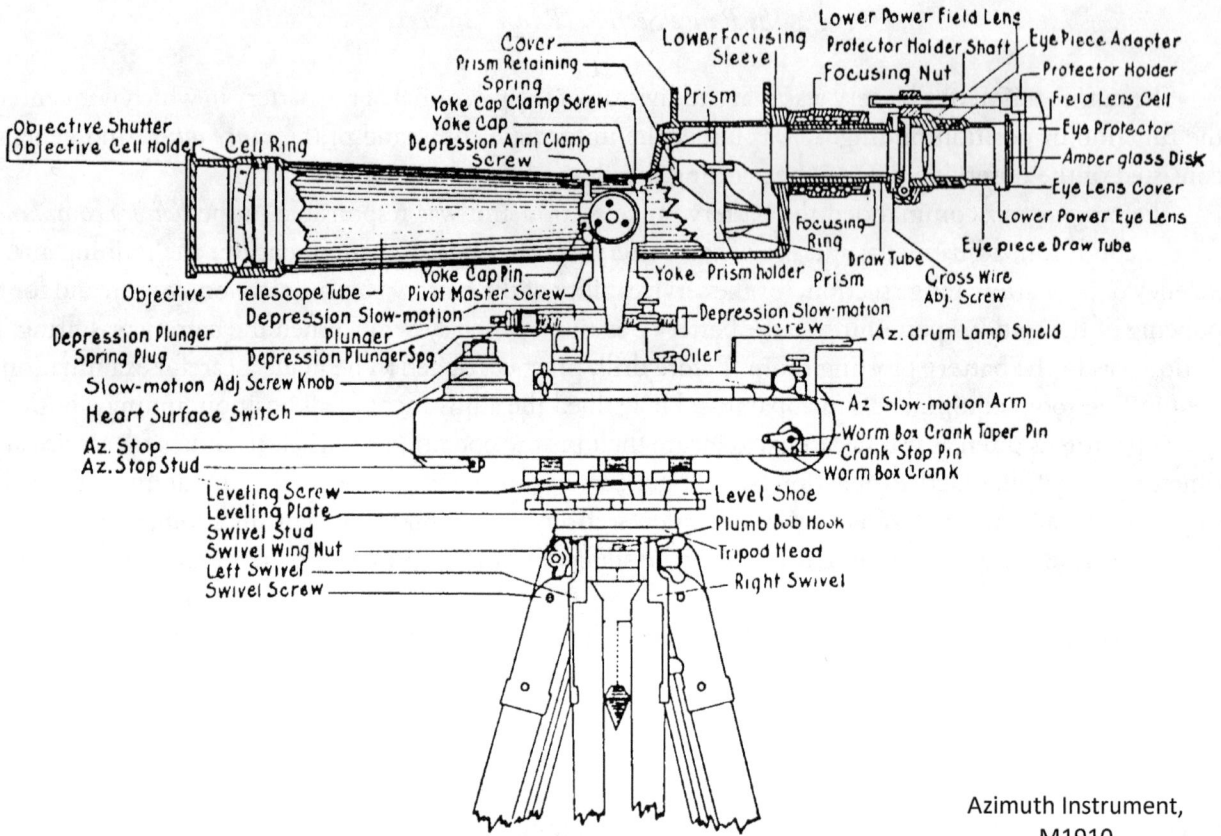

Azimuth Instrument,
M1910

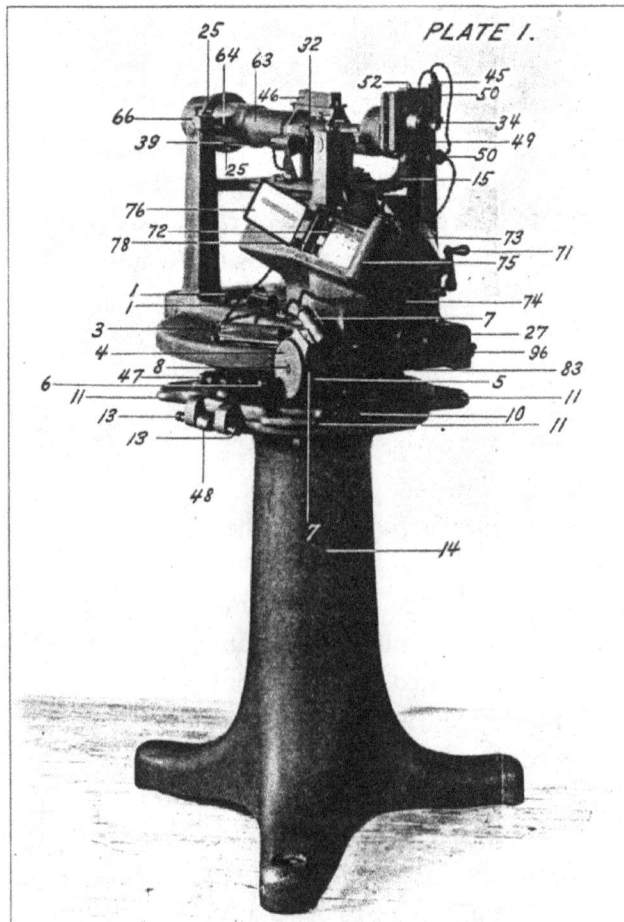

PLATE I.

Swasey Depression
Position Finder Type A-II

to the plotting room or halted his instrument long enough for the azimuth to be read and transmitted by his *reader*. If he was reporting deviations from the target and had no reader, he recorded his own data. In order to insure the identification of the splashes of the shots fired by his battery, the *observer* had to be informed from the battery of the instant of firing and of the time of flight of the projectile or of the expiration of the time of flight.

The *spotting reader* functioned under the direction of the *observer*. Upon arrival at his station he performed such duties as are directed by his *observer*, tested the functioning of his communication with the plotting room, and reported to his *observer*. When the target had been assigned, tracking started, and the impacts of shots occurred, the *reader* read, transmitted to the proper *spotting board assistant*, and recorded the data determined from observation of the splashes.

Plotting room detail (Plotter, chief of section)-

When the plotting room was equipped with a Whistler-Hearn or with a 110 degree plotting board (see next section), the plotting room detail was composed of the following personnel (the numbers refer to the numbered designation of each member of the detail)

The *plotter* was chief of the plotting room detail and as such was responsible to the *range officer* for the adjustment, condition, and serviceability of the plotting room apparatus; for the training and efficiency of the plotting room detail; and for the condition and policing of the plotting room. He received the reports from the observation stations, from the spotting stations, and from the various members of the plotting room detail; and reported to the *range officer*, "Sir, range section in order," or reported such defects as he was unable to remedy without delay. He was responsible for the orientation, adjustment, and use of the plotting board. During drill, practice, or action, the *plotter* plotted on the plotting board the points representing the positions of the target at times of observation (plotted points) and determined the set-forward points.

Plotting room personel were (this varied with the type of plotting board used):

No. 1, *angular travel device operator* (case II) or gun arm azimuth reader (case III).

No. 2, *primary arm setter.*

No. 3, *secondary arm setter.*

No. 4, *range correction board operator.*

No. 5, *set-forward device operator.*

No. 6, *percentage corrector operator.*

No. 7, *deflection board operator.*

No. 8, *assistant deflection board operator* (used only with the deflection board M1)

No. 9, *fire adjustment board operator,* (an officer, if available).

No. 10, *spotting board operator* (Where airplane observation of the fall of shots was provided, the use of the spotting board was unnecessary. In such cases Nos. 10, 11, and 12 became unnecessary and were eliminated. With some spotting boards only one assistant was necessary.)

Nos. 11 and 12, *assistant spotting board operators.*

No. 13, *data transmission device operator* (If the battery was not equipped with a data transmission device, No.13 was eliminated. Where the data transmission system MS was used, four operators were required).

Nos. 14 and 15, *recorders* (Recorders in such numbers as were necessary to insure complete and accurate records for the purpose of drill and target practice analyses were provided. Nos. 14 and 15 were provided as regularly assigned members of the plotting room detail. When they are not required for recording purposes, they were given other duties. Operators of instruments recorded their own data when practical).

A coast artillery mortar battery plotting room with a Whister-Hearn plotting board
(Battery Anderson, Fort Monroe, VA, circa 1914, National Archives)

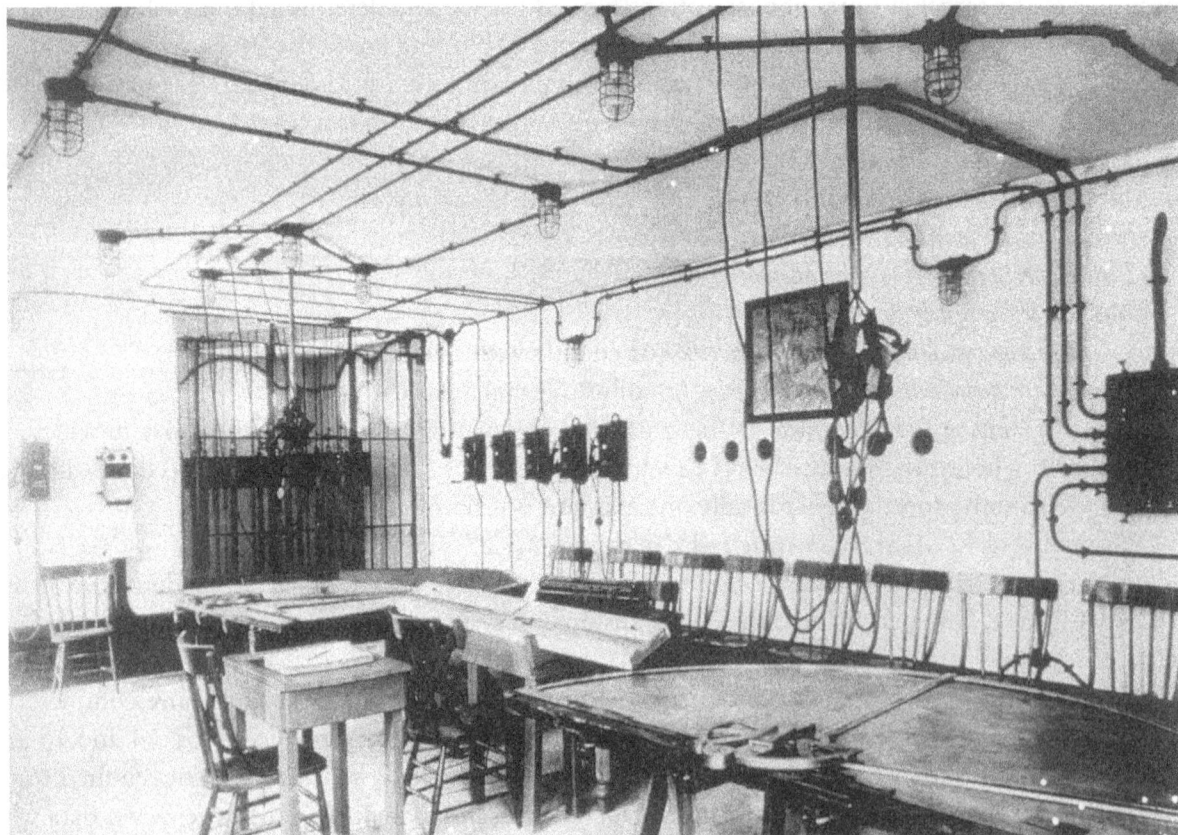

Plotting Room at Battery Baird, Fort Sherman, Harbor Defenses of the Panama Canal, as completed (NA).

Where the plotting room was equipped with the Cloke plotting board (M1923), the plotting room detail was composed of the following personnel (the numbers refer to the numbered designation of each member of the detail):

Plotter.

Platen operator.

No. 1, *angular travel device operator* (case II only).

No. 2, *plotting arm setter.*

No. 3, *relocating arm setter.*

Nos. 4 to 15, inclusive [same as plotting room detail, (a) above].

Rapid-fire batteries (6-inch or less)—In a rapid-fire battery not provided with a plotting room, where corrections to firing data are usually made and applied in the battery commander's station, the range section ordinarily was combined with additional personnel. The combined unit consisted of the following personnel:

Observer, battery commander's (azimuth).

Observer, self-contained range finder.

Observer, spotting.

Reader, battery commander's.

Reader, self-contained range finder.

Reader, spotting (often unnecessary).

Range correction ruler (or *range percentage corrector) operator.*

Deflection board operator.

Fire adjustment board (or *over/short adjustment chart) operator.*

Transmission device or *display board operator* (if necessary).

Telephone operators, one for each phone.

Recorder.

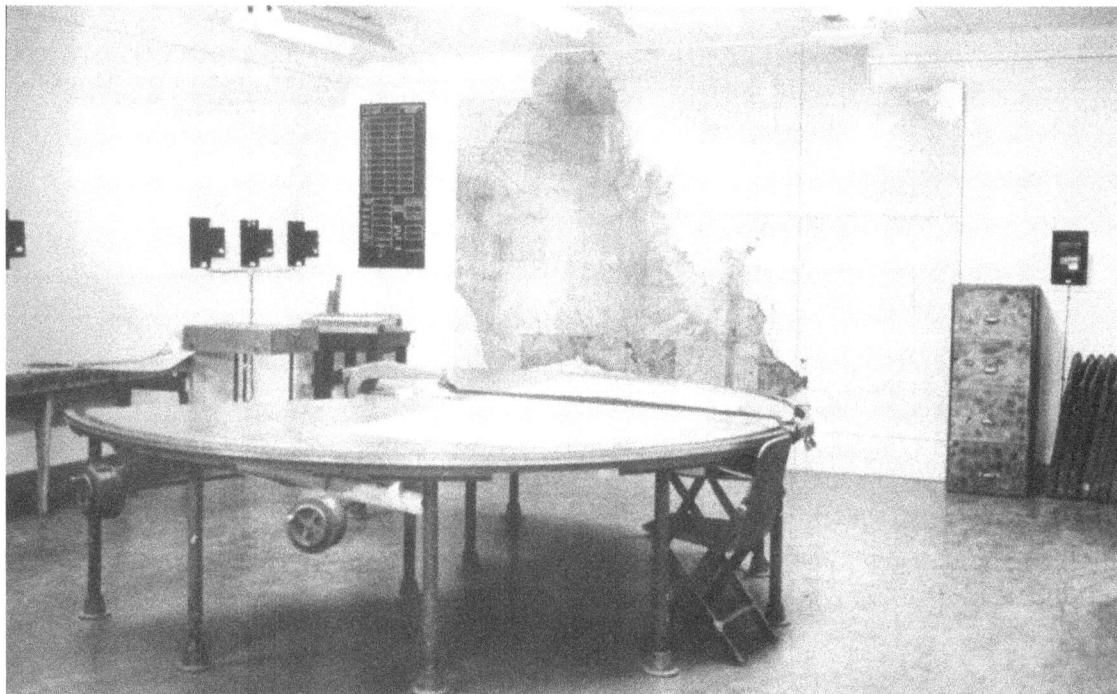

Plotting Room at Battery 405, Fort Hase, Harbor Defenses of Kaneohe Bay, Hawaii, 1945
with a 110° plotting board (NA, courtesy Glen Williford)

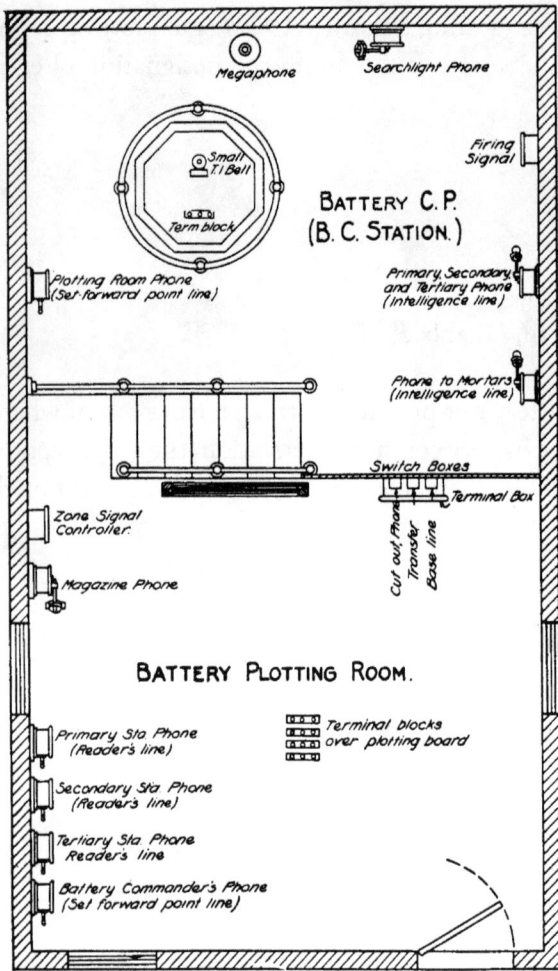

Battery Commander's Station & Plotting Room, interior arrangement, circa 1910s

Beginning with the new 16 inch battery construction in the 1920s, the plotting rooms were located in separate structures which were covered with earth and accessible only from the rear. As the new casemated large caliber batteries were built in the late 1930s, the plotting room and in many cases the radio switchboard room were placed in separate buried structures apart from the main gun battery and magazines, usually behind the batteries from the seashore. These structures were variously called "plotting and spotting rooms" or "plotting and switchboard rooms," abbreviated as P.S.R.

Entrance to the PSR For Battery 134, Fort Emory, HD San Diego

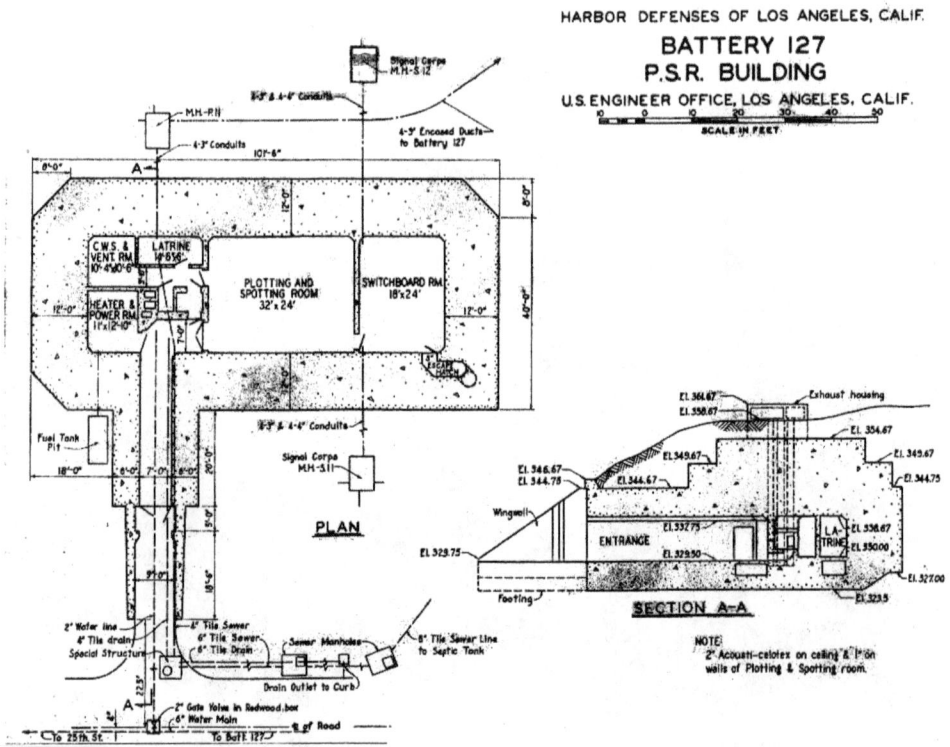

Plan for the PSR of Battery Bunker (BCN 127), Harbor Defenses of Los Angeles

Terminal Box | Base Line Switch Box | Cut Out Switch Box | B' Tel | B" Tel | BC Tel | Gun Tel | (14)

ANGULAR TRAVEL COMPUTER (1)

WIND COMPONENT INDICATOR

(5) **SET FORWARD RULER**

PLOTTING ROOM
Showing Apparatus and Assignment of Plotting Room Detail

RANGE CORRECTION BOARD (4)

DEFLECTION BOARD (8)

Asst. Plotter — Platen Operator — **CLOKE PLOTTING BOARD** — Plotter — (3) Relocating Arm Setter — (2) Plotting Arm Setter

RANGE PERCENTAGE CORRECTOR (6) (7)

MANNING DETAIL — ORIENTATION CHART — DIAGRAM OF WATER AREAS

FIRE ADJUSTMENT BOARD (9)

(11) (12) (10) **SPOTTING BOARD**

DATA TRANSMISSION DEVICES
Range | ELEV ZONE AZ. | Zone Signal Indicator (13)

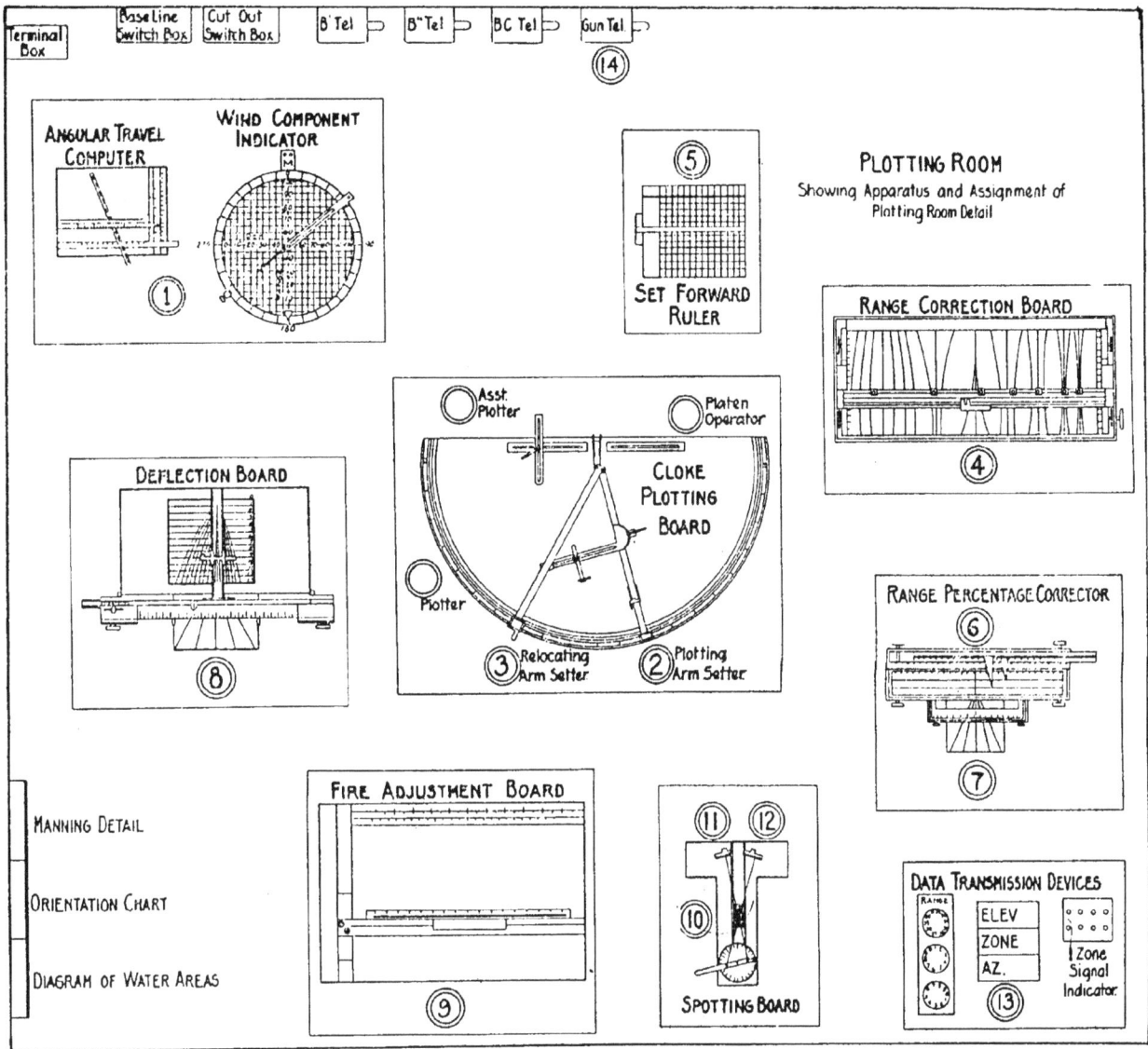

Equipment and stations in a plotting room of the 1920's

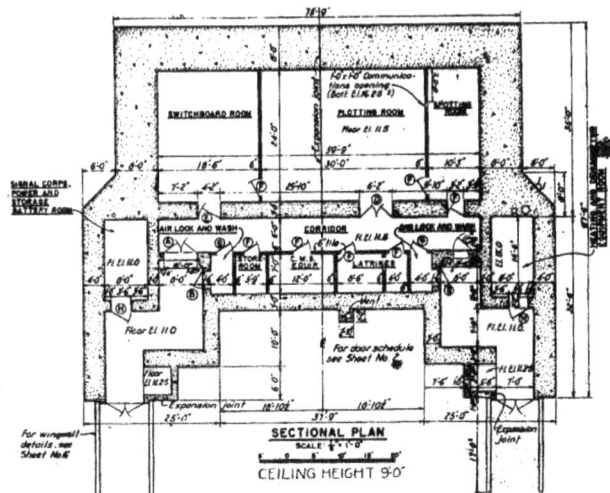

SWITCHBOARD ROOM — PLOTTING ROOM — AIR LOCK AND WASH — CORRIDOR — AIR LOCK AND WASH — SIGNAL CORPS POWER AND STORAGE BATTERY ROOM — LATRINE

SECTIONAL PLAN
CEILING HEIGHT 9'-0"

PSR for Battery Smith (BCN 118), Harbor Defenses of the Chesapeake

Coast Artillery Plotting Boards 1904-1945

Fredrick M. Baldwin

The most important piece of equipment used in seacoast artillery range and position finding was the plotting board, with its mechanical solution to the trigonometric problem of locating a distant point—the target.

Surveyors have long used the angle-side-angle relationship to determine inaccessible distances. Coast artillery plotting boards were based on this method, the concept of a mechanical analog solution to trigonometric problems which would be both accurate and relatively fast.

With the extended base line thus available, more accurate ranges were obtained than with the single station instruments such as the depression position finder (limited to the height of site) or the self-contained range finders with their base line lengths of 9 or 15 feet.

Over the years, three basic types of plotting boards were developed for use with the horizontal base system:

- Whistler-Hearn plotting board
- Cloke plotting and relocating board
- 110° plotting board

In general, the operation using the 110° board was as follows: Each base end station was manned by an observer and a reader, the reader being connected by phone line to the arm setter for his station. The observer tracked the target with his observing instrument. The time-interval bells [often] rang three strokes every set number of seconds as follows—ring, 3 count pause, 2 rings. At the first ring, the reader read the azimuth in whole degrees, i.e.: '242 point' to the arm setter who then moved his arm to that setting. In the meantime the observer continued to track the target. However, at the last stroke of the 3 rings, he would momentarily stop tracking, the reader would read the hundredths of the azimuth from the hundredths dial and repeat this to the arm setter who would set this on his arm index. At this time he would report 'set' to the plotter.

When both arms had been set, which was generally simultaneous, the plotter placed the point of the target marker (targ) at the intersection of the arms, and pressed the button of the targ, making an indentation on the plotting paper. He then called 'clear' and the arm setters moved their arms to one side.

The plotter, once he had three points plotted, used the appropriate set forward scale to locate the setforward point, at which time the gun arm was moved to this point, now marked by the targ and the range read for case II, and both range and azimuth for case III.

This uncorrected data then went to the appropriate correction devices before going to the guns. This was repeated for each time interval.

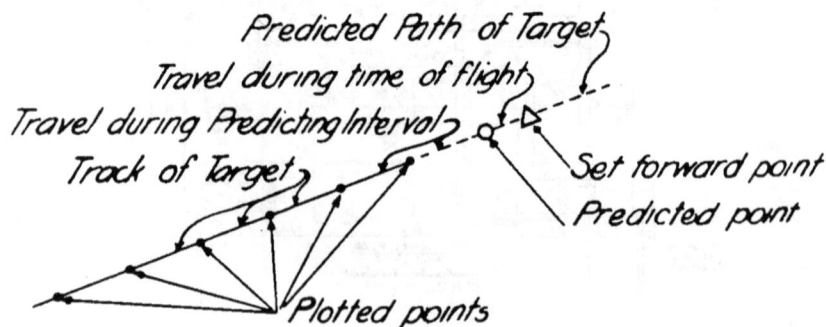

Plotting Board information

Whistler-Hearn Plotting Board M1904

The Whistler Hearn plotting board was the older type used for fixed seacoast artillery installations. It could be used either for two station (horizontal base) or single-station (vertical or self-contained base). The standard scale of the board was 300 yards per inch, with 450 yards per inch for special applications. The base line arm was installed parallel to the edge of the board on the diameter of the azimuth circle. The primary arm was pivoted at the center of the azimuth circle. The secondary arm center could be moved along the base line arm in either direction to its proper location. The directing point was represented by the gun arm center, around which it pivoted, its location being adjustable by offsets parallel to and perpendicular to the base line. Range was read from the gun arm and azimuth from the gun arm circle, not from the outer azimuth circle. A coupler, equal in length to the offset of the secondary arm center from the primary arm center, was used to keep the secondary arm parallel to the auxiliary arm, which was used to set the secondary azimuth. The azimuth circle of each board was oriented for the particular installation where it was to be used.

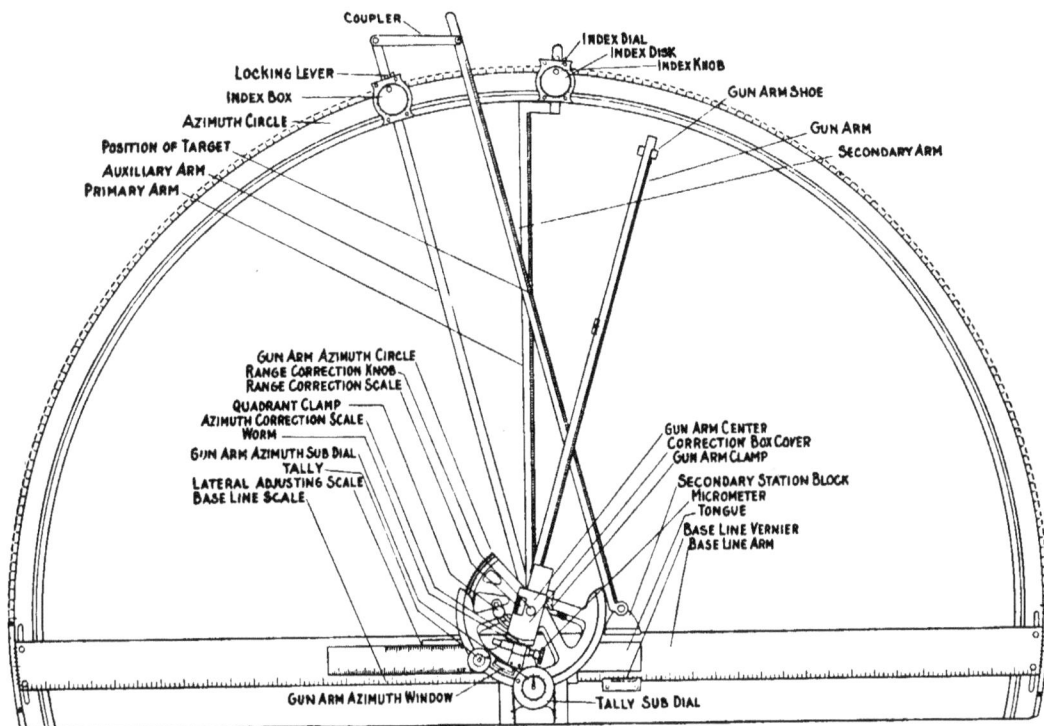

The Whistler-Hearn plotting board M1904

Cloke Plotting & Relocating Board M1923

The Cloke board was developed for use with mobile seacoast artillery installations. The scales of the board were 300, 600, 750, and 1500 yards per inch depending upon the location and range of the battery. This was a truly universal hoard, having an adjustable azimuth circle as well as means to locate the directing point and base end stations on a movable platen. The Cloke board was unique in that the target location was at the center of the board at the pivot point of the two arms, and the base line and directing point were represented on a platen that moved over the board, with the directing point being plotted to obtain the target course. In operation, the platen was held against the base line stop until the arms were set, at which time the platen clamp lever was locked and the platen moved out on the plotting arm until it intersected the relocating arm, at which time the directing point push-button was depressed making a mark on the plotting paper. The platen was then unlocked and returned to the base line stop. The arms were then cleared to allow the plotter to complete his work. Prediction and course plotting were similar to that used on the other boards. Range and azimuth were read from the relocating arm.

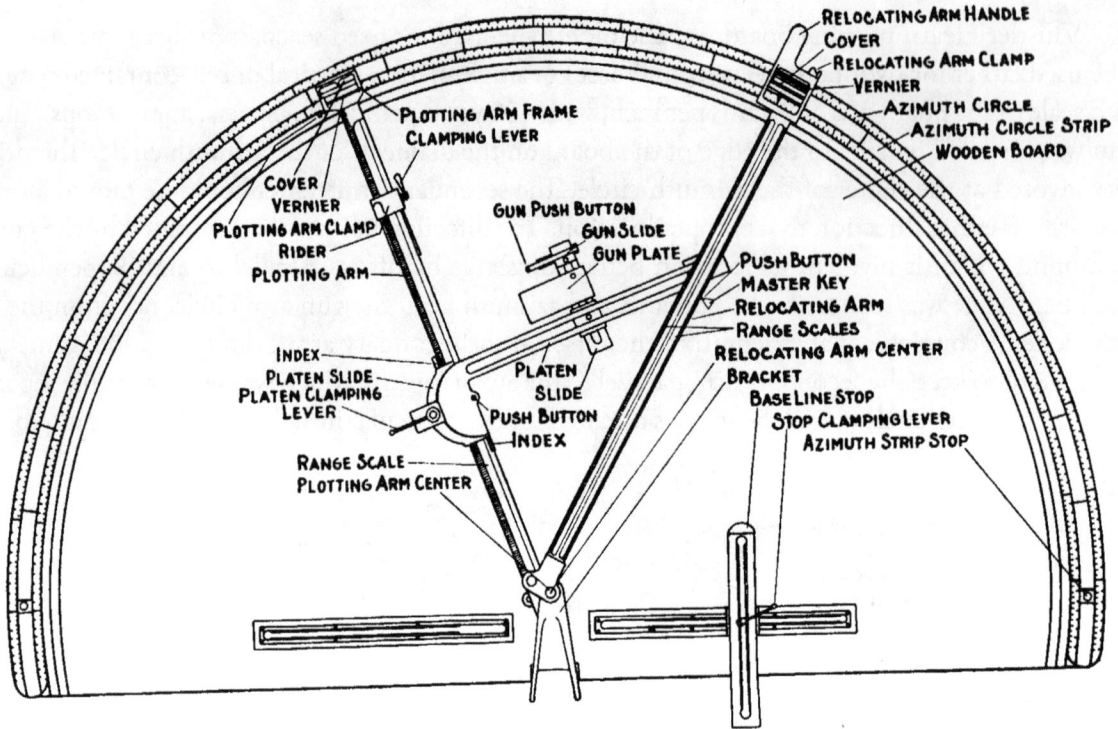

The Cloke plotting and relocating board M1923

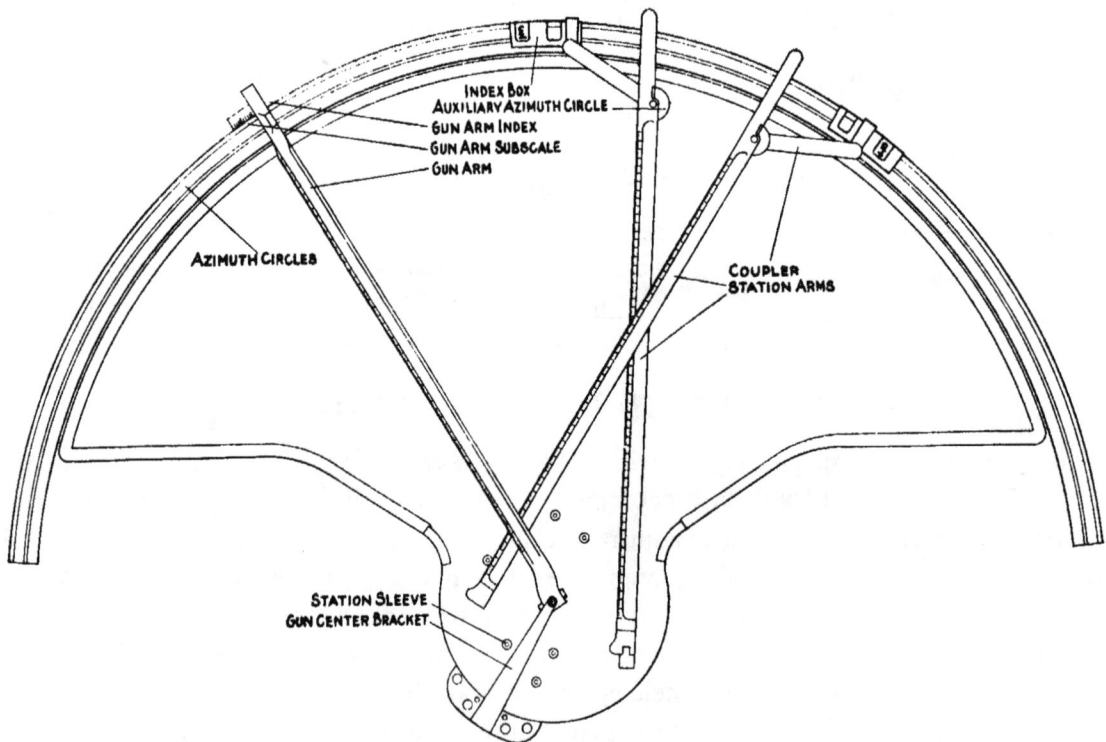

The 110° plotting board

110° Plotting Board

The 110° Board, M1915, 1918, M3 and M4 were for use with all types of fixed seacoast artillery installations. Like the Whistler-Hearn, the 110° board provided means for target location by either the single or two station method. The center of the board was the directing point location and the base end stations were represented by permanent station-sleeves positioned in proper relation to the directing point. The scale of the board was similar to the Whistler Hearn, except for the M3 and M4 boards which were special for certain long range installations. The M4 board at Battery Harris, a 16-inch battery located at Fort Tilden, HD New York, was in the order of 800 yards per inch, resulting in arms some six feet long and a board diameter of some 12-14 feet. The arms on these boards were equipped with couplers equal in length to the station offsets from the directing point. Range was read from the gun arm and azimuth from the outer azimuth circle. Station changes were made by moving the appropriate arm to the new station-sleeve and changing to the proper length coupler. The azimuth circle was not adjustable, being fixed for the particular installation.

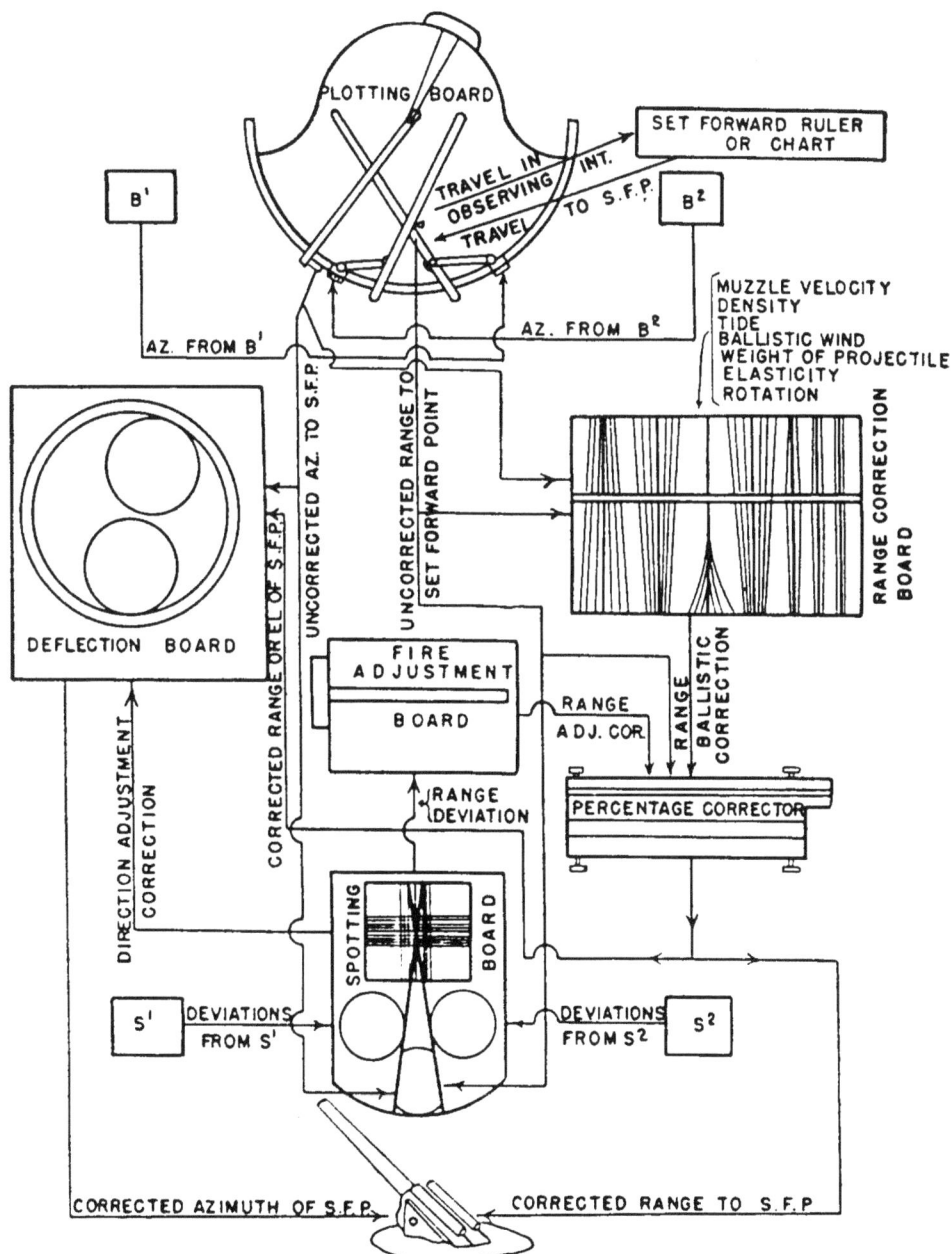

Fire Control Data Flow Chart featuring a 110° plotting board

Function of Position Finding Systems

(derived from U.S. Army Field Manual 4-15 [1940])

The equipment found in the plotting room of any one seacoast artillery battery differed from that in almost any other battery. There were several reasons for this. Not all batteries had the same instruments; the requirements of the various batteries were different depending on caliber and range; and finally, minor modifications were made in the systems to suit the preferences of the battery officers.

Obviously it would be impossible to describe the operation of all possible combinations of the numerous instruments and improvised instruments that were in use without burying the main idea in a mass of detail. This section will describe the operation of one system.

The system described here is for a typical circa-1940 battery of fixed guns of major caliber, and the plotting room equipped with the 110 degree plotting board; range correction board M1; percentage corrector M1; deflection board M1; spotting board M2; and fire adjustment board M1. The base end observers and the spotting observers are equipped with azimuth instruments, M1910A1.

ACTION BEFORE TARGET WAS ASSIGNED
PREPARATION FOR CALCULATION OF DATA

Before the target was assigned, all preparations for the calculation of firing data were made. The *range officer,* having maintained his equipment in adjustment by frequent tests, made a quick survey of the most important features.

All communication lines were tested by the men who operated them.

The meteorological message was received by the man designated for that duty and was recorded on a form drawn up by the *range officer.* Air temperature was taken from the message at once, but the proper ballistic density and ballistic wind to use were not determined until the target was assigned and its range determined.

The tide message was received from the tide station and recorded.

All possible information as to weights of projectiles was received from the *battery executive officer.* By coordination within the battery, this information was in the hands of the *range officer* well in advance of the time the projectiles were to be used.

A record of all available information as to the action of the powder on hand in the battery was maintained by the *range officer.* This included information as to its previous performance—reduced to standard temperature—together with powder tag markings, current temperature, and other information. The *range officer* determined beforehand the muzzle velocity to be assumed with any combination of ammunition likely to be used, and insofar as possible the *battery commander* gave both the *range officer* and the *battery executive officer* advance warning as to the combination that would be ordered.

ACTION WHEN TARGET WAS ASSIGNED—PRELIMINARY STEPS.

The assignment of the target to the battery by the *group commander* was followed by its assignment to the various elements of the battery by the *battery commander.* This assignment conveyed the *battery commander's* decision as to the ammunition, the observation and spotting stations to be used, the case of pointing, and the method of tracking used.

The following steps were taken immediately by the plotting room detail;

(1) The plotting board was made ready for tracking, using the stations and method of tracking ordered.
(2) The required communication set-up was arranged by proper manipulation of the switchboxes.
(3) The spotting board was made ready for spotting, using the stations ordered.

(4) The operators of the range correction board, percentage corrector, and deflection board turned to the charts corresponding to the ammunition ordered.

(5) The *operator of the range correction board* made notation of the muzzle velocity, the height of site (tide), the weight of projectile, and the temperature (elasticity) curves to be used.

The *observers* designated to track the target, and the *gun pointers* when case II pointing was ordered, identified the target and brought their instruments or guns to bear on it. Each then reported, "_____ on target." As each *observer* reported on target, a battery officer commanded "track," without waiting for other observers to report.

The *observers* followed the target. If vertical base tracking was ordered, the *observer* tracked in azimuth as described, and in addition he tracked in range by keeping the horizontal hair of his instrument at the waterline of the target. At the third stroke of the bell, he held both the azimuth reading and the range reading stationary long enough for the *reader* to transmit them to the plotting room.

APPROXIMATE DATA.

As soon as the plotting room received the first readings from the observation stations, the position of the target was plotted, and the *plotter* called out, "Approximate data." He then read off the range to the plotted point, and the *gun arm operator* read off the azimuth to that point, both loud enough to be heard by all in the plotting room.

Using this approximate range, the ballistic wind zone was selected and the ballistic wind and ballistic density were taken from the meteorological message by the *range officer.* The ballistic density and rotation curves to be used on the range correction board were noted. The direction and speed of the ballistic wind were set on the deflection board.

The *deflection board operator* turned the deflection board to the approximate azimuth. This determined the range and lateral components of the ballistic wind for use on the range correction board and on the chart of the deflection board itself. The chart was turned to the approximate range, the pointer was brought to the curve corresponding to the lateral wind component, and a first (approximate) reading sent to the guns (in case III pointing) which were traversed to the target.

The *range correction board operator* turned his chart to the approximate range, took the range component of the ballistic wind from the deflection board, and determined an approximate ballistic correction. This he transmitted to the *percentage corrector operator.*

The *percentage corrector operator* turned his range scale to the approximate range and transmitted the range to the guns except when ranges were set on the guns in angular units (as elevations), in which case he read off an approximate elevation which he sent to the guns. After receiving the ballistic correction he sent to the guns a second approximate range (or elevation) if it differed materially from the first.

The spotting board was set to the approximate range and azimuth.

CORRECTED DATA.

After the course of the plotted points was steadied so that prediction was possible, the *plotter* made a prediction. The range and the azimuth of the set-forward point were called out for all to hear. All other data were transmitted in tones as low as reliable transmission would permit. The outstanding characteristic of a well-trained range section was that it was quiet.

The *percentage corrector operator* set the range to the set-forward point on his board, used the ballistic correction already set, and if necessary converted this range into elevation. He transmitted the firing range or the firing elevation to the guns immediately or held it for transmission on signal, according to the method used in that particular battery.

The *deflection board operator* set the range and the azimuth of the set-forward point on his board, using the lateral component of the ballistic wind that showed on the wind component indicator after the azimuth was set. If pointing was by case III, the corrected azimuth was read from the board by the *assistant operator*. If pointing was by case II, travel was computed by a device built into the board. This device was operated by the *assistant* who computed the deflection and transmitted it immediately to the guns.

The *range correction board operator* set the range to the set-forward point on his board, followed the curves, and gave a new ballistic correction to the percentage corrector operator when it changed by a tenth of 1 percent.

SPOTTING SYSTEM METHOD OF OPERATION.

Each *spotting observer* tracked the target, keeping the vertical hair of his instrument on the observing point. When the splash occurred, tracking was stopped immediately, and the angular deviation of the splash, as marked with the pointer in the azimuth scope, was then read from the deflection scale in the instrument and was transmitted to the spotting board. *Axial observers* observed on the center of the splash; *flank observers* observed on the edge of the splash nearest the battery.

To assist the *spotting observers* in identifying the splash, a stop watch kept at the spotting board was started when the shot was fired and, when the time of flight elapsed, the warning "Splash" was called out to the *observers*.

The angular deviations observed by the *spotting observers* were set into the spotting board. Range deviations in percentages were read off the spotting board and transmitted to the *range adjustment board operator*, among other operations, for spotting correction.

Chain of Command, US Army Harbor Defenses (1940)

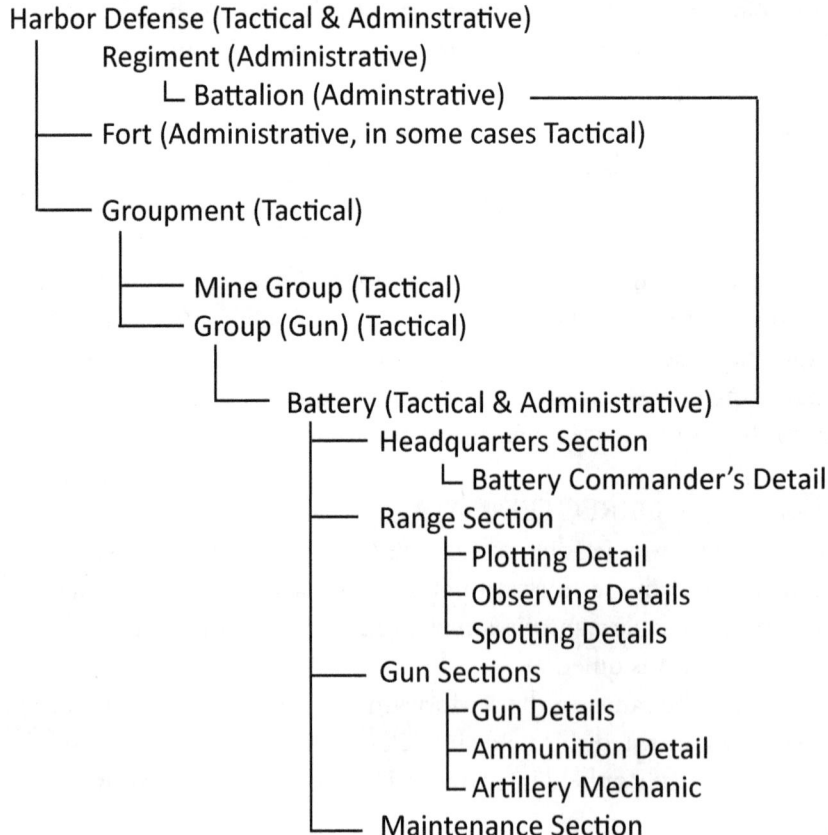

```
Harbor Defense (Tactical & Adminstrative)
    │   Regiment (Administrative)
    │       └ Battalion (Adminstrative) ──────────────┐
    ├── Fort (Administrative, in some cases Tactical)  │
    │                                                  │
    └── Groupment (Tactical)                           │
            │                                          │
            ├── Mine Group (Tactical)                  │
            └── Group (Gun) (Tactical)                 │
                    │                                  │
                    └── Battery (Tactical & Administrative) ─┘
                            ├── Headquarters Section
                            │       └ Battery Commander's Detail
                            ├── Range Section
                            │       ├ Plotting Detail
                            │       ├ Observing Details
                            │       └ Spotting Details
                            ├── Gun Sections
                            │       ├ Gun Details
                            │       ├ Ammunition Detail
                            │       └ Artillery Mechanic
                            └── Maintenance Section
```

Gun Battery Organization and Duties II: Firing Sections (1940)

Firing Sections (Battery Executive Officer, Emplacement Officer)

The battery was divided into firing sections which were commanded by the *battery executive officer*. He might also have served as one of, if not the only, *emplacement officer*. The firing section outlined below is for one gun of a 12-inch or 14-inch disappearing gun. Each primary-caliber gun had its own firing section. Hence for a two gun battery of 12-inch or 14-inch disappearing guns, there would be two firing sections. The sections were charged with the preparation, pointing, loading, and firing of the guns. Other gun types had similar arrangements, but with differing numbers of men.

Gun squad (gun commander, chief of section and gun squad)
The manning details of the gun squad is listed below. A diagram that shows the formation and post positions of these men during the firing procedure is shown on the following page.

Gun pointer
Range setter
Chief of breech
Range display board operator
Deflection (azimuth) display board operator
Azimuth recorder
Range recorder
Breech detail (*cannoneers* 1-4)
Elevating detail (*cannoneers* 5, 6)
Traversing detail (*cannoneers* 7, 8)
Tripping detail (*cannoneers* 9, 10)
Truck detail (*cannoneers* 11, 12)
Powder serving detail (*cannoneers* 13-18)
Sponge detail (*cannoneer* 19)
Rammer detail (*cannoneers* 20-22, other *cannoneers* assisted as needed)

Ammunition squad (chief of ammunition)
The ammunition squad was responsible for the preparation of the projectiles and powder bags in the galleries and magazines and for the subsequent transportation to the gun platform. The crew was divided into two details, the powder detail and the projectile detail. Upon arriving at their posts, the *chief of ammunition* would forward the current information on the types of projectiles available and the current condition of the powder. Projectiles were fuzed, if necessary, and placed on the loading tables. Storage cans containing powder bags were readied for opening.

Powder detail
(*cannoneers* 23-31)
Projectile detail
(*cannoneers* 32 -41)

Artillery
Mechanic

Chief of
Ammunition

Chief of Breech

Range Recorder

Defl.(Az.) Recorder

Range Display
Board Operator

Defl.(Az.) Display
Board Operator

Range Setter

Gun Pointer

Gun Commander

Chief of Section

40"

30"

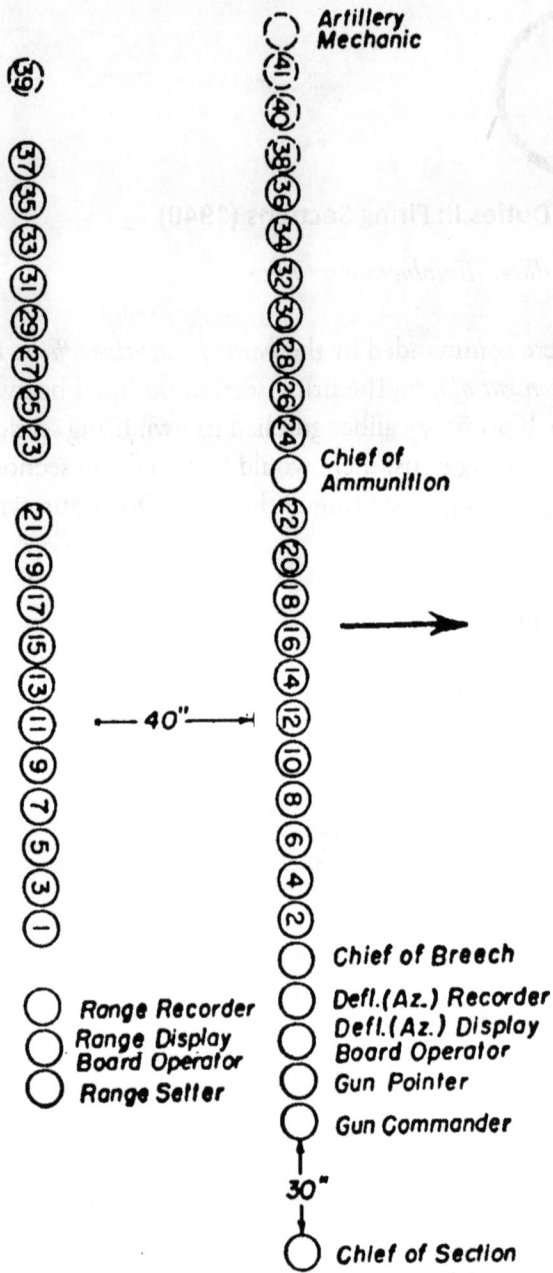

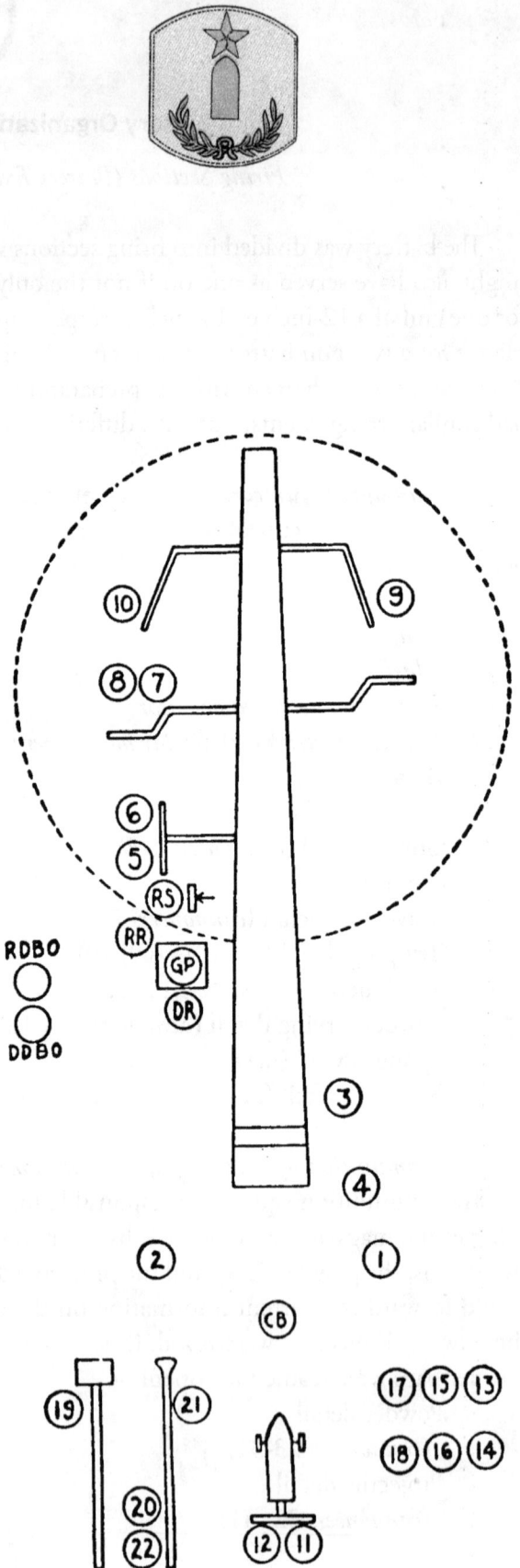

Formation of the firing section for 12-inch and 14-inch guns on disappearing carriages (left).

Posts of the gun squad for 12-inch and 14-inch guns on disappearing carriages (right).

RDBO

DDBO

RS

RR

GP

DR

CB

Maintenance Section- (senior NCO, chief of section)

The maintenance section was responsible for the battery mess, supply, battery repair facilities, and transportation.

Mess sergeant	Supply sergeant
Cooks	Battery clerks
Assistants	Unassigned personnel

Chief artillery mechanic -An *artillery mechanic* was nominally assigned to each gun. He was to perform any minor adjustments or repairs to the guns or other mechanisms in the emplacement, assisted by members of the gun section, if need be. He was also in charge of the battery storerooms and the supplies—equipment, tools, oil, paint, and cleaning materials. He was assigned to the maintenance section, though he stood in with the gun section in the parade assembly.

The Firing Drill

On the day of the gun drill the men were sent to their assigned positions by sections. The observing and spotting details were sent to their respective stations. All base end stations assigned to the battery would have been manned in an actual battle, but only some may have been manned for practice drill if the target was to be in a certain area. A single tactical battery had a battery commander, one range section, one plotting room, two or more observation and spotting details, a firing section for each gun in the battery, and a maintenance section. When the men arrived at the emplacement, the sections formed up on the battery parade.

The following commands were then given by the *battery commander* and/or section chiefs. To make it a little more concise and a bit easier to follow, the actions of both the range section and the gun sections have been combined and only a very general discription of the actions is given.

Details- At this command, the sections came to attention.

Detail!—A gun squad receives its instructions at Battery Osgood's 14-inch disappearing carriage gun in the early 1920s. Note the line of projectiles, the lone set of drill powder bags, and the swab for the breech.

Posts- The men of the battery then headed to their stations.

Examine ... equipment (gun, instrument, scope, etc.) The men would examine the instruments and equipment at their stations. The gun squads would examine their pieces and insure they were in working order. The ammunition squads would examine and prepare the projectiles and powder bags for use, as well as forwarding current information on these munitions to the plotting room detail. All stations would check and establish communication lines. The plotting room detail would make sure each of its assigned instruments was in order and establish the proper communications links. In addition, they would then receive a number of messages including the meteorological message, the tide message, the weights of the projectiles available, and powder information, which would be used to determine the correct firing data.

Report- Each man would then report to his chief whether or not his equipment was in order. In large gun squads, it was necessary to "call off," that is the chief would call off each station and get a response. If all was in order and ready, the chiefs would report to the section chiefs, who would in turn report to the *battery commander* or the appropriate officers. The battery was now ready to receive its target assignment. At some point during this period, the time interval (TI) bell system would be activated. A TI bell was located at all the stations; the battery commander's station, all four BESs, the plotting room and at the gun platform. The bell would ring (for example—3 rings, one second apart) simultaneously at all locations every 20 seconds. This was plotting interval for tracking the target and was used to calculate when to fire the gun. On drill, the battery ran by the TI bell.

Track Target- The target was assigned to the battery by the *group commander.* Once the *battery commander* received the target assignment, he could then determine what projectiles and powder charges would be used and which observing details would be used for determining the range. He also decided which pointing case— I, II, or III—would be used. The various details were then informed of these decisions.

As soon as the observing details located the target, the *observer* kept his azimuth scope trained on the center of the target, then notified his section chief. Once the observing details reported in as "on target," the *battery commander* would give the command "Track" to the range section. At the last ring of the TI bell, the *observer* would pause tracking long enough for the *reader* to report the current azimuth over the telephone to the appropriate plotting board *arm setter.* The *plotter* would determine the coordinates of the target at 20 second intervals on the plotting board (tracking). After at least three points were determined, the information could be used to determine a predicted point, the place the target would be in the time interval from the time the observation was made to when the gun was fired, otherwise known as the "dead time." The plotting room detail now had enough information to determine firing coordinates for the "set forward point"—the position of the target at the point of impact. The plotting room detail also corrected these coordinates and forwarded them to the gun squad *recorders* and *board operators,* in the proper firing intervals (1 to 2 minutes, depending on the efficiency of the squads) who would also write them down on the range board. After firing began the spotting details would also add information, and the corrections from the spotting board were also added to the determination of the firing coordinates. They would continue this process until receiving the command "Cease Tracking."

The gun squad would begin to point the gun according to the azimuths and elevations listed on the board. Hand signals were often used to signal left, right, raise, and lower the proper increments. Some gun carriages were electrically powered for this task, but could also be set manually with hand cranks. As the proper settings were reached, the range setter indicated "range set," and the gun pointer indicated "azimuth set." The ammunition squad also brought out the first projectiles and one set of powder bags to the gun platform. Everything was in readiness for the next command.

Commence Firing- This command from the *battery commander* committed the gun to firing.

Load- This command was usually given by the *gun commander*. Upon receiving the command, the breech was opened, the gun tube swabbed, the projectile rammed home, the powder bags rammed home, the breech closed, and the primer inserted in the firing lock in the breech.

Trip- At this command the gun counter weight was released (tripped) and the gun was raised to firing position— "in battery."

Fire (or Relay)- The gun was fired on the last ring of the appropriate TI bell interval, or by command. This should have occurred momentarily after the gun was tripped "in battery," to minimize the time the gun was exposed above the parapet. If the gun was not ready for some reason, the command "Relay" was given to set the gun for firing in the next TI. Commands 6 through 9 were then repeated until the command "Cease Firing" was given. A crack crew could fire these guns at 1 minute intervals. Misfires had a special procedure that was followed.

Cease Firing- The gun squad halted firing on this command, but would continue to track the target with the gun sight.

New Target-Track- On this command the range section would receive a new target from the *battery commander*. Tracking would begin again, and the procedure would start again from command 5 and continue until a "Cease Fire" command was again received.

Cease Tracking- At this point the engagement was over. The group commander had indicated to the BC that there were no more targets. The plotting room detail, observing details, spotting details, and the gun squad all quit tracking.

Replace Equipment- All equipment and materials were replaced for storage by each detail. The gun was cleaned and "bedded down." The ammunition squad secured the battery magazines. All instruments were cleaned and prepared for their next use.

Close Stations- Stations were inspected, closed and secured. All section chiefs were required to keep record of material used during the firing and report these to the battery commander and the battery clerk for recording. Information as to the number of rounds fired, shells used, powder charges used, hits and misses, etc. were all noted in the emplacement record book.

Battery Dismissed- All sections were dismissed by the battery commander at the end of the "engagement." The section chiefs then took over to secure their areas.

(This page and the next page) Gun drill at one of the 14-inch disappearing gun located at Fort Hughes, Caballo Island, in Manila Bay, the Philippines, showing the crew loading a shell and powder bags. The six photo sequence was taken by Col. George Ruhlen, Jr. in the late 1930s (from the Ruhlen Collection, Fort MacArthur Museum, San Pedro, CA.)

Gun drill with Battery Osgood's 14-inch disappearing carriage gun—(top) everything is ready, (middle) load! (bottom) fire! (from the Fort MacArthur Museum photograph collection.)

References

Coast Artillery School Publications, *Fire Control and Position Finding for Seacoast Artillery*, The Bookshop, Fort Monroe, VA (no date).

Drill Regulations for Coast Artillery, GPO, Washington, D.C. (1914)

Field Manual (Army Coast Artillery) FM 4-5, *Seacoast Artillery Organization and Tactics,* War Department, GPO, Washington, DC (1940)

Field Manual (Army Coast Artillery) FM 4-15, *Seacoast Artillery Fire Control and Position Finding,* War Department, GPO, Washington, DC (1940)

Field Manual (Army Coast Artillery) FM 4-80, *Service of the Piece, 12-inch and 14-inch guns on Disappearing Carriages*, War Department, GPO, Washington, DC (1940)

Clint C. Hearn. *Fire Control and Direction for Coast Artillery.* U.S. Signal School, 1907.

Danny R. Malone. "Seacoast Artillery Radar 1938-46," *CDSG News* Vol. 3, Issue 5 (Nov. 1989), pp. 1-11, and "Addendum," *CDSG News* Vol. 4, Issue 2 (May 1990), pp. 37-39.

H.L. Morse. "The Evolution of Our System of Position Finding And Fire Control," *The Journal of the United States Artillery*, Vol. 39, No. 2 (March-April 1913), pp. 137-190. (reprint available as a PDF file on the CDSG Publications CD ROM, CDSG Press, www.cdsg.org.)

R.O.T.C. Manual, Coast Artillery, Basic, 9th Edition, Military Service Publishing Company, Harrisburg, PA (1938).

Bolling W. Smith. "Vertical and Horizontal Base Position Finding Systems." *CDSG Journal* Vol. 13, Issue 3 (Aug. 1999) pp. 4-23.

Bolling W. Smith. "Time Interval Systems." *Coast Defense Study Group News* Vol. 6 Issue 2 (May 1992) pp. 42-79.

Bolling W. Smith. "Meteorological and Tide Stations 1890-1917." *Coast Defense Study Group Journal* Vol. 11 Issue 1 (February 1997) pp. 23-36.

Bolling W. Smith. "Radio and Coast Defenses in the Endicott Era 1899-1916." *Coast Defense Study Group Journal* Vol. 11 Issue 2 (May 1997) pp. 31-38.

Bolling W. Smith. "Coast Artillery Telephones 1915-1945." *Coast Defense Journal* Vol. 31 Issue 3 (August 2017) pp. 4-58 and Vol. 31 Issue 4 (November 2017) pp. 4-53.

Bolling W. Smith. "Birth of Seacoast Artillery Fire Control and Direction." *Coast Defense Journal* Vol. 34, Issue 4 (Fall 2020) pp. 12-63.

Technical Manual (Army Coast Artillery), TM 4-305 *Coast Artillery Gunners' Instruction, Fixed Seacoast Artillery, First and Second Class Gunners,* War Department, GPO, Washington, DC (April 10, 1942).

Training Regulations, TR 435-220, *The Battery Command (Fixed),* War Department, GPO, Washington, D.C., (1924).

Training Regulations, TR 435-221, *Fire Control and Position Finding*, War Department, GPO, Washington, D.C., (1926).

Training Regulations, TR 435-270, *Service of the Piece, 12-inch and 14-inch guns on Disappearing Carriages,* War Department, GPO, Washington, D.C., (1924).

Training Regulations, TR 435-290, *The Fire Command,* War Department, GPO, Washington, D.C., (1924).

Training Regulations, TR 435-295, *The Fort Command,* War Department, GPO, Washington, D.C., (1924).

Training Regulations, TR 435-300, *The Coast Artillery Command,* War Department, GPO, Washington, D.C., (1924).

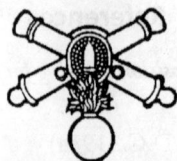

AMMUNITION SERVICE IN AMERICAN SEACOAST DEFENSE BATTERIES

Mark Berhow and Thomas J. Vaughan, Jr.

The new seacoast defense batteries of the 1890s were, for the most part, built with the ammunition storage at a level one story below that of the gun platform. The large degree of protection needed for the magazine and the perceived requirement for the batteries to blend in with the surrounding topography resulted in this arrangement. The projectiles and powder bags had to be moved horizontally from the magazines, vertically up one story to the gun platforms, and then horizontally out to the guns themselves. This was not a problem for the smaller caliber guns (5-inches or less) as these could be carried by the crew. But the larger caliber weapons required mechanical devices. This required a system that was rapid enough not to effect the rate of fire of the guns as well as simple and rugged enough to be run by uneducated enlisted men.

While the problem of horizontal movement was easily solved with the use of overhead trolley rails and wheeled ammunition trucks, the development of the vertical hoisting devices took place over a period of twenty years. In 1890, little thought was given to exactly how the ammunition would be raised from a lower level. Initially cranes or davits with simple block and tackle or differential pulley systems were used both to bring the ammunition up to the loading platforms, and, in the case of the barbette guns, to load the weapons themselves. But, this method proved to be unsatisfactory.

Several engineer officers contributed ideas over the next few years and prototype hoisting systems were installed at a few batteries at various places being built.

In 1896 the Corps of Engineers adopted the "balanced platform hoist," which was essentially a balanced set of dumb waiters in twin shafts built in the battery structure, as the standard hoisting mechanism. This system proved to be too delicate. The wire ropes connecting the platforms stretched, and it became impossible to keep the platforms correctly aligned.

The engineers then turned to continuous chain hoists. Several prototype chain hoists were installed along the East Coast in late 1890s. Two general types of continuous chain hoists were used in large numbers: the Hodges hoist, and the Taylor-Raymond hoist. After 1903 all new 10-inch caliber and larger batteries were fitted with the Taylor-Raymond hoist, and by 1910 many of the older batteries of 10-inch caliber and above had been retrofitted with the Taylor-Raymond hoists. A few batteries (mostly 6-inch, and a few others) retained the Hodges hoist.

Through out the development of modern-era defenses, several batteries were built featuring hoistless designs. These were initially used only under conditions that afforded good protection for the magazines, like the mortar batteries. Most barbette and disappearing gun batteries built 1894-1915 had lower level magazines and vertical hoist mechanisms. Eventually, however, the vertical lift systems were abandoned altogether as the post-1915 seacoast defense batteries were built with the ammunition storage magazines and the gun platforms at the same level, allowing the ammunition trucks to be pushed directly to the guns from the magazines.

Three key definitions: *ammunition supply* is the process of manufacturing and transporting ammunition to a gun emplacement; *ammunition storage* is the care and keeping of ammunition in the magazines; and *ammunition service* is the movement of the ammunition from the emplacement magazine to the gun.

Ammunition Storage

Cartridge Storage: ammunition of 4.7-inch caliber or smaller consisted of complete ("fixed") cartridges composed of shell and powder in a brass casing. These were stored in the boxes they were shipped in and carried to the gun platform by hand.

Projectile Storage: In general, projectiles were stored on their sides in rows along the walls of the shell room with their points toward the wall so that the bases could be inspected from the center path. Each tier of projectiles was usually laid on a pair of timber skids 3-4 inches thick. Beginning in 1908, larger shell rooms allowed for two rows of stacked shells with passageways between the walls and the shell points. Later specialized storage tables were used for a single row of shells.

Powder Storage: Powder bags were usually encased in hermetically sealed metallic cases and then enclosed in wooden boxes or cases for transportation. Early on it was customary to store them vertically (either in or out of the wooden boxes), piling some cases on top of others. In the later, smaller caliber batteries they were stored horizontally, often on specialized storage tables.

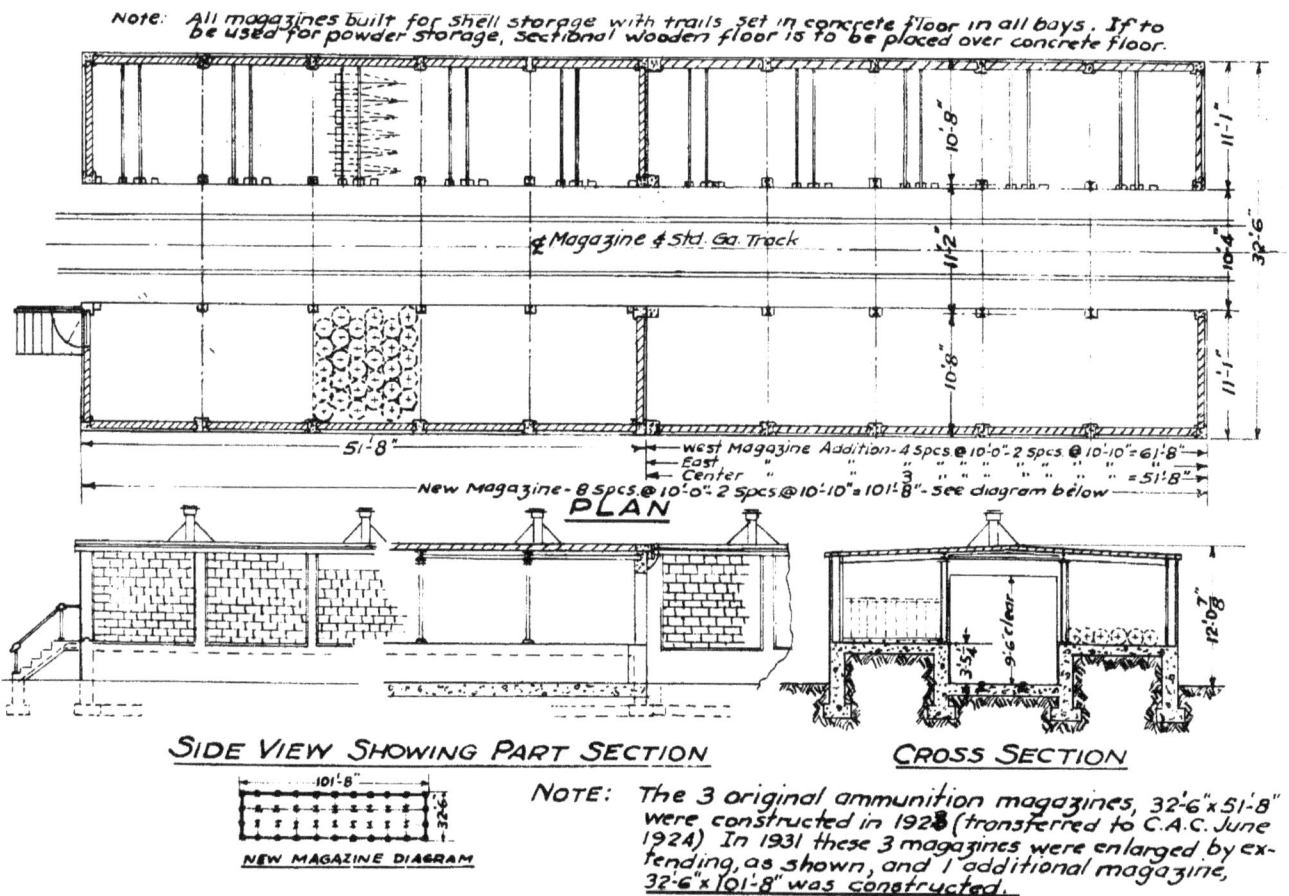

Drawing of one of the detached magazines for Battery Harris, a 16 inch gun battery at Fort Tilden, NY, built in 1923 and modified in 1937. Note the locations of the vertically-stacked powder canisters and the horizontally-stored projectiles. (National Archives)

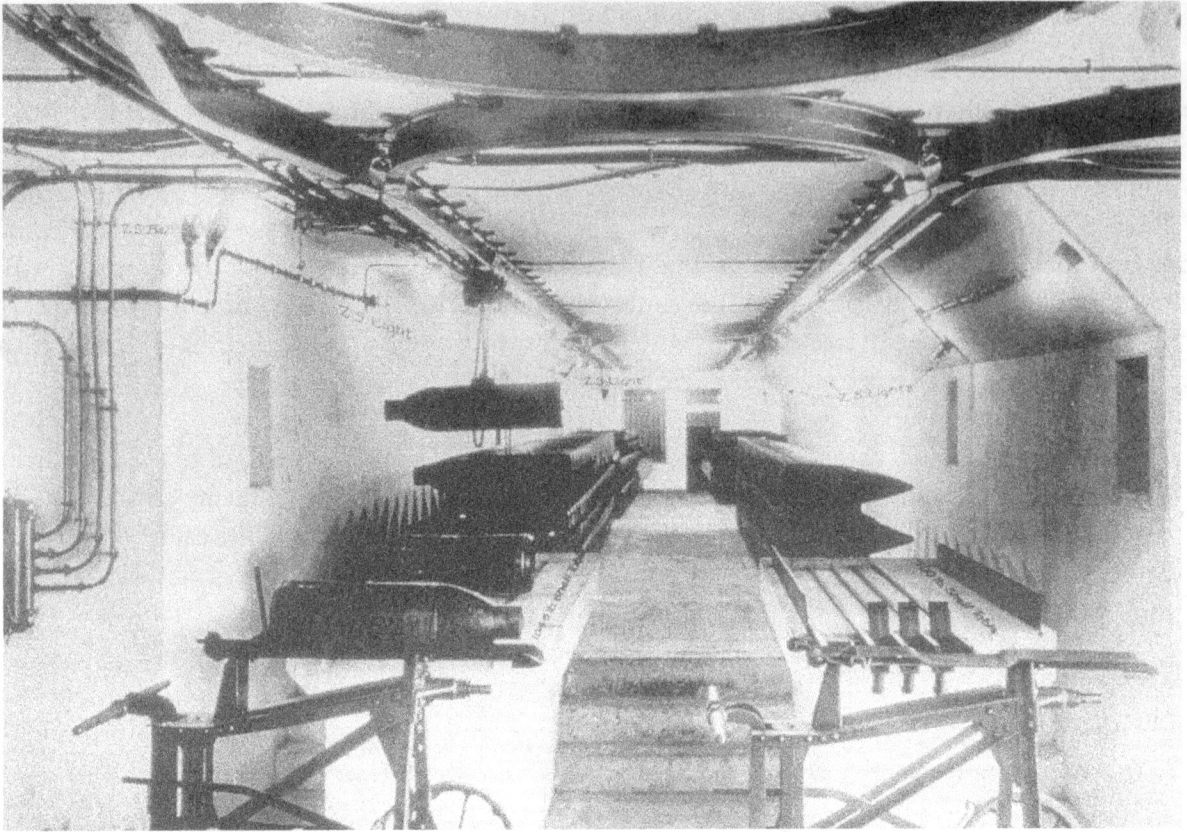

12-inch mortar projectile room at Battery Howard, Fort Sherman, Panama Canal Zone. This battery was completed in 1916 and shows the latest developments in projectile storage. Note the overhead trolleys, the projectile hoist, the projectile trucks, projectile delivery tables, and the two different types of projectiles, armor-piercing (left) and high explosive (right). (National Archives, Still Pictures No. 77-CD-7F-1)

Projectile room (left) and powder room (right) of Battery 405, Oahu, Hawaii (1940s)
(courtesy Glen Williford)

Ammunition Service: Horizontal Movement

Trolley Rail Systems: In nearly all US batteries of 6-inches or larger caliber, an overhead trolley rail system was installed in the magazine level of the battery for moving the projrctiles. These trolleys ran from the rear of the projectile rooms out to the vertical hoists and out to an exterior entrance to the magazine level. If there was more than one projectile room, a simple switching system connected the tracks to the one leading to the hoists/exterior. The trolley mechanism had chains fed through a simple triplex or differential chain fall that was attached to the tongs to lift the projectile from the floor of the magazine and transport it to the vertical hoist receiving table. Some of the later batteries had delivery tables at the ends of the magazines. The projectiles were brought by the trolley to these tables, then rolled onto a shell truck to be pushed to the gun platform. This trolley system remained a standard feature from the early Endicott batteries through the World War II modernization program batteries. It was used to deliver projectiles directly to the gun platform in many of the casemated batteries.

Trucks and Handbarrows: Projectile (i.e. shell/shot) carts were specifically designed for each caliber of shell. Some, such as those for the 6-inch guns, carried up to six projectiles or both a projectile and its powder charge. The carts for the larger caliber guns generally carried one projectile only. They were designed to be the right height to deliver the projectile directly into the breech of a disappearing gun or a 12-inch mortar. Battery personnel pushed these carts from the hoist delivery table to the gun. There were often special storage niches in the parapet walls or traverses to house the trucks when not in use.

A 10-inch disappearing gun loading platform with the crew ready to load the gun. Note (from left to right) the overhang covering the ammunition hoists (with the battery name "Henry Benson" on a sign above), the time range board, a shell on a shot cart, a hand stretcher for the powder bags, the breech swab (propped up at the rear of the emplacement left of the top of the stairs), the gun with the breech open, the rammer (propped up at the rear of the emplacement right of the top of the stairs) and the open cart storage nitches to the right. The 1903 battery plan on the next page resembles this battery very closely. (from the collection of the Puget Sound Coast Artillery Museum).

Upper Level

Truck Recesses

N2 2
WORKING PLATFORM

Ammuntion
Hoist
Wells

Room

Room

Loading Platform

Truck Corridor

Crane

Crane

Lower Level Walk

Lower Level

N2 2
GUN
WELL

Shell
Room

Powder
Room

Accum-
ulater
Room

Guard
Room

Store
Room

Hoist Shafts

Hoist Room

Crane   Trolley Rails   Crane

Lower Level Walk

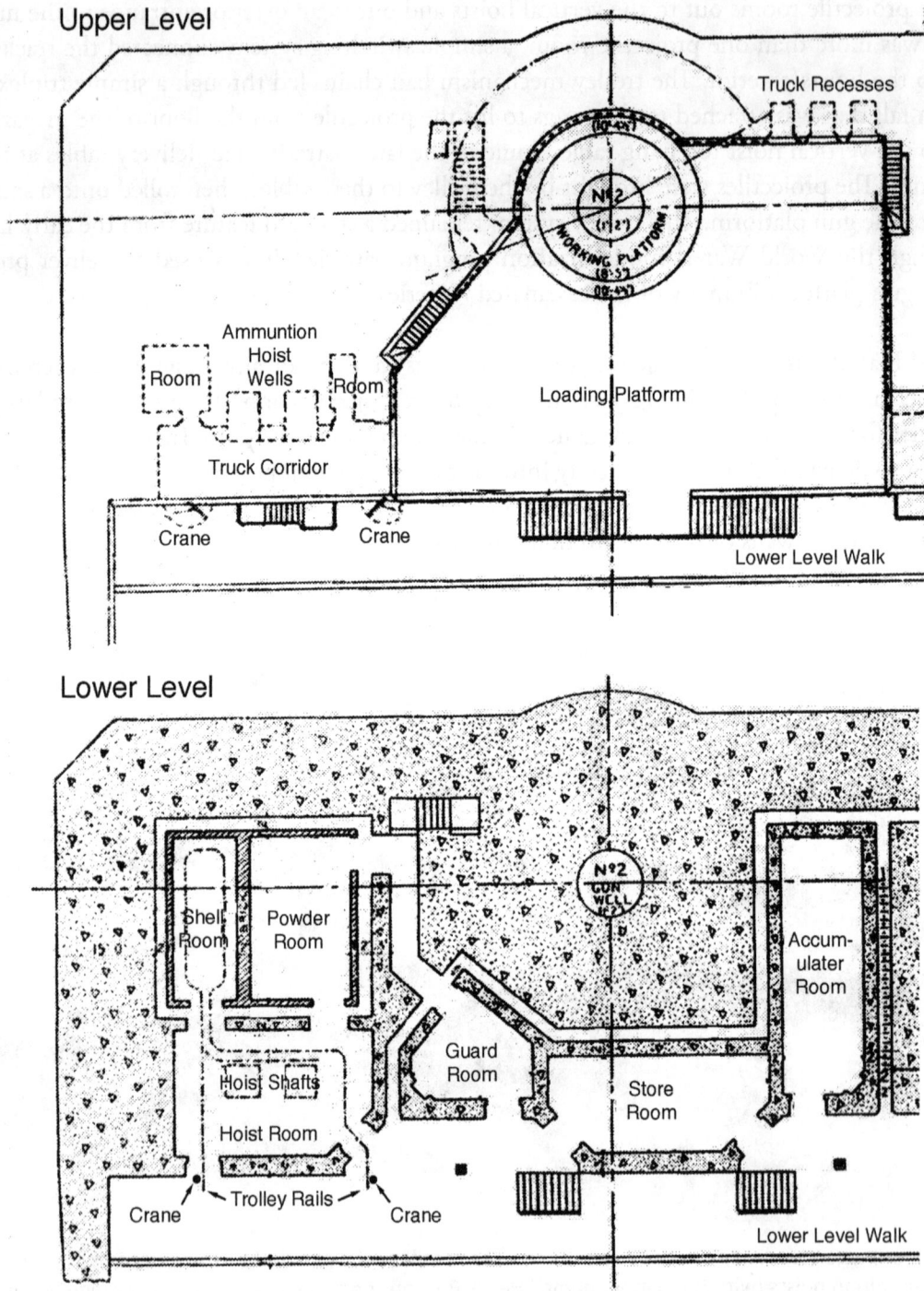

Part of a 10-inch 1903 battery plan (reproduced from Winslow) showing upper level service area
and lower level ammunition storage rooms and trolley rail trace.
See previous page for a photographic view of this battery layout.

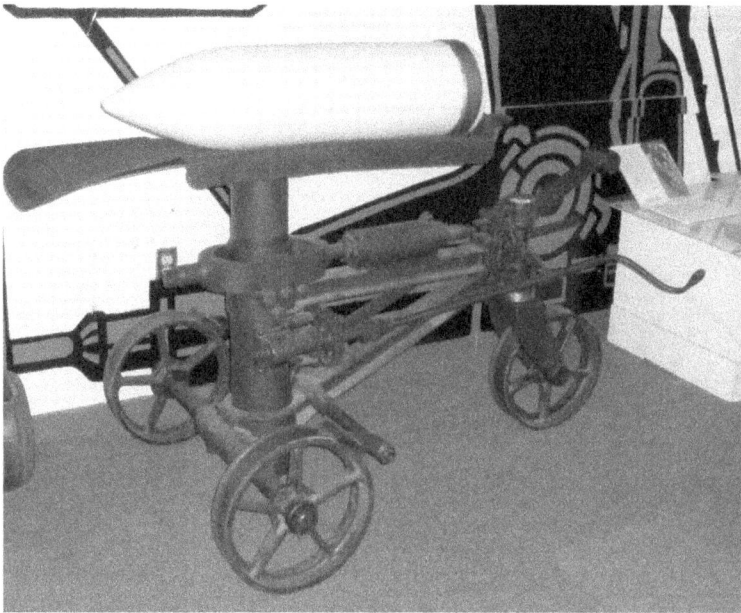

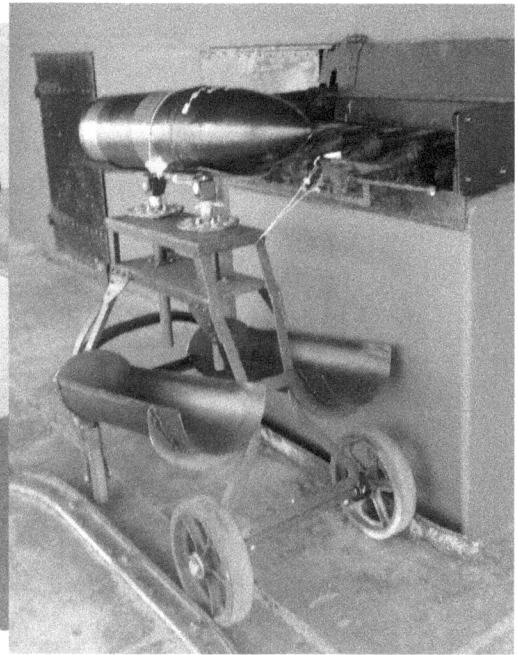

(left) Projectile truck for 10-inch projectiles at Fort Mott State Park visitor center (right) replica shot cart at Fort Casey State Park, WA

Overhead trolley rail and carrier at Battery Calef-Wilkeson, Fort Rosecrans, HD San Diego, CA

Ammunition Passages & Lobbies

The original 1894 style of 8-inch and 10-inch disappearing carriage batteries (for example Battery Jack Adams at Fort Warren, shown on page 307) had an interior ammunition service.

A zigzag passage on two levels connected the magazine with the crane level around the exterior of the loading platform. The zigzag path prevented any projectiles or splinters from entering the magazines. The projectiles were carried on a trolley along a passage leading to the bottom of the 2-story hoist well. The projectiles were lowered to the floor and rolled into the hoist well. They were lifted by a hoist suspended from a trolley attached to the ceiling of the hoist well at the upper level. This trolley was rolled into the 'lobby' (a passage at the upper level with a ceiling level with the ceiling of the hoist well) and out to the crane level around the loading platform. There they were lowered to the ground and rolled to the base of one of the cranes. From there they were hoisted to the loading platform.

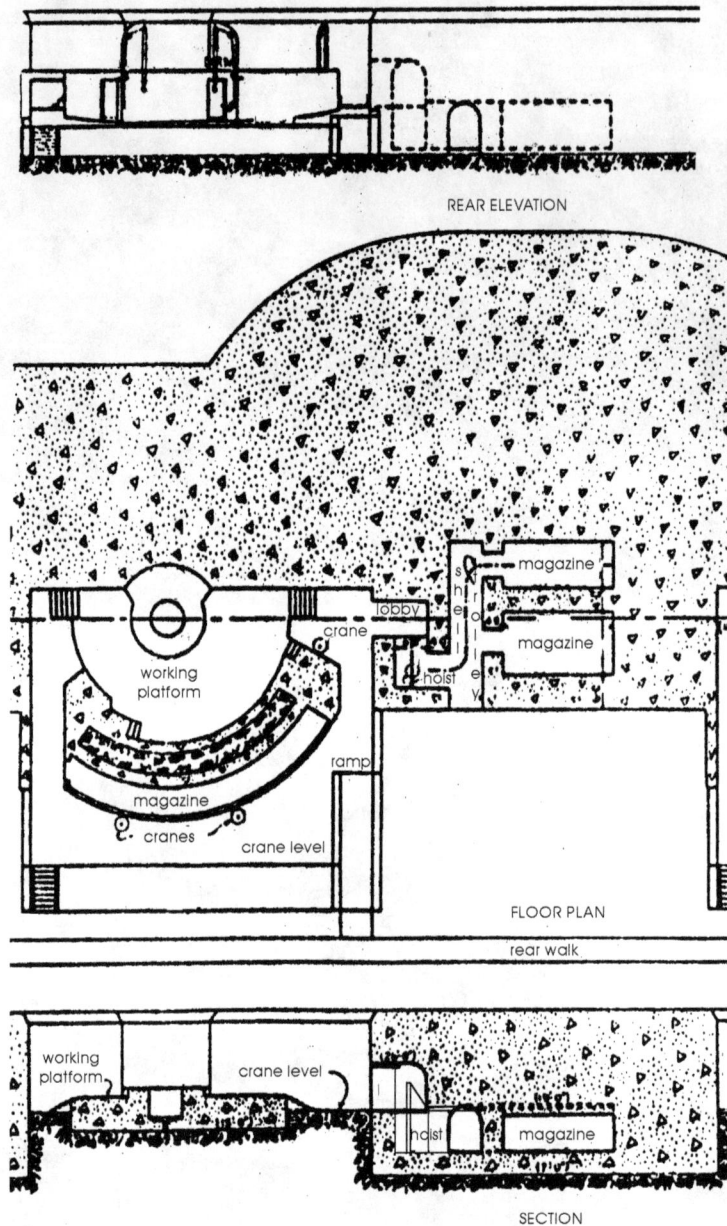

REAR ELEVATION

magazine

lobby

crane

working
platform

magazine

hoist

ramp

magazine

cranes

crane level

FLOOR PLAN

rear walk

working
platform

crane level

hoist

magazine

SECTION

Plans for an early 10 inch disappearing rifle battery showing ammunition service details.

This arrangement brought the shells out very adjacent to the front wall of the emplacement. Early plans were to traverse the gun so that the breech was near the front wall for loading so the crew received maximum protection from the parapet. Later this drill was changed and the gun kept on target while being loaded from the rear of the loading platform to get a faster rate of fire.

When these batteries were modernized, the passages were altered and steps were added. These were used as passageways for manual powder service. The passages still exist in these modernized batteries. Sometimes traces of the original trolleys and/or rings for attaching hoists can be seen in the ceiling which otherwise seem out of place in a stairway.

These passages and lobbies were omitted in the next series of batteries and emplacements that were built with balanced-platform hoists.

Battery Adams at Fort Warren in Boston harbor. This battery shows the details of the simple early ammunition delivery area. The inset to the right of the lower photo shows both the upper bracket and lower socket where the curved metal support for the block and tackle hoist was located.
(All photos by Mark Berhow)

Ammunition Service: Projectile Hoists

Crane (Davit) Hoists

The earliest system of vertical hoists were triplex or differential blocks with chains, attached to the end of a metal crane which was bent at the upper end to extend out 3 to 4 feet. The crane was inserted into a socket some two feet from a wall and held vertical by a bracket and collar attached to the wall. These were located outside to the rear or to one side of the loading platform. The actual hoisting was done by either brute manpower or by a winch. In the case of the earlier model barbette guns of 8-, 10-, and 12-inches, a similar crane was used to hoist the projectile from the truck to the breech. This system was not very popular with the crews and soon other hoisting systems were developed. The cranes often remained as back up hoist systems even after the continuous chain hoists were installed. Few of the crane davits remain at batteries today; more often only the brackets and sockets remain.

Crane davats at Battery Wilhelm (top), Fort Flagler, HD Puget Sound, Battery Bradford (bottom left), Fort Terry, HD Long Island Sound, (MAB 1995 & 1996), and Triplex block at Battery Worth (bottom right) Fort Casey, HD Puget Sound (Dan Rowbottom 2003).

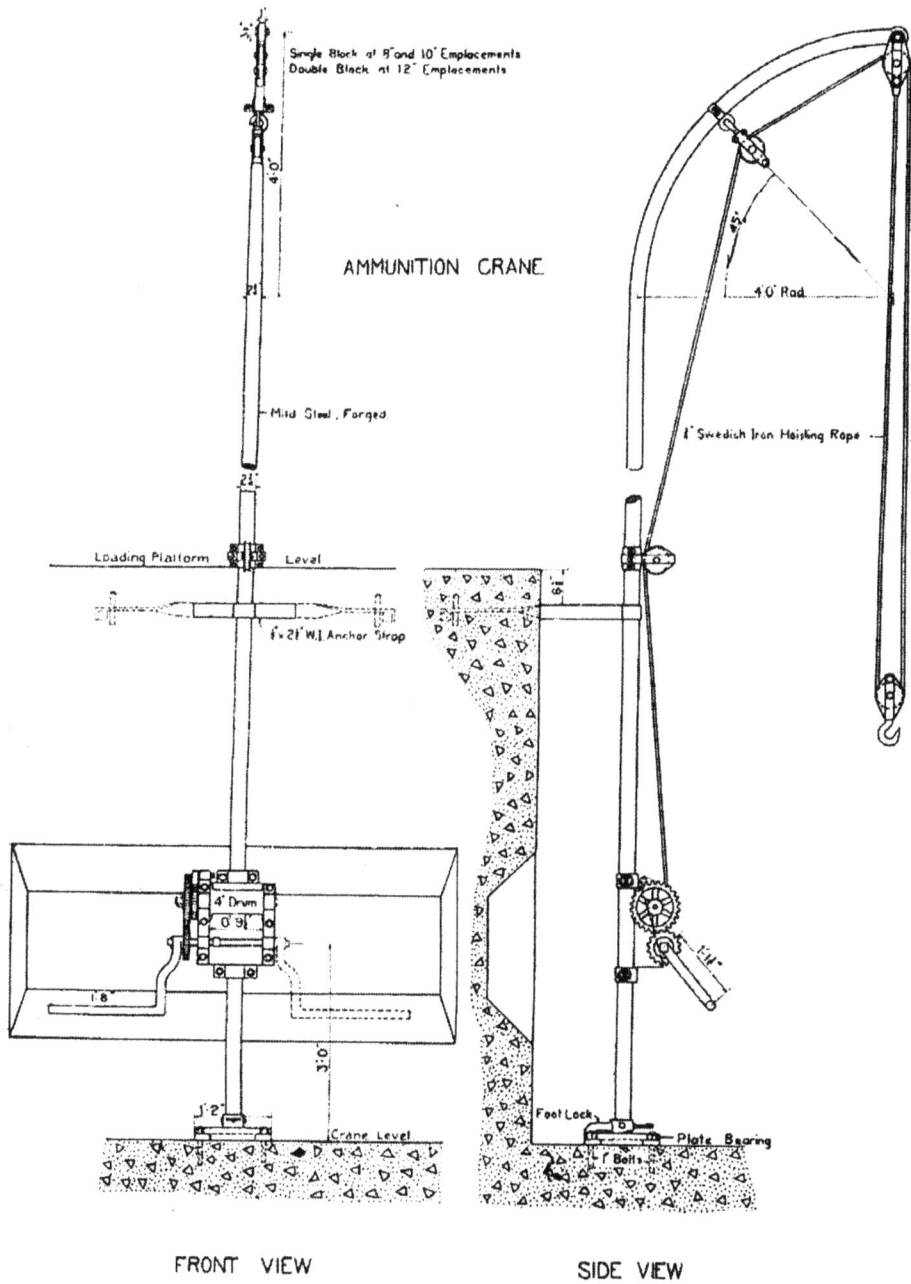

Crane Hoist System (from Winslow).

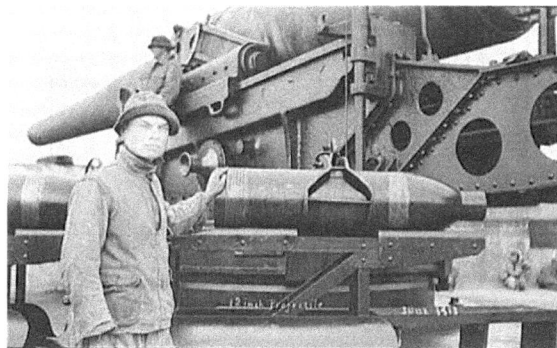

Shell with hoist collar at Battery Saffold, Fort Winfield Scott, HD San Francisco (Golden Gate Natl. Rec. Area)

Balanced Platform Hoists

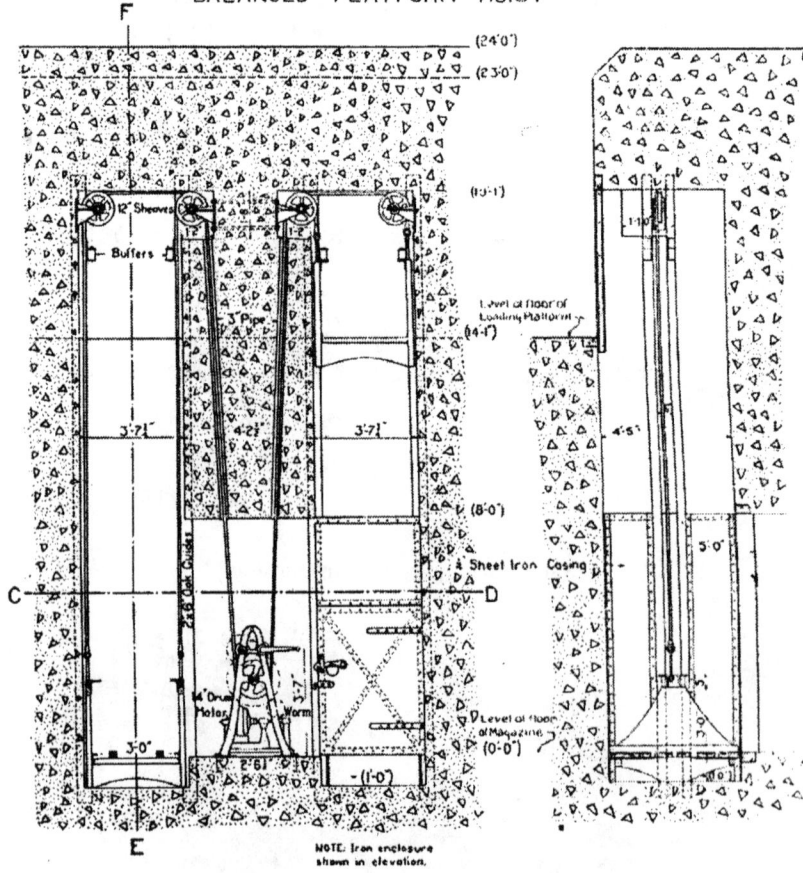

Adapted for general use in 1896, and installed in new batteries built after that date, the hoist consisted of two platforms—about 3 feet by 4 feet—operating in two wells a few feet apart. The two platforms were connected by a wire rope, which passed through a pulley on the upper level system and a winch on the lower level. The length of wire rope was such that when one platform was exactly on the level of the magazine floor, the other would be exactly on the level of the loading platform. The platforms were made of a size that exactly contained a standard ammunition truck. The trucks were loaded in the projectile and powder rooms, and pushed to the balanced platform hoist. At the same time an empty truck was placed on the platform at the loading platform level. The winch then moved the platforms to their opposite positions. This was the first vertical hoist system to be located inside battery structures, with the loading platform egress protected by an overhanging concrete splinterproof.

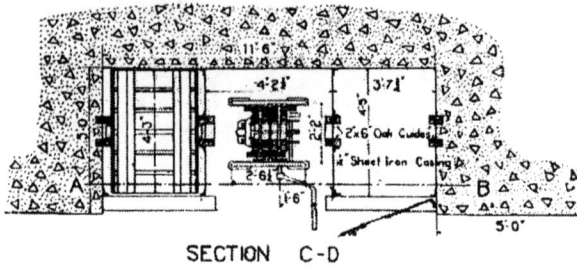

SECTION C-D

BALANCED PLATFORM HOIST

SECTION A-B   SECTION E-F

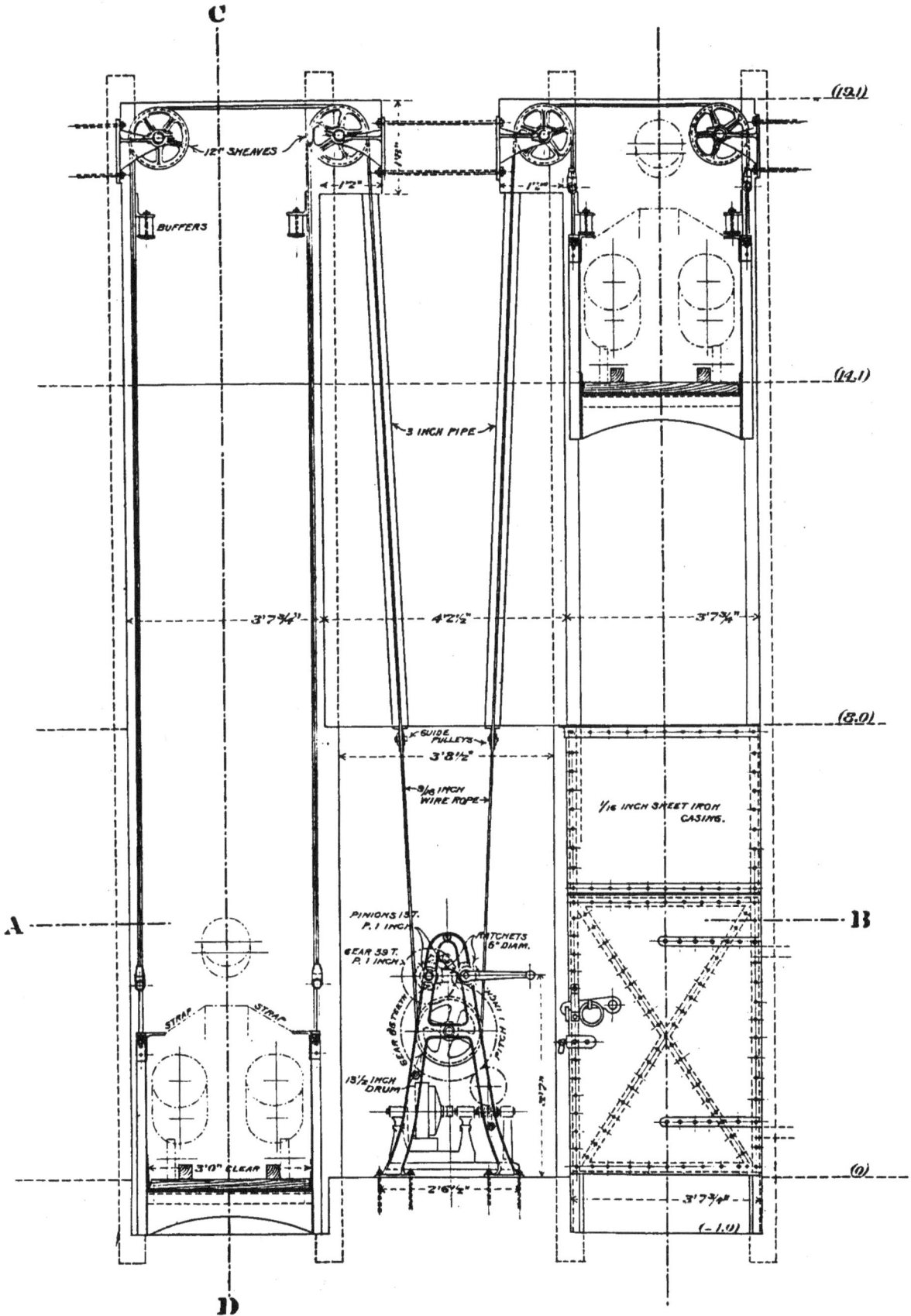

Balanced platform ammunition hoist.

Balanced platform ammunition hoist system at Battery Worth, Fort Casey, WA (MAB)

Balanced platform ammunition hoist at Battery Calef-Wilkeson, Fort Rosecrans, HD San Diego, CA (MAB)

When new, these hoists functioned quite well. But, as the rope stretched with the wear and tear of use, it became impossible to keep the platforms correctly aligned. Often, it was so bad the battery crews went back to the crane hoists. This hoist was superceded by the continuous chain hoists. Some of the balanced platform hoists were physically replaced by the new hoist systems, others were retained and used as powder hoists, and still others were removed and the loading platform level of the well was floored over and used as ammunition truck storage niche.

Hodges Hoists

Because the balanced platform hoists proved unsatisfactory, the engineers turned to several proposals for chain hoists. The first design that was adapted and deployed in significant numbers after 1898 was that of Harry F. Hodges. All chain hoists had two sets of sprockets mounted on axles to which a set of flat link chains were attached in a continuous loop from the lower set of sprockets to the upper set. These were rotated by a hand crank or an electrically powered motor located in bottom of the hoist well. Hodges' hoist had a set of six rigidly fixed cup-shaped arms supported by a bracket. As the chains rotated, the arms passed through openings in the receiving table and picked up the projectiles and powder bags. As they reached the top they rotated over the top sprocket and deposited the ammunition on the arms of the delivery table. Designs were implemented for 6-inch, 8-inch, 10-inch and 12-inch projectiles.

Problems developed with the larger projectiles. If the hoist was operated at too fast a speed, the larger projectiles were rolled onto the delivery table with too much speed to be stopped effectively, and the arms had a tendency to tear powder bags. Most of the Hodges hoists that were retained in batteries after 1910 were hand operated.

Hodges ammunition hoist system for 6-inch projectiles at Battery Carpenter, Fort McKinley, Portland, ME.
Hoist and receiving table (left) delivery table (right) (photos by Mark Berhow).

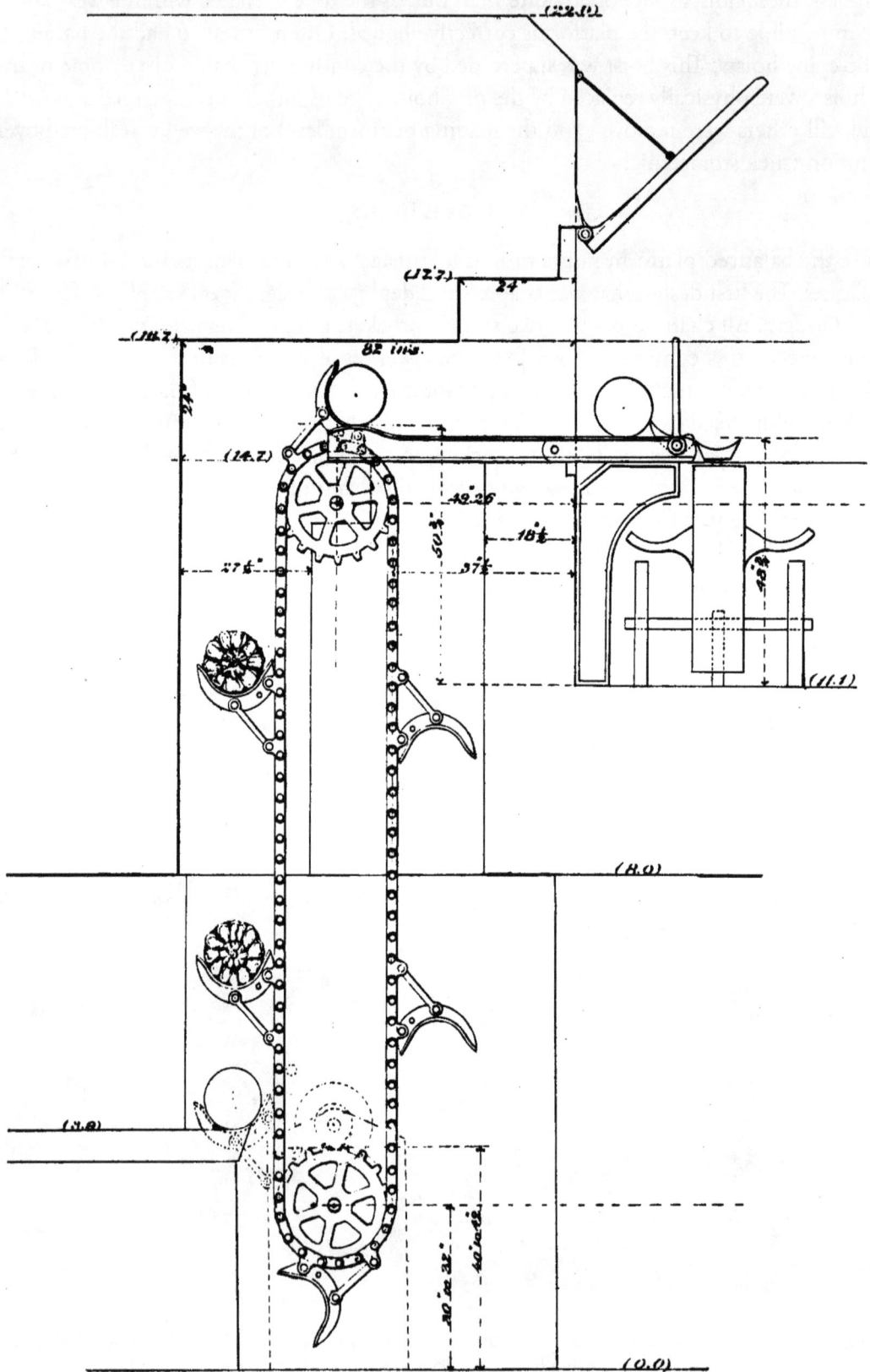

Hodges ammunition hoist.

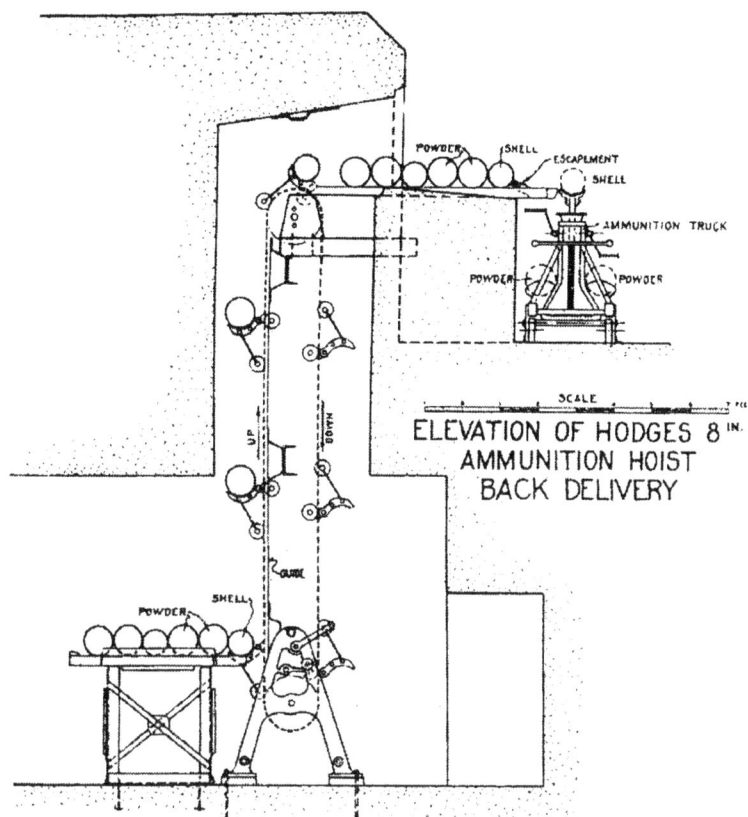

ELEVATION OF HODGES 8 IN.
AMMUNITION HOIST
BACK DELIVERY

Restored Hodges ammunition hoist, back delivery, at Battery New Peck (ex-Gunnison), Fort Hancock, Sandy Hook, NJ. (photo by Mark Berhow).

Other Hoists

As a result of the various vertical lift mechanisms tried out by the army, several batteries had different and sometimes unique hoist systems. Some of the earlier larger caliber batteries operated without hoists, notably those in San Francisco and Boston, which were designed to function without them. Most hoist-equipped 6-inch batteries retained the Hodges design. A few of the larger caliber batteries, mostly on the East Coast, also retained the Hodges hoist. Experimental hoist designs were used in batteries in the harbor Defenses of the Delaware River (Forts Mott, Delaware and DuPont), Baltimore (the 4.7-inch and 5-inch batteries), Cape Fear (the 4.7-inch battery at Fort Caswell), and some of the batteries in the Harbor Defenses of Boston. More details on these batteries can be found in David Hansen's article listed in the sources.

Battery Basinger, a 3-inch battery at Fort Strong, had a hoist of unique design—NOT a Raymond hoist. In each of the two magazines there was a table on which a shipping crate of 4 rounds was placed and the whole thing was hoisted up to an opening in the side of the emplacement. This is one of apparently three 15-pounder (3-inch) batteries to have such hoists—the other two were Battery Allen and Battery Albertis at Fort Delaware.

Raymond Hoists

Robert R. Raymond designed a different chain hoist in 1899. Raymond's hoist had carriers that hung loosely from the chains, and as they passed over the upper axle shaft they tipped forward, rolling the projectiles gently onto the delivery table. The carriers then traveled in the inverted position down the to the bottom of the hoist well, where they were turned right side up scooping up a new load. A hand-operated pan fed the ammunition onto the carrier.

The 5-inch rapid-fire batteries at Boston—Battery Field and Battery Rice—had hand-operated Raymond hoists. These 5-inch hoists did not have a receiving table; the projectiles were placed directly on the carriers by hand. The 1901-type 6-inch rapid fire batteries at Boston—Battery McCook and Battery Whipple—had hand-operated Raymond hoists. These were later removed and the tops of the shafts sealed when the batteries were modernized.

No. 660,998.

R. R. RAYMOND.
HOISTING APPARATUS.
(Application filed Aug. 21, 1899.)

Patented Oct. 30, 1900.

(No Model.)

Fig. 1

Fig. 2

Fig. 3

Fig. 4

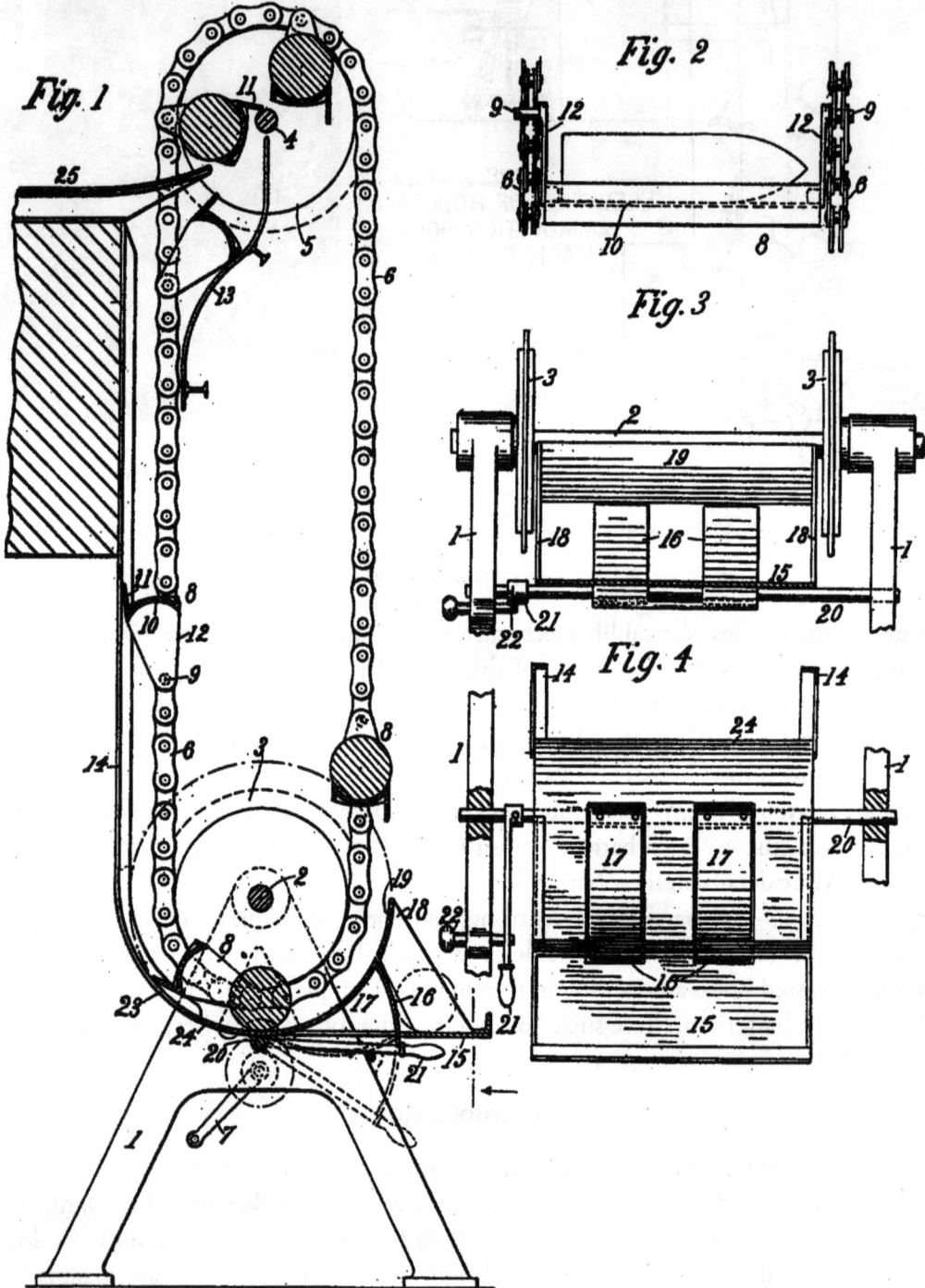

Witnesses:
Raphael Netter
James N. Catlow

Inventor
Robt. R. Raymond
by Robt. F. Gaylord Att'y.

Raymond ammunition hoist

Taylor-Raymond hoist for 8-inch projectiles at Battery Thompson, Fort McKinley, Portland, ME
(Photo by Mark Berhow)

Taylor-Raymond Hoists

In 1902, Harry Taylor submitted a proposal for a chain hoist system that eliminated the tipping movement of the Raymond hoist. It gently pushed the projectiles and powder bags out onto the delivery table and provided a simpler method for picking up projectiles and powder from the receiving table. The Taylor-Raymond hoist was simpler and easier to use than the Hodges hoist, and it was less expensive.

The Engineers adapted the hoist in early 1904 for all new construction. In addition, they ordered that 44 older batteries be retrofitted with the new hoist system. This action, along with other improvements, radically changed the appearance of many of the earlier batteries. By 1910, most of the installations of Taylor-Raymond hoists were complete.

The remains of this hoisting system is the one seen in most pre-1915 American seacoast batteries remaining today.

Delivery table of a Taylor-Raymond hoist for 12-inch projectiles, Battery Wilhelm, Fort Flager, Puget Sound
(Photo by Thomas J. Vaughan, Jr.)

PLAN OF DELIVERY TABLE
T.-R. HOIST

ESCAPEMENT

DELIVERY TABLE

ESCAPEMENT

DOWN

UP

TROLLEY RAIL

MOTOR

RECEIVING TABLE

WINCH

PLAN OF RECEIVING TABLE
T.-R. HOIST

FIG. N° 5
SCALE

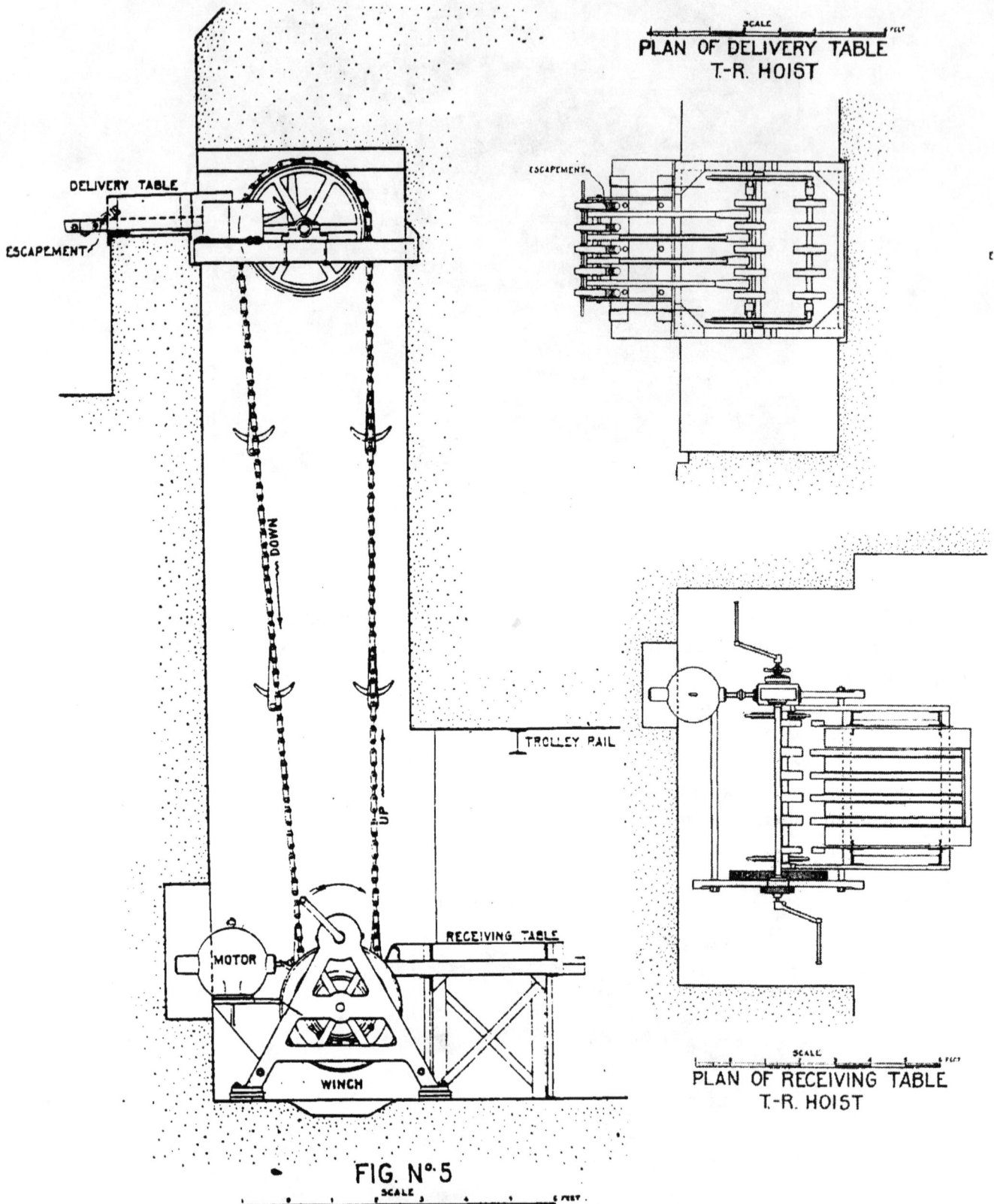

ELEVATION OF TAYLOR-RAYMOND
10IN. AMMUNITION HOIST
BACK DELIVERY

Taylor-Raymond ammunition hoist.

Taylor-Raymond ammunition hoist system for 10-inch projectiles at Battery Harker, Fort Mott, HD of the Delaware

The delivery table at the loading platform level.

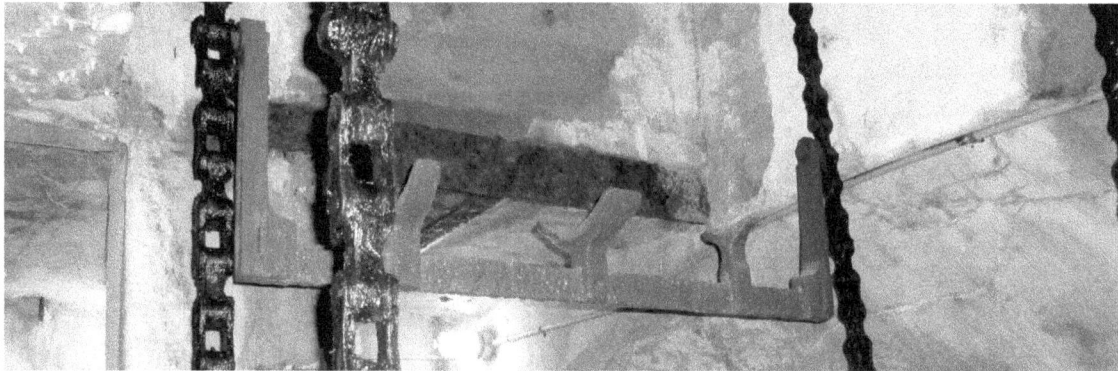

a shell cradle on the chain

Bottom: the winch and receiving table at the magazine level (photos by Mark Berhow 2009).

Ammunition Service: Powder Hoists

In general, powder was carried by hand in the smaller caliber batteries. Larger bags were loaded on a stretcher or handbarrow and transported from the magazine or hoist to the gun. The vertical chain hoists were designed to be used for both powder and shell, but the artillery soon stopped hoisting powder in the same hoist as projectiles. It was apparently because powder charges didn't roll as smoothly as projectiles and tended to jam up on the delivery tables. Subsequently, powder hoists were installed in some large caliber batteries (10-inch, 12-inch, and one 14-inch battery in the Philippines) between about 1908 and 1912. After the installation of the chain hoists, several batteries used the balanced platform hoists for powder service. Apparently all powder hoists were removed prior to World War II and the powder was transported by hand or on a stretcher.

Top of powder hoist (type "A") (left) and shell hoist (Taylor-Raymond) (right),
Battery Kinzie, Fort Worden, Puget Sound (photo by Thomas J. Vaughan, Jr.)

There were three types of powder hoists:

"A" – A *vertical lift type.* Construction details are unclear for this type. It was primarily designed to fit into existing shafts (usually a second balanced-platform hoist shaft in an emplacement where one set of balanced-platform hoists was converted to Taylor-Raymond).

"B" – An *inclined type (with a rectangular shaft).* This type can be described as being similar to a Taylor-Raymond hoist, but set at a 45-degree angle. Instead of shell carriers, simple rods connected the two chains, and canvas slings were arranged to form pockets between the rods. There was a receiving table at the bottom and a delivery table at the top. Powder charges rode up sideways in the pockets in the canvas slings. This appears to have been the preferred type where it could be used.

"C" – An *inclined type (with a round shaft).* A round shaft was drilled through the existing emplacement, a cylindrical steel tube with guide rails was inserted, and a mechanism somewhat similar to a Navy 'dredger' hoist was installed. This was essentially a type of 'conveyor' that transported powder charges from a receiving table at the bottom to a delivery table at the top. A powder charge was placed on the receiving table, oriented parallel to the axis of the shaft, and raised by a pusher plate that ran back and forth. The Type C powder hoist was subsequently considered very dangerous and was removed early (by 1917 in Boston).

Top of type "A" powder hoist built into a old balanced platform hoist well ,
Battery Halleck, Fort Hancock, New Jersey
(Photo by Thomas J. Vaughan, Jr.)

Top of a Taylor-Raymond hoist (right) and a type "A" powder hoist (left) (above)
(detail, below) in Battery Wilhelm, Fort Flagler, Puget Sound (photos by Thomas J. Vaughan, Jr.)

The top of a type "A" powder hoist, Battery Kinzie, Fort Worden, Puget Sound
(photo by Thomas J. Vaughan, Jr.)

Top of a type "B" powder hoist, Battery DeRussy, Fort Monroe, Hampton Roads
(Photo by Thomas J. Vaughan, Jr.)

Top of a type "C" powder hoist, Battery Hitchcock, Fort Strong, Boston
(Photo by Thomas J. Vaughan, Jr.)

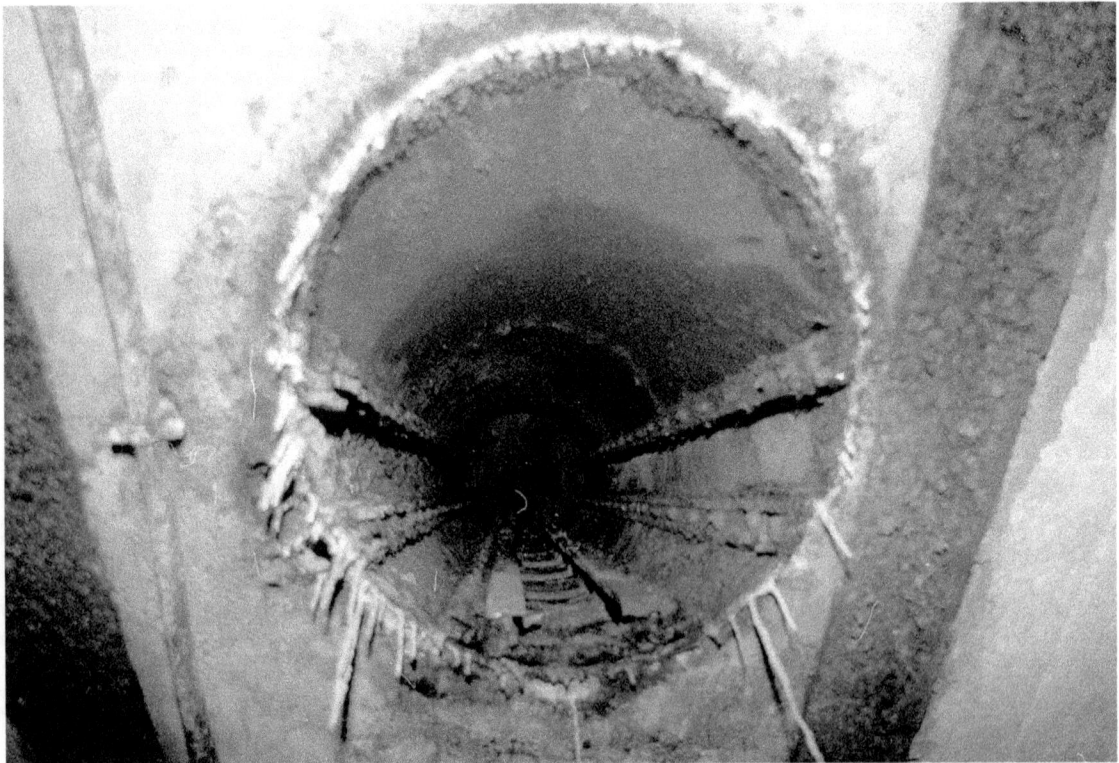

The steel-lined shaft of a type "C" powder hoist, Battery Steele, Fort Terry, Long Island Sound
(Photo by Thomas J. Vaughan, Jr.)

Over the service years of the batteries, ammunition hoists were altered, modified, and removed. That practice has continued to this day with hoist wells being filled in and sealed up for safety reasons, or buried altogether. Few of the ammunition trucks or hoisting mechanisms, either motor or hand-driven, remain today. Many batteries retain their overhead trolley rails.

The Taylor-Raymond hoist system was installed in most of the 10-inch and 12-inch batteries of the Endicott era, and a few 14-inch of the 1910s. The single shaft is usually located at the back of a traverse under an overhang. In most cases the hoists have been completely removed, but often the characteristic delivery table and sometimes the upper axle shaft will remain. The shaft leads down to the receiving room, often a rather spacious area with adaptations in the floor for the receiving table and hoist mechanism.

Batteries that had Hodges hoists tended to have narrower shaft wells and were located in 6-inch batteries designed previous to 1903, for the most part.

There are many seacoast artillery batteries that retain the twin wells characteristic of a balanced platform hoist, even after Taylor-Raymond hoists were installed.

Specific information on the hoisting mechanisms can be found in the Corps of Engineers Reports of Completed Works (RCW) Form 1 for a particular battery. Many of these of various dates from 1920-46 are now available through the CDSG Press in the Annual Conference notes.

Sources

Hansen, David M. " 'With nearly every problem solved,' the development of mechanical ammunition hoists in America's coastal fortifications." *CDSG Journal,* Vol. 12, Issue 4 (Nov. 1998), pp. 4-34.

Winslow, Eben, E. "Ammunition Supply, Service, and Storage," in Notes on *Seacoast Fortification Construction,* Occasional Papers, Engineer School, US Army, No. 61, Government Printing Office, Washington, DC, 1920, pp. 71-99.

Battery Dutton, Fort H.G. Wright, HD Long Island Sound.
Note the crane davits located to the rear of the loading platform and the delivery table for the hoist under the protected traverse to the left.

CONTROLLED MINES IN US ARMY SEACOAST DEFENSES
Mark Berhow

Throughout the modern or "concrete" era of American harbor defenses (1890-1950), mines were considered to be one of the primary harbor defense weapons, yet they have not received the historical attention they deserve. This may be due to the fact that the physical facilities for the mine defenses that remain at the old coast artillery forts were small in comparison with the gun defense structures. The minefields and the mine planters are long gone, and many of the shore facilities have also been destroyed or altered over the years. This article is an introduction to the US Army controlled mine defenses with special illustrated attention to the equipment and facilities used in those defenses until about 1940. It hopefully will serve as a guide for interpretation of the mine defenses and their remaining structures for the historians and visitors to former coast artillery forts.

The written material in this article will be relatively brief, and certain details will of necessity be left out. The mine equipment and procedures changed continually during this time period. This article will discuss the mines, facilities and procedures used during the period 1920-1940, unless specifically noted otherwise. For more complete information on the US Army mine defenses, the reader is encouraged to consult the source material listed at the end of this section. Some of these are currently available in reprint form.

A Brief History of US Army Controlled Mine Use

Submarine mines, or "torpedoes" as they were originally called, were an American invention. Used during the Revolutionary War, several improvements were made during the next 80 years. During the American Civil War (1861-65), mines were an important and effective part of the defense of the Confederacy.

In 1865, the U.S. Army Board of Engineers recommended the use of mines as a part of established harbor defenses. In 1869 the Engineer School was established at Willets Point, New York, and the development of a modern submarine mining system was assigned to that school. Col. Henry L. Abbot was assigned duty as commandant of that school, and over the next 20 years he perfected a submarine mining system for use by the Army. When the Endicott Board published its report on the state of US harbor defenses in 1886, the core of its proposal to revamp the defenses was the use of bouyant submarine mines. Abbot's basic system, with some revisions to the mechanical and electrical equipment, was used by the US Army from the 1880s through World War II.

The US Army began installing mine defense facilities in major US harbors in the 1880s. However, mines were not actually planted, except for the tests run by Abbot, until the Spanish American War in 1898 when mines were planted in 28 US harbors. The lessons learned from that effort resulted in the some changes in mine planting facilities, mostly making them a bit more "user friendly."

In 1901, the submarine mine program, still based out of Fort Totten at Willets Point, was transferred from the Corps of Engineers to the Artillery Corps. The Artillery Corps established its Submarine Mine School (moved to the Coast Artillery School at Fort Monroe, Virginia, in 1908) and its primary mine depot at Fort Totten. By 1903, the construction of mine defense facilities was underway at many U.S. harbors under the "Endicott-era" construction program. In 1904 the first four specially built mine planters were ordered by the Army. Four more planters joined these in 1909. The mine planters were manned by civilians and were intended to travel from harbor to harbor to perform their duties. Mine facilities were also built overseas in Panama, Hawaii, and the Philippines.

During 1911-1912 several organizational changes were made, but equipment shortages plagued the system. Things changed somewwhat with the entry of the US into the World War in Europe. Someharbors had mine fields laid in 1917-18. In 1917 orders were placed for 10 more mine planters and more mines. In addition, the troublesome civilian crew system on the planters was eliminated and the planters were subsequently manned by army personnel, supplemented by the local garrison with each change of station. However, this did not last long as spending cuts in the 1920s eliminated all but 5 of the older mine planters and all but 3 of the new planters. This "quiet era" continued into the 1930s.

In 1937 a new mine planter was ordered, and the mine defenses were brought up to war footing with the construction of new facilities at the harbors which were to continue to have mine defenses. With the build up of US military forces in 1940, the army ordered 16 new planters built to replace its aging fleet. New mines were designed and built, along with a few new facilities to maintain them. During the Second World War years M4 ground mines were planted at several major harbors around the continental United States and maintained there throughout the war.

In 1944, some of the new mine planters were transferred to the navy. In 1947, many of the remaining coast artillery commands were deactivated and the others quickly followed during the next two years. That year, it was decided to transfer all mine duties to the navy's harbor defense units. The navy would experiment with controlled harbor defense mines into the 1950s, when it also discontinued their use.

US Army Controlled Submarine Mines

Submarine mines were basically categorized as controlled or uncontrolled, and as either buoyant or ground. Uncontrolled mines, once laid, were detonated by contact with a passing vessel, friendly or unfriendly. Controlled mines were connected by cables to a control device on shore. Buoyant mines were those that floated and were restrained to an optimal depth in the channel by an anchor, while ground mines rested on the bottom. It should be noted here that for a short time after World War II the US Navy assumed control of former Army controlled minefields. All army mines were the controlled type; the navy "automatic" contact mines and other types of mines are not discussed in this article.

Definition of a Controlled Mine (TM 2160-20, 1930)—A controlled submarine mine, as employed by the United States Army, was a watertight steel case, containing explosive and a firing device, with a means for control of fire by electrical connection to shore. The case was of such size that it was held at a predetermined depth by the length of mooring rope that was attached to the anchor which retained it at the location at which planted.

The United States Army used controlled buoyant mines to defend American harbors during the period from the 1880s until early 1943. Some 4,000 controlled ground mines were used during the period 1943-1945.

Functions of Submarine Mines (TM 2160-20, 1930)

- To effect the destruction or serious damage of hostile vessels which approach within effective range.
- To supplement the offensive action of other weapons in repelling hostile attack.
- To prevent the close approach or entry into a harbor of hostile surface vessels under cover of night, fog, or smoke, when by reason of the invisibility of the ships from shore, other weapons are wholly or partially ineffective.
- To limit or prevent the navigation by hostile naval submarines of channels or water areas.
- To restrict the freedom of maneuver of hostile naval vessels in formation.
- By the morale effect of an unseen threat, to enforce a constant element of caution and uncertainty in the planning and execution of all hostile naval operations within the water areas known or believed to be protected by mine fields.
- To give warning of hostile submarine activities or of the presence of hostile surface vessels.

Controlled mines were used to close such portions of harbor entrances that led to channels required for the use of friendly naval forces and friendly commercial shipping, or included a debauching area required by naval forces. In addition, controlled mine fields were limited to distances not exceeding 10,000 yards from the shore and to depths not greater than 250 feet. The effective range was 2,000 yards from channel mouth inner limit, 8,000 yards from the outermost searchlight locations, with a minimum depth of 20 feet, and a maximum depth of 250 feet. The maximum current in the channel could not exceed 3 knots. The number of mines and their planting depth was determined by the locality to be defended.

Structures, Vessels and Equipment

For each defended harbor, one or more of each of the following was required:
- On-shore structures—
  - The mine casemate which housed the power and control equipment.
  - The mine (torpedo) storehouse for the storage of empty mines and equipment.
  - The loading room for assembling and loading the mines.
  - The magazine for the storage of explosives.
  - The cable tank for storage of cables under water.
  - The mine wharf for loading the mines on the planter.
  - Trackage (tramway) connecting the mine storehouse, magazine, cable tank, and mine wharf.
  - Mine group commander's station with plotting room
  - Base end stations.
- Mine planting vessels—including the regular and auxiliary mine planters, distribution box boats, and motor mine yawls.
- Position finding equipment—including the base end observing instruments, mine group commander's telescope, plotting board, predicting device, stop watch, telephones, and telephone lines.
- Searchlights—for the illumination of the mine field at night. These may or may not have been separately assigned to the mine command.
- Rapid-fire guns for protecting the mine field.

The War Department policy was that the controlled mines, along with all essential accessory items required to plant and operate them, were to be stored locally, contiguous to the projects they were to be used in, in readiness for immediate use and in quantities sufficient to plant the mine field authorized by the project and provide maintenance of these projects after planting. In practice, most authorized projects did not have all the materials needed to plant a complete mine field at the sites they were to defend.

COMPONENTS OF THE CONTROLLED MINE SYSTEM

Mines

Mines proper were composed of the casing, explosive (gun cotton or dynamite before 1912, trinitrotoluene (TNT) thereafter) compound plug with mine transformer, moorings, and anchors. The unit for planting, called a "group," consisted of 19 mines planted at intervals of 100 feet across the waterway to be defended, with submergence as required by tactical considerations. The mines in each group were numbered from left to right as viewed by an observer on the inshore side of the group. One group of mines thus defended 1,900 feet (18 intervals plus 50 feet to each flank). An appropriate number of mine groups to cover the length and breadth of the waterway provided the underwater defense. The groups of mines were numbered from left to right looking offshore, beginning with the most advanced line, if the project included two or more lines of mines. A spherical mine used during the years 1880 to 1943 is illustrated here, though there were a number of variations.

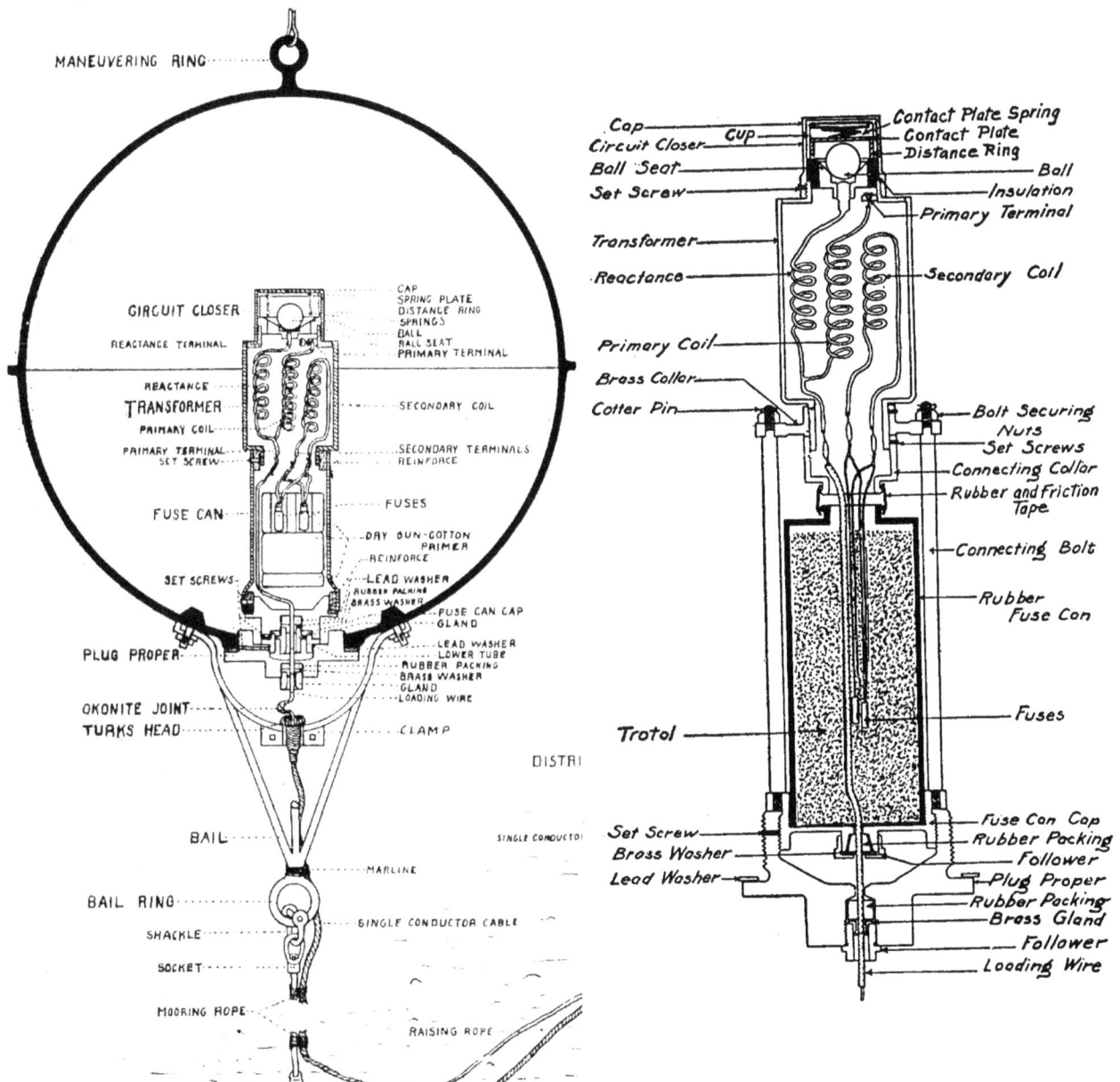

No. 32 buoyant controlled US Army mine (left) and the interior detail of a trotol (trinitrotoluene explosive) M1910 mine compound plug with rubber fuze can (right). (Illustrations adapted from TM 2160-20, 1930)

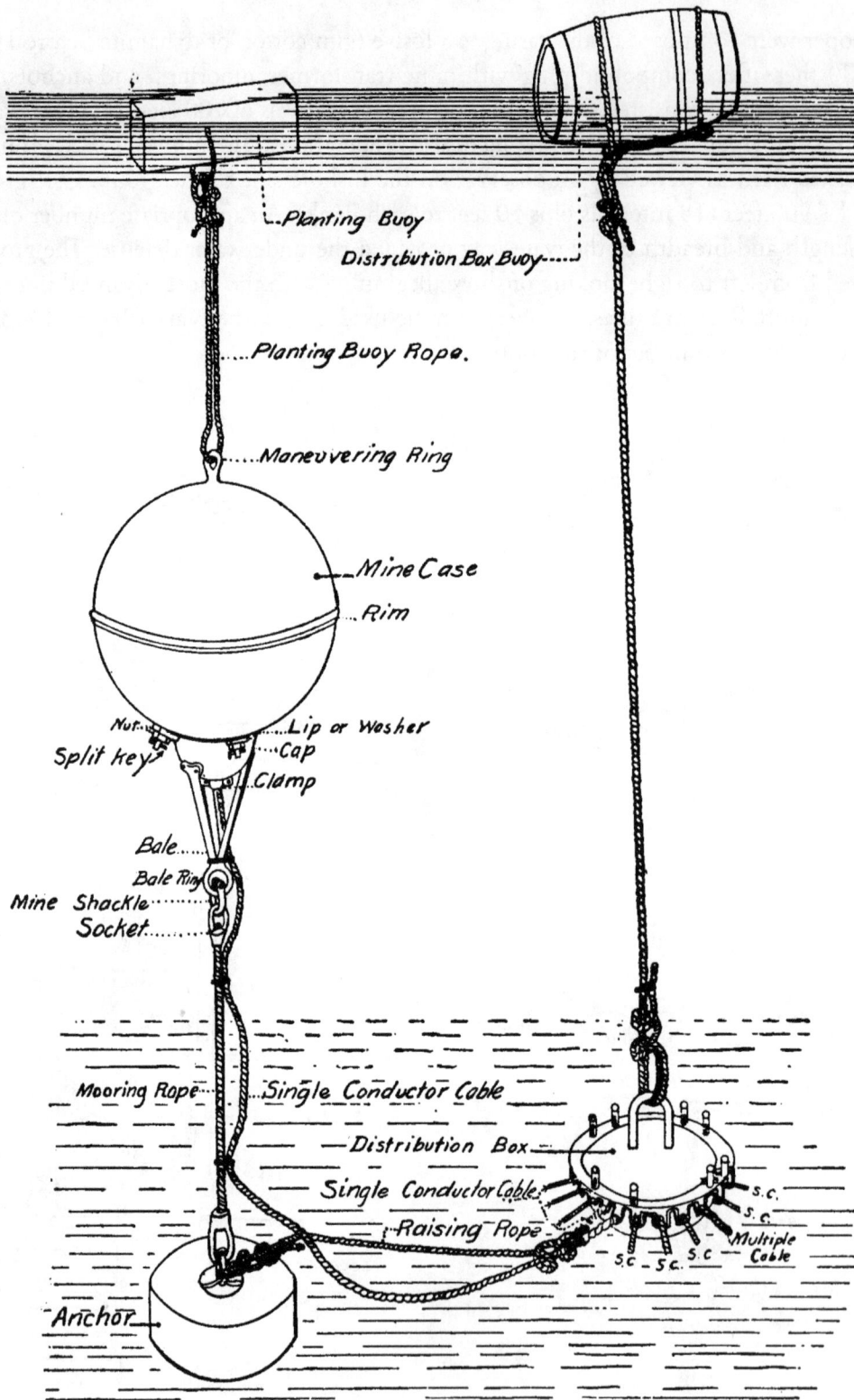

Planting Buoy

Distribution Box Buoy

Planting Buoy Rope.

Maneuvering Ring

Mine Case

Rim

Nut

Lip or Washer

Split key

Cap

Clamp

Bale

Bale Ring

Mine Shackle

Socket

Mooring Rope

Single Conductor Cable

Distribution Box

Single Conductor Cable

Raising Rope

S.C.

S.C.

S.C.

S.C.

S.C.

S.C.

Multiple Cable

Anchor

US. Army controlled mine and distribution box as deployed

The mines were delineated by a numerical designation indicating displacement capacity. During this period there were two basic mine shapes: the spherical mine and the sphero-cylindrical mine.

Size, Weight, and Buoyancy of Mine Cases (TM 2160-20, 1930)

Mine case number, shape	length spherical portion (inches)	length cylindrical portion (inches)	weight empty (pounds)	cap and plug (pounds)	TNT charge (pounds)	assembled loaded mine (pounds)	displacement sea (pounds)	bouyant effort
32 (spherical)	32	0	241	71	100	412	635	223
40 (spherical)	40	0	401	71	200	762	1,241	479
40 (cylindrical)	40	20.4	550	71	200	821	1,242	421
42 (cylindrical)	32	26.9	643	71	200	914	1.435	521
42 (cylindrical corr.)	32	26.9	496	71	200	767	1,480	713
43 (cylindrical)	32	30.4	693	71	200	964	1,540	576
46 (cylindrical)	32	42	859	71	200	1,130	1,886	756
48 (cylindrical)	32	50.6	982	71	200	1.253	2,143	890
48 (cylindrical corr.)	32	50.6	711	71	200	982	2,239	1,257
50 (cylindrical)	40	25.5	946	71	200	1,217	2,428	1,211

Mines at Boston harbor, 1920s (NA; Al Schroeder Collection, MHI)

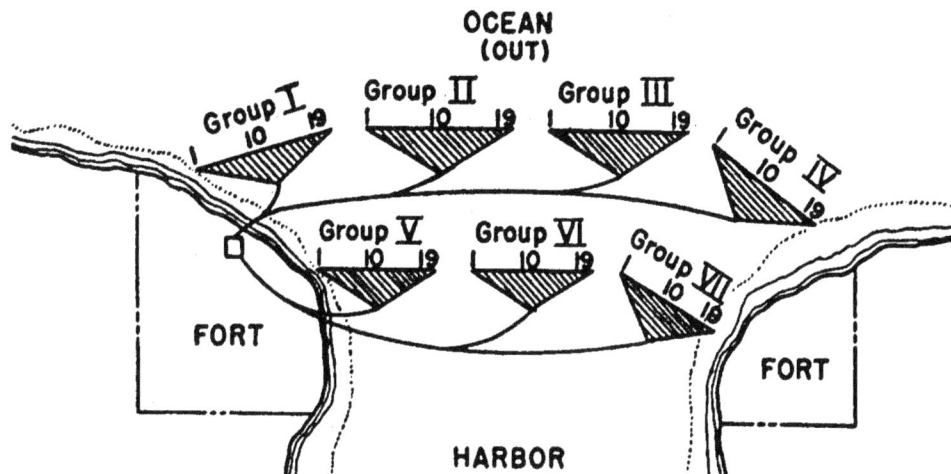

Mine group numbering system

The Grand Mine Group

The basic "grand" mine group arrangement for controlled submarine mine fields was developed by Col. Henry Abbot in the 1880s. In the 1880s the largest insulated cable that could be manufactured in quantity contained 7 single conductors. Abbot's grand group consisted of 7 groups, of three mines, each group connected to a junction box. The mines were deployed about 100 yards apart. In the junction box, the three wires of individual mines were connected and a single conductor cable was connected to one of the 7 slots in the distribution box. The 7-conductor cable from the distribution box was then connected to a cable terminal vault on the shore, which was in turn connected to a panel in the mine casemate. Activation from shore of one of the seven units of a group resulted in the activation (or detonation) of all three mines hooked to the activated conductor.

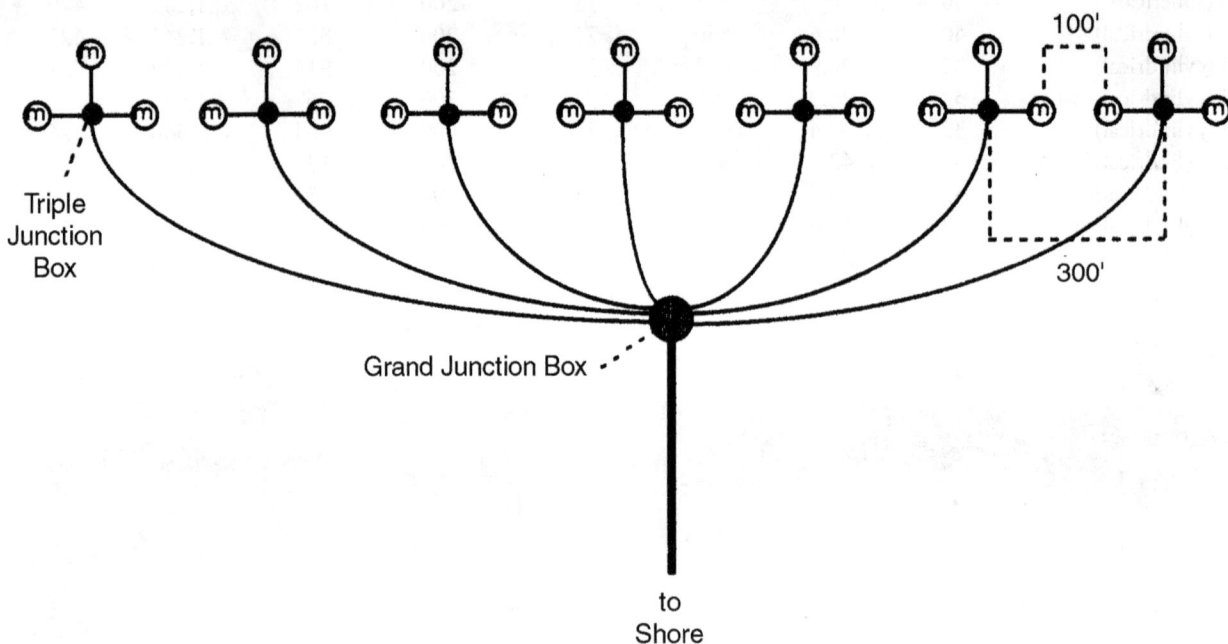

Abbott "Grand" Mine Group (1874)

Later, as cable with more conductors became available, the junction box was eliminated and the grand group consisted of 19 mines individually attached to a distribution box. A 20-conductor cable (the 20th cable was used for telephone communication between the ship and the mine casemate) then connected the distribution box to the cable vault and the mine casemate on the shore (see next page for illustration). By the late 1932, the 20-conductor cable for each grand group had been replaced with a single-conductor cable. The 19 mine group was controlled by two 20-position telephone-type selector switches, one at each end of the cable (at the casemate and in the distribution box) allowing for the selective control of each mine in the grand group.

The number of grand groups deployed depended on the width and length of the channel being defended. Abbot's protocol called for a minimum of three lines of defense, but this may not have always followed. During peacetime, the mines were deployed only for practice, then retrieved, unloaded, and stored ashore. Mines were only actually deployed in defenses for long periods during the Spanish-American War and World War II. When deployed, mines had to be periodically retrieved to repair damage caused by being at the mercy of the elements, tides, currents and passing ships.

100'

Distribution Box

Shoreline

No. 1
60
Searchlight
Position

Cables

M' M'
P P

Double Mine
Primary Base End
Stations and
Mine Plotting Rooms

Cable Terminal Hut

CRF

Rapid-Fire
Gun Battery
(Coincidence
Range Finder
Station)

M" M"

Mine
Casemate

Double Mine
Secondary Base End
Stations

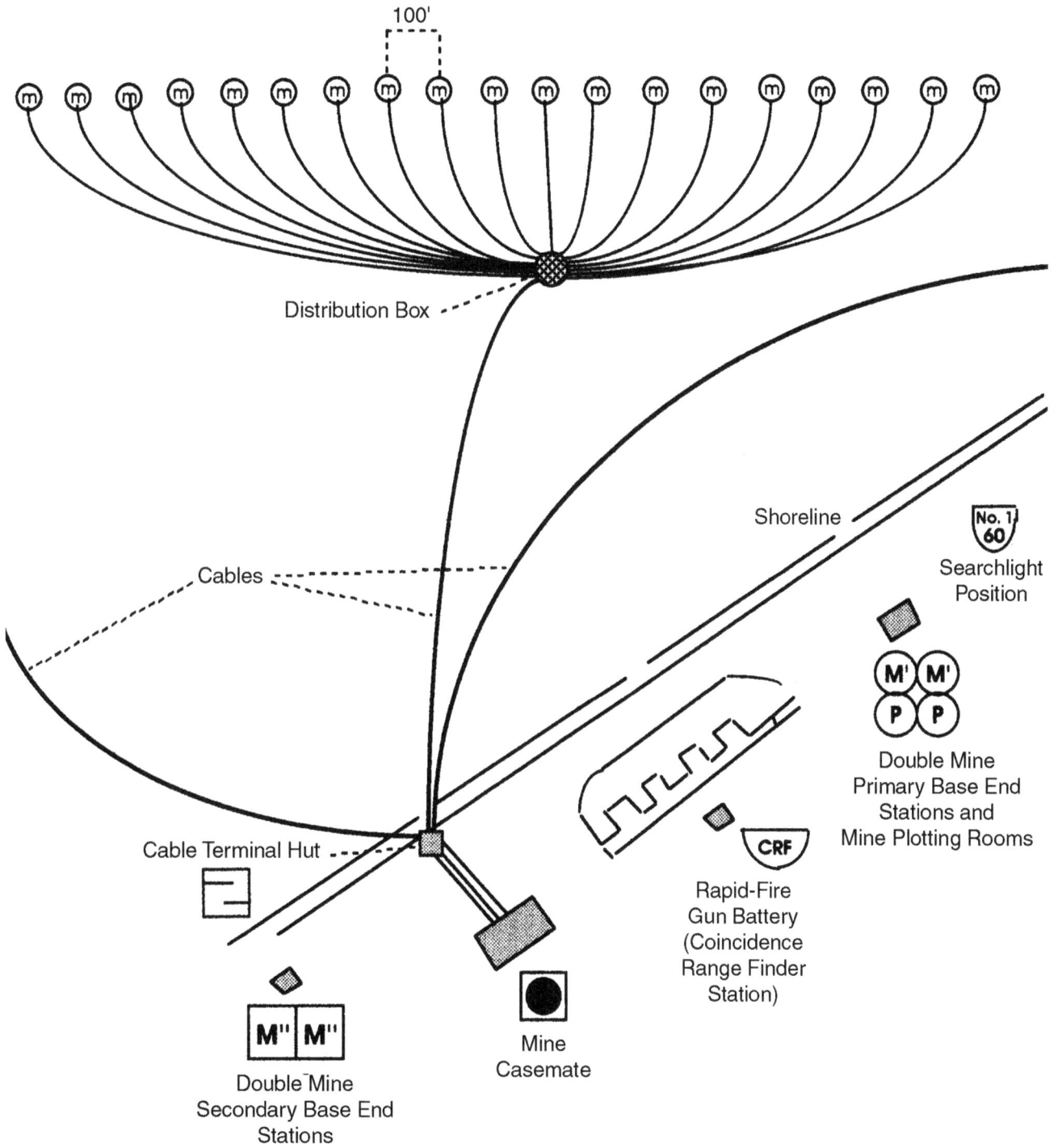

The "Grand" Mine Group circa 1930-1943
with shore fire control facilities (with map symbols for the shore structures)

Control Devices—The 1930 single-conductor system consisted of 19 mines, 19 lengths of single-conductor cable leading from the group back to the distribution box which contained the selector and accessories, the single conductor cable leading from the distribution box to the terminal hut on shore, the lead-covered cable running in a conduit from the terminal hut to the mining casemate, and the instruments and wiring included on the operating board. In the older 19-conductor system, an operating board with its master block and 19 individual mine blocks was used to control each mine in the group.

Power and Switchboard Equipment—The power system included a direct current (DC) generator (either directly connected to a stationary gasoline engine, or belt-driven by an oil engine), an 80-volt storage battery, motor-generating sets to provide alternating current (AC) for firing, a casemate transformer, a power panel to control direct and alternating current, and an interupter panel for the single conductor system.

Army controlled mines used direct current for operation, supervision, and signaling, and used alternating current for all firing. When a mine was fired, either by observation or contact, alternating current was sent though the selectors to the mine. Without the alternating current, the mines could not be detonated.

Methods of fire of the army controlled mines (in order of tactical consideration):

Delayed-Contact Fire—The mines were activated and upon contact with a vessel, the mine was manually detonated after a set (hand-timed) delay. This was to insure an explosion directly under the bottom of the ship.

Contact Fire—The mines were activated to detonate upon contact. This was only to be used if delayed-contact fire could not be used.

Observation Fire—The mines were fired by command from shore. The target would be tracked, and when located directly over the position of a mine, that mine would be detonated by signal from the casemate. This was used only when conditions would not allow for contact fire.

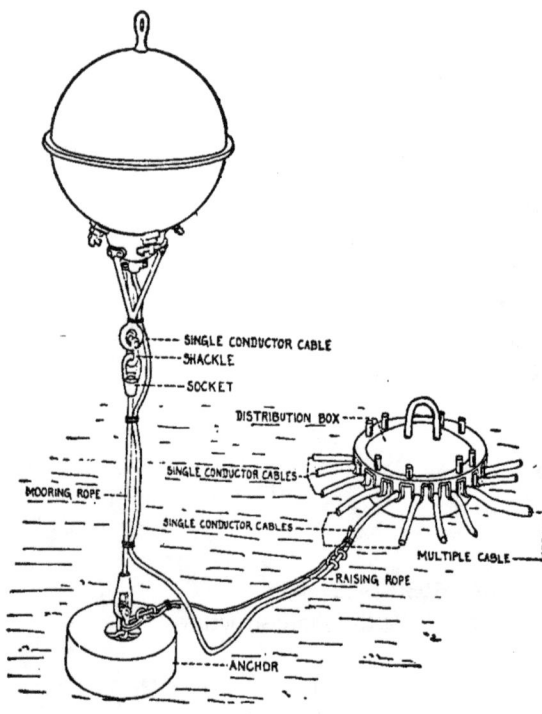

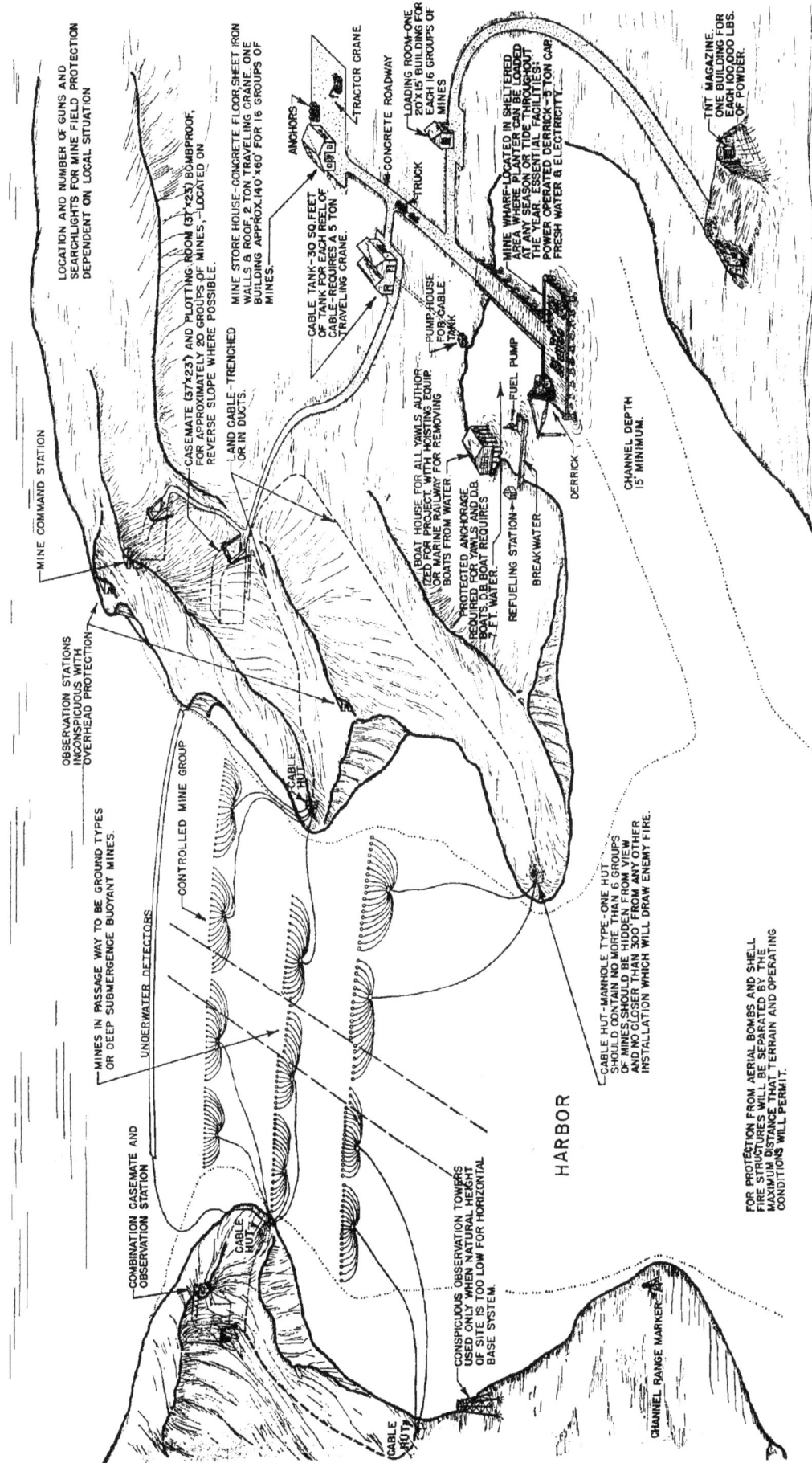

LOCATION AND NUMBER OF GUNS AND SEARCHLIGHTS FOR MINE FIELD PROTECTION DEPENDENT ON LOCAL SITUATION

TRACTOR CRANE

ANCHORS

CONCRETE ROADWAY

TRUCK

LOADING ROOM—ONE 20'x15' BUILDING FOR EACH 16 GROUPS OF MINES

CASEMATE (37'x23') AND PLOTTING ROOM (37'x23') BOMBPROOF, FOR APPROXIMATELY 20 GROUPS OF MINES,—LOCATED ON REVERSE SLOPE WHERE POSSIBLE.

MINE STORE HOUSE—CONCRETE FLOOR, SHEET IRON WALLS & ROOF, 2 TON TRAVELING CRANE. ONE BUILDING APPROX. 140'x60' FOR 16 GROUPS OF MINES.

CABLE TANK—30 SQ. FEET OF TANK FOR EACH REEL OF CABLE—REQUIRES A 5 TON TRAVELING CRANE.

LAND CABLE—TRENCHED OR IN DUCTS.

PUMP HOUSE FOR CABLE TANK

MINE COMMAND STATION

MINE WHARF—LOCATED IN SHELTERED AREA WHERE PLANTER CAN BE LOADED THE ANCHOR, OR TIDE, THROUGHOUT THE YEAR ESSENTIAL FACILITIES: POWER OPERATED DERRICK—5 TON CAP FRESH WATER & ELECTRICITY.

TNT MAGAZINE. ONE BUILDING FOR EACH 100,000 LBS. OF POWDER.

FUEL PUMP

DERRICK

BOAT HOUSE FOR ALL YAWLS AUTHOR-IZED FOR PROJECT WITH HOISTING EQUIP OR MARINE RAILWAY FOR REMOVING BOATS FROM WATER

PROTECTED ANCHORAGE REQUIRED FOR YAWLS AND D.B. BOATS. D.B. BOAT REQUIRES 7 F.T. WATER.

REFUELING STATION

BREAKWATER

CHANNEL DEPTH 15 MINIMUM.

OBSERVATION STATIONS INCONSPICUOUS WITH OVERHEAD PROTECTION

MINES IN PASSAGE WAY TO BE GROUND TYPES OR DEEP SUBMERGENCE BUOYANT MINES.

UNDERWATER DETECTORS

CONTROLLED MINE GROUP

COMBINATION CASEMATE AND OBSERVATION STATION

CABLE HUT

CABLE HUT

CABLE HUT—MANHOLE TYPE—ONE HUT SHOULD CONTAIN NO MORE THAN 6 GROUPS OF MINES, SHOULD BE HIDDEN FROM VIEW AND NO CLOSER THAN 300' FROM ANY OTHER INSTALLATION WHICH WILL DRAW ENEMY FIRE.

HARBOR

CONSPICUOUS OBSERVATION TOWERS USED ONLY WHEN NATURAL HEIGHT OF SITE IS TOO LOW FOR HORIZONTAL BASE SYSTEM.

CABLE HUT

CHANNEL RANGE MARKER

FOR PROTECTION FROM AERIAL BOMBS AND SHELL FIRE STRUCTURES WILL BE SEPARATED BY THE MAXIMUM DISTANCE THAT TERRAIN AND OPERATING CONDITIONS WILL PERMIT.

Illustration from FM 4-6, 1942

The Mine Flotilla: Mine Planters, DB Boats, and Yawls

The mine flotilla consisted of a mine planter, a distribution box boat (DB boat, also known as a L-boat after the numerical designation on the side of the boat), and two to four yawls. There were rarely enough mine planters for one to be assigned to each harbor that had mines. Often one or two mine planters were assigned to a major harbor. They would be called on to perform service in other harbors or in the laying and maintenance of the military underwater cable systems. Smaller harbors without assigned mine planters would be visited by a planter for practice, and during wartime after it had performed its duty at its primary station. Many harbor defenses used vessels that were jury-rigged to serve as makeshift mine planters and DB boats due to the lack of a sufficient number of actual army mine planters.

Army mine planters were custom built after 1904. They had wide decks to hold assembled mines, and the necessary booms and davits for loading the assembled mines from the wharf and planting the mines out in the channel. The four mine planters ordered in 1904 were 150 ft. in length, 32 ft. beam, 14 ft. draft and 447 tons in displacement The 1940 series of mine planters were 188 ft. in length, 37 ft. beam, 12 ft. 6 in. draft and displaced 1,320 tons. The ships had a wartime crew of two officers, 6 warrant officers, and 41 men. Usually the fleet would plant one group of mines at a time.

Distribution box boats were smaller boats that held the distribution box while it was being attached to the cables as the mines were being planted. The DB boat had a large boom to hoist an assembled distribution box over the bow and lower it into the water.

The mine yawls were typical small craft which were used to ferry ropes, cables, etc., from the mine planter to the DB boat or shore as needed.

(left) US Army Mine Planter Abbott in Boston Harbor, 1940s.
(above) US Army Mine Planter Baird at Fort Warren, HD Boston, 1940s (NA, Al Schroeder Collection, MHI, Carlisle)

Mine flotilla: detail of a mine planter dropping a mine (top), a distribution box boat (L-boat) (middle), and a mine yawl (M-boat) (bottom). Photos taken in the Boston harbor area circa 1938 (William Hale, Lynn, MA).

US Army Mine Planter General Franklin J. Bell, Columbia River 1936
(Edwin Bartcher; Marshall Hanft Collection)

US Army Mine Planters, DB Boats, and Yawls

Dates indicate the years the vessels were in mine planting service. A number of the 1942 class are still afloat.

(MP-Mine Planter, MPB- Mine Planter Battery)

Name	Years of Service	Name	Years of Service
Class of 1904		Class of 1937	
General Henry J. Hunt	1904-24	Colonel Ellery J. Niles (MPB 4)	1937-65
General Henry J. Knox	1904-24		
Colonel George Armistead	1904-35	Class of 1942	
Major Samuel J. Ringgold	1904-22	MP-1 General Henry Knox (MPB 8)	1942-46
		MP-2 General Henry J. Hunt (MPB 9)	1942-45
Class of 1909		MP-3 Col. George Armistead (MPB 10)	1942-45
General John M. Schofield (MPB 6)	1909-47	MP-4 General Samuel J. Mills (MPB 11)	1942-58?
General Royal T. Frank (MPB 1)	1909-22	MP-5 Lt. William J. Sylvester (MPB 12)	1942-45
General E.O.C. Ord (MPB 6)	1909-46	MP-6 Brig. Gen. Henry L. Abbott (MPB 13)	1942-58?
General Samual M. Mills	1909-21	MP-7 Maj. Gen. William P. Randolph (MPB 14)	1942-51
		MP-8 Col. John P. Story (MPB 15)	1942-44
Class of 1917		MP-9 Maj. Gen. Arthur Murray (MPB 16)	1942-45
General William M. Graham (MPB 3)	1917-46	MP-10 Maj. Gen. Erasmus Weaver (MPB 17)	1942-49
		MP-11 Col. Samuel Ringgold (MPB 18)	1942-51
Cable Ships 1918		MP-12 General Royal T. Frank (MPB 19)	1942-49
Joseph Henry (built 1909)		MP-13 Col. Alfred A. Maybach (MPB 20)	1942-51
Cyrus W. Field (built 1901)		MP-14 Col. Horace F. Spurgeon (MPB 21)	1943-49
		MP-15 Col. Charles W. Bundy (MPB 22)	1943-44
Class of 1918		MP-16 Col. George W. Ricker (MPB 23)	1943-44
General Absalom Baird (MPB 2)	1919-46		
General J. Franklin Bell (MPB 7)	1919-46	U.S. Junior Mine Planters (10)	
General William M. Graham	1919-41	Distribution box boats L-1 through L-112	
Colonel George F.E. Harrison	1919-42	Motor mine yawls M-1 to M-489	
General Edmund Kirby	1919-21	Several other Auxilliary Vessels over 50 gross tons	
General William P. Randolph	1919-24		
General John P. Story	1919-24	Charles D. Gibson, "The Army's Mine Planter Service" *Journal of*	
Colonel Albert Todd	1919-21	*America's Military Past* Vol. 28, No. 3, pp 52-71.	
Colonel Garland N. Whistler	1919-21	Clay, Steven E. *U.S. Army Order of Battle 1919-1942*, Combat Studies	
Colonel John V. White	1919-21	Institute Press, Ft. Leavenworth, KS 2010. Vol. 2, Chap. 27	
		Army Mine Planters and Cable Ships.	

Mine Planting

This is a very short summary of the actions taken in planting a mine group, circa 1930, using the single conductor cable, selector switches, and the 19-mine group. Many of the details, especially those involved in the circuit testing, will not be discussed. See the following pages for illustrations. For more complete information on this process refer to the 1930 submarine mining manual.

The channel where the mines were to be planted was surveyed and the approximate locations of distribution boxes were determined. A group could be completely laid and planted in one day, or parts could be laid out in advance. For instance, the locations of the distribution box and the positions of some, usually mines 4, 7, 10, 13 & 16, or all of the individual mines could be marked with anchors and buoys before the actual laying of the mine field.

For the actual mine planting, the mine planter and DB boat were loaded at the mine wharf with the equipment to plant a complete group of mines. The personnel at the mine casemate tested all the DC and AC power equipment and internal circuit connections. The DB boat headed to the location of the distribution box for the mine group being planted. The planter, meanwhile, proceeded to a point near the mine casemate, and the group shore cable was connected to the mine casemate via the cable terminal hut. The planter then proceeded toward the DB location playing out the shore cable as it proceeded. If necessary, sections of cable could be joined with the assistance of the DB boat using special junction boxes.

The DB boat either moored to an existing anchor at the location of the distribution box, or deployed an anchor and then moored. Once the DB boat was moored to the DB anchor, the planter approached and transferred the end of the main cable to the DB boat. Using a telephone, the DB boat crew established communications with the mine casemate via the shore cable. The cable end was attached to the distribution box, but not connected. The crews were now ready to plant the mines of the group.

The planter then headed to the location of the number 10 mine, in the center of the group, some several hundred yards ahead of the location of the distribution box. Soundings had been taken before planting to determine the exact depth so the mines could be rigged at the depot to float at the correct depth. The mine, anchor, and cable were prepared for planting: the cable end was attached and connected to the mine, the mooring rope, raising rope, and buoy were attached to the mine and anchor, and the mine attached to the triplex block. The mine and anchor were then swung over the side of the planter. The heaving lines were attached to the DB end of the cable and the heaving lines passed to the DB boat directly or via a yawl. At the command "let go," the mine, anchor and cable were dropped into the channel. The cable end was brought to the DB boat with the heaving lines and the cable connected and secured to the appropriate switch selector in the distribution box aboard the DB boat.

The planter repeated this procedure for mines No. 9 through No. 1 in a straight line using mine yawls attached to the buoys for line of sight alignment. Then mines No. 11 through 19 were planted. Upon the completion of the planting, the mine planter and the yawls cleared away from the area of the mine group. On the DB boat, after receiving the "all clear" signal from the planter and the mine casemate, the telephone was disconnected from the shore cable. The shore cable was then connected to the selector for a series of tests run by the mine casemate, including a dry test of the mine circuit, an operating board test of the selector circuit, and a manual operating board test. After a set period of time to conduct the tests, the cable was disconnected from the selector and reconnected to the telephone to check for the results of the tests. If the okay was given, the shore cable was reconnected to the selector, the distribution box secured, and then lowered by the DB boat into the channel. This procedure was repeated for each group to be planted to complete the authorized project minefield.

Each mine's location was then determined by taking sightings on each mine. A small boat crew would hold a broom stick over the mine buoy, the mine base stations would sight on the stick, and the plotting room would plot the location on a master map. In action, this map would be placed on the plotting board to determine which mines would be fired as the target was tracked.

When the field was retrieved, the groups were taken up in essentially the reverse order in which they were laid—the distribution box was retrieved first, and the shore cable disconnected from the selector. Then the cable to the No. 19 mine was disconnected and taken up by the planter as it approached the location of the mine. The retrieving rope was grappled and the mine and anchor brought up and disconnected, etc., for all the mines in the group. Finally, the planter took up the shore cable, and returned to the wharf to unload.

A MINE PLANTER READY TO L
ONE GROUP.

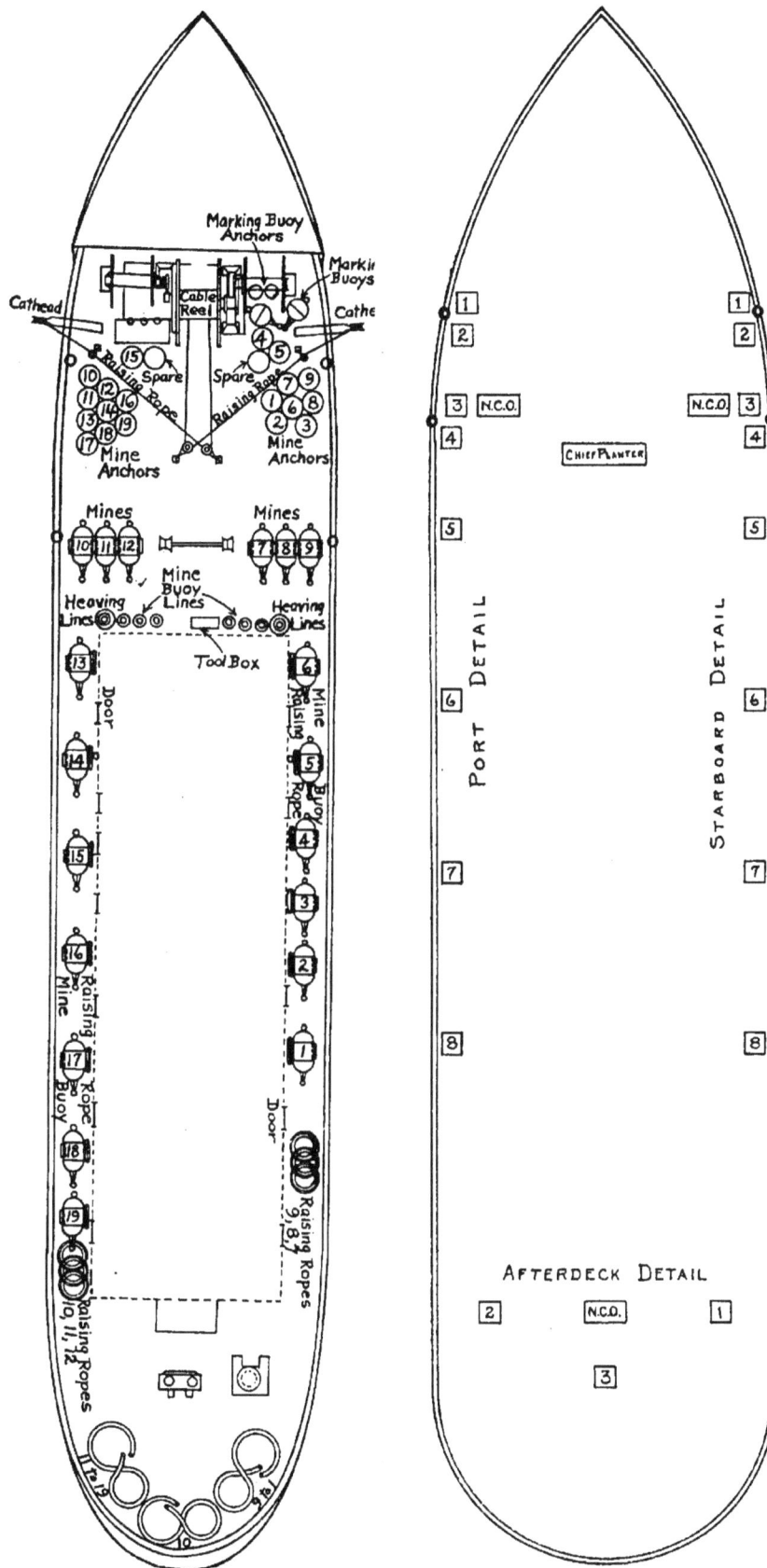

(Left) Arrangement of material on deck of a 172-foot mine planter, (Right) Posts of mine planting details.
Planting jobs for starboard and port details— [1] & [2] anchor davit men; [3] & [4] mine davet men;
[5] heaving line men; [6], [7] & [8] cable men. Aft Deck Detail—[1], [2] & [3] cable men.
(from TM 2160-20, 1930)

PLAN

Mine Planter

Bridge

No. 8 Anchor on Davit

← Tide

Line of Sight

2000# Anchor

Mine Cable #9

2000# Anchor

D.B. Mooring Rope

Mine Cable #10

Mine Cable #12

Mine Yawl

Anchor No. 9 Mine

← Tide

Mine Cable

Mine Cable

Mine Cable

Mine Yawl

Anchor No. 10 Mine

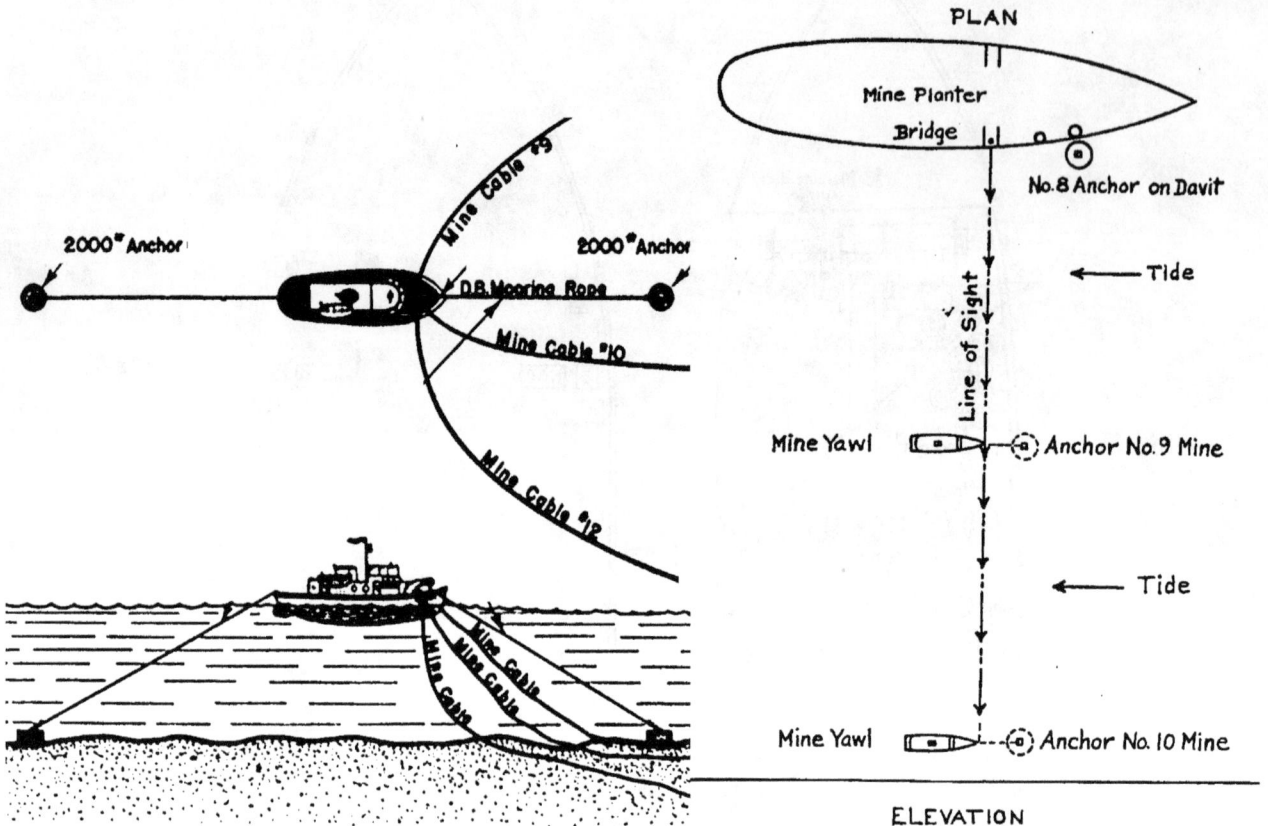

Proper distribution boat anchoring (upper left), the proper method for aligning mines (upper right) and mine planter approach patterns (below).
From FM 4-6, 1942.

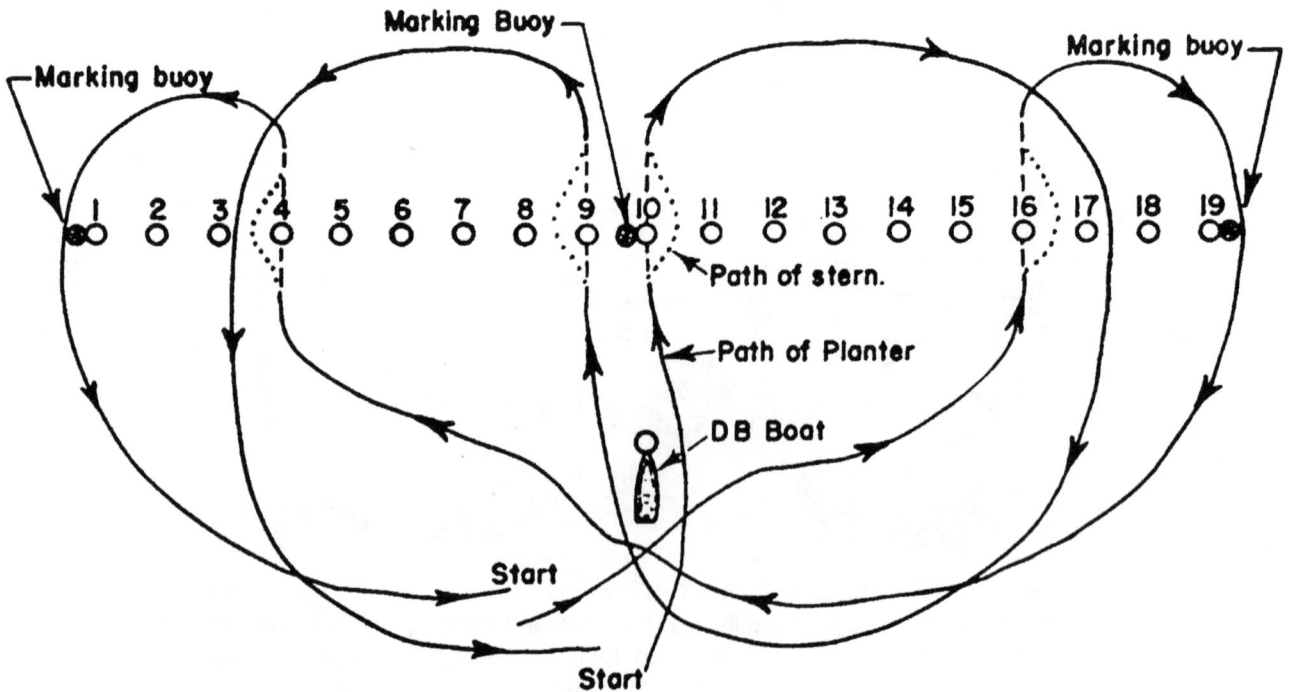

ELEVATION

Bridge ⊔

No. 8 Anchor

← Tide

Tide →

Marking buoy

Marking Buoy

Marking buoy

1 2 3 4 5 6 7 8 9 10 11 12 13 14 15 16 17 18 19

Path of stern.

Path of Planter

DB Boat

Start

Start

AFT DAVIT
FORWARD DAVIT
DIFFERENTIAL PULLEY
(TRIPLEX BLOCK)
MINE BUOY
TRIPPING HOOK
ANCHOR
MOORING ROPE
RAISING ROPE

Figure "8" aft deck
DB end of cable –135'
Heaving line
Aft detail
Raising rope
Mine end of cable

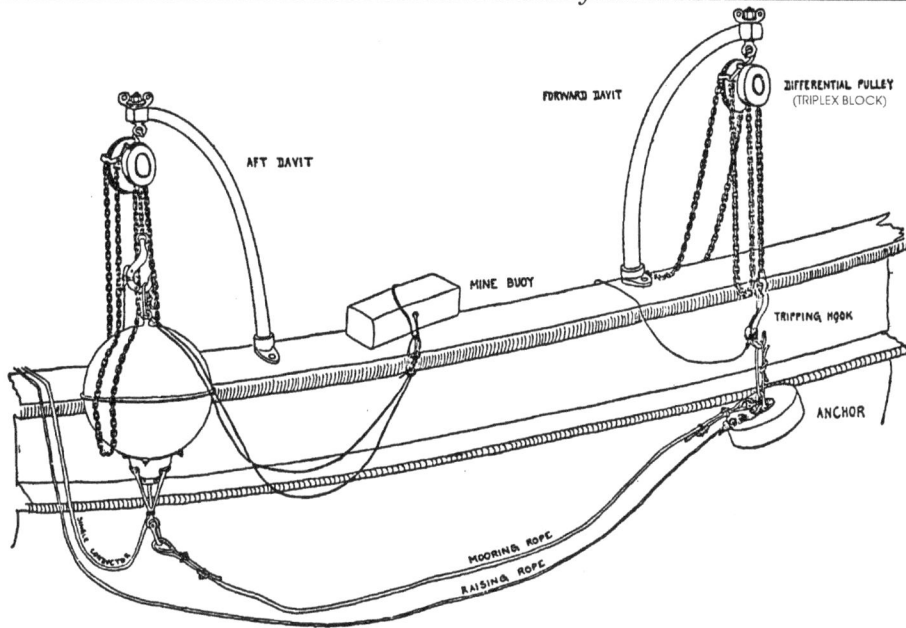

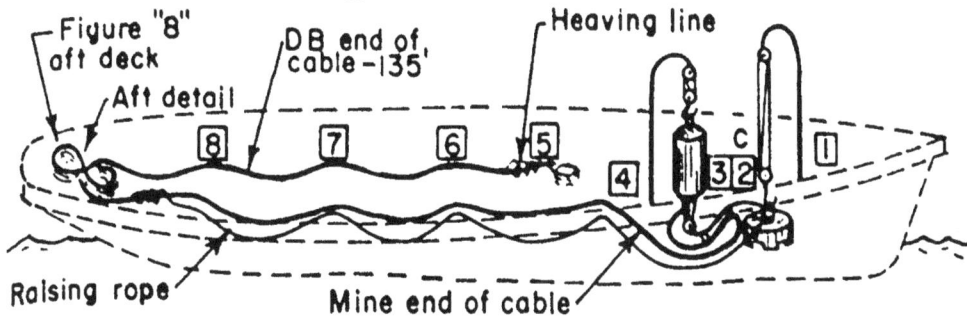

① At instant of passing DB-boat.

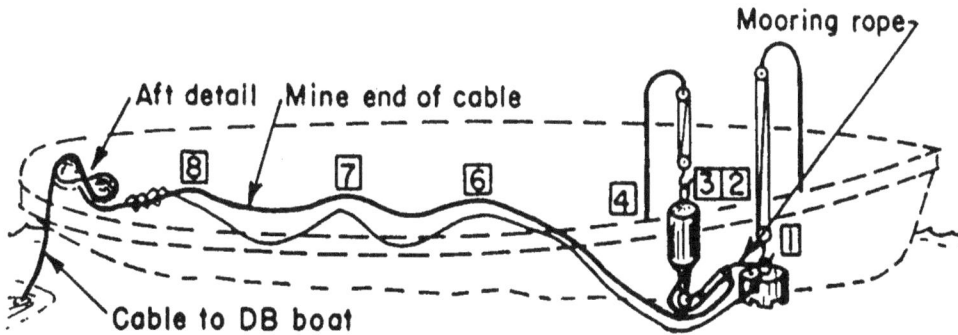

Mooring rope
Aft detail   Mine end of cable
Cable to DB boat

② Between DB-boat and line of mines.

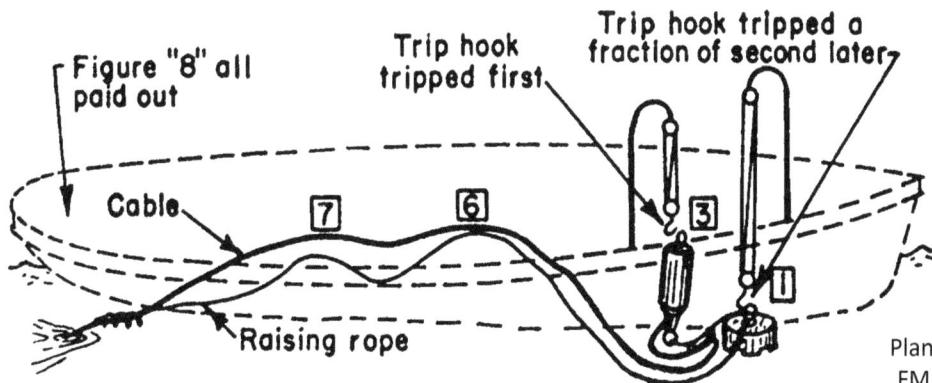

Figure "8" all paid out
Trip hook tripped first
Trip hook tripped a fraction of second later
Cable
Raising rope

Planting mines
FM 4-6, 1942

③ Instant after "let go."

MINE BUOY, WOOD REMOVE BUOYS AFTER ENTIRE FIELD IS PLANTED AND TESTED.

BOWLINE KNOT

BOWLINE KNOT

MEASURING LINE 5/8" MANILA ROPE OR EQUIVALENT.

SUBMERGENCE MARKS

CYLINDRICAL MINE CASE

CLAMP SECURES TURKSHEAD TO MINE CASE CAP

SINGLE CONDUCTOR MINE CABLE.

SHACKLE
THIMBLE
12" MIN. CLIP 30"
6"

5' MINIMUM

MOORING ROPE

CLAMP

MIN. 5', MAX 10', DEPENDING ON LOCAL HYDROGRAPHY

12" MIN CLIP

THIMBLE

SHACKLE, THIMBLE & CLIPS

SHACKLE
RING
SHACKLE
2000 LB ANCHOR

TO DISTRIBUTION BOX →

RAISING ROPE

DISTRIBUTION BOX BUOY

REMOVE BUOYS AFTER ENTIRE FIELD IS PLANTED AND TESTED.

BOWLINE ON STANDING LINE

BOWLINE ON RUNNING LINE

MANILA LINE OR WIRE ROPE. LENGTH = DEPTH AT HIGH WATER PLUS ALLOWANCE FOR LOCAL CURRENT CONDITIONS.

200' 200' 200'

D.B. RAISING ROPE

DISTRIBUTION BOX BOAT ANCHOR, 2000 LBS. OR HEAVIER, WITH MOORING ROPE, LENGTH OF WHICH DEPENDS ON LOCAL CONDITIONS. NORMALLY DEPTH OF WATER PLUS 100 FEET.

RAISING ROPE.

DISTRIBUTION BOX

SHORE CABLE

SHACKLE
RING
SHACKLE

SINGLE CONDUCTOR MINE CABLES.

Tide at time soundings are taken.

Mean low water line.

S = Submergence.

L = Length to cut mooring rope.

D = Depth of water at mean low water.

C = Length of mine case.

E = Length of mooring rope ends for attaching clips = 2 ft.

$$L = D + 2E - S - C \quad \text{(In feet)}$$

NOTE:

    For rocky or hard sand bottoms deduct height of anchor.

    For soft sand or mud bottom make no allowance for height of anchor, which usually is buried.

S

C

E 2 ft.

D

L

E 2 ft.

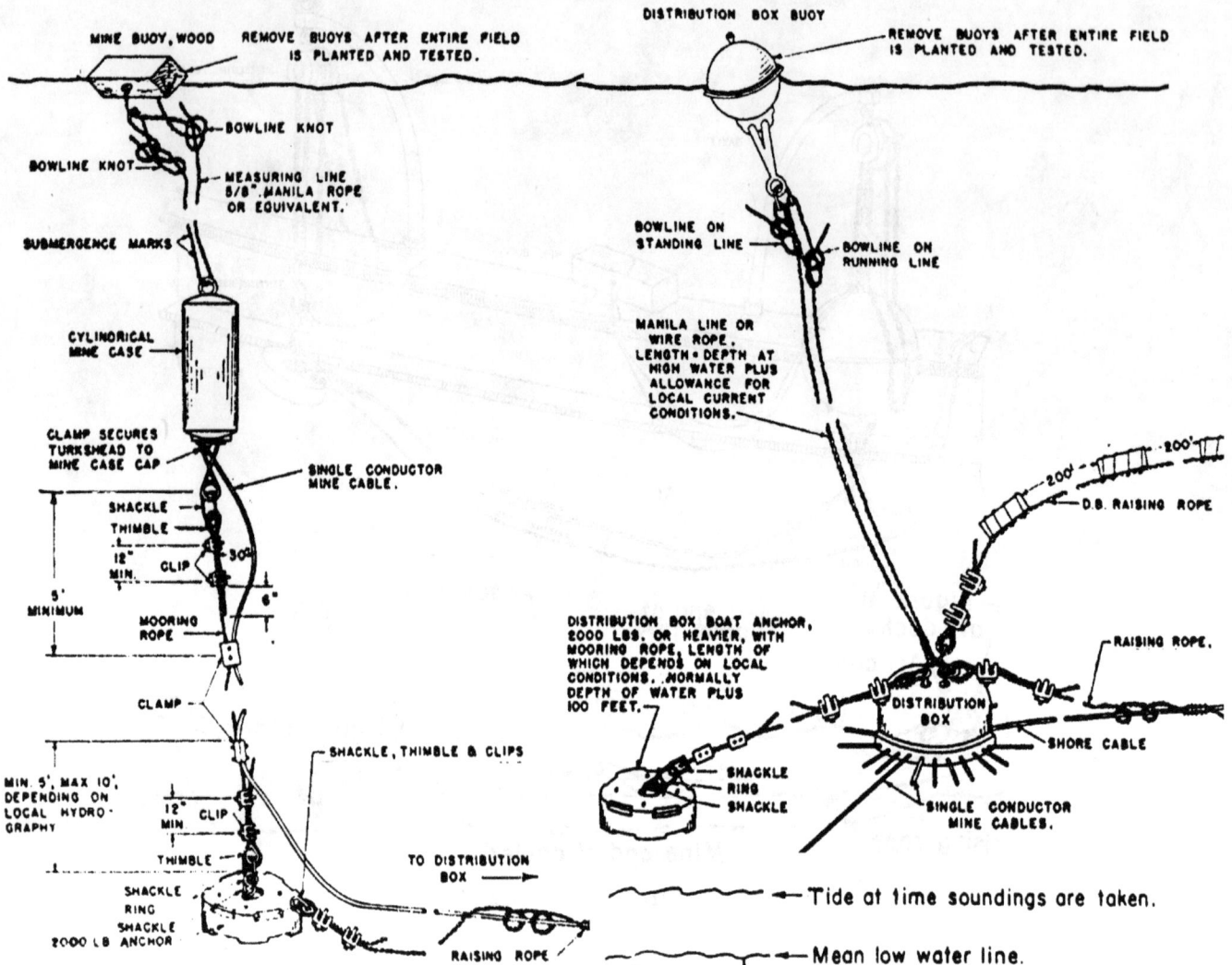

Planted mine (upper left), planted distribution box (upper right) and determination of submergence depth (below). Illustrations from FM 4-6, 1942.

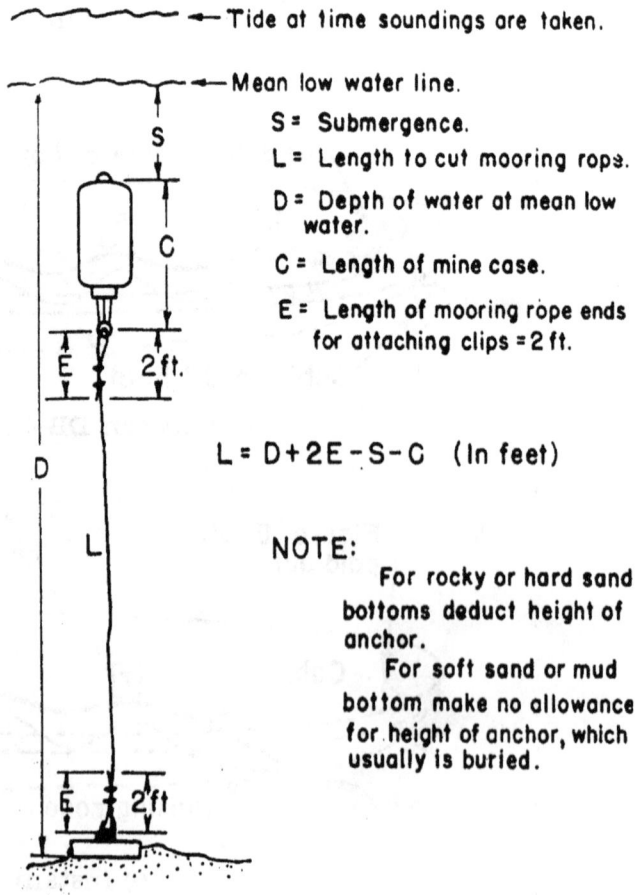

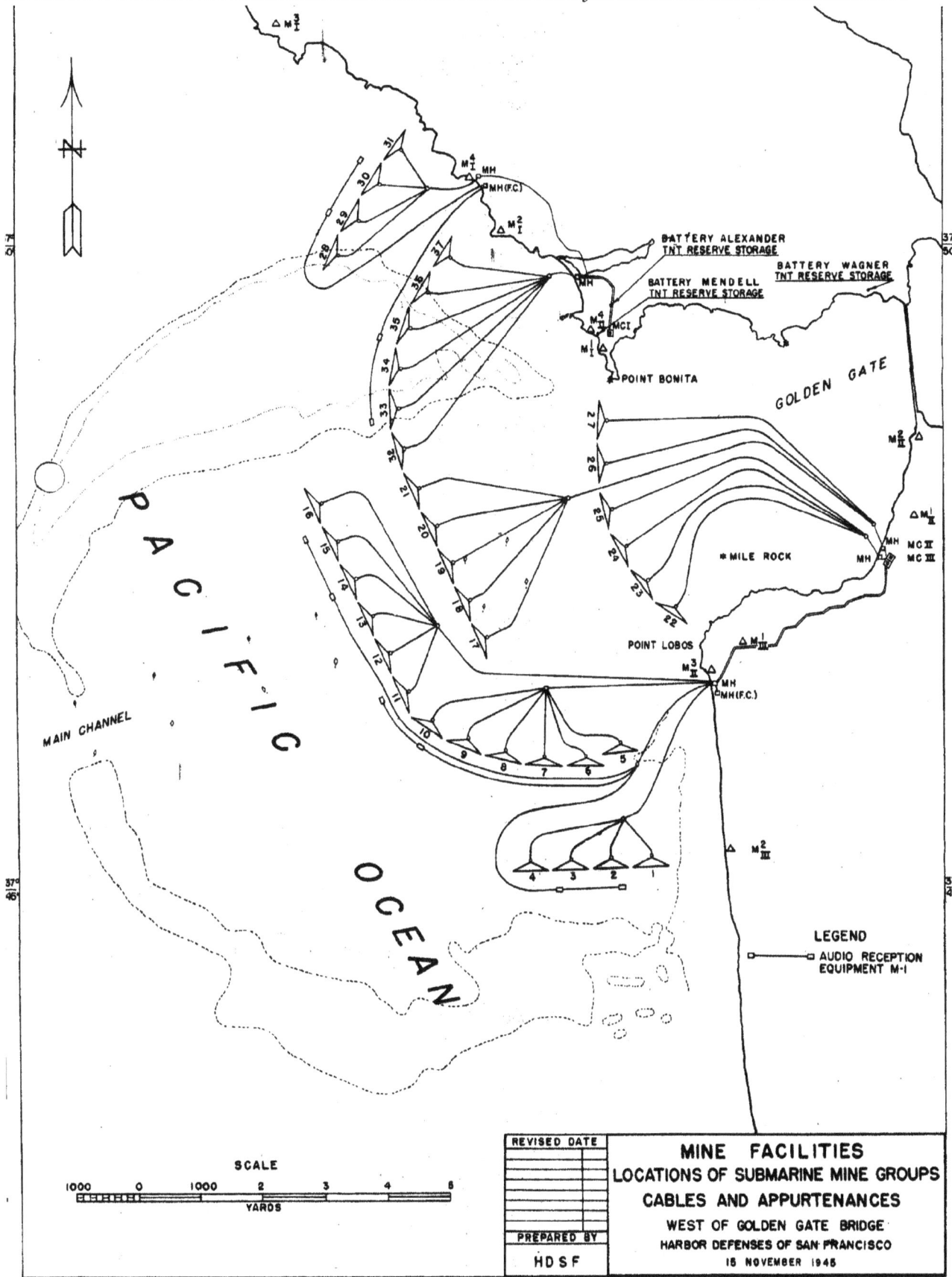

Mine Defenses of the San Francisco Bay, 1945
Exhibit 2, Annex D, Underwater Defense Project, Supplement to Harbor Defense Project, San Francisco

Mine Shore Facilities

The equipment for the entire minefield was stored ashore during peacetime, disassembled without explosives or fuses. They were only to be deployed during times of war. The shore facilities included all the buildings and equipment necessary to store, assemble, test, transfer, and repair the mines, cables, anchors, and distribution boxes. In addition all the buildings and equipment for controlling and detonating the mines was also located ashore. Remaining examples of mine shore facilities can be seen today at Fort Hancock (Sandy Hook Unit, Gateway National Recreation Area, New Jersey), Fort Wadsworth (Gateway National Recreation Area, Staten Island, New York), Fort Stevens (Fort Stevens State Park, Oregon), Fort Totten (City of New York), Fort Winfield Scott (Golden Gate National Recreation Area, California), Fort Baker (Golden Gate National Recreation Area, California), Fort Dade (Egmont Key State Park, St. Petersburg, FL) and Fort Wetherill (Fort Wetherill Park, Jamestown, RI), among others. Fort Wetherill's mine facilities are in especially fine shape. The State Game Department has adapted the interior of the buildings for their use, while the exteriors of the buildings have been restored.

Time line for the development of the San Francisco Bay mine facilities:

1883	Mine depot built on Alcatraz Island
1889	Mine casemates built at Fort Mason & Alcatraz Island
1891	Mine depot built on Yerba Buena Island
1890s	Mine casemates built at Cavallo Point (Fort Baker), and Mortar Hill (Angel Is., Fort McDowell)
1897	Mine casemate at Quarry Point (Angel Island, Fort McDowell)
1907-10	Mine depot built at Fort Winfield Scott
1908	Mine casemate built at Fort Barry
1909	Mine casemate built at Fort Baker
1912	Mine casemate built at Baker Beach (Fort Winfield Scott)
1937-41	Mine depot built at Fort Baker
1943	New mine casemate built at Baker Beach (Fort Winfield Scott)

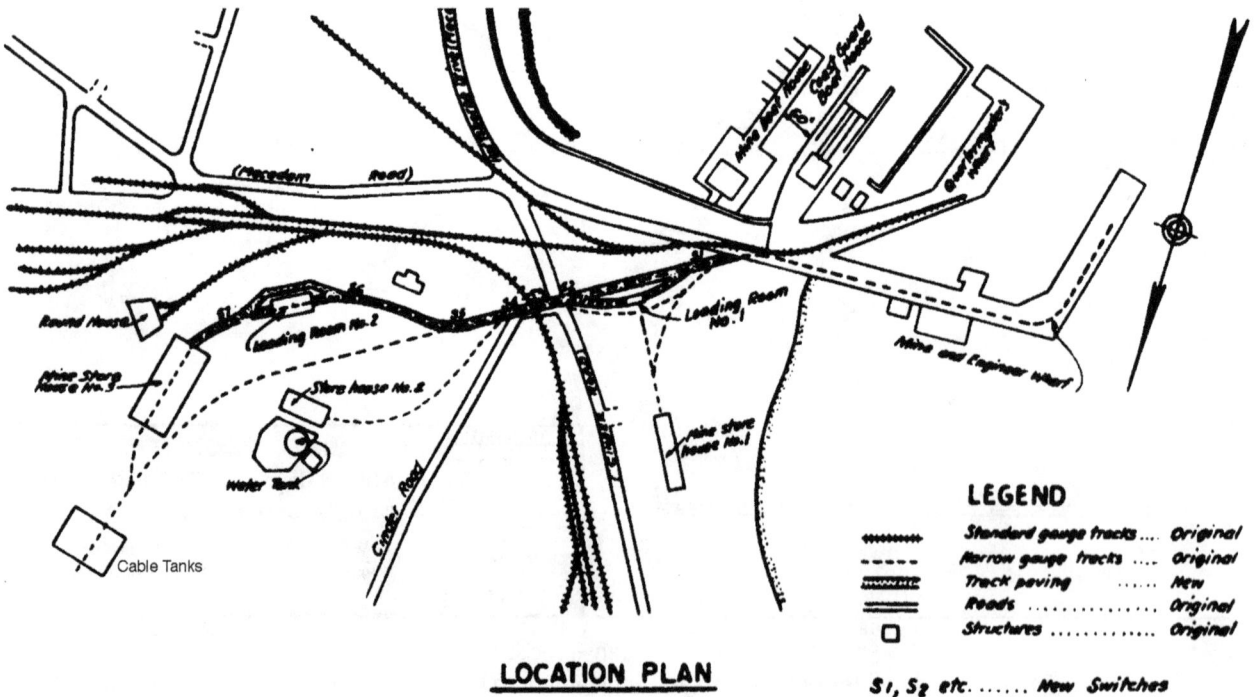

LEGEND

	Standard gauge tracks	Original
	Narrow gauge tracks	Original
	Track paving	New
	Roads	Original
□	Structures	Original

LOCATION PLAN

S1, S2 etc. .... New Switches

Mine Depot, Fort Hancock, HD New York, 1940s

The Mine facilities at Fort Wetherill, Harbor Defenses of Narragansett Bay, RI

Mine Storehouse (left) and Mine Storehouse, Cable Tank, and Loading Room (right), Fort Wetherill, RI 2011 (MAB)

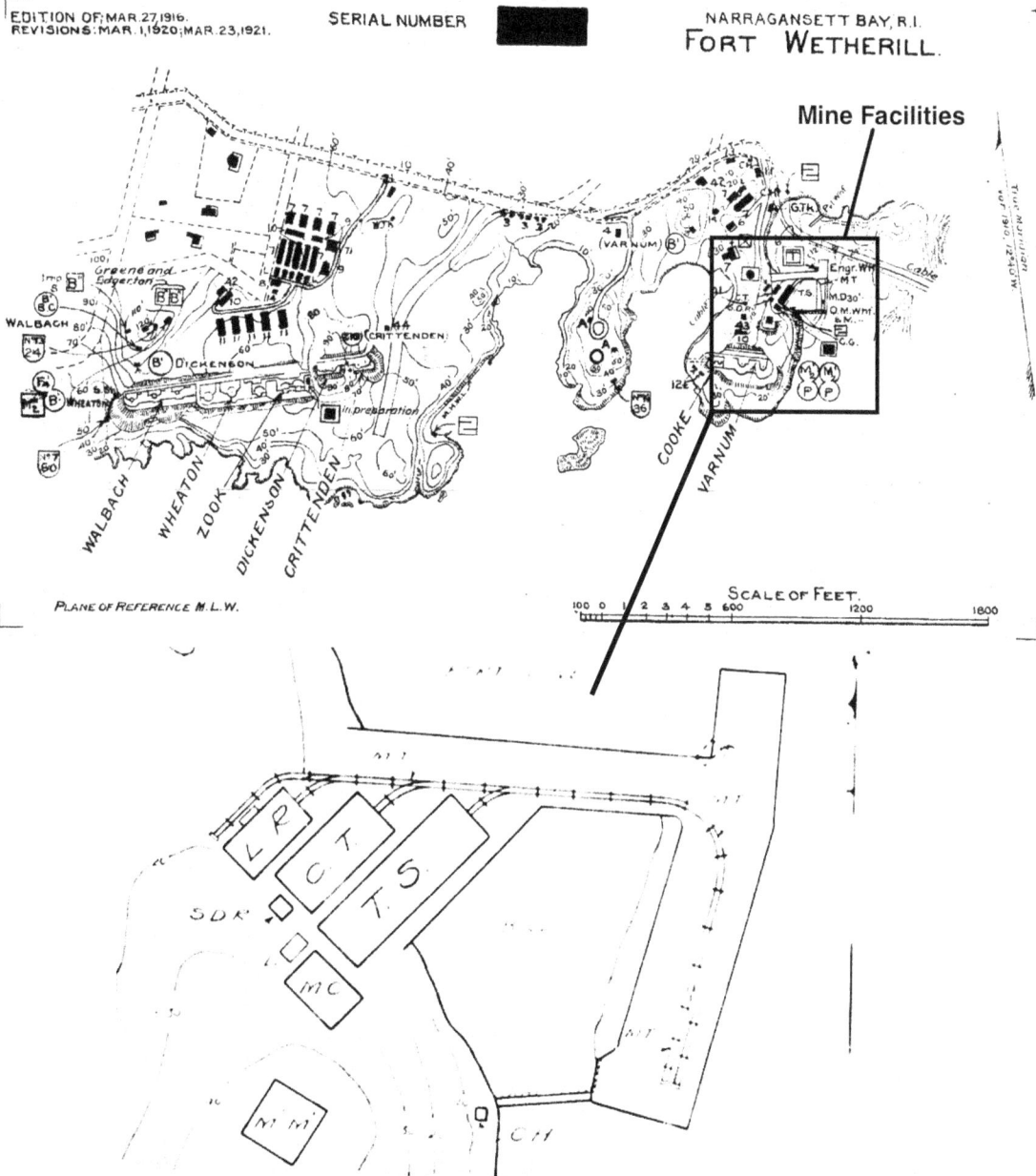

ARMY MINE BASE
LITTLE CREEK, VA.
GASOLINE & FUEL
OIL SYSTEMS

Mine
Pier

Flotilla
Pier

FUEL OIL DISPENSING UNIT

GASOLINE DISPENSING UNIT

Marginal   Pier

Boat
House

Officer's Mess

2" SUCTION LINE

PROP. LINE
MON. 1

15,000 GAL. FUEL OIL
STORAGE TANK.
(UNDERGROUND)

1500 GAL. GASOLINE
STORAGE TANK
(UNDERGROUND)

Loading
Rooms

U.S.E.D. MON
"A-1943"

P.L. MON.
5 & 6

U.S.E.D.
D-194

U. S. COAST GUARD

RESERVATION

14

15

16

17

18

19   20

Mine Storage
Building

12

HUB

SENTRY HOUSE
PROP. LINE MON. 2

Mine Cable Tank
Building

12

storage

3" FILL LINE

Magazines

12

Mag

Mag

U.S.E.D. MON.
"B-1943"

FUEL OIL PUMP
HOUSE (REPORTED
SEPARATELY)

RAILROAD   TRACK

3" FILL LINE TO
PENN. R. R.
(ABANDONED)

11.9

12

10.1

SENTRY HOUSE

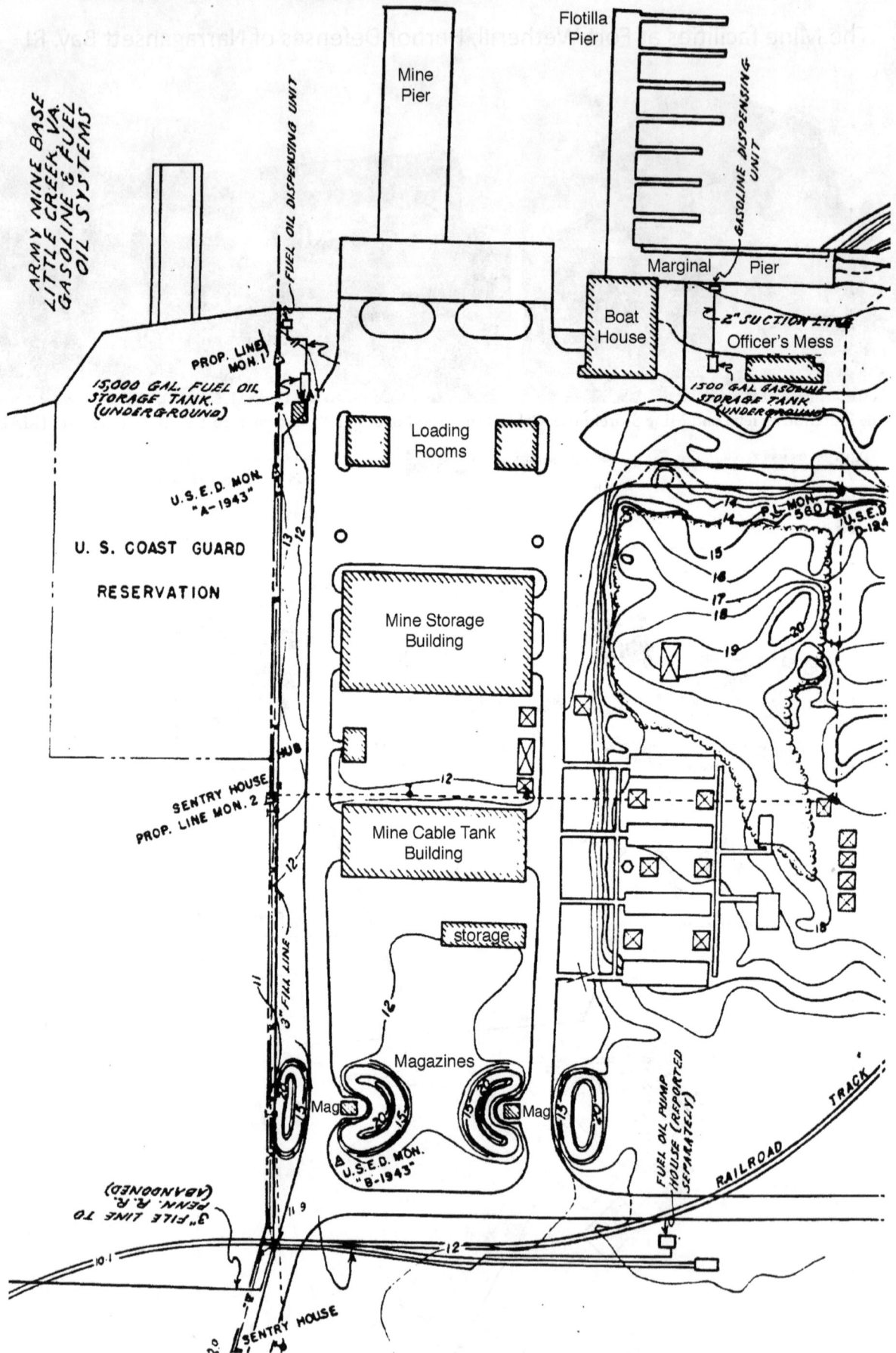

Mine Depot, Little Creek Military Reservation, HD Chesapeake Bay, VA, 1940s

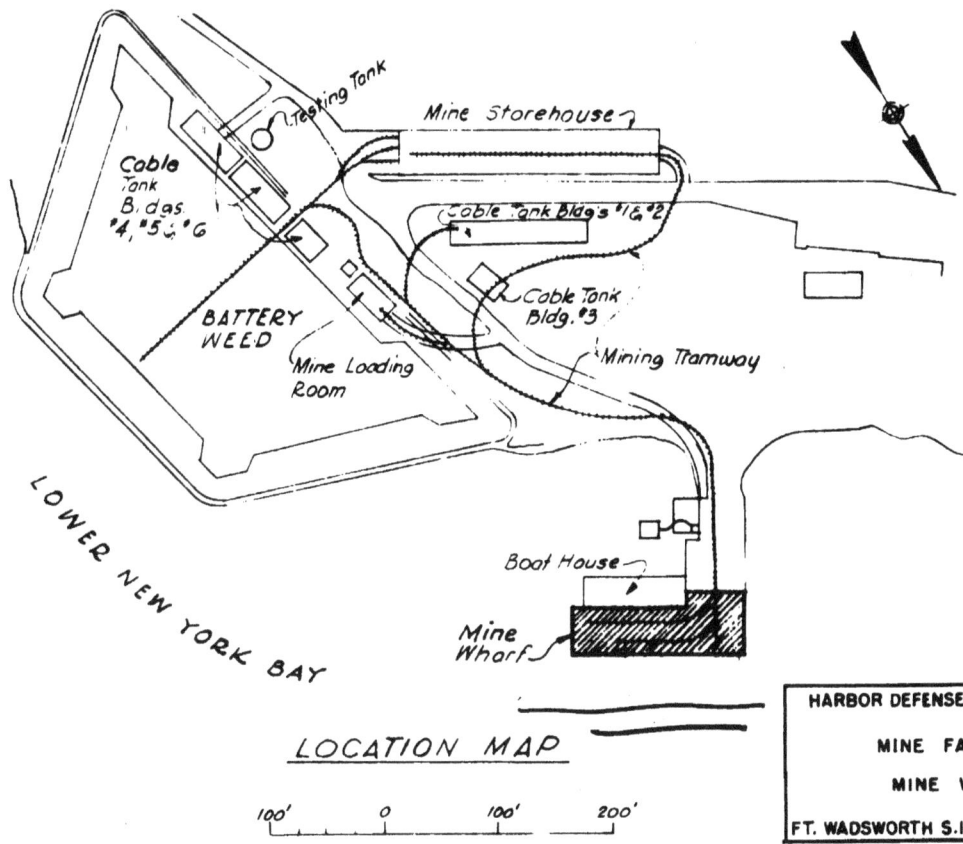

Testing Tank

Mine Storehouse

Cable Tank Bldg's #1 & #2

Cable
Tank
Bldgs
#4, #5, #6

Cable Tank
Bldg. #3

BATTERY
WEED

Mining Tramway

Mine Loading
Room

LOWER NEW YORK BAY

Boat House

Mine
Wharf

LOCATION MAP

100'   0   100'   200'

HARBOR DEFENSES OF NEW YORK
MINE FACILITIES
MINE WHARF
FT. WADSWORTH S.I.   NEW YORK

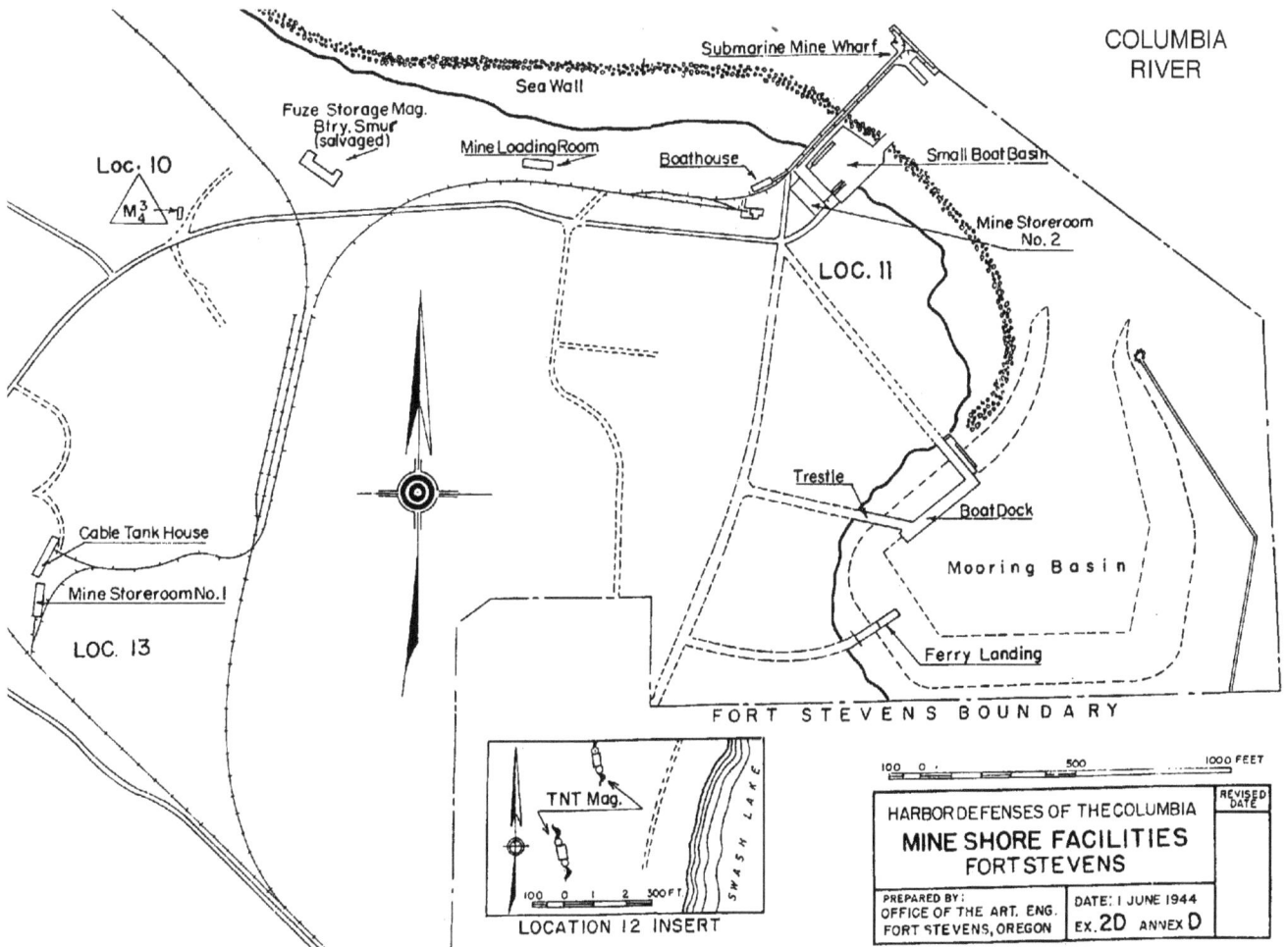

COLUMBIA
RIVER

Submarine Mine Wharf

Sea Wall

Fuze Storage Mag.
Btry. Smur
(salvaged)

Mine Loading Room

Boathouse

Small Boat Basin

Loc. 10

M$^3_4$

Mine Storeroom
No. 2

LOC. 11

Cable Tank House

Trestle

Mine Storeroom No. 1

Boat Dock

Mooring Basin

LOC. 13

Ferry Landing

FORT STEVENS BOUNDARY

100 0   500   1000 FEET

TNT Mag.

SMASH LAKE

100 0   1   2   300 FT.

LOCATION 12 INSERT

HARBOR DEFENSES OF THE COLUMBIA	REVISED DATE
MINE SHORE FACILITIES FORT STEVENS	
PREPARED BY: OFFICE OF THE ART. ENG. FORT STEVENS, OREGON	DATE: 1 JUNE 1944 EX. 2D ANNEX D

Mine (Torpedo) Storehouses

All the disassembled mine equipment was stored in these buildings—empty mine casings, anchors, compound plugs, circuit closers, shackles, mooring sockets, distribution boxes, and the various tools and equipment. These buildings had racks and shelves, movable overhead hoists with blocks and tackle for moving the equipment, ropes, chains, tools, and appliances. These buildings were one to two stories high ,and the building sizes varied greatly from post to post.

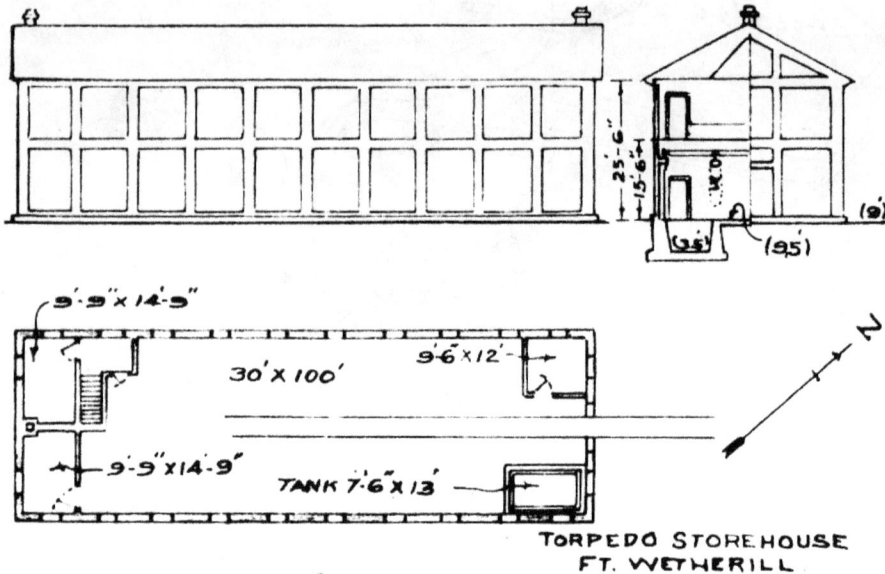

TORPEDO STOREHOUSE
FT. WETHERILL

PANAMA CANAL FORTIFICATIONS
— NAOS ISLAND —
Mine Storehouse

Torpedo Storehouses
(Top) Fort Warren, 1941 (cable tank to left, storehouse right) (NA, Al Schroder Collection, MHI, Carlisle)
(bottom) Interior, Fort Warren, HD Boston, 1920s(?) (NA, Al Schroeder Collection, MHI, Carlisle)

Torpedo Storehouses:
(Top) Interior of storehouse at Fort Winfield Scott 1920s
Fort Wetherill, HD Narrangansett Bay 2003, (MAB),
(Middle) Manchester State Park, WA (across from Fort Ward), 1996 (MAB)

Cable Tanks

The mine cables were stored in large fresh water tanks to prevent deterioration. Each concrete tank had to be capable of holding 10,000 feet (3 tons, pre WW I, later 20,000 ft., 5 tons) of cable on a reel. The cable tanks were usually enclosed in a building to control evaporation and were equipped with overhead cranes with hoists and tackle. The cables were initially tested as well as repaired in this facility.

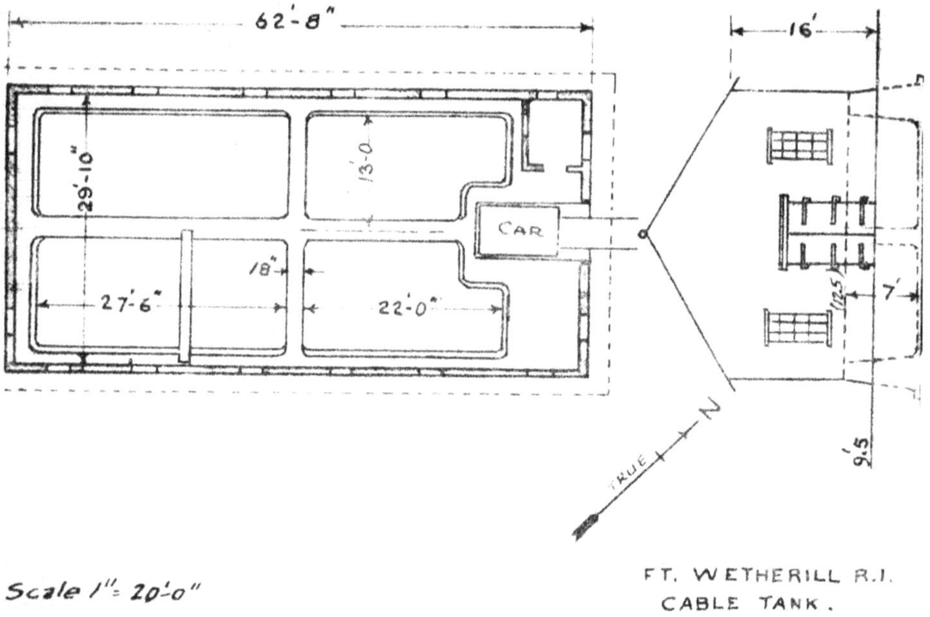

62'-8"
16'
29'-10"
13'-0
CAR
18"
27'-6"
22'-0"
7'
TRUE N
Scale 1" = 20'-0"
FT. WETHERILL R.I.
CABLE TANK.

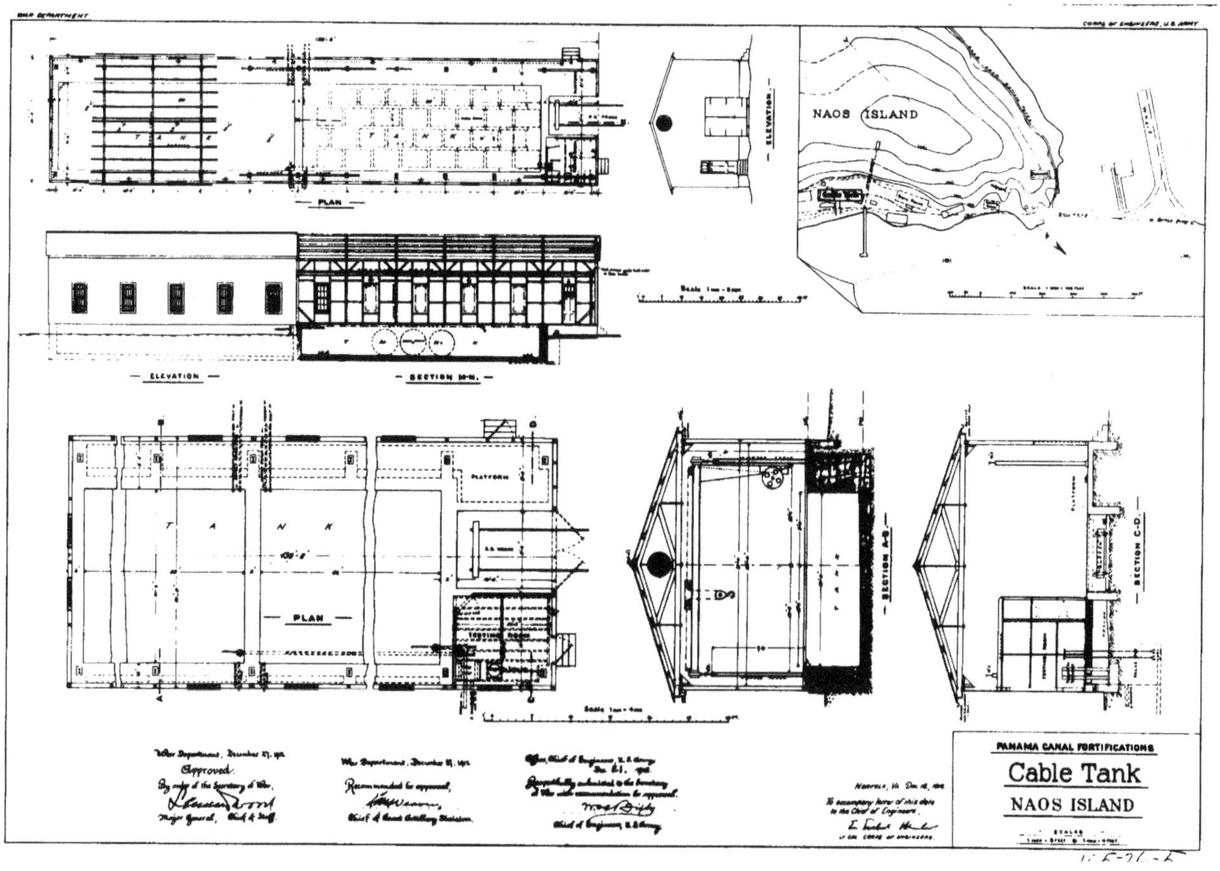

NAOS ISLAND
PLAN
ELEVATION
PLATFORM
PANAMA CANAL FORTIFICATIONS
Cable Tank
NAOS ISLAND

~PLAN~

SECTION A-A   SCALE: 1"=20'
    N.S.

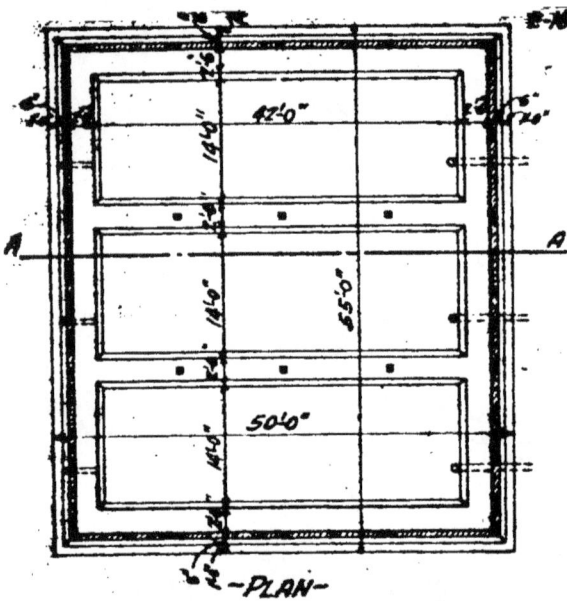

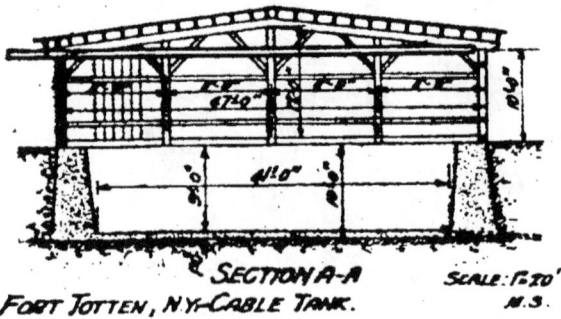

FORT TOTTEN, N.Y.-CABLE TANK.

(Top) Mine cable tank building, Naos Island (Fort Grant), HD Balboa, Panama Canal Zone (MAB 1999)
((Middle left) Foundations of the cable tank facility, Fort Hancock (Sandy Hook), HD New York. (Photo by Mark Berhow 1995) (Bottom) Cable tank building at Fort Baker, HD San Francisco, (photo by Mark Berhow 1997)

Loading Rooms

Loading rooms were where the mine components were assembled and tested after all the necessary equipment and materials were brought from the storehouses, cable tanks, and magazines. The detachment was under the command of the chief loader, a noncommissioned officer. Typically the facility was composed of two rooms, one for loading and assembling mines, and one for loading and assembling the compound plugs (the detonating devices). The loading room often had a small cable test tank and overhead hoists. Mines were loaded with explosives, fitted with their compound plugs, connected to their cables, tested for watertightness and electrical circuit integrity, and then sent out to the wharf. Mines for a complete group were assembled before being sent out to the wharf. Cables were cut and prepared for attachment and laid out in transportable figure "8"s. A service dynamate room for temporary storage of the dynamite to be used was often located adjacent to the loading room, as shown in the drawing below.

Magazines

The explosives used in the mines, dynamite and gun cotton before 1912, trinitrotoluene (TNT) after, were stored in magazine facilities at the post. Only a fraction of what was needed for the complete project was kept at the post, the rest was shipped from a US Arsenal as needed. Sometimes there were specially built magazines just for the mine compound plug detonators near the loading room. Fuses were kept in a separate room away from the bulk explosives.

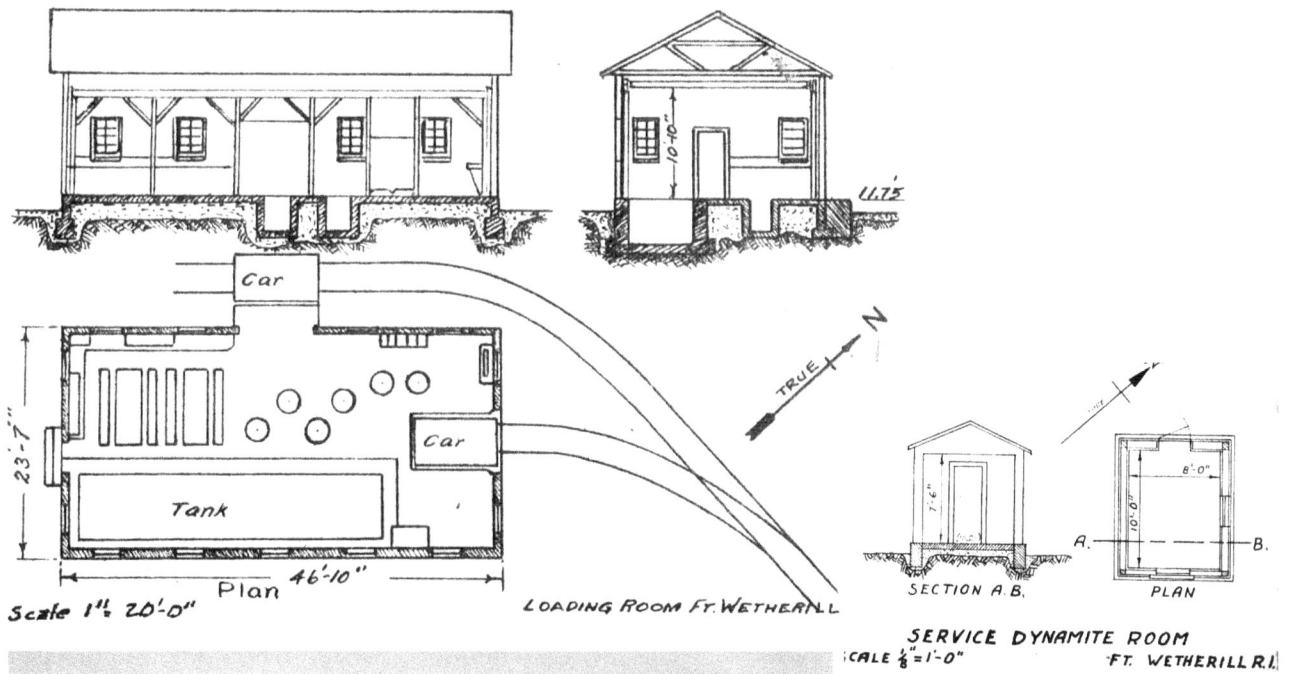

Loading Room,
Fort Hancock
HD New York
(Sandy Hook Unit, Gateway NRA, NJ).
Photo by Mark Berhow 1997.

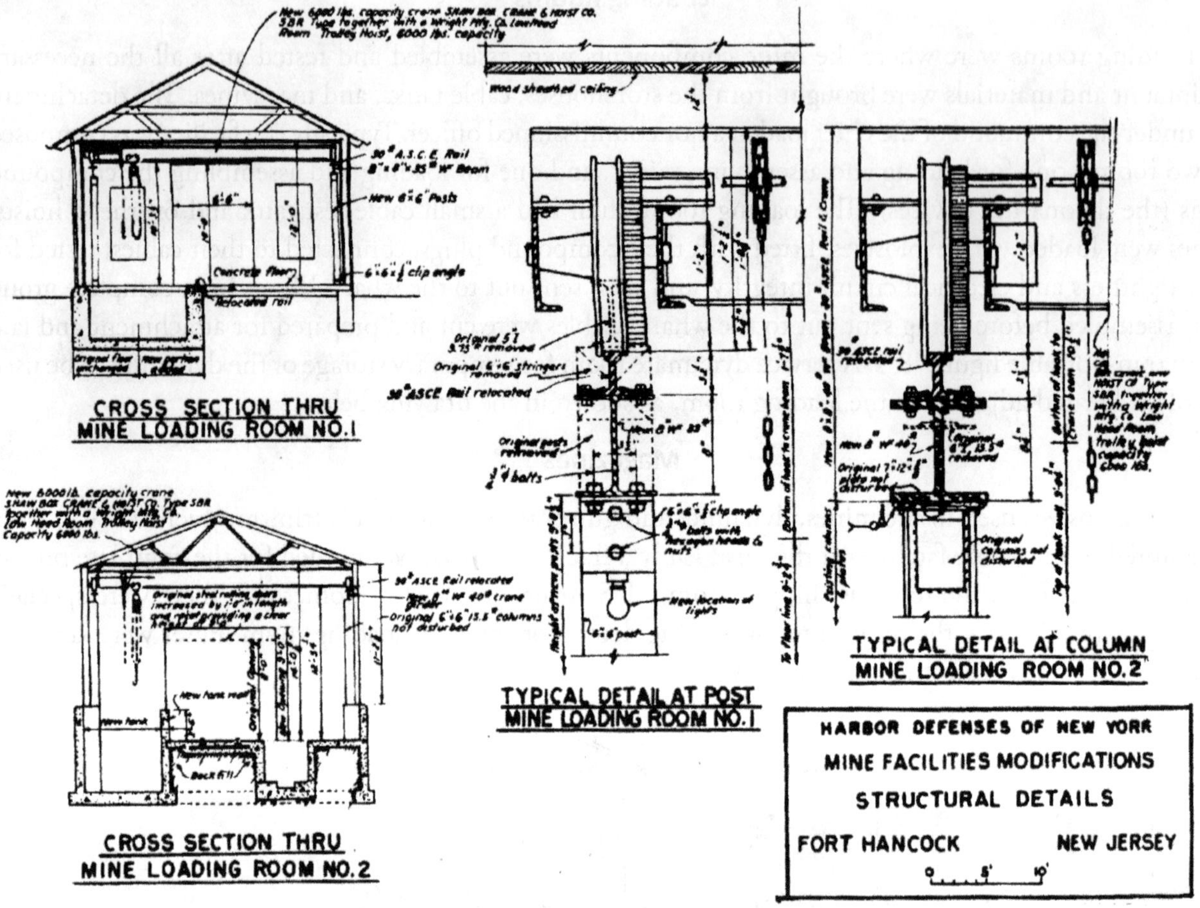

CROSS SECTION THRU
MINE LOADING ROOM NO. 1

CROSS SECTION THRU
MINE LOADING ROOM NO. 2

TYPICAL DETAIL AT POST
MINE LOADING ROOM NO. 1

TYPICAL DETAIL AT COLUMN
MINE LOADING ROOM NO. 2

HARBOR DEFENSES OF NEW YORK
MINE FACILITIES MODIFICATIONS
STRUCTURAL DETAILS
FORT HANCOCK   NEW JERSEY

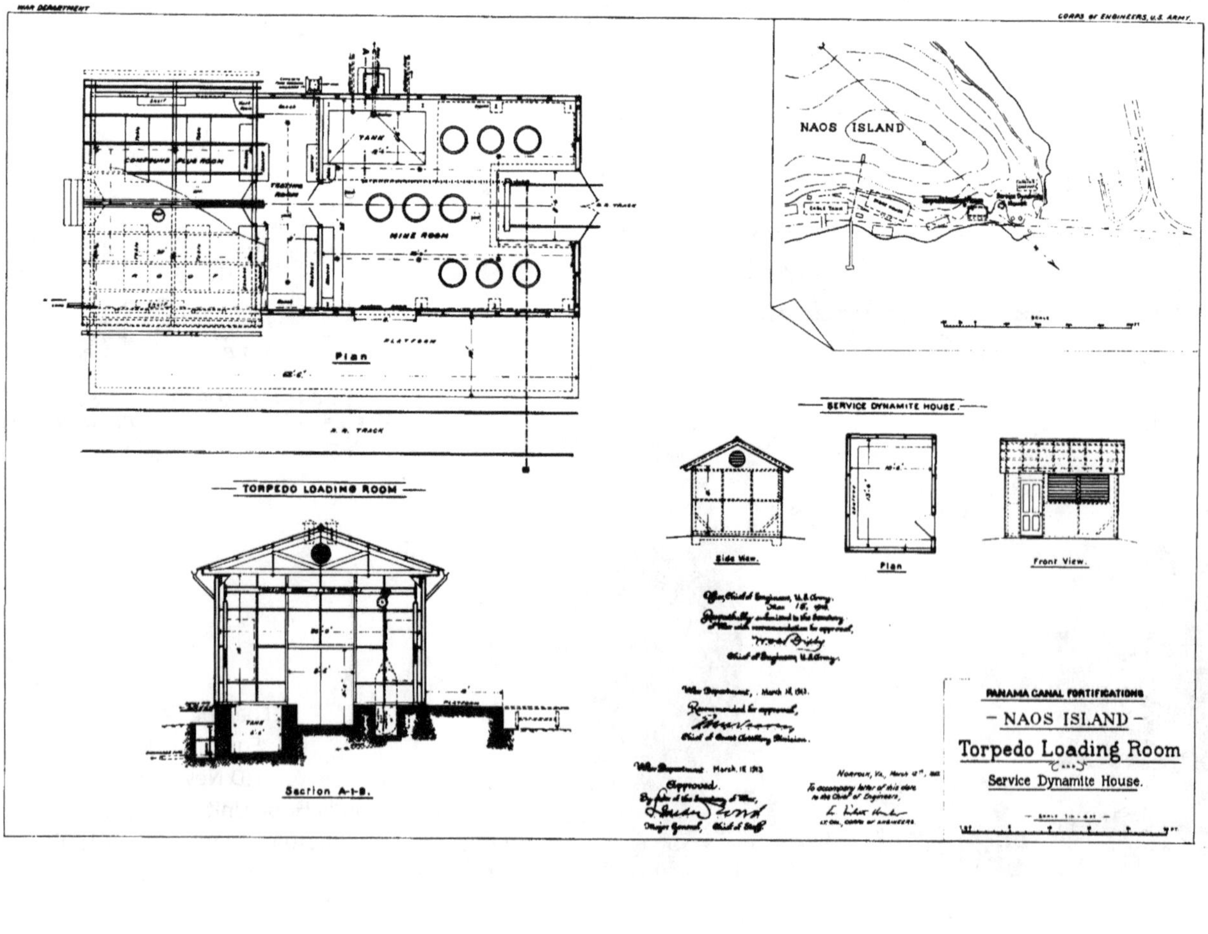

TORPEDO LOADING ROOM

Plan

Section A–1–B.

NAOS ISLAND

SERVICE DYNAMITE HOUSE

Side View.   Plan   Front View.

PANAMA CANAL FORTIFICATIONS
— NAOS ISLAND —
Torpedo Loading Room
Service Dynamite House.

Mine Tramway

All mine depot facilities were connected by a rail system. If the post had a regular railroad system, the tramway may have been incorporated using the same gauge rail as the railroad. If not, typically a 3-foot gauge rail system was installed connecting the mine wharf with the loading room, torpedo storehouses, and the cable tanks. The flat cars could be manually pushed from facility to facility. If a hill was involved, a hoisting engine was usually installed.

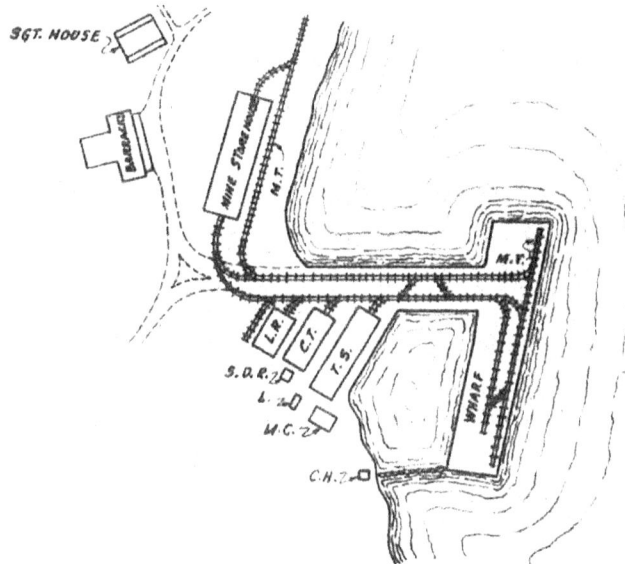

TRAMWAYS
FORT WETHERILL
SCALE : 1" : 200'
W.A.T.

Mine equipment on tramway cars on the mine wharf at Fort Warren, HD Boston.
(NA, Al Schroeder Collection, MHI Carlisle)

Cable Terminal Huts

The mine connections from the shore to the mine case-mate were laid in underground tunnels. These were (usually) permanent installations. The tunnels ended at small concrete structures of a variety of designs, called cable terminal huts, boxes, or vaults. The shore cables from the mine groups were brought up to the shore and routed to the mine casemate through these vaults.

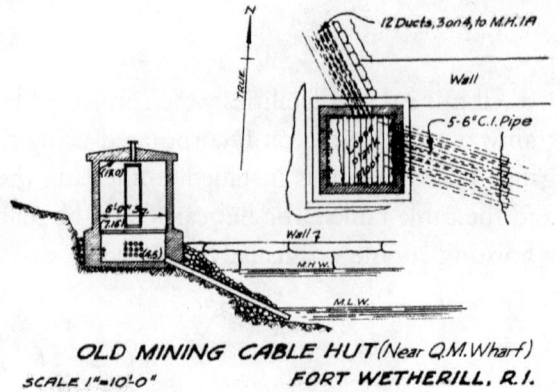

OLD MINING CABLE HUT (Near Q.M. Wharf)
SCALE 1"=10'-0"   *FORT WETHERILL, R.I.*

Mine Wharf

A typical mine wharf was a pier with a large square work surface area and a heavy loading derrick. The planter would load the mines, cables, and distribution boxes from the wharf, which was connected by the tramway to the loading room. Usually mine wharves had boathouses for storing the smaller yawls and their equipment. Ideally, a post would have a separate wharf for mine activities. Quartermaster and engineer wharves would be used for other activities so as not to interfere with the mine planting, but often the mine wharf was used for other post shipping activities as well.

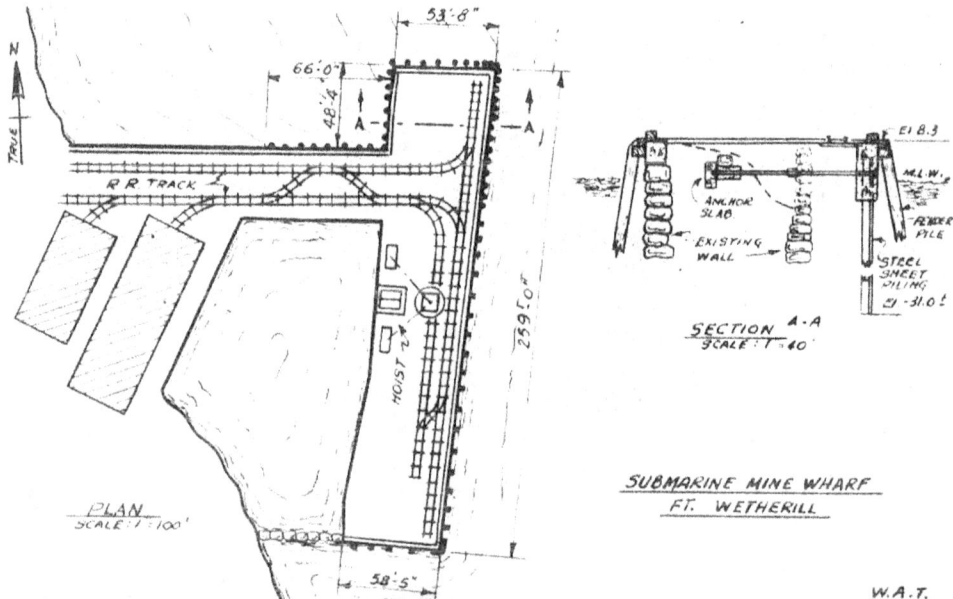

SUBMARINE MINE WHARF
FT. WETHERILL

Mine Wharf, Fort Wetherill, HD Narrangansett Bay), Photo by Mark Berhow, 2003

Mine Casemates

This structure held the firing controls for a mine field or portion thereof. It was the most heavily protected structure of the mine defense facilities. In the 1880s, mine casemates were sometimes located inside a reinforced casemate of an older fortification, hence the name. In the 1890s, mine casemates were located in poorly ventilated concrete vaults built into hillsides. The mine casemates held the control (operating) boards which held the master switches (boxes) for the individual mines, and the batteries that powered the system. After the turn of the century, the mine casemates were larger structures built of brick or concrete with windows for better ventilation. The structure was divided into separate rooms for the operating board, power equipment, storage batteries, and a ready room. Casemates were redesigned again in the 1910s with the construction of roomier wood frame buildings which were protected by concrete reinforced hillsides to the oceanside. The men manning the casemate were under the command of the casemate electrician.

Equipment in 1930s casemate included a 5-kw gasoline generator set (or a Hornsby-Akroyd oil engine and generator), storage batteries, motor generators, the casemate transformer, power panel, interrupter panel, and operating boards.

The minefield could be maintained in different modes of activation from the control panel in the mine casemate: OFF—kept the mines inactive and allowed all vessels to pass without harm; TEST— provided a low voltage DC current through all components of the minefield to confirm its electrical integrity; SIGNAL— which allowed an indication in the electrical lights in the casemate when a mine's ball switch was "tripped" by contact with a passing vessel; CONTACT FIRE— a passing vessel caused a pulse of high voltage AC current to detonate the mine; DELAYED CONTACT FIRE— A tripped mine would be detonated under the vessel by a high voltage AC current after a timed delay; and OBSERVATION (or COMMAND) FIRE—a selected mine was detonated by high voltage AC current by throwing a switch in the mine casemate on command.

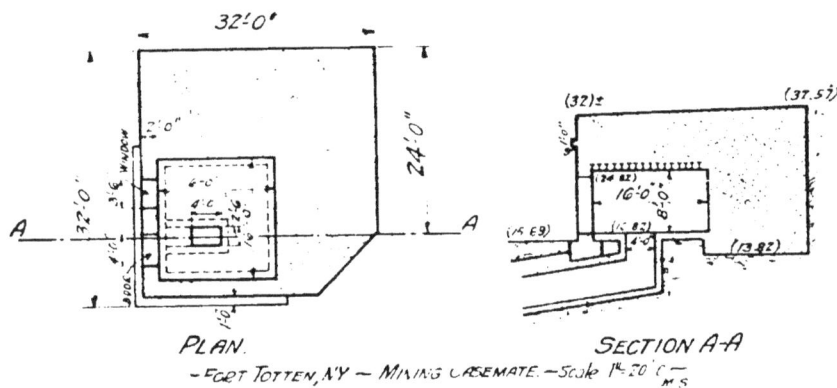

(Above) Early (1890s) type mine casemate (Fort Totten)

Entrance to 1890s
Mine Casemate,
Fort Washington,
HD Potomac River
(Alex Holder, 1990)

The Mining Casemate at the Middle Point Reservation

Greg Hagge
(from *CDSG News* Vol. 3, Iss. 5 (Nov. 1989) p. 21)

The purpose of the casemate was to house the power plant, battery, power distribution board and mine operating or firing panels for the Rich Passage mine defense project. The power plant consisted of a Hornsby-Akroyd, one-cylinder, 4-stroke kerosene oil engine. This engine would drive a 5-kilowatt DC generator by means of a belt. The DC power thus generated could be used to light the casemate, power the mine operating panels, and charge the storage battery. Because the oil engine required from 10 to 20 minutes to preheat the vaporizer before the engine could be started, it was important to keep the storage battery fully charged for immediate use. The storage battery was the normal source of power to the mine operating panels.

In the early part of this century, a storage battery consisted of numerous glass or rubber boxes called "jars." Each was one foot square and 18 inches tall, with thick lead positive and negative plates suspended in sulfuric acid and with a basically open top. The storage battery room held 40 two-volt cells (jars) for a

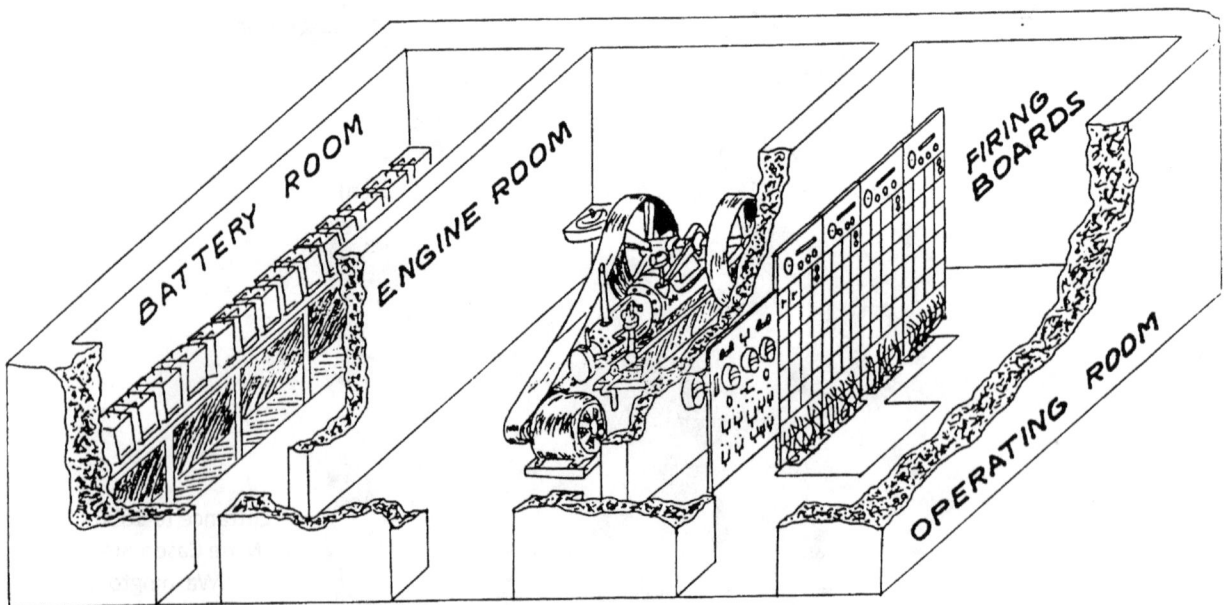

Mine casemate at Middle Point, Harbor Defenses of Puget Sound (illustration by Greg Hagge)

total of 80 volts DC. These large, glass-cased cells would be connected in series to form the storage battery. The cells would be placed on sand-filled trays on shelves along the walls of the storage battery room. The sand trays would contain any acid that spilled or leaked from the cells during operation or maintenance.

The operating room was the heart of the controlled mine system. The main power panel had connections for AC and DC power. Two DC-powered motor-generators would produce the AC power needed to complete the mine firing process. The power panel could select the source of DC power from the storage battery, the casemate generators or external post power. The AC power came from the casemate AC motor-generator, or from external AC power where provided.

The mine operating panels had an upper block with a bell and 3 testing lights, a smaller master block with the DC power test switch, and the AC power firing switch. Nineteen mine blocks, each with a test switch, mine switch, power switch, and automatic switch with its solenoid plunger and release switch controlled each individual mine in the nineteen-mine "grand group" or line. The cables from the mines were brought up through the floor and connected to the operating panel. The DC power would be used to test the circuit of the mine for grounds and arm the firing device as well as the sensor that would indicate when the mine was disturbed by a passing ship. The AC power would be applied to actually fire the mine. By using this system of AC and DC power, the mines could be tested and monitored with absolute safety against accidental explosions while planting and maintaining the mine field. This was of great importance in a controlled mine field as the idea was to avoid the indiscriminate destruction of friend and foe alike which could happen with contact mines.

The Middle Point mine casemate was tuned over to the Coast Artillery in 1905, but was abandoned in 1910 in favor of new facilities located across Rich Passage at Fort Ward, Bainbridge Island. The "new" mine facilities at Fort Ward proper have been destoyed or altered by development. However, visitors can still see the old casemate and the torpedo storehouse at Middle Point, now part of Manchester State Park.

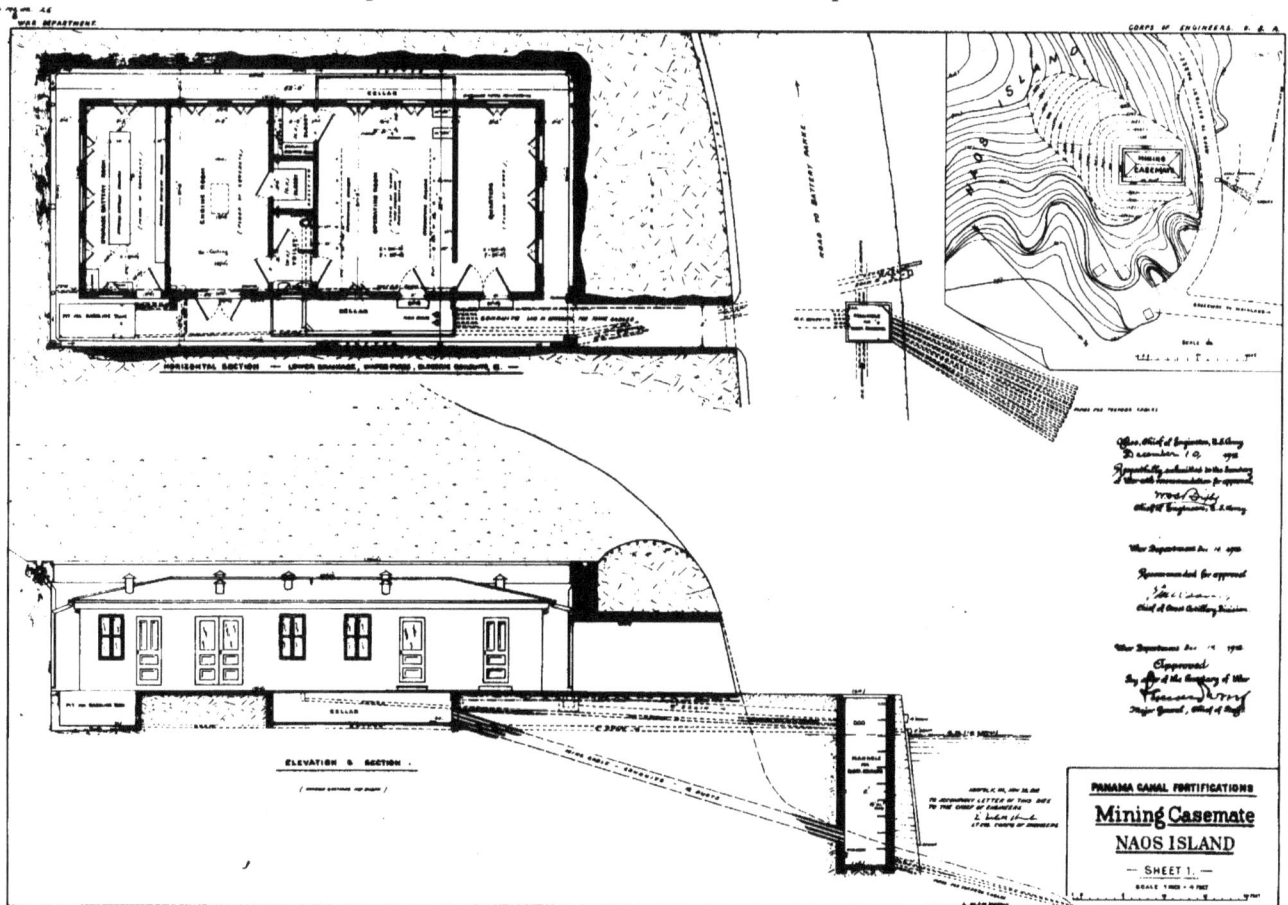

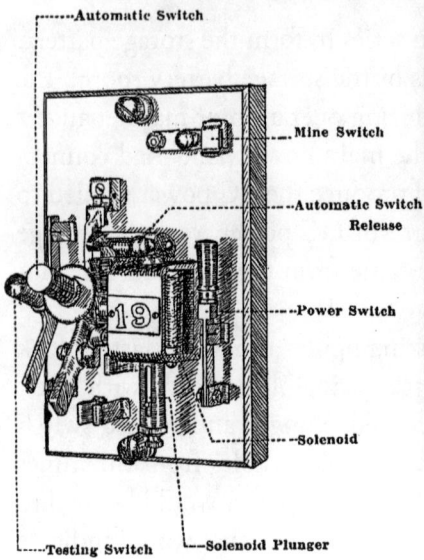

Mine block switch (1910s)

Mine casemate operating room, Fort Wint, HD Subic Bay, 1909, NARA

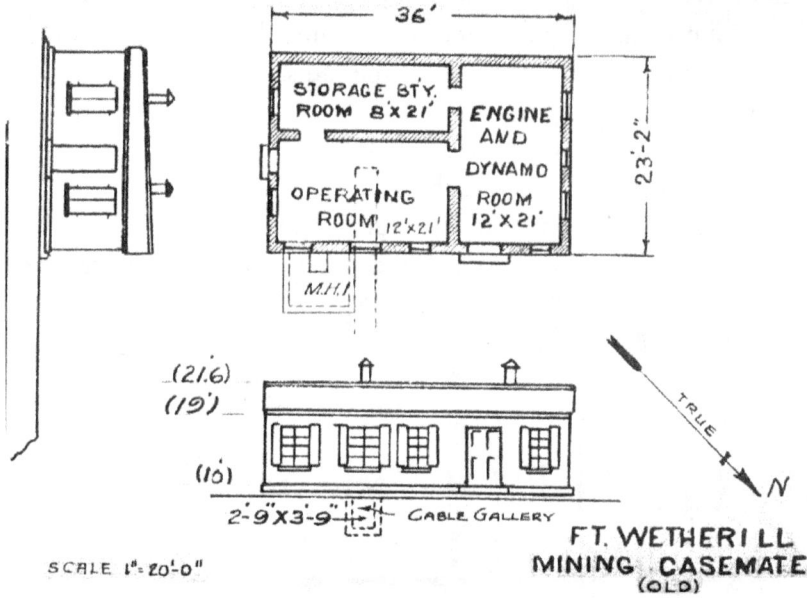

FT. WETHERILL
MINING CASEMATE
(OLD)

SCALE 1"=20'-0"

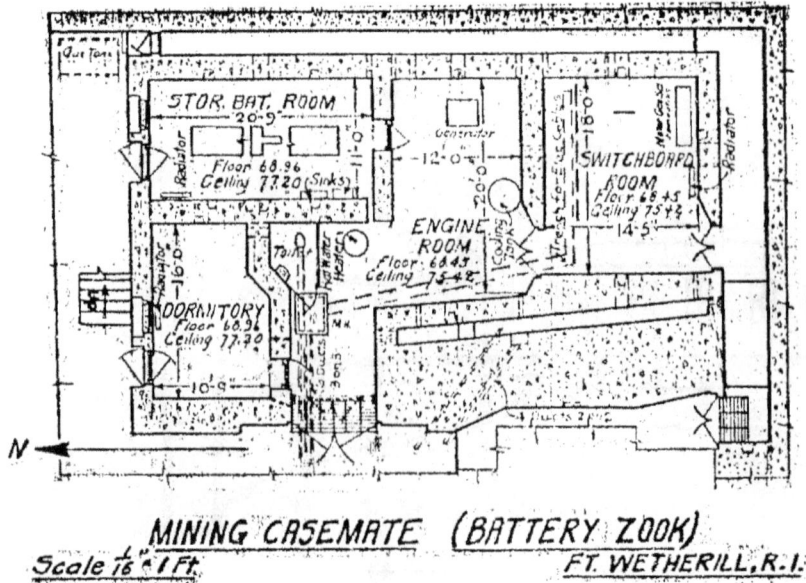

MINING CASEMATE (BATTERY ZOOK)
Scale 1/16"= 1 Ft.   FT. WETHERILL, R.I.

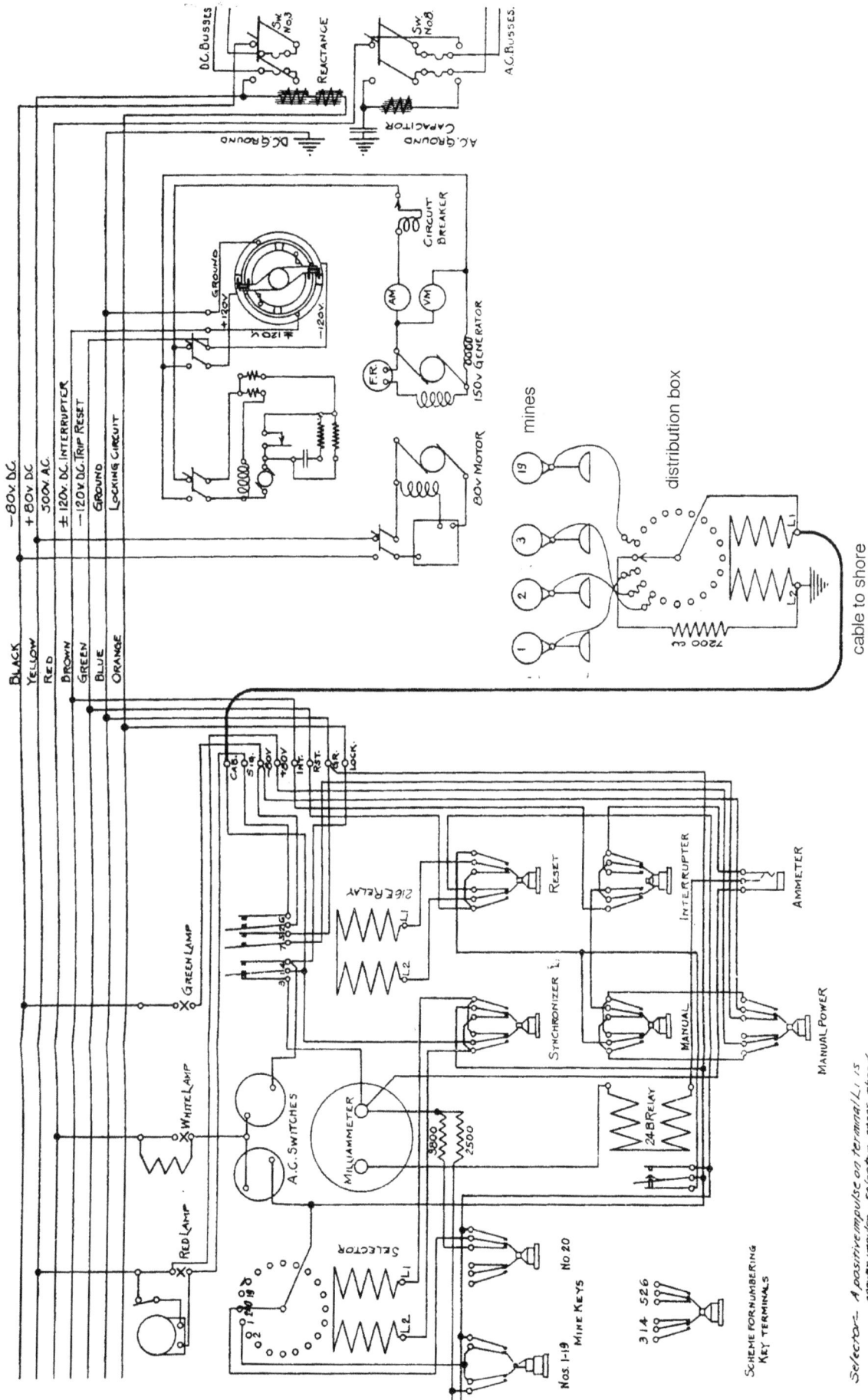

Mine circuit diagram (TM 2160-20, 1930)

Mine Casemates: Fort Michie, HD Long Island Sound (Alex Holder, 1997); Fort Strong, HD Boston (Alex Holder, 1992); Fort Stevens, HD Columbia River (Mark Berhow 1996); Fort Barry, HD San Francisco (Mark Berhow 1997)

Controlled Mine Fire Control

Mine Primary Stations: The fire control system used for determining mine locations during planting and for observation fire was eventually (post 1900) very similar to that used for the big guns (see section "Fire Control and Gun Drill" for a discussion of the fire control and position finding procedures and equipment). As the ranges used were relatively short (4,000 to 8,000 yards) and the channel covered by the command (usually) was restricted, it was decided sometime in the early 1900s to only have one baseline for a mine command. However, two base end stations were constructed at each end of the baseline so that two targets could be tracked at the same time, hence the usual arangement of double mine primary stations and double mine secondary stations. The mine primary was the station of the mine commander (group or battery, depending on the size of the minefield and the year). The stations had a DPF and/or an azimuth scope, telephone connections, and a time interval bell. The observation stations were often a part of structure that contained the large plotting room (with two plotting boards) and may have had other rooms as well.

Double Mine Primary Station, Fort McKinley, HD Portland, ME (Alex Holder, 1989).
This structure has since been destroyed

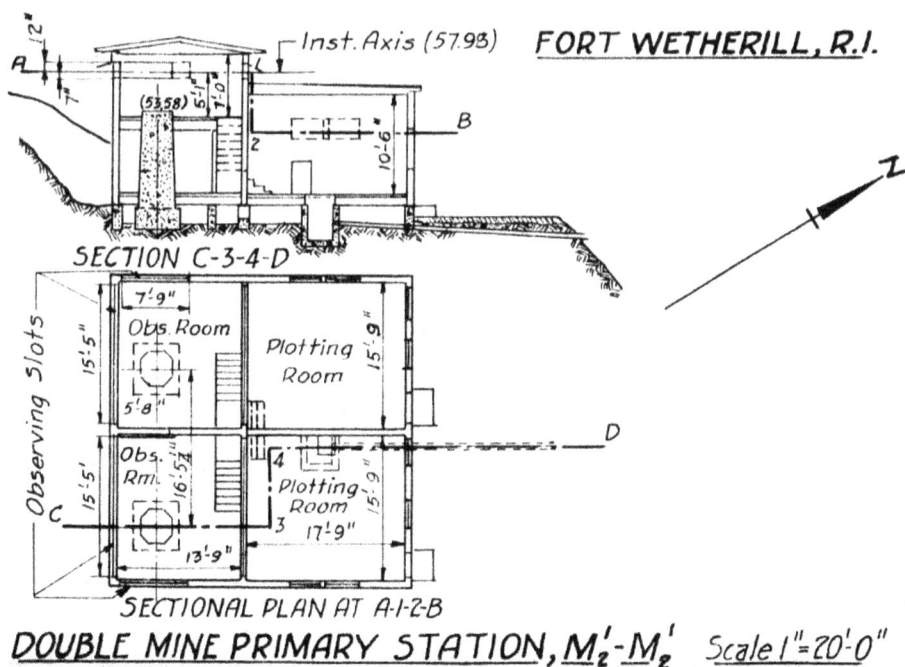

FORT WETHERILL, R.I.

SECTION C-3-4-D

SECTIONAL PLAN AT A-1-2-B

DOUBLE MINE PRIMARY STATION, M_2'-M_2' Scale 1"=20'0"

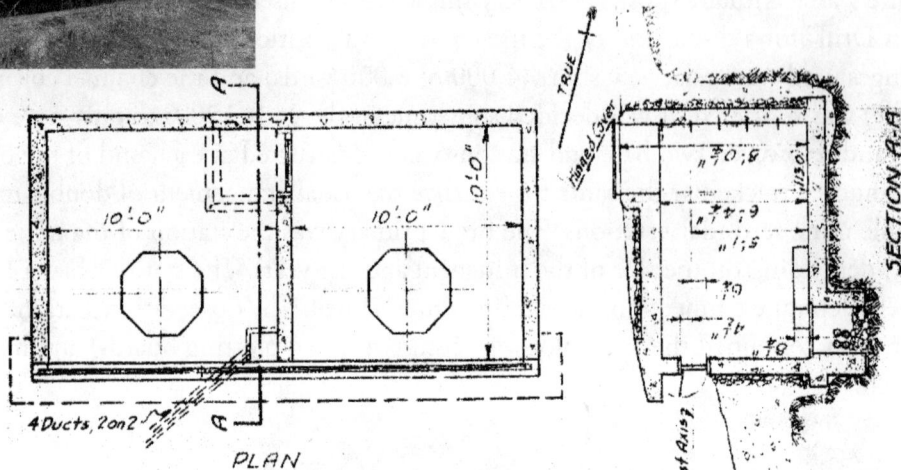

DOUBLE MINE SECONDARY STATION (M₂″-M₂″)
FORT WETHERILL, R.I.
Scale ⅛″=1'-0"

Mine Secondary Stations: Similar to the mine primary stations, but without the plotting room.

Mine Tactical Organization

The tactical organization of the mine command changed considerably over the years, as with all tactical organization in the Coast Artillery Corps. The description given below is from the late 1930s, and is given as an example. Under the overall direction of the harbor defense commander, the large gun batteries were in tactical units called groups. The mine commands (unless unusually large or separated) were under the overall command of the mine group commander. He had under his charge the various mine batteries—the mine operations battery, the mine planter (considered administratively a battery by 1940), the rapid-fire gun batteries assigned to protect the mine field, and the searchlights assigned to the mine field. Not all searchlights in a harbor defense were assigned to the mine group; many were assigned to the heavy caliber gun groups and used jointly by all the elements in the harbor defense.

The mine group was the tactical unit charged with the operation of the controlled mine defense of a harbor, and comprised the personnel, submarine mine equipment, structures, vessels, and equipment necessary to plant, operate, and to protect all or part of the authorized controlled mine project.

MINE GROUP (1930) Mine group commander
    Mine group staff—operations officer, communications officer, mine property officer
        Mine group detachment—command detail, fire control detail, communications detail, message center detail, electrical specialist (NCO)
        Mine property detachment—clerks
        Rapid-fire gun batteries

MINE BATTERY (1930)—(Captain)
    HQ Section
    Maintenance Section (senior NCO)
    Loading and Property Section (loading officer, chief loader)
        Loading Details
        Mine Property Detail

Range Section (range officer, plotter(s) observer(s))
    Plotting Details
    Observing Details
Casemate Section (casemate officer, casemate electrician, asst. casemate electrician,
    power operator, telephone operators)
Planting Section (mine planter commander, executive officer, coxswains,
    planter crew (master, mates, engineers)
    HQ Detachment, deck dept., engine dept., steward's dept.
    Planting Detachment (chief planter)
    D.B. boat crew
    Mine yawl crews

Sources and References

Charles H. Bogart, *Controlled Mines, A History of Their Use by the United States* Weapons and Warefare Monograph No. 50, Bennington, VT 1986 (copies can be ordered from Ray Merriam, 218 Beech St., Bennington, VT 05201).

Merle T. Cole, "U.S. Army Mine Planter Service, 1901-1929," *CDSG News,* Vol. 6, No. 4, (1992) pp. 56-58.

William H. Dorrance, "U.S. Army Mine Planters," *CDSG News,* Vol. 6, No. 3 (1992) pp. 39-41.

Frank T. Hines and Franklin W. Ward, *The Service of Coast Artillery,* Goodenough & Woglam Co., NY 1910

Charles D. Gibson, "The Army's Mine Planter Service" *Journal of America's Military Past* Vol. 28, No. 3, pp 52-71.

U.S. War Department, *Manual for Submarine Mining,* Edition of 1912, GPO, Washington, DC

U.S. War Department, *Drill Regulations, Coast Artillery,* Edition of 1914, GPO, Washington, DC

U.S. War Department, AR 90-70: *Mine Group* (April. 22, 1927) GPO, Washington, DC.

U.S. War Department, AR 90-150: *Army Mine Planters, General* (December 15, 1927) GPO, Washington, DC.

U.S. War Department, AR 90-155: *Assignment, Control, and Use of Army Mine Planters* (April. 30, 1927) GPO, Washington, DC.

U.S. War Department, AR 90-160: *Small Boats* (May 2, 1927) GPO, Washington, DC.

U.S. War Department, AR 610-10: *Army Mine Planter Service* (February 26, 1927) GPO, Washington, DC.

U.S. War Department, AR 610-100: *Warrent Officer Army Mine Planter Service-Clothing Allowance* (April. 20, 1925) GPO, Washington, DC.

U.S. War Department, TM 435-316: *The Battery, Submarine Mine* (December 30, 1926) GPO, Washington, DC.

U.S. War Department, TR 435-315: *The Mine Group* (December 30, 1926) GPO, Washington, DC.

U.S. War Department, TR 435-317: *The Army Mine Planter* (December 30, 1926) GPO, Washington, DC.

U.S. War Department, TR 435-330: *Tactical Employment of Harbor Defense Searchlights and Illuminating Devices* (December 31, 1929) GPOe, Washington, DC.

U.S. War Department, TR 1160-20: *Repair and Testing of Submarine Mine Cable* (April 2, 1928) GPO, Washington, DC.

U.S. War Department, TM 2160-20: *Submarine Mining* (October 15, 1930) GPO, Washington, DC.

U.S. War Department, TM 4-220: *Controlled Submarine Mine Material* (April 27, 1942) Government Printing Office, Washington, DC.

U.S. War Department, FM 4-6: *Seacoast Artillery—Tactics and Technique of Controlled Submarine Mines,* May 1, 1942, GPO, Washington, DC.

K.L. Waters, "The Army Mine Planter Service," *Warship International* (1985) No. 4, pp. 400-411.

E.E. Winslow, Chapter XIV "Submarine Mining" in: *Notes on Seacoast Fortification Construction,* Occasional Papers, Engineer School, U.S. Army, No. 61, GPO, Washington, DC. 1920 (reprint available from the CDSG Press)

Robert D. Zink, "Controlled Submarine Mining in the United States," *CDSG Journal,* Vol. 9, No. 4 (1995) pp. 42-54.

SEARCHLIGHTS IN AMERICAN SEACOAST DEFENSES

Mark Berhow

This section will summarize the key details of the use of searchlights in American seacoast defenses 1900-1945. While some mention will be made of searchlights in antiaircraft defense, this article will not cover that subject in any detail.

Although the concept is as old as fire, the practical problem of illuminating a harbor area during the night was tied to the development of the electric light. As soon as electric generators became available, the US Army began experimenting with electric arc lights as a way of illuminating harbor approaches during the night. The Engineer School established at Willett's Point shortly after the Civil War began using a small electrical searchlight in the mid-1870s. The range was low, 150 to 200 yards at the most.

The 1886 Endicott Report called for the purchase of 200 lights to be installed in the seacoast defenses of the United States, but with no real provision for how it was to be accomplished. The development of the electric light continued into the 1890s. In 1893 a 60-inch German-manufactured searchlight was displayed at the Chicago World's Fair. This light was purchased by the US government and installed at the Coast Artillery School at Fort Monroe for drills and experiments. The war with Spain in 1898 provided funds for the first large scale purchase of some 40-odd lights, mostly 30-inch, with a few 24-inch and 36-inch.

60-inch fixed searchlight and shelter at Fort Casey, Harbor Defenses of Puget Sound, 1920s
(Puget Sound Coast Artillery Museum)

Still, the problem of night illumination remained more or less a theoretical one until the Army got serious about the effort with a special appropriation in the 1901 fortification act for the purchase and installation of electrical searchlights and the electric plants to run them. The Board of Engineers made a careful study of the number of searchlights needed and recommended the installation of 344 lights. The next year, the Board of Engineers recommended the purchase of 60-inch searchlights, just then being produced, which had a range of 6,000 to 8,000 yards. They recommended that they be placed as near the shore as possible, but that their positions be as concealed as possible. Tests and joint maneuvers with the navy occurred during 1904-05.

By the time of the Taft Board Report in 1906, the importance and use of searchlights was of a high priority. The Board recommended the installation of 36-inch and 60-inch searchlights and their electrical plants at all defended harbors, after their proper location was determined by trial. A few years later, it was realized that the 36-inch searchlight was not powerful enough for use and the 60-inch light became the only one suitable.

The War Department then gathered together a number of portable 36-inch searchlight units which were combined into a "searchlight brigade" under the nominal command of Capt. W.C. Davis, C.A.C. The brigade was first mobilized in Portland, ME. This unit assessed the number of lights required and the best possible locations for all the defended harbor areas. Regular, but small appropriations were made over the next few years for the purchase and installation of the searchlights recommended by the various harbor defense projects.

Many of the searchlights installed during the period 1901-1917 were fixed, that is located in a structure of some sort for concealment during the day, with their electrical generator. The Army also examined mobile searchlight models as well; the searchlights and electrical generators made to be carted by trucks and set up at a predetermined position. Over the years, the mobile searchlights became more reliable, durable, and rugged. By the late 1930s, the Coast Artillery switched to using mobile searchlights and replaced fixed searchlights where at all possible. During World War II, the US seacoast defenses used mostly mobile searchlights.

General Operation Principles

The US Army searchlights used an electric arc as the source of light. The arc is formed between a positive and a negative carbon rod (see figure next page). Initially composed of solid carbon with a copper wire core, the later rods were composed of the rare earth elements cerium and lanthanum which gave a higher intensity light. The actual light source is a incandescent ball of vapor formed in the high-intensity arc held in the crater of the positive carbon. This is generated by a high current which vaporizes the core of the positive carbon, which is automatically moved forward and rotated to maintain a consistent crater and keep the arc at the focal point of the mirror. The higher the current, or the smaller the carbon, the more intense the light. The "high intensity" lamp carbons were achieved with the better thinner carbons noted above and cooling was provided with blasts of air. This was called the "Sperry" lamp. A reflecting parabolic mirror is located behind the arc, which reflects the light forward in an almost parallel pattern, which produces a well-defined narrow beam of light.

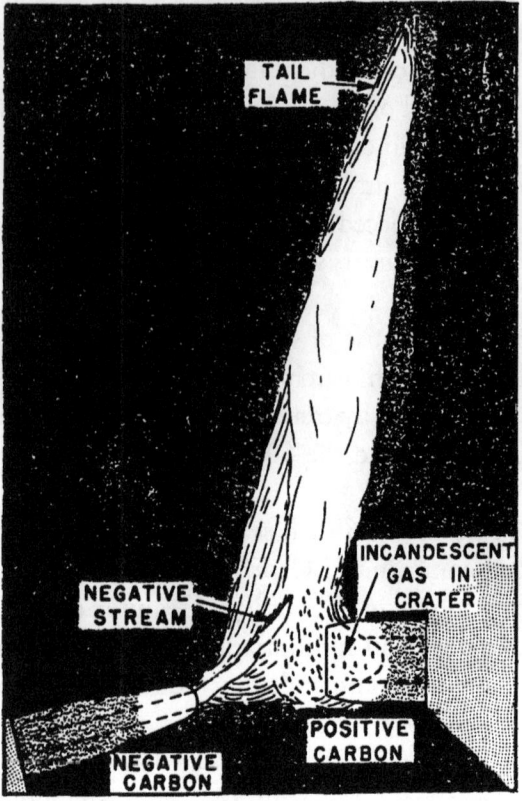

the electric arc

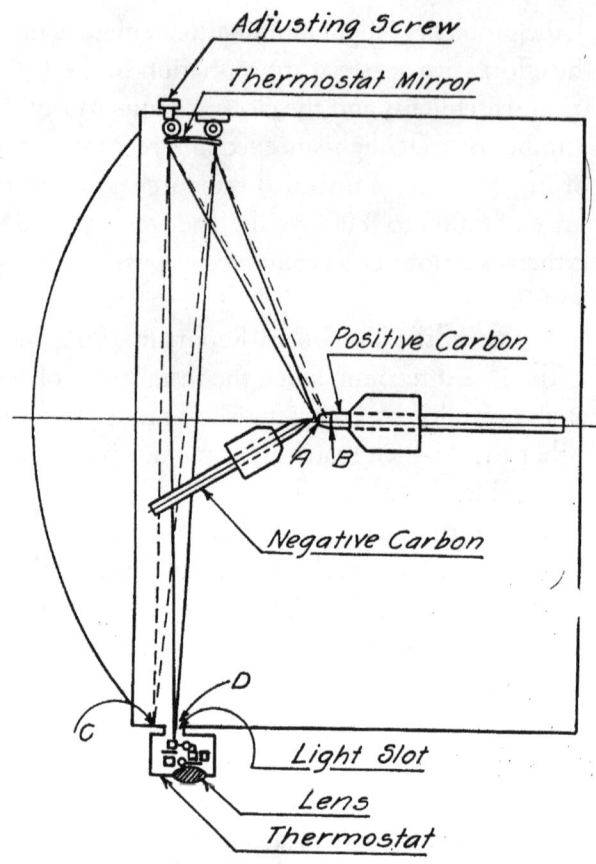

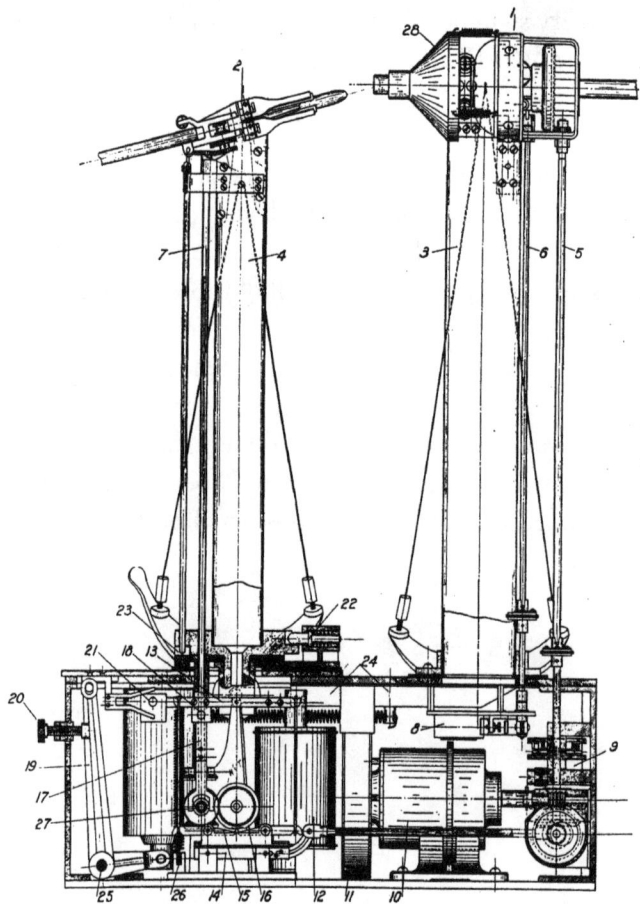

Carbon feeding & rotating mechanism in a Sperry light

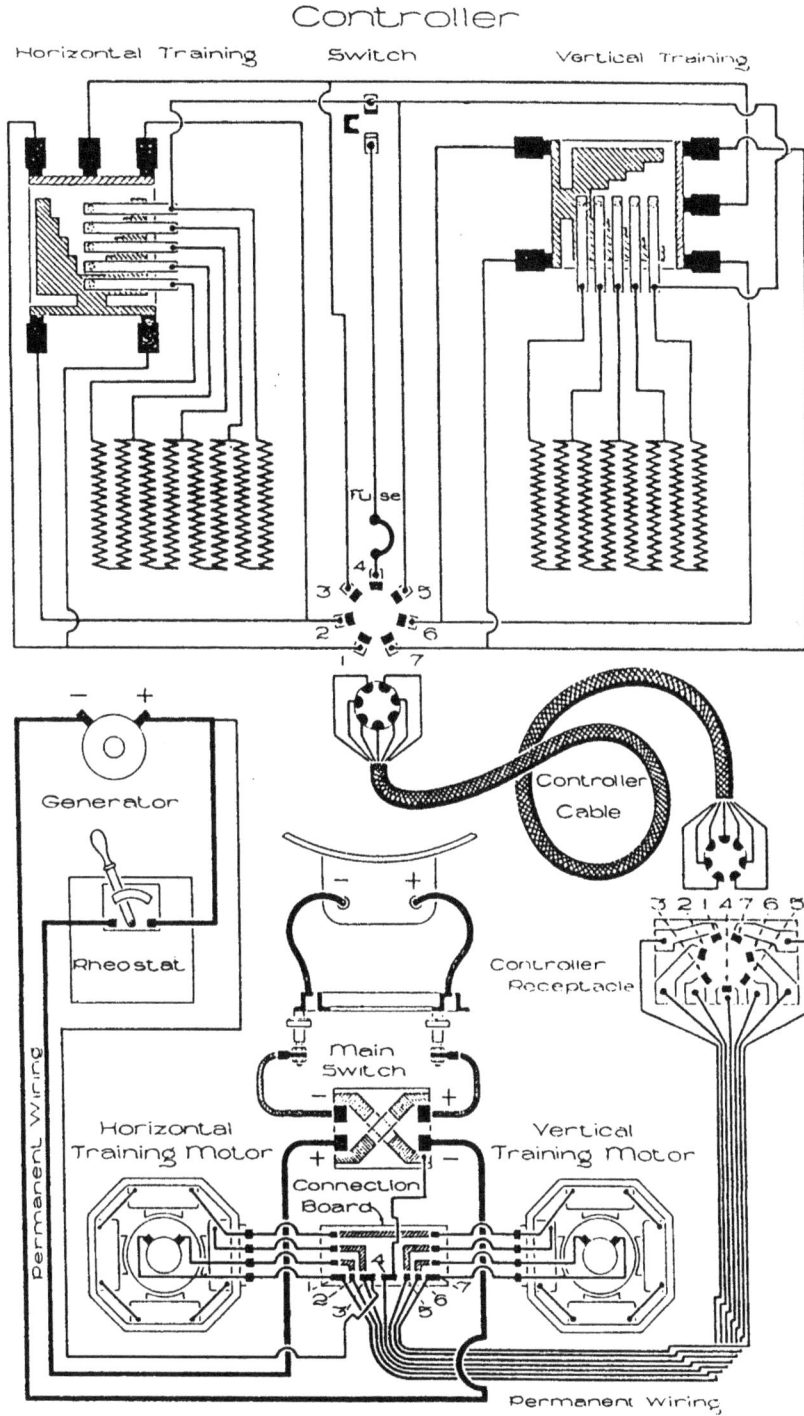

Controller

Horizontal Training   Switch   Vertical Training

Fuse

3 4 5
2 6
1 7

Generator

Rheostat

Controller
Cable

Controller
Receptacle

3 2 1 4 7 6 5

Permanent Wiring

Horizontal
Training Motor

Main
Switch

Vertical
Training Motor

Connection
Board

Permanent Wiring

Projector

Controller operation

30-inch light at Fort H.G. Wright, 1900s (National Archives)

Fixed Searchlight Projectors

The basic design of the searchlight remained the same over the years, with a number of refinements in electrical equipment, motorized parts, controls, etc., and the gradual shift from the heavier fixed searchlights to the lighter mobile searchlights. The earlier fixed searchlights were of 24-inch, 30-inch, 36-inch and 60-inch mirror diameters, and were composed of the lamp, (rheostat, carbons, and feeding and rotating mechanisms providing the arc), the mirror, the drum housing, a series of glass strips across the front to prevent flickering in the wind (the "front door") which could be opened or closed, a turntable and trunnions on the drum, supported by arms, motorized to alter the drum's elevation and azimuth, electrical cabling, a controller for power, azimuth and elevation (located some distance away from the searchlight itself), and the electrical generating plant. By the later part of the 1910s, the 60-inch light became the one of general use, though some of the smaller units remained in used into the 1920s.

Classification of Searchlights (1909)

Initially there were four tactical "types" of lights:

Sentinel beams—fixed beams across a channel that any entering ship would have to cross.
Searching lights—swung back and forth over a specified area to detect an enemy ship
Illuminating lights—fixed on a particular target after it was detected by the searching or sentinel lights.
Blinding lights—to be concentrated on the pilot house of incoming targets.

The sentinel beams and blinding lights were soon dropped, and all searchlights were used as either searching or illuminating.

36-INCH SEARCHLIGHT AND CONTROLLER.

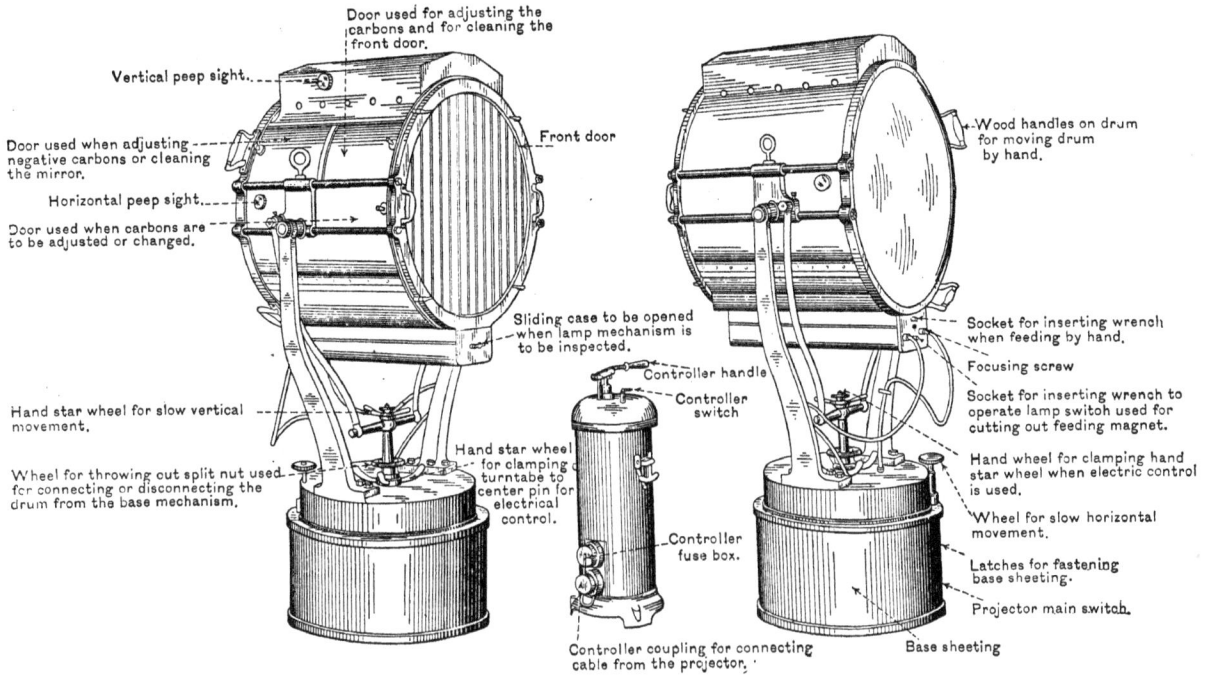

Door used for adjusting the carbons and for cleaning the front door.

Vertical peep sight.

Door used when adjusting negative carbons or cleaning the mirror.

Horizontal peep sight.

Door used when carbons are to be adjusted or changed.

Front door

Sliding case to be opened when lamp mechanism is to be inspected.

Hand star wheel for slow vertical movement.

Wheel for throwing out split nut used for connecting or disconnecting the drum from the base mechanism.

Hand star wheel for clamping turntable to center pin for electrical control.

Controller handle

Controller switch

Controller fuse box.

Controller coupling for connecting cable from the projector.

Wood handles on drum for moving drum by hand.

Socket for inserting wrench when feeding by hand.

Focusing screw

Socket for inserting wrench to operate lamp switch used for cutting out feeding magnet.

Hand wheel for clamping hand star wheel when electric control is used.

Wheel for slow horizontal movement.

Latches for fastening base sheeting.

Projector main switch.

Base sheeting

36-inch Searchlight

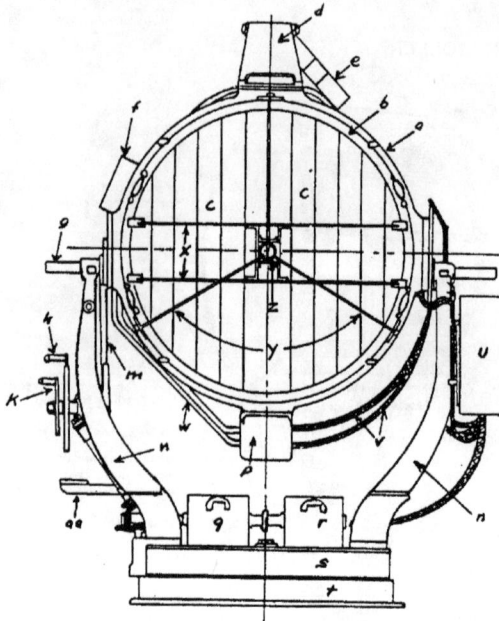

FIG. 80-a—Front View, 60-inch Fixed Light

a—Drum
b—Door
c—Glass Strips
d—Ventilating Hood
e—Muffler
f—Thermostat Instrument Box
g—Drum Trunnions
h—Traversing Handwheel
k—Elevating Handwheel
m—Elevating Rack
n—Right and Left Arms

p—Lamp Trough
q-r—Training Mechanism Boxes
s—Turntable
t—Base
u—Contactor Panel Box
v—Power Leads
w—Conduits
x—Reenforcing Strips
y—Braces
z—Carbon Tube
aa—Operator's Platform

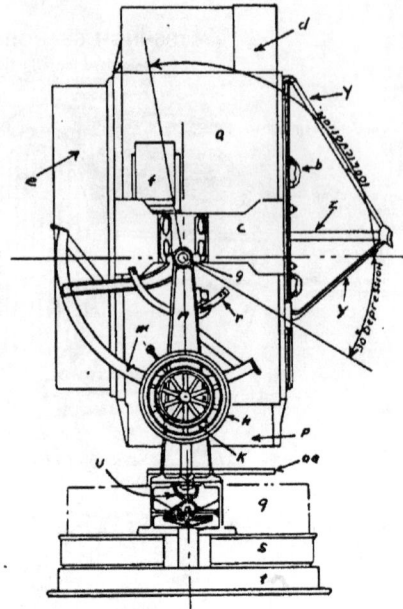

FIG. 80-b—Side View, 60-inch Fixed Light

a—Drum
b—Door Handles
c—Side Plate
d—Ventilating Hood
e—Mirror Frame and Dome
f—Thermostat Instrument Box
g—Drum Trunnion
h—Traversing Handwheel
k—Elevating Handwheel
m—Elevating Rack

n—Right Arm
p—Lamp Trough
q—Training Mechanism Box
r—Elevating Scale Arm
s—Turntable
t—Base
u—Training Gears
y—Braces
z—Carbon Tube
aa—Operator's Platform

60-inch Searchlight

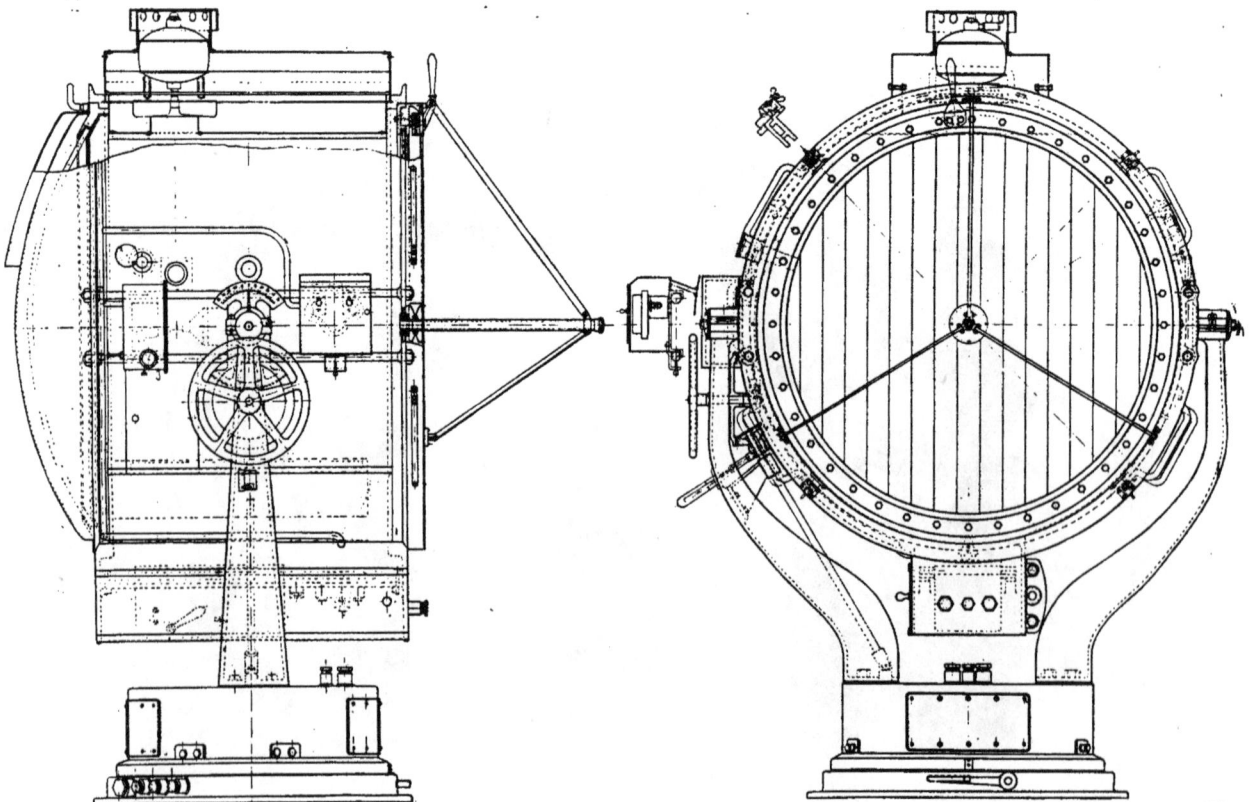

FIG. 4.
SEARCHLIGHT DRUM AND EXTERIOR MECHANISM OF THE SPERRY SEARCHLIGHT LAMP.

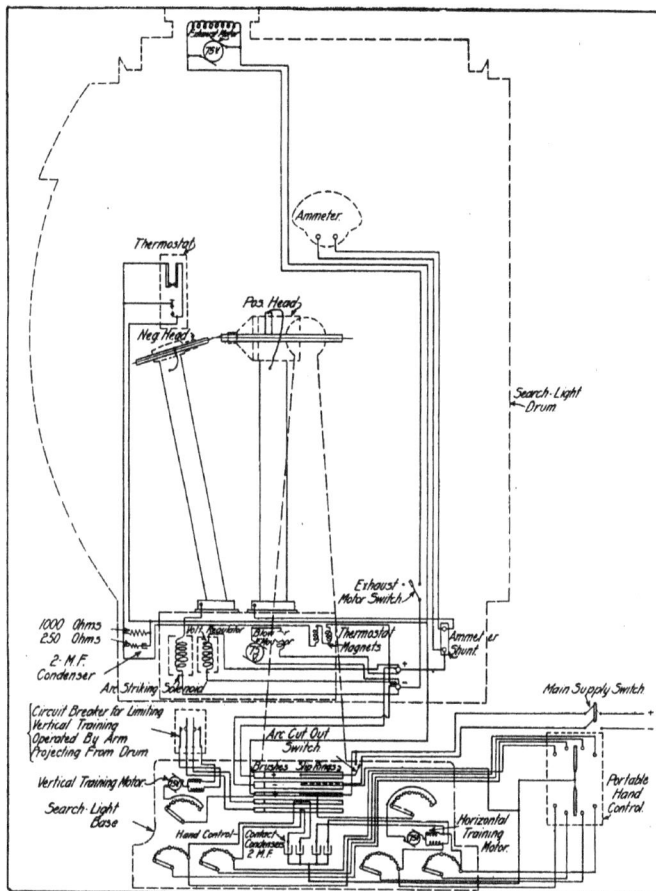

60-inch fixed Sperry searchlight

Searchlights in a given area were generally grouped under one commander, either a gun command or a mine command, or (rarely in seacoast defense) as a separate searchlight command.

The Location of Searchlights (circa 1910)

1) Located with consideration to effectively search the approaches to the harbor area to locate and identify the enemy. The lights were located in advance of the outside line of the primary armament at a distance 1/2 of the effective range of the light.

2) To illuminate continuously with minimum interference a number of targets simultaneously in the defended area. The lights were dispersed in series along the defended area on the flanks of the observation stations and gun batteries.

3) Lights were also placed to illuminate potential landing areas for raiding parties.

4) Lights were positioned to cover mine fields and channels leading to the inner harbors against torpedo boats and mine sweepers.

5) Duplicate lights were set up to provide relief to primary lights that were disabled.

The siting of searchlights was largely based on range. In general, they were located as close to the water's edge as possible to maximize their effective range of between 8,000 and 15,000 yards. They were sited to illuminate the area for effective use of guns and mines. The searchlights were placed in reference to the positions of the base end stations of the horizontal position finding system. The lights had to be located high enough to give a good beam above the base end station, if possible, which tended to be about 60 feet above the high water mark. On the Pacific coast this did not tend to be much of a problem, but the low lying Atlantic and Gulf coasts required elevated platforms or "disappearing" towers for searchlights. Also, consideration was given to where the next searchlight was located, to overlap the illumination arcs. Some lights were co-located so that one could be a backup in case the other malfunctioned or was knocked out by enemy action. Only one of these was used at a time to prevent confusion caused by crossing beams. The development of portable searchlights expanded the tactical positioning in that the searchlights did not have to be on the fixed government reservations, but could be deployed temporarily anywhere they were required.

Fixed searchlights were generally provided with a shelter. The shelters functioned to protect the searchlight from the elements, to house the electrical generator, and to provide concealment for the light when it was not in use. In many places searchlights were mounted on small cars which were moved back and forth on a few feet of track from a recessed shelter to the operating position. Searchlights were also housed in fake water towers, or beach cottages and the like, which had faces that could be opened for use. Other searchlights on the tops of hills were lowered vertically into pits with a movable roof, or they rolled back down an inclined plane. The disappearing towers mentioned above also functioned as a method of concealment. This all became moot with the adaptation of the mobile light, since the light could simply be moved from the exposed position when not in use.

Searchlight tower and shelter

Bascuole "disappearing" seachlight tower

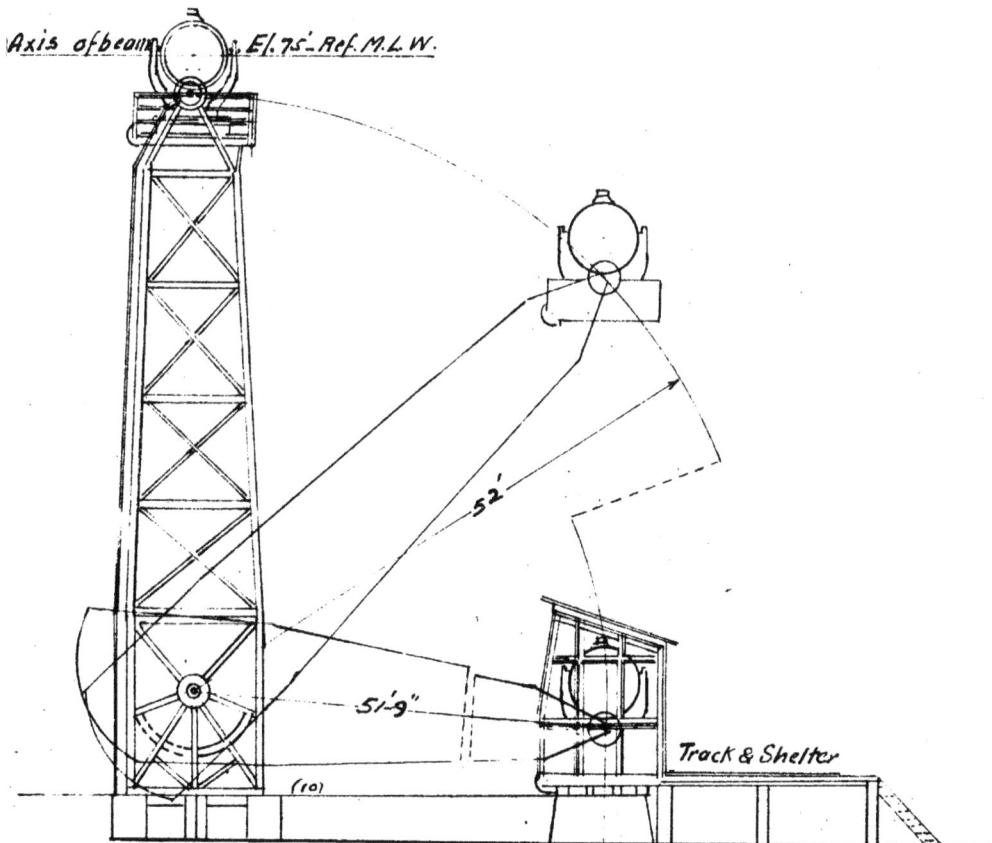

Axis of beam : El. 75' Ref. M.L.W.

52'

51'9"

(10)

Track & Shelter

REPORT of COMPLETED WORKS
SEEA COAST FORTIFICATIONS
(Searchlights)
Corrected to   1922

COAST DEFENSES of PENSACOLA
FORT PICMENS, FLA.
SEARCHLIGHTS Nos 8/C
60 INCH - FIXED
Scale - 1"- 15'
0   5   10   15'

Fixed Searchlight No. 18 in its shelter, FortWorden, HD Puget Sound
(Puget Sound Coast Artillery Museum)

Fixed Searchlight No. 18 deployed showing power controls, Fort Worden, HD Puget Sound
(Puget Sound Coast Artillery Museum)

Searchlight shelters: (clockwise from top left) Fort Totten (Alex Holder 1986), Top of Disappearing-type at Fort Andrews (Alex Holder 1988), interior of disappearing-type, Fort Terry (Mark Berhow 2000), Combination searchlight shelter and power plant structure in the Harbor Defenses of Kodiak (Alex Holder 1996), shelter for searchlight #18 Fort Worden (MAB 2010)

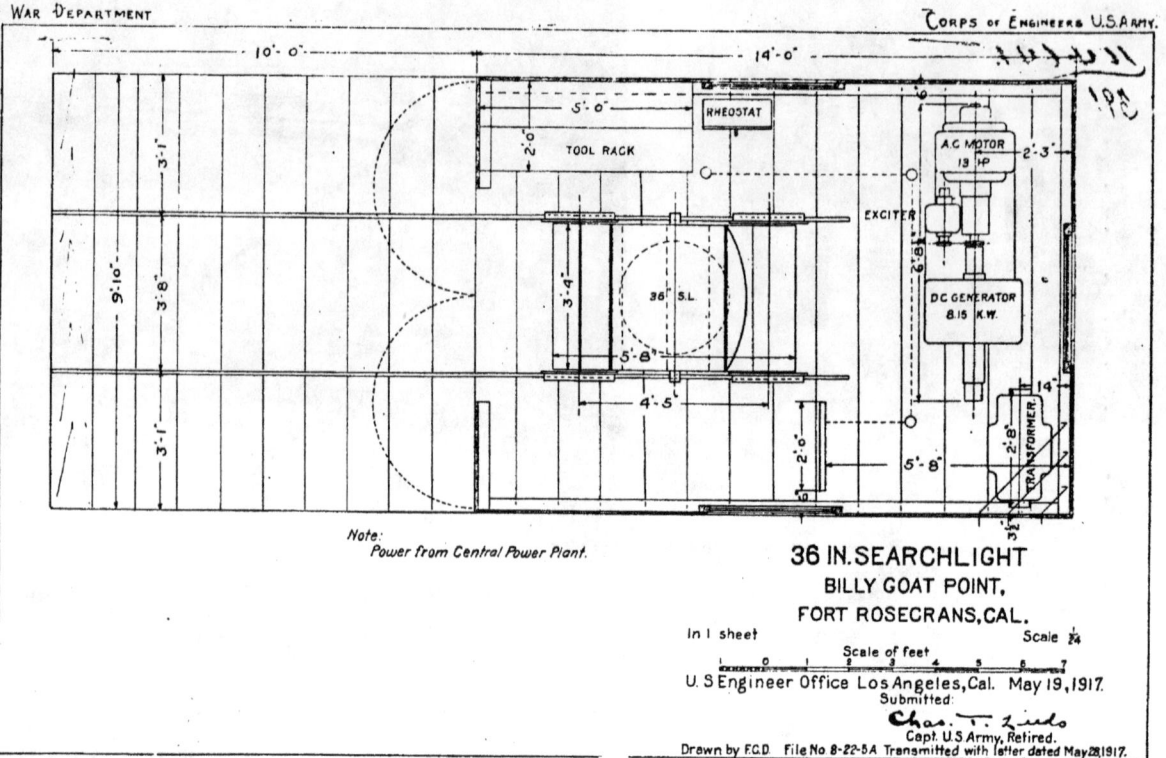

WAR DEPARTMENT

CORPS OF ENGINEERS U.S. ARMY.

RHEOSTAT

TOOL RACK

A.C. MOTOR 13 HP

EXCITER

36" S.L.

D.C. GENERATOR 8.15 K.W.

TRANSFORMER

Note:
Power from Central Power Plant.

36 IN. SEARCHLIGHT
BILLY GOAT POINT,
FORT ROSECRANS, CAL.

In 1 sheet   Scale ¼
                Scale of feet

U.S Engineer Office Los Angeles, Cal. May 19, 1917.
Submitted:
Chas. T. Leeds
Capt. U.S. Army, Retired.
Drawn by F.C.D. File No. 8-22-5A Transmitted with letter dated May 28,1917.

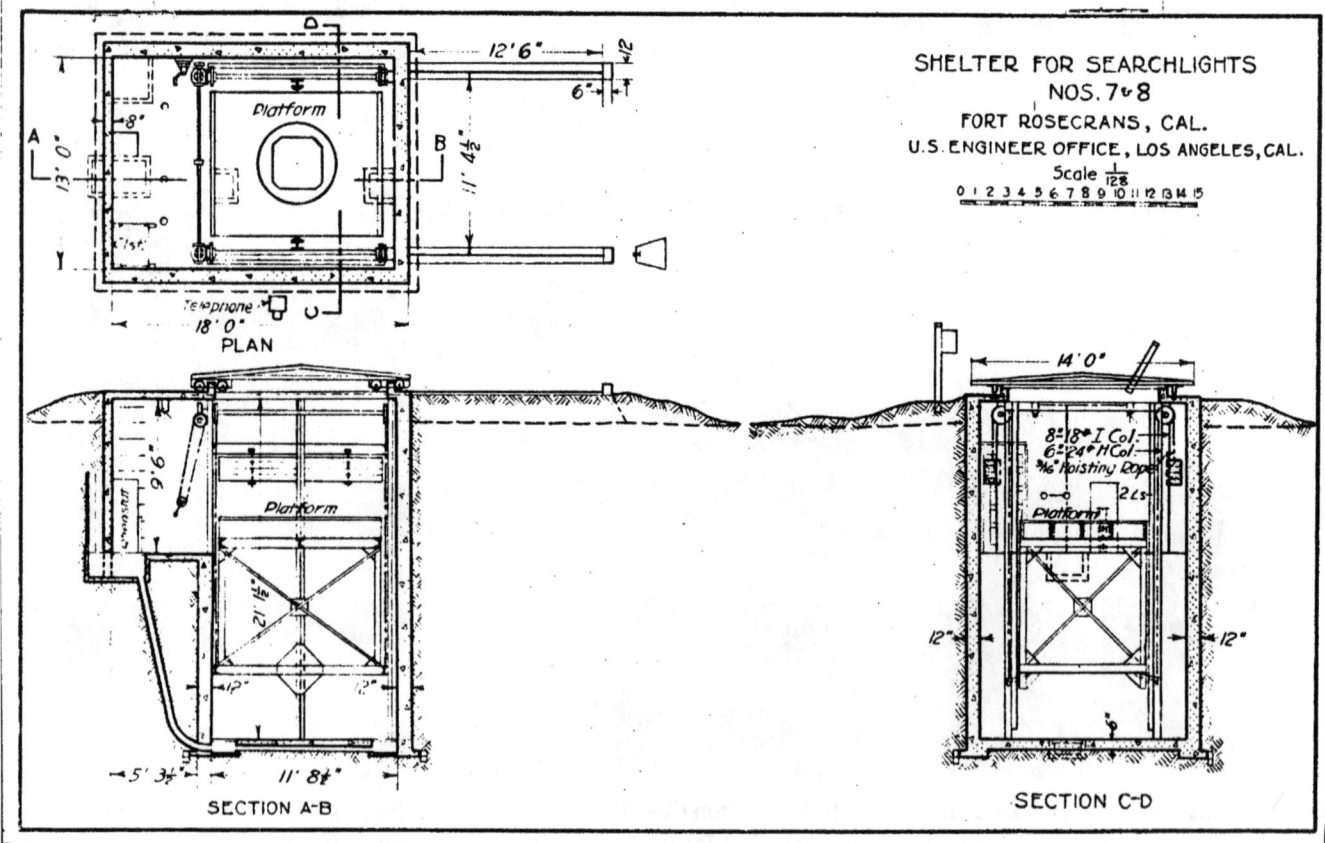

SHELTER FOR SEARCHLIGHTS
NOS. 7 & 8
FORT ROSECRANS, CAL.
U.S. ENGINEER OFFICE, LOS ANGELES, CAL.
Scale 1/128
0 1 2 3 4 5 6 7 8 9 10 11 12 13 14 15

PLAN

Platform

Telephone

SECTION A-B

Platform

SECTION C-D

Platform

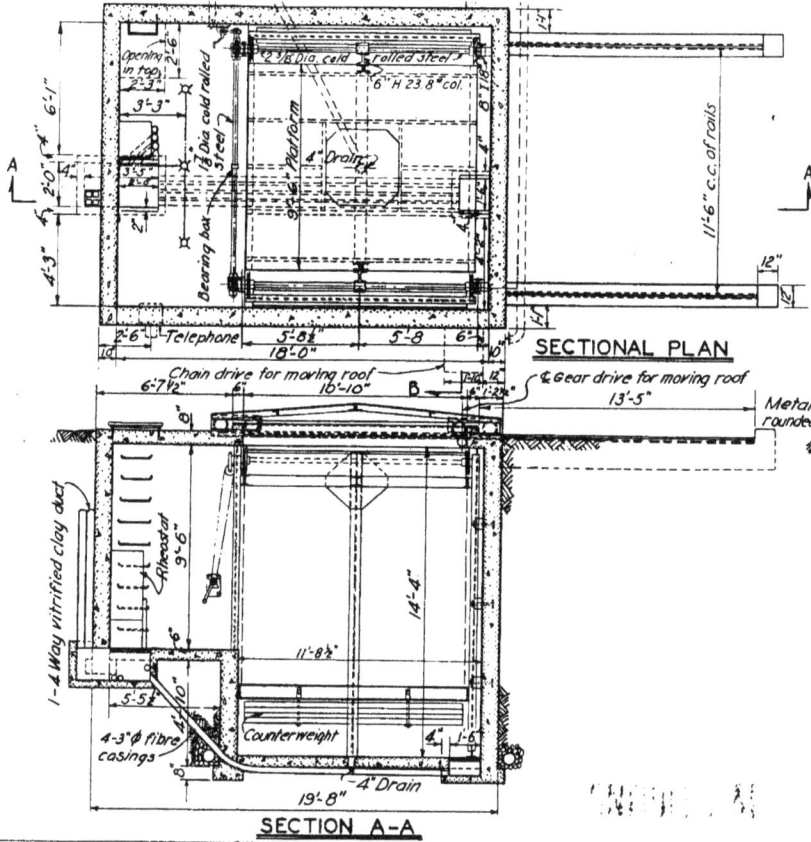

HARBOR DEFENSES OF SAN DIEGO, CALIF.

SEARCHLIGHT SHELTER NO. 15

U.S. ENGINEER OFFICE, LOS ANGELES, CALIF.

SCALE IN FEET

CONFIDENTIAL

SECTIONAL PLAN

SECTION A-A

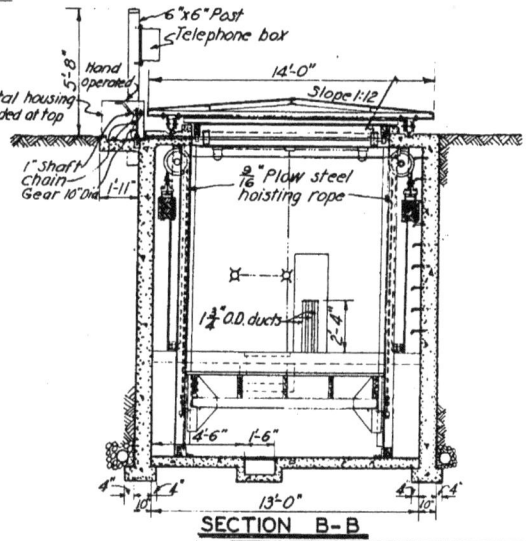

SECTION B-B

HARBOR DEFENSES OF SAN DIEGO, CALIF.

POWER PLANT SHELTER
FOR SEARCHLIGHT NO. 15

U.S. ENGINEER OFFICE, LOS ANGELES, CALIF.

SCALE IN FEET

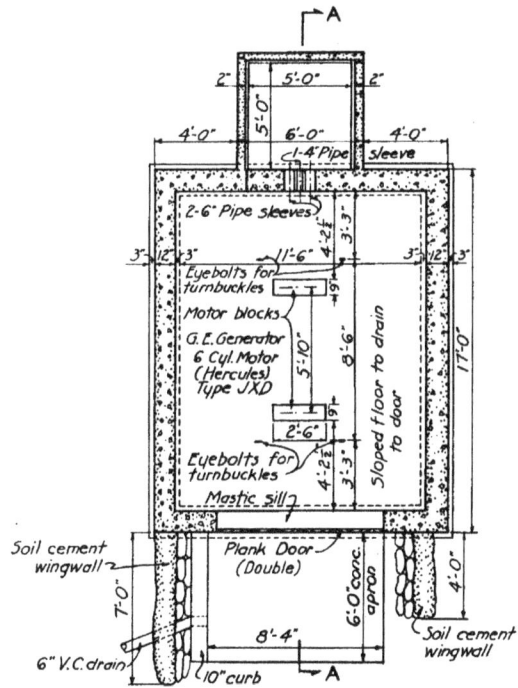

SECTIONAL PLAN C-C

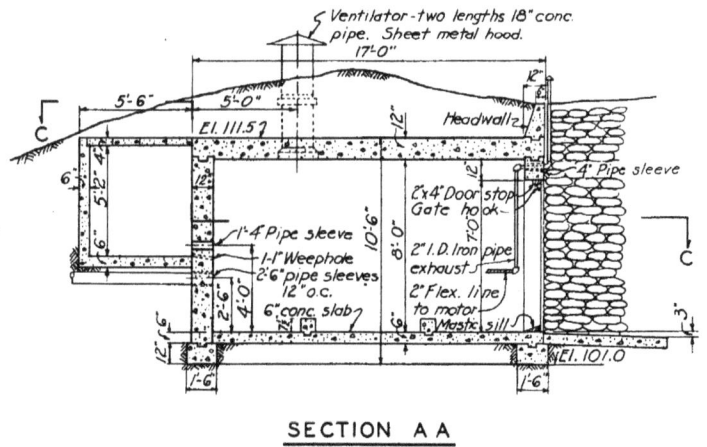

SECTION A A

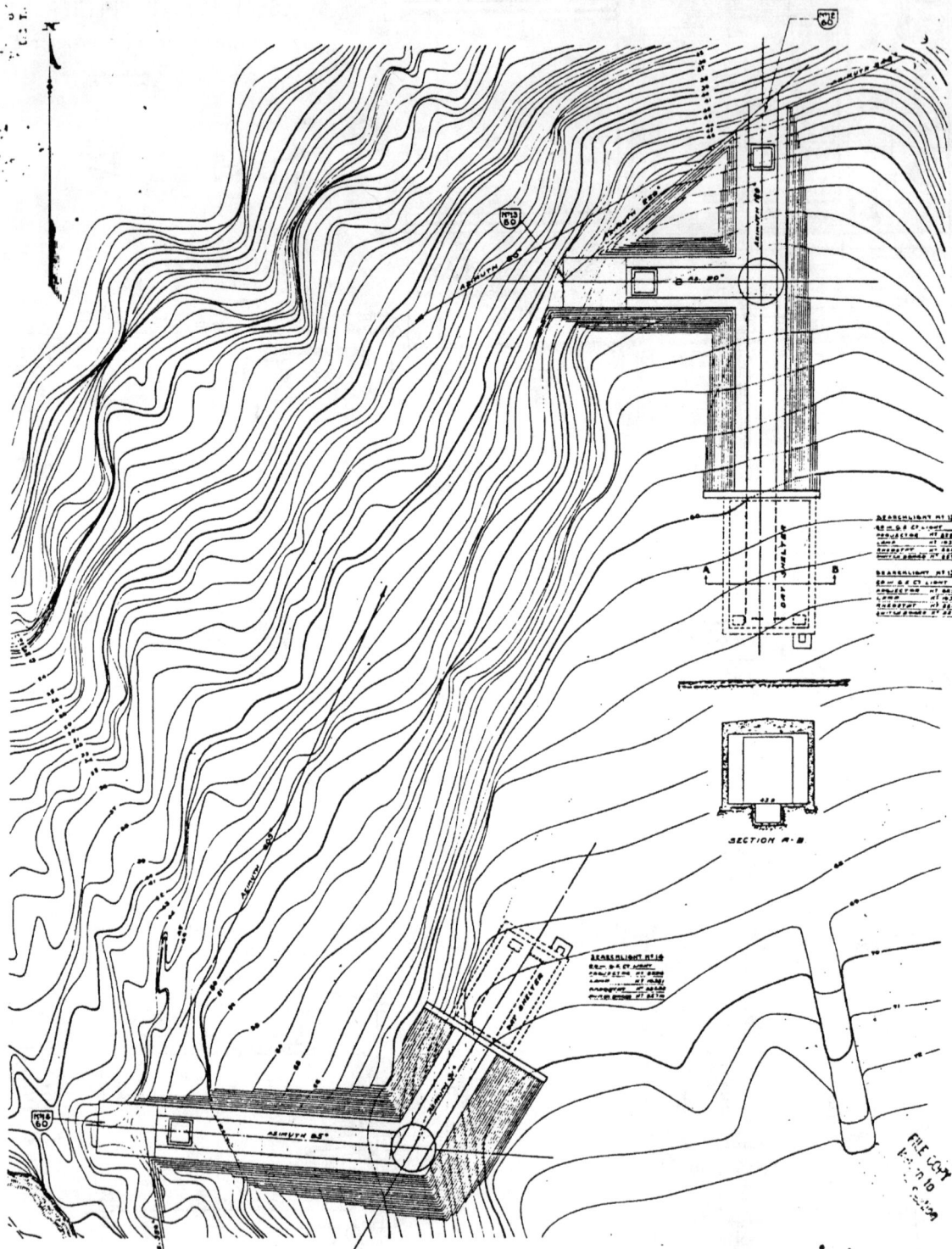

DEFENSES OF PHILIPPINE ISLANDS
ENTRANCE TO MANILA BAY
SEARCHLIGHT INSTALLATIONS
SEARCHLIGHTS
FORT FRANK CARABAO ISLAND

Scale 1"=10 feet

U.S. Engineer Office, Manila, P.I.
Nov. 14, 1913.

Respectfully submitted to the Chief of Engineers;
approval recommended.

Designation of Searchlights

Searchlights were designated by consecutive numbers facing seaward from right to left. Mobile searchlights were designated in the same manner; predetermined locations were numbered consecutively from right to left facing seaward and searchlights assigned to that position were given that number.

Early on (1910s) searchlights were assigned as either searching or illuminating lights and as either fort, fire, or mine command lights. Later this "permanent" assignment was dropped and lights were assigned based on tactical demands. Certain searchlights were assigned to search their respective areas. Once targets were spotted, that searchlight, or another, was designated to illuminate that target. Other lights would pick up the searching job, and once the target passed out of the range of one illuminating light, another would be assigned to it.

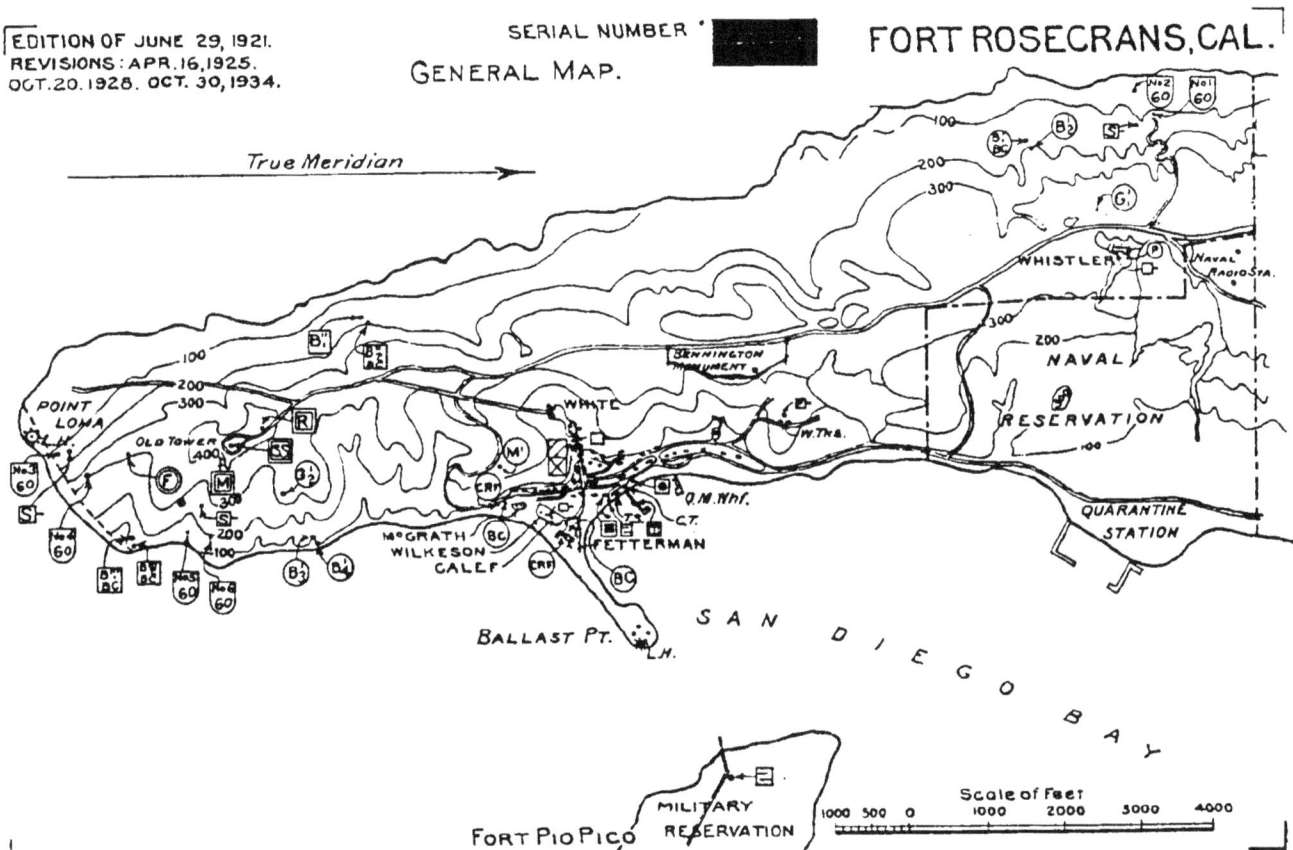

Fort Rosecrans military reservation 1934. Fixed searchlight locations are indicated by the number "60" (the size of the searchlight) and the individual tactical numbers 1-6.

Searchlight Squad

In 1914 the searchlight squad consisted of: a controller operator, a searchlight operator, an assistant, a watcher, an engineer, and a telephone operator. By 1937 this was simplified somewhat to a light commander (NCO), a controller operator, a light operator, and a power plant operator. Mobile searchights may have been assigned more men for drivers and to assist in the set up of the equipment.

Mobile Searchlights

Though initially horse drawn, the development of the motorized vehicle, especially during the WW I years, brought the mobile searchlight to the forefront of tactical use. Initially the mobile motor unit consisted of two trucks, one mounting the 25 kW generator, the other mounting the searchlight. Later, the development of lighter searchlights and a rubber wheeled trailer resulted in the elimination of one of the trucks. Additional searchlights were developed for anti-aircraft work, and eventually the standardized units of either General Electric or Sperry manufacture of the M1941 & M1942 designs were used during WW II for both AA and HD defense.

36-inch seachlight on a trailer

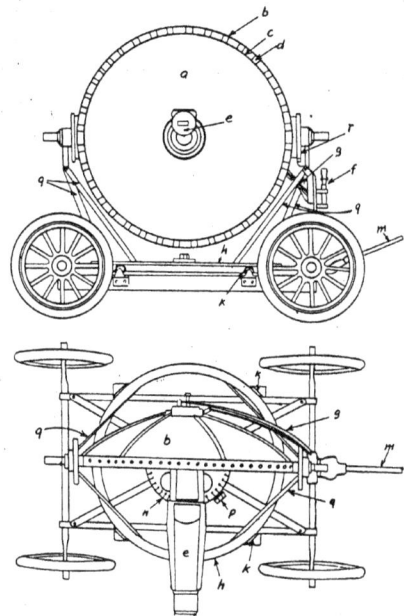

Fig. 90—Sketch of Mobile Searchlight in Operating Position (Open Type)

a—Reflecting Mirror
b—Mirror Frame
c—Mirror Retaining Ring
d—Mirror Retaining Lugs
e—Lamp Mechanism
f—Power Switch
g—Power Leads
h—Turntable

k—Traversing Rollers
m—Maneuvering Handle for Hand Towing
n—Azimuth Circle Plate
p—Azimuth Pointer
q—Trunnion Supports
r—Elevation Clamp

M1934 Mobile Searchlight (left)
M1933 Mobile Searchlight (right)

1. Lamp control mechanism box.
2. Elevation zero reader.
3. Azimuth zero reader.
4. Hand controller socket.
5. Arc and elevation control box.
6. Ventilating fan motor.
7. Lamp unit.
8. Arc view peep sight.
9. Ground-glass finder.
10. Front drum.
11. Mirror.
12. Positive carbon.
13. Azimuth control box.

M1941 General Electric Mobile Searchlight (left)
M1941 Sperry Mobile Seachlight (right)

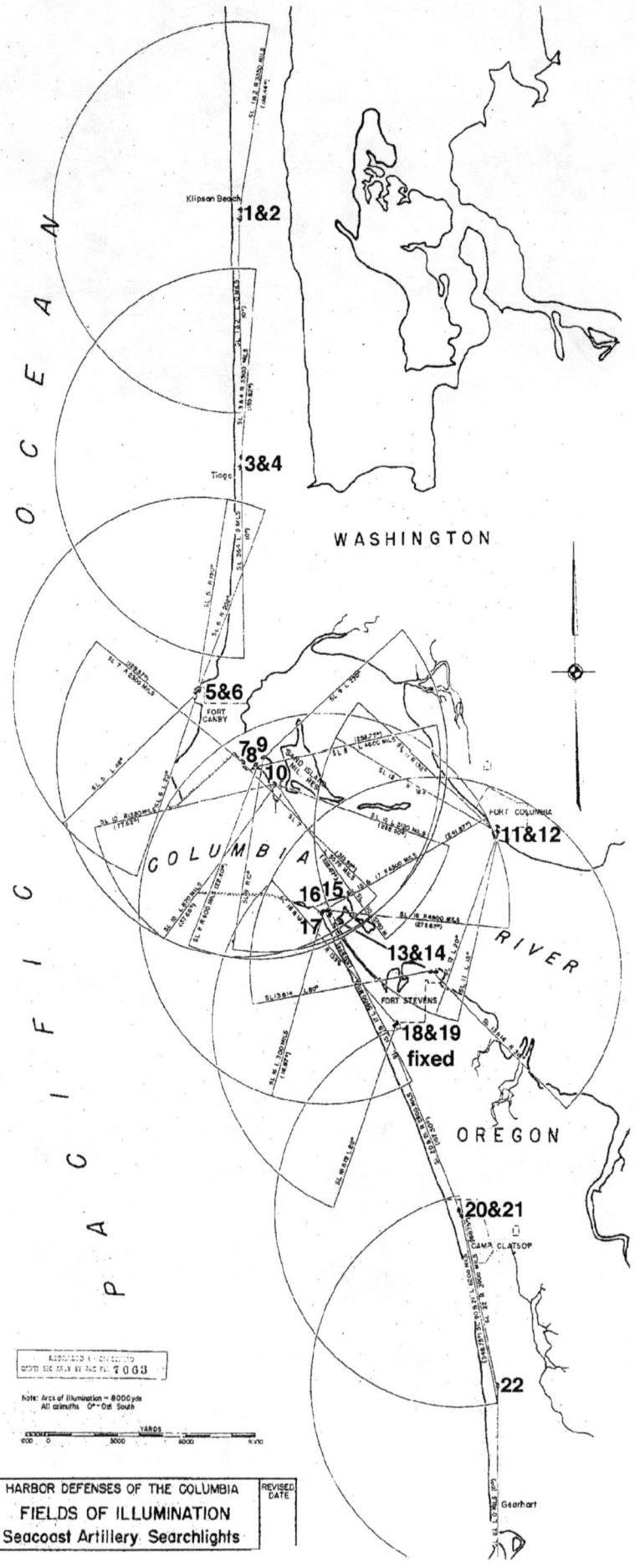

Searchlight on display at Fort Stevens
State Park, OR

Fields of illumination for the searchlights assigned to the Harbor Defenses of the Columbia, 1944 (National
Archives).

Antiaircraft Searchlights

Antiaircraft artillery regiments were formed in the years following WW I and assigned to the Coast Artillery Corps. They had mobile searchlights, some that were specially designed for AA use. The searchlights were usually deployed as searchlight batteries. These were augmented with sound ranging devices for detecting the approach of aircraft by sight and sound.

Searchlights and sound locators of the 64th Coast Artillery Regiment on parade

M1933 Mobile AA Searchlight

Antiaircraft Searchlights

Antiaircraft artillery regiments also fielded more sophisticated equipment as WWII progressed. One element of the Coast Artillery Corps that would see action after the opening of the war was antiaircraft searchlights. These searchlight units were usually deployed as searchlight batteries. These were simple and without the mobile range finders used in detecting the approach of aircraft based upon sound.

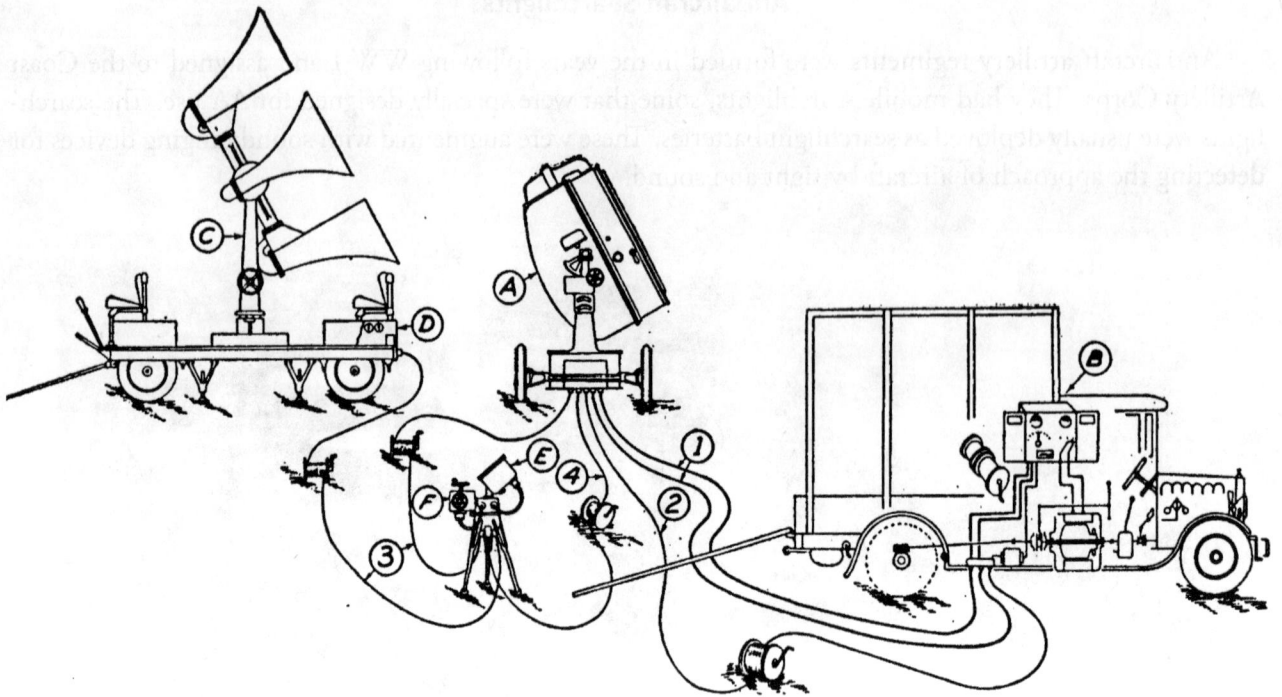

M1933 Searchlights on Cadillac truck, Fort MacArthur 1930s (Fort MacArthur Museum)

TACTICAL EMPLOYMENT OF SEARCHLIGHTS

From: COAST ARTILLERY FIELD MANUAL FM 4-5,
SEACOAST ARTILLERY ORGANIZATION AND TACTICS, 1940

SECTION I
GENERAL

MEANS OF ILLUMINATION.—Three means of securing illumination of water areas and targets at night are searchlights, star shells, or airplane flares. Successful defense against naval attack or raid at night will depend largely upon the efficiency and correct tactical use of searchlights. Star shells (not standard) and airplane flares are supplementary to searchlights and extend the range of effective illumination beyond the limited range of the searchlights.

SECTION II
MISSIONS, TYPES, AND FUNCTIONS

MISSION.—The primary mission of searchlights is illumination of hostile naval targets during periods of darkness. In addition to their primary mission they may be employed to search water areas, to search or illuminate beaches, to serve as barrier lights, and to serve as a means of signal communication. However, their employment for purposes other than illumination of targets should be reduced to a minimum.

TYPES.—Searchlights are classified according to type as fixed and mobile.
a. *Fixed searchlights* include those mounted in fixed positions or capable of but limited movement for purposes of protection or concealment.
b. *Mobile searchlights* include types which are transported on special motor trucks with power units either integral with the truck or transported or towed by a separate vehicle, and a type designated *portable* for which no special motor vehicle is provided.

FUNCTIONS.—Searchlights are classified according to functions as barrier, searching, and illuminating. Functions of searchlights are not rigid. Any light may on occasion perform functions other than those pertaining to its normal assignment.
a. *Barrier lights* are those used in certain exceptional cases to detect passage of vessels toward or through a channel or harbor entrance.
b. *Searching lights* are those used to search a water area for the purpose of detecting presence of vessels therein.
c. *Illuminating lights* are those used to illuminate a target in order that it may be tracked and fired upon.

SECTION III
EMPLACEMENT OF SEARCHLIGHTS

DETERMINATION OF NUMBER.—a. Determination of exact number and location of searchlights required for any particular locality is a local problem which can be solved most satisfactorily by actual tests with mobile lights. It will depend on local hydrography and topography, particularly on configuration of the shore line as related to character, extent, and location of the armament; on number and grouping of horizontal base lines; on location of base-end stations with relation to other elements of the defense and to channels of approach, and on general atmospheric conditions in the area. In general the number of lights should be sufficient to provide two lights at each end of each separate group of horizontal base lines for use primarily as illuminating lights for the armament served by that group of base lines, and a sufficent number of additional lights as determined by local conditions to provide for searching and barrier missions.

b. Installation in pairs.— In general in each important position a pair of lights instead of a single light should be installed. Where the interval between adjacent ends of two adjacent groups of base lines is relatively small, a total of two (or three, depending upon requirements for searching lights) lights instead of four lights (one pair for each group of base lines) will often be sufficient for the interval, since with a proper communication system any of these lights should be able to serve either of the two groups of base lines.

HEIGHTS OF SITE.—a. *Minimum elevation.*—A searchlight in its operating position should be not less than 40 feet above the water in order that usefulness of the light may not be limited by effect of curvature of the earth. A minimum height of 60 feet is desirable to facilitate observation.

b. Maximum elevation.—(1) The disadvantage of too great a height of site is the unilluminated or dead space between the end of the beam and the shore within which presence of a vessel may not be detected when the light is being used at its outer ranges. Lights normally used for barrier or searching lights should not be sited at heights greater than about 100 feet in order that the diffused light from the beam may serve to eliminate any dead space in illumination of the water throughout the entire length of the beam.

(2) Lights normally used for illuminating lights may be sited at heights as great as 300 or 400 feet. Siting illuminating lights at different heights within permissible limits facilitates searchlight maneuvering by increasing the number of lights that may be used to cover an area without mutual interference.

NECESSITY FOR ADVANCED LOCATIONS.—a. *Barrier lights* when required normally will be located on flanks of entrances they are designed to protect. This may result in their location in positions exposed to attack. It may be practicable in some cases to locate a barrier light so as to enfilade a narrow channel or water approach.

b. Searching and illuminating lights should insofar as practicable be located to cover every part of assigned areas, including water areas suitable for use of small boats used in landing attacks although not navigable for larger vessels, and including suitable landing places. Range of seacoast artillery greatly exceeds searchlight range. This precludes practicability of covering with searchlights the entire zone of its effective fire. It also demands that full advantage be taken of advanced positions for lights, whenever available, to the front or toward a line of approach on a flank of the guns. In general, on account of dead space created in front of the guns, lights should not be installed in rear of the guns they serve unless the lights are at a considerably higher elevation than the guns.

LOCATION WITH REFERENCE TO OTHER ELEMENTS.—a. *Difficulty of observing through beam:*—A target though well illuminated cannot be seen through the length of a searchlight beam and cannot be seen very clearly through the width of the beam. The ideal arrangement therefore is to have the controller operator illuminate the target with the side of the beam on which he and all observers and gun pointers are stationed. It follows that if possible an illuminating light should be on a flank of all elements it serves.

b. Flank locations; difference in elevation.—Based on the same fundamentals given in *a*. above, the greater the difference between elevation of the light and the observers, the better the observation of the target. Experience has shown that, where practicable, the light should not be less than 150 yards from the flank of the nearest fire-control station or battery in order that diffused light from the beam may not interfere with observation, and that the difference in elevation of the light and of any observer who is to observe in its beam should be not less than 20 feet. Where practicable, an illuminating light should be sited higher than the elements it serves, since, with this relative position, when the beam is held so that only its lower elements rest upon the target, it interferes least with observation. This consideration should not however be permitted to govern to the extent of requiring the light to be installed on a conspicuous structure or to be unduly withdrawn to the rear.

c. Illumination of foreground.—Location of the light should be such that illumination of the foreground does not indicate other elements of the defense and that fire directed at the light or at the other elements will not endanger both.

d. Illumination of mine field.—Location of lights should provide for effective illumination of the mine field and of water areas containing nets, booms, and other underwater barriers subject to enemy attack.

INSTALLATION OF LIGHTS IN PAIRS.—*a. Mutual support.*—Lights normally should be installed in pairs so that in case one light is put out of action for any reason, whether by enemy action or for purposes of recarboning or readjusting, the other light will be readily available for use. While two lights should be located with a view to being capable of mutually acting one for the other, they must be sufficiently separated to insure that both may not be destroyed by a single shot or bomb. As a rule lights installed in such a manner cover a greater water area than a single light and tend to reduce dead space.

b. Economy of installation and operation.—Installation of lights in pairs furthermore serves to reduce cost of installation, since a single structure can house generating sets for both lights, and similarly it reduces personnel overhead required for operation of these sets.

PROTECTION.—The most effective protection against enemy gunfire or air attack is afforded by concealment of lights when not in action and by concealment of power plants. The most satisfactory cover and concealment for fixed lights are secured by—

a. Mounting searchlights on a tower that can be lowered vertically into a prepared well when not in use, this well being so constructed that when covered its position cannot be located readily by visual or photographic means.

b. Mounting light on a car which can be run into prepared cover when not in use.

c. Mounting light on a bascule mount or swinging tower that can be lowered under trees or other cover affording concealment from observation from the water and from the air.

d. Withdrawing mobile lights during the day from operating positions to other positions affording better concealment, and by frequent changes of operating position.

e. Locating power plants of fixed lights in well concealed bombproof structures, and mobile plants in woods or other positions where concealment and cover may be secured both against aerial and terrestrial observation.

BEACH ILLUMINATION.—*a. Secondary missions of searchlights.*—While effective illumination of naval targets is the primary consideration in location of searchlights, adequate provision must be made where necessary for support of beach defense by providing for searching and illuminating landing places. In determining location of lights in any defended area, either within or without a harbor defense, thought should be given to illumination of these landing places.

b. Locations.—In many cases this is practicable if the light is located on a projecting point of land from which the shore on either flank can be illuminated. In some instances advantage can be taken of small islands near the shore for positions on which a searchlight can illuminate both water to the front and shore line to rear and flanks.

LOCATION WITH REFERENCE TO CONTROLLER AND SEARCHLIGHT.—Ordinarily each fixed light is operated from a control station by means of a distant electric controller. The controller is located in such fire-control station as affords an excellent view of the water area normally to be covered by the searchlight. In the location of the light, the position of the control station is important because of desirability of minimizing length of both telephone line and controller cable. It is also desirable that the light be not more than a few hundred yards from the controller and on its flank, as otherwise it may be difficult in case many lights are in action for the controller operator to recognize his own beam.

MOBILE SEARCHLIGHTS.—Mobile searchlights are employed outside harbor defenses with tractor-drawn and railway artillery, and to supplement fixed lights within the harbor defenses. Where suitable roads are available their mobility is an advantage in that alternate positions may be readily prepared and occupied, and isolated or distant stations used where fixed lights would be difficult to maintain. When plans contemplate their employment in a harbor defense, positions and communication lines should be prepared and maintained at all times.

SECTION IV
SEARCHLIGHT DIRECTION AND CONTROL

GENERAL.—a. *Searchlight direction,* includes determination as to when each light will be put in action, who will control it, and the general nature of employment, including the decision as to whether it is to be used as a barrier, searching, or illuminating light, and the particular water area or ship formation to be covered.

b. Searchlight control includes the issue of necessary instructions to operators of the light to put it into or out of action and assignment of a mission which, in the case of illuminating lights, includes assignment of the target. Searchlight control is exercised by the officer who communicates directly with the controller operator or with the operator of the light in case the controller is not used.

PHASES INVOLVED.—Two phases are involved in direction and control of searchlights used by seacoast artillery. The first phase extends to the time targets are definitely located and assigned to groupments or groups and involves use of barrier and searching lights only; the second phase introduces use of illuminating lights. Control of such lights as may be used to support beach defenses during the second phase will be exercised by the senior tactical commander directly charged with support of beach defense, or by such subordinates as he may designate. The following applies in the direction and control of searchlights:

a. In general, direction of all searchlights that cover a given area must be centralized in the senior tactical commander who can observe the area. This is necessary to insure that the number of lights in use shall be maintained at the minimum consistent with the tactical situation, and to obviate interference between lights in action.

b. During the first phase, the officer or officers exercising searchlight direction will also exercise searchlight control of such lights as may be put into action. Exercise of such control which will involve special signal communication will be facilitated by a carefully prepared plan of illumination.

c. During the second phase, direction and control of such barrier lights and searching lights as may be maintained in action for intelligence purposes normally will continue to be exercised by the officer who exercised these functions during the first phase. This officer may relinquish control of an illuminating light or lights to a subordinate commander when the—

(1) Subordinate commander is in better position to observe the water area concerned.

(2) Subordinate commander is directing all or major part of the action.

(3) Engagement for any reason assumes such a character that centralized control is impracticable.

CONTROL.—*a.* To facilitate searchlight control the navigable water area is divided into subareas, each designated by a suitable name, which may be the same as the subareas into which the battle area is divided.

b. Looking seaward, fixed searchlights of a harbor defense are designated consecutively from right to left by numerals. Sites selected for occupation by mobile searchlights are designated consecutively from right to left by prime numerals. A mobile searchlight, when occupying a given site, is designated by the number of that site.

c. Commanders exercise their functions of searchlight control through searchlight officers who utilize a jack set or cordless switchboard over which direct telephone communication may be had with each searchlight and each controller operator.

CONTROLLER OPERATOR.— In addition to executing commands of the searchlight officer, he is responsible that the:

a. Target assigned to his light is properly illuminated and that in searching the beam is combed for targets. He should be provided with an observing instrument to enable him to follow his target effectively and to discover targets in his beam.

b. Azimuth scales of the controller box and searchlight always read the same. The controller operator will cause the searchlight operator to mesh the light with the controller mechanism at an azimuth designated by him at the command PREPARE FOR ACTION, and will check the azimuth at appropriate intervals thereafter.

c. Approximate azimuth desired should always be reached before the light goes into action. When the light is ordered into action and given a mission, the controller operator will first traverse it *dark* or, if using hand control, cause the searchlight operator to traverse the light *dark* to the approximate azimuth of the target to be covered or area to be searched.

d. Searchlight officer is informed when the carbon of his light is down to about 5 inches and when in following a target his light gets within 10° of its limit of traverse.

e. Hand control is started without command from the searchlight officer and without delay in case electrical control goes out of order while the light is in action. The controller operator will give the necessary commands for hand control to the searchlight operator to continue to search or follow by hand, and then notify the searchlight officer that electrical control is out of order and that he is maneuvering the light by hand.

SEARCHLIGHT TELEPHONE OPERATOR.—The searchlight telephone operator is stationed at the light. He receives the commands of the controller operator and the searchlight officer. To insure closer coordination, the searchlight telephone operator, keeping his head set on, will operate the traversing handwheel when hand control has been ordered.

COMMANDS.—Commands and their meanings employed in operation and control of searchlights are as follows:

a. Prepare for action.—The light will be put in its operating position, the power plant started, the light and telephone tested, and the personnel will take their posts.

b. In action.—The designated searchlight will be put in operation. Thus EIGHT, IN ACTION signifies that the searchlight occupying searchlight position 8 is to be put in operation.

c. Out.—The designated searchlight immediately will be put out, thus placing the light in the same condition as STAND BY. Thus TWO, OUT signifies that fixed searchlight No. 2 is to be put out of operation, but the power plant is to be kept running, and if the position of the target is known, the light will be kept trained upon it.

d. Rest.—The designated light, if in action, will be put out and the power plant shut down. Personnel except telephone operators will fall out, but will remain in the vicinity.

e. Stand by.—The designated light will be kept ready to go into action immediately and, if the position of the target is known, the light will be kept trained upon it.

f. Hand control.—The designated searchlight will be operated by means of the hand controller. ELECTRICAL CONTROL means that the designated searchlight will be controlled by means of the electrical controller.

g. Azimuth—.—The designated searchlight will be set at the designated azimuth, thus TWO, AZIMUTH FOUR ZERO.

h. Search.—The designated searchlight will be used to search its entire area, thus TWO, SEARCH. If it is desired to search a certain subarea the command is — SEARCH —, thus TWO, SEARCH LYNNHAVEN.

i. Search right (left).—The particular function upon which the designated light may be engaged will be discontinued and it will be used to search right (left) until ordered to halt.

j. Follow.—The beam of the designated searchlight will be kept on the target even if the latter passes out of the area which the light has been ordered to search, thus TWO, FOLLOW.

k. Cover.—The searchlight designated first will be used to cover the target being illuminated by another searchlight, thus EIGHT, COVER TWO.

l. Focus, spread, contract, right, left, raise, lower, stop,— These are the commands used to accomplish the objects indicated by them.

m. Slower, slow, fast, faster.—The rate of searching may be regulated by these commands.

n. Elevate.—At this command the beam is raised 30° and held there until further orders.

o. Two controller, two light.—When it is necessary to distinguish between the telephone operator on the controller telephone and the one on the light telephone, the commands TWO CONTROLLER or TWO LIGHT will be used.

SEQUENCE AND REPETITION OF COMMANDS.—*a.* When the searchlight officer calls a controller operator and gives a command, the controller operator will—

(1) Answer to his designation.
(2) Repeat the command.
(3) Receive the "OK" from the searchlight officer.
(4) Call his light and execute the command.

He will not execute the command or call his light before he has repeated the command to the searchlight officer and received "OK."

b. Operation at light.—As the men at the light can see little in its beam, commands such as SEARCH, COVER, and FOLLOW are not sent to the light. If the operation is by hand, all the searchlight operator needs to know is in what direction to traverse his light, how slowly, and when to stop. The controller operator therefore does not transmit commands from the searchlight officer *verbatim* to the light, but translates them into proper commands for the light.

TECHNIQUE OF OPERATION.—*a. Hand operations.*—If normal electrical control goes out of order or gives a traverse too uneven for searching or illuminating, hand operation will be used. It is practicable to bring hand operation to a point where the controller can move his light by voice nearly as effectively as by his controller mechanism. Hand operation may be made especially effective when used against slowly moving targets at outer ranges. Hand operation is mainly a matter of teamwork between the controller operator and searchlight operator. Common sense and close cooperation between the two men are the requirements. The searchlight telephone operator takes post at the traversing handwheel and operates it.

He concentrates on attaining the exact rate of traverse desired by the controller. The controller operator determines the approximate azimuth, orders it set by hand, and follows this with the command IN AC-TION, thus, from Controller to light, TWO LIGHT, APPROXIMATE AZIMUTH 50, IN ACTION. The controller operator also coaches the searchlight operator constantly, giving the necessary commands to keep the beam in the position desired.

b. Interference.—A searchlight which is put in action before it is approximately at the azimuth where it is to be used may interfere with other beams or disclose a friendly patrol vessel. Therefore it is invariably traversed dark to the desired approximate azimuth. This procedure has the further advantage of saving time, as the controller operator starts traversing his light while receiving the command and sends his commands to the light while traversing with a timing calculated to bring the light to the desired azimuth and into action simultaneously.

SECTION V
TACTICAL EMPLOYMENT

GENERAL.—a. *Use of other reconnaissance agencies.*— While it is very important that capabilities and probable lines of action of the enemy be analyzed before an attack begins, it is equally important that the enemy be kept in ignorance of defensive dispositions. Naval patrols may be depended upon to report the presence of enemy vessels. However, the command must build up a reconnaissance organization, utilizing such means as patrol boats, subaqueous sound ranging, radio direction finding, and distant illumination by airplane flares when available.

b. General use of searchlights.—As searchlights have a limited range, their use as a means of discovering the presence of the enemy near the limits of their range may do more harm than good. They can be seen for great distances, marking limits of defenses and serving as aiming points and aids to navigation for the enemy. Their use should be limited to situations where the enemy, already discovered, is so close that searchlights no longer aid him but on the contrary blind him and also illuminate enemy ships sufficiently to permit shore batteries to open fire. Movements of enemy vessels having been followed by the means already mentioned above, the guns, observing instruments, and lights should be kept laid as closely as possible on their targets in order that when the time to open fire arrives, lights may be put in action and rapid and concentrated fire opened from all available armament with the least possible delay. No light should be put into action until its use is required for accomplishment of a specific mission. The minimum number of lights necessary for accomplishment of the mission should be employed.

c. Use of searchlights on approach of enemy vessels.—On the approach of enemy vessels the commander exercising searchlight direction may put into action a minimum number of lights, preferably those most distant from important elements of the command, with a view to picking up or illuminating targets until a decision is made as to assignment of targets to tactical units.

d. Use of searchlights as reconnaissance agencies .—In those cases where searchlights are the only reconnaissance agencies available, they should be employed for this purpose to the minimum extent consistent with gaining the necessary information, and in any case searchlights performing reconnaissance missions normally should be operated intermittently.

EFFECTIVE RANGE.—*a.* Effectiveness of a searchlight beam is dependent upon atmospheric conditions; fog, mist, rain, snow, or smoke reducing effectiveness in proportion to their respective densities. Searchlights are most effective when the air is free from moisture and smoke, and on dark nights when the moon and stars are not visible. Effectiveness of observation further depends upon the size, color, and course of the target, upon the position of the observer with respect to the target and to the light, upon the background, and upon the skill of the observer and of the searchlight control operator.

b. Normal effective range of a searchlight under favorable conditions does not greatly exceed 8,000 yards. The range is greater in localities having unusually favorable atmospheric conditions and for lights sited at heights of 150 feet or greater. Further, searchlight range may be materially increased by using two lights to illuminate a target because of the advantage of increased light intensity and two planes of illumination.

BARRIER LIGHTS.—While targets are more easily picked up when passing through a fixed beam than when passing through a moving beam, use of barrier lights is rarely justified because of the valuable information that the fixed beam gives to the enemy both as to the location of defense elements and range to the shore line. In addition to giving information to the enemy, fixed beams have the disadvantage of making an approach to the shore from a direction outside the beam more difficult to detect.

SEARCHING LIGHTS.—a. A searching light is used to search through a definite and well-defined sector, the limits of which have been determined and prescribed. As these sectors may correspond to subdivisions of the water area as used for purposes of target assignment and identification, their limits may be determined by well-defined features such as islands or, where this is impracticable, by limiting azimuths from each light.

b. Width of sector.—(1) In the use of a searching light it is important that consideration be given to the width of the assigned sector, angular speed of the beam while searching the sector, and the procedure employed in searching the sector.

(2) An appreciable time is required for the eye to receive an impression and considerably more time is required to confirm the impression. A beam may be traversed at what appears to the observer to be a slow angular speed, yet the linear velocity of the end of the beam may be so great that the observer will not detect the target.

(3) A searching light should not be assigned to a sector of more than 60°. For traversing a sector of this size approximately 6 minutes will be required. Assuming the beam to be traversed continuously across the sector and back again, approximately 12 minutes will elapse between successive illuminations of areas near either flank of the sector so that a vessel moving at 30 knots could, by selecting the proper instant, come in through the entire zone of effective illumination without being illuminated by the light.

c. Fundamentals covering employment.—It follows that in the case of searching lights—

(1) The number of lights should be large enough so that, in searching, the sector to be covered by each light shall be as small as practicable and preferably less than 60°.

(2) The angular speed of traversing the beam should not exceed about 10° traverse per minute of time.

(3) The procedure used in traversing the light should be such as to render it difficult for an enemy to predict probable movements of the beam. As a rule the light should not be traversed continuously across the sector in the same direction, and no simple system easily understood by the enemy should be employed.

(4) A searching light should not be put into action until such time as reasonable assurance exists of probable presence of the enemy within the approximate range of searchlight illumination.

(5) A searching light may at any time be used for an illuminating light when a target has been detected.

ILLUMINATING LIGHTS.—Limitations upon the number of targets that can effectively be illuminated simultaneously may result in the necessity for use at night of concentrated fire on targets that can be given proper and simultaneous illumination. Where it is practicable for two vessels in a given formation to be illuminated, the two flank vessels, or the leading and the rear vessels, depending upon the formation, should be illuminated to simplify identification of targets by distant observers.

BLINDING EFFECT.—*a.* A searchlight beam directed upon an enemy vessel will interfere seriously both with those directing the movement of the vessel and with the fire direction of the vessel armament. This has special application when an enemy attempts a run-by at night.

b. The difficulty of seeing through a searchlight beam has been mentioned. By a skillful use of harbor defense searchlights, our own vessels leaving a harbor by night may be protected against enemy observation. Similarly, if information is received of the approach of enemy destroyers or submarines at night for an attack upon vessels at anchor, their observation can be seriously obstructed by crossing searchlight beams in advance of the anchorage.

Sources

Artillery Board, *Searchlights, Artillery Notes No. 9,* Fort Monroe, VA March 15, 1903.

Baird, C.W. and E. P. Noyes, "Searchlights," *J. U. S. Artillery*, Vol. 47, No. 1, (Jan.-Feb. 1917) pp. 1-23.

Board of Engineers, *Subject: Searchlights required in defense of coast of U.S., Mimeograph No. 39,* Board of Engineers, Army Building, NY March 8, 1901., 1st Supplement, Dec 1, 1902, 2nd Supplement Jan. 3, 1903, 3rd Supplement Sept. 14, 1904.

Coast Artillery School, *Coast Artillery Weapons and Material,* Coast Artillery School, Fort Monroe, VA, 1929.

General Electric Company, *Instructions for Installing and Operating Searchlight Projectors, U.S. Government Type,* No. 9033, GE Co., Schenectady, NY, Apr. 20, 1896.

Gibson, A. "The Sperry Searchlight," *J. U.S. Artillery,* Vol. 46 (1916) pp. 37-46.

Hines, F.T. and F.W. Ward, "Searchlights," *The Service of Coast Artillery,* Goodenough & Woglam, Co., NY, 1910, pp. 402-413.

Patterson, M.L., "Searchlights for Military Use," *J. U.S. Artillery,* Vol. 53 (1920) pp. 383-396.

School Board, *Searchlight, Artillery Notes No. 31*, Coast Artillery School Press, Fort Monroe, VA 1908.

School Board, *Searchlight (revision of Artillery Notes No. 31), Artillery Notes No. 32,* Coast Artillery School Press, Fort Monroe, VA 1912.

Waldron, A.E., and J.L. Hall, *Remarks on Search-Light Projectors for Coast Defense Service,* Number 13, Occasional Papers, Engineer School of Application, United States Army, Washington Barracks, Wash. DC, 1904.

War Department, *Handbook for the Use of Electricians in the Care and Operation of Electrical Machinery and Apparatus of the U.S. Seacoast Defenses,* GPO, Wash. DC, 1902.

War Department, Office of the Adjutant General, "Searchlights—Night Drill," *Drill Regulations for Coast Artillery, United States Army,* GPO, Wash. DC, 1914, pp. 121-124.

War Department, "Searchlights," *Technical Manual 4-210, Seacoast Artillery Weapons,* Army Field Printing Plant, Coast Artillery School, Fort Monroe, VA, Oct. 15, 1944, pp. 172-183.

War Department, *Training Regulations No. 435-330, Tactical Use of Searchlights in Harbor Defense,* GPO, Wash. DC, June 30, 1925.

War Department, *Training Regulations No. 435-75, Searchlight Battery, Antiaircraft Artillery,* GPO, Wash. DC, May 20, 1924.

War Department, FM 4-5, Seacoast Artillery Organization and Tactics, US GPO, Wash. DC, 1940.

Winslow, E.E., "Night Illumination of Battlefields," *Notes on Seacoast Fortification Construction,* Number 61, Occasional Papers, Corps of Engineers, US Army, GPO, Wash. DC, 1920, pp. 369-406.

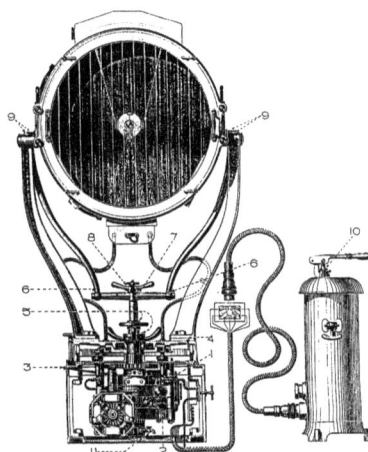

Seacoast Artillery Radar, 1938-46

Danny R. Malone

The Coast Artillery Corps expressed an interest in the development of a workable electronic ship detection device in 1937. The Coast Artillery requested a set that would give advance warning as well as provide azimuth and range data for non-visual engagement by shore batteries. The first attempts utilized infrared type detection, but by 1938, radar had proved superior during trials and the first set was developed. The development and production of this seacoast artillery radar was at the time deferred due to the priorities of the late 1930s of seeing bomber aircraft as the greatest threat to the United States.

SCR-268

In 1937 the Signal Corps had taken over the development of radar and produced the ancestor of all U. S. radar—the SCR-268 (SCR stands for Signal Corps Radio/Radar)(Fig.1). This radar was a large set that utilized a bedspring-type antenna and operated on a long wavelength of 150 centimeters at 205 mc frequency. It was mounted on a 4-wheel trailer for mobility, weighed over 14 tons, and could be emplaced and made operational within two hours. Its chief function was to provide elevation and azimuth for direction of the 60-inch searchlights utilized for anti-aircraft defense. It was operated by a crew of three personnel and gave range, azimuth, and elevation (not height) of aircraft targets. The SCR-268 was capable of a 360 degree search, and the antenna was capable of searching from +15 to +85 degrees elevation. To accomplish this the set required a 13.2 kilowatt power supply.

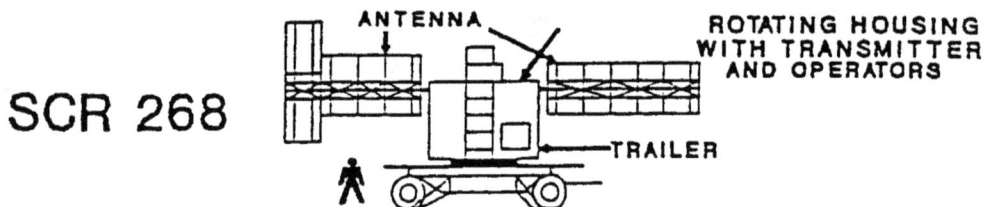

Figure 1.

The set produced a effective range of 24 miles (although 60 miles was achieved on occasion), but was credited with an poor showing in azimuth and elevation accuracy. It achieved a 4 degree average error, as opposed to the 1 degree expected. Post war, the set was credited with being able to furnish reliable range on aircraft flying 10,000 feet or higher at distances up to 40,000 yards. The principal failings of the SCR-268, however, were the poor performance against low flying (500 feet or below) aircraft and the excessive interference of ground or water conditions that created clutter on the scope. Both of these were a direct result of the long wavelength operation and low frequency utilized. An additional restraint was the use of a simple linear oscilloscope (A-scope) presentation (Fig. 2) instead of the later Plan Position Indicator, which is still used today. This A-scope resulted in a less accurate presentation that was difficult for an operator to interpret.

Despite these problems, the SCR-268 remained in service throughout WW-II, serving until 1944 as an anti-aircraft fire control radar and short-range early-warning radar. In 1944, there were 1200 still in service as searchlight control equipment. Later in the war, the SCR-268 was modernized with a PPI type scope, mechanically aided tracking, and operation on a shorter wavelength to become the improved SCR-516.

Two additional radar sets were developed at this same time for early warning of approaching aircraft. These were the SCR-270 (Fig. 3) and SCR-271 (Fig. 4). The SCR-270 was a mobile (calling it "transportable" would have been a better description due to the time required to emplace it) and the SCR-271 which was a fixed version of the same radar.

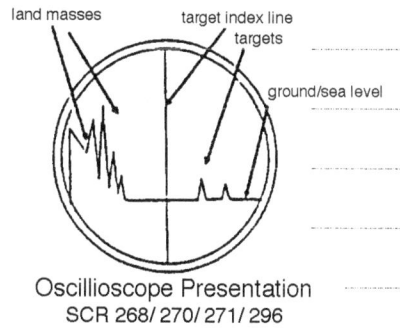

Osccillioscope Presentation
SCR 268/ 270/ 271/ 296

Figure 2.

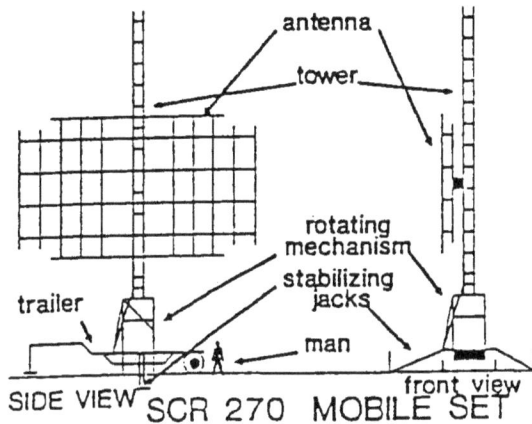

SIDE VIEW SCR 270 MOBILE SET front view

Figure 3.

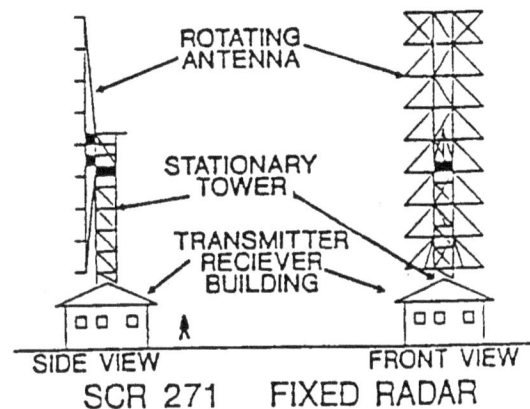

SCR 271 FIXED RADAR

Figure 4.

Both operated on the long wavelength of 300 centimeters at 100 MC (refered to as MHz today) and utilized the same oscilloscope presentation of the SCR-268. While achieving the credible detection range of 150 miles against aircraft, they suffered from poor reliability, poor low-level (below 500 feet) performance, and the inaccuracy and difficulty of interpreting the oscilloscope presentation. Wartime improvements, such as the addition of a PPI indicator (SCR-539) helped, but the long wavelength resulted in interference caused by ground clutter and weather features which severely reduced their effectiveness. This rendered them almost unusable in certain areas (such as Panama), and they were of questionable value in other areas. Despite these limitations and their later proven vulnerabilities to jamming, they were produced under the principle that they were better than nothing. By mid 1942, the United States coast was guarded by over 250 radar sets, consisting of two SCR-271s, seventy SCR-270s, and one hundred ninety four SCR-268s. The majority of these sets were allocated to the Pacific Coast. While the personnel of the period complained of the deficiencies of these early sets, it is noteworthy that the attack on Pearl Harbor was spotted, and tracked, by one of these early SCR-270s at a distance of 132 miles.

SCR-296A

With the production and deployment of the air defense radar the army now turned to modernizing the seacoast defenses, whose requirement in 1937 generated the research that resulted in the development of radar. Because of its lower priority, the new seacoast artillery radar was not further considered until late 1940. Once adequate quantities of the SCR-268, SCR-270, and SCR-271 types had been produced and successfully deployed, the Signal Corps began development in earnest during 1941. A prototype was obtained from the contractor and modified with lobe switching which enabled it to track a target. The set was tested, the modification approved, and a production order was placed for 20 service sets. This became the SCR-296 (Fig. 5). The SCR-296 was a surface search radar which had been developed from the navy's FC shipboard set. It was utilized for fire control and operated with a 40 cm wavelength at a 700 MC frequency.

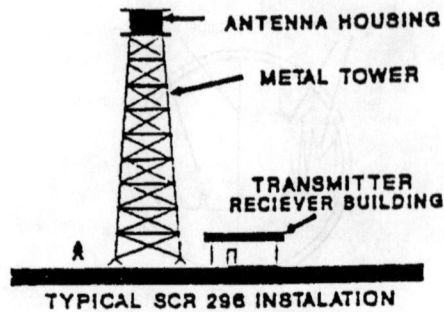

TYPICAL SCR 296 INSTALATION

Figure 5.

The SCR-296A was the standard World War II fire control radar utilized by the U.S. Coast Artillery for engaging surface targets. It was authorized for issue to all modern batteries of 6-inch and larger on a basis of one set per battery. This set was basically identical to the navy Mark III or FC set utilized on ships. The function of the radar was, during periods of poor visibility, to provide the range and azimuth of the target vessel to the plotting room of the battery. In addition, it could be used to provide such data to additional batteries as required or to give spotting corrections to other batteries if advantageously sited. The set was normally installed with: a prefabricated metal lattice tower containing the antenna; an operating room containing transmitter/receiver, indicating panels, power panels, and communication devices; and two 25 kw gasoline generators. Normally the operating room was in a prefabricated metal building (HO-2-A) and the generators (Fig 6b) in two other prefabricated metal buildings (HO-l-A). (Fig 6a) The components weighed 18,943 lbs. (operating building), 31,777 lbs. (antenna), 9,580 lbs. (antenna housing), 6,420 lbs. (radar components), 16,225 lbs. (accessories), and 13,450 lbs. (2 generators and houses). This resulted in a total weight, in action, of 96,388 lbs., not including the IFF equipment. Variations were possible and, in high risk areas, the operating room could be in bombproof locations such as the battery plotting room bunker, as at Battery Hatch (Oahu Island, Hawaii), or located within the metal or concrete multi-story fire control towers. In the instances of tower locations, the antenna was located on the roof. (fig 16 a-b)

The first limitation on the system was the necessity for the proper location of the antenna. As radar at this time was believed limited to line of sight, it followed that generally the higher the antenna location the greater the range. (fig 7) As the antenna was not capable of elevation or depression during operation, it was possible to site the antenna at too high a location which resulted in an excessive non-scanned, or dead" area close in to the set (fig 8). Experience with the set showed that the antenna became less effective, due to decreased maximum range, at heights below 100 feet above sea level and created too much "dead zone at heights above 500 feet.

To provide for this height requirement, the obvious answer was to utilize natural heights such as hills, bluffs etc. However, at many sites on the Atlantic and Caribbean U.S. coasts, no such sites were available. For these low-lying sites, a prefabricated steel tower (fig 6a, fig 16a, fig 19) was issued with the set. This tower was supplied in 25-, 50-, 75-, and 100-foot heights to accommodate the needs of each particular site. This allowed for a minimum of 100 feet to be reached; however, contemporary texts stated that a minimum of 150 feet above sea level was necessary to achieve the minimum effective range necessary to control heavy caliber gunfire. The absolute minimum was necessary to prevent nearby ground clutter with the radar transmissions.

As mentioned, many expedients were utilized to gain the necessary height. The antennas were mounted on the tops of existing steel or concrete fire control towers (fig 16b), with a short (25 feet) lattice tower on top of the hill formed by the battery plotting and switchboard room (PSR), for example as at Battery Hatch, or directly on the battery commander's station, such as at Battery 302 (Oahu Island, Hawaii), which was at a considerable height above sea level.

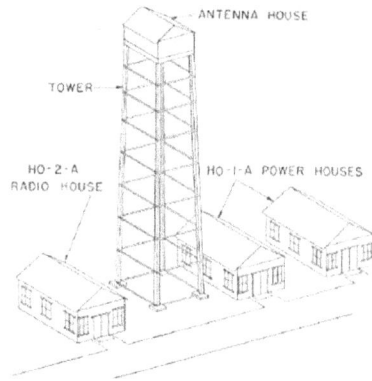 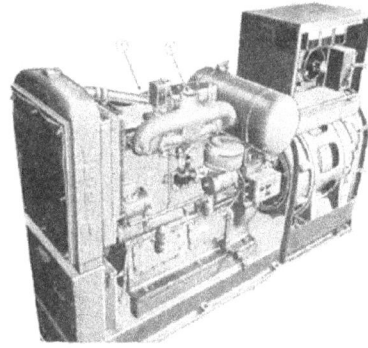

Figure. 6a.   Figure 6b.
from War Department, FM 4-95, "Service of The Radio Set 3CR 296,"15 Sept 1943.

Table 7

Height	Radar Horizon
100 ft.	25,000 yds
150	30,000
200	35,000
250	40,000
300	42,000
350	46,000
400	50,000
450	52,000

Data forTable 7 from Qakes, Pauline M., "U.S. Army in World War II, The Signal Corps, The Test," Office of the Chief of Military History, United States Army, Washington, D.C., 1957.

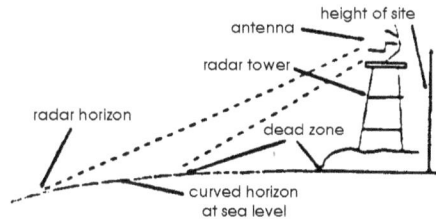

Figure 8

While these higher sites gave longer range, the arrangement of the radar was further restricted by the maximum length of the transmission cable between the antenna and the transmitter/ receiver that was 175 feet. This restriction was due to the technology of the period that was unable to prevent signal (electronic power) loss at greater distances. The ideal solution was, when possible, to make the distance even shorter that to minimize this loss, make the signal stronger, and result in better performance. This was the reason that some transmitter/receivers were mounted in the towers and on BC and plotting room locations.

Along with the height requirements for antenna site selection, it was necessary to locate the site in such a position that no obstructions, such as hills, towers, or buildings, were located within 8-10 degrees of the flanks of the sweep area or within 8 to 10 degrees below the antenna sweep area. Any obstructions could cause reflection of the side lobes of the transmissions and cause the radar to falsely show the target to be located midway between the actual target and the obstruction.

Mounted on top of the tower was a platform on which was mounted the antenna (fig 9). This antenna was used for both transmission and reception. It was a directional antenna formed of five curved ribs on which was mounted a metal screen reflector. Located in the center, running horizontally across the antenna, were the dipoles. The dipoles and the reflector worked together for both transmitting and receiving. The whole apparatus was approximately 6 feet long by 6 feet wide. This antenna was capable of being remotely controlled in azimuth and had an electronic data transmission device, which indicated its position on indicators located in the operating room. This action was accomplished by a pair of selsyn motors, which operated much like a modern TV antenna rotator. The antenna did not continuously rotate as in modern radar, but was directed at the target and in receiving functioned as a modern radio direction finder. It was preset in elevation and adjustment of this was not possible during operation. These restrictions resulted

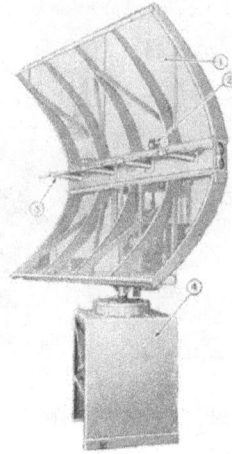

Figure 9.
from War Department, FM 4-95, "Service of The Radio Set SCR 296," 15 Sept 1943.

in an ability to search only a narrow sector (the beam had only a 9 degree width in azimuth and 11 degrees height in elevation) and limited its effectiveness when used as an early warning set. The antenna was housed in a round wooden housing shaped much like a contemporary water tower. A ladder was provided for personnel access, and an electric winch was available for hoisting major components to and from the ground. Because of the danger presented by the powerful concentrated radio waves of the antenna (much like the modern microwave oven) safety interlock switches were positioned on the platform access trapdoor which shut down the power on its opening. Frequently this resemblance to a water tower was exploited by disguising the antenna housing, as was done at Battery 301 (Oahu, Island, Hawaii) by including a dummy water measuring device on the outside.

The radar system normally operated on post (commercial) AC power, requiring 20 kw for efficient operation. In isolated locations and as emergency back-up each set was provide with two PE 84C portable gasoline powered generators (Fig 6b.), each capable of producing 25 kw of 120 volt, 60 cycle, AC current. A standard location such as at Charleston had these components located in prefabricated metal buildings designated HO 1-A (fig 6a). Each set weighed 5000 pounds, and was supplied with gasoline by a 1000-gallon underground gasoline tank which was located some distance from the building. As each generator was capable of supplying more than the required 15 kw necessary for operation of the radar and its auxiliary systems, this resulted in a primary and two alternate emergency operating systems, each capable of independently operating the radar.

The power requirement resulted in the third limitation on SCR-296A location, that each of the two 25 kw gasoline generators could be located no further than 150 feet from the operating room of the radar (fig. 19). This restriction was a result of the fact that any further cable length would result in excessive power loss to the transmitter/receiver. This fact, coupled with the transmission cable length restriction resulted in a compact system that was concentrated in an area no larger than 300 feet. Such compactness, resulted in making the system more vulnerable to enemy shelling and bombing and, as a result, many stations were placed in bunkers or tunnels to reduce vulnerability.

The most important components of the radar were located in the operating room. In this area was located the main unit which contained the radar transmitter, receiver, power control panel, high voltage rectifier, and operating controls (fig 15). This area represented the "heart" of the system. The most important area of the operating room contained the operating equipment. This consisted of a table, on which were located a bank of six instrument boxes which allowed the system to acquire and determine the range and azimuth of the target (fig 10). These consisted of, from left to right: the rectifier (1), the azimuth indicator (2), the azimuth oscilloscope (3), the azimuth control unit (4), the range oscilloscope (5), and the range-measuring unit (6).

Figure 11a.

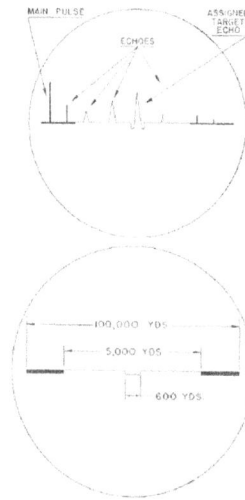

Figure 11b.

Figure 10.
from War Department, FM 4-95, "Service of The Radio Set SCR 296," 15 Sept 1943.

The standard manning crew for the SCR-296A operating room consisted of five personnel. The first was the NCO who functioned as chief of the section. The others were the range operator, the azimuth operator, the range reader, and the azimuth reader. The operating room required a minimum of nine cable pairs for telephonic communication. It was equipped with 4 telephones. It had a command telephone—to communicate with the battery commander—an azimuth reader's phone, and range reader's phone, all connected through the fire control switchboard. In addition it had a time-interval signal and, if equipped, an electronic data transmission system. The fourth telephone, the post line, went through the post switchboard and was intended for non-tactical messages.

In operation, the range and azimuth (rough bearings) were transmitted telephonically from the HDCP SCR-582/682A radar to the range and azimuth readers in the SCR-296A operating room. This coaching was necessary due to the poor target discrimination of the radar and its narrow beam that meant that the antenna had to be pointed almost exactly at the target.

At this time both the azimuth and range readers repeated the data and directed their respective operators to traverse left or right (azimuth) or to increase or decrease the range until the indicated settings appeared on the indicators. At this time, the readers gave the command "steady." At this time the target would have been within the scan of the radar beam and appeared as a peak in the line representing the radar sweep signal in the range oscilloscope (fig 11a)

The range operator, seated at the table, then observed the target signal on this range oscilloscope and adjusted the signal. When the set was operating, the 5-inch wide screen of the oscilloscope electrically generated a horizontal line which represented time. This line was 4 inches in length and had a noticeable "U" shaped dip in the center, which represented a 600 yard area. The total length of the complete line represented a 100,000 yard sweep line (Fig 11b) . The center "U" notch could, by use of the "image spread" knob, be expanded to a width of 4 inches to represent a 5000 yard search area, while the center notch expanded to represent an expanded presentation of the same 600 yard area. After adjusting the signal strength dial to bring the target echo to a height of between 3/4 to 1 inch in height, the Range Operator rotated the range handwheel, located on the Range Unit (fig 10, item 6), to move the target echo into the notch. Once adjusted he continued to rotate the handwheel to the left or right to keep the target in the notch. At this time he called "On Target."

After the range operator had "locked onto" the target the azimuth operator turned the handwheel located on the azimuth control unit (Fig 10, item 4) to match the height of the two "V" shaped pips (Fig 12c) which appeared in the azimuth oscilloscope. These pips *only* appeared *after* the range operator had centered the target in the notch and represented this target. These two "V"s were one of the operating secrets of the SCR-296A and were a function of the "lobe switching" mode. If the antenna transmitted from the center,

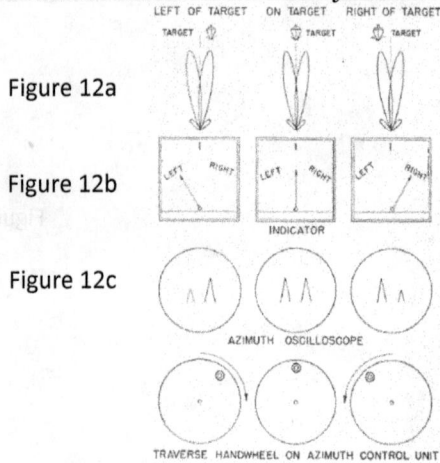

Figure 12a

Figure 12b

Figure 12c

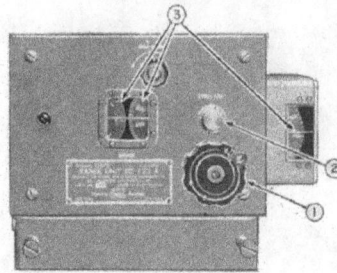

Figure 13.

Figure 14.

from War Department, FM 4-95,"Service of The Radio Set SCR 296," 15 Sept 1943.

it functioned as a simple direction finder. The "lobe switching" allowed the antenna to alternately transmit a number of radar pulses slightly off center. This allowed the target reflection to be received at slightly different angles (fig 12a) and increased the accuracy of the azimuth measuring.

The operator then adjusted the handwheel on the Azimuth Control Unit to the left or right, which rotated the antenna, and caused the target to be centered between the two lobes (fig 12a). The correct adjustment was indicated when the two "V"s visible on the azimuth oscilloscope assumed an equal height (fig 12c). As an aid to proper adjustment, a meter (fig 12b) was provided on the azimuth indicator that indicated the target position. When the operator had the needle centered, the target was properly located. While this method was more reliable than utilization of the azimuth oscilloscope, its use was limited. This meter could not be utilized if another SCR-296A was operational within its sweep area or if natural interference (land mass, towers, etc., were on the edges of the beam as the meter would indicate a false target location midway between the actual target location and the interference.

Once the target was positioned, both operators continued to utilize their respective handwheels to track it. Late in the war, some sets were equipped with "aided tracking." This was a system which, when the range and azimuth change rate had been established, automatically accomplished the tracking. During this operation the operators had to use their adjustment wheel only for minor adjustments and it greatly increased the accuracy of the range and azimuth determination.

After stabilizing the tracking, the range reader then observed the range measuring unit (fig 10, item 6) and read the dials (Fig 13, item 3). Proceeding from left to right, the first square window, two dials with a common indicator line running through the center. The left-hand dial made a 360 degree revolution for every 100,000 yards, the maximum range of the set. It was numbered from 1 to 10 and each number represented 10,000 yards. The dial on the right of the indicator window made a revolution for each 10,000 yards and was marked 0000 to 10,000. Each number represented 1000 yards and each was further subdivided into 100-yard increments indicated by 10 short lines. Moving on, the next knob upper (fig 13, item 2) was utilized for "zeroing" the range knobs with the range tracking wheel (Fig 13, item 1) located below it. Attached to the extreme right side of the unit was a square box. This contained the final range indicator, and consisted of a dial that rotated once for each 1000 yards. It was marked with numbers from 1 to 10. Each number represented a 100-yard increment and each was further subdivided by 5 small lines which represented 20-yard intervals. This enabled the reader to interpret the range to the nearest 5 yards.

For azimuth determination the azimuth reader observed the azimuth indicator (Fig 10, item 2). It contained two round dials located in a common face (Fig 14). The index lines consisted of two "Y" shaped pointers. The stem indicated the reading point. The dial on the left turned one rotation for each 360 degrees of antenna rotation and was numbered at each 10 degree mark. Each 10 degrees was subdivided by 5 short lines into 2 degree increments. The right dial turned one complete revolution for each 10 degrees

Figure 15 from War Department, FM 4-95, "Service of The Radio Set SCR 296," 15 Sept 1943.

and was marked with numbers from 1 to 10. Each degree was subdivided by 20 short lines which allowed the azimuth to be interpreted to the nearest 0.05 degree.

In operation both operators ceased tracking the target as the time-interval bell signal from the battery plotting room rang. The readers then reported the dial reading, interpolated to the nearest 5 yards (range) or 0.05 degrees (azimuth) telephonically to the plotting room. If the set was equipped with the aided tracking equipment, it was not possible to stop the tracking and the reader's job was made more difficult by having to read the dials while they were in motion. As they became available, some sets were equipped with an electrical data transmission which gave continuous range and azimuth information to dials in the plotting room. In those batteries equipped with the Gun Data Computers M-l or M-8, the data transmission was connected directly to the computer. This eliminated both the "dead time" delay and probability of human error associated with the telephonic procedure, as well as freeing the plotting room from the delay necessary in waiting for the next time interval bell. In modern terms, the plotting room was supplied with "real time" data. Despite these advantages, the telephonic time-interval method was retained and practiced for emergency use.

In addition to the basic set, the SCR-296A was later provided with a RC 136 IFF system. This consisted of a smaller antenna mounted near the main antenna (Fig 16c), as well as a separate transmitter/receiver in the operating room. In action, it functioned as a separate transmitter receiver, which activated a Mark III transponder fitted to friendly ships. When activated, this transmitter caused a series of coded pips to appear near targets on the main radar, thus allowing the identification of friendly ships. As with all new devices this worked better in theory than practice, and if ships were closer than 1200 yards in range or separated by less than 20 degrees of azimuth, the system was incapable of identification. As each battery was assigned its own radar, it was necessary to equip the sets with a selection of four different magnetrons.

Figure 16a   Figure 16b   Figure 16c

Figure 16a from Thompson, George R., and Harris, Dixie R, Figure 16b from War Department, FM 4-96, "Service Of The Radio Set SCR 582î," 9 Nov 1943. Figure 16c from Orman, Lt Col L. M. , "Radar, A Survey," Coast Artillery Journal, Vol LXXXIX #2 (Mar-Apr, 1946).

When installed, they varied the transmitting frequency between 680 and 720 MC (MHz), and prevented interference between sets operating in close proximity.

According to contemporary manuals, the SCR-296A was given credit for being able to track the shell splashes and thus be used for fire adjustment. However, postwar assessment stated that due to the poor resolution of the set, it was impossible to distinguish between separate shell bursts and such an adjustment was impossible. This lack of resolution and azimuth discrimination also meant that the SCR-296A operator was only able to separate targets that were more than 275 yards apart and more than 12 degrees apart in azimuth.

With all of these failings, the SCR-296A was credited with the ability to track and give warning of surface targets at a range of over 100,000 yards. This figure represents the maximum range. Postwar, the SCR-296A was considered to be able to give reliable range on large targets (battleships and cruisers) at 40,000 yards, while smaller vessels (destroyers) could only be tracked at ranges of 20,000 yards. Such a limitation restricted its utility with the 16-inch weapons, which had a range of 44,900 yards. The SCR-296A served throughout the war as the standard coast artillery fire control radar. (For site plans see Figure 19 and Figure 31).

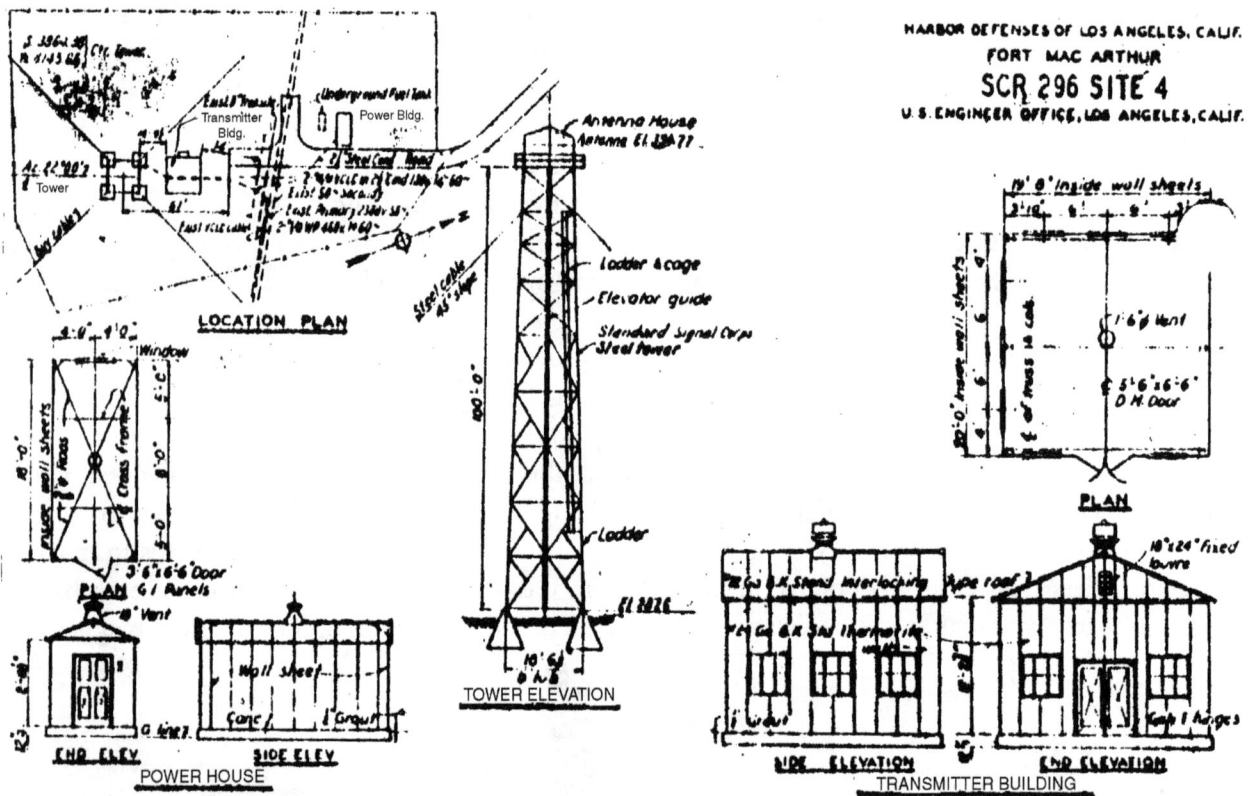

Figure 19. Site plan for the SCR 296A radar facility for a 6-inch gun battery at Fort MacArthur, CA (NARA)

SCR-582/682A

While the SCR-296A was acceptable as a fire control radar, the coast artillery had a second mission for which, due to its limitations, it was unsuitable. This was a need for surveillance radar which would give early warning, and be able to track multiple targets, as well as show their relationship to natural features within a defended harbor. The answer to this problem emerged late in December 1941, as the XT-3. This first set was tested between 27 and 30 December at Fort Dawes, Boston harbor.

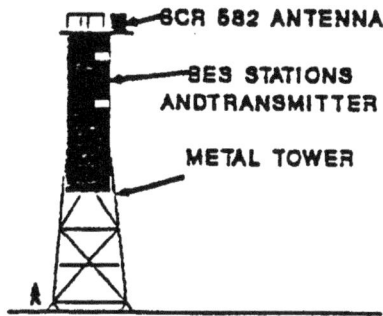

TOWER TYPE SCR 582 INSTALATION

Figure 17

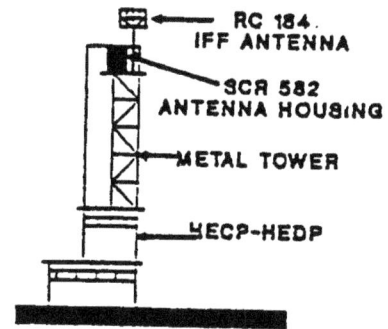

RADAR AND IFF ON HECP-HEDP

Figure 18

The set was then standardized as the SCR-582 (fig. 17). It differed from the SCR-296A in operating in the 10.7 cm wavelength microwave band at 3000 m/c. It presented its information on a Plan Position Indicator (PPI) scope that made the presentation more clear and easier to interpret by less highly skilled operators. This combination enabled the SCR-582 to track multiple targets and thus enabled such organizations as the HECP/HDCP to function during periods of reduced visibility. The SCR-582 antenna was housed in a cylindrical plywood cover and mounted on towers, similar to the SCR-296A, but utilized a smaller, 4-foot diameter dish-shaped antenna which constantly rotated. The results were excellent for the period, and against surface targets a maximum range of 90,000 yards was possible with the maximum effective range being listed postwar as 35,000 yards, with a range accuracy of + or - 25 yards, and an azimuth error of + or - 2 degrees. An additional advantage was that the entire installation weighed less than 2 tons when installed and required only 2.5 kilowatts of power.

The SCR-582 had a large primary scope (PPI) (fig 21) which was graduated for up to 90,000 yards range and covered the 360 degree area of the radar coverage. The second smaller scope was called a Precision Position Indicator (P3I) or "B" scope for short (fig 22).

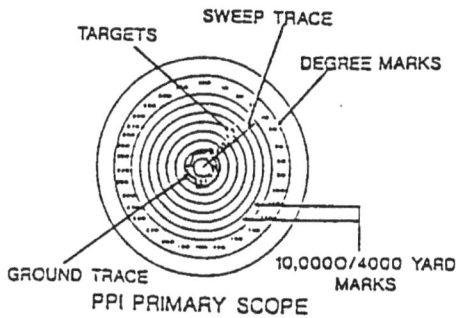

PPI PRIMARY SCOPE

Figure 21

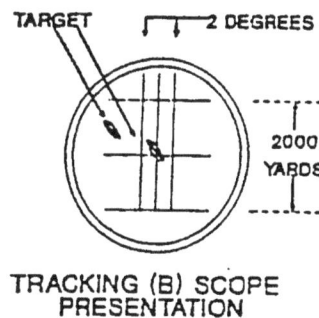

TRACKING (B) SCOPE PRESENTATION

Figure 22.

This B scope was utilized to magnify a section of the radar search area by narrowing the presentation down to an area of 4000 yards in range and 40 degrees in azimuth. The scope was marked with a rectangular scale that clarified the target's position into smaller subdivisions. This magnification allowed both a greater accuracy and provided the additional capability of separating and tracking multiple targets that might appear as a single target on the 90,000-yard scope. This additional scope also ensured that the harbor remained under constant surveillance on the main (PPI) scope. An additional capability was that the P3I scope could be located up to 1,000 yards from the main radar and that up to four P3I/B scopes could be operated from the primary SCR-582. Such an arrangement gave the HECP and HDCP commanders the ability to track several specific targets while maintaining the continuous surveillance of the entire harbor area. Additionally a device called a VG remote indicator was utilized as one of the additional scopes. This device was a projection type remote unit that presented the information shown on the main PPI scope by utilizing lenses and mirrors to project the display on the undersurface of a 30-inch ground glass screen. This display was usually equipped with an outline map of the local area and was utilized for maintaining a plot of target courses. It had the great advantage of being able to be seen in a bright room, unlike the PPI/P3I scopes that needed almost total darkness to be seen. This was an obvious advantage to a HDCP/HECP commander. In addition, the P3I was credited with being able to be utilized as an emergency fire control radar with a limited ability to spot both target and shell splashes; due to its surveillance capabilities and its smaller antenna the SCR-582 was frequently mounted on existing structures, such as steel or concrete fire control towers or HECPs (fig 18) such as at Charleston, SC. In the latter case the structure also mounted an associated Identification Friend or Foe (IFF) SCR-184 antenna (fig 23b). This device transmitted a coded signal to the target, which triggered a radio transponder, which then caused a coded signal to appear on the radar operator's scope and thus identify the ship or aircraft as friendly or hostile.

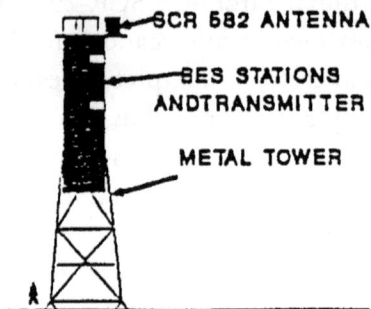

TOWER TYPE SCR 582 INSTALATION

Figure 23a

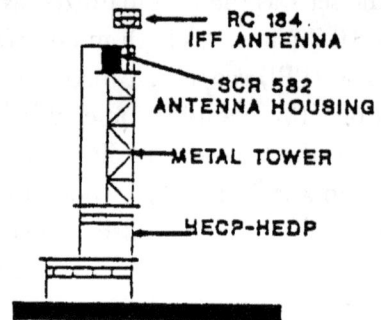

RADAR AND IFF ON HECP-HEDP

Figure 23b

While designed as a surface search radar, the SCR-582 very early in its career, while in use in North Africa, demonstrated a credible aircraft warning capability by being able to track low-flying aircraft between 500 and 1000 feet at a range of 40,000 yards. As a result the SCR 582 was modified with the ability to tilt its antenna. This was possible to preset manually between 0 and + 25 degrees on both sets, but after beginning operation could not be changed, thus lessening the radar's ability to perform both air and surface warning simultaneously. Its success in this form led to it being made mobile by being mounted on three 2-1/2 ton trucks to become the SCR-682A (fig 23a). The mobile SCR-682A unit operated on the 10 cm wavelength, weighed 7 tons, and operated from a 30 foot tower. It was capable of being emplaced and made operational in 3 to 5 hours. In addition to its mobility and antenna modification, the SCR-682A operated with a power output about 5 times greater than its parent SCR-582. This resulted in a maximum range of 240,000 yards, with a postwar quoted effective range of 35,000 yards against surface targets. The SCR-682A was credited postwar with having the same azimuth accuracy as the SCR-582 but considerably superior range accuracy. In addition the power requirement was reduced to 2.4 kilowatts The SCR-682A did not (in 1944) utilize the B or spotting scope but had four separate range settings for the primary PPI scope (fig 24). It was capable of presenting its information on the PPI at 10,000, 40,000, 160,000, and 240,000 yards ranges. The electrically produced rings on the PPI divided the scope into coarse and fine range markers, which varied with the main range settings. The 10,000 yard was subdivided into 2000 (coarse) and 500(fine) , the 40,000 into l0,000 (coarse) and 2000 (fine), the 160,000 into 40,000 (coarse) and l0,000 (fine) yards while at the maximum 240,000 yard setting there was only a 40,000 coarse setting with no fine adjustment. According to the manuals of the period this system was not accurate enough for fire control and was (in 1944) for surveillance purposes only. Initially the SCR-582 was authorized for fixed harbor defense installations and the SCR-682A was for mobile batteries (such as the 155 mm guns). This situation was changed due to the low numbers of SCR-582s produced (55 total) and the low serviceability

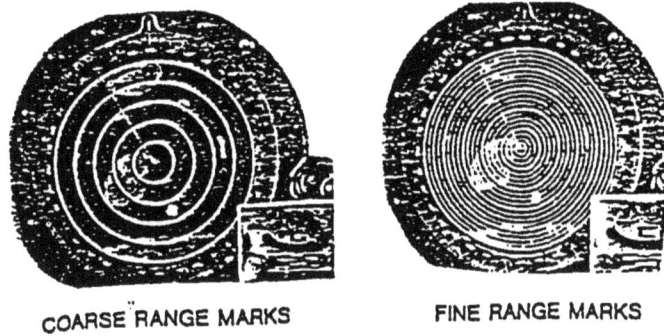

COARSE RANGE MARKS   FINE RANGE MARKS

PPI SCR 682

Figure 24

rate from which the set suffered. This situation was made worse by a failure to procure adequate repair parts for the SCR-582. The result was the wartime use of the mobile SCR-682A set being modified to use the B scope and VG scope, which enabled it to be used in lieu of the "official" SCR-582. The parts shortage and low serviceability rate combined to cause the SCR-582 to be declared obsolete and officially replaced by the SCR-682A by 1946 (see fig 25 for site layout, see fig 31 for overall site layout).

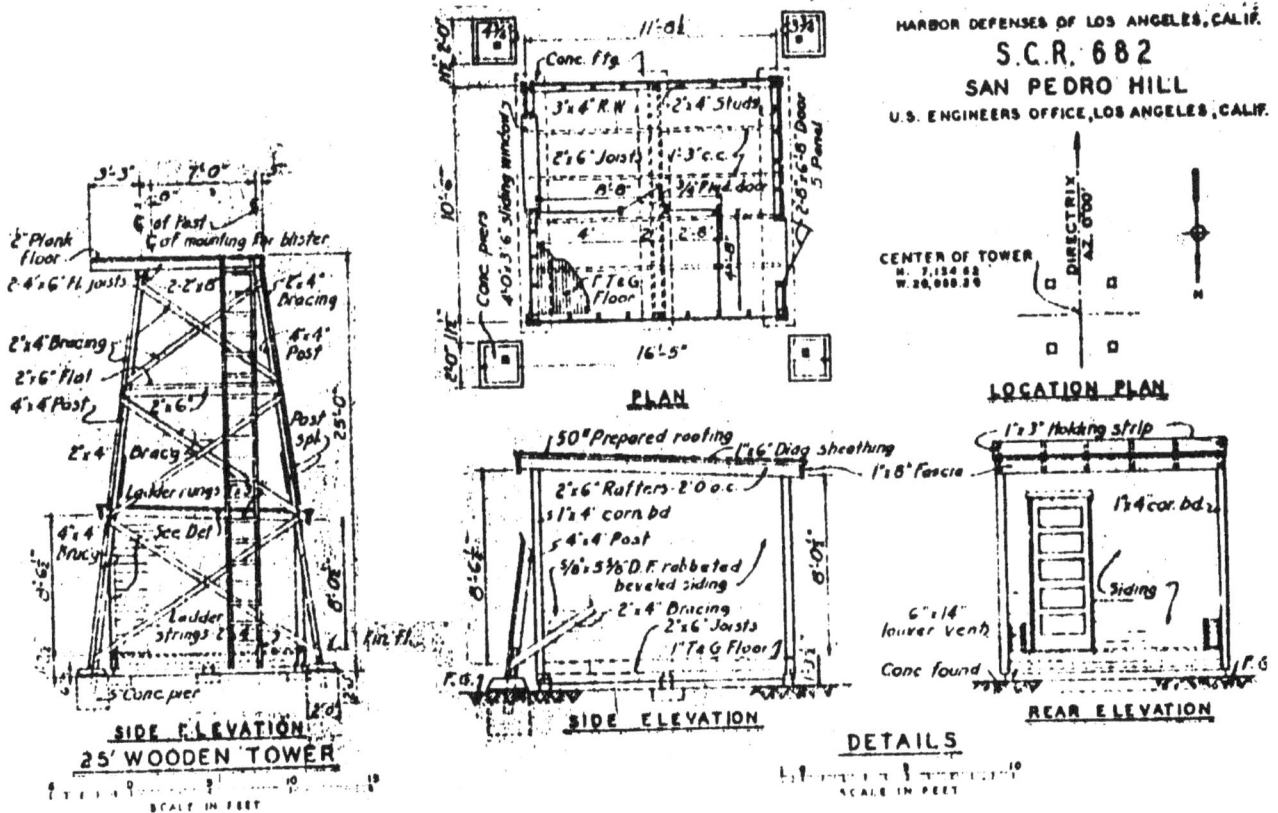

Figure 25. Site layout for SCR 682A facility at San Pedro Hill, Harbor Defenses of Los Angeles (NARA).

SCR-547

Before and throughout most of the war the anti-aircraft, both 3-inch and 90-MM mobile batteries, were dependent for the determination of aircraft altitude and the resulting slant range on the optical height finder M-l. This equipment was basically a stereoscopic rangefinder, which when properly adjusted on the target, automatically transmitted the data (range) to the M-9 gun director (computer) which integrated its own azimuth reading, and produced data which caused all 4 guns of the battery to directed at the target. This system was of less accuracy than desired and, as early as 1938, the possibility of utilizing radar for this mission was discussed. Due to higher priorities it was not until 1941 that the system was produced under the nomenclature SCR-547.

The coast artillery had began to install anti-motor torpedo boat batteries consisting generally of two fixed 90 mm (M-3) and two mobile 90 mm (M1A1) guns. These had as their primary mission the defense of harbors against small rapidly moving and maneuvering targets. As these batteries had as their secondary mission anti-aircraft defense, they were initially equipped with both the M-9 directors and the M-1 height finder. The heightfinder was utilized by the AMTB batteries to provide optical range input when firing on surface targets. As such areas frequently were obscured by fog, as well as the probable target's normal utilization of smokescreens to prevent detection, this system was of limited utility. The Coast Artillery recognized a requirement for radar fire control for such batteries. As the batteries were to be equipped for both antiaircraft and surface gunfire (azimuth/early warning), neither the SCR-268, which was unable to depress and track surface targets, nor the SCR-296A, due to the previously described weaknesses, were suitable and a substitute was necessary. When the SCR-547 became available, it was utilized to provide the necessary radar range and bearing.

The SCR 547 (fig 26) was a continuous wave set which operated on a ten centimeter wavelength. The equipment, which weighed 6 tons, was mounted on a two-wheel trailer. The radar consisted of a rotating central housing on each side of which was positioned a large circular solid parabolic antenna which was capable of elevation. Two operators were stationed on the mount, one of which operated the A scope presentation of the oscilloscope, while the other kept the antenna pointed at the target by observing it through a sighting telescope. A third operator maintained telephonic communication with the battery command post. The set was equipped with power-aided tracking which when engaged assisted in holding a standard traverse rate to maintain target contact. As stated above, the set required three operators and required 4.6 kilowatts of power for operation. The latter was provided by an accompanying truck-mounted generator.

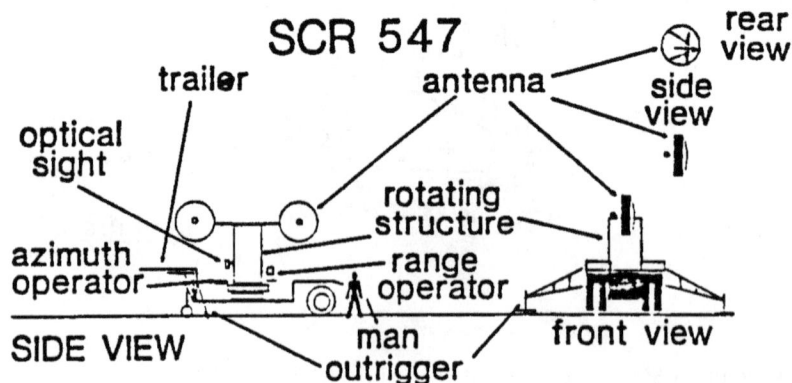

Figure 26

During daylight firing against surface targets the antenna of the set was optically directed at the target in the same manner as that for firing on aircraft. Upon acquisition of the target in the A type scope of the range operator, the target signal was adjusted to place it in the step of the beam (fig 27) . This was accomplished during the Full Sweep (wide beam) operation. When this was completed, the operator switched to Precision Sweep (narrow beam) operation which gave a magnified portion of the full sweep beam and allowed a more precise adjustment of the target echo which provided a more precise range determination.

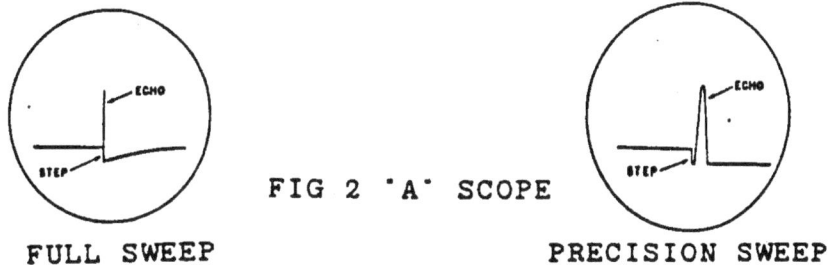

FIG 2 "A" SCOPE

FULL SWEEP   **PRECISION SWEEP**

Figure 27

This system provided for daylight, clear visibility only, and for inclement weather or night operation the SCR-547 was modified by the battery mechanic to semi-permanently attach a test oscilloscope (supplied with the set) which resulted in giving the azimuth operator a second A scope presentation. During blind firing the azimuth operator moved the antenna until the target echo reached maximum height on the scope. This method while not as accurate as the optical method, was credited in contemporary texts with being able to give accuracy readings of + or minus 10 mills. Once the azimuth operator located the target, the range operator performed his normal operation to obtain a range readout. It should be noted that contemporary texts state that this method placed a severe strain on the azimuth operators vision and recommended that he and the range operator switch positions at 20 minute intervals.

The set was capable of tracking bomber-sized aircraft out to a distance of 20,000 yards and gave a range reading with an accuracy of + or minus 25 yards. It was limited, however, by being unable to track targets closer than 100 yards in range. It was unable to be utilized in weather conditions that restricted optical tracking. Despite being used in combat, the system was replaced, as much as possible, upon availability of the superior SCR-584 (see below) which combined tracking, height finding, and early warning in one set.

SCR-584/598

Both the SCR-582 and SCR-682A were effectively employed in the Mediterranean and Pacific theaters in independent organizations called coast artillery surface warning batteries. They were used in forward areas to warn of air or surface attack and to give convoys navigational assistance in dangerous areas. During WW II, the SCR-296A and SCR-582 were primarily used for heavy gun batteries while the SCR-682A was authorized for coast artillery mobile batteries of 155 mm. While these radar sets performed adequately, the Coast Artillery Corps developed several other radar sets to perform functions for which the SCR-296A and SCR-582 were unsuitable. The first requirement was voiced in 1941 when the Chief of Coast Artillery recognized that while he had weapons and searchlights capable of illuminating what was perceived as one of the greatest threats of the period, the motor torpedo or PT boat, but the SCR-268, which was the standard searchlight control radar, was incapable of operating at zero degree elevation and thus was unusable.

The first result was the development of the SCR-598, which was a modification of the SCR-296A with shorter range, but improved accuracy against small, fast, rapidly maneuvering targets. As an interim set the SCR-584 (fig 28), which had been produced for directing the 90 mm anti-aircraft guns, was modified to enable it to be used for surface fire control in the mobile 155 mm batteries. Additionally this set was standard equipment for the 90 mm AMTB, and the mobile 90 mm coast artillery batteries. This modified

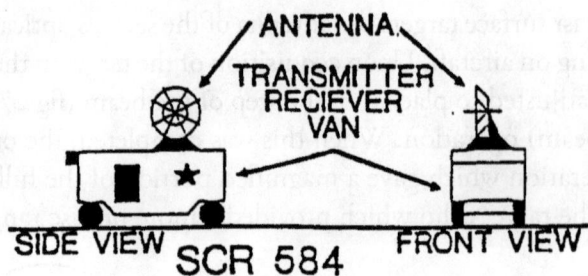

Figure 28

SCR-584 was used with the 155 mm mobile batteries to protect the forward island bases of the Pacific. The SCR-584 was housed in a mobile van and differed from the SCR-296/582/682 family in utilizing a parabolic antenna mounted directly on the van roof instead of on a tower. It operated on a 10 centimeter wave length, and due to operation with a narrow transmission beam and certain wartime modifications (N2 gate) it not only was able to screen out the surface clutter caused by wave action, but also virtually immune to enemy jamming by the aluminum strips (window).

The Post-War Position Finding Radar: the MPG-1

The final result of the research on coast artillery radar fire control was produced too late for service during WW II, but when introduced postwar, corrected all of the previous systems deficiencies. By all accounts, this radar, which was standardized as the AN/MPG-1 (fig 29), was one of the best sets produced for Coast Artillery service. The MPG-1 operated on still a shorter wavelength of 3 centimeters, and was mounted in a single semi-trailer which was towed by 6-ton tractor. The trailer, which functioned as the operating shelter, contained the radar consoles, the antenna, a portable generator, test equipment, and spare parts. The antenna consisted of a constantly rotating 8-foot unprotected array, on a 17-foot collapsible tower. The entire equipment weighed 11 tons and required 10 kilowatts of power.

The shelter contained the transmitter and two 7-inch scopes (fig 30a and 30b). The first was the PPI and the second was the tracking or "B" Scope. The PPI scope was capable of operating in two range scales, 30,000 yards and the maximum 80,000 yards. This scope was marked in 10 degree increments to give a 360 presentation on the defended harbor.

The PPI was also divided into 8 concentric circles which represented 4000 or 10,000 yard range marks depending on the range scale in use. As in the earlier series this was used for surveillance, and when the target range decreased to 28,000 yards or less, the image was transferred to the B Scope. This second 7-inch scope magnified the target presentation that represented an area 2000 yards wide and 10 degrees in azimuth. The B scope was divided by electronically produced lines, three vertical and three horizontal. The horizontal lines were located at the top middle and bottom of the scope and represented 1000-yard range deflection from the center or target locator line. The vertical lines were closely spaced in the center of the scope and represented 1 degree azimuth deflection on either side of the center or target locator line. The operator tracked the target, keeping the blip at the intersection of the center horizontal and vertical lines of the B scope, and the range and azimuth were read off dials on the console. This presentation was accomplished in such a manner as to be in the same plane as the target signals so that it was automatically aligned with the target signals on the main scope. The smooth tracking of the target was made easier by the inclusion of mechanically aided tracking. The set was equipped with two tracking rates, a fast rate for fast close targets such a PT type craft, and a slower rate for larger, slower, or distant targets.

In addition, the MPG-1 had a third scope utilized for spotting the shell bursts (fig 30c). This spotting scope was the prime difference between the MPG-1 and the earlier SCR-296/582 families. The spotting scope gave an identical presentation to that of the tracking B scope. The spotting operator utilized a joystick to move a transparent plate with 5 vertical and 3 horizontal lines to position the center on the shell splash which enabled him to read the deviation in range and azimuth of the shot.

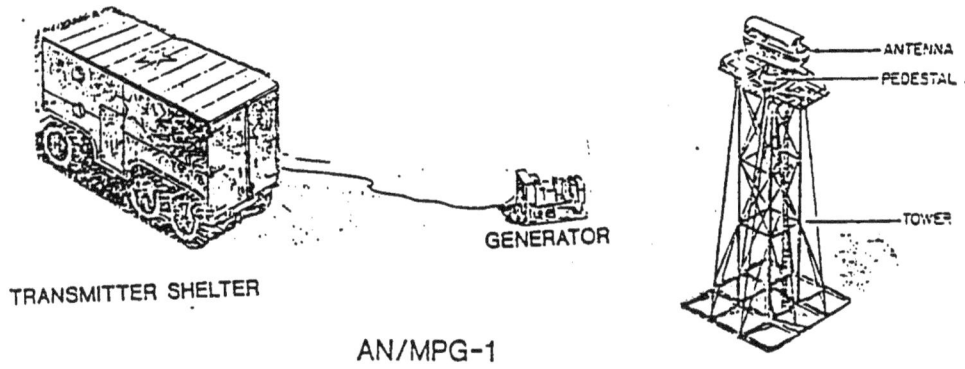

AN/MPG-1

Figure 29.

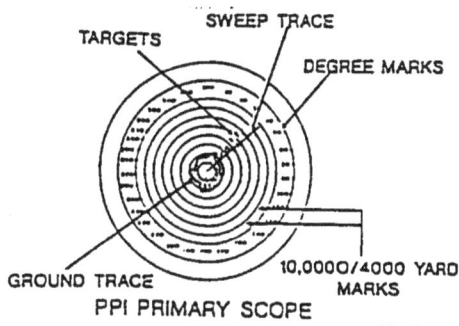

PPI PRIMARY SCOPE

Figure 30a

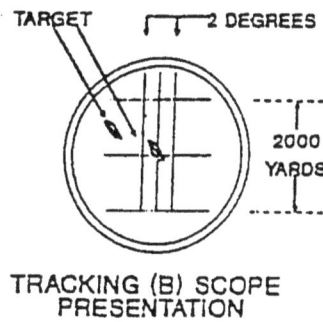

TRACKING (B) SCOPE PRESENTATION

Figure 30b

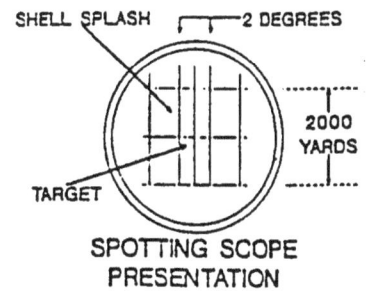

SPOTTING SCOPE PRESENTATION

Figure 30c

Utilizing these improvements, the AN-MPG-1 system achieved an accuracy of being within + or - 10 yards error in range data and within 0.05 degrees error in azimuth data. It was credited postwar with a reliable range of 50,000 yard against large (battleship or heavy cruiser) targets. A fixed version, the AN-FPG-1, was developed for use with the heavy fixed installations such as the 16-inch batteries. Other than the lack of mobility, the primary change was that the tracking B scope and the spotting scope were modified to allow use at a range of up to 50,000 yards. This modification allowed the armament to be used at maximum range, something not possible with the earlier SCR-296/682 and MPG-1 systems.

HARBOR DEFENSES OF THE COLUMBIA
FORT CANBY
LOCATION MAP

SCALE IN FEET

SERIAL NO.
CONFIDENTIAL
EDITION OF 1 MAY 1946
REVISED:

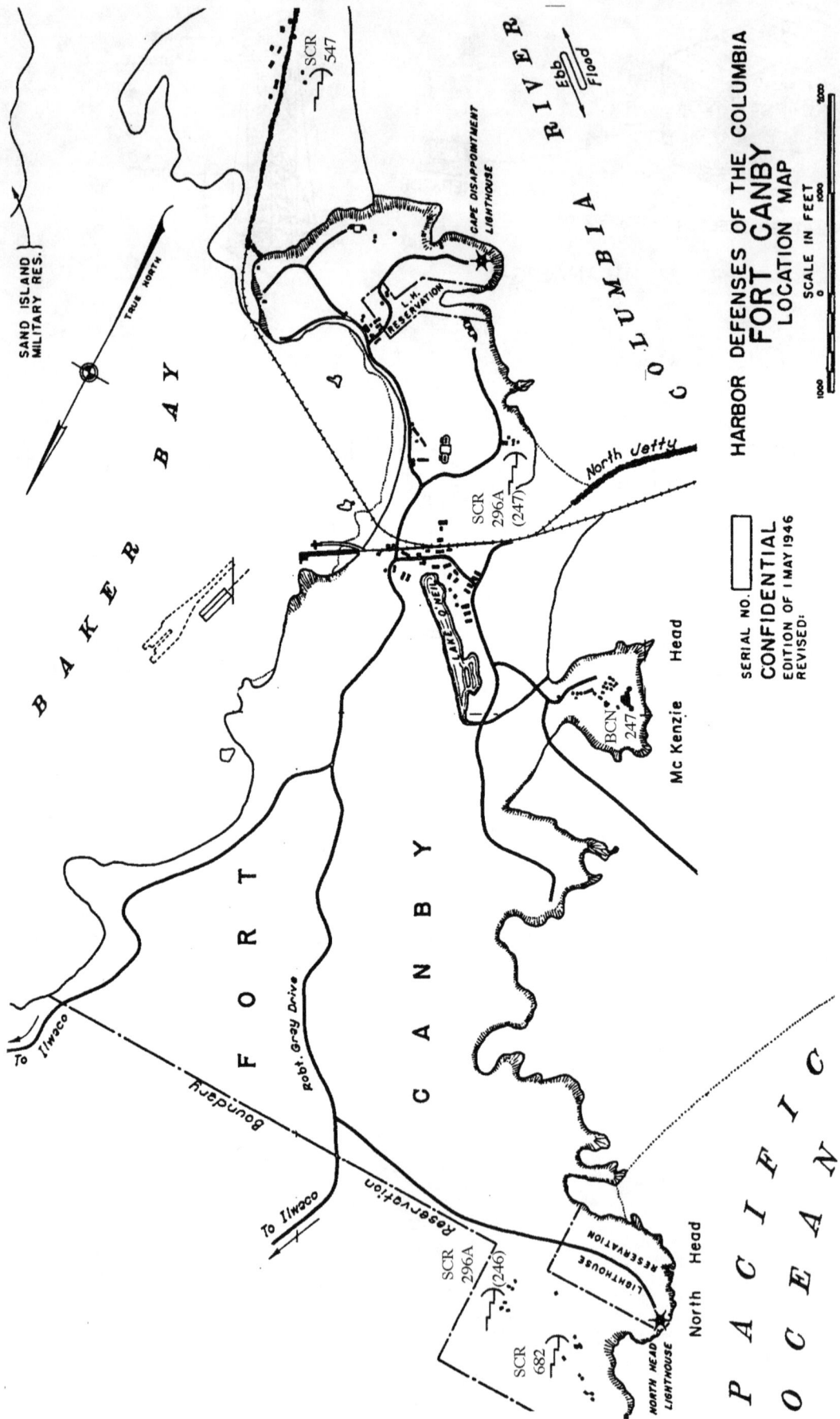

Figure 31. Location of radar units, Fort Canby, Washington, 1946. Four radar sets were located on this reservation, the SCR-682 for the Harbor Defenses of the Columbia and the SCR-296A for Battery 246, Fort Columbia, on North Head, the SCR-296A for Battery 247 on Cape Disappointment and the SCR-547 for the 90 mm guns of the AMTB battery on a jetty sand spit. Note that the SCR-296A for Battery 247 is located on Cape Disappointment, some distance away from the battery itself on McKenzie Head.

Selected Bibliography

Clement, A. W., "Seacoast Artillery Radar," *Coast Artillery Journal* Vol. 91 (May-June, 1948).

Orman, Lt Col L. M., "Radar, a Survey," *Coast Artillery Journal* Vol. 89 #2 (Mar-Apr, 1946).

Orman, Lt Col L. M., "Identification of Friend or Foe," *Coast Artillery Journal* Vol. 89 #6 (Nov-Dec 1946).

Terret, Dulany., *The Signal Corps, The Emergency,* U.S. Army in World War II, Office of the Chief of Military History, United States Army, Washington, D.C. 1956.

Thompson, George R., and Harris, Dixie R, and Qakes, Pauline M., *The Signal Corps, The Test,* U.S. Army in World War II, Office of the Chief of Military History, United States Army, Washington, D.C. 1957.

Thompson, George R., Harris, Dixie R, *The Signal Corps, The Outcome,* U.S. Army in World War II, Office of the Chief of Military History, United States Army, Washington, D.C. 1966.

War Department, FM 4-95, *Service of the Radio Set SCR-296,* 17 Sept 1943.

War Department, FM 4-96, *Service of the Radio Set SCR 582,* 9 Nov 1943.

War Department, FM 4-97, *Service of the Radio Set SCR 682A,* 17 Aug 1944.

War Department, TM 11-1366, *Radar Set AN/MPG-1 and Radar Set AN/FPG-1.,* 15 March, 1945.

Coast Artillery Electrical Power Plants

Bolling W. Smith

This article is a revised, illustrated version of Bolling Smith's "Emplacement Powerplants," article originally published in the *Coast Defense Study Group Journal* Vol. 7, Issue 3, August 1993, pp. 45-50.

Several illustrations are from Lorimer D. Miller's 1917 article: "The 25 Kw Gasoline Driven Generating Set Used in the Coast Defenses of the United States", *The Journal of the United States Artillery*, Vol. 48 No.1, July-August, 1917, pp. 54-79.

By 1900, electricity had become a vital necessity for the Coast Artillery. It was used to traverse and elevate some of the large guns, to light emplacements, to operate ammunition hoists, to power searchlights, to control submarine mines, and for communications, in addition to standard garrison uses. Even when the power came from storage batteries, these had to be charged from generators. The requirement that coast defenses be self-contained resulted in power rooms being included in most batteries and mining casemates, and separate searchlight powerhouses were constructed.

Because direct current (D.C.) technology developed faster than alternating current (A.C.), and because D.C. was necessary to charge storage batteries and operate searchlights, the Engineers preferred D.C. power until the final batteries of the Taft era were built, when the ability to utilize commercial A.C. power in Los Angeles led to the first A.C. powered emplacements. Even in WW-II, D.C. plants were installed to power searchlights.

Since power plants were less than totally reliable, and were at any rate subject to hostile fire, it was deemed essential that any system have a separate reserve, in the event the primary source was disabled. Initially, steam generating plants were generally used in the emplacements, but the Hornsby-Akroyd oil engine was also extensively purchased, especially during the Spanish-American War, when a wide variety of power plants, steam and kerosene, were hastily purchased and installed, often in small powerhouses quickly constructed behind the emplacements rather than inside them. Kerosene engines were particularly well suited to locations where abundant supplies of good water for steam plants were not readily available. Banks of storage batteries were used as reserves, periodically charged from the primary plant, and used for short periods, when it was uneconomical to operate the engine and generator. Since both steam and kerosene engines required some time to start up, this was an important factor. Although these power plants were hardly standardized, they generally were of small capacity, in the range of 5 kilowatts (kW), and were usually intended only to provide power for lighting, and that at a level that a decade later would be considered totally inadequate.

The Hornsby-Akroyd oil engine was a four-cycle kerosene-fueled engine built by the De La Vergne Refrigerating Machine Company, New York, N.Y. It was manufactured in a wide range of sizes, from 1 $1/4$ to 125 horsepower (HP). It featured large flywheels to help maintain speed regularity. In 1911, 28 Hornsby-Akroyd engines remained in place in the coast defenses, with an average rating of 16 HP, producing approximately 10 kW of electric power, usually through a belt driven D.C. generator. Al`though figures are not available as to how many engines were purchased, it is reasonable to assume that many of the smaller engines, over a dozen years old, would have been replaced by 1911. This engine was well suited for small applications, where a steam plant would not have been practical, especially in mining casemates. The kerosene engine could be started up more quickly than a steam engine, but still had a startup time of 8 to 18 minutes, depending on the size of the engine.

The Hornsby-Akroyd oil engine. http://vintagemachinery.org/mfgindex/imagedetail.aspx?id=5986

In the early emplacements, the power room and storage battery rooms were well protected, in the interior of the emplacement; so much so that their location, coupled with the poor ventilation of the early emplacements, resulted in so much condensation that the electrical equipment deteriorated rapidly. The solution seemed to be to bring the plants out into temporary structures outside of the emplacements, where adequate ventilation was possible, and to replace the plants in their protected locations in the event of hostilities. In the inevitable nature of things, however, the old power rooms quickly became utilized for other functions, and had a hasty return been decided on, there would have been no place to return them to. This problem was finally resolved around 1903, when it became the policy to design all new emplacements with power rooms which were both well ventilated and well protected.

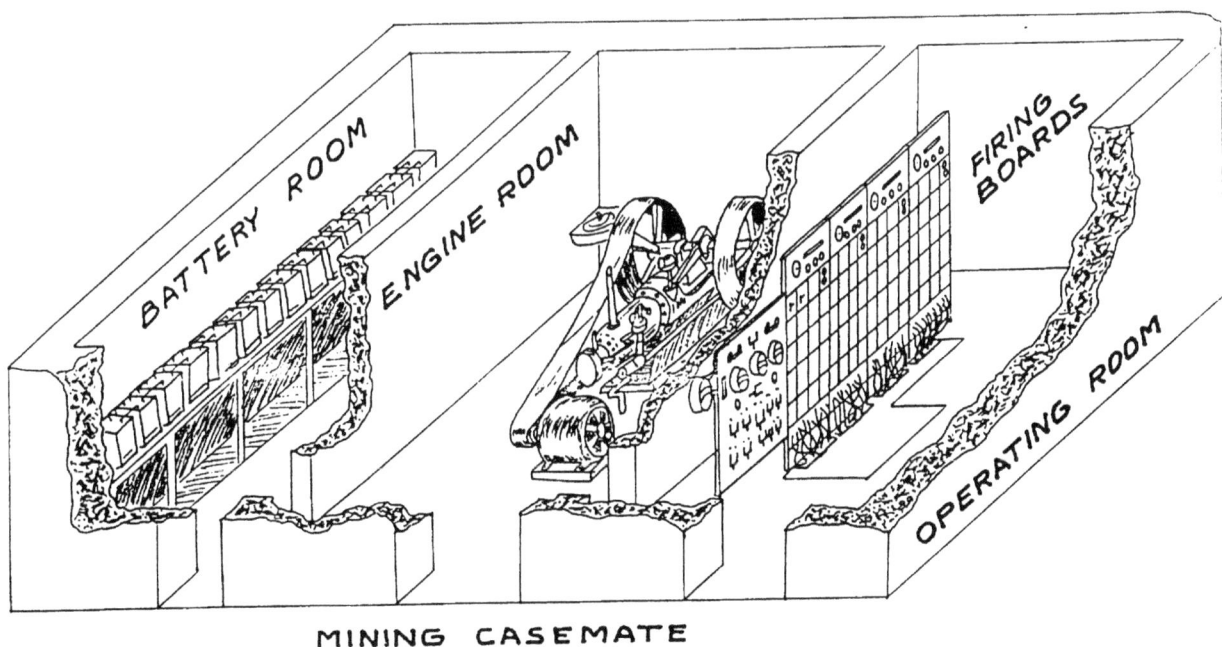

MINING CASEMATE

A Hornsby-Akroyd oil engine used for the water pumping system at Fort St. Phillip near Buras, Louisiana

In 1906, a sub-committee of the Taft Board recommended central, steam powered, D.C. power plants for coast defenses. Smaller D.C. oil or gasoline power plants were recommended as reserves for individual or close groupings of searchlights or emplacements, depending on local conditions, replacing the storage batteries whose expense and maintenance had finally proven too great. Realizing Congress would be slow to appropriate the funds for the central plants, Col. Fredrick V. Abbot, an assistant to the Chief of Engineers, determined to install the reserves first, and assigned two officers to find a satisfactory internal combustion engine to power these generators. The two fuels available were gasoline and kerosene, frequently referred to as gas and oil. Kerosene was the preferred fuel, as there was less danger of fire, a not inconsiderable advantage in combat. Additionally, the Coast Artillery had experience with the Hornsby-Akroyd oil engines, which were already in service. Offsetting this advantage, however, were significant disadvantages. Kerosene engines required considerable preheating before starting, and even then were often difficult to start. Any kerosene engine adopted, then, must be quick starting. Gasoline engines, on the other hand, also had drawbacks, in addition to their more volatile fuel. In particular, speed regulation was a problem with early gasoline engines. Diesel engines had not progressed sufficiently at this date to even be considered.

Two oil engines were evaluated, the Mietz and Weiss 10 kW engine, and the De La Vergne 25 kW engine. The Mietz and Weiss was a two cylinder, hot chamber engine, requiring ten minutes preheating before starting. Even then, despite the use of compressed air, it could never be started without great difficulty. On one occasion, a steam line was run into one cylinder in an unsuccessful attempt to start the engine, presenting the interesting picture of an internal combustion engine operated from a steam boiler. This engine was installed in Fort Banks, Mass., but was never really satisfactory. The second oil engine tested was ordered from the De La Vergne Machine Company, maker of the Hornsby-Akroyd engine. The four cylinder engine was under construction for over a year, and when completed was sent to Fort Monroe for testing. It had several unique features, including placing the generator between the two pairs of cylinders, so that it could be run at half capacity by using only two of the cylinders, thereby saving fuel. By the time it was finished, however, an improved gasoline engine had been developed, and proved so satisfactory that oil engines were no longer seriously considered.

FIG. 3.   10 Kw. KEROSENE SET
(MIETZ & WEISS, NEW YORK CITY)

2096

FIG. 4.   25 Kw. KEROSENE SET
(DE LA VERGNE MACHINE COMPANY, NEW YORK CITY)

2097

The first gasoline engine evaluated was built in 1906 by the Westinghouse Company of Schenectady, N.Y., an experienced firm. The General Electric Company (G.E.), which did not build engines at this date, supplied the generator. One of these models had already been installed in Fort Wool, Va., during the 1905 maneuvers. The three-cylinder engine was comparatively crude, and several changes were made, including replacing the single throttle with a throttle for each cylinder in an effort to improve regulation. An unusual system of cooling the water was used; the fresh water was in an sealed loop, and flowed from the engine to an inner tank, surrounded by an outer jacket, through which flowed salt or other impure water. This improved model was installed at Fort Standish, Mass., but it had several insurmountable defects. The poor regulation obtained with three cylinders showed that four cylinders were necessary, and the need for twice as many water pumps was a major drawback. About the same time, a 25 Kw set was purchased from the Westinghouse Machine Company, East Pittsburgh, Pa. This engine was installed in Fort Revere, Mass., where it proved an excellent engine, but unsuited to the particular conditions of coast defense, since it required ten feet of headroom to remove the cylinders and the normal height in the emplacements was only seven feet. The floor at Fort Revere was excavated two feet to allow the installation, but in too many emplacements the water table was too close to floor level to allow this solution.

FIG. 1. 25 KW. GASOLINE SET
(WESTINGHOUSE COMPANY, SCHNECTADY, N. Y.)   2094

In June 1907, a new engine appeared which seemed to be the answer. The five-cylinder "Dock" engine, invented by Mr. H. Dock and manufactured by the New York Safety Steam Power Company, Hope Valley, R.I., seemed ideal. Light and compact, it offered excellent regulation, and promised durability and efficiency. The search seemed over, and specifications for bids on sixty engines were issued, tailored to the Dock engine. Everything seemed settled until the General Electric Company submitted a bid for an engine exactly like the five cylinder Dock engine, but 20 per cent cheaper, and in addition, offered a four-cylinder version at a further ten per cent reduction. The problem was that the makers of the Dock engine had pro-

FIG. 2. 25 KW. GASOLINE SET
(WESTINGHOUSE MACHINE COMPANY, EAST PITTSBURGH, PENN.)

FIG. 5. 25 KW. "DOCK" GASOLINE SET
(NEW YORK SAFETY STEAM POWER COMPANY, HOPE VALLEY, R. I.)

duced a working prototype, while G.E. had never built a four or five cylinder engine, and certainly could not submit one in operation within the thirty days required by the specifications.

In late 1907, however, events took an unexpected turn. The Chief of Engineers decided to give General Electric an order for ten four cylinder engines, despite their failure to win the contract. The makers of the Dock engine, in the meanwhile, despite the success of their prototype, were unable to build an acceptable second engine. Six engines were built and found unsatisfactory before the factory was destroyed by fire and the company went bankrupt. Ironically, all the rejected models were the product of established manufacturers, while in the end, the four cylinder by unproven G.E. was adopted as the standard, and between 1908 and 1917, 270 of these sets were installed in coast defenses.

FIG. 7 ACCESSORIES TO THE GENERAL ELECTRIC COMPANY'S 25 KW. GASOLINE SET

As issued, the G.E 25 kW generating set consisted of a gasoline engine, direct coupled to a D.C. generator, and a radiator. Both engine and radiator could fit through a standard three-foot emplacement doorway, and the cylinders could be removed within a seven foot headroom. The first fifteen engines were required to be semi-transportable, and the weight limit of 4000 pounds meant that the base had to be made of bronze, but subsequent engines had cast iron bases. These sets proved to be durable, virtually foolproof, and suitable to a wide range of climates. The only complaint was their noise level, and several minor changes were made to make them quieter. The gasoline engine, designated Type GM-12, was a vertical cylinder, four cycle model, with maximum output of 54 HP at the two hour overload rate and a continuous rating of 43 HP. The speed varied from 560 rpm at no load to 575 rpm at full load. The cylinders were 7 1/4 inch diameter x 7 1/2 inch stroke. Fuel was supplied from a 370 gallon tank buried about two feet below ground level, outside of the emplacement. The use of a fuel pump avoided the fire hazard of a gravity feed system; as soon as the engine stopped, the gasoline would run back down into the outside tank. A hand pump was used for starting the engine. The carburetor was equipped with a resistance heating unit to prevent carburetor icing when starting in cold weather, and with a valve to draw preheated air from the crankcase, thereby also passing the products of combustion out through the cylinders and the exhaust, so that they would not leak out into the power room. The throttle, located between the carburetor and the "T" cylinder heads, was controlled by the fly-ball governor. These parts were responsible for the excellent regulation of the engine, which proved better than previously thought possible from a gasoline engine.

FIG. 6. 25 Kw. GASOLINE SET—IGNITION SIDE
(GENERAL ELECTRIC COMPANY, SCHNECTADY, N. Y.)

FIG. 10. CARBURETOR SIDE

FIG. 16. CENTRAL POWER PLANT, FORT H. G. WRIGHT, NEW YORK 2109

FIG. 17. CENTRAL POWER PLANT, FORT H. G. WRIGHT, NEW YORK, SHOWING SWITCHBOARD

The ignition utilized a low voltage A.C. magneto, a make-and-break mechanism, a step-up transformer, and a high voltage distributor. Since the magneto would not produce enough current to operate the spark plugs until the magneto was up to speed, dry cells were supplied, along with a switch to cut them in for starting. The recommended method of starting involved the operator turning the engine over by means of a crank attached to the flywheel. An alternate method of instantaneous starting was supplied, whereby a blank black powder cartridge was exploded in one cylinder. This was not only effective, even if the dry cells were discharged, but required less effort than hand starting, so much so that some operators got into the habit of using this method normally, at least until orders were issued to restrict its use to emergencies.

Radiators were not usually installed in stationary engines; the hot water being normally allowed to run off. In fortification power plants, however, radiators were necessary to allow the reuse of the cooling water, often a scarce commodity. The radiators were not installed in the same room as the engines, due to their noise, and to the 10,000 cubic feet per minute of air delivered by the fan, powered by a 3 HP electric motor. One modification to reduce the noise level involved a cut in resistor to reduce the speed of the fan from 1150 rpm to 950 rpm when maximum cooling capacity was not needed. The water pump was one problems area, and several different designs were tried to improve longevity and reduce noise. The normal coolant was water, and if the local water was too impure, rainwater or distilled water could be used. For colder locations, instructions for an alcohol-water antifreeze mixture were issued, but this mixture was not recommended for general use and seemed to be little used, as artillery drills were not normally held when the temperature was below freezing.

The engine was direct coupled to the General Electric MPC Type D.C. generator. With the exception of a few 220 volt units, all the generators were rated 115 volts at 560 rpm. They were compound wound to give the same voltage at either full or no load, and had a two-hour overload capacity of 33.3 kW. The continuous rating was 27 kW at 43 HP, of which 2 kW was needed for the radiator fan, giving the usable output of 25 kW. The engine and generator were mounted on a rectangular concrete foundation, at least two feet thick, and approximately 59 inches x 26 inches, with eight mounting bolts. The fan and radiator foundation, at least six-inches thick, was approximately 60 1/2 inches x 34 inches and narrower at the fan motor end, with six mounting bolts. The G.E. 25 kW Gasoline Generating Set remained in use until after the passing of the Coast Artillery, but due to larger power requirements and improvements in diesel engines, the primary source of emplacement power in the WW-II generation of fortifications was the diesel engine and A.C. generator. The G.E. 25 kW set, however, remained in service in large numbers through WW-II, and the number of foundations for engines and radiators remaining today is proof of the vital role they played for over forty years.

Most of the larger coast artillery posts received a central power plant which had coal fueled steam boilers to drive electrical generators. In general all coast artillery posts were connected to commercial electrical grids for lighting and general electrical uses by 1917. The cental power plant was a back up to the commercial power, and to the various emplacement and searchlight power plants on the post.

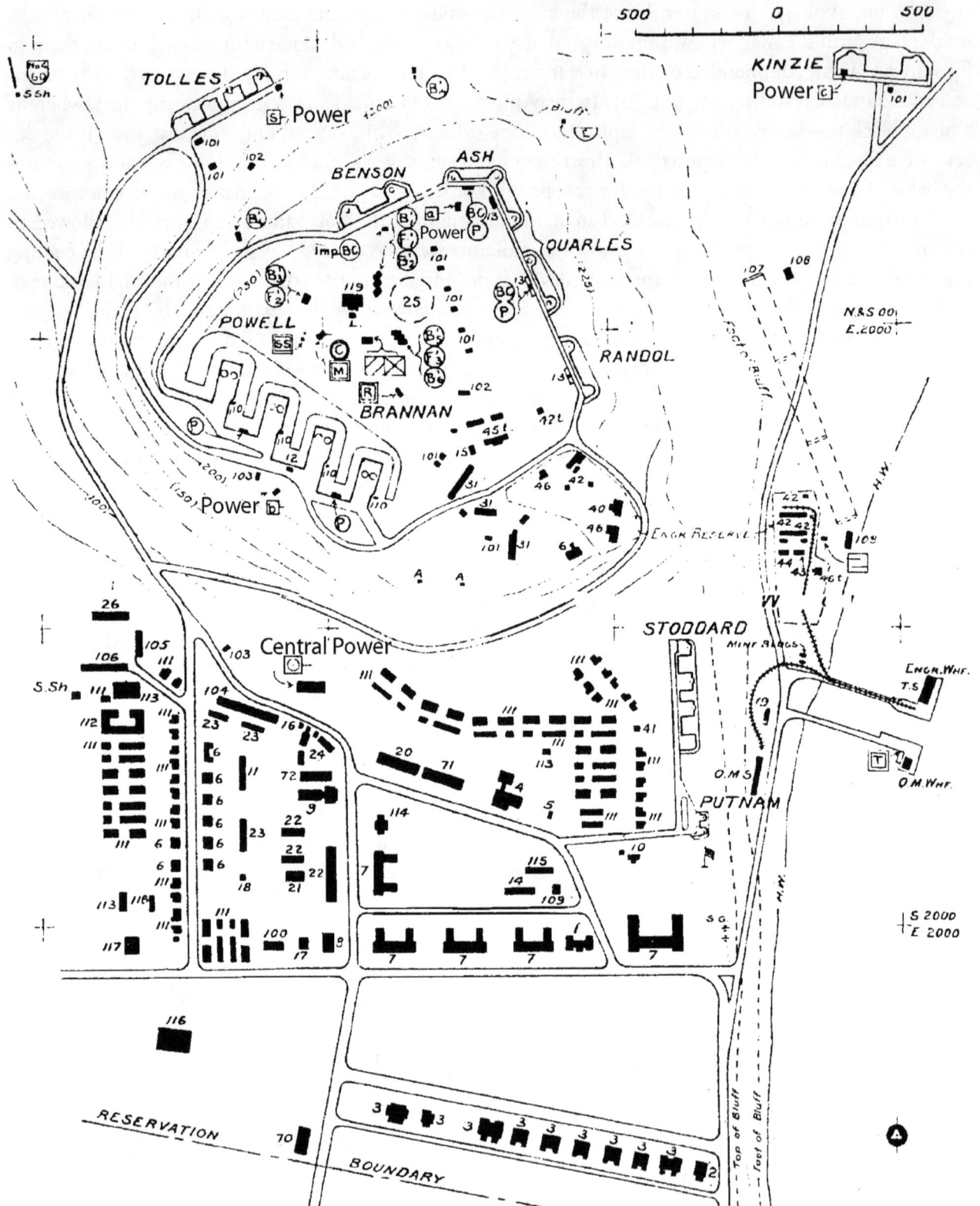

Detail map of part of the Fort Worden Reservation, Harbor Defenses of Puget Sound, showing the locations of five of the six power plants on the site (square symbols with a "o" or a letter inside and a dash on the outside)— the Central Power Plant, the "a" power plant at Battery Ash, the "b" power plant at Batteries Brannon and Powell, the "c" power plant at Battery Kinzie and the "s" power plant at Battery Tolles.

Central power plant at Fort Worden, Harbor Defenses of the Puget Sound (NARA)

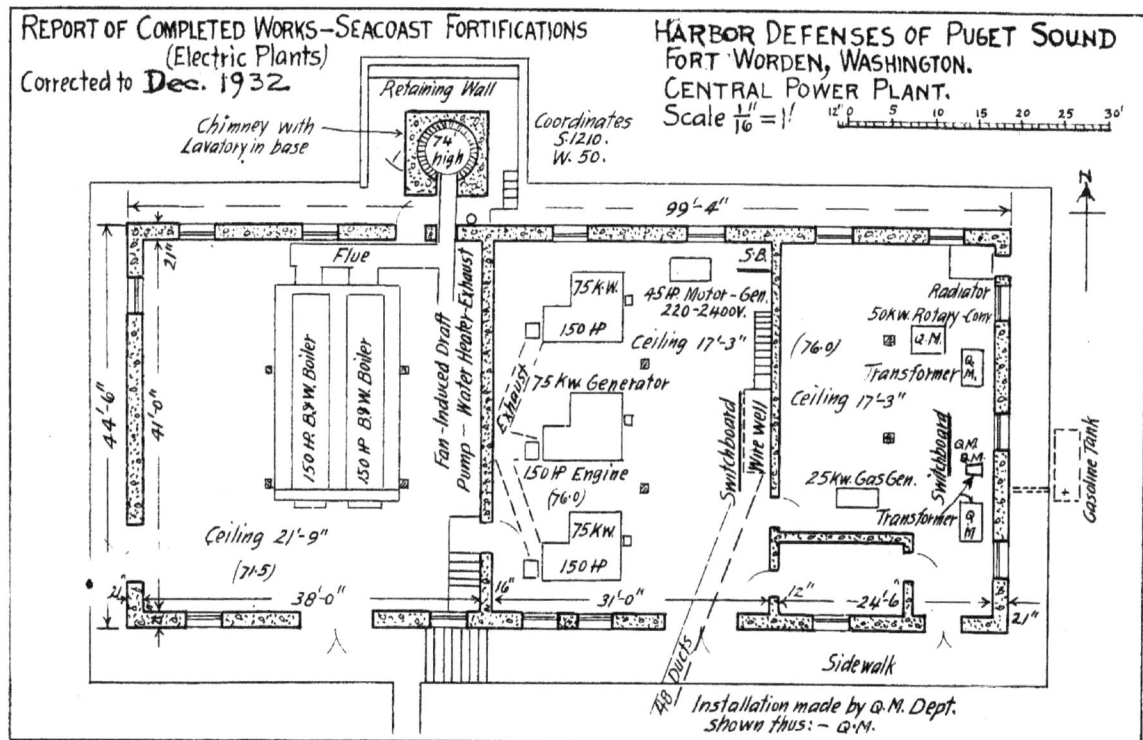

Interior views central power plant at Fort Worden (NARA)

Battery Ash powerplant at Fort Worden (NARA)

Power plant for searchlights No. 21 & 23 (old SL No. 2 & 3) at Fort Worden (NARA)

The powerplant for searchlight 18 (old SL No. 4) at Fort Worden (NARA)

Power control for searchlight No. 18 (old SL No. 4) at Fort Worden (NARA)

Restored power plant (above) and cooling radiators (below)
Battery Osgood-Farley, Fort MacArthur Museum, San Pedro, CA 2016 (Mark Berhow)

Fuse boxes for the electrical system at Battery Osgood-Farley, 2016 (Mark Berhow)

Interior lighting for shell magazine at Battery Osgood-Farley 2016 (Mark Berhow)

Power in Seacoast Fortification Modernization Program of 1940-45

The standard 200 Series 6-inch batteries built during WW-II contained three 150 HP Worthington diesel engines, each driving a Westinghouse 460 Volt 125 kW 3 phase A.C. generator, while the larger, 100 Series 16-inch batteries each contained three 340 HP Worthington Model CC-6 diesel engines and three Westinghouse 460 Volt 375 kW 3 phase A.C. generators. There were several 100 Series batteries, however, which had non-standard power plants, including Battery Gray (#107), Fort Church, R.I., which used diesel engines to drive D.C. generators. The previously built 16-inch batteries had significantly smaller power plants. The plan for the power facility of Battery Richmond C. Davis at Fort Funston, CA, the first casemated 16-inch battery built in 1938-40, is shown below.

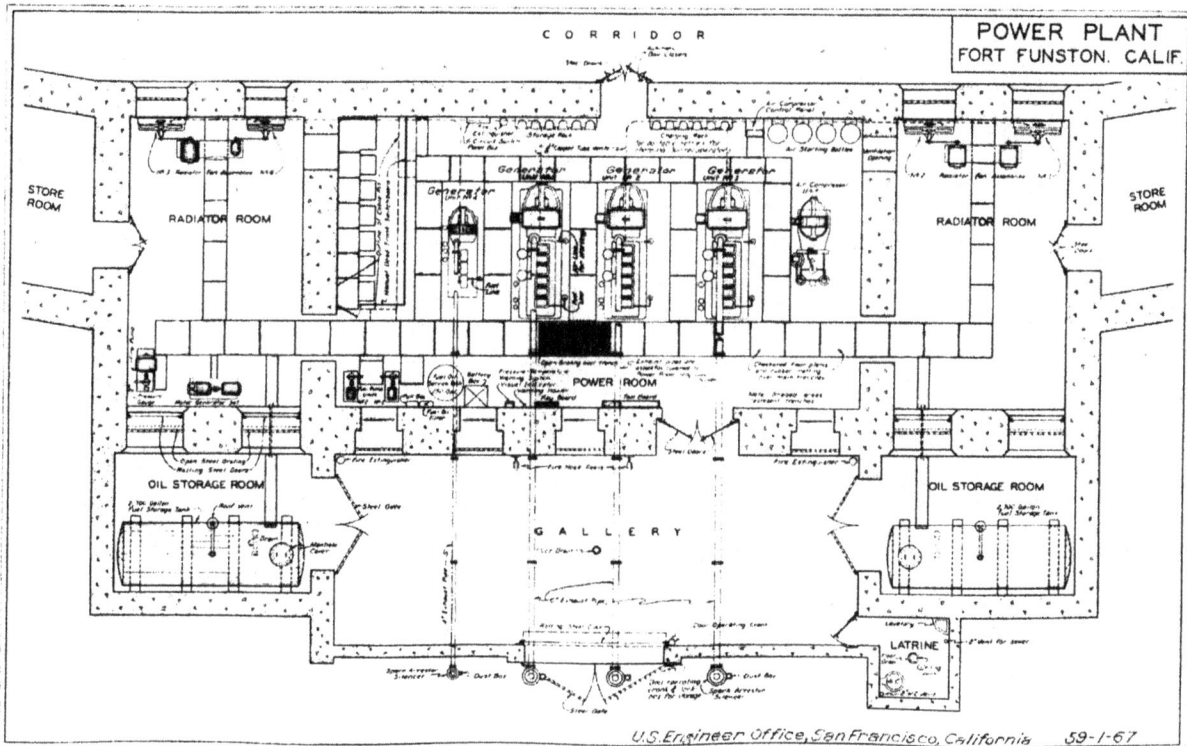

The exisiting 12-inch and 16-inch batteries that were casemated during the WW II years generally received a new power plant system as part of the casemating process.

Chronology of Power Plant Design Procurement and Installation Matters 1940-1942.

This document, dated July 21, 1942, was found in Archives II, RG 77, Entry 1006, D.F. 662, Box 14. No source was given, but the context and the presence in RG 77 seem to make it clear that it was from the office of the chief of engineers.

Item No. 1: September 11, 1940

The modernization program was authorized by the secretary of war in a letter to the chief of engineers, July 27, 1940, with amendments. This provided for construction of 19 new and extensive modification of 17 existing major-caliber batteries, and 26 secondary-caliber batteries, to include necessary fire control elements in the continental United States and overseas bases.

Responsibility for battery power plant design, procurement, and installation was assigned to the chief of engineers by AR 100-20, February 10, 1936, as it had been previously. The basic plan was to use available designs and minimize development to allow expeditious accomplishment. DC power plants were being procured by the district engineer, Philadelphia, when it was learned that the Ordnance Department was trying to procure AC motors for seacoast guns. Procurement by the engineers was stopped pending necessary data from the Ordnance Department to redesign the power equipment.

Item No. 2: 1941-1942

The chief of ordnance recommended an increase in power with conversion to AC to increase the speed of operations and fire power of the batteries. The chief of coast artillery concurred, with the general understanding that the changes would not materially delay the modernization program.

Earlier 16-inch batteries had 90-100 kW diesel generators, but now new estimates ran as high as 1000 kVA per gun. This required larger power rooms. The chief of engineers was unable to obtain definite data on the type of power equipment required until sufficient progress was made by the Ordnance Department, and the construction program suffered from the inability to complete power room designs.

Item No. 3: January 21, 1942

Circular No. 17, January 21, 1942, assigned the design, procurement, and installation of power equipment to the chief of ordnance, pending revision of AR 100-20, February 10, 1936.

Engineer requirements for lighting, etc., of major-caliber batteries were approximately 50 kVA, a minor amount compared with the requirements under consideration by the Ordnance Department for gun operation.

Item No. 4: 1940-1942

The construction program was progressed by engineer field forces as much as possible in the absence of complete data on the structural requirement to accommodate the new power equipment.

Because many construction contracts were already in force, it became increasingly difficult to avoid costly delays or even terminations of contracts. The Ordnance Department issued various sketch drawings on the basis of which, with supplementary consultations, it was possible to issue enough instruction to engineer field forces to continue construction at a reduced rate. These instructions were generally type plans approved by the representatives of the chief of ordnance. Resources resorted to keep field construction progressing were:

    a. Temporary omission of entire areas of construction, principally power room areas.
    b. Temporary omission of floor slabs, manholes, trenches, engine bases, etc.

As the ordnance development program progressed, structural changes to accommodate revisions in power plants, gun mounts, and ammunition service were requested. Sometimes, as many as three conflicting requests for one item were received. The chief of engineers cooperated as much as possible, consistent with minimizing confusion on the part of field construction forces and maintaining maximum progress. The 1942 policy as related to the chief of ordnance was in the 3rd indorsement, July 17, 1942, "CE 662 SPEEF, Subject: 16" barbette carriages arrangement of power equipment in fortifications," paragraph 1, which stated it had become increasingly clear that Ordnance Department power plant development and procurement difficulties would make minor structural changes unavoidable. There followed a summary of the general procedure which was to be followed to achieve the desired results in the field:

Radical structural changes in approved designs issued as the basis for construction contract documents could not be undertaken.

Minor structural changes required to accommodate changes in power, armament, or communications equipment were practicable for uncompleted construction and, when requested, would be incorporated in type plans from time to time with the understanding on the part of all concerned that the changes were not applicable to completed construction.

1. Lubricating oil cooler.
2. Lubricating oil filter.
3. Fuel oil filter.
4. Lubricating oil pump.
5. By-pass valve.
6. Safety control unit.
7. Force feed lubricator.
8. Outlet water manifold.
9. Air intake manifold.
10. Air cleaner.
11. Exhaust manifold.
12. Inlet water manifold.

FIGURE 4. Left side, Diesel engine for a 16-inch gun battery.

1. Governor.
2. Air starting manifold.
3. Master air valve.
4. Control wheel.

FIGURE 3. Right side, Diesel engine for a 16-inch gun battery.

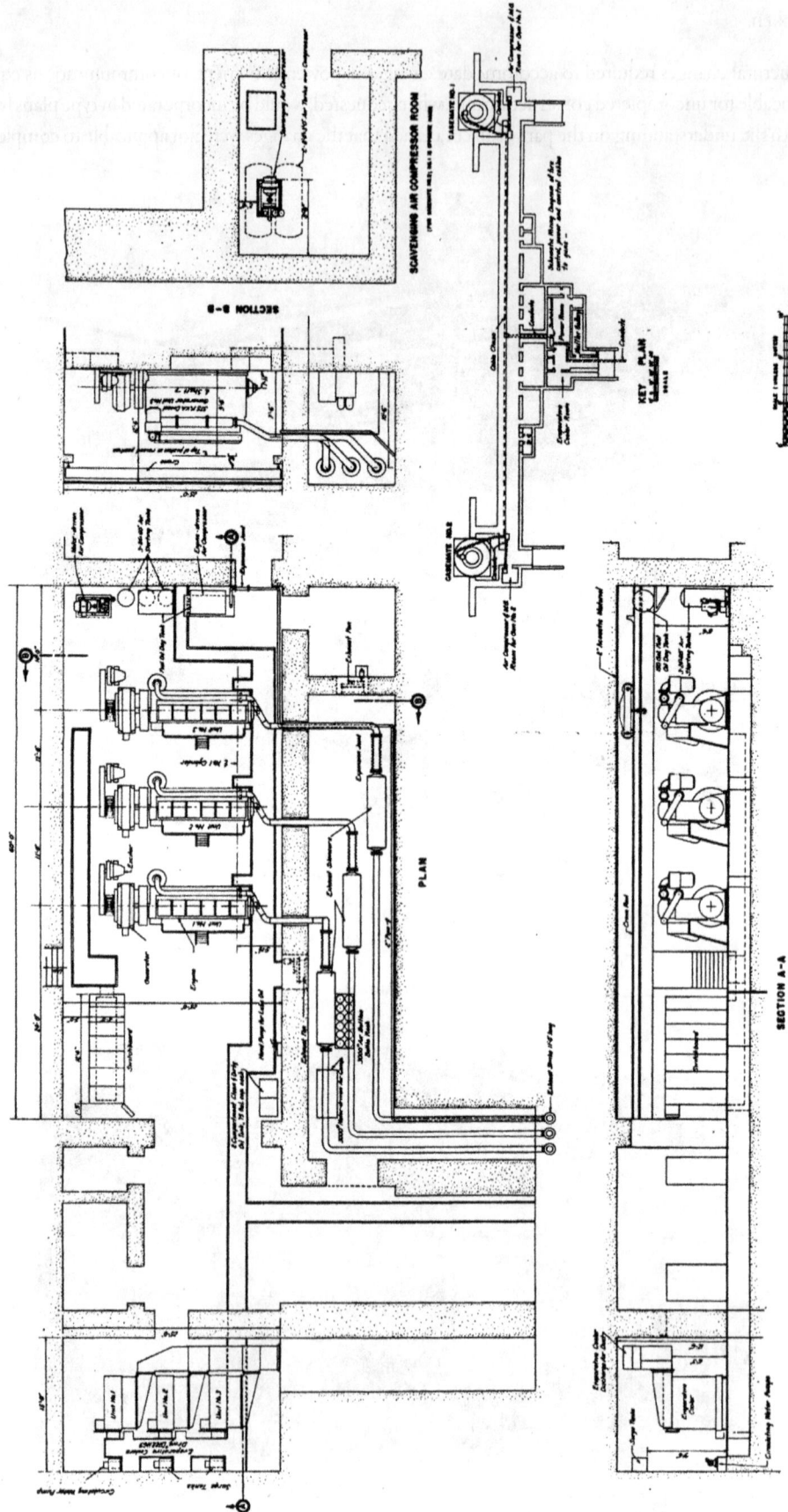

Equipment layout, M1 power plant for 16-inch gun battery.

Diesel electric plants for 6-, 12-, and 16-inch Gun Emplacements, Coast Artillery Training Bulletin, Vol. 4, No. 7, July 1945.

M1 power plant for 16-inch battery. *War Department, SNL 20, 1943.*

Power plant for Battery Steele, Harbor Defenses of Portland, Maine (NARA)

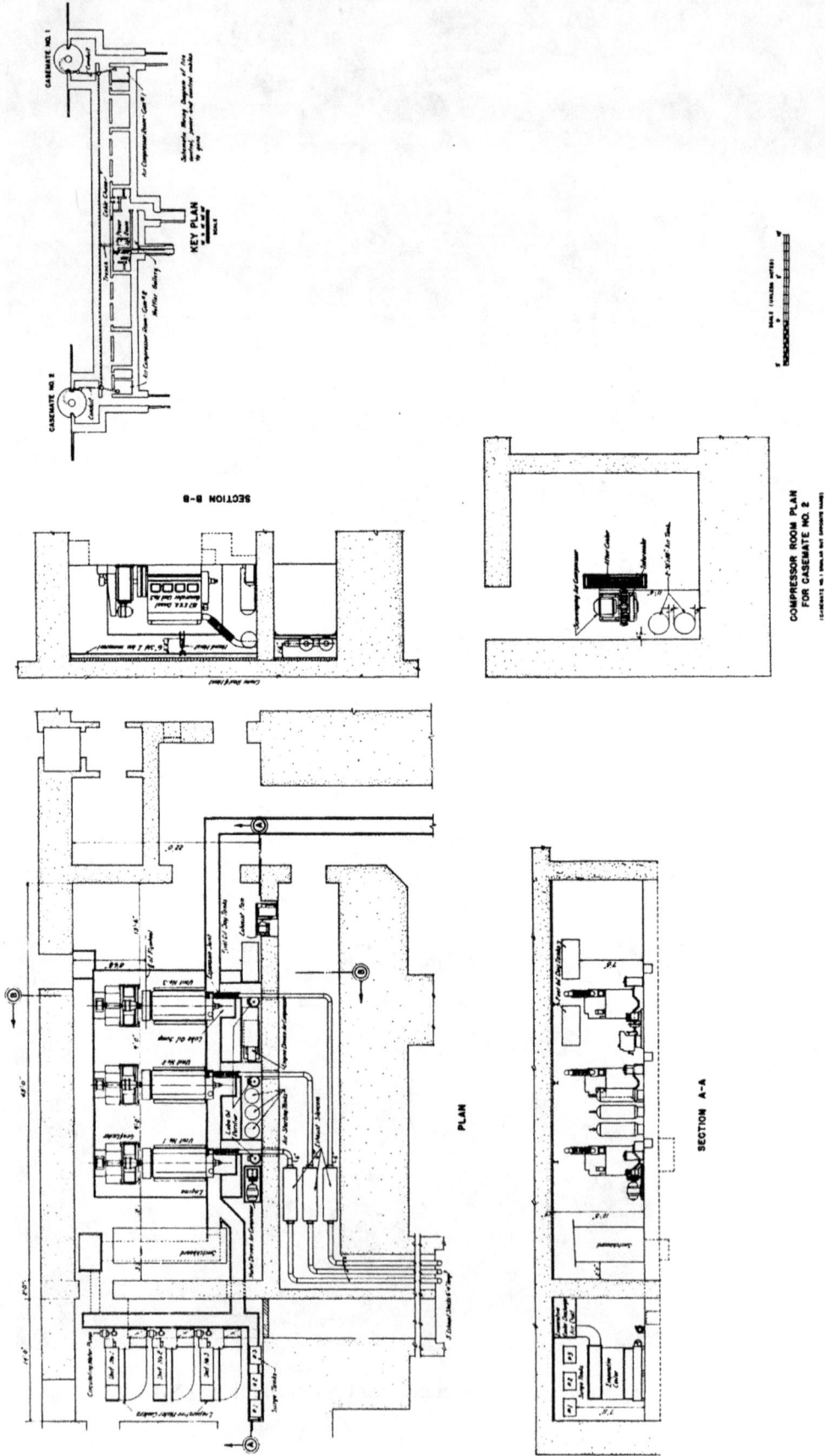

Equipment layout, M2 power plant for 12-inch gun battery.

Diesel electric plants for 6-, 12-, and 16-inch Gun Emplacements, Coast Artillery Training Bulletin, Vol. 4, No. 7, July 1945.

1. Control lever.
2. "Stop" push button.
3. Air cock.
4. Fuel oil filter.
5. Air valve.
6. Fuel oil plunger pump.
7. Lubricating oil pump.
8. Pressure adjustment screw.

FIGURE 8. Right side, Diesel engine for a 12-inch gun battery.

1. Air trap.
2. Water outlet pipe.
3. Safety alarm.
4. By-pass valve.
5. Lubricating oil cooler.
6. Water inlet pipe.

FIGURE 9. Left side, Diesel engine for a 12-inch gun battery.

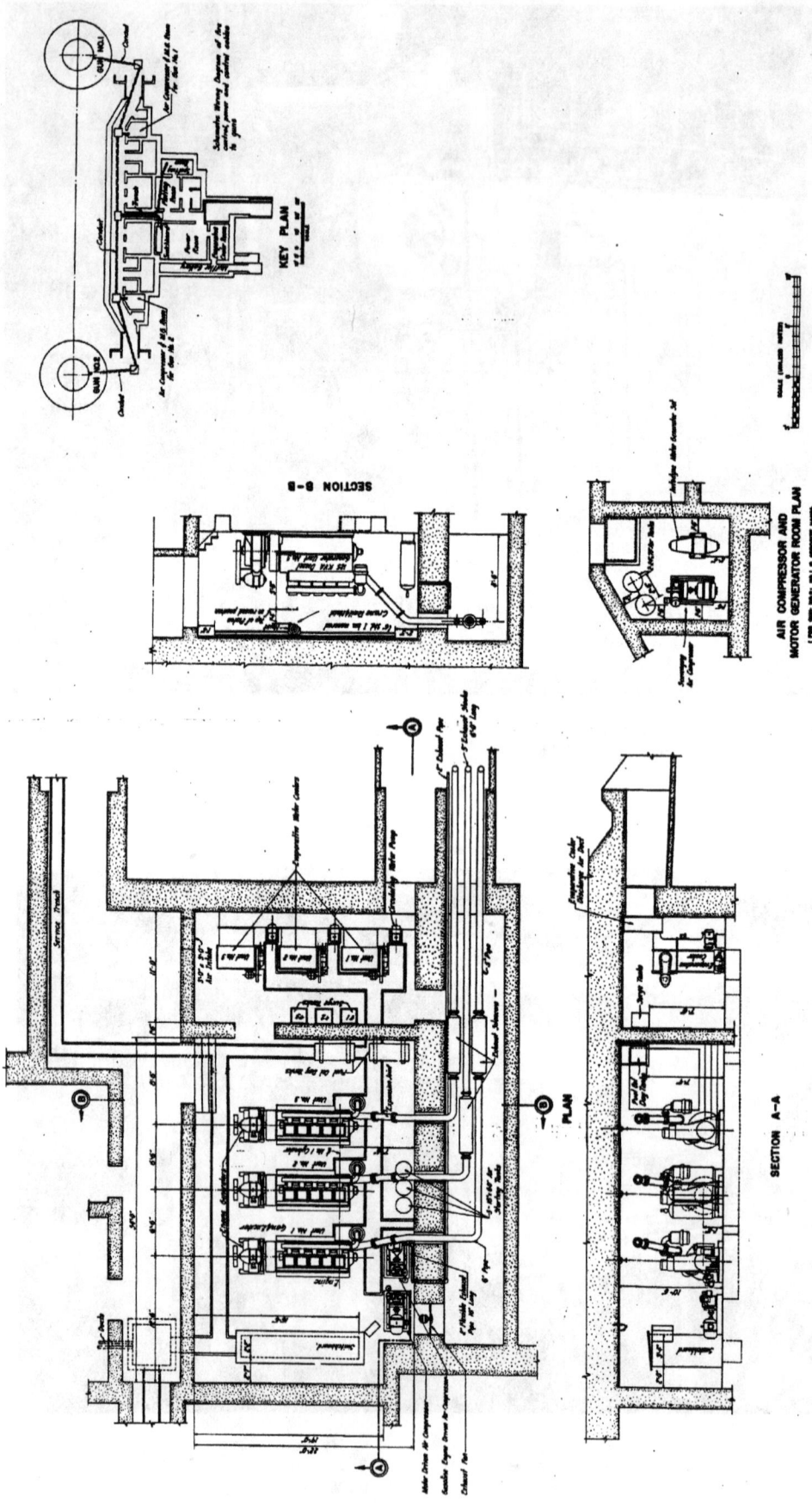

Equipment layout for standard 6-inch gun battery.

Diesel electric plants for 6-, 12-, and 16-inch Gun Emplacements, Coast Artillery Training Bulletin, Vol. 4, No. 7, July 1945.

1. Overspeed linkage.
2. Governor.
3. Air starting manifold.
4. Control wheel.

FIGURE 1. Right side, Diesel engine for a 6-inch gun battery.

1. Lubricating oil filter.
2. Lubricating oil cooler.
3. Inlet water manifold.
4. Exhaust manifold.
5. Air intake manifold.
6. Outlet water manifold.
7. Air cleaner.
8. Crankcase breather.

FIGURE 2. Left side, Diesel engine for a 6-inch gun battery.

Power plant for Battery 241, San Pedro, CA 2016 (Mark Berhow)

Power plant for Battery 241, San Pedro, CA 2016 (Mark Berhow)

Obviously, the frequency of such revisions had to be kept to a minimum to avoid confusion in field operations. The field offices were permitted to make minor changes or to meet local conditions with the understanding that none of these changes would interfere with machinery arrangements shown in approved type plans. When minor field office changes were approved, the Ordnance Department was furnished copies of the field drawings.

As a final note, since almost all batteries built before the 1930s utilized D.C. power, it was not possible to use transformers to change the generated voltage to meet specific requirements. As a result, emplacements commonly contained motor-generators, direct coupled sets with a D.C. motor driving a D.C. generator, which would generate power at the desired voltage. They were also common in switchboard rooms. In some instances, motor-generators were used to convert A.C. power to D.C., such as when the commercially available power was A.C. and the battery required D.C. Before the development of semiconductor rectifiers, this was a standard technique. In summary, emplacement power plants evolved from steam to kerosene to gasoline to diesel, although not always in a steady progression in individual emplacements.

Power plants constituted a major element in the coast defenses, and they are today one of the most common remains of our Coast Artillery history. Some sites, like Fort MacArthur, Fort Moultrie, Fort Stevens and Fort Columbia even offer historic power plants for inspection.

Sources

Hope, Offnere, "Power Installation and Operation for Coast Artillery Posts", *The Journal of the United States Artillery*, Vol. 36 No. 3, November-December 1911, pp. 231-241.

Miller, Lorimer D., "The 25 Kw Gasoline Driven Generating Set Used in the Coast Defenses of the United States", *The Journal of the United States Artillery*, Vol. 48 No. 1, July-August, 1917, pp. 54-79.

Smith, Bolling W. "Emplacement Powerplants," *Coast Defense Study Group Journal* Vol. 7, Issue 3, August 1993, pp. 45-50. (this article is a revised version of this original article).

U.S., Army, Corps of Engineers, *Engineer Mimeographs* No.13 (1896), No. 41 (1901) and No. 75 (1903), (Washington, D.C.).

U.S., Army, Corps of Engineers, *Instruction Book No. 1, Installation, Care, and Operation of 25 Kw. Gasoline-Electric Generating Sets, G.E. Co. Type GM-12*. (Washington, D.C.: GPO, 1916).

U.S., Army, Corps of Engineers, Reports of Completed Works, Miscellaneous, Various Dates.

U.S., Artillery School, *Notes on Oil Engines*, Artillery Notes No. 12, (Ft. Monroe, Va.: Artillery School Press, 1903).

U.S., Artillery School, *Notes on the Oil Engine, Generator and Storage Battery*, (Ft. Monroe, Va.: Artillery School Press, 1901).

U.S., National Coast Defense Board, "...Coast Defenses of the United States and the Insular Possessions" (Washington, D.C.: GPO, 1906) pp. 16, 35, 36.

U.S., War Department, *Handbook for the Use of Electricians in the Operation and Care of Electrical Machinery and Apparatus of the U.S. Seacoast Defenses*, (Washington, D.C.: GPO, 1902).

U.S., War Department, Diesel electric plants for 6-, 12-, and 16-inch gun emplacements, *Coast Artillery Trainin Bulletin*, Vol. 4, No. 7, July 1945.

Winslow, Eben Eveleth, *Notes on Seacoast Fortification Construction*, Engineer School-Occasional Paper No. 61. (Washington, D.C.: GPO 1920), pp. 269-302.

Coast Artillery Organization
A Brief Overview

Bolling W. Smith & William C. Gaines

Artillery Organization

The organization of American coast artillery is a complex story. Until World War I, coast artillery meant seacoast artillery; the World War brought additional functions, especially antiaircraft artillery.

Seacoast artillery was always caught in an organizational conundrum. The other combat arms served with the field army, organized in regiments of fixed strength. Seacoast artillery, however, primarily served at separate harbors, manning varying numbers of batteries with widely differing sizes and number of guns. No uniform system of manning would suit all the different harbors.

One obvious answer was to give the coast artillery a flexible, or task, organization, tailoring the different forces to the varying requirements. Different size units could man different batteries, and the number of batteries could vary for each harbor. However, in the event of war, if there was no serious threat against the coastline, the coast artillery had no intention of being left out of the struggle, or the laurels that would result from victory. To the extent that an organizational structure fit the requirements of harbor defense, it was ill suited to field service. In addition, a regimental structure fostered unit identity and esprit de corps. This dilemma had no clear resolution, and over the next half-century the War Department opted for different organizational solutions, resulting in frequent organizational changes.

In partial response, the coast artillery developed a twin-track chain of command. One track was administrative, responsible for housing, clothing, feeding, instructing, and disciplining personnel, as well as for keeping of the individual and unit records for which the army has always been known. The authority and responsibility of both the tactical and administrative chains of command were carefully defined in army regulations.

The tactical chain of command was concerned solely with the control of men and materiel in action. Sometimes the tactical grouping coincided with the administrative organization, and sometimes it did not. This conflict between roles and organizational systems haunted the coast artillery to the very end. This paper, however, shall be limited to the administrative organization.

The Old Regiments

With the first potential struggle between the new republic and a foreign power in the 1790s, formation of a seacoast artillery organization was authorized. This, termed the Corps of Artillerists and Engineers, was charged with constructing and manning the "First" System of fortifications to guard the new nation's harbors and seaports. By the beginning of the 19th century, this corps had expanded into two regiments of Artillerists and Engineers, and in 1808 a light artillery regiment was constituted in part to help man a new "Second System" of seacoast fortifications that largely replaced the First System. At the beginning of the second war with Great Britain in 1812, two additional regiments of artillery were formed. During the War of 1812, these were amalgamated into the Corps of Artillery.

In 1821, the army was reduced, and the Light Artillery, the Ordnance, and the Corps of Artillery were consolidated into four regiments of artillery, each having nine firing companies, lettered A through I. One company in each regiment was designated a light battery. Each company was commanded by a captain, while a tenth captain in each regiment performed ordnance duty. In 1832, when the Ordnance Corps was created, the ordnance captains were separated from the artillery and transferred to the new corps, but artillery lieutenants continued regimental ordnance work under four-year details until the Act of July 5, 1838, at the onset of the Second Seminole War, completed the severance of the Ordnance Department from the artillery. The light batteries in each regiment were not "horse artillery," whose men were all mounted. Not until 1836 was one battery in each regiment authorized mounts, making them horse artillery.

The Act of July 5, 1838, also authorized the addition of a tenth lettered company, K, to each regiment, and with the beginning of the War with Mexico, the Act of March 3, 1847, added Companies L and M to each of the four regiments. In both of these conflicts, especially the Second and Third Seminole Wars, the artillery companies functioned more as infantry, a role that would be repeated into the 20th century.(1)

Concurrent with the reorganization of the artillery in the early 1820s, a Third System of seacoast fortification was begun, resulting in the construction or rebuilding of more than three dozen fortifications to protect the nation's harbor and seaports, and the planning of several dozen more before the outbreak of Civil War in 1861. Before the Civil War, the four regiments were most often posted in the numerous seacoast fortifications and rotated on an irregular basis between the southern and northern coasts.

With the onset of the Civil War, the 5th Regiment of U.S. Artillery was constituted. During that war, the regimental structure was primarily administrative rather than tactical, as the artillery companies were chiefly employed individually in support of infantry and cavalry brigades. As part of the field armies, they were rarely formed in tactical organizations larger than a battalion and were generally single companies attached to the various infantry brigades and divisions of the army. Of the 60 companies of Regular Army artillery in the Civil War, only four remained in the seacoast defenses, the reminder serving with the armies in the field. Heavy artillery regiments of the state militias were used to man the seacoast fortifications, as well as those those surrounding the capital, until late in the war, when they were used as infantry.

Following the Civil War, many companies returned to their primary prewar role of manning the seacoast forts. The Act of July 28, 1866, designated the unmounted, or "foot" companies of the regiments as batteries. The mounted batteries were redesignated May 20, 1871, as "light batteries," and the foot companies were commonly termed "heavy" batteries.(2)

In the aftermath of the Civil War, the forts of the Third System were alarmingly vulnerable, and Congress had no interest in financing their replacement, even if there had been any consensus on what to replace them with, which there was not. The principal post-Civil War army functions, fighting Indians and occupying the conquered South, demanded little in the way of artillery. The five regiments of artillery each had 12 batteries, A to M, omitting J. Two of these 12 batteries were designated horse or light artillery and the remainder as heavy batteries. In general, "light" meant field artillery and "heavy" meant seacoast. However, the artillery materiel differed only in size, and techniques were sufficiently similar that heavy batteries could be used as light artillery when needed.

The "heavy batteries," however, had a dual role. In theory, they were trained as both seacoast artillery and as infantry, but several factors tended to emphasize their infantry role. The few serviceable seacoast guns actually mounted did not justify a large force, practice ammunition for the large guns was expensive in a time of austerity, and lastly, no foreign threat seemed sufficiently dangerous to tie down a large portion of the small peacetime army in harbor defenses at the expense of other, more pressing demands.

The individual batteries of the artillery regiments were scattered over many posts, essentially infantry, with rudimentary seacoast artillery training and a handful of light batteries mixed in. Complaints by officers that these troops were unready to actually serve the seacoast armament were increasingly aired in popular and professional publications. Zealous artillery officers saw remarkable improvements in the artillery being adopted in European armies, and pondered how to make the artillery ready for the time when Congress would loosen the purse strings and begin to equip our defenses with modern armaments. These officers insisted that the artillery should be artillery in more than name only, and they debated how best to organize the artillery to make it most effective.

Endicott Era

In the decades following the Civil War, as navies adopted steam, armor, and rifled guns, American seacoast defenses became so weak as to be an absolute embarrassment to the country. Finally, in 1885, Congress was moved to action, or at least to the discussion of action. Congress directed President Grover Cleveland to appoint a board of officers and civilians, chaired by Secretary of War William C. Endicott, to report what fortifications or other defenses were necessary to defend America's harbors.

Given the unsettled state of technology, the Endicott Board recommended guns on a variety of mountings. The recommendations were enormous and arguably unrealistic, covering 27 locations at a cost of over $126 million, including armament, floating batteries, submarine mines, and torpedo boats. Initial Congressional response was not enthusiastic, and appropriations only averaged about a half million dollars a year until 1896 - after all, the army was only then completing development of its first generation of breechloading weapons.

During the end of the 1880s, the artillery began to move slowly toward the emerging technology. As they agitated for increased spending on armament and fortifications, artillery officers attempted to remake the seacoast artillery into the technically advanced body of troops required to serve the new guns and their accessories. Civil War-era Rodman cannon, whether smoothbore or rifled, required little sophistication, and fire control changed little in half a century. However, as new, breechloading guns and mortars began to appear in the mid-1890s, it was obvious that their greatly increased range would require new methods and devices for fire control. In addition, forward-thinking officers could see the need for searchlights and electrical power, all requiring greater expertise of a more technical nature.

These questions, of course, were not to be settled by the artillery alone, but also by Congress and the executive branch, who were vigorously debating the role, form, and function of the entire military establishment.

Congress was not quick to reorganize the army, but funding for fortifications increased as the conflict with Spain broke out. In 1898, Congress increased the strength of the artillery by two regiments, the 6th and 7th, and authorized 14 batteries for each of the seven regiments.(3) Since there was little need for heavy artillery with the mobile army, the regular artillery largely manned siege and field artillery during the brief Spanish-American War. Following that war, heavy batteries were retained in the captured defenses of Cuba and Puerto Rico, while both heavy and light batteries served in the field against the Philippine Insurrectionists.

The military incompetence exposed by the Spanish-American War, combined with the panic that swept the seaboard, fueled demands for change. The start of the new century saw sweeping changes in the organization of the army, and one of the branches most affected was the artillery.

1901 Corps Organization

In 1901, Congress became convinced that the coast artillery should have a task organization, and so it abolished the artillery regiments and created an artillery corps. Reflecting the divergence in methods and equipment since the old muzzleloading days, the corps was composed of 30 companies of field artillery and 126 companies of coast artillery. Coast artillery companies were numbered 1 to 126 and field artillery batteries from 1 to 30, but they were united in one branch of service, under a single chief of artillery. Eighty-two existing heavy batteries were designated coast artillery companies, and 44 new companies were created by splitting existing ones, after which both units filled out their ranks with new recruits.

At the head of the Artillery Corps was a chief of artillery, with the rank of brigadier general. Although more of an inspectorate than a command position, the importance of this position can hardly be overstated. For the first time, there was a systematic attempt to regularize the equipment and training of the artillery. Further, seacoast defense planning, until now almost entirely the province of the Corps of Engineers, was now largely the responsibility of the artillery.

Coast Artillery Companies were units of men, and batteries were the fixed emplacements with their guns. Each company was authorized 109 enlisted men, including non-commissioned officers. Although the secretary of war was empowered to fix the strength of individual companies as needed, it does not appear he took advantage of this authority.(4)

Depending on its assignment, one company might man more than one battery. Companies were assigned to forts, which were administrative units commanded by the senior officer present.(5) The forts at each harbor were grouped into artillery districts (the occasional isolated small harbor was included in a larger artillery district), which were the primary tactical and administrative units, with the bulk of the staff and logistical services. The artillery district commander was analogous to a regimental commander in the infantry or cavalry; in turn, he was subordinate to the general who commanded all the troops in the territorial department.(6) The administrative relationship of these troop units was not markedly different from other army units, and was spelled out in detail in army regulations. Over the next half-century, the administrative organization was repeatedly changed, as the army struggled to balance the conflicting requirements and find the most efficient organization.

As the new batteries were created and manpower grew, the new officer positions were filled by West Point graduates, transfers from other branches, and appointments from civilian life. Officers who had served in the Volunteers during the recent war took advantage of the opportunity to join the Regular Army. Unlike the infantry and cavalry, the Artillery Corps continued to be all white, reflecting both its level of technical sophistication and the prevailing views of the capabilities of blacks.

1907 Organization, Refining the Corps Organization

Within five years, technological advances had become even more rapid, further separating field artillery from seacoast. No longer was an officer considered qualified to command both light and heavy artillery. The techniques were now too different, and specialization was required.

As a result, in 1907, Congress split field artillery and coast artillery into separate branches, creating a separate Coast Artillery Corps (CAC). Artillery officers had to opt for either coast or field. The responsibilities of the seacoast artillery had grown since 1901. America was beginning to construct harbor defenses to protect its newly acquired overseas possessions, and additional troops were required to man them. In addition, companies were assigned to plant and control the submarine mines. In an attempt to keep up with these expanded responsibilities, Congress authorized an increase in the Coast Artillery Corps to 170 companies, and the army again struggled to create and fill the new companies. In 1908, the chief of artillery became chief of coast artillery, severing his connection with the field artillery.

Administratively, the change made little difference except that officers now served solely in one branch or the other. All the coast artillery defending a given harbor remained in an artillery district, commanded by a field grade CAC officer, who assigned tactical command roles to his officers as additional duties.

These were in many ways the golden years of the coast artillery. Finally united in their own branch, they enjoyed a reputation as the most scientific of the combat arms, and recruitment was aided by the prospect of relatively settled garrison duty, safer and more comfortable than duty in the field. As the expenditures recommended by the Endicott and Taft Boards finally began to create modern defenses for the country with the technical accessories now necessary, and as barracks accommodations finally caught up with the expanding force, the principal problem the CAC faced was finding enough men to man all the guns now in place.

Although the corps did not muster its authorized strength, had it done so, it would still have been well short of enough to man the guns on a war footing. The chief of coast artillery calculated he needed three shifts, or reliefs, for every gun, with fewer reliefs for the searchlights. The men who planted the mines could man the rapid-fire guns defending their minefields, but the number of men needed to fully man the batteries was extraordinary. The chief of coast artillery proposed a method to deal with this shortage: the most sophisticated units, such as mine planters and searchlight crews, would have to be manned at full peacetime strength by Regular Army units, as would the territorial garrisons, who were too isolated to count on reinforcement in the event of war. In addition, the regular army would supply one relief for batteries defending home harbors. The remainder of the reliefs would be supplied by state troops.

Since raw troops could not be substituted for trained men in the event of sudden hostilities, the chief attempted to persuade the states to form and train coast artillery companies within their National Guard or militia. Some states, most notably New York, responded well, and the army put considerable effort into training these citizen soldiers. The results for many units were encouraging, but the strength of the coast artillery militia was never nearly enough to man the guns on a wartime footing. This problem was actually never solved, despite the continued efforts of successive chiefs of coast artillery.(7)

In 1913, the defenses of individual harbors were renamed coast defenses, as in "the Coast Defenses of San Francisco." Coast artillery districts were retained, but the term now had a different meaning. Three continental coast artillery districts were created within the geographic departments: The North Atlantic Coast Artillery District (Portland to the Delaware, but the Delaware was soon transferred to the South Atlantic District) and the South Atlantic Coast Artillery District (Baltimore to Galveston) in the Eastern Department, and the Pacific Coast Artillery District (San Diego to Puget Sound) in the Western Department. The Philippine and Hawaiian Departments and the Canal Zone included all troops within their limits, including the coast defenses.(8) This had the additional advantage of creating more positions for coast artillery colonels. Over the coming years, as the organization of the army changed, the number and composition of coast artillery districts was adjusted accordingly.

Repeated Renumbering 1916-17

On June 3, 1916, the years of persistence finally paid off. No doubt with a view to the World War then consuming Europe, Congress voted a large increase in the CAC. The strength of the corps was to be increased by about a third, to 29,469 enlisted men, over the next five years, in annual increments ending with that of July 1, 1920. The foreign defenses would get 6,352 men - 1,533 for Oahu, 2,470 for the Philippines, and 2,349 for Panama. The remaining 23,117 men would defend the continental United States. Based on the 1908 formula, the militia was to provide 17,329 men, but the 10,860 men actually in the militia coast artillery, while substantially stronger than the year before, was still considerably short of the number required.(9)

On July 24, 1916, coast defense commanders were authorized to subdivide their manpower into companies as they deemed fit, with the companies to be redesignated as companies in their fort, as "1st Company, Ft. Flagler." A year later, in furtherance of this end, the companies were again renumbered, this time within each coast defense, as "13th Company, Coast Defenses of Puget Sound."(10)

This introduced great confusion into the historical record. No longer were units created by the War Department; they were now created or abolished by coast defense commanders. To further the confusion, the new companies were not given additional designations in the CAC-wide serially numbered system. Their only designation was within their fort from mid-1916 until August 1917 and from August 1917 within their coast defense. Lastly, as companies were assigned to regiments being organized for service in France, their place was taken by new companies. Thus, if the 3rd Co., Long Island Sound, were assigned to a regiment being formed, the 18th Company, LIS, could be redesignated the 3rd Co., LIS (II). In some cases, three or four companies were created with the same designation, kept separate only by Roman numerals.

As a result, the exact disposition of the coast artillery companies between 1916 and 1922 cannot be fully delineated. We will outline the general changes, while Appendix I, The Coast Artillery in World War I, lists what has been uncovered about specific companies during this period.

During 1916, 24 new coast artillery companies were constituted and organized. In some cases, these companies were formed by disbanding Regular Army companies that had been in existence before the 1916 reorganization. (11)

World War I

Effective May 1, 1917, just after America entered the World War, the number of continental coast artillery districts was increased to five: the North Atlantic Coast Artillery District (Portland to Narragansett Bay), the Middle Atlantic Coast Artillery District (Long Island Sound to Chesapeake Bay), the South Atlantic Coast Artillery District (The Cape Fear to Galveston), the South Pacific Coast Artillery District (San Diego to San Francisco) and the North Pacific Coast Artillery District (The Columbia and Puget Sound). Overseas, the coast defenses of the Panama Canal were under the Panama Coast Artillery District, while the coast defenses of Hawaii and the Philippine Islands were under their respective departments.(12)

America's entry into World War I caused widespread changes throughout the entire army, but it primarily impacted the CAC in four ways. First, the timetable for the previously authorized increase in manpower was accelerated. Eight companies of the 1917 augmentation had been created when the United States entered the war on April 6, 1917, at which time the remaining 1917 augmentation was immediately ordered. Further augmentations were authorized later in 1917 and in 1918. After the declaration of war, another 71 companies were organized in April, May, June, and July. Only one company of Regular Army coast artillery was constituted between August and December 1917, in October. In December 1917, another 27 Regular Army companies were constituted and organized. Before the war ended, nearly 276 new companies were constituted and organized in the CAC, in addition to the National Guard companies called into federal service. (See Appendix I)

Secondly, National Guard coast artillery was called into federal service. Between July 1917 and February 1918, 171 state National Guard coast artillery companies were called into federal service. Legally, these units lost their National Guard identity when federalized, and as personnel were transferred into and out of the units, they began to lose their state identities.

Thirdly, the CAC largely devoted its efforts to training tens of thousands of men in heavy land artillery. When it entered the war, the United States had neither the men nor the materiel to provide the large force of heavy artillery necessary for warfare in France. Since the field artillery was hard pressed to provide the lighter, more mobile, artillery needed, the Coast Artillery Corps was the logical source of men to man the heavy artillery, especially since the dominance of the Royal Navy insured that the German navy would not seriously threaten the American seacoast. In addition, the only heavy artillery in this country was the fixed seacoast defenses. Spurred no doubt by fear of the consequences of sitting out the war at home, safe and uninvolved, the CAC largely devoted its efforts to training men in the new techniques of heavy land artillery, with such advice, manuals, and equipment as our more experienced allies could provide. Meanwhile, the army removed many guns from fixed fortifications, to be mounted on railway and wheeled carriages for use in France. However, most of these railway guns did not arrive in France before the Armistice, and in their absence, we utilized French and British heavy weapons. Had the war lasted a year or two longer, American-built heavy artillery might have made a significant impact.

To man these guns, a provisional brigade of three coast artillery regiments, each with 12 firing batteries, was formed. Initially numbered the 6th, 7th, and 8th Provisional Regiments, CAC, they were renumbered in February 1918 as the 51st, 52nd, and 53rd Artillery Regiments, CAC. Beginning in December 1917 and continuing into 1918, Regular Army and National Guard companies were combined, creating the 54th through the 71st Artillery Regiments, CAC. Selected batteries from the 51st, 52nd, and 53rd Artillery Regiments were used to organized the Provisional Howitzer Regiment, which was later designated the 44th Regiment. In a July 1918 reorganization, the 51st, 52nd, and 53rd Regiments were reduced in size to six firing batteries; the excess batteries were used to organize the 42nd and 43rd Regiments. To meet the need for higher units to command and control theses regiments, additional brigades were created.

A total of 57 regiments were constituted in the CAC at coast artillery forts. The first regiments were created by assigning entire companies, either National Guard or Regular Army. Later regiments were formed by assignment of individuals. Of these regiments, 34 arrived overseas, where they were grouped into 11 brigades; the rest were either organizing or awaiting transportation when the war ended. (See Appendix II, Coast Artillery Regiments – WWI.)(13)

Lastly, the evolving nature of warfare created a new weapon, the airplane, and therefore the requirement for weapons to combat it. Again, the job went to the Coast Artillery Corps, which at least had experience in firing at moving targets. Before the war ended, ten gun and six machine-gun battalions were organized, of which seven gun and five machine-gun battalions made it to France. The battalions, assigned to army corps, consisted of a headquarters battery and four lettered batteries. During the last month of the war, the American antiaircraft forces were reorganized to better correspond to the French organization. The gun battalions were abolished and numbered sectors were formed, with a headquarters and supply battery and numbered firing batteries. These sectors were assigned as field army troops. This organization was short lived, as the AA units soon began to return to the United States.(14)

At the end of WWI, the CAC, with an enlisted strength of 147,000, was much more varied than it had been two years before, with heavy artillery batteries, regiments, and brigades with or destined for the field armies; and antiaircraft and trench mortar units for specialized roles. In the U.S. and its possessions, gun, mortar, mine, and searchlight companies remained organized into coast defenses, for harbor defense.

1919-1921

Many of the companies created during WWI, whether for regiments destined for overseas service or for the coast defenses, were demobilized during the six weeks that followed the signing of the armistice. This demobilization was so precipitous, especially in the coast defenses, that in many cases new companies had to be reconstituted in 1919 and 1920 to provide even a semblance of garrisons for the coast defenses.(15)

Most of the stateside regiments intended for France were discharged in December 1918, the remainder in January 1919. The regiments in France were largely demobilized as soon as possible after return to the states, typically by March 1919. However, 12 regiments that had served in France were retained in active service at the end of the war for training purposes. This number was shortly reduced to 10: 42nd, 43rd, 52nd, and 53rd Artillery, CAC (Ry) formed the 30th Brigade, posted at Camp Eustis, VA. The 31st Brigade, consisting of the 55th, 57th, and 59th Artillery, CAC (TD), was initially posted at Fort Winfield Scott, CA, before moving to Camp Lewis, WA, in the later months of 1919. The 39th Brigade, at Camp Jackson, SC, was composed of three more tractor-drawn regiments: the 44th, 51st, and 56th Artillery. CAC (TD).(16)

On October 5, 1920, the five CA districts were numbered (1st, 2nd, 3rd, 4th, and 9th) to correspond to the numbered corps areas, the coast artillery districts coinciding with the corps area boundaries. Of the nine corps areas in the continental United States, five, the I, II, III, IV, and IX Corps Areas, had coastal boundaries, while the other four, V, VI, VII, and VIII Corps Areas, were in the Zone of the Interior. The I Corps Area and the 1st Coast Artillery District encompassed New England. The II Corps Area and 2nd Coast Artillery District comprised the states of New York, New Jersey, Pennsylvania, and Delaware. The III Corps Area and 3rd Coast Artillery District was made up of the states of Maryland, Virginia, and about half of North Carolina. The IV Corps Area and the 4th Coast Artillery District encompassed part of North Carolina, as well as the states of South Carolina, Georgia, Florida, Alabama, Mississippi, Louisiana, and Texas. The IX Corps Area and the 9th Coast Artillery District covered the states of California, Oregon, Washington, and the Territory of Alaska. In Hawaii, the Panama Canal Zone, and the Philippines, the harbor defenses were designated administratively as separate Coast Artillery districts within the Hawaiian, Panama, and Philippine Departments.(17)

Continental coast artillery regiments were all assigned to corps areas. The coast artillery district commander was responsible to the corps commanders for the "instruction, training, and tactical employment of all coast artillery troops within his district." On the other hand, the district commanders were not charged with any duties pertaining to administration and supply.(18)

The Coast Artillery Corps had seized the opportunity presented by the shortage of heavy field artillery to secure an important role in the World War. In the future, however, the Field Artillery could not be expected to view this with favor, and admittedly, it blurred the distinction between field and coast artillery. On August 1, 1921, the War Department ruled that the CAC would furnish all artillery necessary for land and coast fortifications, and all AA, railway, and trench mortar artillery for use either with fortifications or with the armies in the field. This was not, however, the final word on the subject. The general order went on to say that nothing would prohibit "the organization within the coast artillery of such mobile units as may be needed in land or coast fortifications or the employment of such units with field armies whenever or wherever conditions of combat indicate the desirability of such employment." Thus, the door was clearly left open for the CAC to again supply heavy field artillery for use in the field.(19)

Reductions in the size of the army ordered in 1920 and 1921 took a major toll on the units that had been retained for training at the end of the World War. In early summer 1921, the 31st Brigade (TD) and the 57th Artillery were inactivated and their personnel transferred to the 55th and 59th Artillery Regiments, which were transferred later that summer to Hawaii and the Philippine Islands, respectively. By August 1921, the 39th Brigade (TD) and the 44th and 56th Artillery Regiments were also inactivated and the 51st Artillery was transferred to Camp Eustis, VA, where what remained of the mobile units retained in the continental United States were posted. The 42nd, 43rd, and 53rd Artillery, CAC (Ry), Regiments and the 1st Bn, 52nd Artillery, CAC (Ry), were also inactivated in August. The 2nd and 3rd Bns, 52nd Artillery (Ry), were retained at Camp Eustis in active status, although at reduced strength levels.

Although the size of the CAC had been reduced substantially in the post-war economy, efforts were made to continue training for the missions acquired during the World War. The overseas and insular possessions required garrisons strong enough to withstand attack until relief could arrive. Consequently Hawaii, the Philippine islands, and the Panama Canal Zone, having the largest coast artillery garrisons, became the principle training locations for all of the CAC's assigned missions: seacoast artillery, antiaircraft artillery, and mobile railway and tractor-drawn artillery. For that reason, the 60th Artillery Bn (AA) was shipped to the Philippines in 1921 and the Hawaiian Railway Bn was constituted and organized in Hawaii in 1922. As emphasis on antiaircraft training continued to increase, in 1921 the War Department authorized three AA battalions for service in the United States: 1st, 2nd, and 3rd Artillery Bns, CAC (AA), each consisting of a headquarters detachment, a searchlight battery, a gun battery, and a machine gun battery. In Hawaii, an antiaircraft regiment, the 64th Artillery (AA), with eight firing batteries in addition to the usual headquarters, service, and battalion headquarters batteries, was also constituted and organized in 1921. These AA units were created by inactivating existing seacoast artillery organizations and transferring their personnel to the new units.

1922-24 Renumbering Again

In May 1922, the Hawaiian Railway Bn and the Hawaiian AA Regiment were redesignated the 41st Artillery Bn and the 64th Artillery (AA). At the same time, the 1st through the 3rd AA Bns were renumbered the 61st through the 63rd Artillery Bns (AA).(20)

At the end of 1921, there were some 274 CAC companies. Of these however, 188 were active and 86 were inactive. In an attempt to restore the esprit de corps and historic lineage of the companies that in some cases traced back over one hundred years, in 1922 the 170 companies organized prior to the 1916 reorganization had their pre-1916 serially numbered designations restored, and those active companies organized following the 1916 reorganization were designated the 171st – 198th Companies, CAC. Twenty-five pre-1916 companies had been assigned as batteries of regiments formed during the World War and demobilized at the end of the war. These units were reconstituted and consolidated with active companies then serving in the coast defenses and the consolidated company assigned the pre-1916 designation held by the reconstituted company. In addition, the regimental headquarters and firing batteries retained as active or inactive units following the war were also designated as companies of the CAC, typically between 199 and 274. (See Appendix III, 1922 Station List.)(21)

Under War Department orders issued in 1922, Regular Army CAC companies were numbered between 1 and 300; National Guard companies were numbered between 301 and 500, and Organized Reserve companies between 501 and 950.

Coast defense commanders were instructed to divide the men assigned to their command into the number of companies that corresponded to the number of first sergeants allotted to the command, with one headquarters company per coast defense. Where coast defense commands contained only caretaking units, those men would be assigned to the headquarters battery, which would be kept active, although at reduced strength.(22)

In 1922, another 45 companies were inactivated, but later in the year, 14 companies of Philippine Scout Coast Artillery, numbered 275 through 288, were organized in the Harbor Defenses of Manila and Subic Bays, followed by the 289th Co. in January 1923.

1924, Return of the Regiments

Meanwhile, the War Department and the Office of the Chief of Coast Artillery continued to reevaluate the organization of the seacoast artillery. While the current organizational system was well suited to the demands of harbor defense, the World War had shown that regimental organization was necessary for units serving with the mobile army. In addition, a regimental organization was expected to improve morale by enhancing unit identification. After much thought and discussion, the decision was made to reorganize the Coast Artillery Corps, once again. The antiaircraft, railway, and tractor-drawn units had always been organized into regiments and battalions. Now, the seacoast artillery would also be organized into regiments.

Effective July 1, 1924, the serially numbered company designations were abolished and the sole designations held after July 1, 1924, by the CAC were as batteries of the various regiments and battalions in existence from the World War and immediate postwar period, and those regiments constituted in 1924. These regiments would be of four different types: mobile tractor-drawn 155 mm gun (three regiments), mobile railway artillery (two regiments), mobile antiaircraft artillery (five regiments), and harbor defense (sixteen regiments). The 16 newly constituted harbor defense regiments (1st-16th) were assigned to the domestic and territorial coast defenses. In addition to the first 16 harbor defense regiments, the companies of Philippine Scouts were organized into the 91st and 92nd CA Regiments, Philippine Scouts.

The initial assignments for the 16 harbor defense regiments were:

1st CA:	Panama Canal Department
2nd CA:	Panama Canal Department
3rd CA:	CD of Los Angeles, San Diego, and the Columbia
4th CA:	Panama Canal Department
5th CA:	CD of Southern New York
6th CA:	CD of San Francisco
7th CA:	CD of Sandy Hook and the Delaware
8th CA:	CD of Portland and Portsmouth
9th CA:	CD of Boston
10th CA:	CD of Narragansett Bay and New Bedford
11th CA:	CD of Long Island Sound
12th CA:	CD of Chesapeake Bay
13th CA:	CD of Pensacola, Charleston, Key West, and Galveston
14th CA:	CD of Puget Sound
15th CA:	Hawaiian Department
16th CA:	Hawaiian Department

Implementation of the new order resulted in the 1st-7th harbor defense regiments adopting and continuing the historic lineage of the seven artillery regiments that had been abolished in the 1901 reorganization. As most batteries of the old regiments had become companies of coast artillery in the 1901 reorganization, the War Department directed that those companies formed from batteries of the old regiments be assigned to the newly constituted coast artillery regiments as far as was practicable. In reality, the batteries were normally transferred "less personnel and equipment." This generally meant that the unit and its equipment did not move, but were redesignated. It was said that the only thing that actually moved was the flag, although unit records were also transferred.

However, by June 1924, of the 289 coast artillery companies, only 144 were active. The new regiments were made up of units, now termed batteries instead of companies, assigned to numbered battalions in each regiment. The number of batteries per regiment varied from seven to eleven; one was the headquarters battery and the rest were lettered batteries. Again, the number of active batteries equaled the number of first sergeants allotted, plus one reduced-strength caretaking battery. One or more batteries in each coast defense command would be a general utility battery; the remainder of the batteries would man gun, mine, or searchlight batteries.

The War Department had previously organized the National Guard coast artillery into regiments, and this was extended to the Organized Reserve. The number of assigned batteries and battalions in the National Guard and Organized Reserve varied considerably due to their allocation by state. The Organized Reserve regiments were organized into brigades to facilitate mobilization. Regimental numbers between 1 and 100 were reserved for Regular Army regiments; between 195 and 299 were reserved for National Guard regiments; and Organized Reserve units were numbered between 500 and 699. When additional Organized Reserve regiments were created in 1930, they were numbered between 900 and 999.(23)

1925-1941

At the start of 1925, the 18 harbor defense regiments contained 18 HQ batteries and 60 active lettered batteries, an average of a little more than three active firing batteries per regiment. The two railway regiments were in much the same condition, while the six AA regiments did a little better. The three tractor-drawn regiments were the strongest, averaging six lettered batteries per regiment.(24)

On June 9, 1925, coast defense commands were renamed harbor defense commands, to describe their true role more accurately, and to emphasize that the coast artillery was to defend key locations, rather than the entire coastline.(25)

The 1920s and 1930s were a period of continual retrenchment; many units were deactivated in whole or in part, and in army terminology, demobilized, inactivated, reorganized, redesignated, or reassigned. These terms, which had precise meanings within the War Department, often meant that a unit existed for years in name only, with no actual personnel assigned. As an example, a regiment might be listed on active status, but only two or three of its batteries would actually be active. The rest would be inactive, and the regimental manpower would be no greater than a weak battalion.

In addition, units might be reassigned less personnel and equipment, units were transferred between the Regular Army, Organized Reserves, and National Guard, and inactive units were moved between organizations. It is not easy to keep track of which units existed in reality at any given time. This reached a crescendo in World War II, when units appeared and disappeared like props in a magician's show.

The turf battle between the CAC and the field artillery continued, and at the close of 1927, all divisional, corps, and general headquarters artillery (except for railway and antiaircraft) was assigned to the field artillery. The 155 mm tractor-drawn coast artillery units were now dedicated to seacoast defense, no longer charged with supporting the field armies as they had in WWI. Additionally, the CAC was relieved of the need to train on and maintain the larger howitzers of the siege artillery. While this threatened the CAC role in any future war, it did simplify training, allowing the CAC to focus its meager resources on seacoast and antiaircraft defense. Shortly thereafter, all trench mortar units were transferred to the field artillery.(26)

The headquarters batteries for regular army HD regiments were assigned to the headquarters post of a harbor defense. Some of the smaller harbor defenses received a detachment from a regular regiment assigned to one of the larger harbor defenses. For example, batteries of the 3rd Coast Artillery, headquartered at the Harbor Defenses of Los Angeles, were assigned to the Harbor Defenses of the Columbia River and the Harbor Defenses of San Diego, neither of which received its own regiment until 1940. The regular army AA, TD, and Ry units were headquartered either at a large coast artillery post or at some other army location that had space for storing and firing armament, such as Fort Eustis, VA, and Fort Sheridan, IL.

National Guard HD regiments were headquartered at National Guard armories in their representative states. In the case of large cities, like New York, the entire regiment might be based at one armory. In other cases, the different batteries would be assigned to armories at towns and cities around the state, usually near the coast. The batteries held regular training programs during the year at their armories and the regiment as a whole usually conducted a 2-week summer camp at a coast artillery post. The same was true for the National Guard TD regiments. National Guard AA regiments were not necessarily based in states anywhere near the coast and they could hold their summer camps wherever appropriate training facilities could be found.

The Organized Reserve, consisting largely of cadres of volunteer officers and senior NCOs, were often headquartered at a federal facility, such as a post office, in cities around the nation. These units varied greatly in manpower and training activities. Some OR regiments did not have enough personnel to conduct regular training, but other OR regiments built up enough recruits to hold summer camp training exercises and win performance awards.

At this time, regiments in the army were developing distinctive regimental crests and insignia. In 1919, unique unit coats of arms were added to the regimental colors (flags) in place of the arms of the United States, thus making the colors distinctively individual, while the retention of the eagle showed the Federal nature of the organization. Regimental insignia, based on the regimental crest, were designed, approved, and worn on the uniform, in addition to many unofficial uses, such as on stationery, pictures, etc. As the coast artillery in 1920 had few regiments, coats of arms had been designed for a number of the coast defenses. After 1924, as new regular regiments were formed, their regimental designs were approved by the War Department. When all harbor defense troops were assigned to regiments, the regimental insignia superseded the coast defense insignia. Many Organized Reserve regiments also received approved arms and insignia.(27)

The overall picture of the organization of the Coast Artillery Corps remained more or less stable for a dozen years after 1925. By 1930, most regular harbor defense units in the continental United States were reduced to mere skeletons and nearly all of the continental harbor defenses were in caretaker status. The territorial defenses were maintained near full peacetime strength, and most of the training was concentrated there. A few new units were added and some units transferred or inactivated. Other units swapped armament, were reassigned or inactivated during the late 1920s and 1930s. For example: The 17th Coast Artillery regiment was designated for the Hawaiian Department in 1926, but was never activated. As time went by, the antiaircraft role grew, and four more antiaircraft units were added (66th-69th).

Five additional National Guard units were added in the late 1920s. The 1st and 4th Coast Artillery Regiments were converted to mixed harbor defense and antiaircraft in 1932, while the 2nd Coast Artillery Regiment was sent to Fort Monroe and the 12th Coast Artillery Regiment was inactivated. The 59th and the 92nd Coast Artillery Regiments switched armament, and several National Guard units were moved to different states or to non-coast artillery assignments. Some Organized Reserve units were transferred to the regular army for other uses, complicating the history of these units.(28)

Eventually, the harbor defense units were standardized as type-A, B, and C regiments; type-D regiments were in reality only separate battalions. The units' organization was governed by the tactical requirements of the harbor defenses. By the late 1930s, a number of National Guard cavalry organizations were reclassified as coast artillery, the best known being the 200th CA, which served on Bataan. In 1940, as new Regular

Army coast artillery regiments were activated and National Guard regiments were called into federal service, the initial type assigned to some regiments was altered. In most cases, the original 16 harbor defense regiments were reduced in size. When war came, the various elements of the regiments were increased in size and inactive units were activated, but the number of elements remained constant.

All harbor defense regiments were made up of a regimental HQ and HQ Battery, a searchlight battery, band, and a varying number of battalions, each of which contained an HQ and HQ Battery or Detachment and three firing batteries.(29)

> *Type-A regiments* contained three battalions, with a total authorized strength in 1940 of 1911 officers and men.
> *Type-B regiments* contained two battalions; the regimental HQ and HQ Battery and medical detachment were reduced in comparison to type-A regiments. The total authorized strength in 1940 was 1370 officers and men.
> *Type-C regiments* contained four battalions; the regimental HQ and HQ Battery and the medical detachment were increased in comparison to type-A regiments. The total authorized strength in 1940 was 2320 officers and men.
> *Type-D regiments* consisted of a single battalion with a small regimental HQ and HQ Battery, later designated as a HQ Detachment. The authorized strength of a type-D regiment was 802 officers and men.

Finally, on July 1, 1935, Congress authorized an additional 5,918 men for the CAC, an increase of almost 50 percent. Since these increases were all in the grade of private, with no increases in officers, no new units were formed, but existing units were significantly augmented. In addition, the National Guard CAC was increased to 12,960.(30)

In 1940, the Selective Service Act resulted in the first peacetime draft, and a rapid build-up of all Regular Army coast artillery regiments. Existing National Guard coast artillery regiments were called into federal service, originally for a year's training.(31)

The harbor defense assignments were readjusted and six additional Regular Army harbor defense regiments were established (18th-23rd). New antiaircraft regiments were authorized (70th-78th) and a few other mobile regiments, such as the 56th.(32)

In 1941, the National Guard regiments federal service was extended indefinitely. During the same time, regiments assigned to the Organized Reserve were activated. Using their original officers and senior NCOs as cadre, the regiments were built up using volunteers and draftees. These Organized Reserve regiments, however, were redesignated and assigned new regimental numbers.

By 1940, antiaircraft regiments were either mobile or semi-mobile, the primary difference being the number of motor vehicles assigned to the regiment. AA regiments had one gun battalion, which included a searchlight battery, and one machine-gun battalion. During World War II, when first 37 mm and then 40 mm automatic weapons became available, the MG battalion became an automatic weapons battalion.(33) In the spring of 1942 most AA regiments received a third battalion that became the regimental automatic weapons battalion and the former MG battalion was reequipped with 90 mm guns.

World War II

At the beginning of the war, the coast artillery, seacoast and AA, was largely grouped in regiments, with a few separate battalions and batteries for special assignments. The seacoast regiments were assigned to harbor defenses and to temporary harbor defenses, including smaller ports that had not previously been defended and some whose defenses had been abolished between the world wars. Many of the AA regiments were grouped into brigades.

In 1941, the army was reorganized into defense commands—territorial agencies within the continental limits of the United States that coordinated, prepared, and initiated the employment of army forces and installations against enemy action within their boundaries. Four continental defense commands were established on March 17, 1941: The Northeastern, Southern, Central, and Western Defense Commands. The Northeastern Defense Command was designated the Eastern Theater of Operations on December 20, 1941, and renamed the Eastern Defense Command on March 20, 1942. The Western Defense Command was additionally designated the Western Theater of Operations on December 11, 1941. The Eastern Defense Command absorbed the Central Defense Command on January 15, 1944, and the Southern Defense Command on March 1, 1945. On October 23, 1943, the Western Defense Command's status as a theater of operations was terminated and on November 1, 1943, Alaska was separated from the Western Defense Command, becoming a separate theater of operations.

Each defense command was divided into sectors, which were in turn divided into sub-sectors and/or local sectors. Sectors, usually one or more harbor defenses, were the wartime equivalent of coast artillery districts. Sub-sectors were generally the wartime equivalent of harbor defenses. Some sub-sectors could contain more than one harbor defense. The Florida sub-sector of the Southern Sector contained the Harbor Defenses of Key West as well as the Temporary Harbor Defenses of Miami Beach, Fort Lauderdale, Jacksonville, and Tampa. A local sector was a subdivision of a sub-sector, and might or might not have harbor defense units, but might have bases for the support of the harbor defenses.(34)

In a major reorganization of the army, on March 9, 1942, the harbor defenses were placed under Army Ground Forces and the Antiaircraft Command was established. Each coastal defense command had an antiaircraft command (the Eastern, Southern, and Western AA Commands).(35)

On December 24, 1942, a major reorganization of the army's AA units was initiated. The AA regimental organization had proven insufficiently flexible, and in response, the War Department split the regiments into separate battalions. Where it was desirable to combine battalions, primarily AA, they were assigned to groups, which were flexible headquarters that could control a variable number of battalions.

By early 1944, a similar reorganization was undertaken with regard to HD and TD regiments. Harbor defense battalions were generally stationed in fixed continental defenses, Panama, Alaska, and Hawaii. Tractor-drawn units reinforced the harbor defenses, established and maintained temporary harbor defenses, and defended overseas bases, largely in the Pacific and Caribbean. Antiaircraft groups and separate battalions were initially assigned to the continental U.S. and to Panama and Hawaii, but soon AA units were widely assigned to mobile armies in theaters around the globe.

While not all units went through the same transitions at the same time, there are general patterns. At the start of the war, coast artillery units and posts swelled with an influx of new men and units. As the war moved away from American shores, the number of seacoast artillery units and their strength declined sharply. By early 1943, general service personnel were being replaced by limited service troops, releasing the general service personnel for overseas service. A number of units were disbanded, their personnel augmenting other harbor defense units. The remaining harbor defense regiments were broken up into separate HD battalions in October 1944. Excess personnel were often organized into field artillery battalions. In 1945, the remaining coast artillery units were assigned as batteries of newly created harbor defenses.

As new fixed-gun batteries were completed at harbor defenses, and the threat to the temporary harbor defenses in the continental U.S. waned, the need for tractor-drawn units at home decreased. On the other hand, the need for mobile coast defense units to defend newly taken islands in the Pacific was growing. The tractor-drawn regiments, like the harbor defense regiments, were broken up into separate battalions in 1944. Some of these battalions were formed into groups, primarily for overseas service, while others remained as separate battalions.

Beginning in 1943, coast artillery (AA) battalions, groups, and brigades were redesignated "Antiaircraft Artillery" (AAA) battalions, groups, or brigades. Headquarters batteries of the AA regiments became headquarters batteries of AAA groups, and subsequently, additional groups were created. The former AA regiments were broken up into separate AAA gun, automatic weapon, and searchlight battalions. These battalions were renumbered and assigned to groups, which were often assigned to brigades.(36)

The entire picture is much more complicated than can be fully given in a brief account. Early in the war, some battalions from coast artillery regiments were assigned to separate missions, both in and out of the continental United States. These battalions were often redesignated, while the parent regiment recreated new battalions out of recruits, bearing the old designation.

The situation in Hawaii was particularly complex. Separate HD and AA batteries were created with troops hastily forwarded to the island early in 1942, and all types of units were formed, disbanded, redesignated, and transferred to the war zone with dizzying rapidity.(37)

A number of tractor-drawn battalions saw duty, guarding American-held islands in the Pacific. These units seldom saw action as seacoast artillery, but some were used against land targets as heavy field artillery, pre-war policy notwithstanding. Similarly, AA battalions were deployed in considerable numbers in both the European and Pacific Theaters. The general absence of an aerial threat from the Luftwaffe resulted in many of the AA units, primarily automatic-weapon battalions, in the ETO being employed in a ground support role. In the Pacific, there was a greater threat of Japanese attack, and consequently, less diversion of the AA units to other roles.

As the odds of even isolated naval raids on the U.S. coast diminished, the manpower of the harbor defenses was further reduced, and units were disbanded or inactivated, their personnel converted to other roles. By the end of the war, only a few coast artillery batteries were left to watch over the fixed defenses. Ironically, the remaining batteries were grouped into harbor defenses, continuing the cyclic nature of coast artillery organization.

Postwar 1946-1972

In 1946, the demobilization of the army was well underway, and the skeleton seacoast artillery units were again relegated to caretaking status. In 1950, army artillery was once again combined into a single branch. One aspect of this reorganization was the establishment of the Army Antiaircraft Artillery Command (ARAACOM) for continental defense, followed by the re-activation (and in some cases re-designation) of the army antiaircraft artillery battalions. Sixty-six Regular Army antiaircraft battalions, World War II-era coast/antiaircraft artillery battalions de-activated at the end of the war, were reactivated by ARAACOM during 1950-1953, followed by National Guard AAA battalions. The Regular Army and the National Guard contained automatic weapon, gun, and later missile battalions. Antiaircraft battalions were assigned to combat divisions, as well as to continental defense.

The regimental artillery organization reappeared in 1957 when the combat arms regimental system (CARS) was adopted to provide a regimental structure that would perpetuate unit history and tradition, while allowing the flexibility of a battalion organization. Battalions were listed as elements of historic regiments, but the regiments exercised no command or supply functions, and the battalions were given assignments without regard to the other battalions in the regiment. With the creation of the regiments, the identity of the separate AA battalions disappeared.

On June 20, 1968, air defense artillery became a branch of the army, separate from the field artillery. Over the next few years, 24 air defense artillery regiments were formed, a mixture of old coast artillery regiments and new regiments (or battalions) formed during 1942-1945. These ADA regiments were numbered in parallel to regiments in the field artillery, field artillery and ADA regiments having the same numbers. (38)

In summary, the initial Corps of Artillerists and Engineers became a regimental structure that again gave way to a corps organization and in 1821 again became a regimental structure. This was converted to a corps of serially numbered companies in 1901; a regimental structure was subsequently reactivated in 1924, followed by the conversion to a group/battalion organization during World War II, ending with the redesignation of the remaining active harbor defense batteries as elements of the harbor defense command, e.g. Battery A, HD of New York. The AAA battalions were reactivated in the 1950s; a regimental air defense artillery echelon was reestablished in 1957, ending with the creation of a separate air defense artillery branch. This progression of changes, along with the creation, renumbering, consolidation, and transfer, with and without personnel and equipment, of so many units, makes it very difficult to trace the heritage of the coast artillery units.

Footnotes

1. Francis B. Heitman, *Historical Register and Dictionary of the United States Army*, Vol. I, GPO, 1903, pp. 50-64. S.C. Vestal, "Field Service of the Coast Artillery in War," *Coast Artillery Journal (CAJ)*, Vol. 56, No. 3 (March 1922), pp. 199-220.

2. Vestal, "Field Service of the Coast Artillery in War."

3. War Department General Order (WDGO) 13, Feb. 13, 1901.

4. Army Reorganization Act (31 Stat. 748), February 2, 1901. WDGO 9, Feb. 6, 1901. WDGO 48, May 31, 1902.

5. War Department, *Drill Regulations for Coast Artillery (Provisional) 1906*, GPO, 1906, p. 13.

6. WDGO 81, June 13, 1901. WDGO 80, June 4, 1903.

7. Bolling W. Smith, "In Time of Peace, Prepare for War: The Peacetime Strength and Distribution of the U.S. Coast Artillery, 1901-1937," *CD Journal (CDJ)*, Vol. 17, No. 1 (Feb. 2003), pp. 15-25.

8. WDGO 9, Feb. 6, 1913. WDGO 14, Feb. 19, 1913.

9. *Annual Report of the Secretary of War*, Annual Report of the Chief of Coast Artillery (ARCAC), 1916, pp. 1163-65, 1174.

10. WDGO 31, July 24, 1916. WDGO 28, July 26, 1917.

11. Historical Section, Army War College, *Order of Battle of the United States Land Forces in the World War, American Expeditionary Forces (Order of Battle, AEF)*, 3 Vols., GPO, 1937-1949, pp. 1201, 1209, 1211-12, 1221-22, 1227.

12. WDGO 46, April 24, 1917. "Army List and Directory," May 20, 1917, GPO, 1917, pp. 34-36.

13. Historical Section, Army War College, *Order of Battle of the United States Land Forces in the World War: Zone of the Interior, Directory of Troops*, Vol. 3, Pt. 3, GPO 1988, pp. 1136-42.

14. James A. Sawicki, *Antiaircraft Artillery Battalions of the US Army*, Vol. I (Dumfries, VA: Wyvern Pub., 1991), pp. 4-6. GHQ, AEF GO 205, Nov. 14, 1918. Army War College, *United States Army in the World War*, Vol. 16 (General Orders, GHQ, AEF), GPO, 1948, pp. 534-35.

15. Historical Section, Army War College, *Order of Battle, AEF*, Vol. 3, Pt. 2, 1937-49, pp. 1142-1230.

16. R. Ernest Dupuy, "With the 57th in France," Pt. VII, *Our Army*, Jan. 1930, p. 35. "Historical Data and Station and Movement List, 51st Coast Artillery (Tractor Drawn) Regiment" and "56th Coast Artillery Regiment," Organizational Records Section, Military Personnel Records Section, National Personnel Records Center, National Archives and Records Administration, St. Louis, MO, (Hereafter ORS, MPRU.) Janice E. McKinney, *Air Defense Artillery*, Army Lineage Series, GPO, 1985, pp. 159-278. O.C. Warner, "Movement of the 52nd Coast Artillery (Ry) from Fort Eustis to Fort Story Va. and return May 1929," *CAJ*, Vol. 71, No. 2 (August 1929), pp. 152-59.

17. WDGO 62, Oct. 5, 1920.

18. Ibid.

19. WDGO 36, Aug. 1, 1921.

20. WDGO 20, May 15, 1922.

21. War Department, "1922 Station List." Historical Branch, War Plans Division, General Staff, "Outlines of Histories of Regiments, United States Army," May 1, 1921.

22. WDGO, 21, May 18, 1922.

23. Barnes, H.C., Lt. Col., "A Regimental Organization for the Coast Artillery Corps," *CAJ*, Vol. 60, No. 4 (1924), pp. 293-99. "Active Coast Artillery Batteries of the Regular Service," *CAJ*, Vol. 62, No. 1 (Jan. 1925), pp. 74-77. "Coast Artillery National Guard Organizations," *CAJ*, Vol. 62, No. 3 (Mar. 1925), p. 257. "Coast Artillery Reserve Units," *CAJ*, Vol. 63, No. 5 (Nov. 1925), pp. 494-495. WDGO 8, Feb. 27, 1924.

24. "Active Coast Artillery Batteries of the Regular Service."

25. WDGO 13, June 9, 1925.

26. WDGO 22, Dec. 31, 1927. WDGO 2, March 1, 1928.

27. Robert E. Wyllie, "Coats of Arms and Badges of the Coast Artillery Corps," *CAJ*, Vol. 59 (Aug. 1923), pp. 123-42.

28. E.M. Harris, "Harbor Defense Regiments and Battalions," and "Mobile Seacoast Artillery Regiments and Battalions," typescript, August 1949, NARA, College Park, MD. "The Reorganization and New Training Objective of the Coast Artillery Corps," *CAJ*, Vol. 72, No. 1 (Jan. 1930), pp. 1-11. Mark Berhow, "Caretaker Status in the Coast Artillery, 1912-1948," *CDJ*, Vol. 14, No. 4 (Nov. 2000), pp. 48-57.

29. War Department Tables of Organization 4-61,; 4-71, and 4-81, Feb. 1, 1940.

30. ARCAC, 1935, pp. 1-3.

31. E.M. Harris, "Harbor Defense Regiments and Battalions," and "Mobile Seacoast Artillery Regiments and Battalions."

32. "Officer's Station List, August 1, 1940," *CAJ*, Vol. 83, No. 4 (Jul.-Aug. 1940), pp. 393-400.

33. Tables of Organization 4-11, Dec. 15, 1938, Nov. 1, 1940.

34. William C. Gaines, "Glossary of Terms Related to Coast Artillery Operations in World War II," *CDSG Journal*, Vol. 11, No. 4 (Nov. 1997), pp. 58-60. For more details, see William C. Gaines, "Joint Army and Navy Coast Defense Commands: 1927-1945," *CDSG Journal*, Vol. 12, No. 1 (Feb. 1998), pp. 45-57.

35. Executive Order 9082, Feb. 28, 1942. WD Cir 59, Mar. 2, 1942.

36. Shelby L. Stanton, *World War II Order of Battle: An Encyclopedic Reference to U.S. Army Ground Forces from Battalion through Division, 1939-1946* (Rev. Ed.) (Mechanicsburg, PA: Stackpole Books, 2006). Sawicki, *Antiaircraft Artillery Battalions of the U.S. Army.*

37. William C. Gaines, "Antiaircraft Defense of Oahu 1916-1945," *CDJ*, Vol. 15, No. 2 (May 2001), pp. 22-67. William C. Gaines, "Railway Artillery on Oahu, 1922-1944," *CDJ*, Vol. 16, No. 3 (Aug. 2002), pp. 22-58.

38. Sawicki, *Antiaircraft Battalions of the U.S. Army.* K.R. Lamison and John W. Wike, "Combat Arms Regimental System," *Army Information Digest,* Vol. 19, No. 9 (Sept. 1964), US Army AG Publications Center, Baltimore, MD. Janice E. McKenney, *Air Defense Artillery.*

Coast Artillery Companies 1901-1924

Bolling W. Smith

The following listing of serially numbered companies of coast artillery gives their locations and changes in designations from their creation in 1901 until 1924, when the separate company numbers were finally abolished. Redesignations were effective in the year shown. The year shown for locations indicates that the company was at the new post in that year, but could have arrived there during the previous year. Units in parentheses in the headings indicate the units from which the companies were formed, either by redesignation of lettered companies of the seven artillery regiments in 1901, or by splitting existing companies of coast artillery.

1st Company (A/1st Artillery)
  1901 – Fort Dade, FL
  1902 – Fort De Soto, FL
  1907 – Fort Armstrong, TH
  1908 – Fort Levett, ME
  1910 – Fort McKinley, ME
  1916 – Temp., Fort Sam Houston
  1916 – 1st Company, Fort McKinley, ME
  1917 – 1st Company, CD Portland
  1917 – A/6 Provisional Artillery Regiment
  1918 – A/51st Artillery Regiment, CAC
  1922 – 1st Company, CAC (additional designation)
  1924 – A/51st CA Regiment (TD), Fort Eustis, VA

2nd Company (B/1st Artillery)
  1901 – Fort Trumbull, CT
  1903 – Fort H.G. Wright, NY
  1916 – 1st Company, Fort Ruger, TH
  1917 – 11th Company, CD Oahu
  1921 – 4th Company, CD Honolulu
  1922 – 2nd Company, CAC
  1924 – B/1st CA Regiment (HD), inactive

3rd Company (C/1st Artillery)
  1901 – Fort Moultrie, SC
  1902 – Fort Getty, SC
  1903 – Fort Moultrie, SC
  1909 – Fort Hamilton, NY
  1916 – 2nd Company, Fort Hamilton, NY
  1917 – 2nd Company, CD Southern New York
  1922 – C/59th Artillery Regiment, CAC
  1922 – 3rd Company, CAC (additional designation)
  1924 – C/1st CA Regiment (HD), inactive

4th Company (D/1st Artillery)
  1901 – Jackson Barracks, LA
  1901 – Fort St. Philip, LA
  1902 – Jackson Barracks, LA
  1907 – Fort DuPont, DE
  1911 – Fort Mott, NJ
  1915 – Fort Mills, PI
  1916 – 13th Company, Fort Mills, PI

  1916 – Fort Wint, PI
  1917 – 13th Company, CD Manila & Subic Bays
  1917 – Fort Mills, PI
  1922 – 4th Company, CAC
  1924 – D/1st CA Regiment (HD), inactive

5th Company (F/1st Artillery)
  1901 – Fort Screven, GA
  1907 – Fort Williams, ME
  1916 – 2nd Company, Fort Williams, ME
  1917 – H/6th Provisional Artillery Regiment, CAC
  1918 – H/51st Artillery Regiment, CAC
  1922 – 5th Company, CAC (additional designation)
  1924 – F/43rd Artillery Regiment (Ry), inactive

6th Company (G/1st Artillery)
  1901 – Fort Monroe, VA
  1916 – 2nd Company, Fort Monroe, VA
  1917 – 6th Company, CD Chesapeake Bay (I)
  1917 – E/60th Artillery Regiment, CAC
  1921 – A/1 AA Battalion, CAC
  1922 – A/61st Artillery Battalion (AA), CAC
  1922 – 6th Company, CAC (additional designation)
  1924 – A/61st CA (AA) Regiment, Fort Monroe, VA

7th Company (H/1st Artillery)
  1901 – Fort Barrancas, FL
  1907 – Fort Banks, MA
  1916 – 2nd Company, Fort Banks, MA
  1917 – K/6th Provisional Artillery Regiment, CAC
  1917 – K/51st Artillery Regiment (Ry), CAC
  1917 – 6th Battery, Howitzer Regiment, CAC
  1918 – D/44th Artillery Regiment, CAC
  1922 – 7th Company, CAC (additional designation)
  1924 – HQ/1st CA Regiment (HD), Fort DeLesseps, CZ

8th Company (I/1st Artillery)
  1901 - Fort Morgan, AL
  1907 – Fort Preble, ME
  1910 – Fort McKinley, ME
  1916 – Fort Amador, CZ
  1916 – 9th Company, Fort Grant, CZ

1917 – 9th Company, CD Balboa
1922 – 8th Company, CAC
1924 – A/1st CA Regiment (HD), inactive

9th Company (L/1st Artillery)
1901 – Fort Barrancas, FL
1907 – Fort Warren, MA
1909 – Mine company
1916 – 1st Company, Fort Warren, MA
1917 – 7th Company, CD Boston
1921 – 11th Company, CD Long Island Sound
1921 – HQ/2nd AA Battalion, CAC
1922 – 9th Company, CAC (additional designation)
1924 – HQ/62nd CA Regiment (AA), Fort Totten

10th Company (M/1st Artillery)
1901 – Fort Moultrie, SC
1902 – Fort Getty, SC
1903 – Manila, PI
1904 – SF Presidio, CA
1913 – Ft De Russy, TH
1916 – 1st Company, Fort De Russy, TH
1917 – 9th Company, CD Oahu
1921 – 2nd Company, CD Honolulu
1922 – 10th Company, CAC
1924 – E/1st CA Regiment (HD), Fort Randolph, CZ

11th Company (N/1st Artillery)
1901 – Fort Morgan, AL
1901 – Key West Barracks, FL
1907 – Fort Schuyler, NY
1911 – Fort Mills, PI
1916 – 9th Company, Fort Mills, PI
1917 – 9th Company, CD Manila & Subic Bays
1918 – Fort Hughes, PI
1922 – 11th Company, CAC
1924 – G/1st CA Regiment (HD), Fort Randolph, CZ

12th Company (O/1st Artillery)
1901 – Fort Clark, TX
1902 – Fort HG Wright, NY
1916 – 2nd Company, Fort HG Wright, NY
1917 – E/7th Provisional Artillery Regiment, CAC
1918 – E/52nd Artillery Regiment, CAC
1918 – 1st Battery, Howitzer Regiment, CAC
1918 – E/81st Artillery, CAC
1918 – E/44th Artillery, CAC
1921 – Inactivated
1922 – 12th Company, CAC (additional designation)
1923 – E/44th CA Regiment (TD), inactive

13th Company (B/2nd Artillery)
1901 – Fort Monroe, VA
1910 – Fort Mills, PI
1912 – SF Presidio, CA
1913 – Fort Miley, CA
1916 – 1st Company, Fort Miley, CA

1917 – 18th Company, CD San Francisco
1918 – E/40th Artillery Regiment, CAC
1922 – 13th Company, CAC
1924 – B/2nd CA Regiment (HD), inactive

14th Company (C/2nd Artillery)
1901 – Fort Warren, MA
1902 – Fort Screven, GA
1907 – Fort Greble, RI
1909 – Mine company
1916 – 1st Company, Fort Greble, RI
1917 – 7th Company, CD Narragansett Bay
1919 – 6th Company, CD Narragansett Bay
1922 – 14th Company, CAC
1924 – C/2nd CA Regiment (HD), inactive

15th Company (D/2nd Artillery)
1901 – Fort Barrancas, FL
1916 – Fort Randolph, CZ
1916 – 3rd Company, Fort Randolph, CZ
1917 – 9th Company, CD Cristobal
1922 – 15th Company, CAC
1924 – D/2nd CA Regiment (HD), inactive

16th Company (E/2nd Artillery)
1901 – Fort Fremont, SC
1902 – Fort Moultrie, SC
1905 – Fort Moultrie, SC
1909 – Mine company
1915 – Fort Sherman, CZ
1916 – 5th Company, Fort Sherman, CZ
1917 – 5th Company, CD Cristobal
1922 – 16th Company, CAC
1924 – E/2nd CA Regiment (HD), Fort Sherman, CZ

17th Company (G/2nd Artillery)
1901 – Havana, Cuba
1902 – Santiago, Cuba
1904 – Fort Washington, MD
1913 – Mine company
1915 – Fort Mills, PI
1916 – 4th Company, Fort Mills, PI
1917 – 4th Company, CD Manila & Subic Bays
1922 – B/62nd Artillery (AA), CAC
1922 – 17th Company, CAC (additional designation)
1924 – B/62nd CA Regiment (AA), Fort Totten, CZ

18th Company (H/2nd Artillery)
1901 – Havana, Cuba
1902 – Cienfuegos, Cuba
1903 – Fort Schuyler, NY
1911 – Fort Mills, PI
1916 – 12th Company, Fort Mills, PI
1917 – 4th Company, CD Manila & Subic Bays
1922 – 18th Company, CAC
1924 – H/2nd CA Regiment (HD), Fort Sherman, CZ

19th Company (I/2nd Artillery)
1901 – Havana, Cuba
1902 – Santiago, Cuba
1904 – Fort Caswell, NC
1909 – Mine company
1916 – 1st Company, Fort Caswell, NC
1917 – 1st Company, CD Cape Fear
1922 – 19th Company, CAC
1924 – I/2nd CA Regiment (HD), inactive

20th Company (K/2nd Artillery)
1901 – Havana, Cuba
1904 – Fort Barrancas, FL
1916 – Temp., Del Rio, TX
1916 – 2nd Company, Fort Barrancas, FL
1917 – 2nd Company, CD Pensacola
1922 – 20th Company, CAC
1924 – K/2nd CA Regiment (HD), inactive

21st Company (L/2nd Artillery)
1901 – Havana, Cuba
1902 – Cienfuegos, Cuba
1903 – Fort McHenry, MD
1906 – Fort Howard, MD
1915 – Fort Randolph, CZ
1916 – 1st Company, Fort Randolph, CZ
1917 – 7th Company, CD Cristobal
1922 – 21st Company, CAC
1924 – L/2nd CA Regiment (HD), Fort Sherman, CZ

22nd Company (M/2nd Artillery)
1901 – Havana, Cuba
1904 – Fort Barrancas, FL
1916 – 1st Company, Fort Barrancas, FL
1917 – 5th Company, CD Pensacola
1922 – 22nd Company, CAC
1924 – A/2nd CA Regiment (HD), inactive

23rd Company (N/2nd Artillery)
1901 – Havana, Cuba
1902 – Fort McKinley, ME
1915 – Fort Mills, PI
1916 – 10th Company, Fort Mills, PI
1917 – 10th Company, CD Manila & Subic Bays
1918 – Fort Frank, PI
1922 – 23rd Company, CAC
1924 – F/2nd CA Regiment (HD), inactive

24th Company (O/2nd Artillery)
1901 - Havana, Cuba
1903 – Fort Preble, ME
1916 – Temp., Plattsburg Barracks, NY
1916 – 2nd Company, Fort Preble, ME
1917 – 8th Company, CD Portland (I)
1917 – A/54th Artillery, CAC
1919 – Disbanded, Camp Devens, MA

1922 – Reconstituted and consolidated
        with 1st Company, CD Portland,
        as 24th Company, CAC
1924 – G/2nd CA Regiment (HD), Fort Sherman, CZ

25th Company (A/3rd Artillery)
1901 – Fort Santiago, Manila, PI
1903 – Fort Miley, CA
1916 – 2nd Company, Fort Miley, CA
1917 – 19th Company, CD San Francisco
1918 – C/18th Artillery Regiment, CAC
1918 – 19th Company, CD San Francisco
1922 – 25th Company, CAC
1924 – A/3rd CA Regiment (HD), Fort MacArthur, CA

26th Company (B/3rd Artillery)
1901 – Fort Flagler, WA
1916 – 1st Company, Fort Flagler, WA
1917 – 13th Company, CD Puget Sound
1922 – 26th Company, CAC
1924 – B/3rd CA Regiment (HD), Fort MacArthur, CA

27th Company (D/3rd Artillery)
1901 – Manila, PI
1903 – Fort Baker, CA
1904 – SF Presidio, CA
1913 – Fort Winfield Scott, CA
1916 – 7th Company, Fort Winfield Scott, CA
1917 – 7th Company, CD San Francisco (I)
1917 – 4th Separate AA Company
1922 – 27th Company, CAC
1924 – D/3rd CA Regiment (HD), Fort Rosecrans, CA

28th Company (E/3rd Artillery)
1901 – SF Presidio, CA
1904 – Camp McKinley, Oahu, Honolulu, HI
1905 – Fort Rosecrans, CA
1912 – Mine company
1914 – Temp., San Ysidro, CA
1915 – Fort Rosecrans, CA
1916 – 1st Company, Fort Rosecrans, CA
1917 – 1st Company, CD San Diego
1922 – 28th Company, CAC
1924 – E/3rd CA Regiment (HD), Fort Columbia

29th Company (G/3rd Artillery)
1901 - Manila, PI
1901 – SF Presidio, CA
1913 – Fort Winfield Scott, CA
1914 – Temp., Tecate, CA
1915 – Fort Winfield Scott, CA
1916 – 9th Company, Fort Winfield Scott, CA
1917 – 9th Company, CD San Francisco, CA
1922 – C/62nd Artillery Regiment (AA), CAC;
        12th Company, CAC (additional designation)
1924 – C/62nd CA Regiment (AA), Fort Totten, NY

30th Company (H/3rd Artillery)
1901 – Manila, PI
1901 – San Diego Barracks, CA
1904 – Fort Worden, WA
1915 – Fort Rosecrans, CA
1916 – Fort Worden, WA
1916 – 1st Company, Fort Worden, WA
1917 – 1st Company, CD Puget Sound
1922 – E/62nd Artillery Regiment (AA), CAC,
    30th Company, CAC (additional designation)
1924 – E 62nd CA Regiment (AA), Fort Totten, NY

31st Company (I/3rd Artillery)
1901 – Manila, PI
1903 – Fort Caswell, NC
1916 –Temp., Del Rio, TX
1916 – 1st Company, Fort Caswell, NC
1917 – I/8th Provisional Artillery Regiment, CAC
1917 – L/53rd Artillery, CAC
1918 – E/53rd Artillery, CAC
1922 – 12th Company, CAC (additional designation)
1924 – C/3rd CA Regiment (HD), inactive

32nd Company (K/3rd Artillery)
1901 – Manila, PI
1901 – Fort Lawton, WA
1902 – Fort Liscum, AK
1903 – Fort Baker, CA
1916 – Temp., Monterrey, CA
1916 – 3rd Company, Fort Baker, CA
1917 – 12th Company, CD San Francisco
1918 – A/18th Artillery Regiment, CAC
1918 – 12th Company, CD San Francisco
1922 – F/62nd Artillery (AA), CAC;
    32nd Company, CAC (additional designation)
1924 – F/62nd CA Regiment (AA), Fort Totten, NY

33rd Company (L/3rd Artillery)
1901 – Manila, PI
1901 – Fort Canby, WA
1902 – Stevens, OR
1903 – Fort Columbia, WA
1907 – Mine company
1916 – Fort Mills, PI
1916 – 14th Company, Fort Mills, PI
1917 – 14th Company, CD Manila & Subic Bays,
    Fort Frank, PI
1918 – Fort Mills and Fort Hughes, PI
1922 – D/62nd Artillery Regiment (AA), CAC;
    33rd Company, CAC (additional designation)
1924 – D/62nd CA Regiment (AA), inactive

34th Company (M/3rd Artillery)
1901 – Fort Stevens, OR
1907 – Mine company
1916 – 1st Company, Fort Stevens, OR
1917 – 1st Company, CD Columbia, WA

1922 – 34th Company, CAC
1924 – F/3rd CA Regiment (HD), inactive

35th Company (N/3rd Artillery)
1901 – Fort Monroe, VA
1908 – Fort Mills, PI
1910 – Fort Monroe, VA
1916 – 1st Company, Fort Monroe, VA
1917 – 1st Company, CD Chesapeake Bay
1922 – 35th Company, CAC
1924 – G/3rd CA Regiment (HD), inactive

36th Company (O/3rd Artillery)
1901 - Manila, PI
1903 – Fort Moultrie, SC
1910 – Fort Du Pont, DE
1911 – Fort Mott, NJ
1916 – Fort Mills, PI
1916 – 5th Company, Fort Mills, Fort Wint, PI
1917 – 5th Company, Fort Mills, Fort Mills, PI
1917 – 5th Company, CD Manila & Subic Bays
1922 – 36th Company, CAC
1924 – HQ/3rd CA Regiment (HD), Fort MacArthur, CA

37th Company (A/4th Artillery)
1901 – Fort Washington, MD
1905 – Fort McKinley, ME
1908 – Mine company
1916 – 3rd Company, Fort McKinley, ME
1917 – 13th Company, CD Portland
1920 – 5th Company, CD Portland
1921 – HQ & CT/2nd Battalion (AA) , CAC,
    Fort Totten, NY
1922 – Service/62nd Artillery Battalion, AA) CAC;
    37th Company, CAC (additional designation)
1924 – Service/62nd CA Regiment (AA), Fort Totten, NY

38th Company (C/4th Artillery)
1901 – Fort Caswell, NC
1903 – Manila, PI
1904 – SF Presidio, CA
1913 – Fort Winfield Scott, CA
1916 – 4th Company, Fort Winfield Scott, CA (I)
1917 – 1st Company, Fort MacArthur, CA;
    1st Company (I), CD Los Angeles;
    3rd Separate AA Company
1918 – Fort Totten, NY
1919 – Disbanded Fort Totten, NY
1922 – Reconstituted and consolidated
    with 2nd Company, CD Los Angeles,
    as 38th Company, CAC, Fort MacArthur
1924 – C/4th CA Regiment (HD), Fort Amador, CZ

39th Company (D/4th Artillery)
1901 – Fort McHenry, MD
1907 – Fort DeSoto, FL
1910 – Fort Morgan, AL

1914 – Temp., Brownsville, TX
1916 – 1st Company, Fort Morgan, AL
1917 – 1st Company, CD Mobile Bay
1922 – 39th Company, CAC
1924 – D/4th CA Regiment (HD), Fort Amador, CZ

40th Company (E/4th Artillery)
1901 – Fort Howard, MD
1915 – Fort Grant, CZ
1916 – 5th Company, Fort Grant, CZ
1917 – 5th Company, CD Balboa
1922 – 40th Company, CAC
1924 – E/4th CA Regiment (HD), inactive

41st Company (G/4th Artillery)
1901 – Fort Monroe, VA
1916 – Temp., Fabens, TX
1916 – 2nd Company, Fort Monroe, VA
1917 – 2nd Company, CD Chesapeake Bay
1922 – 41st Company, CAC
1924 – G/4th CA Regiment, Fort Amador, CZ

42nd Company (H/4th Artillery)
1901 – Fort Mott, NJ
1911 – Fort Mills, PI
1916 – 11th Company, Fort Mills, PI
1917 – 11th Company, CD Manila & Subic Bays
1922 – 42nd Company, CAC
1924 – H/4th CA Regiment (HD), inactive

43rd Company (I/4th Artillery)
1901 – Fort Trumbull, CT
1901 – Fort Terry, NY
1916 – 1st Company, Fort Terry, NY
1917 – 12th Company, CD Long Island Sound
1918 – 8th Company, CD Long Island Sound
1922 – 43rd Company, CAC
1924 – I/4th CA Regiment (HD), Fort Amador, CZ

44th Company (K/4th Artillery)
1901 – Fort Hunt, VA
1914 – Fort Sherman, CZ
1916 – 1st Company, Fort Sherman, CZ
1917 – 1st Company, CD Cristobal
1922 – 44th Company, CAC
1924 – K/4th CA Regiment (HD), inactive

45th Company (L/4th Artillery)
1901 – Fort Du Pont, DE
1907 – Fort Monroe, VA
1908 – Fort Du Pont, DE
1915 – Fort Grant, CZ
1916 – 3rd Company, Fort Grant, CZ
1917 – 3rd Company, CD Balboa
1922 – 45th Company, CAC
1924 – A/4th CA Regiment (HD), Fort Amador, CZ

46th Company (M/4th Artillery)
1901 – Fort Strong, MA
1916 – Temp., Fort Sam Houston, TX;
        3rd Company, Fort Strong, MA
1917 – C/6th Provisional Artillery Regiment, CAC
1918 – C/51st Artillery Regiment, CAC
1918 – C/43rd Artillery Regiment, CAC
1922 – 46th Company, CAC
1924 – I/4th CA Regiment (HD), inactive

47th Company (N/4th Artillery)
1901 – Fort Hunt, VA
1916 – 1st Company, Fort Hunt, VA
1917 – 1st Company, CD Potomac
1922 – Consolidated with 2nd Company,
        CD Potomac, as 47th Company, CAC
1924 – F/4th CA Regiment (HD), inactive

48th Company (O/4th Artillery)
1901 – Fort Hancock, NJ
1916 – 1st Company, Fort Hancock, NY
1917 – 1st Company, CD Sandy Hook
1922 – 48th Company, CAC
1924 – HQ/4th CA Regiment (HD), Fort Amador, CZ

49th Company (A/5th Artillery)
1901 – Hamilton, NY
1901 – Fort Columbus, NY
1902 – Fort Williams, ME
1916 – Plattsburg Barracks, NY;
        3rd Company, Fort Williams, ME
1917 – G/6th Provisional Artillery Regiment, CAC
1918 – G/51st Artillery Regiment, CAC;
        D/57th Artillery Regiment, CAC
1919 – Fort Winfield Scott, CA
1920 – Camp Lewis, WA
1921 – Disbanded
1922 – Reconstituted and consolidated
        with HQ Company, CD Puget Sound,
        as 49th Company, CAC, Fort Worden, WA
1924 – A/5th CA Regiment (HD), inactive

50th Company (B/5th Artillery)
1901 – Fort Wadsworth, NY
1909 – Fort Wint, PI
1911 – Fort McKinley, ME
1915 – Fort Levett, ME
1916 – 1st Company, Fort Levett, ME
1917 – 9th Company, CD Portland
1922 – 50th Company, CAC
1924 – B/5th CA Regiment (HD), inactive

51st Company (C/5th Artillery)
1901 – Fort Hamilton, NY
1909 – Fort Mills, PI
1911 – Fort McKinley, ME
1916 – 2nd Company, Fort McKinley, ME

1917 – 12th Company, CD Portland;
    C/54th Artillery Regiment, CAC
1919 – Camp Devens, MA
1919 – Disbanded
1922 – Reconstituted and consolidated
    with 4th Company, CD Portland
    as 51st Company, CAC
1924 – C/5th CA Regiment (HD), inactive

52nd Company (E/5th Artillery)
1901 - Fort Columbus, NY
1901 – Fort Hancock, NJ
1903 – Fort Rodman, MA
1916 – 1st Company, Fort Rodman, MA
1917 – 1st Company, CD New Bedford
1922 – 52nd Company, CAC
1924 – E/10th CA Regiment (HD), Ft Rodman, MA

53rd Company (G/5th Artillery)
1901 - Fort Wadsworth, NY
1916 – 1st Company, Fort Wadsworth, NY
1917 – 5th Company, CD Southern New York
1922 – 53rd Company, CAC
1924 – G/5th CA Regiment (HD), inactive

54th Company (H/5th Artillery)
1901 – Fort Totten, NY
1906 – Depot Torpedo Company
1907 – Mine company
1909 – Fort Mills, PI
1911 – Fort Wadsworth, NY
1916 – 2nd Company, Fort Wadsworth, NY
1917 – 6th Company, CD Southern New York
1922 – 54th Company, CAC
1924 – E/5th CA Regiment (HD), inactive

55th Company (I/5th Artillery)
1901 - Fort Hancock, NJ
1909 – Fort Mills, PI
1911 – Fort DuPont, DE
1913 – Fort DeRussy, TH, Mine
1916 – 2nd Company, Fort DeRussy, TH
1917 – 10th Company, CD Oahu
1921 – 3rd Company, CD Honolulu
1922 – 55th Company, CAC
1924 – F/5th CA Regiment (HD), inactive

56th Company (L/5th Artillery)
1901 – El Morro, San Juan, PR
1904 – Fort Wadsworth, NY
1911 – Fort Hancock, NJ
1915 – Fort Jay, NY
1916 – Fort Hancock, NJ; Temp., Columbus, NM;
    6th Company, Fort Hancock, NJ
1917 – 6th Company, CD Sandy Hook
1922 – 56th Company, CAC
1924 – D/5th CA Regiment (HD), inactive

57th Company (M/5th Artillery)
1901 - Fort Wadsworth, NY
1905 – Torpedo company
1907 – Fort Wint, PI, mine company
1909 – SF Presidio, CA
1913 – Fort Winfield Scott, CA
1914 – Temp., San Ysidro, CA, Mine
1915 – Fort Winfield Scott, CA
1916 – 3rd Company, Fort Winfield Scott, CA
1917 – 3rd Company, CD San Francisco
1922 – 57th Company, CAC
1924 – HQ/5th CA Regiment (HD), Fort Hamilton, NY

58th Company (N/5th Artillery)
1901 - Fort Monroe, VA
1905 – Torpedo company
1908 – Mine company
1916 – 3rd Company, Fort Monroe, VA
1917 – 3rd Company, CD Chesapeake Bay
1922 – 58th Company, CAC
1924 – A/12th CA Regiment (HD), Fort Monroe, VA

59th Company (O/5th Artillery)
1901 – San Cristobal, San Juan, PR
1904 – Fort Andrews, MA
1916 – 1st Company, Fort Andrews, MA
1917 – 1st Company, CD Boston (I)
1922 – 59th Company, CAC
1924 – A/9th CA Regiment (HD), Fort Banks, MA

60th Company (A/6th Artillery)
1901 – SF Presidio, CA
1905 – Torpedo company
1908 – Mine company
1913 – Fort Winfield Scott, CA
1916 – 2nd Company, Fort Winfield Scott, CA
1917 – 2nd Company, CD San Francisco
1918 – A/50th Artillery Regiment, CAC;
    2nd Company, CD San Francisco
1922 – 60th Company, CAC
1924 – A/6th CA Regiment (HD), Fort Winfield Scott

61st Company (B/6th Artillery)
1901 – USAT *Meade* enroute Manila to San Francisco,
1901 – SF Presidio, CA
1902 – Fort Baker, CA
1914 – Temp., Calexico Company, CA
1915 – Fort Baker, CA
1916 – 2nd Company, Fort Baker, CA
1917 – 11th Company, CD San Francisco (I);
    C/1st AA Battalion
1918 – 24th AA Battery, 1st AA Sector
1919 – 11th Company, CD San Francisco
1922 – 61st Company, CAC
1924 – B/6th CA Regiment (HD), inactive

62nd Company (C/6th Artillery)
1901 – USAT *Meade* enroute Manila to San Francisco
1901 – SF Presidio, CA
1901 – Fort Mason, CA
1902 – Fort Worden, WA
1916 – 2nd Company, Fort Worden, WA
1917 – 2nd Company, CD Puget Sound
1922 – 62nd Company, CAC
1924 – C/6th CA Regiment (HD), inactive

63rd Company (E/6th Artillery)
1901 – USAT *Meade* enroute Manila to San Francisco
1901 – Alcatraz Is, CA
1902 – Fort Casey, WA
1909 – Fort Worden, WA
1916 – Temp., Fort George Wright, WA;
        3rd Company, Fort Worden, WA
1917 – 3rd Company, CD Puget Sound
1922 – 63rd Company, CAC
1924 – E/6th CA Regiment (HD),
        Fort Winfield Scott, CA

64th Company (F/6th Artillery)
1901 – USAT *Meade* enroute Manila to San Francisco
1901 – SF Presidio, CA
1901 – Alcatraz Is, CA
1902 – Fort Miley, CA
1911 – SF Presidio, CA
1913 – Fort Winfield Scott, CA
1916 – 8th Company, Fort Winfield Scott, CA
1917 – 8th Company, CD San Francisco
1922 – 64th Company, CAC
1924 – F/6th CA Regiment (HD), inactive

65th Company (H/6th Artillery)
1901 – USAT *Meade* enroute Manila to San Francisco
1901 – Fort McDowell, CA
1902 - SF Presidio, CA
1913 – Fort Winfield Scott, CA
1916 – 6th Company, Fort Winfield Scott, CA
1917 – 6th Company, CD San Francisco
1918 – C/40th Artillery Regiment, CAC;
        6th Company, CD San Francisco
1922 – 65th Company, CAC
1924 – H/6th CA Regiment (HD), inactive

66th Company (I/6th Artillery)
1901 – Camp McKinley, Honolulu, HI
1904 – San Francisco Presidio, CA
1908 – Fort Barry, CA
1914 – Temp., Tecate, CA
1916 – 1st Company, Fort Barry, CA
1917 – 15th Company, CD San Francisco
1918 – D/40th Artillery, CAC;
        15th Company, CD San Francisco
1922 – 66th Company, CAC
1924 – I/6th CA Regiment (HD), inactive

67th Company (K/6th Artillery)
1901 – Camp McKinley, Honolulu, HI
1904 – SF Presidio, CA
1913 – Fort Winfield Scott, CA
1916 – 10th Company, Fort Winfield Scott, CA
1917 – 10th Company, CD San Francisco;
        B/1st AA Battalion
1918 - B/1st AA Sector
1919 – 10th Company, CD San Francisco
1922 – 67th Company, CAC
1924 – K/6th CA Regiment (HD),
        Fort Winfield Scott, CA

68th Company (L/6th Artillery)
1901 – USAT *Meade* enroute Manila to San Francisco
1901 – Fort Baker, CA
1913 – Fort Kamehameha, TH
1916 – 1st Company, Fort Kamehameha, TH
1917 – 1st Company, CD Oahu
1921 – 1st Company, 1st CD Honolulu
1922 – 68th Company, CAC
1924 – D/6th CA Regiment (HD), inactive

69th Company (M/6th Artillery)
1901 – Fort Monroe, VA
1916 – Temp., El Paso, TX;
        4th Company, Fort Monroe, VA
1917 – 4th Company, CD Chesapeake Bay
1922 – 69th Company, CAC
1924 – G/6th CA Regiment (HD), inactive

70th Company (N/6th Artillery)
1901 – USAT *Meade* enroute Manila to San Francisco
1901 – SF Presidio, CA
1912 – Fort Mills, PI
1916 – 7th Company, Fort Mills, PI
1917 – 7th Company, CD Manila & Subic Bays
1918 – 7th Company, CD Manila & Subic Bays,
        Fort Wint, PI
1922 – 12th Company, CAC
1924 – Service/65th CA Regiment (AA), inactive

71st Company (O/6th Artillery)
1901 – USAT *Meade* enroute Manila to San Francisco
1901 – SF Presidio, CA
1902 – Fort Casey, WA
1916 – 1st Company, Fort Casey, WA
1917 – 9th Company, CD Puget Sound
1922 – 71st Company, CAC
1924 – HQ/6th CA Regiment (HD),
        Fort Winfield Scott, CA

72nd Company (A/7th Artillery)
1901 – Fort Greble, RI
1907 – Fort Screven, GA
1908 – Mine company
1916 – 1st Company, Fort Screven, GA

1917 – 1st Company, CD Savannah
1922 – 72nd Company, CAC
1924 – A/7th CA Regiment (HD), Fort Hancock, NJ

73rd Company (B/7th Artillery)
1901 – Buffalo, NY
1902 – Fort Monroe, VA
1916 – Fort Grant, CZ; 6th Company, Fort Grant
1917 – 6th Company, CD Balboa
1922 – 73rd Company, CAC
1924 – B/7th CA Regiment (HD), Fort Hancock, NJ

74th Company (D/7th Artillery)
1901 – Fort Williams, ME
1907 – Fort Screven, GA
1916 – Temp., Del Rio, TX;
         4th Company, Fort Screven, GA
1917 – 2nd Company, Fort Screven, GA;
         M/8th Provisional Artillery Regiment, CAC
1918 – M/53rd Artillery Regiment, CAC;
         F/53rd Artillery Regiment, CAC
1922 – 74th Company, CAC
1924 – D/7th CA Regiment (HD), Fort Hancock, NJ

75th Company (E/7th Artillery)
1901 – Fort Preble, ME
1907 – Fort Moultrie, SC
1908 – Fort Morgan, AL
1913 – Fort Kamehameha, TH
1916 – 2nd Company, Fort Kamehameha, TH
1917 – 2nd Company, CD Oahu
1921 – 2nd Company, CD Pearl Harbor
1922 – 75th Company, CAC
1924 – E/7th CA Regiment (HD), Fort DuPont, DE

76th Company (F/7th Artillery)
1901 – Fort Banks, MA
1907 – Fort Barrancas, FL
1909 – Fort Hancock, NJ
1916 – 2nd Company, Fort Hancock, NJ
1917 – 2nd Company, CD Sandy Hook
1922 – 76th Company, CAC
1924 – F/7th CA Regiment (HD), inactive

77th Company (G/7th Artillery)
1901 – Fort Warren, MA
1907 – Fort Barrancas, FL
1916 – Temp., Del Rio, TX;
         3rd Company, Fort Barrancas, FL
1917 – 3rd Company, CD Pensacola
1919 – 3rd Company, CD Pensacola absorbed
         by 5th Company, CD Pensacola
1922 – Reconstituted and consolidated
         with 3rd Company, CD Key West,
         as 77th Company, CAC, Fort Crockett, TX;
         B/60th Artillery Battalion (AA), CAC;
           77th Company, CAC (additional designation)
1924 – B/60th CA Regiment (AA), Fort McKinley, PI

78th Company (H/7th Artillery)
1901 – Fort Adams, RI
1907 – Fort Moultrie, SC
1916 – 2nd Company, Fort Moultrie, SC
1917 – H/8th Provisional Artillery Regiment, CAC
1918 – H/53rd Artillery Regiment, CAC;
         E/42nd Artillery Regiment, CAC
1922 – 78th Company, CAC
1924 – HQ/7th CA Regiment (HD), Fort Hancock, NJ

79th Company (I/7th Artillery)
1901 – Fort Adams, RI
1907 – Fort Monroe, VA
1907 – Fort Caswell, NC
1915 – Fort Michie, NY
1916 – 1st Company, Fort Michie, NY
1917 – 23rd Company, CD Long Island Sound
1918 – 18th Company, CD Long Island Sound;
         15th Company, CD Long Island Sound (II)
1922 – 79th Company, CAC
1924 – C/7th CA Regiment (HD), inactive

80th Company (K/7th Artillery)
1901 – Fort Schuyler, NY
1907 – Key West Barracks, FL
1916 – 1st Company, Key West Barracks, FL
1917 – 1st Company, CD Key West
1922 – C/60th Artillery Battalion (AA), CAC;
         80th Company, CAC
1924 – C/60th CA Regiment (AA), Fort McKinley, PI

81st Company (L/7th Artillery)
1901 – Fort Slocum, NY
1906 – Fort Schuyler, NY
1911 – Fort Du Pont, DE
1914 – Fort Grant, CZ
1916 – 1st Company, Fort Grant, CZ
1917 – 1st Company, CD Balboa
1922 – 81th Company, CAC
1924 – G/7th CA Regiment (HD), inactive

82nd Company (N/7th Artillery)
1901 – Fort Totten, NY
1916 – 1st Company, Fort Totten, NY
1917 – 6th Company, CD Eastern New York (I)
1919 – 1st Company, CD Eastern New York;
         absorbed by 2nd Company, CD ENY
1922 – Reconstituted and consolidated
         with SL/2nd AA Battalion, CAC,
         as A/62nd Artillery Battalion (AA), CAC,
         with additional designation
         as 82nd Company, CAC
1924 – A/62nd CA Regiment (AA), Fort Totten, NY

83rd Company (49th Company)
1901 – Fort Columbus, NY
1901 – Fort Hamilton, NY

1902 – Fort Revere, MA
1911 – Fort Strong, MA
1916 – Temp., Plattsburg Barracks, NY;
       4th Company, Fort Strong, MA
1917 – 8th Company, CD Long Island Sound (I);
       C/55th Artillery, CAC
1922 – 83rd Company, CAC (additional designation)
1924 - C/55th CA (TD)
1925 – I/64th CA Regiment (AA), Fort Shafter, TH

84th Company (54th Company)
1901 – Fort Hamilton, NY
1916 – Temp., Fort Jay, NY
1916 – 3rd Company, Fort Hamilton, NY
1917 – F/8th Provisional Artillery Regiment, CAC
1918 – F/53rd Artillery Regiment, CAC;
       F/42nd Artillery Regiment (Ry), CAC
1922 – 84th Company, CAC (additional designation)
1924 - F / 42nd CA Regiment (Ry), inactive

85th Company (53rd Company)
1901 - Fort Wadsworth, NY
1903 – Manila, PI
1904 – Fort Casey, WA
1916 – 2nd Company, Fort Casey, WA
1917 – 10th Company, CD Puget Sound
1922 – 85th Company, CAC
1924 – D/14th CA Regiment (HD), Fort Casey, WA

86th Company (57th Company)
1901 - Fort Wadsworth, NY
1911 – Fort Mills, PI
1916 – 6th Company, Fort Mills, PI
1917 – 6th Company, CD Manila & Subic Bays,
       Fort Wint, PI
1918 – 6th Company, CD Manila & Subic Bays,
       Fort Mills, PI
1922 – 86th Company, CAC
1924 – E/15th CA Regiment (HD), inactive

87th Company (81st Company)
1901 – Fort Totten, NY
1901 - Fort Slocumb, NY
1916 – 2nd Company, Fort Grant, CZ
1917 – 2nd Company, CD Balboa
1922 – 87th Company, CAC
1924 – B/65th Regiment (AA), Fort Amador, CZ

88th Company (2nd Company)
1901 – Fort Trumbull, CT
1901 – Fort Mansfield, RI
1911 – Fort Terry, NY
1916 – 2nd Company, Fort Terry, NY
1917 – C/7th Provisional Artillery Regiment, CAC
1918 – C/52nd Artillery Regiment, CAC
1922 – 88th Company, CAC (additional designation)
1924 – C/52nd CA Regiment (Ry), Fort Eustis, VA

89th Company (69th Company)
1901 – Fort Monroe, VA
1901 – Fort Banks, MA
1909 – Fort Williams, ME
1916 – Temp., Plattsburg Barracks, NY;
       4th Company, Fort Williams, ME
1917 – F/6th Provisional Artillery Regiment, CAC
1918 – F/51st Artillery Regiment, CAC;
       C/57th Artillery Regiment, CAC
1921 – Disbanded
1922 – Reconstituted and consolidated
       with 3rd Company, CD San Diego,
       as 89th Company, CAC, Fort Rosecrans, CA
1924 – F/65th CA Regiment (AA), inactive

90th Company (39th Company)
1901 – Fort McHenry, MD
1904 – Fort McKinley, ME
1911 – Fort Mills, PI
1916 – 8th Company, Fort Mills, PI
1917 – 8th Company, CD Manila & Subic Bays
1922 – 90th Company, CAC
1924 – A/16th CA Regiment (HD), Ft DeRussy, TH

91st Company (4th Company)
1901 – Jackson Barracks, LA
1914 – Temp., Brownsville, TX
1915 – Fort Kamehameha, TH
1916 – 3rd Company, Fort Kamehameha, TH
1917 – 3rd Company, CD Oahu
1922 – 91st Company, CAC
1924 – B/15th CA Regiment (HD),
       Fort Kamehameha, TH

92nd Company (28th Company)
1901 - SF Presidio, CA
1904 – Honolulu, TH
1905 – Fort Flagler, WA
1916 – 2nd Company, Fort Flagler, WA
1917 – 14th Company, CD Puget Sound
1922 – 92nd Company, CAC
1924 – C/14th CA Regiment (HD), inactive

93rd Company (34th Company)
1901 – Fort Stevens, OR
1916 – Temp., Vancouver Barracks, WA;
       3rd Company, Fort Stevens, OR
1917 – 3rd Company, CD Columbia
1922 – 93rd Company, CAC
1924 – F/14th CA Regiment (HD), inactive

94th Company (26th Company)
1901 – Fort Flagler, WA
1916 – Temp., Fort Lawton, WA
1916 – 3rd Company, Fort Flagler, WA
1917 – 15th Company, CD Puget Sound
1919 – Absorbed by 13th and 15th Cos CD Puget Sound

1922 – Reconstituted and consolidated
        with 7th Company, CD Puget Sound,
        as 94th Company, CAC
1924 – B/14th CA Regiment (HD), inactive

95th Company (48th Company)
1901 - Fort Hancock, NJ
1906 - Torpedo company
1915 – Fort Mills, PI
1916 – 2nd Company, Fort Mills, PI
1917 – 2nd Company, CD Manila & Subic Bays
1918 – Fort Drum, PI
1919 – Fort Mills, PI
1922 – 95th Company, CAC
1924 – F/15th CA Regiment (HD), inactive

96th Company (67th Company)
1901 – Camp McKinley, Oahu, TH
1901 – Fort Warren, MA
1914 – Fort Revere, MA
1916 – 1st Company, Fort Revere, MA
1917 – 1st Company, CD Boston (I);
        A/55th Artillery Regiment, CAC
1922 – 96th Company, CAC (additional designation)
1924 – A/55 CA Regiment (TD), Fort Shafter, TH

97th Company (68th Company)
1901 – Fort Adams, RI
1916 – 1st Company, Fort Adams, RI
1917 – 6th Company, CD Narragansett Bay
1919 – 4th Company, CD Narragansett Bay
1921 – 1st Company, CD Narragansett Bay
1922 – 97th Company, CAC
1924 – A/10th CA Regiment (HD), inactive

98th Company (51st Company)
1901 – Fort Hamilton, NY
1916 – 1st Company, Rockaway Beach, NY
1917 – 9th Company, CD Southern New York
1921 – Disbanded
1922 – Reconstituted and consolidated
        with 1st Company, CD Southern New York,
        as 98th Company, CAC; Fort Mills, PI
1922 – HQ/59th Artillery Regiment (TD), CAC
1924 – HQ/59th CA Regiment (TD), Fort Mills, PI

99th Company (8th Company)
1901 – Fort Morgan, AL
1907 – Fort Moultrie, SC
1908 – Fort Morgan, AL
1914 – Fort Mills, PI
1916 – 16th Company, Fort Mills, PI
1917 – 16th Company, CD Manila & Subic Bays
1922 – 99th Company, CAC
1924 – B/16th CA Regiment (HD), inactive

100th Company (43rd Company)
1901 – Fort Terry, NY
1908 – Mine company
1916 – 4th Company, Fort Terry, NY
1917 – 14th Company, CD Long Island Sound (I);
        HQ/56th Artillery Regiment, CAC
1919 – Camp Jackson, SC
1921 – Disbanded
1922 – Reconstituted and consolidated
        with 3rd Company, CD Long Island Sound,
        as 3rd Company, CD Long Island Sound (II)
1924 – G/11th CA Regiment (HD),
        Fort H.G. Wright, NY

101st Company (82nd Company)
1901 – Fort Totten, NY
1915 – Fort Jay, NY
1916 – 2nd Company, Fort Totten, NY
1917 – G/7th Provisional Artillery Regiment, CAC
1918 – G/52nd Artillery Regiment, CAC;
        3rd Battery, Howitzer Regiment;
        A/44th Artillery Regiment, CAC
1922 – 101st Company, CAC (additional designation)
1924 – A/44th CA Regiment (AA), inactive

102nd Company (38th Company)
1901 – Fort Caswell, NC
1907 – Fort Adams, RI
1916 – Temp., Eagle Pass, TX;
        5th Company, Fort Adams, RI
1917 – 5th Company, CD Narragansett Bay (I)
1919 – 1st Company, CD Narragansett Bay
1922 – 102nd Company, CAC
1924 – B/10th CA Regiment (HD), inactive

103rd Company (40th Company)
1901 – Fort Howard, MD
1916 – Temp., Fort Bliss, TX
1916 – 2nd Company, Fort Howard, MD
1917 – 2nd Company, CD Baltimore (I)
1918 - Disbanded
1922 – Reconstituted and consolidated
        with 10th Company, CD Baltimore,
        as 103rd Company, CAC
1922 – Fort Monroe, VA
1924 – E/12th CA Regiment (HD),
        Fort Washington, MD

104th Company (37th Company)
1901 – Fort Washington, MD
1908 – Mine company
1913 – Fort Armstrong, TH
1916 – 1st Company, Fort Armstrong, TH
1917 – 8th Company, CD Oahu
1921 – 1st Company, CD Honolulu
1922 – 104th Company, CAC
1924 – HQ/16th CA Regiment (HD), Fort DeRussy, TH

105th Company (29th Company)
- 1901 – SF Presidio, CA
- 1910 – Fort Ruger, TH
- 1916 – 2nd Company, Fort Ruger, TH
- 1917 – 12th Company, CD Oahu
- 1921 – 5th Company, CD Honolulu
- 1922 – 105th Company, CAC
- 1924 – C/16th CA Regiment (HD), Fort Ruger, TH

106th Company (32nd Company)
- 1901 – Fort Lawton, WA
- 1902 – Camp Skagway, AK
- 1902 – Fort Flagler, WA
- 1908 – Fort Worden, WA
- 1914 – Mine company
- 1916 – 4th Company, Fort Worden, WA
- 1917 – 4th Company, CD Puget Sound
- 1922 – 106th Company, CAC
- 1924 – HQ/14th CA Regiment (HD), Fort Worden, WA

107th Company (75th Company)
- 1901 – Fort Preble, ME
- 1911 – Fort Williams, ME
- 1914 – Fort Preble, ME
- 1916 – 1st Company, Fort Preble, ME
- 1917 – E/6th Provisional Artillery Regiment, CAC
- 1918 – E/51st Artillery Regiment, CAC;
  E/43rd Artillery Regiment, CAC
- 1922 – 107th Company, CAC (additional designation)
- 1924 – E/43rd CA Regiment (Ry), inactive

108th Company (74th Company)
- 1901 – Fort Williams, ME
- 1903 – Manila, PI
- 1904 – Fort Worden, WA
- 1916 – 5th Company, Fort Worden, WA
- 1917 – 5th Company, CD Puget Sound
- 1922 – 108th Company, CAC
- 1924 – A/14th CA Regiment (HD), inactive

109th Company (72nd Company)
- 1901 – Fort Greble, RI
- 1916 – 2nd Company, Fort Greble, RI
- 1917 – B/6th Provisional Artillery Regiment, CAC
- 1918 – B/51st Artillery Regiment, CAC
- 1922 – 109th Company, CAC (additional designation)
- 1924 - B/51st CA Regiment (TD), Fort Eustis, VA

110th Company (79th Company)
- 1901 – Fort Adams, RI
- 1910 – Fort Greble, RI
- 1916 – Temp., Plattsburg Barracks, NY;
  3rd Company, Fort Greble
- 1917 – 8th Company, CD Narragansett Bay
- 1919 – 6th Company, CD Narragansett Bay;
  5th Company, CD Narragansett Bay
- 1922 – 110th Company, CAC
- 1924 – C/10th CA Regiment (HD), inactive

111th Company (1st Company)
- 1901 – Fort Dade, FL
- 1901 – Fort DeSoto, FL
- 1916 – Fort Mills, PI; 15th Company, Fort Mills, PI
- 1917 – 15th Company, CD Manila & Subic Bays
- 1922 – 111th Company, CAC
- 1924 – D/16th CA Regiment (HD), inactive

112th Company (45th Company)
- 1901 – Fort DuPont, DE
- 1916 – Temp., Del Rio, TX
- 1916 – 4th Company, Fort DuPont, DE;
- 1917 - 4th Company, CD Delaware (I)
- 1918 – A/60th Artillery Regiment, CAC
- 1922 – 112th Company, CAC
- 1924 – HQ/12th CA Regiment (HD),
  Fort Monroe, VA

113th Company (47th Company)
- 1901 – Fort McHenry, MD
- 1902 – Fort Hancock, NJ
- 1916 – Temp., Fort Jay, NY;
  3rd Company, Fort Hancock, NJ
- 1917 – 3rd Company, CD Sandy Hook
- 1922 – 113th Company, CAC
- 1924 – B/9th CA Regiment (HD), inactive

114th Company (80th Company)
- 1901 – Fort Schuyler, NY
- 1901 – Fort Slocumb, NY
- 1915 – Fort Wadsworth, NY
- 1916 – 3rd Company, Fort Wadsworth, NY
- 1917 – E/8th Provisional Artillery Regiment, CAC
- 1918 – E/53rd Artillery Regiment, CAC;
  7th Battery, Howitzer Regiment, CAC;
  E/51st Artillery Regiment (TD), CAC
- 1922 – 114th Company, CAC (additional designation)
- 1924 – E/51st CA Regiment (TD), inactive

115th Company (30th Company)
- 1901 – San Diego Barracks, CA
- 1904 – Fort Rosecrans, CA
- 1914 – Temp., San Ysidro, CA
- 1916 – 2nd Company, Fort Rosecrans, CA
- 1917 – 2nd Company, CD San Diego (I)
- 1918 – A/65th Artillery, CAC
- 1919 – Disbanded Camp Lewis, WA
- 1922 – Reconstituted and consolidated
  with 2ndCompany, CD San Diego,
  as 115th Company, CAC
- 1924 – I/14th CA Regiment (HD), inactive

116th Company (5th Company)
- 1901 – Fort Screven, GA
- 1915 – Fort Grant, CZ
- 1916 – 4th Company, Fort Grant, CZ
- 1917 – 4th Company, CD Balboa

1922 – 116th Company, CAC
1924 – A/65th CA Regiment (AA), Fort Amador, CZ

117th Company (3rd Company)
1901 – Fort Moultrie, SC
1901 – Fort Fremont, SC
1902 – Fort Getty, SC
1903 – Fort Moultrie, SC
1905 – Fort Fremont, SC
1907 – Fort Adams, RI
1916 – 2nd Company, Fort Adams, RI
1917 – K/7th Provisional Artillery Regiment, CAC
1918 – K/52nd Artillery Regiment, CAC;
      B/42nd Artillery Regiment, CAC
1922 – 117th Company, CAC (additional designation)
1924 – B/42nd CA Regiment (Ry), inactive

118th Company (6th Company)
1901 – Fort Monroe, VA
1916 – 5th Company, Fort Monroe, VA
1917 – HQ & Supp Company, 8th
      Provisional Artillery Regiment, CAC
1917 – HQ/1st Separate Artillery Bde, CAC
1918 – HQ/1st Separate Artillery Bde (Ry), CAC
1922 – 118th Company, CAC (additional designation)
1924 – HQ/30th CA (Ry) Brigade, inactive

119th Company (42nd Company)
1901 – Fort Delaware, DE
1902 – Fort Mott, NJ
1903 – Temp., Worlds Fair, St Louis, MO
1905 – Fort Washington, MD
1906 – Fort Mott, NJ
1910 – Fort Washington, MD
1914 – Fort Sherman, CZ
1916 – 4th Company, Fort Sherman, CZ
1917 – 3rd Company, CD Cristobal
1922 – 119th Company, CAC
1924 – HQ/1st Battalion,
      65th CA Regiment (AA), inactive

120th Company (46th Company)
1901 – Fort Strong, MA
1905 – Torpedo company
1916 – 1st Company, Fort Strong, MA
1917 – 9th Company, CD Boston
1921 - Disbanded
1922 – Reconstituted and consolidated
      with 1st Company, CD Boston,
      as 120th Company, CAC
1924 – C/9th CA Regiment (HD), Fort Duvall, MA

121st Company (11th Company)
1901 – Key West Barracks, FL
1910 – Fort Screven, GA
1916 – 2nd Company (I), Fort Screven, GA
1917 – 3rd Company, CD Savannah;
      C/61st Artillery Regiment, CAC

1919 – Disbanded at Camp Upton, NY
1922 – Reconstituted and consolidated
      with 3rd Company, CD Savannah,
      as 121st Company, CAC
1924 – HQ/13th CA Regiment (HD), Fort Barrancas

122nd Company (52nd Company)
1901 – Fort Columbus, NY
1901 – Fort Hamilton, NY
1902 – Key West Barracks, FL
1909 – Fort Hamilton, NY
1916 – Temp., Columbus, NM;
      6th Company, Fort Hamilton, NY
1917 – L/7th Provisional Artillery Regiment, CAC
1918 – I/52nd Artillery Regiment, CAC;
      C/42nd Artillery Regiment, CAC
1922 – 122nd Company, CAC (addl designation)
1924 – C/42nd CA Regiment (Ry), inactive

123rd Company (55th Company)
1901 – Fort Hamilton, NY
1916 – 5th Company, Fort Hamilton, NY
1917 – 3rd Company, CD Southern New York
1922 – 123rd Company, CAC
1924 – HQ/8th CA Regiment (HD), Fort Preble, ME

124th Company (50th Company)
1901 – Fort Wadsworth, NY
1901 – Fort Constitution, NH
1905 – Fort Andrews, MA
1916 – 2nd Company, Fort Sherman, CZ
1917 – 2nd Company, CD Cristobal
1922 – 124th Company, CAC
1924 – D/65th CA Regiment (AA), Fort Sherman, CZ

125th Company (12th Company)
1901 – Fort Clark, TX
1902 – Fort Trumbull, CT
1906 – Fort Michie, NY
1907 – Fort Terry, NY
1916 – 4th Company, Fort Kamehameha, TH
1917 – 4th Company, CD Oahu
1921 – 4th Company, CD Pearl Harbor
1922 – 125th Company, CAC
1924 – A/15th CA Regiment (HD),
      Fort Kamehameha, TH

126th Company (33rd Company)
1901 – Fort Canby, WA
1902 – Fort Worden, WA
1916 – 6th Company, Fort Worden, WA
1917 – 6th Company, CD Puget Sound
1919 – Absorbed by the 5th Company, CD Puget Sound
1922 – Reconstituted and consolidated
      with 12th Company, CD Puget Sound,
      as 126th Company, CAC
1924 – H/14th CA Regiment (HD), inactive

127th Company
    1907 – Fort Fremont, SC
    1908 – Mine company
    1911 – Fort Crockett, TX, Mine company
    1913 – Fort Crockett, TX (No longer mine company)
    1916 – Temp., Marathon, TX;
        2nd Company, Fort Crockett, TX
    1917 – 2nd Company, CD Galveston
    1922 – HQ/60th Artillery Battalion (AA), CAC;
        127th Company, CAC (additional designation)
    1924 – HQ/60th CA Regiment (AA), inactive

128th Company (4th Company)
    1907 – Fort McHenry, MD
    1911 – Fort Crockett, TX
    1914 – Brownsville, TX
    1915 – Fort Crockett, TX
    1916 – 1st Company, Fort Crockett, TX
    1917 – 1st Company, CD Galveston
    1922 – A/60th Artillery Battalion (AA), CAC;
        128th Company, CAC (additional designation)
    1924 – A/60th CA Regiment (AA), Fort McKinley, ME

129th Company (14th Company)
    1908 – Fort Adams, RI
    1910 – Mine company
    1916 – 3rd Company, Fort Adams, RI
    1917 – 3rd Company, CD Narragansett Bay
    1922 – 129th Company, CAC
    1924 – D/10th CA Regiment (HD), inactive

130th Company (102nd Company)
    1908 – Fort Adams, RI
    1916 – Temp., Eagle Pass, TX;
        4th Company, Fort Adams, RI
    1917 – I/7th Provisional Artillery Regiment, CAC
    1918 – I/52nd Artillery Regiment, CAC;
        A/42nd Artillery Regiment, CAC
    1922 – 130th Company, CAC (additional designation)
    1924 – A/42nd CA Regiment (Ry), inactive

131st Company (2nd Company)
    1908 – Fort HG Wright, NY, Mine company
    1916 – 4th Company, Fort HG Wright, NY
    1917 – 2nd Company, CD Long Island Sound
    1922 – 131st Company, CAC
    1924 – H/11th CA Regiment (HD),
        Fort H.G. Wright, NY

132nd Company (52nd Company)
    1908 – Fort Trumbull, CT, Mine company
    1911 – Fort HG Wright, NY, Mine company
    1916 – 3rd Company, Fort HG Wright, NY
    1917 – 1st Company, CD Long Island Sound
    1922 – 132nd Company, CAC
    1924 - I/11th CA Regiment (HD),
        Fort H.G. Wright, NY

133rd Company (100th Company)
    1908 – Fort Terry, NY, Mine company
    1916 – 3rd Company, Fort Terry, NY
    1917 – 13th Company, CD Long Island Sound;
        A/56th Artillery Regiment, CAC
    1919 - Camp Jackson, SC
    1921 – Disbanded at Camp Jackson, SC
    1922 – Reconstituted and consolidated
        with 11th Company, CD Long Island Sound,
        as 133rd Company, CAC
    1924 – HQ/11th CA Regiment (HD),
        Fort H.G. Wright, NY

134th Company (125th Company)
    1908 – Fort Michie, NY
    1912 – Fort HG Wright, NY
    1916 – Temp., Eagle Pass, TX
    1916 – 1st Company, Fort HG Wright, NY
    1917 – A/7th Provisional Artillery Regiment, CAC
    1918 – A/52nd Artillery Regiment, CAC
    1922 – 134th Company, CAC (additional designation)
    1924 – A/52nd CA Regiment (Ry), inactive

135th Company (54th Company)
    1908 – Fort Totten, NY, Mine company
    1916 – 2nd Company, Fort Totten, NY
    1917 – 3rd Company, CD Eastern New York
    1921 – 1st Company, CD Long Island Sound
    1922 – 135th Company, CAC
    1924 – A/11th CA Regiment (HD), inactive

136th Company (95th Company)
    1908 – Fort Hancock, NJ, Mine company
    1916 – 4th Company, Fort Hancock, NJ
    1917 – 4th Company, CD Sandy Hook
    1922 – 136th Company, CAC
    1924 – D/9th CA Regiment (HD), inactive

137th Company (86th Company)
    1908 – Fort Hancock, NJ, Mine company
    1916 – 5th Company, Fort Hancock, NJ
    1917 – 5th Company, CD Sandy Hook
    1922 – 137th Company, CAC
    1924 – E/9th CA Regiment (HD), inactive

138th Company (56th Company)
    1908 – Fort Mott, NJ
    1910 – Mine company
    1911 – Fort Mills, PI, Mine company
    1916 – 1st Company, Fort Mills, PI
    1917 – 1st Company, CD Manila and Subic Bays
    1922 – 138th Company, CAC (additional designation)
    1924 – C/43rd CA Regiment (Ry), inactive

139th Company (18th Company)
    1908 – Fort DuPont, DE, Mine company
    1916 – 2nd Company, Fort DuPont, DE

1917 – 2nd Company, CD Delaware
1922 – 139th Company, CAC
1924 – B/12th CA Regiment (HD), Fort Monroe, VA

140th Company (21st Company)
1908 – Fort Howard, MD, Mine company
1916 – 1st Company, Fort Howard, MD
1917 – 1st Company, CD Baltimore
1920 – 9th Company, CD Chesapeake
1921 – HQ/1st AA Battalion, CAC
1922 – HQ/61st AA Battalion, CAC;
      140th Company, CAC (additional designation)
1924 – HQ/61st CA Regiment (AA), Fort Monroe, VA

141st Company (31st Company)
1908 – Fort McHenry, MD
1913 – Fort Strong, MA, Mine company
1916 – Temp., Fort Sam Houston, TX
1922 – 9th Company, CD Long Island Sound;
      141st Company, CAC
1924 – B/11th CA Regiment (HD), inactive

142nd Company (104th Company)
1908 – Fort McHenry, MD, Mine company
1912 – Fort Mills, PI
1916 – 3rd Company, Fort Mills, PI
1917 – 3rd Company, CD Manila & Subic Bay
1922 – 142nd Company, CAC
1924 – E/42nd CA Regiment (Ry), inactive

143rd Company (17th Company)
1908 – Fort Washington, MD
1913 – Fort Kamehameha, TH
1916 – 5th Company, Fort Kamehameha, TH
1917 – 5th Company, CD Oahu
1921 – 5th Company, CD Pearl Harbor
1922 – 143rd Company, CAC
1924 - C/15th CA Regiment (HD),
      Fort Kamehameha, TH

144th Company (16th Company)
1908 – Fort Moultrie, SC, Mine company
1915 – Fort Grant, CZ
1916 – 8th Company, Fort Grant, CZ
1917 – 8th Company, CD Balboa
1922 – 144th Company, CAC
1924 – HQ/65th CA Regiment (AA),
      Fort Amador, CZ

145th Company (36th Company)
1908 – Fort Moultrie, SC
1916 – Temp., Del Rio, TX
1916 – 3rd Company, Fort Moultrie, SC
1917 – 3rd Company, CD Charleston
1922 – 145th Company, CAC
1924 – C/13th CA Regiment (HD), inactive

146th Company (70th Company)
1908 – SF Presidio, CA, Mine company
1909 – Fort Wint, PI
1912 – Fort HG Wright, NY
1916 – Temp., Eagle Pass, TX;
      5th Company, Fort HG Wright, NY;
      3rd Company, CD Long Island Sound
1917 – C/56th Artillery Regiment, CAC
1919 – Camp Jackson, SC
1921 – Disbanded at Camp Jackson, SC
1922 – Reconstituted and consolidated
      with 8th Company, CD Long Island Sound,
      as 146th Company, CAC
1924 – K/11th CA Regiment (HD),
      Fort H.G. Wright, NY

147th Company (60th Company)
1908 – SF Presidio, CA, Mine company
1913 – Fort Winfield Scott, CA
1914 – Temp., San Ysidro, CA
1916 – 1st Company, Fort Winfield Scott, CA
1917 - 1st Company, CD San Francisco
1918 – HQ/18th Artillery Regiment, CAC;
      1st Company, CD San Francisco
1922 – 147th Company, CAC
1924 – F/10th CA Regiment (HD), inactive

148th Company (61st Company)
1908 – Fort Baker, CA
1916 – 1st Company, Fort Baker, CA
1917 – 14th Company, CD San Francisco;
      D/1st AA Battalion, CAC
1918 – 25th AA Company (Separate), CAC
1919 – 14th Company, CD San Francisco
1922 – 148th Company, CAC
1924 – C/11th CA Regiment (HD), inactive

149th Company (63rd Company)
1908 – Fort Casey, WA
1916 – 3rd Company, Fort Stevens, OR
1917 – 3rd Company, CD Columbia
1922 – 149th Company, CAC
1924 – E/14th CA Regiment (HD), inactive

150th Company (62nd Company)
1908 – Fort Ward, WA, Mine company
1916 – 1st Company, Fort Ward, WA
1917 – 16th Company, CD Puget Sound
1922 – 150th Company, CAC
1924 – G/14th CA Regiment (HD), Fort Worden, WA

151st Company (9th Company)
1908 – Fort Revere, MA
1910 – Fort Andrews, MA
1916 – 2nd Company, Fort Andrews, MA
1917 – I/6th Provisional Artillery Regiment, CAC

1918 – I/51st Artillery Regiment, CAC;
   C/44th Artillery, CAC
1922 – 151st Company, CAC
1924 - C/44th CA Regiment (TD), inactive

152nd Company (7th Company)
1908 – Fort Banks, MA
1916 – 2nd Company Fort Banks, MA
1917 – 13th Company, CD Boston;
   HQ/55 Artillery, CAC
1921 – HQ/55 Artillery Regiment (TD), CAC
1922 – 152nd Company, CAC (additional designation)
1924 – HQ/55th CA Regiment (TD), inactive

153rd Company (59th Company)
1908 – Fort Andrews, MA
1916 – Temp., Plattsburg Barracks, NY
1916 – 3rd Company Fort Andrews, MA
1917 – L/6th Provisional Artillery Regiment
1918 – L/51st Artillery Regiment, CAC;
   C/51st Artillery Regiment, CAC
1922 – 153rd Company, CAC (additional designation)
1924 - C/51st CA Regiment (TD), inactive

154th Company (37th Company)
1908 – Fort McKinley, ME, Mine company
1916 –4th Company, Fort McKinley, ME;
   Temp Fort Sam Houston, TX
1917 – 14th Company, CD Portland
1919 – Absorbed by 13th Company, CD Portland
1922 – Reconstituted and consolidated
   with 6th Company, CD Portland,
   as 154th Company, CAC
1924 - A/8th CA Regiment (HD), inactive

155th Company (90th Company)
1908 – Fort Williams, ME, Mine company
1916 – 1st Company Fort Williams, ME
1917 – 3rd Company, CD Portland
1922 – 155th Company, CAC
1924 – B/8th CA Regiment (HD), inactive

156th Company (120th Company)
1908 – Fort Constitution, NH, Mine company
1916 – 1st Company, Fort Constitution, NH
1917 – 3rd Company, CD Portsmouth
1919 – 3rd Company, CD Portsmouth
   absorbed by 2nd Company,
   CD Portsmouth
1922 – 1st and 2nd Companies consolidated
   and redesignated 156th Company, CAC
1924 – E/8th CA Regiment (HD),
   Fort Constitution, NH

157th Company (58th Company)
1908 – Fort Wadsworth, NY, Mine company
1912 – Fort Terry, NY

1916 – 3rd Company, Fort Terry, NY
1917 – 15th Company, CD Long Island Sound
1918 – 9th Company, CD Long Island Sound
1921 – 7th Company, CD Long Island Sound
1922 – 157th Company, CAC
1924 – D/11th CA Regiment (HD), inactive

158th Company (65th Company)
1908 – SF Presidio, CA
1913 – Fort Winfield Scott, CA
1917 - HQ & Supply/1st AA Battalion, CAC
1919 – Disbanded
1922 – Reconstituted and consolidated
   with 4th Company, CD Los Angeles,
   as 158th Company, CAC
1924 – C/12th CA Regiment (HD), Fort Monroe, VA

159th Company (66th Company)
1908 – Fort Barry, CA
1909 – San Francisco Presidio, CA
1910 – Fort Ruger, TH
1916 – 3rd Company, Fort Ruger, TH
1917 – 13th Company, CD Oahu
1921 – 6th Company, CD Honolulu
1922 – 159th Company, CAC
1924 – E/16th CA Regiment (HD), inactive

160th Company (10th Company)
1908 – SF Presidio, CA
1909 – Fort Stevens, OR
1915 – Fort Rosecrans, CA
1916 – Fort Stevens, OR;
   2nd Company, Fort Stevens, OR
1917 – 2nd Company, CD Columbia
1922 – 160th Company, CAC
1924 – K/14th CA Regiment (HD), inactive

161st Company (68th Company)
1908 – Fort Barry, CA
1916 – Temp., Presidio Monterrey, CA;
   2nd Company, Fort Barry, CA
1917 - 16th Company, CD San Francisco
1922 – 161st Company, CAC
1924 – E/11th CA Regiment (HD), inactive

162nd Company (20th Company)
1908 – Key West Barracks, FL, Mine company
1910 – Fort Dade, FL, mine company
1916 – 1st Company, Fort Dade, FL
1917 – 1st Company, CD Tampa Bay
1922 – 162nd Company, CAC
1924 – A/13th CA Regiment (HD), Fort Barrancas, FL

163rd Company (15th Company)
1908 – Fort Barrancas, FL, Mine company
1910 – Fort Pickens, FL, Mine company
1916 – 1st Company, Fort Pickens, FL

1917 – 1st Company, CD Pensacola
1922 – 163rd Company, CAC
1924 – B/13th CA Regiment (HD), Fort Pickens, FL

164th Company (22nd Company)
1908 – Jackson Barracks, LA, Mine company
1914 – Temp., Brownsville, TX
1916 – Temp., Del Rio, TX;
          1st Company, Jackson Barracks, LA
1917 – 1st Company, CD New Orleans
1922 – 164th Company, CAC
1922 – Fort Howard, MD
1924 – D/12th CA Regiment (HD), Fort Howard, MD

165th Company
1907 – Fort Monroe, VA, Mine company
1908 – Fort Totten, NY, Mine company
1916 – 4th Company, Fort Totten, NY
1917 – 1st Company, CD Eastern New York
1919 – Absorbed by 5th Company,
          CD Eastern New York
1921 – Disbanded at Fort Totten, NY;
          reconstituted and consolidated
          with Gun/2nd AA Battalion, CAC,
          and redesignated B/62nd Artillery
          Battalion (AA), CAC, with additional
          designation as 165th Company, CAC
1924 – G/62nd CA Regiment (AA), inactive

166th Company
1907 – Fort Monroe, VA
1916 – Temp., Fort Oglethorpe, GA
1916 – 7th Company, Fort Monroe, VA
1918 – 7th Company, CD Chesapeake Bay
1922 – 166th Company, CAC
1924 – D/12th CA Regiment (HD), inactive

167th Company
1907 – Fort Monroe, VA
1909 – Fort Totten, NY
1916 – Temp., Columbus, NM;
          5th Company, Fort Totten, NY
1917 – 5th Company, CD Eastern NY
1919 – Absorbed 6th Company,
          Eastern New York, and
          redesignated 2nd Company,
          CD Eastern New York
1921 – Disbanded at Fort Totten, NY;
          AAMG/2nd Battalion (AA) CAC
1922 – Reconstituted and consolidated
          with AAMG/2nd Battalion (AA),
          CAC, redesignated C/62nd Artillery
          Battalion (AA), CAC, with additional
          designation of 167th Company, CAC
1924 – H/62nd CA Regiment (AA), inactive

168th Company (69th Company)
1907 – Fort Monroe, VA
1916 – Temp., Fort Oglethorpe, GA;
          8th Company, Fort Monroe, VA
1917 – 8th Company, CD Chesapeake Bay;
          HQ/60th Artillery Regiment, CAC
1921 – AAMG /1st AA Battalion, CAC
1922 – C/61st AA Battalion, CAC;
          168th Company, CAC (additional designation)
1924 – E/61st CA Regiment (AA), Fort Monroe, VA

169th Company
1907 - Fort Monroe, VA
1908 – Mine company
1916 – 9th Company, Fort Monroe, VA
1917 – 9th Company, CD Chesapeake Bay
1919 – Absorbed by 3rd Company, CD Chesapeake Bay
1922 – Reconstituted and consolidated with
          11th Company, CD Chesapeake Bay,
          as 169th Company, CAC
1924 – G/12th CA Regiment (HD), inactive

170th Company (19th Company)
1908 – Fort Morgan, AL, Mine company
1914 – Temp., Brownsville, TX
1915 – Fort Moultrie, SC
1916 – 1st Company, Fort Moultrie; SC
1917 – 1st Company, CD Charleston
1922 – 170th Company, CAC
1924 – D/13th CA Regiment (HD), Fort Moultrie, SC

171st Company
1918 - 7th Company, CD Portland, Fort Preble, ME
1922 – 171st Company, CAC
1924 - C/8th CA Regiment (HD), inactive

172nd Company (59th Company, CAC)
1917 – HQ Company, NE Department
1917 – 2nd Company, CD Boston
1922 – 172nd Company, CAC
1924 - HHB/ 9th CA Regiment (HD), Fort Banks, MA

173rd Company
1917 – Artillery Engineer Company, Fort Adams, RI
1918 – 2nd Company, CD Narragansett Bay
1922 – 173rd Company, CAC
1924 – HQ/10th CA (HD), Fort Adams, RI

174th Company
1922 - 7th Company, CD Narragansett Bay
1924 - G/10th CA Regiment (HD), inactive

175th Company
1922 - 10th Company, CD Long Island Sound
1924 - F/11 CA Regiment (HD), inactive

176th Company
1917 - 7th Company, Fort Hamilton, NY
1917 - 4th Company (I) CD Southern New York
1918 – A/59th Artillery Regiment, CAC
1922 – 176 Company, CAC (additional designation)
1924 – A/59th CA Regiment (TD), Fort Mills, PI

177th Company
1917 - 4th Company, Fort Wadsworth, NY
1917- 7th Company, CD Southern New York
1919 – Fort Hamilton, NY
1922 – 177th Company, CAC
1924 - F/9th CA Regiment (HD), inactive

178th Company
1917 – 7th Company, Fort Hancock, HY
1917 – 7th Company, CD Sandy Hook
1922 – 178th Company, CAC
1924 – G/9th CA Regiment (HD), inactive

179th Company
1917 – Det of CD Tampa
1917 – El Morro, PR
1918 – 1st Company, San Juan
1919 – 3rd Company, CD Delaware
1922 – 179th Company, CAC
1924 – H/13th CA Regiment (HD), inactive

180th Company
1917 - 5th Company, Fort Moultrie, SC
1917 – 2nd Company, CD Charleston
1922 – 180th Company, CAC
1924 – K/13th CA Regiment (HD), inactive

181st Company
1918 - 2nd Company, (II) CD Savannah,
    Fort Screven, GA
1922 – 181st Company, CAC
1924 – F/13th CA Regiment (HD), inactive

182nd Company
1917 - 2nd Company, Key West Barracks, FL
1917 – 2nd Company, CD Key West
1922 – 182nd Company, CAC
1924 – E/13th CA Regiment (HD),
    Key West Barracks, FL

183rd Company
1917 - 3rd Company, Fort Crockett, TX
1917 – 3rd Company, CD Galveston
1922 – 183rd Company, CAC
1924 – G/13th CA Regiment (HD),
    Fort Crockett, TX

184th Company
1917 – 6th Company, Fort Kamehameha, TH
1917 – 6th Company, CD Oahu
1921 – 6th Company, CD Pearl Harbor
1922 – 184th Company, CAC
1924 – D/15th CA Regiment (HD), inactive

185th Company
1917 – HQ Company, Fort Kamehameha, TH
1917 – 7th Company, CD Oahu, Fort Armstrong, TH
1917 - Fort Kamehameha, TH
1918 – Fort Armstrong, TH
1918 – Fort Kamehameha, TH
1919 – Fort Armstrong, TH
1919 – Fort Kamehameha, TH
1921 – 7th Company, CD Pearl Harbor
1022 – 185th Company, CAC
1924 – HQ/15th CA Regiment (HD),
    Fort Kamehameha, TH

186th Company
1917 – 4th Company, Fort Ruger, TH
1917 – 14th Company, CD Oahu
1921 – 7th Company, CD Honolulu
1922 – 186th Company, CAC
1924 – F/16th CA Regiment (HD), inactive

187th Company
1917 – 17th Company (II), Fort Mills, PI
1917 – 17th Company, CD Manila and Subic Bays
1917 – Fort Hughes, PI
1918 – Fort Mills, PI
1922 - 187th Company, CAC
1924 – F/43rd CA Regiment (Ry), inactive

188th Company
1917 - 18th Company, Fort Mills, PI
1917 – 18th Company, CD Manila and Subic Bays
1917 – Fort Hughes, PI
1918 – Fort Mills, PI
1918 – Fort Drum, PI
1918 – Fort Mills, PI
1919 – Fort Drum, PI
1922 – 188th Company, CAC
1924 – I/13th CA Regiment (HD), inactive

189th Company
1917 - 19th Company, Fort Mills, PI
1917 – 19th Company, CD Manila and Subic Bays
1917 – Fort Frank, PI
1917 – Fort Mills, PI
1918 – Fort Frank, PI
1919 – Fort Drum, PI
1922 – 189th Company, CAC
1924 – E/53rd CA Regiment (Ry), inactive

190th Company
1917 - 20th Company, Fort Mills, PI
1917 – 20th Company, CD Manila and Subic Bays
1917 – Fort Frank, PI

1917 – Fort Mills, PI
1922 – 190th Company, CAC
1924 – F/53rd CA Regiment (Ry), inactive

191st Company
1917- 21st Company, Fort Mills, PI
1917 - 21st Company, CD Manila and Subic Bays
1917 - Inactivated
1922 – 191st Company, CAC
1924 – D/44th CA Regiment (TD), inactive

192nd Company
1917 - 4th Company, Fort Sherman, CZ
1917 – 4th Company, CD Cristobal
1922 – 192nd Company, CAC
1924 – H/65th CA Regiment (AA), inactive

193rd Company
1917 – 1st Company, Fort DeLesseps, CZ
1918 – 6th Company, CD Cristobal
1922 – 193rd Company, CAC
1924 – G/65th CA Regiment (AA), inactive

194th Company
1916 – 7th Company, Fort Grant, CZ
1917 - 7th Company, CD Balboa
1922 – 194th Company, CAC
1924 – F/65th CA Regiment (AA), inactive

195th Company
1916 – 2nd Company, Fort Randolph, CZ
1917 – 8th Company, CD Cristobal
1922 – 195th Company, CAC
1924 – C/65th CA Regiment (AA), Fort Randolph

196th Company
1917 - 5th Company, Fort DuPont, DE
1917 – 5th Company, CD Delaware
1917 – 1st Trench Mortar Battery
1922 – 196th Company, CAC (additional designation)
1924 – D/8th CA Regiment (HD), inactive

197th Company
1917 – Sound Ranging Battery No 1
1922 – 197th Company, CAC (additional designation)
1924 – Sound Ranging Battery No 1, Fort Eustis, VA

198th Company
1918 – HQ Company, 39th Arty Bde
1922 - HHB, 39th Artillery Brigade, CAC
1924 – HHB, 39th CA (TD) Brigade, inactive

199th Company
1921 – A/ Hawaiian Railway Battalion
1922 – A/41st Arty Battalion (Ry), CAC
1924 - A/41st CA Battalion (Ry),
          Fort Kamehameha, TH

200th Company
1921 – B/ Hawaiian Railway Battalion
1922 – B/41st Arty Battalion (Ry), CAC
1924 - B/41st CA Battalion (Ry),
          Fort Kamehameha, TH

201st Company
1918 – HQ/42nd Artillery Regiment (Ry)
1922 - HHB/42nd Artillery Regiment, CAC (Ry)
1924 – HQ/42nd CA Regiment (Ry), inactive

202nd Company
1918 – Supply/42nd Artillery Regiment (Ry)
1921 – Service/42nd Artillery Regiment (Ry)
1922 – 202nd Company, CAC (additional designation)
1924 – Service/42nd CA Regiment (Ry), inactive

203rd Company
1917 – 4th Company, Fort Hamilton, NY
1917 – M/7th Provisional Artillery Regiment, CAC
1918 – D/42nd Artillery Regiment CAC (Ry)
1922 - 203rd Company, CAC (additional designation)
1924 – D/42nd CA Regiment (Ry), inactive

204th Company
1918 – HQ/43rd Artillery Regiment CAC
1922 – 204th Company, CAC (additional designation)
1924 – HQ/43rd CA Regiment (Ry), inactive

205th Company
1918 – Supply/43rd Artillery Regiment (Ry)
1921 – Service/43rd Artillery Regiment (Ry)
1922 – 202nd Company, CAC (additional designation)
1924 – Service/43rd Regiment (Ry), inactive

206th Company
1917 – Detachment, 3rd Company, CD Sandy Hook
1918 – C/57th Artillery Regiment, CAC
1918 – A/43rd Artillery Regiment ,CAC
1922 – 206th Company, CAC (additional designation)
1924 – A/43rd CA Regiment (Ry), inactive

207th Company
1916 - 21st Company, NYNG
1917 – 21st Company, CD Sandy Hook
1918 – D/57th Artillery Regiment, CAC
1918 – B/43rd Artillery Regiment, CAC
1922 – 207th Company, CAC (additional designation)
1924 – B/43rd CA Regiment (Ry), inactive

208th Company
1917 – 5th Company, Fort McKinley, ME
1917 – 15th Company, CD Portland
1917 – D/6th Provisional Artillery Regiment
1918 – D/51st Artillery Regiment, CAC
1918 – D/43rd Artillery Regiment (Ry), CAC
1922 – 208th Company, CAC (additional designation)
1924 – D/43rd CA Regiment (Ry), inactive

209th Company
  1918 – HQ Co/Provisional Howitzer Regiment, CAC
  1918 – HQ Co/81st Artillery Regiment (TD), CAC
  1918 – HQ Co/44th Artillery Regiment (TD), CAC
  1921 – HQ/44th Artillery Regiment (TD), CAC
  1922 – 209th Company, CAC (additional designation)
  1924 – HQ/44th CA Regiment (TD), inactive

210th Company
  1918 – Supply/Provisional HowitzerRegiment
  1918 – Supply/ 81st Artillery Regiment (TD), CAC
  1918 – Supply/44th Artillery Regiment (TD), CAC
  1921 – Service/ 44th Artillery Regiment (TD), CAD
  1922 – 210th Company, CAC (additional designation)
  1924 – Service/44th CA Regiment (TD), inactive

211th Company
  1921 – HQ&CT/1st Battalion, 44th Art. (TD), CAC
  1922 – 211th Company, CAC (additional designation)
  1924 – HQ&CT/1st Battalion,
      44th CA Regiment (TD), inactive

212th Company
  1921 – HQ&CT/2nd Battalion,
      44th Artillery (TD), CAC
  1922 – 212th Company, CAC (additional designation)
  1924 – HQ&CT/2nd Battalion,
      44th CA Regiment (TD), inactive

213th Company
  1921 – HQ&CT/3rd Battalion,
      44th Artillery (TD), CAC
  1922 – 213th Company, CAC (additional designation)
  1924 - HQ&CT/3rd Battalion,
      44th CA Regiment (TD), inactive

214th Company
  1917 – 6th Company, Fort Totten, NY
  1917 – H/7th Provisional Artillery Regiment
  1918 – H/52nd Artillery Regiment (Ry), CAC
  1918 – B/44th Artillery Regiment (TD), CAC
  1922 - 214th Company, CAC (additional designation)
  1924 – B/44th CA Regiment (TD), inactive

215th Company
  1917 – 2nd Company, Fort Schuyler, NY
  1917 – F/7th Provisional Artillery Regiment
  1918 – F/51st Artillery Regiment (Ry), CAC
  1918 - F/44th Artillery Regiment (TD), CAC
  1922 - 215th Company, CAC (additional designation)
  1924 – F/44th CA Regiment (TD), inactive

216th Company
  1917 – 2nd Company, Fort Mott, NJ
  1917 – 8th Company, CD Delaware
  1917 – HQ & Supply/6th Provisional
      Artillery Regiment, CAC
  1918 – HQ & Supply Company,
      51st Artillery Regiment (TD), CAC
  1921 – HQ/51st Artillery Regiment (TD), CAC
  1922 – 216th Company, CAC (additional designation)
  1924 – HQ/51st CA Regiment (TD), Fort Eustis, VA

217th Company
  1918 – Supply/51st Artillery Regiment, CAC
  1921 – Service/51st Artillery Regiment, CAC
  1922 - 217th Company, CAC(additional designation)
  1924 – Service/51st CA Regiment (TD),
      Fort Eustis, VA

218th Company
  1921 – HQ&CT/1st Battalion,
      51st Artillery Regiment (TD), CAC
  1922 – 218th Company, CAC (additional designation)
  1924 - HQ&CT/1st Battalion,
      51st CA Regiment (TD), Fort Eustis, VA

219th Company
  1921 – HQ&CT/2nd Battalion,
      51st Artillery Regiment(TD), CAC
  1922 – 219th Company, CAC (additional designation)
  1924 - HQ&CT/2nd Battalion,
      51st CA Regiment (TD), inactive

220th Company
  1921 – HQ&CT/3rd Battalion,
      51st Artillery Regiment (TD) CAC
  1922 – 218th Company, CAC (additional designation)
  1924 - HQ&CT/3rd Battalion,
      51st CA Regiment (TD), inactive

221st Company
  1916 – 4th Company, Fort Andrews, MA
  1917 – 5th Company, CD Boston
  1917 – M/6th Provisional Artillery Regiment, CAC
  1918 – M/51st Artillery Regiment (Ry), CAC
  1918 – D/51st Artillery Regiment (TD), CAC
  1922 – 221st Company, CAC (additional designation)
  1924 – D/51st CA Regiment (TD), inactive

222nd Company
  1917 – 3rd Company, Fort Tilden, NY
  1917 – G/8th Provisional Artillery Regiment, CAC
  1918 – G/53rd Artillery Regiment (Ry), CAC
  1918 – F/51st Artillery Regiment (TD), CAC
  1922 – 222nd Company, CAC (additional designation)
  1924 – F/51st CA Regiment (TD), inactive

223rd Company
  1917 – HQ & Supply/7th Provisional
      Artillery Regiment, CAC
  1918 – HQ/52nd Artillery Regiment (Ry), CAC
  1922 – 223rd Company, CAC (additional designation)
  1924 – HQ/52nd CA Regiment (Ry), Fort Eustis, VA

224th Company
1921 – Service/52nd Artillery Regiment, CAC
1922 – 224th Company, CAC (additional designation)
1924 – Service/52nd CA Regiment (Ry),
Fort Eustis, VA

225th Company
1917 – 6th Company, Fort Terry, NY
1917 – B/7th Provisional Artillery Regiment
1918 – B/52nd Artillery Regiment (Ry), CAC
1922 – 225th Company, CAC (additional designation)
1924 – B/52nd CA Regiment (Ry), inactive

226th Company
1917 – 6th Company, Fort HG Wright, NY
1917 – D/7th Provisional Artillery Regiment
1918 – D/52nd Artillery Regiment (Ry), CAC
1922 – 226th Company, CAC (additional designation)
1924 – D/52nd CA Regiment (Ry), Fort Eustis, VA

227th Company
1916 – 1st Company, Fort Washington, MD
1917 – I/8th Provisional Artillery Regiment
1918 – I/53rd Artillery Regiment (Ry), CAC
1918 – E/52nd Artillery Regiment (Ry), CAC
1922 – 227th Company, CAC (additional designation)
1924 – E/52nd CA Regiment (Ry), Fort Eustis, VA

228th Company
1917 - 3rd Company, Fort Washington, MD
1917 – K/8th Provisional Artillery Regiment
1918 – K/53rd Artillery Regiment, (Ry), CAC
1918 – F/52nd Artillery Regiment (Ry), CAC
1922 – 228th Company, CAC (additional designation)
1924 – F/52nd CA Regiment (Ry), Fort Eustis, VA

229th Company
1917 – HQ & Supply Co/8th Provisional
Artillery Regiment, CAC
1918 – HQ/53rd Artillery Regiment (Ry), CAC
1922 – 229th Company, CAC (additional designation)
1924 - HQ/53rd CA Regiment (Ry), inactive

230th Company
1921 – Service/53rd Artillery Regiment (Ry), CAC
1922 – 230th Company, CAC (additional designation)
1924 – Service/53rd CA Regiment (Ry), inactive

231st Company
1917 – Detachments, Fort Howard, MD
1917 – A/8th Provisional Artillery Regiment, CAC
1918 – A/53rd Artillery Regiment, CAC
1922 – 231st Company, CAC (additional designation)
1924 – A/53rd CA Regiment (Ry), inactive

232nd Company
1917 – 11th Company, Fort Monroe, VA

1917 – B/8th Provisional Artillery Regiment, CAC
1918 – B/53rd Artillery Regiment (Ry), CAC
1922 – 232nd Company, CAC (additional designation)
1924 – A/53rd CA Regiment (Ry), inactive

233rd Company
1917 – 12th Company, Fort Monroe, VA
1917 – C/8th Provisional Artillery Regiment, CAC
1918 – C/53rd Artillery Regiment (Ry), CAC
1922 – 223rd Company, CAC (additional designation)
1924 – C/53rd CA Regiment (Ry), inactive

234th Company
1917 – 10th Company, Fort Monroe, VA
1917 – D/8th Provisional Artillery Regiment, CAC
1918 – D/53rd Artillery Regiment (Ry), CAC
1022 – 234th Company, CAC (additional designation)
1924 – D/53rd CA Regiment (Ry), inactive

235th Company
1916 – 5th Company, Mass NG CAC
1917 – 20th Company, CD Boston
1918 – Supply/55th Artillery Regiment (TD), CAC
1921 – Service/55th Artillery Regiment (TD), CAC
1922 – 235th Company, CAC (additional designation)
1924 – Service/55th CA Regiment (TD), inactive

236th Company
1921 – HQ Det& CT/1st Battalion/55th
Artillery Regiment, CAC
1922 – 236th Company, CAC (additional designation)
1924 - HQ Det& CT/1st Battalion,
55th CA Regiment (TD), Fort Shafter, TH

237th Company
1921 – HQ Det & CT/2nd Battalion,
55th Artillery Regiment, CAC
1922 – 237th Company, CAC (additional designation)
1924 - HQ Det & CT/2nd Battalion,
55th CA Regiment (TD), Fort Ruger, TH

238th Company
1921 – HQ Det & CT/3rd Battalion,
55th Artillery Regiment, CAC
1922 – 238th Company, CAC (additional designation)
1924 - HQ Det & CT/3rd Battalion, 55th
CA Regiment (TD), Fort Kamehameha, TH

239th Company
1919 – B/55th Artillery Regiment (TD), CAC
1922 – 239th Company, CAC (additional designation)
1924 – B/55th CA Regiment (TD), Fort Shafter, TH

240th Company
1919 – D/55th Artillery Regiment (TD), CAC
1922 – 240th Company, CAC (additional designation)
1924 – D/55th CA Regiment (TD), Fort Ruger, TH

241st Company
1919 – E/55th Artillery Regiment (TD), CAC
1922 – 241st Company, CAC (additional designation)
1924 – E/55th CA Regiment (TD), Fort Ruger, TH

242nd Company
1919 – F/55th Artillery Regiment(TD), CAC
1922 – 242nd Company, CAC (addl designation)
1924 – F/55th CA Regiment (TD), Fort Ruger, TH

243rd Company
1922 – G/55th Artillery Regiment (TD), CAC
1922- 243rd Company, CAC (additional designation)
1924 – G/55th CA Regiment (TD),
        Fort Kamehameha, TH

244th Company
1922 – H/55th Artillery Regiment (TD), CAC
1922- 244th Company, CAC (additional designation)
1924 – H/55th CA Regiment (TD),
        Fort Kamehameha, TH

245th Company
1922 – I/55th Artillery Regiment (TD), CAC
1922- 245th Company, CAC (additional designation)
1924 – I/55th CA Regiment (TD),
        Fort Kamehameha, TH

246th Company
1918 – Detachments/1st Company, CD S. New York
1918 - 43rd Company, CD Southern New York
1918 - 19th Company, CD Southern New York
1919 – 1st Company, (III) Southern New York
1919 – HQ/ 59th Artillery Regiment (TD), CAC
1921 – Fort Lewis, WA; Fort Mills, PI
1922 – 246th Company, CAC
        (additional designation), Ft. Hamilton, NY
1924 – F/8th CA Regiment (HD), inactive

247th Company
1921 – Service/59th Artillery Regiment (TD), CAC
1922 – 247th Company, CAC
1924 – Service/59th CA Regiment (TD), inactive

248th Company
1921 – HQ&CT/1st Battalion,
        59th Artillery Regiment, CAC
1921 - Fort Lewis, WA; Fort Mills, PI
1922 - 248th Company, CAC (additional designation)
1924 – HQ&CT/1st Battalion,
        59th CA Regiment (TD), inactive

249th Company
1921 – HQ&CT/2nd Battalion,
        59th Artillery Regiment, CAC
1921 - Fort Lewis, WA; Fort Mills, PI
1922 - 249th Company, CAC (additional designation)

1924 – HQ&CT/2nd Battalion,
        59th CA Regiment (TD), inactive

250th Company
1921 – HQ&CT/3rd Battalion,
        59th Artillery Regiment, CAC
1921 - Fort Lewis, WA; Fort Mills, PI
1922 - 250th Company, CAC (additional designation)
1924 – HQ&CT/3rd Battalion,
        59th CA Regiment (TD), inactive

251st Company
1919 – A/59th Artillery Regiment (TD), CAC
1919 – Fort Winfield Scott, CA
1920 – Fort Lewis, WA
1921 – Fort Mills, PI
1922 – 4th Company, CD Southern New York
1922 – 251st Company, CAC (additional designation)
1924 – G (SL)/8th CA Regiment (HD), inactive

252nd Company
1919 – B/59th Artillery Regiment (TD), CAC
1922 – 252nd Company, CA (additional designation)
1924 – B/59th CA Regiment (TD), Fort Mills, PI

253rd Company
1919 – C/59th Artillery Regiment (TD), CAC
1919 – Fort Winfield Scott, CA
1920 – Fort Lewis, WA
1921 – Fort Mills, PI
1922 – 253rd Company, CAC (additional designation),
        Fort Hamilton, NY
1924 – C/59th CA Regiment (TD), Fort Mills, PI

254th Company
1919 – D/59th Artillery Regiment (TD), CAC
1919 – Fort Winfield Scott, CA
1920 – Fort Lewis, WA
1921 – Fort Mills, PI
1922 - 254th Company, CAC
1924 – D/59th CA Regiment (TD), Fort Mills, PI

255th Company
1919 – E/59th Artillery Regiment (TD), CAC
1919 – Fort Winfield Scott, CA
1920 – Fort Lewis, WA
1921 – Fort Mills, PI
1922 – 255th Company, CAC (additional designation)
1924 – E/59th CA Regiment (TD), Fort Mills, PI

256th Company
1919 – F/59th Artillery Regiment (TD), CAC
1919 – Fort Winfield Scott, CA
1920 – Fort Lewis, WA
1921 – Fort Mills, PI
1922 – 256th Company, CAC (additional designation)
1924 – F/59th CA Regiment (TD), Fort Mills, PI

257th Company
  1917 - 13th Company, Fort Monroe, VA
  1917 – 5th Company, CD Chesapeake Bay
  1921 – B/1st AA Battalion, CAC
  1922 – B/61st Artillery Battalion (AA), CAC
  1922 – 257th Company, CAC (additional designation)
  1924 – B/61st CA Regiment (AA), Fort Monroe, VA

258th Company
  1921 – HQ Det& CT/2nd AA Battalion, CAC
  1922 – HQ/62nd Artillery Battalion (AA), CAC
  1922 – 258th Company, CAC (additional designation)
  1924 – HQ/62nd CA Regiment (AA),
      Fort Totten, NY

259th Company
  1921 – HQ/3rd AA Battalion, CAC
  1922 – HQ/63rd Artillery Battalion (AA), CAC
  1922 – 259th Company, CAC (additional designation)
  1924 – HQ/63rd CA Regiment (AA), inactive

260th Company
  1921 – HQ/3rd AA Battalion, CAC
  1922 – HQ/63rd Artillery Battalion (AA), CAC
  1922 – 259th Company, CAC (additional designation)
  1924 – HQ/63rd CA Regiment (AA),
      Fort Winfield Scott, CA

261st Company
  1921 – B/3rd AA Battalion, CAC
  1922 – B/63rd Artillery Battalion (AA), CAC
  1922 – 261st Company, CAC (additional designation)
  1924 – B/63rd CA Regiment (AA),
      Fort Winfield Scott, CA

262nd Company
  1921 – C/3rd AA Battalion, CAC
  1922 – C/63rd Artillery Battalion (AA), CAC
  1922 – 262nd Company, CAC (addl designation)
  1924 – C/63rd CA Regiment (AA), inactive

263rd Company
  1921 – HQ/Hawaiian AA Regiment, CAC
  1922 – HQ /64th Artillery Regiment (AA), CAC
  1922 – 263rd Company, CAC (additional designation)
  1924 – HQ/64th CA Regiment (AA), Fort Shafter, TH

264th Company
  1921 – Service/Hawaiian AA Regiment, CAC
  1922 – Service/64th Artillery Regiment (AA), CAC
  1922 – 264th Company, CAC (additional designation)
  1924 – Service/64th CA Regiment (AA), Fort Shafter

265th Company
  1921 – HQ&CT/1st Battalion, Hawaiian
      AA Regiment, CAC
  1922 – HQ&CT/1st Battalion, 64th Artillery
      Regiment (AA), CAC

  1922 – 265th Company, CAC (additional designation)
  1924 – HQ&CT/1st Battalion, 64th CA
      Regiment (AA), Fort Shafter, TH

266th Company
  1921 – A/Hawaiian AA Regiment, CAC
  1922 – A/64th Artillery Regiment (AA), CAC
  1922 – 266th Company, CAC (additional designation)
  1924 – A/64th CA Regiment (AA), Fort Shafter, TH

267th Company
  1921 – B/Hawaiian AA Regiment, CAC
  1922 – B /64th Artillery Regiment (AA), CAC
  1922 – 267th Company, CAC (additional designation)
  1924 – B/64th CA Regiment (AA), Fort Shafter, TH

268th Company
  1921 – C/Hawaiian AA Regiment, CAC
  1922 – C /64th Artillery Regiment (AA) CAC
  1922 – 268th Company, CAC (additional designation)
  1924 – C/64th CA Regiment (AA), Fort Shafter, TH

269th Company
  1921 – D/Hawaiian AA Regiment, CAC
  1922 – D/64th Artillery Regiment (AA), CAC
  1922 – 269th Company, CAC (additional designation)
  1924 – D/64th CA Regiment (AA), Fort Shafter, TH

270th Company
  1921 – E/Hawaiian AA Regiment, CAC
  1922 – E/64th Artillery Regiment (AA), CAC
  1922 – 270th Company, CAC (additional designation)
  1924 – E/64th CA Regiment (AA), Fort Shafter, TH

271st Company
  1921 – F/Hawaiian AA Regiment, CAC
  1922 – F/64th Artillery Regiment (AA) CAC
  1922 – 271st Company, CAC (additional designation)
  1924 – F/64th CA Regiment (AA), Fort Shafter, TH

272nd Company
  1921 – G/Hawaiian AA Regiment, CAC
  1922 – G/64th Artillery Regiment (AA), CAC
  1922 – 272nd Company, CAC,
      (additional designation)
  1924 – G/64th CA Regiment (AA), Fort Shafter, TH

273rd Company
  1921 – H/Hawaiian AA Regiment, CAC
  1922 – H/64th Artillery Regiment (AA), CAC
  1922 – 273rd Company, CAC (additional designation)
  1924 – H/64th CA Regiment (AA), Fort Shafter, TH

274th Company
  1921 - HQ&CT/Gun Battalion, Hawaiian
      Antiaircraft Regiment, CAC, Fort Ruger, TH

1922 - HHD&CT/2nd Battalion,
   64th Artillery Regiment (AA), CAC,
   Fort Shafter, TH
1924 - HHD&CT/2nd Battalion,
   64th CA Regiment (HD), Fort Shafter, TH

275th Company
1922 - 275th Company, CAC (PS)
1924 – B/92nd CA Regiment (PS), Fort Mills, PI

276th Company
1922 – 276th Company, CAC (PS)
1924 – F/91st CA Regiment (PS), Fort Frank, PI

277th Company
1922 – 277th Company, CAC (PS)
1924 – C/91st CA Regiment (PS), Fort Mills, PI

278th Company
1922 - 278th Company, CAC (PS)
1924 – HQ/92nd CA Regiment (PS), Fort Mills, PI

279th Company
1922 – 279th Company, CAC (PS)
1924 – D/91st CA Regiment (PS), Fort Hughes, PI

280th Company
1922 - 280th Company, CAC (PS)
1924 – A/92nd CA Regiment (PS), Fort Mills, PI

281st Company
1922 – 281st Company, CAC (PS)
1924 – HQ/91st CA Regiment (PS), Fort Mills, PI

282nd Company
1922 – 282nd Company, CAC (PS)
1924 – A/91st CA Regiment (PS), Fort Mills, PI

283rd Company
1922 – 283rd Company, CAC (PS)
1924 – B/92nd CA Regiment (PS), Fort Mills, PI

284th Company
1922 – 284th Company, CAC (PS)
1924 – B/91st CA Regiment (PS), Fort Mills, PI

285th Company
1922 – 285th Company, CAC (PS)
1924 – G/91st CA Regiment (PS), Fort Wint, PI

286th Company
1922 – 286th Company, CAC (PS)
1924 – E/91st CA Regiment (PS), Fort Mills, PI

287th Company
1922 - 287th Company, CAC (PS)
1922 – G/59th Artillery Regiment (TD), CAC
1924 – C/92nd CA Regiment (PS), Fort Mills, PI

288th Company
1922 - 288th Company, CAC (PS)
1922 – H/59th Artillery Regiment (TD) CAC
1924 – D/92nd CA Regiment (PS), Fort Mills, PI

289th Company
1922 - 289th Company, CAC (PS)
1924 – F/92nd CA Regiment (PS), Fort Mills, PI

Sources

American Society of Military Insignia Collectors. "Historic Backgrounds Coast Artillery 1938."

Chief of Coast Artillery to Commanding Officer CD of Tampa, April 18, 1921, "Prior History of Company Units, CAC," Entry 370, File 0004, General Historical File Box 396, Records of United States Coast Artillery Districts and Defenses 1901-1942, RG 392, NARA, College Park, MD.

McKinney, Janice E. *Air Defense Artillery* [Army Lineage Series]. GPO, 1985.

U.S. Army War College, Historical Section. *Order of Battle of the United States Land Forces in the World War (1917-19): Zone of the Interior.* GPO, 1949, Vol 3.

U.S. War Department. *Army List and Directory.* GPO, 1901-1916.

U.S. War Department. General Order No. 21, May 18, 1922.

U.S. War Department. General Order No. 8, February 27, 1924.

U.S. War Department, Historical Branch, War Plans Division. *Outlines of History of Regiments United States Army.* GPO, 1921.

The Coast Artillery in WW I

When war erupted in Europe in 1914, the U.S. Army was ill prepared for a European conflict. The CAC especially was greatly under strength with, only 701 officers and 19,321 EM organized into 170 separate companies, distributed among 24 domestic and seven overseas harbors. These personnel were supposed to man half of the gun and mortar batteries and all of the mine defenses in the continental defenses, and all of the overseas gun and mortar batteries and mine defenses, on a peacetime basis. To actually accomplish this mission however, called for 1,312 officers and 30,309 enlisted personnel. Thus, the CAC was short 10,000 men in November 1914, stretched thin as it was in the process of deploying 49 of its 170 companies to the new coastal fortifications in the Panama Canal Zone, Hawaii, and the Philippine Islands.

The War Department formed the high-ranking Board of Review to consider the defenses of the nation. Among the board's recommendations was a substantial increase in the strength of the CAC, some 11,000 additional enlisted men over the next four years. The first increment was to be about 2,000 men in 1916. The nationís entry into the World War resulted in the full implementation of the CAC augmentation, enabling the creation of 105 new companies in 1916 and 1917. Although authorized 30,009 men, there were still only 28,527 coast artillerymen serving in the nationís coast defenses at the end of October 1917. To this must be added the some 5,000 officers and enlisted men of the Provisional CA Brigade in France.

Recognizing the need for additional troops, and in response to Allied requests for more artillery on the Western Front, an additional 14,500 ìreplacementî enlisted troops were authorized for the CAC, bringing its authorized strength to just over 45,000. During the course of the war, 47,386 National Army draftees were sent to coast artillery posts in the Continental U.S. There, many of them were assigned to the regiments of coast artillery sent to, or intended for, France in 1918. When the war ended in November 1918, there were 35,015 CAC troops in the A.E.F. and another 26,272 in the US awaiting shipment to France. In addition, 6,478 officers and men served the insular possessions and Panama Canal Zone, while 34,308 officers and men served in the continental coast defenses.

The following 18 companies were constituted and organized in 1916. In some cases, these companies were formed by disbanding Regular Army Coast Artillery companies that had been in existence before the 1916 reorganization. Their initial designations and subsequent redesignations 1917-1919 are noted below:

    4th Co, Ft. Andrews; M/6 Prov Rgt; M/51 Rgt; D/51 Rgt
    2nd Co, Ft. Dade; 2nd Co, CD Tampa
    1st Co, Ft. Dupont; 1st Co, CD Delaware
    3rd Co, Ft. Dupont; 3rd Co, CD Delaware
    1st Co, Ft. Hamilton; 1st Co,(I), CD Southern NY
    4th Co,(I), Ft. Hamilton; Demob Jun 1917, Rockaway Beach
    3rd Co, Ft. Howard; 3rd Co, CD Baltimore
    4th Co, Ft. Howard; 4th Co, CD Baltimore
    2nd Co, Jackson Barracks; 2nd Co, CD New Orleans
  17th Co,(I), Ft. Mills; Demob Jun 1917
    2nd Co, Ft. Morgan; 2nd Co, CD Mobile
    4th Co, Ft. Moultrie; 4th Co, CD Charleston
    3rd Co, Ft. Screven; 3rd Co, CD (I) CD Savannah; C/61 Rgt
    3rd Co, Ft. Stevens; 3rd Co, CD Columbia R
    1st Co, Ft. Strong; 9th Co, CD Boston
    2nd Co, Ft. Strong; 10th Co, CD Boston
    1st Co, Ft Washington; 3rd Co, (II) CD Potomac
    2nd Co, Ft. Washington; Demob July 1917, Ft. Wool

Eight more companies were organized in 1917, before the US entered WW I. Their initial designations and subsequent redesignations 1917-1919 are noted below:

> 3rd Co, Ft. Barry; 17th Co, CD San Francisco
> 3rd Co, Ft Miley (HQ Co)
> 3rd Co, Ft. Rosecrans
> 4th Co, Ft. Rosecrans
> 4th Co, Ft. Winfield Scott
> 6th Co, Ft. Terry; B/7 Prov Rgt
> 6th Co, Ft. Totten
> 6th Co, Ft. H.G. Wright

With entry into World War I, the entire augmentation authorized in 1915 was immediately ordered, and further augmentations were authorized later in 1917 and in 1918. In the 1917 augmentation after the declaration of war, 71 companies were organized in April, May, June, and July. While some would be retained in the coastal defenses through the end of the war, many were reassigned to one of the CAC regiments formed for service in France. Their initial designations and subsequent redesignations during the period 1917-1919 are noted below, as are Demobilizations prior to November 1918:

April 1917
> 1st Co, Ft. Fremont; HQ Co, Ft. Screven; 2 Co,(II), Ft Screven; 2nd Co,(I), CD Savannah
> 5th Co, Ft. Mckinley; C/6th Prov Rgt; C/51 Rgt
> 6th Co, Ft. Mckinley; 16th Co,(I), CD Portland; C/54 Rgt

May 1917
> 1st Co, Ft. Columbia; 4th Co, CD Columbia
> Hq Co, Jackson Barracks; 5th Co, Jackson Barracks; 5th Co, CD New Orleans
> 10th Co, Ft. Monroe; D/8 Prov Rgt; D/53 Rgt; D/51 Rgt; D/43 Rgt
> 11th Co, Ft. Monroe; B/8 Prov Rgt
> 3rd Co, Ft. Rosecrans; 3rd Co, CD San Diego
> 4th Co, Ft. Rosecrans; 4(I) Co, CD San Diego

June 1917
> 4th Co, Ft. Barrancas; 4th Co, CD Pensacola
> 3rd Co, Ft. Caswell; 3rd Co, CD Cape Fear
> 2nd Co, Ft. Constitution; 4th Co, CD Portsmouth
> 3rd Co, Ft. Constitution; 5th Co, CD Portsmouth
> 3rd Co, Ft. Crockett; 3rd Co, CD Galveston
> 3rd Co, Ft. Dade; 3rd Co, CD Tampa
> 1st Co, Ft. Delesseps; 6th Co, CD Cristobal
> 5th Co, Ft. Dupont; 1st Trench Mortar Btry
> 6th Co, Ft. Dupont; 6th Co, CD Delaware
> 7th Co, Ft. Grant; 7th Co, CD Balboa
> 10th Co, Ft. Grant; 10th Co, CD Balboa
> 11th Co, Ft. Grant; 11th Balboa
> 4th Co (II), Ft. Hamilton; M/7 Prov Rgt; M/52 Rgt; D/42 Rgt
> 7th Co, Ft. Hamilton; 4th Co, CD Southern NY
> 7th Co, Ft. Hancock; 7th Co, CD Sandy Hook
> 2nd Co, Ft. Hunt; 2nd Co, CD Potomac
> 3rd Co, Jackson Barracks; 3rd Co, CD New Orleans
> 4th Co, Jackson Barracks; 4th CD New Orleans
> 12th Co, Ft. Monroe; C/8 Prov Rgt; C/53 Rgt
> 13th Co, Ft. Monroe; 5th Co, CD Chesapeake Bay
> 3rd Co, Ft. Morgan; 25th Co, CD Mobile Bay

4th Co, Ft Morgan; 4th Co (I), CD Mobile Bay; B/1 Trench Mortar Bn
1st Co, Ft. Mott; 7th Co, CD Delaware
2nd Co, Ft Mott; Hq&Sup/ 6 Prov Rgt; Hq/51 Rgt
5th Co, Ft. Moultrie; 2nd Co, CD Charleston
2nd Co, Ft Randolph; 8th Co, Cristobal
4th Co, Ft. Randolph; 10th Co, CD Cristorbal
1st Co, Rockaway Beach; 9th Co, CD Southern NY
3rd Co, Rockaway Beach; G/8 Prov Rgt; G/53 Rgt; 8 Bty/How Rgt; F/51 Rgt
2nd Co, Ft. Rodman; 2nd Co, CD New Bedford
1st Co, Ft. Schuyler; F/7 Prov Rgt; F/52 Rgt; 2 Bty/How Rgt; F/44 Rgt
2nd Co, Ft. Schuyler; 4th Co, CD Eastern NY
4th Co (II), Ft. Screven; 4th Co (I), CD Savannah
4th Co, Ft Sherman; 4th Co, CD Cristobal
1st Co, Ft. Smallwood; 5th (I), CD Baltimore; E/58 Rgt
7th Co, Ft. Terry; 16th Co, CD L.I.S.
7th Co, Ft. Totten; 2nd Co, CD Eastern Ny
4th Co, Ft. Wadsworth; 7th Co, CD Southern Ny
5th Co, Ft. Wadsworth; 8th Co, CD Southern Ny
3rd Co, Ft. Washington 4th Co (I), CD Potomac; Demob Jan 1918 Ft Washington
7th Co, Ft. H.G. Wright; 4th Co, CD L.I.S.

July 1917
Hdqtrs Co, Ft. Adams; 1st Co, CD Narra Bay
Arty Eng Co, Ft. Adams; 2nd Co, CD Narra Bay
5th Co, Ft. Andrews; 6th Co, CD Boston
4th Co, Ft. Casey; 12th Co, CD Puget Sound
Hdqtrs Co, Ft Constitution; 1st Co, CD Portsmouth
Arty Eng Co, Ft Constitution; 2nd Co, CD Porstsmouth
4th Co, Ft. Greble; 4th Co, CD (I) Narra Bay
6th Co, Ft Kamehameha; 6th Co, CD Oahu
Hdqtrs Co, Ft. Kamehameha; 7th Co, CD Oahu
2nd Co, Key West Barracks; 2nd Co, CD Key West
3rd Co, Ft. Macarthur; 3rd Co, CD Los Angeles; Demob Feb 1918 Ft Macarthur
4th Co, Ft. Macarthur; 4th Co, CD Los Angeles
17th Co (II), Ft. Mills; 17th Co, CD Manila & Subic Bays
18th Co, Ft Mills; 18th Co, CD Manila & Subic Bays
19th Co, Ft. Mills; 19th Co, CD Manila & Subic Bays
20th Co, Ft. Mills; 20th Co, CD Manila & Subic Bays
21st Co, Ft Mills; 21st Co, CD Manila & Subic Bays; Demob Nov 1917
4th Co, Ft. Ruger; 14th Co, CD Oahu
5th Co (Hq), Ft. Williams; 1st Co(I), CD Portland; Hq Co/ 54 Rgt
6th Co, Ft. Williams; 2nd Co, CD Portland
7th Co, Ft. Worden; 7th Co, CD Puget Sound
8th Co, Ft. Worden; 8th Co, CD Puget Sound

In July 1917 a provisional brigade of three coast artillery regiments was formed for service in France. Assembled at Fort Adams, RI, this brigade was shipped to France in August to man railway and other mobile artillery in support of the Allied armies. The three regiments were formed by transferring companies from Atlantic coastal fortifications. The regiments went through several reorganizations during 1918. These are reflected in the lists above. Additionally, the companies manning the coast defenses were again redesignated in August 1917, this time as elements of the coast defense to which they were assigned, rather than to specific forts in that coast defense. Only one company of RA CAC was constituted between August and December 1917, in October (5th Co(I), CD Mobile; B/1 Trench Mortar Bn). The next wave of expansion came in December 1917, when another 27 companies of RA CAC companies were constituted and organized.

During the World War, nearly 276 new companies were constituted and organized in the CAC, in addition to the 171 NG companies. While RA and NG troops formed the cadres for many of the new companies, their ranks were filled out with National Army (NA) draftees.

August 1917 saw the expansion of the CAC as a substantial part of the NG CAC companies was mobilized. In a few instances, NG coast defense commands were called into federal service as early as July, while those of several states were not called until early January 1918.

1st California NG CD Command, CAC (Mobilized Aug 1917)
    1st Co, Redes 21st Co, CD San Francisco, Demob March 1918
    2nd Co, Redes 22nd Co, Cd San Francisco, Demob March 1918
    3rd Co, Redes 23rd Co, Cd San Francisco, Demob April 1918
    4th Co, Redes 24th Co, Cd San Francisco, Demob April 1918
    5th Co, Redes 25th Co, Cd San Francisco, Demob January 1918
    6th Co, Redes 26th Co, CD San Francisco; B/67 Rgt
    7th Co, Redes 27th Co, CD San Francisco, D/18 Rgt
    8th Co, Redes 6th Co, CD San Diego; B/65 Rgt
    9th Co, Redes 29th Co, CD San Francisco; F/40 Rgt
    10th Co, Redes 30th Co, CD San Francisco; Sup/67 Rgt
    11th Co, Redes 25th Co, CD San Francisco, Demob March 1918
    12th Co, Redes 28th Co, CD San Francisco; D/67 Rgt

2nd California NG CD Command, CAC (Mobilized Aug 1917- Feb 1918)
    13th Co, Redes 7th Co, CD San Diego; B/25 Rgt
    14th Co, Redes 8th Co, CD San Diego; F/25 Rgt
    15th Co, Redes 5th Co, CD San Diego; Demob October 1918
    16th Co, Redes 5th Co, Cd Los Angles; D/ 2 AA Bn
    17th Co, Redes 6th Co, Cd Los Angeles; Demob January 1918
    18th Co, Redes 6th Co (Ii), CD San Diego; E/25 Rgt
    19th Co, Redes 7th Co (I), CD Los Angeles; Demob January 1918
    20th Co, Redes 8th Co (I), CD Los Angeles; Demob January 1918
    21st Co, Redes 9th Co (I), CD Los Angeles; Demob January 1918
    22nd Co, Redes 10th Co (I), CD Los Angeles; Demob January 1918
    23rd Co, Redes 11th Co (I), CD Los Angeles; Demob January 1918
    24th Co, Redes 12th Co (I), CD Los Angeles; Demob January 1918

Connecticut NG CD Command, CAC (Mobilized Aug 1917)
    1st Co, Redes 32nd Co, L.I.S. (Ft. Terry); 7th Co, CD L.I.S.
    2nd Co, Redes 33rd Co, L.I.S. (Ft. Terry); 19th Co, CD L.I.S.
    3rd Co, Redes 26th Co, L.I.S. (Ft. Terry); Sup/56 RGT
    4th Co, Redes 27th Co, L.I.S. (Ft. Terry); F/56 RGT
    5th Co, Redes 34th Co, L.I.S. (Ft. Terry); 23th Co, CD L.I.S.
    6th Co, Redes 35th Co, L.I.S. (Ft. Terry); 14th Co, CD L.I.S.
    7th Co, Redes 36th Co, L.I.S. (Ft. Terry); 16th Co, CD L.I.S.
    8th Co, Redes 28th Co, L.I.S. (Ft. HG Wright); 28th Co, CD L.I.S.
    9th Co, Redes 29nd Co, L.I.S. (Ft. HG Wright); 29th Co, CD L.I.S.; E/56 RGT
    10th Co, Redes 30th Co, L.I.S. (Ft. HG Wright); 5th Co (I), CD L.I.S.
    11th Co, Redes 37th Co, L.I.S. (Ft. Terry); B/56 Rgt
    12th Co, Redes 31st Co, L.I.S. (Ft. HG Wright); 6th Co (I), CD L.I.S.
    13th Co, Redes 38th Co, L.I.S. (Ft. Terry); D/56 Rgt

District of Columbia NG CD Command, CAC (Mobilized Jan 1918)
    1st Co, Redes D/60 Rgt
    2nd Co, Redes 5th Co (I), CD Potomac (Ft. Washington)

Florida NG CD Command, CAC (Mobilized Aug 1917)
    1st Co, Redes 3rd Co, CD Key West
    2nd Co, Redes 4th Co, CD Tampa
    3rd Co, Redes 6th Co, CD Pensacola

Maine NG CD Command, CAC (Mobilized Aug 1917)
    1st Co, Redes 7th Co, Ft. Williams; 27th Co (I), CD Portland
    2nd Co, Redes 2nd Co, Ft. Levett; 2nd Co, CD Portland; D/ 54 Rgt
    3rd Co, Redes 7th Co, Ft. Mckinley; 24th Co, CD Portland; E/54 Rgt
    4th Co, Redes 1st Co, Ft. Baldwin; 29th Co (I), CD Portland; D/54 Rgt
    5th Co, Redes 8th Co, Ft. Williams; 8th Co (I), CD Portland; E/54 Rgt
    6th Co, Redes 3rd Co, Ft. Levett; 23rd Co, CD Portland
    7th Co, Redes 9th Co, Ft. Williams; 19(I) CD Portland; B/54 Rgt
    8th Co, Redes 3rd Co, Ft. Preble; 21st Co, CD Portland; Sup/54Rgt
    9th Co, Redes 8 Co, Ft. Mckinley; 25th Co (I), CD Portland; F/54 Rgt
    10th Co, Redes 4th Co, Ft. Preble; 21st Co, CD Portland
    11th Co, Redes 9th Co, Ft. McKinley; 26th Co (I), CD Portland; E/54 Rgt
    12th Co, Redes 10th Co, Ft. Mckinely; 28th Co, Portland
    13th Co, Redes 1st Ft. Lyon; 10th Co (II), CD Portland; E/54 Rgt

Maryland NG CD Command, CAC (Mobilized Jan 1918)
    1st Co, Redes 5th Co (II), CD Baltimore
    2nd Co, Redes 6th Co, CD Baltimore, 75 men assigned to F/58 Rgt
    3rd Co
    4th Co (3rd & 4th Cos inactivated and personnel assigned to form 1st Trench Mortar Battery).

Massachusetts NG CD Command, CAC (Mobilized Aug 1917)
    1st Co, Redes 16th Co, CD Boston (Ft. Revere)
    2nd Co, Redes 17th Co, CD Boston (Ft. Revere)
    3rd Co, Redes 18th Co (I), CD Boston (Ft. Strong); F/55 Rgt
    4th Co, Redes 19th Co (I), CD Boston (Ft. Banks); D/55 Rgt
    5th Co, Redes 20 Co (I), CD Boston (Ft. Andrews); Sup/55 Rgt
    6th Co, Redes 21st Co, CD Boston (Ft. Strong)
    7th Co, Redes 22nd Co, CD Boston (Ft. Banks)
    8th Co, Redes 23rd Co, CD Boston (Ft. Andrews)
    9th Co, Redes 24th Co, CD Boston (Ft. Heath)
    10th Co, Redes 25th Co, CD Boston (Sprngfield Arsenal)
    11th Co, Redes 26th Co, CD Boston (Ft. Andrews)
    12th Co, Redes 27th Co, CD Boston (Sprngfield Arsenal)

New Hampshire NG CD Command, CAC (Mobilized Jun 1917)
    1st Co, Redes 4th Co, Ft. Constitution; 9th Co, CD Portsmouth
    2nd Co, Redes 5th Co, Ft. Constitution; 6th Co, CD Portsmouth
    3rd Co, Redes 6th Co, Ft. Constitution; 7th Co, CD Portsmouth
    4th Co, Redes 7th Co, Ft. Constitution; 8th Co, CD Portsmouth

New Jersey NG CD Command, CAC (Mobilized Feb 1918)
    1st Co, Redes 11th Co, CD Delaware (Ft. Mott); 5th Co, CD Delaware
    2nd Co, Redes D/2nd Trench Mortar Bn

New York NG CD Command, CAC, 13th Coast Defense Command (Mobilized Jan 1918)
    1st Co, Not Mobilized
    2nd Co, Redes 17th Co, CD Southern NY
    3rd Co, Redes E/59 Rgt
    4th Co, Redes Sup/59 Rgt
    5th Co, Redes 20th Co, CD Southern NY
    6th Co, Broken up, personnel assigned to 59 Rgt

7th Co, Redes 22nd Co, CD Southern NY

8th Co, Redes 23rd Co, CD Southern NY; 28 men to 40th & 28 men to 23rd Cos, CD SNY

9th Co, Redes 24th Co, CD Southern NY

10th Co, Redes 25th Co, CD Southern NY

11th Co, Redes 26th Co, CD Southern NY

12th Co, Redes F/59 Rgt

New York NG CD Command, CAC, 9th Coast Defense Command (Mobilized Jul 1917)

13th Co, Redes 13th Co, CD Southern NY (Ft Hancock)

14th Co, Redes 14th Co, CD Southern NY (Ft Hancock)

15th Co, Redes 15th Co, CD Southern NY (Ft Hancock)

16th Co, Redes 16th Co, CD Southern NY (Ft Hancock)

17th Co, Redes 17th Co, CD Southern NY (Ft Hancock)

18th Co, Redes 18th Co, CD Southern NY (Ft Hancock); 44 men to 57th Rgt

19th Co, Redes 19th Co, CD Southern NY (Ft Hancock); 109 men to B/57 Rgt

20th Co, Redes 20th Co, CD Southern NY (Ft Hancock)

21st Co, Redes 21st Co, CD Southern NY (Ft Hancock)

22nd Co, Redes 22nd Co, CD Southern NY (Ft Hancock); Large levy to 57 Rgt

23rd Co, Redes 23rd Co, CD Southern NY (Ft Hancock)

24th Co, Redes 24th Co, CD Southern NY (Ft Hancock); 104 men to B, D, E, & F/37 Rgt

New York NG CD Command, CAC, 8th Coast Defense Command (Mobilized as separate units, 1917-18)

26th Co, Redes 28th Co, CD Southern NY

27th Merged

28th Merged

29th Merged

30th Redes 29th Co, CD Southern NY; 54 men to the 35th Co, CDSNY

32nd Merged

33rd Co, Redes 30th Co, CD Southern NY

34th Co, Redes 2nd Co, Rockaway Beach; 10th Co, CD Southern NY; 10th Co,
       Eastern NY mobilized and designated as 2nd Co, Rockaway Beach in August 1917.

35th Redes 31st Co, CD Southern NY

36th Merged

North Carolina NG CD Command, CAC (Mobilized Sept 1917)

1st Co, Redes 7th Co, CD Cape Fear (Ft. Caswell)

2nd Co, Redes 8th Co, CD Cape Fear (Ft. Caswell)

3rd Co, Redes 3rd Co, CD Cape Fear (Ft. Caswell)

4th Co, Redes 4th Co, CD Cape Fear (Ft. Caswell)

5th Co, Redes 5th Co, CD Cape Fear (Ft. Caswell)

6th Co, Redes 6th Co, CD Cape Fear (Ft. Caswell)

Oregon NG CD Command, CAC (Mobilized Jan 1918)

1st Co, Redes 13th Co, CD Columbia River (Ft. Stevens)

2nd Co, Redes 14th Co, CD Columbia River (Ft. Stevens)

3rd Co, Redes 15th Co, CD Columbia River (Ft. Stevens)

4th Co, Redes 16th Co, CD Columbia River (Astoria)

5th Co, Redes 5th Co, CD Columbia River (Ft. Canby)

6th Co, Redes 6th Co, CD Columbia River (Ft. Stevens)

7th Co, Redes 7th Co, CD Columbia River (Ft. Columbia)

8th Co, Redes 8th Co, CD Columbia River (Ft. Stevens)

9th Co, Redes 9th Co, CD Columbia River (Ft. Stevens)

10th Co, Redes 10th Co, CD Columbia River (Ft. Columbia)

11th Co, Redes 11th Co, CD Columbia River (Ft. Stevens)

12th Co, Redes 12th Co, CD Columbia River (Ft. Canby)

Rhode Island NG CD Command, CAC (Mobilized Aug 1917)
 1st Co, Redes 9th Co, CD Narrgansett Bay (Ft. Adams)
 2nd Co, Redes 28th Co, CD Boston (Ft. Standish)
 3rd Co, Redes 10th Co (I), CD Narrgansett Bay (Ft. Wetherill)
 4th Co, Redes 13th Co, CD Narrgansett Bay (Ft. Getty)
 5th Co, Redes 19th Co, CD Narrgansett Bay (Ft. Getty)
 6th Co, Redes 20th Co, CD Narrgansett Bay (Ft. Greble)
 7th Co, Redes 21th Co, CD Narrgansett Bay (Ft. Wetherill)
 8th Co, Redes 22th Co, CD Narrgansett Bay (Ft. Greble)
 9th Co, Redes 29th Co, CD Boston (Ft. Standish)
 10th Co, Redes 30th Co, CD Boston (Ft. Standish)
 11th Co, Redes 11th Co, CD Narrgansett Bay (Ft. Wetherill)
 12th Co, Redes 12th Co, CD Narrgansett Bay (Ft. Kearney)
 13th Co, Redes 31th Co, CD Boston (Ft. Warren)
 14th Co, Redes 14th Co, CD Narrgansett Bay (Ft. Kearney)
 15th Co, Redes 15th Co, CD Narrgansett Bay (Ft. Greble)
 16th Co, Redes 16th Co, CD Narrgansett Bay (Ft. Wetherill)
 17th Co, Redes 17th Co, CD Narrgansett Bay (Ft. Greble)
 18th Co, Redes 18th Co, CD Narrgansett Bay (Ft. Wetherill)
 19th Co, Redes 3rd Co, CD New Bedford (Ft. Rodman)
 20th Co, Redes 32th Co, CD Boston (Ft. Standish)

South Carolina NG CD Command, CAC (Mobilized Aug 1918)
 1st Co, Redes 6th Co, CD Charleston (Ft. Moultrie)
 2nd Co, Redes 7th Co, CD Charleston (Ft. Moultrie)
 3rd Co, Redes 8th Co, CD Charleston (Ft. Moultrie)
 4th Co, Redes 9th Co, CD Charleston (Ft. Moultrie)
 5th Co, Redes 10th Co, CD Charleston (Ft. Moultrie)

Texas NG CD Command, CAC (Mobilized Aug 1917)
 1st Co, Redes 4th Co, CD Galveston (Ft. Crockett)
 2nd Co, Redes 5th Co, CD Galveston (Ft. Crockett)
 3rd Co, Redes 6th Co, CD Galveston (Ft. Crockett)
 4th Co, Redes 7th Co, CD Galveston (Ft. Crockett)
 5th Co, Redes 8th Co, CD Galveston (Ft. Crockett)

Virginia NG CD Command, CAC (Mobilized Feb 1918)
 1st Co, Redes elements of 42nd Division
 2nd Co, Redes elements of 42nd Division
 3rd Co, Redes 10th Co, CD Chesapeake Bay (Ft. Story)
 4th Co, Redes 8th Co, CD Chesapeake Bay (Fishermans Is.)
 5th Co, Not mobilized
 6th Co, Redes 6th Co, CD Chesapeake Bay (Ft. Monroe)
 7th Co, Redes 11th Co, CD Chesapeake Bay (Ft. Monroe)
 8th Co, Redes 12th Co, CD Chesapeake Bay (Ft. Monroe)
 9th Co, Not mobilized

Washington NG CD Command, CAC (Mobilized Jan 1918)
 1st Co, Redes 17th Co, CD Puget Sound (Ft. Worden)
 2nd Co, Redes 18th Co, CD Puget Sound (Ft. Worden)
 3rd Co, Redes 19th Co, CD Puget Sound (Ft. Worden)
 4th Co, Redes 20th Co, CD Puget Sound (Ft. Flagler)
 5th Co, Redes 21st Co, CD Puget Sound (Ft. Worden)
 6th Co, Redes 22nd Co, CD Puget Sound (Ft. Worden)
 7th Co, Redes 23rd Co, CD Puget Sound (Ft. Worden)
 8th Co, Redes 24th Co, CD Puget Sound (Ft. Worden)
 9th Co, Redes 25th Co, CD Puget Sound (Ft. Casey)

10th Co, Redes 16th Co, CD Puget Sound (Ft. Flagler)
11th Co, Redes 27th Co, CD Puget Sound (Ft. Flagler)
12th Co, Redes 28th Co, CD Puget Sound (Ft. Casey)

The decision made in the early winter of 1917/1918 to send additional regimental-sized coast artillery organizations to France created a need for additional îreplacementsî to replace the personnel transferred to the regiments. The 11 initial regiments constituted in December 1917 and January 1918 were formed by simply transferring existing companies manning fixed batteries in the coast defenses. The resulting shortage of men trained in the use of the seacoast armament resulted in subsequent regiments being organized by individual transfers.

The new companies organized mostly in December 1917 received their designations as elements of their assigned coast defense and replaced the companies that were in the process of training and preparing for deployment to France. These newly organized companies were:

December 1917
    19th Co (II), CD Boston (Ft. Banks)
    26th Co (I), CD Narragansett Bay (Ft. Adams)
    27th Co (I), CD Narragansett Bay (Ft. Adams)
    28th Co (I), CD Narragansett Bay (Ft. Greble)
    30th Co (I), CD Narragansett Bay (Ft. Adams)
    31st Co, CD Narragansett Bay (Ft. Getty)
    32nd Co, CD Narragansett Bay (Ft. Getty)
    31st Co, CD San Francisco (Presidio)
    32nd Co, CD San Francisco (Presidio)
    33rd Co, CD San Francisco (Presidio)
    34th Co, CD San Francisco (Presidio)
    35th Co, CD San Francisco (Presidio)
    36th Co, CD San Francisco (Presidio)
    37th Co, CD San Francisco (Presidio)
    38th Co, CD San Francisco (Presidio)
    39th Co, CD San Francisco (Presidio)
    40th Co, CD San Francisco (Presidio)
    41st Co, CD San Francisco (Presidio)
    42nd Co, CD San Francisco (Presidio)
    43rd Co (I), CD San Francisco (Presidio)
    44th Co (I), CD San Francisco (Presidio)
    45th Co (I), CD San Francisco (Presidio)
    46th Co (I), CD San Francisco (Presidio)
January 1918
    9th Co, CD Cape Fear (Ft. Caswell)
    10th Co, CD Cape Fear (Ft. Caswell)
    11th Co, CD Cape Fear (Ft. Caswell)
    12th Co, CD Cape Fear (Ft. Caswell)
February 1918
    13th Co, CD Cheasapeake Bay (Ft. Monroe)
    17th Co, CD Columbia River
    18th Co, CD Columbia River
    4th Co (II), CD Delaware
March 1918
    1st Co (II), CD Boston
    13th Co, CD Boston
    20th Co (II), CD Boston
    29th Co (II), CD Boston
    14th Co, CD Chesapeake Bay

April 1918
    8th Co, CD Delaware
    5th Co, CD Charleston
    11th Co, CD Charleston
    12th Co, CD Charleston
    Hq Co, CD Charleston
    15th Co, CD Chesapeake Bay
    16th Co, CD Chesapeake Bay
    19th Co, CD Columbia River
    20th Co, CD Columbia River
    21st Co, CD Columbia River
    22nd Co, CD Columbia River

May 1918
    23rd Co, CD Columbia River

July 1918
    4th Co, CD Boston
    17th Co, CD Chesapeke Bay

August 1918
    5th Co, CD Boston
    9th Co, CD Delaware
    10th Co, CD Delaware
    11th Co (II), CD Delaware

November 1918
    6th Co (II), CD Columbia
    7th Co (II), CD Columbia

Sources

Historical Section, Army War College. *Order of Battle of the United States Land forces in the World War, American Expeditionary Forces,* 3 Vol. GPO, 1937-1949.

Records of the Office of the Chief of Coast Artillery, Archives II, NARA College Park, MD

Kelton, R.H.C. ìField Service of the Coast Artillery in the World War.î *Coast Artillery Journal,* Vol. 4, No. 4 (April 1922), pp, 295-309.

Coast Artillery Regiments in World War One

*	Never Organized		
#	Regiment assembled at San Francisco from several posts, March 1918		
%	Not demobilized		
PoE	Port of Entry		

RGT	ORGANIZED	STATIONS IN US	DEMOBILIZED
6th Prov	Ft Adams, July 1917	PoE Hoboken	51st CA Rgt
7th Prov	Ft Adams, July 1917	PoE Hoboken	52nd CA Rgt
8th Prov	Ft Adams, July 1917	PoE Hoboken	53rd CA Rgt
How Rgt	AEF, Mar 1918		44th CA Rgt
1-14	*		*
15	Ft Crockett, Oct 1918		Ft Crockett
16	*		*
17	Ft Monroe, Oct 1918		Ft Monroe
18	Ft W Scott, Oct 1918		Ft W Scott
19	Ft MacArthur, Oct 1918		Ft MacArthur
20	Ft Crockett, Oct 1918		Ft Crockett
21	Ft Pickens, Nov 1918		Ft Pickens
22	*		*
23	*		*
24	*		*
25	Ft Rosecrans, Oct 1918		Ft Rosecrans
26	Ft Screven, Nov 1918		Ft Screven
27	Ft Stevens, Oct 1918	Cp Eustis	Cp Eustis
28	Ft Strong, Nov 1918	Ft Revere	Ft Revere
29	Ft Williams, Nov 1918		Ft Williams
30	Ft HG Wright, Nov 1918	Cp Eustis	Cp Eustis
31	Ft Hancock, Oct 1918		Cp Eustis
32	Ft Hamilton, Oct 1918		Cp Eustis
33	Ft Strong, Sept 1918		Cp Eustis
34	Ft Totten, Oct 1918		Cp Eustis
35	Ft DuPont, Nov 1918		Ft DuPont & Cp Meade
36	Ft Moultrie, Sept 1918	Cp Eustis, Cp Stuart	Ft Monroe
37	Ft Hancock, Sept 1918	Cp Eustis, Cp Stuart	Ft Hancock
38	Ft Hamilton, Sept 1918	Cp Eustis, Cp Stuart	Ft Hamilton
39	Ft Worden, Sept 1918	Cp Upton, Cp Grant	Cp Grant
40	Ft W Scott, Sept 1918	Cp Upton, Cp Grant	Presidio of SF
41	Ft Monroe, Oct 1918		Ft Monroe
42	AEF, Aug 1918	Cp Stuart, Cp Eustis	%
43	AEF, Aug 1918	Cp Hill, Cp Eustis	%
44	AEF, Aug 1918 from How Rgt	Cp Mills, Ft Totten	%
45	Cp Eustis, July 1918	Cp Stuart, PoE N News, Cp Mills	Cp Dix
46	Cp Eustis, July 1918	PoE N News, Cp Mills	Cp Dix
47	Cp Eustis, July 1918	Cp Stuart, PoE NNews, Cp Eustis	Cp Eustis
48	Cp Eustis, July 1918	Cp Stuart, PoE Nnews	Cp Grant
49	Cp Eustis, July 1918	Cp Stuart, PoE NNews, Cp Merritt	Cp Grant
50	Cp Eustis, July 1918	Cp Stuart, PoE NNews	Cp Grant

RGT	ORGANIZED	STATIONS IN US	DEMOBILIZED
51	AEF, Feb 1918, from 6th Prov	Cp Mills, Ft Hamilton, Cp Jackson	%
52	AEF, July 1917, from 7th Prov	Cp Stuart, Cp Eustis	%
53	AEF, Mar 1919, from 8th Prov	Cp Stuart, Cp Eustis	%
54	CD Portland, Jan 1918	PoE Hoboken	Cp Devens
55	CD Boston, Dec 1917	Cp Merritt, PoE Hoboken, Cp Mills, Ft HG Wright, Ft W Scott, Cp Lewis	%
56	Ft HG Wright, Dec 1917	PoE Hoboken, Cp Mills, Ft Schuyler, Cp Jackson	%
57	Ft Hancock, Jan 1918	PofHoboken, Cp Merritt, Ft W Scott, Cp Lewis	%
58	Ft Totten, Feb 1918	PoE Hoboken	Cp Upton
59	Ft Hamilton, Jan 1918	Cp Upton, Ft W Scott, Cp Lewis	%
60	Ft Monroe, Feb 1918	Cp Stuart, PoE NNews, Cp Merritt	Ft Washington
61	Ft Moultrie, May 1918	Cp Eustis, Cp Stuart, PoE NNews	Cp Upton
62	CD San Francisco, Jan 1918	Cp Mills, PoE Hoboken, Cp Stuart, Cp Eustis	Cp Eustis
63	CD Puget Sound, Dec 1917	Cp Mills, PoE Hoboken, Cp Mills	Cp Lewis
64	CD Tampa, Jan 1918	Cp Upton, PoE Hoboken, Cp Stuart, Cp Eustis	Cp Eustis
65	Ft Stevens, Dec 1917 #	San Francisco, Cp Merritt, PoE Hoboken, Cp Dix	Cp Lewis
66	CD Narragansett, Mar 1918	PoE Boston	Cp Upton
67	Ft W Scott, May 1918	Cp Mills, PoE Hoboken, Presidio of SF	Presidio of SF
68	Ft Terry, June 1918	PoE Boston, Cp Mills, Ft Wadsworth	Ft Wadsworth
69	CD Puget Sound, May 1918	Cp Mills, PoE Philadelphia	Cp Eustis
70	Ft Hamilton, June 1918	Ft Wadsworth, PoE Hoboken, Cp Merritt	Cp Sherman
71	CD Boston, May 1918	PoE Boston, Cp Merritt	Cp Devens
72	CD Portland, June 1918	Pof E Montreal, Cp Upton	Cp Grant
73	Ft Banks, July 1918	Cp Mills, PoE Hoboken, Cp Mills	Cp Devens
74	Ft Schuyler, June 1918	Cp Upton, PoE Hoboken, Cp Mills, Ft Totten	Ft Totten
75	Ft Moultrie, Sept 1918	Cp Merritt, PoE Hoboken, Cp Stuart	Cp Grant

CAMPS

Cp Devens, MA, Cp Dix, NJ, Cp Eustis, VA, Cp Grant, IL, Cp Hill, VA, Cp Jackson, SC, Cp Lewis, WA Cp Merritt, NY, Cp Mills, NY, Cp Sherman, OH, Cp Stuart, VA, Cp Upton, NY

From: *Order of Battle, Zone of the Interior*, Vol. 2, Pt. 3, CMH, pp. 1136 - 1142

COAST ARTILLERY CORPS REGIMENTS 1924-1941

Abbreviations: RA = regular army, RAI = regular army, inactive, NG = national guard, OR = organized reserve, PS = Philippine Scouts, HD = harbor defense artillery, RY = railway artillery, AA = antiaircraft artillery, TD = tractor drawn artillery, inact = inactivated, redes = redesignated, react = reactivated, assign = assigned, trans = transferred, conv = converted, org = organized, disb = disbanded, CAR = coast artillery regiment, reg = regiment, Bn = battalion.

*CONUS RA, 200-series NG, and 600-series RAI regiments to be mobilized or activated under 1938 Mobilization Plan.

See heraldry section for photos of insignia. This list is as accurate as possible for the regiments existing between 1924-1941. The picture becomes much more complex after 1942. For more detailed information see other references cited in this section and:

William C. Gaines "Coast Artillery Organizational History, 1917-1950, Part I, Coast Artillery Regiments 1-196," *Coast Defense Journal* Volume 23, Issue 2, (May 2009) pp. 4-51.

William C. Gaines "Coast Artillery Organizational History, 1917-1950 Part II, Coast Artillery Regiments, OR and AUS," *Coast Defense Journal* Volume 23, Issue 3 (August 2009) pp. 70-93.

Reg #	Year Org	Initial Station	Army	Type Cadre	Notes; inactivated/disbanded
1	1924	Ft DeLessups, PCZ	RA	HD	Orig. 1st Artillery 1821, inact Nov 1944
2*	1924	Ft Sherman, PCZ	RA	HD	Orig. 2nd Artillery 1821, trans Ft Monroe 1932, inact Oct 1944
3*	1924	Ft MacArthur, CA	RA	HD	Orig. 3rd Artillery 1812, inact Oct 1944
4	1924	Ft Amador, PCZ	RA	HD	Orig. 4th Artillery 1821, inact Nov 1944
5*	1924	Ft. Hamilton, NY	RA	HD	Orig. 5th Artillery 1861, inact Apr 1944 Cp Rucker, AL
6*	1924	Ft W. Scott, CA	RA	HD	Orig. 6th Artillery 1898, inact Oct 1944
7*	1924	Ft Hancock, NJ	RA	HD	Orig. 7th Artillery 1898, inact Apr 1944 Ft L. Wood, MO
8*	1924	Ft Preble, ME	RA	HD	inact Apr 1944 Cp Shelby, MS
9*	1924	Ft Banks, MA	RA	HD	inact Apr 1944 Cp Hood, TX
10*	1924	Ft Adams, RI	RA	HD	inact Apr 1944 Cp Forrest, TN
11*	1924	Ft H.G. Wright	RA	HD	inact Apr 1944 Ft L. Wood, MO
12	1924	Ft Monroe, VA	RA	HD	inactivated & trans to PCZ 1932, not reactivated as CA
13*	1924	Ft Barrancas, FL	RA	HD	inact Aug 1944
14*	1924	Ft Worden, WA	RA	HD	inact Oct 1944
15	1924	Ft Kamehameha	RA	HD	inact Aug 1944
16	1924	Ft DeRussey, HI	RA	HD	inact Aug 1944
17	(1926)	Hawaiian Dept	RAI		Constituted as an inactive unit, never activated, disb 1944
18	1940	Ft Stevens, OR	RA	HD	inact May 1944 Cp Breckenridge, KY
19	1940	Ft Rosecrans, CA	RA	HD	625th CAR redes 19th CAR 1940, inact Oct 1944
20	1940	Ft Crockett, TX	RA	HD	two Bn, disb Aug 1944
21	1940	Ft DuPont, DE	RA	HD	disb Oct 1944
22	1940	Ft Constitution	RA	HD	614th CAR redes 22nd CAR 1940. disb Oct 1944
23	1940	Ft Rodman, MA	RA	HD	616th CAR redes 23rd CAR 1940, One Bn, inact Apr 1944 Ft Hood TX
24	1942	Ft. HG Wright, NY	RA	HD	1 Bn, to Newfoundland, inact Sep 1944 Cp Standish, MA
30	1942	Ft. Lewis WA	RA		redes Jul 1944 Cp Robinson AK
31	1943	Cp Pendleton, VA	RA	HD	redes Apr 1943 Key West, FL
35	1943	Ft. Brooke, PR	RA	HD	redes Nov 1944
36	1943	Puerto Rico	RA	HD	redes Nov 1944 Panama
39	1943	Dutch West Indies	RA	HD	disb May 1944 Dutch West Indies
40	1942	Alaska	RA		inact Dec 1944
41	1921	Ft Kamahemeha	RA	RY	Inact 1931, react 1942, redes HD 1943, inact May 1944
42*	1918		RAI	RY	Inact assign Org Reserve 1921, not activ, disb Jun 1944
43*	1918		RAI	RY	Inact assign Org Reserve 1921, not react, disb Jun 1944
44*	1918		RAI	TD	Inact assign Org Reserve 1921, redes 54th CAR 1941
46	1943	Cp Pendleton, VA	RA	TD	disb Apr 1944 Cp Shelby, MS
47	1943	Cp Pendleton, VA	RA	TD	disb Feb 1944, Cp Pickett, VA
48	1942	San Francisco, CA		SL	inact Jan 1944 Ft Ray AK

49	1942	Los Angeles, CA		SL	inact May 1944 Cp Barkley, AL
50	1942	Cp Pendleton, VA	RA	TD	disb Jan 1944, Ft Devens, MA
51	1918	Ft. Eustis, VA	RA	TD	Inact 1931, react 1938? Puerto Rico 1940, inact Jun 1944
52*	1917	Ft. Eustis, VA	RA	RY	moved to Ft Hancock 1930s, inact May 1943
53	1917		RAI	RY	react 1942, inact Jun 1944 Cp Pendleton
54	1941	Cp Wallace, TX	RA	TD	44th CAR redes to 54th CAR 1941, inact Apr 1944 Ft Ord CA
55	1917	Ft Kamahemeha	RA	TD	inact Jun 1944
56	1918	Ft. Cronkhite, CA	RAI	TD	demob 1921, 506th redes 56th 1941, inact Feb 1944
57*	1918	Ft Monroe?	RAI	TD	demob 1921, reconst RAI 1926, react 1941, inact May 44, HI
58	1942	Chile	RA	TD	disb Jun 1944
59	1918	Ft Mills, PI	RA	TD	conv to HD 1930, surrendered 1942, inact Apr 1946
60	1922	Ft McKinley, PI	RA	AA	surrendered 1942, inact Apr 1946
61	1921	Ft Monroe, VA	RA	AA	trans to Ft Sheridan, IL 1920s, inact Aug 1943 Great Britain
62	1922	Ft Totten, NY	RA	AA	inact Mar 1943 Italy
63	1921	Ft W. Scott, CA	RA	AA	trans to Ft MacArthur 1930, inact Dec 1943 Seattle, WA
64	1921	Ft Shafter, HI	RA	AA	inact Dec 1943
65	1924	Ft Amador, PCZ	RA	AA	Inact 1932, react 1938?, inact Apr 1943 Ft Ord, CA
66	1926		RAI	AA	Inact1926, react 1942 Ft Bragg NC, inact Nov 1943 Puerto Rico
67	1926		RAI	AA	Inact1926, react 1941 Ft. Bragg, NC, inact Jun 1944 Italy
68	1926		RAI	AA	Inactive 1926, react 1939 Ft. Williams, ME, inact Jun 1944 Italy
69	1926	Ft Crockett, TX	RA	AA	Inactive 1926, Activated 1930 Aberdeen, MD, inact Sep 1943 San Diego, CA
70	1939	Ft Monroe VA	RA	AA	917th CAR redes 1940, inact Nov 1943 South Pacific
71	1941	Ft Story, VA	RA	AA	504th CAR redes 71st, inact Sep 1943 Washington DC
72	1939	Ft Randolph, PCZ	RA	AA	inact Sep 1943
73	1939	Ft Amador, PCZ	RA	AA	inact Dec 1943
74	1941	Ft Monroe, VA	RA	AA	503rd CAR redes 74th, inact Apr 1944 Italy
75	1940	Ft Lewis, WA	RA	AA	509th CAR redes 75th, inact Feb 1945 Ft Bliss, TX
76	1941	Ft Bragg NC	RA	AA	502nd CAR redes 76th, inact Nov 1943 South Pacific
77	1941	Ft. Bragg NC	RA	AA	505th CAR redes 77th, inact Nov1943 South Pacific
78	1941	Cp Haan, CA	RA	AA	517th CAR redes 78th, inact Feb 1944 Attu, AK
79	1941	Ft Bliss, TX	RA	AA	inact Sep 1943 Great Britain
82	1940	Ft Randolph, PCZ	RA	AA	inact Sep 1943
83	1940	Ft Amador, PCZ	RA	AA	inact Sep 1943
84	1942	Ft Read, Trinidad	RA	AA	disb Feb 1944 Cp Stewart, GA
85	1942	Cp Davis, NC	RA	AA	One Bn, inact Sep 1943 Norfolk, VA
86	1942	Cp Haan, CA	RA	AA	inact Jan 1943
87	1942	Panama	RA	AA	inact Dec 1943 Ft Bliss, TX
88	1942	Panama	RA	AA	inact Sep 1943
89	1942	Washington DC	RA	AA	inact Sep 1943
90	1942	Cp Stewart, GA	RA	AA	inact May 1944 North Africa
91	1924	Manila Bay, PI	PS	HD	redes TD 1930, surrendered 1942
92	1924	Manila Bay, PI	PS	HD	surrendered 1942
93	1941	Cp Davis, NC	RA	AA	inact Dec 1943 Hawaii
94	1941	Cp Davis, NC	RA	AA	inact May 1943 South Pacific
95	1941	Cp Davis, NC	RA	AA	inact Dec 1943 Hawaii
96	1941	Cp Davis, NC	RA	AA	inact Dec 1943 Hawaii
97	1941	Ft Kamehameha	RA	AA	inact Dec 1943
98	1941	Schofield Barracks	RA	AA	inact Dec 1943
99	1941	Cp Davis, NC	RA	AA	inact Dec 1943 Cp Stewart, GA
100	1941	Cp Davis, NC	RA	AA	disb Apr 1943 Cp Stewart, GA
196	1942	Ft Amador, PCZ	RA	AA	inact Sep 1943
197	1922	Concord, NH	NG	AA	Fed 1940, inact Apr 1943 New Guinea
198	1924	Wilmington, DE	NG	AA	Orig. 1 Del. Vol. Inf. 1861, Fed 1940, inact Mar 1943 S. Pac
200	1925	Raeford, NC.	NG	AA	Trans to Deming, NMNG 1930s, Fed 1940, Tr. PI 1941 surrendered 1942, inact Apr 1946
201	1940	Puerto Rico	NG	AA	Fed 1940, redesig Apr 1941
202	1924	Chicago, IL	NG	AA	Orig. IL 6 Inf 1920, Fed 1940, inact Sep 1943 Bremerton, WA

203	1924	Aurora, MO	NG	AA	Orig. MO 2 Inf 1890, Webb City, MO Fed 1940, inact Jan1944 Alaska
204	1940	Sheveport, LA	NG	AA	Fed 1940, inact Sep 1943 San Diego, CA
205	1940	Olympia, WA	NG	AA	Fed 1940, inact Sep 1943 Santa Monica, CA
206	1923	Marianna, AR	NG	AA	Orig. 141 MG Bn 1917, Fed 1940, disb Mar 1944 Ft Bliss, TX
207	1925?	New York, NY	NG	AA	trans to Infantry 1940?, inact Apr 1943 Cp Edwards, MA
208	1941	West Hartford, CT	NG	AA	inact May 1943 New Guinea
209	1940	Buffalo, NY	NG	AA	inact Oct 1943 Italy
210	1941	Detroit, MI	NG	AA	inact Feb 1944 Adak, AK
211	1924	Boston, MA	NG	AA	Orig. 1777 MA inf unit 45 MA Vol 1862, Fed 1940, inact sep 1943 San Francisco, CA
212	1921	New York, NY	NG	AA	Orig. NY 11 Inf 1847, Fed 1940, inact 1943 Seattle, WA
213	1922	Allentown, PA	NG	AA	Orig. 4 Inf PA 1874, Fed 1940, inact Apr 1944 Italy
214	1933?	Washington, GA	NG	AA	264th CAR conv to 214th, inact Nov 1943 South Pacific
215	1940	Mankato, MN	NG	AA	conv from infantry, Fed 1940, disb Mar 1944 Ft Bliss, TX
216	1940	St. Paul, MN	NG	AA	conv from infantry, Fed 1940, inact Sep 1943, San Francisco
217	1940	St. Cloud, MN	NG	AA	Fed 1940, inact Sep 1943 Oakland, CA
240*	1923	Portland, ME	NG	HD	Orig 1 vol Militia 1854, Fed1940, inact Oct 1944
241*	1923	Boston, MA	NG	HD	Orig MA 1 inf 1878 , Fed 1940, inact Oct 1944
242*	1927	Bridgeport,CT	NG	HD	Orig. militia 1739, Fed 1940, inact Sep 1943
243*	1924	Providence, RI	NG	HD	Fed 1940, inact Oct 1944
244*	1924	New York, NY	NG	HD	9 NY St. Mil. 1859 conv to TD, Fed 1940, inact May 1944 Cp Pendelton, VA
245*	1924	Brooklyn, NY	NG	HD	64 Inf NY 1812, Fed1940, inact Oct 1944
246*	1923	Richmond, VA	NG	HD	Fed 1940, inact Apr 1944
248*	1924	Aberdeen, WA	NG	HD	CAC Res 1909, Bn 1924-1935, Fed 1940, inact May 1944 Cp Barkley, TX
249*	1923	Salem, OR	NG	HD	Fed 1940, inact Oct 1944
250*	1923	San Francisco, CA	NG	TD	CAC Res 1909, Fed 1940, inact May 1944 Cp Gruber, OK
251	1924	San Diego, CA	NG	AA	conv to TD, Fed 1940, inact Mar 1944 South Pacific
252*	1924	Wilmington, NC	NG	TD	conv to ?, inact Apr 1944 Ft Jackson, SC
253	1940	Puerto Rico	NG	TD	Fed 1940, inact 1946
260	1924	Washington, DC	NG	HD	redesig AA 1929, Fed 1940, inact Sep 1943 Seattle, WA
261*	1940	Jersey City, NJ	NG	HD	Dover, DE 1940, Fed 1940, redes Bn Jan 1941
263*	1925	Greenwood, SC	NG	HD	Beauford Artillery 1776, Fed 1940, inact Oct 1944
264*	1925?	GA	NG	HD	conv to 214th CAR 1941, disbanded
265*	1923	Jacksonville, FL	NG	HD	Fed 1940, disb Jul 1944 Alaska
369	1924	New York, NY	NG	AA	conv from 369 Inf, inact Jun 1942 Hawaii
428	1943		RA	AA	conv from Inf, disb May 1944 South Pacific
501	1925	Boston, MA	OR	AA	reconst & act 1942 at Cp Haan, CA, inact Sep 1943 Benica CA
502	1925	New York, NY	OR	AA	reconst & act May 1942 Ft Sheridan IL, inact Sep 1943 Patterson NJ
503	1925	Williamsport, PA	OR	AA	reconst & act May 1942 Ft Lewis WA, inact Dec 1943 Ft Glenn AK
504	1925	Chattanooga, TN	OR	AA	reconst & act Jul 1942, Cp Hulen TX, inact Jan 1943
505	1925	Fort Monroe, VA	OR	AA	reconst & act Jul 1942 Cp Edwards, MA, inact Mar 1944 Italy
506	1925	Rock Island, IL	OR	AA	reconst & act Jun 1942, Cp Edwards, MA, inact Jan 1943
507	1925	Ft Leavenworth, KS	OR	AA	trans to Iowa mid 1930s, reconst & act Aug 1942 Cp Haan CA, inact Sep 1943, Long Beach CA
508	1925	El Paso, TX	OR	AA	Pittsburgh, PA 1940, reconst & act Sep 1942 Cp Edwards MA, inact Jul 1943 Italy
509	1925	Seattle, WA	OR	AA	reconst & act Dec 1942 Ft Bliss TX, inact Jan 1943
510	1925	Chester, PA	OR	AA	Philadelphia, PA 1940, reconst & actDec 1942 Ft Sheridan IL, inact Jan 1943
511	1925	Ft Monroe, VA	OR	AA	Cleveland, OH 1940, reconst & actNov 1942 Cp Haan CA, inact Jan 1943
512			OR	AA	unorganized 1940, reconst & act Jun 1942 Ft Bliss TX, inact Jan 1943
513	1925	New York, NY	OR	AA	Buffalo, NY 1940, reconst & act Sep 1942 Ft Bliss TX, inact Jan 43
514	1925	Schenectady, NY	OR	AA	reconst & act Mar 1942 Cp Davis, NC, inact May 1943
515	1924	Topeka, KS	OR	AA	reconst & act Dec 1941 Luzon Is PI, surrendered 1942
516	1925	Schuylkill Arsn, PA	OR	AA	demobilized 1933
517	1925	Presidio of SF	OR	AA	conv to 78th CAR 1940

518	1925	Presidio of SF	OR	AA	unorganized 1940
519	1925	Los Angeles, CA	OR	AA	disb 1943
520	1924		OR	AA	dropped circa 1926
521	1925	New York, NY	OR	AA	East Orange, NJ 1940, disb 1943
522	1925	New York, NY	OR	AA	disbanded 1933 Buffalo, NY
523	1925	Erie, PA	OR	AA	Pittsburgh, PA 1940, disb 1943
524	1925	Atlanta, GA	OR	AA	disb 1943 Decatur, GA
525	1925?	Indianapolis, IN	OR	AA	Charleston, WV 1940
526	1925	Rockford, IL	OR	AA	Detroit, MI 1940, disb 1943
527	1925	Des Moines, IA	OR	AA	St. Louis, MO 1940, disb 1943
528	1925?	Minneapolis, MN	OR	AA	demobilized 1933
529	1925	Seattle, WA	OR	AA	Portland, OR 1940, disb 1943
530	1925	New York, NY	OR	AA	disb 1942
531	1925	LaCrosse, WI	OR	AA	Chicago, IL 1940, disb 1943
532	1925	East St. Louis, IL	OR	AA	Springfield, IL 1940
533	1925	New York, NY	OR	AA	disb 1943
534	1925	Raliegh, NC	OR	AA	Columbia, SC 1940, disb 1943
535	1922	Louisville, KY	OR	AA	Indianpolis, IN 1930, disb 1943
536	1925	Detroit, MI	OR	AA	disb 1943
537	1925	Minneapolis, MN	OR	AA	disb 1943
538	1921	Topeka, KS	OR	AA	unorganized 1925, disb 1943
539	1925	New York, NY	OR	AA	
540	1922	Birmingham, AL	OR	AA	unorganized 1940, disb 1943
541	1933?	Lexington, KY	OR	AA	unorganized 1925
542	1923	Portland. ME	OR	AA	unorganized 1925, disb 1943
543	1925	New London, CT	OR	AA	Manchester, NH 1940
544	1925	New Orleans, LA	OR	AA	Hartford, CT 1940, disb 1943
545	1923	Jackson, MS	OR	AA	unorganized in 1925
546 to 574 regiments org 1928			OR	AA	redesignated 901 to 929 regiments 1928-30
591	1928		OR	AA	redesignated 945th 1928
592	1928		OR	AA	redesignated 946th 1928
597	1928		OR	AA	RA inactive 1933, disbanded 1943
601*	1925	Boston, MA	OR	RY	Bridgeport, CT 1940
					reconst & act AA Feb 1942 Ft Bliss TX, inact Sep 1943 Philadephia
602*	1925	New York, NY	OR	RY	reconst & act AA Feb 1942 Ft Bliss TX, inact Sep 1943, New York
603*	1925	Chester, PA	OR	RY	Philadelphia, PA 1940
					reconst & act AA Mar 1942 Culver City CA, inact Apr 1943
604*	1925	Presidio SF, CA	OR	RY	Salt Lake City, UT 1940
					reconst & act AA Mar 1942 Ft Bliss TX, inact Sep 1943 New York
605	1925	Seattle, WA	OR	RY	disbanded prior to 1938
					reconst & act AA Mar 1942 Cp Stewart GA, inact Jun 1943 Boston
606*	1925	Boston, MA	OR	TD	reconst & act AA Jun 1942 Cp Edwards MA, inact Jan 1943
607*	1925	New York, NY	OR	TD	reconst & act AA Jun 1942 Cp Hulen TX, inact Jan 1943
608	1925	Presidio SF, CA	OR	TD	conv to 56th CAR 1941
					reconst AA Jun 1942 Cp Hulen TX, inact Jan 1943
609	1925	Ft Monroe, VA	OR	TD	disb prior to 1938,
					reconst & act AA Dec 1942 Cp Edwards MA, inact Jan 1943
610	1925	Los Angeles, CA	OR		reconst & act AA Dec 1942 Cp Davis, NC, inact Jan 1943
611	1925		OR		reconst & act AA Dec 1942 Ft Bliss TX, inact Jan 1943
612	1925		OR		reconst & act AA Sep 1942 Cp Stewart GA, inact Jan 1943
613	1925	Portland, ME	OR	HD	unorganized 1940, reconst AA Apr 1942 Cp Davis NC, inact Jan 1943
614*	1925	Portland, ME	OR	HD	trans RAI before 1938, conv to 20th CAR 1940
					reconst & act AA Apr 1942 Panama, inact Dec 1943 Cp Stewart, GA
615	1925	Boston, MA	OR	HD	Wilmington, DE 1940
					reconst & act AA Apr 1942 Panama, inact Sep 1943

616*	1925	Providence, RI	OR	HD	trans RAI before 1938, conv to 23rd CAR 1940
618	1925	New London, CT	OR	HD	Elizabeth, NY 1940, disb 1944
619	1925	New York, NY	OR	HD	
620	1925	New York, NY	OR	HD	
621*	1925	Wilmington, NC	OR	HD	trans RAI prior to 1938, disb 1944
622*	1925?	Washington, DC	OR	HD	inactive 1925, trans RAI before 1938, disb 1944
623	1925	Atlanta, GA	OR	HD	Jacksonville, FL 1940
624	1933?	Oklahoma City, OK	OR	HD	listed as unorganized 1925
625*	1924	Los Angeles, CA	OR	HD	trans RAI before 1938, conv to 19th CAR 1940
626*	1925	Los Angeles, CA	OR	HD	trans RAI before 1938, disb 1944
627*	1925	Presidio SF, CA	OR	HD	trans RAI before 1938, disb 1944
628*	1925	Seattle, WA	OR	HD	Presidio SF, CA 1940, disb 1944
629	1925	Seattle, WA	OR	HD	Portland, OR 1940
630*	1925	Seattle, WA	OR	HD	trans RAI before 1938, disb 1944
631	1924	HD E. NY	OR	HD	not active, disbanded 1946
632	1924	HD Baltimore	OR	HD	not active, disbanded 1946
633	1924	HD Potomac	OR	HD	not active, disbanded 1946
634	1924	HD Cape Fear	OR	HD	not active, disbanded 1946
635	1924	HD Savannah	OR	HD	not active, disbanded 1946
636	1924	HD Jacksonville	OR	HD	not active, disbanded 1946
637	1924	HD Tampa	OR	HD	not active, disbanded 1946
638	1924	HD Mobile	OR	HD	not active, disbanded 1946
639	1924	HD New Orleans	OR	HD	not active, disbanded 1946
653	1925	New York, NY	OR	AA	disbanded prior to 1938?
701	1942	Ft Totten, NY	RA	AA	organized in 1933?, inact Apr 1943 Newport RI
901	1930	Worchester, MA	RAI	AA	
902	1930	Boston, MA	OR	AA	
903	1930	Hartford, CT	OR	AA	
906	1930	Portland, ME	RAI	AA	disbanded 1943
908	1930	New York, NY	RAI	AA	disbanded before 1938
909	1930	New York, NY	RAI	AA	disbanded 1933
910	1930	New York, NY	RAI	AA	disbanded 1943
913	1930	Washington, DC	OR	AA	
916	1930	Richmond, VA	OR	AA	
917	1930	Roanoke, VA	RAI	AA	conv to the 70th CAR 1940
925	1930	Jacksonville, FL	OR	AA	
932	1930	Columbus, OH	OR	AA	disbanded 1943
933	1930	Cincinnati, OH	RAI	AA	disbanded 1943
938	1930	Cincinnati, OH	RAI	AA	
945	1930	Detroit, MI	OR	AA	disbanded 1943
946	1928				demobilized 1933
950	1930	Lansing, MI	OR	AA	
951	1930	Chicago, IL	RAI	AA	
955	1930	Duluth, MN	OR	AA	disbanded 1943
958	1930	St. Louis, MO	OR	AA	
960	1930	Topeka, KS	RAI	AA	
969	1930	San Antonio, TX	OR	AA	disbanded 1943
970	1930	Texas	OR	AA	unorganized 1940
972	1930	Dallas, TX	RAI	AA	disbanded 1943
973	1930		RAI	AA	unorganized 1940
974	1930	Denver, CO	RAI	AA	disbanded 1943
975	1930	Los Angeles, CA	OR	AA	disbanded 1943
976	1930	Los Angeles, CA	OR	AA	disbanded 1943
977	1930	Los Angeles, CA	OR	AA	disbanded 1943
979	1930	Presidio SF, CA	OR	AA	demobilized 1933

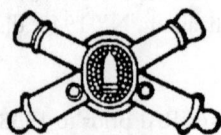

ORGANIZATION OF THE COAST ARTILLERY CORPS,1939

From *R.O.T.C. Manual, Coast Artillery, Basic, 11th Edition (Revised)*
Military Service Publishing Company, Harrisburg, PA (1939)

GENERAL ARTILLERY ORGANIZATION

Introduction

Of the manifold types of weapons with which a modern army is equipped, the most important is that one which is generically termed a gun (fire-arm). In its most general meaning, a gun may be defined as a tubular machine, closed at one end, in which the expansive force of gas is utilized to propel a projectile in a definite direction; but there are so many different kinds of guns that the term is not often used in this comprehensive sense. It is more usual to classify guns as small arms, which have an interior diameter of less than one inch, and cannon, which have an interior diameter of one inch or more.

Originally, the term artillery was applied to all devices for propelling missiles through the air, so when firearms were introduced they were included with the other missile-throwing weapons. To distinguish them from the mechanical weapons, all guns, whatever their size, batteries are designated according to their functions as headquarters batteries, gun batteries, howitzer batteries, mortar batteries, machine-gun batteries, submarine mine batteries, searchlight batteries, service batteries, ammunition batteries, or sound-ranging batteries.

A **headquarters battery** is assigned organically to each battalion, regiment, and brigade. It is organized for purposes of command, intelligence, reconnaissance, observation, administration, signal communication, liaison, fire direction, and supervision of supply. Its strength, composition, and organization depend upon the functions of the unit to which it is assigned.

Gun, howitzer, mortar, and machine-gun batteries are organized primarily for the delivery of fire. Each is equipped with pieces of like type and caliber and has personnel and equipment necessary for command, maneuver, signal communication, observation, and delivery of fire. Each normally operates as a part of a battalion, and each is given a permanent alphabetical designation within the regiment.

Submarine mine batteries are organized for the installation, operation, and maintenance of controlled submarine mine fields in the defense of harbors. Their strength and organization depend upon the extent of the mine fields and upon the armament with which they may be provided.

Searchlight batteries are organized primarily for the operation of searchlights used in the illumination of targets at night.

A **service battery** is assigned organically to each regiment of mobile artillery. It is organized for the supply and baggage transport of the regiment.

Ammunition batteries are organically a part of an ammunition train and are organized for the transport and supply of ammunition.

Sound-ranging batteries are organized primarily for the purpose of determining ranges by observation on sources of sound.

Combat trains, which are analogous to batteries, are organically a part of some battalions. They are organized for the purpose of furnishing a mobile reserve of ammunition for the battalions and a means of transporting ammunition from an ammunition distributing point to the batteries. In time of peace, battalion combat trains may be combined with battalion headquarters batteries.

Battalion. The battalion is primarily a tactical unit. It consists of a headquarters and headquarters battery, a battalion combat train (except in certain units of seacoast artillery and of general headquarters reserve artillery), and either three batteries, as in light artillery, or two batteries, as in medium and heavy artillery. Battalions normally operate as part of a regiment and are given permanent numerical designations within the regiment.

Groups and Groupments. It is sometimes necessary to distribute batteries and battalions so that tactical control from battalion and regimental headquarters becomes difficult or even impossible. In such cases, it is customary to form temporary organizations known as groups and groupments from among units having a common tactical mission. Commanders and staffs are obtained from the headquarters of the organic units from which the groups and groupments are assembled. Groups and groupments are not organized unless the normal organization is inadequate or unsuitable. A group consists of two or more batteries assembled from different battalions. A groupment consists of two or more battalions, groups, or larger tactical units assembled from different organic units.

Regiment. The regiment is both an administrative and a tactical unit. It consists of a headquarters and headquarters battery, a band, a service battery, an attached chaplain, an attached medical detachment, and either three 2-battery battalions, as in medium and heavy artillery, or two 3-battery battalions, as in light artillery. Regiments of fixed artillery have a variable organization which depends upon the amount of seacoast artillery available. Regiments of field artillery normally operate as part of a brigade; regiments of antiaircraft artillery may operate as part of a brigade; regiments of seacoast artillery are not organized into brigades. A regiment is given a permanent numerical designation.

Ammunition Train. An ammunition train is assigned organically to each infantry division, corps, and field army. In divisions and corps, the train is an integral part of the artillery brigade. Ammunition trains are organized to provide a mobile reserve of ammunition and to transport ammunition. Each consists of a train headquarters and ammunition batteries of a number and type which depend upon the type of train (division, corps, or army).

Brigade. A brigade is the largest organic artillery unit. Its functions are primarily tactical, and it consists of a headquarters and headquarters battery, three or more regiments, and, in division and corps artillery, an ammunition train. In addition, a corps artillery brigade or a brigade of heavy artillery from general headquarters reserve artillery includes a sound and flash battalion for observation and an attached ordnance company for ordnance repairs.

ORGANIZATION OF THE COAST ARTILLERY CORPS

The R.O.T.C. Manual, Coast Artillery, Basic, 11th Edition (revised)
The Military Service Publishing Company, Harrisburg, PA 1939

In the Army of the United States, the service of artillery is divided between two arms known as the Field Artillery and the Coast Artillery Corps. When the two arms were first separately organized, the Field Artillery served mobile weapons and the Coast Artillery Corps served fixed weapons; but the recent rapid development of armament and the introduction of new types have modified the original distinction. At present, the Field Artillery may be defined as that arm of the military service which mans and operates all mobile artillery designed and equipped primarily for use against relatively immobile targets; and the Coast Artillery Corps may be defined as that arm of the military service which mans and operates all artillery designed primarily for use against mobile targets. Thus, with some modification both arms may make use of weapons of the same type and caliber.

Coast Artillery Corps

The Coast Artillery Corps includes all fixed artillery, all antiaircraft artillery, all railway artillery, all tractor-drawn artillery especially assigned for coast-defense purposes, all controlled submarine mine installations, and all subaqueous sound-ranging installations, together with the searchlights, power plants, communications, trains, and other accessories necessarily incident to the maintenance and tactical employment of these weapons. It consists of units of the Regular Army, of the National Guard, and of the Organized Reserves. Its mission is the attack of enemy naval vessels by means of artillery fire and submarine mines and the attack of enemy aircraft by means of fire from the ground.

The Chief of Coast Artillery

Supervision of all activities of the Coast Artillery Corps devolves upon the Chief of Coast Artillery, under the Chief of Staff of the Army. He does not exercise any command of troops. His duties pertain solely to supervision of coast artillery training, development, assignment, and other activities, to the formulation of tactical doctrines and mobilization plans, to cooperation with other arms and services, and to advice of the Chief of Staff on matters pertaining to the Coast Artillery Corps. He is assisted by a staff consisting of an executive, a personnel section, a materiel and finance section, an organization and training section, and a plans and project section.

Coast Artillery Districts

All coast artillery troops within corps areas embracing sections of the seacoast are organized into coast artillery districts, which have territorial limits and numerical designations coinciding with those of the corps areas. The district headquarters consisting of a district commander and such staff as may be assigned by the War Department, is normally at or near the headquarters of the corps area of which it forms a part. The district commander commands all coast artillery troops stationed within the territorial limits of the district, including the coast artillery units of the Organized Reserves and those of the National Guard when in the service of the United States.

In time of war, the sea frontier of the United States will be organized into frontier commands, sectors, and subsectors. When such commands come into existence, the command of all coast artillery troops located within them passes automatically from the coast artillery district commander to the frontier, sector, or subsector commander, as the case may be. The coast artillery district commander then becomes available for assignment to duty by the frontier commander.

Unit Organization

For purposes of administration and training and for the purpose of facilitating tactical employment outside of harbor defenses, coast artillery troops other than subaqueous sound-ranging units are organized into batteries, battalions, and regiments. A brigade organization is provided for antiaircraft and tractor-drawn artillery. Subaqueous sound-ranging troops are organized into batteries only.

The organization of all the various kinds of artillery is very similar, since it provides in all cases for the command, service, and tactical employment of artillery weapons. Differences in the details of the organization necessarily exist because of the differences in caliber, tactical use, mobility, and means of transport of the various weapons.

Coast artillery regiments of the Regular Army are designated by number from 1 to 100; those of the National Guard are numbered from 101 to 300; and those of the Organized Reserves are numbered from 301 upward. Brigades of the Regular Army are numbered from 1 to 50, those of the National Guard from 51 to 150, and those of the Organized Reserves from 151 upward.

Antiaircraft Artillery

The assignment of antiaircraft artillery to the elements of armies in the field is as follows:

(1) To each corps: 1 antiaircraft artillery regiment (3 inch guns).

(2) To each field army: 1 antiaircraft artillery brigade (3 inch or 105-mm guns).

(3) To general headquarters reserve: all available antiaircraft artillery not allotted to armies and corps. These units are used in covering establishments, sensitive points, and important places in the communications zone and as a reserve for reinforcing armies, and equipped of equipment, independent corps, and cavalry operating directly under general headquarters. Reinforcement of subordinate units may be by brigade or by regiment.

(4) For overseas possessions and for the zone of the interior, such fixed and mobile antiaircraft artillery brigades and regiments are provided as defense projects may require.

(5) For frontier defense, brigades and regiments are assigned according to the special requirements as determined by defense plans.

Railway and Tractor-Drawn Artillery

Railway and tractor-drawn artillery units are normally allotted to frontier, sector, subsector, and harbor-defense commands in accordance with the requirements of the situation. Their distribution is based upon the needs of the localities to be defended, and they are generally assigned by battalion or regiment. In coastal operations, battalions are usually so widely separated that direct control by regimental commanders becomes impracticable. At the same time, battalions from different regiments may be so located that they may readily be grouped under a single commander. For these reasons, battalions will frequently be formed in groupments, as is usual with fixed artillery in harbor defenses.

In the defense of unfortified harbors, mobile seacoast artillery is organized and emplaced under the same principles that govern the organization and emplacement of artillery in a harbor defense. Such groups and groupments are formed as the mission, local hydrography, and tactical considerations may demand. When large amounts of mobile artillery are assembled in a particular locality, tactical and administrative considerations may warrant their formation by the sector commander into harbor defenses.

When assigned to harbor defenses, railway and tractor-drawn artillery units are absorbed in the harbor-defense organization and become an integral part of the command. They are assigned to positions by the harbor-defense commander so that their fire best supports or supplements that of the fixed armament and to groups and groupments or forts as the organization may require.

Fixed Antiaircraft Artillery

The organization of antiaircraft artillery regiments assigned to fixed armament is made to conform to the requirements of each locality provided with such armament. They are given a flexible organization which can be adapted to the requirements of the fixed armament or to employment with the field forces without extensive reorganization. The gun defense may be organized into one or more battalions, the number of batteries may vary, and the size of each battery conforms to the number of guns and to the character and amount of the equipment. The organization of the machine gun units will conform more closely to that prescribed for mobile units, because their employment may require changes of position and alterations of concentration of fire to meet special situations or the tactics of the enemy.

Brigades

There are two types of brigades, the antiaircraft and the 155-mm gun, tractor-drawn brigades.

The Antiaircraft Brigade. The antiaircraft brigade consists of a headquarters and headquarters battery, and three regiments. Its staff consists of nine officers in peace and war. (Executive, adjutant, intelligence officer, plans and training officer, supply officer, communications officer, munitions officer, assistant communications officer and two aides).

The 155-mm Gun Tractor-Drawn Brigade. This brigade consists of a headquarters and headquarters battery and three regiments. Its staff is organized as shown for the antiaircraft brigade.

Antiaircraft Regiments

This regiment is composed of a headquarters and band, headquarters battery, service battery and two battalions. The 1st battalion is a gun and the 2d battalion, a machine gun battalion. The staff consists of the executive, adjutant, intelligence officer, plans and training officer, supply officer (commands the service battery and is included in the totals of that organization), communications officer, munitions officer. (See Plate 1.)

a. Antiaircraft regimental headquarters battery. The headquarters battery consists of a battery headquarters, a maintenance section and an operation section which is made up of a regimental headquarters detail, intelligence detail, plans and training detail and communication detail.

b. Antiaircraft regimental service battery. This battery consists of a battery headquarters section, a regimental section (divided into a personnel detail and a supply officers' detail), a battalion section (with a battalion detail for the two battalions) and a maintenance section.

c. Antiaircraft gun battalion. The first battalion, gun, consists of a headquarters battery and combat train, searchlight battery and (in war-time) three and (in peace), two batteries. (See Plate 2.)

(1) *The headquarters and headquarters battery and combat train.* This organization includes the battalion headquarters. The staff consists of the executive, adjutant, intelligence officer, plans and training officer, communication officer, reconnaissance officer, and the munitions officer who commands the combat train. The battalion supply officer commands the battalion section of the service battery and is included in the totals of the service battery.

The headquarters battery proper consists of the battery headquarters section; an operations section (consisting of a battalion headquarters detail, intelligence detail, and a communication detail; combat train (war time) consisting of a headquarters and three sections; and a maintenance section.

(2) *Searchlight battery.* This battery consists of a headquarters section, maintenance section and two (peace) and three (war) platoons. Each platoon consists of five sections with a platoon headquarters and a communication detail.

(3) *Antiaircraft gun battery.* This battery is organized into (a) battery headquarters consisting of a battery headquarters section, command detail, range detail and communication detail; (b) a firing section consisting of four gun sections and machine gun and executive officers' detail, and (c) a maintenance section.

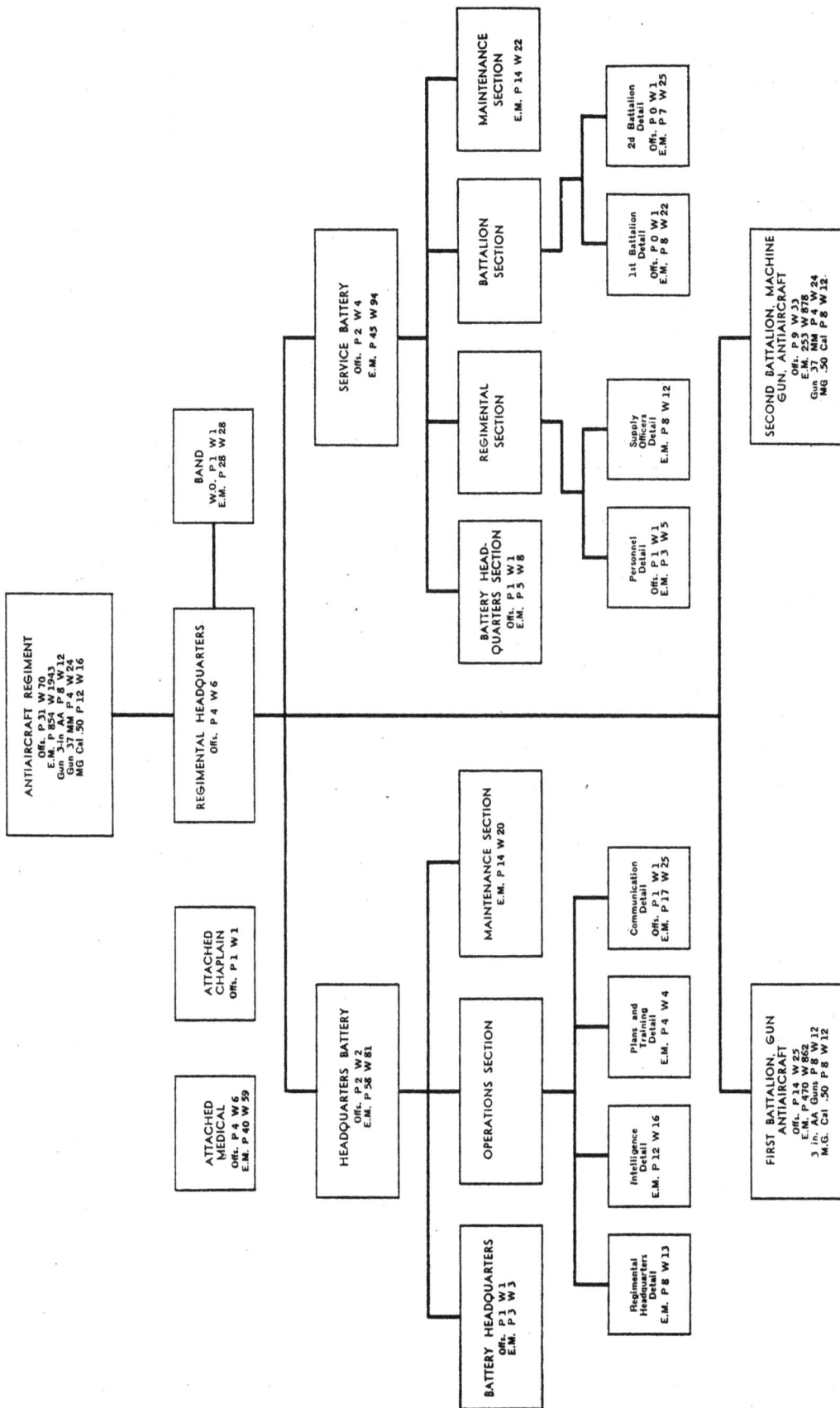

Coast Artillery Antiaircraft Regiment

ANTIAIRCRAFT REGIMENT
Offs. P 31 W 70
E. M. P 854 W 1943
Gun 3-in AA P 8 W 12
Gun 37 MM P 4 W 24
MG Cal .50 P 12 W 16

REGIMENTAL HEADQUARTERS
Offs. P 4 W 6

ATTACHED CHAPLAIN
Offs. P 1 W 1

BAND
W.O. P 1 W 1
E.M. P 28 W 28

ATTACHED MEDICAL
Offs. P 4 W 6
E.M. P 40 W 59

HEADQUARTERS BATTERY
Offs. P 2 W 2
E.M. P 58 W 81

SERVICE BATTERY
Offs. P 2 W 4
E.M. P 45 W 94

BATTERY HEADQUARTERS
Offs. P 1 W 1
E.M. P 3 W 3

OPERATIONS SECTION

MAINTENANCE SECTION
E.M. P 14 W 20

BATTERY HEADQUARTERS SECTION
Offs. P 1 W 1
E.M. P 5 W 8

REGIMENTAL SECTION

BATTALION SECTION

MAINTENANCE SECTION
E.M. P 14 W 22

Regimental Headquarters Detail
E.M. P 8 W 13

Intelligence Detail
E.M. P 12 W 16

Plans and Training Detail
E.M. P 4 W 4

Communication Detail
Offs. P 1 W 1
E.M. P 17 W 25

Personnel Detail
Offs. P 1 W 1
E.M. P 3 W 5

Supply Officers Detail
E.M. P 8 W 12

1st Battalion Detail
Offs. P 0 W 1
E.M. P 8 W 22

2d Battalion Detail
Offs. P 0 W 1
E.M. P 7 W 25

FIRST BATTALION, GUN ANTIAIRCRAFT
Offs. P 14 W 25
E.M. P 470 W 862
3 in. AA Guns P 8 W 12
M.G. Cal .50 P 8 W 12

SECOND BATTALION, MACHINE GUN, ANTIAIRCRAFT
Offs. P 9 W 33
E.M. P 251 W 878
Gun 37 MM P 4 W 24
MG .50 Cal P 8 W 12

Coast Artillery Antiaircraft Gun Battalion

ANTIAIRCRAFT GUN BATTALION
Offs. P 14 W 25
E.M. P 470 W 862
Gun 3 inch AA P 8 W 12
MG .50 Cal. P 8 W 12

- HEADQUARTERS BATTERY AND COMBAT TRAIN
  Offs. P 2 W 3
  E.M. P 56 W 108
  - Battery Headquarters Section
    Offs. P 1 W 1
    E.M. P 2 W 2
  - Operations Section
    - Battalion Headquarters Detail — E.M. P 6 W 9
    - Intelligence Detail — E.M. P 6 W 8
    - Plans and Training Detail — E.M. P 2 W 3
    - Communication Detail — Offs. P 0 W 1, E.M. P 15 W 27
  - Combat Train
    - Headquarters — Offs. P 1 W 1, E.M. P 2 W 5
    - 1st Section — E.M. P 12 W 12
    - 2d Section
    - 3d Section
  - Maintenance Section — E.M. P 11 W 18

- ANTIAIRCRAFT SEARCHLIGHT BATTERY
  Offs. P 3 W 4
  E.M. P 154 W 250
  Searchlight P 10 W 15
  - Battery Headquarters Section — Offs. P 1 W 1, E.M. P 7 W 8
  - Platoon — Offs. P 1 W 1, E.M. P 66 W 73, Searchlights P 5 W 5
    - Platoon Headquarters — Offs. P 1 W 1, E.M. P 4 W 7
    - Communication Detail — E.M. P 7 W 11
    - Searchlight Section
      - Sound Locator Squad — E.M. P 6 W 6
      - Searchlight Squad — E.M. P 5 W 5
    - Searchlight Section
    - Searchlight Section
    - Searchlight Section
    - Searchlight Section
  - Platoon (War)
  - Maintenance Section — E.M. P 15 W 23

- ANTIAIRCRAFT GUN BATTERY
  Offs. P 3 W 4
  E.M. P 130 W 168
  Gun 3-in AA P 4 W 4
  MG Cal .50 P 4 W 4
  - Battery Headquarters
    - Battery Headquarters Section — Offs. P 1 W 1, E.M. P 4 W 5
    - Command Detail — E.M. P 6 W 8
    - Range Detail — Offs. P 1 W 1, E.M. P 20 W 21
    - Communication Detail — E.M. P 7 W 8
    - Machine Gun and Executive Officers Detail — Offs. P 1 W 2, E.M. P 13 W 20
  - Firing Section
    - Gun Section — E.M. P 16 W 21, Gun AA P 1 W 1
    - Gun Section
    - Gun Section
    - Gun Section
  - Maintenance Section

- ANTIAIRCRAFT GUN BATTERY

- ANTIAIRCRAFT GUN BATTERY (War)

d. *Antiaircraft machine gun battalion.* The organization of this battalion consists of a headquarters and headquarters battery and combat train, a machine gun battery, and one (peace) or three (war) 37 -mm batteries. The command and staff consists of a battalion commander, an executive officer, adjutant, communication officer (commands headquarters battery); plans and training officer; supply officer (member of service battery); and the munitions officer who commands the combat train. (See Plate 3.)

(1) *Headquarters, headquarters battery and combat train.* This organization consists of battalion headquarters (4 officers in war time); a battery headquarters section; an operations section which consists of a battalion headquarters detail, intelligence detail, plans and training detail, and a communication detail; a combat train (war); and a maintenance section.

(2) *Antiaircraft machine gun battery.* This organization consists of a battery headquarters; an operations section, consisting of a command detail and a communication detail; a maintenance section, and two (peace) or three (war) platoons. Each platoon (.50 caliber machine gun) has a headquarters and range section and two sections of two squads each.

(3) *Antiaircraft 37 -mm gun battery.* This, battery consists of (a) a battery headquarters section; (b) an operations section, consisting of command detail and communication detail; (c) a maintenance section and (d) two (peace) or four (war) platoons, each consisting of a platoon headquarters, a range section, and two gun sections.

155-mm Gun Tractor-Drawn Regiments

The peace organization of this regiment consists of a regimental headquarters and band, headquarters battery, service battery and two battalions. In time of war it has an additional battalion. Its war time regimental staff consists of the executive, the adjutant, intelligence officer, the plans and training officer, supply officer, communications officer, munitions officer, reconnaissance officer assistant communication officer and assistant plans and training officer. The supply office commands the service battery and is included in the totals of that organization. (See Plate 4.)

a. *Regimental headquarters battery, 155-mm gun.* This battery consists of (1) a battery headquarters; (2) operations section organized into a regimental headquarters detail, plans and training detail, intelligence and liaison detail, reconnaissance detail and a communication detail; and (3) a maintenance section.

b. *Regimental service battery, 155-mm gun.* The battery comprises (1) a battery headquarters section; (2) the regimental section, consisting of a personnel detail and a supply officer's detail; (3) three battalion sections and a (4) maintenance section.

c. *155-mm gun, tractor-drawn battalion.* The battalion consists of a headquarters and headquarters battery, a combat train, and two batteries. At war strength, the battalion is commanded by a major. The staff consists of the executive, adjutant, plans and training officer, intelligence officer, liaison officer, reconnaissance officer, communication officer, and the supply officer, who commands the battalion section of the service battery and is included in the totals of that battery.

(1) *155-mm gun tractor-drawn battalion headquarters battery.* The battery consists of (a) a battery headquarters section; (b) an operations section comprising a battalion headquarters detail, a plans and training detail, an intelligence and liaison detail, a reconnaissance detail and a communication detail; and (c) a maintenance section.

(2) *155-mm gun tractor-drawn battalion combat train.* This train consists of (a) headquarters section; (b) two platoons of two sections each and a maintenance section.

(3) *155-mm gun tractor-drawn battery.* The battery consists of (a) a battery headquarters section; (b) an operations section, consisting of a command post detail, a communication detail and a reconnaissance detail; (c) the firing battery comprising two gun platoons and a machine gun detail. A gun platoon consists of two gun sections and platoon detail. There are four 155-mm guns in the battery.

ANTIAIRCRAFT MACHINE GUN BATTALION
Off. P 9 W 33
E.M. P 255 W 878
Gun 37-MM P 4 W 24
MG Cal .50 P 8 W 12

ANTIAIRCRAFT 37-MM GUN BATTERY
War

ANTIAIRCRAFT 37-MM GUN BATTERY
War

ANTIAIRCRAFT 37-MM GUN BATTERY
Off. P 3 W 7
E.M. P 101 W 203
Gun 37-MM P 4 W 8

ANTIAIRCRAFT MACHINE GUN BATTERY
Off. P 3 W 6
E.M. P 130 W 203
MG Cal .50 P 8 W 12

HEADQUARTERS BATTERY AND COMBAT TRAIN
Off. P 1 W 2
E.M. P 22 W 66

37-MM Gun Platoon
War

37-MM Gun Platoon
War

37-MM Gun Platoon
War

Gun Section
E.M. P 8 W 9

Range Section
E.M. P 17 W 18

37-MM Gun Platoon
Off. P 1 W 1
E.M. P 36 W 41
Guns 37-MM P 2 W 2

Platoon Headquarters
Off. P 1 W 1
E.M. P 3 W 5

Maintenance Section
E.M. P 14 W 15

Operations Section

Battery Headquarters Section
Off. P 1 W 2
E.M. P 3 W 5

Communication Detail
Off. P 0 W 1
E.M. P 8 W 9

Command Detail
E.M. P 4 W 6

Platoon
War

Platoon

Platoon
Off. P 1 W 1
E.M. P 53 W 55

Section

Section
E.M. P 19 W 19
MG Cal .50 P 2 W 2

Platoon Headquarters and Range Section
Off. P 1 W 1
E.M. P 15 W 17

Squad

Squad
E.M. P 9 W 9
MG Cal .50 P 1 W 1

Maintenance Section
E.M. P 9 W 14

Operations Section

Battery Headquarters
Off. P 3 W 5
E.M. P 3 W 5

Communications Detail
Off. P 0 W 1
E.M. P 8 W 13

Command Detail
E.M. P 4 W 6

Maintenance Section
E.M. P 1 W 11

Combat Train
E.M. P 0 W 20
War

Communication Detail
E.M. P 13 W 19

Plans and Training Detail
E.M. P 1 W 1

Operations Section

Intelligence Detail
E.M. P 3 W 5

Battery Headquarters Section
Off. P 1 W 1
E.M. P 1 W 3

Battalion Headquarters Detail
E.M. P 3 W 7

Coast Artillery Antiaircraft Machine Gun Battalion

Coast Artillery 155-mm Tractor-Drawn Gun Regiment

Railway Artillery Regiments

The peace-time organization of this regiment consists of a headquarters and band, headquarters battery, service battery and two battalions. The war time organization has an additional battalion. The staff consists of the executive, adjutant, intelligence, plans and training officer, supply officer (commands service battery and is included in the totals of that organization), communication officer, reconnaissance officer, railway and munitions officer. (See Plate 5.)

a. *Regimental headquarters battery, railway artillery.* The battery consists of (1) a battery headquarters; (2) the regimental section, subdivided into a regimental headquarters detail, intelligence detail, plans and training detail, communication detail, reconnaissance detail and a railway and munitions detail; and (3) a maintenance section. The organization of this battery is the same for both types of railway regiments. b. *Regimental service battery, railway artillery.* The battery has (1) a battery headquarters section, (2) a regimental section, consisting of a personnel detail and a supply officer's detail; (3) a maintenance section and (4) three battalion sections.

c. *Railway artillery battalion* (except those for 12-inch and 14-inch guns, see below.) The battalion consists of a headquarters battery and two batteries. It is commanded by a major, and his staff consists of an adjutant, plans and training officer, intelligence officer, communication officer, reconnaissance officer, railway and munitions officer, and a supply officer who commands the battalion section of the service battery and is included in the totals of the service company.

(1) *Battalion headquarters battery, railway battalion.* The battery consists of (a) a battery headquarters section; (b) an operations section, made up of a battalion headquarters detail, intelligence detail, plans and training detail, communication detail, reconnaissance detail, railway and munitions detail; and (c) a maintenance section. This organization is used by both types of railway regiments.

(2) *Railway artillery battery* (except those for 12-inch and 14-inch guns, see below). The battery has (a) a battery headquarters, consisting of a command detail, range section, and communication section; (b) a firing section consisting of four gun sections and a machine gun and executive officer's detail. This battery is armed with four 7-inch, 8-inch, or 10-inch guns or with 12-inch mortars. It is also provided with four 50 caliber machine guns AA, Browning.

The only difference in the organization of railway artillery regiment for 12-inch and 14-inch guns and the other railway regiments is in the number of guns in the firing battery, two instead of four, and the number of enlisted men in the battalion and gun batteries.

Harbor Defense Regiments

There are three types of harbor defense regiments known as "A", "B", and "C". All three have the same command and staff. The colonel commands the regiment and his staff consists of an executive, adjutant, intelligence officer, plans and training officer, supply officer, searchlight officer, communication officer, assistant adjutant and assistant plans and training officer. In the type "A" regiment its organization in war time consists of a regimental headquarters and band, a headquarters battery and three battalions of three batteries each. In peace it has one less battalion. The type "B" regiment is organized in the same manner except that in war and in peace it has two battalions of three batteries each. The type "C" regiment is the same except that in war time it has three battalions of four batteries each and in two battalions of three batteries each. (See Plates 6, 7, and 8.)

a. *Regimental headquarters battery, harbor defense types "A", "B," and "C."* This battery has (1) a harbor defense section, consisting of (a) an administrative section composed of a headquarters, supply and maintenance sections; (b) a tactical section containing an intelligence, plans and training, and command post sections; (c) technical section, consisting of an artillery engineer, searchlight, and communication sections; and (2) a battery section consisting of a headquarters and a maintenance section.

RAILWAY ARTILLERY REGIMENT (Excepting 12-inch and 14-inch Guns)
Offs. P 34 W 68
E.M. P 853 W 1687
MG Cal .50 P 26 W 36
Guns 7, 8, 10-inch or 12-inch mortar P 16 W 24

BAND
WO P 1 W 1
E.M. P 28 W 28

ATTACHED MEDICAL
Offs. P 2 W 3
E.M. P 18 W 37

ATTACHED CHAPLAIN
Off. P 1 W 1

REGIMENTAL HEADQUARTERS
Offs. P 4 W 6

HEADQUARTERS BATTERY
Offs. P 2 W 3
E.M. P 46 W 81

SERVICE BATTERY
Offs. P 4 W 5
E.M. P 39 W 81

RAILWAY ARTILLERY BATTALION (Excepting 12-inch and 14-inch Guns)
Offs. P 12 W 18
E.M. P 370 W 499
Gun, 7, 10-inch or 12-inch Mortars
MG Cal .50 P 10 W 10
P 8 W 8

RAILWAY ARTILLERY BATTALION (Excepting 12-in and 14-inch guns)

RAILWAY ARTILLERY BATTALION (Excepting 12-in and 14-inch guns) — War

First Battalion Section
Offs. P 1 W 1
E.M. P 10 W 17

Second Battalion Section

Third Battalion Section — War

Maintenance Section
E.M. P 8 W 15

Battery Headquarters
Offs. P 1 W 1
E.M. P 5 W 6

Regimental Section

Regimental Headquarters Detail
E.M. P 6 W 30

Intelligence Detail
E.M. P 3 W 8

Plans and Training Detail
E.M. P 1 W 2

Communication Detail
Off. P 1 W 1
E.M. P 19 W 31

Maintenance Section
E.M. P 7 W 16

Reconnaissance Detail
E.M. P 4 W 5

Railway and Munitions Detail
Off. P 0 W 1
E.M. P 1 W 3

Battery Headquarters Section
Off. P 1 W 1
E.M. P 7 W 8

Regimental Section

Personnel Detail
Off. P 1 W 1
E.M. P 2 W 3

Supply Officer Detail
E.M. P 2 W 4

Railway Artillery Battalion Headquarters Battery
Offs. P 4 W 8
E.M. P 62 W 131

Battery Headquarters Section
Offs. P 1 W 1
E.M. P 4 W 4

Operations Section

Maintenance Section
E.M. P 7 W 12

Battalion Headquarters Detail
E.M. P 4 W 5

Intelligence Detail
E.M. P 1 W 7

Plans and Training Detail
E.M. P 2 W 4

Communication Detail
Off. P 0 W 1
E.M. P 17 W 28

Reconnaissance Detail
E.M. P 5 W 9

Railway and Munitions Detail
Off. P 0 W 1
E.M. P 22 W 62

Railway Artillery Battery

Railway Artillery (Excepting 12-inch and 14-inch guns)
Offs. P 4 W 5
E.M. P 154 W 184
Gun P 4 W 4
MG P 4 W 4

Battery Headquarters

Command Detail
Off. P 2 W 2
E.M. P 7 W 12

Range Section
E.M. P 23 W 29

Communications Section
E.M. P 9 W 14

Maintenance Section
E.M. P 8 W 13

Firing Section

Machine Gun and Executive Officers' Detail
Off. P 2 W 3
E.M. P 11 W 12

Gun Section
E.M. P 24 W 26

Gun Section

Gun Section

Gun Section

Gun Section

Coast Artillery Railway Artillery Regiment (except for 12-in & 14-in)

COAST ARTILLERY REGIMENT, HARBOR DEFENSE
Type "A"
Offs. P 30 W 69
E.M. P 820 W 1583

ATTACHED MEDICAL
Offs. P 4 W 6
E.M. P 27 W 45

ATTACHED CHAPLAIN
Off. P 1 W 1

REGIMENTAL HEADQUARTERS
Offs. P 5 W 10

BAND
WO P 1 W 1
E.M. P 28 W 28

Headquarters Battery
Types "A", "B", and "C"
Offs. P 3 W 5
E.M. P 128 W 172

Coast Artillery Battalion
Harbor Defense, Type "A" or "B"
Offs. P 11 W 18
E.M. P 332 W 461

Coast Artillery Battalion
Harbor Defense, Type "A" or "B"
War

Harbor Defense Section

Administrative

Tactical

Technical

Battery Section

Headquarters
E.M. P 8 W 9

Supply
E.M. P 2 W 6

Maintenance
E.M. P 6 W 10

Intelligence
E.M. P 1 W 6

Plans & Training
E.M. P 4 W 5

Command Post
E.M. P 7 W 11

Artillery Engineer
E.M. P 25 W 33

Searchlight
E.M. P 33 W 40

Communications
E.M. P 13 W 15

Headquarters
Off. P 3 W 5
E.M. P 6 W 6

Maintenance
E.M. P 23 W 31

Coast Artillery Harbor
Defense Battalion Headquarters
Battery, Types "A", "B" and "C"
Offs. P 2 W 6
E.M. P 20 W 41

Coast Artillery Harbor
Defense Battery Types
"A", "B", "C", and "D"
Offs. P 3 W 4
E.M. P 104 W 140

Coast Artillery Harbor
Defense Battery Types
"A", "B", "C", and "D"

Battalion Headquarters Detail
E.M. P 4 W 9

Intelligence Detail
E.M. P 3 W 5

Battery Headquarters Section
Offs. P 1 W 1
E.M. P 1 W 4

Operations Section

Maintenance Section
E.M. P 5 W 9

Plans and Training Detail
E.M. P 2 W 5

Communication Detail
Offs. P 0 W 1
E.M. P 5 W 9

Command Detail
Offs. P 1 W 1
E.M. P 5 W 7

Range Section
Offs. P 1 W 1
E.M. P 22 W 26

Battery Headquarters

Communica-tion Section
E.M. P 1 W 4

Firing Section

Maintenance
E.M. P 8 W 9

Executive Officers' Detail
Offs. P 1 W 2
E.M. P 2 W 2

Gun Section
E.M. P 33 W 46

Gun Section

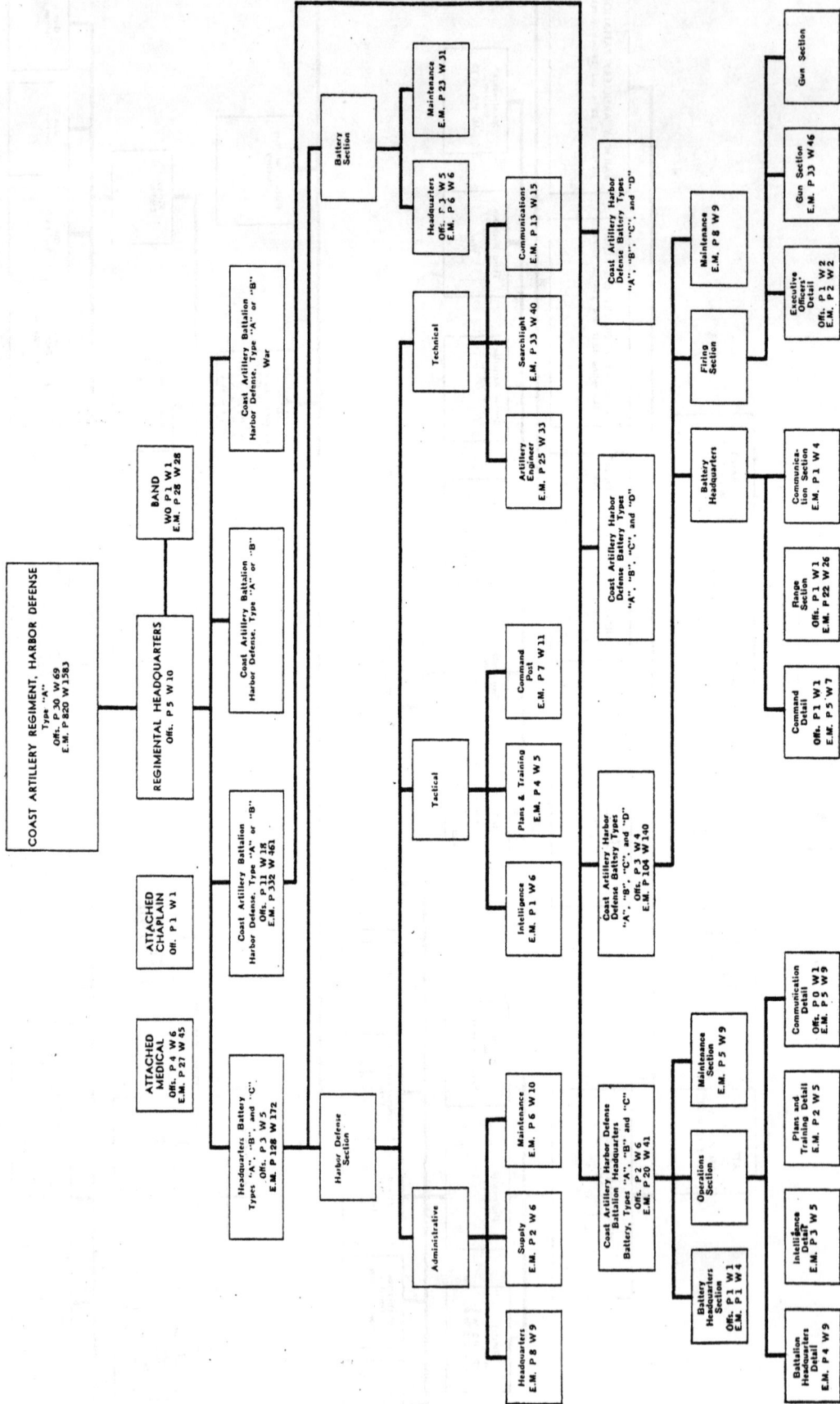

Coast Artillery Harbor Defense Regiment, Type "A"

```
                        ┌─────────────────────────────┐
                        │   COAST ARTILLERY REGIMENT,  │
                        │   Harbor Defense Type "B"    │
                        │     Offs. P 30  W 51         │
                        │     E.M.  P 820  W 1122      │
                        └─────────────────────────────┘
```

ATTACHED MEDICAL	ATTACHED CHAPLAIN	REGIMENTAL HEADQUARTERS	BAND
Offs. P 4 W 6	Offs. P 1 W 1	Offs. P 5 W 10	WO P 1 W 1
E.M. P 27 W 45			E.M. P 28 W 28

Headquarters Battery	Coast Artillery Harbor Defense Battalion	Coast Artillery Harbor Defense Battalion
Offs. P 3 W 5	Offs. P 11 W 18	
E.M. P 128 W 172	E.M. P 332 W 461	
(Same as in Type "A" Regiment)	(Same as in Type "A" Regiment)	

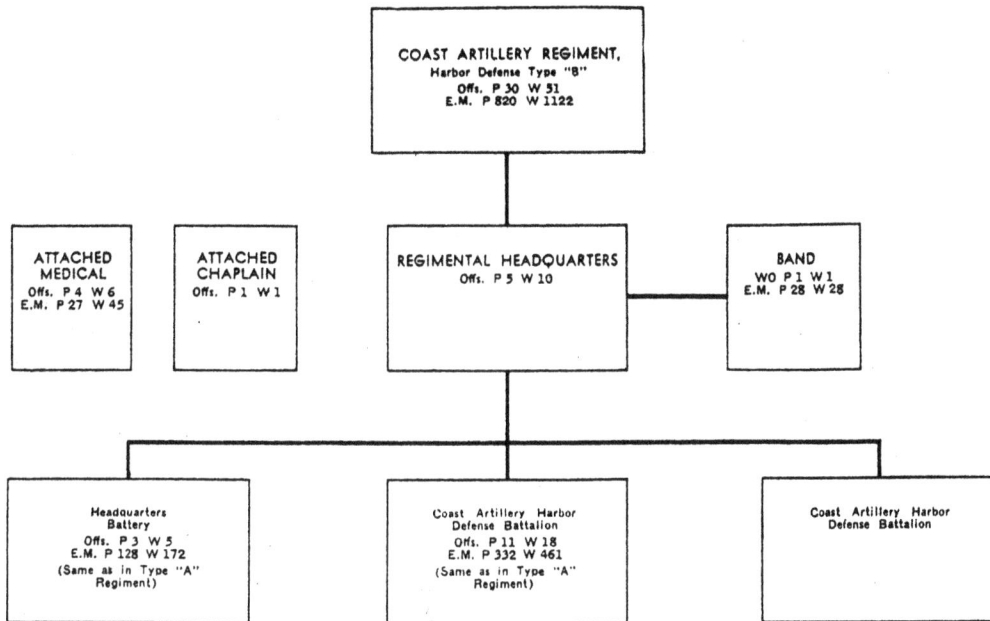

Coast Artillery Harbor Defense Regiment, Type "B"

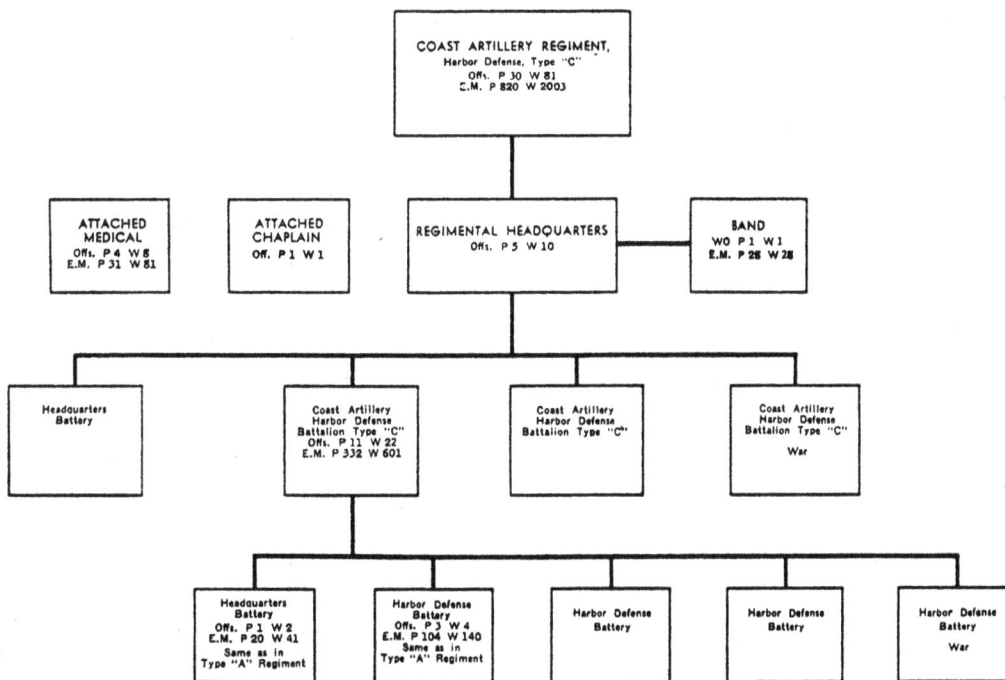

```
                        ┌─────────────────────────────┐
                        │   COAST ARTILLERY REGIMENT,  │
                        │  Harbor Defense, Type "C"    │
                        │     Offs. P 30  W 81         │
                        │     E.M.  P 820  W 2003      │
                        └─────────────────────────────┘
```

ATTACHED MEDICAL	ATTACHED CHAPLAIN	REGIMENTAL HEADQUARTERS	BAND
Offs. P 4 W 8	Off. P 1 W 1	Offs. P 5 W 10	WO P 1 W 1
E.M. P 31 W 81			E.M. P 28 W 28

Headquarters Battery	Coast Artillery Harbor Defense Battalion Type "C"	Coast Artillery Harbor Defense Battalion Type "C"	Coast Artillery Harbor Defense Battalion Type "C"
	Offs. P 11 W 22		
	E.M. P 332 W 601		War

Headquarters Battery	Harbor Defense Battery	Harbor Defense Battery	Harbor Defense Battery	Harbor Defense Battery
Offs. P 1 W 2	Offs. P 3 W 4			
E.M. P 20 W 41	E.M. P 104 W 140			War
Same as in Type "A" Regiment	Same as in Type "A" Regiment			

Coast Artillery Harbor Defense Regiment, Type "C"

The 63rd Coast Artillery (AA) Regiment assembled on the parade ground at Fort MacArthur, CA, in 1939.
Note the presence of the band, 11 lettered batteries with their guidions, and the regiment's antiaircraft equipment
behind. The regiment commander, Col. E.A. Stockton, stands out center in front of the regimental flags.
(Fort MacArthur Museum)

Coast Artillery Harbor Defense Separate Battalion, Type "D"

COAST ARTILLERY SEPARATE BATTALION.
Harbor Defense
Type "D"
Offs. P 17 W 34
E.M. P 414 W 850

BATTALION HEADQUARTERS
Offs. P 5 W 10

HARBOR DEFENSE BATTERY — War

HARBOR DEFENSE BATTERY — War

HARBOR DEFENSE BATTERY

HARBOR DEFENSE BATTERY

HARBOR DEFENSE BATTERY
Offs. P 3 W 4
E.M. P 104 W 140
Same as in Type "A" Regiment

HEADQUARTERS BATTERY. HARBOR DEFENSE
Type "D"
Offs. P 3 W 4.
E.M. P 102 W 150

Battery Section

Harbor Defense Section

Technical
Maintenance E.M. P 16 W 26
Headquarters Offs. P 3 W 4 E.M. P 5 W 6
Communication E.M. P 9 W 14
Searchlight E.M. P 31 W 39
Artillery Engineer E.M. P 17 W 28

Tactical
Command Post E.M. P 6 W 9
Plans and Training E.M. P 3 W 5
Intelligence E.M. P 1 W 3

Administrative
Maintenance E.M. P 3 W 7
Supply E.M. P 4 W 4
Headquarters E.M. P 7 W 9

b. *Harbor defense battalion, types "A," and "B."* This type of battalion consists of a (1) battalion headquarters; (2) a headquarters battery; and (3) three firing batteries. It is commanded by a major whose staff consists of an adjutant, intelligence officer, plans and training officer, and communication officer.

c. *Harbor defense battalion, type "C."* This battalion is similar to types "A" and "B" except that in time of peace it has three firing batteries and in time of war four firing batteries.

(1) Battalion headquarters battery, harbor defense, types "A," B," and "C." The battery consists of a (a) battery headquarters, section, (b) an operation section, composed of a battalion headquarters detail, intelligence detail, plans and training detail, and communications detail; and (c) a maintenance section.

(2) *Harbor defense battery, types "A,"B," "C," and "D."* The battery has (a) a battery headquarters, consisting of a command detail, range section, communication section; (b) a firing section consisting of two gun sections and executive officers detail; and (c) a maintenance section.

d. *Separate harbor defense battalion, type "D."* This battalion consists of a battalion headquarters, a headquarters battery and in peace three and in war five firing batteries. The firing battery is of the same type as found in the other harbor defense organizations. The battalion is commanded by a lieutenant colonel and has a staff consisting of an executive, adjutant, intelligence officer, plans and training officer, supply officer, searchlight officer, communication officer, assistant adjutant, and assistant plans and training officer. (See Plate 9).

(1) Separate harbor defense battalion headquarters battery, type "D." This battery has a harbor defense section, composed of (a) an administrative section which has a headquarters section, a supply section, and a maintenance section; (b) a tactical section which consists of an intelligence section, plans and training section, and command post section; (c) a technical section consisting of an artillery engineer section, searchlight section, and communication section. It also has a battery section with a headquarters and a maintenance section.

GUN BATTERY, FIXED ARTILLERY

The Gun Battery

The gun battery is a complete administrative and fire unit and includes both materiel and personnel It may on occasion be employed as a tactical unit, in which case the battery commander exercises functions of fire direction. It may pertain to fixed, tractor-drawn, railway, or antiaircraft artillery; and its weapons may be guns, howitzers, mortars, or machine guns. Antiaircraft searchlight batteries, submarine mine batteries, and subaqueous sound-ranging batteries are analogous to gun batteries. Each battery manning cannon, antiaircraft machine guns, antiaircraft searchlights, or submarine mines receives a permanent alphabetical designation, beginning with Battery A in each regiment.

Comparison of Batteries. The following 1937 table compares the battery organization and the strength of the various sections during peacetime, more men were to be added on wartime footing:

	HD	TD	Ry	AA Gun	AA MG	AA SL
Commissioned personnel:						
Battery commander	1	1	1	1	1	1
Executive	1	1	-	1	-	-
Assistant executive	1	2	2	1	-	-
Range officer	1	-	-	1	-	-
Reconnaissance officer	-	1	1	-	-	-
Reconnaissance & communication officer	-	-	-	-	1	-
Railway officer	-	-	1	-	-	-
Platoon commanders	-	-	-	-	3	3
Total commissioned	4	4	4	4	4	4
Enlisted personnel:						
Battery headquarters section	5	5	5	7	4	1
Operations section:						
Command-post detail	1	6	7	8	7	-
Range detail	22	-	24	21	-	-
Reconnaissance detail	-	-	24	-	-	-
Communication detail	-	17	17	7	9	-
Firing platoons	-	-	-	-	138	-
Firing sections	68	-	114	-	-	-
Firing battery	-	86	-	100	-	-
Platoons	-	-	-	-	-	216
Maintenance section	8	16	22	28	23	23
Total enlisted	104	184	100	109	181	210

Fixed Battery

A battery of fixed artillery consists of a number of permanently emplaced cannon of the same caliber and characteristics, grouped with the object of concentrating their fire on a single target and of being commanded by a single individual, together with all personnel, structures, installations, and equipment provided for their protection, command, and service. The number of cannon vanes from one to four, depending principally upon the caliber, which may vary from 3 to 16 inches. The permanent emplacement of each fixed battery is given an individual name by the War Department, as Battery Ruggles, by which it is known and distinguished from other emplacements.

Submarine Mine Battery

A submarine mine battery is an administrative and fire unit of the harbor defense and is employed for the installation, maintenance, and operation of a controlled mine field. It is divided into a battery headquarters section, an operations section (containing a command-post detail and a range detail), a casemate section, a loading and property section (consisting of loading, cable, explosive, and maintenance details), a planting section (consisting of mine planter, distribution-box boat, and small boat details), and a maintenance section. In addition, fire-control and gun sections may be included for manning certain of the mine defenses, or the loading and planting sections may have an additional assignment as gun sections. The battery officers include a battery commander, an executive, a casemate officer, and a loading officer.

Sound-Ranging Battery

Subaqueous sound-ranging batteries are formed as required for employment either within or without harbor defenses. No prescribed organization has been adopted, and the details of organization would be determined by local requirements and the equipment available. Each subaqueous sound-ranging battery receives a permanent numerical designation, beginning with *1st Sound-Ranging Battery.*

Battery Subdivisions

Battery headquarters section. Under the direct supervision of the battery commander, the battery headquarters section handles all the administrative work of the battery.

Operations section. The command-post detail, under the direct supervision of the battery commander, establishes and operates the battery command post and assists the battery commander in the conduct of fire. In railway batteries, it makes such reconnaissance of roads, gun positions, observing stations, and spotting stations as may be necessary and prepares such maps and orientation data as may be required.

Range detail. The range detail, under the command of the range officer, installs and maintains all fire-control and position-finding equipment, mans the position-finding, observing, and spotting stations, furnishes the data for determining the range and direction to targets, determines the positions of impacts, operates the plotting and computing devices used to determine the firing data, and records and transmits data at prescribed intervals. In tractor-drawn batteries, the duties of the range detail are included in those of the reconnaissance detail. Machine gun batteries have no range detail.

The reconnaissance detail, under the command of the reconnaissance officer, reconnoiters routes and positions, performs all topographical operations, assists in the preparation of maps and charts, and establishes observation posts.

The communication detail is responsible for the establishment of all means of signal and fire-control communication available to the battery. Its duties include the installation and maintenance of the battery telephone system, operation of the battery panel station and establishment and operation of the battery message center. The usual means of communication within the battery is the telephone, although the megaphone is frequently used for short distances. Visual signaling and messenger service are provided to insure communication in case of disruption of the telephone service. In addition electrical and mechanical devices are used in fixed and antiaircraft artillery for the transmission of data and signals.

Maintenance section. The maintenance section is charged with the procurement, storage, care, and issue of supplies, rations, clothing, and equipment and with the care and operation of all transportation provided for the battery.

Firing section. The firing section of a battery is subdivided into gun sections, one gun section for each gun assigned to the battery, except in machine gun batteries, where a gun section consists of two gun squads.

An antiaircraft machine gun detail, provided in each gun battery, is a part of the firing section. Equipped with four antiaircraft machine guns, it is responsible for providing protection against low-flying enemy aircraft.

Duties of Battery Officers

Battery commander. The battery commander, normally a captain, is responsible for the efficiency of the battery, for the maintenance of all armament, accessories, and fire-control equipment in serviceable condition and for the readiness of the battery for war service. He is charged with the basic training of the individuals comprising the battery and with the administration, subsistence, and supply of the battery. He conducts the fire of the battery In practice and in action, and he keeps up to date the gun or emplacement book and other records prescribed by regulations and orders. When the battery is not operating as part of a battalion (or group), when communication with the battalion (or group) command post is disrupted, and when battery commander's action is ordered, he exercises the functions of fire direction usually exercised by the battalion (or group) commander and conducts the battery action as nearly as possible in conformity with the particular tactical plan which is adapted to the existing tactical situation.

He keeps himself fully informed as to the ammunition supply of the battery and reports expenditures and deficiencies. When necessary, he reconnoiters and selects the exact positions for the battery, position-finding stations, observing stations, and spotting stations and the routes thereto. He makes frequent inspections of the battery and provides for the replacement of personnel, supplies, ammunition, and equipment. He keeps a permanent record of daily attendance at artillery instruction.

Executive. The battery executive, normally a first lieutenant, commands the firing section and is responsible to the battery commander for its training and efficiency. He makes all necessary preparations to expedite the opening of fire. He is charged with the care and operation of the materiel of the firing section, with the protection of the personnel and materiel at or near the gun positions, with the reliability of the communication system at his post, with the resupply of ammunition, and with the replacement of casualties in the firing section. All other officers assigned to the gun sections are assistants to the battery executive.

Range, reconnaissance, and communication officers. The range officer of fixed and antiaircraft gun batteries, the reconnaissance officer of railway and tractor-drawn batteries, and the reconnaissance and communication officer of antiaircraft machine gun batteries perform similar duties, the title indicating in each case the primary functions. The details of these duties depend to a great extent upon the character of the armament which is assigned to the battery. In general, this officer commands the range or reconnaissance detail and the communication detail and is responsible to the battery commander for their training and efficiency. He supervises the installation, care, and operation of the battery communications and of the fire-control and time-interval equipment with which the battery may be provided. He accompanies the battery commander on reconnaissance and makes the necessary surveys and computes the data for orientation of instruments and guns. He supervises the determination of the firing data and all corrections thereto.

Railway officer. The railway officer of a battery of railway artillery is in charge of the motor transportation and the rolling stock of the battery. Under the battery commander, he is directly in charge of the make-up of all trains and of train movements. He maintains a file showing the classes, number, and types of all guns and cars assigned to the battery and loading diagrams showing the axle loads, over-all dimensions, distances between wheels and trucks, maximum degree of track curvature for which the materiel is designed, the over-hang on curves, and end-view clearance diagrams showing the maximum contour around the mounts and cars. He also maintains a file showing the trackage and gauge of all railways in his locality, clearance diagrams of all structures having obstructions or clearance limitations, location of all curves greater than 9° maximum grade of tracks, and the weight of rails and kind of ballast.

The Harbor Defense

Composition. A harbor defense is an administrative and tactical unit provided for the defense of a harbor or other water area. Administratively, the harbor defense is composed of one or more forts, at which the elements of the harbor defense are located. Tactically, the harbor defense embraces one or more groupments, or groups, of seacoast artillery, usually supplemented by submarine mine defenses, and means for local defense against attack from sea, air, or land, together with the personnel, materiel, and accessories required for their administration and tactical employment.

The armament may include fixed and mobile seacoast artillery, fixed and mobile antiaircraft artillery, fixed and mobile searchlights, and controlled submarine mines, the amount and kind of each depending upon strategical and tactical considerations. The armament is manned by coast artillery troops, but the garrison of a harbor defense normally includes troops from the administrative, technical, and supply services and may include troops from other combat arms. Harbor defenses are designated by the name of the harbor, city, locality, or water area they defend.

Administrative Organization. The administrative organization of a harbor defense is based entirely upon the fort as the administrative and supply unit. The fort commanders as such, enter the tactical chain of command only in the event of loss of communication with the harbor defense or in the immediate defense of their forts. When all of the elements of the harbor defense are located at a single fort, the administrative functions of the fort are absorbed in those of the harbor defense, and the fort organization, as such, is omitted.

Tactical Organization. The proper tactical organization of a harbor defense is essentially a matter of efficiency in administration, supply, and tactical employment. Each harbor defense is a separate problem and is organized to meet its particular requirements. In general, the organization contemplates grouping together the armament having fields of fire which cover the same, or adjacent, water areas, the actual location of the armament being a secondary consideration. Where there are two or more water approaches to the defended area, or where for any reason it is necessary to group part of the armament for the defense of one normal zone and part for another, or where the number of groups is such that they cannot be advantageously controlled by one individual, it is necessary to organize two or more groupments. Where the number of groups is not large and there is but a single approach to the defended area, the echelon of groupment may be omitted.

Normally, a harbor defense consists of a headquarters, a harbor defense section, two or more groupments or separate groups, and such administrative, technical, supply, and tactical units as may be assigned.

Headquarters. The harbor defense headquarters includes the harbor defense commander, a tactical staff, and an administrative staff. The senior coast artillery officer present for duty is the harbor defense commander. Often he commands both the harbor defense and a regiment. He organizes his staff from the available officers and assigns the other officers to appropriate stations and duties. He may also act as fort commander at the fort at which his headquarters and command post are established, in which case his staff functions also as the fort staff. He controls all seaward, landward, and antiaircraft defense with which his command is charged. In some cases, the landward or the antiaircraft defense or both may be provided by agencies not under his command. He organizes appropriate groupments and designates their commanders, and he coordinates the action of groupments by assigning normal and contingent zones, by prescribing missions, and by other appropriate instructions.

The tactical staff consists of an executive, a plans and training officer, an intelligence officer, a communication officer and a searchlight officer. The administrative staff consists of an adjutant, an artillery engineer, an ordnance officer, a chemical officer, a disbursing officer, a quartermaster, a surgeon, and a chaplain. Training, administration, and technical and staff functions of the harbor defense and routine duties connected with the exercise of these functions are performed at the harbor defense headquarters. Functions of tactical command are exercised from the harbor defense command post.

The Fort

Composition. A fort is an area within a harbor defense wherein are located harbor defense elements capable of offensive action against hostile war craft and which is organized to provide for such action and for its own protection and administration. It is primarily an administrative command, designed to provide a centralized control over the administrative and technical details pertaining to the personnel and materiel located thereat. Under exceptional circumstances, it may be employed as a tactical unit and its organization provides for the possibility of such use.

The amount, types, and calibers of the armament, the submarine mine installations and the auxiliary installations and equipment at a fort depend chiefly on the location of the fort with reference to the zones and areas defended by the harbor defense of which it is a part, and it may include fixed, mobile, and antiaircraft artillery. All the seaward defenses of the harbor may be contained in a single fort or they may be distributed among several forts and may be augmented by mobile artillery in temporary emplacements.

The composition of a fort varies with local conditions and is not prescribed. It may contain one or more tactical commands or it may contain only parts of one or more tactical commands. The troops may consist of any or all classes of troops contained in a harbor defense garrison. Forts are designated by the names of persons who rendered distinguished service to the government during their public careers.

Organization. The fort organization consists of a headquarters, a fort detachment, and such tactical units as may be specifically placed under the command of the fort commander. The fort detachment is a subdivision of the harbor defense section and is similar in organization and duties to the harbor defense detachment. The fort is organized to provide for the administration and technical control of the personnel and materiel located thereat, to facilitate the employment of tactical units by their respective commanders, and to permit employment of the fort as a tactical entity in the event of disruption of communication with the harbor defense or in the event of a landing attack made under conditions requiring the use of the fort troops for its defeat. When communication with other elements is disrupted, the senior line officer at the fort directs the action of all elements located thereat until normal tactical relations are re-established.

Headquarters. The fort headquarters includes a fort commander and an administrative staff. The tactical command exercised by the fort commander is normally restricted to that of a groupment, group, or battery, depending upon his position in the normal chain of command, other defense elements at the fort being commanded by the commanders of the tactical units to which such elements pertain in the normal chain. There is normally, therefore, no tactical staff provided for the fort headquarters, and when one is required it may be formed from the staff of the fort commander's normal tactical command and its duties and organization will then be similar to those of the harbor defense tactical staff to the extent required by the local situation. The fort administrative staff consists of an adjutant, an artillery engineer, an ordnance officer, a quartermaster, and a surgeon; and their duties are similar to those of corresponding staff officers of the harbor defense.

The Groupment

Composition. A groupment in a harbor defense is a tactical command composed of two or more groups, which cover the same or adjacent areas with their fields of fire, together with the personnel, materiel, and accessories required for its employment as a unit. The details of its composition vary in different harbor defenses but it normally consists of a number of gun, howitzer, and mortar groups and a mine group: It ordinarily includes such fixed and mobile armament and controlled mine fields as cover the same general defensive zone and can be advantageously handled in fire action by a single individual. In addition, the groupment may be provided with elements for local defense. Groupments are designated by the name of some geographical feature with which the groupment is by location identified.

Organization. The groupment consists of a headquarters, a groupment detachment, two or more groups; and such additional tactical units as may be assigned or attached to the groupment. The headquarters includes the groupment commander and a staff consisting normally of a plans and training officer, an intelligence officer, a communication officer, and a searchlight officer. The number of staff officers depends on local conditions, and when the groupment commander is also a fort commander appropriate fort and groupment staff duties are consolidated.

The groupment detachment is a subdivision of the harbor defense section and consists of the enlisted personnel necessary to assist the groupment commander and his staff in the exercise of their functions. Its organization and duties are, in general, similar to those of the harbor defense detachment.

The Gun Group

Composition. A gun group in a harbor defense is a tactical unit composed of two more rifle, howitzer, or mortar batteries which cover the same general water area with their fields of fire, together with the personnel and installations necessary for the employment of the group as a unit. In the organization of a gun group, fire direction is the governing consideration, and fire direction is facilitated when all the weapons of the group are of similar types and characteristics. However, it is frequently impracticable to obtain this condition without a prohibitive sacrifice of other desirable factors involved in fire direction, and in some cases such a group cannot be formed with the armament available. Hence, in general, a group composed of weapons that may suitably be employed against similar targets is considered to satisfy tactical requirements. Groups may be composed of fixed batteries, of mobile batteries, or of a combination of both. Groups are designated by number in each harbor defense.

Organization. The gun group consists of a headquarters, a group detachment, and two or more batteries. The headquarters includes the group commander and a staff consisting of a plans and training and intelligence officer, a communication and searchlight officer, and such other commissioned assistants as may be made necessary by special conditions. The group detachment is a subdivision of the harbor defense section and normally includes command-post, intelligence, plans and training, communication, and searchlight details. The functions of the group staff and the duties of the group detachment are, so far as applicable, similar to those of the groupment staff and groupment detachment.

The Mine Group

Composition. A mine group is a tactical and technical unit for the employment of controlled submarine mines provided for the defense of a given water area, together with the armament, structures, personnel, equipment, and vessels necessary for the planting, operation, and protection of the mine field.

Organization. The mine group consists of a headquarters, a mine group detachment, two or more batteries, and one or more mine planters. The group headquarters includes the group commander and a staff consisting of a plans and training and intelligence officer, a communication and searchlight officer, and a mine property officer. The mine group detachment is a subdivision of the harbor defense section and, with the addition of a mine property detail, is similar in organization and duties to a gun group detachment.

Duties of the Tactical Staff

The duties and responsibilities of corresponding staff officers of the several echelons are similar, although they vary in scope and in detail with the size and functions of the unit. In the higher echelons, assistants to the staff officers may be required; in the lower echelons, the duties of two or more staff positions may be assigned to a single officer.

Executive. The executive is the principal assistant and adviser to the unit commander. He acts as chief of staff, directs and coordinates the various sections 'of the staff, and acts for the commander in his absence.

Intelligence officer. The intelligence officer (S-2) is responsible for the collection, collation, and dissemination of all military intelligence; obtains and distributes maps; acts as agent of the artillery information service; prepares and maintains situation maps; establishes intelligence observation posts; supervises the unit scouts; and is responsible for the training of the intelligence detail.

Plans and training officer. The plans and. training officer (S-3) prepares all programs and schedules for the training of the unit, all plans for its tactical employment, all field orders, all charts and maps to accompany field orders, all plans for movements and all visibility, dead-space, and operation maps; keeps the war diary and a record of all operations; establishes the command post; and is responsible for the training of the plans and training detail.

Reconnaissance officer. The reconnaissance officer makes reconnaissance of routes and positions and special reconnaissance as directed, is in charge of topographical operations, supplies the intelligence and plans and training officers with information for situation and operation maps, assists the plans and training officer in the preparation of maps and charts, establishes and, maintains observation posts for plotting and spotting is in charge of plotting and spotting personnel, and is responsible for the training of the reconnaissance detail.

Communication officer. The communication officer supervises and coordinates the training of the communication personnel in the various units of the organization, establishes and maintains the signal communications of the unit, establishes and maintains the message center, is responsible for the care and maintenance of the signal equipment, assists in the administration of the headquarters battery, and is responsible for the training of the communication detail.

Munitions officer. The munitions officer is responsible for the requisition, receipt, and distribution of ammunition (including pyrotechnics) and for keeping the ammunition records and reports, keeps himself informed concerning ammunition supply throughout the unit, and in a battalion, commands the combat train or munitions detail.

Liaison officer. The liaison officer represents the artillery commander with a supported or other designated unit, is normally stationed at the command post of the designated unit, keeps the artillery commander informed of the artillery requirements and the situation confronting the designated unit, and advises the commander of the designated unit as to the nature of artillery assistance that may be expected.

Searchlight officer. The searchlight officer is responsible for the care, operation, maintenance, and minor repair of all searchlights and searchlight power plants under the control of the commander and advises the commander in matters connected with the tactical employment of the searchlights.

Mine property officer. The mine property officer is responsible for the serviceability of all mine storerooms, cable tanks, wharves, boat houses, and mine boats other than mine planters.

Duties of the Administrative Staff

Adjutant. The adjutant (S-1) is in charge of routine administrative matters, correspondence, assignment of quarters and billets, and postal and welfare services; commands the band; and is responsible for the training of all specialists not included in the various staff sections.

Supply officer. The supply officer (S-4) is responsible for the requisition, receipt, and distribution of rations, equipment, and supplies other than ammunition; receives, coordinates, and forwards all requisitions for supplies; and provides storage facilities for supplies kept on hand. The regimental supply officer commands the service battery; battalion supply officers command the battalion sections of the service battery.

Artillery engineer. The artillery engineer advises the commander on technical questions pertaining to the repair, maintenance, and operation of the systems of fire control, communication, searchlights, power plants and lines of power distribution, and is accountable for and maintains records of all property issued by the Corps of Engineers and the Signal Corps for the fixed armament.

Ordnance officer. The ordnance officer advises the commander on technical questions pertaining to the repair, maintenance, and operation of the armament; has supervision over the ordnance repair shops of the command; and is accountable for and maintains records of all property issued by the Ordnance Department for the fixed armament.

Disbursing officer. The disbursing officer is charged with the administration of all functions pertaining to the Finance Department (except the audit of property accounts) and with the disbursement of and accounting for public funds.

Surgeon. The surgeon commands the hospital and coordinates and supervises all Medical Department activities of the unit.'

Quartermaster. The quartermaster is responsible for the maintenance, operation and supply of all administrative water, animal-drawn, and motor transportation, for the storage and issue of rations, forage, fuel, clothing, and other supplies pertaining to the Quartermaster Department, and for the upkeep of buildings and grounds, and is accountable for and maintains records of all property issued by the Quartermaster Department.

Chemical officer. The chemical officer is charged with all chemical warfare activities including gas-proofing of plotting rooms, inspection of gas masks, and instruction of troops in defense against gas.

CATAGORIES OF DEFENSE FOR COAST ARTILLERY OPERATIONS IN WW II

Compiled and edited by William C. Gaines

There were two groups of defense categories, tactical categories designated by letters, and readiness categories designated by numbers. Although there was a relationship between the two, they were designed to meet two distinct needs. The lettered categories, A, B, C, D, E, and F, were set down by the War Department in accordance with the provisions of JAAN 1935, and were based upon expectancy and intensity of enemy assault as determined by military intelligence. They were applied to the specific coastline or military area concerned in general.

The numbered categories, I, II, and III, were determined by the tactical command on the scene and were applied to the sub-sector, harbor defense, or specific armament within a harbor defense. While these conditions could be applied to a defense command or sector, attacks of that scope would be highly unlikely.

The use of the numbered categories obviated the need for detailed or lengthy orders to meet an emergency or attack. As an example, it was only necessary, when the Japanese air raid on Pearl Harbor came, for the 2nd Coast Artillery District to send the three word message to Colonel Ruhlen commanding the Harbor Defenses of the Delaware at Fort Miles: "Condition II Immediately." This brief message of alert was sufficient to set in motion a state of readiness that called for all harbor defense and antiaircraft artillery observation stations and communications to be manned on a constant basis. Further, if ordered, at least one major caliber battery and one secondary battery were to be manned (along with the necessary searchlights at night) and ready to fire.

The use, however, of one category classification did not necessarily demand the use of another. For example: Category E represented the greatest danger for the military area to which it was applied. The application of Category E, however, did not put all elements in Condition I. It is obvious that a high state of readiness can be maintained for only a limited time and should be used carefully if the military personnel strength is not to be dissipated before the attack is developed. On the other hand, Category A represents a condition in which the area is determined to be free from attack. It is possible, however, that an isolated enemy raid on an installation would throw a harbor defense or the armament concerned into Condition I.

For example: If an enemy force in strength were operating in the waters off Florida and Georgia, having occupied an island in the Bahamas as a base, the eastern seaboard would no doubt be thrown into Category E, but no defense would necessarily be put into Condition I. However, no harbor defense along coastline of the Carolinas, Georgia, and Florida would be less prepared for action than that demanded by Condition II.

If a coastline was declared free from attack and in Condition A, as in the case of the Eastern Seaboard in 1944, a raid by an enemy submarine or a surface raider disguised as a merchantman would surely throw a harbor defense into Condition I if they appeared in a harbor entrance.

Tactical Categories of Defense

(Abstracted from JAAN 1935)

Category A - Coastal frontiers (sea frontiers and defense commands) that would probably be free from attack, but for which a nominal defense must be provided for political reasons in sufficient strength to repel raids by submarines, by surface vessels operating by stealth or stratagem, or isolated raids by aircraft operating chiefly for morale effect.

Category B - Coastal frontiers that may be subject to minor attacks.

Category C - Coastal frontiers that in all probability would be subject to minor attack. Under this category, the coastal defense area should be provided in general with the means of defense, both army and navy, required to meet the following enemy naval operations: those incident to controlling the sea, those against shipping, and minor attacks against land areas. The harbor defenses should be fully manned and air support arranged. Long range air reconnaissance would be provided if practicable. If sufficient forces were available, outposts would be established outside of harbor defenses along the sensitive areas of the shore-line. The inner mine barrages would be established; a full inshore patrol and complete control of shipping would, as a rule, be instituted; and certain outer mine barrages and defensive sea areas may be established, and a limited off-shore patrol instituted.

Category D - Coastal Frontiers that may be subject to major attack. Under this category, the coastal defense areas should, in general, be provided with the means for defense, both army and navy, required to meet enemy naval operations preliminary to joint operations. All available means of defense would generally find application, and a stronger outpost and a more extensive patrol, inshore and off shore, than for Category C, would be required. Under this category, certain defensive sea areas and maritime control areas would be established. In addition, an antiaircraft gun and machine gun defense of important areas outside of harbor defenses should be organized; general reserves should be strategically located so as to facilitate prompt reinforcement of the frontiers; and plans should be developed for the defense of specific areas likely to become theaters of operations. Long range air reconnaissance would be provided and plans made for use of the GHQ air force.

Category E - Coastal frontiers that in all probability would be subject to major attack. Under this category, in additions to the measures required for Category D, there would be required generally the concentrations of the troops necessary to defend the area against a serious attack in force together with additional naval forces to provide intensive inshore and off shore patrols. Defensive sea areas and maritime control areas would be established. Air defense would be provided for as in Category D. All or a part of the GHQ Air Forces might be ordered to the threatened area to operate either under direct control of Army GHQ or that of the army commander of the theater of operations or frontier.

Category F - Possessions beyond the continental limits of the United States which might be subject to either minor or major attack for the purpose of occupation, but which cannot be provided with adequate defense forces. The employment of existing local forces and local facilities would be confined principally to the demolition of those things it was desired to prevent falling into enemy hands.

Readiness Categories of Defense

Condition One - Maximum readiness for action. In the harbor defenses, all stations, communications, and armament were manned in accordance with existing plans; antiaircraft troops similarly man observations posts and communications, and hold gun and searchlight crews at or in the immediate vicinity of their armament; supporting infantry maintain observation and patrol elements in accordance with plans and hold reserves in readiness. Coast artillery units were not able to continue on Condition One indefinitely with the personnel usually available; hence Condition One was ordered for brief periods only, in general, not exceeding six hours in any one day. When this condition was ordered on initial activation, immediate readiness for action would be the first objective. The movement of supply elements and setting up messes and camps would be secondary.

Condition Two - The state of readiness that could be maintained indefinitely. Harbor defense and antiaircraft observation stations and communications were manned continuously, if necessary with reduced personnel or in reduced amounts, but at least one station per battery. Command posts of all echelons were operated continuously, with sufficient personnel for current requirements. Armament, equipment, and personnel not actually at battle stations would be kept in such readiness that Condition One might be assumed within three minutes during daylight hours and within five minutes during hours of darkness.

Condition Three - Minimum readiness appropriate to war or emergency conditions. At least one secondary battery in each harbor defense, with necessary searchlights, would be maintained in Condition Three; at least half of the total number of antiaircraft batteries, both fixed and mobile, would be maintained in Condition Two; at least one additional observation post per groupment would maintain continuous observation, harbor entrance control posts were operated continuously, and communications would be manned at command posts down to groupment and antiaircraft regiments. The remaining armament and personnel off duty would be able to assume Condition One within one hour.

Classification of Material

As a minimum, for each active lettered battery organization in each harbor defense, one complete battery or one complete unit of submarine mine material, together with accessories required to render it effective as a combat unit was classified as Class A. The initial classification of seacoast batteries in the continental United States was prescribed by the War Department in the harbor defense projects. The harbor defense commander concerned was then authorized to raise the classification of a battery from B or C to A by assigning the materiel to an organization primarily for the purpose of training or mobilization. Likewise, the same authority could return batteries to any classification not lower than that initially prescribed by the War Department.

Class A - Materiel which was assigned to an organization in a primary capacity for the purposes of regular and frequent training, together with all installations required to make that materiel effective. This materiel was maintained at all times in such condition as to permit its preparation for service by a full strength manning party in not more than twenty-four hours.

Class B - Materiel which was not assigned to an organization for frequent and regular training but which was important to the performance of the mission of a harbor defense. For this class the maintenance year was divided into the active and inactive seasons. The active season extended for approximately six months, as designated by the harbor defense or department commander, while the inactive period extended throughout the remainder of the year. During the active season, all materiel was placed in operating condition, fully assembled (with some exceptions) and maintained in such condition as to permit its being prepared for service by a full strength manning party in not more than twenty-four hours. During the inactive season, materiel was maintained in such a manner as to permit its being prepared for service by a full strength manning party within seventy-two hours.

Class C - Materiel which was not assigned to an organization for regular and frequent training, and was no longer considered vital to the performance of the mission of the harbor defenses, but which was still capable of furnishing some fire support. Maintenance of this class was ordinarily such that more than seventy-two hours would be required to restore the armament fully to an active condition. Such restoration, however, normally could be accomplished by a full strength manning party in less than fifteen days.

BARRACKS, BAKERIES, AND BOWLING ALLEYS
NON-TACTICAL STRUCTURES AT AMERICAN COAST ARTILLERY FORTS

Mark Berhow, Joel Eastman, and Bolling Smith

This section deals with a topic often neglected in studies on American coast defenses—non-tactical structures. These were the buildings that housed the men and supplies necessary to run and maintain a fort, which were small towns unto themselves. The text describes the permanent and temporary buildings constructed at coast artillery forts in the twentieth century. Photographs illustrate standard structures and a gazetteer lists forts where these buildings survive.

Barracks, Administration Building and Officer's Quarters at Fort Columbia, Washington
(Mark Berhow 2008)

Large brick and stone forts built previous to 1860 were usually designed to be self-contained, with quarters and storehouses located within the defensive walls. However, it was difficult to maintain a garrison in these cramped quarters for long, and non-tactical structures were soon built outside of the walls. After the Civil War most of the non-tactical structures were constructed outside of the original forts, and planners were attempting to organize the buildings around a parade ground similar to the posts established on the western frontier.

When the harbor defenses of the United States were modernized in 1890-1910, a new system of defensive works were created. Rather than the compact defenses of the early systems, the modern forts consisted of tactical and non-tactical structures spread over hundreds of acres of land. The U.S. Army Corps of Engineers selected the locations, purchased additional land, sited, designed, and constructed the tactical structures—batteries, mine facilities, observation stations, plotting rooms and searchlight shelters. The Quartermaster Corps (up to 1941) sited, designed and constructed most of the non-tactical structures—barracks, officer's quarters, administration buildings, storehouses, recreation buildings, and other structures. The quartermasters created a landscape plan to utilize the land efficiently while at the same time creating an aesthetically attractive post laid out in tradition patterns. The Quartermaster Corps used the same standard building plans and layouts at coast artillery forts as it did at cavalry, field artillery, and infantry forts—but it was usually more challenging to implement a traditional plan at the unique locations of coast artillery forts than it was at sites in the interior.

Layouts of Non-Tactical Structures, 1890-1917

The center of the non-tactical area of a coast artillery fort was the parade ground. Officer's quarters were sited on one side of the parade ground, while barracks were placed on the other, at a lower elevation, if possible. The administration building or harbor defense headquarters was given a prominent location on the parade ground, as were the commanding officer's quarters, the flagpole, and bandstand. Non-commissioned officers quarters were located off of the parade ground proper, as was the post exchange, gymnasium, bowling alley, hospital, guardhouse, bakery, fire station, chapel, library, officers club, and theater. Many forts in isolated areas had cemeteries. Good existing examples of an 1890s coast artillery posts are Fort Worden, Harbor Defenses of the Puget Sound, and Fort Columbia, Harbor Defenses of the Columbia River, both now Washington State parks.

BARRACKS 21.-27.-26.-10.-9.   OFF. QRS. 3.-4.-5.-6.; 22-23.-24.-25.
LOOKING EAST FROM GUARDHOUSE.

Fort Worden, 1913, barracks to the left, officers row to the right. (NARA)

Middle and bottom photos, soldiers on parade at Fort Worden (Greg Hagge photo collection)

Officers Quarters, Fort Flagler (Greg Hagge photo collection)

Barracks, Fort Flagler (Greg Hagge photo collection)

NCO quarters, hospital, hospital steward's quarters, QM commissary building, Fort Flagler

Although the parade ground was used as a general athletic field, tennis and handball courts, and baseball fields were also built in open areas of the fort. Storehouses, commissary, workshops, and stables were usually centered near the quartermaster wharf. A system of permanent roads served the entire fort, and the streets were usually named. Railroads and tramways were built during the construction of the forts, and these lines often continued to be used. Forts (eventually) had their own water, sewer, telephone, and electrical systems. If municipal water and commercial power services were available, the army used them, but at many sites the engineers built their own water and electrical plants and distribution systems. Sewer pipes ran into the ocean.

Mine Storehouse. 23.- Post Exchange. 27 ~ Barracks. 9 - Headquarters. 4 - Officers Quarters 3 - 2.: 1.
Fire Apparatus House. 22.
Taken Jan 1913.

Fort Columbia 1913 (NARA)

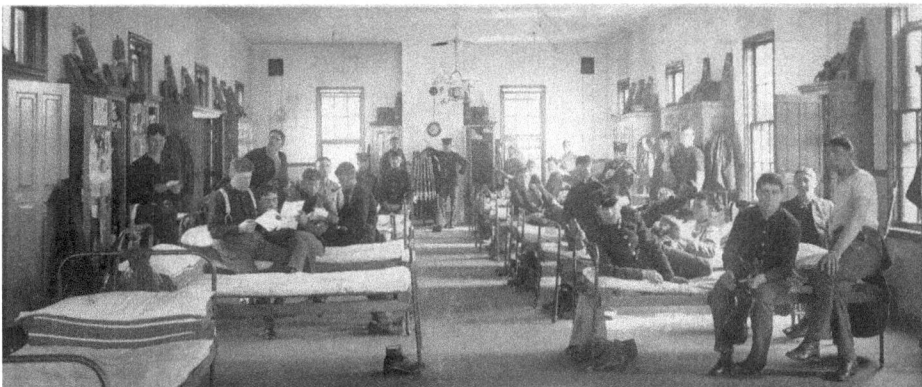

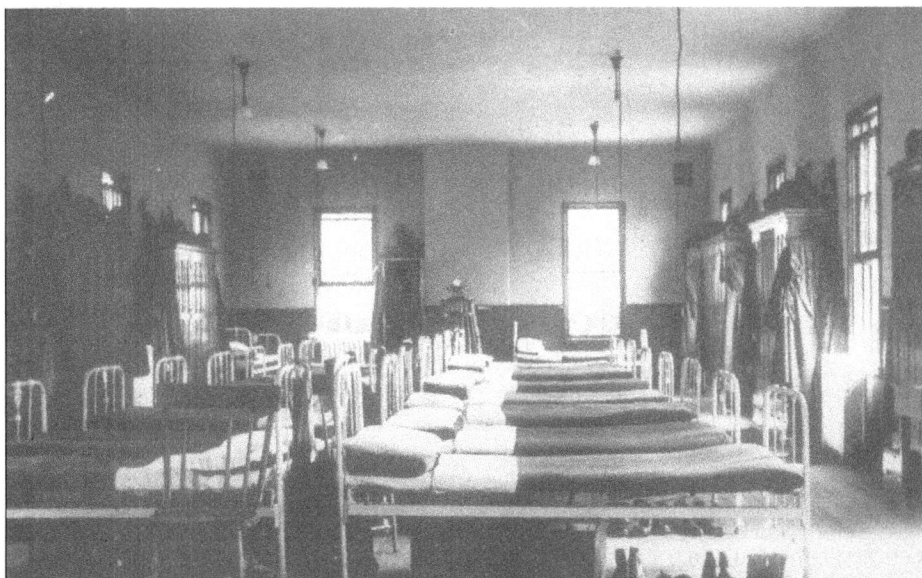

middle and bottom photos, Interior of barracks, Fort Columbia, WA 1910s (Greg Hagge photo collection)

Top and middle photos, mess hall in barracks at Fort Columbia (Greg Hagge photo collection)

Officers and family on porch steps, Fort Casey, Christmas 1913 (Greg Hagge photo collection)

Barracks mess hall, Fort Worden (Greg Hagge photo collection)

Barracks recreation/sitting room, Fort Worden (Greg Hagge photo collection)

Funeral at Fort Worden cemetary (Greg Hagge photo collection)

Baseball game at Fort Flagler (Greg Hagge photo collection)

Ice houses, and in northern areas, ice ponds, were also built to provide refrigeration for food in the years before electrical cooling became available. Systems for the disposal of garbage and rubbish were also created. Garbage and combustible waste were burned in crematoria, while non-combustible materials were disposed of in landfills or dumped into the ocean. The major fuel at forts was coal, and a system of unloading, transporting, and storing the fuel was developed, usually relying on mule-drawn wagons.

1938 photo of the Fort Worden reservation at Point Wilson.
Admiralty Head is visable in the distance across the entrance to the Puget Sound (NARA)

Bird's eye view of the Fort Worden garrison area 1937 (NARA)

Permanent Non-Tactical Structures at Coast Artillery Forts

Most of the non-tactical structures at the forts constructed during the Endicott-Taft period were designed to be permanent structures. These wood-frame buildings were built on stone foundations with slate roofs, sided with local brick, clapboard, or stucco. The Quartermaster Corps architect's office created standard plans for all types of buildings. Those designed at the turn-of-the-century—when most Coast Artillery forts were constructed—were of Colonial Revival style with elements of Queen Anne style in the officers quarters. As the century progressed, new styles were adopted, such as Italianate and Spanish Revival, and these styles were used when additional buildings were constructed. Store houses and pumping plants used more practical industrial or utilitarian styles.

The interiors of buildings were finished with wood floors, plaster walls with wood trim, and pressed metal ceilings. All structures where officers and men lived or worked had electricity, running water and flush toilets. Each barracks was designed to house a company or battery of 100 men and was self-contained with its own kitchen, dining room, day room, barber shop, and tailor shop. Sleeping quarters were on the second floor, while the lavatory and latrine were located in the basement in northern climates. In the south, separate lavatory and latrine buildings were sometimes built. Large forts had double barracks–two 100 man barracks built end-to-end–which functioned as two separate barracks. Forts which served as the headquarters post for a harbor defense usually had a band barracks.

Officer's quarters varied in size and elaborateness depending upon the rank of officer for whom the building was intended. The Commanding Officer's Quarters was usually the largest and most elaborate of the officer's quarters, and it was placed, if possible, on the highest and most prominent location on the parade ground. Other senior officers were assigned single quarters, while the majority of the quarters were double quarters for two families. Large forts had a Bachelor Officer's Quarters with its own mess. Non-Commissioned Officer's quarters were usually double sets.

Recreation was considered important by the Army at the turn-of-the-century. It was believed that it not only maintained physical fitness, but also promoted competitiveness which made the men more effective in combat. Every large fort was provided with a gymnasium and bowling alley, as well as athletic fields, and handball and tennis courts.

Maintenance, supply and transportation required numerous permanent buildings. Carpenter and plumbing shops, Quartermaster and Commissary storehouses, a stable and wagon sheds, power and pumping plants, coal storage and wood sheds all usually were located in the same area, often near the Quartermaster wharf.

Listed below are non-tactical structures and facilities that were built at many Coast Artillery forts during the Endicott period.

Administration Building
Artillery Engineer Storehouse
Athletic Field
Bakery
Barracks, single and double
Barracks, band
Basketball Court
Baseball Field
Blacksmith Shop
Bowling Alley
Carpentry Shop
Cemetery
Coal Storage
Commissary Storehouse
Dump
Electrical Substation
Flag Pole
Fire Apparatus Building
Garages
Garbage Crematory
Gasoline pump and tank
Greenhouse
Guard House
Gymnasium
Harbor Defense Headquarters Building
Handball Courts
Hospital
Hospital Steward's Quarters
Ice House
Library and Reading Room
Non-Commissioned Officers Club

Officers Club
Oil House
Ordnance Machine Shop
Ordnance Storehouse
Plumbing Shop
Post Theater, 1938
Post Exchange
Power (electrical) Plant
Pumping (water) Plant
Quartermaster Storehouse
Quartermaster Wharf
Quarters, Officers, single and double
Quarters, Bachelors Officers, and mess
Quarters, Non-Commissioned, single & double
Quarters, Married Enlisted Men
Quarters, Firemen, double
Salute Gun
Scale House
Sentry Box
Service (enlisted men's) Club
School House
Stable
Swimming Pool
Teamsters Office
Tennis Courts
Veterinary Office
Wagon Shed
Water Tank
Well Shelter
Wood Shed
Work Shops

Parade at Fort Stevens, OR

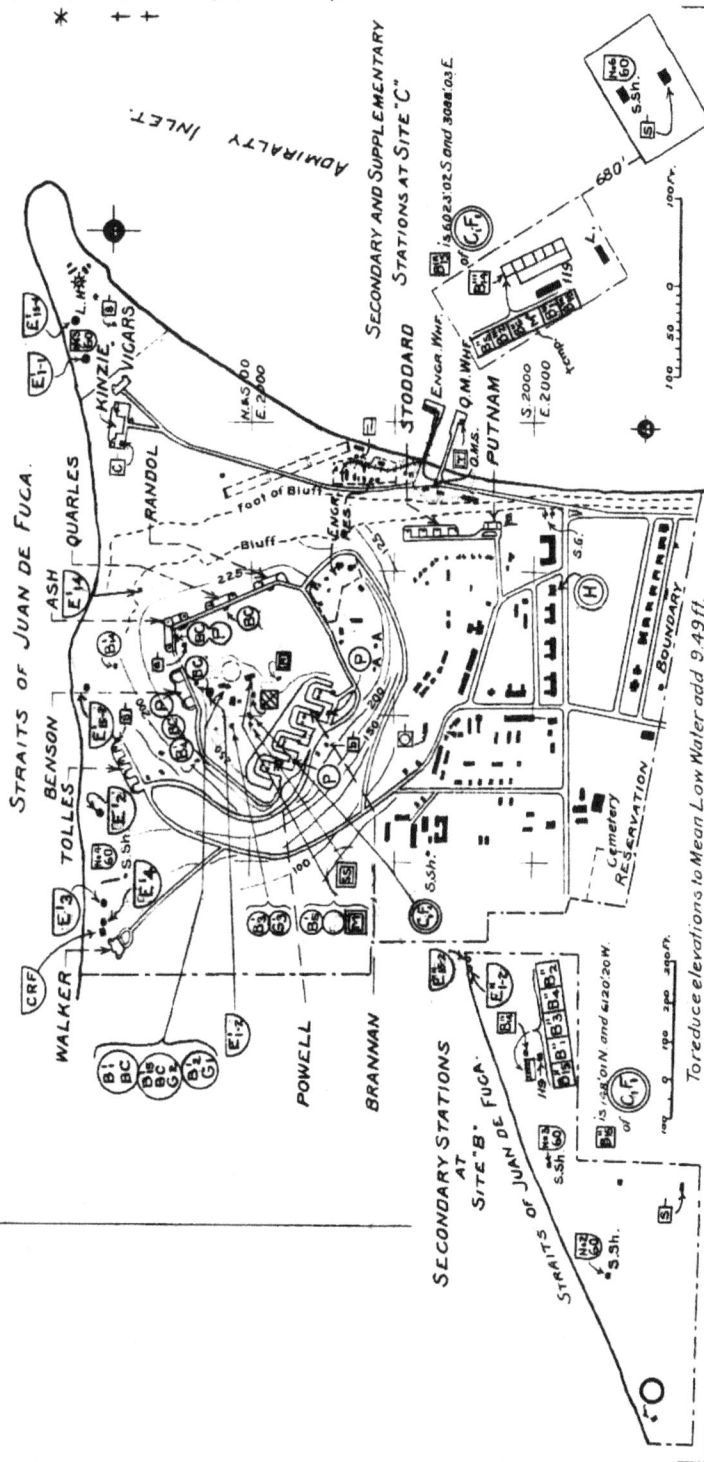

Fort Worden reservation, 1936 (NARA) for symbol key please see section on blueprints and maps

EDITION OF MAY 7, 1921.
REVISIONS: APR. 8, 1925,
NOV. 2, 1928, MAY 22, 1936.

SERIAL NUMBER

CONFIDENTIAL

PUGET SOUND
FORT WORDEN - D1 & 2
POINT WILSON
Scale of Feet
500   0   500

LEGEND
1. ADMINISTRATION BLDG.
2. COMDG. OFFICERS QRS.
3. OFFICERS QRS.
4. HOSPITAL.
5. HOSPITAL STWD'S. QRS.
6. N.C. OFFICERS QRS.
7. BARRACKS.
8. GUARD HOUSE.
9. POST EXCHANGE.
10. BRICK CASTLE.
11.
12.
13. CONCRETE ST. HO.
15. MILITIA STORE HOUSE.
14. ARTY. ENGR. ST. HO.
16. SCALE HOUSE.
17. FIRE STATION.
18. CIV. EMPL. QRS.
19. ARTY. ENGR. CABLE HO.
20. Q.M. OFFICE.
21. COMMISSARY ST. HO.
22. Q.M. & COMY. ST. HO.
23. Q.M. STOREHOUSE.
24. Q.M. WORK SHOP.
25. Q.M. RESERVOIR.
26. Q.M. STABLE.
27. Q.M. QUARTERS.
31. ORDNANCE STOREHOUSE.
40. ENGR. OFFICE.
42. ENGR. STOREHOUSE.
43. ENGR. BLACKSMITH SHOP.
44. ENGR. CARPENTER SHOP.
45. ENGR. GARAGE.
46. ENGR. QUARTERS.
70. SERVICE CLUB.
71. OFFICER'S CLUB.
31. ORD. STORE HO.
100. BAKERY.
101.
102.
103. OIL HOUSE.
104. COAL SHED.
105. WAGON SHED.
106. WAGON SHED AND
     TEAMSTER'S QRS.
107. RIFLE BUTTS.
108. TARGET SHED.
109. ARTILLERY ENGR.
110. TELEPHONE BOOTH.
111. CANTONMENT BLDGS.
112. AUTOMOBILE SCHOOL.
113. GARAGE.
114. BAND BARRACKS.
115. OFFICERS BOWLING ALLEY.
116. HANGAR.
117. GENERATOR HO.
118. STORE HOUSE.
119. DORMITORY
119. GREEN HOUSE.
120. PORT. S/L BLDG.

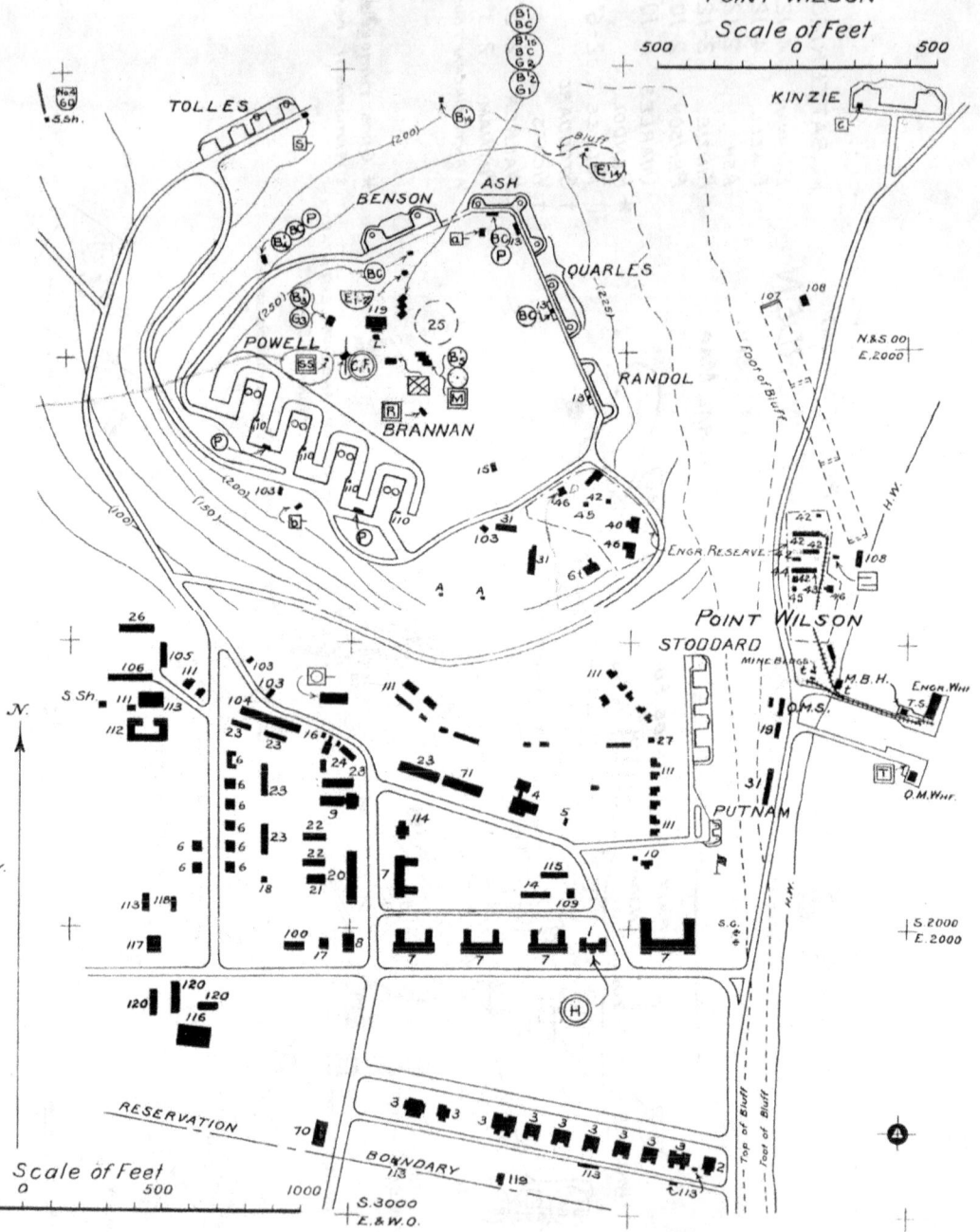

Scale of Feet
500   0   500   1000

Quartermaster Buildings: Records and Plans

from an article by Bolling Smith in the *Coast Defense Journal* Vol. 16, No. 2 pp. 29-42.

The Quartermaster Corps kept careful historical records on the buildings and structures for which it was responsible. These records were transferred to the Corps of Engineers along with the responsibility for construction and maintenance. At Archives II in College Park, MD, Entries 393 and 394, RG 77 (Records of the Chief of Engineers), contain many of these records, arranged generally in alphabetical order by post. Entry 393 contains the records of "active" posts, 1905-1942, while the much smaller Entry 394 contains the records of "abandoned" posts, 1905-1924.

The "Historical Record of Buildings" described individual structures. The term building was used in the broadest sense, and included wharfs, manure pits, tennis courts, and even statues.

The first such forms in the record, dating from 1905, are un-numbered. They contain information on two buildings, one on each side of the 10 x 12 card-stock form. Filed by post building number, the forms list the construction date, materials, and equipment (to include wash basins, showers, urinals, screen doors, and wall lockers), as well as an annual list of expenditures for repairs. Perhaps most valuable, the forms normally displayed a 4 x 5-inch black and white photograph of the structure. While some of these photographs are dark and some have faded, many are extremely sharp, showing gleaming new buildings, or in some cases, failing remnants from the last century. Occasionally, a terse notation will be found to the effect that "Structure is underground, hence no photograph" (in the case of a reservoir), "No photography permitted" (in the case of a magazine), or "Structure burned before photograph could be taken." As a group, however, these photographs provide an unparalleled glimpse of the actual appearance of these forts almost 100 years ago. With the buildings mostly gone, these photographs are our best information on how they actually looked.

In 1913, the forms were designated Form 173 a. By 1921, the forms, now 173 A, were enlarged to 10 x 14 inches and covered only one structure. On the reverse a grid pattern was provided for a simple plan of the structure. Plans were drawn for some structures; others had blueprints pasted on, while still others were blank. In 1924, the form was renumbered 117, but otherwise remained relatively unchanged (see the example of the form 173A for the Commanding Officer's Quaters at Fort Worden, HD Puget Sound, WA on the following page).

These forms cover buildings built until the Corps of Engineers assumed responsibility days before the United States entered World War II. Both permanent and temporary buildings are included, as well as a number of civilian structures taken over by the army when the land on which they stood became part of a military post.

Quarters, which included barracks and houses for officers and NCOs, along with the buildings that most closely represented the service, such as administration buildings, guardhouses, post exchanges, and theaters, tended to be attractive buildings. Built in a number of styles at different posts, they were intended to create an atmosphere of attractive order.

Supporting buildings, on the other hand, tended to be more utilitarian. Without the stylistic embellishments of the buildings that served to represent the army, the supporting structures were normally simple frame buildings, although the prevalence of galvanized tin structures is surprising.

One of the most important entries on the "Historical Record of Buildings," was the OQMG (Office of the Quartermaster General) plan number, which was normally listed, at least for 1891-1917 buildings. These plan numbers, in turn, lead to another valuable source. From 1891 through 1917, the quartermasters built most structures to numbered standard plans. As these plans were updated, letter suffixes were added. Thus, for example, standard plans No. 120 were for a double set of officers' quarters, and standard plans No. 120-E were for duplex lieutenant's quarters.

WAR DEPARTMENT.
Q.M.C. Form No. 117 (Old No. 173A)
Revised June 28, 1939

Post Plan No. ....

O.Q.M.G.: Plan No. ....145-A.....   Building No. ...3...

Place ......**Fort Worden, Washington**......
Designation of building ......**Field Officer Quarters**...... Capacity...**1 Field Off.**
Total cost, $..**13,521.36**.... Date completed ...**April 15, 1904**....
Material: Walls......**Frame**......Foundation......**Stone**......
Roof ......**Slate**......Floors......**Wood**......
Total floor area above basement, square feet......3228......
Size: Main building ...**37' x 53'**... Wings.. - - .. Basement..1135.
a. ...**Boiler, Automatic Oil Burner**... Height of first floor above
    (How heated)   ground...4'
b. ......**Steam**...... How lighted ......**Electricity**......
    (Type of heat)   Water connections ......**Yes**......
c. ...**Automatic Oil Water Heater**... Sewer connections ......**Yes**......
  (Type of domestic hot water heater) Gas connections ... - - ...

METERS INSTALLED
(Give quantity and capacity)

COOKING RANGES INSTALLED
(Give quantity and size)

Coal ... - - ... Gas ... - - ...
Gas ... - - ... Electric **2(1-AC 115V)(1-AC 220**
Electric **1-4 Burner, 1 Oven** Oil ... - - ...
Oil ... - - ... Steam ... - - ...
Steam ... - - ... Water ... - - ...

REFRIGERATORS INSTALLED
(Give quantity and size)

Gas ... - - ...
Electric **1-6 ft. cap.**
Ice ... - - ...

Approval of Secretary of War
as required by A. R. 30-1435
(Give date and File Number)

ADDITIONS AND INSTALLATIONS

(Below enter chronologically all modifications, additions, introductions of water, sewer, lights, heating, etc.)

DATE		COST	DATE	COST
1/6/41	Total amount expended up to and including F.Y. 1940, for Maintenance, Repair and Alteration	13,460.75		

INSTRUCTIONS.—"a" State whether heated from central heating or by individual heating plants, stoves, furnaces, or fireplaces.
"b" State whether steam, vapor, hot water, or hot air.
"c" State whether gas, coal, oil, or central heating plant.

See reverse side of form.

16—6868

The Cartographic Branch at Archives II contains "Standard Plans of Army Post Buildings (Received from Quartermaster Office) 1891-1917." These are hundreds of standard plans prepared by the Office of the Quartermaster General. A notebook lists the plans by number and suffix, with the number of sheets prepared. To request them, merely specify RG 77, PI NM-19, "Standard Plans of Army Post Buildings 1891-1917," with the plan number and letter suffix desired. The number of sheets varies from one to more than a dozen, and averages around eight or nine for larger structures. The plans are in ink on linen, usually about 24 x 37 inches. For most buildings there are front, rear, and side elevations, and plans for each floor. These plans show structural details, as well as plumbing, heating, and lighting fixtures. The remaining sheets show smaller details, such as doors, windows, coal chutes, and furniture such as cupboards and dressers. The scale for the elevations and plans is usually $1/4$ inch = 1 foot, while the scale for the details varies from $1/2$ inch to 3 inches = 1 foot.

THE PUGET SOUND FORTS TODAY:
EX-COAST ARTILLERY RESERVATIONS AS WASHINGTON STATE PARKS

Near Port Townsend, Washington, are three coast artillery posts—now Washington State Parks—that retain a superb collection of tactical and non-tactical structures, as well as some truly rare examples of coast artillery weapons. Port Townsend is located on the NE corner of the scenic Olympic Peninsula. Located on the head of land at the north end of the town is Fort Worden State Park, possibly the premier publicly-owned, intact Endicott-era reservation in the US. Across Admiralty Inlet is Fort Casey State Park with its impressive main battery row and the only two 10-inch disappearing guns in the United States. The third big Puget Sound Endicott fort, Fort Flagler State Park, is located at the tip of Marrowstone Island some 3 miles to south (but a 35 mile drive by car). A visit to Forts Worden, Casey, and Flagler provides a very comprehensive overview of the components of the early modern American coastal defense systems. The parks are located some 60 miles north of downtown Seattle, and it takes 2-3 hours to drive there depending on which route, and the number of ferries, taken.

The photographs and maps provided here give an overview of the features of Forts Worden and Casey as they exist today. Nearly all the salient features found in the turn of the century American coast defenses forts can be seen. The garrison building collection at Fort Worden is superb and the seacoast artillery piece collection at Fort Casey is unmatched by any other site in the US, only exceeded by the collection found on Corregidor Island in the Philippines.

Officer's Row, Fort Worden (MAB 2010)

Fort Worden State Park, Port Townsend, WA

The Fort Worden reservation was obtained by the State of Washington in 1957. The state was fortunate in obtaining the entire reservation intact with most of its buildings in good shape. The state has turned most of the buildings into a conference center, renting out the barracks and other building space to groups and organizations. The park rents out the officer's quarters as vacation housing, so one can even stay here and really get feeling for the officer's life, if planned far enough in advance as these rentals are very popular. The state developed a large camping area on the sand spit leading to Point Wilson. In the late 1990s, the state began an impressive effort to repair and restore the exteriors of these buildings.

Inspection of the maps included in this article show that many of the key buildings present in 1921 still remain today. The main parade ground area, all the entertainment facilities, the quartermaster store houses, the hospital, a band barracks, even the post cemetery are all there for inspection. You may not be able to get into all the buildings due to their use, but they are there to see and photograph. These buildings are being maintained and will be around for the foreseeable future.

All of Fort Worden's gun batteries remain and are in very good shape—two 12-inch mortar batteries, two 3-inch rapid fire batteries, two 6-inch disappearing batteries, a 5-inch battery, a magnificent late design 12-inch disappearing battery out by the lighthouse, and the fantastic main gun line of 10- & 12- inch gun batteries on the top of the hill. One of the disappointments with Worden is the lack of intact fire control and communication structures on the post. Many were built of wood and have been destroyed over the years. Others, along with some nice old searchlight shelters, can be found using old maps and poking around on the trails in the park. All in all, a visit to Fort Worden State Park will give an excellent overall view of the site, and the tactical and non-tactical structures that made up an American coast artillery fort.

Guardhouse, Fort Worden (MAB 2010)

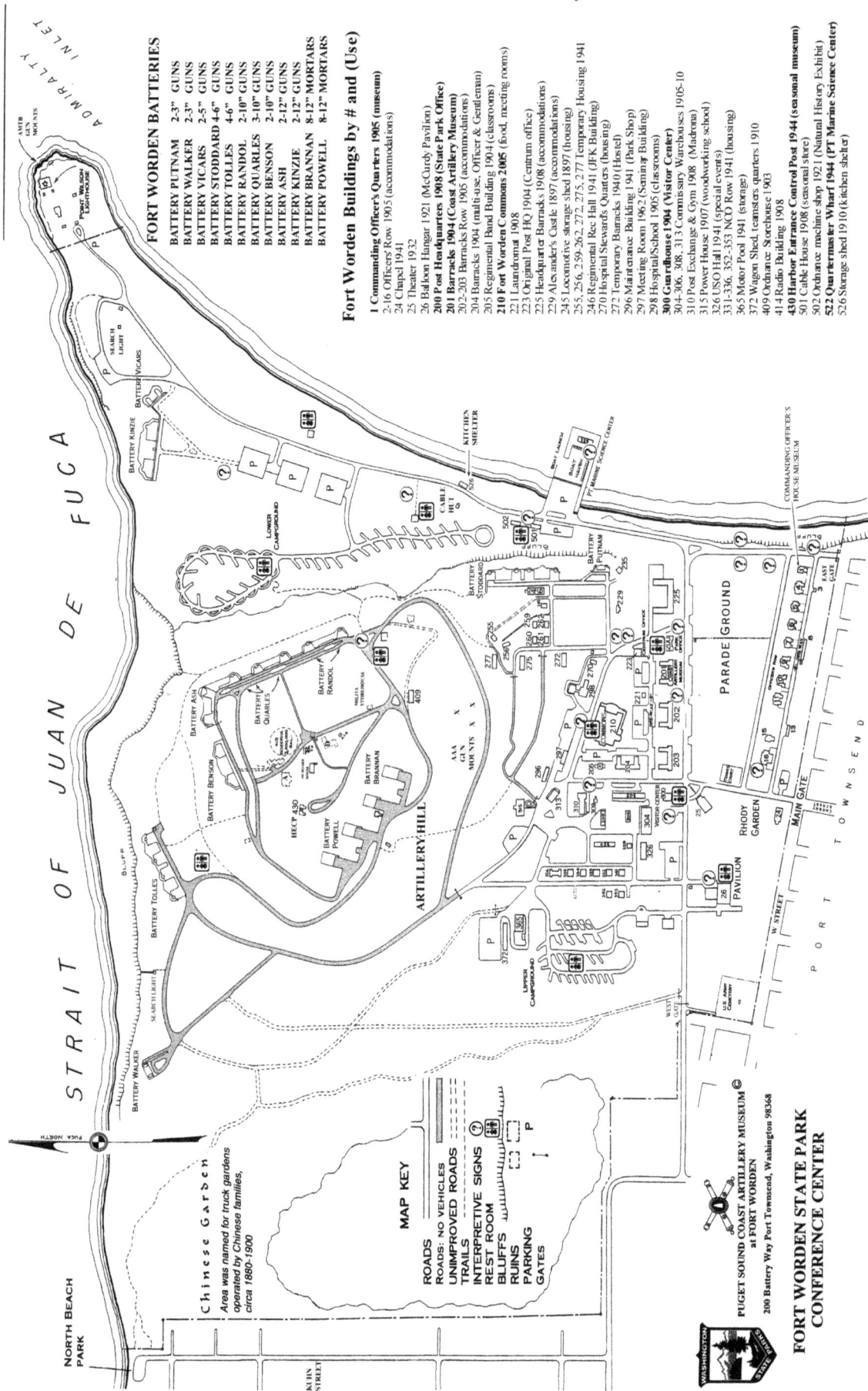

FORT WORDEN BATTERIES

BATTERY PUTNAM	2-3" GUNS
BATTERY WALKER	2-3" GUNS
BATTERY VICARS	2-5" GUNS
BATTERY STODDARD	4-6" GUNS
BATTERY TOLLES	4-6" GUNS
BATTERY RANDOL	2-10" GUNS
BATTERY QUARLES	3-10" GUNS
BATTERY BENSON	2-10" GUNS
BATTERY ASH	2-12" GUNS
BATTERY KINZIE	2-12" GUNS
BATTERY BRANNAN	8-12" MORTARS
BATTERY POWELL	8-12" MORTARS

Fort Worden Buildings by # and (Use)

1 Commanding Officer's Quarters 1905 (museum)
2-16 Officers' Row 1905 (accommodations)
24 Chapel 1941
25 Theater 1932
26 Balloon Hangar 1921 (McCurdy Pavilion)
200 Post Headquarters 1908 (State Park Office)
201 Barracks 1904 (Coast Artillery Museum)
202-203 Barracks Row 1905 (accommodations)
204 Barracks 1904 (multi-use, Officer & Gentleman)
205 Regimental Band Building 1904 (classrooms)
210 Fort Worden Commons 2005 (food, meeting rooms)
221 Laundromat 1908
223 Original Post HQ 1904 (Centrum office)
225 Headquarter Barracks 1908 (accommodations)
229 Alexander's Castle 1897 (accommodations)
245 Locomotive storage shed 1897 (housing)
255, 256, 259-262, 272, 275, 277 Temporary Housing 1941
246 Regimental Rec Hall 1941 (JFK Building)
270 Hospital Stewards Quarters (housing)
272 Temporary Barracks 1940 (Hostel)
296 Maintenance Building 1941 (Park Shop)
297 Meeting Room 1962 (Seming Building)
298 Hospital/School 1905 (classrooms)
300 Guardhouse 1904 (Visitor Center)
304-306, 308, 313 Commissary Warehouses 1905-10
310 Post Exchange & Gym 1908 (Madrona)
315 Power House 1907 (woodworking school)
326 USO Hall 1941 (special events)
331-336, 352-353 NCO Row 1941 (housing)
365 Motor Pool 1941 (storage)
372 Wagon Shed, teamsters quarters 1910
409 Ordnance Storehouse 1903
414 Radio Building 1908
430 Harbor Entrance Control Post 1944 (seasonal museum)
501 Cable House 1908 (seasonal store)
502 Ordnance machine shop 1921 (Natural History Exhibit)
522 Quartermaster Wharf 1944 (PT Marine Science Center)
526 Storage shed 1910 (kitchen shelter)

MAP KEY

ROADS
ROADS: NO VEHICLES
UNIMPROVED ROADS
TRAILS
INTERPRETIVE SIGNS
REST ROOM
BLUFFS
RUINS
PARKING
GATES

Chinese Garden

Area was named for truck gardens operated by Chinese families, circa 1880-1900

PUGET SOUND COAST ARTILLERY MUSEUM ©
at FORT WORDEN
200 Battery Way Port Townsend, Washington 98368

**FORT WORDEN STATE PARK
CONFERENCE CENTER**

NORTH BEACH PARK

KUHN STREET

Fort Worden State Park, 2010 (Puget Sound Coast Artillery Museum)

Company Barracks, Fort Worden, 2010 (MAB)

Double Company Barracks, Fort Worden, 2010 (MAB)

Band Barracks, Fort Worden, 2010 (MAB)

Double Officer's Quarters, Fort Worden, 2010 (MAB)

Large Double Officer's Quarters, Fort Worden, 2010 (MAB)

Commanding Officer's Quarters, Fort Worden, 2010 (MAB)

Batchelor Officer's Quarters, Fort Worden, 2010 (MAB)

Fort Administration Building, Fort Worden, 2010 (MAB)

Headquarters Building, Harbor Defenses of Puget Sound, Fort Worden, 2010 (MAB)

Non-Commissioned Officer's Quarters, Fort Worden, 2010 (MAB)

Hospital, Fort Worden, 2010 (MAB)

Theater, Fort Worden, 2010 (MAB)

WW II Chapel, Fort Worden, 2010 (MAB)

Post Exchange and Gymnasium, Fort Worden, 2010 (MAB)

WW II USO Building, Fort Worden, 2010 (MAB)

Cemetary, Fort Worden, 2010 (MAB)

Balloon Hanger (modified for a theater), Fort Worden, 2010 (MAB)

Centrral Powerhouse, Fort Worden, 2010 (MAB)

Quartermaster Building, Fort Worden, 2010 (MAB)

Quartermaster Commissary Storehouses, Fort Worden, 2010 (MAB)

Wagon Shed, Fort Worden, 2010 (MAB)

Main gun line, Fort Worden, 2010 (MAB)

Ordnance Storehouse, Fort Worden, 2010 (MAB)

Harbor Entrance/Defense Command Post, Fort Worden, 2010 (MAB)

About the only thing you cannot see at Fort Worden is examples of the actual artillery pieces themselves. For that you merely need to ride the ferry from downtown Port Townsend over to Keystone Landing on Whidby Island and visit Fort Casey State Park.

WW II Mobilization Barracks at Fort Casey, now a university camp, 2010 (MAB)

Fort Casey State Park, Whidby Island, WA

Fort Casey did not fare as well as Fort Worden in non-tactical building survival. Many buildings, including the barracks, were torn down in 1938. New barracks were built in 1941-42. After deactivation in 1953, the reservation was divided in half. Most of the garrison area went to Seattle Pacific University, which uses the area today as an extension campus. Some additional land went into private ownership. The rest of the reservation, including all but one of the batteries, went to the state which developed a state park.

The impressive fortification area was built on a low bluff overlooking Admiralty Inlet. The trees and shrubs have been kept under control and the area looks much the same as it did when the army was there. The main parking lot overlooks the impressive main gun line, seven emplacements in a row for 10-inch guns on disappearing carriages. At the center of attention, but to the north end of the main gun line, are two actual M1895MI (the model designation) 10-inch guns on M1901 disappearing carriages. These were retrieved from Battery Warwick, Fort Wint, on El Grande Island in Subic Bay, the Philippines, brought over by the State of Washington in 1963. One really can not appreciate the size of these weapons until one sees them up close and in person—and in a battery emplacement.

Next to the main gun line is a four emplacement 6-inch gun battery and next to that is a two gun 3-inch rapid fire battery—Battery Trevor with its guns. The state retrieved two 3-inch M1903 guns on M1903 pedestal mounts from Fort Wint which were re-installed in Battery Trevor. Fort Casey is doubly blessed with four examples of two types of rare American seacoast artillery. Also open for inspection, and in very fine shape, are a number of fire control structures, the observation stations which housed the optical instruments and communication equipment used in preparing the data for range and direction of the fire of the guns. In front of the battery area are a number of interesting searchlight housing structures.

Battery Worth, Fort Casey, 2010 (MAB)

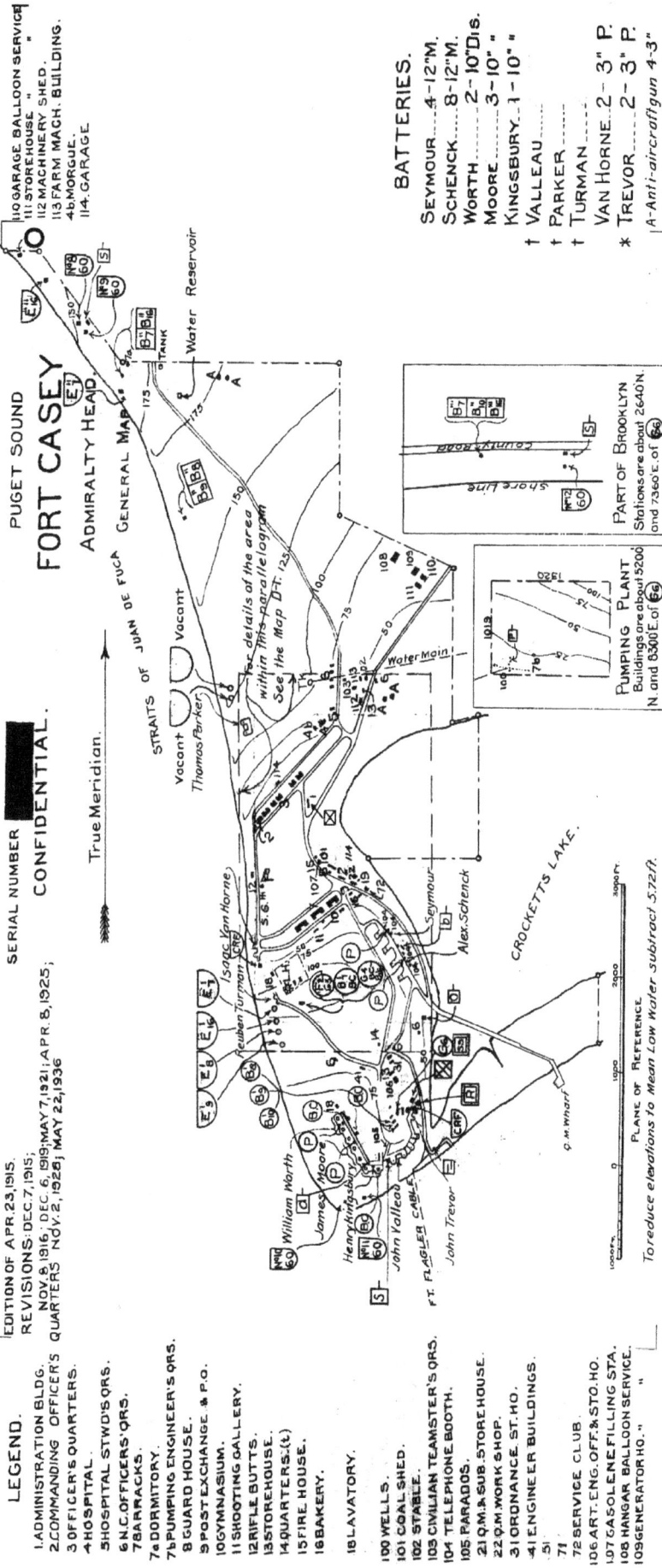

Fort Casey Reservation 1936 (NARA)

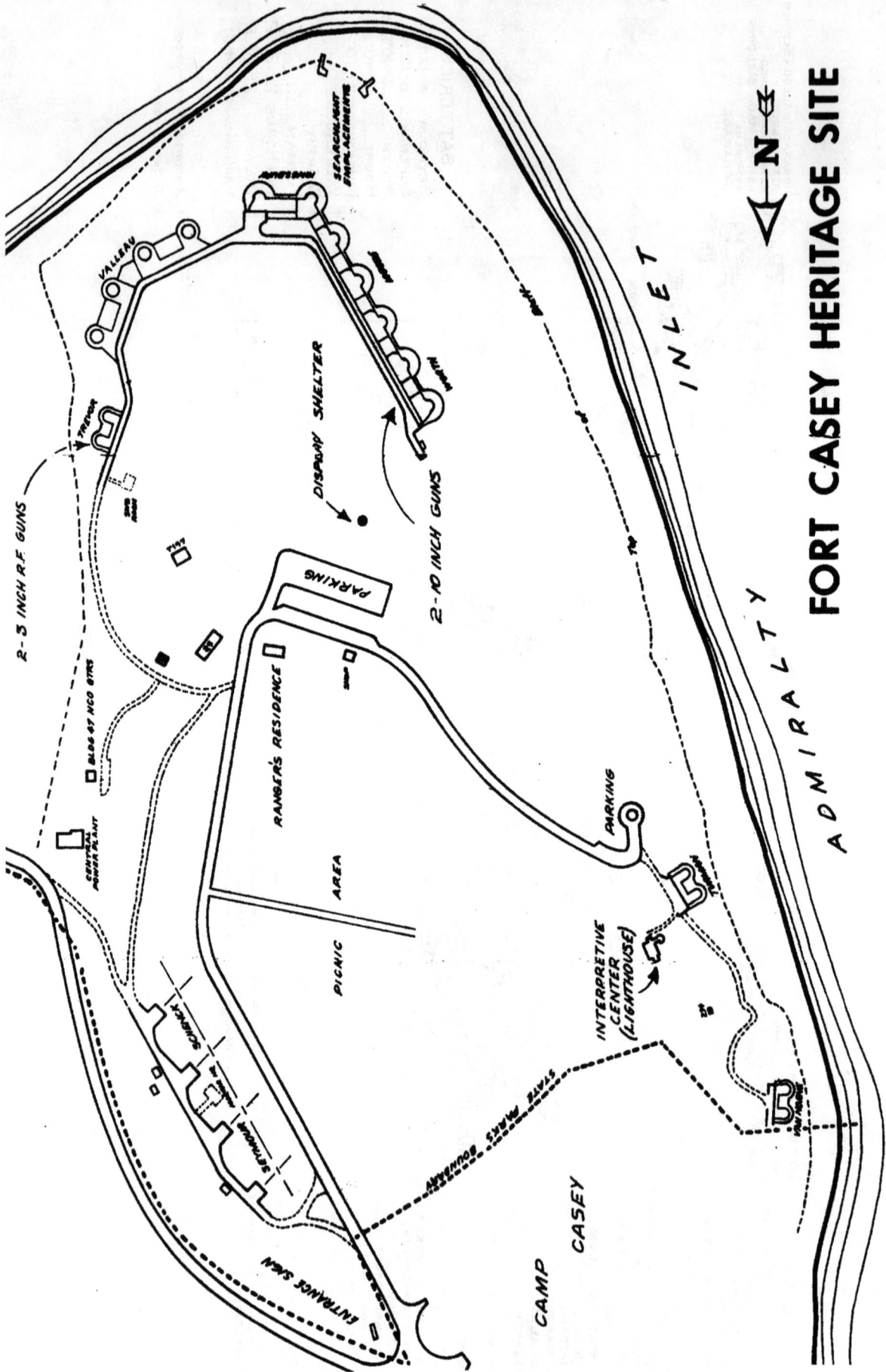

FORT CASEY HERITAGE SITE

Fort Casey State Park (WSPR)

Parade Ground, Fort Casey, 2010 (MAB)

Fire Control Stations, Fort Casey, 2010 (MAB)

Battery Worth, Fort Casey, 2010 (MAB)

The 10 inch guns on disappearing carriages at Battery Worth, Fort Casey, 2010 (MAB)

The 10 inch guns on disappearing carriages at Battery Worth, Fort Casey, 2010 (MAB)

3- inch pedestal mount guns at Battery Trevor, Fort Casey, 2010 (MAB)

Restored magazine and hoist at Battery Worth, Fort Casey, 2010 (MAB)

Fort Flagler State Park, Nordland, WA

Fort Flagler also lost a number of buildings in 1938, including its barracks and officer's quarters. A number of WW II Mobilization barracks were built in the 1941. The state of Washington received the entire original reservation which was then made into a state park in the 1960s. All the batteries remain, though some are in danger of being undermined by substantial erosion along the bluff at the north tip of the island. The main line of two 10-inch barbette batteries and the 12-inch "altered gun-lift" battery are unique. Two 3-inch guns (a M1903 army pedestal and a navy pedestal mount) have been remounted in Battery Wansboro. A number of searchlight shelters and fire control stations also remain relatively intact. The remaining post buildings include the hospital, hospital steward's quarters, a few storehouses, and the WW II barracks. The state has developed camping and picnicking areas, but the bulk of the reservation has been left undeveloped. Recently, the state has begun the clearing of bush and vegetation around the batteries and military roadways in the park, improving both the maintenance of these structures and visual access.

Left to right, Hospital Stewards Quarters, NCO Quarters, Hospital, NCO Quarters, and Quartermaster Building, Fort Flagler, 2010 (MAB)

PUGET SOUND
FORT FLAGLER
MARROWSTONE POINT.
GENERAL MAP

Caretaking Status.

SERIAL NUMBER
CONFIDENTIAL.

EDITION OF APR.23,1915.
REVISIONS: DEC.7,1915; NOV.8,1916; DEC. 6,1919;
MAY.7,1921; APR.8,1925; NOV.2,1928; MAY 22,1936

LEGEND.
1 ADMINISTRATION BLDG.
2 COMMANDING OFFICER'S QUARTERS
3 OFFICER'S QUARTERS.
4 HOSPITAL.
5 HOSPITAL STWD'S.QRS.
6 N.C.OFFICERS QRS.
7 BARRACKS.
7a DORMITORY.
8 GUARD HOUSE.
9 POST EXCHANGE.
10 POST OFFICE.
11 ARTILLERY ENGR. ST.HO.
12 FIRE HOUSE.
13 BAKERY.
14 BLACKSMITH SHOP.
15
16 LAVATORY.
17 CARPENTER SHOP.
18 OIL HOUSE.
19 FUEL SHED.
100 TELEPHONE BOOTHS.
101 RIFLE BUTTS.
102 ROOT HOUSE.
103 STABLE.
104 WAGON SHED.
105 HARNESS ROOM.
106 SHOOTING GALLERY.
21 Q.M.STOREHOUSE.
22 COMMISSARY ST.HO.
31 ORDNANCE ST.HO.
41 ENGR. DEPT. BLDGS.
113 TEAMSTERS QRS.
70
71 BOWLING ALLEY.
72 SERVICE CLUB.
107 CANTONMENT 16 BLDGS
108 PLUMBING SHOP.
109 TOOL HOUSE.
110 PAINT SHOP.

BATTERIES.
BANKHEAD...4-12"M.
WILHELM...2-12"N.Dis.
† RAWLINS...
REVERE...2-10"N.Dis.
† GRATTAN...
† CALWELL...
† LEE...
DOWNES...2-3"P.
WANSBORO.2-3"P.
A-Anti aircraft gun-2-3"
† Armament removed

Fort Flagler Reservation 1936 (NARA)

Restored Hospital Doctor's Quarters, Fort Flagler, 2010 (MAB)

The other Washington coast artillery fort parks are Fort Ward State Park (about half of the old reservation, the rest including the garrison area is privately owned), Fort Ebey State Park, Salt Creek County Recreation Area & Striped Peak State Nature Reserve (Camp Hayden), Manchester State Park (Middle Point Military Reservation), Goat Island Game Reserve (Fort Whitman), Fort Columbia State Park, and Cape Disappointment State Park (Fort Canby). A visit to the Washington State "fort" parks is a superb opportunity for those historians wishing to see examples of 1890 to 1945-era American seacoast fortifications in all their glory—batteries, buildings, and weapons.

Battery 131, Camp Hayden (Salt Creek Recreation Area), 2010 (MAB)

Cantonments and Temporary Non-Tactical Buildings

In addition to permanent structures many temporary buildings were constructed over time at Coast Artillery forts. These were wood-framed and wood-sided structures built on posts or frost walls, rather than permanent foundations. They were sometimes built utilizing available materials or lumber purchased with funds collected from officers, NCOs, or enlisted men—or by using or converting a building originally on the site or used during the construction of the fort. Such buildings were used as chapels, libraries, officers clubs, NCO and enlisted men's clubs, enlisted men's housing, temporary barracks, officers quarters, and offices.

However, the bulk of the temporary buildings at these forts were built during World War I and World War II. When the United States entered the Great War, the Quartermaster Corps created standardized designs for temporary wooden buildings which could be quickly and inexpensively constructed at training camps opened all over the country. The designs—called the 600 series buildings—included barracks, lavatories and latrines, mess halls, officers quarters, and theaters. Some of these buildings were constructed at coast artillery posts. Most typically the coast artilley posts received a theater, where performances were staged and movies shown to entertain the troops. These 1930s-era "War Department Theaters" often became Service Clubs for enlisted men.

World War I Cantonment Area, Fort MacArthur, CA, 1992 (MAB)

When World War II broke out in Europe in 1939, the Quartermaster Corps created new standardized designs for temporary "mobilization" buildings to be built at a new generation of training camps as the country mobilized to defend itself. The designs—designated the 700 series buildings—included 300 hundred different types of buildings, and they were much more elaborate and better built than the 600 series.

Although temporary, these structures were designed with a two-year life span since the length of the mobilization period was unknown. They were built to standard construction specifications, wood framing 36 inches on center, drop wood siding and rolled asphalt roofing, double-hung windows, central heating,and running water. In general, the toilets, wash-basins, and showers were located in the barracks rather than in a separate latrine. One new type that was built at all large World War II sites, including coast artillery forts and reservations, was a chapel designed to be used by all religious denominations, as part of an effort to support and motivate soldiers of all faiths.

The buildings were initially unpainted, but in 1940 the War Department ordered that they be painted. Immediately after the Pearl Harbor attack, all the buildings were painted khaki or other camouflage colors. Large numbers of these buildings were constructed at permanent Coast Artillery forts in 1940 and 1941. The most common ones built were Chapels, Company Administration and Storehouses, Barracks, Mess Halls, Recreation Buildings, Garages, Storehouses, Officers Quarters and Mess, and Theaters. These sturdy structures housed the men of the National Guard who were called to active duty in the fall of 1940 and the draftees used to expand Coast Artillery Regiments thereafter.

Mobilization buildings built at existing harbor defense forts were sited on a pragmatic basis where space permitted. If land was available, the buildings were grouped to serve a battery (company) of men: two 63 man barracks, a mess hall, a recreation building, and battery administration/storehouse building. If land was scarce, buildings were fitted in wherever space was available, even on the parade ground. After mobilization began, but before Pearl Harbor, some forts commissioned landscape development plans to make the forts as aesthetically pleasing as possible after the addition of the mobilization structures by building new streets and adding scrubs and trees. On December 1, 1941, the Corps of Engineers took over the responsibility for building non-tactical structures from the Quartermaster Corps.

Many Mobilization buildings are still in use today at active Army and National Guard posts and a few survive at coast artillery forts. Listed below are the typical mobilization buildings that would be found at a large coast artillery fort.

Administration Building	Motor Pool
Barracks	Nurses Quarters and Mess
Company Administration/Storehouse	Officers Quarters and Mess
Fire Station	Post Chapel
Garage	Post Exchange
Garbage Grinder Building	Radio Shelter
Hospital	Recreation Building
Hospital Barracks	Storehouse
Infirmary	Theater
Mess Hall	

Post Chapel (left) and 700 series Mobilization Cantonment Construction (right) 1941, Fort McKinley, ME
(from the collection of Joel W. Eastman)

WW II Mobilization Barracks at Fort MacArthur, CA 1992 (MAB)

Administration and Storehouse Building, 700 series Mobilization Cantonment Construction (1941) Fort McKinley, Great Diamond Island, HD Portland, ME (from the collection of Joel W. Eastman)

Barracks 700 series Mobilization Cantonment Construction (1941) Fort McKinley, Great Diamond Island, HD Portland, ME (from the collection of Joel W. Eastman)

World War II Coast Artillery Forts and Reservations:

As the US entered World War II a whole new generation of forts and batteries were being built. However, given the rush to complete construction, few of the new harbor defense sites were formally named. Instead, the sites were referred to as "military reservations," and the local names of the sites were used to designate them, such as "Jewell Island Military Reservation."

Once the US entered the war, a new type of temporary non-tactical structure was adopted, called "Modified Theater of Operations" (MTO) buildings. Designed to be used in war zones, the buildings were modified somewhat for use in the Zone of the Interior, the United States, but they were truly temporary. These structures had wood framing 48 inches on center with fiberboard sheathing and 15 pound rolled felt siding held on with wooden battens. Like World War I temporary designs, separate barracks and latrines were adopted, and they were heated by magazine stoves or space heaters. Standard plans were prepared for hundreds of building types. In April 1942, the War Department ordered that Modified Theater of Operations buildings were to be built at all new camps, including new harbor defense sites, and for the expansion of existing camps, posts and stations. In October 1942, the War Department replaced the MTO designs with new Theater of Operations (TO) plans based on the Mobilization designs (700 series), with additional bracing, improved insulation and ventilation, and a variety of siding types other than tar paper.

Modified Theater of Operations buildings built to support new batteries completed after Pearl Harbor also followed a pragmatic approach in siting. Barracks and latrines to serve AMTB batteries were built as close to the batteries as possible given the land available and the nature of the terrain. Some groups of buildings visible from the sea were laid out to appear to be summer cottages. At large reservations with several batteries and large numbers of buildings, landscape plans were developed which included parade grounds, flag poles and named streets. The only examples that survive today are ones that were sold after the war, moved and turned into houses and cottages. Listed below are the types of TO buildings that would be found at a large World War II Coast Artillery fort or military reservation.

Administration
Barracks
Fire Station
Infirmary
Lavatory and Latrine

Mess Hall
Officers Quarters and Mess
Post Exchange
Recreation Building
Warehouse

Officer's Quarters—Modified Theater of Operations Cantonment Construction (1942) Jewell Island Military Reservation, HD Portland, ME (from the collection of Joel W. Eastman)

A Gazetteer of Remaining Coast Artillery Forts

Fort Columbia State Park, Chinook, WA

Fort Columbia, of the harbor defenses of the Columbia River, is an excellent example of a small Endicott-era fort. Built at the turn of the last century, the post was turned over pretty much intact to the state of Washington in the late 1950s, which developed it into a historical park. The barracks, two officers quarters, the administration building, two NCO quarters, the hospital, the hospital stewards quarters, the quartermaster building, the fire station, the guardhouse, and the ordnance storehouse remain, along with four gun batteries, two mine casemates, a power house, and several fire control stations. In 1994, the state obtained and relocated two rare surviving 6-inch guns on the WW II shielded mounts and re-installed them in Battery 246, adding to the state's impressive collection of surviving American coast artillery weapons. The state has also recently completed an extensive repair program on the buildings both outside and inside.

Barracks, Administration Building, and Officer's Quarters, Fort Columbia (Andrew Berhow 2008)

The Golden Gate National Recreation Area, National Park Service,
The Harbor Defenses of San Francisco, CA

The collection of tactical and non-tactical structures which remain from the Coast Artillery forts and Army reservations around the entrance to San Francisco bay are truly one of the finest groups of military architecture in the United States. All of these reserves are open to the public and are being preserved and managed as public park and trust lands. Several districts of historical buildings are on the National Historic Landmarks list. The remaining buildings and structures span the years of military use from the 1850s to the 1970s. Forts Baker, Barry, Cronkhite, Winfield Scott, Mason, and McDowell, as well as the Presidio itself, all have an extensive collection of structures. The Fort Baker parade ground buildings have been recently extensively restored for use as lodge and meeting facilities.

Double Company Barracks, Fort Winfield Scott (MAB 2015)

Harbor Defense Headquarters, Fort Winfield Scott (MAB 2015)

Fort Baker Parade (MAB 2015)

Double Officers Quarters, Fort Barry (MAB 2011)

Barracks, Fort Cronkhite (MAB 2015)

The buildings range from elegant officer's quarters and enlisted men's barracks to large warehouses, storehouses, administrative offices, airplane hangers, medical facilities, and cavalry stables. The remaining seacoast defenses structures include two Third System fortifications, a Civil War-era battery, a unique dynamite battery, dozens of Endicott-era batteries, batteries built during the World War II era, and Nike missile defenses of the 1950s and 1960s.

The establishment of the early modern era coast artillery garrisons led to major building collections which remain from the 1910s at Fort Winfield Scott (next to the Presidio), and Forts Baker and Barry on the Marin Headlands. Fort Cronkhite, north of Fort Barry on the Marin Headlands, is one of the few remaining (partially) intact WW II-era posts and retains a number of mobilization buildings from the period 1940-1945. Major military building collections exist at Fort Mason, the Presidio of San Francisco, Fort Winfield Scott, Fort Baker, Fort Barry, Fort Cronkhite, and Fort McDowell (now Angel Island State Park) dating from the 1850s and on through WW II.

These excellent sites are well documented by brochures and booklets available from the various book stores located around the Golden Gate National Recreation area, including: *Map and Guide to the Seacoast Fortifications of the Golden Gate, Official Map and Guide to the Presidio of San Francisco, the Presidio of San Francisco, a Self-Guided Tour to its Architecture; Nike Missile Site SF-88, Alcatraz—Island of Change; Fort Point—Sentry at the Golden Gate;* and the *Golden Gate National Recreation Area—Guide to the Parks.*

Fort MacArthur Reservations, Los Angeles (San Pedro), California

Fort MacArthur, which defended Los Angeles harbor, was the only complete defense facility built on the continental US during the "Taft-era" construction (1910s). The "mission-revival" style buildings around the parade ground are intact—although it is off limits to the general public as it is used by Los Angeles

Post Exchange and Barracks (above) and Officers Quarters (below) Fort MacArthur (MAB 1995)

AFB personnel. It retains the original barracks, guardhouse, post exchange, and officer's quarters, as well as a number of buildings built in the 1930s and 1940s. The main fortification areas were on separate tracts of land and are now City of Los Angeles parks—Angels Gate and White Point.

Angels Gate Park retains examples of both WW I and WW II mobilization building areas. Battery Osgood-Farley, home of the Fort MacArthur Museum, is quite possibly the most fully preserved early modern American seacoast battery in the United States.

Fort Rosecrans, San Diego, California

A number of Army buildings, including barracks, quartermaster buildings, and officer's quarters, remain in the garrison area of old Fort Rosecrans. The entire reservation, less the acreage set aside for the Cabrillo National Monument, was transferred to the Navy in the late 1950s. The main Fort Rosecrans garrison area is now part of the submarine base and is generally off limits to civilians. Cabrillo National Monument has control of a few searchligh shelters and some base end stations, but few military buildings.

Barracks, Fort Roscrans, 1992 (MAB)

Fort Crockett, Galveston, Texas

This was the main Army garrison post for the Galveston area forts, including Fort San Jacinto and Fort Travis. The post had to be completely rebuilt after the 1900 huricane. The buildings were made much more substantial. The post was used through World War II. A large number of building remain on he old reservation which is used by a variety of national, state, local and private entities. The buildings under the care of NOAA have been renovated for new uses.

Barracks, Fort Crockett, 2008 (MAB)

Fort Barrancas, Pensacola, Florida

This was the main Army garrison post for the Pensacola area forts, including Fort Pickens and Fort McRee. Many of the post structures were built during the early 1900s and remain today. However, the garrison area was transferred to the Navy in the 1950s, and may not be open to the public.

Fort Caswell, Cape Fear, North Carolina

The Baptist State Convention of North Carolina purchased the entire reservation in 1949. The Baptists had purchased a reserve that contained over 70 buildings. Since then the organization has refurbished most of the buildings and put them to use as classrooms, residence rooms and halls, auditoriums, administrative offices, and maintenance facilities. The overall appearance of the buildings and batteries remain unchanged, making this a first-class existing example of a turn of the century coast artillery post. The Assembly is not open to the general public and permission must be sought to visit Fort Caswell.

Hospital and quarters, Fort Caswell, NC

The Harbor Defenses of Chesapeake Bay:
Fort Monroe, Fort Story, & Camp Pendleton

Fort Monroe

These three fine posts contain a variety of interesting buildings and structures. Fort Story and Camp Pendelton are active Army facilities and entry to certain areas may be restricted. Fort Monroe was offically closed as a active military post in 2011. Parts of the reservation were designated as Fort Monroe National Monument in November 2011.

Fort Monroe is the "cradle" of the Coast Artillery Corps; it was the location of the office of the Chief of the Coast Artillery and the Coast Artillery School. Fort Monroe retains a large collection of quarters, barracks, administration, school, supply, and storage buildings, etc. Many of these buildings have been remodeled and used for other purposes over the years. The magnificent Third System fort, one of the largest built by the United States, is also in good shape. The modern-era batteries, however, have not fared as well. About half of the batteries built at Monroe have been destroyed to make space for additional construction; the other half have had their earthen covers removed for the most part. Development of the site will be ongoing in the years ahead. If you have an interest in American coast artillery, this post is a must see.

Coast Artillery School Building (MAB 2019)

Coast Artillery Library (MAB 2019)

Across Hampton Roads from Fort Monroe is Fort Story, a sprawling reservation with a number of interesting features for the coast artillery enthusiast. Most of the historic period buildings were built as cantonment, temporary, or mobilization structures during either World War I or World War II. Sadly many of them have either been torn down or modified. Further down the coast, past Virginia Beach, is Camp Pendleton, a Virginia National Guard reservation. This post has a fine collection of WW II-era mobilization buildings.

Fort DuPont State Park, Delaware City, Delaware

This post has been used for a number of years by the state of Delaware as a record storage facility, a civil defense center, and a mental hospital. A portion by the water is now a state park. More than half of the buildings that were at this post still remain, including the barracks, a number of officer's quarters, the administration building, storehouses, and more, which are stilled used by the state and are off limits to photography. The site has all but one of its batteries remaining and a large natural area along the shore of the Delaware River. A ferry ride from nearby Delaware City takes you to both Fort Delaware State Park on Pea Patch Island and on over to Fort Mott State Park in New Jersey.

Administration Building, Fort DuPont (MAB 2009)

Fort Mott State Park, Salem, New Jersey

A supurb coast artillery park with an active program of restoration and interpretation. Fort Mott State Park features a number of preserved buildings, including a restored fire control tower, a restored administration building, and a newly restored Peace Magazine. The park contains a fine collection of unique Endicott-era batteries.

Peace Magazine, Fort Mott (MAB 2009)

Gateway National Recreation Area, the Harbor Defenses of New York

The Fort Hancock reservation, now the Sandy Hook unit of Gateway National Recreation Area, contains the largest and most diverse set of tactical and non-tactical structures of the early modern era, 1890-1920. A number of interesting defenses were built here, from the uncompleted Third System fort, a unique gun-lift battery, an early mortar battery, through a large collection of Endicott-era gun batteries, post WW I batteries, and a Nike missile battery. The Army Proving Grounds were located here until 1920. Consequently, the Army built a large number of buildings and facilities at this site, most of which remain today. The contrast between the yellow-brick coast artillery buildings and the red-brick ordnance department buildings is particularly interesting. The reservation, except for a section used by the Coast Guard, was turned over to the National Park Service in the mid-1970s. The Park Service has worked to stabilize these structures over the past few years and seek adaptive reuses for them. One barracks has been converted into a research center, and they are seeking tenants to fix and repair the officer's quarters. While not in as good shape as the collection of buildings in the Golden Gate National Recreation Area, the Sandy Hook collection preserves the essence of a large coast artillery post and has a number of unique features not found at any other post.

20 inch Rodman and NCO Quarters, Fort Hancock (MAB) 1997

Parade Ground, Fort Hancock, 1997 (MAB)

Harbor Defense Headquarters, Fort Hancock, 1997 (MAB)

Officers Quarters, Fort Hancock, 1997 (MAB)

Firehouse and commisary buildings, Fort Hancock, 1997 (MAB)

Fort Wadsworth, located at the Narrows on Staten Island, has a good collection of brick garrison structures of various dates and structures. However, many of these buildings are still in use by various government and defense agency and are not open to general public access. The highlight here is the Third System works of Fort Tomkins and Battery Weed. Several of the modern era batteries have been partially buried by the Navy. Across the Narrows in Brooklyn is Fort Hamilton, which is still an active Army post. Fort Hamilton also has an excellent collection of brick garrison structures but their modern use may restrict access to many of them. About half of the Third System work and all of the modern era batteries have been destroyed for road construction.

Officers Quarters, Fort Hamilton, 1997 (MAB)

Fort Tilden, located at Rockaway Point, has a few remaining structures. Many of the buildings at this post were mobilization construction built during WW I and WW II, and most of these buildings have since been removed.

Fort Totten, New York, New York

Occupied by various Army and New York government units until just recently, this post has been generally off limits to the public. The military has just recently left and the City of New York is deciding what to do with reservation. This reservation has an excellent collection of non-tactical buildings, all in pretty good shape, plus the unfinished Third System fort and the partially overgrown Endicott-era batteries. It also has an excellent collection of early mine shore facilities. Permission may still be required to visit the fort as it is not generally open to the public. That may change, as one of the functions being considered by the City of New York is green-space parkland.

Officer's Club, Fort Totten

Fort Terry (Plum Island), Long Island Sound, NY

The Island was eventully transferred to the U.S. Department of Agriculture which used the location for it infectious animal disease research. The buildings have been variously used and or neglected over the past 50 years, but a large number still remain. The Animal Disease Center is scheduled to be moved to a location in Kansas and the fate of Plum Island and the remains of Fort Terry have yet to be determined.

Double Company Barracks, Fort Terry, 2003 (MAB)

Fort H.G. Wright, Fisher's Island, NY

Fishers Island has long been the private retreat of the residents there. The lands utilized for the harbor defense have been returned to private owners for the most part. Most of the officers quarters and few other buildings that remain have been tuned into private residences or other commercial uses. The island community does not encourage visitors.

Officers Quarters, Fort H.G. Wright, 2003 (MAB)

Fort Andrews (Peddock's Island), Boston Harbor Islands Park, Boston, MA

Bordering sheltered Hingham Bay, the 134 acres of Peddock's Island comprise the most diverse island in Boston Harbor, encompassing four spit-connected, forested drumlin hills which include the abandoned, 30+ buildings of circa 1900 Fort Andrews, a colony of 47 circa1900 summer cottages, and a nature preserve. Peddock's Island is accessible only by boat or helicopter, served in season by Bay State ferries, Cruise Lines from Long Wharf, Boston; Hewitt's Cove, Hingham and the George's Island water taxi or by private boat. The MDC acquired Peddock's in 1970. The Fort has been deemed eligible for inclusion on the National Register of Historic Places. Fort Andrews is the only circa1900 military reservation in the Boston area to retain the preponderance of its historic permanent core of buildings. In 2012 several buildings were demolished and removed however several others were stabilized for future restoration and use.

Headquarters Building, Fort Andrews, 2006 (MAB)

Barracks, Fort Andrews, 2006 (MAB)

Fort Preble, South Portland, Maine

Fort Preble has most of its turn of the century buildings, although a number of them have been minimally altered and new buildings have been constructed among them. The buildings on the parade ground look much as they did at the turn of the century—the administration building, hospital and barracks. Behind these are the guard house (with an addition), fire station and bakery (which have been minimally altered). On the hill above the parade ground are four officers quarters, one of which has been turned into a hospitality center. Below the officers quarters is a World War II mobilization building, a quartermaster storehouse (with a wood frame addition), a cable tank, mine storehouse, ordnance machine shop and ordnance storehouse. Fort Preble is now the campus of Southern Maine Community College and open to the public.

Officers Quarters, Fort Preble, 2005 (MAB)

Fort Williams, Cape Elizabeth, Maine

Most of the buildings at Fort Williams have been demolished, but a few excellent structures survive—a Captain's Quarters, Bachelor Officers Quarters, Artillery Engineer's Storehouse, Militia Store House, Fire Station, Gun Shed (for 155 mm guns), fire station, the central power house, and transformer house. The Portland Head Light Museum devotes a room to the history of the fort, an interpretive center has been created in emplacement two of Battery Blair, and other surviving structures have interpretive signs.

Band Stand, Fort Williams, 2005 (MAB)

Fort Levett, Cushing Island, Maine

Several brick storehouses and officer quarters remain at this post along with the stable, the fire station, and the magnificent hosptial building. The barracks have all been demolished. Fort Levett is owned by a private residential association and permission must be obtained to visit the site.

Hospital, Fort Levett, 2005 (MAB)

Fort McKinley, Great Diamond Island, Portland, Maine

Fort McKinley has an excellent collection of turn of the century brick buildings, including its administration building, officers quarters, non-commissioned officers quarters, barracks, guard house, hospital, bakery, fire station, post exchange, bowling alley, quartermaster storehouses, ordnance storehouse, ordnance shop, work shop, stable, water pumping station, stand pipes, central power house, school, and quartermaster wharf. The fort is on the national register of Historic Places and so the exterior of the buildings must be maintained. Fort McKinley is a private condominium association, but volunteers offer tours by appointment to small parties during the summer and maintain a small museum.

Barracks, Fort McKinley, 2005 (MAB)

Barracks, Fort McKinley, 2005 (MAB)

Officer's Quarters, Fort McKinley, 2005 (MAB)

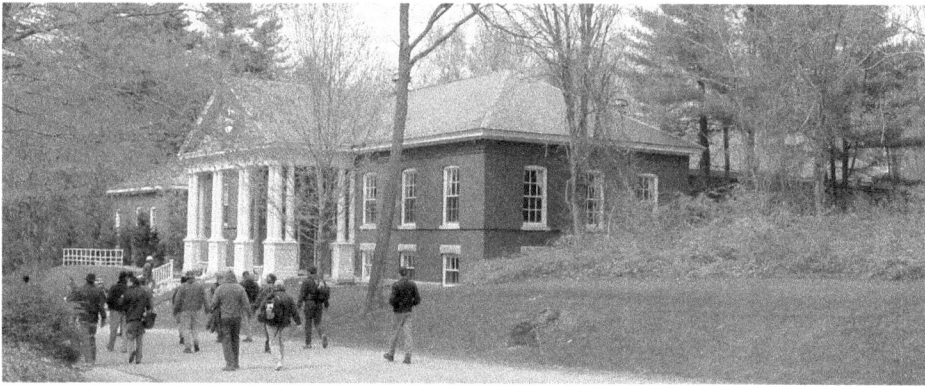
Post Exchange, Fort McKinley, 2005 (MAB)

Bowling Alley, Fort McKinley, 2005 (MAB)

Gymnasium, Fort McKinley, 2005 (MAB)

NCO Quarters, Fort McKinley, 2005 (MAB)

Administration Building, Fort McKinley, 2005 (MAB)

Power Plant, Fort McKinley, 2005
(MAB)

Stable, Fort McKinley, 2005 (MAB)

Bakery, Fort McKinley, 2005 (MAB)

QM Storehouse,
Fort McKinley, 2005 (MAB)

Work Shop,
Fort McKinley, 2005 (MAB)

Ordnance Storehouse,
Fort McKinley, 2005 (MAB)

INSIGNIA OF THE COAST ARTILLERY CORPS

By Greg Hagge, Bolling Smith, and Mark Berhow

Coast Artillery insignia on overpass at the entrance to Fort Winfield Scott at the Presidio of San Francisco

There are several distinct phases in the evolution of uniforms and accessories used by the Coast Artillery Corps. Generally speaking, these changes are driven by the periodic modernization of the army's uniform regulations as a whole. With few exceptions the personnel of the CAC wore the same regulation uniform and insignia as the rest of the army with the appropriate branch insignia applied. The following sections will outline general insignia and uniforms examples and changes as they relate to the Coast Artillery only.

At the turn of the 19th century, there was no difference between field and garrison artillery uniforms and insignia. This changed with the introduction of the Artillery Corps and the specialized insignia for officers in March of 1902. The previous 1901 pattern crossed cannon insignia with a red enameled disc at the center was modified to distinguish between field and coast artillery officers. Field artillery would have a wagon wheel device in the red disc and Coast Artillery would have a projectile at the center. Enlisted men of both branches of artillery used a stubby crossed cannon insignia patterned after the 1895 artillery insignia. The addition of numbers below the cannons identified the numbered independent companies/batteries. For a short time both field and coast artillery enlisted men used the same insignia. Later, the Field artillery was again reorganized into lettered batteries, thus leaving the numbered insignia exclusively for the coast artillery. The cannon pins were finished in black for the olive drab field uniform and gold for the dress blue coat. Another difference is that only the dress uniform used a cap insignia. Cap insignia was attached to the front of the cap with a thumbscrew. The olive drab service cap was not provided with cap style pins. The 1901 "fat" style cannon insignia was only manufactured for the first 126 companies, the black finish made only as a pin back.

The insignia for many branches of the army were redesigned in 1904. The coast artillery received a new "thin" style cannon device in black and gold finish. This time both service and dress cap badges were provided, and they were manufactured for company numbers up to 170. At different times reorganization changed the designation of companies. The numbers were sometimes broken off to accommodate these changes.

The officer's cannon insignia was also redesigned in the "thin" style. This was used through the 1920s. Officer's insignia did not have company numbers attached. The only devices attached to the officer's cannon insignia were to denote service in staff positions. These were small devices attached to the bottom of the insignia for Quartermaster, Ordnance, Mines, Chaplain, Adjutant General, Commissary and a few others.

In 1908 a new type of enlisted insignia was introduced for the service uniform—the collar disc. Troubles with manufacturing prevented the distribution of the disc until about 1910. This pattern disc is known as "Type-I" discs in the insignia collector's jargon. It was about one inch in diameter finished black. This type of insignia has a screw stud on the back with a round thumbnut for attachment. The service branch (i.e.: artillery, infantry, etc) device was worn on the left collar and the "US" disc on the right (at first the "US" and branch device were worn in sets on each side just as the previous insignia were). The background of the disc was a scored crosshatch design. Coast Artillery was distinguished from Field artillery by placing the crossed cannon device above center to accommodate placing the company number below the cannons, just

Officer insignia   Enlisted insignia

1902 style

1902 style

1901 style

1904 style

1904 style

1904 style

Coast Artillery Insignia 1901, 1902 & 1904 types: officers (left) enlisted men (right).

Type "I" collar disk   Type "I" collar disk (with company #)   Officers' collar insignia
                                              (1917-18)

like the former collar pins. Field artillery was organized in regiments of lettered batteries. Their left collar insignia had cannons centered on the disc with the letter below. A variation on this was provided for NCO staff. This field artillery insignia had the letter below with the regimental number above the cannons. Normally the "US" disc has the regimental number below the "US", when this number is used. The relevance to Coast Artillery is that enlisted men of the CAC field regiments of the First World War used field artillery collar insignia. Officers of these field regiments wore CAC cannon pins with numbers attached above the projectile device. The projectile device was added to the enlisted collar insignia in 1917, but seldom used much before 1919. Around 1920 the background pattern was changed to an open crosshatching with tiny shield shaped "dots" in the open spaces, all in very small detail.

In 1924 the Army reorganized the CAC from a mix of independent companies and regiments entirely into regiments. This was accompanied by the introduction of regimental distinctive unit insignia (DUI) for the uniforms. The DUI was worn on the shoulder strap by officers and on the lapel below the branch insignia by enlisted men (behind the disc on standing collars). The colorful pins were intended to identify each regimental organization and inspire esprit de corps. Each of the regular Army and National Guard regiments had authorized designs registered with the Institute of Heraldry. The designs, based on the unit's heraldic crest, often reflected historical themes and geographic elements (see following sections on unit flags and heraldry). There are many variations of these insignia to be found—some through authorized changes in the designs and some through manufacturer's errors, die variations and different types of screw or pin attachments. Some of these organizations exist today and use the same DUI design, but most have long ago been deactivated or have drastically changed the design (See section on heraldry following).

As noted above, the 1924 regulations required all buttons and insignia to be gold for the service uniform, so the collar insignia changed as well. This new background design in gold finish is known as "Type-II" collar brass. Officers began wearing a large cannon insignia with regimental numbers on the shirt collar (formerly very small black devices) for field service. A new smaller design with the regimental numbers above the projectile device was worn on the lapels of the service coat. This style insignia was used with minor variations until the CAC was disbanded.

About 1930 a new style enlisted collar disc was introduced. This was of two-piece construction consisting of a flat smooth brass disc with a branch device attached by a screw fitting from the rear. Any branch of service could be assembled with this disc. This is known as "Type-III" brass. It is still in use to the modern era. As before the "US" disc had the regimental number under the "US". The left disc had the crossed cannon and projectile device with the battery letter below. A variation on this included a battalion number above the device. Another unofficial addition was "AA" over the cannons for officers as well as enlisted men's collar brass. This was mostly a National Guard phenomenon. In 1943, as a wartime measure the insignia was redesigned for economy of materials. The brass was replaced by stamped and plated steel. The new brass plated insignia is of one-piece construction. Also at this time the DUIs were commonly made of silver. There was no shortage of this metal; "sterling" was not the "high class" model, it was the substitute!

Type "II" collar insignia

Type "III" collar insignia

Cloth Insignia of the CAC

The most prolific type of cloth insignia specific to coast artillery is enlisted men rating insignia of the 1900-1930s. The great variety of ratings and their associated insignia is too large and complex to treat here, but a set of examples from 1908 are illustrated on the next page. They were manufactured in bright colors for the dress uniform and in drab for the service uniform. The purpose of the specialized insignia was to show the specialty and level of qualification of the individual soldier. It was necessary to qualify as a first class gunner before other specialty qualifications could be earned. Some of these ratings also meant extra pay. The competition was tough and successful achievement could mean a chance to be selected for advanced training at the Fort Monroe enlisted specialty schools. The cutting edge technologies of master gunner (civil engineering skills), electrical and power plant operations (steam boilers and electric generators), radioman and other highly technical skills were on the curriculum.

Enlisted rank chevrons are another category of cloth insignia with many variations over the years. Field and coast artillery used the same chevrons as well as some of the same rating insignia. Some of the names for the grades were different for coast artillery, sergeant major junior grade being one of them. The dress chevrons were red artillery branch color facing cloth on dark blue uniform cloth. Service uniforms used full color on olive drab background at first, and then changed to drab or gray on olive drab background. In the 1920s the service chevrons were all mounted on a dark blue background. Some of the rating insignia was red on blue, some drab on blue. These details changed regularly and again are too complex to treat in this article.

There were only a few unit-specific or department-specific CAC shoulder sleeve insignia (SSI) "patches." The red triangle of the 55th CAC (below left) and the white triangle of the 59th CAC (below, right) used in the early 1920s are the most noteworthy. Some locations, especially overseas, wore a local department SSI beginning in the late 1920-1930s, the Hawaiian, Panama Canal Zone and the Philippine departments being the most important. WW II command SSI such as the Alaskan Defense Command (ADC) is another example. State side CAC organizations generally did not wear SSI until about 1942 when the insignia were developed for the continental defense, coastal, and AA commands.

Shoulder patch insignia of the Coast Artillery in World War II.

A Note on Collecting Insignia

When contemplating collecting any of these or other types of insignia, be aware that reproductions and "re-strikes" abound. Learn the differences between modern insignia and "old" types. If your interest is in assembling insignia from a specific era, there are many things to look for. The accompanying illustrations may help in selecting insignia for period uniforms and displays more carefully.

Coast Artillery Enlisted Men Sleeve Insignia

The insignia shown here are for dress uniforms from the 1908 Quartermaster uniform catalog in use c. 1904-1917. Note that the titles and designs varied considerably over the years.

COAST ARTILLERY.
(Par. 89, 90, G. O. 169 W. D. 1907. & Par. III, G. O. 15 W. D. 1908.)

SERGEANT MAJOR,
SENIOR GRADE.

MASTER ELECTRICIAN.

ENGINEER.

ELECTRICIAN SERGEANT
FIRST CLASS.

ELECTRICIAN SERGEANT,
SECOND CLASS.

MASTER GUNNER.

CHIEF MUSICIAN.

SERGEANT MAJOR,
JUNIOR GRADE.

CHIEF TRUMPETER.

PRINCIPAL MUSICIAN.

FIRST SERGEANT.

DRUM MAJOR.

SERGEANT.

COMPANY QUARTER-
MASTER SERGEANT.

CORPORAL.

LANCE CORPORAL.

FIREMAN.

COOK.

MECHANIC.

SERVICE IN WAR.

SERVICE.

GUN COMMANDER.

GUN POINTER.

CASEMATE ELECTRICIAN.

OBSERVER, FIRST CLASS.

OBSERVER, SECOND CLASS.

PLOTTER.

CHIEF PLANTER.

CHIEF LOADER.

FIRST CLASS GUNNER.

SECOND CLASS GUNNER.

FIRST CLASS GUNNER.

SECOND CLASS GUNNER

GUN OR MORTAR COMPANY.

MINE COMPANIES

Coast Artillery Enlisted Men Sleeve Insignia

1918-1930s

RATED POSITIONS IN THE COAST ARTILLERY

CASEMATE
ELECTRICIAN

OBSERVER
1st CLASS

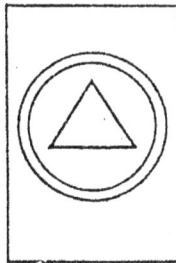

OBSERVER
2nd CLASS

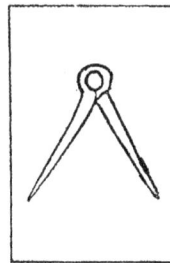

PLOTTER

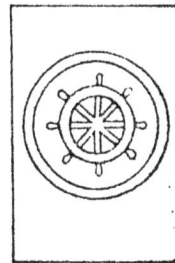

COXSWAIN

CHIEF
PLANTER

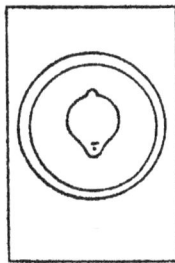

CHIEF
LOADER

GUN
COMMANDER

GUN
POINTER

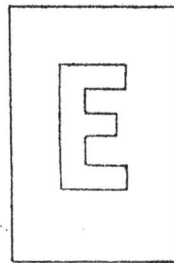

EXCELLENCE

Key references on Army insignia:

Leon W. LaFramboise, *History of the Artillery, Cavalry, and Infantry Branch Service Insignia* (Watson Publishing Co., Steeleville, MO, 1976)

William K. Emerson, *U.S. Army Soldiers and Their Chevrons: An Illustrated Catalog and History from the Revolutionary War to the Present* (R. James Bender Publishing, San Jose, CA 2013).

1902   1904   regimental # (1917)   Antiaircraft artillery

Line drawings of officer collar or cuff insignia

Chaplain   Commissary   Ordnance   Quartermaster   Judge Advocate

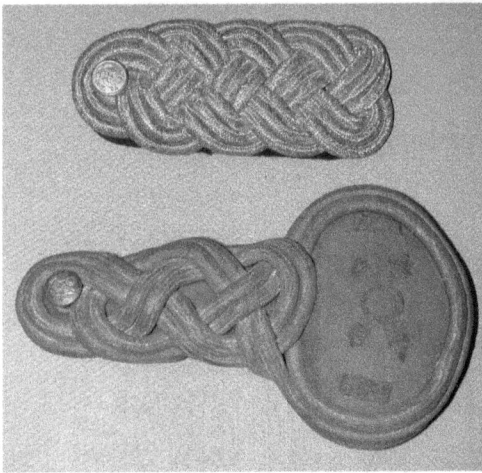

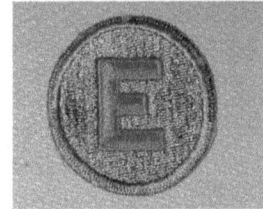

Left to right; officer's epaudetts (1900s), master gunner dress sleeve insignia, "E" sleeve patch for Excellence, sewn dress sleeve insignia

Enisted man 1930s-era type III collar insignia

Triangle insignia on the equipment of the 59th Coast Artillery Regiment

HERALDRY OF THE COAST ARTILLERY CORPS

The adoption of arms and badges for organizations of the Army was formally approved towards the end of 1919. Previously regiments were authorized and encouraged to obtain such insignia, but no official use was made of them, neither did the War Department exercise any control or supervision over the designs, and the result was a great variety, many defying the laws of heraldry, and a remarkable number containing historical inaccuracies.

In 1919 the War Department authorized the Supply Division of the Chief of Staff to use regimental arms on the colors in place of the arms of the United States, thus making the color truly regimental in character, instead of being a national emblem as it had previously been. The retention of the eagle showed the Federal nature of the organization, but the remainder of the design applied only to that particular unit which reflected the traditions, ideals, wars, battles, and other incidents connected to that unit's history. In addition a distinctive insignia was developed from an element of the arms for use as a marker and an emblam to be worn on the uniform.

A coat of arms, in the ordinary acceptation of the phrase, consists essentially of a *shield*, with the most important accessories being the *crest* and *motto*.

The *shield* consists of a base metal (gold or silver) and one or more solid colors on which are placed designs to illustrate the history of the unit.

The *crest* was formerly worn on the helmet and, whenever practicable, was so shown. Due to the manner in which the arms and crest were placed on the regimental color, the helmet was often omitted, but on drawings, stationery, etc., it was used to support the crest, thus avoiding the appearance of a crest suspended in midair. The heraldic wreath typifed the torse of cloth or silk formerly used to fasten the crest to the helmet, and was always shown. It was placed between the helmet and the crest, or as the support of the crest if the helmet was omitted. The mantling was an accessory of the helmet. It symbolized the mantle formerly worn over the knight's armor, and was always the principal color of the shield, lined with the principal metal; and the same rule holds true for the wreath.

The *motto* was placed on a scroll or ribbon, usually below the shield, but occasionally elsewhere, there being no fixed rule about its placement or color.

In 1919 the Coast Artillery Corps had few regiments, so coats of arms were designed for the various coast defense commands and a small number were authorized distinctive unit insignia.

In 1924, the regimental structure was returned to the entire Coast Artillery Corps and each new regiment was required to produce a coat of arms design for approval. Throughout the 1920s and 1930s, War Department policy dictated that the organizational colors would not be issued until a coat of arms was approved by the Chief of Staff, or after 1924, the Adjutant General.

The first distinctive insignia for uniforms was produced for the 51st Coast Artillery Regiment in 1924, followed by the 55th. Most existing active Regular Army, National Guard, and Organized Reserve regiments eventually had a coat of arms approved. This policy was continued into the early years of World War II, until it was more or less suspended in 1943. A color guide to the DUIs for the Coast Artillery Corps regiments is included at the end of this section.

Coast Artillery, Antiaircraft Artillery, Air Defense Artillery Distinctive Insignia Catalog, compiled by the American Society of Military Insignia Collectors (ASMIC), 526 Lafayette Avenue, Palmerton, PA 18071-1621.

Sawicki, James P., *Antiaircraft Battalions of the U.S. Army,* Volumes 1 & 2, Wyvern Publications, Dumphries, VA, 1991.

Stanton, Shelby L., *Order of Battle, US Army in World War II,* Presidio Press, Novato, CA, 1984.

Wyllie, Robert E. "Coats of Arms and Badges of The Coast Artillery Corps" *Coast Artillery Journal* Volume 59 (August 1923) pp. 123-142.

COAT OF ARMS OF THE THIRD ARTILLERY REGIMENT

Text adapted from James P. Sawicki "Antiaircraft Battalions of the U.S. Army," Wyvern Publications, Dumphries, VA, 1991;
for definitions of heraldic terms consult this reference.

Shield:	*Or,* on a chevron *gules* above an imperial Chinese dragon of the like armed *azure* three mullets argent, on a chief of the second two pallets of the fourth an arrow in fess counterchanged.
Crest:	Out of a mural crown *or* masoned *gules* a garland—the dexter branch cactus, the sinister palm—proper encircling a sun in spendor argent.
Motto:	*Non Cedo Ferio* (Yield Not, Strike).
Symbolism:	Scarlett is used for artillery. The two white stripes on the scarlett chief, the colors of the campaign streamers for the War of 1812, commemorating the participation of several companies of the regiment. The arrow alludes to the Indian wars. The chevron and stars indicate service in the Civil War. The stars also refer to the numerical designation of the regiment. The dragon represents service in China; the claws and teeth are blue to indicate that elements of the regiment served in the China Relief Expedition as infantry. The mural crown, cactus, and palm signify the regiment's participation in the Mexican War and elements of the regiment in the Philippine Insurrection. The sun in its glory commemorates the laurels earned by the regiment during its days of glory.
Distinctive Insignia:	An adaption of the crest and motto of the coat of arms (right).

FLAGS FOR THE COAST ARTILLERY

Bolling W. Smith

Exerpted from *CDSG Journal,* Vol. 10, Issue 2 (May 1996) pp. 85-91

In the U.S. Army, flag is a general term, while a color was originally the flag of an unmounted unit and a standard the flag of a mounted unit. Over time, this distinction inevitably became more complicated. Although the sizes of colors and standards changed, colors were always larger than standards. The staff for a color was termed a pike, while the staff for a standard was called a lance.

Organizational Colors and Standards

Units that were issued a silken national color or standard were also issued a regimental color or standard. They were the same size, of the same material, with the same fringe and carried on identical pikes or lances. The restrictions on the use of the silken national flags also applied to the organizational flags.

The 1923 regulations describe the regimental colors and standards, the fabric being red for the artillery. In place of the coat of arms of the United States, a regiment having an approved coat of arms displayed it on a shield on the breast of an eagle, with the regimental motto on a scroll in the eagle's beak. The regimental crest was above the eagle's head. A regiment having an approved badge, but no coat of arms, replaced the crest with the badge, while the eagle's breast would remain feathered. In either case, a scroll below the eagle carried the unit's designation. The cord was scarlet and gold.

In 1931, the regulations were expanded to cover regiments in the National Guard and Organized Reserves. Guard regiments with approved coats of arms would display them the same as active army regiments, while the crest would be that approved for regiments of their state. In the case of a regiment allocated to more than one state, each state crest would be displayed above the eagle's head. In the case of a regiment in the Organized Reserves with an approved coat of arms, the crest would be the image of the Lexington Minute Man.

(Left) the colors of the 248th Coast Artillery Regiment
(Right) Retiring the regimental colors of the 240th Coast Artillery Regiment, 1950 (H. Lawrence)

Guidons

Guidons are small, swallow-tailed flags, normally carried by company size units. The 1923 regulations give the first description found of coast artillery guidons. They were scarlet, of course, with a yellow coast artillery device. In yellow above the device was the regimental number, and below it was the battery designation. The regimental headquarters battery was indicated by "HQ," the service battery by "S," and a battalion headquarters battery and combat train by the number of the battalion. In the case of companies not assigned to a regiment, the device was raised on the guidon to maintain balance. The guidons were of wool bunting, swallow-tailed, 2 feet 3 inches on the pike or lance, 3 feet 5 inches on the fly, and forked 15 inches.

By 1944, reflecting the changing organization of the Coast Artillery Corps, the regulations provided for the number over the device to be either a regiment or separate battalion. For separate batteries the number was placed below the device, which was then raised enough to "preserve the symmetry of the guidon."

Certain coast artillery officers, by virtue of their rank or position, were entitled to personal flags. In the 1904 regulations, general officers were authorized boat flags of scarlet bunting, 4 feet 9 inches by 3 feet. Rank was indicated by white stars placed on the center line of the flag. The chief of artillery used the general officer's flag appropriate to his rank. By Change No. 66 to Army Regulations (C.A.R. No. 66), December 31, 1917, a similar flag, 26 inches by 18 inches, was issued to be flown from an automobile while engaged in official duties.

In 1908, the Quartermaster Department adopted specifications for a boat flag for artillery district commanders. It was to be of the best scarlet duck bunting, with the coast artillery device in yellow bunting sewn in the center, 18 by 24 inches for small boats and launches and 27 by 36 for larger boats. In the regulations dated April 15, 1917, this flag was specified for only coast defense commanders, while district commanders were authorized an automobile flag, 18 by 26 inches, of the same design. On the last day of 1917, however, C.A.R. 66 restricted the use of the boat flag to district commanders, while authorizing coast defense commanders to fly the same pennant as post commanders. District commanders who were general officers flew the general officer's flags instead of the district commander's.

In the 1923 regulations, no automobile flags were authorized for commanders who were not general officers, and district commanders lost any distinctive flag, while coast defense commanders continued to use the two sizes of boat flags. The smaller, or launch, flag was lengthened to 26 inches, the same size as the earlier automobile flag. The boat flag for general officers remained unchanged, while the automobile flag gained a fringe of knotted yellow silk 1 1/2 inches wide. The automobile flag for chief of coast artillery, however, now contained the corps device, surrounded by an appropriate number of white stars.

Coast Artillery Vessels

The coast artillery operated a substantial number of vessels, the great majority of which were involved in the planting of submarine mines. A flag for these mine boats is described in Quartermaster Department specifications approved in 1905. The flag was to be of white bunting, four by six feet, made of two, 24 inch wide strips joined with a horizontal seam. Dyed on the flag was the design, a blue mine superimposed on two white crossed cannon, with the words "SUBMARINE" above and "DEFENSE" below, in white, in the form of an arc. The background of the flag was dyed scarlet. Although this flag does not appear in the 1910 or 1917 regulations, it does appears in the October 1917 *National Geographic Magazine*. In the 1923, 1931 and 1944 regulations, the lettering was omitted, and it is called a mine planter flag.

COAST DEFENSE COATS OF ARMS

Exerpted from "Coats of Arms and Badges of The Coast Artillery Corps"
by Colonel Robert E. Wyllie, C. A. C. *Coast Artillery Journal* Volume 59 (August 1923) pp. 123-142.
and other notes in the *Coast Artillery Journal* 1928 & 1929.

Coast Defenses of Portland the shield is divided horizontally, the upper half red, the lower silver. On the red is a silver star of five Points, and on the lower half is a pine tree in natural colors. The star bas a double significance; it symbolizes the Pole Star, this being the most northerly defense on the Atlantic Coast, while its five points represent the five forts, Williams, McKinley, Levett, Preble and Baldwin, This star is placed on a background of artillery red. The pine tree is the old emblem of the State of Maine, and appears on the coat of arms of that State. The crest is a phoenix, and is taken from the arms of the city of Portland, whose history it well typifies. In this case the body and head are purple, the wings gold. The flames are always shown in natural colors. The motto for these defenses is *Terrae Portam Defendamus,* the translation being "We defend the land gate (or port)." Defendamus is the motto of the Coast Artillery School, and Terrae Portam is a latin rendering of Portland. (Shield adopted by the 8th Coast Artillery Regiment.)

Coast Defenses of Boston a shield of artillery red, on which is the Mayflower under full sail in natural colors. The crest consists of a wreath of the colors on which is a dexter arm, embowed, habited gray with white ruff grasping a staff with the flag of Bunker Hill attached, all proper. The motto is *Prima Libertatio Acie,* "In the first line of battle for liberty." (Shield adopted by the 9th Coast Artillery Regiment)

Coast Defenses of Narragansett Bay. The shield is red on the upper half for artillery, and blue below taken from the arms of Rhode Island, which has a gold anchor on a blue shield. The dividing line between these two colors is embattled to show fortifications. On the red is the gold sundial-compass of Roger Williams (Date 1638) the founder of the Colony of Rhode Island and Providence Plantations. In the lower half is a gold fleur-de-lis, to commemorate the services of the French during the Revolution. The crest is two gold crossed cannons supporting the anchor of Rhode Island in blue. The motto of the state, *Hope,* is also used by the Coast Defenses. The anchor, the device of Rhode Island, is the symbol of stability.

Coast Defenses of Long Island Sound the shield is gold, and bears a blue diagonal stripe, known as a bend, on which are three silver towers. The bend with its towers represents the line of three forts, Wright, Michie and Terry, placed diagonally across the entrance of the Sound. On each side of the bend is a narrow parallel stripe of black, symbolizing the iron defenses. The crest is the head of a fish hawk in natural colors, which bird abounds in that vicinity. The motto is a command to the enemy, *Stop.* (Crest adopted by the 11th Coast Artillery Regiment.)

Coast Defenses of Sandy Hook has a shield of artillery red on the upper half and gold below, the line between the two being embattled. On the red and rising out of the embattlements is the Statue of Liberty in gold, and in the lower half is the Sandy Hook lighthouse placed between two bursting shells. The lighthouse and shells are black, while the flames from the shells are in the natural color of fire. The crest is a gold panther, breathing fire, placed on the battlements of a red tower. The motto of these defenses is *Obscurata lucidior,* and refers to the incident when the darkening of the lighthouse furthered the light of liberty in the country. A supporter for these arms to be used in all cases except on the colors. When Hudson explored New York Bay and the river which bears his name in 1609, his ship, the "Half Moon," was anchored in the Horse-shoe near Sandy Hook, in commemoration of which the shield of these defenses is displayed in front of the "Half Moon." (Crest adopted by the 7th Coast Artillery Regiment.)

Coast Defenses of Chesapeake Bay has for its base the arms of Lord Delaware, the first Governor of the Colony of Virginia. His arms consisted of a silver shield bearing a jagged black stripe placed horizontally across the centre known to heralds as a fess dancetty. To this is added a red cross, symbolic of the landing of the first settlers at Cape Henry in 1607, their first act being to erect a cross and offer thanks for their safe arrival. The crest is a hand in a gauntlet of silver mail grasping a gold trident, which commemorates the battle between the Monitor and the Merrimac in Hampton Roads. The mailed hand grasping the trident of Neptune, the god of the seas, fittingly symbolizes that supremacy. The motto is *Portam Primam Defendo* "I defend the first gateway."

Coast Defenses of Pensacola is based entirely on the defense of Fort Pickens during the Civil War. That fortification was the only place within the territorial boundaries of the Confederacy over which the Stars and Stripes flew during the whole of the Civil War. This was a specially meritorious incident and the War Department has commemorated it by permitting the use the eagle in gold on a shield of artillery red. The crest is an arm clothed in Union blue, while a gold flaming torch of liberty is held erect in the hand. The motto is *Fides ultra finem* "Faithfulness beyond the end."

Coast Defenses of San Francisco has a shield of purple on the upper half, with gold below. On this is a charge known in heraldry as a pile, an inverted triangle having the base coincident with the top line of the shield and the apex very near the bottom. The upper half of the pile is gold, the lower blue. In the center of the pile is a red demi-sun. This combination represents the setting sun seen through the Golden Gate of San Francisco Harbor. The crest is a grizzly bear, the emblem of California, in black. (Crest adopted by the 6th Coast Artillery Regiment)

Coast Defenses of Puget Sound has a shield of artillery red and on it are five horizontal stripes of gold. At each side of the shield is a semi-circular piece (flaunch) of ermine. Ermine is represented in heraldry by black tails, very much conventionalized, on a white or silver background, and in this case it recalls the fur trade and the positions of the ermine on the shield indicate the straits across which are placed five fortifications which bar the way to an invader. The red is not only for artillery, but, in connection with the gold of the bars, commemorates the Spaniard who discovered the Straits in question. The crest is a full-faced sun, known as a "sun-in-splendor." This is always shown with rays issuing from the entire perimeter, alternately straight and wavy, the straight rays denoting the light received from the sun, while the wavy rays represent the heat. A human face is depicted on the sun itself. This was the crest of Lieutenant Peter Puget, Royal Navy, one of Captain Vancouver's officers, for whom the Sound was named.

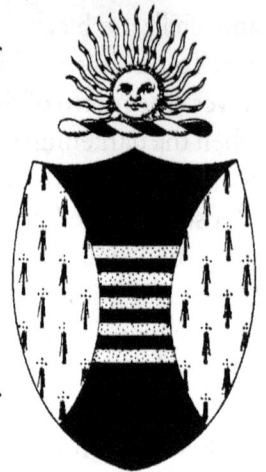

Coast Defenses of Cristobal has a shield of artillery red, and in the centre a medieval vessel known as a caraval in gold. In the upper corner is a silver portcullis, the barred gateway used in the middle ages at the entrance of castles. The red and gold together again make the Spanish colors to commemorate the discoverers of this part of the continent. It was in 1502 that Columbus skirted this coast and landed near the Chagres River, which is indicated by the caraval. The portcullis is symbolical of the canal, which when open forms a passage between the two oceans, but when closed by these defenses bars the way to the enemy.

The crest of this command is unique, the arm of a pirate, tattooed with skull and crossbones, having on the upper arm a sleeve of white and green with crimson cuff and gold buttons, the hand brandishing a pirate's cutlass in black.

The motto is *Nullius pavit occursum* "He fears no encounter" and can be considered as referring both to the old buccaneers and to the present defenders of the canal.

Coast Defenses of Balboa has a red shield. On it is a gold chevron sprinkled with red hearts. Above the chevron are two portcullises in gold and below is an old type gold cannon, placed vertically and on its summit a garland of Holy Ghost orchids. The gold and red of the chevron form the Spanish colors. The hearts are an allusion to Amador, the principal fort of these defenses, named after the first President of Panama. The portcullis has the same significance as in the arms of the Coast Defenses of Cristobal, two are used in this case to represent the two sets of locks at the Pacific end of the canal. The cannon is for artillery, and its garland is formed of orchids which are said to grow only on the Isthmus. Like the crest of its neighbor at the Atlantic end of the canal, that of these defenses is based on the old buccaneers; an arm in a blue rolled up sleeve, tattooed on the forearm with skull and crossbones, and holding a smoking pistol of 17th century type. The motto of the command is *Strength, Loyalty, Valor.*

Until March, 1921, there was but one coast defense command in Hawaii, known as the Coast Defenses of Oahu, but it was then split into the *Coast Defenses of Pearl Harbor* and *Coast Defenses of Honolulu.* The arms of these two are designed to show their common origin. In each case the shield is gold, surrounded by a border of eight horizontal stripes (or bars), silver, red, blue, silver, red, blue, silver, red, commencing at the top. The interior line of the border is embattled to show fortifications. The eight stripes are taken from the old Hawaiian flag and arms and signify the eight islands of the group.

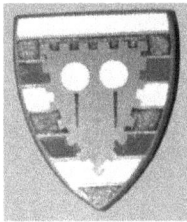

Coast Defenses of Pearl Harbor has two upright black sticks, each surmounted by a silver ball, placed on the gold. These are known as tabu sticks, and were formerly placed in front of the entrance to the king's palace, etc., every[thing] behind the sticks being "tabu" to the common man. The crest of the Coast Defenses of Pearl Harbor is the Helmet of King Kamehameha the Great in red and gold, placed on a garland of palm branches. The principal fort of these defenses is named after that monarch, while red and yellow were the royal colors. The palm branches symbolize victory. The motto is *"Defenders of Pacific Pearls."*

In place of these tabu sticks the *Coast Defenses of Honolulu* uses two ancient Hawaiian spears, crossed like the letter "X," red in color. Crossed spears were used in the old times immediately at the door of the King's tent. Diamond Head, known by reputation all over the world, is used as a crest in red to denote both artillery and the color of the soil at Fort Ruger, which is located at that famous extinct volcano. The motto is the well known Hawaiian word *Kapu,* which means "Keep out." (Shield adopted by the 16th Coast Artillery Regiment.)

Coast Defenses of Manila and Subic Bays is based on the arms of the Philippine Islands, although different meanings are attached to the devices used. The arms of the Philippines are red in the upper half, blue in the lower. On the red is the Spanish castle in gold, below is a silver seahorse grasping a sword. The whole is set upon the shield of the United States, so that the latter forms a border. These defenses omit the border and have reversed the shield, putting the blue on top, and separating the blue and red by a wavy line, the heraldic way of indicating water, which, in this case, consists of the two bays defended.

The seahorse in the upper half is denuded of his sword and represents the island of Caballo, on which Fort Hughes is located. The castle below represents Corregidor, or Fort Mills, the principal fortification, and it is placed between two croziers of gold, symbolizing the monk (El Fraile, Fort Drum) and the nun (La Monja) of the legend, familiar to all who have served in Manila. The crest of these defenses is a carabao's head, full face in the natural colors, and represents Fort Frank on Carabao Island. The motto is *Corregidor omnia vigilat* (Corregidor guards all).

Coat of Arms for the Harbor Defenses of Charleston

Coast Artillery Journal Vol. 69 (1928) p. 349.

Shield: *Gules* a palmetto tree proper.
Crest: On a wreath *or* and *gules,* a dexter arm, embowed, habited in the Continental artillery uniform (blue with red cuffs and yellow buttons) grasping the Fort Moultrie flag (blue with a white increscent in dexter chief and the word "LIBERTY" also in white along lower edge of proper).
Motto: Let's not fight without a flag.

These arms commemorate the repulse of the British fleet in 1776. The shield is red for artillery and on it the palmetto tree of South Carolina, adopted by that state as its emblem because Fort Moultrie in 1776 was constructed largely of palmetto logs. The crest is to recall Sergeant Jasper's exploit, when he replaced the flag which had been shot down off the parapet, and the motto is the remark attributed to him at the time.

Harbor Defenses of New Bedford

Coast Artillery Journal Vol. 68 (1928) p. 451.

The Coat of Arms for the Harbor Defenses of New Bedford bears on a *Shield: Gules,* an arm embowed brandishing a harpoon proper.

The City of New Bedford from its earliest days was known as the "Whaling City" which accounts for the arm and harpoon on the shield.

Coat of Arms for the Harbor Defenses of Eastern New York

Coast Artillery Journal Vol. 68 (1928) p. 347.

Shield: Ermine on a chevron *vert* a mine case between
two Engineer castles *agent.*
Crest: On a wreath of the colors a dexter arm in armor, embowed proper
charged with a mullet *gides* grasping in the naked hand a sword *argent*
hilted *or.*
Motto: *Sic Vis Pacem, Para Bellum.*

Fort Totten was originally the site of the Engineer School of Application, later the Coast Artillery School of Submarine Defense. The Chevron in green, the school color, with its charges, shows its history. It is now the seat of the Artillery District shown by the crest, the star indicating a general officer, the arm with sword the power of command.

Coat of Arms of the Harbor Defenses of Portsmouth

Coast Artillery Journal Vol. 68 (1928) p. 537.

Shield: Gyronny of eight *azure* and *gules,* a three-bastioned fort voided *argent.*
Crest: On a wreath of the colors a ship *gules* flagged proper in stocks *argent,*
from the seal of the State of New Hampshire.
Motto: We Are One.

The field is taken from one of the two earliest New Hampshire flags known to exist, that of the Second New Hampshire Regiment of the Continental Army in the Revolutionary War. (This flag bears in the upper comer next to the staff, eight triangles, alternately red and blue, so arranged as to form two crosses, one upright and the other diagonal.) The field commemorates the capture, on December 14-15, 1774, of Fort William and Mary (now Fort Constitution) by the Colonial Americans of New Hampshire, the first American victory of the Revolutionary War. The three bastions of the fort is used as a charge, represent the three forts of the harbor defenses, Fort Constitution New Hampshire, at chief, Fort Foster, Maine, dexter base, Fort Stark, New Hampshire, sinister base. The fact that the three forts are represented as bastions joined together by curtain walls so as to form a single fort signifies their union in the Harbor Defenses of Portsmouth and the close cooperation of the three in the common defense of Portsmouth.

The motto, "We are One," taken from the old flag mentioned in connection with the field, also alludes to this union and cooperation.

The ship on the stocks, used as a crest, is taken from the seal of the State of New Hampshire, of which seal it is the most prominent feature. Its significance lies in the fact that the Harbor Defenses of Portsmouth defend the only port in the state. Its tincture, red, is that of the Coast Artillery Corps, the combatant arm manning the defenses.

Coat of Arms of the Harbor Defenses of Southern New York

Coast Artillery Journal Vol. 69 (1928) p. 73.

Shield:	*Vair,* three bars *gules,* jessant from the middle one a demilion saliant, ragardant *or.*
Crest:	On a wreath of the colors *(argent* and *azure)* a beaver couchant proper.
Motto:	*Volens et Potens.*

The crest is the beaver of New York, the only charge on the original arms of New Netherlands adopted in 1623, and now on the seal of New York City. The shield symbolizes the battle of Long Island, August 27, 1776, which took place near the present Fort Hamilton. The color of the field is vair, a fur, which is said to come from an animal called Varus, the back of which is blue, the belly white. Tradition relates that a Hungarian general displayed his cloak made of varus fur as an ensign to rally his men and succeeded in turning defeat into victory. Similarly, Washington, after the battle of Long Island, by a masterly retreat across the East River, rendered the British victory fruitless. The three bars represent the three enemy forces under Grant and Cornwallis and the British fleet. The lion in a springing position issuing from the center bar symbolizes the piercing of Cornwallis' command by the American brigade under General Stirling.

Coat of Arms of the Harbor Defenses of the Delaware

Coast Artillery Journal Vol. 69 (1928) p. 161.

Shield:	Azure, three lions' heads earsed *or,* 2 and 1.
Crest:	On a wreath of the colors a griffin's head earsed *azure,* beaked and eared *or.*
Motto:	*Semper Paratus.*

The history of this region shows that it was colonized and occupied by the Swedish, Dutch, and English, who are shown on these arms by the three lionsheads, each of those countries having a gold lion on their coat of arms. The color blue is common to all three flags and also to the flag of the United States. The griffin's head is taken from the crest of Lord Delaware for whom the state, river, and defenses were named.

Coat of Arms of the Harbor Defenses of Galveston

Coast Artillery Journal Vol. 69 (1928) p. 495.

Shield:	*Gules,* a ship under sail, in chief a mullet, both *argent.*
Crest:	On a wreath of the colors a cotton boll proper.

The State of Texas is shown by the lone star, the shipping from the port of Galveston by the ship and by the crest, cotton being the principal product of the port.

Coat of Arms of the Harbor Defenses of the Potomac

Coast Artillery Journal Vol. 69 (1928) p. 264.

Shield: *Gules.* two bars, *argent.* in chief three mullets of the like.
Crest: On a wreath of the colors an eagles head earsed *sabled.* armed *or.*
Motto: *Exitus acta probat.*

The shield in this case is the same as in the coat of arms of the Washington family, with colors reversed. The crest is taken from the arms of the Digges family, the original owners of the land on which Fort Washington is now located.

Coat of Arms of the Harbor Defenses of Baltimore

The shield is the Coat of Arms of the Calvert family, to which Lord Baltimore, the founder of Maryland, belonged. This now forms the 1ˢᵗ and 4ᵗʰ quarters of thearms of the State of Maryland. The chief commemorates the writing of the Star-Spangled Banner by Francis Scott Key during the battle of Fort McHenry, September 13, 1814. The flag at that time had fifteen stars and fifteen strips. The embattled partition line is for the defense of the fortress.

The translation of the motto is "with song and deed."

The crest features a soldier in the uniform of 1812.

Courtesy of the U.S. Army Institute of Heraldry.

Coat of Arms of the Harbor Defenses of Key West

Coast Artillery Journal Vol. 69 (1928) p. 437.

Shield: *Gules* a tower *or* in chief a key fesswise of the like.
Motto: *Quod Habemus Defendemus.*

The shield is red for artillery, Key West being the strategic key to the Gulf of Mexico is represented by a key and tower. The motto means, "We defend what we have."

Coat of Arms for the Harbor Defenses of San Diego

Coast Artillery Journal Volume 70 (January 1929) page 72.

Shield: *Azure,* a pile raguly *or.*
Crest: On a wreath of the colors *or* and *azure* and anchor proper (grayish) behind an eight point mullet of rays *or.*
Motto: *Paratus* (Prepared).

The blue shield and the yellow pile are symbolic of the blue ocean and the yellow land of Point Loma. The place was first visited by the Spaniards, Cabrillo in 1542, and the edges of the pile are made raguly (ragged) as the Spanish flag at that time bore a cross.

The crest symbolizes the hardest fought battle of the Mexican War in California, near San Diego, at San Pasquale, December 6, 1846. General Stephen W. Kearny commanded the Americans, consisting of one company of the First Dragoons, a few sailors sent by Commodore Stockton from San Diego, and a volunteer company from San Diego. The anchor commemorates Stockton's sailors, and Kearny's Dragoons wore on their helmets the eightpointed gold star of rays.

Coat of Arms for the Harbor Defenses of Los Angeles

Coast Artillery Journal Volume 70 (Feb. 1929) pp. 151-152

Shield: Parti per fess wavy *gules* and *azure,* in chief two angels habited of the second and *argent* and winged *or* proper and in base two keys in saltire of the fourth and third.
Crest: On a wreath of the colors (*or* and *gules*) a crescent *gules.*
Motto: *Nosotros Los Defenderemos.*

The escutcheon combines San Pedro (Los Angeles Harbor) and the City of Los Angeles, both of which are defended by the guns of Fort MacArthur, and is an excellent example of "canting" heraldry. Los Angeles being represented by the two angels and of symbolic heraldry, San Pedro being represented by the keys of St. Peter. Los Angeles Harbor or San Pedro Harbor is in the lee of Point Fermin, which was a point of note with the early explorers. Cabrillo in 1542 named it "Bahia de los Humos," and it appears on the charts of Vizcanio, 1603, under the name of "Ensenada De San Andres." In 1734 the Spanish Admiral Gonzales gave it the name of San Pedro, which name continues in use today. It was a regular loading and unloading place for vessels from the date of the founding of the Pueblo of Los Angeles in 1781.

The motto refers to both the port and the city "We shall defend them." On account of the Spanish origin of the community in which the harbor Defenses are situated, the motto, "Nosotros los defenderemos," is in Spanish.

The crest pertains to the first organization to garrison these Harbor Defenses in 1917, the 38th Company, Coast Artillery Corps, known in 1812 and the years following as Capt. S.B. Archer's Company.

Other Coast Artillery Coats of Arms

Harbor Defenses of Savannah

Harbor Defenses of Cape Fear

The Coast Artillery School

Coast Artillery Regimental Insignia 1924-1942

prepared by Mark Berhow, Greg Hagge, and Bob Capistrano, 2006/2022
and the American Society of Military Insignia Collectors (www.asmic.org)

COAST ARTILLERY CORPS

COAST ARTILLERY SCHOOL

BARRAGE BALLOON SCHOOL

Entry key:
Insignia
Variant(s)
Regiment # Cadre Type Year Organized
Insignia authorized/approved (For more unit history see CAC Organization section)

Army Cadre Designation: RA = regular army	RAI = regular army, inactive	NG = national guard
OR = organized reserve	PS = Philippine Scouts	
Type of regiment: HD = harbor defense artillery	RY = railway artillery	AA = antiaircraft artillery
TD = tractor drawn artillery		

Only units with insignia that were officially approved (or worn) during the period 1920-1942 are shown.

1st RA HD 1924
1925/1926

2nd RA HD 1924
1924/1926

3rd RA HD 1924
1925/1926

4th RA HD 1924
1928/1929

5th RA HD 1924
1925/1925

6th RA HD 1924
1924/1924

7th RA HD 1924
1924/1924

8th RA HD 1924
1924/1924

9th RA HD 1924
1924/1924

10th RA HD 1924
1926/1926

11th RA HD 1924
1924/1925

12th RA HD 1924
1924/1925

13th RA HD 1924
1924/1924

14th RA HD 1924
1924/1925

15th RA HD 1924
1924/1925

16th RA HD 1924
1922/1923

18th RA HD 1940
1940/1940

19th RA HD 1940
(ex-625th CA)

20th RA HD 1940
1940/1940

21st RA HD 1940
1941/1941

22nd RA HD 1940
(Ex-614th CA)

23rd RA HD 1940
(Ex-616th CA)

41st RA RY 1921
1923/1923?

42nd RAI RY 1918
1937/1937

43rd RAI RY 1918
1937/1937

44th RAI RY 1918
1937/1937
(to 54th 1941)

51st RA TD 1918
1922/1923

52nd RA RY 1917
1923/1923?

53rd RAI RY 1917
1942?

54th RA TD 1941
(44th redesig. to 54th)

55th RA TD 1917
1922/1923

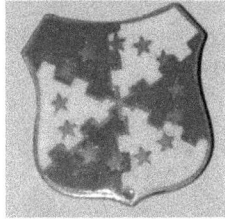

56th RA TD 1941
(Ex-506th)

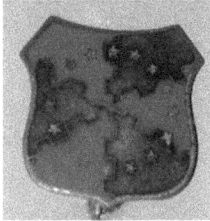

NIGHT HIDES NOT

57th RA TD 1918
1931/1931

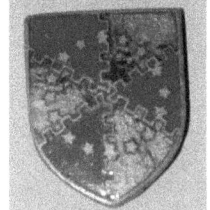

VETO

59th RA TD 1918
1923/1923?

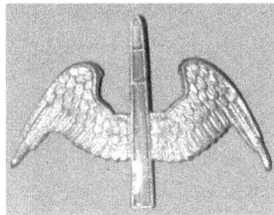

DEFENDIMUS

60th RA AA 1922
1924/1924

COELIS IMPERAMUS

61st RA AA 1921
1923/1923

62nd RA AA 1922
1923/1923

63rd RA AA 1921
1923/1923

AMOR PATRIAE

64th RA AA 1921
1922/1923

65th RA AA 1924
1925/1925

67th RAI AA 1926
1931/1936

68th RAI AA 1926
1941/1941

OLAM

69th RAI AA 1926
1930/1930

70th RA AA 1940
(Ex-917th)

TUEBOR

71st RA AA 1940
(Ex-504th)

UNDIQUE VENIMUS

72nd RA AA 1940
1941/1941

73rd RA AA 1940
1940/1941

74th RA AA 1940
(Ex-503rd)

FULMEN JOVIS JACIMUS

75th RA AA 1940
(Ex-509th)

76th RA AA 1940
(Ex-502nd)

WE RISE TO DEFEND

77th RA AA 1940
(Ex-505th)

78th RA AA 1940
(Ex-517th)

79th RA AA 1941
1941/1942?

83rd RA AA 1940

90th RA AA 1942

91st PS HD 1924
1928/1928

92nd PS HD 1924
1937/1937

94th RA AA 1941
1941/1942

99th RA AA 1941
1941/1942

100th RA AA 1941

197th NH NG AA 1922
1927/1927

198th DE NG AA 1924
1933/1934

200th NM NG AA 1925
1926/1941

202nd IL NG AA 1924
1925/1925

203rd MO NG AA 1924
1925/1926

204th LA NG AA 1940
1942/1943

205th WA NG AA 1940
1942/1943

206th AR NG AA 1923
1930/1930

207th NY NG AA 1925?
1928/?

209th NY NG AA 1940
1941/1941

211th MA NG AA 1924
1923/1923

212th NY NG AA 1921
1927/1927

213th PA NG AA 1922
1932/1932

214th GA NG AA 1933?
1940/1940 (Ex-264th)

215th MN NG AA 1940
(Conv from 205th Inf)

216th MN NG AA 1940
(Conv from 206th Inf)

217th MN NG AA 1940
1942/1942

240th ME NG HD 1923
1929/1929

241st MA NG HD 1923
1924/1928

242nd CT NG HD 1927
1928/1928

243rd RI NG HD 1924
1927/1927

244th NY NG HD 1924
1936/?

245th NY NG HD 1924
1925/1925

246th VA NG HD 1923
1932/1932

248th WA NG HD 1924
1934/1934

249th OR NG HD 1923
1928/1928

250th CA NG HD 1924
1925/1925

251st CA NG AA 1924
1928/1928

252nd NC NG TD 1924
1929/1929

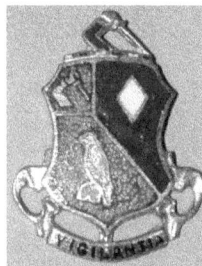

260th DC NG AA 1924
1928/1928

261st NJ NG HD 1940
1941/1942 also DENG

263rd SC NG HD 1925
1935/1935

264th GA NG HD 1925?
1931/1932

265th FL NG HD 1923
1928/1928

369th NY NG AA 1924
Ex-369th Inf

501st OR AA 1925
1925/1925

502rd OR AA 1925
1925/1925 (to 76th)

503rd OR AA 1925
1926/1926 (to 74th)

504th OR AA 1925
1926/1926 (to 71st)

505th OR AA 1925
1933/1933 (to 77th)

506th OR AA 1925
1929/1938 (to 56th)

506th RA AA 1942

507th OR AA 1925
1927/1929

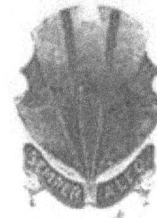

508th OR AA 1925
1932/1933

509th OR AA 1925
1929/1938 (to 75th)

510th OR AA 1925
1926/1927

511th OR AA 1925
1928/1928

513th OR AA 1925
1938/1939

514th OR AA 1925
1925/1925

516th OR AA 1925
1927/1927

517th OR AA 1925
1938/1938 - *to 78th*

519th OR AA 1925
1930/1931

521st OR AA 1925
1927/1927

529th OR AA 1925
1935/1935

535th OR AA 1933?
1936/1937

542nd OR AA 1930?
1931/1931

522nd OR AA 1925
1927/1927

530th OR AA 1925
1928/1928

536th OR AA 1925
1928/1928

543rd OR AA 1925
1930/1930

523rd OR AA 1925
1925/1927

531st OR AA 1925
1925/1925

537th OR AA 1925
1938/1939

544th OR AA 1925
1925/1925

524th OR AA 1925
1929/1930

532nd OR AA 1925
1928/1928

538th OR AA 1933?
1938/1939

545th OR AA 1925
1925/1925

526th OR AA 1925
1929/1929

533rd OR AA 1925
1925/1926

539th OR AA 1925
1925/1926

601st OR RY 1925
1925/1925

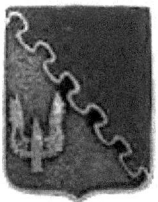

527th OR AA 1925
1934/1935

534th OR AA 1925
1926/1940

540th OR AA 1925
1930/1930

602nd OR RY 1925
1929/1929

603rd OR RY 1925
1926/1926

604th OR RY 1925
1926?

FACEMUS

604th RA AA 1942

FACEMUS

605th OR RY 1925
1931/1932

606th OR TD 1925
1931/1937

607th OR TD 1925
1924/1924

608th OR TD 1925
1932/1932

614th OR HD 1925
1931/1931 (to 20th)

615th OR HD 1925
1926/1927

616th OR HD 1925
1934/1934 (to 23rd)

619th OR HD 1925
1925/1925

620th OR HD 1925
1933/1934

PERCUTIMUS ATQUE DELEMUS

621st OR HD 1925
1925/1925

622nd OR HD 1926?
1928/1928

625th OR HD 1925
1939/1939 (to 19th)

SIEMPRE ADELANTE

626th OR HD 1925
1928/1928

627th OR HD 1925
1927/1929

628th OR HD 1925
1936-1939

630th OR HD 1925
1933/1934

903rd OR AA 1930
1934/1935

906th RAI AA 1930
1931/1931

SURSUM IN ADVERSUM

908th RAI AA 1930
1930/1931

ALWAYS ALERT

909th RAI AA 1930
1930/1930

910th RAI AA 1930
1933/1939

913th OR AA 1930
1931/1931

916th OR AA 1930
1932/1939

917th RAI AA 1930
1935/1935 (to 70th)

932nd OR AA 1930
1932/1932

933rd RAI AA 1930
1931/1932

938th RAI AA 1930
1931/1931

945th OR AA 1930
1938/1939

950th OR AA 1930
1938/1939

955th OR AA 1930
1932/1932

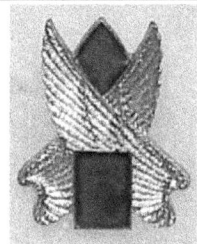

960th RAI AA 1930
1936/1937

969th OR AA 1930
1933/1935

970th OR AA 1930
1933/1933

972nd RAI AA 1930
1935/1935

974th RAI AA 1930
1936/1938

975th OR AA 1930
1930/1939

976th OR AA 1930
1935/1936

977th OR AA 1930
1930/1930

Compiled from the following sources:

American Society of Military Insignia Collectors (ASMIC), comp., *Coast Artillery, Antiaircraft Artillery, Air Defense Artillery Distinctive Insignia Catalog,* Palmerton, PA, 1999.

Sawicki, James P., *Antiaircraft Artillery Battalions of the U.S. Army,* 2 Vols., Wyvern Publications, Dumphries, VA, 1991.

Stanton, Shelby L., *World War II Order of Battle: An Encyclopedic Reference to U.S. Army Ground Forces from Battalion through Division, 1939-1946* (Revised Edition), Stackpole Books, Mechanicsburg, PA, 2006.

UNIFORMS OF THE COAST ARTILLERY, 1895-1945

Greg Hagge and Mark Berhow

Private Glenn F. Knight, Battery E, 213th Coast Artillery Regiment at Camp Pendleton, VA, in 1940
This was a typical uniform of the coast artillerymen in the 1930s (Glen B. Knight collection)

In 1895 the uniform regulations of 1881 were in effect with various amendments. The artillery uniforms were the standard pattern of the 1880-90s. The dress uniform was the thigh length frock coat in dark blue with sky blue trousers and the spiked helmet. The daily duty wear was the 1885 pattern, five-button sack coat with the same trousers and the new 1895 pattern service cap. The spiked helmet of foot troops (field artillery was mounted and used a plumed helmet with red cords) and the black leather belt were the same as used by the infantry. The only difference between infantry uniforms and artillery was the red trim and cannons on the eagle helmet plate. In addition to the frock coat, officers had a dress coat similar to the navy style. It was a waist length form fitting jacket with mohair trim and a standing collar. Although very plain, it was sharp looking, with artillery insignia on the collar and shoulder straps to indicate rank.

When troops started to arrive at the newly opened coast artillery posts in the mid to late 1890s, this was the uniform of the day. When artillery was reorganized into independent companies, the new company number was applied to the shield of the helmet plate.

In 1902 a new uniform regulation was adopted. This is the uniform regulation most commonly associated with the Coast Artillery Corps. The new enlisted men's dress jacket was still dark blue with red trim, but the style was updated to the cut of the current men's clothing fashion. The jacket had shoulder straps and a standing collar. The short skirt was vented at the sides (later dropped) and sported new pattern buttons (smooth faced with the new national eagle motif) and all new gilded insignia. The trousers were still sky blue and retained the late 1890s flared cuff. Black high top shoes completed the outfit, and brown leather belt and accessories were worn for dress occasions. The full dress ensemble was also adorned with a snazzy red breast cord with two big "waffles." From head to toe, the new uniforms were very colorful and impressive to look at. The Coast Artillery Corps was supposed to use up the remaining stocks of the old pattern five-button sack coat. This may have made some economic sense, but the troops had other ideas. About 1905 a new version of the sack coat began to appear in mass around the army. This up-dated version of the sack coat had six buttons, but was otherwise about the same as the old pattern. Interestingly this sack coat seems to have been a privately-purchased item, not a regulation issue coat.

The 1902 officer's uniforms were also updated in detail but largely remained similar to the 1881 style. The big change in the new regulations was in the field uniforms for all ranks. Some experimenting had been done with khaki uniforms in the late 1890s. In 1902 both the officers and enlisted men had new pattern four pocket jackets with a narrow stand and fall collar. Both coat and trousers were olive drab wool with brown leather accessories. Field equipment of brown leather and tan colored woven web began to be adopted. Eventually web material replaced much of the leather equipment. The coast artilleryman was equipped as infantry for field service. The 1889 pattern gray field or garrison hat remained in use with the olive drab bell crown cap for the service uniform. This bell crown cap became a symbol of the modern Army.

In 1906 the service coat was modified by changing to gusseted balloon pockets, and in 1910 the coat was slightly redesigned returning to patch pockets and adding a standing collar. With small modifications this coat was used through 1927.

In 1912 the uniform regulations were revised. Among other details the buttons and headgear were redesigned.

Soldier in the 1902/05 dress uniform with red breast cord.

Men of the 63rd Co at Fort Worden, 1910s. Most are wearing the service uniform, c. 1913. Note the old 1902 crush cap and the new 1912 version, left and center. Note that some of the coats have the 1906 pattern gusseted pockets, as well as patch pockets. Two variations on the denim pullover top and trouser are shown..

Two men of the 63rd Company, Fort Worden, on guard duty in their service uniforms with accoutrements, c. 1914. It was considered stylish to turn up the shirt collar under the coat. The M1905 bayonet and scabbard show well here.

Two men of the 63rd Company, c. 1914. This is the dress blue uniform worn daily. The breast cords and belt would convert this to full dress. The man on the left has the M1902 cap, on the right is the M1912 cap.

The bell crown cap gave way to a round flat topped style. The stiffening was removed except in the front. This gave the cap a distinct slant to the rear. This form of the service cap would remain in use for over 80 years. The field or garrison hat changed style and color to what we know as the "Smoky the Bear" hat in olive drab. This hat also has remained in use to the present era for drill instructors. General use of the garrison hat ceased in 1942. The uniform buttons changed to a rimed type with a ruled background, retaining the national eagle motif. Field equipment continued to evolve, changing from khaki to an unusual olive green hue.

The year 1917 had a significant change in Army uniforms—the old "blue" uniformed army passed into history. The blue dress uniform was dropped that year, never to return as a standard "required uniform" for enlisted men. Officers were also no longer required to have the dress blues or official formal occasion uniforms. The dress blue uniform remained as a privately purchased item.

(left) A corporal of the 63rd Company c. 1915. Note the "Type I" discs on his collar and the "first class gunner" insignia on the left sleeve.

If the commanding officer asked his men to attend a party in dress blues, they had better be wearing them! Enlisted men on special duty at embassies or the War Department staff also were required to acquire dress blues.

Enlisted men portraits circa 1910-1917 (Bolling Smith Collection)

Officers of the 49th CAC in France, January 1919. With a change of headgear to the full brimmed campaign hat, or the billed service "saucer" cap, this would be the look of the 1920s. Note the officer at center is wearing the "Marksman-A" shooting badge.

Col. Edward J. Cullen, CAC, Lt. Col. J.H. Pirie, CAC, and Maj. O. Krupp, CAC, of the Development Division, CATC, Ft. Monroe, VA, during the firings at Battery Kingman in June and July 1919. NARA Still Pictures, 111-SC-WWI, SC Photo 60773.

Soldiers at Fort Columbia, WA in the mid-1910s. The men are in service uniforms. (WSPR)

Men of the 63rd Co, Btry Kinzie, Fort Worden, c. 1910. All are wearing the denim coveralls.

Drill and Work Detail Clothing

The occupational work of the Coast Artillery was a messy business. The nickname of "cosmoline soldiers" was well founded. For the enlisted men manning and maintaining the big guns it was a constant routine of drill and cleaning the equipment. The uniform of the day for drill was the field service olive drab wool uniform as described above. To preserve this clothing—and protect it from the worst of the oil and grime—denim work coveralls were provided. These would be worn over the service uniform while performing duties at the emplacements. These denim work clothes are what is seen in use in nearly every photograph of troops serving the guns. The first variety was rust brown denim bib overalls with a pull over top and a round toped full brim "Daisy May" hat. The pullover top, or jacket, could have any number of pockets and a wide variety of placket configurations including a full button front. Later patterns were made in blue denim, which seems to have completely replaced the brown by about 1918. The blue work clothing remained in use for "fatigue" through about 1943 for all branches of the army. After WW I, there seems to have been a Quartermaster (QM) specification for the purchase of the denim work clothing. Before c. 1917 the brown and blue work clothing seems to have been an "off the shelf" purchase, not having a QM tag. Known examples and the wide variety in details tend to support this observation.

Ft Worden men manning the 6-inch guns of Battery Stoddard. The bib overalls show well here.

Enlisted men in work detail coveralls at Fort Worden, WA, photo from the mid-1910s, Note rank chevrons on sleaves of several of the men. (Puget Sound Coast Artillery Museum)

Officer and enlisted men in work detail coveralls at Fort Worden, WA, photo dated 1924/25. Lt. Oliver Hazen is standing in the bottom row, third from the left, wearing a campaign hat and the Sam Brown belt with the M1910 jacket. (Puget Sound Coast Artillery Museum).

Officers of the 59th Coast Artillery Regiment, C.A.C., Fort Mills, the Philippines, 1937, Col George Ruhlen, commanding (seated fith from left)

In 1922 the officer's uniform acquired a much-loved accessory, the "Sam Brown" belt with shoulder strap (which remained in use through 1942 when some frills of the uniforms were dropped in favor of more practical fashions).

Only minor alterations in uniforms occurred up until 1924. At this time the brass insignia was changed from gold for dress uniforms and black for field service to gold for all uniforms. This dressed up the olive drab coat.

The next major change was in 1927. The standing collar gave way to an open necked design with flat lapels. Some minor adjustment were made over the next ten years for trends in style, but this basic pattern would remain in use into modern times, the major changes being in material and color. Except in detail and color the officers and enlisted men's uniforms were fairly similar in style. Up to 1927 the officers coat and trousers were the same color. After this time the trousers would be changed to a shade of gray and the coat to a darker shade of olive green/brown, introducing what became known as "pinks and greens". The enlisted uniform colors remained about the same.

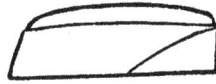
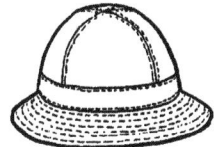

Garrison Cap

Field Hat

Service Cap

Service Hat

M1927 Officers' Service Coat, Wool

1920s-1930s uniforms
(left) Soldier in work dungarees over service uniform,
(middle) sergent in service uniform, men in brown & blue work deniums,
(right) officer in service uniform
(Greg Hagge photo collection)

In 1943 another detail was changed—the black silk/wool necktie was replaced with a tan necktie. This basic uniform (1927 pattern) remained in use until 1957. The current Class-A dress coat is basically the same cut but in a different color. The Second World War did not bring many changes in uniforms for the Coast Artillery. The officers and men tended to use the regulation service uniform for duty/off duty and the various field uniforms and accessories for daily wear. The M1941 short waist jacket was very popular. Being "stateside" and not in combat, the CAC attire did not change much from the pre-war times. The M1917A1 "dish pan" style helmet eventually was replaced with the new M1 "steel pot." The field gear was replaced with the new M1944 pattern in dark olive green. Generally the supply system cycled out the old material, as new material became available. The next major change of style for the army was the general introduction of the M1944 field uniform, the "Ike" jacket. This was intended to be a combat dress, styled on the British field uniform. State side troops received this outfit along with the rest of the army.

For more information on U.S. Army uniforms consult William K. Emerson, *Encyclopedia of United States Army Insignia and Uniforms* (University of Oklahoma Press, Norman, OK, 1996).

Officer at Fort Worden, WA, in the 1930s, wearing a M1927-style jacket and a Sam Brown belt (Puget Sound Coast Artillery Museum)

Coast Artillery soldiers in WW II-era service uniforms and coats with a M1910 azimuth scope at Fort Story, VA in the 1940s. (Shawn Welch collection)

Capt. Harry J. Harrison in a service uniform at Fort MacArthur in 1941 (Fort MacArthur Museum collection)

1sr, 2nd, 3rd, 4th, and 9th Coast Artillery District shoulder patches (1940s) (Greg Hagge Collection)

Colonel Fulton Q.C. Gardner, CAC, at the Coast Artillery School, Fort Monroe, VA, c. 1935. The officer-style coast artillery insignia are on his lapels and the coast artillery school distictive unit insignias (DUIs) are on his shoulder straps above his rank insignia. (courtesy the Gardner family)

Private Norman E. Hope, Harbor Defenses of Portsmouth, NH, 1942. Typical enlisted man dress uniform 1930-1945. Note the circular coast artillery collar insignia on the upper jacket lapel and the distictive insignia for the 22nd Coast Artillery Regiment on the lower lapel. He is wearing the Eastern Defense Command patch on his sholder, which were worn stateside after 1941. (courtesy Norman E. Hope)

European Theater of Operations, circa 1945.
Among the mix of styles here, most men are wearing the short-waisted M1944 "Ike" jackets. The fourth man from the left is dressed in the pre-war style "pinks and greens." The second man from the right is dressed in a service uniform that was similar to those worn at the stateside coast artillery posts.

LEGEND & LORE OF THE COAST ARTILLERY CORPS

In those years before the beginning of World War I, the Army was relatively small and the officers who served in the various arms—the infantry, the cavalry, the artillery and the other specialized branches— saw and/or corresponded with each other frequently. This enabled an efficient transfer of stories and other gossip. The men who served in the Coast Artillery Corps quickly began to develop an "esprit de corps" and a distinctive identity for their Corps. They scrambled to develop an "esprit" that was distinctive from that of the older branches, especially the rival Field Artillery Corps. They wrote coast artillery songs, they developed coast artillery punch, told coast artillery stories, and developed a coast artillery military etiquette. Here are a few examples to enjoy!

COAST ARTILLERY PUNCH
A recipe found by Elliot Deutsch.

Ingredients:
1 quart Rum
1 quart Sherry or Gin
1 pint Brandy
1-2 quarts Tea
1/4 to 1/2 pounds Sugar
2 Oranges
2 Lemons
1 quart Champagne

Instructions:
Mix rum, sherry, brandy, tea, and sugar.
Juice the oranges and lemons, grate lemon peels and add.
Chill with ice.
Just before serving, add champagne.
Enjoy immediately.

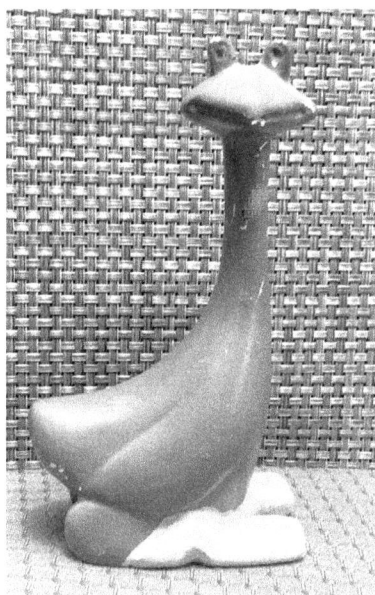

THE OOZLEFINCH
MYTHICAL MASCOT OF THE COAST ARTILLERY CORPS
a compendium prepared by Mark A. Berhow

By end of World War I the Oozlefinch was firmly ensconced as the mascot of the Coast Artillery Corps. The Oozlefinch was a fictitious bird born in stories told at the Fort Monroe Officer's Club during the early 1900s, and of which there is only one "official" image, a plaster statue that sat in a glass case for many years in the "Gridiron Room" of the officer's club. This happened during the early years of the Coast Artillery Corps, following the merging of all the Army's artillery regiments and batteries into a "Corps of Artillery" in 1901 and the separation of the units that manned the heavy fixed seacoast artillery weapons and units that manned the light mobile artillery units into the Coast Artillery Corps and Field Artillery Corps, respectively in 1907.

The Oozlefinch was a part what made up the distinctive "esprit" of the Coast Artillery Corps. Its story was spread by the close-knit group of officers who served in the Corps though conversation, letters, and most importantly for us latter day historians, military newspapers and professional journals. The Oozlefinch spent much of his time in silent contemplation (after all, he was a statue). Occasionally during the 1930s and early 1940s he would stir and speak out on the issues of the day, which were then recorded in the Fort Monroe newspapers and in the *Coast Artillery Journal.* For example: "What I Mean, We gotta Keep Advancin'," by Oozlefinch, *Coast Artillery Journal,* Volume 72, No. 1, Jan. 1930, pp. 49-52; and "This Man's Army" by Private Oozlefinch, *Coast Artillery Journal,* Volume 72, No. 4, Apr. 1930, pp. 366-368.

His image was used for informal insignia, signs, lapel pins, paperweights, and earrings. Apparently, some of the Coast Artillery antiaircraft units used his image as an insignia during their service overseas in World War II. The Oozlefinch went on to become the patron mascot of the army missilemen of the 1950s.

The story of the Oozlefinch is best told by those who were there in their own words. One of the members of the group of officers at Fort Monroe who developed the Oozlefinch story was E. R. Tilton, who was then assigned to the Quartermaster Corps. His story of the origins of the Oozlefinch and its association with the infamous "Gridiron Club" follows, exerpted from "History of the 'Oozlefinch'," By Colonel E. R. Tilton, originally printed in the *Liaison* newsletter on June 21, 1919, and reprinted in *Coast Artillery Journal* Volume 69, No. 1, July 1928, pp. 60-63.

The Origin of the Oozlefinch

A number of years ago, I think it was about 1905, a certain officer of the Artillery Corps who was more or less famed for his sayings (then Captain H. M. Merriam), spoke often about the existence of the "Oozlefinch." When questioned about this bird, he was rather close about describing either its appearance or its habits, or where it could be found. All that he ever disclosed was that "the Oozlefinch was a bird which flew tail foremost to keep the dust out of its eyes."

Any naturalist, even a nature faker, having this much of a description to work on, would probably assume that the eyes of the bird were of such prominence that it had to fly in the manner described to protect them. Hence the eyes must be important, probably large and prominent and not otherwise protected, an assumption which proved to be correct. A little while before Christmas in the year mentioned above, Mrs. Tilton, while shopping in Hampton, came across the present "Oozlefinch" in a small shop, and being struck with the prominent eyes of the animal, bought him. I then took the bird over to the Fort Monroe Club and let him perch behind the bar. He, under the loving care of Keeney Chapman, retained his place behind the bar for many years.

The bird was almost lost several times, but when a shavetail lieutenant in the Coast Artillery School tried to steal him away, he was enclosed in a glass cage for safe-keeping.

Early in 1908, the construction of the present Coast Artillery School was begun. The Torpedo School at Fort Totten, New York, was moved to Fort Monroe and consolidated with the Artillery School. When the consolidation took place, General (then Major) R. P. Davis came to Monroe as director of the combined schools and as President of the Artillery Board. The Board then consisted of Major Davis, Captain F. W. Coe (now *(June 1919)* Chief of Coast Artillery), Captain H. J. Hatch, and Lieutenant Halsey Dunwoody, Secretary. I was then Constructing Quartermaster and Captain Curtis G. Rorebeck was Post Quartermaster. During the building of the School the offices of the Artillery Board were in the second front of casemates. The office of the Constructing Quartermaster adjoined those of the Artillery Board in the same front. It was then the custom in those days to adjourn to the Club, not far away, after the labors of the day.

The "Oozlefinch" awoke from his sleep of several years, being aroused by the noise of the constant shaking of the dice box by members of the Artillery Board and the two Quartermasters. He insisted on joining the festivities and the location of his glass cage was changed from the bar to the mantel shelf of the second room from the bar. (In those days the bar was in the west end of the second front in the bastion under the flag staff.) This room became famous, not only from the fact that the "Oozlefinch" lived there, but because the sessions of the Artillery Board were held there every afternoon until long after retreat, winter and summer. The "Oozlefinch" with his all seeing eyes, took in all the work of the Board, and was so deeply interested in its proceedings that it practically became a member, and he never missed a meeting.

This room became known, eventually, as the "Gridiron Room" and the members who gathered there formed what afterwards became known throughout Fort Monroe as "the Club." The membership was limited, and woe betide the unfortunate who passed through the room to get a quiet drink, by himself, at the bar beyond. He generally had to pay toll, and was then allowed to proceed on his way.

It is a fact that the proceedings of the Board on Artillery matters of import in those days were discussed in that room. The present "Drill Regulations for Coast Artillery" saw the first light of day therein. It was natural for the "Oozlefinch" to absorb all the knowledge which was there, and he became the emblem of the "Gridiron Club" as well as a full-fledged member. (He has not a single feather on him.)

The Coat-of-Arms of the "Gridiron Club" came to life about this time, and after a course in Heraldry I designed the Coat-of-Arms and it was adopted. The Coat-of-Arms created quite a sensation amongst the non-initiated, and the secrets of its composition were never divulged to outsiders. There is no reason, now, why the heraldic story of the Coat-of-Arms should not be given to the Coast Artillery World.

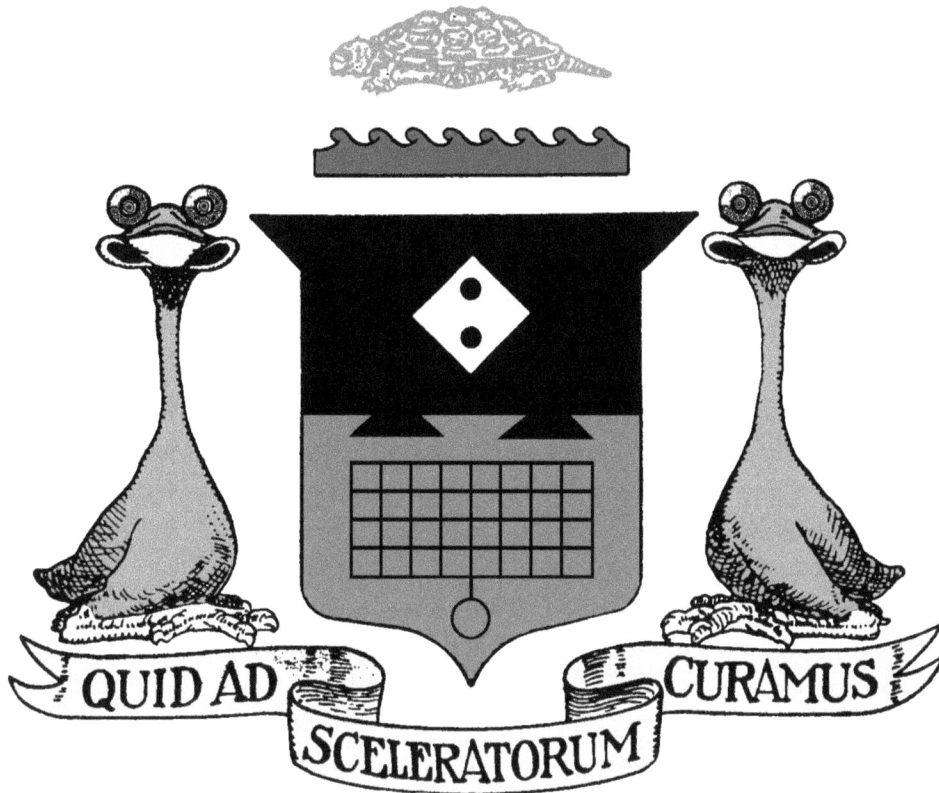

QUID AD SCELERATORUM CURAMUS

The body of the shield *"parti per fess, dovetailed"* indicated the general woodenness, not of the Artillery Board and the other members of the "Gridiron Club" but of the passing throng who paid not their toll cheerfully in passing through the Sanctum to the bar. *"Gules and Sable":* The color of the shield, red and black—red for the Artillery and black, in mourning for those who lost at dice by throwing the lowest spots. *"In honor, a deuce spot of dice, lozenged, proper":* The honor point of the shield was given to the lowest marked dice, as it was the one which most frequently appeared to some members, the law of probability not withstanding. *"in nombril a gridiron sable":* The lower half of the shield is given over to the memory of those who did not belong to the "Gridiron Club," who were constantly roasted by it, both when present and, I regret to say also, when absent.

The motto: *"Quid ad sceleratorum curamus,"* was the result of much thought and research. A visit to the Artillery School Library and a consultation with the then librarian disclosed the fact that there was no English-Latin dictionary in the Library, but they had a Lexicon which might serve that purpose. After an exhaustive examination of this lexicon for the Latin equivalent of the good old English word "Hell," the word "sceleratorum" was found. This word means the "place of the damned," which is as near as the ancient Romans came to the word desired. So the motto literally translated means: "What in Hell do we care!"

The supporters, *"two Oozlefinches, regardant, proper,"* were a natural selection, "regardant" meaning looking, or better, all seeing, with the great eyes that this bird has to protect while in flight in the manner described.

The crest, *"a terrapin, passant dexter proper,"* was selected owing to the great number of these animals, cooked to perfection by Keeney and served with great pomp to the members of the Artillery Board on occasions of state. This was always accompanied by libations of *"red top,"* red top being a now obsolete drink made in Champagne in France and once transported into the United States, in times gone by, that now seem almost prehistoric.

The wavy bar over which the terrapin is passing, represents the adjacent waters of the Chesapeake, the natural habitat of this animal.

This Coat-of-Arms that appeared so mystifying to the outside, was once stolen by a civilian gentleman from the Klondike who visited the Fort Monroe Club as a guest. He having been roasted by the "Gridironers" for his tales of great wealth, took this petty means of revenge. The Coat-of-Arms was recovered from his room in the Chamberlin Hotel, by strategy.

I think that on the back of the Coat-of-Arms appear the names of the original members of the "Gridiron Club," but of this I am not, now, sure. In any event, it is a fitting memorial to those days in Fort Monroe when work was intermingled with joys and pleasures, and which produced as great results as can be expected even today, when the mind of the Artillerist is given great activity by absorption of cold tea, coca cola, and such other concoctions.

To say that the Oozlefinch's origin was concocted in some sort of alcoholic haze may be a real understatement, as noted in this description of the habits of famous bird. (From: "Meet Mr. Oozlefinch," *The Recoil*, the Fort Monroe R.O.T.C. Camp newspaper, June 29. 1923, reprinted in the *Coast Artillery Journal*, Volume 59, No. 5, October 1923, p. 368.)

During his captivity he has responded in diverse ways to his environment. At the time of capture his mouth was able to imbibe a quart per minute, and his protruding eyes were capable of discerning an approaching glass or bottle at an incredible distance in any direction whatever. During the three years that he was perched behind the club's bar he never refused a drink and in that period his mouth trebled in capacity and his eyes became greatly enlarged and increased in penetrating power. The neck, already of great length, elongated at the rate of one-eighth inch per minute, increasing the time of transit of the "likker" and prolonging the delightful taste.

These changes in the features of the mascot caused great concern among the club members. Although he always paraded his only feather with great gusto and abandon, his constant inebriated condition was a reflection upon the club, and so somebody devised the plan of turning his mind from drink to gambling.

In 1918 he was moved from the bar to his present position on the mantel, and was placed in a glass case to prevent his likkeriferious instincts leading him from the straight and narrow path, and also to secure him against the attentions of visitors who sometimes attempted to lure him away.

The Artillery Board met in the same room with Oozlefinch. He witnessed the compilation of the present Coast Artillery Drill Regulations. Once imbued with the spirit of the Artillery he forsook dice and "likker" forever. His service to the Corps as tactical adviser and builder of morale has been inestimable.

This description of the Oozlefinch comes from a 1930 article. (From: "What I Mean, We gotta Keep Advancin'," by Oozlefinch, *Coast Artillery Journal*, Volume 72, No. 1, Jan. 1930, pp. 49-52.)

The Oozlefinch is that strange bird known to the Coast Artillery for his peculiar appearance and habits. One of his most peculiar habits is flying backwards to keep the dust out of his eyes. We don't know whether this is due to the prominence of his eyes or the absence of feathers. While he flies faced to the rear, the direction of his movement is always to the front. Where he picked up his rough language we can't imagine. It probably came from association with some rough Coast Artillerymen who came back from service overseas after the war and gathered at the Club hoping for something that was not there. These days, the Oozlefinch maintains a morose silence. No longer are the affairs of the Coast Artillery discussed in his presence. The tinkle of feminine laughter has replaced other kinds of tinkle. The rattle of the dice box is gone. The Oozlefinch can't talk in this atmosphere.

The Transplanted Oozlefinch

During World War I there was a big controversy as another Oozlefinch was spotted among the rolling stock of the Railway Artillery Reserve of the A.E.F. and adopted as its mascot. This lead to some interesting correspondence, some fine (?) poetry, and a further development of the Oozlefinch lore including his distinctive cry "gazook-gazoo." (From: "The Railway Artillery Reserve, American E. F." by Col. H. C. Barnes, C.A.C., Coast Artillery Journal, Volume 71, No. 1, July 1929, pp. 1-33.)

HEADQUARTERS RAILWAY ARTILLERY RESERVE

From: The Commanding General   France, 24 October, 1918
To: The Chief of Coast Artillery, Washington, D.C.
Subject: Distinctive Insignia for Railway Artillery Reserve, American E. F.

General:

1. I have the honor to send you herewith, a design of the "Pochoir," (stencil) which you will readily understand is the distinctive mark painted upon all transportation belonging to the Railway Artillery Reserve, American Expeditionary Force.

2. This design, which is the combined effort of all the genius contained at the present time in the Railway Artillery Reserve, is intended to represent as well as can be recalled by memory, the Oozlefinch, a rare bird which you will recall was incarcerated in a cage in one of the card rooms at the Fort Monroe Club. You will recall that this bird is a *sui generis,* and believed to be the only one in captivity, hence, after much reflection, I have concluded to adopt it as the emblem of the Railway Artillery Reserve, American Expeditionary Force, being symbolic of the only Railway Artillery Reserve known to exist in our service. You will note from examination of the accompanying drawing, that the Oozlefinch is very proud of himself. He wears a trench helmet, perhaps uselessly, but with effect. He has not many feathers, but in order to give a coquettish appearance, he has his left foot cocked up in the air. On his foot you will notice a wrist watch, which indicates 7:30. This is the hour for all hands in the Railway Artillery Reserve to begin work. On his right leg, he wears a *plaque d'identité,* which all chic soldiers are now supposed to wear in France. You will further notice that he is perched upon a section of rail, symbolic of the Railway Artillery Reserve, being surrounded by epis, which permits him to fire in any direction. The design is placed upon a white polygon, surrounded by red, suggestive of the Coast Artillery Corps, and having many sides, is supposed to be an allusion to the capabilities of the Coast Artillery Corps officers, who, in France, perform any duty but that pertaining to the Artillery Corps.

3. The motto of this design is *"Abandonné en France, sans ami,"* which you will readily interpret, "Abandoned in France without friends."

4. I ask you to accept again the continuance of my highest esteem and beg to remain,

Very sincerely,
William Chamberlaine
(Commanding General, R.A.R., A.E.F.)

Office of the
CHIEF OF COAST ARTILLERY
Washington, D.C.

November 19, 1918

From: Chief of Coast Artillery.
To: Commanding General, R.A.R., A.E.F.
Subject: Distinctive mark for Railway Artillery.

1. The "Pochoir" of the Railway Artillery Reserve which you so kindly forwarded, arrived in perfect condition and at present adorns the wall of my office where I can gaze upon it with all the admiration and perfect understanding that it has awakened in me.

2. Feeling that such a work of art and genius should be embodied in the archives of the great war, I turned it over for a day to the Military Intelligence Bureau who, wishing to show their undying appreciation and gratitude for such an unprecedented honor, submitted the following information, which is, to my mind, both interesting and instructive:

The OOZLEFINCH, a rare and almost extinct bird having but one feather, which it displays with great pride and gusto. This bird lives entirely on "hopes," which it forages from promises, rumors, mimeographs, and unconfirmed orders.

While ordinarily of happy disposition it has been noted that lately the OOZLEFINCH has been plunged at times into the depths of despair, despondency and desolation, which is doubtless caused by the fact that it is unable to ascertain if the hour is 7:30, symbolic of the time at which all hands in the R. A. R. commence work, is a.m. or p.m.

The chief enjoyment of the OOZLEFINCH is to sneak off to an artillery park and their to listen to an M. T. S. calling its mate, repeating to himself all the while, "gazook-gazoo"—which, when translated, means, "I DIDN'T KILL A SINGLE BOCHE 'CAUSE OUR POWDER DIDN'T COME."

3. As I remarked several times since its arrival, the "Pochoir" is not only interesting and instructive but inspiring, as can be well shown by the following "Ode to the OOZLEFINCH," written by an Ordnance officer after a short glance at the wonderful bird, and anything that can inspire an Ordnance officer is indeed a thing to be marveled at and its glory should be sung in every publication from the COAST ARTILLERY JOURNAL to the *Police Gazette*.

ODE TO THE OOZLEFINCH

O   ffensives, are his dotage, advancing foot by foot
O   riented, so to shoot *"Dans tous les azimuths"*;
Z   ealous and Resourceful, and just 'twixt me and you
L   eave it to our warlike bird, is the word at G. H. Q.
E   very time he flaps his wings, the big guns go into action
F   iring from a railway track to get the proper traction.
I   n this war he was so young, Ah, yes it was too bad,
N   ever once could he flap his wings when one feather's all he had;
C   ome what may, however, this one fact is a cinch,
H   ere's to the R. A. R., by gosh, and its little—

F. W. Coe
Major General
Chief of Coast Artillery

HEADQUARTERS RAILWAY ARTILLERY RESERVE
American Expeditionary Forces, France

6 January 1919

From: The Commanding General
To: The Chief of Coast Artillery, Washington, D.C.
Subject: Distinctive Insignia for Railway Artillery Reserve, American E. F.

1. The Railway Artillery Reserve is highly gratified to know its "pochoir," the Oozlefinch, properly placed upon the wall of the Headquarters of its parent Coast Artillery Corps, is receiving its due measure of respect and admiration.

2. It is a thought, perhaps, that the merits of this rare, noble and almost extinct bird are not fully appreciated, and that this failure is the result of a lack of sympathetic understanding on the part of your Military Intelligence Bureau.

3. It is true upon his arrival on the scene of action "over here," he had little else but "hopes" upon which to live, and further, that these "hopes" were only born and kept alive by his own foraging from promises, rumors, mimeographs, and unconfirmed orders—the latter being given at times "By direction." However, being a wise bird, and knowing that "hope deferred maketh the heart sick," he bestirred himself and foraged, not only from "promises, rumors, etc.," but from other more tangible sources, with the result that as time passed and his development proceeded, he subsisted on a substantial fare of big guns and plenty of ammunition—"hopes" serving but to spur him on to added activities.

4. In fleeting moments of the past, shadows of despair may have clouded the countenance of the Oozlefinch, customarily so contented, yet expressive of punctilious pride and martial bearing; but this mental desolation, if in truth it ever really existed, other than in the minds of the uninitiated, did not reflect any indecision as to whether 7:30 is a.m. or p.m. Only one who views the struggle at long range from beyond the seas could fail to know that in the land of France, the military day contains but one 7:30.

5. In the days of his early development it is said that he did derive satisfaction from repeating to himself the refrain "Gazook-Gazoo." However, in his later days, it was noticed that this expression was entirely eliminated from his vocabulary. The fact that his foraging produced results in powder, as well as other necessities, and that evidence was not lacking to prove more than a "Single Boche" succumbed to his marksmanship would indicate that possibly your translation of the above refrain is made with a reasonable degree of accuracy. In this connection it might be interesting to record the fact that, among other Boche casualties which are to be credited to the Oozlefinch, he proudly plumes himself on having brought down one Boche airplane and having captured the occupants thereof.

6. The Oozlefinch of the Railway Artillery Reserve was born in time of strife and tribulation; but over this handicap he proved his mastery. After ten active months on the battlefield, he sent his last token to the Hun at 10:57, 11 November, 1918, and now that his part in this great war is over, he will fly back to his home with the Coast Artillery Corps and there reflect with satisfaction upon the fact that he deposited within the enemy lines five million five hundred sixty-eight thousand pounds of cold and convincing steel.

7. During the coming years of enervating peace, his faithful adherents will proudly call to mind his record which may be briefly stated as follows:

He left his home for "over there," among the first to go,
      Hell bent to give the cruel Hun a rocky road to hoe.
At first he fed on "hopes," which he foraged, by direction
      From orders, rumors, promises—all lacking confirmation.

The song he sang: "Gazook-Gazoo," in those days of hoped deferred,
      May indicate that despondency then gripped our sacred bird.
We'll translate it into English, which may assist you some:
      "I haven't killed a single Boche, 'cause my powder hasn't come."

His innate wisdom later on showed this to be wrong,
      So he set his mind and heart and soul to find another song.
He saw the fields he foraged in were barren, save for "hope,"
      So he looked about, in his wise young way, for fields of better dope.

He found these fields, and he foraged there—guns and powder galore.
      And then his troops went to the front. Believe me, the Boche were sore.
For months and months our sacred bird harassed and harried the foe,
      And did his bit to make for the Hun a rocky road to hoe.

Counter-battery fire! Interdiction fire! He was always in the fray;
      And his last he fired at three minutes of 'leven on that good armistice day.
So now he the song he sings to himself is bright and jolly and gay,
      As he sends his troops, with hollers and whoops, back on their homeward way.

His faithful adherents will never forget him,
      To do so would be absurd,
And they dedicate here, the following ode,
      To their grand, resourceful bird:

O ffensives were his food and drink, he wanted but to shoot,
O riented, so to fire *"Dans tous les azimuths";*
Z ealous and resourceful, and just 'twixt me and you
"L eave it to our warlike bird," was the answer over sea.
E very time he flapped his wings, the big guns got into action
F iring from a railway track, to get the proper traction.
I n this war he was so young, 'twas wondered what he'd do,
N ot only did he do his bit, but he did it straight and true.
C ome all of you and join us. Here's a fact that is a cinch—
H appy is the R. A. R. to toast its—

                      William Chamberlaine,
                      Brigadier General, U. S. A.

This toast to the Oozlefinch comes from the same article that carried the above correspondence:

A TOAST TO THE OOZLEFINCH

1. You may sing of the EAGLE, that splendid bird
            Who dwells by the sounding sea.
    Whose head is bald, but whose eyes are keen,
            Whose sentinel form is but seldom seen
                Keeping watch o'er the Land of the Free.
    Long—too long—has he watching perched
            And slumbering seemed—perchance,
    Till he stretched his wings, as in days of yore,
            for a flight which no eagle had ventured before,
                To the battle-scarred land of France!

2. You may raise, Civilians, your hats to him,
            While soldiers from every land—
    Conscript, volunteer, veteran grim—
            Stand to salute when his myriads trim
                Pass—then strike hand to hand!
    Yet, Bulldog Tommy and Poilu bold—
            Ye who at naught would flinch—
    And sons of the Roman Legions old,
            In our EAGLE'S BROOD you may not have been told,
                Of his fledgling—the OOZLEFINCH.

3. Drawn by his snorting steeds of steel—
            Like Titan of the ancient days—
    Beneath the tread of each chariot wheel
            The strong earth trembles, as dreading to feel
                The burden that on her he lays.
    Aloud when the storm of his cannon breaks,
            With the blast of each thunderbolt thrust
    The foe's proud citadel shudders and quakes
            While each stricken rampart totters and shakes
                And crumbles to rubbish and dust.

4. A stout trench helmet is worn by him,
            Protecting his massive pate
    (Full of trig. functions and logarithms);
            His watch he wears on a nether limb,
                He is ready early or late.
    Though nobody loves him, he does not cry,
            But dances—his lone feathered tail
    He proudly raises, he cannot fly
            To the rear, or retreat—he would scorn to try,
                For he marches forward by rail.

5. You may sing of the bird-men, reckless and free,
     Of Infantry, fearless and strong,
But what of the Railway Artillery
     When the field guns fail, when the enemy
          Has held his position too long?
And, comrades all, though with honors few
     And with glory small, yet our part
We each of us did, though none others knew
     To the BRAVE OLD FLAG how we were true
          Both in deeds and thoughts of the heart.

6. So "HOW!" Here we drink to the Oozlefinch,
     And the Railway Artillery.
They did their duty—it wasn't a "cinch."
     With "Bière de Châlons" our friendship we'll clinch,
          While we toast them and VICTORY!
And here's to our General, gallant and wise,
     And to Mailly and Haussimont
And THAT LAIR whence the Oozlefinch shall rise
     And smite our foeman betwixt the eyes
          Should they e'er dare their faces to show!

ABRAHAM B. COX, Captain, Ordnance, U. S. A.

The Transplanted Oozlefinch may have been a slightly different species of Oozlefinch as noted below. (From: "Meet Mr. Oozlefinch," *The Recoil,* the Fort Monroe R.O.T.C. Camp newspaper, June 29. 1923, reprinted in the *Coast Artillery Journal,* Volume 59, No. 5, October 1923, p. 368.)

In 1918 a strange bird with distinct Oozlefinch attributes was observed to roost at night upon the rolling stock of the Railway Artillery Reserves, A. E. F. This bird was equipped with trench helmet and wrist watch. He rapidly made friends with the personnel who named him the "Transplanted Oozlefinch" and adopted him as their mascot. Fort Monroe was somewhat disturbed at the first news of the discovery of a second Oozlefinch but the name "Transplanted Oozlefinch" was taken as an admission of the superiority of the original. Its eyes were not unduly prominent, nor did it fly backwards. In spite of the lack of prominence of eyes, he was capable of observing everything within many kilometers.

An artist fortunately made a drawing of the Transplanted Oozlefinch and incorporated it in the emblem of the R. A. R. He is depicted perched on a section of rail (symbolic of the R. A. R.) with his left foot cocked in the air, and wearing a wrist watch on his left leg.

After having fired 5,568,000 lbs. of steel into German territory the T. O. disappeared one dark night from his perch on an ammunition car and has never been seen since, but his memory is preserved to posterity by the R. A. R. emblem. We may still rejoice that the original Oozlefinch, the one and only "Coastillery Bird" remains in captivity as the celebrated mascot of the Coast Artillery Corps.

E. R. Tilton thought the Transplanted Oozlefinch was the "spirit" of the Oozlefinch overlooking his men in France during their service there. (From: "History of the 'Oozlefinch'," By Colonel E. R. Tilton, *C. A., Liaison,* June 21, 1919, reprinted in *Coast Artillery Journal* Volume 69, No. 1, July 1928, pp. 60-63.)

It is a good thing that the Coast Artillerymen who were fortunate enough to cross the seas and go to war remembered the existence of the "Oozlefinch" (though not his shape) and took him as their sacred standard, as Napoleon did his Eagles. The "Oozlefinch" never crossed the ocean to France in person. His spirit led the Coast Artillerymen who went over, and it would be by all means proper to bestow upon him the required number of Silver Chevrons indicating his war service, and it is to be hoped that he will wear them with the same feelings of devotion to duty which causes those of us who stayed at home to wear them. Wound stripes, they are sometimes called, wounds to personal feelings and professional ambitions. Perhaps the Chief of Coast Artillery might be influenced to grant the "Oozlefinch" some special type of war chevrons, say two gold and two silver, on April 6, 1919, to indicate the bird's influence on the fortunes of the Coast Artillery Corps for service "over here" in the body and service "over there" in the spirit. He is certainly deserving of it.

The Missile Oozlefinch

In 1946 the Coast Artillery School was closed at Fort Monroe and its functions moved to Fort Winfield Scott at the Presidio of San Francisco as a part of the Artillery School and the School of Mines. The Oozlefinch was moved to San Francisco as well. This school was soon closed, and the Coast Artillery was officially abolished as a separate branch of the Army in 1950, all of its remaining units being integrated into a unified Artillery arm as antiaircraft artillery.

Some of the officers who served with the Coast Artillery Corps during World War II were involved with the growing air defense artillery function during the early 1950s. Somebody had retained the glass encased Oozlefinch statue after the close of the Fort Scott Coast Artillery School. And, in due course, the Oozlefinch was brought to the air defense artillery training grounds at Fort Bliss, Texas, where the Army Antiaircraft Defense Command (ARAACOM), was training its crews for the deployment of the new the surface to air NIKE missiles in its role as a continental defense system.

The Oozlefinch made the transition to the development of a new "esprit de corps" for the missilemen, who were bringing with them the roots from the old Coast Artillery Corps. The commanding general of the center was proclaimed "Chief Oozlefinchling I" and authorized to speak during the Oozlefinch's many absences to the missile ranges. The Oozlefinch became the guardian of those who trained at Fort Bliss and a new "club" sprang up. (selected exerpts from: "History of the Oozlefinch," Fort Bliss, circa 1958, reprinted in the *Fort MacArthur Alert* Volume 6, Issue 3 (July 1994) pp 1-3.)

In 1946 the Oozlefinch finally became restless at Fort Monroe, and as all his friends began to depart to be replaced by individuals of various branches, he decided to move to Fort Scott, California, where the Seacoast Artillery Branch of the Artillery School and School of Mines were activated. When these schools were closed, about 1948, the Oozlefinch retired to some unknown cloister where he turned his eyes inward and engaged in deep meditation over the events of the times and need for modernization of the Artillery.

On 6 July 1956, the Oozlefinch, legendary featherless bird of the Coast Artillery Corps, awakened from his sleep of several years, tucked a Nike in the crook of his nude left leg, and traveling by ways known only to himself, arrived at Fort Bliss, Texas, the home of the antiaircraft and Guided Missile Center—there to become the guardian of all missilemen.

Since, as it is well known, the Oozlefinch always flies backwards to keep dust, trivia and other inconsequentia out of his eyes, the Nike is always positioned at the correct attitude.

He charged himself, in addition, with particular care for Very Important Visitors to the Air Defense Center and specifically, not only to protect such visitors from the long-winded, technical briefings and orientations to which they are subjected, but to accord them suitable recognition as "Oozlefinchlings" for their punishment.

To reward both these visitors and others, the amazing bird created the Ancient and Honorable Order of the Oozlefinch, directed its incorporation under the Laws of the State of Texas, and from time to time approved the "awarding" of degrees to those deemed worthy of this honor. Among the degrees are: Master, First Class, Gunner, Apprentice, 24 Hour Expert Oozlefinchling, and Charitable Oozlefinchling.

These degrees, carrying various qualifications as prerequisites for award, all require that the recipient be physically present at the Air Defense Center for induction. The Oozlefinch has also authorized still another degree, the coveted "Oozlefinchling, Old Timer." This degree is bestowed upon persons who qualify by virtue of their association with the bird long before he took over his present job of protecting the men who man the missiles, as well as their dedication and faithfulness to the spirit of the Oozlefinch. This degree can be awarded to persons who are prohibited by age, space, or other ills, from journeying in person to the shrine of the Oozle at the U.S. Army Air Defense Center. One of the first "Old Timer" degrees was awarded to Capt. Ellis C. Baker, who retired shortly after World War I after service with the 42d Railway Artillery Regiment. It was a letter from Captain Baker to General Wood, promptly relayed to the Oozlefinch, of course, recalling the captain's association with the awkward angel of the artillerymen during World War I, which prompted the establishment of the "old Timer" degree.

The original Oozlefinch is likely somewhere in a army cabinet either at Fort Bliss or Fort Sill. It was remembered fondly by the old officers of the Coast Artillery Corps and the men who have been trained at the Fort Bliss missile ranges and is still part of U.S. Army artillery legend today.

Fort Monroe's Coast Artillery School Library bookplate features two crossed Oozlefinches

Ex Libris (from ex librīs, Latin for 'from the books (or library)'), also known as a bookplate, is a printed or decorative label pasted into a book, often on the front endpaper, to indicate ownership. (Wikipedia)

The "Dragon Lady" from the "Terry and the Pirates" comic strip and the Oozlefinch from Coast Artillery Corps OCS Class 29 classbook, Fort Monroe, VA, 1944

THE
COSMOLENE ~
C.M.T.C.

IT COVERS EVERYTHING AT FORT MAC ARTHUR

YE RYME OF YE TYME-RAYNGE BOARDE
By MAJ. FRED M. GREEN, C. A. C.
(With usual apologies to T. Coleridge)

An earlier version of this poem, circa 1910 or so, was reprinted in the *CDSG Journal* Volume 9, Issue 1 (1995) and in the previous editions of this reference guide. A later version appeared in the *Coast Artillery Journal* Volume 73 (1930) pp. 495-498 with additional stanzas and a running commentary in a "Old English" lyric, which has been edited here. It describes fairly closely the proceedures involved in fire control in the early 1910s and voices the frustration that many soldiers must have often felt. The comment "I've just changed to blue" refers to the old army dress uniforms which were blue, the change to olive-drab dress uniforms occured in the mid-1910s. The name "Erasmus" in the last stanza of the poem refers to Major General Erasmus D. Weaver, Chief of the Coast Artillery in the 1910s.

(**Original**) AUTHOR'S NOTE: *Seacoast gun batteries formerly plotted at fifteen-second intervals. No predictions were made on the plotting board; the correction for range-change during the time of flight was computed on the Pratt range board, and combined with the ballistic correction.*

In 1913 it was required that set-forward points be plotted on the plotting board, a graphic time-range relation be maintained on a blackboard at one side of the emplacement, and a T-square and a stop watch be used to determine the proper corrected range for the instant of fire. At the command "Trip," this range was predicted a suitable number of seconds.

These requirements were heartily disliked by most officers. Under this system the inherent errors of prediction were made manifest, and the system was unjustly blamed for their existence. The emplacement pattern of time-range board was difficult to operate, and many delays and personnel errors resulted from its use. Also, the earliest boards were built without any protection from rain, and became inoperative in wet weather due to blurring of the chalk lines and figures.

To penalize unnecessary exposure of the piece to hostile observation and fire, a percentage deduction was made from the final figure of merit for each second the piece remained in battery after the lapse of a maximum allowable period—say ten seconds. I cannot now guarantee the accuracy of these figures but I can vouch for the possibility of the result indicated below. Each practice was supervised by an umpire detailed for that purpose.

Argument

A soldier was mourning
by the sea-beach,

and seized another, a passing
soldier, to whom he
proposed to relate his
tale.

Part the First

Upon the seashore, sad and gray—
    Upon the sandburred strand—
There stood a dismal soldier
    With a piece of chalk in hand.

This saddened, weary cosmoline,
    He hove a mournfull sigh
And reaching out a skinny claw
    -grabbed a passer-by.

The passer-by he waxed full woe;
    "Hold off your hand," quoted he,
"By your oily, greasy denim pants
    Now why do you stop me?

The passer-by pleads a
previous engagement,

but cannot escape

and so runs an absence.

The aged cosmoline commences
his recital,

notes the time-range board,
and shows the confusion
engendered thereby,

and the difficulties of interpolation,
to which the operator succumbed.

He shows further how
his valiant captain, doubtless
a wily man in the
ways of war, avoids the
need for interpolating.

He relays how the
range section labored
lugubriously

and were discomforted by
the requirement of prediction
when the course was
sinuous,

and of the lamentable
register of the plotter at
this circumstance,

"First call has sounded for parade—
    I've just changed into blue;
I need must run like hell—
    Do you not stand it too?"

He holds him with his greasy hand:
    "Our battery fired—" quoth he;
"Hands off— let go my dress-coat sleeve,"
    And so his hand drops he.

He holds him with his glitterying eye.
    His victim must stande stille,
While lo! Assemble for parade
    Rings clearly from the hill.

"The target sped across the bay-
    The Umpire he was there,
Our service practice for to see,
    To judge just what was fair.

"Our captain gazed with deep disgust
    Upon the time-range board;
The writer who posted ranges there
    Showed plainly he was floored.

"The curve ran up-the curve ran down
    the curve ran all around-
The range-board man he beat his breast
    And rolled upon the ground.

"Our captain was a hardy soul:
    Quoth he, 'Now what the hell?
'Goddam these silly lines of chalk
    We'll fire on the bell'

Part the Second

"Aloft, behind the battery,
    gun-shy brain-squad sat-
They sought to trak the target
    And to guess where it was at;

"And when they could locate the thing—
    The orders they were striked—
The plotter cried, 'Now clear away,
    I'm going to predict!'

"But when he studied over its track
    He thought it was bewitched;
Loudly he raved, and cursed, and swore,
    And he even sonofabitched.

and how he was again baffled
by each change in
course of the ship.

And how sorrows multiplied,

and how The captain, doubtless
constrained by the colonel
and other senile warriors,

was driven to extreme
measures; and how, just as
the gun been tripped
into battery,

The whole system been tripped
up by failure of the time-
range board,

thus leaving the gun in
battery without data, so yet
the penalty runs.

Yet further delays are
caused by a chance shower
of rain, thus making the
board all

wet in every sense, and the
practice is irretrievably
ruined before a single
shot

had been fired.

Despite all interruption, the
cosmoline will shrive himself, for yet he had been full

"At length the course was straightened out—
　　Its wriggles he was learning—
Anon cried one from up above,
　　'The goddamned target's turning!'

Part the Third

"Anon upon the time-range board
　　We see a course once more,
But—mark how yon deflections jump
　　From two-point-five to four!

"The afternoon is waning fast,
　　The captain he is tiring—
'Oh, damn the data!' loud he yells:
　　'Attention! Commence firing!'

"The shell was seated from the truck,
　　The powder home was rammed,
The breach was closed, the peace was tripped—
　　And *then,* the T-square jammed!

"The stop-watch stopped! He can't predict!
　　When did the last bell ring?
The T-square's stuck—he drops his chalk—
　　'To hell with this damn thing!'

"Above the lofty parapet
　　The muzzle comes in sight;
Above yon parapet it stays
　　'Till ranges shall come right.

"Yet lo! upon the fatal board
　　Strange characters appear:
A 'reading lost,' 'corrected range';
　　The chalk begins to smear

"As rain-drops small run down the wall—
　　And also down our lines—
The figures blur upon the arnie
　　We can't tell sevens from nines—

"So consequently we have lost,
　　Before our first shot went,
All of our figures of merit
　　With a penalty hundred percent!"

Part the Fouth and Last

"Pass in review"—parade is over—
　　The men will soon be in;
"Stand fast," commands the cosmoline,
　　"I must confess my sin!

to bursting of grief and woe.
He recounts the hardships
and privations
inseparable

from military service,
and the shock to his finicky
sensiblilities at
seeing

a Model 1912 belt over a
blue uniform.

And how the patient soldier

finally lost his patience

over the time-range boards,

and did scoff and jear
villionously there at.

He shows his erudition
by quoting the equation of
an inclined straight line

not passing through ye
origin, as set forth in
divers Godlye workes on
Analytique Goemetrie.

He philosophizes bitterly,
and, disparing of better
things, considers putting
in

for a transfer to the doughs,
as did many a poor wyte
in that sad time.

"For years I've patiently endured
    A gun-mount weird and strange;
"I've used all sorts of godamn ways
    For identifying out the range;

"I've cut the grass; I've trimmed the sod;
    I've shined electric lights;
I've taken visitors 'round, by god,
    And showed them all the sights;

"With gaudy full dress uniform
    Of scarlet, gold, and blue,
I've worn a woven pea-green belt
    Of an appalling hue;

'I've peeled the spuds; I've scrubbed the floor
    I've gladly shined my gun;
I've dusted out behind the door;
    I've risen before the sun;

"All these—and more—have I endured
    Without a single growl,
But of contentment I am cured—
    I'm going to make Rome howl!

"I'm damned if I will go to war,
    And join in battle's hell,
Armed with some chalk, some cotton waste,
    And a nickel-plated bell!

"It makes me sore to drill with them—
    It fills my pants with pain
To think how chalk-line systems fail
    If it should chance to rain.

"And when I face my maker
    And the Pearly Gates I see,
I want some better last words than,
    'y is mx plus b';

"Prediction is uncertain
    In this world, I always find;
Perhaps they'll sail in a strait line-
    Perhaps they ain't that kind.

"A T-square's not my weapon,
    And a doughboy bunch I'll find;
Instead of chalk lines, I prefer
the good old skirmish line

"Where they don't fight with erasers
    And a little tinkling bell;
May the devil snatch Erasmus' board
    And burn it up in hell!"

There are many coast artillery cement batteries that show the effects of modifications made during their service years, sometimes sympathetically and sometimes not. This poem records one engineer's attitude to the many changes. The author, W. L. Marshall, had charge of the construction of the defense of southern and eastern New York beginning in 1900; he was later Chief of Engineers 1908 - 1910.

-oOo-

TREATMENT OF GUN EMPLACEMENTS
PARAPHRASED AND SLIGHTLY REMODELED FROM A BRILLIANT WESTERN POET BY W. L. MARSHALL, AN EXASPERATED DAUBER OF CEMENT

I take a little gravel,
And I take a little tar;
With various ingredients,
Imported from afar.

then -

I hammer it and roll it,
And when I go away,
I think they have a 'placement
That will last for many a day.

but -

I must come with picks and smite it,
To lay a water main;
And then I call the workmen,
To put it back again.
To run a hoist or cable,
I must take it up once more;
And then I put it back again,
Just where it was before.

but again   -

I must take it up for conduits,
To run the telephone;
And then I put it back again,
As hard as any stone.
I must take it up for wires,
To feed the 'lectric light,
And then must put it back again,
Which is no more than right.

Oh! the platform's full of furrows,
There are patches everywhere;
I would like to swear about,
But 'tis seldom that I dare.
They are "very handsome" 'placements;
A credit to the Corps?
We are always digging of them up,
Or putting down some more.

Coast Artillery Songs

In keeping with efforts to keep the "esprit de corps" of the Coast Artillery Corps alive, here are reproductions of three "Coast Artillery Songs." The first was published during World War I in a "US Army Song Book" dated 1918 and based on a popular melody. The second comes from a soldier who was stationed at Fort Monroe during WW I. The third, an original composition, was published in 1934. These songs were a part of the efforts in the Corps to build a distinctive tradition for itself. Comments by Mark Berhow.

Coast Artillery Song

Air: "The Son of a Gambolier"

1. Oh, they said the Coast Ar-til-ler-y would nev-er go to war;___ And all that they were fit___ for was to hang a-round the shore___ But when in France they need-ed men to shoot the tens and twelves, Why they ca-bled to the Pres-i-dent to send our loy-al selves.

2. When Bri-tish Tom-mies took the field to stop the bar-brous Hun,___ They found their light ar-til-ler-y was beat-en gun for gun,___ So Mar-shal French got on the wire and quick-ly told the king That the Gar-ri-son Ar-til-ler-y would be the on-ly thing.

3. So lim-ber up the six-es and the tens and oth-er ones,___ And brack-et on the O. T. line un-til you get the Huns___ There may be ma-ny plans and schemes, to set this old world free, But you'll find in ev-'ry one a part for Coast Ar-til-ler-y.

CHORUS

Then its home boys home, it's___ home that we would be,___ Its home boys home when the na-tion shall be free.___ We're in this war un-til it ends, and Ger-man-y will see That the end of all the Kai-ser's hopes is the Coast Ar-til-ler-y.

Coast Artillery Marching Song

(Tune: "One Keg of Beer for the Four of Us")

Enlisted in the army, turned down the Field,
        Almost joined the doughboys —am glad I didn't yield
Assigned to the Coast, I'm as happy as can be,
        For now I am a member of the COAST ARTILLERY.

Chorus:
Glorious, Glorious, We'll make our Uncle Sam victorious,
        Load her up with shell and we'll give the Kaiser hell,
                As we blast the bloody Germans out of France.

On to Monroe, then to France,
        Limber up the big boys and make the Boches dance,
We'll clear the way for our gallant Infantry,
        For we are the gunners of the Coast Artillery.

(Repeat chorus)

Says von Hindenburg to Kaiser Bill,
        "Dam that artillery, it never will be still,
They're shooting like the devil and it's very plain to me,
        That we're up against the Gunners of the Coast Artillery."

(Repeat chorus)

Black Jack Pershing, he says, says he,
        "Send along another bunch of Coast Artillery,
They'll blast us a path through the lines of the huns,
        So bring along the mortars and the twelve inch guns."

    Submitted by Donald G. Cronan, who was given a copy by his cousin, whose father, Lt. Herbert Scholz, had it while stationed at Fort Monroe during World War One. Reprinted in the CDSG Journal, *Volume 12, Issue 1 (1998).*

The Coast Artillery Song, 1934

This commentary was published in the Coast Artillery Journal, *Volume 77, Number 5 (September-October 1934) page 376; the score itself was published on pages 348-351.*

In the preceding issue of the COAST ARTILLERY JOURNAL it was predicted that we would be able to make full pronouncements concerning the long awaited official Coast Artillery song. For once we are able to make good on a promise, and the song is reproduced elsewhere in this issue. We hope that it will speak for itself and will prove a boon to the Coast Artillery Corps for years to come.

It may be of interest to our readers to know something about the history of the project and the trials and tribulations through which it has passed. From the inception of the idea the pathway has not been strewn with roses. It has encountered some adverse criticism, it has met with procrastination, delay, inertia and a host of other enemies all of them bent upon delaying its progress or strangling it in infancy. The task that confronted Mohammed in moving the mountain was no greater in comparison than the difficulties confronting those whose task it was to bring the plan to a full fruition.

It will be recalled that other arms of the service have a distinctive song, the best known of these being "The Caissons go Rolling Along" and "O'er the Broad Missouri." It was felt that, in this respect, the Coast Artillery Corps should be on an equal footing with its sister arms. Several attempts have been made to bring out a Coast Artillery song; the best known of these compositions was one which made its appearance during the World War but this, like all the others, was open to the serious objection that the lyrics had been adapted to a well known and popular musical score; therefore, at best this song was only half "Coast Artillery."

At a meeting of the Executive Council of the Association held in the early part of 1933 it was decided to offer cash prizes for the lyrics and music of a Coast Artillery song. A cash prize always stimulates human endeavor and it was hoped that something worthwhile would result. It was pointed out at the time that a musical hit to catch and hold the popular fancy and favor is more the result of accident than design. This is borne out by the fact that of the thousands of compositions produced annually only a very small percentage strike a responsive cord and are elevated to any degree of popularity. The announcement of the contest brought forth 18 contributions. Some of these consisted of lyrics only, others of lyrics adapted to popular musical scores. Several consisted of music only. A committee was appointed by the President of the Association to make recommendations to the Council for the award. This committee formulated the following hypotheses as outlining in a broad general way the requirements to which the winning number should conform.

a. It should be an inspiring military march that would instinctively quicken the pulse, raise the head, expand the chest and cause tired soldiers to pick tip their feet with less effort.

b. It should be an original and distinctive Coast Artillery production. (Lyrics adapted to existing popular airs did not seem to fill the bill.)

c. The score was considered to be more important than the lyrics, for the reason that it is easier to fit words to music than vice versa.

The words as published elsewhere in this issue are subject to change. It is not unlikely that they can be revised to more nearly portray the record of glorious achievements of the Coast Artillery. Will some poet laureate come to the front with his ideas as to what they should be? The JOURNAL will publish all helpful suggestions.

All entries, without name of author, were submitted to the committee on award. Each member of the committee reached his conclusion and recommendation independently of the others. Subsequently at a meeting of the committee the merits and demerits of the few outstanding numbers were considered. In this the committee had the benefit of the advice and suggestions of a nationally known musician. As a result of all of this painstaking effort the committee recommended to the Executive Council of the Association that the first prize be given to Messrs. J. H. Hewett and Arthur H. Osborn, the latter is a lieutenant of the New York National Guard. Their composition is dedicated to Major General John W. Gulick, former Chief of Coast Artillery under whose regime the project took definite form. It may not be amiss to record that the co-authors have produced a number of song hits among the better known of these are the following:

The Princeton Cannon Song.

The Guard of Old Nassau.

The Mummy March and a Brigade March, dedicated to the Coast Artillery Brigade, N.Y.N.G.

We are sorry that there is no second prize, therefore the best we can do is to give honorable mention, and all the distinction that this carries, to the composition entitled "The Coast Artillery," words by Major Edward B. Dennis and melody by Mr. Kurt Freier.

The committee on award desires to make due and grateful acknowledgment to Captain W. J. Stannard, leader of the army band, and to Major R. J. Hernandez, Editor of the *Quartermaster Review,* for their painstaking efforts, valuable assistance and helpful suggestions in assisting the Coast Artillery song to emerge from its cocoon and to develop the full statute of an accomplished fact. To Major Hernandez belongs the credit for the excellent band arrangement now in course of production. This we hope will become nationally popular; certainly it should be included among the repertoire of all Regular Army, National Guard and R.O.T.C. bands.

CRASH ON ARTILLERY!

Words and music by J. F. Hewett and A. H. Osborn

(Dedicated to Major General John W. Gulick)

Crash, on! -- with your guns, boys --- Let ev - 'ry- -- shell tell ------ Push on! --- to the end boys -- Let your guns give --- 'em Hell ------.' - Go on, and fight, on --- for e - ver -- No mat-ter

(Gun effect)

2nd Stanza

Who backs the infantry when the fighting begins?
Who cracks the enemy with a fire that wins?
Who brings the planes down in flames, boys, o'er land
    and o'er sea?
Ever and forever, it's the Coast Artillery.

3rd Stanza

Drink to the flag, boys, of the grand old C.A.
Here's how! to the men who will fight come what may.
And here's a toast to the gunners of each battery
Here's health! to the General, and the Coast Artillery.

BLUEPRINTS, EXHIBITS, AND SITE PLANS
PERIOD MAPS OF US ARMY COAST/HARBOR DEFENSE INSTALLATIONS, 1900-45

Mark Berhow

Reference material used in this section:

Hines, Franklin and Frank Ward, T*he Service of Coast Artillery,* Goodenough and Woglom Company, NY (1910), reprinted by the Coast Defense Study Group Press, Bel Air, MD (1997), p. 65.

War Department, United States Army, Office of the Chief of Staff, *Coast Artillery Drill Regulations,* Document No. 474, USGPO, Washington, DC, 1914, pp. 182-183.

War Department, Office of the Chief of Engineers, *Confidential blueprints for fortifications plats issued by the Office of the C. of E.,* National Archives, Cartographic Branch, College Park, MD (4 sheets, symbols and abbreviations, 195 plats covering 28 coast defenses), maps dated 1921-22. A copy of the compete set of these maps is available through the Coast Defense Study Group. The set is approximately 300 pages, in the negative format (white lines–on–black background), sharp and clear enough to easily read all numbers, symbols and writing. (Unfortunately, there is not such a set of maps available for the WW II-era fortifications, though copies of some are in the collections of CDSG members.)

War Department, *General Orders No. 114,* Wash. D.C., May 23, 1907, "Abbreviations and conventional signs . . ." [reprinted in the *CDSG News* Vol. 3, Issue 3 (May 1989) pp. 18-19.]

War Department., Office of the Chief of Staff, Coast Artillery Division, *Artillery Bulletin No. 51 (Serial No. 102)* Subject: "Symbols and abbreviations for harbor charts and fortification index maps.", Wash. D.C., Feb . 4, 1913. [reprinted in the *CDSG News* Vol. 3, Issue 3 (May 1989] p. 17.

War Department, Office of C. of E., Letter: 061.2b-F248, To: Recorder, Coast Artillery Board, Fort Monroe, VA, Subject: Confidential Blue Prints, August 9, 1922. 2 indorsements and list of blueprints (11 pages).

War Department, Office of C. of E., File 323.5, "Symbols and Abbreviations for Use on Fire Control and other Drawings," June 18, 1927 (9 sheets) National Archives, Pacific Southwest Region, Laguna Nigel, CA, RG 77, Box 46, Folder H-5.

War Department, *Technical Regulations, No. 1050-5, Symbols for Seacoast Defense Fire-Control Maps, Diagrams, and Structures.,* US GPO, Washington, DC, May 10, 1939 (8 pages).

War Department, *FM 4-155 Coast Artillery Corps Field Manual, Reference Data (Seacoast Artillery and Antiaircraft Artillery),* US GPO, Washington DC, 1940.

War Department, Annexes to Harbor Defense Projects, various locations and dates (National Archives RG 407, Records of the Adjudent Generals Office, Entry 366).

Site maps, site plans, annex exhibits, confidential blueprints, D-series maps—these are all terms that have been used to describe various maps which depict sites used by the U.S. Army, at one time or another, in connection with harbor defense fortifications and fire control. CDSG members have circulated copies of many such maps among themselves for many years; lately they have been handed out at the CDSG conferences, and they now often accompany articles in the *CDSG Journal.* These maps have been keys to ferreting out the identification of the various remaining structures during site visits, yet there is some confusion over where these maps come from, what their cryptic symbols mean, and even what they are called.

As new harbor defense construction was proposed, planned and carried out by the U.S. Army Corps of Engineers, the engineers prepared a large number of maps and plans to go with those projects as needed. Once the works were completed, the Corps of Engineers maintained and updated maps of the various military reservations and sites. These maps were kept as part of the records of the various Corps of Engineer district offices around the country. Copies were turned over to qualified parties in the army, such as the Coast Artillery Corps, the Quartermaster Corps, etc. In due course, the records of the Corps of Engineers and the other branches of the army have been turned over to National Archives. Copies of a number of interesting maps, dated from the 19th and the first half of the 20th centuries, have been duplicated from the holdings in the National Archives by historians and researchers.

Most maps of harbor defense installations are located in the Cartographic Branch of the National Archives. This section was just recently relocated to new facility in College Park, MD. This branch also has maps from the modern or "concrete" era of American harbor defenses, but for a number of reasons, many of the more frequently seen maps have come from other National Archives holdings. This article will discuss three concrete-era (1890-1950) map formats most frequently seen by the members of the CDSG—the Confidential Blueprint map series (1900-1935), the confidential or secret blueprint map series (1940-1948) and the exhibits from the annexes to harbor defense projects (1940s). An integral part of this article is the reproduction of a symbol key dated 1909 and two symbol and abbreviation keys dated 1921 and 1940.

After 1902 an optical system for fire control based on trigonometric principles was developed for more precisely aiming coast artillery guns. The structures that were built to house the optical and communication elements of this system were often numerous and small in relation to the other major buildings on a military reservation, and many required a detailed description making it complicated to label them on a map, so a set of map symbols was developed to indicate the fire control structures. As these fire control structures were built in the years following 1905, they were incorporated into the maps on which the Corps of Engineers recorded the location of all the structural elements of the fortifications in the seacoast defenses.

As new construction was finished, maps were revised and updated by the Corps of Engineers. One such series of maps was reproduced as negatives from a master positive in blueprint style, which meant maps were composed of white lines on a blue or dark background. As they were classified "confidential" by the War Department, they became known as "confidential blueprints." A number of these confidential blueprints have been found in various cartographic and textual Corps of Engineers records in the National Archives. As they were updated frequently, maps of different dates provide a snapshot of what structures were on a given military reservation on a given date.

The confidential blueprint series of maps has general maps of each defended harbor, and general maps of each of the forts and military reservations in the harbor defense. If it was warranted, larger scale maps of parts of some forts were also included. These were labeled "D" for "detail" and followed in series, D-1, D-2, D-3, etc., as required. These maps show the location of batteries, various components of the fire control and communication system, mine facilities, and all the post buildings. Identification of each structure was shown by name, symbol, abbreviation, or number.

These maps were also used by the Artillery Corps, later the Coast Artillery Corps, as a convenient way of keeping track of the structures under their control. Keys to these fire control map symbols began appearing in coast artillery manuals, such as drill regulations, training regulations, and later field manuals. Most importantly for modern day historians, a complete update of these maps was performed during the years 1920-1922, just after the major construction projects of the Endicott and Taft programs were completed, before some of the smaller harbor defense areas were eliminated. On July 12, 1922, the Coast Artillery Board at Fort Monroe requested a complete list and set of these maps for their records. In August, 1922, the corps complied with a set of maps. Several maps were missing from the Board's collection, and by November, the engineers had sent the missing maps for San Francisco. The other missing maps were explained as having been deleted for being obsolete. The 1922 collection contains about 290 maps of 29 harbor areas.

It is from this collection that many of the maps used by the CDSG have come from. A complete set of the maps collected in 1922 has been found in the National Archives. The set includes the correspondence, the list, and a key to the map symbols and abbreviations for these maps.

An important thing to note is the tactical fire control organization, which varied during the course of the history of the Coast Artillery.

In 1909, each battery was under the immediate command of the officer stationed at the battery commander's station (BC). Mine commanders manned their posts at the mine primary (M') station. In the defended harbor areas, called the Coast Defense Command, batteries were grouped into Fire Commands, each under the overall command of the fire commander stationed at the fire command primary station (F'). The Fire Commands were then grouped together by geographical areas under the command of the officer in command of that entire sector of the coast defense. This command was initially called the Battle Command, but later was changed to the Fort Command. This officer was stationed at the primary fort command station (C').

In 1925, this chain of command was changed slightly. Individual gun batteries were assigned to a gun group (G). Forts (F) were also used as tactical commands. All forts and/or groups were under the Harbor Defense (H), which was under the overall command of the harbor defense commander. This was outlined in an article in the *Coast Artillery Journal,* which is partially reprinted in this article. Later an additional tactical organization, the groupment (C), was added below the Harbor Defense. It was composed of two or more groups.

Many of the maps in this series, some initially drawn in the first decade of the 1900s, were updated over the years up to the late 1930s. The dates are noted in the series in the upper left hand corner of the maps. Many of the later versions of the maps have been converted to positives, that is black lines on white background.

A 1909 and a 1921 map symbol key are reproduced on the following pages, along with a partial article reprint from the 1925 article in the *Coast Artillery Journal,* and example confidential blueprint maps of the entrance to Puget Sound and two maps of Fort Flagler, all dated 1921. The rest of these maps are available from the CDSG.org website and on the CDSG ePress CD ROM.

1909 SYMBOL CHART

Primary Station of a Battery	Ⓑ′
Secondary Station of a Battery	B″
Supplementary Station of a Battery	B‴
Battery Commander's Station	ⒷⒸ
Battle Commander's Station	ⒸⒸ
Emergency Station of a Battery	Ⓔ
Primary Station of a Fire Command	Ⓕ′
Secondary Station of a Fire Command	F″
Supplementary Station of a Fire Command	F‴
Illuminating Light	I
Primary Station of a Mine Command	Ⓜ′
Secondary Station of a Mine Command	M″
Supplementary Station of a Mine Command	M‴
Double Primary Station of a Mine Command	Ⓜ′✳Ⓜ′
Double Secondary Station of a Mine Command	M″✛M″
Meteorological Station	Ⓜ
Separate Observing Room	Ⓞ
Separate Plotting Room	Ⓟ
Post Telephone Switchboard	⊠
Searchlight	Ⓢ
Searchlight 36 inch	③⑥
Searchlight 60 inch	⑥⓪
Tide Station	Ⓣ
Signal Station	S S
Wireless Station	W S

SYMBOLS AND ABBREVIATIONS—1921 CONFIDENTIAL BLUEPRINTS

Name	Abbr.	symbol	Sta. w/o roof
Fort Commander's Station	C	Ⓒ	
Primary Station, Fire Command	F'	Ⓕ'	
Secondary Station, Fire Command	F"	F"	
Supplementary Station, Fire Command	F'''	F'''	
Primary Station of a Battery	B'	Ⓑ'	
Secondary Station of Battery	B"	B"	
Supplementary Station of a Battery	B'''	B'''	
Battery Commander's Station	BC	Ⓑ.C.	
Primary Station, Mine Command	M'	Ⓜ'	
Secondary Station, Mine Command	M"	M"	
Supplementary Station, Mine Command	M'''	M'''	
Double Primary Station, Mine Command	M'-M'	(M'+M')	
Double Secondary Station Station, Mine Command	M"-M"	M"+M"	
Separate Plotting Room	P.	Ⓟ.	
Separate Observing Room	O.	Ⓞ.	
Self-contained Horizontal Base	C.R.F.	C.R.F.	
Emergency Station	E.	Ⓔ.	
Spotting Station	Sp.	Ⓢ.	
Meteorological Station	Met.	M.	
Tide Station	T.	T.	
Searchlight (30, 60, etc., relates to the size of the lights)	S.	No 60	No 36
Controller Booth	C.B.	●	
Watchers Booth	W.	⊕	
Signal Station	S.S.	SS	
Radio Station	R.	R	
Cable Terminal	C. Ter.	⊟	
Post Telephone Switchboard	P.S.B.	⊠	
Mining Casemate	M.C.	■	

Name	Abbr.	symbol	Sta. w/o roof
Loading Room	L.R.	▣	
Switchboard Room	S.W.B.	◨	
Central Powerhouse	C.P.H.	◉⊢	
Powerhouse (and Searchlight Powerhouse)	P.H.	□⊢	
Combined Stations, in same room		B.C	
Combined Stations, in communicating rooms		F' B	B.C / P.
Combined C and F' Station in same room		F'	
Differentiation of auxiliary plants		a⊢ b⊢ c⊢ etc.	

Abbreviations

Cable Gallery	C.Gal.
Cable Tank	C.T.
Cable Hut (commercial cable)	C.H.
Coast Guard Station	C.G.S.
Engineer Wharf	Engr. Whf.
Gasoline Tank	G.Tk.
Guard House	G.H.
Latrine	L.
Lighthouse	L.H.
Lighthouse Wharf	L.H.Whf.
Magazine	Mg.
Mining Boathouse	M.B.H.
Mining Derrick	M.D.
Mining Tramway	M.T.
Ordnance Machine Shop	O.M.S.
Mine Wharf	M.Whf.
Private Wharf	Pvt.Whf.
Radio (commercial station)	Rad.
Railway Wharf	Ry.Whf.
Saluting Battery	Sl.B.
Searchlight Shelter	S.Sh.
Service Dynamite Room	S.D.R.
Steamship Wharf	S.S.Whf.
Sunset Gun	S.G.
Tide Gauge (not a Tide Station)	T.G.
Torpedo Storehouse	T.S.
Tower	Tw.
Water Tank	W.Tk.
Weather Bureau	W.B.

Additional Symbols and Abbreviations	Abbr.	Symbol
Pumping Plant	P.P.	
Radio Powerhouse	R.P.H.	
Searchlight Powerhouse	S.P.H.	
60 inch Searchlight No. 7	S.$^{60}_{7}$	
Coincidence Rangefinder	C.R.F.	
Quartermaster Wharf	Q.M.Whf.	

Subscripts for use in both Legend and on Face of Plat are—

Imp.	Improvised. (for temporary fire control structures only.)	B'' imp.	
p.	Portable. (Principally used for portable searchlights etc.)	S$^{36}_{p2}$	
s.	Superseded. (for abandoned buildings, etc.)	24s.	
t.	Temporary. (For all uses except fire control structures.)	19t.	

Datum Point—location indicated by intersection of lines or by dot at end of arrow.

Triangulation Station.

Intersection Point.

Benchmark.

Lighthouse.

Such other topographic signs as may be necessary will be taken from the Engineer Field Manual (Professional Papers, Corps of Engineers, No. 29) pages 74 to 97.

System of Numbering (for buildings)

1. Administration Building
2. Commanding Officer's Quarters
3. Officer's Quarters
4. Hospital
5. Hospital Steward's Quarters
6. Non-commissioned Officer's Quarters
7. Barracks
8. Guard House
9. Post Exchange

10 to 19 and 100 to 199	Post Buildings
20 to 29 and 200 to 299	Quartermaster Buildings
30 to 39 and 300 to 399	Ordnance Buildings
40 to 49 and 400 to 499	Engineer Department Buildings
50 to 59 and 500 to 599	Signal Corps Buildings
60 to 69 and 600 to 699	Reserved for future requirements
70 to 79 and 700 to 799	Religious and Social Buildings
80 to 89 and 800 to 899	Government Buildings not under War Dept. Control
90 to 99 and 900 to 999	All Private Buildings (Private dwellings, stores, contractor's buildings and buildings purchased with the land but not assigned to public use.)

Notes: Maneuver buildings were classed as post buildings.

CHANGES IN DESIGNATIONS RELATED TO COAST DEFENSES
Coast Artillery Journal Vol. 63 (1925) pp. 172-73.

Paragraph V of General Orders 13, War Department, June 9, 1925, was of such vital interest to Coast Artilleryman that it was published in order to be available for ready reference:

1. To the end that the designations of units comprising the fortifications of the United States may be more truly descriptive, and that they may more nearly conform to the terms used in other branches of the service, the following changes therein are made:

    a. The principal harbor defense tactical and administrative unit, heretofore designated the "Coast Defense Command," will hereafter be known as the "Harbor Defense," the commanding officer of such a unit will be called the "Harbor Defense Commander," and his staff the "Harbor Defense Staff."

    b. The unit heretofore designated the "Fort Command" will hereafter be known as the "Fort."

    c. Units heretofore designated the "Fire Command" and the "Mine Command" will be known as the "Group."

2. It is not practicable to revise and republish all orders, Training Regulations, Army Regulations, and other War Department publications in which the designations of these units appear; nor is it considered practicable to issue detailed changes in such publications to effect these corrections, as the instances where the designations are used are so numerous as to render such action uneconomical.

3. In the first column of the tabulation in paragraph 5 are enumerated old designations and in the second column the corresponding new designations of units referred to in paragraph 1; and whenever the terms in the first column now appear in any publication issued by the authority of the War Department, the corresponding items shown in the second column will be substituted therefor.

4. Hereafter in all official correspondance between officers and individuals in and under the War Department, in all orders, bulletins, circulars, regulations, and other official publications issued by the authority of the War Department, where mention in made of any of the units referred to in the tabulation in paragraph 5, the designations shown in the second column will be used.

5. Old and New Designations

Old Designation	*New Designation*
Coast Defenses of	Harbor Defenses of
These coast defenses	This harbor defense
This coast defense	This harbor defense
This coast defense command	This harbor defense
Fixed defenses	Harbor defense (includes both fixed and mobile armament)
Fort command	Fort
Fire command	Group
Mine command	Group
	Harbor defense command post
Fort commander's station or command post.	Fort command post
	Group command post (first group, second group, etc.)
Battery commander's station or command post.	Battery command post
Primary station, fire command	Primary station (first group, second group, etc.)
Primary station, mine command	Primary station (first group, second group, etc.)
Double primary station, mine command	Double primary station (first group, second group, etc.)
Secondary station, fire command	Secondary station (first group, second group, etc.)
Secondary station, mine command	Secondary station (first group, second group, etc.)
Double secondary station, mine command	Double secondary station (first group, second group, etc.)
Supplementary station, fire command	Supplementary station (first group, second group, etc.)
Supplementary station, mine command	Supplementary station (first group, second group, etc.)
Separate observing room	Observation post

ENTRANCES TO PUGET SOUND

Scale 200,000

CONFIDENTIAL

SERIAL NUMBER

EDITION OF APR 23, 1915.
REVISIONS DEC 7. 1915;
NOV 8, 1916; DEC. 6, 1919; MAY 7, 1921.

SARATOGA PASSAGE AND ADMIRALTY INLET ARE CONNECTED 13 MILES SOUTH OF THIS POINT BY DEEP NAVIGABLE WATER

SARATOGA PASSAGE.

CAMANO ISLAND.

PT. POLNELL

CRESCENT HARBOR

FORBES PT.
OAK HARBOR

WHIDBEY ISLAND

PENN COVE

BLOWERS BLUFF

WATSAK PT

MITCHELL BLUFF

FORT CASEY, ADMIRALTY HEAD.

LAGOON PT.

FORT FLAGLER
MARROWSTONE PT

Shield's Springs

ADMIRALTY INLET

PORT TOWNSEND BAY

Lt. Pt HUDSON

POINT WILSON
FORT WORDEN

MIDDLE

POINT

PT. PARTRIDGE

Light buoy

TRUE MERIDIAN.

FIDALGO Is.

(W) IKA Is.
FORT WHITMAN
GOAT ISLAND

(HOPE)

SKAGIT!

DECEPTION
HO. PUGET

DECEPTION PASS

N

PUGET SOUND

FORT FLAGLER

GENERAL MAP

SERIAL NUMBER 124

CONFIDENTIAL.

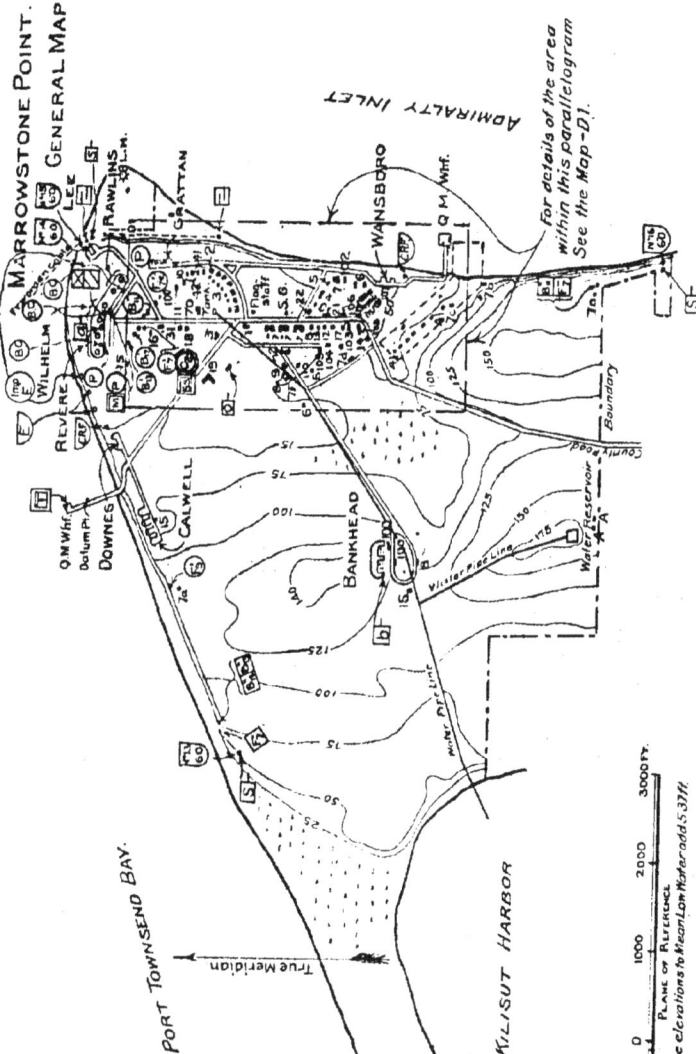

EDITION OF APR. 23, 1915.
REVISIONS: DEC. 7, 1915; NOV. 8, 1916; DEC. 6, 1919
REVISIONS BLDG. MAY 7, 1921.

MARROWSTONE POINT.

ADMIRALTY INLET

PORT TOWNSEND BAY.

KILISUT HARBOR

True Meridian

For details of the area
within this parallelogram
See the Map-D1.

PLANE OF REFERENCE.
0 1000 2000 3000 FT.
To reduce elevations to Mean Low Water add 5.37 ft.

LEGEND.
1 ADMINISTRATION BLDG.
2 COMMANDING OFFICER'S QUARTERS
3 OFFICER'S QUARTERS
4 HOSPITAL.
5 HOSPITAL STWD'S. QRS.
6 N.C.OFFICERS' QRS.
7 BARRACKS.
7a DORMITORY.
7b POST BLACKSMITH'S QRS.
7c POST PLUMBER'S QRS.
7d TEAMSTERS' QRS.
8 GUARD HOUSE.
9 POST EXCHANGE.
10 POST OFFICE.
11 ARTILLERY ENGR. ST. HO.
12 FIRE HOUSE.
13 BAKERY.
14 BLACKSMITH SHOP.
15 CAMP KITCHENS.
16 LAVATORY.
17 CARPENTER SHOP.
18 OIL HOUSE.
19 FUEL SHED.
100 TELEPHONE BOOTHS.
101 RIFLE BUTTS.
102 ROOT HOUSE.
103 STABLE.
104 WAGON SHED.
105 HARNESS ROOM.
106 SHOOTING GALLERY.
21 Q.M. STOREHOUSE.
22 COMMISSARY ST. HO.
31 ORDNANCE ST. HO.
41 ENGR. DEPT. BLDGS.

70 CHAPEL.
71 BOWLING ALLEY.
72 SERVICE CLUB.

107 CANTONMENT 25 BLDGS.
108 PLUMBING SHOP.
109 TOOL HOUSE.
110 PAINT SHOP.

BATTERIES.
BANKHEAD... 4-12" M.
WILHELM... 2-12" N. DIS.
RAWLINS...
REVERE... 2-10" N. DIS.
GRATTAN...
CALWELL...
LEE...
DOWNES... 2-3" P.
WANSBORO... 2-3" P.
A - Anti-aircraft gun - 2-3"

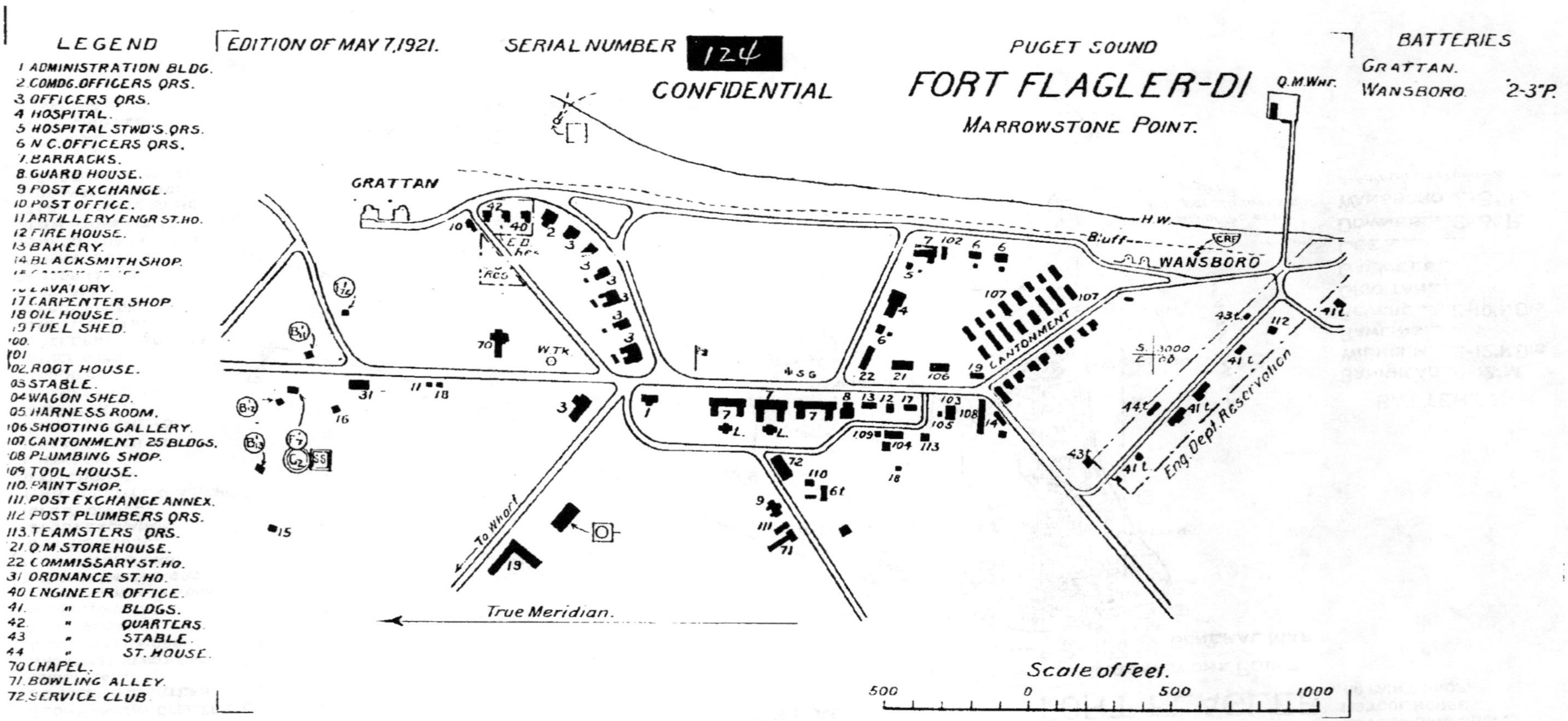

LEGEND
1 ADMINISTRATION BLDG.
2 COMDG. OFFICERS QRS.
3 OFFICERS QRS.
4 HOSPITAL.
5 HOSPITAL STWD'S. QRS.
6 N.C. OFFICERS QRS.
7 BARRACKS.
8 GUARD HOUSE.
9 POST EXCHANGE.
10 POST OFFICE.
11 ARTILLERY ENGR ST. HO.
12 FIRE HOUSE.
13 BAKERY.
14 BLACKSMITH SHOP.
LAVATORY.
17 CARPENTER SHOP.
18 OIL HOUSE.
19 FUEL SHED.
100.
101
102 ROOT HOUSE.
103 STABLE.
104 WAGON SHED.
105 HARNESS ROOM.
106 SHOOTING GALLERY.
107 CANTONMENT 25 BLDGS.
108 PLUMBING SHOP.
109 TOOL HOUSE.
110 PAINT SHOP.
111 POST EXCHANGE ANNEX.
112 POST PLUMBERS QRS.
113 TEAMSTERS QRS.
21 Q.M. STOREHOUSE.
22 COMMISSARY ST. HO.
31 ORDNANCE ST. HO.
40 ENGINEER OFFICE.
41 " BLDGS.
42 " QUARTERS.
43 " STABLE.
44 " ST. HOUSE.
70 CHAPEL.
71 BOWLING ALLEY.
72 SERVICE CLUB.
EDITION OF MAY 7, 1921.
SERIAL NUMBER 124
CONFIDENTIAL
PUGET SOUND
FORT FLAGLER-DI
MARROWSTONE POINT.
BATTERIES
GRATTAN.
WANSBORO 2-3"P.
Q.M. Whf.
GRATTAN
WANSBORO
H.W.
Bluff
CANTONMENT
Eng. Dept. Reservation
To Wharf
True Meridian.
W.Tk.
S.G.
Scale of Feet.
500
0
500
1000

By the beginning of World War II, several key events had occurred which brought out an entirely new series of maps. The new Harbor Defense Modernization Program was begun in 1940 to completely replace the older defenses. The existing fire control structures were re-assigned or incorporated into the new construction program. The new construction brought with it a new series of confidential or secret blueprints. These maps are not as often seen by CDSG members, as not as many of them have been located and reproduced from the archives.

More often, we see maps that were used as exhibits for the various annexes to the harbor defense projects from the mid 1930s and the 1940s. A harbor defense project was a written document which described all existing and projected harbor defense elements, including structures. Annexes updated the projects. The annexes produced between1944-1948 cover much of the new harbor defense construction then underway with a detailed description and a set of maps that showed where these new structures were located, the field of fire of the guns, radar coverage, etc. The maps are particularly useful for the precise location of the 1940s-era fire control elements, both planned and built. Many of these annexes have been located and reproduced from the archives. Many of these maps have their own symbol key located on one or more of the general maps.

A more general key for these map symbols from FM 4-155, published in 1940, is reproduced in the following pages, along with copies of the confidential blueprint maps of the Columbia River and Fort Columbia dated 1944. Also reproduced is Exhibit 1A (the location of elements) from the Annex to the harbor defense project for Los Angeles, July 1944.

SYMBOLS
FM 4-155, REFERENCE DATA 1940
(SEACOAST ARTILLERY AND ANTIAIRCRAFT ARTILLERY)

TABLE *C.-Symbols for seacoast artillery fire-control maps, diagrams, and structures*
Part 1.—Basic symbols

Name	Abbreviation	Symbol
Harbor defense command post	H D C P	(H)
Groupment command post	Gpmt C P	(C)
Fort command post	Ft C P	(F)
Gun group command post	G C P	(G)
Mine group command post	M C P	(M)
Seacoast battery command post	B C P	(BC)
Harbor defense observation station	H D O P	△ H
Groupment observation station	Gpmt O P	△ C
Fort observation station	Ft O P	△ F
Gun group observation station	G O P	△ G
Mine group observation station	M O P	△ M
Battery observation station	B O P	△ B
Emergency observation station	E O P	△ E
Antiaircraft observation post	A A O P	A A △
Battery spotting station	S O P	△ S
Separate observation station	O P	△ O

Name	Abbreviation	Symbol
Operations and plotting room	O P R	
Plotting room	P	
Self-contained base range-finder station	R F	
Magazine	Mg	
Shellroom	S Rm	
Temporary or improvised fire-control structures	Imp	
Mine casemate	M C	
Mine loading room	L R	
Searchlight, 60-inch seacoast	S L	
Searchlight, seacoast, other than 60-inch	S L	
Antiaircraft searchlight	A A S L	
Searchlight shelter	S Sh	
Searchlight powerhouse	S P H	
Searchlight controller booth	C B	
Data booth	Data B	
Watchers booth	W Bth	
Meteorological station	M E T	

Name	Abbreviation	Symbol
Tide station	Td	
Signal station	S S	
Fire Control switchboard room	F S B	
Post telephone switchboard room	P S B	
Combined fire-control & post telephone S B room	F S B P S B	
Cable terminal	C Ter	
Powerhouse	P H	
Radio powerhouse	R P H	
Central powerhouse	C P H	
Pumping plant	P P	
Datum point		OR
Triangulation station		OR
Intersection point		Black Beacon
Benchmark	B M	BM 1232
Lighthouse	L H	

Part 2.-Numbers for harbor defense installations.—a. In harbor defense, seacoast artillery installations of each type are numbered consecutively from right to left, facing the center of the field of fire of the harbor defense. Antiaircraft installations pertaining to the harbor defense may be numbered in any convenient sequence.

b. Groupments, gun groups, mine groups, batteries, and all installations functioning directly under the harbor defense commander, such as harbor defense observation stations, searchlights, and underwater listening posts, are numbered consecutively, each type in a separate series, beginning with number 1. These numbers normally are shown as subscripts to the letter included in the appropriate symbol. Exceptions are included among the examples that follow.

Name	Abbreviation	Symbol
Harbor defense observation station	$HDOP_3$	
Fort observation station	$FtOP_3$	
Antiaircraft observation post	$AAOP2$	
Magazine or shell room	Mg 2 or S Rm 2	

c. Groupment, group, and battery observation and spotting stations assigned to a unit are numbered consecutively within the unit, each type in a separate series, beginning with number 1. These numbers are shown as superscripts to the letter included in the appropriate symbol, the unit number remaining as the subscript.

Name	Abbreviation	Symbol
Groupment observation station	$Gpmt_2OP_2$	
Gun group observation station	G_2OP_1	
Mine group observation station	M_2OP_1	
Battery observation station	B^1_1OP	
Spotting station	S^1_3OP	
Emergency observation station	$E_2{}^1OP$	
Temporary or improvised fire control structures	$B_3{}^2$ Imp.	

d. In certain cases it is desirable to show additional information regarding an installation, such as its size and whether fixed, portable, or mobile. Such information is placed either in the symbol or to the right thereof.

Name	Abbreviation	Symbol
60-inch seacoast searchlight; fixed, portable or mobile.	SL 2F (P or M)	2F(P or M)
Seacoast searchlight other than 60-inch	SL$^{36}_{3P}$	36'
Antiaircraft gun battery or composite battery, fixed or mobile.	A A No. 2 (F or M)	

e. Where two stations are combined in one room, the symbols are superimposed one upon the other, and the letters representing each station are inclosed in the combined symbol.

Name	Abbreviation	Symbol
Combined groupment command post and fort command post.	Gpmt Ft Cp	
Combined battery observation and spotting station.	$B^2_1 S^2_1$ O P	
Combined group command post and battery command post.	$G_1 B_2$C P	
Combined battery command post and battery observation station.	B_2C P B^2_2 O P	

f. Where stations are adjacent in the same structure, the symbols are tangent to each other and are arranged to show the relative location, as:

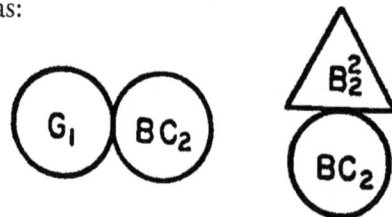

g. Where communication may be had by voice through a passage, door, window, or voice tube, the symbols are left open at the point of contact, as:

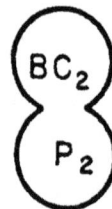

Part 3.—Communications symbols for use on harbor defense fire-control charts and diagrams.

Telephone cable (numerals indicate number of pairs and gage)	26-19
Speaking tube	
Mechanical data transmission line	
Electrical data transmission line	
Searchlight controller line	
Zone signal and magazine telephone line	
Firing signal line	
Time interval bell line	
Submarine cable (numerals indicate number of pairs and gage)	50-19

Part 4.-Abbreviations

Cable gallery	C Gal
Cable tank	C T
Cable hut (commercial cable)	C H
Coast Guard station	C G S
Engineer wharf	Engr Whf
Gasoline tank	G Tk
Guardhouse	G H
Latrine	L
Lighthouse wharf	L H Whf
Mine boathouse	M B H
Mine derrick	M Drk
Mine tramway	M Tmy
Mine wharf	M Whf
Ordnance machine shop	O M S
Private wharf	Pvt Whf
Radio (commercial station)	Rad
Railway wharf	Ry Whf
Saluting battery	Sl B
Service dynamite room	S D R
Steamship wharf	S S Whf
Quartermaster wharf	Q M Whf
Superseded (for abandoned buildings, etc.)	24 s
Temporary (for all uses except fire-control structures)	19 t
Sunset gun	S G
Tide gage	T G
Torpedo storehouse	T S
Tower	Tw
Water tank	W Tk
Weather bureau	W B

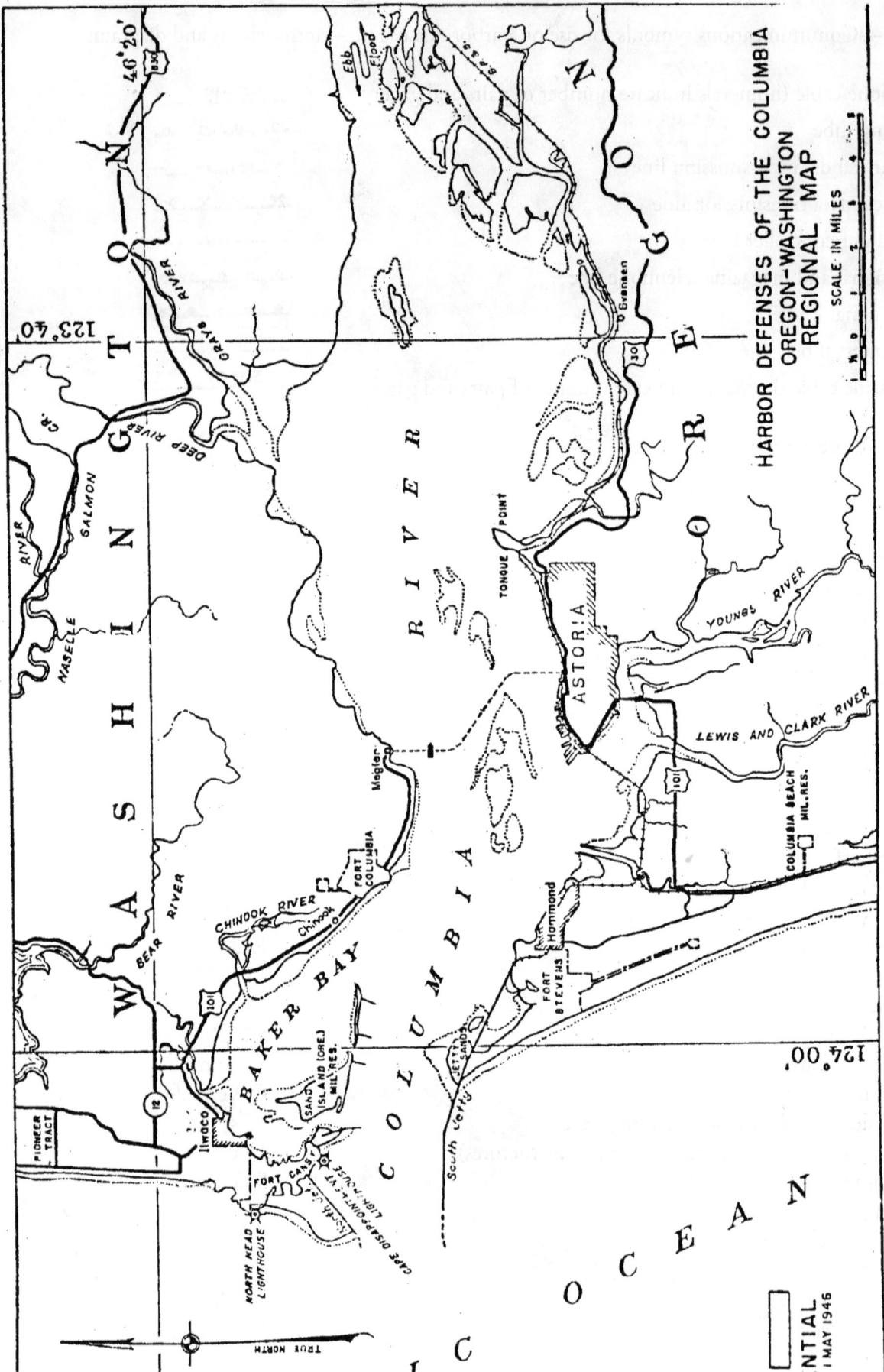

HARBOR DEFENSES OF THE COLUMBIA
OREGON-WASHINGTON
REGIONAL MAP

SCALE IN MILES

WASHINGTON

OREGON

COLUMBIA RIVER

ASTORIA

BAKER BAY

PACIFIC OCEAN

PIONEER TRACT

Ilwaco

NORTH HEAD LIGHTHOUSE

FORT CANBY

CAPE DISAPPOINTMENT LIGHTHOUSE

SAND ISLAND (ORE.) MIL. RES.

South Jetty

FORT STEVENS

Hammond

COLUMBIA BEACH MIL. RES.

LEWIS AND CLARK RIVER

YOUNGS RIVER

TONGUE POINT

FORT COLUMBIA

Megler

CHINOOK RIVER

Chinook

BEAR RIVER

DEEP RIVER

GRAYS RIVER

SALMON RIVER

NASELLE RIVER

Gvensen

TRUE NORTH

ENTIAL
1 MAY 1946

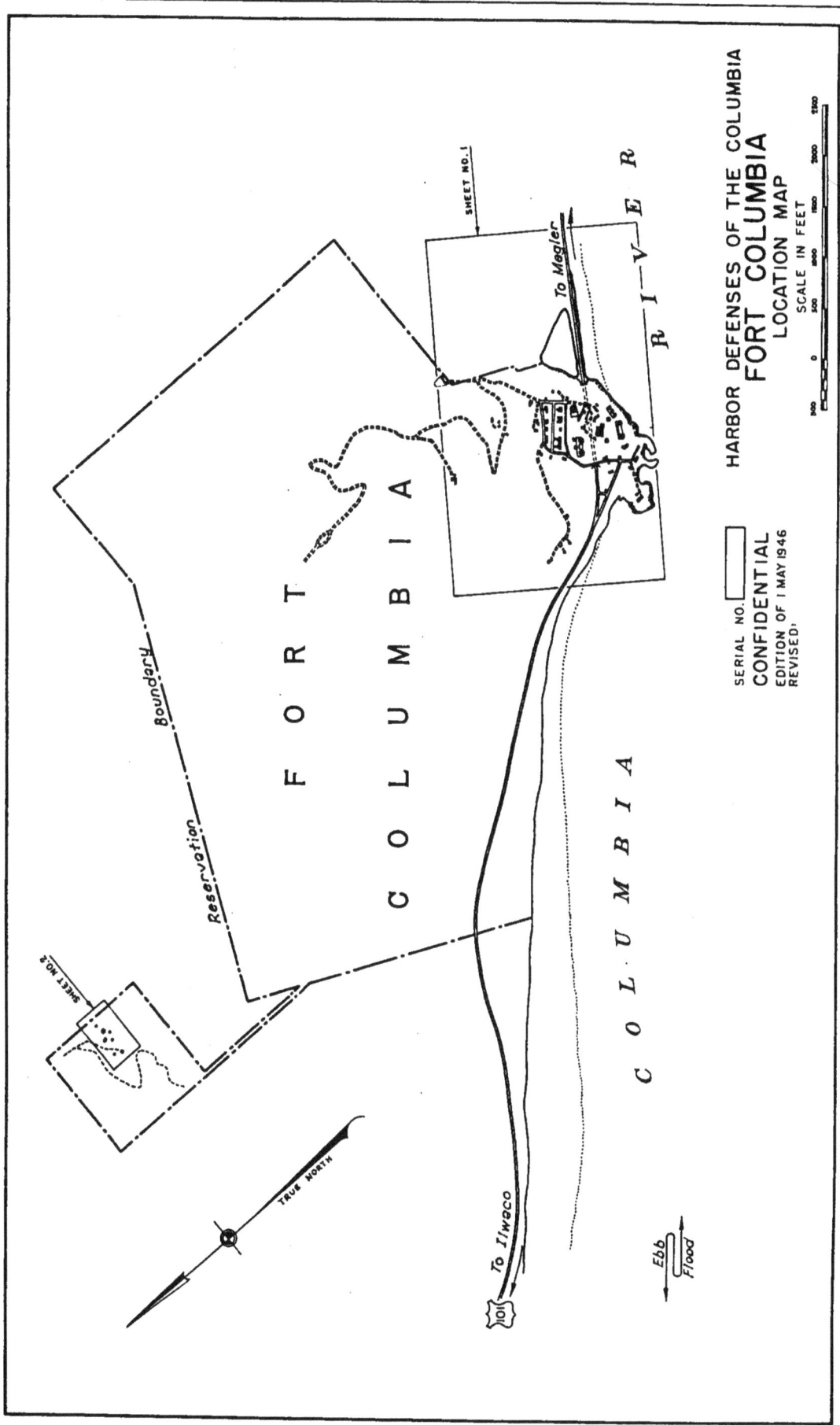

HARBOR DEFENSES OF THE COLUMBIA
FORT COLUMBIA
LOCATION MAP

SCALE IN FEET

SERIAL NO.
CONFIDENTIAL
EDITION OF 1 MAY 1946
REVISED:

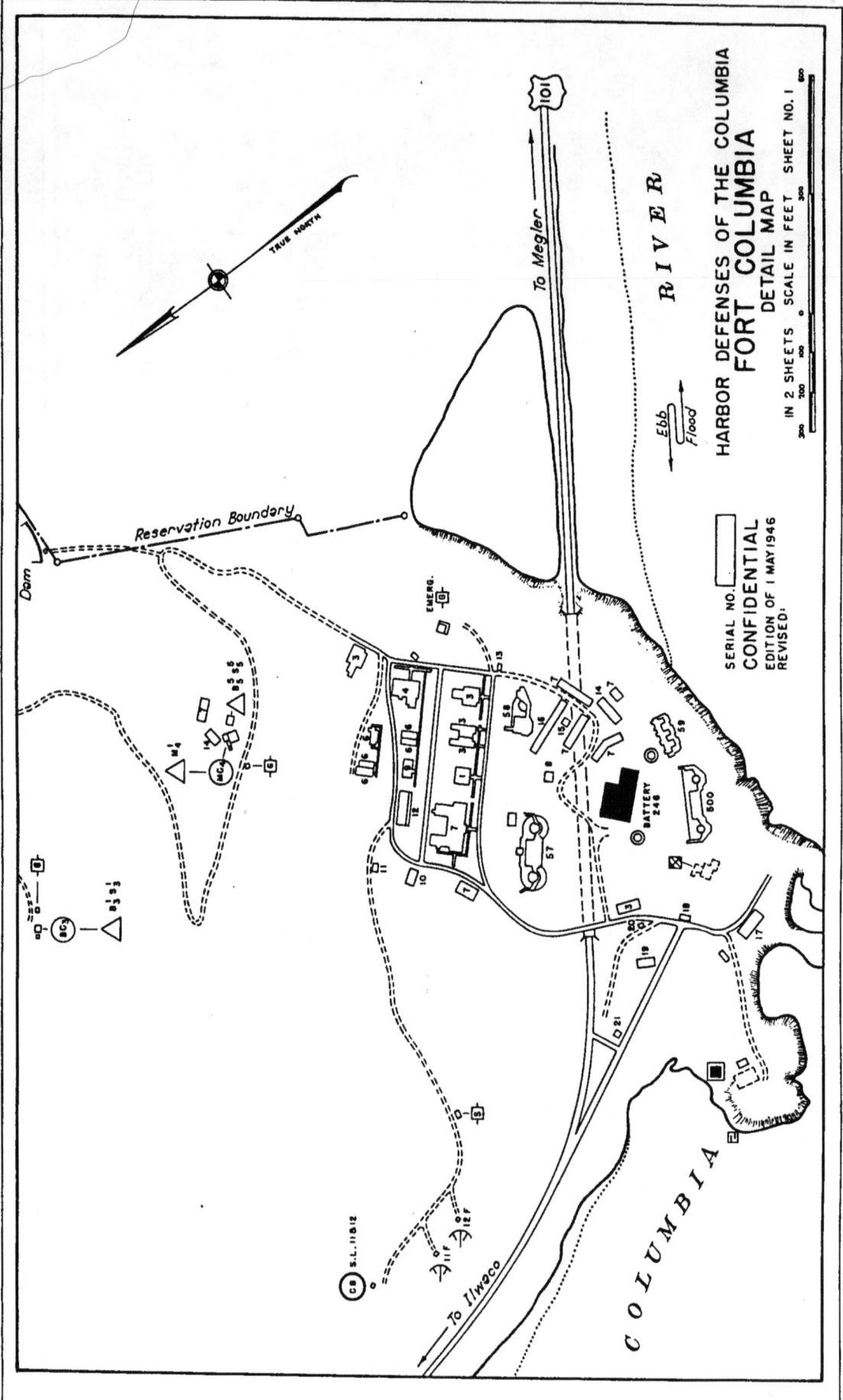

HARBOR DEFENSES OF THE COLUMBIA
FORT COLUMBIA
DETAIL MAP

IN 2 SHEETS   SCALE IN FEET   SHEET NO. I

CONFIDENTIAL
EDITION OF 1 MAY 1946
REVISED:

SERIAL NO.

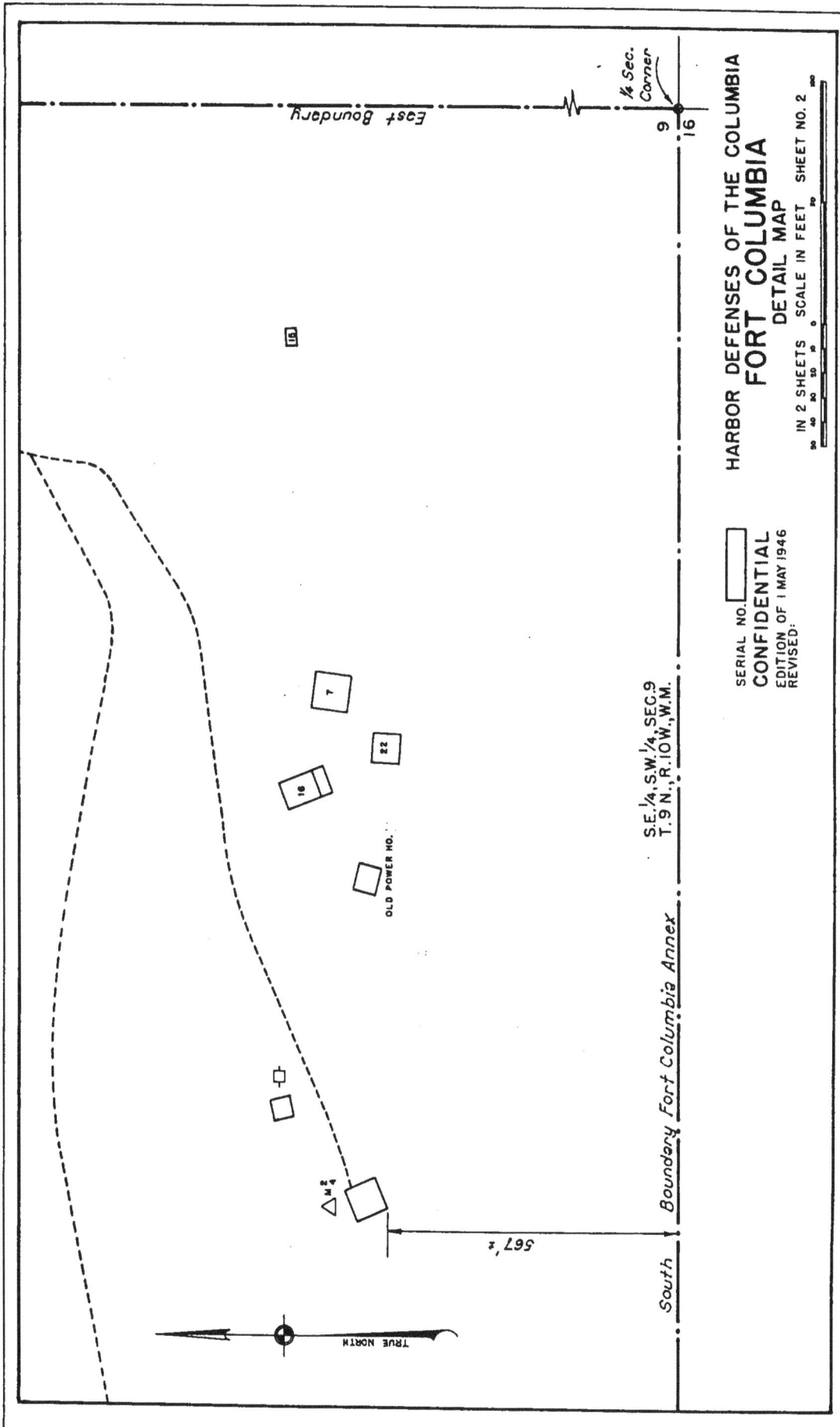

HARBOR DEFENSES OF THE COLUMBIA
FORT COLUMBIA
DETAIL MAP
IN 2 SHEETS   SCALE IN FEET   SHEET NO. 2

SERIAL NO.
CONFIDENTIAL
EDITION OF I MAY 1946
REVISED:

S.E.¼,S.W.¼,SEC.9
T.9 N., R.IOW.,W.M.

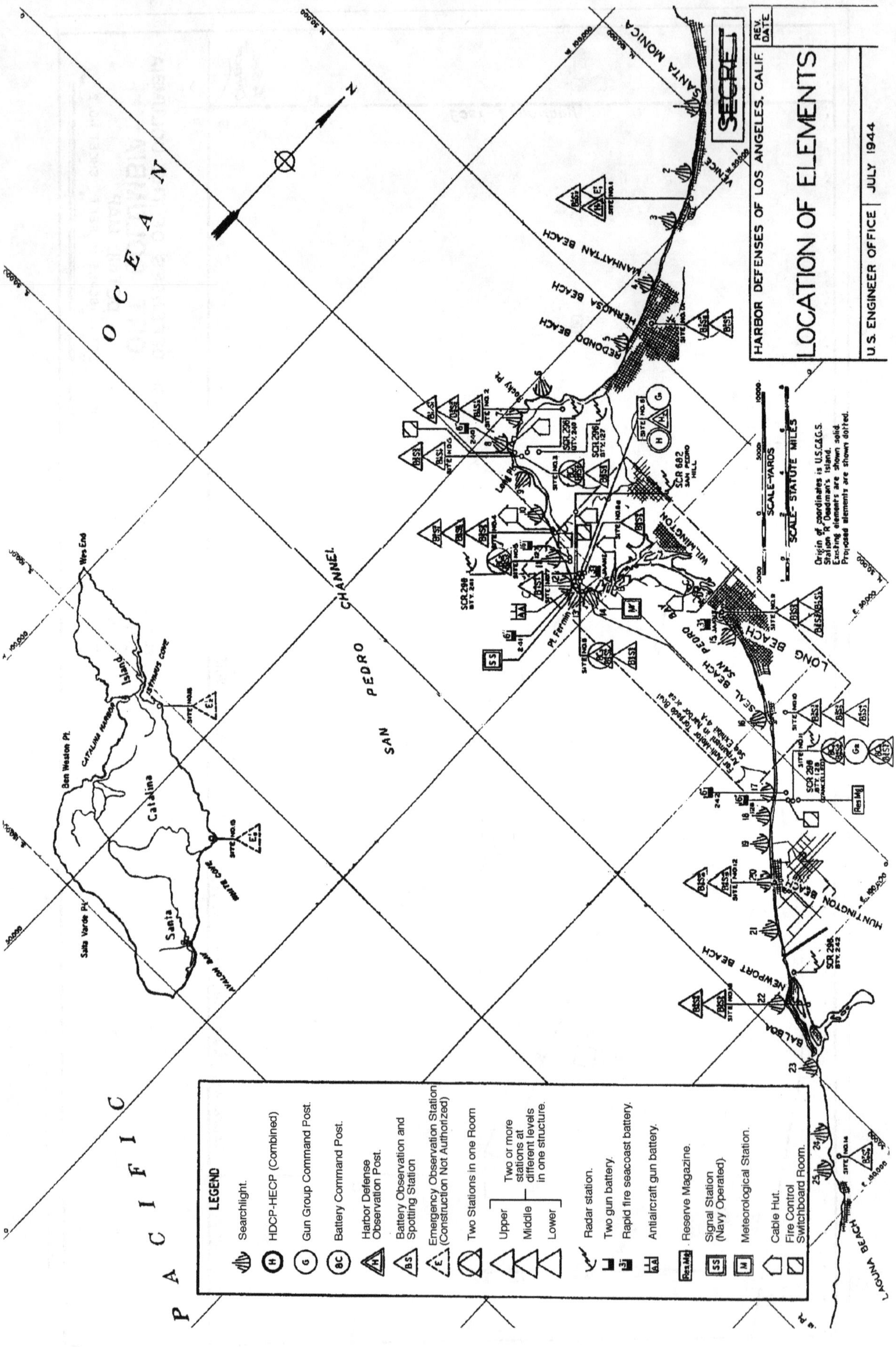

SECRET
REV. DATE
HARBOR DEFENSES OF LOS ANGELES, CALIF.
LOCATION OF ELEMENTS
U.S. ENGINEER OFFICE
JULY 1944
PACIFIC OCEAN
SAN PEDRO CHANNEL
SANTA MONICA
VENICE
MANHATTAN BEACH
HERMOSA BEACH
REDONDO BEACH
Pt. Fermin
WILMINGTON
SAN PEDRO BAY
LONG BEACH
SEAL BEACH
HUNTINGTON BEACH
NEWPORT BEACH
BALBOA
LAGUNA BEACH
Santa Catalina Island
West End
Ben Weston Pt.
Salta Verde Pt.
SCR 682 SAN PEDRO HILL
SCALE-YARDS
SCALE-STATUTE MILES
Origin of coordinates is U.S.C.&G.S. Station "R" Deadman's Island.
Existing elements are shown solid.
Proposed elements are shown dotted.
LEGEND
Searchlight.
HDCP-HECP (Combined)
Gun Group Command Post.
Battery Command Post.
Harbor Defense Observation Post.
Battery Observation and Spotting Station
Emergency Observation Station (Construction Not Authorized)
Two Stations in one Room
Upper
Middle
Lower
Two or more stations at different levels in one structure.
Radar station.
Two gun battery.
Rapid fire seacoast battery.
Antiaircraft gun battery.
Reserve Magazine.
Signal Station (Navy Operated)
Meteorological Station.
Cable Hut.
Fire Control Switchboard Room.

IMPORTANT SOURCE DOCUMENTS FOR
THE STUDY OF MODERN US HARBOR DEFENSES

This is an annotated list of some of the source documents used by researchers in the study of the modern US Army seacoast defenses (1886-1947). This is by no means complete. A brief description of what these sources are given besides just giving a title. Original copies of the sources listed can be found at either the National Archives (www.nara.gov) or at the Institute of Military History Library (carlisle-www.army.mil/usamhi/) at Carlisle Barracks, PA. Some of these sources can also be found in the holdings of major city and university libraries which are designated repositories for Federal records.

A more complete bibliography can be found in *Defending America's Coasts 1775-1950* (EP-870-1-57), USACE, Historical Division, Office of the Chief of Engineers, GPO, Washington, DC, 1997. An excellent guide to doing research in the National Archives and Records Administration holdings is *Environmental Cleanup at Former and Current Military Sites: A Guide to Research* (EP-870-1-64), USACE, Historical Division, Office of the Chief of Engineers, GPO, Washington, DC, 2001). Copies of these books can be requested from the USACE Office of History in Alexandria, VA, or they can be downloaded as PDF files from the Office's publications page—www.hq.usace.army.mil/history/pubs.htm.

The Coast Defense Study Group has reprinted a number of these documents and offer them for sale as printed books of as downloadble PDF files. A catalog can be found on the cdsg.org website.

Key Reports—Modern Era

Report of the Board on Fortifications or other Defenses Appointed by the President of the United States under the Provisions of the Act of Congress Approved March 3, 1885, House Executive Document No. 49, 49th Congress, 1st session (2 volumes, GPO, Washington D.C., 1886). This is the start of the modern era of the United States coastal defenses. This board report, known as the **Endicott Board Report,** reviewed the state of American harbor defenses in 1885 and recommended a completely new system of harbor defenses based on the emerging technology of rifled breech loading cannons and armor. The major harbors of the United States were ranked in terms of military and economic importance. When Congress began to appropriate money for the construction of the new defenses in the late 1880s, this report was cited as the authorization source for the harbors to be defended. Reprinted as hardcover book with a separate set of the figures by the CDSG Press.

Report of the National Coast Defense Board . . . on the Coast Defenses of the United States and the Insular Possessions, Senate Document No. 248, 59th Congress 1st Session (GPO, Washington D.C., 1906) This report, known as the **Taft Board Report,** surveyed the progress in harbor defenses made since the Endicott Report and recommended a number of technical improvements. It also recommended the fortification of key harbors in the newly acquired overseas possessions such as Hawaii and the Philippines in addition to those of the Panama Canal. Note: the fortification of entrances to the Panama Canal was authorized in the Spooner Act of 1902. Included with the reprint of the Endicott Report by the CDSG Press.

Report of the Board of Review of the War Department to the Secretary of War (November 26, 1915) on the Coast Defenses of the United States, the Panama Canal, and the Insular Possessions, House Document No. 49, 64th Congress, 1st Session (G.P.O., Washington D.C., 1916). This report marks the beginning of the transition from the disappearing guns as the major weapon in American harbor defenses to the longer range 12- and 16-inch barbette carriage guns of the post World War I era. Included with the reprint of the Endicott Report by the CDSG Press.

Memo "Harbor Defenses," War Plans Division, Office of the Assistant Chief of Staff, 8 Mar 1923, War Plans Division 1105. 18 pages. "This study, approved 17 Apr 1923, remained the basic definition of War Department policy with respect to Harbor Defenses until the eve of WW II." (Note 2, page 45 in: *Guarding the United States and its Outposts, the US Army in WW II, The Western Hemisphere,* by Conn, Engleman & Fairchild, Center of Military History, US Army, GPO, Washington, DC, 1989). Reprinted by the CDSG in the *Coast Defense Journal* Volume 36, Issue 2 (Spring 2022).

Memo "Harbor Defenses in the Continental United States," War Plans Division, Office of the Assistant Chief of Staff, 6 Aug 31, AG 660.2 (9-30-31) War Plans Division 1105-55, 9 pages. This memo resulted in the creation of the Harbor Defense Board. Reprinted by the CDSG in the *Coast Defense Journal* Volume 36, Issue 2 (Spring 2022).

Documents pertaining to the 1940 Modernization Program for the Continental United States. This is some of the correspondence during study of Harbor Defense Board that was used as the blueprint for the 1940 Harbor Defense Modernization Program (also known as the World War II Construction Program). The July 27, 1940 letter recommending the building of "twenty-seven 16-inch batteries and fifty 6-inch batteries at major harbors around the continental United States," is a key document. Many of these documents were reprinted by the CDSG in the *Coast Defense Journal* Volume 36, Issue 2 (Spring 2022). Additional overseas construction (Alaska, Canada, Panama, Hawaii, and the Caribbean) were added later, some as separate programs. Other letters were the official army authorization for various parts of the program.

US War Department, Letter: "Abandonment of Harbor Defense Posts no longer required for Sea Coast Defense," Harbor Defense Board, President, AG 602 (7-27-40) M, 6 pages, Exhibit A (12 pages), Exhibit B (2 pages) (National Archives RG 407, Entry 360, Class. 602, Box 958).

US War Department, Letter: "Modernization of Harbor Defense Projects, Continental United States," AGO, AG 660.2 (9-16-40) M-WPD-M 26 Sept. 1940.

US War Department, Letter: "Revision of Anti-Aircraft Annexes, Harbor Defense Projects," AGO, AG 660.2 (10-23-40) M-OCCA 1 Nov. 1940.

US War Department, Letter: "Surface Craft Detectors, SCR 296," AGO, AG 413.68 (5-4-42) WC to the CG WDC 3 Nov. 1941.

US War Department, Letter: "Defense of Harbors Against Motor Torpedo Boats," AGO AG 660.2 (12-12-42) MSC-E to the CG WDC, 21 Feb 1942.

US War Department, Letter: "Radio Set SCR-582 Central Surveillance Detector," AGO AG 413.44 (12-12-42) MC to the CG WDC 13 May 1942.

US War Department, Letter: "Revision of Underwater Defense Projects," AGO, AG 660.3 (9-29-42) OB-S-E 20 Oct. 1942.

US War Department, Letter: "Defenses of Harbors Against Motor Torpedo Boats," AGO, AG 660.2 (4-12-43) OB-S-SPDDO-M, 17 Sept. 1943.

Annual Reports (Congressional Records)

These are often cited in *CDSG Journal* articles. These reports are part of the public record and the ones from the War Department are of particular interest to modern day researchers. These can be found at all major public libraries that are Fereral record repositories. Some of the reports are separately bound. Several of these reports pertaning to seacost artillery are offered as PDF copies by the CDSG ePress.

Annual Report of the War Department
Annual Report of the Chief of Engineers
Annual Report of the Chief of Ordnance
Annual Report of the Chief of Artillery, Chief of Coast Artillery

Army Record Books from Harbor Defense Installations

The National Archives Record Group 392 includes four classes of record books that are from coast defenses installations. These were the official record books maintained at the various sites, with information on commanders, names of sites, and fire control systems. They are: Battery Emplacement Books, Fort Record Books, Fort Record Book Files (supplements to the Fort Record Books), Mine Command Record Books.

Histories of the Defense Commands during WW II

These typescript documents were prepared by the defense commands following WW II. They contain a wealth of information on Coast Artillery troop deployments on the east, gulf and west coasts during the war. Never formally published, only a few copies of each have survived in the military libraries. PDF copies are available from the CDSG ePress.

History of the Eastern Defense Command
History of the Southern Defense Command
History of the Western Defense Command

Coast Artillery Drill Regulations, U.S. Army, 1898, 1906, 1909, 1914

Drill regulations for the coast artillery appeared after the development of fire control procedures beginning in 1894. Several book bound editions were printed. PDF copies are available from the CDSG ePress.

Manual for Submarine Mining, 1898, 1912, 1926, 1930

A manual of procedures for the handing of submarine mines was also developed 1890s. It was revised several times as well. The 1930 manual was TM 2160-20 (see TMs, below). PDF copies are available from the CDSG ePress.

Training Regulations (separate series, TR 435 coast artillery) 1920s

These individually printed chapters replaced the book-bound "Drill Regulations" in the 1920s. These "TRs" were designed to be assembled into binders and could be "custom assembled" based on the interests of the individual officer or enlisted man. PDF copies of some are available from the CDSG ePress.

Field Manuals, 1940-1945

In 1940, the training regulations were overhauled, redone and printed as Field Manuals (FM) and Technical Manuals (TM), all part of the Army Manual system. What follows is a list of FMs from FM 4-155 (1940) and a list of revisions printed in the mid 1940s from the Army Ordnance Museum at Aberdeen Proving Grounds, Maryland. A fairly complete set of original FMs and TMs are housed at Carlisle Barracks. PDF copies of some of these titles are available from the CDSG ePress.

Some abbreviations include: SOP- service of the piece, IMUC-instructions for mounting, using and caring for, AA- antiaircraft artillery, D-description.

Coast Artillery Field Manuals- Seacoast Artillery

FM 4-5	Organization and Tactics
FM 4-6	Tactics and Technique of Controlled Submarine Mines, Buoyant
FM 4-7	Tactics and Technique of Controlled Submarine Mines, Ground
FM 4-10	Gunnery
FM 4-15	Fire Control and Position Finding
FM 4-19	Examinations for Gunners
FM 4-20	Formations, Inspections, Service, and Care of Material (retitled "Firing Preparations, Safety Precautions, Care and Service of Material" in 1943)

FM 4-24	Service of the Piece (SOP), 155 mm Gun M1
FM 4-25	SOP, 155 mm Gun M1917 & M1918
FM 4-29	Service of the Seacoast Searchlight
FM 4-30	Service of the Gun Data Computer M1
FM 4-32	Service of the Base-End Data and Gun Data Transmission Systems
FM 4-35	SOP, 14-inch Gun M1920MII on Railway Mount M1920
FM 4-40	SOP, 12-inch Mortar, Railway Artillery
FM 4-45	SOP, 12-inch Gun Railway Mount M1918, Railway Artillery
FM 4-48	SOP, 8-inch Gun, MkVI, Mod. 3A2, on Barbette Carriage M1
FM 4-49	SOP, 8-inch Gun, MkVI, Mod. 3A2, on Railway Mount M1A1
FM 4-50	SOP, 8-inch Gun, Railway Artillery
FM 4-51	SOP, Operation and Care of Railway Artillery Equipment
FM 4-55	SOP, 12-inch Mortar (Fixed Armament)
FM 4-60	SOP, 12-inch gun, Barbette Carriage
FM 4-61	SOP, 12-inch Gun, Casemated Mount
FM 4-65	SOP, 10-inch Gun, Disappearing Carriage
FM 4-70	SOP, 6 inch Gun, Disappearing Carriage
FM 4-74	SOP, 6-inch Gun M1903A2 or M1905A2, on BC M1 or M2 and 6-inch Gun M1 on BC M3 or M4
FM 4-75	SOP, 6-inch Gun, Barbette Mount M1900
FM 4-80	SOP, 12-inch and 14-inch Disappearing Carriages
FM 4-85	SOP, 16-inch Gun and Howitzer
FM 4-86	SOP, 16-inch Gun Casemated Mount
FM 4-90	SOP, 3-inch Rapid-Fire Mount
FM 4-91	SOP, 90 mm Gun M1, on Mount M3
FM 4-97	Service of the Radio Set SCR 682-A

Coast Artillery Field Manuals- Antiaircraft Artillery

FM 4-100	AA: Organization and Tactics (1943)
FM 4-105	Organization and Tactics, Antiaircraft Artillery (1940)
FM 4-110	Gunnery, Fire Control, and Position Finding, Antiaircraft Artillery Gun
FM 4-111	Position Finding and Fire Control, Antiaircraft Artillery Searchlights
FM 4-112	Gunnery, Fire Control and Position Finding, AA automatic weapons
FM 4-115	Operation of Material and Employment of Personnel, AA Searchlight Units
FM 4-119	AA: Examination for Gunners
FM 4-120	Formations, Inspections, Service and Care of Material, AAA
FM 4-121	Fire Control, Guns
FM 4-125	SOP, 3-inch Antiaircraft Artillery Gun
FM 4-126	AA: SOP—90 mm AA Gun on M1A1 Mount
FM 4-127	AA: SOP—90 mm Gun M2, on Mount M2
FM 4-128	AA: SOP—4.7-inch AA gun
FM 4-130	SOP, 105 mm Antiaircraft Artillery Gun
FM 4-135	Marksmanship and SOP, Antiaircraft Artillery Machine Gun
FM 4-140	SOP, 37 mm Antiaircraft Artillery Gun
FM 4-142	Service of Height Finders M1 & M2
FM 4-143	Service of Height Finder SCR-547
FM 4-144	AA: Service of Radio Set SCR-584
FM 4-146	AA: Service of Radio Set SCR-545
FM 4-150	Exam for Gunners

FM 4-155	Reference Data (Seacoast Artillery and Antiaircraft Artillery) (1940)
FM 4-155	SOP, Caliber .50 AA Machine Gun
FM 4-160	Coast Artillery Training
FM 4-176	Service of Radio Set SCR-268
FM 4-181	Employment of Barrage Balloons
FM 4-182	Barrage Balloon Technique
FM 4-183	AA: Barrage Balloon Control
FM 4-184	AA: Barrage Balloon Site Installations
FM 4-187	AA: Barrage Balloon, Service of the Balloon and Balloon Equipment, Very Low Altitude
FM 4-188	Service of the Balloon and Balloon Equipment, Very Low Altitude
FM 4-191	Barrage Balloon, Service of Cable Armament, Low Altitude
FM 4-192	AA: Barrage Balloon, Service of the Cable Armament, Low Altitude
FM 4-192?	AA: Barrage Balloon, Gas Generation, Use, Purification, and Service of Hydrogen Generator
FM 4-196	Barrage Balloon, Rigging and Fabric Repair
FM 4-198	Barrage Balloon, Reference Data

Technical Manuals, 1930s-1945

These manuals were usually for specific items of equipment or supply. This list is derived from the listing of manual changes obtained from the Army Ordnance Museum. This list only has those entries pertaining to fixed (and some mobile) seacoast artillery material. PDF copies of some of these titles are available from the CDSG ePress.

TM 4-205	Coast Artillery Ammunition
TM 4-210	Seacoast Artillery Weapons
TM 4-220	Controlled Submarine Mine Material
TM 4-234	Antiaircraft Artillery Target Practice
TM 4-235	Seacoast Artilley Target Practice
TM 4-237	Radio-Controlled Target (JR) Boat
TM 4-238	Coordinate Conversion Tables
TM 4-245	Preservation and Care of Seacoast Defense Material
TM 4-305	CA Gunners Instruction, Fixed Seacoast Artillery
TM 4-310	Gunners Instruct., Fixed Seacoast Art., Expert
TM 4-315	Gunner's Instruct., Mobile Seacoast Artillery, 1st & 2nd Class Gunners
TM 4-320	Coast Artillery Gunner's Instruction

Selected TM 9s

TM 9-235	37 mm AA Material
TM 9-252	40 mm AA Material
TM 9-252	40 mm Auto, Gun M1 & 40 mm AA gun Car. M2
TM 9-345	155 mm Gun Material M1917, M1918 & Modifications (1942)
TM 9-370	90 mm Antiaircraft Gun Material, retitled - 90 mm Gun M1A1 on Mount M1A1
TM 9-371	90 mm Gun M1 and 90 mm Antiaircraft Gun Mount T2E1
TM 9-372	90 mm Gun M2 and 90 mm Antiaircraft Gun Mount M2
TM 9-373	90 mm Gun M1 and 90 mm Gun Mount M3

TM 9-421	3-inch Seacoast Material (1942)
TM 9-424	6-inch Seacoast Material: Gun M1900 mounted on Barbette Carriage M1900
TM 9-428	6-inch Seacoast Material: Gun M1903A2 & M1905A2 mounted on Barbette Carriage M1, retitled Gun 6-in M1903A2 & M1905A2, Carriage Barbette M1, Gun 6-inch M1, Carriage M3, power plant M4
TM 9-429	6-in Seacoast Material: Gun T2 and Barbette Carriages M3 & M4, retitled Gun 6-in M1903A2 & M1905A2, Carriage Barbette M2, Gun 6-in M1, Carriage M4, power plant M4
TM 9-442	8-inch Seacoast Material: Gun MkVImod3A2, Barbette Carriage M1
TM 9-451	12-inch Gun M1895M1A4 on Barbette Carriage M1917, Power Plant M2 (1945)
TM 9-452	12-inch Seacoast Material: Power Plant and Auxiliary Equipment for Barbette Carriage M1917
TM 9-456	12-inch Seacoast Material, Mortar M1890M1/Carriage M1986M1 & MII 1942
TM 9-457	12-inch Seacoast Material, Mortar M1908 mounted on Mortar Carriage M1908
TM 9-458	12-inch Seacoast Material, Mortar M1912 mounted on Mortar Carriage M1896MIII
TM 9-463	8-inch Gun MkVImod3A2 on railway mount M1A1
TM 9-471	16-inch Seacoast Gun Material: Gun MkIIMI on Barbette Carriage M4
TM 9-471-1	16-inch Seacoast Gun Material: Gun MkIIMI on Barbette Carriage M5
TM 9-472	16-inch Seacoast Gun Material: Gun M1919MII, MIII on Barbette Carriage M1919
TM 9-472-1	16-inch Seacoast Gun Material: Gun MkIIMI on Barbette Carriage M2 & M3
TM 9-472	Power Plant M1: for 16-inch Gun MkIIMI on Barbette Carrige M4 & M5

TM 9-500 (U500) - 541(U540) various Telescope Mounts

TM 9-U542	Range Finder Mount M62
TM 9-U543	Range Quadrant M3 & M8
TM 9-U544	Range Quadrant M4, M5, & M6
TM 9-U545	Range Quadrant M10C, D
TM 9-U546	Elevatin Quadrant M1
TM 9-U548	Azimuth Instrument M1910M1
TM 9-U549	Azimuth Instrument M1918A2, M1
TM 9-U550	Azimuth Instrument M2A1
TM 9-U551	Observation Instrument A.A.B.C. M1
TM 9-U552	Bracket Fuze Setter M1916M2
TM 9-U556	Computing Sight M13
TM 9-U559	Telescope Mount M69
TM 9-U560	Depression Position Finder M1, M2, M2A1
TM 9-U561	Spotting Board M3
TM 9-U562	Plotting Board M3 & M4
TM 9-U563	Plotting and Relocating Board M1
TM 9-616	Generating Unit M5 & M6
TM 9-617	Generating Unit M18
TM 9-618	Generating Unit M7, M7A1, M15, M15A1
TM 9-623	Hieght Finder M1 & M1A1

TM 9-624	Height Finder M2
TM 9-627	Cable System M12 and off-carriage components of the remote control system M14 (for 6-in Barbette Carriages M2 & M4)
TM 9-659	Directors M5, M5A1, & M6
TM 9-659-1	Directors M5A2, & M5A3
TM 9-1345	Maintenance—155 mm Guns, M1917, M1917A1, & M1918M1; Carriages M1917, M1917A1, M1918, M1918A1, M2 & M3; limbers, M1917, M1917A1, M1918, M1918A1, & M3
TM 9-1370A	Maintenance—90 mm Gun M1 and Mounts M1 & M1A1; Gun & Upper Carriage
TM 9-1370B	Maintenance—90 mm Gun M1 and Mounts M1 & M1A1; Lower Carriage
TM 9-1371	Maintenance—90 mm Gun M1 and Mount M3
TM 9-1401	Repair of Submarine Mine Cases

TM 9-1500s & 1600s Maintenance—sights, telescopes, quardants, range finders, sighting systems, periscopes, generating units, height finders, fuze setters, remote control systems, gun data computers, directors, data transmission systems, azimuth insturments, depression position finders, sound locators, etc. Examples:

TM 9-1570	Plotting Boards for Seacoast Artillery (1942)
TM 9-1585	Range Finders, Short Base Coincidence Types
TM 9-1647	Data Transmission Systems for Seacoast and Railway Artillery
TM 9-1653	Data Transmission System M6 (for 90 m AA gun Mount M1)
TM 9-1675	Azimuth Instrument M1910 & M1910A1
TM 9-1680	Azimuth Instrument M1918 & M1918A2
TM 9-1685	Depression Position Finder M1907
TM 9-1695	Depression Position Finder M1
TM-9-2005	Ordnance Material-General, Dec. 1943 (in 7 volumes)
TM 9-2300	Standard Artillery and Position Finding Equipment, Feb. 1944 (Artillery Material and Associated Equipment)
TM-9-2305	Fundamentals of Artillery Weapons, Sept. 1947
TM 9-2601	Instruction Guide: Elementary Optics & Application to Fire Control Instr.
TM 9-2681	Instruction Guide: Plotting Boards M3 & M4
TM 9-2682	Instruction Guide: Spotting Board M3
TM 9-2683	Instruction Guide: Plotting Board M5

F190-W1   Data Transmission Systems

Ordnance Department Documents

This list was derived from Ordnance Department Document (ODD) No. 1467 "List of Blanks, Pamphlets, etc." The first section is a list of the headings found in the ODD No. 1467, the second section is a list of those ODDs that pertain to Coast Artillery. For a more complete list, the researcher is referred to ODD 1467. Again, a fairly complete collection of these documents is in the holdings of Carlisle Barracks and at the National Archives in RG-287 (Records of the Government Printing Office). PDF copies of some of these titles are avaialble from the CDSG ePress

Note: gun carriage titles in this listing indicate only the carriage model. The full title carries the designation of the gun models that were mounted in these carriages.

Abbreviations: IMUC-Instructions for Mounting, Using and Caring for; DU-Description and Use; (years printed, R = reprint) for carriage abbreviations, see the Emplacment Guide section.

List of topics in ODD 1467:

1-400	Property returns, invoices, receipts, vouchers, requisitions
401-600	Specifications, Instructions to bidders, contracts, bonds
800-1033	Target firing, etc.
1034-1177	Inspection and Proof reports
1185-1273	Progress Reports
1274-1358	Civil Service Blanks
1369-1379	Personal Reports of Officers
1417-1422	Bonds
1467-1651	Ordnance Office Forms
1656-2200	Pamphlets, etc., descriptive of guns, carriages, & other Ordnance materials
2229-2512	Forms for specific Ordnance Stations
2534-2787	Miscellaneous
3020-4020	Additional Specifications and Inspection Reports
1-33, 320-364	War Department Standard Forms

Titles from ODDs 1656-2200—Pamphlets, etc., descriptive of guns, carriages, & other Ordnance material:

1006	Self-Propelled Caterpillar MkII for 155 mm Gun M1918MI (1920)
1022	IMUC 6-inch BC Model of 1900 (1920) see also #1688
1023	IMUC 10-inch M1900, DC M1901 (1907, 1920R) see also #1694
1024	IUC 12-inch Mortar Carriages M1896MI & M1896MII, revised (1916)
1040	Reloading Outfits for .30 cal. gallery . . . M1919 (1920)
1060	3-inch AA Gun Mount M1917 (superceded #1808) (1926)
1067	AA Truck Mount M1917 for the 75 mm Field Gun of 1916 (1921)
1074	155 mm Motor Carriage M1920 (1922)
1075	IMUC 12-inch Mortar Carriage Model of 1896 MI, MII
1078	"Caterpillar" Adapters for 8-in Howizters and 155 mm Carriages (1922)
1091	Proof Firing of Artillery Material (1905, 1906, 1908, 1916, 1919)
1110	Artillery Cart M1918 & .... (1922)
1607	Plotting Boards for Mobile Artillery M1905 & M1906
1647	IMUC 12-inch DCLF M1901 (190?)
1654	Periscopic Azmuth Instrument M1918 (1918)
1655	Azimuth Instruments M1918 (1918)
1656	Azimuth Instrument, Warner & Swasey M1910 DIU (1907, 1911, 1912, 1917R)
1657	IMUC Azimuth Instrument, M1900 & M1900M1 (1905, 1906, 1909, 1907, 1912, 1917R)
1658	Ammunition, Blank Inst for Prep & Use (1908, 1914, 1915)
1659	Handbook of 3-inch Gun Material ((1905, 1906, 1908, 1911, 1912, 1915, 1917)
1660	Handbook of 3.2-inch field battery (1902, 1908, 1911, 1914, 1917R)
1661	Handbook of 5-inch siege gun battery (1903, 1914, 1917R
1662	Handbook of Material for 7-inch Siege Howitzer Battery (1900, 1914. 1917R)
1663	Descrip. Pratt Range Board M1905 (1908, 1915, 1917)

1705A	Supplement to Form #1705 (1914)
1706	12-inch Mortar Dummy M1912 (1912)
1707	IMUC 12-inch Mortar Carriage M1908 (1912, 1919)
1708	IMUC 15-Pdr Dummy Carriage M1912 (1912)
1709	IMUC 12-inch Mortar M1912, Carriage Model of 1896MIII (1913, 1917R)
1710	IMUC 10-inch Dummy DCLF M1912 (1912)
1711	IMUC 6-inch gun M1905, M1908, DCLF M1905MII (1914, 1917R)
1712	IMUC 14-inch M1910, DC Model of 1907 and 1907MI (1910, 1912, 1917R)
1712A	Supplement to 1712 (1924)
1713	IMUC 6-inch M1908MIII, BC Model of 1910 (1914, 1917R, 1921)
1714	IMUC 3.6- & 7-inch Field Mortar M1895, 7-in M1892, 3.6-in M1890 (1915, 1917R)
1715	Equipment from the Cavalry Equipment Board M1912 (1912, 1914, 1918R)
1716	IMUC 16-inch M1895, DCLF Model of 1912 (1917)
1717	Equipment Infantry M1910 (1912, 1914, 1917R)
1718	Equipment Infantry M1910 Description (1912, 1914, 1917R)
1719	Equipment, Horse & Officers & Enlisted Men (1905, 1908, 1917R)
1720	Explosives Regulations for Transportation (1914)
1721	Instructions for Loading Projectiles with Explosive D (1906, 1907, 1911, 1913, 1915, 1917)
1722	IMUC 12" M1895M1, BC Model of 1917 (1917)
1723	Handbook on High Explosives and Shell Loading (1918)
1724	Rules Covering Handling and Storage of Explosives and Ammo (1918)
1725	Methods of Analyzing Amatol and Its Constituents (1919)
1727	Fuzes for Use . . . (1895, 1904, 1905, 1906, 1908, 1913, 1917, 1924R)
1738	Crusher Gages, Tools and Accessories for Determining Pressures in Cannon (1902, 1903, 1908, 1912, 1915, 1917R, 1922)
1741	Grenades, Rifle and Hand (1911, 1918)
1741A	Hand Grenades IDU(1917, 1918)
1742	Chemical Arms now in Use by the German Army (1917)
1743	Grenade V.B. rifle, mark 1 (1917)
1744	Stokes Trench Mortar mk 1 & Shell mk 1, 3-in (1917, 1918, 1921)
1745	Handbook of the 6" Trench Mortar Mark 1, (1919, 1921)
1749	IMUC 4.72-inch 40 cal. R-F Armstrong, BC Armstrong (1898, 1904, 1908, 1911, 1917R)
1750	IMUC 4.7-inch 45 cal. R-F Armstrong, BC Armstrong (1898, 1904, 1908, 1911, 1917R)
1751	IMUC 4.7-inch 50 cal. RF Armstrong BC, Armstrong (1898, 1904, 1908, 1911, 1912R, 1917R)
1752	IMUC 6-inch R-F gun Armstrong, BC Armstrong (1903, 1908, 1917R,)
1753	IMUC 6-inch R-F Armstrong, BC Armstrong (1903)
1755	Gun Making in the United States by Birnie (1914)
1756	IMUC Driggs-Seabury 15-lb, R-F Guns & Their M1898 Masking Parapet Mounts (1904, 1907, 1911, 1914, 1916, 1919)
1757	Gun, Gatling M1895 & M1900 (1896, 1908)
1759	Gun, Machine Colt Automatic (1901, 1905, 1906, 1907, 1916, 1917R)
1760	Hotchiss 2 Pounder Mountain Gun Handbook (1903)
1761	2.95-inch Vickars-Maxim Mountain Gun Material & Pack Outfit (904, 1912, 1916)

1762   1-pdr. Maxim Q.F. Gun on Field Carriage (1903)

1763   IMUC 2.24-in (6 pounder) R-F and Their Parapet Mounts (1903, 1908, 1915,1916)

1765   Desc, Instr 5 & 6 -inch R-F Guns (1903, 19007, 1911, 1916, 1917R)

1766   IMUC 15 lb. R-F, BC Model of 1902 (1905, 1908, 1913, 1916, 1917R)

1768   3-inch Wrought Iron Saluting Gun and Mount (1905, 1907)

1769   Gatling Gun M1895, M1900 & M1903 (1905, 1906, 1910, 1917R)

1770   Maxim Machine Gun 0.30 Calibur M1904 Wheeled Mount M1905 (1906, 1908, 1913, 1915, 1916, 1917R)

1771   Handbook of Gun Material 4.7-in M1906 (1910, 1914, 1917)

1772   Desc. 3-inch gun M1903, Ped Mount M1903 & Breech (1912, 1914)

1773   Handbook 3.8-inch gun (1917)

1774   IMUC 1.457-inch subcaliber guns (1917)

1775   Vickers Machine Gun M1915 with Pack Outfits & Accessories (1917)

1776   Descrip Subcaliber Rifle and Target Device for DCs (1917)

1778   High Explosives for Bursting Charges of Projectiles, Methods (1911)

1779   Handbook of 6-inch Howitzer Material (1913, 1917)

1780   Handbook of 4.7-inch Howitzer Material M1908 (1913, 1916, 1917)

1781   Handbook of 3.8-inch Howitzer Material M1915 (1916, 1917R)

1782   Handbook of 4.7-inch Howitzer Material M1913 on Pedestal Mount M1915 (1916, 1918)

1783   IMUC 4.7-inch Howitzer Railway Mount M1917 (1917)

1784   Handbook of Motorized 4.7-inch Gun Material (1918)

1792   Inspectors Manual for Inspection of Powder & Explosives (1918)

1793   Manuals for Inspectors of Ordnance (1886, 1904, 1906, 1909, 1917)

1793   Supplement: Hydropneumatic Recoil Mechanisms, Inspec. Man. (1919)

1794   Description of the Wind Component Indicator (1904, 1906, 1908, 1917R)

1795   IMUC Instruments for Fire Control Systems for Coast and Field Artillery (1906, 1909, 1912, 1916, 1917R, 1918R)

1796   Handbook of Fire Control Equipment for Field Artillery (1916, 1917)

1797   Handbook of Range-Finders 70 cm & 80 cm (1915)

1798   Handbook for the 8-inch Howitzer Material M1917 (Vicars Mk VI) (1918)

1798   Rocking Bar Sights for 8-inch Howitzer carr. M1917 (Vicars) (1922)

1800   Methods for Testing in Toluol Plant Operations (1918)

1803   Instructions for Using and Repairing Dudgeon's Hydraulic Jacks (1902, 1904, 1907, 1909R, 1917)

1805   French Detonating Fuze M1899 (1917)

1806   Explosive Projectiles Fired with a Reduced Charge from 75 mm (1917)

1807   French Fuze Percussion Detonating 24/31 type 1 (1917)

1808   IMUC 3-inch gun, M1917 Antiaircraft Mount (1917, 1919, 1926)

1812   Machines, Tools, Supplies for Ordnance Repair Shops List (1905, 1907, 1908, 1909, 1911, 1912, 1916, 1918R)

1814   Descr. & Inst. for Care & Oper.Firing Magnetos Types GA & MA (1917)

1815   Handbook for British 75 mm Gun Material M1917 (1918)

1816   IMUC 8-inch Gun RY Mount M1918 (1918)

1817   75 mm Gun M1897 MI (French) (1918)

1818   155 mm Gun M1918 (Filloux) (1918)

1819   75 mm Gun M1916 (1916)

1820	Description of 12-inch Mortar and Instructions for Its Care (1900, 1904, 1907, 1913, 1917, 1919R)
1849	List of Parts for 75 mm Gun Material (1917)
1850	Notes on the French 75 mm Gun Material (1917)
1851	DICO Schneider 155 mm Howitzer M1915 (1917)
1861	Handbook of Ordnance Data (1918, 1919R)
1862	The Story of the 75 (mm Field Gun) by W.N. Dickerson (1920)
1863	History of 155 mm Artillery Projects (1920)
1864	Ordnance Schools During the Period of War (1919)
1865	History of Rifles, Revolvers, and Pistols (1920)
1866	Automatic Pistol M1911 .45 cal. (1912, 1914, 1917R)
1868A	Paints and Markings for Field Artillery Ammunition (1898, 1902, 1903, 1904, 1914, 1917)
1868B	Paints for Seacoast Projectiles (1918)
1869	Oils, Paints, and Materials for Cleaning and Maintaining Seacoast Guns, Carriages, Sights and Position Finding Instruments (1898, 1908, 1909, 1917)
1870	Smokeless Powder for Small Arms and Cannons (1910, 1913, 1917)
1872	Seacoast Artillery Ammunition and Instr. Prep., Care & Use (1905, 1906, 1907, 1912, 1915, 1917)
1873	Lewis Depression Position Finders M1898 Type A & B (1902, 1909, 1913, 1918R)
1874	Rafferty Depression Position Finder, Type B (1907)
1874	History of 4.7 in Artillery by Anna Dee (1919)
1875	Swasey Depression Position Finder Types A, AI, AII (1904, 1905, 1908, 1911, 1913, 1915, 1917R)
1876	Lewis Depression Position Finders M907 (1909, 1915, 1917, 1918)
1877	Distinctive Colors for Projectiles (Chart) (1904)
1879	List of Ordnance and Ordnance Stores (1908, 1910, 1914, 1916, 1917, 1918)
1880	IMUC Drill Primer Outfit (1907)
1881	Primers for Use in Service Cannon (1904, 1906, 1908, 1915, 1917)
1886	Price list of Ordnance Stores . . . Miscellaneous (1906)
1887	Price List of Articles of Submarine Mine Equipment (1906, 1917)
1888	Smokeless Powder and Other Explosives ... Regulations for Care & Test (1902, 1907, 1912, 1914, 1917)
1889	Price List of 3-inch Gun Material (1907, 1910, 1912, 1914, 1917)
1890	Price List of Machine Gun and Automatic Gun Equipment .... (1907, 1917, 1919R)
1891	Price List of 2.95-inch Mountain Gun V.M. (1916)
1892	Price List of Ammunition and its Components (1909, 1914, 1917)
1893	Price List of Subcalibur Guns, ... Cartridges ... Pertaining to Seacoast Artillery (1909, 1915, 1917)
1894	Price list of 8, 10, 12, and 14 -inch Guns and Carriages, etc. (1909, 1914, 1917)
1895	Price List of 12 inch Mortars, etc. (1909, 1917)
1896	Price List of Seacoast Guns and Carriages 2.24 to 6-inch Incl. (1909, 1914, 1918)
1897	Price List of Seacoast Targets .... (1909, 1914, 1917)
1898	Price List of Fire Control Instruments .... (1910, 1917)
1899	Price List of United States .30 Caliber Rifle M1917... (1918)
1900	Price List of 4.7-inch Material M1906 (1915, 1917)
1902	Price List of 4.7-inch Howitzer Material (1915, 1917)

1964   Staff Observation Car . . . M1918 (1918)
1965   Small Arms and Ordnance Equipment (1915, 1918)
1966   Directions for the Care and Repair of Leather Equipment (1911)
1970   Ordnance Department Supply and Allowance Tables (1917)
1971   Gun Material Table (1917)
1972   Reconnaissance Car Mk II . . . M1918 (1918)
1973   Machine Gun Car Mk II . . . M1918 (1918)
1974   3-inch Field Gun Trailer M1918 (1918)
1975   4-ton Shop Trailer . . . M1918 (1918)
1976   Machine Gun Car, Mk III . . . M1918 (1918)
1977   35-ton Tank Mk 8 (1918)
1979   Artillery Wheeled Tractor M1918 (1918)
1980   AA Machine Gun Trailer M1918 ... (1918, 1920)
1981   Heavy Ordnance Mobile Repair Shop (1919)
1982   Bench Reloading Tools, Instr. . . . for Small Arms Cartridges (1901)
1984   List of Hand Reloading Tools for Gallery Practice Ammunition (n.d.)
1985   12-inch Gun M1918 on Sliding RY Mount M1918 (1918)
1986   IMUC&D 1 Pdr, 75 mm Subcaliber Guns in Bore . . . . (1906, 1910, 1914)
1986   IMUC 1 pdr & 75 mm Subcaliber Guns in Bore . . . (1906, 1907, 1910, 1914, 1917R)
1987   IMUC&D 75 mm Subcalibur Tubes in 12-inch Mortar Carriage (1904)
1987   IMUC 12-inch Mortar Carriage M1918 (1919)
1988   Hand Tools for Reloading Gallery Practice Cartridges M1903 Rifle (1905)
1989   Tools for Engineer Power Plant Repairs, List (1914)
1990   Decapping & Cleaning Tools for Small Arms Cartidges (1907, 1909, 1912, 1917R)
1991   Coast Artillery Targets and Accessories (1908, 1909, 1913, 1917)
1992   Small Arms Targets and Equipment of Target Ranges (1910, 1914, 1918)
1993   Bench Reloading Tools for Ball Cartridges M1898 & M1905 (1908)
1994   Mobile Artillery Targets and Accessories (1909, 1910, 1914, 1917, 1918R)
1995   6-ton Special Tractor M1917 (1918)
1996   5-ton Artillery Tractor M1917 (1918)
1997   3-ton Truck Chassis MB-1917 (1918)
1998   Target Range Pocketbook for Use with M1903 .30 cal. US Rifle (1908, 1917R)
1999   2-ton Truck Chassis Nash Models 4017-A, -F, -L, (1918)
2000   Observation (Seacoast) Telescope M1908 (1911, 1917)
2001   IMUC 14-inch gun M1909, Turret Mount M1909 (1917)
2002   Ammunition Truck Body M1918 ...(1918)
2003   10-ton Artillery Tractor M1917 (1918, 1919)
2004   Artillery Supply Truck Body M1918 (1918)
2006   20-ton Artillery Tractor M120 HP Holt (1918)
2007   Light Repair Truck Comprising . . . M1918 (1918)
2008   3-inch AA Mobile Gun Mount M1918 (1918)
2009   Artillery Tractor 2 1/2 ton M1918 (1918)
2011   10-ton Trailer M1918 (1918)
2012   1-ton Truck Chassis M1918 (1918)
2015   Handbook of Spade and Firing Platform for 5" Gun Carr. M1917, 6" M1917A &B (1919)
2016   240 mm Howitzer Material M1918 (Schneider) (1918)

2017	155 mm Howitzer Material M1918 (Schneider) Motorized (1918)
2019	Ammunition for 3-inch AA guns M1917 & M1918 (1919)
2020	IMUC 12-inch RY Mount M1918 (1919)
2021	12-inch guns M1895 & M1895MI Breech & Firing Mech. for 12-in RY Carr. M1918 (1919)
2023	Railway Gun Car M1918 (1919)
2024	Railway Cars M1918MI & MII (1919)
2025	Narrow-Gauge Equipment for 12-in RY mortar mount M1918 (1919)
2026	Handbook of Ammunition Car for RY Batteries (1919)
2028	IMUC 12-inch RY Mount M1918
2030	History of District Offices (1920)
2032	History of Trench Warfare Material (1920)
2033	Field Artillery Material AA (1920)
2034	Railway Guns, etc., Vol. 1 & II (1922)
2036	Service Markings of Ammunition and Ammo Containers (1921, 1926)
2038	Field Service, General Duties & Functions (1921)
2042	American Seacoast Material, June 1922
2043	Elements of Mobile Carriage Design (1922)
2044	Descrip & IU Aberdeen Chronograph (1921)
2045	Design Manual Note No. 3, Artillery (Heat Treatment of Castings)(1923)
2046	Elements of Mobile Carriage Design (Supplement) (1923)
2048	Design of Steel Castings (1923)
2049	St. Chamond Recoil Mechanism for 75 mm Gun Carriage M1916MI (1924))
2050	Selection of Material for Guns (1940)
xxxx	Ammunition for Heavy Artillery (Coast Artillery Board) (n.d.)

Journal of the United States Artillery (1892-1922)
& Coast Artillery Journal (1922-1948).

These publications were the professional journals for the artillery and coast artillery. They have a wide variety of news, history, and technical articles on US Coast Artillery. Complete sets are hard to find, but some of the larger libraries have some copies. Issues often turn up at used book sales. The CAJ was originally published by the Coast Artillery School at Fort Monroe, Virginia, and later by the United States Coast Artillery Association. The Journal also published a series of 12 *Gunner's Instruction Pamphlets,* which covered all aspects of antiaircraft, fixed and mobile artillery and submarine mining. In 1948 the Journal was renamed the *Antiaircraft Journal,* which was last published in 1954.

Books, Texts, etc.

Coast Artillery School Publications, Fort Monroe, VA. Here are a few titles I have seen: Coast Artillery Training Bulletins (CATBs). These often went on to become Field Manuals; Army Extension Courses; Coast Artillery Weapons and Material Special Text No. 25, (1933 Edition); Fire Control and Position Finding for Seacoast Artillery (various dates).

Occasional Papers, Engineer School, U.S. Army, G.P.O., Washington D.C. A series of publications by the Engineer School. Most notable in this series related to harbor defense is: *Notes on Seacoast Fortification Construction* by Col. Eben E. Winslow (Occasional Papers No. 61, Engineer School, U.S. Army, G.P.O., 1920). This was one of the first comprehensive publications published on building modern seacoast artillery fortifications. The book had a number of drawings of different coast artillery emplacements. This book was reprinted by the CDSG Press in 1992.

Engineer Mimeograph Series. Another useful series, copies were printed and distributed to various offices, but few were preserved. The collections of these in the Archives have been separated from their drawings and figures. PDF copies of this series is available from the CDSG ePress.

The Service of Coast Artillery, by Frank T. Hines and Franklin W. Ward (Goodenough & Woglom Co., New York, 1910) is also an excellent period introductory reference on American Coast Artillery. The CDSG Press reprinted this book in 1997

Ordnance and Gunnery, by Ormond M. Lissak, John Wiley & Sons, NY, 1908 & 1915
Ordnance and Gunnery, by Lawerence L. Bruff, John Wiley & Sons, NY, 1896 & 1902.
Ordnance and Gunnery, by William H. Tschappat, John Wiley & Sons, NY,
Ordnance and Gunnery, by Earl McFarland, John Wiley & Sons, NY, 1929 & 1932.
Elements of Ordnance, by Thomas J. Hayes, John Wiley & Sons, NY 1938.
   These books were used as text books in the US Military Academy.

Historical Register and Dictionary of the United States Army from its Organization September 29, 1789 to March 2, 1903, by Francis B. Heitman. This has been reprinted several times recently.

Military Service Publishing Company, Harrisburg, Pennsylvania. This company printed a number of texts used by ROTC program and Army Extension Courses. They can be found at used book stores. Examples include: *The R.O.T.C. Manual, Coast Artillery, Basic & Advanced, The Coast Artillery Corps: a Complete Manual of Technique and Material,* and *Seacoast Artillery, Basic Tactics and Techniques.*

Reports of Completed Batteries and Reports of Completed Works

Matthew L. Adams

(Originally published in the *CDSG Journal* Volume 12, Issue 2 (May 1998) pp. 64-68)

Reports of completed batteries (RCBs) and reports of completed works (RCWs) were forms used by the Corps of Engineers to document seacoast fortifications and other buildings related to coast defense. RCBs were in use from 1900 until 1919. RCWs were in use from 1919 until the coast artillery was disbanded in 1950. A time line of the development of RCBs and RCWs is presented at the end of this report. A nearly complete set of RCBs and RCWs for the various harbor defenses are available from the CDSG ePress.

   The antecedent of the RCB was the armament report and sketches described in Sections V and VI, Circular No. 2, Office of the Chief of Engineers (OCE), 1896.(1) The armament report summarized the guns and carriages received at the post and whether the armament was mounted or unmounted. The armament sketches contained a general drawing of the work (of each tier in multi-tiered forts) with each emplacement marked and numbered properly. The sketches also indicated the type of platform, its construction, whether the platform was serviceable or not, and whether the gun was mounted or not. If the gun was mounted, the type of gun and whether it was serviceable or not was also noted. Some of these details can be traced back to Army Regulations in force in 1863.

   Circular No. 2, 1896, was issued in response to the increased clerical load in the Office of the Chief of Engineers from the increased fortification construction activity. In excess of 30 batteries were either completed or under construction and many more were planned at the time this circular was issued (the first would not be transferred to the artillery until a few months after this circular was issued in March).

   When the Spanish-American War started in 1898, fortification construction increased markedly over 1896 levels as Congress appropriated substantial sums for defense. By October 1900, over 125 batteries had been transferred to the artillery. Trying to distill the operational readiness of each harbor defense from the monthly operations reports, armament reports, and armament sketches described in Circular No. 2,

OCE, 1896 (which included statements of financial accounts), must have become increasingly difficult for the staff at the Office of the Chief of Engineers. To make the task of assessing operational readiness easier, the RCB was designed. The RCB form was printed in Circular No. 30, OCE, October 12, 1900.

The major strength of the RCB over earlier forms was its tabular format; a concise summary of the operational readiness of the harbor defenses in any engineer district. In filling out this form, works transferred prior to 1890 (i.e. prior to the completion of any Endicott batteries) only required the first two columns to be completed. The RCB was also to be used for reporting completed range-finding stations, cable tanks, mine casemates, and torpedo storehouses. For each engineer district, the RCB was to be current to October 1, 1900, and it was intended that new works would be added at the bottom of the list as they were transferred to the artillery. Importantly, this was a monthly report.

By order of Circular No. 18, OCE, September 22, 1903, annual armament reports were discontinued. In its place an expanded form of RCB was introduced. In addition to adding the official name of each battery to the RCB, the individual number and name of the manufacturer of each gun or mortar and carriage was to be recorded with the number of the emplacement each gun or mortar and carriage was mounted in (see Army Regulations of 1901, Paragraph 407, as amended by G.O. No. 82, Hq of the Army, AGO, 1902). A number of other minor items were to be recorded as well.

In contrast to its predecessors, these RCBs were to be submitted annually rather than monthly. The shift in frequency was undoubtedly intended to reduce unnecessary paperwork. By the time Circular No. 18, OCE, 1903, was issued, approximately 250 batteries had been transferred to the artillery. Only minor changes were made to the RCB from 1903 until RCWs were created in 1919.

Reports of completed works (RCWs) were created by a circular letter issued by the chief of engineers, Eben E. Winslow, on January 30, 1919.(2) It prescribed that the annual RCBs were no longer required and that all data that the RCBs were intended to provide would be submitted on seven forms referred to as reports of completed works. Furthermore, new forms need only be submitted whenever changes in works made the old forms obsolete, compared to the annual submission of RCBs.

A brief description of the content of each of those seven forms, which changed remarkably little over the next 30 years, follows:

Form	1	all important data relating to an individual battery
Form	2	details of fire control and torpedo structures
Form	3	details of mine wharfs and tramways
Form	4	details of searchlights (a separate sheet for each light)
Form	5	details of electric plants
Form	6	existing Engineer Dept structures of permanent or semi-permanent nature
Form	7	a blueprint of the battery

Four copies of each RCW were supposed to be made; one for each of the following engineer offices: district, military department, divisional, and Office of the Chief of Engineers.(3)

The main difference in the RCW compared to the RCB was the separation of data onto different forms. For example, the details of different electric plants in a harbor defense may have been listed under fort sub-headings in an RCB, which could have been scattered over several pages of a 20+ page RCB from a larger harbor defense. The RCW, on the other hand, consolidated all electrical plants onto a series of Form 5s. This allowed for easier comparison and assessment of material present at each harbor defense.

Reports of completed batteries (RCBs) and reports of completed works (RCWs) were the end products of an increasing clerical load on the staff at the Office of the Chief of Engineers. In comparison to armament reports and sketches of the 1890s, the RCBs of the early 1900s allowed the staff to distill much more easily the operational readiness of fortifications at each harbor defense across the country by listing individual batteries and armament in a table. The RCWs extended this idea by subdividing the information recorded in RCBs into categories (general battery and armament information, fire control structures, searchlights, etc.). RCWs also allowed for greater detail in documenting different elements of coast defenses than the RCB. Both are essential documents in the study of modern U.S. coast defenses, 1890-1950.

Chronological History

1863.

The roots of the RCBs can be found in Army Regulations. Paragraphs 57 and 58, Article XI, of the 1863 Army Regulations state:

57. At all posts with fixed batteries, the position of every gun, mounted or to be mounted, will have its number, and this number be placed on the gun when in position.

58. For every such work a post-book of record will be kept, under the direction of the commander of the post, in which will be duly entered—the number of each gun mounted, its calibre, weight, names of founder and its inspector, and other marks; the description of its carriage and date of reception at the post; where from; and the greatest field of fire of the gun in its position.

1864.

G.O. No. 42, Hq of the Army, 1864, expanded paragraph 57 by specifying the manner in which the guns were to be numbered, i.e. from right to left in a regular series (including unoccupied platforms) starting from the first gun on the left of the main entrance looking out.(4) This is the origin of numbering guns in Endicott and later batteries from right to left facing the enemy.(5)

1896.

Circular No. 2, OCE, 1896, promulgated on March 21, describes the form of armament reports and sketches, as well as the form for the monthly report of operations.

1900.

Circular No. 30, OCE, 1900, promulgated on October 12, describes the original report of completed batteries (RCB).

1903.

Circular No. 18, OCE, 1903, promulgated on September 22, expanded the RCB to include the battery name, and the serial numbers and manufacturers of the guns and carriages, as well as additional information related to fire control stations. It also simplified the format of monthly reports as promulgated in Circular No. 2, OCE, 1896, and Circular No. 30, OCE, 1900.

1904.

Circular No. 11, OCE, 1904, required the addition of information pertaining to motors on carriages to the RCB.

1905.

Circular No. 3, OCE, 1905, required the addition of information pertaining to electrical plants and searchlights to the RCB.

1906.

Circular No. 56, OCE, 1906, dealt with tide gauges and datum points in the RCB.

1909.

Circular No. 39, OCE, 1909, promulgated on November 23, required the addition of information pertaining to projectile and powder hoists to the RCB.

1919.

Circular Letter, 660B-F, issued January 30, 1919, eliminated RCBs and replaced them with seven forms called RCWs.

1921.

Circular Letter (Forts No. 8), 660B-F, issued August 19, 1921. Slight modifications to Forms 4 and 5 of RCW.

1935.

Circular Letter (Forts No. 1, Fiscal Year 1935), 660B- F, issued January 16, 1935. Slight revision and reissue of all RCW forms.

Appendix

The following provides a brief guide to the location in the National Archives and Records Administration (NARA). At the time of this writing, RCBs and RCWs are housed at NARA-I in downtown Washington, D.C.

The following is a list of correspondence file numbers for RCBs (1903-1919) for various harbor defenses in the continental United States, in geographic order:

RG 77, Office of the Chief of Engineers, General Correspondence, 1894-1923, Document File, Entry 103:

Harbor Defense	File No.
Kennebec R./Portland, ME/Portsmouth	48917
Boston	49447
New Bedford/Narragansett Bay	48906
Eastern Long Island Sound	48813
New York (Eastern and Southern)	49105/79501
Delaware R.	49192
Baltimore	48833
Washington, D.C.	49187
Hampton Roads	48916
Wilmington, N.C.	48853
Charleston/Port Royal Sound	48927
Savannah	48827
Key West/Tampa Bay	49231
Pensacola	48855
Mobile	48905
New Orleans	48944
Galveston	66628
San Diego/Los Angeles	48904
San Francisco	49068
Columbia R.	49002
Puget Sound	48920
Guantanamo Bay, Cuba	66283
Honolulu	79705
Manila/Subic Bay	66629
Panama Canal	?????

RCWs can be found in RG 77, Office of the Chief of Engineers, Geographic File, 1918-1945, Classified Files, Entry 1011. Look under file no. 600.914 (the number is derived from the War Department decimal system in general use after 1918) under the harbor defense in which you are interested. Please note that the RCWs for Portland are not located with these records, and have not been located as yet.

Author's note:

I have not seen any RCBs prepared between October 1, 1900 and August 31, 1903. They may be located elsewhere in the Office of the Chief of Engineers correspondence files, or perhaps they were destroyed as redundant when RCBs created as the result of the issuance of Circular No. 18, 1903, were received in Washington, D.C.

Endnotes

1. Circular No. 2, 1896, also describes the form of the monthly report of operations, which were summarized and published at the end of each fiscal year in the Annual Report of the Chief of Engineers between 1896 and 1902.

2. See File No. 660, Box 1, Harbor Defense Geographic Files, 1918-1945, Unclassified Correspondence, Entry 1008, RG 77, Records of the Office of Chief of Engineers.

3. The RCWs located in the Office of the Chief of Engineers are the ones researchers most commonly access.

4. The tangible evidence of the preceding Regulations and this General Order can be found on drawings held in the Cartographics Branch of the National Archives in Record Group 77 (Records of the Office of the Chief of Engineers). A number of armament sketches dated post Civil War can be identified for various posts around the country where the emplacements are numbered right to left, as well as the condition of the emplacement.

5. Refer to Paragraph 407 of the Army Regulations of 1901, as well as G.O. No. 82, Hq of the Army, AGO, 1902.

Annexes and Supplements to Harbor Defense Projects

In the National Archives Record Group 407, Records of the Adjutant General's Office, 1917-, Entry 366, are filed the various Harbor Defense Projects of the 1930s and 1940s. A Harbor Defense Project was a written document which described all existing and projected harbor defense elements, including structures, first prepared in 1932-33. Annexes to Harbor Defense Projects were published to update these projects during 1934-37. These projects and their annexes were utilized by the Local Board Modernization Program Harbor Defenses to prepare their Proceedings 1940-42. Finally Supplements to the Harbor Defense Projects were prepared 1943-44 and updated 1945-46 which superceded the previously issued Annexes and Local Board Proceedings. The supplements detail the progress on the construction of the new 1940s modernization program defenses with descriptions and a set of maps that showed where these new structures were located, the field of fire of the guns, radar coverage, etc. The supplements provide extensive detailed information on all tactical and physical aspects of the harbor defenses on the date of the annex, both existing and proposed, and a number of exhibits detailing the locations of elements.

The supplements follow a format consisting of the following items:

Annex A- Armament.
Annex B- Fire Control.
Annex C- Seacoast Searchlights.
Annex D- Underwater Defense.
Annex E- Anti-Aircraft Artillery (Harbor Defense).
Annex F- Gas Defense.
Annex G- Equipment.
Annex H- Real Estate Required.

These supplements contain specific information on each of the items cited above and contain a number of detailed maps and drawings. Exhibits included fields of fire for the major batteries, tactical organization charts, ammunition allowances, location maps, site maps, cable routings, and various appendices. These provide an excellent place to find very specific information on a given harbor defense during the World War II years.

PDF copies of several of the Projects and Annexes between 1932 and 1940 for 18 major continental United States harbor, the four Alaskan harbors and the Panama Canal Zone defenses are available from the CDSG ePress. The ollection includes copies of all the 1944-45 Supplements to the Harbor Defense Projects.

BASIC RESEARCH IN TEXTUAL RECORDS AT THE NATIONAL ARCHIVES: COAST ARTILLERY FORTS FROM THE ENDICOTT-TAFT ERA THROUGH WWII

Bolling Smith & Glen Williford

There are different reasons to consult the records of the National Archives – subjects may be topical, such as submarine mining, or geographic, such as a particular fort or harbor defense. Those just starting to do research on a geographic basis can find the archives confusing. Entirely too much time is spent re-inventing the wheel - searching for records that someone else has either seen or knows where to locate. It is hoped this article will decrease this, and encourage the sharing of information found in the archives.

This is by NO means a complete guide to the resources of the National Archives. It is only intended as an introduction, a guide to the first level of basic research. Starting and ending dates given for different entries are not absolute, and the files often contain documents dated before or after the years given.

It should also be noted that one of the most important record groups, RG 392, Records of the Coast Artillery Districts and Defenses, has been transferred from the Washington, D.C., archives to the various regional archives.

General

The National Archives largely contains the records of Federal agencies. While both their library and holdings contain some published works, the National Archives is primarily a repository for unpublished documents, divided into several hundred "record groups." For example, the records of the U.S. Army are contained in many different record groups, including those for the Office of the Chief of Ordnance, the Office of the Chief of Engineers, and two for the Adjutant General's Office.

These record groups are subdivided into "entries," sometimes called "series." To assist researchers, there are two types of listing aids at the archives. "MLRs" list each entry in a record group and give a brief description of the contents, showing the size of the entry and where it is located on the shelves at the archives. One complication is that some record groups, for reasons never explained, have more than one entry with the same number, so the descriptive title of the entry and its shelf location are important. In addition, descriptions of the entries are often less than complete, and sometimes so cryptic as to be meaningless.

The second source is the "finding aids." Compiled by the archives to aid researchers and often containing better information on the content of the entries, these may list the contents of an entry box by box. However, finding aids do not cover every entry of every record group, and for some large entries, the finding aids amount to many dozens of notebooks.

At the downtown Washington, D.C., archives, the MLRs and finding aids are in the archivist's office. For military records, this is on the first floor, behind the elevators. The archivists are also located in this room, and can be conveniently consulted.

At the College Park archives, MLRs and finding aids are kept in the glassed-in room off the textual research room on the second floor. For further help, the archivists are in Room 2400, which also contains MLRs and finding aids, sometimes more extensive than those in the research room. A researcher must receive a pass and be escorted to Room 2400, although an appointment is usually not necessary.

Archives I, Washington, D.C. - c. 1894 - 1917

During this period (roughly the Endicott-Taft era), records on forts were usually kept in the general record files of the different departments and branches. Each department used a numerical file number to create and identify separate files. Unlike the later military decimal system, the numbers themselves do not indicate what the file contains, and there is no relation between adjacent numbers or between numbers assigned by different departments. File numbers were simply assigned chronologically as a topic or need arose. Topics are mixed – a number assigned for "New 3-inch battery to be built at Fort Mott" may follow a number about granting leave to a clerk somewhere and precede a file concerning major civil engineering work. Therefore, browsing just does not work. You need to locate a finding aid, index, or some reference giving the precise number of the file needed and you must then request the box that has that file in it. Within the file the individual letters or enclosures (designated by a slash: file 37455/1, 37455/2, etc.) are usually arranged chronologically.

Multiple files may exist on a single subject. Files begun before a fort was named and listed simply by the geographic location (i.e. Great Gull Island or Clark's Point) may not be consolidated with files created after the fort was named (i.e. Forts Michie or Rodman).

At times files were set up in "waves." Some new appropriation or initiative would create a new generation of fire control stations, mine casemates, or searchlights – and the engineer office would allocate 28 new numbers, one for each harbor defense. They did not go back and use the last number for that port and that item. Often after 3 - 5 years without a new entry, a number was not continued as an active file, and a new number was created if the topic came up again. Thus there may be four or five major files on subjects like "position finding," "submarine mining," or "electrification."

The hierarchy of how the information was organized also varies. Some topics were the natural concern of the harbor defense. For example, general defense plans and projects were always created and filed for the "Defense" (e.g. Portland), and not an individual fort (there is no defense project for just Fort Williams). On the other hand, individual gun batteries were appropriated, designed, approved for construction, and built. Thus there is no file for "Batteries in Portland, Maine," or even for "Batteries at Fort Williams," but rather there is a file for each individual battery – Battery DeHart (and from the above example – "1897 battery for 10-inch guns at Fort Williams"), etc. Less frequent were topics handled only at the regional level and the service as a whole – though you can sometimes find a summary report in a specific file – "status of modifications for chain hoists," that has an entry for each and every location. Thus to try and cover the information on a fort you also need to look up the individual batteries or topics about the fort, and also look a level or two higher into the district or region of the command responsible. Fortunately, the index cards generally recognize this and multiple entries of a file exist on numerous cross-referencing cards.

One key to using the index cards is to be creative. Files on batteries are indexed by the battery name, but files created before the battery was named may be filed under the fort name. Similarly, other names for forts should be checked, such as the name of the geographic location. In addition, the cards for the harbor defense should be consulted. As an example, for Battery Bradford, on Fort Terry, the index cards for "Bradford," "Terry," "Plum Island," "Long Island Sound," and "New London" (the original name of the harbor defense) should be requested.

These record systems usually have file card indexes, large collections of up to several hundred boxes of file cards. Arranged alphabetically, they list a topic, and then have listed on them (at first handwritten, later typewritten) a short (sometimes cryptic) description, sometimes the name of the author, date, and file/letter number. For some forts there may be 30 double-sided file cards holding hundreds of records. These can be used to identify specific records to look at – or also as a shortcut for more productive browsing. If file #17652 shows up a dozen times in 1898 - 1901 it is likely a major file with specific letters about a key battery or subject you are interested in. Often 10 - 12 file numbers will provide a good coverage of a fort.

It does take a little knowledge and sometimes some dead-ends to catch on to the alphabetical protocol. Forts and batteries are listed by name – and not under "F" for fort or "B" for battery. However, all specific gun sizes are listed numerically under "G" for guns or "M" for mortars.

At times files were destroyed or discarded – sometimes a note was inserted in the file system, other times not. Within files individual letters and enclosures are often missing – in other words the numbers "skip." Oversized maps or plans and often photographs were usually, but not always, removed and filed separately. However, they are usually noted on the document "map enclosure 110-19-4," and as this same numerical system is used at the cartographic archives at College Park, often a map or plan may be found in Cartographics with this number. Photographs do not seem to have fared as well – the location of many remains a mystery.

Archives II, College Park, MD - c. 1917 - 1945

Archives II generally contains military records dating from the adoption of the War Department Decimal System, around the start of World War I. This system, whereby files are grouped by topic, described in more detail in a following section. While this facilitates research on particular topics, it means that files concerning different topics on a particular fort are not located together.

There are, however, exceptions to this classification system, most notably the Records of the Office of the Chief of Artillery/Coast Artillery, 1901 - 1917, in RG 177; the files of the Construction Division in the Records of the Office of the Chief of Engineers; and most importantly the harbor defense files in the Records of the Office of the Chief of Engineers. The first of these is arranged by file number, the last two alphabetically, either by fort name or harbor defense.

Whether in Archives I or II, keep in mind that the researchers who preceded you may not have put the records back exactly in their proper order. Before you conclude that a file or an enclosure does not exist, be sure to check the box to make sure it was not re-filed out of order.

Specific Record Groups

Record Group 77 – Office of the Chief of Engineers

Probably the most important records for coast artillery posts were those kept by the Corps of Engineers, who built and maintained the tactical structures – batteries, mine structures, fire control stations, searchlights, radar, and communications buildings. Because much of this involved detailed plans for weapons and fire control, the engineers routinely documented much of the decision-making process that led to the construction of the tactical structures. In addition, in December 1941, the Construction Division was transferred from the QMC to the Corps of Engineers, who took over responsibility for constructing and maintaining non-tactical structures. Because of this, many of the records of these buildings passed to the engineers, and are therefore filed in RG 77 rather than RG 92.

Archives I

The primary series of engineer records is Entry 103, General Correspondence, 1894 – 1924, arranged by file number. Entry 99 is the name and subject index cards.

Entry 220 is four boxes of binders on harbor defenses maintained in the Office of the Chief of Engineers. The binders, generally one for each harbor defense, contain a cumulative record of engineer activity at the forts until the early 1920s. This key resource should not be overlooked.

Entry 225 contains 15 boxes of miscellaneous papers dealing with harbor defense. The most important are numerous plans by the Board of Engineers for the artillery, mine, searchlight, and land defense of different harbor defenses. In addition, there are reports of completed batteries (RCBs) dating between 1910 and 1919 for most harbor defenses. See http://members.aol.com/vexillarii/Main.html for a listing of the contents of this entry.

Archives II

The key files of the Corps of Engineers dealing with fortification after WWI are arranged geographically in two entries, thankfully separate from the corps' other records. Entries 1007 (classified) and 1009 (unclassified) contain the Harbor Defense Geographic Files. The most interesting files tend to be in 1007 (including the RCW files), but both entries should be searched carefully. The files are organized alphabetically by harbor defense, as listed in the RG 77 finding aids at Archives II.

Since the Construction Division and its responsibilities for non-tactical buildings and systems – barracks, water systems, wharves, etc… - was transferred to the Corps of Engineers in December 1941, the vast bulk of the records compiled and preserved by the QMC are now located in RG 77. There are two primary types of records. Entry 391 contains Construction Completion Reports, 1917 - 43. These often very detailed narrative reports of QMC construction or repair projects were submitted by constructing quartermasters. While they do not exist for all forts or all projects, they do contain a wealth of information on the projects that are covered, often including B&W photos. Entry 391 is arranged alphabetically, and the individual box numbers for each fort are shown in the finding aid.

Entries 393 and 394 contain the Historical Records of Buildings at army posts. The *Coast Defense Journal*, Vol. 16, No. 2 (May 2002), pp. 29 - 42, describes these forms in more detail - they are large cards or sheets with data on each non-tactical building, usually with a photograph and often a floor plan. These are the best single source for non-tactical structures. Entry 393 covers posts still active in 1942, while Entry 394 covers those abandoned by that date. The files are arranged alphabetically by fort, and the finding aid lists the boxes for each fort.

Record Group 92 – Office of the Quartermaster General

The QM Corps (QM Department before 1912) were responsible for non-tactical building construction before December 1941, as well as functions analogous to public utilities, such as roads, fences, wharves, water and electric supply, as well as transportation and provision of uniforms, food, and fuel. Many of these records are vital to understanding the history of the forts and of structures such as barracks. All types of watercraft were also an important quartermaster function.

Archives I

Entry 89 contains correspondence, 1890 – 1914, by file number. Entry 88 contains the record cards and Entry 84 the name and subject index to Entry 89. These documents relate to the wide range of quartermaster responsibilities.

Record Group 94 – Adjutant General's Office

The Adjutant General's Office was the record keeper for the army, as well as managing a number of specific functions, such as personnel. Because they kept records of much military correspondence, the AGO files are relatively large.

Archives I

Entry 25 contains correspondence between 1890 and 1917, arranged by file number. Entry 27 contains the name and subject index cards. Entry 61 contains the Returns of Army Departments, 1818 – 1916. Entry 66 contains Returns of Military Organizations, from the "early 1800s" to 1916. An interesting series is Entry 280, the reports of the National Land Defense Board, 1907 – 1915. Entry 281 is the index to Entry 280.

Record Group 407 – Adjutant General's Office, 1917 - Present

Archives II

The records of the Office of the Adjutant General after 1917 are contained in Record Group 407, at Archives II. The "Central Decimal Files" are in Entry 37, which is divided into eight entries by letter suffixes, such as 37H, which contains classified and unclassified correspondence between 1926 and 1939. Other entries in the 37 series are for different dates, for bulky files, and in some cases for separating classified and unclassified files. These entries are arranged by the decimal system, but decimal file 662, "Gun and Mortar Batteries," is subdivided by location, and may contain numerous documents, some of which are duplicated in other record groups and some of which are not. The interwar correspondence of the War Plans Division found here is particularly interesting.

Harbor defense annexes and supplements of the 1930s and 40s are filed in several places, but the most complete set is found in Entry 366, filed alphabetically. The box numbers for each harbor defense are shown in the finding aid.

Record Group 156 – Office of the Chief of Ordnance

These records, by their nature, are not primarily arranged by location. However, by use of the indexes, files relating to specific forts or harbor defenses can be found.

Archives I

The primary source for the Endicott-Taft era is Entry 28, correspondence between 1894 and 1913, which is arranged by file number. A smaller series, Entry 29, contains additional correspondence between 1910 and 1915. Entry 26 contains the name and subject index to Entry 28 and some of Entry 29. Entry 34 indexes the portion of Entry 29 between 1912 and 1915; however, it is located at Archives II. Appendix I to the archives finding aid has an abbreviated file number list. While only a small portion of the total files are represented, it is a good place to start compiling interesting numbers to have pulled, and may save an entire call cycle at the reference room. Most master file numbers for coast artillery forts seem to be listed in this finding aid.

Archives II

Entry 36 contains general correspondence between 1915 and 1941. Entry 39 is smaller and contains confidential correspondence between 1917 and 1940. Like other records of the period, these entries are arranged in the War Department Decimal System, by subject rather than by location. Entry 34 contains the name and subject indexes to Entry 36 in three subseries, divided by date. Entry 38 contains the name and subject index to Entry 39.

RG 156 (Records of the Chief of Ordnance), Entry 712, at Archives II, College Park, MD, consists of 13 boxes of 3 x 5 cards. Each seacoast gun and carriage has a separate card, filed by model and serial number. In addition, many other pieces of equipment also have cards. The last entries on the cards were made around 1946.

The Ordnance Department tracked the location and/or disposition of each gun and carriage. The omissions, if any, are few. (The cards include those for a number of one-of-a-kind items, such as the 10-inch breechloading mortar, along with equipment from plotting and ammunition cars to star gauges.) The cards are organized by size and model year, often subdivided by manufacturer. Items that had been modified are usually filed separately; for instance, M1895 guns are separate from M1895MIs, while M1895A1s and M1895MIA1s are filed together. Much space is devoted to railway guns, carriages, and such equipment, as well as a large number of trench mortars. Other cards seem to indicate the numbers of certain items ordered during 1941 or 1942, and there are cards filed by harbor defense, which seem to show what remained in 1946.

As always, the accuracy is not guaranteed. There was much correspondence between the Ordnance Office and the forts, discussing the serial numbers and model designations of various pieces of ordnance.

It seems, however, that the Ordnance Office records usually proved correct. In a minority of instances, the first date on the card is July 1, 1917, which seems to reflect the date the particular cards were started, rather than the date the item was acquired.

The cards show the locations and movements of the items, but with little or no comment on the reasons behind the moves. A card may note that a gun was shipped from Fort Monroe to Watervliet Arsenal, but not indicate why. In other instances, especially when a gun was relined, the card may contain a brief notation to this effect.

The Ordnance Department devoted many man-hours to maintaining this file, and it is still invaluable for tracking specific pieces of ordnance or resolving discrepancies on RCWs.

Record Group 165 – War Department Special Staffs and Boards

Archives I

Entry 165 has the records of special army boards, usually those containing representatives of more than one department. Entries 518 and 519 hold the correspondence of the Board to Revise the Report of the Endicott Board (the Taft Board), 1905 - 1906. While much is of a general nature, specific proposals and recommendation for individual harbor defenses are in these holdings. For the Endicott Board, similar records are in RG 77, Entry 485. For both the RG 77 and RG 165 records the entries are small (only a few inches of material), so it is best to just request the entire entry and look for the harbor of interest rather than try to locate a finding aid.

Record Group 177 – Offices of the Chiefs of Arms

Archives II

The general correspondence records of the Office of the Chief of Artillery (1901 - 1907) and Chief of Coast Artillery (1907 - 1918) are arranged by file number in Entry 4, with subject index cards in Entry 1. In addition, a file-by-file list of the contents, searchable by computer, with box numbers, has been prepared. Anyone interested can request it by email at bollingsmith@hotmail.com. There are separate files for fire control, submarine mining, and artillery defense plans for most harbor defenses, as well as other files specific to particular locations.

The correspondence of the Office of the Chief of Coast Artillery after approximately 1918 is in Entry 8, arranged in the War Department Decimal System. Entry 7 contains the alphabetical index to Entry 8. Of particular interest in Entry 8 are the inspection reports filed between decimal 333 and 333.14. These contain a number of different types of reports, but each category is filed by harbor defense, and they constitute a valuable record.

There is considerable overlap between different decimal files, and in some cases, different levels of command filed the same inspection under different decimal file numbers. Each decimal file, in turn, is divided into files for each harbor defense, which were assigned numbers in alphabetical order. For instance, /15 indicates reports for the HD of Manila and Subic Bay. In addition to the designated harbor defenses and forts, reports are included for some Army Reserve and National Guard regiments, as well as for some corps area reserve components. \

In general, these inspections are divided into five groups, decimal file numbers 333, 333.1, 333.12, 333.13, and 333.14.

Decimal file 333 is relatively thin and largely contains the results of "annual surveys" by higher commands. These often mention the inspection only briefly, with excerpts of particularly significant findings and some descriptions of corrective action taken.

Decimal file 333.1 includes overall annual inspections by the chief of coast artillery, departmental (later corps area) and coast artillery district commanders, and their staff, to include the Inspector General's

Department. These cover a wide range of topics, from appearance of personnel to condition of buildings and materiel to shoefitting tests. Recommendations could be as significant as that recommending that Fort Greble, RI, be disposed of. (The recommendation was not approved.)

Decimal file 333.12 deals largely with tactical inspections by coast artillery district commanders. Considerable emphasis was placed on tactical problems used for training exercises in the harbor defenses. Also included in this file are technical inspections, which concentrated on tactical equipment. Some inspections under this file were excruciatingly detailed, such as one 1919 inspection of Fort Kamehameha, which cited with disapproval, among many other things, "1 man with safety pin on shirt to replace button," and "Company Commander saluted inspector."

Decimal file 333.13 includes training inspections. These evaluated training programs, both for assigned units and for such outside units as ROTC, CMTC, National Guard, and Organized Reserve. In a few instances (Manila and Subic Bays, 1931), inspections in this file are essentially tactical.

Decimal file 333.14 contains the largest groupings, the Signal Corps inspections. These annual inspections contained a detailed narrative report and a lengthy (often more than 20-page) form listing EVERY single Signal Corps item on the post. Items such as telephones are listed, as are hand tools, tacks, and even, in one instance, a teakettle. The full form was not filled out every year; inspectors were allowed to submit only the pages with changes. In general, the entire form was usually filled out around every five years, while the number of pages submitted in the intervening years varied considerably. The post commander submitted his response to the inspection report, as did everyone in the chain of command.

When a commander or an inspector recommended changes or additional equipment, the correspondence could be lengthy, although often ending with the statement that no money was available. These reports describe the actual condition of the coast defense communications systems, and are a useful antidote to manuals that only describe the ideal systems. Obviously, these are of particular value to those interested in technical equipment.

WAR DEPARTMENT DECIMAL FILE SYSTEM
Bolling Smith

Just before World War I, the army adopted a decimal system for its files. The initial manual was published 1914, based on the Dewey Decimal System then in use for libraries. Although some changes were made, in general the system remained relatively consistent. The 1943 edition of the *War Department Decimal File System,* which replaced the editions of 1914 and 1918, is over 400 pages, including listings and index, and contains the clear statement "THE ADDITION OF NEW NUMBERS OR DECIMAL SUBDIVISIONS IS PROHIBITED." (Emphasis in the original). Decimal subdivisions other than those in the 1943 manual do exist in earlier files, and probably in later ones also. For example, radar was not even mentioned in the 1943 edition.

Classification of documents was subjective. Different departments might classify documents differently depending on their particular interests, or perhaps simply depending on who made the classification. Also, documents might contain subject matter relevant to several classifications. The archives contain many cards indicating documents had been refiled under different numbers.

Nonetheless, a general description of the system may be of some assistance to researchers. Files were divided into nine general classes:

000-General
100-Finance and accounting
200-Personnel
300-Administration
400-Supplies, services, and equipment
500-Transportation
600-Buildings and grounds
700-Medicine, hygiene, and sanitation
800-Rivers, harbors, and waterways

The alphabetical index to the 1943 edition is 301 pages long, and relatively complete ("white mice...454.7"). Those interested in earlier files may wish to consult an earlier edition, as some numbers were changed. The listings below *from the 1943 edition* pertain to coast defense. In general, references to subjects not related to coast defense have been deleted.

320: ORGANIZATION OF THE ARMY
321-ARMS OF SERVICE AND DEPARTMENTS (Organization and reorganization, duties, functions, policies, etc.) (Coast Artillery Corps, etc.)
322-ORGANIZATIONS AND TACTICAL UNITS (Regiments, batteries, commands, etc.)
323-GEOGRAPHICAL DIVISIONS
    323.3-Military departments and divisions (forts, etc.)
353-TRAINING
    353.1-Target practice
    353.4-Firing (Ballistics, circuits for seacoast, proof firing, salvos and salvo points, etc.)
370: EMPLOYMENT, OPERATION, AND MOVEMENT OF TROOPS
    370.3-Land and marine tactics
    370.31-Submarine mine defense
    370.32-Harbor and coast defense
453-RAILWAY EQUIPMENT
    453.1-Railway cars for guns
    453.2-Railway cars for howitzers or mortars
470: AMMUNITION, ARMAMENT, AND OTHER SIMILAR STORES
    470.2-Dummy armament (drill cartridges, dummy cannon, carriages, cartridges, powder bags, projectiles, etc.)
    470.3-Searchlights and outfits
    470.4-Turrets
    470.5-Armor plate and navy armor.
471-AMMUNITION
    471.3-For seacoast guns
    471.31-parts and accessories. (Accessories, balancer sets, cable, carbons, clutches, controllers, curtains, mirrors of glass or metal obturators, projectors, shutters, stops, etc.)
    471.5-Powder and powder charges. (6-pounder-16-inch gun; maneuver, smokeless, black, saluting, igniting, and shrapnel powder, etc.)
    471.8-Parts and accessories
    471.81 Components. (Bands and caps for projectiles, covers, plungers, tracers, etc.)

471.82-Fuzes, adapters, and boosters

471.83-Primers

471.87-Containers for ammunition.

471.88-Ammunition service

471.9-Miscellaneous

471.93-Ammunition for antiaircraft guns.

471.942-Rockets for antiaircraft guns.

472-CANNONS OR GUNS

472.1-Mobile guns (Subcaliber, siege guns, 12 and 16-inch guns, and over.)

472.2-Howitzers (12 and 16-inch, and over.)

472.3-Seacoast guns (Subcaliber, and 3 to 16-inch, and over)

472.4-Mortars

472.5-Automatic and machineguns.

472.6-Navy guns for submarine chasers and other purposes.

472.8-Parts and accessories.

472.81 Parts (Blocks, breech mechanisms, castings and forgings, firing mechanisms, hoops, jackets, lanyards, safety devices, sleeves, tubes, etc.)

472.82.Accessories (Cleaning devices, covers, rammers, sights, sponges, staves, subtarget gun machines, etc.)

472.9-Miscellaneous cannon

472.93-Antiaircraft guns

473-CARRIAGES, MOUNTS, AND TRIPODS

473.1-For mobile guns

473.2-For howitzers

473.3-For seacoast guns

473.4-Mortar carriages

473.5-Mounts for machine guns.

473.6-Mounts for navy guns.

473.8-Parts and accessories.

473.81-Fixed parts (base ring, chassis, cradle, scales, shields, top carriage, racer, etc.)

473.82-Mechanisms

473.83-Electrical equipment, circuits, conduits, magnetos, telephone, terminal boxes, wiring, etc.

473.84-Spare parts.

473.85-Sights for carriages

473.86-Caissons, covers, limbers, etc.

473.864-Observation towers.

473.865-Covers for carriages.

473.866-Platforms, etc.

473.87-Braces and brackets for firing materials, fuze setters, lamps, range finders, sights, telautographs, telephones, etc.

473.88-Carriers and containers (armament chests, gun carriages, howitzer carriages, mortar carriages, etc.)

473.9-Carriages for miscellaneous guns (For antiaircraft guns, subcaliber guns, etc.)

476-LAND AND SUBMARINE MINES AND EQUIPMENT.

476.1-Equipment for mining

476.2-Cables and equipment

476.3-Supplies and tools (Boat telephones, cable cutters, drying ovens, supply list, etc.)

477-SUBMARINE NETS AND SIMILAR APPLIANCES (Anti-submarine devices, automatic torpedo destroyers, destroyers, etc.)

600: BUILDINGS AND GROUNDS

600.05-Name or designation; naming (Naming posts)

620: BARRACKS AND QUARTERS

621-BARRACKS FOR ENLISTED MEN

623-QUARTERS FOR NONCOMMISSIONED OFFICERS

625-QUARTERS FOR COMMISSIONED OFFICERS

630: POST BUILDINGS

631-ADMINISTRATION, DRILL, AND RECREATION BUILDINGS

660: FORTIFICATIONS

660.1-Foreign

660.2-Coast and antiaircraft defense, and land defense works

660.3-Defense sea areas, mine areas.

660.4-Protective installations (Bombproofing, gasproofing, etc.)

661-FIELD FORTIFICATIONS

662-GUN AND MORTAR BATTERIES

663-MINE STRUCTURES

664-CABLE STRUCTURES

665-FIRE CONTROL INSTALLATIONS

665.01-Reports on, reports and progress.

665.02-Fire control diagrams

665.1-Stations.

665.11-Field of view from

665.12-Protection for

665.2-Commanders' and other stations

666-SEARCHLIGHT STRUCTURES

United States Coast Defense Sites after 1950

Joel Eastman, Bolling W. Smith and Mark A. Berhow

A number of coast defense sites (both from the pre- and post-1890 eras) had been "abandoned" by the Army as active defenses as early as 1928 (for example those at the Mississippi River, Mobile Bay, Tampa Bay, Savannah, Port Royal Sound, Cape Fear River, Potomac River, and Kennebec River). Many of these reservations were reclaimed for use during WW II. A new program of seacoast defense construction was begun in 1940. With the tide of World War II shifting toward the Allies after 1942 and the demands for the production of war material for the mobile Army, Navy, and the Air Corp, the implementation of the 1940 Modernization Program for harbor defenses was slowed. While the construction of structures could keep pace with the original plan, the manufacture of weapons and their accessories could not. In response to these pressures, the 1940 Program was scaled back. By the war's end, the modernization program had completed nearly 200 modern and modernized batteries in the continental United States at a cost of $220 million. This represented about one-half the number of gun installations projected in the 1940 Program, but it was still the most powerful set of coast defenses in the nation's history.

The development in military tactics and technology during World War II brought about massive changes in the concept of coast defense. Air power and naval forces replaced breech loading rifles, reinforced concrete and submarine mines. By the end of the war, all except a few 90 mm batteries were placed on caretaking status, the weapons of the older batteries were being scrapped, while work on a few new batteries continued. By 1949, all remaining army coast artillery was deactivated and the last of the big guns were being scrapped (except for a few special cases). In 1950, the remaining harbor defense commands were disbanded and the Coast Artillery Corps was abolished as a separate branch. Its remaining units, all anti-aircraft artillery, were combined with the Field Artillery. After 150 years of being one of America's military prime missions, the building and manning of permanent coastal fortifications was no longer a part of American military policy. Meanwhile, the responsibility for a limited harbor defense mission was transferred to the U.S. Navy. The seacoast forts and reservations were now available for other purposes. The Army retained some and the Navy took others for training and for use in its new role of harbor defense.

The first large-scale transfer of harbor defense properties from the Army began in 1947 and continued through the mid-1950s. These first rounds of deposals were those reservations that could not be converted to other uses by the various branches of the military. First, other federal agencies had an opportunity to claim all or portions of the former coast artillery sites. Those not transferred were turned over for disposition to the General Services Administration (GSA), who offered them to state, county, and local governments, and finally to private citizens. Before selling or transferring these former forts, the Army either returned to its depots all usable equipment or auctioned items in lots to the public or destroyed such items as powder on site. Many of the smaller, independent plots of land which had been leased or purchased for fire control and searchlight positions were returned to original owners or otherwise sold if the original owners did not want to pay for the Army's "improvements" to their lands.

Another round of general military reservation closures began in the early 1970s throughout the Department of Defense to reduce basing costs. Several large former harbor defense sites, including some of the military reservations around San Francisco, New York, and Pensacola. Given the large size and value of the properties in these three area, Congress passed several laws that directed the ownership of these former forts to be transferred to the National Park Service (NPS) to form parts of three new recreation areas—the Golden Gate, the Gateway, and the Gulf Islands. Several base closure commissions in the 1990s recommended the closure and transfer of other former harbor defense sites including the Presidio of San Francisco, Fort Wadsworth on Staten Island, and Fort Trumbull in New London. The latest round of closures in the late 2000s brought the closure of Fort Monroe, Virginia, parts of which were designated a national historic monument in 2011. Of the all the original forts and military reservations used by the United States for harbor and coastal defenses, only a handful remain in military hands in 2013 — Fort Story, VA, Fort Hamilton, NY, a large part of Fort Rosecrans, CA, Fort Kamehameha, HI, Fort Hase, HI and a few other sites.

Many of these former coast artillery sites are now administered by other national agencies (such as the National Park Service), state agencies, and local governments as parks and recreation areas, since they inevitably have scenic river or ocean views. Depending on how diligently the GSA protected the sites, and the length of time it took to dispose of them, some sites and structures survived in excellent condition, while others suffered at the hands of salvagers and vandals. Sadly, both the local and national government agencies sometimes destroyed historic structures, either because they were considered dangerous and unsightly or expensive to maintain or because they simply did not fit the recreational concept of these new parks.

The National Park Service continues to struggle to preserve and interpret the many sites it has acquired, often hampered by limited funding and by the perception that while the First, Second, or Third System forts were historic, the more modern concrete fortifications were not. Some state and local parks also undertook efforts to preserve and interpret their sites, but none had the resources to undertake full preservation, restoration, and interpretation of all the structures in their care, especially the Endicott-Taft and WWII batteries, buildings and fire control systems. The almost complete scrapping of disappearing guns, mortars, and barbette guns from these modern concrete batteries also lessened the interest in interpreting these sites, as they were perceived to be beyond any restoration effort. For almost a generation, the concrete fortifications survived, if at all, in a state of benign neglect. In more recent years, as the Endicott-Taft and WWII batteries have aged, and as interest in coast defenses has risen, efforts have been made to preserve and even restore the fortifications and other structures that have survived. Outstanding examples of these efforts are the Golden Gate National Recreation Area's Fort Baker, the City of Los Angeles' Fort MacArthur Museum, Oregon's Fort Stevens State Park, Washington State's Fort Worden, Fort Casey and Fort Columbia State Parks, Pinellas County's (FL) Fort DeSoto County Park, Fort Sumter National Monument's Fort Moultrie, New Jersey's Fort Mott State Park, The City of Boston's Fort Warren, Maine's Fort Knox State Park, and Rhode Island's Fort Adams State Park.. At the beginning of the 21st Century, new efforts are underway to preserve, restore, and interpret the United States' seacoast defense legacy at several national, state, and local parks around the United States, raising the hope that steady decline and disappearance of these structures will be slowed.

Today, among the best-preserved seacoast fortifications are the remaining Third System (and earlier) works, which played prominent roles in the American Civil War as military objectives, fortified strong points, and prisons. Only a few Third System forts have been destroyed; the vast majority has been turned into historic parks, and they are protected as historic landmarks from future destruction. While these brick and stone forts are generally protected, the cost of upkeep is high and many have deferred maintenance needs. For example, the major problems facing the forts around New Orleans in repairing and stablizing the damage resulting from Hurricane Katrina in 2005 is still ongoing in 2013.

While many of the Endicott-Taft batteries are now located within parks, they have not been accord the same level of protection or care as earlier forts. Most are considered to be, at the worst, a legal liability or at best, an eyesore to the park. Park administrations have built on top of them, fenced them in, buried them, and destroyed them in an effort to remove these structures from interfering with the park's primary mission of providing recreation space. If the owners of the site don't inflict damage to them, then vandals do by smashing doors and windows, dumping trash, and setting fires so after a while they have turned these structures into derelicts. Those structures that have been left alone have suffered from nature's own attack, particularly those located in northern states where the freeze-thaw cycle works on the concrete and brick, and everywhere the rusting of rebar and other metal work has hastened deterioration. Vegetation and tree roots have also played their part, causing some structures to collapse. Finally, as seacoast defenses the effect of the ocean's waves and spray have reduced some structures to rubble. While most gun emplacements have been constructed in such a way to resist these attacks, those supporting wood and cement-plaster tactical structures have largely collapsed, and even brick structures have been damaged or destroyed by vandals. Non-tactical structures, particularly officers' quarters, have survived at many parks and government-owned sites through adaptive reuse, but at some former posts such structures have been completely removed, leaving only the more stubborn concrete emplacements.

Many World War II-era concrete batteries and tactical structures survive in very good condition due to their relatively recent construction and to the high quality of their cement work. However, many have been sealed, some buried, a few demolished, and other used as foundations for homes. Only a few are open for public visitation, and only one World War II battery, the Golden Gate Recreation Area's Battery Townsley on the Marin Headlands is being restored. High economic values for oceanfront property have been key driver in the destruction of many of these World War II structures, even though the structures themselves were in excellent condition. In some instances, adaptive reuse has given these structures a new lease on life, although with varying results as far as historic preservation. Steel radar towers were taken down soon after the war and steel observation towers are being removed as they age. Some concrete towers have been demolished as eyesores and attractive nuisances. While these towers are important sites, their days are numbered, and substantial infusions of money will be required to preserve many examples. World War II theater-of-operations buildings were almost all removed after the war, while many mobilization buildings still survive, particularly on military controlled sites, but even those are being demolished after 50 years of use.

Public interest in the history of American coast defenses has grown substantially during the last 60 years as people discover this fascinating aspect of American military history from visiting former military posts turned into public parks and lands. The publication of *Seacoast Fortifications of the United States: An Introductory History*, by E.R. Lewis in 1970 was a pivotal event, giving the public and park personnel a well-documented interpretive history of American coast defenses. A group of coast defense enthusiasts held the first national conference in 1978, and they organized the Coast Defense Study Group (CDSG) in 1985. The CDSG's annual conferences, *Journal, Newsletter,* web site, and reprints of key coast defense books have played important roles in fostering interest in the history of American coast defense and assisting both the public and park staffs in understanding the fascinating history of these defenses and to interpret their surviving elements. Finally, the very nature of these fortifications have played a role in growing public interest since they were designed to withstand the impact of naval artillery, these massive structures have been able to withstand both the natural climate and economic development longer than other military features from the same periods. The mere fact that these structures which incorporated the leading edge of technology of their times are still standing 60 to 150 years after their effective use ended, draws the public to study them and to seek to understand they purpose and history. It is hoped that this interest will translate into efforts to preserve and restore these sites for current and future generations.

A SHORT HISTORY OF THE
COAST DEFENSE STUDY GROUP
Terrance McGovern

Today, the Coast Defense Study Group (CDSG) is a tax-exempt corporation dedicated to the study of seacoast fortifications. The CDSG promotes and encourages the study of coastal defenses, primarily but not exclusively those of the United States of America. This study of coast defenses and fortifications includes their history, architecture, technology, strategic and tactical employment, and evolution. The goals of the CDSG include educational study, technical research and documentation, interpretation of sites, and providing assistance to others interested in the study of coast defense. Our key missions include the preservation of coast defense sites, equipment, and records, and conducting chartable activities that promote this preservation. We endeavor to achieve these goals by publishing the *Coast Defense Journal* and the *CDSG Newsletter* on a quarterly basis and by holding annual conferences at various U.S. harbor defenses and special tours to coast defense sites around the world. We have also formed two special committees further our goals. The first is the *CDSG Press* to reprint key books on coast defenses and provide copies of past *CDSG Conference and Tour Notes* to allow our members access to these important documents. Secondly, we have created the *CDSG Fund* to allow our members to make charitable gifts for the preservation of coast defense sites, equipment, and records for use by current and future generations. Finally, we seek to educate non-members about coast defenses through our own website at cdsg.org.

As with all volunteer organizations, these activities and services did not appear overnight. It took 25 years of effort and the input of many of our roughly 500 members. The roots of the CDSG can be traced back to the two people that founded the organization -- Bob Zink and Glen Williford at the annual meeting of the Council of Abandoned Military Posts (CAMP) in Charleston, South Carolina in 1977. CAMP's focus on a wide range of military history resulted in its annual meetings being a "marathon" of field tours that allowed for only limited time to explore each site visited. Bob and Glen lamented that a more ideal tour would be a group that would focus on visiting and photographing coast defenses. At that time they could identify a handful of folks with similar interests, and thought it would be good to have their own occasional meetings and regular communications forum to help share information. It was from these discussions that the idea to form an organization to study coast artillery separate from CAMP was born.

Bob Zink took the lead in gathering a small group for a first meeting in July 1978 in New York, dubbed "St. Babs I" (after St. Barbara, the patron saint of artillerists). Attending were Glen Williford, Bob Zink, Ed Jerue, Nelson Lawry, Charlie Robbins, Tom Hoffman (Sandy Hook NPS), and Russ Gillmore (Fort Hamilton Museum). Also, attending part of meeting was Terry Hahn, Jack Fein and Ray Lewis. This meeting established a basic format for our annual conferences, with field trips during the day to coast defense sites and discussions about fortifications at night. Also created at the meeting was the "Coast Artillery Routed Exchange" (CARE) -- the organization's first embryonic "newsletter." This was basically a single large envelope of various coast artillery documents that was mailed from member to member. Each person would in turn take out the old material they had offered to share, copy or note the information added by other members, and try to place something new in the packet. This service lasted several years, with the addition of Dave Cameron and David Hansen to the routing.

The next St. Babs meeting was held six years later in 1984, when the group again visited the harbor defenses of New York. At that meeting the membership was expanded and an organizational structure was adopted. From this point on we began having regular meetings each year and started the publication of a more formal newsletter, *CDSG News*, under the editorship of Charlie Kimbell. The Coast Defense Study Group was informally organized and named during the last evening's discussion at the third "St. Babs" gathering held at Fort Monroe, Virginia in 1985. Appointed directors at this time were Bob Zink and Glen Williford. By the Portland, Maine meeting in 1986, the CDSG was up and running. At the Boston, Massachusetts meeting in 1988 the editorship of the *CDSG News* was transferred from Charlie Kimbell to Bob Zink. At the same time, Elliot Deutsch took over the duties of membership secretary and treasurer from Charlie Kimbell. In 1989, we held our first "overseas" tour to the defenses of Bermuda. The group was formally organized with the adoption of its first by-laws in January 1989 that established a three-person board of directors with rotation via direct election of one director each year. The first elected directors were Alex Holder, Mike Kea, and Terry McGovern.

The 1990s saw a steady growth in our membership -- total membership at the beginning of the decade was 143, and by 1999 it had risen to 420. At the 1990 annual conference at the defenses of Baltimore/Washington, the editorship of the quarterly newsletter was transferred from Bob Zink to Joel Eastman. Our first tour to the "Mecca" of surviving American coast artillery, Corregidor Island in the Philippines, was conducted in 1991 with 25 members attending. We also visited the defenses of Oahu upon our return flight to the USA. The editorship of the newsletter was transferred in 1992 to Bolling W. Smith, making him the fourth editor of our quarterly publication.

The Coast Defense Study Group, Inc. was incorporated as a 501(c)(3) Maryland non-profit corporation on August 23, 1993 and our current by-laws were issued at that time. In becoming a charitable organization, the CDSG formed the CDSG Fund to allow members to make tax-deductible donations, and for the CDSG to make tax-exempt gifts to organizations that promote the goals of the CDSG. Also in 1993, the *CDSG News* changed its name to the *CDSG Journal*. We visited another key site for American coast defense by touring the harbor defenses of the Panama Canal in 1993. The new by-laws also saw the creation of several new committees to further the goals of the CDSG. A preservation committee was formed to promote the protection of coast defense sites and equipment. The CDSG Press was created to reprint hard to locate coast artillery manuals. The Nominations and Audit Committees were also established.

CDSG Press' first reprint, *Notes on Seacoast Fortification Construction* by Col. Eben E. Winslow, was published in 1994. In 1995, the journal was split into two parts with B.W. Smith editing the *CDSG Journal* and Mark Berhow producing the *CDSG Newsletter*, which focused on current events and organization news. The year 1997 saw the formation of the Project and Representative Committees to help channel our members' interests in developing new activities and for our members to interface with local coast defense sites. The CDSG ended the decade by establishing a website in 1998 to provide online information about American coast artillery and the CDSG, and has published three original works: *American Seacoast Defense: A Reference Guide*, edited by Mark Berhow in 1999, *Artillerists & Engineers*, by Arthur C. Wade in 2011 and *WWII Harbor Defenses of San Diego* by H.R. Everett in 2021.

The turn of the new century saw the maturing of our membership and leveling off of the group's growth. Elliot Deutsch, after many years of service, transferred the finance duties to Terry McGovern and a few years later the membership function to Glen Williford. In 2000, the *CDSG Journal* changed its name to the *Coast Defense Journal* to indicate a broader coverage of the subject. The CDSG organized its first ever tour to a purely foreign coast defense site in 2002 with a visit to Vladivostok, Russia. The CDSG has held an annual conference every year since 1985, and with the 2004 meeting in Charleston and Savannah, the group had officially visited all major defended harbor in the continental United States. The CDSG has continued to hold annual conferences on a rotating basis to the former harbor defenses in the US so by 2022 we will have completed our second round of conferences to each of the 21 defended harbors. Special tours have taken CDSG members to fortifications in Spain, Finland, France, Norway, Canada, Switzerland, Singapore, and Sweden. One key activity over the last 35 years has been the accumulation of nearly a terabyte of scanned archival material on range of coast defenses and ordnance which is made available to our members through the CDSD ePress.

The CDSG has continued to expand its preservation and interpretation efforts by providing technical, networking, and grants to former coast defense sites and to protect artifacts. The CDSG Fund has now made grants totaling almost $100k to other organizations working to preserve and interpret former coast defense sites or artifacts. The CDSG Representative program has grown to include regional CDSG leaders to coordinate local representative's outreach to the owner of the former coast defenses sites. Recently, the www. CDSG.org website was complete redone using new software and expanded content. Also, the CDSG's primary publication – American Seacoast Defenses a Reference Guide – was revised into an expanded 3rd Edition.

The future of the CDSG is depends on volunteer members to organize and run our annual conferences and special tours to various harbor defense locations around the world. Our publications are also based on our volunteer members writing or producing the quarterly journal, organizational newsletter, or reprint books. The CDSG looks to each of our members to volunteer his or her time to staff our committees and activities. After 40 years, the CDSG has become an established organization with a small but supportive membership. We have established ourselves as a key source of information on American seacoast defenses and have actively enhanced the availability of information and documentation on the subject. We continue to seek ways to support preservation and restoration efforts at former seacoast defenses sites around the U.S.

GLOSSARY OF TECHNICAL TERMS, 1860
Derived from: *A Legacy in Brick and Stone* by John Weaver, 2001
Bulwark & Bastion by James R. Hinds & Edmund Fitzgerald, 1981

Abatis (ab ah TEE). An obstacle made from tree branches and brush that have been sharpened on the side facing the attacker.

Advanced Work. An independent secondary work being within cannon range of the main fortification.

Angle of Defense. The angle fanned by one face and it's opposite flank.

Angle of The Flank. The angle made by a curtain and flank.

Angle of The Shoulder. The angle formed by a face and flank of a bastion.

Angle of Traverse. The angle through which a gun can be rotated. The pintel location, the embrasure size, and the carriage design determine this angle.

Approaches. Trenches dug by an attacker, generally parallel or at a slight angle to the curtains of a fort. Also referred to as parallels.

Arrow. A work placed at the salient of a glacis with a caponier or communication back to the covered way.

Banquette (bahn KET). The raised earthen or masonry platform behind the parapet on which riflemen can stand when firing.

Banquette Slope. The earthen or masonry slope leading from the banquet to the covered way or terraplein. Riflemen could use this slope for reloading their weapon while shielded from direct fire by the parapet.

Barbette (bar BET). The platforms behind the parapet on which guns are mounted, such that they can fire over the wall. Guns are said to be mounted en barbette when they are mounted on these platforms.

Barbette Tier. The top tier of a fort where guns are mounted en barbette rather than en casemate.

Bartizan. A small masonry turret hanging out from a wall, usually at a salient angle, and supported on corbels. Bartizans were used in Spanish fortifications as sentry boxes.

Bastion (BAST yun). A projection in the fort wall, usually at a corner, which allows flanking fire along the wall. A full bastion has two faces and two flanks; a demibastion has one face and one flank.

Bastion Head. A field work with two faces and two converging flanks.

Bastionet. A very small full bastion. Sometimes casemated.

Batardeu. A solid masonry barrier 7 or 8 feet thick crossing the entire breadth of the ditch opposite the flanked angles of the bastion. It served to keep portions of a ditch flooded.

Battery On A Cavalier. A battery in which the platforms for the guns or mortars were raised above the natural level of the ground. Also called raised batteries, these were sometimes made in front of the first parallel, especially if the fort under attack were on higher ground.

Battery. A group of gun emplacements, either within a fort or more often in the outworks of a fort. Also, a group of guns commanded by a single officer.

Berm. A ledge between the edge of the ditch and the base of the parapet, which served to keep the earth from sliding into the ditch.

Blind. Two sticks joined together with two spars about 4 feet wide, used to shelter against a cross fire in siege operations.

Blockade. An attack on a fortification solely by closing it on and denying its defenders provisions and supplies.

Blockhouse. A structure of heavy timber or masonry. A blockhouse may be a stand-alone structure to guard a particular location, it may be used at a salient or along the curtain of a fort in lieu of a bastion, or it may be a central structure surrounded by an earthwork. A blockhouse often was a multistory structure with the second level having a larger trace than the first. The overlap would contain machicolations – openings to fire down on an attacker.

Bombardment. An attack by firing bombs or shells into a fort to destroy its buildings.

Bombproof. A structure protected from indirect fire; a place of refuge during a siege.

Bonnet. An outwork covering the salient angle of a ravelin.

Breach. An opening created in a wall, generally during a siege. A breach was generally made by cannon fire or a mine.

Breastwork. In field fortification, a low parapet without a banquette, chiefly intended for protection against fire, or any low defensive work mainly designed as a protection from fire and not as an obstacle.

Bulwark. A circular work, originally of timber and earth, later of stone, erected to keep early siege guns out of range of town walls, or to shield and protect gates.

Caltrop Or Crow's Foot. A four pronged obstacle, usually of iron, sometimes used against cavalry. In later field fortifications, boards with nails driven through were often substituted.

Canister Shot. Containers filled with iron balls, usually ranging in size from three-fourths inch to two inches in diameter. On firing, the canister splits open discharging the shot in a shotgun-like pattern.

Capital. A line bisecting a salient angle.

Caponier (cap un YAIR). A small, vaulted outwork designed to provide flanking fire along a wall, generally located in the ditch. A caponier may or may not be attached to the scarp wall. Also, a defensive passageway, generally located in a ditch or closing a ditch. Caponiers had loopholes, howitzer embrasures, or both.

Carnot Wall (CAR noh wall). A detached scarp; a scarp wall, which is separated from the ramparts of a fort, usually by a chemin de ronde.

Carronade. A short-barrel cannon used for flanking fire, similar to a flank howitzer. Carronades were originally designed for shipboard use, but were also used for antipersonnel missions in forts.

Casemate (CASE mate). A vaulted, bombproof room of masonry construction, sometimes called a gunroom. It provides overhead protection to gunners, and allows tiers of guns to be stacked. Casemates may also provide firing positions for small arms, and may be used for quarters, kitchens, storage, and other sundry functions.

Castle. A tall fort. The term generally refers to a fort with multiple tiers of guns.

Cavelier. A work within a fort's enciente raised 10 or 12 feet higher than the rest of the works, and often used within a bastion or at its gorge.

Center of a Bastion. The point where the curtains would intersect if extended into the bastion.

Chamade. A signal made by the besieged consisting of the beating of drums on the rampart next to the point of the attack and indicating a desire to capitulate.

Chamber. A room excavated at the end of a mine into which the powder would be placed.

Chemin de Ronde (SHEH min deh RON dah). A pathway around the interior of the scarp wall of a detached-scarp fort, serving as a firing platform for riflemen.

Cheval-de-Frise. An obstruction consisting of pointed poles extending radially through a central axel. These were often used to block a breach in a wall, or as a barrier to a cavalry charge.

Circumvallation. A continuous entrenchment, often a line of redans, facing away from a besieged fortification and helping to prevent its relief. It was rarely used.

Citadel (SIT ah dell). A stronghold inside the scarp of a fort, which serves as a defensive barracks and a last line of defense.

Command. To overlook a work or a plain. A work or location is said to command another work or location if it stands above it.

Cordon. A projection, often of rounded stone, at the junction of the scarp and the exterior slope. Its primary function was to protect the masonry from weather.

Counterfort. A buttress on the earthen side of a wall, generally either the scarp or the parade wall, to increase its strength.

Counterguard. An outwork erected before the bastions to shelter their faces from breaching batteries placed on the covered way. Counterguards were also used before ravelins. The counterguard consisted of two faces joined to forming a salient angle.

Countermine. A tunnel going outward from the fort, used as a defense against mining the fort walls. A countermine generally had areas in which explosives could be placed to cave in the mine of a siege force or to bring down a bastion of the fort if the bastion falls to an attacker (see also, Listening Galleries).

Counterscarp Gallery. Casemated areas behind the counterscarp wall, allowing a cross-fire into the ditch. Counterscarp galleries generally contained howitzers at the corners, with loopholes in the remaining walls. Access is from the ditch or a tunnel under the ditch.

Counterscarp. The wall across the ditch from the scarp. The counterscarp served as a revetment as well as an obstruction to entry of the ditch by making it a significant drop to the level of the ditch.

Countervallation. Either a continuous entrenchment similar to the circumvallation, but facing the besieged fortification, or a series of separate entrenchments doing the same. This served to prevent a breakout.

Coup de Main (KOO dah MIN). An assault of a fortification by storm, rather than by siege. A coup de main would allow the possession of a fort in a very short time period compared to that required by a siege.

Coverface. An earthen or earth-and-masonry outwork designed to protect the masonry of the fort from siege guns. A coverface often mounted cannon en barbette. The "casemated coverface" at Fort Monroe was an all-masonry casemated water battery that increased the firepower of the fort.

Covert (KUV ert) *Way, or Covered Way.* The area above the counterscarp, covered by the guns of the fort but protected from view by an attacker.

Crenel. An open-topped embrasure at the top of a wall. A crenel is defined by two adjacent merlons.

Crochets (croe SHAYS). Angled or curved paths around a traverse that spans a covert way, generally wide enough for only one soldier abreast to pass.

Crownwork. A very elaborate outwork consisting of two demibastions joined by curtains to a central bastion.

Cunette. An open-topped drain, often in the center of a dry ditch. A cunette was used to keep a dry ditch free from water.

Curtain Angle. The angle formed by the curtain and the flank of a bastion.

Curtain. The portion of the scarp between bastions

Dead Space. An area right below the parapet that the guns could not be depressed to cover.

Decagon. A ten sided fortification.

Defensive Barracks. Barracks built to serve as a keep, usually of masonry construction and loop holed.

Demibastion (DEM mee BAST yun). A half-bastion; a bastion with only one face and one flank.

Demigorge. At the rear of a bastion or outwork, the line between the capital and the flank or face.

Demilune. Literally, "half moon." A demilune is a semicircular outwork, generally used for land defense. The term is used interchangeably with ravelin, but generally ravelin implies a salient while a demilune implies a semicircular face.

Detached Bastion. A bastion separated from the scarp wall, often by a ditch or chemin-de-ronde. A detached bastion was usually connected to the main work by a caponier or gallery.

Detached Scarp. A masonry scarp in advance of an earthen rampart, usually separated by a chemin-de-ronde. This allowed the destruction of the scarp wall during a siege without damaging the mass of the fort. See also Carnot Wall.

Detached Work. An independent work beyond the outworks of a fort, generally located outside of cannon range from the main work. The "Martello Towers" at Key West would be detached works to Fort Taylor.

Ditch. A low area, either wet or dry, which inhibits the passage of the enemy to the scarp wall. The ditch is located between the scarp and counterscarp wall. A wet ditch is usually referred to as a moat.

Double Sap. When a parapet was thrown up on both sides of a sap it was called a double sap. Such saps were often used close to the enemy work.

Drawbridge. A bridge, generally over the ditch, that controls entrance to the work and can be raised or pivoted to deny access.

Elevated Battery. A battery in which the platforms for the guns or mortars were laid on the natural level of the ground, and having a parapet raised above it with earth from the ditch. See battery.

Embrasure (em BRAY Shure). An opening in the scarp, the counterscarp wall, or the parapet which allows a cannon to fire through it.

Empty Bastion. A hollow bastion.

En Barbette. The practice of mounting cannon such that they fire over a wall rather than through an opening (embrasure) in that wall.

En Casemate. The practice of mounting cannon in a casemate such that they can fire through an embrasure.

Enceinte (awn SENT). The body or mass of a fort, inside the ditch.

Enneagon. A nine sided fortification.

Epagon. A seven sided fortification.

Escalade. An attack made by climbing over the parapet of a fort, usually using ladders.

Esplanade. The level space separating a citadel of a fortress from the town, or, an open area inside a fortification used for drilling troops, a parade.

Exterior Crest. The crest of the exterior slope.

Exterior Side of a Work. The side facing the enemy, also, an imaginary line drawn from one salient of a bastion to that of another, a front.

Exterior Slope. The slope at the base of a parapet, between the cordon and the superior slope. The exterior slope is usually the steepest slope on the exterior of the parapet.

Face. The side or sides of the fort which open onto the water, where the primary coastal guns will be located. On a bastion, the face is the wall of the bastion, which is nearly parallel to the curtain.

Facines. Bundles of sticks, which were commonly used for revetments, etc. Usually they were about 6 or 8 inches in diameter. Extra long ones, about 16 feet long, were called saucissions.

Faussbray. A low rampart around the outside of the enceinte.

Fire. The fire of guns was usually described in terms of the way it struck the target. Direct fire hit in front; enfilading fire took a target in flank; oblique fire, at an angle; reverse fire, from behind. Sometimes fire was described in terms of the trajectory of the projectiles. Thus we hear of the high angle fire of mortars, of cross fire, of plunging fire and of ricochet fire. Finally, fire may be classified according to the direction it is delivered, i.e. to the front, on the oblique, to the flanks.

Flank Howitzer. A howitzer designed to fire the length of the ditch, usually from the flank of a bastion. A flank howitzer generally fired canister shot or grapeshot.

Flank, or Cheek. The side. On a bastion, the flank is the portion which is most nearly perpendicular to the curtain of the fort. The flank of a bastion provided the primary fire along the curtain.

Flat Bastion. A bastion situated on a curtain wall between rather than at one of its angles.

Flying Sap. When fire from the defenders of a work was light the attackers could speed up their approaches by placing many gabions along the trench before filling them. This was called a flying sap.

Fortress. A fortification with a civilian population; a fortified city.

Fougass. An early version of the land mine. A stone fougass resembled a modern claymore mine. Engineers dug a 6-foot shaft in a slope inclined to the horizontal at an angle of about 45 degrees. At the bottom they placed a charge of, say, 55 pounds of powder. Above this charge they put a strong wooden shield and three or four cubic yards of loose stones not weighing less than 11 pound apiece. A wire running through a hose or tube activated the firing device. Other versions of the fougass included boxes of live shells over a powder charge or simply powder charges.

Fourneau. See chamber.

Fraises: A row of sharpened logs placed horizontally or angled downward. These were generally imbedded in an earthen rampart of a fort to inhibit escalade.

Front: The side of a fort. On an unbastioned fort, the front is measured from corner to corner. On a bastioned fort, it is measured from the salient of a bastion to the salient of the opposing bastion, including the curtain wall between the bastions.

Fuze. A means of igniting the bursting charge of a shell or bomb, usually consisting of a piece of wood hollowed out and filled with a powder composition timed to burn at a certain rate. It was cut to the proper length, driven into the shell before firing and ignited from the muzzle blast.

Gabion (GAB ee on): A bundle or open-topped cylinder filled with materials – often branches or stones – used to form a barrier. Gabions were used in temporary (such as siege) works and to fill breaches in a permanent fortification.

Gallery: A passageway, generally masonry or masonry revetted. A gallery differed from a caponier in that a gallery has no defensive mission and therefore has neither loopholes nor embrasures.

Garrison Carriage. See siege and garrison carriage.

Genouillere. The interior elevation of a parapet where an embrasure had been cut through was called a genouillere. This part of the parapet protected the lower portion of the gun carriage.

Glacis (glah SEE): The gentle slope beyond the outworks, cleared of all obstructions, which an attacker must traverse to reach the fort.

Gorge Angle. In polygonal fortifications, the angle formed by the junction of the gorge with one of the flanks.

Gorge Bastion. In polygonal fortification a bastion or bastionet at the angle of the gorge and a flank.

Gorge: The side or sides of the fort away from the water, where land defense is the primary concern. On a bastion, the gorge is the narrow portion of the bastion toward the parade.

Grenade. A hollow 3-inch iron sphere filled with powder and thrown by grenadiers or others after the ignition of its fuze. Also, a similar projective designed to explode on impact and furnished with vanes.

Grillage: A web of timbers, often placed in perpendicular layers, used as the foundation of a fort. Cedar was the most popular wood to use to form a grillage.

Guard Room. A room near the entrance of the fort, often in a gate tower, where the guard was stationed. Often there was a cell for prisoners adjacent to the guardroom. Sometimes the guardroom might have a loop-holed wall covering the passage from the main gate.

Half Bastion. See demibastion.

Half Moon. See demilune, a ravelin.

Hexagon. A six sided work.

Hornwork: An elaborate outwork, consisting of two demibastions connected by a curtain.

Hot-Shot Furnace or Shot Furnace: A furnace for heating shot. A masonry structure, it consisted of a fire chamber below rails on which the shot would travel. A loading mechanism sat at one end and an unloading mechanism at the other. Hot shot was used to set fire to wooden ships, sails, and rigging. In several cases hot shot was used against a town to set it afire.

Howitzer: A short-barreled cannon. Flank howitzers generally were used for antipersonnel missions, firing canister shot. Seacoast howitzers (or siege howitzers) generally utilized a high angle of fire (between a gun and a mortar).

Hurdles. Barriers about 3 feet high and 2 broad used to stop up breaches in besieged works.

Hurter. A piece of timber 6 inches square placed to prevent the wheels of gun carriages from damaging the parapet of a siege battery or field work, also, on the seacoast carriage. The pieces on either end of the chassis to keep the recoiling part of the carriage from running off.

Ichnography. A plan of the horizontal characteristics of a work.

In Advance: Toward an attacker, or toward the likely route of an attacker.

In Battery: A gun in firing position, i.e., at the front of the lower carriage.

Insult. An open assault on a work.

Interior Crest. The crest of a parapet's superior slope, the highest point of the parapet.

Interior Side of a Fortification. The imaginary line from the center of one bastion to the center of the next.

Interior Slope. The vertical or near-vertical slope from the crest of the parapet to the banquette, or the terraplein if there is no banquette. This slope is generally masonry or wood.

Interior Works. After a fort's scarp has been breached, it was sometimes possible for the defender to erect an earthwork, sealing off the endangered area. A usual interior work was a retrenchment of a bastion, which had been breached, that is an earthwork thrown up across the gorge.

Investment: The surrounding of a work in preparation for a siege, preventing resupply or reinforcement of the work. The time period of a siege is generally counted as beginning when the work is invested.

Keep, or Safety Redoubt. In permanent fortification, a blockhouse in ravelin or behind a bastion to which troops could retire if it were overrun. In field works, a similar structure of timber and earth. Usually, these keeps were square or cross-shaped. The sides were protected by two thicknesses of 12-inch thick lumber. The interior dimensions had to be at least 9 feet high and 20 feet wide. Loopholes were usually about 3 feet apart. Keeps were usually ditched.

Land Front: The front of a fort facing the land, where a siege would take place. In the Third System, the land front was generally designed in the style of Vauban with protection of the masonry and levels of defense.

Level of Site (Plane of Site). The original ground level.

Line of Defense. The line from a salient to its opposite flank.

Lines. Chains of field fortifications, either continuous entrenchments or works placed at intervals.

Listening Gallery: A tunnel extending beyond the scarp of a fort that is used to detect the sounds of the mining operations of a siege force.

Lodgement. A foothold gained by attackers in some part of a work.

Loophole: A narrow opening in a wall which allows a rifle or musket to fire through it.

Lunettes. In permanent fortification, works built on both sides of a ravelin, with one of their faces flanked by the ravel in and the other by the bastion, also, a ravelin-like work erected beyond the second ditch. In field works, a work with two faces and two parallel flanks, open at the gorge. It was also called a bastion head when the flanks were diverging, not parallel.

Machicoulis Gallery: A protruding gallery that allowed firing through the floor, thus protecting the area along the scarp without the use of a bastion or caponier. The overhang of blockhouses often served as machicoulis galleries.

Magazine: The place for storage of powder inside a fort. The main (or storage) magazine would store the bulk of the powder, and day-use (or service) magazines would be secondary storage depots. Magazines were carefully designed to prevent sparks and to provide a dry atmosphere for the powder.

Magistral Line. A line formed by tracing the line of the scarp's cordon around the fort, the base line from which the other parts of a permanent fortification were laid out.

Main Gateway. The principal entrance to a fortified work, usually wide enough for wagons and often arched over.

Mantlet. A movable parapet made of strong planks about 4 feet long, 3 feet wide and mounted upon two wheels. It served as a shelter for sappers.

Martello Tower: A round tower of three or more floors mounting heavy artillery en barbette on the top level (generally one to three cannon) and infantry-defense weapons (howitzers, carronades, and small arms) on the lower levels. Based on a design in Corsica, which successfully defended a harbor, the entrance was always on the second floor. This nomenclature was extended to encompass any tower-like small fort, whether round or rectangular and with or without outworks.

Mask. Anything that hid a battery or work or shielded it from fire.

Merlon: A rectangular projection from the top of a wall. Crenels and merlons alternate on a crenellated wall.

Militia Artillery: A group of reserve soldiers who were trained in artillery drill. Militia artillery were used to supplement the regular garrison in most Third System forts.

Mine: An underground excavation passing under a work. A mine could be filled with explosives that would be detonated to cause a breach in the defenses.

Moat: Another name for the ditch, usually implying a wet ditch.

Morro Castle (El Morro). Any of several fortifications such as those of San Juan, Puerto Rico, Santiago de Cuba, or Havana, Cuba, dating between the Sixteenth and Eighteenth Centuries. Generally triangular in form, these works are built upon a headland (morro) overlooking a channel or entrance to a harbor. Usually these forts have embrasures in exposed walls facing the channel, while the front facing the landside resembles a hornwork, is ditched and furnished with a covered way and glacis. These works belong to the Italian School.

Octagon. An eight-sided fortification.

Ondecagon. An eleven sided work.

Orillion. A rounded section of bastion at the shoulder serving to cover the retired and lower flank from oblique fire.

Orthography. A drawing of the vertical characteristics of a work.

Outworks: Works located within or beyond the ditch, i.e., works outside the enceinte of the fort.

Palisade: A wooden wall of sharpened stakes. On an earthwork fort, the palisade was generally located at the center of the ditch, shielded from artillery by the counterscarp.

Pan Coupe: A feature of an unbastioned fort, consisting of a wall that connects two curtains. It follows the trace of the gorge of a bastion in a bastioned fort, replacing a sharp salient with two broader salients.

Parade: The open area in the center of a fort, usually used for drilling troops, for barracks area, etc.

Parados: An earthen traverse located on the parade of a fort, parallel to the scarp. A parados prevented reverse fire on the ramparts and contained shot and shell passing over the ramparts.

Parallel: See Approaches.

Parapet: The low wall along the top of the rampart, generally masonry or masonry-revetted earth, which protects the artillery and the banquette. The highest point of the rampart.

Pas de Souris. See stairs.

Petard. A type of gun resembling a brass pot which was fastened to a strong square plank with an iron hook to fix it against a gate or palisade in order to break it down.

Pintel: The pin around which a gun carriage rotates. A front-pintel carriage for a gun mounted in a casemate generally has the pintle at the narrowest part (throat) of the embrasure. A center-pintel carriage, sometimes used on a barbette emplacement, allows a full 360-degree traverse of the gun.

Place d'Armes (PLAHSS deh ARMS): A gathering place for counterattacking forces. A place d'armes can be within a fort or a part of the outworks of a fort. A salient place d'armes is located beyond the salient of a bastion or pan coupe; a reentering place d'armes is located at the midpoint between bastions or pans coupe.

Place. A fortified town.

Platform. A floor of strong planks laid upon joists or sleepers, and usually rectangular, upon which guns or mortars might be placed, also the level, earthen base for this floor.

Plongee. In embrasures, the slope of the sole or bottom.

Polygonal Fort: A fort with the trace of a polygon, as opposed to a bastioned fort or a fort with a rounded trace. In the early Third System, a regular pentagon was used but a five-sided truncated hexagon became the most common trace later in the system.

Portcullis (port cue LEE): A wooden or iron gate which bars access but not sight, often on the inner side of a sally port. This allows the attacker to be trapped within the sally port while under fire from the defenders.

Postern (POST ern): The "back door" of a fort. Posterns are secondary entrances leading through the ramparts and either into or over the ditch. Posterns often connect to exterior batteries or other outworks.

Priest-Cap or Swallowtail. In field fortification, a redan-like work with two salient angles.

Profile. A vertical section through a fortification.

Rampart (RAM part): The area above the casemates, including the parapet and the banquette.

Ravelin (RAVeh lin): A V-shaped outwork, usually with a closed back. Ravelins were placed in advance of either a curtain or the gorge of a fort, giving protection to the scarp and providing additional batteries.

Re-enterings. Those angles formed by the junction of the faces and flanks or flanks and curtains which have the apex pointing inward.

Redan (reh DAHN): A small, V-shaped, open back outwork placed in advance of either a curtain or the gorge of a fort, protecting the scarp and providing additional batteries.

Redoubt (reh DOOT): An unbastioned fort some distance from the main fort which is designed to form the first line of defense and provide early warning of an attack. Also, a small, unbastioned fort.

Reentering: An angle pointing toward the fort; an angle of greater than 90 degrees, such as where a bastion meets a curtain.

Relief. Height of the interior crest of a rampart or parapet above the bottom of the ditch.

Retired Flank. Whenever the flank of a bastion was drawn back so as to be covered against enfilade by the shoulder of the bastion it was said to be retired. See orillion.

Revet: To face an embankment, usually with masonry or wood, to sustain it.

Revetment: A depressed area sustained by masonry or wood, with the surrounding earth providing protection.

Ricochet Fire. A technique involving loading guns with light charges and elevating them 10 or 12 degrees so as to send the shot over the parapet and bounce it along the rampart.

Salient Bastion. In polygonal fortification, the bastion or bastionet at the junction of two faces.

Salient: An angle pointing away from the fort; an angle of less than 90 degrees, the "point."

Sally Port: The main entrance or entrances to a fort, often including layers of defense against an invader. The sally port is generally located in the gorge wall.

Sap Roller. In sieges, a roller made of two large concentric gab ions 6 feet in length; the outer one 4 feet in diameter, the inner one 2 feet, 8 inches. The intervening space was filled up with pickets of wood so make them musket proof. The roller was used to protect the sappers when they were at work.

Saps. In sieges, a sap was a trench by which the attackers could move closer to a hostile fortification. Saps were usually dug in zigzag patterns to prevent the enemy from raking them so effectively with his guns. See double sap & flying sap.

Saucisson. Either a long fuse of cloth or leather filled with powder for firing a mine or a very long fascine used in the erection of batteries and repair of breaches.

Scarp, or Scarp Wall: The perimeter wall of the fort.

Seacoast Front: The front of a fort facing the water, designed to control passage and/or dominate a harbor. In the Third System, the seacoast front was generally designed in the style of Montalembert with unprotected masonry providing multiple tiers of guns.

Seacoast Gun Carriage. A type of carriage having the barrel or tube on a top carriage which was in turn mounted on rails along which it could recoil. The rails or chassis pivoted on a pintle to allow easy traversing of the piece from side to side. This type of carriage, a part of the French Gribeauval System, was first introduced into the United States in the 1790's.

Second Covered Way. A covered way beyond the second ditch.

Second Ditch. A ditch, usually flooded, located beyond the glacis.

Sector of Fire. An arc covered by the fire of a gun.

Sector Without Fire. The blind area in front of a salient angle.

Shell: A projectile fired from a cannon or mortar that has a hollow interior, generally filled with explosives.

Ship-of-the-line, or capital ship: A large warship. A ship-of-the-line was generally a three-masted sailing ship with three of four decks of cannon. These large ships were the principal warships of a fleet prior to the use of ironclads.

Shot: A solid iron projectile fired from a cannon. Shot had far more momentum than shell, but did its damage through momentum only.

Shoulder Angle. In polygonal fortification, the angle formed by the junction of a face and flank.

Shoulder Bastion. In polygonal fortification, a bastion or bastionet at the intersection of a face and a flank.

Shoulder of a Bastion. The junction between the face of a bastion and its adjacent flank was said to form its shoulder.

Shoulder: The place on a bastion where the face and the flank meet. The shoulder of a bastion is generally its widest point.

Siege and Garrison Carriage. A heavy two wheeled carriage for mounting siege or garrison guns. This carriage, like a field carriage, rested on its wheels and trail when in battery, its trail being raised and secured to a two-wheeled limber for movement. It was, however, much more massive in construction.

Siege: The protracted taking of a fort, beginning with investment and continuing through the breaching and ultimate reduction of the work. By the time of the Third System, it was understood that ultimately no fort could hold out indefinitely to a siege.

Sole. The bottom of an embrasure.

Sortie or Sally. A sudden attack by a portion of the besieged garrison on the enemy and his siege works.

Stairs. Steps of masonry were often made at the gorges of works and at the salient and re-entering angles of the counterscarp.

Star Fort. A fort formed by joining equal sized salients around a circular figure. Such a fort had alternate salient and re-entering angles. This type of fort should not be confused with pentagonal, bastioned works.

Stockade. Either a palisade of tree trunks or a small picket work made from tree trunks 9 to 12 inches in diameter and about 12 feet long. This defense was often used to close the gorge of a redan or lunette so as to make it more secure against infantry.

Sunken Battery. A battery made by excavating the interior about 3 feet deep and erecting a low parapet from it in front. Sunken batteries took only half as much work as elevated batteries.

Superior Slope: The slope that reaches the crest of the parapet. The superior slope is usually the shallowest slope of a parapet.

Tablette. The flat coping stone that surmounted the top of the scarp.

Talus. The forward slope of a parapet, or the rampart slope.

Tambour. A loop holed stockade or timber wall with two faces, made to cover the gorge of a bastion or an entrance.

Tenailles (ten EYE): Low, detached outworks, generally placed in the ditch, usually between bastions. Tenailles usually have casemated demibastions at the ends with an earthen mass between the casemates. This mass is sometimes revetted with masonry.

Tenaillions. Works on either side of the ravelin with both faces parallel or almost parallel to the ravelin's faces.

Tennaille Trace. Works using this trace were similar to star forts except that the salients were of unequal size, usually large and small alternately.

Terraplein (TARE a pleh): The area atop the ramparts behind the banquette and parapet, supporting the barbettes.

Terraplein Parade. The interior level space within the parapets of a field fortification, and shielded by them.

Throat: The narrowest part. On a bastion, the narrow portion between the flanks, nearest the parade. On an embrasure, the narrowest portion of the opening.

Totten embrasure: An embrasure reinforced with iron, developed by J. G. Totten. These embrasures provided a smaller opening and a wider angle of traverse of the cannons than European embrasures. Totten embrasures usually, but not always, utilized Totten Shutters.

Totten shutters: Iron doors which close an embrasure between cannon firings. This intricate system was developed by J. G. Totten.

Tower Bastion: A bastion of reduced size, often of significant height.

Trace: The line defining the exterior of the enceinte, or body, of a fort. The trace is generally referred to by the geometric figure that defines it.

Traverse (trah VERSE): 1. An earthen or masonry-revetted earthen mound which breaks the open area of the terraplein, separating men and guns to eliminate enfilading fire and minimizing damage from exploding shells. Traverses also allow defense of portions of the ramparts when other portions have been stormed, and often house bombproofs and magazines. 2. The arc of stone on which the rear wheels of guns travel. The individual stones are generally referred to as traverse blocks. The iron rail that the wheels travel on is generally referred to as a traverse rail.

Traverse Circle. A circular or semicircular track upon which a seacoast carriage revolved on its pintle.

Traverse Platform. See seacoast carriage, traverse circle.

Trous-de-Loop, or Loop Trap Holes. Rows of pits, either conical or pyramid shaped, with a strong stake in the center of each. The pits were used as an obstacle against cavalry and were either 2- or 8 feet deep so that they would be useless as cover. Each hole was usually 6 feet in diameter at the top and 18 inches at the bottom.

Truck Carriage. A low garrison or naval gun carriage made of two wooden sides in which the barrel rested. A transom and front and rear axels joined the sides or brackets, the carriage had four small trucks or wheels. During the Seventeenth, Eighteenth and early Nineteenth Centuries such carriages were sometimes used in forts or taken from ships to use in a siege.

Water Battery: A water battery is a group of cannon facing seaward and firing as a unit. A battery was generally commanded by a single officer. Use of the term water battery often implies a separate, detached work housing a group of cannon.

Work: A fortification. This is the broad term that applies to forts, fortresses, towers, redoubts, batteries, etc.

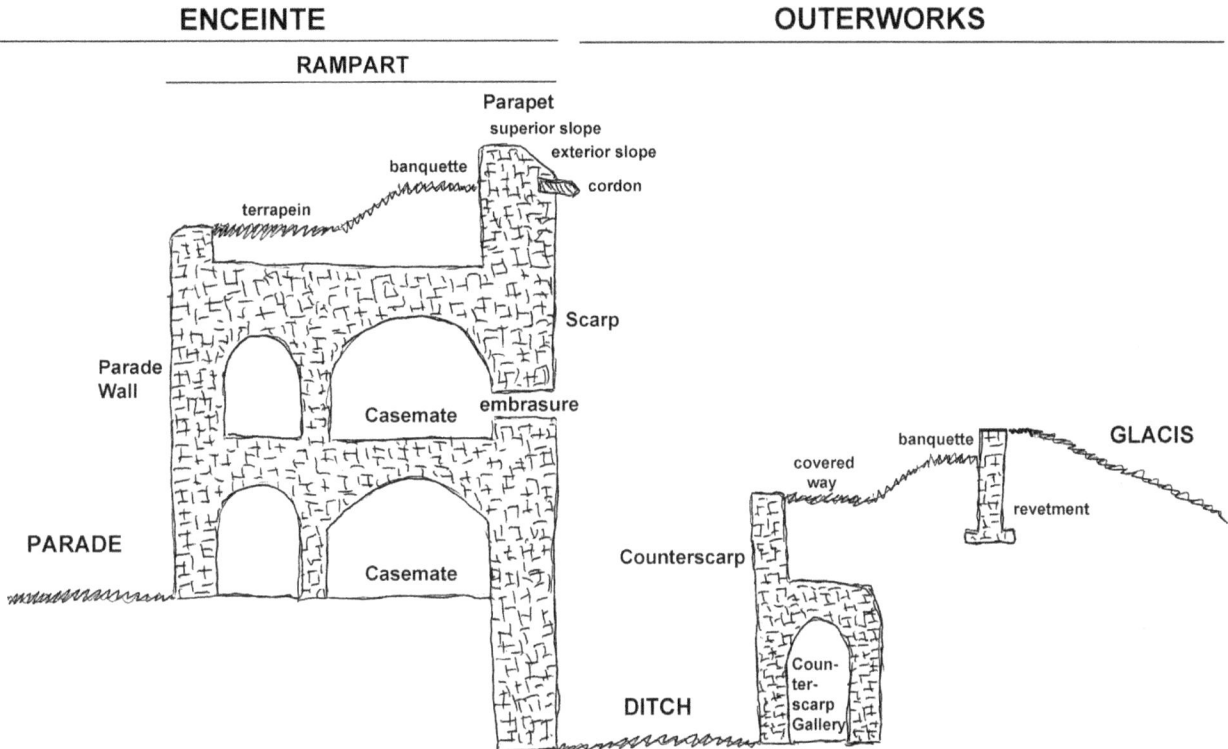

Cross section of a fortification illustrating some of the terms

Governer's Island, New York, Signal Corps Photo 1930s
Fort Jay is a classic Second System "star" fort with four bastions and a ravelin on the right side.
Castle Williams (top) is Second System four tier rounded casemated work.

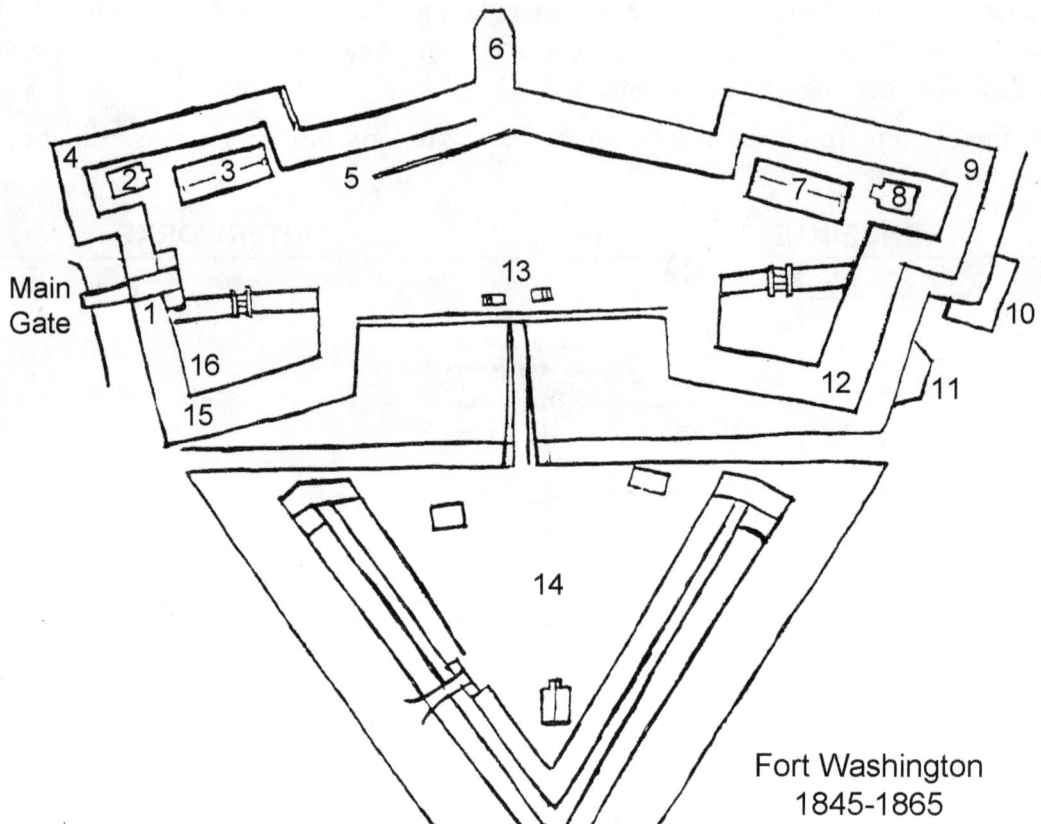

Fort Washington
1845-1865

1. Gate Tower
2. Powder Magazine
3. Officer's Quarters
4. North Bastion
5. Ramp Curtain
6. Bastionet
7. Enlisted Men's Barracks
8. Powder Magazine

9. South Bastion (with Postern)
10. Counterscarp Gallery
11. Mortar Battery
12. South Demibastion (with casemates in lower level)
13. Wall (with Water Gate)
14. Ravelin (water battery)
15. North Demibastion
16. Fort's well

Fort Washington 1930s (Signal Corps photo, NARA)

NARRANGANSETT
BAY

↑
North

BRENTON
COVE

7

1

3

3

10

3

14

1

10

7

14

2

9

9

1

1

2

11

11

2

5

5

6

4

6

6

7

7

13

13

12

8

12

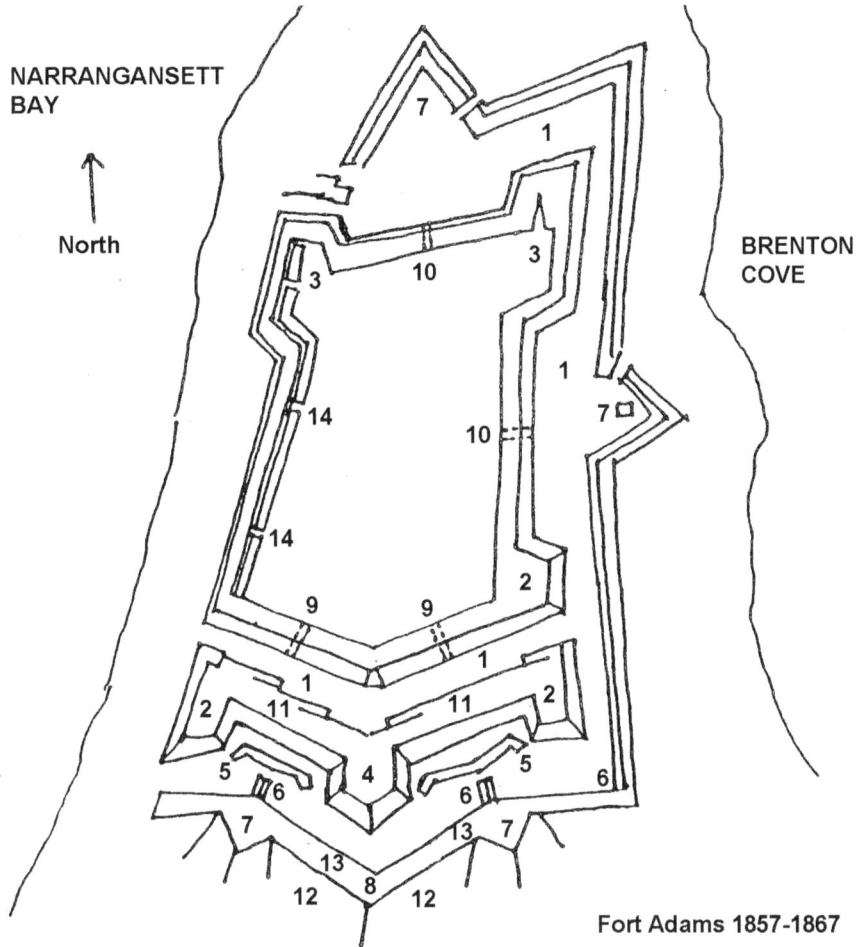

Fort Adams 1857-1867

1. Ditch
2. Demibastions
3. Hollow Bastions
4. Full Bastions
5. Tennailes
6. Caponiers
7. Re-entering Places of Arms

8. Banquette
9. Posterns
10. Gates
11. Crownwork (with Rifle Galleries)
12. Glacis
13. Covered Way (above Counterscarp Gallery)
14. Bridges to second-tier Casemates

Fort Adams 1932 (Signal Corps photo, NARA)

Fort Adams Advance Redoubt 1932 (Signal Corps photo, NARA)

1.

Fort Monroe, 1930s (Signal Corps photo, NARA)
This large fort has seven bastions and a encircling wet moat. The outer works outside the moat at the top of this photo consisted of a revetted glacis with a ravelin at top center and a redoubt at top left (both filled with more modern artillery emplacements in this photo). A water battery that was located across the moat on the upper right of this photo has been removed, but three emplacements remain at the ravelin on the right end of the outerworks.

Fort Point, San Francisco, California, a four tier casemated work with two bastions, with rifle slits facing the rear.

Fort Schuyler, NY, 1930s (Signal Corps photo, NARA) with its hornwork outerwork to the left.

Fort Tomkins and Battery Richmond, NY1930s (Signal Corps photo, NARA)
Battery Richmond (later renamed Battery Weed) is a four tier casemated work with bastions.
Fort Tomkins (left) has a barbette seafront tier and was surrounded on three sides by a dry ditch and a glacis.
In this photo, the glacis on the lower (southern) side has been removed for the concrete battery built later.

GLOSSARY OF TERMS, 1914
From: *Service of the Coast Artillery* by F.T. Hines and F.W. Ward, 1910
and *Coast Artillery Drill Regulations, United States Army, 1914,* War Department,
Office of the Chief of Staff, Washington, Government Printing Office

Absolute Deviation. The distance measured in a straight line from the center of the target to the point of impact.

Accumulator Room. A room in the battery provided for a storage battery.

Aeroscope. A device used in the meteorological station and the fire, mine, and battery primary stations for the indication of the azimuth of the wind in in degrees, the velocity of the wind and the density of the atmosphere by reference numbers. In the latter stations it may also contain a dial indicating the height of tide.

Aiming. The operation, with the aid of a sight, of giving a cannon the direction and elevation necessary to hit the target.

Air Spaces. The galleries and narrow spaces around interior rooms to facilitate ventilation and assist in keeping the rooms dry.

All-Round-Fire. Fire delivered through the entire circumference of an azimuth circle.

Ammunition Hoist. The device by means of which ammunition is raised to the loading platform. Separate hoists are used for projectiles and powder, or the latter is served by hand.

Ammunition Recess. The space built in the parapet wall at loading platform level for the temporary storage of ammunition.

Ammunition Shoes. Any type of slipper made entirely of a non-metallic substance.

Ammunition Truck. A steerable three-wheeled truck, with suitable space for carrying a complete charge from the delivery or reserve table to the breech of cannon.

Ammunition. A general term applied to projectiles, explosives used for propelling projectiles, explosives used for filling projectiles, primers used for discharging guns and mortars, and fuses used for exploding projectiles. When the projectile, propelling charge, and primer are held permanently together by a. metallic case inclosing the powder and primer, or otherwise, in condition to be handled as a unit in loading, the ammunition is called "fixed ammunition." When the projectile, propelling charge, and primer are not so held together, but are handled separately in loading, the ammunition is called "separate-loading ammunition."

Anemometer. An instrument used in the meteorological station to determine the velocity of the wind in miles per hour.

Aneroid Barometer. A watch-shaped instrument used in the meteorological station to determine the pressure or density of the atmosphere.

Angle danger. See Danger angle.

Angle of departure. The angular elevation of the line of departure above the line of sight. Quadrant angle of departure is the angular elevation of the line of departure above the horizontal plane through the muzzle of the gun in the firing position.

Angle of depression. The angular depression of the line of sight below the horizontal plane.

Angle of Fall. The angle of fall is the angle which the tangent to the trajectory at the point of impact makes with the line of shot. In practical gunnery the angle of fall is often expressed as a slope, i.e., 1 on 10, meaning that for one foot of drop in vertical height the projectile would travel ten feet horizontally.

Angle of impact. The angle between the line of impact and the tangent to the surface at the point of impact. It is the complement of the angle of incidence.

Angle of incidence. The angle between the line of impact and the normal to the surface at the point of impact.

Angle of jump. The angular elevation of the line of departure above the position of the axis of the bore at the time the piece was pointed. In determining the sight or quadrant elevation to be used, this angle must be subtracted algebraically from the angle of departure given in the range table; the angle of jump differs for different guns, carriages, and ranges, and is determined by experiment.

Angle of Position (or Depression). The angle between the line of sight and a horizontal plane through the axis of the trunnions.

Angle of splash. See danger angle.

Angular Elevation or Depression. The angular elevation or depression of the target includes the depression due to the curvature of the earth. It is sometimes called position angle.

Angular Velocity. The ratio of the angular travel or motion of a body, to the time consumed in describing the angle.

Approaches. The water area over which the enemy may be expected. In fortification, roadways entering the battery parade.

Apron. That portion of the superior slope of a parapet or the interior slope of a pit and the interior slope of a mortar pit designed to protect the elopes against blast. Also called blast slope.

Armament. Guns and mortars of various sizes and powers, including their carriages. In the coast artillery service armament is classified as primary, intermediate, and secondary.

Armor-Piercing Projectiles. Shot and shell designed to penetrate heavy side and turret armor of war vessels.

Armstrong Guns. The built-up gun construction of Great Britain, the germ of which is to be found in the coiled welded system of Sir William Armstrong, introduced in 1852.

Armstrong Rapid-Fire Gun. A seacoast cannon usually 40 to 50 calibers long and of 6-inch and 4.72-inch caliber. The latter is usually called "four point seven. "

Artillery Garrison. The personnel assigned to duty at a coast artillery fort.

Artillery. The name given to all firearms discharged from carriages ashore in contradistinction to "small arms," which are discharged from the hands. It also denotes the particular troops employed in the service of such fire- arms. Artillery in the U. S. military service is known as coast artillery and field artillery, the latter being classified as light artillery, horse artillery, siege artillery, and mountain artillery.

Atmosphere Board. A device pertaining to the equipment of the meteorological station. A graphic table by means of which the reference numbers to be recorded on the dial of the aeroscope indicator can be determined from readings of the barometer and thermometer.

Automatic Firing. A term applied to one of three methods of exploding submarine mines, i.e., where the apparatus is so arranged that the mine explodes upon contact.

Auxiliary Horizontal Base System. When either the battery primary or secondary station becomes inoperative, the battery commander's station is used as a base end station in place of the disabled station.

Axial Vent. A vent in which the opening is parallel with the axis of the bore. It is the type of vent used in modern seacoast cannon.

Axis of cannon or axis of bore. The central line of the bore.

Axis of trunnions. The central line of the trunnions.

Azimuth (of a point). In coast artillery usage, the horizontal angle measured in a clockwise direction from the south line through the observer & position to the line from the observer to the point. For example, the azimuth of a point B from A is the angle (measured clockwise from the south) between the north and south line through A and the line from A to B. The north point has an azimuth of 180°.

Azimuth Difference. The difference between two azimuths of a point as read from two other points, as, for example, the difference between the azimuth of the target as read from the primary station and the azimuth of the target as read from the directing gun or point of a battery.

Azimuth Instrument. An instrument for determining azimuths. It is sometimes used as a position-finding instrument in secondary stations.

Azimuth of a Point. In coast artillery, the horizontal angle measured in a clockwise direction from south to a line from the observer to the point. For example, the azimuth of a point R from A is the angle (measured clockwise from the south) between the north and south line through A and the line from A to B. The north point has an azimuth of 180 degrees.

Azimuth Setter. The member of a mortar detachment who lays the mortar in azimuth.

Backlash. The play between a screw and its nut where the latter is loosely fitted. A reverse movement of any part of a mechanical gear, caused by irregularities, without moving other connecting parts.

Bale. That part of an assembled submarine mine used for the attachment of the mooring cable to the mine case; also provided for the protection of the joint where the single conductor cable enters the mine.

Ballistic Board. See RANGE BOARD. A device used to determine the total range corrections to be applied to the range found on the plotting board.

Ballistic Tables. Tables of data used in connection with ballistic formula in the solution of ballistic problems; certain functions used in the various formulas are previously calculated for the certain velocities and then tabulated.

Ballistics. That branch of the science of artillery which treats of the motion of projectiles. Interior ballistics treats of the motion within the gun. Exterior ballistics treats of the motion outside of the gun.

Banquette. The step between the truck and loading platform.

Bar and Drum Sight. An open sight used on rapid-fire guns.

Barbette Carriage. See CARRIAGE OR MOUNT.

Barbette. A mound of earth or a platform on which guns are mounted to fire over a parapet. In field fortification this level is distinguished by the name of Banquette Tread.

Barometer. See ANEROID BAROMETER and MERCURIAL BAROMETER.

Base Fuse. A firing device inserted in the base of cored shot and armor- piercing shell to ignite the bursting charge. Base fuses are used when the point of shell requires great strength, as for penetrating armor.

Base line. A horizontal line the length and direction of which have been determined. This line is used in position finding, especially for long ranges; the stations at its ends are called "observing stations." It can be called "right-handed" or "left-handed, " depending on whether the secondary station is to the right or left of the primary from the point of view of a person facing the field of fire. The base end observing stations are called primary, secondary, or supplementary.

Base ring. The metal ring which is bolted to the concrete of the emplacement and which supports the weight of the gun or mortar carriage.

Base-end Station. An observing station located at either end of a base line, designed to contain an azimuth instrument or depression position finder. Base-end stations are designated as primary, secondary, or supplementary.

Battery Commander. The senior artillery officer present for duty at a battery. The battery commander exercises both administrative and tactical command. As an administrative officer he is responsible that every effort is made to keep the battery supplied with the proper equipment, implements and ammunition. He keeps a record of the daily attendance at drill and instruction, with names of absentees, reason and authority for such absence. As a tactical officer he is responsible that the personnel of his battery is efficient in drill, in practice and in action, that the equipment and fire-control installation provided for his battery are in serviceable condition and that no permanent modifications are made therein without proper authority. He is required to see that the officers and men of his battery are instructed in the care, preservation and use of artillery material, and that records are kept and reports rendered as prescribed in orders and regulations. He is authorized to modify the manual of the piece within the limits pre- scribed by the drill regulations and to make temporary modifications or changes in the fire-control installation, provided such modifications or changes permit of prompt return to their original condition. Permanent changes may be made in the provisional installations upon the approval of the district commander, the changes to be reported to the War Department. Changes in the standard installation can only be made with the approval of the War Department. He should be encouraged by his superiors to improvise devices and methods that will simplify the fire-control system or increase the efficiency of his command. He is responsible that the plotting-room details for the fire and battle commanders' stations are kept in practice during the period of indoor instruction. He should inspect his battery and make a test of the fire-control system thereof weekly. In battle command or fire command drill or action he exercises limited fire control, acting on orders received from higher commanders. When "battery commander's action" is ordered he exercises independent fire-control and fights his battery in accordance with his own judgment. In cases of emergency he acts without waiting for orders, in accordance with instructions previously given.

Battery Commander's Station. An observing station at or near the battery, usually in rear of the center traverse. The new type is a combined observing room, emergency instrument room, plotting room, etc., at the rear of the traverse.

Battery Commander's Walk. The elevated walk leading from the battery commander's station along the rear of the battery.

Battery Emplacement Book. A loose-leaf record book provided for each battery, in which all important data relating to the emplacements, guns, fire- control equipment, etc., of a battery are kept.

Battery Field of Fire. The area covered by the armament of a battery; it covers that portion of the fire area which can be most conveniently defended by the battery in question.

Battery Manning Table. A table containing a list of names detailing the personnel of a battery to their posts.

Battery parade. The area in rear of the emplacements where the gun or pit sections form.

Battery. The entire structure erected for the emplacing, protection and service of one or more guns or mortars, together with the guns and mortars so protected. The guns of a battery are of the same size and power, and are grouped with the object of concentrating their fire on a single target and of their being commanded directly by a single individual. Normally a battery of the primary armament consists of two guns or two pits of mortars. Under exceptional circumstances a single gun with its fire- control service may constitute a battery.

Bevel Wheel. A wheel having teeth cut on a bevel or conical surface called a face; where the inclination of the face is 45 degrees it is called a miter wheel. Two of these wheels with teeth engaged at right angles to each other form a bevel or miter gear.

Blasting-Gelatine. The most powerful of the detonating and disruptive explosives. It is only suitable for purposes of demolition.

Blending Powder. The process of mixing powders of the same or different lots so as to obtain charges of uniform characteristics.

Blending Room. A dry, well-lighted and ventilated room in which several charges of the same lot of smokeless powder are taken from the powder storage cases and thoroughly mixed and blended.

Blending. The process of mixing powders of the same or different lots so as to obtain charges of uniform characteristics.

Blind Shell. A shell which does not explode upon impact or when intended to do so.

Boat Telephone. A simple type of telephone used in communicating over cable between boats at distribution boxes in the mine field and the mining casemate.

Bomb-Proof. A term applied to military structures of such immense thickness and strength that shells can not penetrate them.

Bomb. A missile which also receives the names of bombshell and shell. The use of the word is practically obsolete in gunnery.

Booth or Recess. Any recess or construction for the accommodation of telautograph, telephone, etc.

Bore-Plug. See CLINOMETER REST.

Bore-Sighting. In coast artillery, the process by which the line of sight and axis of the bore prolonged are caused to converge on a point at or beyond mid-range.

Bore. The interior of a cannon forward of the front face of the breechblock. It is composed of the gas check seat, the powder chamber, the centering slope, the forcing slope, and the rifled portion called the "main bore." The length of bore is the distance from the front face of the breechblock proper (not the mushroom head) when in position to the face of the muzzle measured along the axis of the bore.

Bound. The path of a shot comprised between two grazes.

Bourrelet. That part of a projectile between the main body and the head, which includes the beginning of the ogive.

Brackets. Metal supports for telautograph cases, etc.

Breech Bushing. That part of the breech which contains the threaded and slotted sectors of the breech recess.

Breech Detail. A detail of the gun detachment charged with the duty of opening and closing the breech.

Breech mechanism. The breechblock, obturating device, firing mechanism, and all parts used in operating the breechblock of a cannon.

Breech recess. The opening in a cannon which receives the breechblock.

Breech reinforce. The part of a cannon in front of the breech and in rear of the trunnions band.

Breech, face of. The rear plane of a cannon perpendicular to the axis of the bore.

Breech. The mass of metal behind the plane of the rear section of the bore of a cannon, the section being taken at right angles to the axis of the bore.

Breechblock. The metal plug which closes the breech of a cannon.

Brown Prismatic Powder. A brown gunpowder of translucent celluloid appearance, in the form of a six-sided prism with a hole in the center.

Building up Charges. The operation of preparing charges of brown prismatic, nitro-cellulose or spero-hexagonal powder, by properly placing the grains in silk bags.

Built-up Cannon. Types of cannon in which the parts are built up of either cylindrical forgings or a single forging wrapped with a rectangular or ribbon form of wire.

Buoy. A floating object moored to the bottom, used for temporarily marking the positions of mines, junction and distribution boxes, channels, and target positions.

Bursting Charge or Shell Filler. The charge of explosive required for bursting a projectile ; it may be poured in loose or by melting.

Button Drill Primer. A form of primer so called from the fact that the head of the friction wire is formed in the shape of a button.

Cake Powder. Gunpowder which has become lumpy from having absorbed moisture.

Caliber. A name given the minimum diameter of the bore of a firearm; it is the diameter of the main bore in inches measured at the top of diametrically opposite lands, or minimum diameter of the rifled portion of the bore. Also used to express the length of cannon; e.g., a 12-inch gun 42 calibers long would be 42 feet long ; a 6-inch gun 52 calibers long would be 26 feet long.

Calibration. The operation of adjusting the range scale so that the range reading at any particular elevation of the gun will indicate the true distance to the center of impact of a group of shots fired from that particular gun and mount at that elevation with the standard velocity and under normal atmospheric conditions. It is desirable to calibrate the guns of a battery under the same atmospheric conditions, although this is not absolutely necessary. It is absolutely necessary that uniform ammunition be used for calibration firing of all guns of a particular battery. When the individual guns of a battery are calibrated the battery is calibrated, for the centers of impact of a series of shots from each gun under normal atmospheric conditions will coincide at the point indicated by any range setting. When guns of a battery "shoot together" (that is, give the same range for the same range setting) they may be fired on the same data, but are not calibrated unless the range under normal atmospheric conditions is that indicated by the range setting. It is not feasible to determine by actual firing all the points of a range scale, and therefore it is assumed that the gun is calibrated when a range scale constructed from a computed range table is adjusted on the gun so as to give the proper setting for a mid-range.

Cannister. A projectile with a thin wall inclosing a number of small steel balls but without a bursting charge, the case being ruptured by the shock of discharge as the projectile leaves the gun. Designed for use against infantry at short range.

Cannon Powder. A term applied to large-grained black or brown gun- powder to distinguish it from rifle, mortar or mammoth powder. It is used as a base in the manufacture of prismatic powders.

Cannon. A general term for artillery weapons and firearms not carried nor fired in the hands, from which projectiles are thrown by the force of expanding powder gases. Guns are long (generally from 30 to 50 calibers) have flat trajectories, and are used for low-angle fire (less than 15 degrees), with high velocities (from 2000 to 3000 f.s., about). Mortars are short (generally about 10 calibers), and are used for high-angle fire (from 45 degrees to 70 degrees), with low velocities (from 550 to 1300 f.s., about). Howitzers are intermediate between guns and mortars. The term "piece" is used when referring to a cannon of any class. Cannon of the United States land service are classified according to their use into coast, siege, field, and mountain. Built-up cannon are made by shrinking forgings (jacket and hoops) over an inner tube. Wire-wound cannon are made by winding wire under tension around a tube; a jacket and hoops may be shrunk over the wire-wound tube.

Cannonade. The act of discharging shot and shell from cannon for the purpose of destroying an enemy. To discharge cannon. Also written cannonry.

Cannoneer. An artilleryman engaged in the firing, or one who manages or assists in managing a cannon.

Canopy. The projecting roof over delivery tables of ammunition hoists of modernized gun batteries.

Cap-square. That part of a gun or mortar carriage which fits over the trunnion and holds the trunnion in the trunnion bed.

Capital. As applied to an emplacement - the line through the pintle center bisecting the arc of the interior crest. As applied to a battery - the perpendicular to the line of pintle center at its middle point.

Capped Projectile. A projectile having a soft iron cap over its point to give stability to the point when commencing penetration and to give the armor an initial pressure at the point of penetration.

Carriage or Mount. The means provided for supporting a cannon. It includes the parts for giving elevation and direction, for taking up the recoil on discharge, and for returning the piece to the firing position. They are classified as follows: 1. Fixed. A mount provided for guns and mortars in permanent works and not designed to be moved from place to place. 2. Movable or Wheeled. A carriage or mount provided with wheels for ready transportation of the piece mounted thereon. Guns of the movable armament are mounted on this type of carriage. COAST CARRIAGES. Those used for coast artillery cannon. They may be divided into four classes, depending upon the nature of cover afforded by the emplacements: a. Barbette. Where the gun remains above the parapet for loading and firing. b. Disappearing. Where the gun is raised above the parapet for firing, and recoils under cover for loading. c. Masking mount. Where the gun remains above the parapet for loading and firing but can be lowered below the level of the crest for concealment. d. Casemate. Where the gun fires through a port. RAPID-FIRE GUN CARRIAGES (except the 6-inch on disappearing carriage) are constructed so that the gun recoils in a sleeve and returns to the loading position immediately after firing. Note. If the carriage can be traversed so that the gun may be fired in all directions it is said to have all-round fire. If it can not be traversed so that the gun can be fired in all directions, it is said to have limited fire.

Carriage, fixed. A mount provided for guns and mortars in permanent works and not designed to be moved from place to place.

Carriage, movable (wheeled mount). A carriage or mount provided with wheels for transportation of the piece mounted thereon.

Carriages, seacoast. Those used for coast artillery cannon. They may be divided into four classes, depending upon the nature of cover afforded by the emplacements. (a) Barbette: where the gun remains above the parapet for loading and firing. Barbette carriages are used for guns of 3-inch or greater caliber. The Pedestal mount is a type of barbette carriage used for guns up to 6 inches in caliber. (b) Disappearing: where the gun is raised above the parapet for firing and recoils under cover for loading. This mount is used for guns of 6-inch or greater caliber. (c) Masking parapet mount: where the gun remains above the parapet for loading and firing but can be lowered below the level of the crest for concealment. This mount is also called the balanced pillar mount and is used for guns up to 5 inches in caliber. (d) Casemate. Where the gun fires through a port.

Cartridge Bags. Bags made of a special quality of silk and sewed with silk thread; used to contain the powder charge for cannon. This material burns rapidly and completely, thus avoiding the danger of a smoldering residue in the powder chamber.

Cartridge case. A container in which powder is sealed for shipment and storage.

Cartridge Extractor. That part of a breech-loading gun which ejects the empty cartridge case from its seat in the bore.

Cartridge Room. A room of the magazine for the storage of cartridges.

Cartridge. A complete load of fixed ammunition (projectile, powder, and primer) as used in small arms.

Case. The charge holder of a submarine mine. Case I. In gunnery, a class of gun pointing where direction and elevation are both given by the sight. Case II. In gunnery, a class of gun pointing where direction is given by the sight and elevation by the range scale on the carriage. Case III. In gunnery, a class of gun pointing where direction is given by the azimuth scale and elevation by quadrant or by the range scale on the carriage. With mortars this Case is used exclusively. It is used with guns, (1) in firing by battery, (2) when the position of the target is known but the target is not visible from the gun on account of smoke, fog, etc., (3) in predicted firing by piece when for any reason direction cannot be accurately given by the sight.

Casemate Battery. A storage battery in the mining casemate provided as a means of furnishing direct current for the *mine system.*

Casemate Electrician. A specially qualified member of a mine command assigned to the care and operation of the mining casemate.

Casemate Officer. A coast artillery officer stationed at the mining case- mate who controls its operation.

Casemate. A bombproof chamber, usually of masonry, in which cannon were placed to be fired through embrasures or portholes; or one capable of being used as a magazine. See MINING CASEMATE. This also refers to an emplacement surrounded by a re-enforced concrete structure. These were used for batteries of two 12 and 16-inch barbette guns (and in one case two 8-inch barbette guns) built or rebuilt during the period 1936-1945.

Cast-iron Shot. Projectiles made of cast iron, used in service target practice.

Cathead. A projecting piece of timber or iron with a pulley at the head, attached to either side of the bow of mine planters, by which anchors and mine cases are hoisted or lowered.

Center of Gravity Band. A painted band one-half a caliber wide on gun projectiles, and six inches wide on mortar projectiles. The center of gravity of the projectile is the center of the band. It also indicates where the shot tongs are placed.

Center of Gravity of Cannon. The center of gravity of a cannon is near the intersection of its axis and the axis of the trunnions. The preponderance is the excess or moment of weight in rear of the axis of the trunnions over that **of the front, or the converse.**

Center of Gravity. That point in a body, or system of bodies rigidly connected, upon which the body or system will balance itself in all positions, though acted upon by gravity.

Center of Impact. The mean point of impact, or the mean of all the hits. When a projectile strikes the target a number of times, it is the mean trajectory.

Center of the target. As used in coast artillery practice, the point from which deviations are measured. (See Deviation).

Center Pintle. A term applied to a seacoast carriage when its axis of rotation is approximately through the center, that is, when it traverses about a point at its center.

Centering slope. The conical part of the bore between the powder chamber and the forcing slope. It is for the purpose of bringing the axis of the projectile in line with the axis of the bore.

Centrifugal Fuses. Firing devices, the action of which depend upon centrifugal force; inserted either in the base or point of an armor or deck-piercing shell to ignite the bursting charge.

Chamber. See SHOT OR POWDER CHAMBER.

Charge (or powder) section. One of the component parts of a charge when the charge is made up of two or more separate parts.

Charge. The charge consists of the powder and the projectile (propelling charge). Also the explosive placed in the cavity of a projectile (bursting charge). The powder for all large cannon, to include 4.7-inch guns, is enclosed in silk or serge bags and is separate from the projectile. In guns of greater caliber than six inches it is put up in two or more sections or bags. For smaller calibers the projectile and powder are not separate; such ammunition is called "fixed."

Chase. That part of a cannon in front of the trunnion band.

Chassis. That part of a gun carriage upon which the top carriage moves backward and forward. The chassis carries recoil rollers and the top carriage rests upon these rollers. .

Chief Loader. A non-commissioned officer in charge of the loading of submarine mines. Insignia: Red mine case within a yellow circle, all to be of cloth.

Chief of Ammunition Service. A non-commissioned officer in charge of the magazines, galleries, and service of ammunition for a gun battery, or a mortar emplacement.

Chief of Coast Artillery. A general officer of coast artillery charged with the duty of keeping the War Department advised of the efficiency of the coast artillery personnel and material, and making such recommendations as will tend to promote its efficiency; to confer with and advise the chiefs of bureaus in all matters relating to coast artillery; to correspond direct with command- ants of artillery service schools and the president of the Artillery Board on questions of a technical character not involving matters of command, discipline, administration or the status or interests of individuals; to make recommendations as to the instruction, examination, promotion, assignment, special duty, and transfer of coast artillery officers and men, as well as methods and courses for their instruction ; to issue direct to coast artillery officers bulletins and circulars on technical matters. He is a member of the General Staff Corps and the Board of Ordnance and Fortifications.

Chief Planter. A non-commissioned officer in charge of the service of a mine planter. Insignia: Red mine case within a yellow circle, all to be of cloth.

Chord of the trajectory. The straight line joining the extremities of the trajectory, i.e., the straight line from the muzzle of the gun (in the firing position) to the point of splash.

Chronograph. An instrument for measuring the velocity of projectiles.

Chronometer. An instrument for accurately measuring time.

Circle. A plane figure bounded by a curved line, called its circumference, which returns into itself, and which is everywhere equally distant from a point within it called a center. A circumference is divided into 360 equal parts, each of which is known as a degree, each degree into 60 parts, each of which is known as a minute; each minute into 60 parts, each of which is known as a second. In the coast artillery service degrees are divided into hundredth, and the minutes and seconds eliminated.

Circuit-Closer. A device by which submarine mines are fired electrically by the vessel closing the circuit.

Cleaner. An artilleryman charged with the care of armament out of service.

Clinometer rest. A device inserted in the muzzle of a gun for the purpose of supporting a clinometer; also, called "bore plug" or "bore rest."

Clinometer. An instrument for measuring vertical angles with great accuracy; for example, the inclination of the axis of the bore to the horizontal.

Coast Artillery Board. A board, consisting of such number of coast artillery officers as the War Department may direct, to which may be referred from time to time all subjects pertaining to the coast artillery upon which the War Department or the Chief of Coast Artillery may desire the board's opinion and recommendations.

Coast Artillery Company. An integral part of the Coast Artillery Corps, assigned to some portion of the armament. The strength is fixed in orders; the present authorized personnel of a gun company of the primary armament is, Captain, 1 First Lieutenant, 1 Second Lieutenant, 1 First Sergeant, 1 Quartermaster Sergeant, 8 sergeants, 12 Corporals, 2 Mechanics, 2 Musicians, 2 Cooks, and 81 Privates. The number of privates may be varied in companies depending upon the armament to which it is assigned; this is particularly the case in companies of the mine defense.

Coast Artillery Fort. The coast defenses at any military post and the garrison assigned thereto. Its command devolves upon the senior regular coast artillery officer present.

Coast Artillery Garrison. The personnel, to include regular coast artillery, coast artillery reserves, and coast artillery supports, assigned to a coast artillery fort.

Coast Artillery Material. Coast artillery material is classified as armament, range equipment, power and light equipment, and submarine defense equipment.

Coast Artillery Militia. Troops of the organized militia organized as coast artillery for the purpose of supplementing the regular coast artillery in time of war.

Coast Artillery Reserves. Militia troops organized as coast artillery for the purpose of supplementing the regular coast artillery.

Coast Artillery Supports. Infantry or other troops assigned to coast artillery forts to support the artillery in repelling land attacks in the immediate' vicinity of the fortifications.

Coast artillery supports. Troops of the mobile army assigned to coast artillery forts to repel land or landing attacks in the immediate vicinity of the fortifications.

Coast Defense Officer. A coast artillery officer assigned to duty on the staff of division or department commanders to act in an advisory capacity with respect to matters pertaining to the efficiency of coast artillery material and to the drill, instruction and employment of coast artillery in connection with coast defense generally.

Coast Defense. The military and naval dispositions and operations necessary to resist a naval attack on any part of the coast line.

Collar. A device made of wood, placed upon the chase of a gun to make the diameter equal to that of the body of the piece, to enable it to be rolled with facility.

Colors of the Coast Artillery Corps. The national color is of silk 5 feet 6 inches fly, 4 feet 4 inches on the pike, which is 9 feet long, including spearhead and ferrule; the union is 2 feet 6 inches long, with stars embroidered in white silk on both sides of the union; the edges are trimmed with knotted fringe of yellow silk 1/2 inches wide; there are two cords 8 feet 6 inches long, having tassels, and composed of red, white, and blue silk strands. The official designation of the artillery district is engraved on a silver band placed on the pike. The corps color, of the same dimensions as the national color, is of scarlet silk, having embroidered upon it in colors the official coat of arms of the United States, of suitable size. Below the coat of arms, in the middle, embroidered in yellow silk, are two cannon, crossed; also a scroll embroidered in yellow silk and bearing the inscription, "U. S. Coast Artillery Corps," embroidered in red silk; the edges are trimmed with knotted fringe of yellow silk 1\2 inches wide; cord and tassels same size as those of the national color, but of red and yellow silk strands. Both sides of the color are embroidered alike.

Colors on Projectiles. See PAINTS ON PROJECTILES.

Colors. The silken national, and district or regimental flags carried by regiments or battalions of engineers, troops comprising coast artillery districts, and regiments of infantry. The word is used in contradistinction to standards, which are smaller in dimension and are carried by regiments of cavalry, and regiments of field artillery. They are carried in battle, campaign, and on all occasions of ceremony at district or regimental headquarters where two or more companies of the district or regiment participate. Each coast artillery post, where two or more companies of coast artillery are stationed, is furnished with a service color, which is the national color made of bunting or other suitable material, but in all other respects similar to the silken national color. Garrisons of coast artillery posts other than district headquarters may use them upon all occasions.

Combination Electric Friction Primer. A primer combining the principles of friction and electric primers, the electric features being modified.

Combination Fuse. A fuse inserted in the point of shrapnel which ignites the bursting charge either upon impact or at the completion of the set time interval; it contains both a time and percussion fuse, thus increasing the chance of bursting.

Common Electric Primer. A primer whose action depends upon the ignition of a small charge of fulminate fired by means of a platinum bridge heated to incandescency by an electric current.

Common Friction Primer. A primer which operates solely by friction.

Communications. Means of transmitting orders or messages through the tactical chain of artillery command. In the case of mobile troops, it includes all routes, such as roads, railroads, etc., by which an army communicates with its base, or by which several parts of an army communicate with each other.

Composite Artillery Type Telephone. A special type of telephone provided for permanent artillery communications.

Computer. A member of the fire-control section who operates a range or deflection board.

Cone of Dispersion. See ONE OF SPREAD.

Cone of Spread. The imaginary cone containing the diverging bullets or fragments upon the explosion of a shrapnel shell. This cone is very long.

Conical. Round and tapering to a point.

Console. A bracket whose projection is not more than half its height; any small bracket.

Contact Firing. The electrical firing of a submarine mine when it is struck by an enemy's ship.

Converted Gun. A smooth-bore gun in which a tube containing rifling has been inserted.

Cordage. Ropes and cord collectively. See chapter on CORDAGE.

Cored Shot. A projectile the center of which is partly hollowed. It can be filled with a bursting charge and used the same as a shell.

Corrected Range. The fictitious range which determines the elevation to be given the gun, in order to hit the target.

Corridor or Truck Corridor. The elevated passageway in rear of the traverse connecting adjacent gun emplacements at the loading platform level.

Corridor wall. The traverse wall along the corridor.

Corridor. The uncovered passageway in rear of a traverse connecting two adjacent emplacements.

Counter-recoil buffers. Devices on gun and mortar carriages for the purpose of reducing the shock due to the return of the piece to the firing position.

Counterweight Well. The pit in the front end of a gun platform for the reception of the counterweight of a disappearing carriage.

Counterweight. The weight used in bringing a gun on a disappearing carriage to the firing position. The pit in the gun platform for the reception of the counterweight is called the counterweight well.

Cover Post. Positions for the members of a mortar detachment at the command "Take Cover."

Cradle. A device employed for transporting heavy guns a short distance.

Crane. A mechanical device for raising ammunition by means of differential or other blocks.

Critical Dimension. A term used in connection with powder grains. It is the dimension or thickness of the web between the perforations in a multi-perforated grain. Also called least dimension.

Cross Fire. Cross fire is where the projectiles from guns in different positions cross one another at a particular point.

Crow's Nest. The name commonly applied to the observing station located in the parapet or traverse.

Crusher Gauge. A device inserted in the mushroom head of the breech- block, or in the bottom of the bore, to determine the maximum pressure of the bore. Commonly called pressure gauge.

Curvature of the Earth. A term applied in gunnery to define the amount of bending of the water surface of the earth from the normal due to the earth's shape. This bending of the water surface causes the target to be on a lower level than that of the gun, i.e., the apparent difference of level between the axis of the gun and mean low water as the position of the target is increased in proportion to the range.

Cut-off Jack Set. A form of telephone bridged on lines to primary stations and booths to enable communication officers to talk direct to any primary station in the fire command.

Cylindrical. Having the form of a cylinder; uniformly circular.

Cylindro-Conical. Having the form of a cylinder the forward portion of which is conical.

Danger angle. (Also called angle of splash). The angle which the tangent to the trajectory at the point of splash makes with the plane containing the point of splash and parallel to the horizontal plane through the muzzle of the piece in the firing position.

Danger space. The horizontal distance within which a target of a given height would be hit by a projectile. The danger space varies with the range the flatness of the trajectory, the height of the target, and the height of the gun above the target. The maximum range which is all danger space is called the danger range.

Data line. A telephone line used for the transmission of data. (See Intelligence line.)

Datum Point. A fixed point, the azimuth and range of which, from one or more observing stations, have been accurately determined. Such point or points are used in proving the accuracy of range and position finding instruments.

Deflection Board. A device for determining the algebraic sum of the deflection corrections for wind, drift and travel of target during the time of flight and the predicted interval. It is used to determine the reference numbers for the deflection scale of the sight in Cases I and II, and the azimuth correction reference number in Case III; and, for mortars, the corrected azimuth.

Deflection Scale. A scale provided on sights, graduated in degrees and hundredths for the purpose of obtaining and applying corrections for deviation.

Deflection. The angle between the plane of sight and plane of departure; it is usually expressed as a reference number, and is set off on the sight deflection scale.

Delayed Automatic Firing. A term applied to one of the three methods of exploding submarine mines, i.e., where the apparatus is so arranged that the mine is exploded when a signal is given which indicates that it has been struck.

Delivery table. The hoist table from which the projectiles are delivered to the trucks.

Density of loading. The mean density of the whole contents of the powder chamber. It is the ratio of the weight of the powder charge to the weight of a volume of distilled water at the temperature of maximum density (39.2° F.) which will fill the powder chamber. The formula for computing it is -- d (density of loading) = 27.68 w/V in which w is equal to the weight of the powder in pounds and V the volume of the chamber in cubic inches.

Depression Position-Finder. A telescopic instrument used in the primary and secondary stations of a fire-control base line, to read either vertical or horizontal angles. When used in the first instance it is a depression-position finder, while in the second it is used the same as an ordinary azimuth instrument. Objects when viewed from an elevation appear under different angles of depression according to their distance from the point or points of observation; this fact is taken advantage of in the vertical base system and is the principle upon which depression-position range finders are constructed. The depression-position range finder solves, mechanically, the problem of determining one side of a vertical right triangle, having given a side and two adjacent angles. The given, or base side, is the distance above sea level of the axis about which the telescope is elevated or depressed. The lower angle is constant and equal to 90 degrees. The depression angle varies with the distance of the observed target.

Detonation. The practically instantaneous combustion or decomposition of a disruptive explosive of high order.

Deviation at the Target. If from the target a line be drawn perpendicular to the plane of direction intersecting the plane containing the line of shot, the length of this perpendicular is the "deviation at the target."

Deviation, azimuth. The difference between the azimuths from the directing point of the battery to the center of the target and to the point of splash at the instant the projectile strikes.

Deviation. As used in coast artillery practice, deviations are either the horizontal distances of the points of splash from the center of the target, or the rectilinear coordinates of those distances. Deviations are measured in a plane passing through the water line of the target and parallel to the horizontal plane through the muzzle of the piece in the firing position. (a) Absolute deviation. The shortest distance between the center of the target and the point of splash. (b) Lateral deviation. The distance between the plane of direction and the plane of splash measured (right or left) from the center of the target and perpendicular to the plane of direction. (c) Longitudinal deviation. The perpendicular distance (over or short) of the point of splash from the vertical plane passing through the center of the target and perpendicular to the plane of direction. (d) Mean lateral deviation. The algebraic mean of the lateral deviations of a series of shots. (e) Mean longitudinal deviation. The algebraic mean of the longitudinal deviations of a series of shots. (f) Mean absolute deviation. The algebraic mean of the absolute deviations of a aeries of shots. (g) Range deviation. The difference between the range to the target (at the instant the projectile strikes) and the range to the point of splash. The range deviation is equal to the longitudinal deviation when the lateral deviation is zero.

Difference Chart. A graphic device constructed upon geometric principles by which information as to range and azimuth of a target from one point the primary station is converted mechanically into similar information with respect to any other point, as the directing gun or point of a battery.

Directing point. A point at or near the battery for which relocation is made at the plotting room. It is the point for which the gun center of the plotting board is adjusted. When the pintle center of a gun is taken as the directing point, such gun is called the "directing gun."

Disappearing Carriage. A gun carriage so constructed that it will carry its gun to a firing position above the parapet and upon discharge carry it back to the original loading position behind the parapet.

Displacement of any Point. The horizontal distance in yards of that point from the directing point.

Displacement. The horizontal distance from the vertical axis of the position finder of the battery primary station to the pintle center of the directing gun, or some other directing point. In a marine sense the word implies the quantity of water displaced by the hull of a ship, the weight of the displaced liquid being equal to that of the displacing body.

Distribution Box. A cast-iron case through which the cables of a group of submarine mines are distributed from the multiple-core cable which runs to the mining casemate.

Double Primary Station. A building so constructed as to contain under one roof two primary stations with their plotting rooms.

Double Secondary Station. A building so constructed as to contain under one roof two secondary stations.

Drift. The divergence of the projectile from the plane of departure due to its rotation, its ballistic character and the resistance of the air. It is generally in the direction of rotation, except for extreme elevations of high-angle fire, in which case it may be opposite to the original direction of rotation. For the United States service rifled guns it is to the right. It may be expressed either in yards or angular measure.

Driggs-Schroeder Rapid-Fire Gun. A rapid-fire gun of 2.24-inch caliber, commonly called 6-pounder. They are distinguished by two models of breech mechanism, namely, screw-block and drop-Week.

Driggs-Seabury Rapid-Fire Gun. A rapid-fire gun of 2.24 and 3-inch caliber, commonly called 6 and 15-pounders respectively.

Drill Primer. A primer provided for reloading, used for drill purposes.

Dunnite. See EXPLOSIVE "D."

Dynamite. A detonating and disruptive explosive of high order used for filling submarine mines and demolitions of all kinds.

Earthworks. In fortification, a general term for all military constructions, whether for attack or defense, in which the material employed is chiefly earth.

Ecrasite. A high explosive compound of foreign manufacture.

Elasticity. That property by which bodies recover their former shape and volume after having yielded to some force. Elasticity of steel permits its extension to a certain limit, beyond which a permanent set would take place. This limit is known as the limit of elasticity.

Electric Mines. Submarine mines fired by electric current; they are of two classes, controllable and non-controllable.

Electrician Detachment. A detachment consisting of the electrician sergeants and necessary assistants detailed from the enlisted personnel, charged with the care and preservation of the electrical installations, etc., of a fort.

Electrician Sergeant. A non-commissioned staff officer. Electrician sergeants are of two classes. The first class are charged with the immediate supervision, care, and operation of a division of the electrical installations, including searchlights and power plants when necessary, in addition to the duties prescribed for electrician sergeants, second class. Any duty in connection with the electrical installations, including the mechanical work of repairing electrical apparatus and the care and operation of searchlights. The second class are charged specifically with the care, repair and maintenance of the electrical installations, including lines and means of communication, as well as any duty in connection with the electrical installations, including mechanical work necessary in repairing the electrical apparatus, and the care and operations of searchlights and small power plants.

Elevating band. A band around a gun near the breech to which are attached the elevating- arms. By means of the elevating gearing, the elevating arm gives elevation to the gun.

Elevation Setter. The member of a mortar detachment who lays the mortar in elevation.

Elevation table. A table of ranges with corresponding quadrant elevations for a direct-fire gun on a mount provided with an elevation device graduated in ranges. The quadrant elevations tabulated in the elevation table are the angles of departure of the range table corrected for curvature of earth, height of sight, and jump.

Elevation. The inclination in a vertical plane given to the axis of the bore in pointing a gun; the angular elevation of the axis of the bore above the line of sight is the sight elevation; the angular elevation of the axis of the bore above the horizontal is the quadrant elevation.

Elliptical. Oblong with rounded ends.

Elongated Projectiles. The modern type of coast artillery projectiles.

Emergency Position Finder. A self-contained horizontal position finder, or a depression position finder constructed for use on low sights and sufficiently accurate for emergency purposes. It is operated from the observing station (crow's nest) at a battery in case the regular instrument or station is destroyed.

Emergency Station. A range-finding station provided for use in case a regular station of a permanent base line has been destroyed. These stations are so located that they can be used to replace the primary or secondary station of one or more base lines. The term is also applied to the observing station at the battery, where an emergency depression position-finder is mounted.

Emergency System. A system of position-finding, used in an emergency. It employs a self-contained position-finder located at the battery, with or without a plotting board.

Emplacement book. A book containing all necessary data concerning the battery.

Emplacement Officer. A coast artillery officer in immediate charge of the emplacements of a gun battery or one pit of a mortar battery. He is the battery commander's assistant at the guns to which he is assigned. He is responsible to the battery commander for the condition of the emplacement and material, as well as for the efficiency of its service. Upon arrival at the emplacement before a drill, practice or action, he makes a careful inspection of all parts of the guns and carriages, the equipment and implements to be used, giving special attention to the elevating and traversing devices as well as those for running the piece in and from battery; recoil cylinders to see that they arc properly filled with the right amount and kind of oil, and that the plugs are properly inserted; obturators to see that they are properly adjusted; pads in serviceable condition. In case of firing to see that the throttling and buffer valves are properly set and locked; to see that the motor generator, motors and controllers, firing attachments, firing batteries and circuits are in order; that the sights, subscales of azimuth circles are in adjustment; that sponges and rammers are of proper kind and gauge and in serviceable condition; that suitable vessels are provided for the water necessary for sponging out the bore; that all necessary charts are on hand; and that ammunition hoists are in working order. As the efficiency of a battery depends primarily on the correctness of adjustment of each of the devices mentioned above, the necessity of a thorough inspection can not be too strongly recommended. Emplacement officers of mortar batteries are charged with the corresponding duties in so far as they pertain to a mortar battery.

Emplacement. That part of a battery pertaining to the position, protection and service of one gun or mortar, or a group of mortars.

Endurance of Cannon. The life of a cannon or the number of times a piece is capable of being fired before relining is necessary. In the case of heavy guns their life is assumed to be approximately 250 service shots.

Energy of Recoil. An expression denoting the work done in recoil of a gun when fired. The recoil may be reduced by decreasing the weight of the projectile, decreasing the muzzle velocity, or by increasing the weight of the gun.

Energy of the projectile. The energy stored up in the projectile by the force of the expanding gases generated by the explosion of the powder charge. It is expressed usually in foot-tons. The formula for computing it is -- $E = WV2/ (4480g)$ -- in which W is the weight of the projectile in pounds, V its velocity in feet per second, and g the acceleration due to gravity (mean value 32.16). V may be taken as the velocity at any instant and the energy remaining at that instant can be determined from the formula.

Enfilade Fire. Fire which rakes a fighting line, the gun being on the prolongation of the line. In naval or fortress engagements fire delivered on the stern or bow of a ship so that the projectiles rake the whole length of the deck.

Engineer. An enlisted specialist of the noncommissioned staff Coast Artillery Corps who, under the artillery engineer, is placed in charge of one or more power plants at a coast artillery fort.

Enlisted specialists. Noncommissioned staff officers of the Coast Artillery Corps who are assigned to technical duties at coast artillery forts. The various grades are master electrician, engineer, electrician sergeant first class, electrician sergeant second class, master gunner, and fireman.

Equalizing pipe. A pipe connecting the front ends of two recoil cylinders for the purpose of equalizing the pressure therein.

Erosion. The gradual enlargement and scoring of the bore due to the action of powder gases on the metal of the lands and grooves.

Error, probable. The probable error of a gun in any direction is that error which is as likely to be exceeded as not in the case of any single shot of a series fired with the same elevation and azimuth settings. This is equivalent to saying that in the long run 50 per cent of all shots fired with the same elevation and azimuth settings will have an error less than the probable error.

Error. As used in coast artillery practice, errors are either the horizontal distances of the points of splash from the center of impact, or the rectilinear coordinates of those distances. Errors are measured in a plane passing through the center of impact and parallel to the horizontal plane through the muzzle of the piece in the firing position. (a) Absolute error. The shortest distance between the center of impact and the point of splash. (b) Lateral error. The distance between the plane of splash and a plane through the center of impact parallel to the plane of direction, measured (right or left) from the center of impact and perpendicular to the plane of direction. (c) Longitudinal error. The perpendicular distance (over or short) of the point of splash from a vertical plane passing through the center of impact and perpendicular to the p lane of direction. (d) Mean absolute error. The arithmetical mean of the absolute error of a series of shots. (e) Mean lateral error. The arithmetical mean of the· lateral errors of a series of shots. (f) Mean longitudinal error. -The arithmetical mean of the longitudinal errors of a series of shots.

Exploder. An electrical machine operated by hand, used to fire electric fuses and primers.

Explosive "D," or Dunnite. A shell filler the exact ingredients of which are secret. It is a high explosive compound. It is not fusible and shells are filled by compression. It is the least sensitive to shock of all the explosives used in the service.

Explosive Compound. An explosive whose ingredients are united chemically. Nitro-glycerine and guncotton are explosive compounds.

Explosive House. A structure in which high explosives are stored on a military post; frequently referred to as a magazine. Authorities do not agree as to the most suitable type. Some advocate light wooden structures, the debris of which in case of explosion would be thrown a comparatively short distance; others recommend a building of corrugated iron with asphaltum floors, making the house fireproof. Structures of this characte: are usually protected by earthworks.

Explosive Mixture. An explosive whose ingredients are mixed mechanic- ally. Gunpowder is an explosive mixture.

Explosive. Any substance by whose decomposition or combustion, gas is generated with great rapidity. Military explosives consist of solids or liquids which, through the application of heat or shock, are susceptible of being converted suddenly into gases through chemical reactions.

Exterior crest. The line of intersection of the superior and exterior slopes.

Exterior slope. The outer slope of the battery.

Extreme Range. The greatest accurate range obtained by a projectile in its flight. For example, the extreme range of the 16-inch breech-loading rifle, model 1895, is approximately 21 miles. The extreme range against armor for the 12-inch breech-loading rifle, model 1900, is approximately 13,000 yards, or 1\ miles.

Face of the Breech. The rear terminal plane of the gun perpendicular to the axis of the bore.

Face of the Muzzle. The front terminal plane of the gun perpendicular to the axis of the bore.

Faces of the Rimbases. The end planes of the rimbases perpendicular to the axis of the trunnions.

Faces of the Trunnions. The end planes of the trunnions perpendicular to their axis.

Field of fire. The area covered by the armament of a battery, or with reference to a single gun, it is the area covered by that gun.

Fifteen-Pounder. Term applied to a 3-inch rapid-fire gun. It denotes the proper weight of projectile for the piece.

Fillet. That portion of metal, filling the re-entrant angle formed by two surfaces not tangent, thus avoiding a sharp corner. The term is also applied to the metal cut away in removing the sharp edge formed by the intersection of two surfaces.

Fire area. The area covered by the armament of a fire command.

Fire control diagram. A diagram showing the assignment of batteries to fire or mine commands, the division of fort commands into fire and mine commands, the assignment of searchlights, and the system of communications for the tactical chain of command in any particular coast defense command.

Fire control installation. The materiel as installed, which is employed in the fire control or fire direction of any unit, is called the "fire-control installation" for that unit.

Fire control symbols. The symbols are used in fire-control diagrams and for other purposes.

Fire control. Fire control is the exercise of those tactical functions connected with the concentration and distribution of fire, including the assignment and identification of targets. The exercise of those tactical functions which determine: (a) The objective of fire. (6) The volume and concentration of fire, (c) The accuracy of fire. The term fire-control system includes the means employed in fire-control, the scheme of its installation and method of its use. The material as installed, which is employed in the fire-control of a battery or district, is called the fire-installation for that battery or district. Installations are either standard or provisional. Fire-control material may be classified under the following heads: a. Instruments for the observation and location of targets. b. Instruments for the determination of firing data. The personnel employed in fire control is called the fire-control personnel.

Fire direction. Fire direction is the application of the methods and training necessary to secure accuracy of fire. A battery commander exercises fire direction while the fire commander exercises fire control.

Fire discipline. The efficiency of personnel in action, involving accuracy and alertness resulting from organization, drill, and combined practice. It is measured by the length of time required to exercise fire control and fire direction; the time required to assign targets and to fire accurately.

Fire Left. When marked on the deflection scales of telescopic sights indicates minus deflection (muzzle pointed left).

Fire Right. When marked on the deflection scales of telescopic sights indicates plus correction (muzzle pointed right).

Fire. The discharge of firearms and the destruction caused by their projectiles. Artillery fire is classified as direct, curved and high-angle fire. Direct fire is fire with high velocities and angles of elevation not exceeding fifteen degrees. Curved fire is fire with low velocities and angles of elevation not less than fifteen degrees. High angle fire is fire with low velocities and angles of elevation not less than forty-five degrees. Kinds of: (a) Direct fire. Fire with high velocities and with angles of elevation not exceeding 20°. (b) Curved fire. Fire with low velocities and with angles of elevation not exceeding 45°. (c) High angle fire. Fire with low velocities and with angles of elevation above 45°.

Fired Standard Primer. See DRILL PRIMER.

Fireman. An enlisted specialist of the noncommissioned staff Coast Artillery Corps, who under the engineer is assigned to duty for firing boilers, running engines, and other work in a power plant at a coast artillery fort.

Firing Interval. The interval of time between consecutive shots of the same gun or mortar in continuous firing.

Firing Machine or Electric Firer. See EXPLODER.

Fixed Ammunition. When the cartridge case is attached to the projectile, the two together are called fixed ammunition. It is not used in large calibered guns on account of the disadvantage of handling and the difficulty of arranging and preserving.

Fixed Armament. Guns and mortars of various sizes and powers, mounted on stationary carriages.

Fixed Defenses. Defensive works ashore within the line of defense.

Fixed Light. A searchlight intended to demarcate the outer limit of a battle area and illuminate any vessel entering it.

Fixed Mount. A mount or carriage provided for guns and mortars in permanent works and not designed to be moved from place to place.

Flanking Fire. Fire directed along the front of or nearly parallel to the enemy's line.

Foot-Ton. The energy expended or necessary to raise a weight equal to a long ton, or 2,240 pounds, one foot.

Forcing Cone. The part of the bore of a gun immediately in front of the centering slope. It is formed by cutting away the lands so as to decrease their height uniformly from front to rear.

Forcing slope. The part of the bore immediately in front of the centering slope. The rifling begins at the junction of the centering slope and the forcing slope. The tops of the lands at this point are cut down so that-less power is required at first to force them through the copper rotating band. The lands attain their full height at the front end of the forcing slope.

Forcing. As applied to a projectile, forcing is the operation by which a projectile is made to take hold of the grooves of the bore.

Fort Record Book. A permanent confidential record book containing the history of the works, their object, armament, scheme of defense, and all information of value regarding the equipment and installation. It is supplemented by the Fort Record Book Files, in which copies of all confidential papers and maps are kept.

Fortified Point. A general term indicating a city, harbor, anchorage, estuary, or any limited portion of the coast line that is defended by fixed defenses.

Friction Primer. See COMMON FRICTION PRIMER.

Friction. A force acting between two bodies at their surface of contact, so as to resist their sliding on each other. Friction is of three kinds; sliding and rolling friction, which act with solids; and fluid friction, which acts with liquids and gases.

From battery. The position of a gun when withdrawn from its firing position.

Front Pintle. A term applied to a coast carriage where its axis of rotation is at or near its front end, i.e., where it traverses about a point in front of its center.

Frontal Fire. Fire which is directed perpendicularly, or nearly so, to the objects fired at.

Fulminate. A very sensitive explosive compound used in fuses, primers and caps.

Fuse. A device attached to a projectile for the purpose of causing the explosion of the bursting charge either by impact or at the expiration of a certain time of flight. Fuses are classified according to construction, as ring resistance, combination, time, and percussion, centrifugal, and detonating; they are classified according to location in the projectile as point and base.

Gallery. Any passageway covered overhead and at the sides.

Galvanometer. An instrument used for detecting the existence, and determining the strength and direction of an electric current.

Garrison Flag. The national flag, 36 feet fly and 20 feet hoist. Hoisted only on holidays, important occasions and during engagements. In the latter case it is called the battle flag.

Garrison Gin. A lifting tackle used in mechanical maneuvers of coast artillery armament.'

Gas Check Pad. A pad made of asbestos and tallow enclosed in a canvas cover and compressed under heavy pressure. Under the weight of firing the plastic nature of the pad causes it to press outward against the gas check seat and inward against the spindle, forcing the split rings firmly in their seats and completely stopping the passage to the escape of gas.

Gas check seat. That part of the bore of a cannon where the gas check pad rests when the breechblock is closed.

Gas Check. The essential mechanical feature of an obturator which enables it to prevent the escape of gas.

Gear Wheel (or Cog Wheel). A wheel with teeth on the circumference to mesh with a rack, worm ring, or another gear wheel.

General Defense Plan. The scheme of defense formulated prior to an attack. A variety of these plans, based on the character of attack to be expected, should be prepared and issued to the command.

Gravimetric Density. A term which refers to the ratio of the weight of a unit volume of a standard powder to the weight of the same volume of any other powder, i.e., the gravimetric density of a powder is the weight in pounds of a volume of 27.68 cubic inches of the powder not pressed together by its own weight. (27.68 is the number of cubic inches occupied by one pound of water).

Gravity. That force which tends to draw all bodies toward the earth with uniformly increasing velocity. Its mean value equals 32.16 foot-seconds. An example of gravity may be demonstrated as follows: A projectile thrown from a mortar at, say, an angle of 90 degrees, would travel upward until the propelling force under or behind it ceased to exist, it would then take a down- ward course drawn by gravity and strike the surface at a velocity equal to the original initial velocity which it had on leaving the muzzle.

Graze. The point at which a projectile strikes a surface and rebounds onward.

Grooves. In ordnance, the spiral hollow cuts made in the surface of the bore.

Guard Room. A room in the battery, or guard house, set aside for the use of the guard.

Gun Carriage. See CARRIAGE or MOUNT.

Gun Commander. A specially qualified non-commissioned officer in direct charge of a gun section. When assigned in command of mortar pits they are called pit commanders] to ammunition sections, chiefs of ammunition service.

Gun Commander's Range Scale. See RANGE SCALE.

Gun Company. A company assigned to the service of direct-fire guns only.

Gun Cotton. A detonating and disruptive explosive of high order made of unspun cotton waste, used in shells, torpedoes and for demolitions of all kinds.

Gun Differences. Differences in range and azimuth to the target from the gun and from the directing point, due to gun displacement.

Gun Displacement. The horizontal distance in yards from the vertical axis of the directing gun to the pintle center of any other gun of the battery, or from the directing point to the pintle center of any gun of the battery.

Gun levers. Two steel arms on a disappearing carriage which support the gun at one end and the counterweight at the other end. The gun trunnions rest in trunnion beds on the gun levers, and the counterweight is suspended from a steel crosshead which joins the ends of the gun levers. The gun levers are pivoted near their middle upon a gun-lever axle which rests in bronze bushed axle beds in the top carriage.

Gun Lift. See CARRIAGE or MOUNT.

Gun or piece. A general term applied to any firearm from which a missile is propelled by the force of expanding gas. In a restricted sense, the term "gun" is applied as defined under "Cannon." Any firearm or instrument for throwing projectiles by the expanding force of powder gas, consisting of a tube or barrel closed at one end. In a restricted sense, the term is applied to that class of cannon in which the length of bore is great in comparison with the caliber

Gun Platform. That part of a battery upon which the gun carriage rests.

Gun Pointer. A specially qualified member of a gun section charged with the proper aiming or laying of a gun, or the chief of a mortar detachment who supervises the loading and laying of a mortar. Insignia: Red crossed cannon within a yellow circle, all to be of cloth.

Gun Section. A detail of the enlisted personnel, consisting of a gun commander, a gun detachment, ammunition detachment and reserve. There is one gun section for each piece assigned to a battery of the primary armament for service or drill; for batteries of the secondary armament, the detachment for all of the pieces constitutes one gun section under a single gun commander.

Gunner. A specially qualified enlisted man who has passed the examination in elementary gunnery. They are classified as first- and second-class gunners. Insignia: First Class. In gun or mortar company; red projectile, with red bar below. In mine company; red mine case, with red bar below. Second Class. Same as first class, omitting the bar. All to be of cloth.

Gunner's Quadrant. An instrument usually used in laying mortars to give quadrant elevation by either applying it at the breech or muzzle.

Gunnery Specialists. Specially qualified enlisted men, such as Master Gunners, Master Electricians, Engineers and Electrician Sergeants who have successfully pursued the course of instruction at the Coast Artillery School.

Gunnery. That branch of military science which comprehends the theory of projectiles, and the manner of constructing and using ordinance.

Gunpowder. A black or brown granular explosive mixture of low order, made of niter, charcoal and sulphur.

Hang Fire. A delayed ignition of the powder charge caused either by defective primer or charge. See MISSFIRE.

Harbor Charts. Charts covering the water area of each fortified harbor within the field of fire of the armament at that point. They are made on a scale of 500 yards to the inch and the area covered is marked off in one -inch squares which are numbered consecutively. The outlines of the harbor, depths of water, channels, locations of batteries and stations are accurately indicated. A chart mounted on a suitable board, with a range-scale arm pivoted at the point indicating the station to which it pertains is furnished each Battle, Fire, Mine and Battery Primary Observing Station.

Height of site. The altitude of the axis of the gun trunnions in the firing position above the plane of mean low water.

High Angle Fire. See FIRE.

Hoist room. The room in a battery containing the receiving table of the ammunition hoist.

Hoist. See AMMUNITION HOIST.

Homogeneous. Of the same kind or nature; consisting of similar parts or of elements of like nature.

Hoops. In ordnance, hoops are cylindrical forgings concentric with the tube of built-up cannon, superimposed upon the tube, jacket" or other hoops. The "A" Row are over the jacket, extending from the breech to the trunnions. The "B" Row when used are on "A" Row. The "C" Row are all those hoops in contact with the tube in front of the jacket. The "D" Row are over the front end of the jacket and the "C" Row in front of the trunnions.

Horizontal Angle. An angle measured in the horizontal plane, or whose sides lie wholly in such plane.

Horizontal Base System. The system of range -finding in which the position and range of the target are definitely located by the use of the primary and secondary arms of the plotting board, from azimuth readings taken simultaneously at the primary and secondary observing stations.

Horizontal Plane. That plane which passes through or contains the line of the horizon. The horizontal plane referred to in the artillery position finding service is the horizontal plane containing the water level.

Horizontal Position Finder. The Azimuth instrument.

Horizontal Range. The longitudinal distance from the muzzle of the gun to the point of fall measured in the initial plane. It is the range given in all fundamental range tables.

Horizontal Velocity. The component of the muzzle velocity parallel to the horizontal plane passing through the muzzle of the gun.

Howitzers. Those cannon whose relative length and caliber range between the gun and mortar classes. They are used principally by the mobile army.

Hydraulic Jack. A portable machine for exerting great pressure for lifting or moving a heavy body through a small distance by hydraulic power.

Hygrometer. An instrument for measuring the degree of moisture in the atmosphere.

Hypothetical Targets. Imaginary targets of assumed dimensions, for heavy guns. This target is outlined by two standard pyramidal targets towed 60 feet from center to center with a red streamer suspended from a wire or hemp cord midway between them, the hypothetical target represented being a vertical rectangle 30 by 60 feet. For mortars the hypothetical target is a circle 100 yards in diameter the center of which is indicated by a pyramidal target.

Identification of Target. The act or process of recognizing a target which has been designated.

Igniter or Igniter Charge. A small charge of rifle powder placed in con- tact with the propelling charge to insure the ignition of the latter. See PRIMER CHARGE.

Igniting Primer. Primers used in cartridge cases for subcaliber tubes not provided with percussion firing mechanism. They require for ignition an auxiliary friction or electric primer which is inserted in the vent of the spindle in the same manner as for service firing.

Illuminating Light. A searchlight whose primary function is to follow a target that has been assigned to a fire command.

In Battery. The term used to indicate that a gun is in its proper position for firing.

In Commission. The term used to indicate those batteries to which personnel is assigned.

In Service. The term used to indicate those batteries to which personnel is assigned and at which daily drills are held.

Indication of a Target. Any method employed to designate a target.

Inflammation. The spread of flame over the surface and into the perforations of powder grains.

Initial Pressure. The first or starting pressure. The term is frequently applied in reference to initial tension, or the stress developed in the body of a built-up gun, by the method of fabrication. Initial tension is produced by shrinking over a tube or hoop a heated hoop that will have a slightly smaller diameter when cooled, each hoop compressing the one beneath it.

Initial Velocity of Rotation. The rate of motion at which a projectile is traveling around its longer axis at the instant it leaves the muzzle of the gun.

Initial Velocity of Translation. The rate of motion at which a projectile is traveling in the direction of its flight at the instant it leaves the muzzle of the gun. See MUZZLE VELOCITY.

Initial Velocity. The rate of travel at which a projectile leaves the muzzle of a cannon. Generally called muzzle velocity.

Inside of a Beam. A point is so called when it lies between the battle commander's station and a line passing through the axis of a searchlight beam.

Intelligence Line. A telephone line connecting a primary and secondary station; used for the transmission of general information. The line over which data are sent from the reader at the secondary station to the secondary arm setter at the plotting board in the primary station plotting room is called the data line.

Interior crest. The line of intersection of the interior slope with the superior slope. If there be no interior slope, it is the line of intersection of the interior wall and superior slope.

Interior Slope or Wall. The inner slope or wall of gun parapets or mortar pits. The inner slope of a parapet connecting the interior wall and superior slope

Interior wall. The inner parapet wall.

Intermediate Armament. The armament of a coast artillery fort used to attack lightly armored or unarmored vessels; it may be employed effectively to supplement the primary armament in the attack of armored vessels, or the secondary armament in the defense of the mine fields. It includes the 6-inch, 5-inch and 4.72-inch guns.

Jacket. A cylindrical forging, concentric with and shrunk on the tube; it generally extends from the breech of the gun to a plane beyond the trunnions.

Judgment Firing. See OBSERVATION FIRING.

Jump, Angle of. The angle included between the line of departure and the, axis of the bore when the piece is pointed. Experience has shown that the actual range for a given elevation does not correspond to the computed range. For convenience it is assumed that the entire discrepancy is due to an increase or decrease of the elevation of the gun at the instant the projectile leaves the muzzle, and this small difference in angle is called the jump. Therefore in determining the sight or quadrant elevation to be used to obtain a given range, a correction, which differs for different guns, carriages and ranges, must be applied to the angle of departure given in the range table. It may be determined by experiment. In practice it is included in the necessary range corrections as determined by trial shots.

Junction Box. A device used in splicing cable, its object being to protect the joint and cause the strain to come upon the armor of the cable rather than on the joint itself. The Single function box, small, consists of two rectangular plates of cast iron ^-inch in thickness, 20 inches long and 6 inches wide, united by four ^-inch bolts at the corners (the bolts having square shanks and hexagonal nuts to facilitate clamping) ; the plates are hollowed in the middle to form a chamber which receives the Turks' heads and joint. The ends of the plates are curved to admit the cable ends, and the Turks' heads are clamped to the lower plate by straps and screw bolts, the cavity of the upper plate covering them when bolted in position. This type of box is used in splicing single conductor cable. The Single junction box, large, is similar in construction and is used in splicing multiple cables or cables of more than one core. The Grand junction box, or junction box used as a distribution box when 7 conductor multiple cable is used, consists of two circular plates of cast iron, f-inch in thickness and 21 inches in diameter, united by four 1-inch bolts at the corners. The joints and 7 conductors and multiple cable are clamped in the box in a similar manner to that described for the single junction boxes.

Kentledge. Old cast iron articles which have become unserviceable, such as condemned guns, shot and shell, etc.

Laflin & Rand Firing Machine, or Electric Firer. See EXPLODER.

Land Front. Those portions of the defenses which are provided to repel an attack from the land area in rear of or on the flank of permanent seacoast works.

Lands. In ordnance, the surfaces or ribs of the bore between two adjacent grooves of the rifling.

Landward defenses. Those portions of the defenses which are provided to repel an attack from the land area in rear of or on the flank of permanent seacoast works.

Lanyard. A strong cord to one end of which a brass hook is attached. Used for exploding the friction primer when the piece is to be fired. See SAFETY LANYARD.

Large Caliber Guns. The class of guns included in the primary armament.

Lateral. Of or pertaining to the sides.

Latrine. A closet for soldiers in camps or barracks.

Laying. The operation of giving a gun the direction and elevation necessary to hit the target without the use of a sight.

Le Boulenge Chronograph. See CHRONOGRAPH.

Least Dimension. A term used in connection with powder grains. The least dimension of a grain of powder is the dimension measured between the perforations of a multi-perforated grain over or through which the fire spreads in order to consume the entire grain. See CRITICAL DIMENSION.

Length of the Bore. The distance from the front face of the breechblock when seated, to the face of the muzzle.

Limited Fire. Fire delivered through a restricted circumference of an azimuth circle.

Limits of Fire. The terminating azimuths of the field of fire of a battery,

Line of Collimation. The line in which the optical axis of the telescope should be when properly adjusted. The line of collimation and the line representing the axis of the telescope, when in proper adjustment, coincide.

Line of Defense. In coast artillery, the coast line; it consists of fortified and unfortified portions. The fortified portions are those which include important harbors, cities, roadsteads, estuaries and approaches thereto. The unfortified portions are those lying between or adjacent to the fortified portions. Lines of defense for both the coast guard and coast artillery supports are determined upon and planned in detail in time of peace for each fortification; the works necessary for each line are surveyed and mapped out in detail.

Line of Departure. A line representing the prolongation of the axis of the gun at the instant the projectile leaves the bore; it is therefore tangent to the trajectory at the muzzle. It is sometimes called the line of fire.

Line of direction. The straight line from the muzzle of the gun (in the firing position) to the center of the target at the instant the shot strikes.

Line of fall. The tangent to the trajectory at the point of fall.

Line of Impact. The line tangent to the trajectory at the point of impact.

Line of Shot. The line from the gun to the point of impact.

Line of Sight. A straight line passing through the sights of the piece; at the instant of firing this line passes through the target.

Line of sight. The axis of collimation of the telescope or the straight line passing through the sights of the piece; at the instant of firing this line passes through the center of the target.

Line. A line is that which has length, but not breadth nor thickness. A curve or curved line is a line having no finite portion of a straight line.

Litmus Paper. Paper saturated with blue or red litmus, used in testing for acids or alkalies. It is essential to the service in that it is used for testing powder and explosives to determine whether or not they are deteriorating in storage; which fact is indicated when the blue litmus paper is turned red in the presence of acid fumes which are given off to a greater or less extent when a powder or explosive is deteriorating. The amount of deterioration is indicated by the length of the time required for the paper to change color. Litmus. A dye stuff extracted from certain lichens as a blue amorphous mass which consists of a compound of the alkaline carbonates, with certain coloring matters relating to orcin and orcein. When litmus is used as a dye it is turned red by acids and restored to its blue color by alkalies (common salt is a good one).

Load. A single charge of powder and a single projectile as combined for firing in a gun or mortar.

Loading Platform. That surface upon which the cannoneers stand while loading the piece.

Loading Position. At gun batteries; breech closed, cannoneers at posts for inspection, projectile and powder charges on truck near the delivery table. At mortar batteries; mortars horizontal, breech closed, cannoneers except No. 6, at post of inspection, projectiles on trucks about ten feet in rear of mortars, powder at entrance to pit, No. 6 is at the entrance to the powder magazine.

Loading Room. A room suitably equipped for the loading of submarine mines.

Loading tray. A device used to protect the breech recess while loading the projectile.

Location of a Target. The determination of its range and azimuth from a given point. See RELOCATION OF A TARGET.

Long Roll. A drum alarm signal, when, if practicable, each man goes direct to his post at a run.

Longitudinal. Extending in length; running lengthwise. The longitudinal extent of a gun would be its length from breech to muzzle.

Lot. A term used by manufacturers to designate a certain amount of explosive manufactured at one time. All of the explosive of one lot should possess uniform characteristics.

Lug. A projecting piece to which anything is attached, or against which anything bears, or through which a bolt passes.

Lyddite. A high explosive of British manufacture.

Machine and Rapid-Fire Gun Mounts. Guns of this class used in coast defense are mounted on moving or traveling carriages, and fixed mounts, called rapid-fire mounts, which are disappearing or non-disappearing, recoil or non- recoil.

Machine guns. Guns of one or more barrels using fixed ammunition and provided with mechanism for continuous loading and firing. The mechanism may be operated by manpower or by the force of recoil. Guns in which the force of recoil is used to operate the breechblock are termed "semiautomatic." When this force is used also to load and fire the guns, they are termed "automatic."

Magazine. In a literal sense any place where stores are kept; as a military expression a magazine signifies rooms and galleries for the storage of powder, primers, fuses, etc. Magazines are classified as peace magazines and storage magazines.

Main Bore. That part of the bore in front of the forcing cone.

Maneuvering Rings. Large cast iron rings fastened in the walls of emplacements, designed for holdfasts in mechanical maneuvers.

Manning Party. The personnel assigned to the service of any specific element of the defense.

Manning Table. A list of the names of those who constitute a manning party, with the particular post to which each is assigned.

Marine Obstructions. Sunken hulks, subwater piles, dams, booms, barricades, rope entanglements and any other form of barrier that may delay the enemy in navigating a defended water area.

Mark One. A term used to indicate the first improvement of the original model of a particular type of gun, mortar, etc.

Masking Mount. See CARRIAGE or MOUNT.

Master Electrician. A non-commissioned staff officer charged with the general supervision of the electrical and power installations of an artillery district or post. He assists the artillery engineer in his work and is required to make inspections and tests of electrical plants and installations, and perform such other technical duties as may be necessary.

Master gunner. An enlisted specialist of the noncommissioned staff Coast Artillery Corps who is assigned to duty as assistant to the artillery engineer in connection with the preparation of charts, maps, drawings: range tables, etc., in a coast defense command.

Material Target. A target used for all subcaliber practice with guns. When used as a fixed target it is moored fore and aft, as nearly broadside to the battery firing as possible. For battle, fire and mine command practice two or more targets are used on the same towline, separated by about 100 yards. For record subcaliber practice at moving targets it is towed at the end of a 300- yard towline. The standard material target for use as stated above consists of a buoyant base upon which is mounted a vertical rectangular frame 10x24 feet, covered to within 2 feet of the bottom with white cotton cloth divided into three panels, the middle panel being painted black. The base consists of two parallel flotation sills each of a 10x10 inch timber, surmounted by a 3x10 inch plank nailed thereto; three cross-pieces, a prow and four diagonal braces, the whole being fastened with a bridle for towing.

Maul. A heavy wooden beater or hammer, used in driving stakes, tent pegs, etc. Usually miscalled mallet by inexperienced soldiers.

Maximite. A high explosive shell filler. It is fusible and very suitable for armor piercing shell, which are charged by melting the maximite and pouring it in.

Maximum Ordinate. The vertical distance between the line of sight and the summit of the trajectory.

Maximum Pressure. The greatest pressure in the bore of a firearm. The pressure indicated by the pressure gauge.

Maximum Range. The greatest range obtainable by using the maximum elevation permitted by the carriage.

Mean Lateral Deviation. The arithmetical mean of the lateral deviations of the points of impact of a series of shots, from the center of the target.

Mean Longitudinal Deviation. The arithmetical mean of the longitudinal deviations of all the points of impact of a series of shots, from the center of the target. For example, if six shots were fired and struck as follows: No. 1 50 yards over. No. 2 10 yards short. No. 3 10 yards over. No. 4 30 yards over. No. 5 2 yards over. No. 6 12 yards short. The mean longitudinal deviation would be 114/6 = 19 yards.

Measure of Uniformity. The regularity in the velocity given by a number of consecutive shots. It is calculated by taking the mean observed velocity, and from it deducting the difference in velocity of each shot, and dividing the sum of the differences by the number of shots fired.

Mechanic. A specially qualified artilleryman holding the grade of that name in a coast artillery company.

Mechanics. The science which treats of the nature of forces and of their actions on bodies, either directly or by the agency of machinery.

Medium-Caliber Guns. Guns of 4-inch, 4-7-inch, 5-inch and 6-inch caliber.

Melinite. A high explosive compound of foreign manufacture

Melting and Thawing Explosives. The explosive shell filler maximite is melted by placing it in a copper watertight vessel and immersing it in a boiling water bath, the temperature being kept practically at 212 degrees F. Dyna- mite is thawed by putting the cartridges or sticks of frozen dynamite in a water- tight vessel and immersing it in warm water. If sufficient time is available a better method would be to leave the boxes open for several hours in a warm room or by taking the cartridges out of the boxes and laying them on a shelf in a room at which the temperature is about 70 degrees F., and thus allow the cartridges to thaw out gradually.

Mensuration. The act or process of measuring. That branch of applied geometry which gives rules for finding the length of lines, the areas of surfaces, etc., from certain simple data of lines and angles.

Mercurial Barometer. An instrument used in the meteorological station to determine the pressure or density of the atmosphere.

Meteorological Message. The message sent to fire commanders by a meteorological observer. It includes the barometer and thermometer readings, the atmosphere reference numbers and the velocity and azimuth of the wind.

Meteorological Observer. An enlisted man in charge of the meteorological station.

Meteorological Station. A station containing instruments for obtaining and sending out to the various primary stations data relating to the density of the atmosphere and the velocity and direction of the wind.

Micrometer Caliper or Gauge. A caliper or gauge with a micrometer screw for measuring dimensions with great accuracy.

Micrometer. An instrument, used with a telescope, for measuring minute distances or apparent diameters of objects which subtend minute angles. The measurement given directly is that of the image of the object formed at the focus of the object glass.

Mine Command. Such portions of submarine defenses and rapid-fire guns for the protection thereof as may be efficiently controlled by one man. A mine command consists normally of one or more rapid-fire batteries, with the necessary elements for a complete mine system, including proper installation, control and repair of the mine fields. There may be more than one mine command in a battle area, depending upon the size of the harbor and the mine defenses necessary for the protection thereof.

Mine Commander. A coast artillery officer assigned as such in orders from district headquarters. He exercises both administrative and tactical command of a mine command. As an administrative officer it is his duty to have his command supplied with all material necessary for carrying out the improved scheme of submarine defense. In this he is assisted by a property officer, who is responsible to the mine commander for the material and equipment of the command. Tactically the mine commander is responsible to his battle commander for the condition of the submarine defense material of his command and the efficient service thereof at drill or in action. He is responsible for the instruction of the officers and men of his command in all matters pertaining to the care and use of the mine defense material and rapid-fire guns assigned to his command. In battle -command drill and in action he undertakes to destroy such vessels crossing the mine field as the battle commander may direct. When ordered to exercise independent fire action he fires the mines in accordance with his own judgment. He exercises the same command over rapid-fire batteries assigned to the mine command, as is the case in fire commands. He has control over and is responsible under the battle commander, for, the use of the mine field searchlights.

Mine Company. A company assigned to the service of submarine mines.

Mine Field. The area of water in which submarine mines are planted.

Mine Planter. A seagoing tug 150 feet in length and about 30 feet beam. It has large deck space forward and little rigging. It is equipped with booms, winches, davits, catheads, triplex blocks, etc., necessary in handling and planting assembled mines.

Mine-Field Lights. Searchlights that may be used for searching when no attack on the mine field is anticipated, and to illuminate the channel for the purpose of aiding the entrance of friendly vessels. During an attack on the mine field their function is to deceive the enemy by searching over an area outside the field. The mine field should not be illuminated until the enemy's boats approach one of the fields.

Mine-Field Officer. A coast artillery officer in charge of laying and maintaining the mine field. He is responsible that the planting and loading sections perform their duties correctly and that proper material is used in the assembling and planting of mines; that all precautions for safety are taken in loading, planting and taking up of mines.

Mine. See SUBMARINE MINE.

Mining Casemate. A protected building containing the controlling mechanism of the mine defense.

Minus Correction. When conditions require the use of a range less than the actual range it is termed minus correction.

Minus Deflection. A piece is said to have minus deflection when its axis points to the left of the target.

Misfire. The failure of a powder charge to explode. In case of a misfire in artillery practice the breech will not under any circumstances be opened for ten minutes, nor until the primer has been removed, except when the primer is seated in the cartridge case.

Miter Wheel. See BEVEL WHEEL.

Mortar Battery. The entire structure erected for the emplacement, protection and service of one or more pits of mortars.

Mortar Company. A company assigned to the service of mortars.

Mortar Pit. See PIT.

Mortar. A cannon employed to throw projectiles at high angles of elevation. Their length of bore is small in comparison with the caliber.

Mount. See CARRIAGE or MOUNT.

Mounting Cannon. The mechanical maneuvers necessary to place coast cannon in position.

Movable Armament. Small caliber guns on wheeled mounts, such as the Colt automatic and Gatling machine gun.

Movable Carriage or Wheeled Mount. A carriage or mount provided with wheels for ready transportation of the piece mounted thereon.

Mushroom head. The front part of the De Bange obturator.

Muzzle Velocity. The rate of travel at which a projectile leaves the muzzle of a cannon. It is sometimes called initial velocity.

Muzzle. The front end of a cannon including the mouth of the bore, the face and the swell. The face of the muzzle is the front plane of the gun perpendicular to the axis of the bore.

Nitro-Cellulose Powder. The name applied to a form of smokeless powder used in modern ordnance, in which cellulose (unspun cotton waste) is the base.

Nitro-Glycerine Powder. The name applied to a form of smokeless powder used in modern ordnance, in which nitro-glycerine is the base.

Non-commissioned Staff Officers. The Coast Artillery Corps non-commissioned staff officers consist of sergeants major, senior grade; master electricians, engineers, electrician sergeants, first class; electrician sergeants, second class; master gunners, sergeants major, junior grade, and firemen. They are appointed after due examination, and receive warrants signed by the Chief of Coast Artillery.

Nose. A name sometimes given to the point of projectiles.

Object Glass. The glass in a telescope which is placed at the end of the tube nearest the object.

Oblique Fire. Fire which is directed obliquely to the object fired at.

Observation Firing. A term applied to one of the three methods of exploding submarine mines, i.e., where the time of firing is given from the mine commanders' station.

Observation Telescope. A telescope used in target practice and in action to observe the striking point of shots.

Observer. A member of the fire-control section who is in charge of and uses an observing instrument. Insignia: First Class. Red triangle with a red bar below within a yellow circle. Second Class. Same as first class, omitting the bar. All to be of cloth.

Observing Interval. The time in seconds between two consecutive observations on a target (between two signals of the time interval bell) during tracking. The regular interval for guns is 15 seconds, and for mortars 30 seconds.

Observing Room. The room of a primary station in which the position - finding instrument and necessary accessories are located.

Observing Station. A position constructed in a favorable place for observing the field of fire. A protected position constructed in a parapet or traverse for the purpose of observation, commonly called " Crow's Nest."

Obturating Primer. A primer of any type so constructed as to prevent the escape of powder gas through the vent.

Obturator. A device for preventing the escape of gas. Obturation is the process of preventing the escape of gas. The mushroom head of the breechblock. In gunnery, any device for preventing the escape of gas; the term includes the entire mechanism.

Occult Light. The act of screening or shutting off the beam of a search- light.

Ogive. That curve of the head of a projectile which terminates at the point.

Oil Room. A room in the emplacement for the storage of oil.

Oils. The principal oils used in the coast artillery service are : Hydroline, for filling recoil cylinders. Synovial, for lubricating the breech recess and breechlock of cannon, and general lubrication of the carriage. Light Slushing, for slushing the bore of cannon and all exposed surfaces of the carriage. Lubricant No. 4, for filling grease cups. Linseed (boiled), for use on retraction ropes, mixing paints, etc. Kerosene, for cleaning purposes.

Omniscope. An apparatus used in the Lake type of submarine boats for observation, sighting and steering.

One-Pounder. A rapid-fire gun whose projectile weighs one pound. The caliber 1.457-inch pompom, Vickers-Maxim gun is an example of this type.

Open Sight. See SIGHT.

Opposite Angles. When two lines meet and cross each other four angles are formed, and the opposite angles are equal to each other.

Orders of Fire. 1. Unrestricted Fire. When the only limitation imposed by the fire commander upon the action of a battery is the assignment of a target, the fire is said to be unrestricted. This is the normal fire action of a battery. 2. Restricted Fire. When the range at which to fire, the number of shots, the firing interval, or any other limitation except as to target, is imposed upon the action of a battery, the fire is said to be restricted.

Ordnance Machinist. A civilian expert ordnance machinist, resident at each coast artillery fort.

Ordnance Sergeant. A non-commissioned staff officer charged with the care and preservation of all ordnance property at a coast artillery fort. Chevrons and Insignia: Three bars inclosing a shell and a flame. All to be of black cloth piped with red.

Ordnance. The term applied to artillery armament and the accessories and stores pertaining thereto.

Orientation Table. A table showing the azimuths and distances of various points in a harbor.

Orientation. The process of adjusting an instrument, gun or mortar in azimuth.

Out of Commission. A term applied to armament and fire-control stations that are not in such condition that they could be made ready for service in 24 hours.

Out of Service. A term applied to armament and fire-control stations to which no manning party is assigned but which could be made ready for service in 24 hours.

Outposts. Detachments thrown out from a force for the purpose of protecting it from surprise.

Outside of a Beam. A point is said to be outside of a searchlight beam when it lies on the outer side of a line passing through the axis of the beam as seen from the battle commander's station.

Paints on Projectiles. Projectiles are painted so as to show the material used in their manufacture, their armor-piercing qualities, their center of gravity and their character.

Pantograph. A device used on the plotting board of a fire-command station, to relocate for data for use at any battery in that command.

Parade Slope or Wall. The rear slope or wall of an emplacement.

Parados. A structure in rear of a battery for protection against fire from the rear. It may have interior, superior, and exterior slopes .

Parallax. An apparent displacement of an object observed through a telescope, due to the real displacement of the observer, so that the direction of the object with reference to the observer is changed. In optics, parallax is an apparent displacement of the image upon the cross- wires in a telescope when the eye is moved across the eye-piece. It is due to the non-coincidence of the cross-wires with the focal plane of the objective. Both the image, as formed by the objective, and the cross wires, should lie in the focus of the eye-piece, i.e., in the same plane. The image may be moved back and forth by moving the objective in or out, but the plane of the cross wires is fixed. When the two are brought into the same plane the image is brought upon the cross- wires. To accomplish this the eye- piece should first be focused on the cross-wires so that they appear most distinct, the irregularities of the wires being very apparent. There should be no image visible during this operation, that is, the objective should either be thrown out of focus by turning the focusing knob either all the way out or in; or the tele- scope should be pointed to the sky. The eye-piece should then be moved until the inner and outer limits of distinct vision of the wires are found, and then set at the mean position. The telescope should then be pointed toward the object and moved until the image also comes into focus accurately. If parallax is now found it should be removed by refocusing. The adjustment of the eye-piece will be correct for the same observer regardless of the range and object sighted upon, but it may be necessary to refocus the objective when the range differs materially.

Parapet. That part of a battery which gives protection to the armament and personnel from front fire.

Pawl. A pivoted tongue, sliding bolt or catch, adapted to fall into notches or indental spaces in such a manner as to permit motion in one direction and prevent it in the reverse, as in a windlass.

Penetration of Projectiles. The ability of a projectile to overcome the resisting qualities of an armor plate by completely or partially perforating it. The ability to perform this function depends on the relative merits of the particular projectile, the plate against which it is fired, the striking velocity and the angle of impact.

Percussion Cap. A cap in which the method of explosion is due to a blow; used in fixed ammunition.

Percussion Fuse. A fuse which is armed or prepared for action by the shock of discharge and acts upon impact.

Percussion Primer. The type of primer used in fixed ammunition which is exploded by a blow of the firing pin.

Perforation. This term used with reference to a projectile fired against armor signifies that the projectile passed entirely through the armor.

Personnel. The personal composition of any organized group of officers or men which has for its purpose the accomplishment of some service.

Picric Acid. A detonating and disruptive explosive of high order used as a base for explosives or shell fillers.

Piece. The name applied to any type of cannon, whether gun or mortar. It is also used as a matter of convenience to designate both cannon and carriage when the cannon is mounted.

Pintle center. The vertical axis about which a gun or mortar carriage traverses.

Pit Commander. A non-commissioned officer (gun commander) in charge of a mortar pit.

Pit. That part of a mortar emplacement designated for mounting one or more mortars, usually two or four.

Plane of departure (also called *plane of fire*). The vertical plane containing the line of departure.

Plane of direction. The vertical plane containing the line of direction.

Plane of sight. The vertical plane containing the line of sight.

Plane of splash. The vertical plane containing the chord of the trajectory.

Planting Section. A section of the enlisted personnel of a mine company, consisting of men required afloat. It is divided into a planter detachment and the small boat detachments.

Plotter. A specially qualified enlisted man in charge of the plotting board at a fire-control station. Insignia: Red triangle with a red bar below within a yellow circle, all to be of cloth.

Plotting Board. A device used in the position-finding service to quickly plot to scale the data sent from the position-finding instruments, and in connection with range and deflection boards, to determine the corrected data for firing. It consists essentially of a semi-circular drawing board with a radius of 45 inches, made of well-seasoned lumber. It has mounted upon it the necessary scale arms and gun center to determine the rate of travel of the target both in range and azimuth for use on the range and deflection boards. By means of the gun arm the range and azimuth to the target from the directing point of the battery is found.

Plotting Room. The room in which the plotting detachment works. It is usually located below and communicates with, the instrument room of the battery commander's station, or with the observing room of the primary station.

Plunging Fire. Fire in which the line of departure passes below the horizontal plane.

Plus Correction. When conditions require the use of a range greater than the actual range the correction is called plus correction.

Plus Deflection. A piece is said to have plus deflection when its axis points to the right of the target.

Point Fuse. A firing device inserted in the point of a shell not intended to penetrate; as shrapnel.

Point of fall. The point at which the trajectory again pierces the horizontal plane through the muzzle of the gun.

Point of impact. The point at which the projectile first strikes. When the projectile strikes the water before striking any object, the point of impact and the point of splash are the same.

Point of splash. The point at which the trajectory pierces the surface of the water.

Pointing. The operation of giving the direction and elevation necessary to hit the target. When the sight is used it is called "aiming", when the sight is not used it is called "laying." There are three cases of pointing: Case I. When direction and elevation are both given by the sight. Case II. When direction is given by the sight and elevation by the range scale on the carriage or by quadrant. Case III. When direction is given by the azimuth scale and elevation by quadrant or by the range scale on the carriage.

Position Angle. See ANGULAR ELEVATION.

Position finder. An instrument for locating a target.

Position finding system. The term applied to the system used in determining the range and direction to any target from a battery or station. The system of position finding includes: 1. The horizontal base system, which employs azimuth reading instruments in stations at the ends of a base line, and a plotting board. 2. The depression -position finding system, which employs a depression position finder at a considerable elevation above the sea level, and a plotting board. 3. The emergency system, which ordinarily employs a self-contained instrument located at the battery, with or without a plotting board.

Powder Blast. The force of the powder gas for a short distance in front of the muzzle, which acts destructively on objects close at hand lying within its path.

Powder Cases. Cases in which powder is contained in shipment from arsenals or storage until used. Three types are in common use; the zinc storage case with balata washers; the wooden storage case; and the metallic cartridge case hermetically sealed. The latter is rapidly replacing all the others in the service.

Powder Chamber. The chamber in the bore for the reception of the powder charge; it is usually cylindrical, but frequently conical and sometimes elliptical. It is between the breech recess and the centering slope, which unites it with the forcing cone.

Powder Chart. A graphic chart used to determine the velocity to be expected from a given charge of powder considered as a function of the temperature of the powder.

Powder Chute. In gun emplacements an inclined well or shaft for returning cartridges or dummies to magazine.

Powder Hoist Well. The shaft through which the powder hoist operates.

Powder Hoist. A device for raising powder from the magazine to the loading platform.

Powder Magazine. See MAGAZINE.

Powder Room. See CARTRIDGE ROOM.

Powder. See GUNPOWDER.

Power and Light Equipment. Equipment including engines, dynamos, storage batteries, motors, electric and other kinds of lights, and all material and supplies pertaining thereto.

Power Room. A room in the battery provided for the necessary motor generators, induction motors and switch-boards.

Power Section. A detachment of the enlisted personnel of a mine company consisting of the operators and assistants required at the power plants, searchlights and mining casemate of a mine command.

Power Station or Plant. The principal source of supply of energy, usually electrical, for the power system of the fortifications and stations. The plant consists of a sufficient number of direct connected units to supply all the power needed for the entire installation under conditions of full load.

Predicted Firing. Firing at which guns and mortars are given direction and elevation corresponding to a predicted point.

Predicted Point. A point on the course of a moving target, as indicated on the plotting board at which it is predicted the target will arrive at the expiration of an assumed interval of time. This interval of time is called the predicting interval.

Predicted point. The point at which it is estimated a target will arrive at the end of an assumed interval of time reckoned from the time of the last observation on which the estimate is based. This interval of time is called the "predicting interval."

Predicted Time. The time at which the target should reach a predicted point.

Predicting Interval. See PREDICTED POINT.

Predictor. An accessory of the plotting board used to locate the positions of the predicted and the set-forward points on the plotting board.

Preponderance. The excess (moment) of weight of that part of the piece in rear of the trunnions over that of the front, or the converse. It is measured by the force expressed in pounds necessary to balance the cannon when resting freely on the trunnions.

Pressure Cylinder. A soft copper cylinder used in crusher gauges which is compressed by the explosion of the charge.

Pressure gauge. A gauge placed in cannon to measure the maximum pressure developed during firing.

Pressure Gauge. See CRUSHER GAUGE.

Primary Armament. The armament of a coast artillery fort used to attack the side, turret and deck armor of war vessels, and carry large explosive charges into their interiors. It includes the 8-inch, 10-inch, 12-inch, 14-inch and 16-inch guns, and 12-inch mortars.

Primary Station. The principal station of a base line. See BASE-END STATION.

Primer. A wafer, cap, tube or. other device for communicating fire to the powder charge. There are five classes of primers used in the U.S. Coast Artillery service, namely: Friction, percussion, electric, combination and igniting.

Priming charge or igniter. Small charges of black powder in the ends of powder sections necessary for the ignition of smokeless powder.

Priming Charges. A charge consisting of black powder, quilted in each end of the bag containing the sections of smokeless powder for the purpose of igniting it.

Prismatic Powder. A molded gunpowder hexagonal in shape, with a single round perforation through the center of the grain. It is either brown or black in color, depending upon the color of the charcoal used. The black prismatic powder is made of the ordinary black granulated powder, the "cannon powder" grain being taken as a base.

Probability Factors. A table of factors which, multiplied by the width of a zone containing fifty per cent, of the hits, will give the width of zones containing any other percentage of hits.

Probability of Error. As referred to gunnery, is that particular error in any direction in which it is an even chance will not be exceeded by any shot. It is based upon the rule that when the value of any quantity or element has been determined by means of a number of independent observations, each one liable to a small amount of accidental error, the result of determination will also be liable to some uncertainty. The probable error, therefore, is the quantity, which is such that there is some probability of difference between the determination and the true absolute value of the thing to be determined, exceeding or falling short of it. The probable error of a gun, in any direction, is a distance measured in that direction from the center of impact, of such length that it is an even chance that it will not be exceeded by a single shot, and for which it can be predicted that in the long run 50 per cent, of all shots fired will have a less error.

Probable Zone. The space bounded by two parallel straight lines of such length that by the theory of probabilities 50 per cent, of the points of impact will probably be found.

Profile Board. A thin plate or board having its edge so cut as to represent the outline of an object ; it is used to prove the models of the breech and other exterior parts of a gun.

Progressive Powder. An explosive or propelling agent of low order; for example, the charcoal and nitro-cellulose powders. The explosion of powders of this kind is marked by more or less progression. The mass is ignited at one point and the combustion proceeds progressively over the exterior exposed surfaces and then at right angles to these surfaces.

Projectile. The term applied to a missile thrown from a firearm by an explosive. The principal parts of an armor-piercing projectile are the ballistic cap, the armor-piercing cay, the nose or point, the ogive, the bourrelet, the body, the rotating band, the cavity, the base, the base plug, and the fuse plug. The ballistic cap is for the purpose of reducing the effect of or the retardation due to the resistance

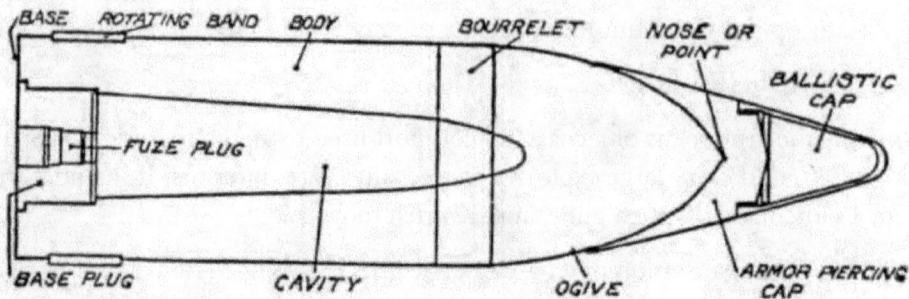

of the air. It consists of a hollow metal cap placed over the armor-piercing cap. The armor-piercing cap is a piece of soft steel placed over the point to prevent the point from bending or breaking on impact against bard-faced armor, and to thereby increase penetration. The lower part of the ogive is turned off to make a cylindrical bearing surface for the front part of the projectile. This surface, called the bourrelet, has a diameter slightly less than the caliber of the gun, but greater than that of the cylindrical portion of the projectile. The rotating band is forced through the rifling of the bore and gives rotation to the projectile. The rotating band also seals the grooves and prevents the escape of gas. When the rifling is worn, due to erosion, broader bands are necessary. Rotation is given to the projectile in order to prevent the projectile from tumbling end over end in the air. The rotation in our service is clockwise, as viewed from the base of the projectile. The base and fuse plugs are arranged to screw to the left, so that the rotation of the projectile to the right may have no tendency to unscrew them. Coast artillery projectiles are cast iron or steel. The service projectiles are the armor-piercing shot, the armor-piercing shell, and the shrapnel. The shot has a thicker wall and contains a smaller bursting charge than the shell. The shrapnel is a projectile which carries a number of bullets to a distance from the gun and there discharges them over an extended area.

Projector. The technical name of a searchlight.

Proof of Gunpowder. A certain test made on separate lots of gunpowder before it is accepted by the War Department.

Proof Plug. See CRUSHER GAUGE.

Protractor. A mechanical instrument used for laying out and measuring angles on paper.

Pyramidal Target. A material target in the form of a pyramid, covered with canvas painted vermilion, divided into rectangles 2 feet wide, which are painted alternately vermilion and white. This pyramid is mounted on a float made of two parallel sills of timber, joined by transoms, two diagonal braces and a prow to which a suitable bridle is attached for towing. This target is used as a fixed target for all trial shots, and in case the material target for heavy guns has not been furnished, two of these targets are towed 60 feet from center to center, with a red streamer suspended from a wire or hemp line at the middle point between them, to represent the material target. For service practice with mortars this target is used to represent the center of a circular hypothetical target 100 yards in diameter. For subcaliber practice with mortars this target without canvas covering but with a flagstaff and flag is used.

Quadrant Angle of Departure. The angle between the line of departure and the horizontal plane through the muzzle. It is obtained from the angle of departure by correcting for the angular elevation or depression of the target, including curvature of the earth.

Quadrant Elevation. The angle between the horizontal and the axis of the bore when the piece is pointed. It is obtained from the angle of departure by correcting for the angular elevation or depression of the target including curvature of the earth and for jump.

Quadrant. The quarter of a circle or the quarter of the circumference of a circle ; an arc of 90 degrees. See GUNNER'S QUADRANT.

Quick-Firing Guns. A British term for rapid-fire guns.

Quickness of Burning. The rapidity with which a grain of powder is consumed. When it is said that the powder is too quick or too slow for a gun, the quickness of burning through the "critical dimension" of the grain is re- ferred to.

Racer. That part of a gun or mortar carriage which rests upon the traversing rollers. On gun carriages the chassis is bolted to the racer, and on mortar carriages the are frames are bolted to the racer.

Rack. A bar or arc, having teeth that engage with those of a gear wheel or worm.

Radial Vent. A vent extending at right angles to the axis of the bore.

Rammer. A rod provided with a graduated brass ring; used for properly seating a projectile in the bore of seacoast cannon.

Ramp. An inclined plane or foot-path, serving as a means of communication from one level to another.

Range Board. A device for obtaining the range corrections which must be made for wind, atmosphere, tide, velocity, and travel of target during the observing interval and time of flight.

Range Deviation. The difference between the range to the target and the range to the point of impact.

Range Difference. The difference in range of a point from any other two points, as, the difference between the range to the target from the directing gun, and the range to the target from any other gun of the same battery.

Range Finder. An instrument for determining the range to a target or object, from some fixed point.

Range Keeper. A specially qualified member of the fire-control section, who operates the time range board and calls out the range to the range setter as often as may be necessary to insure the piece being kept at proper elevation. In restricted fire at a specified interval he keeps the time and indicates to the chief of detachment the proper time to trip the gun.

Range of a Shot. See RANGE.

Range of Ballistic Tables. See THEORETICAL RANGE.

Range Officer. A coast artillery officer in immediate charge of all or a part of the fire-control section. He is stationed at the battery plotting-room. He is responsible to the battery commander for the condition of the material pertaining to the fire- control service, for the instruction of the fire-control personnel and for the efficiency of that service in general. Upon opening the station he should make careful inspection of the equipment, verify the adjustment of the position- finding instruments, plotting-board, etc. After satisfying himself that every- thing is in order and receiving the reports of the chiefs of details, he reports to the battery commander: "Sir, fire-control stations in order," (or reports defects he cannot readily correct). In battle and fire-command drill or action, he receives directly from the fire commander and executes orders, as to the assignment of targets. When direct communication between the fire commander and the battery commander is impracticable he receives directly and executes other orders pertaining to the fire action of the battery. He is responsible to the battery commander for the prompt and accurate transmission to the battery commander of orders received by him from the fire commander. At the close of the drill or action he directs stations to be closed, inspects them and reports to the battery commander, handing him all records pertaining to the work at his station.

Range Rake. An instrument made in the form of an ordinary rake. The main arm is shaped like a gun stock with the cross arm extended at front end. At a convenient distance on the main arm is placed a guide peg representing the rear sight, while on the cross arm pegs are placed at intervals of one-half of an inch to represent points, each point having a value in mils equal to 1/1000 the length of the towline connecting the target with the tug upon which the observations with the range rake are taken. To use the range rake the observer takes position at a point in the center of the stern of the tug and aims the rake so that the line of sight from the guide peg will pass over the center peg on the cross arm and thence to the target. At the instant the splash of the shot occurs, the observer, keeping the rake pointed as above described, sights over the guide peg and in the direction of the splash, observing which peg on the cross arm is in line with the splash. For example, if a shot struck five divisions of the rake beyond the target (or five pegs), assuming the value of each peg at 10 mils, the reading in mils would be 50; if the towline wet measured 295 yards, the shot would have struck 14.75 yards over. This is obtained as follows: 50 X 295 / 1000 = 14.75 yards.

Range Scale. The graduations in yards either on the range scale arc of rapid fire guns, or the quadrant arc of large-caliber guns.

Range Setter. A specially qualified member of the gun section who lays the gun for range.

Range table. A table of the elements of the trajectory of a particular cannon for a standard muzzle velocity and a given projectile. For direct fire guns, with the range as argument, are tabulated angle of departure; change in elevation for 10 yards full range; time of flight; angle of fall; slope of fall; maximum ordinate; striking velocity; perforation of Krupp armor at normal, and 30° from normal, impact; drift; and deflection for 10 miles per hour wind component. The ballistic coefficient used in calculating the table is shown. For indirect fire cannon, with elevations and zones as arguments, are tabulated range; time of flight; drift; angle of fall; maximum ordinate; striking velocity; and perforation of deck steel.

Range-azimuth table. A table of ranges and the corresponding azimuths from a gun to points in the center of the main ship channel or channels. It is kept at the gun and used for firing without the use ***of range-finding apparatus.***

Range. In a limited sense, the horizontal distance from the gun to the target. In a general sense, it is applied to horizontal distances between position finder and target, position finder and splash, gun and splash, etc. The range of a shot is the horizontal distance from the muzzle of the gun in the firing position to the point of splash. (Practically the range is reckoned from the axis of the gun trunnions in the firing position, instead of from the muzzle, but the difference in range is negligible.) The range used in ballistics is the horizontal distance from the muzzle of the gun in the firing position to the point of fall.

Rapid-Fire Gun Carriage. See CARRIAGE or MOUNT.

Rapid-fire gun. A single-barrel breech-loading gun provided with breech mechanism, mounting and facilities for loading, aiming, and firing with great rapidity. The breech mechanism is operated by a single motion of the handle or lever. The smaller calibers use fixed ammunition.

Rate of Fire. The average rate of fire of heavy caliber guns with service charge should be about one shot in forty or fifty seconds.

Rated men. Enlisted men who have passed examinations for the positions and who have been rated by the coast defense commander as gun commanders, gun pointers, observers, plotters, casemate electricians, chief planters, and chief loaders.

Ratings. A particular class or grade to which enlisted men belong. In the coast artillery they may be rated as sergeants major, master electricians, engineers, electrician sergeants, master gunners, firemen, casemate electricians, observers, plotters, chief planters, chief loaders, gun commanders, gun pointers and gunners.

Ready. At gun batteries, a signal given to indicate to the gun pointer that the piece is ready to be fired. At mortar batteries, a signal given to the battery commander that the mortars\are ready to be fired.

Rear Slope. The rear slope to the parade in rear of the battery.

Receiving Table. The hoist table on which ammunition is placed preparatory to raising it to the loading platform level.

Recoil buffers. Devices on gun carriages for the purpose of reducing the shock due to abnormally excessive recoil.

Recoil Cylinder. The hydraulic cylinder attached to the carriage for controlling the recoil of the piece.

Recoil. The backward movement of the gun on firing. Counter recoil is the return of the gun in battery.

Recorder. An enlisted man stationed at each battle- and fire -commander's station who keeps an accurate written record of all orders and communications received and transmitted by the battle or fire commander.

Records of Firing. Data taken during target practice or that relating to the gun, carriage, conditions of loading, laying, etc., which would be of value in connection with future firings.

Rectangular Target. See MATERIAL TARGET.

Reference Numbers. Arbitrary numbers used to avoid "plus" and "minus," "right" and "left," in data for firing. These numbers are used on the graduations of scales on devices of the position-finding equipment to avoid liability of error by the use of one set of numbers instead of two sets. Without reference numbers it would be necessary to have one set of numbers for plus, and another set for minus corrections. For example, if the wind curves on the range board were numbered in both directions from zero, there would be a "plus" 10-mile wind curve and a "minus" 10-mile wind curve. In use it would be comparatively easy to make the mistake of taking the plus 10, instead of the minus 10. The corresponding reference numbers, however, for wind would be 60 for plus 10, and 40 for minus 10 (50 being the zero). In this manner the liability of error is minimized.

Regulations. Under the Constitution of the United States, any rules for the government and regulation of the army made by Congress. Regulations imply regularity and signify fixed forms; a certain method, order or precise determination of functions, rights and duties. It embraces administrative service, system of tactics, and the regulation of service in campaign, garrison and in quarters.

Relay. The command given when mortars are not to be fired as laid, but are to be fired on the next data furnished.

Relocation of a target. A process whereby the range and azimuth of a target from a point may be obtained without observation when the range and azimuth of the target is known from some other point.

Remaining Velocity. The velocity of a projectile at any point of the trajectory.

Reserve Table. A table in a sheltered position for reserve ammunition.

Resistance of the Air. The retarding effect of the air on projectiles during their flight.

Restricted Fire. See ORDERS OF FIRE.

Retardation. The velocity a projectile loses in consequence of a resisting medium.

Retractor. A device for withdrawing the empty cartridge shell rearward from the bore of small -calibered guns.

Reverse Fire. Reverse fire is when the object is fired at from the rear.

Ricochet. The rebound of a projectile along a surface.

Rifle. A cannon or gun with the interior surface of its bore grooved with spiral channels or cuts, thus giving the projectile a rotary motion. If the interior surface of the bore is smooth, the cannon is known as smooth bore.

Rifling. Helical grooves cut in the surface of the bore for the purpose of giving a rotary motion to the projectile. The rib of metal between two adjacent grooves is called a "land." (See Twist of rifling.)

Rimbases. The masses of metal uniting the trunnions of a cannon with the trunnion band.

Ring Resistance Fuse. A base or point fuse used in charged shell and shrapnel. The name is derived from the manner in which the firing pin is maintained in its normal or unarmed position by a brass split-ring spring.

Roadstead. A water area where ships may ride at anchor some distance from the shore. An anchorage off shore.

Rocket. A projectile set in motion by forces residing within itself, usually used for signaling.

Rotating Band. The copper band encircling projectiles near their base for the purpose of giving them angular rotation in passing through the rifling of the bore.

Rotation of Projectile. The act of the projectile turning upon its axis during the time of flight.

Round. A round of ammunition includes a projectile, charge and primer. To fire one round is to discharge one shot from each gun of a battery.

Roving Light. A searchlight intended to search the battle area within the field not covered by fixed lights.

Row. See HOOPS.

Rubber Impression of the Bore. An impression taken on rubber and used to determine the amount of erosion or other irregularities of any portion of the bore.

Safety Lanyard. A safety device attached to seacoast cannon consisting of a lanyard wound on a drum working against the action of a spring and attached to the gun. It is so arranged, by means of a ratchet and pawl, that a pull on the firing lanyard can not be transmitted to the primer until the gun is in battery.

Salvo Fire. Fire concentrated from one or more batteries against a salvo point.

Salvo Point. A point, the azimuth and range of which are known and conspicuously posted in the battery; at which a concentrated fire from one or more batteries may be directed. Certain points in narrow channels are usually selected as salvo points.

Salvo table. A table giving ranges and azimuths of salvo points. ·

Salvo. The simultaneous firing of a single load from each gun or mortar of a battery or from each mortar of a pit. The former is called a "battery salvo" and the latter a "pit salvo."

Screw Box. The breech recess of seacoast cannon.

Searchlight Area. The area of land or water illuminated by a searchlight.

Searchlight Observer*.* A member of a searchlight detachment equipped with a night glass, stationed outside of fixed or roving lights at such distance that he can detect readily any vessel passing into the beam.

Searchlight Officer. A coast artillery officer in charge of the searchlight system covering a battle area, and the manning party assigned thereto. His station is at the battle-command station, or within speaking distance of the battle commander. He is responsible to him for the condition of the search- light material and for the efficiency of the searchlight system. At night drill and in action he stands ready to execute the orders of the battle commander as to searching the battle area. When deemed necessary by the battle commander, the searchlight officer may temporarily take charge of the mine-field lights. He is responsible for the instruction of the personnel assigned to the control and operation of the searchlights, and by frequent inspections sees that they are thoroughly familiar with their duties and the prescribed method of searchlight control. In performing his duties from the battle commander's station he should be constantly on the alert to see that no vessel enters the battle area without being detected either by himself or one of his observers, and should so direct the searchlights as to promptly pick up any vessel entering the harbor.

Searchlight Operator. An enlisted man specially trained in the care and operation of searchlights.

Searchlight Range. The distance at which a target can be illuminated sufficiently for identification and range-finding purposes. The maximum effective illuminating range on a clear night would be approximately 8,000 yards for horizontal-base range-finding system; for purposes of water-lining the target in using the vertical base system the range would be approximately 6,500 yards.

Searchlight Tower. An elevated structure containing the searchlight and its operating mechanism.

Searchlight. A high-power electric arc light, used for night illumination. Searchlights are classified as, Fixed Lights, Roving Lights, Illuminating Lights, Mine Field Lights and Battle Lights. The standard diameters are the 60-inch and the 36-inch lights. A few 30-inch and 24-inch lights are still in use.

Seat of the Charge. The form of that part of the bore of a firearm which contains the charge.

Secondary Armament. The armament of a coast artillery fort used to defend the mine field; to attack lightly armored or unarmored ships, the upper works, etc., and personnel of war ships and defend the inner waters against small boats and landing parties. It includes 3-inch guns, small-caliber guns, and machine guns when on fixed mounts.

Secondary or Auxiliary Power Plant. A reserve unit of power supply usually located at each battery, to furnish it with power in case the central power plant or main source of power supply fails or is put out of action.

Secondary Station. A position-finding station furnished with an authorized range-finding instrument upon which readings are taken simultaneously with those at the, primary station if the horizontal-base system is used; or upon which range and azimuth readings are taken if the vertical -base system is used and it is desired to make observations and obtain data taken from that position.

Sector of Explosion*.* At the moment a cannon is fired, there is a sort of spherical sector of fire formed in front of the piece, called sector of explosion. The extremity of this sector presses against the rear end of the bore while the external portion of it terminates in the air, which it compresses and drives in every direction; the air thus forming a support, the sector reacts with its full force upon the rear end of the bore and causes the recoil of the piece.

Sector. A plane figure part of a circle, enclosed between two radii and the included arc.

Segment. A piece of metal in the form of the sector of a circle, or part of a ring.

Self-Contained Horizontal Position Finder. A position-finding instrument containing a horizontal base line within itself. This base line is from eight to twenty-five feet in length.

Semi-Automatic Machine Guns. Rapid-fire guns in which the force of recoil is used to operate the breechblock, but not to load and fire the piece. See AUTOMATIC MACHINE GUNS.

Separate Observing Room. An observing room which does not adjoin a plotting room or other observing room.

Separate Plotting Room. A room in which the plotting board is located and used; instead of being located in the primary station.

Sergeants Major. Coast artillery corps non-commissioned staff officers assigned to duty as assistants to adjutants in administrative functions. Their tactical duties are the same as those of regimental or battalion sergeants major of infantry and any coast artillery duty pertaining to their proper position.

Service Charge. The maximum quantity of powder that is prescribed to be used in any seacoast cannon.

Serving Table. A table for keeping a supply of projectiles convenient to the breech during loading. It is usually mounted on wheels.

Serving the Vent. The term implies the removing of the old primer from the obturator, inserting a new one, adjusting the slide of the firing attachment, attaching the firing wire or lanyard and cleaning the vent.

Set-Back Point. A point on the course of a target determined in a similar manner as a set-forward point, but in the opposite direction.

Set-forward point. A point on the course of a target in advance of the plotted point at which it is estimated that a target will arrive, at the end of the predicting interval plus the time of flight for the range. It is located (on the plotting board) by laying off from the last plotted position of the target, along the estimated course of the target, a distance equal to the travel of the target during the predicting interval plus the time of flight.

Set-Forward Ruler. A celluloid rule that may be used in the position - finding system for mortars, to determine the set-forward point; it consists of a time-of-flight scale in seconds, a travel scale in yards and a scale giving yards of travel during time of flight plus one minute. It is arranged in the form of a slide rule.

Shears. A form of tackle consisting of two spars lashed together at one point, forming an inverted V.

Shell filler. An explosive used to make up the bursting charge in n projectile.

Shell room or shot room. A room for the storage of projectiles.

Shell tracer. A device attached to the base of a projectile which enables its flight to be followed. In the daytime a smoke (which is visible) is emitted and at night a bright flame.

Shell. A steel or cast iron projectile the center of which is hollowed to be filled with the bursting charge.

Shimose Powder. A high explosive of Japanese manufacture.

Shot Chamber. That part of the bore in which the projectile is seated. It includes part of the centering slope and the rear portion of the forcing cone.

Shot Gallery. A gallery or room in the emplacement for the storage of projectiles.

Shot hoist well. The shaft through which the projectile hoist operates.

Shot hoist. A device for raising projectiles from the hoist room to the loading or truck platform. Sometimes called ammunition hoist.

Shot Room. The room in earlier emplacements for the storage of shot.

Shot Tongs. A mechanical device used to encircle the projectile at the center of gravity to facilitate handling.

Shot. A projectile with a small cavity for explosive; also the firing of a single load from a single gun or mortar.

Shrapnel. A projectile composed of a number of spherical balls enclosed in a cast iron case, with a bursting charge in either point or base to scatter the missiles. The point is armed with a combination fuse. It is distinguished from shell by this point.

Sight Deflection. The horizontal angle between the line of sight and the axis of the piece. It is used to correct for conditions of drift, wind, and movement of target tending to cause deviation.

Sight Elevation. The elevation measured from the line of sight; it is the angle between the line of sight and the axis of the bore when the piece is pointed. It is obtained from the angle of departure by correcting for jump.

Sight standard. The upright on the carriage which supports the sight.

Sight. An instrument by which the gun pointer gives the gun the proper direction for firing. Sights are of two classes, open and telescopic. The former consists of two points which are brought into line with the target by the unaided eye; the latter uses the magnifying power of the telescope and is the standard sight.

Signal Rocket. An ordinary skyrocket used in fortress warfare and exercises, as a means of communication.

Signal Station. A station located if practicable at a height sufficient to give a sky background, from which visual signals are displayed.

Six-Pounder. A rapid-fire gun of 2.24-inch caliber. The name denotes the proper weight for the projectile of the piece.

Small Boats. Launches, cutters, gigs and yawls, use d in connection with submarine mine work, boat drill and transportation.

Small-Caliber Guns. Guns of 3-inch caliber, 6 and 1-pounders.

Smokeless Powder. The name given to nitro-cellulose and nitro-glycerin gunpowder.

Smooth-Bore Cannon. See RIFLE.

Specific Gravity. The ratio of the weight of a body to the weight of an equal volume of water, in the case of solids and liquids; and to an equal volume of air in the case of gases; taken as the standard or unit.

Sphero-Hexagonal Powder. A black gunpowder in the form of a small ball with a six-sided ring around the middle.

Splash, angle of. See Danger angle.

Splash, point of. See Point of splash.

Spline. A rectangular piece fitting grooves like key seats in a hub or shaft, so that while the one may slide endwise on the other, both must revolve together.

Sponge. A swab used for cleaning the chamber and bore of guns and mortars.

Sprocket Wheel. A toothed wheel that engages the links of a chain. Spur Ring. A ring having radial teeth on the circumference. Spur Wheel. A gear wheel having external radial teeth on the circumference.

Staff. For administrative and other purposes, the staff of a coast artillery fort includes an adjutant, surgeon, artillery engineer, ordnance officer, quartermaster and commissary.

Stand Fast. A command at which cannoneers halt until the previous command is repeated. When one member makes a mistake this command is given before the mistake is corrected.

Standards. See COLORS.

Star Gauge. A device for measuring the diameter of the bore of cannon. It is used during the manufacture and when it is necessary to determine if any enlargement of the bore has taken place.

Storage Magazine. A building provided for the storage and preservation of powder or explosives; located so as to be protected from the fire of the enemy.

Storehouse. Every coast artillery fort is provided with one or more store- houses for the care and storage of accessory material.

Storeroom. A room in the emplacement for the storage of necessary material.

Strength. The number composing any military body. Striking Angle. The angle which the line of impact makes with the horizontal plane. It is equal to the angular depression of the point of impact plus the angle between the line of impact and the line of shot. Striking Energy. See ENERGY OF PROJECTILE.

Striking velocity. The velocity of the projectile at the point of impact.

Subcaliber Platform. A steel platform attached to the breech of large- caliber guns upon which the breech detail stands to load the subcaliber tube during practice. After the platform is attached the piece is placed in battery. At batteries not equipped with the prescribed pattern a platform of wood is constructed by lashing heavy planks to the sighting platforms so that they will extend about six feet in rear of the breech; flooring is then nailed to the planks.

Subcaliber Quadrant Scale. Scales used on gun carriages to set the piece in elevation during subcaliber practice in place of the regular service scale. They are constructed for ranges in yards corrected for height of site and curvature, and are used in the same manner as the regular quadrant scale.

Subcaliber Tube. A small caliber gun which is seated in the bore of a gun of larger caliber; used for target practice with ammunition of smaller charges and caliber than the gun in which it is used. In rapid fire guns this device is contained in a dummy projectile. They are classified as 30/100-inch caliber, 1-pounder, and 18-pounder.

Submarine Defense Equipment. Submarine defense equipment includes submarine mines, mobile torpedoes, obstructions and all material pertaining to the placing and service of these means of defense.

Submarine Defenses. Submarine defenses include submarine mines, mobile torpedoes, marine obstructions and submarine boats.

Submarine Mine. A submerged stationary torpedo consisting of an explosive charge and firing device, enclosed in a water-tight steel case, to be fixed in position in a channel which it is desired to close against the passage of an enemy's vessels. They are classified as mechanical, when they are exploded by means of a firing device; and electrical, when exploded by electricity.

Superior slope. The top slope of a parapet or traverse.

Supplementary Station. An auxiliary base-line station used to furnish data in place of the secondary station in case said station is put out of action, or to furnish data over a field of fire not covered by the secondary station.

Surface. The exterior part of anything that has length and breadth; the outside, as the surface of the earth.

Surveyor's Transit. An angle-measuring instrument similar in its essential parts to the azimuth instrument, but of more delicate and complicated design. It may be used to measure both horizontal and vertical angles as well as give the magnetic bearing. Its principal parts are shown by name and detail in chapter on Fire-Control Instruments, etc.

Swell of the Muzzle. The enlargement of the exterior of the gun at the muzzle.

T-I Bell. See TIME-INTERVAL BELL.

Tackle. A purchase formed by reeving a rope through two or more blocks, for the purpose of hoisting.

Tactical Chain of Artillery Command. The combined tactical units of seacoast defense.

Tactical Command. Command at drill and during action.

Tactical Responsibility. Responsibility for all matters affecting the efficiency of a tactical command.

Tactics. The art of drilling troops, and of handling and maneuvering them in the presence of the enemy.

Take Cover. A command which can be given at any time, at which all numbers not designated to remain at their post move at a run to some designated place under cover. As a rule this command is given in mortar batteries only.

Tangent of an Arc. That part of the tangent which, touching the arc at one extremity, is limited by the line passing through the other extremity and the center of the circle.

Tangential Force. A force which acts on a moving body in the direction of the tangent to the path of the body, its effect being to increase or diminish the velocity; distinguished from a normal force, which acts at right angles to the tangent and changes the direction of the motion without changing the velocity. A ricochet shot or a foul tip in base ball, are examples of applied tangential force.

Targ. The piece of metal (or other material) used to indicate the intersection of the arms on the plotting board.

Target or Vessel Tracking. The process whereby successive positions of a moving target are plotted on a chart or plotting board. It includes the observations made by the observers at the position-finding instruments, the plotting of the results of these observations on a plotting board at the same time tracing thereon the plotted track of the course of the target.

Target. The object at which guns or mortars are pointed; as a boat, ship or other object, whether stationary or moving.

Telautograph. An electro-mechanical instrument by means of which the movement of an attached pencil used by a person in writing at one end of the circuit, will automatically trace or reproduce the characters, as written, at the other end.

Telephone. An instrument by means of which a sound produced at one end of a wire is reproduced at the other end. There are two types in use the Service Telephone, used in temporary installations; and the Composite Artillery Type Telephone, used in permanent installations.

Telescopic Sight. A combination sight and telescope used for the aiming of guns at ranges greater than an object could be distinguished with the naked eye. The advantage of telescopic sights are the increased power of vision; large decrease in personal error ; and the great facility and accuracy of aiming a gun at the greater ranges. See SIGHT.

Terrain. The ground, its configuration and natural and artificial diversification. The topographical character of the country, region or tract, as viewed from a military standpoint.

Thawing Explosives. See MELTING AND THAWING EXPLOSIVES.

Theater of Operations. All the territory an army may desire to invade, and all that it may be necessary to defend.

Theoretical Range or Range of Ballistic Tables. The horizontal distance from the muzzle of the gun to that point of the descending branch of the trajectory, called the point of fall, which is at the level of the muzzle of the gun. This definition of the theoretical range is strictly accurate and is conveniently applicable in using quadrant elevation, as with mortar fire; but in using sight elevation it is more convenient, assuming the principle of the rigidity of the trajectory, to define the theoretical range as the distance from the muzzle of the gun to the point of intersection of the trajectory with the line from the muzzle to the target, the distance being measured along this line. The point of intersection is called the point of fall.

Theory of Probability. See PROBABILITY OF ERROR.

Thermometer. An instrument for measuring temperature.

Throttling Bar. A bar in the recoil cylinder to regulate the size of the orifice through which the oil escapes from one side of the piston head to the other.

Throttling pipe. A pipe connecting the rear ends of two recoil cylinders. The throttling and the equalizing pipes are joined by a connecting pipe through which oil flows from one end of the cylinders to the other without passing through the piston heads. The amount of oil which passes through the connecting pipe is controlled by the throttling valve. The recoil of the gun can be controlled to a certain extent by varying the setting of the throttling valve.

Tide Gauge. A mechanical device used to register the height of tide in feet and hundredths above the datum plane (mean low water). It consists essentially of a float connected with an automatic registering device.

Tide Indicator. A device operated electrically to indicate height of tide in all primary stations. It is operated by a controller located in the tide station.

Tide Station. A station at which periodical readings of height of tide are made, recorded and sent to the various primary stations throughout the fire- control command.

Time Fuse. A fuse which ignites the bursting charge at some fixed time after the projectile leaves the muzzle.

Time interval bell or T. I. bell. A bell to indicate the observing interval. Bells ring simultaneously at the emplacements and the observing stations. They are operated by a clock or a motor.

Time interval recorder. The ordinary stopwatch.

Time of Flight Scale. A scale giving the time of flight in seconds for any particular muzzle velocity and projectile, for ranges from one to twelve thousand yards. The term is also applied to the travel of the vessel in yards during the time of flight for any projectile at the range considered. In the case of mortars, the term is applied to the time of flight scale on the predicter, which gives the yards of travel during the time of flight plus one minute.

Time-Interval Bell. A bell with electrical attachment, located in all emplacements, primary and secondary stations, for the purpose of sounding simultaneous signals to indicate the observing interval. Commonly called "T-I Bell."

Time-Interval Recorder. The ordinary stop-watch.

Time-Range Board. A board to show range of target from battery at any instant. It is placed on the emplacement wall and is operated on data from the plotting room.

Tool Room. A room in the battery for the storage of necessary tools and implements.

Top carriage. The top carriage is a part of the gun carriages for guns of 8-inch or greater caliber and for 6-inch guns mounted on disappearing carriages. It consists of the recoil cylinders, the axle bedside frames, and the connecting pipes and transoms.

Torpedo Shell. A deck-piercing shell with an unusually large explosive cavity, fired from mortars for the purpose of carrying a large explosive charge to the decks of war vessels.

Torpedo-Detonating Pierce Fuse. A delayed action fuse used to detonate high-explosive bursting charges in mortar projectiles.

Towing Target. Any target which is capable of being towed behind a boat,

Tracking. The method employed in locating the course of a vessel on the plotting board, by taking simultaneous readings at the two base-end stations at regular intervals, and plotting the location of the target at the instant of each observation. See TARGET OR VESSEL TRACKING.

Trajectory. The curve described by the center of gravity of the projectile in passing from the muzzle of the gun to the point of impact. The lateral travel of the projectile is not proportional to the range, and the trajectory is, therefore, in general, a curve of double curvature, convex to the plane of departure. The trajectory of the range tables ignores the deflection forces, and lies in the plane of departure. Deflections due to wind and drift are considered separately. There is no appreciable error introduced by considering the trajectory as a plane curve lying in a vertical plane. The "principle" of the rigidity of the trajectory assumes that the figure whose outline is composed of the trajectory and its chord, behaves as if it were cut out of cardboard and rotated up and down with the muzzle of the gun as a center. For example, in the following figure, according to this principle, it is assumed that the shape of the trajectory is the same as if the points were in the horizontal plane through the muzzle of the gun and the angle ESO is equal to the angle of fall for a horizontal range equivalent to OS. OPS -Trajectory. F - Point of fall. I -Point of impact. S -Point of splash. W -Center of target. LV -Height of site. OH -Line of departure. OW -Line of direction. AF -Line of fall. BI -Line of impact. OS -Chord of the trajectory. QS -Range of shot. TW -Target. FOS -Angle of depression. ESQ -Angle of splash-danger angle. HOS -Angle of departure. HOF -Quadrant angle of departure. AFO =Angle of fall (range table) for horizontal range OF. BIT -Angle of impact. BIC -Angle of incidence.

Transit Instrument. See SURVEYOR'S TRANSIT.

Travel of Projectile. The distance from the base of a projectile in its seat in the bore to the muzzle of the gun.

Travel of Target. The distance passed over by the target in the time of flight. It is also used to express the distance passed over by the target in an observing interval.

Traverse Slope or Wall. The side slope or wall of the traverse.

Traverse. The structure protecting the armament and personnel from flank fire. In fortification, the structure perpendicular or oblique to the parapet wall, protecting the armament and personnel from flank fire. In gunnery, a term used to indicate the horizontal travel of the piece either to the right or left,

Traversing circle. The metal which is bolted to the concrete and which supports the rear part of the carriage in the front pintle type. Traversing wheels roll on this circle.

Traversing Indicator. A device used by gun pointers to control the traversing of a gun without command.

Traversing rollers. Rollers which rest upon the base ring and which enable the gun or mortar carriage to be given motion right or left.

Tray. See LOADING TRAY.

Trial Shots. Shots fired before practice or action to determine, for guns, the muzzle velocity to be used; for mortars the range and deflection corrections to be applied.

Trinitrotoluol. A detonating and disruptive explosive of high order used as a base for shell fillers.

Tripping. The act of releasing the counterweights of a disappearing carriage, thereby carrying the piece in battery, i.e., moving the top carriage forward so that the muzzle extends over the parapet.

Trolley. A mechanical device for transporting projectiles on horizontally suspended tracks.

Truck platform. If the ammunition trucks run on a different surface from that of the loading platform, this surface is called the "truck platform."

Truck Recess. The spaces built in the parapet wall for the storage of ammunition trucks.

Trunnion Band or Hoop. The hoop around a cannon, of which the trunnions form a part, located at about the center of gravity.

Trunnion-Sight Bracket. A bracket attached to the right trunnion of a gun, which may be used for holding the telescopic sights.

Trunnions. The cylinders which rest in bearing surfaces of the carriage called "trunnion beds." Their axis is perpendicular to the axis of the bore and ordinarily in the same plane; they connect the cannon with the carriage and transmit the force of recoil from one to the other. The faces of the trunnions are the end planes perpendicular to their axis.

Tube. The inner wall of the bore of a built-up gun extending usually from the breech to the muzzle, ordinarily made one piece.

Tug Observer. A coast artillery officer detailed by the post commander for duty aboard a tug engaged in towing and maneuvering targets during subcaliber or service target practice. His principal duties are to supervise the use of the camera or range rake in taking range deviations (overs and shorts) .

Tug Officer. A coast artillery officer in charge of a tug engaged in towing and maneuvering targets during subcaliber or service target practice. The actual maneuvering of targets from the tug is under the supervision of the master of the tug, subject to the orders of the tug officer.

Twist of rifling. The inclination of the grooves to the axis of the gun at any point. When this inclination is constant the twist is uniform; when it increases from the breech to a point near the muzzle it is increasing. Twist is generally expressed in turns per caliber, e. g., one turn in 50 calibers, meaning that the projectile makes one complete rotation in passing over a distance equal to 50 calibers, provided the twist were uniform. In most of the major caliber guns in our service, the twist increases from one turn in 50 calibers to one turn in 25 calibers at a short distance from the muzzle and beyond that point it is uniform.

Unit. A military body acting together. Military units are divided into two classes, i.e., administrative units, which include those necessary for the proper care, housing, clothing, feeding, instruction, disciplining and keeping of military records; and tactical units, which are organized and equipped with a view of their highest efficiency in action.

Unrestricted Fire. See ORDERS OF FIRE.

Velocity of Combustion. The rate of burning of a powder grain.

Velocity of Rotation. The rate of motion of a body around its axis, as a wheel, as distinguished from progressive motion of a body in the direction of a distant point. In gunnery it is the rate of motion of the projectile at any point in the trajectory around its longer axis.

Velocity of Translation. The rate of travel of the projectile in the direction of its flight.

Velocity. As used in coast artillery practice, velocity is the rate of travel of a projectile in feet per second. The velocity at the muzzle is the "initial" or "muzzle" velocity. The velocity at the point of impact is the "striking" velocity. The velocity at any point of the trajectory between the muzzle and the point of impact is the "remaining" velocity at that point. Velocities with guns vary from 2,000 to 3,000 feet per second, with mortars from 550 to 1,300 feet per second.

Vent. A small channel leading from the exterior of the cannon to the powder chamber for the ignition of the powder charge. It is an "axial vent" when it is in line with the axis of the bore. It is a "radial vent" when it is at right angles to the axis of the bore.

Ventilators. The shafts or flues with movable covers for ventilation, leading from interior galleries or air spaces and opening through the superior slope.

Vernier. A small auxiliary scale enabling the measurement of hundredths (or minutes and seconds) in connection with the main scale.

Vertex of an Angle. The point at which the sides of an angle meet.

Vertical Angle. An angle whose sides lie wholly in the vertical plane.

Vertical Base System. The system of range finding in which the azimuth and range are determined by one position-finding instrument, located at either of the (horizontal) base-end stations. The base line in this system extends from the horizontal axis of the trunnions of the instrument vertically to mean low-water level. The vertical base feature of the position-finding instrument is used exclusively in this system.

Vertical Plane. Any plane which passes through a vertical line or contains the line of the plumb bob when freely suspended by a line and allowed to come to rest.

Vessel Charts. Charts, usually blue prints, used in the identification of targets. These charts consist of a system of vessel symbols giving a classification in outline of warships in connection with funnels and masts.

Water Front. That portion of the defenses bearing upon the navigable water areas that may be open to an enemy.

Winch. A machine operated by steam or other motive power, used on mine planters for raising anchors and other heavy weights. It consists of a drum, crank and the necessary gearing arranged for gaining power.

Wind Vane. A device pertaining to the meteorological station, which indicates the azimuth of the wind.

Wind-Component Indicator. A device used in primary stations of the fire-control system to indicate to the operators of the range and deflection boards, the reference numbers corresponding to the range and deflection components of the wind as sent from the meteorological station.

Wireless Station. A station located within the fortification in which wire- less telegraph apparatus is installed.

Worm. A short threaded portion of a shaft, constituting an endless screw formed to mesh with a gear wheel.

Xylol. A colorless oily inflammable liquid, used in the bath for making the 135-degree stability test of nitro-cellulose powder.

Zero. The point from which instruments, etc., are graduated. In communicating numerals, it is a word used to indicate a cipher or naught. For example, 340.30, would be expressed: "Three-four-zero-point-three-zero."

Zone Energy. A term denoting the relative armor-piercing power of different guns. It is estimated by the number of foot-tons per inch of the shot circumference. At the muzzle this power is a maximum, but owing to the resistance of the air it gradually diminishes during flight.

Zone Mort. A French military expression denoting the space in which the projectile has lost its strength, or is spent.

Zone. In mortar firing, the area m which projectiles fall for a given charge of powder, when the elevation is varied between the minimum and maximum. It is also used with reference to other divisions of the defensive area, as "outer defense zone," "inner defense zone," etc.

Zone of Fire. A term synonymous with specified fire areas. In mortar firing it is the particular area in which projectiles fall for a given charge of powder when the elevation is varied between minimum and maximum. The several zones of fire for the 12-inch steel mortars are, 1st Zone, 2,210 to 2,970; 2d Zone, 2,600 to 3,431 ; 3d Zone, 3,070 to 4,030; 4th Zone, 3,631 to 4,800; 5th Zone, 4,429 to 5,940; 6th Zone, 5,520 to 7,476; 7th Zone, 7,027 to 9,250; 8th Zone, 8,758 to 12,019 yards. In the case of rifle fire the defensive area between the guns and 4,000 yards range, is known as the "inner defense zone" between 4,000 and 8,000 yards range, as the "middle defense zone;" and between 8,000 and 12,000 yards range, as the "outer defense zone."

Zone, 50 per cent (sometimes called **probable zone**). The space bounded by two parallel lines within which 50 per cent of the points of impact of all shots fired with the same elevation and azimuth settings will probably lie. The width of the 50 per cent zone in any direction is equal to twice the probable error in that direction. The area common to the 50 percent lateral zone and the 50 per cent longitudinal zone is the 25 per cent rectangle.

Index

This is a brief general index to some of the major subjects covered by this reference guide, use in conjunction with the table of contents.

www.ingramcontent.com/pod-product-compliance
Lightning Source LLC
Chambersburg PA
CBHW062021090426

42811CB00005B/916

9 780974 816739